Studies in Logic
Mathematical Logic and Foundations
Volume 18

Classification Theory for Abstract Elementary Classes

Volume 8
A New Approach to Quantum Logic
Kurt Engesser, Dov M. Gabbay and Daniel Lehmann

Volume 9
Handbook of Paraconsistency
Jean-Yves Béziau, Walter Carnielli and Dov Gabbay, editors

Volume 10
Automated Reasoning in Higher-Order Logic. Set Comprehension and Extensionality in Church's Type Theory
Chad E. Brown

Volume 11
Foundations of the Formal Sciences V: Infinite Games
Stefan Bold, Benedikt Löwe, Thoralf Räsch and Johan van Benthem, editors

Volume 12
Second-Order Quantifier Elimination: Foundations, Computational Aspects and Applications
Dov M. Gabbay, Renate A. Schmidt and Andrzej Szałas

Volume 13
Knowledge in Flux. Modeling the Dynamics of Epistemic States
Peter Gärdenfors. With a foreword by David Makinson

Volume 14
New Approaches to Classes and Concepts
Klaus Robering, editor

Volume 15
Logic, Navya-Nyāya and Applications. Homage to Bimal Krishna Matilal
Mihir K. Chakraborti, Benedikt Löwe, Madhabendra Nath Mitra and Sundar Sarukkai, eds.

Volume 16
Foundations of the Formal Sciences VI. Probabilistic Reasoning and Reasoning with Probabilities.
Benedikt Löwe, Eric Pacuit and Jan-Willem Romejin, eds.

Volume 17
Reasoning in Simple Type Theory. Festschrift in Honour of Peter B. Andrews on His 70th Birthday.
Christoph Benzmüller, Chad E. Brown and Jörg Siekmann, eds.

Volume 18
Classification Theory for Abstract Elementary Classes
Saharon Shelah

Studies in Logic Series Editor
Dov Gabbay

dov.gabbay@kcl.ac.uk

Classification Theory for Elementary Abstract Classes

Saharon Shelah

ISBN 978-1-904987-71-0

College Publications
Scientific Director: Dov Gabbay
Managing Director: Jane Spurr
Department of Computer Science
King's College London, Strand, London WC2R 2LS, UK

http://www.collegepublications.co.uk

Original cover design by orchid creative www.orchidcreative.co.uk
Printed by Lightning Source, Milton Keynes, UK

מקדש לבני האהוב יובב

Dedicated to my beloved son Yovav

Contents

INTRODUCTION TO: CLASSIFICATION THEORY FOR ABSTRACT ELEMENTARY CLASSES

E-53

ABSTRACT

Classification theory of elementary classes deals with first order (elementary) classes of structures (i.e. fixing a set T of first order sentences, we investigate the class of models of T with the elementary submodel notion). It tries to find dividing lines, prove their consequences, prove "structure theorems, positive theorems" on those in the "low side" (in particular stable and superstable theories), and prove "non-structure, complexity theorems" on the "high side". It has started with categoricity and number of non-isomorphic models. It is probably recognized as the central part of model theory, however it will be even better to have such (non-trivial) theory for non-elementary classes. Note also that many classes of structures considered in algebra are not first order; some families of such classes are close to first order (say have kind of compactness). But here we shall deal with a classification theory for the more general case without assuming knowledge of the first order case (and in most parts not assuming knowledge of model theory at all).

For technical reasons the book has been split into two volumes.

§0 INTRODUCTION AND NOTATION

In §2 we shall try to explain the purpose of the book to mathematicians with little relevant background. §1 describes dividing lines and gives historical background. In §5 we point out the (reasonably limited) background needed for reading various parts and some basic

Typeset by $\mathcal{A}\mathcal{M}\mathcal{S}$-TEX

definitions and in §6 we list the use of symbols. The content of the book is mostly described in §2-§3-§4 but §4 mainly deals with further problems and §6 with the symbols used.

Is this a book? I.e. is it a book or a collection of articles? Well, in content it is a book but the chapters have been written as articles, (in particular has independent introductions and there are some repetitions) and it was not clear that they will appear together, see §5(A) for more on how to read them.

§1 Introduction for model theorists

(A) Why to be interested in dividing lines?

Classification theory for first order (= elementary) classes is so established now that up to the last few years most people tended to forget that there are non-first order possibilities. There are several good reasons to consider these other possibilities; first, it is better to understand a more general context, we would like to prove stronger theorems by having wider context, classify a larger family of classes. Second, understanding more general contexts may shed light on the first order one. In particular, larger families may have stronger closure properties (see later). Third, many classes arising in "nature" are not first order ("in nature" here means other parts of mathematics).

Of course, we may suspect that applying to a wider context may leave us with little content, i.e., the proofs may essentially be just rewording of the old proofs (with cumbersome extra conditions); maybe there is no nice theory, not enough interesting things to be discovered in this context; it seems to me that experience has already refuted the first suspicion. Concerning the other suspicion, we shall try to give a positive answer to it, i.e. develop a theory; on both see the rest of the introduction.

In any case, "not first order" does not define our family of classes of models as discussed below. This is both witnessed from the history (on which this section concentrates) and suggested by reflection; clearly we cannot prove much on arbitrary classes, so we need some restriction to reasonable classes. Now there may be incomparable cases of reasonableness and a priori it is natural to expect to be able to say considerably more on the "more reasonable" cases. E.g. we expect that much more can be said on first order classes than on the class of models of a sentence from $\mathbb{L}_{\omega_1,\omega}$.

We are mainly interested here in generalizing the theorems on categoricity, superstability and stability to such contexts, in particular we consider the parallel of Łoś Conjecture and the (very probably much harder) main gap conjecture as test problems.

This choice of test problem is connected to the belief in (a),(b),(c) discussed below (that motivates [Sh:c]).

(*a*) It is very interesting to find dividing lines and it is a fruitful approach in investigating quite general classes of models.

That is, we start with a large family of (in our case) classes (e.g., the family of elementary (= first order) classes or the family of universal classes or the family of locally finite algebras satisfying some equations) and we would like to find natural dividing lines. A dividing line is not just a good property, it is one for which we have some things to say on both sides: the classes having the property and the ones failing it. In our context normally all the classes on one side, the "high" one, will be provably "chaotic" by the non-structure side of our theory, and all the classes on the other side, the "low" one will have a positive theory. The class of models of true arithmetic is a prototypical example for a class in the "high" side and the class of algebraically closed field the prototypical non-trivial example in the "low" side.

Of course, not all important and interesting properties are like that. If F is a binary function on a set A, not much is known to follow from (A, F) not being a group. In model theory introducing o-minimal theories was motivated by looking for parallel to minimal theories and attempts to investigate theories close to the real field (e.g., adding the function $x \mapsto e^x$). Their investigation has been very important and successful, including parallels of stability theory for strongly minimal sets, but it does not follow our paradigm. A success of the guideline of looking for dividing lines had been the discovery of being stable (elementary classes, i.e. $(\mathrm{Mod}_T, \prec)$, [Sh 1]). From this point of view to discover a dividing line means to prove the existence of complementary properties from each side:

(*i*) T is unstable iff it has the order property (recall that T

has the order property means that: some first order formula $\varphi(\bar{x}, \bar{y})$ linearly orders in M some infinite $\mathbf{I} \subseteq {}^{\ell g(\bar{x})}M$ in a model M of T)

(*ii*) T is stable iff $A \subseteq M \models T$ implies $|\mathbf{S}(A, M)|$, the set of 1-types on A for M is not too large ($\leq |A|^{|T|}$).

A case illustrating the point of dividing line is a precursor of the order property, property E of Ehrenfeucht [Eh57], it says that some first order formula $\varphi(x_1, \ldots, x_n)$ is asymmetric on some infinite $A \subseteq M$, M a model of T; it is stronger than the order property (= negation of stability). A posteriori, order on the set of n-tuples is simpler; this is not a failure, what Ehrenfeucht did was fine for his aims, but looking for dividing lines forces you to get the "true" notion.

Even better than stable was superstable because it seems to me to maximize the "area" which we view as being how many elementary classes it covers times how much we can say about them. On the other hand, it has always seemed to me more interesting than $\aleph_0$-stable as the failure of $\aleph_0$-stability is weak, i.e. it has a few consequences. There is a first order superstable not $\aleph_0$-stable class K such that a model $M \in K$ is determined up to isomorphism by a dimension (a cardinal) and a set of reals. This exemplifies that an elementary class can fail to be $\aleph_0$-stable but still is "low": we largely can completely list its models. Such a class is the class of vector spaces over $\mathbb{Z}/2\mathbb{Z}$ expanded by predicates P_n for independent subspaces of co-dimension 2. A model M in this class is determined up to isomorphism by one cardinal (the dimension of the sub-space $V_M = \cap\{P_n^M : n \in \mathbb{N}\}$) and the quotient M/V_M which has size at most continuum (alternatively the set $\{\eta_a : a \in M\}, \eta_a(n) \in \{0,1\}$ and where $\eta_a = \langle \eta_a(0), \eta_a(1), \ldots \rangle$ and $\eta_a(n) = 0 \Leftrightarrow a \in P_n^M$).

Of course, the guidelines of looking for dividing lines if taken religiously can lead you astray. It does not seem to recommend investigation of FMR (Finite Morley Rank) elementary classes which has covered important ground (see e.g. Borovik-Nessin [BoNe94]). This guideline has helped, e.g. to discover dependent and strongly dependent elementary classes, but so far our approach has seemingly not succeeded too much in advancing the investigation.

See more on this in end of §2(B), in particular Question 2.15.

(b) It is desirable to have an exterior a priori existing goal as a test problem.

Such a problem in model theory was Łos conjecture which says: if a first order class of countable vocabulary (= language) is categorical in one $\lambda > \aleph_0$ (= has one and only one model of cardinality λ up to isomorphisms) then it is categorical in every $\lambda > \aleph_0$. At least for me so was Morley conjecture [Mo65] which says that for first order class with countable vocabulary, the number of its models of cardinality $\lambda > \aleph_0$ up to isomorphism is non-decreasing with λ. This motivated my research in the early seventies which eventually appeared as [Sh:a] (with several late additions like local weight in [Sh:a, Ch.V,§4]). Now having introduced "$\aleph_\varepsilon$-saturated models", it seems unconvincing to understand $\dot{I}(\lambda, K)$, the number of models in K of cardinality λ up to isomorphism, for K the class of $\aleph_\varepsilon$-saturated models of a first order class, hence though essentially done then, was not written till much later. Eventually "$\dot{I}(\lambda, T)$ non-decreasing" was done for the family of classes of models of a countable first order theory (which was the original center of interest; see [Sh:c]).

By this solution, there are very few "reasons" for such $K = \text{Mod}_T$ to have many models: being unstable, unsuperstable, DOP (dimensional order property), OTOP (omitting type order property) and deepness (for fuller explanation see after 2.12; see more, characterizing the family of functions $\dot{I}(\lambda, T)$ for countable T in Hart-Hrushovski-Laskowski [HHL00]). So the direct aim was to solve the test question (e.g., the main gap[1]), but the motivation has always been the belief that solving it will be rewarded with discovering worthwhile dividing lines and developing a theory for both sides of each.

The point is that looking at the number of non-isomorphic models and in particular the main gap we hope to develop a theory. Other exterior problems will hopefully give rise to other interesting theories, which may be related to stability theory or may not; this was the point of [Sh 10], in particular the long list of exterior results in the end of its introduction, and the words "classification theory" in the

[1] which says that either $\dot{I}(\lambda, T) = 2^\lambda$ for every ($> |T|$, or large enough) λ or $\dot{I}(\aleph_\alpha, T) \leq \beth_{\gamma(T)}(|\alpha|)$ for every α (for some ordinal $\gamma(T)$); see more in 2.10.

name of [Sh:a]. But, the above point seemingly was slow in being noticed.

Of course, if we consider the family of classes which are "high" by one criterion/dividing line, we expect that with respect to other questions/dividing lines the "previously high ones" will be divided and on a significant portion of them we have another positive theory, quite reasonably generalizing the older ones (but maybe we shall be led to very different theories). E.g. for unstable first order classes [Sh:93] succeeded in this respect: "low ones" are the simple theories and the "high ones" are theories with the tree property (on exciting later developments, see [KiPi98] or [GIL02]).

(c) successful dividing lines will throw light on problems not considered when suggesting them.

The point is that the theory should be worthwhile even if you discard the original test problems. Stability theory is just as interesting for some other problems as for counting number of non-isomorphic models. E.g.

$(*)_1$ the maximal number of models no one embeddable into another.

This sounds very close to counting, so we expect this is to have a closely related answer.

In fact for elementary classes (with countable vocabulary) which have a structure theorem (see 2.10 below), this number is $< \beth_{\omega_1}$, for the others it is very much higher (see more on the trichotomy after 2.12); so the answer to $(*)_1$ turns out to be nicer than the one concerning the number, $\lambda \mapsto \dot{I}(\lambda, T)$.

$(*)_2$ in $\mathfrak{K}$ there are models very similar yet non-isomorphic.

This admits several interpretations which in general have complete and partial solutions quite tied up with stability theory. One is finding $\mathbb{L}_{\infty,\lambda}$-equivalent not isomorphic models of cardinality λ. Stronger along this line are EF_λ-equivalent not isomorphic. Another is that there are non-isomorphic models of T such that a forcing neither collapsing cardinals nor adding too short sequences makes them isomorphic. For non-logicians we should explain that this says in a very strong sense that there are no reasonable invariants, see [Sh 225], [Sh 225a], Baldwin-Shelah [BLSh 464], Laskowski-Shelah [LwSh 489], Hyttinen-Tuuri [HyTu91], Hyttinen-Shelah-Tuuri [HShT 428], Hyttinen-Shelah [HySh 474], [HySh 529], [HySh 602].

$(*)_3$ For which classes K do we have: its models are no more complicated than trees (in the graph theoretic sense say rooted graphs with no cycle)?

This question was specified to having a tree of submodels which is "free" (= "non-forking") and it is a decomposition, i.e., the whole model is prime over the tree. This is answered by stability theory (for Mod_T, T countable)

$(*)_4$ similarly replacing graphs with no cycles by another simple class, e.g., linear orders.

This is very interesting, but too hard at present (see more in Cohen-Shelah [CoSh:919])

$(*)_5$ decidable theories, e.g. we may note that there was much done on decidability and understanding of the monadic theory of some structures (in particular Rabin's celebrated theorem). Those works concentrated on linear orders and on trees. Was this because of our shortcoming or for inherent reasons?

We may interpret this as a call to classify classes, in particular, first order ones by their complexity as measured by monadic logic. This was carried to large extent in Baldwin-Shelah [BlSh 156] for first order classes. Now this seems a priori orthogonal to classification taking number of models as the test question; note that the class of linear orders is unstable but reasonably low for [BlSh 156], whereas any class is maximally complicated if it has a pairing function (e.g. a one-to-one function F^M from $P_1^M \times P_2^M$ into P_3^M while P_1^M, P_2^M are infinite) and there are such classes which are categorical in every $\lambda \geq \aleph_0$. In spite of all this [BlSh 156] relies heavily on stability theory; see [Bl85], [Sh 197], [Sh 205], [Sh 284c]

$(*)_6$ the ordinal κ-depth of a model (Karp complexity).

For a model M and a partial automorphism f of M, $\text{Dom}(f)$ of cardinality $< \kappa$, we can define its κ-depth in M, an ordinal (or ∞) by $\text{Dp}_\kappa(f, M) \geq \alpha$ iff for every $\beta < \alpha$ and subsets A_1, A_2 of cardinality $< \kappa$, there is a partial automorphism f' of M extending f of κ-depth $\geq \beta$ such that $|\text{Dom}(f')| < \kappa, A_1 \subseteq \text{Dom}(f'), A_2 \subseteq \text{Rang}(f')$.

Let

$$\text{Dp}_\kappa(M) = \cup\{\text{Dp}_\kappa(f, M) + 1 : f \text{ a partial automorphism of } M \text{ of cardinality } < \kappa \text{ and } \text{Dp}_M(f) < \infty\}.$$

This measures the complexity of the models and $\text{Dp}_\kappa(T) = \cup\{\text{Dp}(M) + 1 : M \text{ a model of } T\}$ is a reasonable measure of the complexity of T. With considerable efforts, reasonable knowledge concerning this measure was gained by Laskowski-Shelah [LwSh 560], [LwSh 687], [LwSh 871] confirming to some extent the thesis above.

$(*)_7$ categoricity and number of models in $\aleph_\alpha$, in ZF (i.e., with no choice).

See [Sh 840].

You may view in this context the question of having non-forking (= abstracts dependence relations), orthogonality, regularity but for me this is part of the inside theory rather than an external problem

(*d*) non-structure is not so negative.

Now this book predominantly deals with the positive side, structure theory, so defending the honour of non-structure is not really necessary (it is the subject of [Sh:e] though). Still first we may note that finding the maximal family of classes for which we know something is considerably better than finding a sufficient condition. In particular finding "the maximal family ... such that ..." is finding dividing lines and this is meaningless without non-structure results.

Second, this forces you to encounter real difficulties and develop better tools; also using the complicated properties of a class which already satisfies some "low side properties" may require using and/or developing a positive theory.

Last but not least, non-structure from a different perspective is positive. Applying "non-structure theory" to modules this gives representation theorems of rings as endomorphism rings (see Göbel-Trlifaj [GbTl06]; note that the "black boxes" used there started from [Sh:c, VIII]). In fact, generally for unstable elementary class K, we can find models which in some respect represent a pregiven ordered group (see [Sh 800]). This has been applied to clarify in some cases to which generalized quantifiers give a compact logic (see [Sh:e] and more in [Sh 800]).

It may clarify to consider an alternative strategy: we have a reasonable idea of what we look for and we have a specific class or structure which should fit the theory. This works when the analysis we have in mind is reflected reasonably well in the specific case. It may be misleading when the examples we have, do not reflect the complexity of the situation, and it seems to be the case in the problems we have at hand. More specifically, though the "example" of the theory of superstable first order classes stand before us, we do not try to take the way of trying to assume enough of its properties so that it works; rather we try look for dividing lines.
See more on "why dividing lines" in the end of (B) of §2.

(B) Historical comments on non-elementary classes:

Let us return to non-elementary classes. Generally, on model theory for non-elementary classes see Keisler [Ke71] and the handbook [BaFe85]: closer to our interest in the forthcoming book of Baldwin [Bal0x] and the older Makowsky [Mw85], mainly around $\aleph_1$.

Below we present the results according to the kind of classes dealt with (rather than chronologically).
The oldest choice of families of classes (in this context) is the family of class of κ-sequence homogeneous models for a fixed D.

Morley and Keisler [KM67] proved that there are at most $2^{2^{|T|}}$ such models of T in any cardinality. Keisler [Ke71] proved that if $\psi \in \mathbb{L}_{\omega_1,\omega}$ is categorical in $\aleph_1$ and its model in $\aleph_1$ is sequence homogeneous then it is categorical in every $\lambda > \aleph_1$; generalizing (his version of) the proof of Morley's theorem. In [Sh 3] instead of having a monster $\mathfrak{C}$, i.e., a $\bar{\kappa}$-saturated model of a first order T, we have a $\bar{\kappa}$-sequence homogeneous model $\mathfrak{C}$. Let $D = D(\mathfrak{C}) = \{\mathrm{tp}(\bar{a}, \emptyset, \mathfrak{C}) : \bar{a} \in \mathfrak{C}$; i.e., $\bar{a}$ a finite sequence from $\mathfrak{C}\}$; note that $D, \bar{\kappa}$ determines $\mathfrak{C}$ and we look at the class of $M \prec \mathfrak{C}$ (or the class of (D, λ)-homogeneous

$M \prec \mathfrak{C}$). There the stability spectrum was reasonably characterized, splitting and strong splitting were introduced (for first order theory this was later refined to forking). See somewhat more in [Sh 54].

Lately, this (looking at the $\prec$-submodels of a (D, λ)-homogeneous monster $\mathfrak{C}$) has become very popular, see Hyttinen [Hy98], Hyttinen and Shelah [HySh 629], [HySh 632], [HySh 629] (the main gap for $(D, \aleph_\varepsilon)$-homogeneous models for a good diagram D), Grossberg-Lessman [GrLe02], [GrLe0x] (the main gap for good $\aleph_0$-stable (= totally transcendental)), [GrLe00a], Lessman [Le0x], [Le0y] (all on generalizing geometric stability).

We may look at contexts which are closer to first order, i.e., having some version of compactness. Chang-Keisler [ChKe62], [ChKe66] has looked at models with truth values in a topological space such that ultraproducts can be naturally defined. Robinson had looked at model theory of the classes of existentially closed models of first order universal or just inductive theories. Henson [He74] and Stern [Str76] have looked at Banach spaces (we can take an ultraproduct of the spaces, throw away the elements with infinite norm and divide by those with infinitesimal norm). Basically the logic is "negation deficient", see Henson-Iovino [HeIo02].

The aim of [Sh 54] was to show that the most basic stability theory was doable for Robinson style model theory. In particular it deals with case II (the models of a universal first order theory which has the amalgamation property) and case III (the existentially closed models of a first order inductive (= Π^1_2) theory); those are particular cases of (D, λ)-homogeneous models. Case II is a special case of III where T has amalgamation. Lately, Hrushovski dealt with Robinson classes (= case II above). A Ph.D. student of mine in the seventies was supposed to deal with Banach spaces but this has not materialized. Henson and Iovino continued to develop model theory of Banach spaces. Lately, interest in the classification theory in such contexts has awakened and dealing with cases II and III and complete metric spaces and Banach spaces and relatives, see Ben-Yaacov [BY0y], Ben-Yaacov Usvyatsov [BeUs0x], Pillay [Pi0x], Shelah-Usvyatson [ShUs 837].

The most natural stronger (than first order) logic to try to look at, in this context, has been $\mathbb{L}_{\omega_1,\omega}$ and even $\mathbb{L}_{\lambda^+,\omega}$. By 1970 much was known on $\mathbb{L}_{\omega_1,\omega}$ (see Keisler's book [Ke71]); however, if you do not like non-first order logics, look at the class of atomic models of a countable first order T. The general question looks hard. At the early seventies I have clarified some things on $\psi \in \mathbb{L}_{\omega_1,\omega}$ categorical in $\aleph_1$, but it was not clear whether this leads to anything interesting. Then the following question of Baldwin catches my eye (question 21 of the Friedman list [Fr75])

$(*)_1$ can $\psi \in \mathbb{L}(\mathbf{Q})$ have exactly one uncountable model up to isomorphism?
Q stands for the quantifier "there are uncountably many"

This is an excellent question, a partial answer was ([Sh 48])

$(*)_2$ if $\diamondsuit_{\aleph_1}$ and $\psi \in \mathbb{L}_{\omega_1,\omega}(\mathbf{Q})$ has at least one but $< 2^{\aleph_1}$ models in $\aleph_1$ up to isomorphism <u>then</u> it has a model in $\aleph_2$ (hence has at least 2 non-isomorphic models)

Only later the original problem (even for $\psi \in \mathbb{L}_{\omega_1,\omega}(\mathbf{Q})$) was solved in ZFC, see below. It seems natural to ask in this case how many models ψ has in $\aleph_2$, and then successively in $\aleph_n$ (raised in [Sh 48]), but as it was hard enough, the work concentrates on the case of $\psi \in \mathbb{L}_{\omega_1,\omega}$, so ([Sh 87a], [Sh 87b] and generalizing it to cardinals $\lambda, \lambda^+, \ldots$ is a major aim of this book):

$(*)_3$ (a) if $n < \omega, 2^{\aleph_0} < 2^{\aleph_1} < \ldots < 2^{\aleph_n}, \psi \in \mathbb{L}_{\omega_1,\omega}, \dot{I}(\aleph_\ell, \psi) < \mu_{\text{wd}}(\aleph_\ell)$, for[2] $\ell \leq n$ and $\dot{I}(\aleph_1, \psi) \geq 1$ <u>then</u> ψ has a model in $\aleph_{n+1}$ and without loss of generality ψ is categorical in $\aleph_0$

(b) if the assumption of (a) holds for every $n < \omega$ and ψ is for simplicity categorical in $\aleph_0$ <u>then</u> the class Mod_ψ is so-called excellent (see (c))

(c) if $\psi \in \mathbb{L}_{\omega_1,\omega}$ is excellent and is categorical in one $\lambda > \aleph_0$ then it is categorical in every $\lambda > \aleph_0$.

---

[2] $\mu_{\text{wd}}(\aleph_\ell)$ is "almost" equal to $2^{\aleph_\ell}$

Essentially, it was proved that excellent $\psi \in \mathbb{L}_{\omega_1,\omega}$ are very similar to $\aleph_0$-stable (= totally transcendental) first order countable theories (after some "doctoring"). The set of types over a model $M, \mathscr{S}(M)$ is restricted (to not violate the omission of the types which every model of ψ omit). The types themselves are as in the first order case, set of formulas but we should not look at complete types over any $A \subseteq M \models \psi$, only at the cases $A = N \prec M$ or $A = M_1 \cup M_2$ where M_1, M_2 are stably amalgamated over M_0 and more generally at $\cup\{M_u : u \in \mathscr{P}^-(n)\}$, where $\langle M_u : u \in \mathscr{P}^-(n)\rangle$ is a "stable system".

This work was continued in Grossberg and Hart [GrHa89], (main gap), Mekler and Shelah [MkSh 366] (dealing with free algebras), Hart and Shelah [HaSh 323] (categoricity may hold for $\aleph_0, \aleph_1, \aleph_2, \ldots, \aleph_n$ but fail for large enough λ) and lately Zilber [Zi0xa], [Zi0xb] (connected to his programs). Further works on more general but not fully general are [Sh 300], Chapter II (universal classes), Shelah and Villaveces [ShVi 635], van Dieren [Va02] (abstract elementary class with no maximal models). See also the closely related Grossberg and Shelah [GrSh 222], [GrSh 238], [GrSh 259], [Sh 394], (abstract elementary class with amalgamation), Grossberg [Gr91] and Baldwin and Shelah [BlSh 330], [BlSh 360], [BlSh 393]. Lately, Grossberg and VanDieren [GrVa0xa], [GrVa0xb] Baldwin-Kueker-VanDieren [BKV0x] investigate the related tame abstract elementary class including upward categoricity. They prove independently of IV.7.12 that tame a.e.c. with amalgamation has nice categoricity spectrum; i.e. prove categoricity in cardinals $> \mu$ in the relevant cases; in the notation here "tame" means locality of orbital types over saturated model; on IV.7.12, see §4(B) after $(**)_\lambda$. Concerning $\mathbb{L}_{\kappa,\omega}$, see Makkai-Shelah [MaSh 285] (on cateogoricity of $T \subseteq \mathbb{L}_{\kappa,\omega}, \kappa$ compact starting with λ successor), Kolman-Shelah [KlSh 362] ($T \subseteq \mathbb{L}_{\kappa,\omega}, \kappa$ measurable, amalgamation derived from categoricity), [Sh 472] ($T \subseteq \mathbb{L}_{\kappa,\omega}, \kappa$ measurable, only down from successor). See more in the book [Bal0x] of Baldwin on the subject.

Going back, $(*)_3$ deals with $\psi \in \mathbb{L}_{\omega_1,\omega}$, it generalizes the case $n = 1$ which, however, deals with $\psi \in \mathbb{L}_{\omega_1,\omega}(\mathbf{Q})$. On the other hand, $\psi \in \mathbb{L}_{\omega_1,\omega}(\mathbf{Q})$ is not a persuasive end of the story as there are similar stronger logics. Also the proof deals with $\mathbb{L}_{\omega_1,\omega}(\mathbf{Q})$ in an indirect way, we look at a related class K which has also countable models but some first order definable set should not change when extending. So it seems that the basic notion is the right version of elementary extensions. This leads to analysis which suggests the notion of abstract elementary class, $\mathfrak{K}$ with $\mathrm{LS}(\mathfrak{K}) \leq \aleph_0$ which, moreover, is $\mathrm{PC}_{\aleph_0}$ (in [Sh 88], represented here in Chapter I).

Now much earlier Jonsson [Jn56], [Jn60] had considered axiomatizing classes of models. Compared with the abstract elementary classes used (much later) in [Sh 88]=Chapter I, the main[3] differences are that he uses the order $\subseteq$ (being a submodel) on K (rather than an abstract order $\leq_{\mathfrak{K}}$) and assume the amalgamation (and JEP joint embedding property). His aim was to construct and axiomatize the construction of universal and then universal homogeneous models so including amalgamation was natural; Morley-Vaught [MoVa62] use this for elementary class. In fact if we add amalgamation (and JEP) to abstract elementary classes we get such theorems (see I§2, in fact we also get uniqueness in a case of somewhat different character, I.2.17). From our perspective amalgamation (also $\leq_{\mathfrak{K}}=\subseteq$) is a heavy assumption (but an important property, see later). Now, model the-

---

[3]Jonsson axioms were, in our notations, (for a fix vocabulary τ, finite in [Jn56], countable in [Jn60]), K is a class of τ-models satisfying

- (I) there are non-isomorphic $M, N \in K$ in [Jn56]
- $(I)'$ K has members of arbitrarily large cardinality in [Jn60]
- (II) K is closed under isomorphisms
- (III) the joint embedding property
- (IV) disjoint amalgamation in [Jn56]
- $(IV)'$ amalgamation in [Jn60]
- (V) $\cup\{M_\alpha : \alpha < \delta\} \in K$ if $M_\alpha \in K$ is $\subseteq$-increasing
- (VI) if $N \in K$ and $M \subseteq N$ (so $|M| \neq \emptyset$ but not necessarily $M \in K$) and $\alpha > 0$, $\|M\| < \aleph_\alpha$ then there is $M' \in K$ such that $M \subseteq M' \subseteq N$ and $\|M'\| < \aleph_\alpha$ (this is a strong form of the LS property).

Note that for an abstract elementary class $(K, \leq_{\mathfrak{K}})$, if $\leq_{\mathfrak{K}}=\subseteq\restriction K$, then AxIV (smoothness) and AxV (if $M_1 \subseteq M_2$ are $\leq_{\mathfrak{K}}$-submodels of N then $M_1 \leq_{\mathfrak{K}} M_2$) of I.1.2 or II.1.4 and part of AxI become trivial (hence are missing from Jonsson axioms), the others give II, and a weaker form of VI (specifically, for one $\aleph_\alpha$, i.e. $\aleph_\alpha = \mathrm{LS}(\mathfrak{K})^+$, the other cases are proved).

orists have preferred saturated on universal homogeneous and prefer first order classes (Morley-Vaught [MoVa62], Keisler replete) with very good reasons, as it is better (more transparent and give more) to deal with one element than a model. That is, assume our aim is to show that N from our class K is universal, i.e., we are given $M \in K$ of cardinality not larger than that of N and we have to construct an (appropriate) embedding of M into N. Naturally, we do it by approximations of cardinality smaller than $\|M\|$, the number of elements of M. Jonsson uses as approximations isomorphisms f from a submodel M' of M of cardinality $< \|M\|$. Morley and Vaught use functions from a subset A of M into N such that: if $n < \omega, a_0, \dots, a_{n-1} \in A$ satisfy a first order formula in M then their image satisfies it in N. So they have to add one element at each step which is better than dealing with a structure. In fact, also in this book, for a different notion of type, the types of elements continue to play a major role (<u>but</u> we use types which are not sets of formulas over models). So we try to have "the best of both approaches" - all is done over models from K, but we ask existence, etc., only of singletons, for this reason in the proof of the uniqueness of "saturated" models we have to go "outside" the two models, build a third (see V.B.3.18 or II.1.14).

Here we have chosen abstract elementary class as the main direction. This includes classes defined by $\psi \in \mathbb{L}_{\omega_1,\omega}$ and we can analyze models of $\psi \in \mathbb{L}_{\omega_1,\omega}(\mathbf{Q})$ in such context by a reduction. In [Sh 88] = Chapter I Baldwin's question was solved in ZFC. Also superlimit models were introduced and amalgamation in λ was proved assuming categoricity in λ and $1 \le \dot{I}(\lambda^+, \mathfrak{K}) < 2^{\lambda^+}$ when $2^\lambda < 2^{\lambda^+}$. The intention of the work was to prepare the ground for generalizing [Sh 87b]. Note that sections §4,§5 from Chapter I are harder than the parallel in [Sh 87a] because we deal with abstract elementary class (not just $\psi \in \mathbb{L}_{\omega_1,\omega}(\mathbf{Q})$).

Now [Sh 300] deals with universal classes. This family is incomparable with first order and [Sh 155] gives hope it will be easier. Note that in excellent classes the types are set of formulas and this is true even for Chapter I though the so-called materializing replaces realizing a type. In [Sh 300] (orbital)-type is defined by $\le_{\mathfrak{K}}$-mapping. Surprisingly we can still show "λ-universal homogeneous" is equivalent to λ-saturated under the reasonable interpretations (so have to find an element rather than a copy of a model) what was a strong argument for sequence homogenous models (rather than model homogeneous).

In [Sh 576], which is a prequel of the work here we generalize [Sh 88] to any abstract elementary class $\mathfrak{K}$ having no remnant of compactness, see on it below. On Chapter II, Chapter III see later.

I thank the institutions in which various parts of this book were presented and the student and non-students who heard and commented. Earlier versions of Chapter V.A, Chapter V.B, [Sh:e, III], Chapter V.C, Chapter V.D, Chapter V.E were presented in Rutgers in 1986; some other parts were represented some other time. In Helsinki 1990 a lecture was on the indiscernibility from Chapter V.F, Chapter V.G. First version of [Sh 576] was presented in seminars in the Hebrew University, Fall '94. The Gödel lecture in Madison Spring 1996 was on [Sh 576] and Chapter II. The author's lecture in the logic methodology and history of science, Kracow '99, was on Chapter II and Chapter III. In seminars at the Hebrew University, Chapter I was presented in Spring 2002, [Sh 576] was presented in 98/99, Chapter II + Chapter IV were presented in 99/00, Chapter II + Chapter III were presented in 01/02 and my lecture in the Helsinki 2003 ASL meeting was on good λ-frames and Chapter IV.

I thank John Baldwin, Emanuel Dror-Farajun, Wilfred Hodges, Gil Kalai, Adi Jarden, Alon Siton, Alex Usvyatsov, Andres Villaveces for many helpful comments and error detecting in the introduction (i.e. Chapter N).

Last, but not least, I thank Alice Leonhardt for beautifully typesetting the contents of this book.

§2 Introduction for the logically challenged

(This is recommended reading for logicians too, but there are some repetitions of part (A) of §1).

This is mainly an introduction to Chapter II, Chapter III as §1 and the introduction to the others can serve.

> We assume the reader knows the notion of an infinite cardinal <u>but not</u> that he knows about first order

> logic (and first order theories); for reading (most of) the book, not much more is needed, see §5.
>
> Paragraphs assuming more knowledge or are not so essential will be in indented, e.g. when a result is explained ignoring some qualifications and we comment on them in indented text.

(A) What are we after?

This introduction is intended for a general mathematical audience. We may view our aim in this book as developing a theory dealing with abstract classes of mathematical structures that will also be referred to as models. Examples of structures are the field $\mathbb{R}$, any group and any ring. The classes of models we consider are called "abstract elementary classes" or briefly a.e.c. An abstract elementary class $\mathfrak{K}$ is a class of structures denoted by K together with an order relation denoted by $\leq_{\mathfrak{K}}$ which distinguishes for each structure N a certain family $\{M \in K : M \leq_{\mathfrak{K}} N\}$ of substructures (= submodels).

First, rather than giving a formal definition, we will give several examples:

2.1 Examples:

(*i*) the class of groups where the order relation is "being a subgroup".

In this example $\leq_{\mathfrak{K}}$ is simply being substructures. (In the sequel when we do not specify the order relation is means simply to take all substructures).

(*ii*) The class of algebraically closed fields with characteristic zero

(*iii*) the class of rings

(*iv*) the class of nill rings, i.e. ring R such that for every $x \in R$, $x^n = 0$ for some $n \geq 1$

(*v*) the class of torsion R-modules for a fix ring R

(*vi*) the class of R-modules for a fix ring R but unlike the previous cases the relation of $\leq_{\mathfrak{K}}$ is not just being a submodule, it is being a "pure submodule"[4]

---

[4] A left R-module M is a pure submodule of a left R-module N when if $rx = y, x \in N$ and $y \in M$ then $rx' = y$ for some $x' \in M$

(*vii*) the class of rings but $R_1 \leq_{\mathfrak{K}} R_2$ means here: R_1 is a subring of R_2 and if R'_2 is a finitely generated subring of R_2 then $R_1 \cap R'_2$ is a finitely generated subring of R_1

(*viii*) the class of partial orders.

Abstract elementary class form an extension of the notion of elementary class which mean a class of structures which are models of a so-called first order theory. The notion of abstract elementary classes, while more general, does not rely on elementary classes and indeed, for reading this introduction we do not assume knowledge of first order logic.

We will be mainly interested in this book in finding parallel to the "superstability theory" which is part of the "classification theory" (this is explained below; on the first order case see, e.g. [Sh:c], [Sh 200] or other books on the subject, e.g. Bladwin [Bal88]).

Superstability theory can be described as dealing with elementary classes of structures for which there is a good dimension theory; but see on our broader aim below.

A structure M will have a so-called vocabulary τ_M (this is its "kind", e.g. is it a ring or a group). Note that for each class $\mathfrak{K} = (K, \leq_{\mathfrak{K}})$ we shall consider, all $M \in K$ has the same vocabulary (sometimes called language), which we denote by $\tau = \tau_{\mathfrak{K}}$, e.g., for a class of fields it is $\{+, \times, 0, 1\}$ where $+, \times$ are binary functions symbols interpreted in each field as two-place functions and similarly $0, 1$ are individual constant symbols. We may have also relations, (in example (*viii*) the partial order is a relation), note that relation symbols are usually called predicates. The reader may restrict himself to the case of countable or even finite vocabulary with function symbols only. We certainly demand each function symbol to have finitely many places (and similarly for relation symbols).

We try now, probably prematurely, to give exact definitions of some basic notions toward what long term goal we would like to advance, probably it will make more sense after/if the reader continues to read the introduction. (But most of this will be repeated and expanded).

We think that the family of abstract elementary classes $\mathfrak{K}$ (defined in 2.2 below) can be divided, in some ways, so that we can say significant things both on the "low", simple side and on the "high, complicated" side. This sounds vague, can we already state a conjecture? It seems reasonable that a class K with a unique member (up to isomorphism, of course) in a cardinality λ is simple; but what can be the class of cardinals for which this holds? This class is called the "categoricity spectrum of the abstract elementary class $\mathfrak{K}$" (see Definitions 2.2, 2.3 below), we conjecture that is a simple set, e.g. contains every large enough cardinal <u>or</u> does not contain every large enough cardinal. Moreover, this also applies to the so-called superlimit spectrum of $\mathfrak{K}$ (see Definition 2.4). In the "low, simple" case we have, e.g. a dimension theory for $\mathfrak{K}$, and in the "high case" we can prove the class is complicated and so cannot have such a nice theory (this paragraph will be explained/expanded later).

Here we make some advances in this direction.

First, what exactly is an abstract elementary class? It is much easier to explain than the so-called "elementary classes" which is defined using (first order) logic. A major feature are closure under isormorphism and unions.

2.2 Definition. $\mathfrak{K} = (K, \leq_{\mathfrak{K}})$ is an abstract elementary class <u>when</u>

(A)(a) K is a class of structures all of the same "kind", i.e. vocabulary; e.g. they can be all rings or all graphs, τ denote a vocabulary

(b) K is closed under isomorphisms

(c) $\leq_{\mathfrak{K}}$ is a partial order of K, also closed under isomorphisms and $M \leq_{\mathfrak{K}} N$ implies $M \subseteq N, M$ a substructure of N and, of course, $M \in K \Rightarrow M \leq_{\mathfrak{K}} M$

(d) K (and $\leq_{\mathfrak{K}}$) are closed under direct limits, or, what is equivalent, by unions of $\leq_{\mathfrak{K}}$-increasing chains, i.e. if I is a linear order and $M_t (t \in I)$ is $\leq_{\mathfrak{K}}$-increasing with t then $M = \cup\{M_t : t \in I\}$ belongs to K and; morever, $t \in I \Rightarrow M_t \leq_{\mathfrak{K}} M$

(e) similarly to clause (d) inside $N \in K$, i.e., if $t \in I \Rightarrow M_t \leq_{\mathfrak{K}} N$ then $M \leq_{\mathfrak{K}} N$.

Two further demands are only slightly heavier

(B)(f) if[5] $M_\ell \leq_{\mathfrak{K}} N$ for $\ell = 1, 2$ and $M_1 \subseteq M_2$ then $M_1 \leq_{\mathfrak{K}} M_2$

[5]this certainly holds if $\leq_{\mathfrak{K}}$ is defined as $\prec_{\mathscr{L}(\tau(\mathfrak{K}))}$ for some logic $\mathscr{L}$

(g) $(K, \leq_{\mathfrak{K}})$ has countable character, which means that every structure can be approximated by countable ones; i.e., if $N \in K$ then every countable set of elements of N is included in some countable $M \leq_{\mathfrak{K}} N$ (in the book but not in the introduction we allow replacing "countable" by "of cardinality $\leq$ LS($\mathfrak{K}$)" for some fixed cardinality LS($\mathfrak{K}$)).

Not all natural classes are included, e.g. the class of Banach spaces is not, as completeness is not preserved by unions of increasing chains. Still it seems very broad and the question is can we prove something in such a general setting.

2.3 Definition. 1) $\mathfrak{K}$ (or K) is categorical in λ when it has one and only one model of cardinality λ up to isomorphism.
2) The categoricity spectrum of $\mathfrak{K}$, cat($\mathfrak{K}$), is the class of cardinals λ in which $\mathfrak{K}$ is categorical.

A central notion in model theory is elementary classes or first order classes which are defined using so called first order logic (which the general reader is not required here to know, it is explained in the indented text below).

Each such class is the class of models of a first order theory with the partial order $\prec$.

Among elementary classes, a major division is between the so-called superstable ones and the non-superstable ones, and for each superstable one there is a dimension theory (in the sense of the dimension of a vector space). Our long term aim in restricted terms is to find such good divisions for abstract elementary classes, though we do not like to dwell on this further now, it seems user-unfriendly not to define them at all, so for the time being noting that for elementary classes being superstable is equivalent to having a superlimit model in every large enough cardinality; also noting that superstability for abstract elementary classes suffer from schizophrenia, i.e. there are several different definitions which are equivalent for elementary classes, the one below is one of them.

2.4 Definition. Let $\mathfrak{K}$ be an abstract elementary class.
1) We say f is a $\leq_{\mathfrak{K}}$-embedding of M into N when f is an isomorphism of M onto some $M' \leq_{\mathfrak{K}} N$.
2) $\mathfrak{K}_\lambda = (K_\lambda, \leq_{\mathfrak{K}_\lambda})$ where $K_\lambda = \{M \in \mathfrak{K} : \|M\| = \lambda\}$ and $\leq_{\mathfrak{K}_\lambda} = \leq_{\mathfrak{K}} \restriction K_\lambda$.
3) An abstract elementary class $\mathfrak{K}$ is superstable <u>iff</u> for every large enough λ, there is a superlimit structure M for $\mathfrak{K}$ of cardinality λ; where
4) We say that M is a superlimit (for $\mathfrak{K}$) when for some (unique) λ

(a) $M \in \mathfrak{K}$ has cardinality λ

(b) M is $\leq_{\mathfrak{K}}$-universal, i.e., if $M' \in K_\lambda$ then there is a $\leq_{\mathfrak{K}}$-embedding of M' into M, in fact with range $\neq M$

(c) for any $\leq_{\mathfrak{K}}$-increasing chain of models isomorphic to M with union of cardinality λ, the union is isomorphic to M.

5) The superlimit spectrum of $\mathfrak{K}$ is the class of λ such that there is a superlimit model for $\mathfrak{K}$ of cardinality λ.

We shall return to those notions later.

What about the examples listed above? Concerning the strict definition of elementary classes as classes of the form $(\mathrm{Mod}_T, \prec)$ defined below, among the examples in 2.1 the class of algebraically closed fields (example (ii)) is an elementary class since it can be proved that being a sub-field is equivalent to being an elementary substructure for such fields.

In the example (i), the class of models is elementary, i.e., equal to Mod_T: the class of groups, but the order is not $\prec$ but $\subseteq$. This is true also in the examples (iii), rings and $(viii)$, partial orders.
In the example (vi), the class of torsion R-modules is not a first order class as we have to say $(\forall x) \bigvee\limits_{r \in R \setminus \{0\}} rx = 0$ and we really need to use an infinite disjunction. The situation is similar for the class of nill ring (example (iv)). In example (vii), the class of rings with $\leq_{\mathfrak{K}}$ defined using finitely generated subrings not only is the class of structures not elementary by $\leq_{\mathfrak{K}}$ is neither $\prec$ nor $\subseteq$. In the example (vii), R-modules, K is elementary but $\leq_{\mathfrak{K}}$ is different.

Recall[6] the traditional frame of model theory are the so-called elementary (or first order) classes. That is, for some vocabulary τ, and set T of so-called sentences in first order logic in this vocabulary, $K = \mathrm{Mod}_T = \{M : M \text{ a } \tau\text{-structure satisfying every sentence of } T\}$ and $\leq_{\mathfrak{K}}$ being $\prec$, "elementary submodel". Recall that $M \prec N$ if $M \subseteq N$ and for every first order formula $\varphi(x_0, \ldots, x_{n-1})$ in the (common) vocabulary, i.e., from the language $\mathbb{L}(\tau)$ and $a_0, \ldots, a_{n-1} \in M$, $\varphi(a_0, \ldots, a_{n-1})$ is satisfied by M, (symbolically $M \models \varphi[a_0, \ldots, a_{n-1}]$) <u>iff</u> N satisfies this.

Now here an elementary class is one of the form $(\mathrm{Mod}_T, \prec)$, any such class is an abstract elementary class (see below). A different abstract elementary class derived from T is $(\mathrm{Mod}_T, \subseteq)$ but then we should restrict ourselves to T being a set of universal sentences or just Π_2-sentences as we like to have closure under direct limits. For each such T another abstract elementary class which can be derived from it is $(\{M \in \mathrm{Mod}_T : M \text{ is existentially closed}\}, \subseteq)$.

We are <u>not disputing</u> the choice of first order classes as <u>central</u> in model theory but there are many interesting other classes. Most notably for algebraists are classes of locally finite structures and for model theorists are $(\mathrm{Mod}_\psi, \prec_{\mathscr{L}})$ where ψ belongs to the logic denoted by $\mathbb{L}_{\omega_1,\omega}(\tau)$ or just $\psi \in \mathbb{L}_{\lambda^+,\omega}(\tau)$ for some λ where $\mathscr{L}$ is a fragment of this logic to which the sentence ψ belongs; if $\psi \in \mathbb{L}_{\omega_1,\omega}(\tau)$ we may choose a countable such $\mathscr{L}$.

(This logic may seem obscure to non-logicians but it just means that we allow to say $\bigwedge\limits_{i \in I} \varphi_i(x_0, \ldots, x_{n-1})$ where I has at most λ members so enable us to say "a ring is nill, locally finite, etc.", <u>but</u> not "$<$ is a well ordering").

In some sense if we look at classification theory of

[6]we urge the logically challenged: when lost, jump ahead

elementary classes as a building, we note that several "first floors" disappear (in the context of abstract elementary class) but we aim at saving considerable part of the rest (of course not all) by developing a replacement for those lower floors.

We may put in the basement the downward LS theorems (there are small $N \prec M$), it survived. But not so the compactness theorem, even very weak forms like "if $\bar{a} = \langle a_n : n \in \mathbb{N} \rangle, \bar{b} = \langle b_n : n \in \mathbb{N} \rangle$ are sequences of members of M and f_n is an automorphism of M mapping $\bar{a} \restriction n$ to $\bar{b} \restriction n$ then some $\leq_{\mathfrak{K}}$-extension of M has an automorphism mapping $\bar{a}$ to $\bar{b}$". (Note that for "(D, λ)-homogeneous models" (e.g. [Sh 3]) such forms of compactness hold and the point of [Sh 394] is to start investigating classes for which all is nice except that types are not determined by their small restrictions, that is, defining $\mathbb{E}^{\kappa}_{N} = \{(p, q) : p, q \in \mathscr{S}(N)$ and $M \in K_{\kappa} \Rightarrow p \restriction M = q \restriction M\}$, this is, a priori, not the equality ([Sh 394, 1.8,1.9,pg.4]). We lose as well the upward LS theorem (a model have a proper $<_{\mathfrak{K}}$-extension); (those fit the first floor).

Also in abstract elementary classes the roles of formulas disappear. Hence we lose the notion of the type of an element a over a set A inside a model M; so goes the second floor including the "κ-saturated model" (in the traditional sense) down the drain as the types disappear.

What is saved? (I.e. not by definitions but in the positive case of a dividing line which has a non-structure result.) In a suitable sense non-forking amalgamation of models, prime models, a decomposition of a model over a non-forking tree of models (a relative of free amalgamation), and for a different notion of type, being (saturated and) orthogonal, regular and eventually the main gap for the parallel of $\aleph_\varepsilon$-saturated model of a superstable T.

We now try to describe our aim in broad terms; if this seems vague, in (B) below we describe it in a restricted case more concretely. Our aim is to consider a family of classes $\mathfrak{K}$ (all the "reasonable" classes) and try to classify them in the sense of taxonomy, we look for dividing lines among them. This means dividing the family to two, one part are those which are "high", "complicated". Typically we have for each $\mathfrak{K}$ in the "high side" a non-structure result, saying there are many complicated such models $M \in K$ (in suitable sense). Those in the other side, the "low" one have some "positive" theory, we have to some extent understood those models, e.g. they have a good dimension theory.

A reader interested to see more quickly what is done rather than why it is done and what are our hopes should go to (C) below.

A good dividing line of a family of classes is such that we really can say something on both sides, with some being complementary; ideally it also should help us prove things on all K's by division to cases. So it seems advisable to prove the equivalence of an external property (like not having many models) and an internal property (some understanding of models of K). Now clearly such a dividing line is interesting but, of course, there are properties which are interesting for other reasons (see more on this in the end of (A) of §1).

(B) The structure/non-structure dichotomy

More specifically we may ask: which classes have a structure theory? By a structure theory we mean "determined up to isomorphism by an invariant called the dimension or several dimensions or something like that". A non-structure property (or theorem) will be a strong witness that there is no structure theory. So the question is:

2.5 Question: When does a class $\mathfrak{K}$ of models have a structure theory? In particular, each model from $\mathfrak{K}$ is characterized up to isomorphism by a "complete set of reasonable invariants" like those of Steinitz (for algebraically closed fields) and Ulm (for countable torsion abelian groups).

This is still quite vague, and it takes some explanation (and choices) to make it concrete. Instead we shall be even more specific. We shall explain two more concrete questions: categoricity and the main gap and the solution in the known (first order countable vocabulary) case.

Counting the number of models in a class seems very natural and to make sense we have to count them in each cardinality separately. If the reader is not enthusiastic about this counting, some alternative questions lead us to the same place: e.g.: having models which are almost isomorphic but not really isomorphic (see more in $(*)_2$ from §1(B)(c)).

2.6 Definition. For a class K of models and infinite cardinal λ let $\dot{I}(\lambda, K)$ be the number of models in K of cardinality λ up to isomorphism. So for any K it is a function from Card, the class of cardinals to itself; we may write $\mathfrak{K} = (K, \leq_{\mathfrak{K}})$ instead of K.

Now a priori we may get quite arbitrary functions. But it seems reasonable to hope that all our classes $\mathfrak{K}$ will have a simple function $\lambda \mapsto \dot{I}(\lambda, \mathfrak{K})$ and classes with a "structure theory" will have such functions with small values. It seems more hopeful to try to first investigate the most extreme cases (being one and being maximal), considering both our chances to solve and for getting an interesting answer; also we expect the "upper" one to give the important dividing lines. It is most natural to start asking above the spectrum of existence, i.e., being non-zero, i.e., what can be $\{\lambda : \mathfrak{K}_\lambda \neq \emptyset\}$? This had been answered quite completely (see I.1.11,I.1.13), and it seems easier at least from the present perspective.

Considering this, the number one naturally has a place of honor; this is categoricity. Recall K is said to be categorical in λ iff $\dot{I}(\lambda, K) =$ 1.

A natural thesis is

<u>2.7 Thesis</u>: If we really understand when a (reasonable) class is categorical in λ it should have little dependence on λ, ignoring "few, exceptional" cardinals.

[Why? How can we understand why $\mathfrak{K}$ is categorical in λ? We should know so much on the class so that given two models from K of cardinality λ we can construct in a coherent way an isomorphism from one onto the other; but this should work for any other (large enough) cardinal. Also being categorical implies the model is a very simple one, analyzable.

This is, of course, not true for every class of, e.g. if K is the class of $\{(I, <)$: is $<$ well order I, such that if $|I|$ is a successor cardinal then every initial segment has cardinality $< |I|\}$. This class is categorical in $\aleph_\alpha$ iff $\aleph_\alpha$ is a limit cardinal (we could change it to "α even", etc). However, we have to restrict ourselves to "reasonable" classes.]

An antagonist argument against the thesis 2.7 is that for first order T, the class $\{\lambda : T$ has in λ a rigid model, i.e., one without (non-trivial) automorphism$\}$, e.g. can be "any class of cardinals" in some sense, e.g., $\{\aleph_3, \aleph_{762}, \beth_{\omega_3},$ first inaccessibly cardinality$\}$. Essentially any Σ^1_2 class of cardinal (see [Sh 56]).

We may answer that rigidity implies a complicated model so we may have T coding a definition of a complicated class, of cardinals, whereas being categorical implies the models are simple. The antagonist may answer that allowing enough classes of models it would not work, the categoricity spectrum will be weird and probably Łos (see below) has no good enough reasons for his conjecture (of course we can argue till the problem is resolved). We may answer that Łos conjecture implicitly says that first order classes (of countable vocabulary) are "nice", "analyzable". So 2.7 beg the question which classes are reasonable and this book contend that abstract elementary classes are.

Of course, there may be reasonable classes for which "$\mathfrak{K}$ is categorical" depend on simple properties of the cardinal (e.g. being strong limit).

More specifically we may ask: is it true for every (relelvant) $\mathfrak{K}$, either $\mathfrak{K}$ is categorical in almost every λ or non-categorical in almost every λ? Indeed Łos had conjectured that if an elementary class $\mathfrak{K}$ with countable vocabulary is categorical in one $\lambda > \aleph_0$ then $\mathfrak{K}$ is categorical in every $\lambda > \aleph_0$, having in mind the example of algebraically closed fields of a fixed characteristic. A milestone in mathematical logic history was Morley's proof of this conjecture. The solution forces you to understand such $\mathfrak{K}$.

We may ask: Is $\dot{I}(\lambda, \mathfrak{K})$ a non-decreasing function? Of course, this is a question on K but the assumptions are on $\mathfrak{K} = (K, \leq_{\mathfrak{K}})$. This sounds very reasonable as "having more space we have more possibilities". For elementary $\mathfrak{K}$ with countable vocabulary this was conjectured by Morley (for $\lambda > \aleph_0$). It is not clear how to prove it directly so it seemed to me a reasonable strategy is to find some relevant dividing lines: the complicated classes will have the maximal number of models, the less-complicated ones can be investigated as we understand them better. This may lead us to look at the dual to categoricity, the other extreme - when $\dot{I}(\lambda, T)$ is maximal (or just very large).

2.8 Definition. The main gap conjecture for K says that either $\dot{I}(\lambda, K)$ is maximal (or at least large) for almost all λ or the number is much smaller for almost all λ; for definiteness we choose to interpret "almost all λ" as for every λ large enough.

(We cheat a little: see 2.10).

This seems to me preferable to "$\dot{I}(\lambda, K)$ is non-decreasing" being more robust; this will be even more convincing if we succeed in proving the stronger statement:

2.9 The structure/non-structure Thesis For every reasonable class either its models have a complete set of cardinal invariants or its models are too complicated to have such invariants.

This had been accomplished for elementary classes (= first order theories) with countable vocabularies. We suggest that the main gap problem is closely connected to 2.9.

So ideally, for classes $\mathfrak{K}$ with structure for every model M of $\mathfrak{K}$ we should be able to find a set of invariants which is complete, i.e., determines M up to isomorphism. Such an invariant is the isomorphism type, so we should restrict ourselves to more reasonable ones, and the natural candidates are cardinal invariants or reasonable generalizations of them. E.g. for a vector space over $\mathbb{Q}$ we need one cardinal (the dimension = the cardinality of any basis). For a vector space over an algebraically closed field, two cardinals; (the dimension of the vector space and the transcedence degree (= maximal number of algebraically independent elements) of the field, both can be any cardinal; of course, we have also to say what the characteristic of the field is). For a divisible abelian group G, countably many cardinals (the dimension of $\{x \in G : px = 0\}$ for each prime p and the rank of $G/\mathrm{Tor}(G)$ where $\mathrm{Tor}(G)$ is the subgroup consisting of the torsion members of G, i.e. $\{x \in G : nx = 0$ for some $n > 0\}$). For a structure with countably many one-place relations P_n (i.e., distinguished subsets), we need $2^{\aleph_0}$ cardinals (the cardinality of each intersection of the form $\cap\{P_n^M : n \in u\} \cap \{M \backslash P_n^M : n \notin u\}$) for u a set of natural numbers).

We believe the reader will agree that every structure of the form $(|M|, E)$, where E is an equivalence relation, has a reasonably complete set of invariants: namely, the function saying, for each cardinal λ, how many equivalence classes of this cardinality occur. Also, if we enrich M by additional relations which relate only E-equivalent members and such that each E-equivalence class becomes a structure with a complete set of invariants, then the resulting model will have a complete set of invariants. We know that even if we allow such generalized cardinal invariants, we cannot have such a structure theory for every relevant class (e.g. the class of linear orders has no such cardinal invariants). So if we have a real dichotomy as we hope for, we should have a solution of (a case of) the main gap conjecture which says each class K either has such invariant or is provably more complicated.

Let us try to explicate this matter. We define what is a λ-value of depth α by induction on the ordinal α: for $\alpha = 0$ it is a cardinal $\leq \lambda$, for $\alpha = \beta + 1$ it is a sequence of length $\leq 2^{\aleph_0}$ of functions from the set of λ-values of depth β to the set of cardinals $\leq \lambda$ or a λ-value of depth β, and for α a limit ordinal it is a λ-value of some depth $< \alpha$.

An invariant [of depth α] for models of T is a function giving, for every model M of T of cardinality λ, some λ-value [of depth α] which depends only on the isomorphism type of M. If we do not restrict α, the set of possible values of the invariants is known, in some sense, to be as complicated as the set of all models.

This leads to:

2.10 Main Gap Thesis: 1) A class K has a structure theory if there are an ordinal α and invariants (or sets of invariants) of depth α which determines every structure (from K) up to isomorphism.

2) If K fails to have a structure theory it should have

"many" models and we expect to have reasonably definable such invariants.

We can prove easily, by induction on the ordinal α, that

2.11 Observation. The number of $\aleph_\gamma$-values of depth α has a bound $\beth_\alpha(|\tau_K| + |\gamma|)$ where

$$\beth_\beta(\mu) = \mu + \prod_{\varepsilon<\beta} 2^{\beth_\varepsilon(\mu)}.$$

2.12 Corollary of the thesis. *If $\mathfrak{K}$ has a structure theory by the interpretation of 2.10 then there is an ordinal α such that for every ordinal γ, $\mathfrak{K}$ has $\leq \beth_\alpha(|\tau_{\mathfrak{K}}| + |\gamma|)$ non-isomorphic models of cardinality $\aleph_\gamma$.*

It is easy to show, assuming e.g., the G.C.H., that for every α there are many γ's such that $\beth_\alpha(|\omega+\gamma|) < 2^{\aleph_\gamma}$ and even $< \aleph_\gamma$. Thus, if one is able to show that $\mathfrak{K}$ has $2^{\aleph_\gamma}$ models of cardinality $\aleph_\gamma$, this establishes non-structure.

In the case in which the main gap was proved, it turns out that there are only few "reasons" for an elementary class $\mathfrak{K}$ with countable vocabulary to have the maximal number of models:

(*a*) $\mathfrak{K}$ is so called unstable, prototypical example are the class of infinite linear orders and the class of random graphs [formally: in some model from $\mathfrak{K}$ some first order formula $\varphi(\bar{x}, \bar{y})$ with $\ell g(\bar{x}) = m = \ell(\bar{y})$ for every linear order I there is $M \in \mathfrak{K}$ and an m-tuple $\bar{a}_t$ from M for each $t \in I$ such that $\varphi[\bar{a}_s, \bar{a}_t]$ is satisfied in M iff $s <_I t$]

(*b*) $\mathfrak{K}$ has the so called OTOP, it is similar to (a), but the order is defined in a different way, not by a so-called first order formula but by a formula of the form $(\exists \bar{z}) \bigwedge_n \varphi_n(\bar{x}, \bar{y}, \bar{r})$. The prototypical example is straightforward but somewhat cumbersome

(c) it has the DOP, this is harder to define and even to give example too. It means that in some members M of $\mathfrak{K}$, we can define large linear orders by using dimensions

proto-typical example is: for some infinite I and $R \subseteq I \times I, M_{I,R}$ has universe $I \cup \{(s,t,\alpha) : s \in I, t \in I, \alpha < \omega_1$ and $(s,t) \in R \Rightarrow \alpha < \omega\}$ and relation $P^M = \{(s,t,a) : a = (s,t,\alpha)$ for some $\alpha\}$. So R can be defined in $M_{I,R}$ (though is not a relation of M) as $\{(s,t)$: the set $\{x : M_{I,R} \models P(s,t,x)\}$ is uncountable$\}$. But the definition is not first order, it speaks on dimension (actually we can also interpret any graphs). Note that $T = \mathrm{Th}(M_{I,R})$ does not depend on R.

(d) $\mathfrak{K}$ is so called unsuperstable; proto-typical example $({}^{\omega}I, E_n)_{n<\omega}$ where ${}^{\omega}I$ is the set of functions from $\mathbb{N}$ into I and $E_n = \{(\eta,\nu) : \eta,\nu \in {}^{\omega}I$ and $\eta \restriction n = \nu \restriction n\}$

(e) T is deep, proto-typical example is the class of graphs which are trees (i.e. with no cycles).

We return to the more concrete question: the main gap and the thesis 2.9. We can hope that a non-structure theorem should imply $\dot{I}(\lambda, K)$ is large, whereas a structure theorem should enable us to show it is small and even allow us to show it is non-decreasing, and to compute it.

Actually the picture of the "non-structure" side (in the resolved case) is more complicated. In some classes "reasons" (a)-(d) fail but "reason" (e) holds, in this case the members of $\mathfrak{K}$ are essentially as complicated as graphs which are trees (i.e., no cycle); for them we get the maximal number of non-isomorphic models, but we have a "handle" on understanding the models. So, e.g., a result proving this is the following: possibly[7] for

[7] formally: if some (mild) large cardinal exists

some λ we cannot find λ models no one embeddable into the others. For the rest there are stronger results in the inverse direction (e.g. we can code stationary sets modulo the club filter). So it seemed that we end up with a trichotomy rather than a dichotomy. That is, for the question of counting the number of models up to isomorphism the middle family behaves more like the high one: has maximal number. But for the question mentioned above and also for questions of the form: "are there two very similar non-isomorphic models in the class" the middle family behaves like the low (e.g. we can build reasonable invariants when not restricting the ordinal depth). Still there are clear results for each of the three families.

It was (and is) our belief that there is such a theory even for abstract elementary classes and that we should look at what occurs at large enough cardinals, as in small cardinals various "incidental" facts interfere. Notice that a priori there need not be a solution to the structure/non-structure problem or to the spectrum of categoricity problem: maybe $\dot{I}(\lambda, T)$ can be any one of a family of complicated functions, or, worse, maybe we cannot characterize reasonably those functions, or, maybe the question of which functions occur is independent of the usual axioms of set theory.

Now, of course, the aim of classification is not just those specific questions. We rather think and hope that trying to solve them will on the way give interesting dividing lines among the classes. A class K here may have too many models but still we can say much on the structure of its models.

Now the thesies underlining the above is

2.13 Thesis

(a) dividing lines are interesting, and obviously reasonable test questions are a good way to find them (and we try to use test questions of self-interest)

(b) good dividing lines throw light also on questions which seem very different from the original test questions (*c*) in particular, investigating $\dot{I}(\lambda, K)$ (and more profoundly, characterizing the classes with complete set of invariants) is a good way to find interesting dividing lines, but naturally there are other ways to arrive at them and

(*d*) there are measures of complexity of a class (other than $\dot{I}(\lambda, K)$) which lead to interesting dividing lines and some such work was done on elementary classes (see §1).

Behind the discussion above also stands
2.14 Thesis: To investigate classes K it is illuminating to look for each λ, at problems on $\mathfrak{K}_\lambda, \mathfrak{K}$ which is restricted to cardinal λ and

(*a*) to try to prove that the answer does not depend on λ or at least depends just on a small amount of information on λ

(*b*) to discard too small cardinals (essentially to look at asymptotic behaviour)

This seems to be successful in discovering stability (and superstability).

> An illustration is that Rowbottom had defined λ-stable (i.e. $A \subseteq M \wedge |A| = \lambda \Rightarrow |\mathscr{S}(A, M)| \leq \lambda$) but it seems to me only having ([Sh 1]) the characterization of $\{\lambda : T \text{ stable in } \lambda\}$ and the equivalence with the order property and defining "T stable" started stability theory. (Of course, for his aims this was irrelevant).

The rationale is that if the answer is the same for "most λ", this points to a profound property of the class and it forces you to find inherent principles which you may not be so directly led to otherwise. Hence it probably will be interesting even if you care little about these cardinals. A parallel may be that even low dimension algebraic topologists were interested in the solution of Poincare conjecture for dimension ≥ 5. Also the behaviour in too small cardinals may be "incidental". So the class of dense linear order with neither first nor last element and the class of atomless Boolean Algebra or the class of random enough graphs are categorical in $\aleph_0$, but have many complicated models in higher ones. (One may feel these are low theories. This is true by some other criterions, other test problems; in fact, there are dividing lines among the elementary classes for which they are low. Still, for the test questions considered here, provably those classes are complicated, e.g., in a strong sense do not have a set of cardinal invariants characterizing the isomorphism type).

You may wonder:

2.15 Question: Do we recommend dividing lines everywhere? (in mathematics) or is this something special for model theory?

Now dividing lines are meaningful in many circumstances. But on the one hand it is better to list all simple finite groups than to find a dividing line among them. Similarly for the elementary classes categorical in every $\lambda \geq \aleph_0$. On the other hand, surely for many directions there are no fruitful dividing lines. The thesis that appeared here means that for broad front in model theory this is fruitful. (Not everywhere: too strong infinitary logics are out). It seemed that this has been vindicated for stability (and to some extent for simplicity and hopefully for (the family of) dependent elementary classes).

It may be helpful to compare this to alternative approaches in model theory. One extreme position will say that there is a central core in mathematics (built around classical analysis and geometry; and number theory of course) and other areas have to justify themselves by contributing something to this central core. Dealing with cardinals is pointless bad taste, and while some interaction of elementary classes with cardinals had been helpful, its time has passed.

It seemed to me that the criterion and its application leave out worthwhile directions. We all know that some neighboring subjects are just hollow noise and sometimes we are even right. So an excellent witness for a mathematical theory to be worthwhile is its ability to solve problems from others, preferably classical areas or problem from other sciences. Certainly a sufficient condition. What is doubtful is whether it is a necessary condition; we do not agree.

However, even within this narrow criterion, the direct attack is not the only way to look for applications to other areas. Not so seldom do we find that only after developing strong enough theory, deep applications become possible, the history of model theory seems to support this (in particular, lately in works of Hrushovski and Zilber). Looking at large enough cardinals serve as asymptotic behaviour, in which it is more transparent what are the general outlines of the picture.

The reader may wonder how this work is related, e.g. to category theory? universal algebra? soft model theory?. For category theory this work, in short, is closer than classical model theory but still not really close, similarly in category theory each class $\mathfrak{K}$ is equipped with a notion of mapping (rather than $\leq_{\mathfrak{K}}$ being defined from K by some specific logic as in classical model theory). But here we restrict ourselves to embeddings (this is not unavoidable but things are already hard enough without this) and the main difference is that we do not forget the elements.

What about universal algebra? A traditional model[8] theorist definition of model theory is universal algebra logic, so a large part of this work is, by that definition, in universal algebra. I do not see any reason to disagree but still the methods and results are well rooted in the model theoretic tradition.

What about soft model theory? Though our work itself does not need soft model theory, it fits well there (and Chapter I, Chapter IV use infinitary logics hence are not discussed in this part).

First, for many important logics $\mathscr{L}$, for theories $T \subseteq \mathscr{L}(\tau)$ the class $(\mathrm{Mod}_T, \prec_{\mathscr{L}(\tau)})$ or variants are abstract elementary classes (certainly for the logic $\mathbb{L}_{\lambda^+,\omega}$) and by choosing the $\leq_{\mathfrak{K}}$ appropriately also $\mathbb{L}(\mathbf{Q}^{\mathrm{card}}_{\geq\lambda})$; in fact they were the original motivation to look at abstract elementary classes. So if you ask for the part of soft model theory dealing with classification theory or at least investigate categoricity, you arrive here. Also not just varying the logic, but fixing a class Mod_T fits it well.

This work certainly reflects the author's preference to find something in the white part of our map, the "terra incognita" rather than understand perfectly what we have reasonably understood to begin with (which is exemplified by looking at abstract elementary

[8]but no universal algebraist agree

classes on which our maps reflect our having little to say on them, rather than FMR theories or o-minimal theories, cases where we had considerable knowledge and would like to complete it). Anyhow, by experience, there will not be many complaints on lack of generality and broadness.

Note that we would like to get results, not consistency results and allowing definability of well ordering or completeness runs into set-theoretic independence results so restricting ourselves to an abtract elementary class, a framework which excludes well ordering and complete spaces is reasonable. But we shall not really object to cardinal arithmetic assumptions like weak forms of GCH.

In fact, having the non-structure results depend on the universe of set theories is not desirable but is reasonable, as they still witness the impossibility of a positive theory. It is reasonable to adopt this as part of the rules of the games. In some cases, consistency results forbid us to go further (see, e.g. [Sh:93]). But still the positive side should better be in ZFC.

(C) Abstract elementary classes

We now return to the question: With which classes of structures we shall deal? Obviously, "a class of structures" is too general. Getting down to business we concentrate on

⊠ (a) abstract elementary classes

(b) good λ-frames

(c) beautiful λ-frames.

In short, in ⊠(a), see below, we suggest abstract elementary classes (a.e.c.) as our framework, i.e., the family of classes we try to classify; it clearly covers much ground and seems, at least to me, very natural. What needs justification is whether we can say on it interesting things, have non-trivial theorems.

Among elementary (= first order) classes we know which classes have reasonable dimension theory, the so called superstable elementary classes; and we like to understand the case for the family of abstract elementary class . In $\boxtimes(c)$, see below in §3(C), we suggest beautiful λ-frames as our "promised land", as a context where we have reasonable understanding, e.g., have dimension theory, can prove the main gap, etc. (but of course more wide families "on our way" probably will be interesting per se). Now it is very unsurprising that if we assume enough axioms, we shall regain paradise (which means here quite full fledge analog to the so called superstability theory, at least for my taste). Hence the problem in justifying the choice in $\oplus(c)$ is mainly not in pointing to many good properties but have to show that there are enough such frames and/or that it helps prove theorems not mentioning it. On the second, see e.g. 2.20 below. In our context ideally the first means to show that they are the only ones, i.e., the broadest family of abstract elementary class which has so good dimension theory. We are far from this, still we would according to our "guidelines" like at least to get beautiful frames by choosing to consider the classes which fall on the "low" side (in the elementary classes case) by dividing lines (= dichotomies) inside a family of classes which is large and natural, here among abstract elementary classes. That is, the program is to suggest some dividing lines, for the high side to prove the so-called non-structure theorems and for the low side to have some theory. Being always in the low sides we should arrive to beautiful frames.

But most of our work falls under $\boxtimes(b)$, good λ-frames. So it needs double justification: on the one hand we have to show it arises naturally from our program.
[In details, a weak case for "arising naturally" is to start with an abstract elementary classes satisfying some external condition of being "low" like categoricity, and prove that "inside $\mathfrak{K}$" we can find good frames. A strong case is to find a dividing line such that for each low $\mathfrak{K}$ we can find inside it "enough" good frames, and for all other "few". There is another meaning of "arising naturally" which would mean that we have looked at some natural examples and extracted the definition from their common properties; this is not what we mean. We rather try to solve questions on the number of models but of course the first order case was before our eyes as first approximation to the paradise we would like to arrive to.]

On the other hand for such frames, possibly with more assumptions justified similarly we can say something significant.

> In fact, we see good λ-frames essentially as the rock-bottom analogs of the family of elementary classes called superstable mentioned above.

We shall discuss $\boxtimes(a)$ and (b) and (c) in more detail. We start with

$\boxtimes(a)$ abstract elementary classes.

Recall the definition of abstract elementary classes Definition 2.2.

2.16 Explanation: An abstract elementary class is easy to explain (probably much simpler than elementary (= first order) class). Such $\mathfrak{K}$ consists of a class K of structures = models, all of the same "kind", e.g. all rings have the same kind, but a group has a different kind. We express this by saying "all members of K has the same vocabulary $\tau = \tau_K$". E.g., K consists of objects of the form $M = (A^M, F_0^M, F_1^M, Q^M), A^M$ its universe, a non-empty set, F_ℓ^M a binary function on it, Q^M a binary relation. $\mathfrak{K}$ has also an order $\le_{\mathfrak{K}}$ on K, its notion of being a sub-structure (which refines the standard notion). Now $(K, \le_{\mathfrak{K}})$ have to satisfy some requirements: preservation under isomorphisms, $\le_{\mathfrak{K}}$ being an order, preserved by direct limits and also direct limits inside $N \in K$, remembering that our mapping are embedding. Also if $M_1 \subseteq M_2$ are both $\le_{\mathfrak{K}}$-sub-structures of N then $M_1 \le_{\mathfrak{K}} M_2$, and lastly we demand every $M \in K$ has a countable $\le_{\mathfrak{K}}$-sub-structure including any pregiven countable set of elements (or replace countable by a fix cardinality, we ignore this point in the introduction; see II§1).

> Concerning "$M_\ell \le_{\mathfrak{K}} N, (\ell = 1,2), M_1 \subseteq M_2 \Rightarrow M_1 \le_{\mathfrak{K}} M_2$" note that if we define $\le_{\mathfrak{K}}$ as $\prec_{\mathscr{L}}$ for any logic, this will hold.

For elementary classes $\mathfrak{K}$, because of the so-called compactness and Löwenheim-Skolem theorems, the situation in all cardinals is to a significant extent similar.

In particular, if $\mathfrak{K}$ is an elementary class (with countable vocabulary) and λ_1, λ_2 are (infinite) cardinals then there is $M \in \mathfrak{K}$ of cardinality λ_1 iff there is $M \in \mathfrak{K}$ of cardinality λ_2. So recalling that $K_\lambda = \{M \in K : M$ has cardinality $\lambda\}$ and $\mathfrak{K}_\lambda = (K_\lambda, \leq_{\mathfrak{K}} \restriction K_\lambda)$ we have $K_{\lambda_1} \neq \emptyset \Leftrightarrow K_{\lambda_2} \neq \emptyset$. Moreover, any infinite $M \in \mathfrak{K}$ has $\leq_{\mathfrak{K}}$-extension in every larger cardinality. But for abstract elementary classes it is not necessarily true, and even if $(\forall\lambda)K_\lambda \neq \emptyset$ there may be many $\leq_{\mathfrak{K}}$-maximal models, i.e., $M \in K$ such that $M \leq_{\mathfrak{K}} N \Rightarrow M = N$. This (and more) makes the theory very different.

The context of abstract elementary class may seem so general, we may doubt if anything interesting can be said about it; still note that this context does not allow the class of Banach spaces as the union of an increasing chain is not necessarily complete. Certainly a loss. Also the class $(W, \subseteq)$, the class of well orders, is not an abstract elementary class ; (recall I is a well order if it is a linear order such that every non-empty set has a first element). Similarly the class $(K^{\rm fgi}, \subseteq)$ where $K^{\rm fgi}$ = the class of rings (or even integral domains) in which every ideal is finitely generated, is not an abstract elementary class (where $\leq_{\mathfrak{K}}$ is being a subring). However, we get an abstract elementary class when we consider only $K_{\leq n}$ = the class of rings in which every ideal is generated by $\leq n$ elements.

> We may like to replace n by a countable ordinal α, i.e., $K^{\rm fgi}_{\leq\alpha} = \{M \in K : {\rm dp}_M(\emptyset) \leq \alpha\}$; where for a ring M we define dp:$\{u : u \subseteq M$ finite$\} \rightarrow$ the ordinals by ${\rm dp}_M(u) = \cup\{{\rm dp}(w) + 1 : u \subseteq w$ and w is not included in the ideal of M which u generates$\}$. But then we have problems with closure under unions; a reasonable remedy is to have an appropriate $\leq_{\mathfrak{K}}$: $M \leq_{\mathfrak{K}} N$ if M, N are rings and for every finite $u \subseteq M$ we have ${\rm dp}_N(u) = {\rm dp}_M(u)$.
> Why have we restricted ourselves to "countable α"? Only because in clause (g) of Definition 2.2 we have used "countable".

But the family of abstract elementary classes includes all the examples listed in 2.1 in the beginning (of this section, 2).

Also, other abstract elementary classes are $(K, \prec)$ where K is the class of locally finite models of a first order theory T. Another example is $(\mathrm{Mod}_\psi, \prec_{\mathscr{L}})$ where ψ is a sentence from logic $\mathbb{L}_{\lambda^+,\omega}$ with $\mathscr{L}$ the set of subformulas of ψ. Also $(K, \prec)$ where $P \in \tau_{\mathfrak{K}}$ is a unary predicate, T first order and $K = \{M \in \mathrm{Mod}_T : P^M = \mathbb{N}$, the natural numbers$\}$.

A natural property to consider is amalgamation. We say that $\mathfrak{K}$ has the amalgamation property when for any $M_\ell \in \mathfrak{K}, \ell = 0, 1, 2$ and $\leq_{\mathfrak{K}}$-embedding f_1, f_2 of M_0 into M_1, M_2 respectively (this means that f_ℓ is an isomorphism from M_0 onto some $M'_\ell \leq_{\mathfrak{K}} M_\ell$) there are $M_3 \in \mathfrak{K}$ and $\leq_{\mathfrak{K}}$-embeddings g_1, g_2 of M_1, M_2 into M_3 respectively such that $g_1 \circ f_1 = g_2 \circ f_2$. Should we adopt it? Now it is a very important property, we would like to have it, but it is a strong restriction (our prototyical problem, models of $\psi \in \mathbb{L}_{\omega_1,\omega}$ fails it); so we do not assume it, but it will appear as a dividing line.
So the thesis is
2.17 Thesis:

- (a) In the context of abstract elementary classes we can answer some non-trivial questions
- (b) In particular we can say something on the categoricity spectrum
- (c) In the long run a parallel to the main gap will be found.

A reasonable reader may require an example of results. First we quote [Sh 576] represented here in Chapter VI:

2.18 Theorem. *Assume $2^{\aleph_\alpha} < 2^{\aleph_{\alpha+1}} < 2^{\aleph_{\alpha+2}}$ and $\mathfrak{K}$ is an abstract elementary class categorical in $\aleph_\alpha$, in $\aleph_{\alpha+1}$ and has an "intermediate" number of models in $\aleph_{\alpha+2}$, then $\mathfrak{K}$ has at least one model in $\aleph_{\alpha+3}$.*

Note that

2.19 Notation. If $\lambda = \aleph_\alpha$ we let $\lambda^{+n} = \aleph_{\alpha+n}$, so can write this theorem in such a notation, similarly later.

So it is an example for 2.17(a)+(b): not "every function" can occur as $\lambda \mapsto \dot{I}(\lambda, \mathfrak{K})$.

Note that this theorem gives a weak conclusion, but with very weak assumptions. In fact at first glance it seems we are facing a wall: our assumptions are so weak to exclude all possible relevant methods of model theory, in particular all relatives of compactness.

> I.e., we have no compact (even just $\aleph_0$-compact) logic defining our class. Of course, the upward LS cannot be used, it does not make sense: the desired conclusion is a weak form of it. As for the downward Löwenheim Skolem theorem, with only three cardinals available it seems to say very little.
>
> We do not have formulas hence no types and no saturated models. Here we cannot use versions of "well ordering is undefinable" as in previous cases (see Chapter I; if $\aleph_\alpha = \aleph_0$ and $\mathfrak{K}$ is reasonable we have used "no $\psi \in \mathbb{L}_{\omega_1,\omega}(\mathbf{Q})$ defines well ordering (in a richer vocabulary)"; this does not apply in [Sh 576], i.e. Chapter VI even when $\lambda = \aleph_0$ as we demand only $\mathrm{LS}(\mathfrak{K}) \leq \aleph_0$ rather than "$\mathfrak{K}$ is a $\mathrm{PC}_{\aleph_0}$-class"; and we certainly like to allow any $\aleph_\alpha$). Also in general we cannot find Ehrenfeuch-Mostowski models (another way to say well orders are not definable). Also we do not assume the existence of relevant so called large cardinals, e.g. $\mathfrak{K}$ is definable in some $\mathbb{L}_{\kappa,\omega}$, κ a compact or just a measurable cardinal. So indeed no remnants of compactness are available here.

The proof of 2.18 leads us to our second framework, good λ-frames which has a crucial role in our investigations, see below. The main neatly stated result in Chapter II (part (1) of 2.20), Chapter III(part (2) of 2.20) is:

> (omitting a weak set theoretic assumption which will be eliminated in the full version of Chapter VII).

2.20 Theorem. *Assume $\mathfrak{K}$ is an abstract elementary class .*
1) $\mathfrak{K}$ has a member in $\aleph_{\alpha+n+1}$ if ($n \in \mathbb{N}$ and)

- *(a) $n \geq 2$ and $2^{\aleph_\alpha} < 2^{\aleph_{\alpha+1}} < \ldots < 2^{\aleph_{\alpha+n}}$*
- *(b) $\mathfrak{K}$ is categorical in $\aleph_\alpha$ and in $\aleph_{\alpha+1}$*
- *(c) $\mathfrak{K}$ has a model in $\aleph_{\alpha+2}$*
- *(d) $\dot{I}(\aleph_{\alpha+m}, \mathfrak{K})$ is not too large for $m = 2, \ldots, n$.*

2) If (a)-(d) holds for every n then $\mathfrak{K}$ is categorical in every $\aleph_\beta \geq \aleph_\alpha$.

Actually above "$\mathfrak{K}$ having Löwenheim-Skolem number $\leq \lambda$" (rather than $\aleph_0$) is enough.

(D) Toward Good λ-frames (i.e. ⊠(b)):

2.21 Thesis Good λ-frames are a right context to start our "positive" structure theory.

They are a rock-bottom parallel of superstable elementary classes.

Now compared to abstract elementary classes, much more has to be said in order to explain what they are and how to justify them. We describe good λ-frames $\mathfrak{s}$ in several stages. We need several choices to specify our context. Usually in model theory we fix an elementary class $\mathfrak{K}$ and consider $M \in \mathfrak{K}$. Here we concentrate on one cardinal λ, that is, we usually investigate $\mathfrak{K}_\lambda = (K_\lambda, \leq_{\mathfrak{K}_\lambda})$ where $K_\lambda = \{M \in K : M$ has cardinality $\lambda\}$ and $\leq_{\mathfrak{K}_\lambda}$ is defined by $M \leq_{\mathfrak{K}_\lambda} N$ iff $M \leq_{\mathfrak{K}} N, M \in K_\lambda$ and $N \in K_\lambda$. This is not a clear cut deviation, also for elementary classes we sometimes fix λ, and here we usually look at least at $\mathfrak{K}_\lambda$ and $\mathfrak{K}_{\lambda^+}$ together, still the flavour is different. So (the notion "choice" may be seemingly problematic but a better alternative was not found).

2.22 Choice: We concentrate on $\mathfrak{K}_\lambda$, an abstract elementary class restricted to one cardinal.

This seems reasonable because as noted above, transfer from one cardinal to another is central, but in our context quite hard, so we may know various "good" properties only around λ. Also there are $\mathfrak{K}$ which in some cardinals are model theoretically "very simple" but in other (e.g. larger) cardinals complicated, and we may like to say what we can say about $\mathfrak{K}_\lambda$ in λ for which $\mathfrak{K}_\lambda$ is "simple".

2.23 Choice: We concentrate here on $\mathfrak{K}_\lambda$ with amalgamation and the JEP (joint embedding properties).

But is amalgamation not a very strong/positive property? Yes, but amalgamation for models of cardinality λ only is much weaker and its failure in some reasonable circumstances leads to non-structure results, so it can serve as a dividing line. More specifically, we know that if $\mathfrak{K}$ is categorical in $\lambda \geq \mathrm{LS}(\mathfrak{K})$ and $\mathfrak{K}_\lambda$ fails amalgamation and $\mathfrak{K}_{\lambda^+} \neq \emptyset$ then in $\mathfrak{K}_{\lambda^+}$ we have many complicated models (provided that $2^\lambda < 2^{\lambda^+}$; see Chapter I).

2.24 Choice: In $\mathfrak{K}_\lambda$ there is a superlimit model M^* which means that: $M^* \in \mathfrak{K}_\lambda$ is universal, (i.e., any $M' \in \mathfrak{K}_\lambda$ can be $\leq_{\mathfrak{K}}$-embedded into it), has a proper $<_{\mathfrak{K}}$-extension and if M is the union of a $<_{\mathfrak{K}}$-increasing chain of models isomorphic to M^* and M is of cardinality λ, then M is isomorphic to M^*.

Can we give a natural example of a superlimit model? For the abstract elementary class of linear orders, the rational order $(\mathbb{Q}, <)$ is superlimit (in $\aleph_0$). However, this is somewhat misleading as in larger cardinals it is much "harder", in fact, for the abstract elementary class of linear orders there is no superlimit model in $\lambda > \aleph_0$. The abstract elementary class of algebraically closed fields of some fixed character has a superlimit model in every $\lambda \geq \aleph_0$. However, consider the class of $\{(A, E) : E$ an equivalence relation on $A\}$. Easily (A, E) is superlimit in it iff the number of E-equivalence classes as well as the cardinality of each E-equivalence class is the number of elements of A.

Of course, if $\mathfrak{K}$ is categorical in λ then every $M \in \mathfrak{K}_\lambda$ is superlimit (if it is not $\leq_{\mathfrak{K}}$-maximal in which case every $M \in \mathfrak{K}$ has cardinality $\leq \lambda$), but having a superlimit is a much weaker condition and it seems a right notion of generalizing superstability (or, probably, a good first approximation). This may surely look tautological in view of Definition 2.3, but that definition is misleading. There are not few properties which for elementary classes are equivalent to being superstable and we have chosen the existence of superlimit. However, so far the existence of a superlimit model in λ has few consequences.

Why the choice? As this is an exterior way to say that our class is "simple, low"; it is weaker than categoricity and we next demand much more.

> Note that if $\mathfrak{K}$ is an elementary class and $\lambda = \lambda^{\aleph_0} + |\tau_{\mathfrak{K}}|$ or $\lambda \geq \beth_\omega + |\tau_{\mathfrak{K}}|$, then $M \in K_\lambda, M$ is superlimit iff M is saturated and the theory is superstable; see [Sh 868, 3.1].

Now we are very interested in the existence of something like "free amalgamation", which in our context will be called non-forking amalgamation. That is, we are interested in saying when "M_1, M_2 are freely amalgamated over M_0 inside M_3" (all in $\mathfrak{K}_\lambda$). In our main example we have to use a more restrictive notion, having quadruples (M_0, M_1, a, M_3) is non-forking where $M_0 \leq_{\mathfrak{K}} M_1 \leq_{\mathfrak{K}} M_3, a \in M_3 \backslash M_1$. This says that "inside M_3 the element a and the model M_1 are freely amalgamated over M_0". (Mainly in [Sh 576], i.e. Chapter VI, use so called "minimal types", which give rise to such quadruples).

This leads us to define a central notion here: $\mathbf{tp}_{\mathfrak{K}}(a, M, N)$, the "orbit" of $a \in N$ over $M \leq_{\mathfrak{K}} N$. We express (M_0, M_1, a, M_3) is non-forking also as "$\bigcup(M_0, M_1, a, M_3)$" and also as "$\mathbf{tp}_{\mathfrak{s}}(a, M_1, M_3)$ does not fork over M_0" because it is analogous to the non-forking in first order model theory. But this background is not needed, as non-forking is an abstract, axiomatic relation in our context.

> This replaces here the notion of type in the investigation of elementary (= first order) classes. But there the types are defined as $\text{tp}(\bar{a}, A, N) = \{\varphi(\bar{x}, \bar{b}) : \bar{b} \subseteq A, \varphi(\bar{x}, \bar{y})$ is a first order formula and $N \models \varphi[\bar{a}, \bar{b}]\}$. Note: the case A is the universe of $M \leq_{\mathfrak{K}} N$ is not excluded but is not particularly distinguished. In fact, it was unnatural there to make the restriction as there are theorems using our ability to restrict the type to any subset of A (e.g. for inductive proof) and it is im-

portant to have results on any A.

We let $\mathscr{S}_{\mathfrak{K}_\lambda}(M) = \{\mathbf{tp}_{\mathfrak{K}_\lambda}(a, M, N) : M \leq_{\mathfrak{K}_\lambda} N \text{ and } a \in N\}$ be called the set of types over M. The set of axioms (i.e., Definition II.2.1) of good λ-frames expresses the intuition of "non-forking" as a free amalgamation (in fact we are allowed to restrict the non-forking relation to types $\mathbf{tp}_{\mathfrak{s}}(a, M_1, M_3)$ which are, so called basic ones, they should mainly be "dense" enough). We may consider these axioms per se, but we feel obliged to find evidence of their naturality of the form indicated above. So

2.25 Definition. A good λ-frame $\mathfrak{s}$ consists of

(a) an abstract elementary class $\mathfrak{K} = \mathfrak{K}^{\mathfrak{s}}$ and let $\mathfrak{K}_{\mathfrak{s}} = \mathfrak{K}_\lambda$
(b) for $M \in \mathfrak{K}_\lambda$ we have $\mathscr{S}^{\rm bs}_{\mathfrak{s}}(M)$, a subset of $\mathscr{S}_{\mathfrak{K}_\lambda}(M)$ and
(c) a notion of "$p \in \mathscr{S}^{\rm bs}(M_2)$ does not fork over $M_1 \leq_{\mathfrak{K}_\lambda} M_2$" satisfying some reasonable axioms.

How does this help us in proving Theorem 2.20? Relying on the main results of [Sh 576], Chapter VI, we in II§3 prove that there is a good λ^+-frame $\mathfrak{s}$ with $\mathfrak{K}_{\mathfrak{s}} = \mathfrak{K}_{\lambda^+}$. Also in II§3 using a similar theorem from Chapter I for the case $\lambda = \aleph_0$ with a little different assumptions, we get a good $\aleph_0$-frame $\mathfrak{K}$.

We take a spiralic approach: we look at a good λ-frame $\mathfrak{s}$, suggest a question, i.e., dividing lines, if $\mathfrak{s}$ falls under the complicated side we prove a non-structure theorem. If not, we know some things about it and we can continue to investigate it, after we have enough knowledge we ask another question. In II§5 we start with a good λ-frame, gain some knowledge and if there are not enough essentially unique amalgamations we get many complicated models in λ^{++}. If $\mathfrak{s}$ avoids this, we call it weakly successful and understand $\mathfrak{K}_{\mathfrak{s}}$ better. In particular, we define the promised "M_1, M_2 are non-forking amalgamated over M_0 inside M_3", we call this relation NF $= \mathrm{NF}_\lambda$ $= \mathrm{NF}_{\mathfrak{s}}$ and prove that it has the properties hoped for. Listing its desired properties, it is unique. But this has a price: we have to restrict $\mathfrak{K}_{\mathfrak{s}}$ to isomorphic copies of the superlimit models. After assuming $\mathfrak{s}$ fails, another non-structure property we succeed to find for λ^+ another good frame, $\mathfrak{s}^+$ such that $K^{\mathfrak{s}^+}_{\lambda^+} \subseteq K^{\mathfrak{s}}_{\lambda^+}$.

What have we gained? Have we not worked hard just to find ourselves in the same place? Well, $\mathfrak{s}^+$ is a good λ^+-frame and $\dot{I}(\mu, K^{\mathfrak{s}^+}) \leq \dot{I}(\mu, K^{\mathfrak{s}})$ for every $\mu \geq \lambda^+$ and

(∗) for every χ and good χ-frame $\mathfrak{t}$, $K^{\mathfrak{t}}$ has models of cardinality χ^+ and moreover of cardinality χ^{++}.

So this is enough to prove the Theorem 2.20(1), by induction on n.

Let us compare this to [Sh 87a], [Sh 87b]. There in stage n we have some knowledge on models in $\mathfrak{K}_{\aleph_\ell}$ for $\ell \leq n$ but our knowledge decreases with ℓ. Now (all in [Sh 87b]) dealing with $n+1$ we have to consider a question on models of cardinality $\lambda = \aleph_0$, for which our specific tools for $\aleph_0$ (the omitting type theorem and the assumption that $\mathfrak{K}$ is $(\mathrm{Mod}_\psi, \prec)$ where $\psi \in \mathbb{L}_{\omega_1,\omega}$) enable us to have proved a dichotomy, each side implied additional information concerning $\aleph_\ell$ for $\ell \leq n$, again decreasing with ℓ.

[We elaborate: for each $\ell < n$ we can define so called full stable $(\mathscr{P}^-(m), \aleph_\ell)$-systems $\langle M_u : u \in \mathscr{P}^-(m)\rangle$ for $m \leq (n-\ell)$ where $\mathscr{P}^-(m) = \{u : u \subset \{0, \ldots, m-1\}\}$. So our knowledge "decreases" with ℓ: we can handle only systems of lower "dimension". We ask on such systems whether we can find suitable $M_{\{0,\ldots,n-1\}}$, is it weakly unique (up to embedding), is it unique, is there a prime one. We can transfer up a positive property from $(\mathscr{P}^-(m), \aleph_\ell)$ to $(\mathscr{P}^-(m-1), \aleph_{\ell+1})$, and also negative ones if $2^{\aleph_\ell} < 2^{\aleph_{\ell+1}}$. A crucial point is the existence of a strong dichotomy in the cardinality $\aleph_0$, either we have a prime solution or we have $2^{\aleph_0}$ pairwise incompatible ones.

Note that in [Sh 87a], [Sh 87b], we deal with types as in elementary classes (i.e. as set of formulas) but only over models or $\cup\{M_u : u \in \mathscr{P}^-(n)\}$ when $\langle M_u : u \in \mathscr{P}^-(n)\rangle$ is so called stable.]

The proofs of Chapter II seem neater than [Sh 87a], [Sh 87b]: because we are "poorer", we do not have the special knowledge on the first λ. So we do not have to look back, we can forget $\mathfrak{s}$ when advancing to $\mathfrak{s}^+$. This is nice for its purpose but suppose that we have a good λ^{+n}-frame $\mathfrak{s}^n$ for $n < \omega, \mathfrak{s}^{n+1}$ being gotten from $\mathfrak{s}^{+n}$ as above. For this purpose, forgetting the past costs us the future - we cannot say anything on models of cardinality $\geq \lambda^{+\omega}$. This is rectified in Chapter III.

> So in Chapter III we investigate the $\mathfrak{K}_{\mathfrak{s}^{+n}}$ for every n large enough, a priori it is fine to do this for $n \geq 756$, and increasing the number as we continue to investigate. But in spite of this knowledge, considerable effort was wasted on small n, i.e., assuming little on $\mathfrak{s}$, and in III§2-§11 we get the theory of prime, independence, dimension, regular types and orthogonality we like (see, maybe, [Sh:F735] on what we really need to assume).
>
> But for going up we need to deal with $\mathscr{P}^-(n)$-amalgamation - their existence and uniqueness. Then we can go up, see III§12.

§3 On Good λ-frames

This continues §2 and should be "non-logician friendly" too, though it may well be more helpful after some understanding/reading of the material itself.

(A) Getting a good λ-frame

We try below to describe in more details the proof of Theorem 2.20(1) + (2) proved in Chapter II, Chapter III, so we somewhat repeat what was said before in (D) of §2. We have to start by getting good λ-frames. We could have concentrated on the case $\lambda = \aleph_0$ and rely on Chapter I, but as this does not fit the "for non-logicians" we instead rely on [Sh 576], [Sh 603], that is on Chapter VI and the non-structure from Chapter VII, at least the "lean" version.

> For presentation we cheat a little in the non-structure part, saying we prove results like $\dot{I}(\mu^{++}, \mathfrak{K}) = 2^{\mu^{++}}$ when $\mathfrak{K}$ satisfies some "high" property and say $2^{\mu^+} < 2^{\mu^{++}}$. One point is that this relies on using an extra set

theoretic assumption on μ^+: the weak diamond ideal on μ^+ not being μ^{++}-saturated. This is a very weak assumption, it is not clear whether its failure is consistent when $\mu \geq \aleph_1$ and in any case its failure has high consistency strength, that is, if the ideal is μ^{++}-saturated then there are inner models with quite large cardinals. We may eliminate this extra set theoretic assumption as done in the full version of Chapter VII (see later part of the introduction). The second point is we prove only that there are $\geq \mu_{\text{unif}}(2^{\mu^{++}}, 2^{\mu^+})$ many non-isomorphic models in μ^{++}. This number is always $> 2^{\mu^+}$ (recall we are assuming $2^{\mu^+} < 2^{\mu^{++}}$), and is equal to $2^{\mu^{++}}$ when $\mu \geq \beth_\omega$ and conceivably the statement "$2^{\mu^+} < 2^{\mu^{++}} \Rightarrow \mu_{\text{unif}}(2^{\mu^{++}}, 2^{\mu^+}) = 2^{\mu^{++}}$" is provable in ZFC.

Of course, below $\text{LS}(\mathfrak{K}) \leq \lambda$ suffices instead of $\text{LS}(\mathfrak{K}) = \aleph_0$.

So first assume

$\boxdot_1$ $\mathfrak{K}$ is an abstract elementary class, and for simplicity $2^\lambda < 2^{\lambda^+} < \ldots < 2^{\lambda^{+n}} < 2^{\lambda^{+n+1}} < \ldots$, $\mathfrak{K}$ is categorical in λ, λ^+, has a model in λ^{++}, and $\dot{I}(\lambda^{+2}, \mathfrak{K}) < 2^{\lambda^{+2}}$.

We can deduce that $\mathfrak{K}_\lambda$ and $\mathfrak{K}_{\lambda^+}$ have amalgamation.
(Why? Otherwise it has many complicated models in λ^+, λ^{++}, respectively). Now we consider the class $K^{3,\text{na}}_\lambda$ of triples (M, N, a), $M \leq_{\mathfrak{K}_\lambda} N$, $a \in N \backslash M$ with the (natural) order, which is $(M_1, N_1, a_1) \leq (M_2, N_2, a_2)$ iff $a_1 = a_2$ (yes! equal) and $M_1 \leq_{\mathfrak{K}_\lambda} M_2$ and $N_1 \leq_{\mathfrak{K}_\lambda} N_2$.

We may look at them as representing the "orbit (or type of) a over M inside $N, \mathbf{tp}_{\mathfrak{K}}(a, M, N)$", which is not defined by formulas but by mappings, (i.e. types are orbits over M) so if $M \leq_{\mathfrak{K}_\lambda} N_\ell$ and $a_\ell \in N_\ell \backslash M$ then $\mathbf{tp}_{\mathfrak{K}_\lambda}(a_1, M, N_1) = \mathbf{tp}_{\mathfrak{K}}(a_2, M, N_2)$ iff for some $\leq_{\mathfrak{K}_\lambda}$-extension N_3 of N_2 there is a $\leq_{\mathfrak{K}_\lambda}$-embedding h of N_1 into N_3 over M which maps a_1 to a_2, recalling $\mathfrak{K}_\lambda$ has amalgamation.

Why do we consider $K^{3,\text{na}}_\lambda := \{(M, N, a) : M \leq_{\mathfrak{K}_\lambda} N, a \in N \backslash M\}$ instead of $\mathscr{S}^{\text{na}}_{\mathfrak{K}_\lambda}(M) := \{\mathbf{tp}_{\mathfrak{K}_\lambda}(a, M, N) : (M, N, a) \in K^{3,\text{na}}_\lambda\}$? (The types $\mathbf{tp}_{\mathfrak{K}_\lambda}(a, M, N)$ when $a \in M$ are called algebraic (and na stands for non-algebraic) and are trivial, so $\mathscr{S}^{\mathrm{na}}_{\mathfrak{K}_\lambda}(M)$ is the rest.) Now $\mathscr{S}_{\mathfrak{K}_\lambda}(M)$ is very important and for $M_1 \leq_{\mathfrak{K}_\lambda} M_2, p \in \mathscr{S}_{\mathfrak{K}_\lambda}(M_2)$ we can define its restriction to $M_1, p \restriction M_1 \in \mathscr{S}_{\mathfrak{K}_\lambda}(M_1)$, with some natural properties, and this mapping is onto (= surjective) as $\mathfrak{K}_\lambda$ has the amalgamation property. But it is not clear that an increasing sequence of types of length $\delta < \lambda^+$ of types has a bound (when $\mathrm{cf}(\delta) > \aleph_0$), see Baldwin-Shelah [BlSh 862]. For $K^{3,\mathrm{na}}_\lambda$ this holds. That is, if the sequence $\langle (M_\alpha, N_\alpha, a_\alpha) : \alpha < \delta \rangle$ is increasing in $K^{3,\mathrm{na}}_\lambda$, so $\alpha < \delta \Rightarrow a_\alpha = a_0$, then it has a lub: the triple $(\cup\{M_\alpha : \alpha < \delta\}, \cup\{N_\alpha : \alpha < \delta\}, a_0)$.

Some types (and triples) are in some sense better understood: here the ones representing minimal types; where

(∗) $p \in \mathscr{S}^{\mathrm{na}}_{\mathfrak{K}_\lambda}(M)$ is minimal if for every $\leq_{\mathfrak{K}_\lambda}$-extension N of M the type p has at most one extension in $\mathscr{S}^{\mathrm{na}}_{\mathfrak{K}_\lambda}(N)$.

Note that p always has at least one extension in $\mathscr{S}_{\mathfrak{K}_\lambda}(N)$ by amalgamation and we can prove that p has at least one from $\mathscr{S}^{\mathrm{na}}_{\mathfrak{K}_\lambda}(N)$ in our context, and recall that we have discarded the algebraic types, i.e. those of $a \in M$.

It is too much to expect that every $p \in \mathscr{S}^{\mathrm{na}}_{\mathfrak{K}_\lambda}(M)$ is minimal, but what about

3.1 Question: Is the class of minimal types dense, i.e., for every $p_1 \in \mathscr{S}^{\mathrm{na}}_{\mathfrak{K}_\lambda}(M_1)$ there are $M_2 \in \mathfrak{K}_\lambda$ and a minimal $p_2 \in \mathscr{S}^{\mathrm{na}}_{\mathfrak{K}_\lambda}(M_2)$ such that $M_1 \leq_{\mathfrak{K}_\lambda} M_2$ and p_2 extends p_1?

As we are assuming categoricity in λ and λ^+, this is not unreasonable and its failure implies having large $\mathscr{S}^{\mathrm{na}}_{\mathfrak{K}_\lambda}(M)$. Now §3,§4 relying on Chapter VII (earlier: [Sh 603] and part of [Sh 576]) are dedicated to proving that the minimals are dense. (This requires looking more into the set theoretic side but also the model theoretic one; an example of a property which we consider is: given $M_0 <_{\mathfrak{K}_\lambda} M_1$ is there $M_2, M_0 <_{\mathfrak{K}_\lambda} M_2$ such that M_1, M_2 can be amalgamated over M_0 uniquely?).

So we assume the answer to 3.1 is yes that is make the hypothesis:

3.2 Hypothesis. The answer to question 3.1 is yes.

Having arrived here, further investigation shows

(*) $\mathscr{S}^{\rm na}_{\mathfrak{K}_\lambda}(M)$ has cardinality $\le \lambda$.

Now it is natural to define (M_0, M_1, a, M_3) is a non-forking quadruple or $\underset{\mathfrak{s}}{\bigcup}(M_0, M_1, a, M_3)$ iff $M_0 \le_{\mathfrak{K}_\lambda} M_1 \le_{\mathfrak{K}_\lambda} M_3, a \in M_3 \backslash M_1$ and $\mathbf{tp}_{\mathfrak{K}_\lambda}(a, M_0, M_3)$ is minimal. Recalling Candid, we note that having chosen the unique non-trivial extension, we certainly have made the free choice: we have no freedom left on what is $\mathbf{tp}_{\mathfrak{K}_\lambda}(a, M_1, M_3)$! Now we find a good λ-frame $\mathfrak{s}$, with $\mathfrak{K}_\mathfrak{s} = \mathfrak{K}_\lambda$ and $\mathfrak{K}^\mathfrak{s} = \mathfrak{K}[\mathfrak{s}]$ will denote $\mathfrak{K}_{\ge\lambda} = \mathfrak{K} \restriction \{M \in K : \|M\| \ge \lambda\}$ and the set of basic types, is $\mathscr{S}^{\rm bs}_{\mathfrak{K}_\lambda}(M)$ is the set of minimal $p \in \mathscr{S}^{\rm na}_{\mathfrak{K}_\lambda}(M)$. Note that good λ-frame is defined in II§2, existence in our case is proved in II§3.

More accurately, in II§3 we prove in our present context the existence of a good λ^+-frame $\mathfrak{s}$ with $\mathfrak{K}_\mathfrak{s} = \mathfrak{K}_{\lambda^+}$, and we rely on having developed NF_λ in [Sh 576, §8]. But something parallel to [Sh 576, §8] is done in II§6 and described below. Moreover, in Chapter VI this is circumvented at the price of arriving to almost good λ-frame and then by Chapter VII it is even a good λ-frame and it converges with the description here.

We assume here that $\mathfrak{K}_\mathfrak{s} (= \mathfrak{K}^\mathfrak{s}_\lambda)$ is categorical; in the present context this is reasonable (e.g., as otherwise you restrict yourself to $\{M \in \mathfrak{K}_\mathfrak{s} : M \text{ is superlimit}\}$).

(B) The successor of a good λ-frame

Now we look at our good λ-frame $\mathfrak{s}$, and the $\mathfrak{s}$-basic types in this case are the minimal types. But we can forget the minimality and just use the properties required in the definition of a good λ-frame (i.e. we are in Chapter II). Now as $M \in \mathfrak{K}_\mathfrak{s} \Rightarrow \mathscr{S}^{\rm bs}_\mathfrak{s}(M)$ has cardinality $\le \lambda$, we can find $\le_\mathfrak{s}$-increasing chains $\langle M_i : i \le \lambda \times \delta\rangle$ such that for every $i < \lambda \times \delta$ every $p \in \mathscr{S}^{\rm bs}_\mathfrak{s}(M_i)$ is realized in M_{i+1}. It follows that $M_{\lambda\times\delta}$ is determined uniquely up to isomorphisms over M_0 (seemingly, depending on $\mathrm{cf}(\delta) := \mathrm{Min}\{\mathrm{otp}(C) : C \subseteq \delta \text{ unbounded}\}$). In such a case we say that $M_{\lambda\times\delta}$ is brimmed over M_0 and eventually we succeed to prove that the choice of the limit ordinal $\delta(< \lambda^+)$ is immaterial.

(These are relatives of universal homogeneous, saturated models and special models.)

We define $K_{\mathfrak{s}}^{3,\mathrm{bs}}$ as the class of triples (M, N, a) such that $M \leq_{\mathfrak{K}_{\mathfrak{s}}} N$ and $\mathbf{tp}_{\mathfrak{K}_{\mathfrak{s}}}(a, M, N) \in \mathscr{S}_{\mathfrak{s}}^{\mathrm{bs}}(M)$. By the axioms of "good λ-frames" for $(M_1, N_1, a) \in K_{\mathfrak{s}}^{3,\mathrm{bs}}$ and M_2 such that $M_1 \leq_{\mathfrak{s}} M_2$ we can find $M_2' \in \mathfrak{K}_\lambda$ isomorphic to M_2 over M_1 and $N_2 \in \mathfrak{K}_\lambda$, which is $\leq_{\mathfrak{K}}$-above M_2' and N_1 and $\mathbf{tp}_{\mathfrak{s}}(a, M_2', N_2)$ does not fork over M_1. In this case we say $(M_1, N_1, a) \leq_{\mathfrak{s}} (M_2', N_2, a)$, (or use $\leq_{\mathrm{bs}} = \leq_{\mathrm{bs}}^{\mathfrak{s}}$ instead $\leq_{\mathfrak{s}}$).

Having existence is nice, but having also uniqueness is better. So we become interested in $K_{\mathfrak{s}}^{3,\mathrm{uq}}$, the class of $(M, N, a) \in K_{\mathfrak{s}}^{3,\mathrm{bs}}$ satisfying: if $(M_*, N_*, a) \in K_{\mathfrak{s}}^{3,\mathrm{bs}}$ is $\leq_{\mathfrak{s}}$-above (M, N, a), then the way M_*, N are amalgamated over M inside N_* is unique (up to common embeddings).

For the first order case this means "$\mathrm{tp}(N, M \cup \{a\})$ is weakly orthogonal to M"; (i.e., domination).

3.3 Question: 1) (Density) Do we have "$K_{\mathfrak{s}}^{3,\mathrm{uq}}$ is dense in $K_{\mathfrak{s}}^{3,\mathrm{bs}}$ (under $\leq_{\mathfrak{s}}$)"?
2) (Existence) Assume $p \in \mathscr{S}_{\mathfrak{s}}^{\mathrm{bs}}(M)$, can we find a, N such that $(M, N, a) \in K_{\mathfrak{s}}^{3,\mathrm{uq}}$ and $\mathbf{tp}_{\mathfrak{s}}(M, N, a) = p$?

As $\mathfrak{K}_{\mathfrak{s}}$ is categorical, we can prove that density implies existence.

"Have we not been here before?" the reader may wonder. This is the spiral phenomena: in 3.1 we were interested in a different kind of uniqueness. Now we prove that the non-density is a non-structure property and as a token of our pleasure, $\mathfrak{s}$ with positive answer is called weakly successful.

3.4 Hypothesis. The answer to 3.3 is yes, enough triples in $K_{\mathfrak{s}}^{3,\mathrm{uq}}$ exist.

So we have some cases of uniqueness of the non-forking amalgamation. When we (in II§6) close this family of cases of uniqueness, under transitivity and monotonicity we get a four-place relation $\mathrm{NF}_\lambda = \mathrm{NF}_{\mathfrak{s}}$ on $\mathfrak{K}_\lambda$. Working enough we show that $\mathrm{NF}_{\mathfrak{s}}$ conforms reasonably with "M_1, M_2 and are in non-forking ($\equiv$ free) amalgamation over M_0 inside M_3". We justify the definition showing that some natural properties it satisfies has at most one solution (for any good λ-frame).

Now we start to look at models in $K^{\mathfrak{s}}_{\lambda^+}$; in an attempt to find a good λ^+-frame $\mathfrak{s}^+ = s(+)$, a successor of $\mathfrak{s}$. There are some models in $K^{\mathfrak{s}}_{\lambda^+}$; in fact, there is a universal homogeneous one M^* and it is unique so if there is a superlimit $M \in K^{\mathfrak{s}}_{\lambda^+}$ then $M \cong M^*$. Now if $\langle M_i : i < \lambda^+\rangle$ is $\leq_{\mathfrak{K}^{\mathfrak{s}}_{\lambda^+}}$-increasing $M_i \cong M^*$ then $\cup\{M_i : i < \lambda^+\} \cong M^*$ but it is not clear if, e.g., $\cup\{M_i : i < \omega\} \cong M^*$. So we consider another choice of being a substructure in $K^{\mathfrak{s}}_{\lambda^+}$: $M_1 \leq^*_{\lambda^+} M_2$ iff $M_1, M_2 \cong M^*$ and for some $\leq_{\mathfrak{K}}$-representations (also called $\leq_{\mathfrak{K}}$-filtrations) $\langle M^\ell_\alpha : \alpha < \lambda^+\rangle$ of M_ℓ for $\ell = 1, 2$ we have $\mathrm{NF}_{\mathfrak{s}}(M^1_i, M^2_i, M^1_j, M^2_j)$ for every $i < j < \lambda^+$.

> [We say that $\langle M_\alpha : \alpha < \lambda^+\rangle$ is a $\leq_{\mathfrak{K}}$-representation or $\leq_{\mathfrak{K}}$-filtration of $M \in \mathfrak{K}_{\lambda^+}$ when $M_\alpha \in \mathfrak{K}_\lambda$ is $\leq_{\mathfrak{K}_\lambda}$-increasing continuous for $\alpha < \lambda^+$ and $M = \cup\{M_\alpha : \alpha < \lambda^+\}$.]

We would love to understand $\mathfrak{K}_{\lambda^+}$, but this seems too hard, so presently so we restrict ourselves to isomorphic copies of the model we do understand, M^*.

> This conforms with the strategy of first understanding the quite saturated models.

This helps to prove "M^* is superlimit" but with a price: we have to consider the following question.

<u>3.5 Question</u>: Assume $\langle M_i : i \leq \delta\rangle$ is $\leq^*_{\lambda^+}$-increasing continuous, δ a limit ordinal $< \lambda^{++}$ and $i < \delta \Rightarrow M_i \cong M^*$ and $i < \delta \Rightarrow M_i \leq^*_{\lambda^+} N$ and $N \cong M^*$. Does it follow that $M_\delta \leq^*_{\lambda^+} N$?

This is an axiom of an abstract elementary class, so we know that it holds for $(\mathfrak{K}_{\lambda^+}, \leq_{\mathfrak{K}})$ but not necessarily for $\leq^*_{\lambda^+}$. This is another dividing line: if the answer is no, we get a non-structure theorem. If the answer is yes, we call $\mathfrak{s}$ successful.

3.6 Hypothesis. $\mathfrak{s}$ is successful.

We go on and prove that $\mathfrak{s}^+$ is a good λ^+-frame. Well, the reader may wonder: all this work and you just end up where you have started, just one cardinal up? True, but if $\mathfrak{s}$ is a good λ-frame then $K^{\mathfrak{s}}_{\lambda^{++}} \neq \emptyset$, so for a successful $\mathfrak{s}$, applying this to the good λ^+-frame $\mathfrak{s}^+$ we get $\mathfrak{K}_{\lambda^{+3}_{\mathfrak{s}}} \neq \emptyset$. Having "arrived to the same place one cardinal up" is enough to prove part (1) of Theorem 2.20!

More elaborately, under the assumptions of 2.20 there is a good λ^+-frame $\mathfrak{s}_1$ with $\mathfrak{K}^{\mathfrak{s}_1} \subseteq \mathfrak{K}^{\mathfrak{s}}$. Second, if we prove by induction on $k = 1, \ldots, n-1$ that there is a good λ^{+k}-frame $\mathfrak{s}_k$ with $K^{\mathfrak{s}_k} \subseteq K^{\mathfrak{s}_{k-1}}$, the induction step is what we have proved. For $k = n-1$, "$K^{\mathfrak{s}_k}$ has a model in $\lambda^{++}_{\mathfrak{s}_k}$" means that $K_{\lambda^{+n+1}} \neq \emptyset$ as asked for in 2.20(1). All this is Chapter II, so its proof proceeds by "forgetting" the previous $\mathfrak{s}$ when advancing $\mathfrak{s}^+$ and $\lambda^+_{\mathfrak{s}}$. Next assume

$\boxdot_2$ $\mathfrak{s}$ is a λ-good frame, $\dot{I}(\lambda^{+n}, \mathfrak{K}^{\mathfrak{s}}) < 2^{\lambda^{+n}}$ and $2^{\lambda^{+n}} < 2^{\lambda^{+n+1}}$ for $n < \omega$.

We now define by induction on n a good λ^{+n}-frame $\mathfrak{s}^{+n} = \mathfrak{s}(+n)$. Let $\mathfrak{s}^0 = \mathfrak{s}$ and having defined $\mathfrak{s}^{+n}$, it has to be successful by the previous argument so $\mathfrak{s}^{+(n+1)} := (\mathfrak{s}^{+n})^+$ is a well defined good $\lambda^{+(n+1)}$-frame. We can prove by induction on n that $K_{\mathfrak{s}(+n)} \subseteq K$ and $m < n \Rightarrow K^{\mathfrak{s}(+n)} \subseteq K^{\mathfrak{s}(+m)}$.

Note that if $K^{\mathfrak{s}}$ is the class of (A, E) where $|A| \geq \lambda$ and E is an equivalence relation on A then $K^{\mathfrak{s}^{+n}}$ is the class of $(A, E) \in K^{\mathfrak{s}}$ such that E has $\geq \lambda^n$ equivalence classes each of cardinality $\geq \lambda^{+n}$.

(C) The beauty of ω successive good λ-frames

What about part (2) of 2.20, i.e., models in cardinalities $\geq \lambda^{+\omega}$? The connection between $\mathfrak{s}^{+n}, \mathfrak{s}^{+(n+1)}$ is not strict enough. Now though we have $K^{\mathfrak{s}^{+n+1}} \subseteq K^{\mathfrak{s}^{+n}}$, we do not know whether $\leq_{\mathfrak{s}(+n+1)}$ is $\leq_{\mathfrak{K}[\mathfrak{s}(+n)]} \restriction K_{\mathfrak{s}(+n+1)}$ and whether $\mathfrak{K}_{\mathfrak{s}(+n+1)} = K^{\mathfrak{s}(+n)}_{\lambda^{+n+1}}$. We can overcome the first problem. We show that if $\mathfrak{s}$ is so called good$^+$ then $\leq_{\mathfrak{s}(+)} = \leq_{\mathfrak{K}[\mathfrak{s}]} \restriction K_{\mathfrak{s}(+)}$ (and $\mathfrak{s}$ is good$^+$ "usually" holds e.g., if $\mathfrak{s} = \mathfrak{t}^+, \mathfrak{t}$ is good$^+$ and successful, see III§1). In this case $\langle \mathfrak{K}^{\mathfrak{s}^{+n}} : n < \omega \rangle$ is decreasing and even $\langle K^{\mathfrak{s}^{+m}}_{\lambda^{+n}} : m \leq n \rangle$ is decreasing in m, but the orders agree when well defined. The crux of the matter is in the end (III§12, relying on what we prove earlier in Chapter III), to show that for some $\mathfrak{s}^{+\omega}$, $K_{\mathfrak{s}^{+\omega}} = \bigcap\{\mathfrak{K}^{\mathfrak{s}(+n)}_{\lambda^{+\omega}} : n < \omega\}$ and $\mathfrak{s}^{+\omega}$ is so called beautiful, so at last we shall arrive to "the promised land" from $\boxtimes(c)$ from the beginning of §2(C). But this comes only at the very end. In particular before starting we have to know much on the $\mathfrak{K}_{\mathfrak{s}(+n)}$'s. It is enough to prove any of the nice things we like to know on $\mathfrak{K}_{\mathfrak{s}(+n)}$ just for "$n < \omega$ large enough". A priori we may have from time to time to say "if $\mathfrak{s}$ has the desirable properties $(A)_1, \ldots, (A)_{\ell-1}$ then $\mathfrak{s}^{+n}$ has $(A)_\ell$ (as we are assuming all $\mathfrak{s}^{+n} (n < \omega)$ are successful), and so when we prove a desirable property X we prove it for $\mathfrak{s}^{+n}$ when $n \geq n_X$". Originally we were using $n \geq 2$ or $n \geq 3$, but try to use little, say "$\mathfrak{s}$ is weakly successful" (which means n is 0 or 1) and lately try just to finish.

Note also that without loss of generality $\mathfrak{s}$ is type-full, i.e. $\mathscr{S}^{\rm bs}_{\mathfrak{s}}(M) = \mathscr{S}^{\rm na}_{\mathfrak{s}}(M)$, as we can use our knowledge on $\mathrm{NF}_{\mathfrak{s}}$ to define when "$p \in \mathscr{S}^{\rm na}_{\mathfrak{s}}(N)$ does not fork over $M \leq_{\mathfrak{s}} N$" and prove that $\mathfrak{t}$ is a good λ-frame when we define $\mathfrak{t}$ by $\mathfrak{K}_{\mathfrak{t}} = \mathfrak{K}_{\mathfrak{s}}, \mathscr{S}^{\rm bs}_{\mathfrak{t}} = \mathscr{S}^{\rm na}_{\mathfrak{s}}$, and non-forking as above. As we can replace $\mathfrak{s}$ by $\mathfrak{t}$ the "w.l.o.g." above is justified.

Note that the $\mathfrak{K}_{\mathfrak{s}(+n)}$ are categorical, but this is deceptive: $\mathfrak{K}_{\mathfrak{s}(+n)}$ is, but $K^{\mathfrak{s}(+n)}_{\lambda^{+n+1}}$ is not necessarily categorical. So in order to eventually understand the categoricity spectrum in III§2 we sort out when is $\mathfrak{K}^{\mathfrak{s}}_{\lambda^+}$ categorical (for a successful good λ-frame $\mathfrak{s}$).

We define several (variants of) $\mathfrak{s}$ is uni-dimensional, prove the equivalence with "$K^{\mathfrak{s}}$ is categorical in $\lambda^+_{\mathfrak{s}}$" and show that (for successful $\mathfrak{s}$) $\mathfrak{s}$ is uni-dimensional iff $\mathfrak{s}^+$ is uni-dimensional (so this applies to $\mathfrak{s}^{+n}$ and $\mathfrak{s}^{+(n+1)}$ when well defined). So in the case we have chosen, $\mathfrak{s}^+, \mathfrak{s}^{+2}, \ldots$ are uni-dimensional and $K^{\mathfrak{s}(+n)}_{\lambda^{+n}} = K^{\mathfrak{s}}_{\lambda^{+n}}$ so in the beautiful (see below) case it implies categoricity in all $\mu > \lambda$.

We now review Chapter III in more detail. We define and investigate "$\mathbf{J}$ is a set of elements in $N \backslash M$ which is independent over M" in symbols $(M, N, \mathbf{J}) \in K^{3,\rm bs}_{\mathfrak{s}}$. The idea is that if $\langle M_i : i \leq \alpha \rangle$ is $\leq_{\mathfrak{s}}$-increasing, $a_i \in M_{i+1} \backslash M_i$ and $\mathbf{tp}_{\mathfrak{s}}(a_i, M_i, M_{i+1})$ does not fork over M_i for $i < \alpha$, then $(M_0, M_\alpha, \{a_i : i < \alpha\}) \in K^{3,\rm bs}_{\mathfrak{s}}$ and even $(M_0, M', \{a_i : i < \alpha\}) \in K^{3,\rm bs}_{\mathfrak{s}}$ if $M \cup \{a_i : i < \alpha\} \subseteq M' \leq_{\mathfrak{s}} M_\alpha$. But we have to prove that this notion has the expected properties, e.g., the finite character (see III§5).

We know about $(M, N, a) \in K_{\mathfrak{s}}^{3,\mathrm{uq}}$, but also important is $(M, N, a) \in K_{\mathfrak{s}}^{3,\mathrm{pr}}$: the triple is prime, i.e., such that if $(M, N', a') \in K_{\mathfrak{s}}^{3,\mathrm{bs}}$ and $\mathbf{tp}_{\mathfrak{s}}(a, M, N) = \mathbf{tp}_{\mathfrak{s}}(a', M, N')$ then there is a $\leq_{\mathfrak{s}}$-embedding of N into N' over M mapping a to a'. We prove existence in enough cases (mainly for $\mathfrak{s}^+$) and eventually define and investigate also "N is prime over $M \cup \mathbf{J}$" when $(M, N, \mathbf{J}) \in K_{\mathfrak{s}}^{3,\mathrm{bs}}$ and $\mathbf{J}$ is maximal.

Next we develop orthogonality: assume $p_\ell \in \mathscr{S}_{\mathfrak{s}}^{\mathrm{bs}}(M)$ for $\ell = 1, 2$. Then $p_1 \perp p_2$ when: if $(M, N, a) \in K_{\mathfrak{s}}^{3,\mathrm{uq}}$ and $p_1 = \mathbf{tp}_{\mathfrak{s}}(a, M, N)$ then p_2 has a unique extension in $\mathscr{S}_{\mathfrak{s}}(N)$. This means that there is no connection, no interaction between p_1 and p_2. It implies that $(M, N, \{a_i : i < \alpha\}) \in K_{\mathfrak{s}}^{3,\mathrm{bs}}$, i.e., is independent iff for each $j < \alpha, (M, N, \{a_i : i < \alpha, p_j \perp p_i\})$ is independent where $p_i = \mathbf{tp}_{\mathfrak{s}}(a_i, M, N)$. We prove that this behaves reasonably; in particular, is preserved by non-forking extensions. We similarly define $p \perp M$ (when $M \leq_{\mathfrak{s}} N, p \in \mathscr{S}_{\mathfrak{s}}^{\mathrm{bs}}(N)$). Because of the categoricity (and $\mathfrak{s} = \mathfrak{t}^+$) we can prove $K_{\mathfrak{s}}^{3,\mathrm{pr}} = K_{\mathfrak{s}}^{3,\mathrm{uq}}$.

In those terms we can characterize when $(M, N, a) \in K_{\mathfrak{s}}^{3,\mathrm{bs}}$ has uniqueness (i.e., $\in K_{\mathfrak{s}}^{3,\mathrm{uq}}$), under the assumption that there are primes. It holds <u>iff</u> there is a decomposition $\langle (M_i, a_j) : i \leq \alpha, j < \alpha \rangle$ of (M, N), i.e., $M_0 = M, M_\alpha = N, (M_i, M_{i+1}, a_i) \in K_{\mathfrak{s}}^{3,\mathrm{pr}}$ such that $a_0 = a$ and $i \in (0, \alpha) \Rightarrow \mathbf{tp}_{\mathfrak{s}}(a_i, M_i, M_{i+1}) \perp M_0$. We can define regular types such that: for $M \leq_{\mathfrak{s}} N$ and regular $p \in \mathscr{S}_{\mathfrak{s}}^{\mathrm{bs}}(M)$ the dependence relation on $\mathbf{I}_{M,N} = \{a \in N : a \text{ realizes } p\}$ behaves as independence in vector spaces (for others it behaves like sets of finite sequences from a vector space), and regular types are dense (i.e., if $M <_{\mathfrak{s}} N$ then for some $a \in N \backslash M, \mathbf{tp}_{\mathfrak{s}}(a, M, N)$ is regular). So $a \in \mathbf{I}_{M,N}$ depends on $\mathbf{J} \subseteq \mathbf{I}_{M,N}$ iff there are $M_1 \leq_{\mathfrak{K}_{\mathfrak{s}}} N_1$ such that $M \leq_{\mathfrak{K}_{\mathfrak{s}}} M_1, N \leq_{\mathfrak{K}_{\mathfrak{s}}} N_1, \mathbf{J} \subseteq M_1$, the triple $(M, N, \mathbf{J})$ has uniqueness and $\mathbf{tp}_{\mathfrak{s}}(a, M_1, N_1)$ forks over M. It has local character (if $a \in \mathbf{I}_{M,N}$ depends on $\mathbf{J}$ then it depends on some finite subsets of it), monotonicity, transitivity (if $a \in \mathbf{I}_{M,N}$ depends on $\mathbf{J}' \subseteq \mathbf{I}_{M,N}$ and each $b \in \mathbf{J}'$ depends on $\mathbf{J} \subseteq \mathbf{I}_{M,N}$ then a depends on $\mathbf{J}$) and satisfies the exchange lemma. Then we can define (and prove the relevant properties) when "$\{M_i : i < \alpha\}$ is independent over M inside N" and we can deal similarly with "$\langle M_\eta : \eta \in \mathscr{T} \rangle$ is independent inside N" when $\mathscr{T} \subseteq {}^{\omega>}(\lambda_{\mathfrak{s}})$ is closed under initial segments.

We may now consider the main gap in this context (but mostly this is delayed). From some perspective this is ridiculous: $\mathfrak{K}_{\mathfrak{s}}$ is categorical in $\lambda_{\mathfrak{s}}$. But we analyze $\{N : M_* \leq_{\mathfrak{s}} N\}$ for a fixed M_*. (In this still there is some degeneration, but we can analyze models from $\mathfrak{K}^{\mathfrak{s}}_{\lambda^+}$, in this case there is no real difference between what we do and the actual main gap theorem. And if $\mathfrak{s}$ is beautiful, see below, we can do the same for $\mathfrak{K}^{\mathfrak{s}}$).

So if $M \leq_{\mathfrak{s}} N$ (assuming, e.g. $\mathfrak{s}$ is a successful λ-frame with primes, less is needed), we can find a decomposition $\langle N_\eta, a_\nu : \eta \in \mathscr{T}, \nu \in \mathscr{T}\backslash\{<>\}\rangle$ of N which means

⊛ (a) $\mathscr{T} \subseteq {}^{\omega >}(\lambda_{\mathfrak{s}})$ is non-empty closed under initial segments

(b) $N_\eta \leq_{\mathfrak{s}} N$

(c) $\nu \triangleleft \eta \Rightarrow N_\nu \leq_{\mathfrak{s}} N_\eta$

(d) $(N_\eta, N_{\eta\hat{}<\alpha>}, a_{\eta\hat{}<\alpha>}) \in K^{3,\mathrm{pr}}_{\mathfrak{s}}$ if $\eta\hat{}<\alpha>\in \mathscr{T}$

(e) $\{a_{\eta\hat{}<\alpha>} : \eta\hat{} <\alpha>\in \mathscr{T}\}$ is independent in (M_η, N) and is

a maximal such set (with no repetitions, of course)

(f) $N_{<>} = M$,

(g) if $\cup\{N_\eta : \eta \in \mathscr{T}\} \subseteq N' <_{\mathfrak{s}} N, p = \mathbf{tp}_{\mathfrak{s}}(a, N', N) \in \mathscr{S}^{\mathrm{bs}}_{\mathfrak{s}}(N')$

then $p \pm N_\eta$ for some $\eta \in \mathscr{T}$

(h) if $\nu \triangleleft \eta \triangleleft \eta\hat{}\langle\alpha\rangle \in \mathscr{T}$ then $\mathbf{tp}_{\mathfrak{s}}(a_{\eta\hat{}\langle\alpha\rangle}, N_\eta, N_{\eta\hat{}\langle\alpha\rangle}) \perp N_\nu$.

<u>3.7 Question</u>: Is always N prime and/or minimal over $\cup\{N_\eta : \eta \in \mathscr{T}\}$?

The answer is yes iff whenever $\mathscr{T} = \{<>, <0>, <1>\}$ the answer is yes and we then say that $\mathfrak{s}$ have the so-called NDOP. Moreover, its negation DOP is a strong non-structure property: for every $R \subseteq \lambda \times \lambda$ we can find $N_R \in K^{\mathfrak{s}}_{\lambda^{++}}$ and $\bar{a}_\alpha, \bar{b}_\alpha \in {}^{\lambda_{\mathfrak{s}}}(N_R)$ for $\alpha < \lambda$ such that some condition (preserved by isomorphism) is satisfied by $\bar{a}_\alpha\hat{}\bar{b}_\beta$ in N_R iff $(\alpha, \beta) \in R$. Also the NDOP holds for $\mathfrak{s}^+$ iff it holds for $\mathfrak{s}$ when $\mathfrak{s}$ is successful from DOP. We can get $\dot{I}(\lambda^{++}_{\mathfrak{s}}, K^{\mathfrak{s}}) = 2^{\lambda^{++}_{\mathfrak{s}}}$ and more.

* * *

How does all this help us to go up? That is, we assume $\mathfrak{s}^{+n}$ is well defined and successful for every n (equivalently $\mathfrak{s}$ is n-successful for every n) and we would like to understand the models in $\mathfrak{K}^{\mathfrak{s}(+\omega)}$, (so they have cardinality $\geq \lambda^{+\omega}$ and are close to being $\lambda^{+\omega}$-saturated). The going up is done in the framework of stable $\mathscr{P}^{(-)}(n)$-system of models $\langle M_u : u \in \mathscr{P}^-(n)\rangle, \mathscr{P}^-(n) = \{u : u \subset \{0,\dots,n-1\}$; explained below. This is done in III§12 (which should be helpful for completing [Sh 322]).

In short, to understand existence/uniqueness of models (and of amalgamation) in λ, we consider such properties for some n-dimensional systems of models in every large enough $\mu \leq \lambda$. So for $n = 0, 1, 2$ we get the original problems but understanding the n-th case given in λ is intimately connected to understand the $(n+1)$-case for every large enough $\mu < \lambda$. So for $\lambda = \mu^+$ we get a positive property for (μ^+, n) from one for $(\mu, n+1)$.

Why do we need such systems? Consider $\lambda_* \geq \mu_* \geq \lambda_{\mathfrak{s}}$ and we try to analyze models of cardinality $\in [\mu_*, \lambda_*]$ by pieces of cardinality μ_* or $\mu' \in [\mu_*, \lambda_*)$ (in the end we consider $\mu_* = \lambda_{\mathfrak{s}}^{+\omega}$, but most of the analysis is for the case $\lambda_*, \mu_* \in [\lambda_{\mathfrak{s}}, \lambda_{\mathfrak{s}}^{+\omega})$). We can analyze a model M from $\mathfrak{K}$ of cardinality $\lambda_0 \in (\mu_*, \lambda_*]$ by a $\leq_{\mathfrak{K}}$-increasing continuous sequence $\langle M_\alpha : \alpha < \lambda_0\rangle, \mu_* \leq \|M_\alpha\| = \|M_{\alpha+1}\| < \lambda_0$, with $M = \cup\{M_\alpha : \alpha < \lambda_0\}$; so it suffices to analyze $M_{\alpha+1}$ over M_α for each α. We can analyze M_1 over M_0 for a pair of models $M_0 \leq_{\mathfrak{K}} M_1$ of the same cardinality which we call λ_1 when $\lambda_1 > \mu_*$ by an $(\leq_{\mathfrak{K}})$-increasing continuous sequence of pairs $\langle (M_i^0, M_i^1) : i < \lambda_1\rangle$ where $\|M_i^0\| = \|M_{i+1}^0\| = \|M_i^1\| = \|M_{i+1}^1\| < \lambda_1$, and we have to analyze M_{i+1}^1 over $\langle M_i^0, M_i^1, M_{i+1}^0\rangle$ for each i. In the next stage we have $8 = 2^3$ models and have to analyze the largest over the rest. Eventually we arrive to the case that all of them have cardinality μ_*.

In short, we have to consider suitable $\mathscr{P}(n)$-systems $\langle M_u : u \in \mathscr{P}(n)\rangle$ where $\mathscr{P}(n) = \{u : u \subseteq \{0,\dots,n-1\}\}, u \subseteq v \Rightarrow M_u \leq_{\mathfrak{K}^{\mathfrak{s}}} M_v$ and $\|M_u\| = \|M_0\| \in [\mu_*, \lambda_*]$.

We would like to analyze $M_{\{0,\dots,n-1\}}$ over $\cup\{M_u : u \in \mathscr{P}^-(n)\}$ where $\mathscr{P}^-(n) = \mathscr{P}(n)\backslash\{0,\dots,n-1\}$. Such analysis of a "big" system of small models naturally help proving cases of uniqueness, e.g., uniqueness of non-forking-amalgamations suitably defined. So if for μ_* we have positive answers for every n, then this holds for every $\lambda \in [\mu_*, \lambda_*]$.

But we are interested as well in existence proofs. (Note that in the proof we have to deal with uniqueness, existence (and some relatives) simultaneously.) For the existence we need for a given suitable system $\langle M_u : u \in \mathscr{P}^-(n)\rangle$ to complete it by finding $M_{\{0,\dots,n-1\}}$. Well, but what are the suitable systems? Those are defined, by several demands including $u \subseteq v \Rightarrow M_u \leq_{\mathfrak{K}^{\mathfrak{s}}} M_v$ (and many more restrictions which hold if the sequence of approximations chosen above are "fast" enough). We called them the stable ones. For each n, k we can ask on $\mathfrak{s}^{+n}$ some questions on $\mathscr{P}(k)$-systems: mainly versions of existence and uniqueness. A major point is that failure of uniqueness for $\lambda^{+n}, \mathscr{P}(m+1)$ implies failure for $\lambda^{+n+1}, \mathscr{P}(m)$ (using $2^{\lambda^{+n}} < 2^{\lambda^{+n+1}}$). But to get strong dichotomy we have to use systems which have the right amount of brimmness. At last we have a glimpse of "paradise", we can define when $\mathfrak{s}$ is n-beautiful essentially when it satisfies all the good properties on stable $\mathscr{P}(m)$-systems for $m \leq n$. In the end we prove that $\mathfrak{s}^{+n}$ is $(n+2)$-beautiful, i.e. has all the desired properties for $m \leq n+2$ but for this we use $\mathfrak{s}^{n+\ell}$ being successful for $\ell \leq n$.

Having all this we can prove that $\mathfrak{s}^{+\omega}$ has all the good properties (but we have to work on changing the brimmness demands) so is ω-beautiful. This now can be lifted up, in particular $\mathfrak{K}^{\mathfrak{s}(+\omega)}$ has amalgamation and the types $\mathbf{tp}_{\mathfrak{K}[\mathfrak{s}(+\omega)]}(a, M, N)$ are μ-local for $\mu = \lambda^{+\omega}$ (in fact $\mu = \lambda$ is enough) where

(∗) $\mathfrak{K}$ an abstract elementary class with amalgamation, is μ-local when for $M \leq_{\mathfrak{K}} N$ and $a_1, a_2 \in N$ we have: $\mathbf{tp}_{\mathfrak{K}}(a_1, M, N) = \mathbf{tp}_{\mathfrak{K}}(a_2, M, N)$ <u>iff</u> for every $M' \leq_{\mathfrak{K}} M$ of cardinality μ, $\mathbf{tp}_{\mathfrak{K}}(a_2, M', N) = \mathbf{tp}_{\mathfrak{K}}(a_2, M', N)$.

Now for a beautiful $\mathfrak{s}$, in particular we have amalgamation/stable amalgamation, prime models over a triple of models in stable amalgamation. In particular we can prove the main gap. However, here we just present the characterization of the categoricity spectrum (see 2.20(2)) and delay the rest.
On Chapter IV and Chapter VII see §4(B).

§4 Appetite comes with eating

Here we mainly review open questions, Chapter IV, Chapter VII and further relevant works which could have been part of this book but were not completely ready; so decided not to wait because my record of dragging almost finished books is bad enough even without this case. Note that Chapter IV use infinitary logics and most of Chapter VII has largely set theoretic character hence does not fit §2,§3.

But we begin by looking at what has been described so far has not accomplished. (By this division we end up dealing with some issues more than once.)

(A) The empty half of the glass:

(a) Categoricity in one large enough λ:

We have here concentrated on going up in cardinality, (assuming that in ω successive cardinals there are not too many models without even assuming the existence of models of cardinality $\geq \lambda^{+3}$!). We use weak instances of GCH $(2^\lambda < 2^{\lambda^+})$ and prove a generalization of [Sh 87a], [Sh 87b]. But originally, and it still seems a priori more reasonable, probably even more central case should be to start assuming categoricity in some high enough cardinal. There are several approximations in Makkai-Shelah [MaSh 285], Kolman-Shelah [KlSh 362], [Sh 472] using so called "large cardinals".

(Compact cardinals in the first, measurable cardinal in the second and third).

(b) Main Gap:

If we assume that for some "large enough" λ, we do not have "many very complicated models", we expect to be able to show the class is "managable", hence has a structure theory. But the proofs described above, do not do that job. Not only do we usually start with categoricity assumptions, in our main line here we learn whatever we learn only on the $\lambda^{+\omega}$-brimmed models. However, just on the class of models, i.e., on the original $\mathfrak{K}$, we know little. This is not surprising as, e.g. for elementary classes with countable vocabulary, the solution of Łos conjecture predates the main gap considerably.

(*c*) Superstability:

Having claimed that the superstability is a central dividing lines, it is unsatisfactory to arrive at it here from categoricity assumptions only.

That is, the detailed building of apparatus parallel to superstability is built on examples which are mostly categorical. (But if $\psi \in \mathbb{L}_{\omega_1,\omega}$ or $\mathfrak{K}$ is an abstract elementary class which is $\mathrm{PC}_{\aleph_0}$ and $2^{\aleph_0} + \dot{I}(\aleph_1, \mathfrak{K}) < 2^{\aleph_1}$ this is not so: by II§3 there is a good $\aleph_0$-frame $\mathfrak{s}$ whose $\aleph_1$-saturated models belongs to Mod_ψ but $\mathfrak{s}$ is not necessarily uni-dimensional (which is the "internal" form of categoricity)). Probably the main weakness of beautiful λ-frames as a candidate to being the true superstable is the lack of non-structure results which are not "local", in addition to just failure of categoricity.

(*d*) $\aleph_1$-compact structures:

We may like to relax the definition of abstract elementary class to investigate classes of structures satisfying some kind of countable compactness, i.e., any reasonable countable set of demands has a solution. This will include "$\aleph_1$-saturated models" of an elementary class (even with countable vocabulary) also complete metric spaces but those are closer to elementary classes.

What we lose is closure under unions of ω-chains. For elementary classes this corresponds to $\aleph_1$-saturated models (more generally, $\mathrm{LS}(\mathfrak{K})^+$-saturated) and we have stable instead of superstable (the class of complete metric spaces is closer to elementary classes). We have considerable knowledge on them but much less than on superstable ones. In particular, even for elementary classes with countable vocabulary the main gap is not known.

(*e*) Some unaesthetic points in Theorem 2.17

One of them is that from [Sh 576] we get (in II§3) a good λ^+-frame and not a good λ-frame. Second, we use here for simplicity in the non-structure results an extra set theoretic assumption, though a very weak one.

Namely, the weak diamond ideal on λ^+ is not λ^{++}-saturated. The negation of this statement, if consistent, has high consistency strength. In fact, my attempts to derive good λ-frames from [Sh 576] or dealing with weaker versions had delayed Chapter II considerable.

(f) Lack of Counter-examples:

By Hart-Shelah [HaSh 323], Shelah-Villaveces [ShVi 648] there are some examples for the categoricity spectrum being non-trivial. Still in many theorems on dividing lines it is not proved that they are real, i.e., that there are examples.

(g) Natural Examples:

This bothers me even less than clause (f) but for many investigators the major drawback is lack of "natural examples", i.e., finding classes which are already important where the theory developed on the structure side throw light on the special case. (E.g., for simple theories, pseudo finite fields; for $\aleph_0$-stable theories, differentially closed fields of characteristic zero; for countable stable theories, differentially closed fields of characteristic $p > 0$ (and even separably closed fields of charactertistic $p > 0$)).

(B) The full half and half baked:

Some works throw some light on some of the points from (A), in particular Chapter IV, Chapter VI, Chapter VII. Concerning (a), in Chapter IV we assume an abstract elementary class $\mathfrak{K}$ is categorical in large enough μ and we investigate $\mathfrak{K}_\lambda$ for $\lambda < \mu$ which are carefully chosen, specifically we assume

$(*)_\lambda$ (a) $\mathrm{cf}(\lambda) = \aleph_0$ which means $\lambda = \Sigma\{\lambda_n : n < \omega\}$ for some $\lambda_n < \lambda$

(b) $\lambda = \beth_\lambda$ which means that for every $\kappa < \lambda$ not only $2^\kappa < \lambda$ but $\beth_\kappa < \lambda$

where $\beth_\alpha$ is defined inductively by iterating exponentiation, i.e.,

defining inductively $\beth_\alpha = \aleph_0 + \Sigma\{2^{\beth_\beta} : \beta < \alpha\}$

or even

$(**)_\lambda$ $(a) + (b) + \lambda$ is the limit of cardinals λ' satisfying $(*)_\lambda$.

Are such cardinals large? Not in the set theoretic sense (i.e., provably in ZFC there are such cardinals), they are in some sense analog to the tower function in finite combinatorics. Ignoring "few" exceptional μ, a result of Chapter IV is the existence of a superlimit model in $\mathfrak{K}_\lambda$; moreover the main theorem IV.4.10 of Chapter IV says that there is a good λ-frame $\mathfrak{s}$ with $\mathfrak{K}_\mathfrak{s} \subseteq \mathfrak{K}$; the proof uses infinitary logics. Also if the categoricity spectrum contains arbitrarily large cardinals then for some closed unbounded class $\mathbf{C}$ of cardinals, $[\lambda \in \mathbf{C} \wedge \mathrm{cf}(\lambda) = \aleph_0 \Rightarrow \mathfrak{K}$ categorical in $\lambda]$. It seems reasonable that this can be combined with Chapter III, but there are difficulties.

Having IV.4.10 may still leave us wondering whether we have more tangible argument that we have advance. So we go back to earlier investigations of such general contexts. Now Makkai-Shelah [MaSh 285] deal with $T \subseteq \mathbb{L}_{\kappa,\omega}$ categorical in some μ big enough than $\kappa+|T|$ and develop enough theory to prove that the categoricity spectrum in an end-segment of the cardinals starting not too far, but, with two extra assumptions.

First, κ is a strongly compact cardinal, this is natural as our problem is that $\mathbb{L}_{\kappa,\omega}$ lack many of the good properties of first order logic, and for strongly compact cardinals, some form of compactness is regained (even for $T \subseteq \mathbb{L}_{\kappa,\kappa}$), still very undesirable.

Second, we should assume that μ is a successor cardinal, this exhibit that the theory we build is not good enough. Now Kolman-Shelah [KlSh 362] + [Sh 472] partially rectify the first problem: κ is required just to be a measurable cardinal (instead strongly compact), still measurable is not a small cardinal. Moreover, there is an extra, quite heavy price - we deal with the categoricity spectrum just below μ and say nothing on it above so the categoricity spectrum is proved to be an interval instead of an end-segment. A parallel work [Sh 394] replace measurability by the assumption that our $\mathfrak{K}$ an abstract elementary class with amalgamation; a major point there is trying to deal with the theory problem of locality of types (and see Baldwin [Bal0x]). Note that in both works we get amalgamation of $\mathfrak{K}$ below μ.

We address both cases together, assuming only that our abstract elementary class $\mathfrak{K}$ has the amalgamation property below μ. We try to eliminate those two model theoretic drawbacks: starting from a successor cardinal, and looking only below it, in IV.7.12, using Chapter III. For this we prove that suitable cases of failure of non-structure imply cases of $(< \mu, \kappa)$-locality for saturated models (which means if $p \in \mathscr{S}_{\mathfrak{K}}(M), M \in \mathfrak{K}_{<\mu}$ is saturated then $\langle p \restriction N : N \leq_{\mathfrak{K}} M, \|N\| = \kappa \rangle$ determine p). We also show that every $M \in K_N$ is quite saturated, using a generalization of the stability spectrum for linear orders from IV§6.

Finally, we conclude (also for abstract elementary class) $\mathfrak{K}$ with amalgamations assuming enough cases of $2^\lambda < 2^{\lambda^+}$ we can characterize the categoricity spectrum (eliminating earlier restriction to successor cardinals). This is done showing Chapter III applies, so we need the existence of enough λ, such that $\langle 2^{\lambda^{+n}} : n < \omega \rangle$ is strictly increasing.

So we have eliminated the two thorny model theoretic problems and we eliminated the use of large cardinals but we use this weak form of GCH, we intend to deal with it in [Sh 842].

Considering clause (b) from (A), the main gap, it seems far ahead. A more basic short-coming is that in III§12 we get "$\mathfrak{s}^{+\omega}$ is $\lambda_{\mathfrak{s}}^{+\omega}$-beautiful" and "for beautiful μ-frame $\mathfrak{t}$ we can prove the main gap" but this is just for, essentially, the class of $\lambda_{\mathfrak{s}}^{+\omega}$-saturated models.

Concerning (A)(c), superstability, [Sh 842] suggests "$\mathfrak{K}$ is (λ, κ)-solvable" as the true generalization of superstable (remembering superstability is schizophrenic in our context); this is weaker than categoricity and we use this assumption in Chapter IV when not hard. Essentially it means:

$\boxdot$ for some vocabulary $\tau_1 \supseteq \tau_{\mathfrak{K}}$ of cardinality κ and $\psi \in \mathbb{L}_{\kappa^+,\omega}(\tau_1)$, ψ has a model of cardinality $\geq \beth_{(2^\kappa)^+}$ and $([M \models \psi \wedge \|M\| = \lambda \Rightarrow M \restriction \tau$ is superlimit in $\mathfrak{K}]$.

A major justification for the parallelism with superstability is that for elementary classes this is equivalent to superstability.

But in [Sh 842], III§12 needs to be reworked hopefully toward the needed continuation.

We can look at results from [Sh:c] which were not regained in beautiful λ-frames. Well, of course, we are far from the main gap for the original $\mathfrak{K}$ ([Sh:c, XIII]) and there are results which are obviously more strongly connected to elementary classes, particularly ultraproducts. This leaves us with parts of type theory: semi-regular types, weight, **P**-simple[9] types, "hereditarily orthogonal to **P**" (the last two were defined and investigated in [Sh:a, V,§0 + Def4.4-Ex4.15], [Sh:c, V,§0,pg.226,Def4.4-Ex4.15,pg.277-284]). The more general case of (strictly) stable classes was started in [Sh:c, V,§5] and [Sh 429] and much advanced in Hernandes [He92].

Note that "a type q is p-simple (or **P**-simple)" and "q is hereditarily orthogonal to p (or **P**)" are essentially the[10] "internal" and "foreign" in Hrushovski's profound works.

Some years ago [Sh 839] started to deal with this to some extent. No problem to define weight, but for having "simple" types we need to be somewhat more liberal in the definition of abstract elementary class - allow function symbols of infinite arity (= number of places) while preserving the uniqueness of direct limit. In the right form which includes the case of $\aleph_1$-saturated models of a stable theory, we generalize what was known (for elementary classes); see more in 4.9 and before.

---

[9]The motivation is for suitable **P** (e.g. a single regular type) that on the one hand $\mathrm{stp}(a, A) \pm \mathbf{P} \Rightarrow \mathrm{stp}(a/E, A)$ is **P**-simple for some equivalence relation definable over A and on the other hand if $\mathrm{stp}(a_i, A)$ is **P**-simple for $i < \alpha$ then $\Sigma\{w(a_i, A) \cup \{a_j : j < i\}) : i < \alpha\}$ does not depend on the order in which we list the a_i's. Note that **P** here is $\mathscr{P}$ there.

[10]Note, "foreign to **P**" and "hereditarily orthogonal to **P** are equivalent. Now ($\mathbf{P} = \{p\}$ for ease)

(a) $q(x)$ is $p(x)$-simple when for some set A, in $\mathfrak{C}$ we have $q(\mathfrak{C}) \subseteq \mathrm{acl}(A \cup \bigcup p_i(\mathfrak{C}))$

(b) $q(x)$ is $p(x)$-internal when for some set A, in $\mathfrak{C}$ we have $q(\mathfrak{C}) \subseteq \mathrm{dcl}(A \cup p(\mathfrak{C}))$.

Note

(α) internal implies simple

(β) if we aim at computing weights it is better to stress acl as it covers more

(γ) but the difference is minor and

(δ) in existence it is better to stress dcl, also it is useful that $\{F \restriction (p(\mathfrak{C}) \cup q(\mathfrak{C}) : F$ an automorphism of $\mathfrak{C}$ over $p(\mathfrak{C}) \cup \mathrm{Dom}(p)\}$ is trivial when $q(x)$ is p-internal but not so for p-simple (though form a pro-finite group).

Lastly, considering (A)(e), to a large extent this is resolved as a product of redoing and extending the non-structure theory of [Sh 576] in Chapter VII.

In view of I§5 it is natural to weaken the stability demand to $M \in K_{\mathfrak{s}} \Rightarrow |\mathscr{S}^{\rm bs}_{\mathfrak{s}}(M)| \le \lambda^+_{\mathfrak{s}}$ as otherwise we restrict the class of models (i.e. in II§3 getting semi-good frames are introduced and investigated by Jarden-Shelah [JrSh 875]. Concerning clause (A)(f), Baldwin-Shelah [BlSh 862] expands our knowledge of examples considerably. Concerning clause (A)(g) see Zilber [Zi0xa], [Zi0xb].

In [Sh:F709] may try to axiomatize the end of I§5 and connect it to good $\aleph_0$-frames, [Sh:E54] will say more on Chapter II. In Chapter VII we also deal with the positive theory of almost good frame and weak versions of $K^{3,{\rm uq}}_{\mathfrak{s}}$. Also [Sh:F735] will consider redoing Chapter III under weaker assumptions and getting more and [Sh:F782] will continue Chapter IV, e.g. how the good λ-frame from IV§4 fit Chapter III. Also [Sh:F888] will try to continue [Sh:E56], and [Sh:F841] to continue Chapter VII.

(C) The white part of the map:

So we would really like to know

4.1 Problem: What can be the categoricity spectrum Cat-Spec$_{\mathfrak{K}}$ = $\{\lambda : \mathfrak{K}$ is categorical$\}$ for an abstract elementary class ?

This seems too hard at present and involves independence results. Note also that easily (by known results, see [Ke70] or see ([Sh:c, VII,§5]) for any $\alpha < \omega_1$ for some abstract elementary class $\mathfrak{K}$ (with LS($\mathfrak{K}$) = $\aleph_0$) we have: $\lambda \in$ Cat-Spec$_{\mathfrak{K}} \Leftrightarrow \lambda > \beth_\alpha$ (just let $\psi = \psi_1 \vee \psi_2 \in \mathbb{L}_{\omega_1,\omega}(\tau), \psi_1$ has a model of cardinality λ iff $\lambda \le \beth_\alpha$ and ψ_2 says that all predicates and function symbols are trivial).

Considering the history it seemed to me that the main question on our agenda should be

4.2 Conjecture: If $\mathfrak{K}$ is an abstract elementary class then either every large enough λ belongs to Cat-Spec$_{\mathfrak{K}}$ or every large enough λ does not belong to Cat-Spec$_{\mathfrak{K}}$ (provably in ZFC).

After (or you may say if) this is resolved positively we should consider

4.3 Conjecture. 1) If $\mathfrak{K}$ is an a.e.c. with LS($\mathfrak{K}$) = χ then

(a) Cat-Spec$_{\mathfrak{K}}$ includes or is disjoint to $[\beth_\omega(\chi), \infty)$ or even better

(a)$^+$ similarly for $[\lambda_\omega, \infty)$ where $\lambda_0 = \chi, \lambda_{n+1} = \min\{\lambda : 2^\lambda > 2^{\lambda_n}\}, \lambda_\omega = \Sigma\{\lambda_n : n < \omega\}$

probably more realistic are

(b) similarly for $[\beth_{(2^\chi)^+}, \infty)$, or at least

(c) similarly $(\beth_{1,1}(\chi), \infty)$ or at least $(\beth_{1,\omega^\omega}(\chi), \infty)$, see IV§0.

This will be parallel in some sense to the celebrated investigations of the countable models for (first order) countable T categorical in $\aleph_1$.

Further questions are: (recall ⊡ above)

4.4 Question: What can be $\{(\lambda, \kappa) : \mathfrak{K}_\lambda$ is (λ, κ)-solvable, $\lambda >> \kappa >> \mathrm{LS}(\mathfrak{K})\}$?

Question 4.4 seems to us to be more profound than the categoricity spectrum as solvability is a form of superstability. We conjecture that the situation is as in 4.3(c); note that solvability seems close to categoricity and we have a start on it (Chapter IV, [Sh 842]).

Still more easily defined (but a posteri too early for us) is:

4.5 Question: 1) What can be $\{\lambda : \mathfrak{K}_\lambda$ has a superlimit model$\}$?

2) Similarly for locally superlimit (see IV.0.4).

3) For suitable Φ what can be $\{\lambda$: if I is a linear order of cardinality λ then $\mathrm{EM}_{\tau(\mathfrak{K})}(I, \Phi)$ is pseudo superlimit$\}$? see IV.0.5(3).

We conjecture it will be a variant of 4.3 but will be harder and even:

4.6 Conjecture. If $\lambda > \beth_{1,1}(\mathrm{LS}_{\mathfrak{K}})$ (or $\lambda > \beth_{1,\omega}(\mathrm{LS}_{\mathfrak{K}})$, then $\mathfrak{K}$ has a superlimit model in λ iff $\mathfrak{K}$ is $(\lambda, \mathrm{LS}_{\mathfrak{K}})$-solvable.

We now return to (D, λ)-homogeneous models. Of course, for special D's we may be interested in some special classes of models, but not necessarily the elementary sub-models of $\mathfrak{C}$. Of course, parallely to the first order case, the main gap for them is an important problem (e.g. the class of existentially closed models of a universal first order theory is a natural and important case). But the most natural main case seems to me the "$\mathfrak{C}$ is (D,κ)-sequence homogeneous" context:

4.7 Problem: Prove the main gap for the class of (D,κ)-sequence-homogeneous $M \prec \mathfrak{C}$; considering what we know, we can assume $\kappa \geq \kappa(D)$, see [Sh 3] (and §1(B)) and concentrate on $\kappa \geq \aleph_1$ and we would like to prove that

(a) either the number of such models of cardinality $\aleph_\alpha = \aleph_\alpha^{<\kappa(D)} + \lambda(D)$ is small, i.e., $\leq \beth_{\gamma(D)}(|\alpha|)$ for[11] some $\gamma(D)$ not depending on α or the number is $2^{\aleph_\alpha}$ (where $\lambda(D)$ is the first "stability cardinal" of D).

(b) $\gamma(D)$ does not depend on κ.

A parallel of "the main gap for the class of $\aleph_\varepsilon$-saturated models of a first order T" in this context is dealt with in Hyttinen-Shelah [HySh 676], and a parallel to the "main gap for the class of model of a totally transcendental first order T" in Grossberg-Lessman [GrLe0x], and surely there is more to be said in those cases but in the problem above, even the case $\kappa = \aleph_1, \mathfrak{C}$ saturated is not covered.

We hope eventually to find a stability theory for the "countably compact abstract elementary class" strong enough to prove as a special case the main gap for the $\aleph_1$-saturated models of elementary classes (i.e., clause (d) of (A)) as said above maybe [Sh 839] help.

The reader may wonder: if not known for elementary classes why you expect more from a general frame? Of course, we do not know, but:

4.8 Thesis: The better closure properties of the abstract frames should help us, being able to, e.g., make induction on frames.

Hence

4.9 Thesis: Some problems on elementary classes are better dealt

[11] of course, $\beth_{\gamma(D)}(|\alpha|)$ may be $\geq 2^{\aleph_\alpha}$ in which case this says little; this consistently occurs for every $\alpha \geq \omega$. But if G.C.H. holds, and if we ask on $\dot{I}\dot{E}(\lambda, -)$ for the class we get clear cut results

with in some non-elementary contexts (close to abstract elementary class), as if we would like during the proof to consider some derived other classes, those contexts give you more freedom. In particular this may apply to

(*a*) main gap for $|T|^+$-saturated models (the parallel of (D, λ)-sequence-homogeneous above in 4.7 and (d) of (A) and discussion on it in (B))

(*b*) the main gap for the class of models of T for an uncountable first order T.

Note that [Sh 300], Chapter II has tried to materialize this, but that program is not finished.

4.10 Problem: Similar questions for the number of pairwise non-elementarily embeddable (D, λ)-sequence homogeneous models.

In the case of the class of models (not the class of $\aleph_1$-saturated models) for countable first order theories, those two problems were solved together.

There are many other interesting questions in this context. An important one, of a different character is:

4.11 Problem: 1) [Hanf number for sequence homogeneous]

Given a cardinal κ, what is the first λ such that: if T is a complete first order theory, $D \subseteq D(T) = \{\text{tp}(\bar{a}, \emptyset, M) : M \text{ a model of } T, \bar{a} \in {}^{\omega>}M\}$ and there is a (D, λ)-sequence-homogeneous model, then for every $\mu > \lambda$ there is a (D, μ)-sequence homogeneous model.
2) Similarly for $\{\kappa$: in $\mathfrak{K}$ we have amalgamation for models of cardinality $< \kappa$ (and $\kappa \geq \text{LS}(\mathfrak{K}) > \aleph_0\}$).
3) Similarly for $(\mathbb{D}, \lambda)$-model homogeneous models (see V.B§3).

Toward this we may define semi-beautiful classses as in III§12 (or [Sh 87a], [Sh 87b]) replacing the stable $\mathscr{P}^-(n)$-systems by an abstract notion, omitting uniqueness and the definability of types and retaining existence. Semi-excellent classes seem like an effective version of having amalgamation, so it certainly implies it; such properties may serve as what we actually have to prove to solve the problem 4.11 above. We may have to use more complicated frames: say classes $\mathfrak{K}_n$ so that $M \in \mathfrak{K}_n$ is actually a $\mathscr{P}^-(n)$-system of models from $\mathfrak{K}$. (See more in [Sh 842]).

Recall that a class $\mathfrak{K}$ of structures with fixed vocabulary τ is called universal if it is closed under isomorphisms, and $M \in \mathfrak{K}$ if and only if every finitely generated submodel of M belongs to $\mathfrak{K}$. So not every elementary class is a universal class, but many universal classes are not first order (e.g., locally finite groups). This investigation leads (see [Sh 300], Chapter II) to classes with an axiomatized notion of non-forking and much of [Sh:c] was generalized, sometimes changing the context (a case of Thesis 4.9), but, e.g., still:

4.12 Problem: Prove the main gap for the universal context.

4.13 Question: Can we in [Sh 576], i.e. Chapter VI weaken the "categorical in λ^+" to "has a superlimit model in λ^+"?

See on this hopefully [Sh:F888].

4.14 Question: Do we use a parallel of III§12 with existential side for serious effect? (See more in [Sh 842]).

§5 Basic knowledge

(A) What knowledge needed and dependency of the chapters

The chapters were written separately, hence for better or for worse there are some repetitions, hopefully helping the reader if he likes to read only parts of this book.

Chapter III depends on Chapter II and Chapter VI depends somewhat, e.g. on II§1, but in other cases there are no real dependency.

In fact, reading Chapter II, Chapter III requires little knowledge of model theory, they are quite self-contained, in particular you do not need to know Chapter I, Chapter II; this apply also to Chapter II and to Chapter VI. Of course, if a claim proves that the axioms of good λ-frames are satisfied by the class of models of a sentence ψ in a logic you have not heard about, it will be a little loss for you to ignore the claim (this occurs in II§3). Still much of the material is motivated by parallelism to what we know in elementary (= first order) contexts. Let me stress that neither do we see any merit in not using large model theoretic background nor was its elimination an a priori aim, but there is no reason to hide this fact from a potential reader who may feel otherwise.

Also the set theoretic knowledge required in Chapter II, Chapter III is small; still we use cardinals and ordinals of course, induction on ordinals, cofinality of an ordinal, so regular cardinals, see here below for what you need. A priori it seemed that somewhat more is needed in the proof of the non-structure theorems, i.e., showing a class with a so-called "non-structure property" has many, complicated models so cannot have a structure theory. But we circumvent this by quoting Chapter VII, or you can say delaying the proof. That is, we carry the construction enough to give a reasonable argument. So the reader can just agree to believe; similarly in Chapter II and in Chapter VI.

In Chapter VII itself, we rely somewhat on basics of II§1, and in the applications (VII§4) we somewhat depend on the relevant knowledge and for VII§5-§8 we assume the basics of II§2. Also VII§9,§10,§11 are set theoretic, mainly use results on the weak diamond which we quote.

The situation is different in Chapter I. Still you can read §1, §2, §3 of it ignoring some claims but in §4,§5 the infinitary logics $\mathbb{L}_{\omega_1,\omega}(\mathbf{Q})$ and its relatives and basic theorems on them are important.

For Chapter IV you need basic knowledge of infinitary logics and Ehenfeucht-Mostowski models, and in IV§4 (the main theorem) we use the definition of good λ-frame from II§2.

(B) Some basic definitions and notation

We first deal with model theory and then with set theory.

5.1 Definition. 1) A vocabulary τ is a set of function symbols (denoted by G, H, F) and relation symbols, (denoted by P, Q, R) (= predicates), to each such symbol a number of places (= arity) is assigned (by τ) denoted by $\text{arity}_\tau(F)$, $\text{arity}_\tau(P)$, respectively.
2) M is a τ-model or a τ-structure for a vocabulary τ means that M consists of:

- (a) its universe, $|M|$, a non-empty set
- (b) P^M, the interpretation of a predicate $P \in \tau$ and P^M is an $\text{arity}_\tau(P)$-place relation on $|M|$
- (c) F^M, the interpretation of a function symbol $F \in \tau$ and F^M is an $\text{arity}_\tau(F)$-place function from $|M|$ to $|M|$ in the case of arity $0, F^M$ is an individual constant.

3) We agree τ is determined by M and denote it by τ_M. If $\tau_1 \subseteq \tau_2$, M_2 a τ_2-model, then $M_1 = M_2 \restriction \tau_1$, the reduct is naturally defined.
4) The cardinality of M, $\|M\|$, is the cardinality, number of elements of the universe $|M|$ of M. We may write $a \in M$ instead of $a \in |M|$ and $\langle a_i : i < \alpha \rangle \in M$ instead $i < \alpha \Rightarrow a_i \in M$, i.e., $\bar{a} \in {}^{\alpha}|M|$.
5) Let $M \subseteq N$ mean that

$$\tau_M = \tau_N, |M| \subseteq |N|, P^M = P^N \restriction |M|, F^M = F^N \restriction |M|$$

for every predicate $P \in \tau_M$ and for every function symbol $F \in \tau_M$.
6) If N is a τ-model and A is a non-empty subset of $|M|$ closed under F^N for each function symbol $F \in \tau$, <u>then</u> $N \restriction A$ is the unique $M \subseteq N$ with universe A.

5.2 Definition. 1) K denotes a class of τ-models closed under isomorphisms, for some vocabulary $\tau = \tau_K$.
2) $\mathfrak{K}$ denotes a pair $(K, \leq_{\mathfrak{K}})$; K as above (with $\tau_{\mathfrak{K}} := \tau_K$) and $\leq_{\mathfrak{K}}$ is a two-place relation on K closed under isomorphisms such that $M \leq_{\mathfrak{K}} N \Rightarrow M \subseteq N$.
3) f is a $\leq_{\mathfrak{K}}$-embedding of M into N <u>when</u> for some $N' \leq_{\mathfrak{K}} N, f$ is an isomorphism from M onto N'.
4) K is categorical in λ <u>if</u> K has one and only one model up to isomorphism of cardinality λ. If $\mathfrak{K} = (K, \leq_{\mathfrak{K}})$ we may say "$\mathfrak{K}$ is categorical in λ".

5.3 Definition. 1) For a class K (or $\mathfrak{K}$) of τ_K-models

(a) $K_\lambda = \{M \in K : \|M\| = \lambda\}$

(b) $\mathfrak{K}_\lambda = (K_\lambda, \leq_{\mathfrak{K}} \restriction K_\lambda)$

(c) $\dot{I}(\lambda, K) = \dot{I}(\lambda, \mathfrak{K}) = |\{M/\cong : M \in K_\lambda\}|$ so K (or $\mathfrak{K}$) is categorical in λ iff $\dot{I}(\lambda, K) = 1$

(d) $\dot{I}\dot{E}(\lambda, \mathfrak{K}) = \sup\{\mu$: there is a sequence $\langle M_\alpha : \alpha < \mu \rangle$ of members of K_λ such that M_α is not $\leq_{\mathfrak{K}}$-embeddable into M_β for any distinct $\alpha, \beta < \mu\}$. But writing $\dot{I}\dot{E}(\lambda, \mathfrak{K}) \geq \mu$ we mean the supremum is obtained if not said otherwise.

(e) $M \in \mathfrak{K}$ is $(\leq_{\mathfrak{K}}, \lambda)$-universal if every $N \in \mathfrak{K}_\lambda$ can be $\leq_{\mathfrak{K}}$-embedded into it. If $\lambda = \|M\|$ we may write $\leq_{\mathfrak{K}}$-universal. If $\mathfrak{K}$ is clear from the context we may write λ-universal or universal (for $\mathfrak{K}$).

We end the model-theory part by defining logics (this is not needed for Chapter II, Chapter III, Chapter VI and Chapter II except some parts of Chapter V.A).

5.4 Definition. A logic $\mathscr{L}$ consisting of

(a) function $\mathscr{L}(-)$ (actually a definition) giving for every vocabulary τ a set of so-called formulas $\varphi(\bar{x}), \bar{x}$ a sequence of free variables with no repetitions

(b) $\models_{\mathscr{L}}$, satisfaction relation, i.e., for every vocabulary τ and $\varphi(\bar{x}) \in \mathscr{L}(\tau)$ and τ-model M and $\bar{a} \in {}^{\ell g(\bar{x})}M$ we have "$M \models_{\mathscr{L}} \varphi[\bar{a}]$" or in words "$M$ satisfies $\varphi[\bar{a}]$"; holds or fails.

As for set theory

5.5 Definition. 1) A power = number of elements of a set, is identified with the first ordinal of this power, that is a cardinal. Such ordinals are called cardinals, $\aleph_\alpha$ is the α-th infinite ordinal.
2) Cardinals are denoted by $\lambda, \mu, \kappa, \chi, \theta, \partial$ (infinite if not said otherwise).

5.6 Definition. 0) Ordinals are denoted by $\alpha, \beta, \gamma, \delta, \varepsilon, \zeta, \xi, i, j$, but, if not said otherwise δ denotes a limit ordinal.
1) An ordinal α is a limit ordinal if $\alpha > 0$ and $(\forall \beta < \alpha)[\beta + 1 < \alpha]$.
2) For an ordinal α, $\mathrm{cf}(\alpha)$, the cofinality of α, is $\min\{\mathrm{otp}(u) : u \subseteq \alpha$ is unbounded$\}$; it is a regular cardinal (see below), we can define the cofinality for linear orders and again get a regular cardinal.
3) A cardinal λ is regular if $\mathrm{cf}(\lambda) = \lambda$, otherwise it is called singular.
4) If $\lambda = \aleph_\alpha$ then $\lambda^+ = \aleph_{\alpha+1}$, the successor of λ, so $\lambda^{++} = \aleph_{\alpha+2}, \lambda^{+\varepsilon} = \aleph_{\alpha+\varepsilon}$.

Recall:

5.7 Claim. *1) If λ is a regular cardinal, $|\mathscr{U}_t| < \lambda$ for $t \in I$ and $|I| < \lambda$ then $\cup\{\mathscr{U}_t : t \in I\}$ has cardinality $< \lambda$.*
2) λ^+ is regular for any $\lambda \geq \aleph_0$ but $\lambda^{+\delta}$ is singular if δ is a limit ordinal $< \lambda$ (or just $< \lambda^{+\delta}$), and, obviously, $\aleph_0$ is regular but e.g. $\aleph_\omega$ is singular, in fact $\aleph_\delta > \delta \Rightarrow \aleph_\delta$ is singular, but the inverse is false.

Sometimes we use (not essential)

5.8 Definition/Claim. 1) $\mathscr{H}(\lambda)$ is the set of x such that there is a set Y of cardinality $< \lambda$ which is transitive (i.e. $(\forall y)(y \in Y \Rightarrow y \subseteq Y)$ and x belongs to λ.
2) Every x belongs to $\mathscr{H}(\lambda)$ for some x.
So for some purpose we can look at $\mathscr{H}(\lambda)$ instead of the universe of all sets.

§6 Index of symbols[12]

a member of a model

A set of elements of model

$\mathfrak{A}$ a "complicated" model

b member of a model

B set of members of models

$\mathfrak{B}$ a "complicated" model

c member of model (also individual constant)

$\mathbf{c}$ colouring, mainly Chapter VII

C set or elements of models or a club

$\mathscr{C}$ club of $[A]^{<\lambda}$,

$\mathfrak{C}$ a complicated model, or a monster

d member of model

$\mathbf{d}$ expanded I-system, III§12; $\mathfrak{u}$-free rectangle or triangle in Chapter VII

[12] some will be used only in subsequent works; in particular concerning forcing

D diagram; set of $(<\omega)$-types in the first order sense realized in a model, Chapter I, Chapter V.B

$\mathbf{D}$ a function whose values are diagrams, Chapter I, Chapter V.B

$\mathbb{D}$ diagram for model homogeneity, Chapter I, so set of isomorphism types of models, also Chapter V.B

$\mathfrak{D}$ a set of $\mathbb{D}$'s, Chapter V.B

$\mathscr{D}$ filter

$\mathscr{D}_\lambda$ club filter on the regular cardinal $\lambda > \aleph_0$

e element of a model or a club

$\mathbf{e}$ expanded I-system (used in continuations), III§12; $\mathfrak{u}$-free rectangle or triangle in Chapter VII

E a club

$\mathbb{E}$ filter

$\mathscr{E}$ an equivalence relations, (e.g. $\mathscr{E}_M, \mathscr{E}_M^{\rm at}$ in II.1.9 for definition of type and $\mathscr{E}_{\mathfrak{K},\chi}, \mathscr{E}_{\mathfrak{K},\chi}^{\rm mat}$ in V.B§3)

f function (e.g., isormorphism, embedding usually)

$\mathbf{f}$ function (Chapter VII in $(\bar{M}, \bar{\mathbf{J}}, \mathbf{f}) \in K_{\rm u}^{3,\rm qt}$, also in II§5, $(\bar{M}, \bar{\mathbf{f}}), (\bar{M}, \bar{\mathbf{J}}, \mathbf{f})$)

F function symbol

$\mathbb{F}$ amalgamation choice function (Chapter VII also see [Sh 576, §3])

$\mathbf{F}$ function (complicated, mainly it witnesses a model being limit, I§3)

g function

$\mathfrak{g}$ witness for almost every $(\bar{M}, \bar{\mathbf{J}}, \mathbf{f})$ see VII.1.22–VII.1.26

G function symbol

$\Game$ game

h function

$\mathfrak{h}$ witnesses for almost every $(\bar{M}, \bar{\mathbf{J}}, \mathbf{f}) \in K_{\rm u}^{\rm qt}$, see VII.1.22–VII.1.26

H function symbol

$\mathscr{H}$ in $\mathscr{H}(\lambda)$, rare here see 5.8

i ordinal/natural number

I linear order, partial order or index set

$\dot{I}$ $\dot{I}(\lambda, K)$, numbers on non-isomorphic models; $\dot{I}\dot{E}(\lambda, K)$" (see Chapter I), also $\dot{I}(K)$, see Chapter VII

$\mathbf{I}$ set of sequences or elements from a model, in particular: $\mathbf{I}_{M,N} = \{c \in N : \mathbf{tp}_{\mathfrak{s}}(c, M, N) \in \mathscr{S}^{\rm bs}_{\mathfrak{s}}(M)\}$, see Chapter II, Chapter III

$\check{I}[\lambda]$ a specific normal ideal, see I§0, marginal here

$\mathbb{I}$ ideal

$\mathscr{I}$ predense set in a forcing $\mathbb{P}$, very rare here

j ordinal/natural number

J linear order, index set, Chapter I

$\mathbf{J}$ set of sequences or elements from a model

$\mathbb{J}$ ideal

$\mathscr{J}$, predense set in a forcing $\mathbb{P}$, very rare here

k natural number

K class of model of a fix vocabulary $\tau_{\mathfrak{K}}$, K_λ is $\{M \in K : \|M\| = \lambda\}$

$\mathfrak{K}$ is $(K, \leq_{\mathfrak{K}})$, usually abstract elementary class

$K^{3,x}_{\mathfrak{s}}$ for $x = \{$bs,uq,pr,qr,vq,bu$\}$, appropriate set of triples (M, N, a) or $(M, N, \mathbf{I})$, see Chapter II, Chapter III

$K^{3,\rm na}_{\lambda}$ for triples (M, N, a), see Chapter VI

$K^{3,x}_{\mathfrak{u}}$ set of triples $(M, N, \mathbf{J}) \in$ $\mathrm{FR}^{\ell}_{\mathfrak{u}}$, see Chapter VII

ℓ natural number

L language (set of formulas, e.g., $\mathscr{L}(\tau)$ but also subsets of $\mathscr{L}(\tau)$ which normally are closed under subformulas and first order operations), used in Chapter I.

LS Löwenheim-Skolem numbers, mainly $\mathrm{LS}(\mathfrak{K}) =$ $\mathrm{LS}_{\mathfrak{K}}$

$\mathscr{L}$ logic, i.e., a function such that $\mathscr{L}(\tau)$ is a language for vocabulary τ (but also a language mainly $\mathscr{L}$ a fragment of $\mathbb{L}_{\lambda^+,\omega}$, i.e., a subset closed under subformulas and the finitary operations)

$\prec_{\mathscr{L}}$ is used for $M \prec_{\mathscr{L}} N$ iff $M \subseteq N$ and for every $\varphi(\bar{x}) \in \mathscr{L}(\tau_M)$ and $\bar{a} \in {}^{\ell g(\bar{x})}M$ we have $M \models \varphi[\bar{a}] \Leftrightarrow N \models \varphi[\bar{a}]$

$\mathbb{L}$ first order logic and $\mathbb{L}_{\lambda,\kappa}, \mathbb{L}^{\ell}_{\lambda,\kappa}$, see Chapter I so $\varphi(\bar{x}) \in \mathbb{L}_{\lambda,\kappa}$ has $< \kappa$ free variables

$\mathbf{L}$ the constructible universe

m natural number

$\mathbf{m}$ an I-system in III§12

M model

$\mathbf{M}$ complicated object, see VI§3,§4

n natural number

$\mathbf{n}$ an I-system in III§12, for continuation and in Chapter V.F

N model

$\mathbb{N}$ the natural numbers

p type

$\mathbf{p}$ member of $\mathbb{P}$, a forcing condition, very rare here

P predicate

$\mathscr{P}$ power set, family of sets,

$\mathbf{P}$ family of types, Chapter III

$\mathbb{P}$ forcing notion, very rare here

q type

$\mathbf{q}$ forcing condition, very rare here

Q predicate

$\mathbf{Q}$ a quantifier written $(\mathbf{Q}x)\varphi$, see Chapter I, if clear from the context means $\mathbf{Q}^{\mathrm{car}}_{\geq\aleph_1}$

$\mathbf{Q}^{\mathrm{car}}_{\geq\kappa}$ the quantifier there are $\geq \kappa$ many

$\mathbb{Q}$ the rationals

r type

$\mathbf{r}$ forcing condition, very rare here

R predicate

$\mathbb{R}$ reals

s member of I, J

$\mathfrak{s}$ frame

S set of ordinals, stationary set many times

$\mathscr{S}$ $\mathscr{S}_{\mathfrak{K}}(M)$ is a set of types in the sense of orbits, $\mathscr{S}^{\mathrm{bs}}_{\mathfrak{s}}(M)$ the basic types (there are some alternatives to bs)

$\mathbf{S}$ or $\mathbf{S}^{\alpha}_{L}(A, M)$: set of complete (L, α)-types over M, so a set of formulas, used when we are dealing with a logic $\mathscr{L}$, may use $\mathbf{S}^{\alpha}_{\mathscr{L}}(A, M)$

$\mathbb{S}$ or $\mathbb{S}(M)$ is a set of pseudo types, are neither set of formulas nor orbits, but formal non-forking extension (for continuations, see [Sh 842])

t member of I, J

tp type as set of formulas

tp type as an orbit, an equivalence class under mapping

$\mathbf{t}$ type function

$\mathfrak{t}$ frame

T first order theory, usually complete

$\mathscr{T}$ a tree

u a set

$\mathfrak{u}$ a nice construction framework, in Chapter VII

unif in $\mu_{\mathrm{unif}}(\lambda, 2^{<\lambda})$, see I.0.5 or VII.0.4(6)

U a set

$\mathscr{U}$ a set

v a set

V a set

$\mathbf{V}$ universe of set theory

w a set

W a set (usually of ordinals)

$\mathscr{W}$ a class of triples $(N, \bar{M}, \bar{\mathbf{J}})$; see III§7

wd in $\mu_{\mathrm{wd}}(\lambda)$ see I§0, VII§0

WDmId_λ the weak diamond ideal, see I.0.5

x variable (or element)

$\mathbf{x}$ complicated object, in Chapter VII such that is a sequence $\langle(\bar{M}^\alpha, \bar{\mathbf{J}}^\alpha, \mathbf{f}^\alpha) : \alpha < \alpha(*)\rangle$

X set

y variable

$\mathbf{y}$ like $\mathbf{x}$

Y set

$\mathscr{Y}$ a high order variable (see I§3)

z variable

Z set

$\mathbb{Z}$ the integers

<u>Greek Letters</u>:

α ordinal

β ordinal

γ ordinal

Γ various things; in Chapter VI a set of models or types

δ ordinal, limit if not clear otherwise

∂ cardinal

Δ set of formulas (may be used for symmetric difference)

ϵ ordinal

ε ordinal

ζ ordinal

η sequence, usually of ordinals

θ cardinal, infinite if not clear otherwise

ϑ a formula, very rare

Θ set of cardinals/class of cardinals

ι ordinal (sometimes a natural number)

κ cardinal, infinite if not clear otherwise

λ cardinal, infinite if not clear otherwise

$\lambda(\mathfrak{K})$ is the L.S.-number of an abstract elementary class ($\geq |\tau_{\mathfrak{K}}|$ for simplicity), rare

Λ set of formulas, used in Chapter IV, Chapter I

μ cardinal, infinite if not said otherwise

ν sequence, usually of ordinals

ξ ordinal

Ξ a complicated object

π permutation

Π product

ρ sequence, usually of ordinals

ϱ sequence, usually of ordinals

σ a term (in a vocabulary τ)

Σ sum

τ vocabulary (so $\mathscr{L}(\tau), \mathbb{L}(\tau), \mathbb{L}_{\lambda,\mu}(\tau)$ are languages)

Υ ordinal and other objects

φ formula

Φ blueprint for EM-models

χ cardinal, infinite if not said otherwise

ψ formula

Ψ blueprint for EM-models

ω the first infinite ordinal

Ω a complicated object

ANNOTATED CONTENTS

ANNOTATED CONTENT FOR CH.N (E53): INTRODUCTION

[We first explain by examples and then give a full definition of an a.e.c. (abstract elementary class), central in our context, $\mathfrak{K} = (K, \leq_{\mathfrak{K}})$, with K a class of models (= structures), $\leq_{\mathfrak{K}}$ a special notion of being a submodel, it means having only the quite few of the properties of an elementary class (like closure under direct limit). Such a class is $(\text{Mod}_T, \prec)$ with $M \prec N$ meaning "being an elementary submodel"; but also the class of locally finite groups with $\subseteq$ is O.K. Second, we explain what is a superlimit model (meaning mainly that a $\leq_{\mathfrak{K}}$-increasing chain of models isomorphic to it has a union isomorphic to it (if not of larger cardinality). We can define "an a.e.c. is superstable" if it has a superlimit model in every large enough cardinality. For first order class this is an equivalent definition. A stronger condition (still equivalent for elementary classes) is being solvable: there is a $\text{PC}_{\lambda,\lambda}$-class, i.e the class of reducts of some $\psi \in \mathbb{L}_{\lambda^+,\omega}$ which, in large

Typeset by $\mathcal{A}_{\mathcal{M}}\mathcal{S}$-TEX

enough cardinality, is the class of superlimit models; similarly we define being (μ, λ)-solvable. Of course we investigate the one cardinal version (hoping for equivalent behaviour) in all large enough cardinals, etc. We state the problem of the categoricity spectrum and the solvability spectrum. We finish explaining the parallel situation for first order classes and explain "dividing lines".]

(B) The structure/non-structure dichotomy,

[We define the function $\dot{I}(\lambda, K)$ counting the number of non-isomorphic models from K of cardinality λ, define the main gap conjecture, phrase and discuss some thesis explaining an outlook and intention. We then explain the main gap conjecture and the case it was proved and list the possible reasons for having many models. We then discuss dividing lines and their relevance to our problems.]

(C) Abstract elementary classes,

[We shall deal with a.e.c., good λ-frames and beautiful λ-frames. The first is very wide so we have to justify it by showing that we can say something about them, that there is a theory; the last has excellent theory and we have to justify it by showing that it arises from assumptions like few non-isomorphic models (and help prove theorems not mentioning it); the middle one needs justifications of both kinds. In this part, we concentrate on the first, a.e.c., explain the meaning of the definition, discuss examples, phrase our opinion on its place as a thesis, and present two theorems showing the function $\dot{I}(\lambda, \kappa)$ is not "arbitrary" under mild set theoretic conditions.]

(D) Toward good λ-frames,

[We explain how we arrive to "good λ-frame $\mathfrak{s}$" mentioned above, which is our central notion; it may be considered a "bare bone case of superstable class in one cardinal". We choose to concentrate on one cardinal λ, so $K_{\mathfrak{s}} = K_\lambda$. Also we may assume $\mathfrak{K}_\lambda$ has a superlimit model, and that it has amalgamation and the joint embedding property, so only in λ! Amalgamation is an "expensive" assumption, but amalgamation in one cardinal is much less so. This crucial difference holds because it is much easier to prove amalgamation in one cardinality (e.g. follows from having one model in λ (or a superlimit one) and few models in λ^+ up to isomorphism and mild set theoretic assumptions). We are interested in something like "M_1, M_2 are in non-forking (= free) amalgamation over M_0 inside M_3". But in the axioms we only have "an element a and model M_1 are in non-forking amalgamation over M_0 inside M_3, equivalently $\mathbf{tp}_{\mathfrak{s}}(a, M_1, M_3)$ does not fork over M_0", however the type is orbital, i.e. defined by the existence mapping and not by formulas. There are some further demands saying non-forking behave reasonably (mainly: existence/uniqueness of extensions, transitivity and a kind of symmetry). So far we have described a good λ-frame. Now we consider a dividing line - density of the class of appropriate triples (M, N, a) with unique amalgamation. Failure of this gives $\dot{I}(\lambda^{++}, K^{\mathfrak{s}})$ is large if $2^\lambda < 2^{\lambda^+} < 2^{\lambda^{++}}$, from success (i.e. density) we derive the existence of non-forking amalgamation of models in $K_{\mathfrak{s}}$. After considering a further dividing line we get $\mathfrak{s}^+$, a good λ^+-frame such that $K^{\mathfrak{s}^+}_\mu \subseteq K^{\mathfrak{s}}_\mu$ for $\mu \geq \lambda^+$. All this (in Chapter II) gives the theorem: if $2^\lambda < 2^{\lambda^+} < \ldots < 2^{\lambda^{+n}}$, $\mathrm{LS}(\mathfrak{K}) \leq \lambda$, $\mathfrak{K}$ categorical in λ, λ^+, has a model in λ^{+2} and has not too many models in $\lambda^{+2}, \ldots, \lambda^{+n}$ then K has a model in λ^{+n+1}. If this holds for every n, we get categoricity in all cardinals $\mu \geq \lambda$. For the first result (from Chapter II) we just need to go from $\mathfrak{s}$ to $\mathfrak{s}^+$, for the second (from Chapter III) need considerably more.]

§3 Good λ-frames,

(A) getting a good λ-frame,

[We deal more elaborately on how to get a good λ-frame starting with few non-isomorphic models in some cardinals. If $\mathfrak{K}$ is categorical in λ, λ^+ and $2^\lambda < 2^{\lambda^+}$ we know that $\mathfrak{K}$

has amalgamation in λ. Now we define the (orbital) type $\mathbf{tp}_{\mathfrak{K}_\lambda}(a, M, N)$ for $M \leq_{\mathfrak{K}} N, a \in N$. Instead of dealing with $\mathscr{S}_{\mathfrak{K}_\lambda}(M)$, the set of such types, we deal with $K^{3,\mathrm{na}}_\lambda = \{(M, N, a) : M \leq_{\mathfrak{K}_\lambda} N \text{ and } a \in N \backslash M\}$, ordered naturally (fixing a!) The point is of dealing with triples, not just types, is the closureness under increasing unions, so existence of limit. Now we ask: are there enough minimal triples? (which means with no two contradictory extensions). If no, we have a non-structure result. If yes, we can deduce more and eventually get a good λ-frame. Here we consider $K^{3,\mathrm{bs}}_{\mathfrak{s}} = \{(M, N, a) : M \leq_{\mathfrak{K}_{\mathfrak{s}}} N, \mathbf{tp}(a, M, N) \in \mathscr{S}^{\mathrm{bs}}_{\mathfrak{s}}(M)$, i.e. is a basic type$\}$ (this is part of the basic notions of a good λ-frame $\mathfrak{s}$).]

(B) the successor of a good λ-frame,

[We elaborate the use of successive good frames in Chapter II. If $\mathfrak{s}$ is a good λ-frame, we investigate "N is a brimmed extension of M in $\mathfrak{K}_{\mathfrak{s}} = \mathfrak{K}^{\mathfrak{s}}_\lambda$", it is used here instead of saturated models, noting that as $K^{\mathfrak{s}}_{<\lambda}$ may be empty we cannot define saturated models. We now consider the class $K^{3,\mathrm{uq}}_{\mathfrak{s}}$ of triples $(M, N, a) \in K^{3,\mathrm{bs}}_{\mathfrak{s}}$ such that if $M \leq_{\mathfrak{K}} M^+$ then M^+, N can be $\leq_{\mathfrak{K}}$-amalgamated uniquely over M as long as the type of a over M^+ does not fork over M. If the class of uniqueness triples (M, N, a) is not dense (in $K^{3,\mathrm{bs}}_{\mathfrak{s}}$) we get a non-structure result. Otherwise (assuming categoricity in λ, a soft assumption here) we can define $\mathrm{NF}_{\mathfrak{s}}$, non-forking amalgamation of models. We then investigate $K^{\mathfrak{s}}_{\lambda^+}$, more exactly the models there which are saturated. Either we get a non-structure result or our frame $\mathfrak{s}$ is successful and then we get a successor, a good λ^+-frame, $\mathfrak{s}^+ = \mathfrak{s}(*)$. Now $K_{\mathfrak{s}(+)} \subseteq K^{\mathfrak{s}}_{\lambda^+}$, but $\leq_{\mathfrak{s}(+)}$ is only $\subseteq \leq_{\mathfrak{K}^{\mathfrak{s}}} \restriction K_{\mathfrak{s}(+)}$.]

(C) the beauty of ω successive good λ-frames,

[Here we describe Chapter III. Assume for simplicity that letting $\mathfrak{s}^0 = \mathfrak{s}, \mathfrak{s}^{n+1} = (\mathfrak{s}^n)^+$ our assumption means that: each $\mathfrak{s}^{+n}$ is a (well defined) successful good λ^{+n}-frame. We first try to understand better what occurs for each $\mathfrak{s}^n$ (at least when n is not too small). But to understand models of larger cardinalities we have to connect better the situation in the various cardinals, for this we use $(\lambda, \mathscr{P}^{(-)}(n))$-systems of models, particularly stable ones and in general properties for (λ, n) are connected to properties of $(\mu, n+1)$ for every large enough $\mu < \lambda$.]

§4 Appetite comes with eating

(A) The empty half of the glass,

[Here we try to see what is lacking in the present book.]

(B) The full half and half baked,

[Here we review Chapter IV which deals with abstract elementary classes which are catogorical (or just solvable) in some large enough μ. We also review Chapter VII which do the non-structure in particular eliminating the "weak diamond ideal on λ^+ is not λ^{++}-saturated" (but also do some positive theory on almost good λ-frames). We also discuss further works, which in general gives partial positive answer to the lackings in the previous subsection.]

(C) The white part of the map,

[We state conjectures and discuss them.]

§5 Basic knowledge,

(A) knowledge needed and dependency of chapters,

(B) Some basic definitions and notation,

[We review the basic set theory required for the reader and then review the model theoretic notation. Some parts need more - mainly Chapter I, Chapter IV.]

§6 Symbols,

Annotated Content for Ch.I (88r): A.e.c. near $\aleph_1$

I.§0 Introduction

[We explain the background, the aims and what is done concerning the number of models of $\psi \in \mathbb{L}_{\omega_1,\omega}(\mathbf{Q})$ in $\aleph_1$ and in $\aleph_2$; here $\mathbf{Q}$ is the quantifier there are uncountably many. Also several necessary definitions and theorems are quoted. We justify dealing with a.e.c. (abstract elementary classes). The original aim had been to make a natural, not arbitrary choice of the context ($\psi \in \mathbb{L}_{\omega_1,\omega}$ or $\psi \in \mathbb{L}_{\omega_1,\omega}(\mathbf{Q})$?, see [Sh 48]). The net result is a context related to, but different than, the axioms of Jónson for the existence of universal homogeneous models. One difference is that the notion of a submodel is abstract rather than a submodel; this forces us to formalize properties of being submodels and decide which we adopt, mainly AxV, (if $M_1 \subseteq M_2$ are $\leq_{\mathfrak{K}}$-submodels of N then $M_1 \leq_{\mathfrak{K}} M_2$). Another serious difference is the omission of the amalgamation property. So they are more like a class of models of $\psi \in \mathbb{L}_{\omega_1,\omega}$, recalling (as a background) that amalgamation and compactness are almost equivalent as properties of logics but formulas are not involved in the definition here.]

I.§1 Axioms and simple properties for classes of models

[We define the a.e.c. and deal with their basic properties, the classical examples being, of course, $(\mathrm{Mod}_T, \prec), T$ a first order theory, but also $(\mathrm{Mod}_\psi, \prec_{\mathrm{sub}(\psi)}), \psi \in \mathbb{L}_{\lambda^+,\omega}(\tau)$. Surprisingly (but not complicatedly) it is proved that every such class $\mathfrak{K}$ can be represented as a $\mathrm{PC}_{\lambda,2^\lambda}$, i.e. the class of $\tau_{\mathfrak{K}}$-reducts of models of a first order T omitting every type $p \in \Gamma$, where $|\Gamma| \leq 2^\lambda$ and the vocabulary has cardinality $\leq \lambda$. So though a wider context than $\mathrm{Mod}(\psi), \psi \in \mathbb{L}_{\lambda^+,\omega}$, it is not totally detached from it by the representation theorem just mentioned above. A particular consequence is the existence of relatively low Hanf numbers.]

I.§2 Amalgamation properties and homogeneity

[We present (D, λ)-sequence homogeneous and $(\mathbb{D}, \lambda)$-model homogeneous, various amalgamation properties and basic properties, in particular the existence and uniqueness of homogeneous models. Those are important properties but here they are usually unreasonable to assume; we have to console ourselves in proving them under strong assumptions (like categoricity) and after working we get the weak version.]

I.§3 Limit models and other results

[We introduce and investigate (several variants of) "limit models in $\mathfrak{K}_\lambda$", the most important one is superlimit. Ignoring the case "M_* is $<_{\mathfrak{K}}$-maximal", M_* is superlimit in $\mathfrak{K}_\lambda$ means that if $\langle M_i : i \le \delta \rangle$ is $\le_{\mathfrak{K}_\lambda}$-increasing continuous, and $i < \delta \Rightarrow M_i \cong M_*$ then $M_\delta \cong M_*$ and another formulation is "$\mathfrak{K}_\lambda \restriction \{M : M \cong M_*\}$ is a λ-a.e.c.". Note that if $\mathfrak{K}$ is categorical in λ, any $M \in \mathfrak{K}_\lambda$ is trivially superlimit. The main results use this to investigate the number of non-isomorphic models. We get amalgamation in $\mathfrak{K}_\lambda$ if $\mathfrak{K}$ has superlimit (or just so called λ^+-limit) models in $\lambda, 1 \le \dot{I}(\lambda^+, K) < 2^{\lambda^+}$ and $2^\lambda < 2^{\lambda^+}$. We at last resolve the Baldwin problem in ZFC: if $\psi \in \mathbb{L}_{\omega_1,\omega}(\mathbf{Q})$ is categorical in $\aleph_1$ then it has a model in $\aleph_2$. In fact, the solution is in considerable more general context.]

I.§4 Forcing and Categoricity

[We assume $\mathfrak{K}$ is a $\mathrm{PC}_{\aleph_0}$-a.e.c. and it has at least one but less than the maximal number of models in $\aleph_1$, we would like to deduce as much as we can on $\mathfrak{K}$ or at least on some $\mathfrak{K}' = \mathfrak{K} \restriction K'$, which is still an a.e.c. and has models of cardinality $\aleph_1$. Toward this we build a "generic enough" model $M \in \mathfrak{K}_{\aleph_1}$ by an $\le_{\mathfrak{K}}$-increasing ω_1-sequence of models in $\mathfrak{K}_{\aleph_0}$ so define $N \Vdash^{\aleph_1}_{\mathfrak{K}} \varphi(\bar{a})$ for suitable $N \in \mathfrak{K}_{\aleph_0}$, i.e. countable. This is reasonable for φ a formula in $\mathbb{L}_{\omega_1,\omega}(\tau_{\mathfrak{K}})$ or even $\mathbb{L}_{\omega_1,\omega}(\mathbf{Q})(\tau_{\mathfrak{K}})$. Now using $\mathbb{L}_{\omega_1,\omega_1}$ seems too strong. But we can do it over a fix $N \in K_{\aleph_0}$, so $N \le_{\mathfrak{K}} M$. What does this mean? We have a choice: should we fix N pointwise (so adding an individual constant for each $c \in N$) or as a set (so adding a unary predicate always interpreted as N). The former makes sense only if $2^{\aleph_0} < 2^{\aleph_1}$ as is the case in §5, so in the present section we concentrate on the second. By the "not many models in $\aleph_1$" we deduce that fixing N, for a "dense" family of M satisfying $N \leq_{\mathfrak{K}} M \in K_{\aleph_0}$ we have: $(M, N) \Vdash^{\aleph_1}_{\mathfrak{K}}$ decides everything. So we know what type $p_{\bar{a}}$ each $\bar{a} \in M$ realizes in any generic enough M^+ when $M \leq_{\mathfrak{K}} M^+ \in K_{\aleph_1}$. But in general the sequence $\bar{a}$ does not realize the type $p_{\bar{a}}$ in M itself (e.g., this phenomena necessarily occurs if the formula really involves $\mathbf{Q}$). So we say $\bar{a}$ materializes the type in (M, N) and we play between some relevant languages (the logics are mainly $\mathbb{L}^{-1}_{\omega_1,\omega}$ which is without $\mathbf{Q}$, $\mathbb{L}^{0}_{\omega_1,\omega} = \mathbb{L}_{\omega_1,\omega}(\mathbf{Q})$, the vocabulary is $\tau = \tau_{\mathfrak{K}}$ or $\tau^{+0} = \tau_{\mathfrak{K}} \cup \{P\}$, P predicate for N; and more cases). If we restrict the depth of the formulas by some countable ordinal, then the number of complete types is countable. We have to work in order to show that the number of complete $\mathbb{L}_{\omega_1,\omega}(\mathbf{Q})$-types realized in quite generic models in $\mathfrak{K}_{\aleph_1}$ is $\leq \aleph_1$ (recalling that there may be Kurepa trees). We end commenting on further more complicated such results and the relevant logics.]

I.§5 There is a superlimit model in $\aleph_1$

[Here we add to §4 the assumption $2^{\aleph_0} < 2^{\aleph_1}$ hence we prove amalgamation of $\mathfrak{K}_{\aleph_0}$ (or get a non-structure result). Someone may say something like §1-§3 are conceptual and rich, I.§4-§5 are technicalities. I rather think that §1,§2,§3 are the preliminaries to the heart of the matter which is §4 and mainly §5. Assuming properties implying non-structure results in ($\aleph_1$ and) $\aleph_2$ fails, we understand models in $\mathfrak{K}_{\aleph_0}$ and $\mathfrak{K}_{\aleph_1}$ better. In particular we get for countable N that the number of types realized in some generic enough $M \in \mathfrak{K}_{\aleph_1}$, so called $\mathbf{D}(N)$ which $\leq_{\mathfrak{K}}$-extend N, is $\leq \aleph_1$, and we can restrict ourselves to subclasses with strong notion of elementary submodel such that each $\mathbf{D}(N)$ is countable. A central question is the existence of amalgamations which are stable, definable in a suitable sense of countable models trying to prove symmetry, equivalently some variants and eventually uniqueness. The culmination is proving the existence of a superlimit model in $\aleph_1$, though this is more than necessary for the continuation (see II§3).]

I.§6 Counterexamples

[Some of our results (in previous sections) were gotten in ZFC, but mostly we used $2^{\aleph_0} < 2^{\aleph_1}$. We show here that this is not incidental. Assuming $\mathrm{MA}_{\aleph_1}$, there is an a.e.c. $\mathfrak{K}$ which is $\mathrm{PC}_{\aleph_0}$, categorical in $\aleph_0$ and in $\aleph_1$, but fails the amalgamation property. We can further have that it is axiomatized by some $\psi \in \mathbb{L}_{\omega,\omega}(\mathbf{Q})$, and we deal with some related examples.]

Annotated Content for Ch.II (600): Categoricity in a.e.c.: going up inductive steps

II.§0 Introduction

[We present the results on good λ-frames and explain the relationship with [Sh 576] that is Chapter VI and with Chapter I. We then suggest some reading plans and some old definitions.]

II.§1 Abstract elementary classes

[First we recall the definition and some claims. In particular we define types (reasonable over models which are amalgamation basis), and we prove some basic properties, in particular, model homogeneity - saturativity lemma II.1.14 which relate realizing types of singleton elements to finding copies of models. We also define "N is (λ, θ)-brimmed over M", etc., and their basic properties. Then we prove that we could have restricted our class $\mathfrak{K}$ to cardinality λ without any real loss, i.e., any λ-a.e.c. can be blown up to an a.e.c. with LS-number λ and any a.e.c. with LS-number $\leq \lambda$ can be restricted to cardinality λ and as long as we ignore the models of cardinality $< \lambda$, this correspondence is one to one (see II.1.23, II.1.24); reading those proofs is a good exercise in understanding what is an a.e.c.]

II.§2 Good frames

[We introduce the central axiomatic framework called "good λ-frames", $\mathfrak{s} = (K_{\mathfrak{s}}, \leq_{\mathfrak{s}}, \mathrm{NF}_{\mathfrak{s}})$. The axiomatization gives the class $K_{\mathfrak{s}}$ of models and a partial order $\leq_{\mathfrak{s}}$ on it, forming an a.e.c., $\mathfrak{K}_{\mathfrak{s}} = (K_{\mathfrak{s}}, \leq_{\mathfrak{s}})$, a set $\mathscr{S}^{\mathrm{bs}}_{\mathfrak{s}}(M)$ of "basic" types over any model $M \in K_{\mathfrak{s}}$, the ones for which we have a non-forking notion. A (too good) example is regular types for superstable first order theories. We also check how can the non-forking of types be lifted up to higher cardinals or fewer models; but unlike the lifting of λ-a.e.c. in §1 in this lifting we lose some essential properties; in particular uniqueness and existence. We end noting some implications between axioms of good λ-frames.]

II.§3 Examples

[We prove here that cases treated in earlier relevant works fit the framework from §2. This refers to [Sh 576], Chapter I and also to [Sh 87a], [Sh 87b], [Sh 48].]

II.§4 Inside the frame

[We prove some claims used later, in particular stability in λ, sufficient condition for M_δ being $(\lambda, \mathrm{cf}(\delta))$-brimmed over M_0 for a chain $\langle M_i : i \leq \delta \rangle$ and the uniqueness of the $(\lambda, *)$-brimmed model over $M_0 \in K_\lambda$. We deal (for those results but also for later uses) with non-forking rectangles and triangles. An easy (but needed in the end) consequence is that $K^{\mathfrak{s}}_{\lambda^{++}}$ is not empty.]

II.§5 Non-structure or some unique amalgamations

[We prove that we have strong non-structure in $K^{\mathfrak{s}}_{\lambda^{++}}$ or for enough triples $(M_0, M_1, a) \in K^{3,\mathrm{bs}}$ we have unique amalgamation of M_1, M_2 over M_0 when $M_0 \leq_{\mathfrak{K}} M_2 \leq_{\mathfrak{K}} M_3, M_0 \leq_{\mathfrak{K}} M_1 \leq_{\mathfrak{K}} M_3$ and we demand that $\mathbf{tp}(a, M_2, M_3)$ does not fork over M_0. Naturally, we use the framework of [Sh 576, §3] or better Chapter VII and we do the model theoretic work required to be able to apply it. More explicitly, from the non-density of such triples with uniqueness we prove a non-structure theorem in λ^{++}. A major point in proving this dichotomy is to guarantee that $\bigcup_{\alpha<\delta} M_\alpha \in K_{\lambda^+}$ is saturated, when $\delta < \lambda^{++}$ and each $M_\alpha \in K_{\lambda^+}$ is saturated at least when $\langle M_\alpha : \alpha < \delta\rangle$ appears in our constructions. For this we use M_α which is $\leq_{\mathfrak{K}}$-represented by $\langle M^\alpha_i : i < \lambda^+\rangle$ so $M_\alpha = \bigcup_{i<\lambda^+} M^\alpha_i$ and $\langle\langle M^\alpha_i : i < \lambda^+\rangle : \alpha < \lambda^{++}\rangle$ is used with extra promises on non-forking of types, which are preserved in limits of small cofinality. Note that we know that in $\mathfrak{K}^{\mathfrak{s}}_{\lambda^+}$ there is a model saturated above λ but we do not know that it is superlimit.]

II.§6 Non-forking amalgamation in $\mathfrak{K}_\lambda$

[Our aim is to define the relation of non-forking amalgamations for models in K_λ and prove the desired properties promised by the name. What we do is to start with the cases which §5 provides us with a unique amalgamation modulo non-forking of a type of an element, and "close" them by iterations arriving to a (λ, θ)-brimmed extension. This defines non-forking amalgamation in the brimmed case, and then by closing under the submodels we get the notion itself. Now we have to work on getting the properties we hope for. To clarify, we prove that "a non-forking relation with the reasonable nice properties" is unique. A consequence of all this is that we can change $\mathfrak{s}$ retaining $\mathfrak{K}_{\mathfrak{s}}$ such that it is type-full, i.e., every non-algebraic type (in $\mathscr{S}_{\mathfrak{K}_\lambda}(M)$ is basic for $\mathfrak{s}$. (This is nice and eventually needed.)]

II.§7 Nice extensions in K_{λ^+}

[Using the non-forking amalgamation from §6, we define nice models ($K^{\rm nice}_{\lambda^+}$) and "nice" extensions in $\lambda^+(\leq^*_{\lambda^+})$, and prove on them nice properties. In particular K_{λ^+} with the nice extension relation has a superlimit model - the saturated one.]

II.§8 Is $K^{\rm nice}_{\lambda^+}$ with $\leq^*_{\lambda^+}$ a λ^+-a.e.c.?

[We prove that $\mathfrak{K}^{\rm nice}_{\lambda^+} = (K^{\rm nice}_{\lambda^+} \leq^*_{\lambda^+})$ is an a.e.c. under an additional assumption but we prove that the failure of this extra assumption implies a non-structure theorem. We then prove that there is a good λ^+-frame $\mathfrak{t}$ with $\mathfrak{K}_{\mathfrak{t}} = \mathfrak{K}^{\rm nice}_{\lambda^+}$ and prove that it relates well to the original $\mathfrak{s}$, e.g. we have locality of types.]

II.§9 Final conclusions

[We reach our main conclusions (like II.0.1) in the various settings.]

Annotated Content for Ch.III (705): Toward classification theory of good λ-frames and a.e.c.

III.§0 Introduction

III.§1 Good$^+$ Frames

[We define when a good λ-frame is successful (III.1.1) and when it is good$^+$ (III.1.3). There are quite many good$^+$ frames $\mathfrak{s}$: the cases of good λ-frames we get in II§3 all are good$^+$ and further, if $\mathfrak{s}$ is successful good λ-frame (not necessarily good$^+$!) then $\mathfrak{s}^+$ is good$^+$ (see III.1.5, III.1.9). Moreover, if $\mathfrak{s}$ is a good$^+$ successful λ-frame, <u>then</u> $\mathfrak{s}^+ = \mathfrak{s}(+)$ satisfies $\leq_{\mathfrak{s}(+)} = \leq_{\mathfrak{K}[\mathfrak{s}]} \restriction K_{\mathfrak{s}(+)}$ (see Definition III.1.7 and Claim

III.1.8), and we can continue and deal with $\mathfrak{s}^{+\ell} = \mathfrak{s}(+\ell)$, (see III.1.14). We define naturally "$\mathfrak{s}$ is n-successful" and look at some basic properties. We end recalling some things from Chapter II which are used often and add some. We prove locality for basic types and types for $\mathfrak{s}(+)$, see III.1.10, III.1.11. In III.1.21 we show that if $M_1 \leq_{\mathfrak{s}} M_2$ are brimmed and the type $p_2 \in \mathscr{S}^{\mathrm{bs}}_{\mathfrak{s}}(M_2)$ does not fork over M_1 then some isomorphism from M_2 to M_1 maps p_2 to $p_2 \restriction M_1$, similarly with $< \lambda$ types. In III.1.16-III.1.20 we essentially say to what we use on $\mathrm{NF}_{\mathfrak{s}}$, assuming $\mathfrak{s}$ is weakly successful; this is the part most used later.]

III.§2 Uni-dimensionality and non-splitting

[We are interested not only in the parallel of being superstable but also of being categorical, which under natural assumptions is closely related to being uni-dimensional. We now define (the parallel of) uni-dimensional, more exactly some variants including non-multi-dimensionality (in III.2.2, III.2.13). We then note when our examples are like that; we show that $\mathfrak{s}(+)$ satisfies such properties when (even iff) $\mathfrak{s}$ does (III.2.6, III.2.10, III.2.17 and more in III.2.12). Of course we show the close connection between uni-dimensionality and categoricity in λ^+ (see III.2.11). Next we deal with minimal types and with good λ-frames for minimals (III.2.13 - III.2.17). We then look at splitting, relevant ranks and connection to non-forking (from III.2.18 on). We also know what occurs if we make $\mathfrak{s}$ type-full (III.2.7) and we then consider frames where the basic types are the minimal types (III.2.15 - III.2.17). We then recall splitting.]

III.§3 Prime triples

[We define $K^{3,\mathrm{pr}}_{\mathfrak{s}}$, the family of prime triples (M, N, a), the family of minimal triples and "$\mathfrak{s}$ has primes" (Definition III.3.2). We look at the basic properties (III.3.5,III.3.8), connection to $K^{3,\mathrm{uq}}_{\mathfrak{s}}$ (III.3.7) and x-decompositions for x = pr,uq,bs in Definition III.3.3. In particular if $\mathfrak{s}$ has primes then any pair

$M <_{\mathfrak{s}} N$ has a pr-decomposition (see III.3.11). We prove the symmetry for "the type of a_ℓ over $M_{3-\ell}$ does not fork over M_0 wherever $M_{3-\ell}$ is prime over $M_0 \cup \{a_{3-\ell}\}$" (III.3.9, III.3.12); note that the symmetry axiom say "for some $M_{3-\ell} \ldots$".]

III.§4 Prime existence

[We deal with good$^+$ successful λ^+-frame $\mathfrak{s}$. We recall the definition of $\leq_{\mathrm{bs}}$ and variants, and prove that $\mathfrak{s}^+$ has primes (III.4.9). For this we prove in III.4.9 that a suitable condition is sufficient for (M, N, a) to belong to $K^{3,\mathrm{pr}}_{\mathfrak{s}(+)}$, proving it occurs (in III.4.3), and more in III.4.5, III.4.14, III.4.20. We use for it $\leq_{\mathrm{bs}}$ (defined with the variants $<^*_{\mathrm{bs}}, <^{**}_{\mathrm{bs}}$ in III.4.2), the relevant properties in III.4.6. We then investigate more on how properties for $\mathfrak{s}^+$ reflects to $\lambda_{\mathfrak{s}}$, for $\mathrm{NF}_{\mathfrak{s}(+)}$ in III.4.15 also in III.4.13(2). Also we consider other sufficient conditions for III.3.9's conclusion in III.4.13(1). Lastly, III.4.20 deals with the examples.]

III.§5 Independence

[We define $\mathbf{I}_{M,N}$ and define when $\mathbf{J} \subseteq \mathbf{I}_{M,N}$ is independent in (M, N), (see Definition III.5.2). In III.5.4 + III.5.5 + III.5.6 + III.5.8(2) we prove fundamental equivalences and properties, including M_0-based pr/uq-decomposition in N/of N and that "independent in (M, N)" has finitary character. We also define "N is prime over $M \cup \mathbf{J}$" denoted by $(M, N, \mathbf{J}) \in K^{3,\mathrm{qr}}_{\mathfrak{s}}$ (Definition III.5.7). We note existence and basic properties (claim III.5.8). We show embedding existence (III.5.9(1)) and how this implies NF (see III.5.9(2)). We show that "normally" independence satisfies continuity (III.5.10) and reflect from $\mathfrak{s}^+$ to $\mathfrak{s}$ (III.5.11). Using this we prove the basic claims on dimension for non-regular types, (see III.5.12, III.5.13 + III.5.14).

We generalize $K^{3,\mathrm{uq}}_{\mathfrak{s}}$, the class of uniqueness triples (M, N, a), to $K^{3,\mathrm{vq}}_{\mathfrak{s}}$, the class of uniqueness triples $(M, N, \mathbf{J}), \mathbf{J}$ independent in (M, N), Definition III.5.15(1). We then define when $(M, N, \mathbf{J}) \in K_{\mathfrak{s}}^{3,\mathrm{vq}}$ is thick (Definition III.5.15) and prove their basic properties, in particular $K_{\mathfrak{s}}^{3,\mathrm{qr}} \subseteq K_{\mathfrak{s}}^{3,\mathrm{vq}}$ (see III.5.16, III.5.16(3)). When $\mathfrak{s} = \mathfrak{t}^+$ we "reflect" $K_{\mathfrak{s}}^{3,\mathrm{qr}}$ to cases of $K_{\mathfrak{t}}^{3,\mathrm{vq}}$ (see III.5.22). Lastly, every triple in $K_{\mathfrak{s}}^{3,\mathrm{bs}}$ can be extended to one in $K_{\mathfrak{s}}^{3,\mathrm{vq}}$ (with the same $\mathbf{J}$, Claim III.5.24).]

III.§6 Orthogonality

[We define when $p, q \in \mathscr{S}_{\mathfrak{s}}^{\mathrm{bs}}(M)$ are weakly orthogonal/orthogonal, (Definition III.6.2), show that "for every $(M, N, a) \in K_{\mathfrak{s}}^{3,\mathrm{uq}}$..." can be replaced "for some ...", (III.6.3) and prove basic properties (III.6.4, III.6.7), and define parallelism (see III.6.5,III.6.6). We define "a type p is orthogonal/super-orthogonal to a model" (Definition III.6.9, the "super" say preservation under NF amalgamation), prove basic properties (III.6.10), and how we reflect from $\mathfrak{s}^+$ to $\mathfrak{s}$ (see III.6.11 concerning $p \perp q, p \perp M$). Orthogonality helps to preserve independence (III.6.12). We investigate decompositions of tower with orthogonality conditions. If $(M, N, a) \in K_{\mathfrak{s}}^{3,\mathrm{uq}}, M \cup \{a\} \subseteq N' < N$ and $p = \mathbf{tp}_{\mathfrak{s}}(b, N', N)$ then p is weakly orthogonal to M (see III.6.14(1),III.6.14(2)), and decompose such triples by it (III.6.14(2)), look at an improvement (III.6.15(1)) and reflection from $\mathfrak{s}^+$ (in III.6.15(2)), how we can use independence, $K_{\mathfrak{s}}^{3,\mathrm{vq}/\mathrm{qr}}$ and orthogonality (III.6.16, III.6.18, III.6.20, III.6.22). In particular by III.6.20(2) if $(M_n, M_n, \mathbf{J}_n)$ $\in K_{\mathfrak{s}}^{3,\mathrm{uq}}$ for $n < \omega$ and $c \in \mathbf{J}_{n+1} \Rightarrow \mathbf{tp}_{\mathfrak{s}}(c, M_{n+1}, M_{n+2}) \perp M_0$ then $(M_0, \bigcup_n M_n, \mathbf{J}_0) \in K_{\mathfrak{s}}^{3,\mathrm{vq}}$. From pairwise orthogonality we can get independence (III.6.21), and one p cannot be non-orthogonal to infinitely many pairwise orthogonal types (III.6.22).]

III.§7 Understanding $K_{\mathfrak{s}}^{3,\mathrm{uq}}$

[In III.7.2, we define $\mathscr{W}, \leq_{\mathscr{W}}$ (weak form of decompositions of triples from $K_{\mathfrak{s}}^{3,\mathrm{vq}}$) and related objects, in III.7.3 we prove basic properties. In III.7.4 we define $K_{\mathfrak{s}}^{x}$ for $x =$ or,ar,br, decompositions of length $\leq \omega$ of triples in $K_{\mathfrak{s}}^{3,\mathrm{or}}$ with various orthogonality conditions (why of length $\leq \omega$? so that in inductive proof when we arrive to a limit case we are already done). We also define fat, related to thick, (III.5.8(5)) and we prove in III.7.6 some properties. In III.7.5 we define "$\mathfrak{s}$ weakly has regulars" and later, in III.7.18, define "almost has regulars". Existence for $K_{\mathfrak{s}}^{3,\mathrm{or}}, K_{\mathfrak{s}}^{3,\mathrm{ar}}$ (assuming enough regulars) are investigated (in III.7.7, III.7.8). We characterize being in $K_{\mathfrak{s}}^{3,\mathrm{uq}}$ in III.7.9, this is the main result of the section. We then deal with universality and uniqueness for fat uq/vq triples (see III.7.11 - III.7.13). We also deal with hereditary and limits of uq/vq triples in III.7.15, III.7.16.]

III.§8 Tries to decompose and independence of sequences of models

[We define and prove existence of x-decompositions $(\bar{M}, \bar{a})$ with $\mathbf{tp}_{\mathfrak{s}}(a_i, M_i, M_{i+1})$ does not fork over some M_j but is orthogonal to M_ζ when $\zeta < j$ and show that $(M_0, M_\alpha, \{a_i : \mathbf{tp}_{\mathfrak{s}}(a_i, M_i, M_{i+1})$ does not fork over $M_0\}) \in K_{\mathfrak{s}}^{3,\mathrm{vq}}$ and also revisit existence for $K_{\mathfrak{s}}^{3,\mathrm{vq}}$ (see III.8.2, III.8.3, III.8.6). We define and investigate when $\langle M_i : i < \alpha \rangle$ is $\mathfrak{s}$-independent over M inside N with witness $\bar{N} = \langle N_i : i \leq \alpha \rangle$ (see III.8.8 - III.8.18). In III.8.19 we return to investigating $\mathrm{NF}_{\mathfrak{s}}$, prove that it is preserved under reasonable limits and by III.8.21 this holds for $K_{\mathfrak{s}}^{3,\mathrm{vq}}$. We also further deal with $K_{\mathfrak{s}}^{3,\mathrm{vq}}$.]

III.§9 Between cardinals, non-splitting and getting fullness

[We deal mainly with varying $\mathfrak{s}$. We fulfill a promise, proving that a weakly successful good λ-frame $\mathfrak{s}$ can be doctored to be full (see III.9.5 - III.9.6. Also we show that if $\mathfrak{s}$ is a successful λ-good$^+$ frame, <u>then</u> we can define a λ^+-good$^+$ successor $\mathfrak{s}^{\mathrm{nf}}$ with $\mathfrak{K}_{\mathfrak{s}^{\mathrm{nf}}} = \mathfrak{K}_{\mathfrak{s}}$ and $\mathfrak{s}^{\mathrm{nf}}$ is full, i.e. $\mathscr{S}^{\mathrm{bs}}_{\mathfrak{s}(+)} = \mathscr{S}^{\mathrm{na}}_{\mathfrak{s}(+)}$; moreover if $\mathfrak{s}$ is categorical and successful.]

III.§10 Regular types

[We deal mainly with type-full $\mathfrak{s}$. We define regular and regular$^+$ (Definition III.10.2),prove some basic equivalences (III.10.4) and prove that the set of regular types is "dense" (III.10.5). To prove that for regular type p, non-orthogonality, ($p \pm q$) is equivalent to being dominated, ($p \trianglelefteq q$) (in III.10.8), we prove a series of statements on regular and regular$^+$ types (in III.10.6). We prove e.g. that if $\langle M_i : i \leq \delta + 1\rangle$ is increasing continuous, $M_\delta \neq M_{\delta+1}$ then some $c \in M_{\delta+1} \backslash M_\delta$ realizes a regular type over M_δ which does not fork over M_j but is orthogonal to M_{j-1} if $j > 0$, for some j, which necessarily is a successor ordinal (III.10.9(3)) that is, we prove that $\mathfrak{s}$ almost has regulars. Hence weakly has regulars as expected from the names we choose. Using this, we revisit decompositions (III.10.12).]

III.§11 DOP

[We deal with the dimensional order property.]

III.§12 Brimmed Systems

[This is the crux of the matter. We deal with systems $\mathbf{m} = \langle M_u : u \in \mathscr{P}\rangle$, $\mathscr{P}$ usually is $\mathscr{P}(n)$ or $\mathscr{P}^-(n)$, which are "stable", as witnessed by various maximal independent sets. A parameter $\ell = 1, 2, 3$ measure how brimmed is $\mathbf{m}$, presently the central one is $\ell = 3$. We then phrase properties related to such stable system, e.g. the weak (λ, n)-existence say every such $(\lambda, \mathscr{P}^-(n))$-system can be completed to a $(\lambda, \mathscr{P}(n))$-system; the strong (λ, n)-existence property says that we can do it "economically", by a "small M_n". We also define weak/strong uniqueness, weak/strong primeness and weak/strong prime existence. The main work is proving the relevant implications. The culmination is proving that if $\mathfrak{s}$ is ω-successful and $\langle 2^{\lambda_{\mathfrak{s}}^{+n}} : n < \omega\rangle$ is increasing, <u>then</u> all positive properties holds and so can understand, e.g. categoricity spectrum (and superlimit models).]

Annotated Content for Ch.IV (734): Categoricity and Solvability of a.e.c., quite highly

IV.§0 Introduction

[Our polar star is: if an a.e.c. is categorical in arbitrarily large cardinals then it is categorical in every large enough cardinal. We make some progress getting some good λ-frames; and to point to a more provable advancement, confirm this conjecture (and even a reasonable bound on starting) for a.e.c. with amalgamation (as promised in [Sh:E36]). In fact we put forward solvability as the true parallel to superstability.]

IV.§1 Amalgamation in K^*_λ

[We assume $\mathfrak{K}$ is categorical in μ (or less-solvable in μ); and the best results are on λ such that $\mu > \lambda = \beth_\lambda > \mathrm{LS}(\mathfrak{K})$ (i.e. λ is a fix point in the beth sequence) and λ has cofinality $\aleph_0$; we fix suitable $\Phi \in \Upsilon^{\mathrm{or}}[\mathfrak{K}]$. We mostly assume $\mu = \mu^\lambda$.

First we investigate $K^*_\theta = \{M : M \cong \mathrm{EM}(I, \Phi)$ for some linear order I of cardinality $\theta\}$, which is in general not an a.e.c. under $\leq_{\mathfrak{K}}$, but in our μ it is. We investigate such models in the logic $\mathbb{L}_{\infty,\partial}$, particularly when θ is large enough than $\partial, \partial > \mathrm{LS}(\mathfrak{K})$ (mainly $\theta \geq \beth_{1,1}(\partial)$). We get more and more cases when $M \prec_{\mathbb{L}_{\infty,\partial}[\mathfrak{K}]} N$ follows from $M \leq_{\mathfrak{K}} N+$ additional assumptions. An evidence of our having gained understanding is proving the amalgamation theorem IV.1.29: the class $(K^*_\lambda, \leq_{\mathfrak{K}})$ has the amalgamation property. In the end we prove that if $\lambda = \Sigma\{\lambda_n : n < \omega\} < \mu$ each λ_n is as above and $< \lambda_{n+1}$ and is μ as above then $\mathfrak{K}_\lambda$ has a local superlimit model, see IV.1.38, in fact we get a version of solvability in λ, see IV.1.41.]

IV.§2 Trying to Eliminate $\mu = \mu^{<\lambda}$

[In §1 essentially (in the previous section) the first step in our ladder was proving $M \prec_{\mathbb{L}_{\infty,\theta}} N$ for $M \leq_{\mathfrak{K}} N$ from K_μ but we have to assume $\mu = \mu^{<\theta}$. As we use it for many $\theta < \lambda$, the investigation does not even start without assuming $\mu = \mu^{<\lambda}$. We eliminate this assumption except "few" exceptions (i.e., for a given $\mathfrak{K}$ and θ).]

IV.§3 Categoricity for cardinals in a club

[We assume $\mathfrak{K}$ is categorical in unbounded many cardinals. We show that for some closed and unbounded class $\mathbf{C}$ of cardinals, $\mathfrak{K}$ is categorical in μ for every $\mu \in \mathbf{C}$ of cofinality $\aleph_0$ (or $\aleph_1$). This is a weak theorem still show that the categoricity spectrum is far from being "random" (as is, e.g. the rigidity spectrum is by [Sh 56]).]

IV.§4 Good frames

[Assume for simplicity that $\mathfrak{K}$ is categorical in arbitrarily large cardinals μ. Then for every $\lambda = \Sigma\{\lambda_n : n < \omega\}, \lambda_n = \beth_{\lambda_n} > \mathrm{LS}(\mathfrak{K})$ there is a superlimit model in $\mathfrak{K}_\lambda$, and even a version of solvability. Moreover there is a good λ-frame $\mathfrak{s}_\lambda$ such that $K_{\mathfrak{s}_\lambda} \subseteq \mathfrak{K}_\lambda, \leq_{\mathfrak{s}_\lambda} = \leq_{\mathfrak{K}} \restriction K_{\mathfrak{s}_\lambda}$. Other works, in particular Chapter III, are a strong indication that this puts us on our way for proving the goal from §0.]

IV.§5 Homogeneous enough linear orders

[We construct linear order I of any cardinality $\lambda > \mu$ such that there are few $J \in [I]^\mu$ up to an automorphism of I and more. This helps when analyzing EM models using the skeleton I. Used only in §2 and §7. The proof is totally direct: we give a very explicit definition of I, though the checking turns out to be cumbersome.]

IV.§6 Linear orders and equivalence relations

[For a "small" linear order J and a linear order I, mainly well ordered we investigate equivalence relations $\mathscr{E}$ on $\mathrm{inc}_J(I) = \{h : h \text{ embed } J \text{ into } I\}$ which are invariant, i.e., defined by a quantifier free (infinitary) formula, hence can (under reasonable conditions) be defined on every I'. We are interested mainly to find when $\mathscr{E}$ has $> |I|$ equivalence classes; and for "there is a suitable I of cardinality λ". The expected answer is a simple question on λ: is $\lambda > \lambda^{|J|}/D$ for some suitable filter D? but we just prove enough for the application in §7, dealing with the case $\lambda > \lambda^{|J|}/D$ holds for some non-principal ultrafilter on $|J|$.]

IV.§7 Categoricity spectrum for a.e.c. with bounded amalgamation

[Let $\mathfrak{K}$ be a.e.c. categorical in μ (or less, $\Phi \in \Upsilon^{\rm or}_{{\rm LS}[\mathfrak{K}]}[\mathfrak{K}]$, if $\lambda > \mu \geq {\rm cf}(\mu) > {\rm LS}(\mathfrak{K})$ and $\mathfrak{K}_{<\mu}$ has amalgamation. Then for $\mu_* < \mu$, every saturated $M \in \mathfrak{K}$ of cardinality $\in [\mu_*, \mu)$ is μ_*-local, i.e., any type $p \in \mathscr{S}_{\mathfrak{K}}(M)$ is determined by its restriction to model $N \leq_{\mathfrak{K}} M$ of cardinality μ_*. Also $M \in K$ is (χ, μ)-saturated, e.g., if $2^{2^\chi} < \mu$. Then we prove that if $\mathfrak{K}$ is an a.e.c. categorical in a not too small cardinal μ and has amalgamation up to μ or less) <u>then</u> it is categorical in every not too small cardinal. We delay the improvements concerning solvability spectrum and saying more in the case $\mathfrak{K} = ({\rm Mod}_T, \prec_{\mathbb{L}_{\kappa,\omega}})$, where $T \subseteq \mathbb{L}_{\kappa,\omega}$, κ measurable. In all cases we eliminate the restriction of starting with "μ successor" and having the upward directions, too.]

Annotated Content for Ch.V.A (300a): Stability theory for a model

(This chapter will appear in book 2.)

V.A.§0 Introduction

[Introduction and notation.]

V.A.§1 The order property revisited

[We define some basic properties. First a model M has the $(\varphi(\bar{x};\bar{y};\bar{z}),\mu)$-order property (= there are $\bar{a}_\alpha,\bar{b}_\alpha,\bar{c}$ for $\alpha<\mu$ such that $\varphi(\bar{a}_\alpha;\bar{b}_\beta,\bar{c})$ is satisfied iff $\alpha<\beta$) and the non-order property is its negation. Also indiscernibility (of a set and of a sequence), and non-splitting. We then prove the non-splitting/order dichotomy: if M is an elementary submodel of N in a strong enough way related to χ and κ and $\bar{a}\in{}^{\kappa}N$ <u>then</u> either $\mathrm{tp}_\Delta(\bar{a},M,N)$ is definable in an appropriate way (i.e., does not split over some set $\leq\chi$ relevant formulas) <u>or</u> N has (ψ,χ^+)-order for a formula ψ related to Δ. Lastly, we prove that (Δ,χ^+)-non-order implies (μ,Δ)-stability if for appropriate χ,μ. We also define various sets of formulas Δ^x derived from Δ.]

V.A.§2 Convergent indiscernible sets

[For stable first order theory, an indiscernible set $\mathbf{I}\subseteq M$ define its average type over M: the set of $\varphi(\bar{x},\bar{b})$ satisfied by all but finitely many $\bar{c}\in\mathbf{I}$. In general not every indiscernible set $\mathbf{I}$ has an average, so we say $\mathbf{I}$ is (Δ,χ)-<u>convergent</u> if any formula $\varphi(\bar{x},\bar{b})$ where $\varphi\in\Delta$ and $\bar{b}$ is from M, divide $\mathbf{I}$ to two sets, exactly one of which has $<\chi$ members. We prove that convergent sets exists (V.A.2.8) under reasonable conditions (mainly non-order). We also prove that convergent sets contain indiscernible ones. Toward the existence we give a sufficient condition in V.A.2.10 for a sequence $\langle\bar{c}_i : i<\mu^+\rangle$ being (Δ,χ^+)-convergent including the (Δ,χ^+)-non-order property which is easy to obtain.]

V.A.§3 Symmetry and indiscernibility

[We prove a symmetry lemma (V.A.3.1), give sufficient conditions for being an indiscernible sequence (V.A.3.2), and when an indiscernible sequence is an indiscernible set (V.A.3.5), and on getting an indiscernible set from a convergent set.]

V.A.§4 What is the appropriate notion of a submodel

[We define $M \leq^{\kappa}_{\Delta,\mu,\chi} N$ which says that for $\bar{c} \in {}^{\kappa>}N$, the Δ-type which it realizes over M inside N is the average of some (Δ, χ^+)-convergent set of cardinality μ^+ inside M. We give an alternative definition of being a submodel (in V.A.4.4) when M has an appropriate non-order property, prove their equivalence and note some basic properties supporting the thesis that this is a reasonable notion of being a submodel. We then define "stable amalgamation of M_1, M_2 over M_0 inside M_3" and investigate it to some extent.]

V.A.§5 On the non-order implying the existence of indiscernibility

[We give a sufficient condition for the existence of "large" indiscernible set $\mathbf{J} \subseteq \mathbf{I}$, in which $|\mathbf{J}| < |\mathbf{I}|$, but the demand on the non-order property is weaker than in V.A.§2 speaking only on non-order among singletons. Even for some first order T which are unstable, this gives new cases e.g. for $\Delta =$ the set of quantifier free formulas.]

Annotated Content for Ch.V.B (300b): Axiomatic framework

(This chapter will appear in book 2.)

V.B.§0 Introduction

[Rather than continuing to deal with universal classes per se, we introduce some frameworks, deal with them a little and show that universal classes with the $(\chi, < \aleph_0)$-non-order property fit some of them (for suitable choices of the extra relations). In the rest of Chapter II almost always we deal with AxFr_1 only.]

V.B.§1 The Framework

[We suggest several axiomatizations of being "a class of models K with partial order $\leq_{\mathfrak{K}}$ with non-forking and possibly the submodel generated by a subset" (so being a submodel, non-forking and $\langle A \rangle_M^{\rm gn}$ for $A \subseteq M$ are abstract notions). The main one here, AxFr_1 is satisfied by any universal class with $(\chi, < \aleph_0)$-non-order; (see §2). For AxFr_1 if M_1, M_2 are in non-forking amalgamation over M_0 inside M_3 then the union $M_1 \cup M_2$ generate a $\leq_{\mathfrak{K}}$-submodel of M_3. In such contexts we define a type as an orbit, i.e. by arrows (without formulas or logic); to distinguish we write **tp** (rather than tp_Δ) for such types. Also "Tarski-Vaught theorem" is divided to components. On the one hand we consider union existence Ax(A4) which says that: the union of an $\leq_{\mathfrak{K}}$-increasing chain belongs to the class and is $\leq_{\mathfrak{K}}$-above each member. On the other hand we consider smoothness which says that any $\leq_{\mathfrak{K}}$-upper bound is $\leq_{\mathfrak{K}}$-above the union.]

V.B.§2 The Main Example

[We consider a universal class K with no "long" linear orders, e.g. by quantifier free formulas (on χ-tuples), we investigate the class K with a submodel notion introduced in V.A§4, and a notion of non-forking, and prove that it falls under the main case of the previous section. We also show how the first order case fits in and how (D, λ)-homogeneous models does.]

V.B.§3 Existence/Uniqueness of Homogeneous quite Universal Models

[We investigate a model homogeneity, toward this we define $\mathbb{D}_\chi(M), \mathbb{D}_\chi(\mathfrak{K}), \mathbb{D}'_{\mathfrak{K},\chi}$ and define "M is $(\mathbb{D}, \lambda)$-model homogeneous". We show that being λ^+-homogeneous λ-universal model in $\mathfrak{K}$ can be characterized by the realization of types of singletons over models (as in the first order case) so having "the best of both worlds".]

Annotated Content for Ch.V.C (300c): A frame is not smooth or not χ-based

(This chapter will appear in book 2.)

V.C.§0 Introduction

[The two dividing lines dealt with here have no parallel in the first order case, or you may say they are further parallels to stable/unstable, i.e. stability "suffer from schizophrenia", there are distinctions between versions which disappear in the first order case, but still are interesting dividing lines.]

V.C.§1 Non-smooth stability

[This section deals with proving basic facts inside AxFr_1. On the one hand we assume we are hampered by the possible lack of smoothness, on the other hand the properties of $\langle - \rangle^{\mathrm{gn}}_M$ are helpful. These claims usually say that specific cases of smoothness, continuity and non-forking hold. So it deals with the (meagre) positive theory in this restrictive context.]

V.C.§2 Non-smoothness implies non-structure

[We start with a case of failure of κ-smoothness, copy it many times on a tree $\mathscr{T} \subseteq {}^{\kappa \geq}\lambda$; for each $i < \kappa$ for every $\eta \in \mathscr{T} \cap {}^i\lambda$ we copy the same things while for $\eta \in \mathscr{T} \cap {}^\kappa\lambda$ we have a free choice. This is the cause of non-structure, but to prove this we have to rely heavily on §1. If we assume the existence of unions, for any $<_{\mathfrak{s}}$-increasing sequence, i.e. Ax(A4), the non-structure (in many cardinals), is proved in ZFC, but using weaker versions we need more.]

V.C.§3 Non χ-based

[We note some basic properties about directed systems and how much they depend on smootheness. We then define when $\mathfrak{s}$ is χ-based: if $M \leq_{\mathfrak{s}} N$ and $A \subseteq N$ has cardinality $\leq \chi$ then for some M_1, N_1 of cardinality $\leq \chi$ we have $\mathrm{NF}_{\mathfrak{s}}(M_1, N_1, M, N)$ and $A \subseteq N_1$. This is a way to say that

$\mathbf{tp}_{\mathfrak{s}}(N_1, N)$ does not fork over M_1, so being χ-based is a relative of being stable, and when it fails, a very explicit counterexample.]

V.C.§4 Stable construction

[We generalize [Sh:c, IV] to this context. That is we deal with constructions: in each stage we add a "small" set which realizes over what was constructed so far a type which does not fork over their intersection. We define and investigate the basic properties of such constructions.]

V.C.§5 Non-structure from "NF is not χ-based"

[Assuming the explicit failure of "χ-based over models of cardinality χ^+", and using the existence of good stationary subsets S^* of regular $\lambda > \chi^{++}$ of cofinality χ^+, we build a model in $\mathfrak{K}^{\mathfrak{s}}_\lambda$ which codes any subset S of S^* (modulo the club filter) hence get a non-structure theorem. Naturally we use the stable constructions from the previous section, §4 and have some relatives.]

Annotated Content for Ch.V.D (300d): Non-forking and prime models

(This chapter will appear in book 2.)

V.D.§0 Introduction

[Here we deal with types of models (rather than types of single elements). This is O.K. for parallel to some properties of stable first order theories T, mainly dealing with $|T|^+$-saturated models.]

V.D.§1 Being smooth and based propagate up

[By Chapter V.C we know that failure of smoothness and failure of being χ-based are non-structure properties, but they may fail only for some large cardinal. We certainly prefer to be able to prove that faillure, if it happens at all, happens for some quite small cardinal; we do not know how to do it for each property separately. But we show that if $\mathfrak{s}$ is $(\leq \chi, \leq \chi^+)$-smooth and (χ^+, χ)-based and LSP(χ) then for every $\mu \geq \chi$, $\mathfrak{s}$ is $(\leq \mu, \leq \mu)$-based, and $(\leq \mu, \leq \mu)$-smoothed and has the LSP(μ). So it is enough to look at what occurs in cardinality LS$(\mathfrak{K}_{\mathfrak{s}})$ for the non-structure possibility (rather than "for some χ"). We then by Chapter V.C get a non-structure result from the failure of the assumption above. We also investigate when $\mathfrak{K}_{\mathfrak{s}}$ has arbitrarily large models. So being "$(\leq \chi, \leq \chi^+)$-smooth, (χ^+, χ)-based, LSP(χ)" is a good dividing line.]

V.D.§2 Primeness

[We define prime models (over A), isolation (for types of the form $N/M + c$) and primary models. We prove the existence of enough isolated types; the difference with the first order case is that we need to deal with $M <_{\mathfrak{s}} \mathfrak{C}$ even if we start with a singleton. From this we deduce the existence of primary models over $A <_{\mathfrak{s}} \mathfrak{C}$ hence primes.]

V.D.§3 Theory of types of models

[We look at TP(N, M) when $N \cap M, N, M$ are in stable amalgamation. The set of such types is called $\mathscr{S}^{\alpha}_{c}(M)$ if $\langle a_i : i < \alpha \rangle$ list the elements of N. For such types we can define non-forking, stationarization and prove properties parallel to the first order case of stable first order classes.]

V.D.§4 Orthogonality

[For types in $\mathscr{S}^{<\infty}_{c}(M)$ we can define weak orthogonality and orthogonality of types and orthogonality of a type to a model and prove expected claims.]

V.D.§5 Uniqueness of $(\mathbb{D}_{\mathfrak{s}}, \mu)$-primary models

[We prove that the non-forking restriction of an isolated type is isolated. We then prove the uniqueness of the primary model.]

V.D.§6 Uniqueness of $(\mathbb{D}_{\mathfrak{s}}, \mu)$-prime models

[We deal with the uniqueness of prime models and only comment on $\mathfrak{C}^{\text{eq}}$.]

Annotated Content for Ch.V.E (300e): Types of finite sequences

(This chapter will appear in book 2.)

V.E.§0 Introduction

[The investigations in Chapter V.C, Chapter V.D do not suggest a parallel to superstable. For this we have to look at types of singletons, and the picture is more complicated, but a very reasonable parallel exist.]

V.E.§1 Forking over models of types of sequences

[We define when $\mathbf{tp}(\bar{c}, N)$ does not fork over $M \leq_{\mathfrak{s}} N$ even for sequences $\bar{c}$ not enumerating any appropriate $N' <_{\mathfrak{s}} \mathfrak{C}$ and investigate the properties.]

V.E.§2 Forking over sets

[We define when $\mathbf{tp}(\bar{c}, B)$ does not fork over A, show the equivalence and compatibility of several variants; we define when $\mathbf{tp}(\bar{c}, B)$ is stationary over A and investigate the basic properties (including symmetry). Compared to the first order stable case there may be "bad types", e.g. there may be no "small" $A \subseteq B$ such that "$\mathbf{tp}(\bar{c}, B)$ does not fork over A". We also define strong splitting in this context and convergence, independence and parallelism.]

V.E.§3 Defining superstability and $\kappa(\mathfrak{s})$

[We define $\kappa(\mathfrak{s})$, a set of regular cardinals, which replace $\{\theta : \theta = \text{cf}(\theta) < \kappa_r(T)\}$ for stable first order T; (superstability means $\kappa(\mathfrak{s}) = \emptyset$) and get a non-structure theorem for unsuperstable $\mathfrak{s}$. We connect $\kappa(\mathfrak{s})$, the existence of $(\mathbb{D}_{\mathfrak{s}}, \lambda)$-homogeneous model in λ and the behaviour of a directed union of quite homogeneous models. For a regular cardinality θ, we have: $\theta \in \kappa(\mathfrak{s})$ iff there is a $\leq_{\mathfrak{s}}$-increasing sequence $\langle M_i : i \leq \theta \rangle$ of models and $p \in \mathscr{S}^1(M_\theta)$ such that for each $i < \theta$ the type p forks over M_i (but not necessarily $p \restriction M_{i+1}$ forks over M_i!). This is related to the existence of (λ, κ)-brimmed models.]

V.E.§4 Orthogonality

[We generalize the orthogonality calculus to the present context.]

V.E.§5 Niceness of types

[In general here we do not know that not all types behave "nicely". But for some we can translate problems about them to problems of types in $\mathscr{S}^\alpha_c(M)$ from Chapter V.D. This motivates the definition of nice and prenice types over models. The prenice ones behave as in stable theories. But without existence of pre-nice types this is of limited interest. However, there are quite many of them and in particular see §6 below.]

V.E.§6 Superstable frames

[We deal with rank of types. For superstable $\mathfrak{s}$, the rank is $< \infty$ and then we show that every $p \in \mathscr{S}^{<\omega}(M)$ is prenice fulfilling a promise from §5. The notion of rank is less central than in the first order case as "every $p \in \mathscr{S}(M)$ has rank $< \infty$" is not equivalent to $\kappa(\mathfrak{s}) = \emptyset$ but to a failure of a weak version of $\aleph_0 \in \kappa(\mathfrak{s})$.]

V.E.§7 Regular types and weight

[We generalize regular types and weight to this context. We delay dealing with **P**-simple, **P**-hereditarily orthogonal to **P** and $w_{\mathbf{P}}$ to [Sh 839].]

V.E.§8 Trivial regular types

[We deal with trivial regular types, the ones where depending on a set is equivalent to depending on some member.]

Annotated Content for Ch.V.F (300f): The Heart of the Matter

(This chapter will appear in book 2.)

V.F.§0 Introduction

[We show that if $\mathfrak{s}$ falls under the high side of some dividing lines, it has many complicated models. If it falls under the low side, we can find $\mathfrak{s}^+ = \mathfrak{s}(+)$ with a stronger $\leq_{\mathfrak{s}(+)}$ which also satisfies AxFr_1.]

V.F.§1 More on indiscernibility

[In our context and in particular for stable theories we can combine getting indiscernibles and Erdös-Rado theorem. E.g. if M is a model of a (first order complete) stable T and $a_{\{\alpha,\beta\}} \in M$ for $\alpha < \beta < (2^\lambda)^+, \lambda \geq |T|$, then we can find $u \in [(2^\lambda)^+]^{\lambda^+}$ such that $\langle a_{\{\alpha,\beta\}} : \alpha < \beta$ are from $u\rangle$ is indiscernible, not just $\langle a_{\{\alpha\}} : \alpha \in u\rangle$ is 2-indiscernible in M.

The point is that we define when $\langle M_u : u \in [\lambda]^{\leq n}\rangle$ in independent (this applies even to $M_u \prec \mathfrak{C}, \mathfrak{C}$ a model of a stable theory). We prove existence of such systems parallel to Erdös-Rado theorem. We then turn to other cases.]

V.F.§2 Order properties considered again

[We start with non-order for infinitary formulas and get a non-structure result. This will justify the concentration on the case we have the relevant non-order property.]

V.F.§3 Strengthening the order $\leq_{\mathfrak{s}}$

[Assuming enough non-order, we derive from the framework $\mathfrak{s}$ a framework $\mathfrak{s}^+ = \mathfrak{s}(+)$ satisfying AxFr_1^*, the order letting $M \leq_{\mathfrak{s}(+)} N$ mean ($\leq_{\mathfrak{s}}$ and) preservation of the satisfaction of some infinitary universal formulas.]

V.F.§4 Regaining existence of ω-unions

[We investigate and get non-structure from failure of the existence of ω-limits for the new notion of being a sub-model, $\leq_{\mathfrak{s}(+)}$. The main point is investigation in the ranks of a tree of the form $\{f : f \text{ is a } \leq_{\mathfrak{s}}\text{-embedding of } M_n \text{ into } N\}$ ordered by $\subseteq$ where $\langle M_n : n < \omega\rangle$ is $\leq_{\mathfrak{s}}$-increasing. We conclude (in Conclusion V.F.4.9) that non-structure follows from failure of $\mathrm{Ax(A4)}_\theta$ for $\theta = \aleph_0$ but get only $\dot{I}(\mu, K_{\mathfrak{s}}) \geq \mu^+$ for many μ's.]

V.F.§5 Non-existence of union implies non-structure

[This section is complementary to the previous one getting non-structure from non-existence of an $\leq_{\mathfrak{s}(+)}$-upper bound of an $\leq_{\mathfrak{s}(+)}$-increasing continuous δ-chain also when $\theta = \mathrm{cf}(\delta)$ is minimal and $\theta > \aleph_0$. So the counterexample is less easily manipulated, and the rank from §4 is meaningless. But by the amount of existence which follows by the minimality of θ (and free amalgamation of families of models), we know more how to construct non-forking trees of models and this enables us to prove non-structure.]

Annotated Content for Ch.V.G (300g): Changing the framework

(This chapter will appear in book 2.)

V.G.§0 Introduction

[We hope that by repeating the operation $\mathfrak{s} \mapsto \mathfrak{s}^+$ up to some limit ordinal δ we get an $\mathfrak{s}^{+\delta}$ for which we can prove the main gap. Here we take some steps toward this.]

V.G.§1 On the family of $\mathfrak{s}$'s

[We define a natural (partial) order on the family of reasonable frameworks $\mathfrak{s}$, and prove its basic properties. In particular, increasing sequences has a limit.]

V.G.§2 From large enough $\mathrm{rk}^{\mathrm{emb},2}_{\bar{M}}(f, N)$ to every ordinal

[This continues V.F§1. Here we are interested in well-founded trees, and so if we start with a well-founded tree of $\leq_{\mathfrak{s}}$-submodels of N of cardinality $\leq \mu$ which form a tree of large enough rank, then there is a large enough subtree which is "free enough" so we can "blow it up" to larger ordinals. This is applied to the case $\mathrm{rk}^{\mathrm{emb},2}_{\bar{M}}(f, N)$ is large enough.]

Annotated Content for Ch.VI, (E46): Categoricity of an abstract elementary class in two successive cardinals revisited

(This chapter will appear in book 2.)

VI.§0 Introduction

VI.§1 Basic properties

[We look at an a.e.c. $\mathfrak{K}$ with $\mathrm{LS}(\mathfrak{K}) \leq \lambda$, with assumptions as in the abstract and, by Chapter I, deduce amalgamation (in

$\mathfrak{K}_\lambda$ and $\mathfrak{K}_{\lambda^+}$). We define the class $K_\lambda^{3,\mathrm{na}}$ of triples (M, N, a) ordered by $\leq = \leq_{\mathrm{na}}$ representing (orbital) types in $\mathscr{S}^{\mathrm{na}}(M)$ for $M \in K_\lambda$, and start to investigate it, dealing with the weak extension property, the extension property, minimality, reduced triples and types (except for minimality, in the first order case, these hold trivially). Our aims are to have the extension property or at least the weak extension property for all triples in $K_\lambda^{3,\mathrm{na}}$, and the density of minimal triples. The first property makes the model theory more like the first order case, and the second is connected with categoricity. We start by proving the weak extension property under reasonable assumptions and a consequence of having too many types, reminding the Δ-system lemma.]

VI.§2 The extension property and toward density of minimal types

[We deal with triples from $K_\lambda^{3,\mathrm{na}}$. Under "expensive" assumptions (mainly categoricity in λ^+) we prove that all triples have the extension property and that we have disjoint amalgamation in K_λ. We prove the density of minimal triples under the strong assumptions: $K_{\lambda^{+3}} = \emptyset$ and an extra cardinal arithmetic assumption ($2^{\lambda^+} > \lambda^{++}$). Now the assumption $K_{\lambda^{+3}} = \emptyset$ does no harm if we just intend to prove Theorem VI.0.2(1),(2)(a), i.e. $K_{\lambda^{+3}} \neq \emptyset$ but is a disaster if we would like to continue as in Chapter II or try to get an almost good λ-frame from the present assumptions (without $K_{\lambda^{+3}} = \emptyset$), i.e. VI.0.2(2)(b). The reader willing to accept these assumptions may skip some proofs later.]

VI.§3 On UQ from non-density of minimal (assuming weak extensions)

[Assume ($\mathfrak{K}_\lambda$ has amalgamation and) the minimal types are not dense in $K_\lambda^{3,\mathrm{na}}$, we define and investigate UQ, the class of triples of models with unique amalgamation. So we have some positive model theoretic consequences from what is a non-structure assumption. We get some non-structure results relying on Chapter VII.]

VI.§4 Density of minimal types

[We continue §3 getting the promised results, relying on Chapter VII.]

VI.§5 Inevitable types and stability in λ

[We continue to "climb the ladder", using the amount of structure we already have (and sometimes categoricity) to get more. We start by assuming there are minimal types, and show that some minimal types are inevitable. We construct $p_i \in \mathscr{S}(N_i)$ minimal $(i \leq \lambda^+)$ both strictly increasing continuous and with p_0, p_δ inevitable, and then as in the proof of the equivalence of saturativity and model homogeneity, we show N_δ is universal over N_0. We can then deduce stability in λ, so the model in λ^+ is saturated. Then we note that we have disjoint amalgamation in K_λ.]

VI.§6 Density of uniqueness and proving for $\mathfrak{K}$ categorical in λ^{+2}

[We give a shortcut to proving the main theorem by using stronger assumptions (may be useful in categoricity theorems). For this we first look at uniqueness triples. If $\dot{I}(\lambda^{+2}, K) = 1$ and $\dot{I}(\lambda^{+3}, K) = 0$ then for some triple $(M, N, a) \in K_{\lambda^+}$, a is "1-algebraic" over M, i.e. this is a maximal triple. Now first assuming for some pair $M_0 \leq_{\mathfrak{K}} M_2$ in K_λ we have unique (disjoint) amalgamation for every possible M_1 with $M_0 \leq_{\mathfrak{K}} M_1 \in K_\lambda$ (and using stability), we get a pair of models in λ^+ which contradicts the existence of maximal triples. We then rely on Chapter VII to prove that there are enough cases of unique amalgamation.]

VI.§7 Extensions and Conjugacy

[We investigate types. We prove that in $\mathscr{S}(N), N \in K_\lambda$ the following: reduced implies inevitable, and non-algebraic extensions preserve the conjugacy classes for minimal reduced types (so solving parallel to the realize/materialize problem from Chapter I, see in particular Definition I.4.3(5), the discussion in the beginning of I§5 just after I.5.1 and Claim I.5.23).]

VI.§8 Almost good frame

[We prove the main theorem in particular find an almost good λ-frame $\mathfrak{s}$ with $\mathfrak{K}_{\mathfrak{s}} = \mathfrak{K}_\lambda$.]

Annotated Content for Chapter VII (838): Non-structure in λ^{++} using instances of WGCH

(This chapter will appear in book 2.)

VII.§0 Introduction

[In addition to explaining what we are doing, we quote some definitions (and results) on the weak diamond.]

VII.§1 Nice construction framework

[The intention is to build (many complicated) models of cardinality ∂^+ by approximations of cardinality $< \partial$. We give the basic definitions: of $\mathfrak{u}$ being a nice construction framework (consisting of a $(< \partial)$-a.e.c. $\mathfrak{K}_{\mathfrak{u}}$, the class of approximations to the desired $M \in K^{\mathfrak{u}}_{\partial^+}$, classes FR_ℓ of triples $(M, N, \mathbf{J})$ for $\ell = 1, 2$ and some relations on $K_{\mathfrak{u}}$) and of $\mathfrak{u}$-free rectangles and triangles. We define approximations of size ∂, i.e. the class of triples $(\bar{M}, \bar{\mathbf{J}}, \mathbf{f})$ from $K^{\mathrm{qt}}_{\mathfrak{u}}$ and some quasi orders on them. We prove some basic properties and define what is meant by: almost$_\ell$ all such triples has a property; this will many times mean $M = \cup\{M_\alpha : \alpha < \partial\} \in K^{\mathfrak{u}}_{\partial}$ is saturated.]

VII.§2 Coding properties and non-structure

[the coding properties are sufficient conditions on $\mathfrak{u}$ for finding many non-isomorphic models in $K^{\mathfrak{u},*}_{\partial^+}$. They have the form that $\mathfrak{K}_{\mathfrak{u}}$ has strong forms of failure of amalgamation of two members of $\mathfrak{K}_{\mathfrak{u}}$, so of cardinality $< \partial$ over a third using FR_1, FR_2]

VII.§3 Invariant coding

[We deal with some further coding properties; the invariant meaning that the relevant isomorphisms (which we demand does not exist) fix some models setwise rather than pointwise.]

VII.§4 Straight Applications of codings properties

[We mainly deal with theorems using the weak coding property of a suitable $\mathfrak{u}$ derived from an a.e.c. with $\partial_{\mathfrak{u}} = \lambda^+$ when $2^\lambda < 2^{\lambda^+} < 2^{\lambda^{++}}$ so assuming WDmId_∂ is not λ^{++}-saturated. The first case (in §4(A)) deals with the density of minimal types for $\mathfrak{K}_\lambda$ when $\mathfrak{K}$ is categorical in λ, λ^+ and has a medium number of models in λ^{++} and $\mathrm{LS}(\mathfrak{K}) \leq \lambda$; this is promised in VI§4. The second case (in §4(C)) deals with an a.e.c. which is $\mathrm{PC}_{\aleph_0}$ and has a medium number of models in $\aleph_1$ and not too many models in $\aleph_2$ and derive uniqueness of one sided stable amalgamation (promised in Chapter I). The third case (in §4(D)) continues the first, proving the density of uniqueness triples (M, N, a) in $K^{3,\mathrm{na}}_\lambda$ under the same assumptions, as promised in VI§6. The fourth case (in §4(E)) proves the density of uniqueness triples in $K^{3,\mathrm{bs}}_{\mathfrak{s}}$, for $\mathfrak{s}$ a good λ-frame as promised in II§5. In addition, concerning the first case we eliminate the use of "$\mathrm{WDmId}_{\lambda^+}$ is λ^{++}-saturated" by using $\mathfrak{u}$ with the vertical coding property, this is done in §4(B); this redo [Sh 603]. Finally in §4(F) we do the full versions of the theorems, assuming only the relevant cases of the WGCH, but relying on the results of the subsequent sections §5-§8.]

VII.§5 On almost good λ-frames

[We say some basic things on <u>almost</u> good λ-frames $\mathfrak{s}$; they arise in Chapter VI. E.g. we prove that "N is brimmed over M" is unique up to isomorphism over M (i.e. if N_ℓ is $(\lambda_{\mathfrak{s}}, \kappa_\ell)$-brimmed over M for $\ell = 1, 2$ then N_1, N_2 are isomorphic over M). This is a consequence of analyzing full and brimmed $\mathfrak{u}$-free rectangles and triangles for some nice construction framework $\mathfrak{u}$ derived from $\mathfrak{s}$.]

VII.§6 Density of weak versions of uniqueness

[For a good λ-frame, for any $\xi < \lambda^+$ we prove that either $K^{\mathfrak{s}}$ has non-structure in λ^{++} by getting vertical uq-invariant coding, from §3, or prove density for $K^{3,\mathrm{up}}_{\mathfrak{s},\xi}$, a quite weak form of uniqueness of triples, i.e. of a kind of uniqueness for a suitable form of amalgamation. As we like to deal also with almost good λ-frames, we rely on §5. This relates to §4(D),§4(E).]

VII.§7 Pseudo uniqueness

[From existence for $K^{3,\mathrm{up}}_{\mathfrak{s},\xi}$ for $\xi = \lambda^+$ we define $\mathrm{WNF}_{\mathfrak{s}}$, a weak form of the class of quadruples $\langle M_\ell : \ell < 4\rangle$ of models from $K_{\mathfrak{s}}$ with M_1, M_2 amalgamated in a non-forking way over M_0 inside M_3. We prove that $\mathrm{WNF}_{\mathfrak{s}}$ is a weak $\mathfrak{s}$-non-forking relation which respects $\mathfrak{s}$.]

VII.§8 Density of $K^{3,\mathrm{uq}}_{\mathfrak{s}}$

[We try to prove non-structure in λ^{++} from failure of density of $K^{3,\mathrm{uq}}_{\mathfrak{s}}$. By §6 we justify assuming existence for $K^{3,\mathrm{up}}_{\mathfrak{s}}$, so by §7 the relation $\mathrm{WNF}_{\mathfrak{s}}$ is a well defined weak $\mathfrak{s}$-non-forking relation on $\mathfrak{K}_{\mathfrak{s}}$ (respecting $\mathfrak{s}$). So we can define $\mathfrak{u}$ such that $(M_0, N_0, a) \leq^{\ell}_{\mathfrak{u}} (M_1, N_1, a)$ implies $\mathrm{WNF}(M_0, N_0, M_1, N_1)$. We also show that it is enough to show $K^{3,\mathrm{up}}_{\mathfrak{s}} \subseteq K^{3,\mathrm{uq}}_{\mathfrak{s}}$. Now the proof splits to two cases. In the first we assume wnf-delayed uniqueness fails and get vertical coding. In the second we assume wnf-delayed uniqueness holds but density of uniqueness triples fail and get horizontal coding (using the properties of WNF).]

VII.§9 The combinatorial part

[We first quote; central in justifying our results is $\mu_{\rm unif}(\partial^+, 2^\partial)$ which "usually" is 2^{∂^+}, (in VII.9.4). We show that building an appropriate tree $\langle M_\eta : \eta \in {}^{\partial^+ \geq}(2^\partial)\rangle$ is enough (in VII.9.1). We present building $\langle \bar{M}_{\eta\char`^<\alpha>} : \alpha < 2^\partial\rangle$ as above (in VII.9.3); as well as the "universal case", i.e. when $M_\eta(\eta \in {}^\partial 2)$ are pairwise non-isomorphic of $M_{<>}$. Also we deal with the results on having many models in ∂ (when $\emptyset \in$ WDmId$_\partial$) and mention the case in each step $\alpha < \partial^+$ we use, e.g. ∂ substeps.]

VII.§10 Proofs of the non-structure theorems, with choice functions

[This has a somewhat more set theoretic character compared to, and fulfills promises from §2,§3. We prove various coding theorems saying that there are many non-isomorphic models in ∂^+. In particular we prove this for nice construction frameworks in cases in which we need amalgamation choice functions.]

VII.§11 Remarks on pcf

[We prove things in pcf relevant to non-structure in a reasonably self contained way. One is a relative of Hajnal free subset theorem. The main other says that if $2^\lambda < 2^{\lambda^+}$ then one of three cases occurs, each helpful in proof of non-structure and some related results. This is a revised version of part of [Sh 603].]

ABSTRACT ELEMENTARY CLASSES NEAR $\aleph_1$
SH88R

§0 Introduction

In [Sh 48], proving a conjecture of Baldwin, we show that ($\mathbf{Q}$ here stands for the quantifier $\mathbf{Q}^{\rm car}_{\geq \aleph_1}$, there are uncountably many)

$(*)_1$ no $\psi \in \mathbb{L}_{\omega_1,\omega}(\mathbf{Q})$ has a unique uncountable model up to isomorphism

by showing that

$(*)_2$ categoricity (of $\psi \in \mathbb{L}_{\omega_1,\omega}(\mathbf{Q})$) in $\aleph_1$ implies the existence of a model of ψ of cardinality $\aleph_2$ (so ψ has ≥ 2 non-isomorphism models).

Unfortunately, both $(*)_1$ and $(*)_2$ were not proved in ZFC because diamond on $\aleph_1$ was assumed. In [Sh 87a] and [Sh 87b] this set theoretic assumption was weakened to $2^{\aleph_0} < 2^{\aleph_1}$; here we shall prove it in ZFC (see §3). However, for getting the conclusion from the weaker model theoretic assumption $\dot{I}(\aleph_1, \psi) < 2^{\aleph_1}$ as there, we still need $2^{\aleph_0} < 2^{\aleph_1}$.

The main result of [Sh 87a], [Sh 87b] was:

$(*)_3$ if $n > 0, 2^{\aleph_0} < 2^{\aleph_1} < \ldots < 2^{\aleph_n}, \psi \in \mathbb{L}_{\omega_1,\omega}, 1 \leq \dot{I}(\aleph_\ell, \psi) < \mu_{\rm wd}(\aleph_\ell)$ for $\ell \leq n, \ell \geq 1$ (where $\mu_{\rm wd}(\aleph_\ell)$ is usually $2^{\aleph_\ell}$ and always $> 2^{\aleph_{\ell-1}}$, see 0.5 below) <u>then</u> ψ has a model of cardinality $\aleph_{n+1}$

$(*)_4$ if $2^{\aleph_0} < 2^{\aleph_1} < \ldots < 2^{\aleph_n} < 2^{\aleph_{n+1}} < \ldots$ and $\psi \in \mathbb{L}_{\omega_1,\omega}, 1 \leq \dot{I}(\aleph_\ell, \psi) < \mu_{\rm wd}(\aleph_\ell)$ for $\ell < \omega$ <u>then</u> ψ has a model in every infinite cardinal (and satisfies Łos Conjecture), (note that $(*)_3$ for $n = 1$, assuming $\diamondsuit_{\aleph_1}$ was proved in [Sh 48]).

Typeset by $\mathcal{A}_\mathcal{M}\mathcal{S}$-TEX

In $(*)_4$, it is proved that without loss of generality $\mathfrak{K}$ is excellent; this means in particular that K is the class of atomic models of some countable first order T. The point is that an excellent class $\mathfrak{K}$ is similar to the class of models of an $\aleph_0$-stable first order T. In particular the set of relevant types, $\mathbf{S}_{\mathfrak{K}}(A, M)$ is defined as $\{p(x) : p(x)$ a complete type over A in M in the first order sense such that $p \restriction B$ is isolated for every finite $B \subseteq A\}$. <u>But</u> we better restrict ourselves to "nice A", that is A which is the universe of some $N \prec M$ or $A = N_1 \cup N_2$ where N_0, N_1, N_2 are in stable amalgamation or $\cup\{N_u : u \in \mathscr{P} \subseteq \mathscr{P}(n)\}$ for some (so called) stable system $\langle N_u : u \in \mathscr{P}\rangle$; on stable such systems in the stable first order case see [Sh:c, XII,§5]. So types are quite like the first order case. In particular we say $M \in \mathfrak{K}$ is λ-full <u>when</u>: if $p \in \mathbf{S}_{\mathfrak{K}}(A, M), A$ as above, $|A| < \lambda$ implies p is realized in M; this is the replacement of λ-saturated for that context.

Why in [Sh 87a] and [Sh 87b], ψ was assumed to be just in $\mathbb{L}_{\omega_1,\omega}$ and not more generally in $\mathbb{L}_{\omega_1,\omega}(\mathbf{Q})$? Mainly because we feel that in [Sh 48], the logic $\mathbb{L}_{\omega_1,\omega}(\mathbf{Q})$ was incidental. We delay the search for the right context to this sequel. So here we are working in a.e.c., "abstract elementary class" (so no logic is present in the context) which are formally like elementary classes, i.e. $(\mathrm{Mod}_T, \prec), T$ first order but note the absence of amalgamation, still they have closure under union of increasing chains. It is $\mathfrak{K} = (K, \leq_{\mathfrak{K}})$ where $\leq_{\mathfrak{K}}$ is the "abstract" notion of elementary submodel. So if $\mathscr{L}$ is a fragment of $\mathbb{L}_{\infty,\omega}(\tau)$ (for a fixed vocabulary), $T \subseteq \mathscr{L}$ a theory included in $\mathscr{L}$, and we let $K = \{M : M \models T\}, M \leq_{\mathfrak{K}} N$ if and only if $M \prec_{\mathscr{L}} N$, we get such a class; if $\mathscr{L}$ is countable then $\mathfrak{K}$ has L.S. number $\aleph_0$. So the class of models of $\psi \in \mathbb{L}_{\omega_1,\omega}(\mathbf{Q})$ is not represented directly, but can be with minor adaptation; see 3.18(2). Surprisingly (and by not so hard proof), every a.e.c. $\mathfrak{K}$ can be represented as a pseudo elementary class <u>if</u> we allow omitting types, (see 1.9). We introduce a relative of saturated models (for stable first order T) and full models (for excellent classes, see [Sh 87a] and [Sh 87b]): limit models; really several variants of this notion. See Definition 3.3. The strongest and most important variant is "$M \in K_\lambda$ superlimit" which means: M is universal (under $\leq_{\mathfrak{K}}$), $(\exists N)(M \leq_{\mathfrak{K}} N \wedge M \neq N)$ and if $M_i \cong M$

for $i < \delta \leq \|M\|$ and M_i is $\leq_{\mathfrak{K}}$-increasing then $\bigcup_{i<\delta} M_i \cong M$. If we restrict ourselves to δ's of cofinality κ we get (λ, κ)-superlimit. Such M exists for a first order T for some pairs λ, κ. In particular (see more in [Sh 868])

$(*)_5$ for every $\lambda \geq 2^{|T|} + \beth_\omega$, a superlimit model of T of cardinality λ exists if and only if T is superstable (by [Sh 868, 3.1]).

Moreover

$(*)_6$ "almost always"; for $\lambda \geq 2^{|T|} + \kappa, \kappa =$ cf(κ) (for simplicity) we have:
a (λ, κ)-superlimit model exists iff T is stable in λ & $\kappa \geq \kappa(T)$ or $\lambda = \lambda^{<\kappa}$.

But we can prove something under those circumstances: if K is categorical in λ or just have a superlimit model M^* in λ, but the λ-amalgamation property fails for M^* and $2^\lambda < 2^{\lambda^+}$ then $\dot{I}(\lambda^+, K) = 2^{\lambda^+}$ (see 3.8). With some reasonable restrictions on λ and K, we can prove e.g. $\dot{I}(\lambda, K) = \dot{I}(\lambda^+, K) = 1 \Rightarrow \dot{I}(\lambda^{++}, K) \geq 1$, (see 3.11, 3.13).

However, our long term main aim was to do the parallel of [Sh 87a] and [Sh 87b] in the present context, i.e., for an a.e.c. $\mathfrak{K}$ and it is natural to assume $\mathfrak{K}$ is $\mathrm{PC}_{\aleph_0}$, here we prepare the ground.

Sections 4,5 present work toward this goal (§5 assuming $2^{\aleph_0} < 2^{\aleph_1}$; §4 without it). We should note that dealing with superlimit models rather than full ones make problems, as well as the fact that the class is not necessarily elementary in some reasonable logics. Because of the second we were driven to use formulas which hold "generically", are "forced" instead of are satisfied, and "the type $\bar{a}$ materialize" instead of realize and gtp$(\bar{a}, N, M)$ instead of tp$(\bar{a}, N, M)$. We also (necessarily) encounter the case "$\mathbf{D}(N)$ of cardinality $\aleph_1$ for $N \in K_{\aleph_0}$", see 5.2, 5.4(6). Because of the first, the scenario for getting a full model in $\aleph_1$ (which can be adapted to $(\aleph_1, \{\aleph_1\})$-superlimit - see 5.17) does not seem to be enough for getting superlimit models in $\aleph_1$ (see 5.39).

We had felt that arriving at enough conclusions on the models of cardinality $\aleph_1$ to start dealing with models of cardinality $\aleph_2$, will be

a strong indication that we can complete the generalization of [Sh 87a] and [Sh 87b], so getting superlimits in $\aleph_1$ is the culmination of this paper and a natural stopping point. Trying to do the rest (of the parallel to [Sh 87a] and [Sh 87b]) was delayed.
Much remains to be done,

0.1 Problem:

1) Prove $(*)_3, (*)_4$ in our context.
2) Parallel results in ZFC; e.g. prove $(*)_3$ for $n = 1, 2^{\aleph_0} = 2^{\aleph_1}$.
Note that if $2^{\aleph_0} = 2^{\aleph_1}$, assuming $1 \le \dot{I}(\aleph_1, K) < 2^{\aleph_1}$ give really less model theoretic consequences, as new phenomena arise (see §6). See §4 (and its concluding remarks).
3) Construct examples; e.g. (an a.e.c.) $\mathfrak{K}$ (or $\psi \in \mathbb{L}_{\omega_1,\omega}$), categorical in $\aleph_0, \aleph_1, \dots, \aleph_n$ but not in $\aleph_{n+1}$.
4) If $\mathfrak{K}$ is a PC_λ class, categorical in λ, λ^+, does it necessarily have a model in λ^{++}?

See the book's introduction Chapter N on the progress on those problems in particular on [Sh 576], redone here in Chapter VI. The direct motivation for [Sh 576] was that Grossberg asked me (Oct. 1994) some questions in this neighborhood (mainly 0.1(4)), in particular:

> $(*)$ assume $K =$ Mod(T), (i.e. K is the class of models of T), $T \subseteq L_{\omega_1,\omega}, |T| = \lambda, I(\lambda, K) = 1$ and $1 \le I(\lambda^+, K) < 2^{\lambda^+}$. Does it follow that $I(\lambda^{++}, K) > 0$?

We think of this as a test problem and much prefer a model theoretic to a set theoretic solution. This is closely related to 0.1(4) above and to 3.11 (where we assume categoricity in λ^+, do not require $2^\lambda < 2^{\lambda^+}$ but take $\lambda = \aleph_0$ or some similar cases) and 5.27(4) (and see 5.2 and 4.8 on the assumptions) (there we require $2^\lambda < 2^{\lambda^+}, 1 \le I(\lambda^+, K) < 2^{\lambda^+}$ and $\lambda = \aleph_0$).

Problem [Sh 576, 0.1] was stated a posteriori but is, I think, the real problem, it says:

> $(**)$ Can we have some (not necessarily much) classification theory for reasonable non-first order classes $\mathfrak{K}$ of models, with no

uses of even traces of compactness and only mild set theoretic assumptions?

This is a revised version of [Sh 88] which continues [Sh 87a], [Sh 87b] but do not use them. The paper [Sh 88] and the present chapter relies on [Sh 48] only when deducing results on $\psi \in \mathbb{L}_{\omega_1,\omega}(\mathbf{Q})$; it improves some of its early results and extends the context. The work on [Sh 88] was done in 1977, and a preprint was circulated. Before the paper had appeared, a user-friendly expository article of Makowsky [Mw85a] represent, give background and explain the easy parts of the paper. In [Sh 88] the author have corrected and replaced some proofs and added mainly §6. See more in [Sh:F709].

We thank Rami Grossberg for lots of work in the early eighties on previous versions, i.e. [Sh 88], which improved this paper, and the writing up of an earlier version of §6 and Assaf Hasson on helpful comments in 2002 and Alex Usvyatsov for very careful reading, corrections and comments and Adi Jarden and Alon Siton on help in the final stages.

* * *

On history and background on $\mathbb{L}_{\omega_1,\omega}, \mathbb{L}_{\infty,\omega}$ and the quantifier $\mathbf{Q}$ see [Ke71]. On (D, λ)-sequence-homogeneous (which 2.2 - 2.5 here generalized) see Keisler-Morley [KM67], this is defined in 2.3(5), and 2.5 is from there. Theorem 3.8 is similar to [Sh 87a, 2.7] and [Sh 87b, 6.3].

Remark. On non-splitting used here in 5.6 see [Sh 3], [Sh:c, Ch.I, Def.2.6, p.11] or [Sh 48].
We finish §0 by some necessary quotation.

By [Ke70] and [Mo70],

0.2 Claim. *1) Assume that $\psi \in \mathbb{L}_{\omega_1,\omega}(\mathbf{Q})$ has a model M in which $\{\mathrm{tp}_\Delta(\bar{a}, \emptyset, M) : \bar{a} \in M\}$ is uncountable where $\Delta \subseteq \mathbb{L}_{\omega_1,\omega}(\mathbf{Q})$ is countable, <u>then</u> ψ has $2^{\aleph_1}$ pairwise non-isomorphic models of cardinality $\aleph_1$, in fact we can find models M_α of ψ of cardinality $\aleph_1$ for $\alpha < 2^{\aleph_1}$ such that $\{\mathrm{tp}_\Delta(a; \emptyset, M_\alpha) : a \in M_\alpha\}$ are pairwise distinct*

where $\mathrm{tp}_\Delta(\bar{a}, A, M) = \{\varphi(\bar{x}, \bar{b}) : \varphi(\bar{x}, \bar{y}) \in \Delta$ *and* $M \models \varphi[\bar{a}, \bar{b}]$ *and* $\bar{b} \in {}^{\omega>}A\}$.
2) If $\psi \in \mathbb{L}_{\omega_1,\omega}(\mathbf{Q}), \Delta \subseteq \mathbb{L}_{\omega_1,\omega}(\mathbf{Q})$ *is countable and* $\{\mathrm{tp}_\Delta(\bar{a}, \emptyset, M) : \bar{a} \in {}^{\omega>}M$ *and* M *is a model of* $\psi\}$ *is uncountable, then it has cardinality* $2^{\aleph_0}$.

Also note

0.3 Observation. Assume (τ is a vocabulary and)

- (a) K is a family of τ-models of cardinality λ
- (b) $\mu > \lambda^\kappa$
- (c) $\{(M, \bar{a}) : M \in K$ and $\bar{a} \in {}^\kappa M\}$ has $\geq \mu$ members up to isomorphism.

Then K has $\geq \mu$ models up to isomorphisms (similarly for $= \mu$).

Proof. See [Sh:a, VIII,1.3] or just check by cardinal arithmetic. $\square_{0.3}$

Further

0.4 Claim. *1) Assume* λ *is regular uncountable,* M_0 *is a model with countable vocabulary and* $T = \mathrm{Th}_{\mathbb{L}}(M_0)$, $<$ *a binary predicate from* $\tau(T)$ *and* $(P^{M_0}, <^{M_0}) = (\lambda, <)$. *Then every countable model* M *of* T *has an end extension, i.e.,* $M \prec N$ *and* $P^M \neq P^N$ *and* $a \in P^N \wedge b \in P^M \wedge a <^N b \Rightarrow a \in M$.
2) Moreover, we can further demand $(P^N, <^N)$ *is non-well ordered and we can demand* $|P^N| = \aleph_1, (P^N, <^N)$ *is* $\aleph_1$*-like (which means that it has cardinality* $\aleph_1$ *but every (proper) initial segment has cardinality* $< \aleph_1$*); and we can demand* N *is countable.*
3) Moreover, we can add the demand that in $(P^N, <^N)$ *there is a first element in* $P^N \backslash P^M$ *and we can add the demand: in* $(P^N, <^N)$, *there is no first element in* $P^N \backslash P^M$.

Proof. 1),2) Keisler [Ke70].
3) By [Sh 43] and independently Schmerl [Sc76]. $\square_{0.4}$

By Devlin-Shelah [DvSh 65], and [Sh:f, Ap,§1] (the so-called weak diamond).

0.5 Theorem. *Assume that* $2^\lambda < 2^{\lambda^+}$.
1) There is a normal ideal $\mathrm{WDmId}_{\lambda^+}$ *on* λ^+ *and* $\lambda^+ \notin \mathrm{WDmId}_{\lambda^+}$, *of course, (the members are called small set) such that: if* $S \in (\mathrm{WDmId}_{\lambda^+})^+$ *(e.g.,* $S = \lambda^+$*) and* $\mathbf{c} : {}^{\lambda^+ >}(\lambda^+) \to \{0,1\}$, *then there is* $\bar\ell = \langle \ell_\alpha : \alpha < \lambda^+\rangle \in {}^{\lambda^+}2$ *such that for every* $\eta \in {}^{\lambda^+}(\lambda^+)$ *the set* $\{\delta \in S : \mathbf{c}(\eta \restriction \delta) = \ell_\alpha\}$ *is stationary; we call* $\bar\ell$ *a weak diamond sequence (for the colouring* $\mathbf{c}$ *and the stationary set* S*).*
2) $\mu_* = \mu_{\mathrm{wd}}(\lambda^+)$, *the cardinal defined by* $(*)$ *below, is* $> 2^\lambda$ *(we do not say* $\geq 2^{\lambda^+}$*!)*

- $(*)$ (α) *if* $\mu < \mu_*$ *and* $\mathbf{c}_\varepsilon$ *for* $\varepsilon < \mu$ *is as above then we can find* $\bar\ell$ *as in part (1) for all the* $\mathbf{c}_\varepsilon$*'s simultaneously*
  - (β) μ_* *is maximal such that clause* (α) *holds.*

3) $\mu_* = \mu_{\mathrm{unif}}(\lambda^+, 2^\lambda)$ *satisfies* $\mu_*^{\aleph_0} = 2^{\lambda^+}$ *and moreover* $\lambda \geq \beth_\omega \Rightarrow \mu_* = 2^\lambda$ *where* $\mu_{\mathrm{unif}}(\lambda^+, \chi)$ *is the first cardinal* μ *such that we can find* $\langle \mathbf{c}_\alpha : \alpha < \mu\rangle$ *such that:*

- (a) $\mathbf{c}_\alpha$ *is a function from* ${}^{\lambda^+ >}(\lambda^+)$ *to* χ
- (b) *there is no* $\rho \in {}^{\lambda^+}\chi$ *such that for every* $\alpha < \mu$ *for some* $\eta \in {}^{\lambda^+}(\lambda^+)$ *the set* $\{\delta < \lambda : \mathbf{c}_\alpha(\eta \restriction \delta) \neq \rho(\delta)\}$ *is stationary (so* $\mu_{\mathrm{wd}}(\lambda^+) = \mu_{\mathrm{unif}}(\lambda^+, 2)$*).*

See more in VII§0,§9 and hopefully in [Sh:E45].

The following are used in §2.

0.6 Definition. 1) For a regular uncountable cardinal λ let $\check I[\lambda] = \{S \subseteq \lambda$: some pair $(E, \bar a)$ witnesses $S \in \check I(\lambda)$, see below$\}$.
2) We say that (E, u) is a witness for $S \in \check I[\lambda]$ if:

- (a) E is a club of the regular cardinal λ
- (b) $u = \langle u_\alpha : \alpha < \lambda\rangle, a_\alpha \subseteq \alpha$ and $\beta \in a_\alpha \Rightarrow a_\beta = \beta \cap a_\alpha$
- (c) for every $\delta \in E \cap S, u_\delta$ is an unbounded subset of δ of order-type $< \delta$ (and δ is a limit ordinal).

By [Sh 420] and [Sh:E12]

0.7 Claim. *Let λ be regular uncountable.*
1) If $S \in \check{I}[\lambda]$ then we can find a witness $(E, \bar{a})$ for $S \in \check{I}[\lambda]$ such that:

(a) $\delta \in S \cap E \Rightarrow \mathrm{otp}(a_\delta) = \mathrm{cf}(\delta)$
(b) *if* $\alpha \notin S$ *then* $\mathrm{otp}(a_\alpha) < \mathrm{cf}(\delta)$ *for some* $\delta \in S \cap E$.

2) $S \in \check{I}[\lambda]$ iff there is a pair $(E, \bar{\mathscr{P}})$ such that:

(a) *E is a club of the regular uncountable λ*
(b) *$\bar{\mathscr{P}} = \langle \mathscr{P}_\alpha : \alpha < \lambda \rangle$, where $\mathscr{P}_\alpha \subseteq \{u : u \subseteq \alpha\}$ has cardinality $< \lambda$*
(c) *if $\alpha < \beta < \lambda$ and $\alpha \in u \in \mathscr{P}_\beta$ then $u \cap \alpha \in \mathscr{P}_\alpha$*
(d) *if $\delta \in E \cap S$ then some $u \in \mathscr{P}_\delta$ is an unbounded subset of δ (and δ is a limit ordinal).*

§1 Axioms and simple properties for classes of models

1.1 Context. 1) Here in §1-§5, τ is a vocabulary, K will be a class of τ-models and $\leq_{\mathfrak{K}}$ a two-place relation on the models in K. We do not always strictly distinguish between K and $\mathfrak{K} = (K, \leq_{\mathfrak{K}})$. We shall assume that $K, \leq_{\mathfrak{K}}$ are fixed; and usually we assume that $\mathfrak{K}$ is an a.e.c. (abstract elementary class) which means that the following axioms hold.
2) For a logic $\mathscr{L}$ let $M \prec_{\mathscr{L}} N$ mean M is an elementary submodel of N for the language $\mathscr{L}(\tau_M)$ and $\tau_M \subseteq \tau_N$, i.e., if $\varphi(\bar{x}) \in \mathscr{L}(\tau_M)$ and $\bar{a} \in {}^{\ell g(\bar{x})}M$ then $M \models \varphi[\bar{a}] \Leftrightarrow N \models \varphi[\bar{a}]$; similarly $M \prec_L N$ for L a language, i.e. a set of formulas in some $\mathscr{L}(\tau_M)$. So $M \prec N$ in the usual sense means $M \prec_{\mathbb{L}} N$ as $\mathbb{L}$ is first order logic and $M \subseteq N$ means M is a submodel of N.

1.2 Definition. 1) We say $\mathfrak{K}$ is a a.e.c. with L.S. number $\lambda(\mathfrak{K}) =$ LS$(\mathfrak{K})$ if:
Ax 0: The holding of $M \in K, N \leq_{\mathfrak{K}} M$ depend on N, M only up to isomorphism, i.e. $[M \in K, M \cong N \Rightarrow N \in K]$ and [if $N \leq_{\mathfrak{K}} M$ and f is an isomorphism from M onto the τ-model M', $f \restriction N$ is an isomorphism from N onto N' then $N' \leq_{\mathfrak{K}} M'$].

Ax I: if $M \leq_{\mathfrak{K}} N$ then $M \subseteq N$ (i.e. M is a submodel of N).

Ax II: $M_0 \leq_{\mathfrak{K}} M_1 \leq_{\mathfrak{K}} M_2$ implies $M_0 \leq_{\mathfrak{K}} M_2$ and $M \leq_{\mathfrak{K}} M$ for $M \in K$.

Ax III: If λ is a regular cardinal, $M_i (i < \lambda)$ is a $\leq_{\mathfrak{K}}$-increasing (i.e. $i < j < \lambda$ implies $M_i \leq_{\mathfrak{K}} M_j$) and continuous (i.e. for $\delta < \lambda, M_\delta = \bigcup_{i<\delta} M_i$) then $M_0 \leq_{\mathfrak{K}} \bigcup_{i<\lambda} M_i$.

Ax IV: If λ is a regular cardinal and M_i (for $i < \lambda$) is $\leq_{\mathfrak{K}}$-increasing continuous and $M_i \leq_{\mathfrak{K}} N$ for $i < \lambda$ then $\bigcup_{i<\lambda} M_i \leq_{\mathfrak{K}} N$.

Ax V: If $N_0 \subseteq N_1 \leq_{\mathfrak{K}} M$ and $N_0 \leq_{\mathfrak{K}} M$ then $N_0 \leq_{\mathfrak{K}} N_1$.

Ax VI: If $A \subseteq N \in K$ and $|A| \leq \mathrm{LS}(\mathfrak{K})$ then for some $M \leq_{\mathfrak{K}} N, A \subseteq |M|$ and $\|M\| \leq \mathrm{LS}(\mathfrak{K})$ (and $\mathrm{LS}(\mathfrak{K})$ is the minimal infinite cardinal satisfying this axiom which is $\geq |\tau|$; the $\geq |\tau|$ is for notational simplicity).
2) We say $\mathfrak{K}$ is a weak[1] a.e.c. if above we omit clause IV.

Remark. Note that AxV holds for $\prec_{\mathscr{L}}$ for any logic $\mathscr{L}$.

Notation: Let $K_\lambda = \{M \in K : \|M\| = \lambda\}$ and $K_{<\lambda} = \bigcup_{\mu<\lambda} K_\mu$ and $\mathfrak{K}_\lambda = (K_\lambda, \leq_{\mathfrak{K}} \restriction K_\lambda)$ and similarly $\mathfrak{K}_{<\lambda}, K_{\leq\lambda}, \mathfrak{K}_{\geq\lambda}, K_{\geq\lambda}$. Recall $\mathbb{L}$ is first order logic.

1.3 Definition. The embedding $f : N \to M$ is called a $\leq_{\mathfrak{K}}$-embedding if the range of f is the universe of a model $N' \leq_{\mathfrak{K}} M$ (so $f : N \to N'$ is an isomorphism onto).

1.4 Definition. Let T_1 be a theory in $\mathscr{L}(\tau_1), \Gamma$ a set of types in $\mathscr{L}(\tau_1)$ for some logic $\mathscr{L}$, usually first order.
1) $\mathrm{EC}(T_1, \Gamma) = \{M : M$ an τ_1-model of T_1 which omits every $p \in \Gamma\}$. We implicitly use that τ_1 is reconstructible from T_1, Γ. A problem

[1]this is not really investigated here

may arise only if some symbols from τ_1 are not mentioned in T_1 and in Γ, so we may write $\mathrm{EC}(T_1, \Gamma, \tau_1)$, but usually we ignore this point.
2) For $\tau \subseteq \tau_1$ we let $\mathrm{PC}(T_1, \Gamma, \tau) = \mathrm{PC}_\tau(T_1, \Gamma) = \{M : M$ is a τ-reduct of some $M_1 \in \mathrm{EC}(T_1, \Gamma)\}$.
3) We say that K, a class of τ-models, is a PC^μ_λ or $\mathrm{PC}_{\lambda,\mu}$ class <u>when</u> for some T_1, Γ_1, τ_1 we have $\tau \subseteq \tau_1, T_1$ a first order theory in the vocabulary τ_1, Γ_1 a set of types in $\mathbb{L}(\tau_1), K = \mathrm{PC}_\tau(T_1, \Gamma_1)$ and $|T_1| \leq \lambda, |\Gamma_1| \leq \mu$.
4) We say $\mathfrak{K}$ is PC^μ_λ or $\mathrm{PC}_{\lambda,\mu}$ <u>if</u> for some $(T_1, \Gamma_1, \tau_1), (T_2, \Gamma_2, \tau_2)$ as in part (3) we have $K = \mathrm{PC}(T_1, \Gamma_1, \tau)$ and $\{(M, N) : M \leq_{\mathfrak{K}} N$ hence $M, N \in K\} = \mathrm{PC}(T_2, \Gamma_2, \tau')$ where $\tau' = \tau \cup \{P\} \subseteq \tau_2, P$ a new one-place predicate, so $|\tau_\ell| \leq \lambda, |\Gamma_\ell| \leq \mu$ for $\ell = 1, 2$.
If $\mu = \lambda$ we may omit μ.
5) In (4) we may say "$\mathfrak{K}$ is (λ, μ)-presentable" and if $\lambda = \mu$ we may say "$\mathfrak{K}$ is λ-presentable".

<u>1.5 Example</u>: If $T \subseteq \mathbb{L}(\tau), \Gamma$ a set of types in $\mathbb{L}(\tau)$, then $K := \mathrm{EC}(T, \Gamma), \leq_{\mathfrak{K}} := \prec_{\mathbb{L}}$ form an a.e.c. with LS-number $\leq |T| + |\tau| + \aleph_0$, that is, satisfy the Axioms from 1.2 (for $\mathrm{LS}(\mathfrak{K}) := |\tau| + \aleph_0$).

1.6 Observation. Let I be a directed set (i.e. partially ordered by $\leq$, such that any two elements have a common upper bound).
1) If M_t is defined for $t \in I$ and $t \leq s \in I$ implies $M_t \leq_{\mathfrak{K}} M_s$ <u>then</u> $\bigcup\limits_{s \in I} M_s \in K$ and for every $t \in I$ we have $M_t \leq_{\mathfrak{K}} \bigcup\limits_{s \in I} M_s$.
2) If in addition $t \in I$ implies $M_t \leq_{\mathfrak{K}} N$ <u>then</u> $\bigcup\limits_{s \in I} M_s \leq_{\mathfrak{K}} N$.

Proof. By induction on $|I|$ (simultaneously for (1) and (2)).

If I is finite, then I has a maximal element $t(0)$, hence $\bigcup\limits_{t \in I} M_t = M_{t(0)}$, so there is nothing to prove.
So suppose $|I| = \mu$ and we have proved the assertion when $|I| < \mu$. Let $\lambda = \mathrm{cf}(\mu)$ so λ is a regular cardinal; hence we can find I_α (for $\alpha < \lambda$) such that $|I_\alpha| < |I|, \alpha < \beta < \lambda$ implies $I_\alpha \subseteq I_\beta \subseteq I, \bigcup\limits_{\alpha < \lambda} I_\alpha = I$, for limit $\delta < \lambda, I_\delta = \bigcup\limits_{\alpha < \delta} I_\alpha$ and each I_α is directed

and non-empty; this is trivial when $\lambda > \aleph_0$ and obvious otherwise. Let $M^\alpha = \bigcup_{t \in I_\alpha} M_t$; so by the induction hypothesis on (1) we know that $t \in I_\alpha$ implies $M_t \leq_{\mathfrak{K}} M^\alpha$. If $\alpha < \beta$ then $t \in I_\alpha$ implies $t \in I_\beta$ hence $M_t \leq_{\mathfrak{K}} M^\beta$; hence by the induction hypothesis on (2) applied to $\langle M_t : t \in I_\alpha \rangle, M_\beta$ we have $M^\alpha = \bigcup_{t \in I_\alpha} M_t \leq_{\mathfrak{K}} M^\beta$. So by Ax III, applied to $\langle M^\alpha : \alpha < \lambda \rangle$ we have $M^\alpha \leq_{\mathfrak{K}} \bigcup_{\beta<\lambda} M^\beta = \bigcup_{t \in I} M_t$, and as $t \in I_\alpha$ implies $M_t \leq_{\mathfrak{K}} M^\alpha$, by Ax II, $t \in I$ implies $M_t \leq_{\mathfrak{K}} \bigcup_{s \in I} M_s$. So we have finished proving part (1) for the case $|I| = \mu$. To prove (2) in this case note that for each $\alpha < \lambda$, $\langle M_t : t \in I_\alpha \rangle$ is $\leq_{\mathfrak{K}}$-directed and $t \in I_\alpha \Rightarrow M_t \leq_{\mathfrak{K}} N$, so clearly by the induction hypothesis for (2) we have $M^\alpha := \cup\{M_t : t \in I_\alpha\}$ is $\leq_{\mathfrak{K}} N$. So $\alpha < \lambda \Rightarrow M^\alpha \leq_{\mathfrak{K}} N$ and as proved above $\langle M^\alpha : \alpha < \lambda \rangle$ is $\leq_{\mathfrak{K}}$-increasing and obviously it is continuous, hence by Ax IV, $\bigcup_{s \in I} M_s = \bigcup_{\alpha<\lambda} M^\alpha \leq_{\mathfrak{K}} N$. $\square_{1.6}$

1.7 Lemma. *Let $\tau_1 = \tau \cup \{F_i^n : i < \mathrm{LS}(\mathfrak{K}), n < \omega\}$, F_i^n an n-place function symbol (assuming, of course, $F_i^n \notin \tau$).*

Every model M (in K) can be expanded to an τ_1-model M_1 such that:

(*A*) *$M_{\bar a} \leq_{\mathfrak{K}} M$ when $n < \omega, \bar a \in {}^n|M|$ and where $M_{\bar a}$ is the submodel of M with universe $\{F_i^n(\bar a) : i < \mathrm{LS}(\mathfrak{K})\}$*

(*B*) *if $\bar a \in {}^n|M|$ then $\|M_{\bar a}\| \leq \mathrm{LS}(\mathfrak{K})$*

(*C*) *if $\bar b$ is a subsequence of a permutation of $\bar a$, then $M_{\bar b} \leq_{\mathfrak{K}} M_{\bar a}$*

(*D*) *for every $N_1 \subseteq M_1$ we have $N_1 \restriction \tau \leq_{\mathfrak{K}} M$.*

Proof. We define by induction on n, the values of $M_{\bar a}$ and of $F_i^n(\bar a)$ for every $i <$ LS$(\mathfrak{K}), \bar a \in {}^n|M|$ such that F_i^n is symmetric, i.e. preserved under permuting its variables. Arriving to n, for each $\bar a \in {}^nM$ by Ax VI there is an $M_{\bar a} \leq_{\mathfrak{K}} M$ such that $\|M_{\bar a}\| \leq$ LS$(\mathfrak{K})$, $|M_{\bar a}|$ include $\cup\{M_{\bar b} : \bar b$ a subsequence of $\bar a$ of length $< n\} \cup \bar a$ and $M_{\bar a}$ does not depend on the order of $\bar{a}$. Let $|M_{\bar{a}}| = \{c_i : i < i_0 \leq \text{LS}(\mathfrak{K})\}$ and define $F^n_i(\bar{a}) = c_i$ for $i < i_0$ and c_0 for $i_0 \leq i < \text{LS}(\mathfrak{K})$.

Clearly our conditions are satisfied; in particular, if $\bar{b}$ is a subsequence of $\bar{a}$, $M_{\bar{b}} \leq_{\mathfrak{K}} M_{\bar{a}}$ by Ax V and clause (D) holds by 1.6 and Ax IV. $\square_{1.7}$

1.8 Remark. 1) This is the "main" place we use Ax V,VI; it seems that we use it rarely, e.g., in 2.11 which is not used later. It is clear that we can omit Ax V if we strengthen somewhat Ax VI for the proofs above.
2) Note that in 1.7, we do not require that $M_{\bar{a}}$ is closed under the functions $(F^n_i)^{M_1}$. By a different bookkeeping we can have it: renaming $\tau_{1,\varepsilon} = \tau \cup \{F^n_i : i < \text{LS}(\mathfrak{K}) \times \varepsilon, n < \omega\}$ for $\varepsilon \leq \omega$ and we choose a $\tau_{1,n}$-expansion $M_{1,n}$ of M such that $m < n \Rightarrow M_{1,n} \restriction \tau_{1,m} = M_{1,m}$. Let $M_{1,0} = M$, and if $M_{1,n}$ is defined, choose for every $\bar{a} \in {}^{\omega>}(M_{1,n})$ a (non-empty) subset $A^{1,n}_{\bar{a}}$ of $M_{1,n}$ of cardinality $\leq \text{LS}(\mathfrak{K})$ such that $A^{1,n}_{\bar{a}}$ is closed under the functions of $M_{1,n}$ and $M \restriction A^{1,n}_{\bar{a}} \leq_{\mathfrak{K}} M$, let $A^{1,n}_{\bar{a}} = \{c_{\bar{a},i} : i \in [\text{LS}(\mathfrak{K}) \times n, \text{LS}(\mathfrak{K}) \times (n+1))$ and define $M_{1,n+1}$ by letting $(F^m_i)^{M_{1,n+1}}(\bar{a}) = c_{\bar{a},i}$. Let $M_1 = M_{1,\omega}$ be the τ_ω-model with the universe of M such that $n < \omega \Rightarrow M_1 \restriction \tau_{1,n} = M_{1,n}$.
3) Actually $M_{1,1}$ suffices if we expand it by making every term $\tau(\bar{x})$ equal to some function $F(\bar{x})$.
4) Alternatively demand for $n > 0$ that $F^n_i(\bar{a})$ is $F^{|u|}_i(\bar{a} \restriction u), u = \{i < n : a_i \notin \{a_j : j < i\}$.

1.9 Lemma. *1)* $\mathfrak{K}$ *is* $(\text{LS}(\mathfrak{K}), 2^{\text{LS}(\mathfrak{K})})$*-presentable.*
2) There is a set Γ *of types in* $\mathbb{L}(\tau_1)$ *in fact complete quantifier free (where* τ_1 *is from Lemma 1.7) such that* $K = \text{PC}_\tau(\emptyset, \Gamma)$.
3) For the Γ *from part (2), if* $M_1 \subseteq N_1 \in \text{EC}(\emptyset, \Gamma)$ *and* M, N *are the* τ*-reducts of* M_1, N_1 *respectively then* $M \leq_{\mathfrak{K}} N$.
4) For the Γ *from part (2), we have* $\{(M, N) : M \leq_{\mathfrak{K}} N$ *so* $N, M \in K\} = \{(M_1 \restriction \tau, N_1 \restriction \tau) : M_1 \subseteq N_1$ *are both from* $\text{PC}_\Gamma(\emptyset, \Gamma)\}$.

Proof. 1) By part (2) the first half of "$\mathfrak{K}$ is $(\text{LS}(\mathfrak{K}), 2^{\text{LS}(\mathfrak{K})})$-presentable holds". The second part will be proved with part (4).
2) Let Γ_n be the set of complete quantifier free n-types $p(x_0, \ldots, x_{n-1})$

in $\mathbb{L}(\tau_1)$ such that: if M_1 is a τ_1-model, $\bar{a}$ realizes p in M_1 and M is the τ-reduct of M_1, then $M_{\bar{a}} \in K$ and $M_{\bar{b}} \leq_{\mathfrak{K}} M_{\bar{a}}$ for any subsequence $\bar{b}$ of any permutation of $\bar{a}$; where $M_{\bar{c}}(\bar{c} \in {}^m|M_1|)$ is the submodel of M whose universe is $\{F_i^m(\bar{c}) : i < \mathrm{LS}(\mathfrak{K})\}$. Clearly there are such submodels (when $K \neq \emptyset$).

Let Γ be the set of p which, for some n, are complete quantifier free n-types (in $\mathbb{L}(\tau_1)$) which do not belong to Γ_n. By 1.6(1) we have $\mathrm{PC}_\tau(\emptyset, \Gamma) \subseteq K$ and by 1.7 $K \subseteq \mathrm{PC}_\tau(\emptyset, \Gamma)$.

3) Similar to the proof of (2) using 1.6(2).

4) The inclusion $\supseteq$ holds by part (3); so let us prove the other direction. Given $N \leq_{\mathfrak{K}} M$ we apply the proof of 1.7 to M, but demand further $\bar{a} \in {}^nN \Rightarrow M_{\bar{a}} \subseteq N$; simply add this demand to the choice of the $M_{\bar{a}}$'s (hence of the F_i^n's). We still have a debt from part (1).

We let Γ'_n be the set of complete quantifier free n-types in $\tau'_1 := \tau_1 \cup \{P\}$ (P a new unary predicate), $p(x_0, \ldots, x_{n-1})$ such that:

(*) if M_1 is an τ'_1-model, $\bar{a}$ realizes p in M_1, M the τ-reduct of M_1, then

  (α) $M_{\bar{b}} \leq_{\mathfrak{K}} M_{\bar{a}}$ for any subsequence $\bar{b}$ of $\bar{a}$ where $M_{\bar{c}}$ (for $\bar{c} \in |M_1|$) is the submodel of M whose universe is $\{(F_i^m)^{M_1}(\bar{c}) : i < \mathrm{LS}(\mathfrak{K})\}$, where $m = \ell g(\bar{c})$ (and there are such models),

  (β) $\bar{b} \subseteq P^{M_1} \Rightarrow M_{\bar{b}} \subseteq P^{M_1}$ for $\bar{b} \subseteq \bar{a}$.

We leave the rest to the reader (alternatively, use $\mathrm{PC}_{\tau'_1}(T', \Gamma), T'$ saying "P is closed under all the functions F_i^n"). $\square_{1.9}$

By the proof of 1.9(4).

1.10 Conclusion. The τ_1 and Γ from 1.9 (so $|\tau_1| \leq \mathrm{LS}(\mathfrak{K})$) satisfy: for any $M \in K$ and any τ_1-expansion M_1 of M which is in $\mathrm{EC}_{\tau_1}(\emptyset, \Gamma)$

(a) $N_1 \prec_{\mathbb{L}} M_1 \Rightarrow N_1 \subseteq M_1 \Rightarrow N_1 \restriction \tau \leq_{\mathfrak{K}} M$

(b) $N_1 \prec_{\mathbb{L}} N_2 \prec_{\mathbb{L}} M_1 \Rightarrow N_1 \subseteq N_2 \subseteq M_1 \Rightarrow N_1 \restriction \tau \leq_{\mathfrak{K}} N_2 \restriction \tau$

(c) if $M \leq_{\mathfrak{K}} N$ then there is a τ_1-expansion N_1 of N from $\mathrm{EC}_{\tau_1}(\emptyset, \Gamma)$ which extends M_1.

1.11 Conclusion If for every $\alpha < (2^{\mathrm{LS}(\mathfrak{K})})^+, \mathfrak{K}$ has a model of cardinality $\geq \beth_\alpha$ then K has a model in every cardinality $\geq$ LS($\mathfrak{K}$).

Proof. Use 1.9 and the classical upper bound on value of the Hanf number for: first order theory and omitting any set of types, for languages of cardinality LS($\mathfrak{K}$) (see, e.g., [Sh:c, VII,5.3,5.5]). $\square_{1.11}$

1.12 Conclusion: Assume that $\mathfrak{K}$ is an a.e.c., $\mu = |\tau_{\mathfrak{K}}| +$ LS($\mathfrak{K}$) and for simplicity $\tau_{\mathfrak{K}} \subseteq \mu$ or just $\tau_{\mathfrak{K}} \subseteq \mathbf{L}_\mu$, recalling $\mathbf{L}$ is the constructible universe of Gödel. If $\lambda > \mu$ and $\mathfrak{A} \prec (\mathscr{H}(\chi, \in)$ and $\mu + 1 \subseteq \mathfrak{A}$ and $\mathfrak{K} \in \mathfrak{A}$ which means $\{(M, N) : M \leq_{\mathfrak{K}} N$ has universe $\subseteq \mu\} \in \mathfrak{A}$ then:

- (a) $M \in \mathfrak{K} \cap K \Rightarrow M \restriction \mathfrak{A} \leq_{\mathfrak{K}} M$
- (b) if $M \leq_{\mathfrak{K}} N$ so both belongs to K and $M, N \in \mathfrak{A}$ then $M \restriction \mathfrak{A} \leq_{\mathfrak{K}} N \restriction \mathfrak{A}$
- (c) if $\mathfrak{A} \prec \mathfrak{B}$ and $[b <_{\mathfrak{B}} \mu \Rightarrow b \in \mathfrak{A}]$ and $\mathfrak{B} \models$ "$M \in K$" then $M[\mathfrak{B}] \in K$
- (d) similarly for $\mathfrak{B} \models$ "$M \leq_{\mathfrak{K}} N$"
  where
  - $(*)_1$ if $M \in \mathfrak{A}$ then $M \restriction \mathfrak{A}$ is the submodel of M with universe $|M| \cap |\mathfrak{A}|$
  - $(*)_2$ if $\mathfrak{B} \models$ "$M \in \mathfrak{K}$" then $M[\mathfrak{B}]$ is the following τ_K-model:
    - (a) it has universe $\{b \in \mathfrak{B} : \mathfrak{B} \models$ "b an element of the model M"$\}$
    - (b) for any m-place predicate Q of τ, $Q^M = \{\langle b_0, \ldots, b_{m-1}\rangle : \mathfrak{B} \models$ "$M \models Q[b_0, \ldots, b_{m-1}]$"$\}$
    - (c) for any m-place function symbol G of τ, similarly.

Proof. Should be clear. $\square_{1.12}$

1.13 Remark. 1) Clearly $\{\mu : \mu \geq$ LS($\mathfrak{K}$) and $K_\mu \neq 0\}$ is an initial segment of the class of cardinals $\geq$ LS($\mathfrak{K}$).
2) For every cardinal $\kappa (\geq \aleph_0)$ and ordinal $\alpha < (2^\kappa)^+$ there is an a.e.c.

$\mathfrak{K}$ such that: $\mathrm{LS}(\mathfrak{K}) = \kappa = |\tau_{\mathfrak{K}}|$ and $\mathfrak{K}$ has a model of cardinality λ iff $\lambda \in [\kappa, \beth_\alpha(\kappa))$. This follows by [Sh:c, VII,§5,p.432] in particular [Sh:c, VII,5.5](6), because

(a) if a vocabulary of cardinality $\leq \kappa$ and $T \subseteq \mathbb{L}(\tau)$ and Γ a set of $(\mathbb{L}(\tau), < \omega)$-types then $K = \{M : M$ a τ-model of T omitting every $\in \Gamma\}$ and $\leq_{\mathfrak{K}} = \prec \restriction K$ form an a.e.c. (we can use Γ a set of quantifier free types, $T = \emptyset$), with $\mathrm{LS}((\mathfrak{K}, \leq_{\mathfrak{K}}) \leq \kappa$

(b) if $\{c_i \neq c_j : i < j < \kappa\} \subseteq T$ then K above has no model of cardinality $< \kappa$.

3) More on such theorems see [Sh 394].
4) We can phrase 1.12 "for any $\mathfrak{B}$ in appropriate $\mathrm{EC}(T_1, \Gamma_1)$", but the present formulation is the way we use it.

§2 Amalgamation properties and homogeneity

2.1 Context. $\mathfrak{K}$ is an a.e.c.

The main theorem 2.8, the existence and uniqueness of the model-homogeneous models, is a generalization of Jonsson [Jo56], [Jo60] to the present context. The result on the upper bound $2^{2^{\aleph_0 + |\tau|}}$ for the number of D-sequence homogeneous universal-models of cardinality is of Keisler-Morley [KM67]. Earlier there were serious good reasons to concentrate on sequence-homogeneous models, but here we deal with the model-homogeneous case. From 2.13 to the end we consider what we can say when we omit smoothness, i.e. AxIV of Definition 1.2.

2.2 Definition. 1) $\mathbb{D}(M) := \{N/\cong : N \leq_{\mathfrak{K}} M, \|N\| \leq \mathrm{LS}(\mathfrak{K})\}$.
2) $\mathbb{D}(\mathfrak{K}) := \{N/\cong : N \in K, \|N\| \leq \mathrm{LS}(\mathfrak{K})\}$.
3) $D(M) = \{\mathrm{tp}_{\mathbb{L}(\tau_M)}(\bar{a}, \emptyset, M) : \bar{a} \in {}^{\omega>}M\}$.

2.3 Definition. Let $\lambda > \mathrm{LS}(\mathfrak{K})$.
1) A model M is λ-model-homogeneous when: if $N_0 \leq_{\mathfrak{K}} N_1 \leq_{\mathfrak{K}} M, \|N_1\| < \lambda, f$ an $\leq_{\mathfrak{K}}$-embedding of N_0 into M, then some $\leq_{\mathfrak{K}}$-embedding $f' : N_1 \to M$ extends f.
1A) A model M is $(\mathbb{D}, \lambda)$-model-homogeneous if $\mathbb{D} = \mathbb{D}(M)$ and M

is a λ-model homogeneous.
1B) Adding "above μ" means in $\mathfrak{K}_{\geq\mu}$.
2) M is λ-strongly model-homogeneous <u>if</u>: for every $N \in K_{<\lambda}$ such that $N \leq_{\mathfrak{K}} M$ and a $\leq_{\mathfrak{K}}$-embedding $f : N \to M$ there exists an automorphism g of M extending f.
3) M is λ-model universal homogeneous (for $\mathfrak{K}$) <u>when</u>: $\lambda >$ LS($\mathfrak{K}$), every[2] $N \in K_{\mathrm{LS}(\mathfrak{K})}$ is $\leq_{\mathfrak{K}}$-embeddable into M and for every $N_\ell \in K_{<\lambda}$ (for $\ell = 0, 1$) such that $N_0 \leq_{\mathfrak{K}} N_1$ and $\leq_{\mathfrak{K}}$-embedding $f : N_0 \to M$ there exists a $\leq_{\mathfrak{K}}$-embedding $g : N_1 \to M$ extending f (unlike (1), we do not demand that N_1 is $\leq_{\mathfrak{K}}$-embeddable into M; the universal is related to λ, it does not imply M is universal).
4) For each of the above three properties and the one below, if M has cardinality λ and has the λ-property then we may say for short that M has the property (i.e. omitting λ).
5) M is (D, λ)-sequence-homogeneous <u>if</u>:

- (a) $D = D(M) = \{\mathrm{tp}_{\mathbb{L}(\tau_M)}(\bar{a}, \emptyset, M) : \bar{a} \in |M|$, i.e., $\bar{a}$ a finite sequence from $M\}$ and
- (b) if $a_i \in M$ for $i \leq \alpha < \lambda, b_j \in M$ for $j < \alpha$ and $\mathrm{tp}_{\mathbb{L}(\tau M)}(\langle a_i : i < \alpha\rangle, \emptyset, M) = \mathrm{tp}_{\mathbb{L}(\tau_M)}(\langle b_i : i < \alpha\rangle, \emptyset, M)$, <u>then</u> for some $b_\alpha \in M$, $\mathrm{tp}_{\mathbb{L}(\tau_M)}(\langle a_i : i \leq \alpha\rangle, \emptyset, M) = \mathrm{tp}_{\mathbb{L}(\tau_M)}(\langle b_i : i \leq \alpha\rangle, \emptyset, M)$.

5A) In (5) we omit D when $D = \{\mathrm{tp}_{\mathbb{L}(\tau_K)}(\bar{a}, \emptyset, N) : \bar{a} \in {}^n N$ where $n < \omega$ and $M \prec_{\mathbb{L}} N\}$.
6) We omit the "model/sequence", when which one is clear from the context, i.e., if D is as in 2.2(3) $=$ 2.3(5)(a), (D, λ)-homogeneous means (D, λ)-sequence-homogeneous: if $\mathbb{D}$ is as in Definition 2.2(1), $(\mathbb{D}, \lambda)$-homogeneous means $(\mathbb{D}, \lambda)$-model-homogeneous, if not obvious we mean the model version.
7) M is λ-universal <u>when</u> every $N \in K_\lambda$ can be $\leq_{\mathfrak{K}}$-embedded into it. Similarly $(< \lambda)$-universal, $(\leq \lambda)$-universal.

2.4 Claim. *Assume N is λ-model-homogeneous and $\mathbb{D}(M) \subseteq \mathbb{D}(N)$, (and* LS($\mathfrak{K}$) $< \lambda$*, of course).*

---

[2]in fact, $N \in K_{\leq\lambda}$ is O.K. by 2.5(2)

1) If $M_0 \leq_{\mathfrak{K}} M_1 \leq_{\mathfrak{K}} M, \|M_0\| < \lambda, \|M_1\| \leq \lambda$ and f is a $\leq_{\mathfrak{K}}$-embedding of M_0 into N, <u>then</u> we can extend f to a $\leq_{\mathfrak{K}}$-embedding of M_1 into N.
2) If $M_1 \leq_{\mathfrak{K}} M, \|M_1\| \leq \lambda$ <u>then</u> there is a $\leq_{\mathfrak{K}}$-embedding of M_1 into N.

Proof. We prove by induction on $\mu \leq \lambda$ simultaneously that:

$(i)_\mu$ for every $M_1 \leq_{\mathfrak{K}} M, \|M_1\| \leq \mu$ (yes! not $< \mu$) there is a $\leq_{\mathfrak{K}}$-embedding of M_1 into N

$(ii)_\mu$ if $M_0 \leq_{\mathfrak{K}} M_1 \leq_{\mathfrak{K}} M, \|M_1\| \leq \mu, \|M_0\| < \lambda$ then any $\leq_{\mathfrak{K}}$-embedding f of M_0 into N can be extended to a $\leq_{\mathfrak{K}}$-embedding of M_1 into N.

Clearly $(i)_\lambda$ is part (2) and $(ii)_\lambda$ is part (1) so this is enough.

Proof of $(i)_\mu$. If $\mu \leq \mathrm{LS}(\mathfrak{K})$, this follows by $\mathbb{D}(M) \subseteq \mathbb{D}(N)$.
If $\mu > \mathrm{LS}(\mathfrak{K})$, then by 1.10 we can find $\bar{M}_1 = \langle M_1^\alpha : \alpha < \mu\rangle$ such that $M_1 = \bigcup_{\alpha<\mu} M_1^\alpha$ and $\alpha < \mu \Rightarrow M_1^\alpha \leq_{\mathfrak{K}} M_1$ and M_1^α is $\leq_{\mathfrak{K}}$-increasing continuous with α and $\alpha < \mu \Rightarrow \|M_1^\alpha\| < \mu$. We define by induction on α, a $\leq_{\mathfrak{K}}$-embedding $f_\alpha : M_1^\alpha \to N$, such that for $\beta < \alpha, f_\alpha$ extend f_β. For $\alpha = 0$ we can define f_α by $(i)_{\chi(0)}$ which holds as by the induction hypothesis, where $\chi(\beta) := \|M_1^\beta\|$. We next define f_α for $\alpha = \gamma + 1$: by $(ii)_{\chi(\alpha)}$ which holds by the induction hypothesis there is a $\leq_{\mathfrak{K}}$-embedding f_α of M_1^α into N extending f_γ.

Lastly, for limit α we let $f_\alpha = \bigcup_{\beta<\alpha} f_\beta$, it is a $\leq_{\mathfrak{K}}$-embedding into N by 1.6. So we finish the induction and $\bigcup_{\alpha<\mu} f_\alpha$ is as required.

Proof of $(ii)_\mu$. First, assume that $\mu = \lambda$ so we have proved $(ii)_\theta$ for $\theta < \lambda$ and $\|M_1\| = \lambda > \|M_0\|$, so $\mathrm{LS}(\mathfrak{K}) < \mu = \lambda$ hence we can find $\langle M_1^\alpha : \alpha < \mu\rangle$ as in the proof of $(i)_\mu$ such that $M_1^0 = M_0$ and let $\chi(\beta) = \|M_1^\beta\|$. Now we define f_β by induction on $\beta \leq \mu$ such that f_β is a $\leq_{\mathfrak{K}}$-embedding of M_β^1 into N and f_β is increasing continuous in β and $f_0 = f$. We can do this as in the proof of $(i)_\mu$ by $(ii)_{\chi(\alpha)}$ for $\alpha < \mu$.

Second, assume $\|M_1\| < \lambda$. Let g be a $\leq_{\mathfrak{K}}$-embedding of M_1 into N, it exists by $(i)_\mu$ which we have just proved. Let g be onto $N'_1 \leq_{\mathfrak{K}} N$, and let $g \restriction M_0$ be onto $N'_0 \leq_{\mathfrak{K}} N'_1$, and let f be onto $N_0 \leq_{\mathfrak{K}} N$. So clearly $h : N'_0 \to N_0$ defined by $h(g(a)) = f(a)$ for $a \in |M_0|$, is an isomorphism from N'_0 onto N_0. So $N_0, N'_0, N'_1 \leq_{\mathfrak{K}} N$. As $\|M_1\| < \lambda$ clearly $\|N'_1\| < \lambda$ so (by the assumption "N is λ-model-homogeneous", see Definition 2.3(1)) we can extend h to an isomorphism h' from N'_1 onto some $N_1 \leq_{\mathfrak{K}} N$, so $h' \circ g : M_1 \to N$ is as required. $\square_{2.4}$

<u>2.5 Conclusion</u> 1) If M, N are model-homogeneous, of the same cardinality ($>$ LS($\mathfrak{K}$)) and $\mathbb{D}(M) = \mathbb{D}(N)$ <u>then</u> M, N are isomorphic. Moreover, if $M_0 \leq_{\mathfrak{K}} M, \|M_0\| < \|M\|$, <u>then</u> any $\leq_{\mathfrak{K}}$-embedding of M_0 into N can be extended to an isomorphism from M onto N.
2) The number of model-homogeneous models from $\mathfrak{K}$ of cardinality λ is $\leq 2^{2^{\mathrm{LS}(\mathfrak{K})}}$; if in Definition 1.2, AxVI, in the definition of LS($\mathfrak{K}$) we omit $|\tau| \leq$ LS($\mathfrak{K}$), the bound is $2^{2^{\mathrm{LS}(\mathfrak{K})+|\tau(\mathfrak{K})|}}$.
3) If M is λ-model-homogeneous and $\mathbb{D}(M) = \mathbb{D}(\mathfrak{K})$ <u>then</u> M is $(\leq \lambda)$-universal, i.e. every model N (in K) of cardinality $\leq \lambda$, has a $\leq_{\mathfrak{K}}$-embedding into M. So if $\mathbb{D}(M) = \mathbb{D}(\mathfrak{K})$ <u>then</u>: M is λ-model universal homogeneous (see Definition 2.3(3)) iff M is a λ-model-homogeneous iff M is $(\lambda, \mathbb{D}(\mathfrak{K}))$-homogeneous.
4) If M is λ-model-homogeneous <u>then</u> it is λ-universal for $\{N \in K_\lambda : \mathbb{D}(N) \subseteq \mathbb{D}(M)\}$.
5) If M is (D, λ)-sequence-homogeneous, ($\lambda >$ LS($\mathfrak{K}$)) <u>then</u> M is a λ-model homogeneous.
6) For $\lambda >$ LS($\mathfrak{K}$), M is λ-model universal homogeneous <u>iff</u> M is λ-model-homogeneous and ($\leq$ LS($\mathfrak{K}$))-universal.

Proof. 1) Immediate by 2.4(1), using the standard hence and forth argument.
2) The number of models (in K) of power $\leq$ LS($\mathfrak{K}$) is, up to isomorphism, $\leq 2^{\mathrm{LS}(\mathfrak{K})}$ (recalling that we are assuming $|\tau(\mathfrak{K})| \leq$ LS($\mathfrak{K}$)). Hence the number of possible $\mathbb{D}(M)$ is $\leq 2^{2^{\mathrm{LS}(\mathfrak{K})}}$. So by 2.5(1) we

are done.

3),4),5) Immediate. $\square_{2.5}$

2.6 Remark. The results parallel to 2.5(1)-(4) for λ-sequence homogeneous models and $D(M)$ hold, too.

2.7 Definition. 1) A model M has the (λ, μ)-amalgamation property (= am.p., in $\mathfrak{K}$, of course) i̲f̲: for every M_1, M_2 such that $\|M_1\| = \lambda, \|M_2\| = \mu, M \leq_{\mathfrak{K}} M_1$ and $M \leq_{\mathfrak{K}} M_2$, there is a model N and $\leq_{\mathfrak{K}}$-embeddings $f_1 : M_1 \to N$ and $f_2 : M_2 \to N$ such that $f_1 \restriction |M| = f_2 \restriction |M|$. Now the meaning of e.g. the $(\leq \lambda, < \mu)$-amalgamation property should be clear. Always $\lambda, \mu \geq \mathrm{LS}(\mathfrak{K})$ (and, of course, if we use $< \mu, \mu > \mathrm{LS}(\mathfrak{K})$).

1A) In part (1) we add the adjective "disjoint" w̲h̲e̲n̲ $f_1(M_1) \cap f_2(M_2) = M$. Similarly in (2) below.

2) $\mathfrak{K}$ has the (κ, λ, μ)-amalgamation property i̲f̲ every model M (in K) of cardinality κ has the (λ, μ)-amalgamation property. The (κ, λ)-amalgamation property for $\mathfrak{K}$ means just the $(\kappa, \kappa, \lambda)$-amalgamation property. The κ-amalgamation property for $\mathfrak{K}$ is just the (κ, κ, κ)-amalgamation property.

3) $\mathfrak{K}$ has the (λ, μ)-JEP (joint embedding property) if for any $M_1 \in K, M_2 \in K$ of cardinality λ, μ respectively there is $N \in K$ into which M_1 and M_2 are $\leq_{\mathfrak{K}}$-embeddable.

4) The λ-JEP is the (λ, λ)-JEP.

5) The amalgamation property means the (κ, λ, μ)-amalgamation property for every $\lambda, \mu \geq \kappa (\geq \mathrm{LS}(\mathfrak{K}))$.

6) The JEP means the (λ, μ)-JEP for every $\lambda, \mu \geq \mathrm{LS}(\mathfrak{K})$.

Remark. Clearly in 2.7, parts (1), (2) first sentence, (3),(5), the roles of λ, μ are symmetric.

2.8 Theorem. *1) If* $\mathrm{LS}(\mathfrak{K}) < \kappa \leq \lambda, \lambda = \lambda^{<\kappa}, K_\lambda \neq \emptyset$ *and* $\mathfrak{K}$ *has the* $(< \kappa, \lambda)$*-amalgamation property* t̲h̲e̲n̲ *for every model M of cardinality λ, there is a κ-model-homogeneous model N of cardinality λ satisfying $M \leq_{\mathfrak{K}} N$. If $\kappa = \lambda$, alternatively the $(< \kappa, < \lambda)$-amalgamation property suffices.*

2) So in (1) if $\kappa = \lambda$, there is a universal, model-homogeneous model of cardinality λ, provided that for some $M \in K_{\leq\lambda}, \mathbb{D}(M) = \mathbb{D}(\mathfrak{K})$ or just $\mathfrak{K}$ has the LS($\mathfrak{K}$)-JEP.
3) If $\mathfrak{K}$ has the amalgamation property and the LS($\mathfrak{K}$)-JEP, *then $\mathfrak{K}$ has the* JEP.

2.9 Remark. 1) The last assumption of 2.8(2) holds, e.g., if $(\leq \mathrm{LS}(\mathfrak{K}), < 2^{\mathrm{LS}(\mathfrak{K})})$-JEP holds and $|\mathbb{D}(\mathfrak{K})| \leq \lambda$.
2) If for some $M \in K, \mathbb{D}(M) = \mathbb{D}(\mathfrak{K})$ then we can have such M of cardinality $\leq 2^{\mathrm{LS}(\mathfrak{K})}$.
3) We can in 2.8 replace the assumption "$(< \kappa, \lambda)$-amalgamation property" by "$(< \kappa, < \lambda)$-amalgamation property" if, e.g., no $M \in K_{<\lambda}$ is maximal.

Proof. Immediate; in (1) note that if κ is singular then necessarily $\lambda > \kappa$ & $\lambda = \lambda^\kappa = \lambda^{<\kappa^+}$ so we can replace κ by κ^+.

2.10 Remark. Also the corresponding converses hold.

2.11 Lemma. *1) If* LS($\mathfrak{K}$) $\leq \kappa$ *and $\mathfrak{K}$ has the κ-amalgamation property then $\mathfrak{K}$ has the (κ, κ^+)-amalgamation property and even the $(\kappa, \kappa^+, \kappa^+)$-amalgamation property.*
2) If $\kappa \leq \mu \leq \lambda$ and $\mathfrak{K}$ has the (κ, μ)-amalgamation property and the (μ, λ)-amalgamation property then $\mathfrak{K}$ has the (κ, λ)-amalgamation property. If $\mathfrak{K}$ has the (κ, μ, μ) and the (μ, λ)-amalgamation property, then $\mathfrak{K}$ has the (κ, λ, μ)-amalgamation property.
3) If $\lambda_i (i \leq \alpha)$ is increasing and continuous, LS($\mathfrak{K}$) $\leq \lambda_0$ *and for every $i < \alpha, \mathfrak{K}$ has the $(\lambda_i, \mu + \lambda_i, \lambda_{i+1})$-amalgamation property then $\mathfrak{K}$ has the $(\lambda_0, \mu + \lambda_0, \lambda_\alpha)$-amalgamation property.*
4) If $\kappa \leq \mu_1 \leq \mu$ and for every $M, \|M\| = \mu_1$, there is $N, M \leq_{\mathfrak{K}} N, \|N\| = \mu$, then the (κ, μ, λ)-amalgamation property (for $\mathfrak{K}$) implies the (κ, μ_1, λ)-amalgamation property (for $\mathfrak{K}$).
5) Similarly with the disjoint amalgamation version.

Proof. Straightforward, e.g.
3) So assume $M_0 \in K_{\lambda_0}, M_0 \leq_{\mathfrak{K}} M_1 \in K_{\mu+\lambda_0}$ and $M_0 \leq_{\mathfrak{K}} M_2 \in K_{\lambda_\alpha}$

and for variety we prove for the disjoint amalgamation version (see part (5)). By e.g. 1.10 we can find an $\leq_{\mathfrak{K}}$-increasing continuous sequence $\langle M_{2,i} : i \leq \alpha \rangle$ such that $M_{2,0} = M_0, M_{2,\alpha} = M_2$ and $M_{2,i} \in K_{\lambda_i}$ for $i \leq \alpha$.

Without loss of generality $M_1 \cap M_2 = M_0$. We now choose $M_{1,i}$ by induction on $i \leq \alpha$ such that:

(∗) (a) $\langle M_{1,j} : j \leq i \rangle$ is $\leq_{\mathfrak{K}}$-increasing continuous
(b) $M_{1,i} = M_1$ if $i = 0$
(c) $M_{1,i} \in K_{\mu+\lambda_i}$
(d) $M_{2,i} \leq_{\mathfrak{K}} M_{1,i}$
(e) $M_{2,i} \cap M_{1,\alpha} = M_{1,i}$.

For $i = 0$ see clause (b), for i limit take union, for $i = j + 1$ apply the disjoint $(\lambda_j, \mu + \lambda_j, \lambda_i)$-amalgamation to $M_{2,j}, M_{1,j}, M_{2,j+1}$. For $i = \alpha$ we are done. $\square_{2.11}$

2.12 Conclusion. If $\text{LS}(\mathfrak{K}) \leq \chi_1 < \chi_2$ and $\mathfrak{K}$ has the κ-amalgamation property whenever $\chi_1 \leq \kappa < \chi_2$ then $\mathfrak{K}$ has the (κ, λ, μ)-amalgamation property whenever $\chi_1 \leq \kappa \leq \lambda \leq \chi_2, \kappa \leq \mu \leq \chi_2$ and $\kappa < \chi_2$.

* * *

It may be interesting to note that even waiving AX IV we can say something.

2.13 Context: For the remainder of this section $\mathfrak{K}$ is just a weak a.e.c., i.e., Ax IV is not assumed.

2.14 Definition. Let $M \in K$ have cardinality λ, a regular uncountable cardinal $> \text{LS}(\mathfrak{K})$. We say M is smooth if there is a sequence $\langle M_i : i < \lambda \rangle$ with M_i being $\leq_{\mathfrak{K}}$-increasing continuous, $M_i \leq_{\mathfrak{K}} M$ and $\|M_i\| < \lambda$ for $i < \lambda$ and $M = \bigcup_{i<\lambda} M_i$.

2.15 Remark. We can define $S/\mathscr{D}$-smooth, for S a subset of $\mathscr{P}(\lambda)$, $\mathscr{D}$ a filter on $\mathscr{P}(\lambda)$, that is: $M \in K_\lambda$ is $(S/\mathscr{D})$-smooth when for every one-to-one function f from $|M|$ onto λ the set $\{u \in \mathscr{P}(\lambda) : M \restriction \{a : f(a) \in u\} \leq_{\mathfrak{K}} M\} \in D$. Usually we demand that for every permutation f on λ, $\{u \subseteq \lambda$: u is closed under $f\} \in \mathscr{D}$, and usually we demand that $\mathscr{D}$ is a normal $\mathrm{LS}(\mathfrak{K})^+$-complete filter).

2.16 Claim. *Assume that* $\lambda = \lambda^{<\lambda} > |\tau_K|, \mathfrak{K}_{<\lambda}$ *has no maximal member and* $\mathfrak{K}$ *has* $(<\lambda, <\lambda, <\lambda)$*-amalgamation property and* $\mathrm{LS}(\mathfrak{K}) < \lambda$ *or at least assume in the* $(<\lambda, <\lambda, <\lambda)$*-amalgamation demand that the resulting model has cardinality* $< \lambda$. *Then* $\mathfrak{K}_\lambda$ *has a smooth model-homogeneous member.*

Proof. Same proof. $\square_{2.16}$

2.17 Lemma. *If* $M, N \in K_\lambda (\lambda > \mathrm{LS}(\mathfrak{K}))$ *are smooth, model-homogeneous and* $\mathbb{D}(M) = \mathbb{D}(N)$ *then* $M \cong N$.

Proof. By the hence and forth argument, left to the reader (the set of approximations is $\{f : f$ isomorphism from some $M' \leq_{\mathfrak{K}} M$ of cardinality $< \lambda$ onto some $N' \leq_{\mathfrak{K}} N\}$ but note that not for any increasing continuous sequence of approximations is the union an approximation). $\square_{2.17}$

2.18 Remark. It is reasonable to consider

(∗) if $M \in K_\lambda, (\lambda > \mathrm{LS}(\mathfrak{K}))$ is smooth and model-homogeneous and $N \in K_\lambda$ is smooth, $\mathbb{D}(N) \subseteq \mathbb{D}(M)$ then N can be $\leq_{\mathfrak{K}}$-embedded into M.

This can be proved in the context of universal classes (e.g. $AxFr_1$ from Chapter V.B).

2.19 Fact: 1) If $\mathfrak{K}_i = (K_i, <_i)$ is a (weak) a.e.c., i.e. with $\lambda_i = \mathrm{LS}(K_i, \leq_i)$ where $\lambda_i \geq \aleph_0$ for $i < \alpha, i < \alpha \Rightarrow \tau_{K_i} = \tau$ and $K = \bigcap\limits_{i<\alpha} K_i$ and $\leq$ is defined by $M \leq N$ if and only if for every $i < \alpha, M \leq_i N$ then $\mathfrak{K} = (K, \leq)$ is a [weak] a.e.c. with $\mathrm{LS}(\mathfrak{K}) \leq \sum\limits_{i<\alpha} \lambda_i$.

2) Concerning AxI-V, we can omit some of them in the assumption and still get the rest in the conclusion. But for AxVI we need in addition to assume AxV + AxIV$_\theta$ for at least one $\theta = \mathrm{cf}(\theta) \leq \sum_{i<\alpha} \lambda_i$.

Proof. Easy.

2.20 Example Consider the class K of norm spaces over the reals with $M \leq_{\mathfrak{K}} N$ iff $M \subseteq N$ and M is complete inside N. Now $\mathfrak{K} = (K, \leq_{\mathfrak{K}})$ is a weak a.e.c. with $\mathrm{LS}(\mathfrak{K}) = 2^{\aleph_0}$ and it is as required in 2.16.

§3 Limit models and other results

In this section we introduce various variants of limit models (the most important are the superlimit ones). We prove that if $\mathfrak{K}$ has a superlimit model M^* of cardinality λ for which the λ-amalgamation property fails and $2^\lambda < 2^{\lambda^+}$ then $\dot{I}(\lambda, K) = 2^\lambda$ (see 3.8). We later prove that if $\psi \in \mathbb{L}_{\omega_1,\omega}(\mathbf{Q})$ is categorical in $\aleph_1$ then it has model in $\aleph_2$ see 3.18(2). This finally solves Baldwin's problem (see §0). In fact we prove an essentially more general result on a.e.c. and λ (see 3.11, 3.13).

The reader can read 3.3(1),(1A),(1B) ignore the other definitions, and continue with 3.7(2),(5) and everything from 3.8 (interpreting all variants as superlimits).
You may wonder can we prove the parallel to Baldwin conjecture in λ^+ if $\lambda > \aleph_0$; it is

> $\circledast_\lambda$ if $\mathfrak{K}$ is λ-presentable a.e.c. with $\mathrm{LS}(\mathfrak{K}) = \lambda$, categorical in λ^+ then $K_{\lambda^{++}} \neq \emptyset$.

This is false when $\mathrm{cf}(\lambda) > \aleph_0$.

3.1 Context. $\mathfrak{K}$ is an a.e.c.

3.2 Example: Let λ be given and $\mathfrak{K} = (K, \leq_{\mathfrak{K}})$ be defined by

$$K = \{(A, <) : (A, <) \text{ a well order of order type } \leq \lambda^+\}$$

$$\leq_{\mathfrak{K}} = \{(M, N) : M, N \in K \text{ and } N \text{ is an end extension of } M\}.$$

Now

(a) $\mathfrak{K}$ is an abstract elementary class with $\mathrm{LS}(\mathfrak{K}) = \lambda$ and $\mathfrak{K}$ categorical in λ^+

(b) if λ has cofinality $\geq \aleph_1$ then $\mathfrak{K}$ is λ-presentable (see, e.g., [Sh:c, VII,§5] and history there); by clause (a) it is always $(\lambda, 2^\lambda)$-presentable,

(c) $\mathfrak{K}$ has no model of cardinality $> \lambda^+$.

Note that if we are dealing with classes which are categorical (or just simple in some sense), we have a good chance to find limit models and they are useful in constructions.

3.3 Definition. Let λ be a cardinal $\geq \mathrm{LS}(\mathfrak{K})$. For parts 3) - 7) but not 8), for simplifying the presentation we assume the axiom of global choice (alternatively, we restrict ourselves to models with universe an ordinal $< \lambda^+$).
1) $M \in K_\lambda$ is locally superlimit (for $\mathfrak{K}$) if:

(a) for every $N \in K_\lambda$ such that $M \leq_{\mathfrak{K}} N$ there is $M' \in K_\lambda$ isomorphic to M such that $N \leq_{\mathfrak{K}} M'$ and $N \neq M'$

(b) if $\delta < \lambda^+$ is a limit ordinal and $\langle M_i : i < \delta \rangle$ is $\leq_{\mathfrak{K}}$-increasing sequence and $M_i \cong M$ for $i < \delta$ then $\bigcup_{i<\delta} M_i \cong M$.

1A) $M \in K_\lambda$ is globally superlimit if (a) +(b) and

(c) M is universal in $\mathfrak{K}_\lambda$, i.e., any $N \in K_\lambda$ can be $\leq_{\mathfrak{K}}$-embedded into M.

1B) Just superlimit means globally. Similarly with the other notions below we define the global version as adding clause (c) from (1A) and the default version is the global one. (Note that in the local version we can restrict our class to $\{N \in K_\lambda : M \text{ can be } \leq_{\mathfrak{K}}\text{-embedded into } N\}$ and get the global one).
2) For $\Theta \subseteq \{\mu : \aleph_0 \leq \mu < \lambda, \mu \text{ regular}\}$, $M \in K_\lambda$ is locally (λ, Θ)-superlimit if:

(a) as in part (1) above

(b) if $\langle M_i : i \le \mu\rangle$ is $\le_{\mathfrak{K}}$-increasing, $M_i \cong M$ for $i < \mu$ and $\mu \in \Theta$ then $\cup\{M_i : i < \mu\} \cong M$.

2A) If Θ is a singleton, say $\Theta = \{\theta\}$, we may say that M is locally (λ, θ)-superlimit.

3) Let $S \subseteq \lambda^+$ be stationary. $M \in K_\lambda$ is called locally S-strongly limit or locally (λ, S)-strongly limit when for some function: $\mathbf{F} : K_\lambda \to K_\lambda$ we have:

(α) for $N \in K_\lambda$ we have $N \le_{\mathfrak{K}} \mathbf{F}(N)$

(β) if $\delta \in S$ is a limit ordinal and $\langle M_i : i < \delta\rangle$ is a $\le_{\mathfrak{K}}$-increasing continuous sequence[3] in K_λ and $M_0 \cong M$ and $i < \delta \Rightarrow \mathbf{F}(M_{i+1}) \le_{\mathfrak{K}} M_{i+2}$, then $M \cong \cup\{M_i : i < \delta\}$

(γ) if $M \le_{\mathfrak{K}} M_1 \in K_\lambda$ then there is N such that $M_1 <_{\mathfrak{K}} N \in K_\lambda$.

4) Let $S \subseteq \lambda^+$ be stationary. $M \in K_\lambda$ is called locally S-limit or locally (λ, S)-limit if for some function $\mathbf{F} : K_\lambda \to K_\lambda$ we have:

(α) for every $N \in K_\lambda$ we have $N \le_{\mathfrak{K}} \mathbf{F}(N)$

(β) if $\langle M_i : i < \lambda^+\rangle$ is a $\le_{\mathfrak{K}}$-increasing continuous sequence of members of K_λ, $M_0 \cong M, \mathbf{F}(M_{i+1}) \le_{\mathfrak{K}} M_{i+2}$ then for some closed unbounded[4] subset C of λ^+,

$$[\delta \in S \cap C \Rightarrow M_\delta \cong M].$$

(γ) if $M \le_{\mathfrak{K}} M_1 \in K_\lambda$ then there is $N, M_1 <_{\mathfrak{K}} N \in K_\lambda$.

5) We define "locally S-weakly limit", "locally S-medium limit" like "locally S-limit", "locally S-strongly limit" respectively by demanding that the domain of $\mathbf{F}$ is the family of $\le_{\mathfrak{K}}$-increasing continuous sequence of members of $\mathfrak{K}_{<\lambda}$ of length $< \lambda$ and replacing "$\mathbf{F}(M_{i+1}) \le_{\mathfrak{K}} M_{i+2}$" by "$M_{i+1} \le_{\mathfrak{K}} \mathbf{F}(\langle M_j : j \le i+1\rangle) \le_{\mathfrak{K}} M_{i+2}$". We replace "limit" by "limit$^-$" if "$\mathbf{F}(M_{i+1}) \le_{\mathfrak{K}} M_{i+2}$", "$M_{i+1} \le_{\mathfrak{K}} \mathbf{F}(\langle M_j : j \le i+1\rangle) \le_{\mathfrak{K}} M_{i+2}$" are replaced by "$\mathbf{F}(M_i) \le_{\mathfrak{K}} M_{i+1}$", "$M_i \le_{\mathfrak{K}}$

[3] no loss if we add $M_{i+1} \cong M$, so this simplifies the demand on $\mathbf{F}$, i.e., only $\mathbf{F}(M')$ for $M' \cong M$ are required

[4] we can use a filter as a parameter

$\mathbf{F}(\langle M_j : j \leq i\rangle) \leq_{\mathfrak{K}} M_{i+1}$" respectively.
6) If $S = \lambda^+$ then we omit S (in parts (3), (4), (5)).
7) For $\Theta \subseteq \{\mu : \aleph_0 \leq \mu \leq \lambda$ and μ is regular$\}$, M is locally (λ, Θ)-strongly limit if M is locally $\{\delta < \lambda^+ : \mathrm{cf}(\delta) \in \Theta\}$-strongly limit. Similarly for the other notions (where $\Theta \subseteq \{\mu : \mu$ regular $\leq \lambda\}$. If we do not write λ we mean $\lambda = \|M\|$. Let locally (λ, θ)-strongly limit mean locally (λ, θ)-strongly limit.
8) We say that $M \in K_\lambda$ is invariantly strong limit when in part (3) we demand that $\mathbf{F}$ is just a subset of $\{(M, N)/\cong : M \leq_{\mathfrak{K}} N$ are from $K_\lambda\}$ and in clause (b) of part (3) we replace "$\mathbf{F}(M_{i+1}) \leq_{\mathfrak{K}} M_{i+2}$" by "$(\exists N)(M_{i+1} \leq_{\mathfrak{K}} N \leq_{\mathfrak{K}} M_{i+2} \wedge ((M_{i+1}, N)/\cong) \in \mathbf{F})$" but abusing notation we still write $N = \mathbf{F}(M)$ instead $((M, N)/\cong) \in \mathbf{F}$. Similarly with the other notions, so if $\mathbf{F}$ acts on suitable $\leq_{\mathfrak{K}}$-increasing sequence of models then we use the isomorphic type of $\bar{M}{}^\frown\langle N\rangle$.

<u>3.4 Obvious implication diagram</u>: For Θ, S_1 as in 3.3(7) and $S_1 \subseteq \{\delta < \lambda^+ : \mathrm{cf}(\delta) \in \Theta\}$ is a stationary subset of λ^+:

superlimit $= (\lambda, \{\mu : \mu \leq \lambda$ regular$\})$-superlimit

↓

(λ, Θ)-superlimit

↓

S_1-strongly limit

↓ ↓

S_1-medium limit, S_1-limit

↓ ↓

S_1-weakly limit.

3.5 Lemma. *0) All the properties are preserved if S is replaced by a subset and if $\mathfrak{K}$ has the λ-JEP, the local and global version in Definition 3.3 are equivalent.*
1) If $S_i \subseteq \lambda^+$ for $i < \lambda^+$, $S = \{\alpha < \lambda^+ : (\exists i < \alpha)\alpha \in S_i\}$ and

$S_i \cap i = \emptyset$ for $i < \lambda$ then: M is S_i-strongly limit for each $i < \lambda$ if and only if M is S-strongly limit.
2) Suppose $\kappa \leq \lambda$ is regular and $S \subseteq \{\delta < \lambda^+ : \mathrm{cf}(\delta) = \kappa\}$ is a stationary set and $M \in K_\lambda$ then the following are equivalent:

- (a) *M is S-strongly limit*
- (b) *M is $(\lambda, \{\kappa\})$-strongly limit*
- (c) *$M \in \mathfrak{K}_\lambda$ is $\leq_{\mathfrak{K}}$-universal not $<_{\mathfrak{K}}$-maximal and there is a function $\mathbf{F} : K_\lambda \to K_\lambda$ satisfying $(\forall N \in K_\lambda)[N \leq_{\mathfrak{K}} \mathbf{F}(N)]$ such that if $M_i \in K_\lambda$ for $i < \kappa$, $[i < j \Rightarrow M_i \leq_{\mathfrak{K}} M_j], \mathbf{F}(M_{i+1}) \leq_{\mathfrak{K}} M_{i+2}$ and $M_0 \cong M$ then $\bigcup_{i<\kappa} M_i \cong M$.*

2A) If $S \subseteq \lambda^+, \Theta = \{\mathrm{cf}(\delta) : \delta \in S\}$ then M is S-strongly limit iff clause (c) in part (2) above holds for every $\kappa \in \Theta$.
3) In part (1) we can replace "strongly limit" by "limit", "medium limit" and "weakly limit".
4) Suppose $\kappa \leq \lambda$ is regular, $S \subseteq \{\delta < \lambda^+ : \mathrm{cf}(\delta) = \kappa\}$ is a stationary set which belongs to $\check{I}[\lambda]$ (see 0.6, 0.7 above) and $M \in K_\lambda$.
The following are equivalent

- (a) *M is S-medium limit in $\mathfrak{K}_\lambda$*
- (b) *$M \in K_\lambda$ is $\leq_{\mathfrak{K}}$-universal not maximal and there is a function $\mathbf{F}$ from $\bigcup_{\alpha<\kappa} {}^\alpha(K_\lambda)$ to K such that*
  - (α) *for any $\leq_{\mathfrak{K}}$-increasing $\langle M_i : i \leq \alpha \rangle$ if $M_0 = M, \alpha < \kappa, M_i$ is $\leq_{\mathfrak{K}}$-increasing, $M_i \in K_\lambda$, then $M_\alpha \leq_{\mathfrak{K}} \mathbf{F}(\langle M_i : i \leq \alpha \rangle)$*
  - (β) *if $\langle M_i : i < \kappa \rangle$ is $\leq_{\mathfrak{K}}$-increasing, $M_0 = M, M_i \in K_\lambda$ and for $i < \kappa$ we have $M_{i+1} \leq_{\mathfrak{K}} \mathbf{F}(\langle M_j : j \leq i+1 \rangle) \leq_{\mathfrak{K}} M_{i+2}$ then $\bigcup_{i<\kappa} M_i \cong M$.*

Proof. 0) Trivial.
1) Recall that in Definition 3.3(3), clause (b) we use $\mathbf{F}$ only on M_{i+1}; (see the proof of (2A) below, second part).
2) For (c) $\Rightarrow$ (a) note that the demands on the sequence are "local",

$M_{i+1} \leq_{\mathfrak{K}} \mathbf{F}(M_{i+1}) \leq_{\mathfrak{K}} M_{i+2}$, (whereas in part (4) they are "global").
2A) First assume that M is S-strongly limit and let $\mathbf{F}$ witness it. Suppose $\kappa \in \Theta$, so we choose $\delta_\kappa \in S$ with $\mathrm{cf}(\delta_\kappa) = \kappa$ and let $\langle \alpha_i : i < \kappa \rangle$ be increasing continuous with limit δ, $\alpha_0 = 0$, α_{i+1} a successor of a successor ordinal for each $i < \kappa$. We now define $\mathbf{F}_\kappa$ as follows: to define $\mathbf{F}_\kappa(M)$ we define $\mathbf{F}_{\kappa,\alpha}$ for $\alpha \leq \delta$ by induction on $\alpha \leq \delta$. Let:

(a) if $\alpha = 0$ then $\mathbf{F}_{\kappa,0}(M) = M$

(b) if $\alpha = \beta + 1$ then $\mathbf{F}_{\kappa,\alpha}(M) = \mathbf{F}(\mathbf{F}_{\kappa,\beta}(M))$

(c) if $\alpha \leq \delta$ a limit ordinal then $\mathbf{F}_{\kappa,\alpha}(M) = \cup\{\mathbf{F}_{\kappa,\beta}(M) : \beta < \alpha\}$.

Lastly, let $\mathbf{F}_\kappa(M)$ be $\mathbf{F}_{\kappa,\delta}(M)$.

Now suppose $\langle N_i : i \leq \kappa \rangle$ is $\leq_{\mathfrak{K}}$-increasing continuous, $N_i \in K_\lambda$ and $\mathbf{F}_\kappa(N_{i+1}) \leq_{\mathfrak{K}} N_{i+2}$ for $i < \kappa$ and we should prove $N_\kappa \cong M$. Now we can find $\langle M_j : j < \lambda^+ \rangle$ such that it obeys $\mathbf{F}$ and $M_{\alpha_i} = N_i$ for $i < \kappa$; so clearly we are done.

Second, assume that for each $\kappa \in \Theta$, clause (c) of 3.5(2) holds and let $\mathbf{F}_\kappa$ exemplify this. Let $\langle \kappa_\varepsilon : \varepsilon < \varepsilon(*) \rangle$ list Θ so $\varepsilon(*) < \lambda^+$ and define $\mathbf{F}$ as follows. For any $M \in \mathfrak{K}$ choose $M_{[\varepsilon]}$ by induction on $\varepsilon \leq \varepsilon(*)$ as follows: $M_{[0]} = M, M_{[\varepsilon+1]} = \mathbf{F}_{\kappa_\varepsilon}(M_{[\varepsilon]})$ and for ε limit ordinal let $M_{[\varepsilon]} = \cup\{M_{[\zeta]} : \zeta < \varepsilon\}$. Lastly, let $\mathbf{F}[M] = M_{[\varepsilon(*)]}$. Now check.
3) No new point.
4) First note that $(a) \Rightarrow (b)$ should be clear. Second, we prove that $(b) \Rightarrow (a)$ so let $\mathbf{F}$ witness that clause (b) holds. Let $E, \langle u_\alpha : \alpha < \lambda \rangle$ witness that $S \in \check{I}[\lambda]$, i.e.

$(*)_1$ (a) E a club of λ

(b) $u_\alpha \subseteq \alpha$ and $\mathrm{otp}(u_\alpha) \leq \kappa$ for $\alpha < \lambda$

(c) if $\alpha \in S \cap E$ then $\alpha = \sup(u_\alpha)$ and $\mathrm{otp}(u_\alpha) = \kappa$

(d) if $\alpha \in \lambda \backslash S \cap E$ then $\mathrm{otp}(u_\alpha) < \kappa$

(e) if $\alpha \in u_\beta$ then $u_\alpha = u_\beta \cap \alpha$.

We can add

$(*)_2$ (f) if $\beta \in u_\alpha$ then β has the form $3\gamma + 1$.

Let $\langle \alpha_\varepsilon : \varepsilon < \lambda \rangle$ list E in increasing order and without loss of generality $\alpha_0 = 0, \alpha_{1+\varepsilon}$ is a limit ordinal (note that only the limit ordinals of S count).

To define $\mathbf{F}'$ as required we shall deal with the requirement according to whether $\delta \in S$ is "easy", i.e. $\delta \notin E$ so $\delta \in (\alpha_\varepsilon, \alpha_{\varepsilon+1}]$ for some $\varepsilon < \lambda^+$ so after α_ε we can "take care of it", or δ is "hard", i.e. $\delta \in E$ so we use the $\alpha \in u_\delta$.

We choose $\langle e_\delta : \delta \in S \backslash E \rangle$ such that $\delta \in (\alpha_\varepsilon, \alpha_{\varepsilon+1}] \cap S$ implies $e_\delta \subseteq \delta = \sup(e_\delta)$ and $\min(e_\delta) > \alpha_\varepsilon$, $\mathrm{otp}(e_\delta) = \kappa, e_\delta$ is closed and $\alpha \in e_\delta \Rightarrow \alpha = \sup(e_\delta \cap \alpha) \vee (\alpha \in \{3\gamma + 2 : \gamma < \delta\})$. If $\delta \in S \cap E$ let e_δ be the closure of u_δ. Let $\langle \gamma_{\delta,\zeta} : \zeta < \kappa \rangle$ list e_δ in increasing order.

We now define a function $\mathbf{F}'$ so let $\langle M_j : j \le i + 1 \rangle$ be given and let $\alpha_\varepsilon \le i < \alpha_{\varepsilon+1}$. We fix ε so $(\alpha_\varepsilon, \alpha_{\varepsilon+1})$ and now define $\mathbf{F}'(\langle M_j : j \le i + 1 \rangle)$ by induction on $i \in [\alpha_\varepsilon, \alpha_{\varepsilon+1})$ assuming that if $\alpha_\varepsilon \le j' + 1 < i + 1$ then $\mathbf{F}'(\langle M_j : j \le j' + 1 \rangle) \le_{\mathfrak{K}} M_{j'+2}$ and further there is $\bar{N}^{j'+1} = \langle N_{j'+1,\xi} : \xi < \alpha_{\varepsilon+1} \rangle$ such that the following holds:

$(*)_3$ $\bar{N}^{j'+1}$ is $\le_{\mathfrak{K}_\lambda}$-increasing continuous, $M_{j'+1} \le_{\mathfrak{K}} N_{j'+1,0}$ and $N_{j'+1,\xi} \le_{\mathfrak{K}_\lambda} M_{j'+2}$

$(*)_4$ if $\delta \in (S \backslash E) \cap (\alpha_{\varepsilon+1} \backslash \alpha_\varepsilon), j' + 1 = \gamma_{\delta,\zeta}$ (so necessarily $j' + 1 \in (\alpha_\varepsilon, \alpha_{\varepsilon+1}), j' + 1 \in \{3\gamma + 2 : \gamma < \lambda\}, \zeta$ is a successor ordinal) then let $\bar{N}^*_{\delta,j'} = \langle N^*_{\delta,j',\zeta'} : \zeta' \le \zeta \rangle$ be the following sequence of length $\zeta + 1$, $N^*_{\delta,j',\zeta'}$ is $N_{\gamma_{\delta,\zeta'},\zeta'}$ if ζ' is a successor ordinal and is $M_{\gamma_{\delta,\zeta'}}$ if ζ' is limit or zero, and we demand $\mathbf{F}(\langle N^*_{\delta,j',\zeta'} : \zeta' \le \zeta \rangle) \le_{\mathfrak{K}} N_{j'+1,\zeta+1}$

$(*)_5$ if $j' + 1 \in u_\delta$ for some $\delta \in S \cap E$ hence $j' + 1 \in \{3\gamma + 1 : \gamma < \delta\}$ and $\zeta = \mathrm{otp}(u_{j'+1}) < \kappa$ and f_ε is the one-to-one order preserving function from $\zeta + 1$ onto $c\ell(u_{j'+1} \cup \{j' + 1\})$ and ζ' is a successor, <u>then</u> $\mathbf{F}(\langle M_{\alpha_{f_\varepsilon(\zeta')}} : \zeta' \le \zeta \rangle) \le_{\mathfrak{K}} M_{\alpha_\varepsilon + 1}$.

This implicitly defines $\mathbf{F}'$. Now $\mathbf{F}'$ is as required: $M_i \cong M$ when $i < \lambda$, $\mathrm{cf}(i) = \kappa$ by $(*)_4$ when $(\exists \varepsilon)(\alpha_\varepsilon < i < \alpha_{\varepsilon+1})$ and by $(*)_5$ when $(\exists \varepsilon)(i = \alpha_\varepsilon)$. $\square_{3.5}$

3.6 Lemma. *Let T be a first order complete theory, K its class of models and $\le_{\mathfrak{K}} = \prec_{\mathbb{L}}$.*

1) If λ is regular, M a saturated model of T of cardinality λ, <u>*then*</u>

M is $(\lambda, \{\lambda\})$-superlimit.
2) If T is stable, and M a saturated model of T of cardinality λ then M is $(\lambda, \{\mu : \kappa(T) \leq \mu \leq \lambda$ and μ is regular$\})$-superlimit (on $\kappa(T)$- see [Sh:c, III,§3]). (Note that by [Sh:c] if λ is singular and T has a saturated model of cardinality λ then T is stable and $\mathrm{cf}(\lambda) \geq \kappa(T)$).
3) If T is stable, λ singular $> \kappa(T), M$ a special model of T of cardinality λ, $S \subseteq \{\delta < \lambda^+ : \mathrm{cf}(\delta) = \mathrm{cf}(\lambda)\}$ is stationary and $S \in \check{I}[\lambda]$ (see above 0.6, 0.7) then M is (λ, S)-medium limit.

Remark. See more in [Sh 868].

Proof. 1) Because if M_i is a λ-saturated model of T for $i < \delta$, $\mathrm{cf}(\delta) \geq \lambda$, then $\bigcup_{i<\delta} M_i$ is λ-saturated. Remembering the uniqueness of a λ-saturated model of T of cardinality λ we finish.
2) Use [Sh:c, III,3.11]: if M_i is a λ-saturated model of T, $\langle M_i : i < \delta\rangle$ increasing $\mathrm{cf}(\delta) \geq \kappa(T)$ then $\bigcup_{i<\delta} M_i$ is λ-saturated.
3) Should be clear by now. $\square_{3.6}$

3.7 Claim. *1) If $M_\ell \in K_\lambda$ are S_ℓ-weakly limit and $S_0 \cap S_1$ is stationary, then $M_0 \cong M_1$, provided κ has (λ, λ)-JEP.*
2) K has at most one locally weakly limit model of cardinality λ provided K has (λ, λ)-JEP.
3) If $M \in K_\lambda$ then $\{S \subseteq \lambda^+ : M$ is S-weakly limit or S not stationary$\}$ is a normal ideal over λ^+.
Instead "S-weakly limit", also "S-medium limit", "S-limit", "S-strongly limit" can be used.
4) In Definition 3.3 without loss of generality $\mathbf{F}(N) \cong M$ or $\mathbf{F}(\bar{M}) \cong M$ according to the case (and we can add $N <_{\mathfrak{K}} \mathbf{F}(N)$, etc.)
5) If K is categorical in λ, then the $M \in K_\lambda$ is superlimit provided that $K_{\lambda^+} \neq \emptyset$ (or, what is equivalent, M has a proper $\leq_{\mathfrak{K}}$-extension).

Proof. Easy.
1) E.g., let $\mathbf{F}_\ell$ witness that M_ℓ is S_ℓ-weakly limit. We can choose (M^0_α, M^1_α) by induction on α such that: $\langle M^\ell_\beta : \beta \leq \alpha\rangle$ is $\leq_{\mathfrak{K}}$-increasing continuous for $\ell = 0, 1, M^0_\alpha \leq_{\mathfrak{K}} M^1_{\alpha+1}, M^1_\alpha \leq_{\mathfrak{K}} M^0_{\alpha+1}$

and $\mathbf{F}_\ell(\langle M^\ell_\beta : \beta \le \alpha+1\rangle) \le M^\ell_{\alpha+2}$. So for some club E_ℓ of $\lambda^+, \delta \in S_\ell \cap E_\ell \Rightarrow M^\ell_\delta \cong M_\ell$ for $\ell = 0, 1$. But $S_0 \cap S_1$ is stationary hence there is a limit ordinal $\delta \in S_0 \cap S_1 \cap E_0 \cap E_1$, hence $M_0 \cong M^0_\delta = M^1_\delta \cong M_1$ as required. $\square_{3.7}$

3.8 Theorem. *If $2^\lambda < 2^{\lambda^+}, M \in K_\lambda$ superlimit, $S = \lambda^+$ <u>or</u> M is S-weakly limit, S is not small (see Definition 0.5) and M does not have the λ-amalgamation property (in $\mathfrak{K}$) <u>then</u> $\dot I(\lambda^+, K) = 2^{\lambda^+}$, moreover there is no universal member in $\mathfrak{K}_{\lambda^+}$ and $(2^\lambda)^+ < 2^{\lambda^+} \Rightarrow \dot I\dot E(\lambda^+, K) = 2^{\lambda^+}$, that is there are 2^{λ^+} models $M \in K_{\lambda^+}$ no one $\le_{\mathfrak{K}}$-embeddable into another.*

3.9 Remark. 0) So in 3.8, if K is categorical in λ <u>then</u> it has λ-amalgamation.
1) We can define a superlimit for a family of models, i.e., when $\mathbf{N} = \{N_t : t \in I\} \subseteq \mathfrak{K}_\lambda$ is superlimit (i.e., if $\langle M_i : i < \delta\rangle$ is $\le_{\mathfrak{K}}$-increasing, $i < \delta \Rightarrow M_i \in \mathfrak{K}_\lambda, \delta$ a limit ordinal $< \lambda^+, M_\delta = \cup\{M_i : i < \delta\}$ <u>then</u> $\bigwedge_{i<\delta} \bigvee_{t\in I} M_i \cong N_t \Rightarrow \bigvee_{t\in I} M_\delta \cong N_t$ (and the other variants). Of course, the family is $\subseteq K_\lambda$ and is not empty. Essentially everything generalizes <u>but</u> in 3.8 the hypothesis should be stronger: the family should satisfy that any member does not have the amalgamation property. E.g. $\mathbf{N} = \mathfrak{K}_\lambda$, (and we can reduce the general case to this by changing $\mathfrak{K}$). But this complicates the situation, and the gain is not clear, so we do not elaborate this.
2) We can many times (and in particular in 3.8) strengthen "there is no $\le_{\mathfrak{K}}$-universal $M \in K_{\lambda^+}$" to "there is no $M \in K_\mu$ into which every $N \in K_{\lambda^+}$ can be $\le_{\mathfrak{K}}$-embedded" for μ not too large. We need $\neg$ Unif$(\lambda^+, S, 2, \mu)$, (see [Sh:f, AP,§1]).

Proof. Let $\mathbf{F}$ be as in Definition 3.3(5) for M. We now choose by induction on $\alpha < \lambda^+$, models M_η for $\eta \in {}^\alpha 2$ such that:

$\circledast_1$ (*i*) $M_\eta \in K_\lambda, M_{<>} = M$,

(*ii*) if $\beta < \alpha$ and $\eta \in {}^\alpha 2$ <u>then</u> $M_{\eta\restriction\beta} \le_{\mathfrak{K}} M_\eta$

(*iii*) if $i+2 \le \alpha$ and $\eta \in {}^\alpha 2$, <u>then</u> $(\mathbf{F}(\langle M_{\eta\restriction j} : j \le i+1\rangle)) \le_{\mathfrak{K}} M_{\eta\restriction(i+2)}$

(iv) if $\alpha = \beta + 1$ and β non-limit, $\eta \in {}^{\alpha}2$, then $M_{\eta \restriction \beta} \neq M_\eta$

(v) if $\alpha < \lambda$ is a limit ordinal and $\eta \in {}^{\alpha}2$ then:

(a) $M_\eta = \cup\{M_{\eta\restriction\beta} : \beta < \ell g(\eta)\}$ and

(b) if M_η fails the λ-amalgamation property then $M_{\eta\hat{}<0>}, M_{\eta\hat{}<1>}$ cannot be amalgamated over M_η, i.e. for no N do we have:

$M_\eta \leq_{\mathfrak{K}} N \in K$ and $M_{\eta\hat{}<0>}, M_{<\eta\hat{}<1>}$ can be $\leq_{\mathfrak{K}}$-embedded into N over M_η.

For $\alpha = 0, \alpha$ limit, we have no problem, for $\alpha + 1, \alpha$ limit: if M_η fails the λ-amalgamation property - use its definition, otherwise let $M_{\eta\hat{}<1>} = M_\eta = M_{\eta\hat{}<0>}$; for $\alpha+1, \alpha$ non-limit - use $\mathbf{F}$ to guaranteee clause (iii), and then for clause (iv) use clause (γ) of Definition 3.3(5), i.e., 3.3(4).

Let for $\eta \in {}^{\lambda^+}2, M_\eta = \bigcup_{\alpha<\lambda^+} M_{\eta\restriction\alpha}$. By changing names we can assume that

$\circledast_1$ (vi) for $\eta \in {}^{\alpha}2 (\alpha < \lambda^+)$ the universe of M_η is an ordinal $< \lambda^+$ (or even $\subseteq \lambda \times (1 + \ell g(\eta))$ and we could even demand equality).

So (by clause (iv)) for $\eta \in {}^{\lambda^+}2$, M_η has universe λ^+.

First, why is there no universal member in $\mathfrak{K}_{\lambda^+}$? If $N \in K_{\lambda^+}$ is universal (by $\leq_{\mathfrak{K}}$, of course), without loss of generality its universe is λ^+. For $\eta \in {}^{\lambda^+}2$ as $M_\eta \in K_{\lambda^+}$, there is a $\leq_{\mathfrak{K}}$-embedding f_η of M_η into N. So f_η is a function from λ^+ to λ^+. Let $\eta \in {}^{\lambda^+}2$, by the choice of $\mathbf{F}$ and of $\langle M_{\eta\restriction\alpha} : \alpha < \lambda^+ \rangle$ there is a closed unbounded $C_\eta \subseteq \lambda^+$ such that $\alpha \in S \cap C_\eta \Rightarrow M_{\eta\restriction\alpha} \cong M$, hence $M_{\eta\restriction\alpha}$ fails the λ-amalgamation property. Without loss of generality for $\delta \in C_\eta$, $M_{\eta\restriction\delta}$ has universe δ. Now by 0.5, if $\langle (f_\rho, C_\rho) : \rho \in {}^{\lambda^+}2 \rangle$ satisfies that for each $\rho \in {}^{\lambda^+}2$, $f_\rho : \lambda^+ \to \lambda^+$ and $C_\rho \subseteq \lambda^+$ is closed unbounded then for some $\eta \neq \nu \in {}^{\lambda^+}2$ and $\delta \in C_\eta \cap S$ we have $\eta \restriction \delta = \nu \restriction \delta, \eta(\delta) \neq \nu(\delta)$ and $f_\eta \restriction \delta = f_\nu \restriction \delta$.

[Why? For every $\delta < \lambda^+, \rho \in {}^{\delta}2$ and $f : \delta \to \lambda^+$ we define $\mathbf{c}(\rho, f) \in 2$ as follows: it is 1 iff there is $\nu \in {}^{\lambda^+}2$ such that $\rho = \nu \restriction \delta$ & $f =$

$f_\nu \restriction \delta$ & $\nu(\delta) = 0$ and is 0 otherwise. So some $\eta \in {}^{\lambda^+}2$ is a weak diamond sequence for the colouring $\mathbf{c}$ and the stationary set S. Now C_η, f_η are well defined and $S' = \{\delta \in S : \delta$ limit and $\eta(\delta) = \mathbf{c}(\eta \restriction \delta, f \restriction \delta)\}$ is a stationary subset of λ^+, so we can choose $\delta \in S' \cap C_\eta$. If $\eta(\delta) = 0$, then $\mathbf{c}(\eta \restriction \delta, f \restriction \delta) = 0$ by the choice of S' but η witness that $\mathbf{c}(\eta \restriction \delta, f \restriction \delta)$ is 1, standing for ν there. If $\eta(\delta) = 1$ there is ν witnessing $\mathbf{c}(\eta \restriction \delta, f_\eta \restriction \delta) = 1$, in particular $\nu(\delta) = 0$, so $\eta, \nu, \eta \restriction \delta$, are as required.]

Now as $\delta \in S \cap C_\eta \subseteq C_\eta$ it follows that $M_{\eta \restriction \delta} \cong M$ hence $M_{\eta \restriction \delta}$ fails the λ-amalgamation property. Also $M_{\eta \restriction \delta}$ has universe δ as $\delta \in C_\eta$ and $M_{\eta \restriction \delta} = M_{\nu \restriction \delta}$ as $\eta \restriction \delta = \nu \restriction \delta$.

So $f_\eta \restriction M_{\eta \restriction \delta} = f_\eta \restriction \delta = f_\nu \restriction \delta = f_\nu \restriction M_{\nu \restriction \delta}$. So $f_\eta \restriction M_{\eta \restriction (\delta+1)}, f_\nu \restriction M_{\nu \restriction (\delta+1)}$ show that $M_{\eta \restriction (\delta+1)}, M_{\nu \restriction (\delta+1)}$, can be amalgamated over $M_{\eta \restriction \delta}$ contradicting clause (v)(b) of the construction, i.e. of ⊛. So there is no $\leq_{\mathfrak{K}}$-universal $N \in \mathfrak{K}_{\lambda^+}$.

It takes some more effort to get 2^{λ^+} pairwise non-isomorphic models (rather than just quite many).

<u>Case A</u>[5]: There is $M^* \in K_\lambda, M \leq_{\mathfrak{K}} M^*$ such that for every N satisfying $M^* \leq_{\mathfrak{K}} N \in K_\lambda$ there are $N^1, N^2 \in K_\lambda$ such that $N \leq_{\mathfrak{K}} N^1, N \leq_{\mathfrak{K}} N^2$ and N^2, N^1 cannot be $\leq_{\mathfrak{K}}$-amalgamated over M^* (not just N). In this case we do not need "M is S-weakly limit".

We redefine $M_\eta, \eta \in {}^\alpha 2, \alpha < \lambda^+$ such that:

⊛$_2$ (a) $\nu \triangleleft \eta \in {}^\alpha 2 \Rightarrow M_\nu \leq_{\mathfrak{K}} M_\eta \in K_\lambda$:

(b) if $\alpha = 0, M_{<>} = M^*$;

(c) if α limit and $\eta \in {}^\alpha 2$ <u>then</u> $M_\eta = \bigcup_{\beta < \alpha} M_{\eta \restriction \beta}$;

(d) if $\eta \in {}^\beta 2, \alpha = \beta + 1$, use the assumption for $N = M_\eta$, now obviously the (N^1, N^2) there satisfies $N^1 \neq N$ and $N^2 \neq N$, so we can have $M_\eta <_{\mathfrak{K}} M_{\eta\hat{}<1>} \in K_\lambda$, $M_\eta <_{\mathfrak{K}} M_{\eta\hat{}<0>} \in K_\lambda$, such that $M_{\eta\hat{}<0>}, M_{\eta\hat{}<1>}$ cannot be amalgamated over M^*.

[5] we can make it a separate claim

Obviously, the models $M_\eta = \bigcup_{\alpha<\lambda^+} M_{\eta\restriction\alpha}$, for $\eta \in {}^{\lambda^+}2$ are pairwise non-isomorphic over M^* and by 0.3 as $2^\lambda < 2^{\lambda^+}$ we finish proving $\dot{I}(\lambda^+, \mathfrak{K}) = 2^{\lambda^+}$.

Note also that for each $\eta \in {}^{\lambda^+}2$ the set $\{\nu \in {}^{\lambda^+}2 : M_\nu$ can be $\leq_{\mathfrak{K}}$-embedded into $M_\eta\}$ has cardinality $\leq |\{f : f$ a $\leq_{\mathfrak{K}}$-embedding of M^* into $M_\eta\}| \leq 2^\lambda$. So if $(2^\lambda)^+ < 2^{\lambda^+}$, then by Hajnal free subset theorem ([Ha61]), there are 2^{λ^+} models $M_\eta \in K_{\lambda^+} (\eta \in {}^{\lambda^+}2)$ no one $\leq_{\mathfrak{K}}$-embeddable into another.

Case B: Not Case A.

Now we return to the first construction, but we can add

(*vii*) if $\eta \in {}^{(\alpha+1)}2$, then if $M_\eta \leq_{\mathfrak{K}} N^1, N^2$ both in K_λ, then N^1, N^2 can be $\leq_{\mathfrak{K}}$-amalgamated over $M_{\eta\restriction\alpha}$.

As $\{W \subseteq \lambda^+ : W$ is small$\}$ is a normal ideal (see 0.5), (and it is on a successor cardinal) it is well known that we can find λ^+ pairwise disjoint non-small $S_\zeta \subseteq S$ for $\zeta < \lambda^+$. We define a colouring (= function) $\mathbf{c}$:

$\circledast_3$ (*a*) $\mathbf{c}(\eta, \nu, f)$ will be defined iff
for some limit ordinal $\delta < \lambda^+, \eta \in {}^\delta 2, \nu \in {}^\delta 2$
and f is a function from δ to λ^+

(*b*) $\mathbf{c}(\eta, \nu, f) = 1$ iff
the triple (η, ν, f) belongs to the domain of $\mathbf{c}$
(i.e., is as in (a)) and M_η, M_ν have universe δ,
f is a $\leq_{\mathfrak{K}}$-embedding of M_η into M_ν
and for some $\rho, \nu\hat{\ } <0> \triangleleft \rho \in {}^{\lambda^+}2$
the function f can be extended to a
$\leq_{\mathfrak{K}}$-embedding of $M_{\eta\hat{\ }<0>}$ into M_ρ

(*c*) $\mathbf{c}(\eta, \nu, f)$ is zero iff it is defined but is $\neq 1$.

For each ζ, as S_ζ is not small, by simple coding, for every $\zeta < \lambda^+$ there is $h_\zeta : S_\zeta \to \{0, 1\}$ such that:

$(*)_\zeta$ for every $\eta \in {}^{\lambda^+}2, \nu \in {}^{\lambda^+}2$ and $f : \lambda^+ \to \lambda^+$, for a stationary set of $\delta \in S_\zeta$

$$\mathbf{c}(\eta \restriction \delta, \nu \restriction \delta, f \restriction \delta) = h_\zeta(\delta).$$

Now for every $W \subseteq \lambda^+$ we define $\eta_W \in {}^{\lambda^+}2$ as follows:

$\eta_W(\alpha)$ is $h_\zeta(\alpha)$, if $\zeta \in W$ and $\alpha \in S_\zeta$ (note that there is at most one ζ)

$\eta_W(\alpha)$ is zero if there is no such ζ.

Now we can show (chasing the definitions) that

$\circledast_4$ if $W(1), W(2) \subseteq \lambda^+, W(1) \nsubseteq W(2)$, <u>then</u> $M_{\eta_{W(1)}}$ cannot be $\leq_{\mathfrak{K}}$-embedded into $M_{\eta_{W(2)}}$.

This clearly suffices.

Why is $\circledast_4$ true? Suppose $W(1) \nsubseteq W(2)$, let $\zeta \in W(1) \backslash W(2)$ and toward contradiction let f be a $\leq_{\mathfrak{K}}$-embedding of $M_{\eta_{W(1)}}$ into $M_{\eta_{W(2)}}$, so $E = \{\delta : M_{\eta_{W(1)} \restriction \delta}, M_{\eta_{W(2)} \restriction \delta}$ have universe δ and $f \restriction \delta$ is a $\leq_{\mathfrak{K}}$-embedding of $M_{\eta_{W(1)} \restriction \delta}$ into $M_{\eta_{W(2)} \restriction \delta}\}$ is a club of λ^+. Hence by the choice of $\mathbf{c}$ and h_ζ there is $\delta \in E \cap S_\zeta$ such that

$\boxtimes$ $\mathbf{c}(\eta_{W(1)} \restriction \delta, \eta_{W(2)} \restriction \delta, f \restriction \delta) = h_\zeta(\delta)$ and $M_{\eta_{w(1)} \restriction \delta}$ is not an amalgamation base.

Now the proof splits to two cases.

<u>Case 1</u>: $h_\zeta(\delta) = 0$.

So $\eta_{W(1)}(\delta) = 0 = \eta_{W(2)}(\delta)$ and by clause (b) of $\circledast_3$ above, i.e., the definition of $\mathbf{c}$ we have the objects $\eta_{W(1)}, \eta_{W(2)}, f \restriction M_{\eta_{W(1)} \hat{} <0>} = f \restriction M_{\eta_{W(1)} \restriction (\delta+1)}$ witness that $\mathbf{c}(\eta_{W(1)} \restriction \delta, \eta_{W(2)} \restriction \delta, f \restriction \delta) = 1$, contradiction.

<u>Case 2</u>: $h_\zeta(\delta) = 1$.

So $\eta_{W(1)}(\delta) = 1, \eta_{W(2)}(\delta) = 0, \mathbf{c}(\eta_{W(1)} \restriction \delta, \eta_{W(2)} \restriction \delta, f \restriction \delta) = 1$. By the definition of $\mathbf{c}$, we can find ν such that $(\eta_{W(2)} \restriction \delta)\hat{}<0> \trianglelefteq \nu \in {}^{\lambda^+}2$ and a $\leq_{\mathfrak{K}}$-embedding g of $M_{(\eta_{W(1)} \restriction \delta)\hat{}<0>}$ into M_ν.

For some $\alpha \in (\delta, \lambda^+), f$ embeds $M_{\eta_{W(1)} \restriction (\delta+1)} = M_{(\eta_{W(1)} \restriction \delta)\hat{}<1>}$ into $M_{\eta_{W(2)} \restriction \alpha}$ and g embeds $M_{(\eta_{W(1)} \restriction \delta)\hat{}<0>}$ into $M_{\nu \restriction \alpha}$.

As $\eta_{W(2)} \restriction \delta\hat{} <0> \triangleleft \nu \restriction \alpha$ and $\eta_{W(2)} \restriction \delta\hat{} <0> \triangleleft \eta_{W(2)} \restriction \alpha$ by clause (vii) above there are f_1, g_1 and $N \in K_\lambda$ such that

(a) $M_{\eta_{W(2)} \restriction \delta} \leq_{\mathfrak{K}} N$

(b) f_1 is a $\leq_{\mathfrak{K}}$-embedding of $M_{\eta_{W(2)} \restriction \alpha}$ into N over $M_{\eta_{W(2)} \restriction \delta}$

(c) g_1 is a $\leq_{\mathfrak{K}}$-embedding of $M_{\nu \restriction \alpha}$ into N over $M_{\eta_{W(2)} \restriction \delta}$.

So

$(b)^*$ $f_1 \circ f$ is a $\leq_{\mathfrak{K}}$-embedding of $M_{(\eta_{W(1)} \restriction \delta)\hat{}<1>}$ into N

$(c)^*$ $g_1 \circ g$ is a $\leq_{\mathfrak{K}}$-embedding of $M_{(\eta_{W(1)} \restriction \delta)\hat{}<0>}$ into N

$(d)^*$ $f_1 \circ f, g_1 \circ g$ extend $f \restriction \delta : M_{\eta_{W(1)} \restriction \delta} \to N$ (both).

So together we get a contradiction to assumption $(*)_1(d)$. $\square_{3.8}$

3.10 Theorem. *1) Assume one of the following cases occurs:*

$(a)_1$ $\mathfrak{K}$ *is* $\mathrm{PC}_{\aleph_0}$ *(hence* $\mathrm{LS}(\mathfrak{K}) = \aleph_0$*) and* $1 \leq \dot{I}(\aleph_1, \mathfrak{K}) < 2^{\aleph_1}$ *or*

$(a)_2$ $\mathfrak{K}$ *has models of arbitrarily large cardinality,* $\mathrm{LS}(\mathfrak{K}) = \aleph_0$ *and* $\dot{I}(\aleph_1, \mathfrak{K}) < 2^{\aleph_1}$.

Then there is an a.e.c. $\mathfrak{K}_1$ *such that*

(A) $M \in K_1 \Rightarrow M \in K$ *and* $M \leq_{\mathfrak{K}_1} N \Rightarrow M \leq_{\mathfrak{K}} N$ *and* $\mathrm{LS}(\mathfrak{K}_1) = \mathrm{LS}(\mathfrak{K})(= \aleph_0)$

(B) *if* K *has models of arbitrarily large cardinality then so does* K_1

(C) $\mathfrak{K}_1$ *is* $\mathrm{PC}_{\aleph_0}$

(D) $(K_1)_{\aleph_1} \neq \emptyset$

(E) *all models of* K_1 *are* $\mathbb{L}_{\infty,\omega}$*-equivalent and* $M \leq_{\mathfrak{K}_1} N \Leftrightarrow M \prec_{\mathbb{L}_{\infty,\omega}} N$ & $M \leq_{\mathfrak{K}} N$ *and* K_1 *is categorical in* $\aleph_0$ *and* $M_* \in (K_1)_{\aleph_0} \Rightarrow K_1 = \{N \in K : N \equiv_{\mathbb{L}_{\infty,\omega}(\tau_K)} M_*\}$

(F) *if* $\mathfrak{K}$ *is categorical in* $\aleph_1$ *then* $(K_1)_\lambda = K_\lambda$ *for every* $\lambda > \aleph_0$*; moreover* $\leq_{\mathfrak{K}_1} = \leq_{\mathfrak{K}} \restriction (K_1)_{\geq \aleph_1}$.

2) If in (1) we add $\mathrm{LS}(\mathfrak{K})$ *names to formulas in* $\mathbb{L}_{\infty,\omega}$ *(i.e. to a set of representatios up to equivalence) then we can assume each member of* K *is* $\aleph_0$*-sequence-homogeneous. The vocabulary remains countable, in fact, for some countable first order theory* T*, the models of* K *are the atomic models of* T *(in the first order sense) and* $\leq_{\mathfrak{K}}$ *becomes* $\subseteq$ *(being a submodel).*

Proof. Like [Sh 48, 2.3,2.5] (using 2.19 here for $\alpha = 2$). E.g. why, if K is categorical in $\aleph_1$ then $\leq_{\mathfrak{K}_1} = \leq_{\mathfrak{K}} \restriction (K_1)_{\geq \aleph_1}$? We have to prove that if $M \leq_{\mathfrak{K}} N$ are uncountable then $M \prec_{\mathbb{L}_{\infty,\omega}(\tau_K)} N$. But there is $M_* \in K_{\aleph_0}$ such that $K_1 = \{M' \in K : M' \equiv_{\mathbb{L}_{\infty,\omega}} M_*\}$ and $(K_1)_{\aleph_1} = K_{\aleph_1} \neq \emptyset$, so it suffices to prove $M \prec_{\mathbb{L}_{\omega_1,\omega}(T)} N$, so assume this is a counterexample so for some $\varphi(x, \bar{y}) \in \mathbb{L}_{\omega_1,\omega}(\tau)$ and $\bar{a} \in {}^{\ell g(\bar{y})}M, b \in N$ we have $N \models \varphi[b, \bar{a}]$ but for no $b' \in M$ do we have $N \models \varphi[b', \bar{a}]$ and without loss of generality the quantifier depth of $\varphi(x, \bar{y}), \gamma$ is minimal (for all such pairs (M, N)). Let $\Delta_\gamma = \{\psi(\bar{z}) \in \mathbb{L}_{\omega_1,\omega}(\tau_K) : \psi$ has quantifier depth $\leq \gamma\}$ hence $M' \leq_{\mathfrak{K}} N', M' \in K_{>\aleph_0} \Rightarrow M' \prec_{\Delta_\gamma} N'$. Also without loss of generality $\|M\| = \|N\| = \aleph_1$. Now choose $M_\alpha \in K_{\aleph_1}$ by induction on $\alpha < \omega_2$, which is $\leq_{\mathfrak{K}}$-increasing continuous (hence $\prec_{\Delta_\gamma}$ increases) and for each α there is an isomorphism f_α from N onto $M_{\alpha+1}$ mapping M onto M_α, recalling the categoricity. By Fodor lemma for some $\alpha < \beta$ we have $f_\alpha(\bar{a}) = f_\beta(\bar{a})$, so $f_\beta^{-1}(f_\alpha(b))$ contradict the choice of $\varphi(x, \bar{y}), b, \bar{a}$. $\square_{3.10}$

We arrive to the main theorem of this section.

3.11 Theorem. *Suppose $\mathfrak{K}$ and λ satisfy the following conditions:*

(A) *$\mathfrak{K}$ has a superlimit member M^* of cardinality $\lambda, \lambda \geq \mathrm{LS}(\mathfrak{K})$, (if K is categorical in λ, then by assumption (B) below there is such M^*; really invariantly λ^+-strongly limit suffice if (d) of $(*)$ of 3.12(2) below holds, see Definition 3.3)*

(B) *$\mathfrak{K}$ is categorical in λ^+*

(C) (α) *$\mathfrak{K}$ is $\mathrm{PC}_{\aleph_0}, \lambda = \aleph_0$ or*

(β) *$\mathfrak{K} = \mathrm{PC}_\lambda, \lambda = \beth_\delta, \mathrm{cf}(\delta) = \aleph_0$ or*

(γ) *$\lambda = \aleph_1, \mathfrak{K}$ is $\mathrm{PC}_{\aleph_0}$ or*

(δ) *$\mathfrak{K}$ is $\mathrm{PC}_\mu, \lambda \geq \beth_{(2^\mu)^+}$; not useful for 3.11, still it too implies $(*)_{\lambda,\mu}$ in 3.12.*

Then K has a model of cardinality λ^{++}.

3.12 Remark. 1) If $\lambda = \aleph_0$ we can wave hypothesis (A) by the previous theorem 3.10.
2) Hypothesis (C) can be replaced by (giving a stronger theorem):

$(*)_{\lambda,\mu}(a)$ $\mathfrak{K}$ is PC_μ and (*b*) any $\psi \in \mathbb{L}_{\mu^+,\omega}$ which has a model M of order-type λ^+, $|P^M| = \lambda$, has a non-well-ordered model N of cardinality λ

(*c*) $\{M \in K_\lambda : M \cong M^*\}$ is PC_μ (among models in K_λ) and

(*d*) for some $\mathbf{F}$ witnessing "M^* is invariantly λ-strongly limit", that is the class $\{(M, \mathbf{F}(M)) : M \in K_\lambda\}$ is PC_μ (if M^* is superlimit this clause is not required as $\mathbf{F}$ = the identity on K_λ is O.K.)

3) It is well known, see e.g. [Sh:c, VII,§5] that hypothesis (C) implies $(*)_{\lambda,\mu}$ from part (2), see more [GrSh 259].

Proof. By 3.12(3) we can assume $(*)_{\lambda,\mu}$ from 3.12(2).

Stage a: It suffices to find $N_0 \le_{\mathfrak{K}} N_1, \|N_0\| = \lambda^+, N_0 \ne N_1$.

Why? We define by induction on $\alpha < \lambda^{++}$ a model $N_\alpha \in K_{\lambda^+}$ such that $\beta < \alpha$ implies $N_\beta \le_{\mathfrak{K}} N_\alpha$ and $N_\beta \ne N_\alpha$. Clearly N_0, N_1 are defined (without loss of generality $\|N_1\| = \lambda^+$ as $\lambda \ge \mathrm{LS}(\mathfrak{K})$, also otherwise we already have the desired conclusion), for limit $\delta < \lambda^{++}$ the model $\bigcup_{\alpha<\delta} N_\alpha$ is as required. For $\alpha = \beta + 1$, by the λ^+-categoricity, N_0 is isomorphic to N_β, say by f and we define $N_{\beta+1}$ such that f can be extended to an isomorphism from N_1 onto $N_{\beta+1}$, so clearly $N_{\beta+1}$ is as required. Now $\bigcup_{\alpha<\lambda^{++}} N_\alpha \in K_{\lambda^{++}}$ is as required.

Hence the following theorem completes the proof of 3.11 (use $\mathbf{F}$ = the identity for the superlimit case).

3.13 Theorem. *Suppose the following clauses:*

(*A*) *$\mathfrak{K}$ has an invariantly λ-strongly limit member M^* of cardinality λ, as exemplified by $\mathbf{F} : K_\lambda \to K_\lambda$ and $\mathfrak{K}_\lambda$ has the* JEP *(see Definition 3.3)*

(*B*) $\dot{I}(\lambda^+, K_{\lambda^+}) < 2^{\lambda^+}$ *or even just* $\dot{I}(\lambda^+, K^{\mathbf{F}}_{\lambda^+}) < 2^{\lambda^+}$ *(or just* $\dot{I}\dot{E}(\lambda^+, K^{\mathbf{F}}_{\lambda^+}) < 2^{\lambda^+}$ *(see below))*

(*C*) *$\mathfrak{K}$ is a* PC_μ *class, as well as $\mathbf{F}$, i.e., K' is* PC_μ *where K' is a class closed under an isomorphism of $(\tau_{\mathfrak{K}} \cup \{P\})$-models, P a unary predicate such that $K'_\lambda = \{(N, M) : N = \mathbf{F}(M)\}$*

(D) *$\mu = \lambda = \aleph_0$ or $\mu = \lambda = \beth_\delta, \mathrm{cf}(\delta) = \aleph_0$ or $\mu = \aleph_0, \lambda = \aleph_1$ or just $(*)_{\lambda,\mu}(c)$ from 3.12(2)*

(E) *K categorical in λ or at least there is $\psi \in \mathbb{L}_{\omega_1,\omega}(\tau^+)$ such that $(M^*/\cong) = \{M \restriction \tau_{\mathfrak{K}} : M \models \psi, \|M\| = \lambda\}$.*

Then we can find $N_0 \leq_{\mathfrak{K}} N_1, N_0 \neq N_1$ such that $N_0, N_1 \in K^{\mathbf{F}}_{\lambda^+}$, where

3.14 Definition. Assume $\mathbf{F} : K_\lambda \to K_\lambda$ satisfies $M \leq_{\mathfrak{K}} \mathbf{F}(M)$ for $M \in K_\lambda$ or more generally $\mathbf{F} \subseteq \{(M,N) : M \leq_{\mathfrak{K}} N$ are from $K_\lambda\}$ satisfies $(\forall M \in K_\lambda)(\exists N)((M,N) \in \mathbf{F})$ or just $(\forall M \in K_\lambda)(\exists N_0, N_1)[(N_0,N_1) \in \mathbf{F} \wedge M \leq_{\mathfrak{K}} N_0 \leq_{\mathfrak{K}} N_1]$. Then we let $K^{\mathbf{F}}_{\lambda^+} := \{\bigcup_{i<\lambda^+} M_i : M_i \in K_\lambda, \langle M_i : i < \lambda^+\rangle$ is $\leq_{\mathfrak{K}}$-increasing continuous not eventually constant and $\mathbf{F}(M_{i+1}) \leq_{\mathfrak{K}} M_{i+2}$ or $(M_{i+1}, M_{i+2}) \in \mathbf{F}\}$ for $i < \lambda$.

3.15 Remark. 1) As the sequence in the definition of $K^{\mathbf{F}}_{\lambda^+}$ is $\leq_{\mathfrak{K}}$-increasing and the sequence is not eventually constant (which follows if $(M,N) \in \mathbf{F} \Rightarrow M \neq N$), necessarily $K^{\mathbf{F}}_{\lambda^+} \subseteq \mathfrak{K}_{\lambda^+}$.
2) Theorem 3.13 is good for classes which are not exactly a.e.c., see, e.g., 3.18.

Considering $K^{\mathbf{F}}_{\lambda^+}$ we may note that the proofs of some earlier claims give more. In particular (before proving 3.13), similarly to 3.8:

3.16 Claim. *Assume that*

(a) $2^\lambda < 2^{\lambda^+}$

(b) *$\mathfrak{K}$ is an a.e.c. and $\mathrm{LS}(\mathfrak{K}) \leq \lambda$*

(c) *$M \in K_\lambda$ is S-weakly limit, S not small (see Definition 0.5)*

(d) *M does not have the amalgamation property in $\mathfrak{K}$ (= is an amalgamation base)*

(e) *$\mathbf{F}$ is as in 3.14.*

<u>Then</u> $\dot{I}(\lambda^+, K^{\mathbf{F}}_{\lambda^+}) = 2^{\lambda^+}$.

Proof. To avoid confusion rename $\mathbf{F}$ of clause (e) as $\mathbf{F}_1$, and choose $\mathbf{F}_2$ which exemplifies "M is S-weakly limit", i.e., as in Definition 3.3(5). Now we define $\mathbf{F}'$ with the same domain as $\mathbf{F}_2$ by $\mathbf{F}'(\langle M_j : j \le i\rangle) = \mathbf{F}_1(\mathbf{F}_2(\langle M_j : j \le i\rangle))$, and continue as in the proof of 3.8 noting that $\mathbf{F}'$ works as well there.

The sequence of models $\langle M_\eta : \eta \in {}^{\lambda^+}2\rangle$ we got there are from $K^{\mathbf{F}_1}_{\lambda^+}$ (so witness that $\dot{I}(\lambda^+, K^{\mathbf{F}_1}_{\lambda^+}) = 2^{\lambda^+}$) because:

- (∗) if the sequence $\langle M_\alpha : \alpha < \lambda^+\rangle, M_\alpha \in \mathfrak{K}_\lambda$ for $\alpha < \lambda^+$ is $\le_{\mathfrak{K}}$-increasing continuous and $\mathbf{F}'(\langle M_j : j \le i+1\rangle) \le_{\mathfrak{K}} M_{i+2}$ then $\cup\{M_\alpha : \alpha < \lambda^+\} \in K^{\mathbf{F}_1}_{\lambda^+}$.

$\square_{3.16}$

Also similarly to 3.10 we can prove:

3.17 Claim. *Assume $\mathfrak{K}$ is a* $\mathrm{PC}_{\aleph_0}$ *and* $\mathbf{F}$ *a* $\mathrm{PC}_{\aleph_0}$ *is as in 3.14. If* $1 \le \dot{I}(\aleph_1, K^{\mathbf{F}}_{\aleph_1}) < 2^{\aleph_1}$ *<u>then</u> the conclusion of 3.10 above holds.*

Proof of 3.13. (Hence of 3.11). The reader may do well to read it with $\mathbf{F}$ = the identity in mind.

<u>Stage b</u>: We now try to find N_0, N_1 as mentioned in stage (a) above by approximations of cardinality λ. A triple will denote here (M, N, a) satisfying $M, N \cong M^*$ (see hypothesis (A)), $M \le_{\mathfrak{K}} N$ and $a \in N\backslash M$. Let $<$ be the following partial order among this family of triples: $(M, N, a) < (M', N', a')$ if $a = a', N \le_{\mathfrak{K}} N', M \le_{\mathfrak{K}} M', M \ne M'$ and moreover $(\exists N'')[N \le_{\mathfrak{K}} N'' \ \ \& \ \ \mathbf{F}(N'') \le_{\mathfrak{K}} N']$ and $(\exists M'')[M \le_{\mathfrak{K}} M'' \ \ \& \ \ \mathbf{F}(M'') \le_{\mathfrak{K}} M']$. (It is tempting to omit a and require $M = M' \cap N$, but this apparently does not work as we do know if disjoint amalgamation $\mathfrak{K}_{\aleph_0}$ exist).

We first note that there is at least one triple (as M^* has a proper elementary extension which is isomorphic to it, because it is a limit model by clause (A) of the assumption).

<u>Stage c</u>: We show that if there is no maximal triple, our conclusions follows.

We choose by induction on α a triple (M_α, N_α, a) increasing by $<$. For $\alpha = 0$ see the end of previous stage, for $\alpha = \beta + 1$, we can define (M_α, N_α, a) by the hypothesis of this stage. For limit $\delta < \lambda^+, (M_\delta, N_\delta, a)$ will be $(\bigcup_{\alpha<\delta} M_\alpha, \bigcup_{\alpha<\delta} N_\alpha, a)$ (notice $M_\delta \leq_{\mathfrak{K}} N_\delta$ by AxIV of 1.2 and M_δ, N_δ are isomorphic to M^* by the choice of $\mathbf{F}$ and the definition of order on the family of triples). Now similarly $M = \bigcup_{\alpha<\lambda^+} M_\alpha \leq_{\mathfrak{K}} N = \bigcup_{\alpha<\lambda^+} N_\alpha$ are both from $\mathfrak{K}^{\mathbf{F}}_{\lambda^+}$ and the element a exemplifies $M \neq N$, so by Stage (a) we finish.
Recall

⊛ if (M, N, a) is a maximal triple <u>then</u> there is no triple (M', N', a) such that $M' \leq_{\mathfrak{K}} N', M <_{\mathfrak{K}} M', N \leq_{\mathfrak{K}} N', a \in N' \backslash M'$ and $(\exists M'')(M \leq_{\mathfrak{K}} M'' \leq_{\mathfrak{K}} \mathbf{F}(M'') \leq_{\mathfrak{K}} M')$ and $(\exists N'')(N \leq_{\mathfrak{K}} N'' \leq_{\mathfrak{K}} \mathbf{F}(N'') \leq_{\mathfrak{K}} N')$.

<u>Stage d</u>: There are $M_i \cong M^*$ for $i \leq \omega$ such that $[i < j \leq \omega \Rightarrow M_j <_{\mathfrak{K}} M_i], i < \omega \Rightarrow \mathbf{F}(M_{i+1}) \leq_{\mathfrak{K}} M_i$ and $|M_\omega| = \bigcap_{n<\omega} |M_n|$ and note that M_i is λ^+-strongly limit.
This stage is dedicated to proving this statement. As M^* is super-limit (or just strongly limit), there is an $\leq_{\mathfrak{K}}$-increasing continuous sequence $\langle M_i : i < \lambda^+\rangle, M_i \cong M^*$ and $\mathbf{F}(M_{i+1}) \leq_{\mathfrak{K}} M_{i+2}$. (Note that this is true also for limit models as we can restrict ourselves to a club of i's). So without loss of generality $\bigcup_{i<\lambda^+} M_i$ has universe λ^+, M_0 has universe λ.
Define a model $\mathfrak{B}$.
Its universe is λ^+.

<u>Relations and Functions</u>:

(a) those of $\bigcup_{i<\lambda^+} M_i$

(b) R-two place: aRi if and only if $a \in M_i$

(c) P (monadic relation) $P = \lambda$ which is the universe of M_0

(*d*) g, a two-place function such that for each $i, g(i, -)$ is an isomorphism from M_0 onto M_i

(*e*) $<$ (two-place relation) - the usual ordering (on the ordinals $< \lambda^+$)

(*f*) relations with parameter i witnessing $M_i \leq_{\mathfrak{K}} \bigcup_{j<\lambda^+} M_j$ (we can instead make functions witnessing $M \in K$ as in 1.9 (the strong version) and have: each M_i is closed under them))

(*g*) relations with parameter i witnessing each $\mathbf{F}(M_{i+1}) \leq_{\mathfrak{K}} M_{i+2}$ and $M_{i+1} \neq M_{i+2}$ (including $(M_{i+1}, \mathbf{F}(M_{i+1})) \in \mathbf{F}$)

(*h*) if $\mu = \lambda$, also individual constant for each $a \in M_0$.

Let $\psi \in \mathbb{L}_{\mu^+,\omega}$ describe this, in particular for clauses (f), (g) use clause (C) of the assumptions. So ψ has a non-well ordered model $\mathfrak{B}^*, |P^{\mathfrak{B}^*}| = \lambda$ (by clause (D) of the assumption see 3.12(2)+(3)). So let

$$\mathfrak{B}^* \models \text{“}a_{n+1} < a_n\text{” for } n < \omega.$$

Let for $a \in \mathfrak{B}^*, A_a = \{x \in \mathfrak{B}^* : \mathfrak{B}^* \models xRa\}$

$$M_a = (\mathfrak{B}^* \restriction \tau_{\mathfrak{K}}) \restriction A_a.$$

Easily $M_a \leq_{\mathfrak{K}} (\mathfrak{B}^* \restriction \tau_{\mathfrak{K}})$ (use clause (f)) and $\|M_a\| = \lambda$. In fact M_a is superlimit or just isomorphic to M^* if $\mu = \lambda$, as ψ includes the diagram of $M_0 = M^*$, having names for all members, and if $\mu < \lambda$ see assumption (E). So $M_{a_n} \leq_{\mathfrak{K}} \mathfrak{B}^* \restriction \tau_{\mathfrak{K}}, M_{a_{n+1}} \subseteq M_{a_n}$ hence $M_{a_{n+1}} \leq_{\mathfrak{K}} M_{a_n}$ by Ax V. Let $M_n := M_{a_n}$. Let $I = \{b \in \mathfrak{B}^* : \bigwedge_{n<\omega} [\mathfrak{B}^* \models b < a_n]\}$.

Also as for $b \in I, M_b <_{\mathfrak{K}} \mathfrak{B}^* \restriction \tau_{\mathfrak{K}}$ and $M_{b_1} <_{\mathfrak{K}} M_{b_2}$ for $b_1 <^{\mathfrak{B}^*} b_2$, by Ax IV clearly $M_\omega := (\mathfrak{B}^* \restriction (\tau_{\mathfrak{K}})) \restriction \bigcup_{b \in I} A_b$ satisfies $M_\omega \leq_{\mathfrak{K}} \mathfrak{B}^* \restriction \tau_{\mathfrak{K}}$ hence $M_\omega \leq_{\mathfrak{K}} M_n$ for $n < \omega$. Obviously $M_\omega \subseteq \bigcap_{n<\omega} M_n$ and equality holds as ψ guarantee

(∗) for every $y \in \mathfrak{B}^*$ there is a minimal $x \in \mathfrak{B}^*$ such that $y \in M_x$.

As each M_b is isomorphic to M^*, of cardinality λ, also M_ω is.

Stage e: Suppose that there is a maximal triple, then we shall show $\dot{I}(\lambda^+, K) = 2^{\lambda^+}$ and moreover $\dot{I}(\lambda^+, K^{\mathbf{F}}_{\lambda^+}) = 2^{\lambda^+}$, and so we shall get a contradiction to assumption (B).

So there is a maximal triple (M^0, N^0, a). Hence by the uniqueness of the limit model for each $M \in K_\lambda$ which is isomorphic to M^* hence to M^0 there are N, a satisfying $M \leq_{\mathfrak{K}} N \cong M^* \in K_\lambda, a \in N \backslash M$ such that: if $M <_{\mathfrak{K}} M' \leq_{\mathfrak{K}} N' \in \mathfrak{K}_\lambda, N <_{\mathfrak{K}} N', (\exists M'')(M \leq_{\mathfrak{K}} M'' \leq_{\mathfrak{K}} \mathbf{F}(M'') \leq_{\mathfrak{K}} M' \cong M^*)$ and $(\exists N'')(N \leq_{\mathfrak{K}} N'' \leq_{\mathfrak{K}} \mathbf{F}(N'') \leq_{\mathfrak{K}} N' \cong M^*)$ then $a \in M'$. (That is, in some sense a is algebraic over M). We can waive $(\exists N'')(N \leq_{\mathfrak{K}} N'' \leq_{\mathfrak{K}} \mathbf{F}(N'') \leq_{\mathfrak{K}} N' \cong M^*)$ as by the definition of strongly limit there is $N'_* \cong M^*$ such that $\mathbf{F}(N') \leq_{\mathfrak{K}} N'_*$. On the other hand by Stage d

$(*)_1$ for each $M \in K_\lambda$ isomorphic to M^* there are $M'_n (n < \omega)$ such that $M \leq_{\mathfrak{K}} M'_{n+1} <_{\mathfrak{K}} M'_n \in K_\lambda, M'_n \cong M^*$ and $\mathbf{F}(M'_{n+1}) \leq_{\mathfrak{K}} M'_n$ and $\bigcap_{n<\omega} M'_n = M$.

For notational simplicity: $M \in K_\lambda, |M|$ an ordinal $\Rightarrow |\mathbf{F}(M)|$ an ordinal.

Now for each $S \subseteq \lambda^+$ we define by induction on $\alpha \leq \lambda^+, M^S_\alpha$, increasing (by $<_{\mathfrak{K}}$) and continuous with universe an ordinal $< \lambda^+$ such that $M^S_\alpha \cong M^*$ and if $\beta + 2 \leq \alpha$ then $\mathbf{F}(M_{\beta+1}) \leq_{\mathfrak{K}} M_{\beta+1}$. Let $M^S_0 = M^*$ and for limit $\delta < \lambda^+$ and let $M^S_\delta = \bigcup_{\alpha<\delta} M^S_\alpha$; by the induction assumption and the choice of $M^*, \mathbf{F}$ clearly M^S_δ is isomorphic to M^*. For $\alpha = \beta + 1, \beta$ successor let M^S_α be such that $\mathbf{F}(M^S_\beta) <_{\mathfrak{K}} M^S_\alpha \cong M^*$. So we are left with the case $\alpha = \delta + 1, \delta$ limit or zero.

Now if $\delta \in S$ hence $M^S_\delta \cong M^*$, choose $M_{\delta+1}, a^S_\delta$ such that $(M^S_{\delta+1}, M^S_\delta, a^S_\delta)$ is a maximal triple (possible as by the hypothesis of this case there is a maximal triple, and there is a unique strong limit model). If $\delta \notin S$ we choose $M^{S,n}_\delta \in K_\lambda$ for $n < \omega$ (not used) such that $M^S_\delta <_{\mathfrak{K}} M^{S,n+1}_\delta \leq_{\mathfrak{K}} M^{S,n}_\delta$ and $\mathbf{F}(M^{S,n+1}_\delta) \leq_{\mathfrak{K}} M^{S,n}_\delta$ for $n < \omega$ and $M^S_\delta = \bigcap_{n<\omega} M^{S,n}_\delta$ and $M^{S,n}_\delta \cong M^*$; and let $M^S_{\delta+1} = M^{S,0}_\delta$

(again possible as $M_\delta \cong M^*$ and an $(*)_1$ above).
Lastly, let $M^S = \bigcup_\alpha M^S_\alpha$.

Now clearly it suffices to prove that if $S^0, S^1 \subseteq \lambda^+, S^1 \backslash S^0$, is stationary then $M^{S^1} \not\cong M^{S^0}$. So suppose f is a $\leq_{\mathfrak{K}}$-embedding from M^{S^1} onto M^{S^0} or just into M^{S^0}. Then $E^2 = \{\delta < \lambda^+ : M^{S^1}_\delta, M^{S^0}_\delta$ each has universe δ and $i < \lambda^+$ implies $[i < \delta \Leftrightarrow f(i) < \delta]\}$ is a closed unbounded subset of λ^+, hence there is a limit ordinal $\delta \in (S^1 \backslash S^0) \cap E^2$. Let us look at $f(a^{S^1}_\delta)$; as $\delta \in S^1$, $a^{S^1}_\delta$ is well defined, also $a^{S_1}_\delta \in M^{S^1}_{\delta+1} \backslash M^{S^1}_\delta$, as $\delta \in E^2$ it follows that $f(a^{S^1}_\delta) \not< \delta$ hence $f(a^{S^1}_\delta)$ belongs to $M^{S^0} \backslash M^{S^0}_\delta$ but $M^{S^0}_\delta = \bigcap_{n<\omega} M^{S^0,n}_\delta$ (as $\delta \notin S^0$).
Hence for some n, $f(a^{S^1}_\delta) \notin M^{S^0,n}_\delta$. Let $\beta \in (\delta, \lambda^+)$ be large enough such that $f(M^{S^1}_{\delta+1}) \subseteq M^{S^0}_\beta$. But then $f(M^{S^1}_\delta) \leq_{\mathfrak{K}} M^{S^0,n}_\delta \leq_{\mathfrak{K}} M^{S^0}_\beta$ and $f(M^{S^1}_{\delta+1}) \leq_{\mathfrak{K}} M^{S^0}_\beta$ and $a^{S^1}_\delta \notin f^{-1}(M^{S^0,n}_\delta)$.
Now $(f(M^{S^1}_\delta)), f(M^{S^1}_{\delta+1}), f(a^{S^1}_\delta))$ has the same properties as $(M^{S^1}_\delta, M^{S^1}_{\delta+1}, a^{S^1}_\delta)$ because if f is an isomorphism from M' onto $M'' \in K_\lambda$ then we can extend f to an isomorphism from $\mathbf{F}(M')$ onto $\mathbf{F}(M'')$ (i.e., the "invariant"). But $(f(M^{S^1}_\delta), f(M^{S^1}_{\delta+1}), f(a^{S^1}_\delta)) < (M^{S^0,n}_\delta, M^{S^0}_\beta, f(a^{S^1}_\delta))$, contradiction. So we are done. $\square_{3.11}$

3.18 Conclusion. 1) If $\mathrm{LS}(\mathfrak{K}) = \aleph_0, K$ is $\mathrm{PC}_{\aleph_0}$ and $\dot{I}(\aleph_1, K) = 1$, <u>then</u> K has a model of cardinality $\aleph_2$.
2) If $\psi \in \mathbb{L}_{\omega_1,\omega}(\mathbf{Q})$ ($\mathbf{Q}$ is the quantifier "there are uncountably many") has one and only one model of cardinality $\aleph_1$ up to isomorphism <u>then</u> ψ has a model in $\aleph_2$.

Proof. 1) By 3.10 we get suitable $\mathfrak{K}_1$ (as in its conclusion) and by 3.11 the class $\mathfrak{K}_1$ has a model in $\aleph_2$, hence $\mathfrak{K}$ has a model in $\aleph_2$.
2) We can replace ψ by a countable theory $T \subseteq \mathbb{L}_{\omega_1,\omega}(\mathbf{Q})$.
Let L be a fragment of $\mathbb{L}_{\omega_1,\omega}(\mathbf{Q})(\tau)$ in which T is included (e.g., L is the closure of $T \cup$(the atomic formulas) under subformulas, $\neg, \wedge, (\exists x), (\mathbf{Q}x)$; in particular L includes, of course, first order logic). By [Sh 48], without loss of generality T "says" that every formula

$\varphi(x_0, \ldots, x_{n-1})$ of L is equivalent to an atomic formula (i.e., of the form $P(x_0, \ldots, x_{n-1})$, P a predicate) and every type realized in model of T is isolated (i.e., every model is atomic), and T is complete in L. Let

$$\begin{aligned} K = \{M : & M \text{ an atomic } \tau(T)\text{-model of } T \cap \mathbb{L} \text{ and if } M \models P[\bar{a}] \\ & \text{and } (\forall \bar{x})[P(\bar{x}) \equiv \neg(\mathbf{Q}y)R(y, \bar{x})] \in T \\ & \text{then } \{b : M \models R[b, \bar{a}]\} \text{ is countable}\} \end{aligned}$$

$M \leq_{\mathfrak{K}} N$ iff $M \leq^* N$, which means:

(a) $M \prec_{\mathbb{L}} N$
(b) if $M \models P(\bar{a})$ and $\forall \bar{x}[P(\bar{x}) \equiv \neg \mathbf{Q}yR(y, \bar{x})] \in T$ <u>then</u> for no $b \in N \backslash M$ do we have $N \models R[b, \bar{a}]$.

So $\mathfrak{K} = (K, \leq_{\mathfrak{K}})$ is categorical in $\aleph_0$, is an a.e.c. and is $\mathrm{PC}_{\aleph_0}$. Let $\mathbf{F}$ be (see 3.3(8)) such that for $M \in K_{\aleph_0}, N = \mathbf{F}(M)$ iff: $M <^{**} N$ which says $M \leq_{\mathfrak{K}} N \in K_{\aleph_0}$ and if $\bar{a} \in M, M \models P[\bar{a}], \forall \bar{x}[P(\bar{x}) \equiv \mathbf{Q}yR(y, \bar{x})] \in T$, <u>then</u> for some $b \in N \backslash M$ we have $N \models R[b, \bar{a}]$. So $\mathbf{F}$ is invariant.

Note that every $M \in K^{\mathbf{F}}_{\aleph_1}$ is a model of ψ. So 3.13 gives that some $M \in K^{\mathbf{F}}_{\aleph_1}$ has a proper extension in $K^{\mathbf{F}}_{\aleph_1}$.
The rest should be easy, just as in stage (a) of the proof of 3.11. $\square_{3.18}$

<u>3.19 Question</u> 1) Under the assumptions of 3.18(2), can we get $M \in K_{\aleph_2}$, such that: if $M \models P[\bar{a}], \forall \bar{x}[P(\bar{x}) \equiv (\mathbf{Q}y)R(y, \bar{x})] \in T$ then $\{b \in M : M \models R[b, \bar{a}]\}$ has cardinality $\aleph_2$? Note that in the proof of 3.13 we show that no triple is maximal.

3.20 Remark. 1) We could have used multi-valued $\mathbf{F}$ then in the proof above $N = \mathbf{F}(M)$ just means the demand there.
2) To answer 3.19, i.e., to prove the existence of $M \in K_{\aleph_2}$ as above we have to prove:

$(*)_1$ there are $N, N_i \in K^{\mathbf{F}}_{\aleph_1}$ for $i < \omega_1$ and $N \leq_{\mathfrak{K}} N_i$ such that if $N \models P[\bar{a}]$ and the sentence $(\forall \bar{x})(P(\bar{x}) \equiv (\mathbf{Q}y)R(y, \bar{x})]$

belongs to T, then for some $i < \omega_1$ there is $b_* \in N_i \backslash N$ such that $N_i \models R[b, \bar{a}]$.
Clearly

$(*)_2$ the existence of N, N_i as in $(*)_1$ is equivalent to "ψ^* has a model" for some $\psi^* \in \mathbb{L}_{\omega_1,\omega}(\mathbf{Q})$ which is defined from $T, \leq_{\mathfrak{K}}$.

Hence

$(*)_3$ it is enough to prove that for some forcing notion $\mathbb{P}$ in $\mathbf{V}^{\mathbb{P}}$ there are N, N_i as in $(*)_1$.

There are some natural c.c.c. forcing notions tailor-made for this

$(*)_4$ consider the class of triples (M, N, a) such that $M \leq_{\mathfrak{K}} N \in K_{\aleph_0}, \bar{a} \in {}^{\omega>}N, \ell < \ell g(\bar{a}) \Rightarrow a_\ell \notin M$, order as in the proof of 3.13. By the same proof there is no maximal triple.

3) We can restrict ourselves in $(*)_2$ to

$$\{R(y, \bar{a}) : \bar{a} \in {}^{\ell g(\bar{x})}N \text{ and } \bar{a} \text{ realizes a type } p(\bar{x})\}.$$

Also we may demand $i < \omega_1 \Rightarrow N_i = N_0$ and we may try to force such a sequence of models (or pairs) and there is a natural forcing. By absoluteness it is enough to prove that it satisfies the c.c.c.

3.21 Problem: If $\mathfrak{K}$ is PC_λ, K categorical in λ and λ^+, does it necessarily have a model in λ^{++}?

Remark. The problem is proving $(*)$ of 3.12.

3.22 Question: Assume $\psi \in \mathbb{L}_{\omega_1,\omega}(\mathbf{Q})(\tau)$ is complete in $\mathbb{L}_{\omega_1,\omega}(\mathbf{Q})(\tau)$, is categorical in $\aleph_1$, has an uncountable model $M, \bar{a} \in {}^nM$ and $\varphi \in \mathbb{L}_{\omega_1,\omega}(\mathbf{Q})(\tau)$ axiomatizes the $\mathbb{L}_{\omega_1,\omega}(\mathbf{Q})(\tau)$-theory of $(M, \bar{a})$. Is φ categorical in $\aleph_1$?

3.23 Question: Can we weaken the demand on M^* in 3.13 to "M^* is a λ^+-limit model"?

§4 Forcing and categoricity

The main aim in this section is, for $\mathfrak{K}$ as in §1 with $\mathrm{LS}(\mathfrak{K}) = \aleph_0$, to find what we can deduce from $1 \le \dot{I}(\aleph_1, K) < 2^{\aleph_1}$, first without assuming $2^{\aleph_0} < 2^{\aleph_1}$.

We can build a model of cardinality $\aleph_1$ by an ω_1-sequence of countable approximations. Among those, there are models which are the union of a quite generic $<_{\mathfrak{K}}$-increasing sequence $\langle N_i : i < \omega_1 \rangle$ of countable models, so it is natural to look at them (e.g. if $\mathfrak{K}$ is categorical in $\aleph_1$, every model in $K_{\aleph_1}$ is like that). We say on such models that they are quite generic. More exactly, we look at countable models and figure out properties of the quite generic models in $\mathfrak{K}_{\aleph_1}$. The main results are 4.13(a),(f). Note that the case $2^{\aleph_0} = 2^{\aleph_1}$, though in general making our work harder, can be utilized positively - see 4.11.

A central notion is (e.g.) "the type which $\bar{a} \in {}^{\omega>}(N_1)$ materializes in (N_1, N_0)", $N_0 \le_{\mathfrak{K}} N_1 \in K_{\aleph_0}$. This is as the name indicates, the type materialized in N_1^+, which is N_1 expanded by $P^{N_1^+} = N_0$; it consists of the set of formulas forced (in the model theoretic sense started by Robinson) to satisfy; here forced is defined thinking on $(K_{\aleph_0}, \le_{\aleph_0})$ so models in $K_{\aleph_1}$ can be constructed as the union of quite generic $<_{\mathfrak{K}}$-increasing ω_1-sequence. As we would like to build models of cardinality $\aleph_1$ by such sequence, the "materialize" in (N_1, N_0) becomes realized in the (quite generic) $N \in K_{\aleph_1}$; but most of our work is in $K_{\aleph_0}$. This is also a way to express $\mathbf{Q}$ speaking on countable models.

By the hypothesis 4.8 justified by §3, the $\mathbb{L}_{\infty,\omega}(\tau_{\mathfrak{K}})$-theory of $M \in K$ is clear, in particular has elimination of quantifiers hence $M \le_{\mathfrak{K}} N \Rightarrow M \prec_{\mathbb{L}_{\infty,\omega}} N$, but for $\bar{N} = \langle N_\alpha : \alpha < \omega_1 \rangle$ as above we would like to understand (N_β, N_α) for $\alpha < \beta$ (from the point of view of $N, \bar{N}$ is not reconstructible, but its behaviour on a club is). Toward a parallel analysis of such pairs we again analyze them by $\langle L^0_\alpha : \alpha < \omega_1 \rangle$ (similarly to [Mo70]).

4.1 Convention. We fix $\lambda >$ $\mathrm{LS}(\mathfrak{K})$ as well as the a.e.c. $\mathfrak{K}$.

The main case below is here $\lambda = \aleph_1, \kappa = \aleph_0$.

4.2 Definition. For $\lambda >$ LS($\mathfrak{K}$) and $N_* \in K_{<\lambda}$ and μ, κ satisfying $\lambda \geq \kappa \geq \aleph_0, \mu \geq \kappa$ and let

1) $\mathbb{L}^0_{\mu,\kappa}$ be first order logic enriched by conjunctions (and disjunctions) of length $< \mu$, homogeneous strings of existential quantifiers or of universal quantifiers of length $< \kappa$, and the cardinality quantifier **Q** interpreted as $\exists^{\geq\lambda}$. But we apply those operations such that any formula has $< \kappa$ free variables, and the non-logical symbols are from $\tau(\mathfrak{K})$ so actually we should write $\mathbb{L}^0_{\mu,\kappa}(\tau_{\mathfrak{K}})$ but we may "forget" to say this when clear; the syntax does not depend on λ but we shall mention it in the definition of satisfaction.

2) For a logic $\mathscr{L}$ and $A_i, A \subseteq N_*$ for $i < \alpha, \alpha < \lambda$ let $\mathscr{L}(N_*, A_i; A)_{i<\alpha}$ be the language, with the logic $\mathscr{L}$, and with the vocabulary $\tau_{N_*,\bar{A},A}$ where $\bar{A} = \langle A_i : i < \alpha \rangle$ and $\tau_{N_*,\bar{A};A}$ consists of $\tau(K)$, the predicates $x \in N_*$ and $x \in A_i$ for $i < \alpha$ and the individual constants c for $c \in A$. (If $A = \emptyset$, we may omit the A; if we omit N_* then "$x \in N_*$" is omitted, if the sequence of the A_i is omitted then the "$x \in A_i$" are omitted, so $\mathscr{L}()$ means having the vocabulary $\tau(K)$). So $\mathscr{L}(N_*, A_i; A)_{i<\alpha}$ formally should have been written $\mathscr{L}(\tau_{N_*,\bar{A};A})$.

3) $\mathbb{L}^1_{\mu,\kappa}$ is defined is as in part (1), but we have also variables (and quantification) over relations of cardinality $< \lambda$. Let $\mathbb{L}^{-1}_{\mu,\kappa}$ be as in part (1) but not allowing the cardinality quantifier **Q**; this is the classical logic $\mathbb{L}_{\mu,\kappa}$.

4) $(N, N_*, A_i; A)_{i<\alpha}$ is the model N expanded to a $\tau_{N_*,\bar{A};A}$-model by monadic predicates for $N_*, A_i (i < \alpha)$ and individual constants for every $c \in A$.

5) For "$x \in N_*$", "$x \in A_i$" we use the predicates P, P_i respectively, so we may write $\mathscr{L}(\tau + P)$ instead $\mathscr{L}(N_*)$, <u>but</u> writing $\mathscr{L}(N_*)$ we fix the interpretation of P.

Let $\tau^{+\alpha} = \tau \cup \{P, P_\beta : \beta < \alpha\}$ and if $L = \mathscr{L}(\tau^{+0})$, i.e., for $\alpha = 0$ then $L(N)$ means L but we fix the interpretation of P as N, i.e., $|N|$, the set of elements of N.

Let $L(N_*, N_i)_{i \in u}$ where u a set of $< \kappa$ ordinals means the language L in the vocabulary $T \cup \{P, P_i : i \in u\}$ when we fix the interpretation of P as N_* and of $P_{\text{otp}(u \cap \alpha)}$ as N_α.

4.3 Definition. 1) For $N_* \in K_{<\lambda}$ and $\varphi(x_0, \ldots) \in \mathbb{L}^1_{\mu,\kappa}(N_*, \bar{A}; A)$ we define by induction on φ when $N_0 \Vdash^\lambda_{\mathfrak{K}} \varphi[a_0, \ldots]$ holds where

$N_* \leq_{\mathfrak{K}} N_0 \in K_{<\lambda}, a_0, \ldots$ are elements of N_0 or appropriate relations over it, depending on the kind of x_i. Pedentically we should write $(N_0, N_*, \bar{A}; A) \Vdash^{\lambda}_{\mathfrak{K}} \varphi[a_0, \ldots]$; and we may do it when not clear from the context.

For φ atomic this means $N_0 \models \varphi[a_0, \ldots]$. For $\varphi = \bigwedge_i \varphi_i$ this means

$$N_0 \Vdash^{\lambda}_{\mathfrak{K}} \varphi_i[a_0, \ldots] \text{ for each } i.$$

For $\varphi = \exists \bar{x} \psi(\bar{x}, a_0, \ldots)$ this means that for every N_1 satisfying $N_0 \leq_{\mathfrak{K}} N_1 \in K_{<\lambda}$ there is N_2 satisfying $N_1 \leq_{\mathfrak{K}} N_2 \in K_{<\lambda}$ and $\bar{b}$ from N_2 of the appropriate length (and kind) such that $N_2 \Vdash^{\lambda}_{\mathfrak{K}} \psi[\bar{b}, a]$.

For $\varphi = \neg\psi$ this means that for no N_1 do we have $N_0 \leq_{\mathfrak{K}} N_1 \in K_{<\lambda}$ and $N_1 \Vdash^{\lambda}_{\mathfrak{K}} \psi[a_0, \ldots]$.

For $\varphi(x_0, \ldots) = (\mathbf{Q}y)\psi(y, x_0, \ldots)$ this means that for every N_1 satisfying $N_0 \leq_{\mathfrak{K}} N_1 \in K_{<\lambda}$ there is N_2 satisfying $N_0 \leq_{\mathfrak{K}} N_2 \in K_{<\lambda}$ and $a \in N_2 \backslash N_1$ such that $N_2 \Vdash^{\lambda}_{\mathfrak{K}} \psi[a, a_0, \ldots]$.

2) In part (1) if $\varphi \in \mathbb{L}^1_{\mu,\kappa}(N_*)$ we can omit the demand "$N_* \leq_{\mathfrak{K}} N$" similarly below.

3) For a language $L \subseteq \mathbb{L}^1_{\mu,\kappa}(N_*, \bar{A}; A)$ and a model N satisfying $N_* \leq_{\mathfrak{K}} N \in K_{<\lambda}$ and a sequence $\bar{a} \in {}^{\lambda>}N$ the L-generic type of $\bar{a}$ in N is $\text{gtp}(\bar{a}; N_*, \bar{A}; A; N) = \{\varphi(\bar{x}) \in L : N \Vdash^{\lambda}_{\mathfrak{K}} \varphi[\bar{a}]\}$.

4) Let $\text{gtp}^{\lambda}_{L}(\bar{a}; N_*, \bar{A}; A; N)$ where $N_* \leq_{\mathfrak{K}} N \in K_\lambda$ and $L \subseteq \mathscr{L}(N_*, \bar{A}; A)$ be $\{\varphi(\bar{x}) : \varphi \in \mathscr{L}(N_*, \bar{A}; A)$ and for some $N' \in K_{<\lambda}$ we have $N \leq_{\mathfrak{K}} N' \leq_{\mathfrak{K}} N$ and $N' \Vdash^{\lambda}_{\mathfrak{K}} \varphi[\bar{a}]\}$; we may omit $\bar{A}, A$ (and omit λ if clear from the context) and may write $\mathscr{L}$ instead of $L = \mathscr{L}(N_*, \bar{A}; A)$; but note Definition 5.5.

5) We say "$\bar{a}$ materializes p (or φ)" if p (or $\{\varphi\}$) is a subset of the L-generic type of $\bar{a}$ in N.

4.4 Definition. Let $N_i (i < \lambda)$ be an increasing (by $\leq_{\mathfrak{K}}$) continuous sequence, $N = \bigcup_{i<\lambda} N_i$, $\|N_i\| < \lambda$ and $L^* \subseteq \bigcup_{\alpha<\kappa} \mathbb{L}^1_{\infty,\kappa}(\tau^{+\alpha})$.

1) N is L^*-generic, if for any formula $\varphi(x_0, \ldots) \in L^* \cap \mathbb{L}^1_{\infty,\kappa}(\tau_{\mathfrak{K}})$ and $a_0, \ldots \in N$ we have:
$N \models \varphi[a_0, \ldots] \Leftrightarrow$ for some $\alpha < \lambda$, $N_\alpha \Vdash^{\lambda}_{\mathfrak{K}} \varphi[a_0, \ldots]$.

2) The $\leq_{\mathfrak{K}}$-presentation $\langle N_i : i < \lambda \rangle$ of N is L^*-generic <u>when</u> for any

$\alpha < \lambda$ of cofinality $\geq \kappa$ and $\psi(x_0, \dots) \in L^*(N_\alpha, N_i)_{i \in I}$ satisfying $I \subseteq \alpha, |I| < \kappa$ and $a_0, \dots \in N$ we have:

$$N \models \psi[a_0, \dots] \Leftrightarrow \text{ for some } \gamma < \lambda, N_\gamma \Vdash^\lambda_{\mathfrak{K}} \psi[a_0, \dots]$$

and for each $\beta \geq \alpha$, with cofinality $\geq \kappa$, N_β is almost $L^*(N_\alpha, N_i)_{i \in I}$-generic (see part (5)).
3) N is strongly L^*-generic <u>if</u> it has an L^*-generic presentation (in this case, if λ is regular, then for any presentation $\langle N_i : i < \lambda \rangle$ of N there is a closed unbounded $E \subseteq \lambda$ such that $\langle N_i : i \in E \rangle$ is an L^*-generic presentation).
4) We say that $N \in K_{<\lambda}$ is pseudo L^*-generic <u>if</u>

(a) for every $\varphi(\bar{x}) = \exists \bar{y} \psi(\bar{x}, \bar{y}) \in L^*$, if $N \Vdash^\lambda_{\mathfrak{K}} \varphi(\bar{a})$ then for some $\bar{b}$, $N \Vdash^\lambda_{\mathfrak{K}} \psi(\bar{a}, \bar{b})$

(b) for every $\bar{a} \in N$, $\bar{a}$ materializes in N some complete L^*-type.

5) We add "almost" to any of the above defined notions <u>when</u>: for $\Vdash^\lambda_{\mathfrak{K}}$, the inductive definitions of satisfaction works except possibly for $\mathbf{Q}$ (e.g., $N \Vdash^\lambda_{\mathfrak{K}} \exists x \varphi(x, \dots)$ iff for some $a \in N$, $N \Vdash^\lambda_{\mathfrak{K}} \varphi(a, \dots)$).

4.5 Remark. 1) Notice we can choose $N_i = N_0 = N$, so $\|N\| < \lambda$. In particular almost (and pseudo) L^*-generic models of cardinality $< \lambda$ may well exist.
2) Here we concentrate on $\lambda = \aleph_1$ and fragments of $\mathbb{L}^0_{\infty,\omega}$ (mainly $\mathbb{L}^0_{\omega_1,\omega}$ and its countable fragments).
3) There are obvious implications, and forcing is preserved by isomorphism and replacing $N (\in K_{<\lambda})$ by N', $N \leq_{\mathfrak{K}} N' \in K_{<\lambda}$.

There are obvious theorems on the existence of generic models, e.g.,

4.6 Theorem. *1) Assume $N_0 \in K_{<\lambda}, \lambda = \mu^+, \mu^{<\kappa} = \mu, L \subseteq \bigcup_{\alpha < \kappa} \mathbb{L}_{\infty,\kappa}(\tau^{+\alpha})$ and L is closed under subformulas and $|L| < \lambda$. <u>Then</u> there are $N_i (i < \lambda)$ such that $\langle N_i : i < \lambda \rangle$ is an L-generic representation of $N = \bigcup_{i<\lambda} N_i$, (hence N is strongly L-generic).*
2) In part (1), $N \in K_\lambda$ if no N', $N_0 \leq_{\mathfrak{K}} N' \in K_{<\lambda}$ is $\leq_{\mathfrak{K}}$-maximal.

Proof. Straightforward. $\square_{4.6}$

4.7 Remark. 1) If $L = \bigcup_{i<\lambda} L_i, |L_i| < \lambda$, then we can get "$\langle N_i : j < i < \lambda\rangle$ is an L_j-generic representation of N for each $j < \lambda$".
2) When we speak on "complete L-type p" we mean $p = p(x_0, \dots, x_{n-1})$ for some n.

From time to time we add some hypothesis and prove a series of claims; such that the hypothesis holds, at least without loss of generality in the case we are interested in. We are mainly interested in the case $\dot{I}(\aleph_1, \mathfrak{K}) < 2^{\aleph_1}$, etc., so by 3.10, 3.17 it is reasonable to make:

4.8 Hypothesis. $\mathfrak{K}$ is $\mathrm{PC}_{\aleph_0}$, $\leq_{\mathfrak{K}}$ refines $\mathbb{L}_{\infty,\omega}$ and $\mathfrak{K}$ is categorical in $\aleph_0$ and $1 \leq \dot{I}(\aleph_1, K)$ and $\dot{I}(\aleph_1, K^{\mathbf{F}}_{\aleph_1}) < 2^{\aleph_1}$ where $K^{\mathbf{F}}_{\aleph_1}$ is as in Definition 3.14 and is $\mathrm{PC}_{\aleph_0}$ or just $\mathbf{K}^{\mathbf{F}}_{\aleph_1} = \{M \restriction \tau_{\mathfrak{K}} : M \models \psi\}$ for some $\psi \in \mathbb{L}_{\omega_1,\omega}(\mathbf{Q})$ (if $\mathbf{F}$ is invariant, this follows).

4.9 Remark. 0) We can add: every $M \in K_{\aleph_0}$ is atomic (model of $\mathrm{Th}_{\mathbb{L}}(M)$).
1) Usually below we ignore the case $\dot{I}(\aleph_1, \mathfrak{K}) < 2^{\aleph_0}$ as the proof is the same.
2) We can deal similarly with the case $1 \leq \dot{I}(\aleph_1, K') < 2^{\aleph_0}$ where $\mathfrak{K}_{\aleph_1} \subseteq K'_{\aleph_1} \subseteq \{M \in \mathfrak{K}_{\aleph_1} : M$ is strongly L_*-generic$\}$ and K' is $\mathrm{PC}_{\aleph_0}$ (or less: $\{M \restriction \tau_{\mathfrak{K}} : M$ a model of $\psi \in \mathbb{L}_{\omega_1,\omega}(\mathbf{Q})(\tau^*)\}$).
3) Can we use $\mathbf{F}$ a function with domain $K_{\aleph_0}$ such that $M \leq_{\mathfrak{K}} \mathbf{F}(M_0) \in K_{\aleph_0}$ for $M \in K_{\aleph_0}$ without the extra assumptions or even $\mathbf{F} : \{\bar{M} = \langle M_i : i \leq \alpha\rangle$ is $\leq_{\mathfrak{K}_{\aleph_0}}$-increasing continous$\} \to \mathfrak{K}_{\aleph_0}$ such that $M_\alpha \leq_{\mathfrak{K}} \mathbf{F}(M_i : i \leq \alpha\rangle)$? We cannot use the non-definability of well ordering (see 3.10(3)); (as in the proof of (f) of 4.13).

4.10 Claim. *1) If $\bar{a} \in N \in K_{\aleph_0}$ and $\varphi(\bar{x}) \in \mathbb{L}^0_{\infty,\omega}(\tau^{+0})$ (so $\bar{a}$ is a finite sequence)* <u>then</u> *$(N, N) \Vdash^{\aleph_1}_{\mathfrak{K}} \varphi[\bar{a}]$ or $(N, N) \Vdash^{\aleph_1}_{\mathfrak{K}} \neg\varphi[\bar{a}]$ (i.e. P is interpreted as N).*
2) If $(N, N) \Vdash^{\aleph_1}_{\mathfrak{K}} \exists \bar{x} \bigwedge p(\bar{x})$, where $p(\bar{x})$ is a not necessarily complete

n-type ($n = \ell g(\bar{x})$) in L where $L \subseteq \mathbb{L}^0_{\omega_1,\omega}(\tau^{+0})$ is countable, <u>then</u> for some complete n-type q in L extending p we have $(N, N) \Vdash^{\aleph_1}_{\mathfrak{K}} \exists \bar{x} \wedge q(\bar{x})$.

Proof. 1) Suppose not, for each $S \subseteq \omega_1$, we define by induction on α, $N^S_\alpha \in K_{\aleph_0} (\alpha < \omega_1)$, increasing (by $\leq_{\mathfrak{K}}$) and continuous, $N^S_0 = N$ and for limit α, $N^S_\alpha = \bigcup_{\beta<\alpha} N^S_\beta$. For $\alpha = 2\beta + 1$ remember that $(N^S_\beta, \bar{a}) \cong (N, \bar{a})$ because $N = N_0 \leq_{\mathfrak{K}} N^S_\beta$ hence $N_0 \prec_{\mathbb{L}_{\infty,\omega}} N^S_\beta \in K_{\aleph_0}$ hence $(N^S_\beta, \bar{a}) \equiv_{\mathbb{L}_{\infty,\omega}} (N, \bar{a})$ hence they are isomorphic. So (N^S_β, N^S_β) forces ($\Vdash^{\aleph_1}_{\mathfrak{K}}$) neither $\varphi[\bar{a}]$ nor $\neg\varphi[\bar{a}]$. So there are M_ℓ (for $\ell = 0, 1$) such that $N^S_\beta \leq_{\mathfrak{K}} M_\ell \in K_{\aleph_0}$ and $(M_0, N^S_\beta) \Vdash^{\aleph_1}_{\mathfrak{K}} \varphi[\bar{a}]$ but $(M_1, N^S_\beta) \Vdash^{\aleph_1}_{\mathfrak{K}} \neg\varphi[\bar{a}]$. Now if $\beta \in S$ we let $N^S_\alpha = M_0$ and if $\beta \notin S$ we let $N^S_\alpha = M_1$.

Lastly, $M_{2\beta+2} = \mathbf{F}(M_{2\beta+1})$ recalling $\mathbf{F}$ is from 4.8. Let $N^S = \bigcup_{\alpha<\omega_1} N^S_\alpha$. Now if $S(0) \backslash S(1)$ is stationary then $(N^{S(0)}, \bar{a}) \ncong (N^{S(1)}, \bar{a})$. Why? Because if $f : N^{S(0)} \to N^{S(1)}$ is an isomorphism from $N^{S(0)}$ onto $N^{S(1)}$ mapping $\bar{a}$ to $\bar{a}$ then for some closed unbounded set $E \subseteq \omega_1$, we have: if $\alpha \in E$ then f maps $N^{S(0)}_\alpha$ onto $N^{S(1)}_\alpha$, so choose some $\alpha \in E \cap S(0) \backslash S(1)$ and choose $\beta \in E \backslash (\alpha + 1)$. Now $(N^{S(0)}_{\alpha+1}, N^{S(0)}_\alpha) \Vdash^{\aleph_1}_{\mathfrak{K}} \varphi[\bar{a}]$, hence $(N^{S(0)}_\beta, N^{S(0)}_\alpha) \Vdash^{\aleph_1}_{\mathfrak{K}} \varphi[\bar{a}]$, and similarly $(N^{S(1)}_\beta, N^{S(1)}_\alpha) \Vdash^{\aleph_1}_{\mathfrak{K}} \neg\varphi(\bar{a})$, but $f \restriction N^{S(0)}_\beta$ is an isomorphism from $N^{S(0)}_\beta$ onto $N^{S(1)}_\beta$ mapping $N^{S(0)}_\alpha$ onto $N^{S(1)}_\alpha$ and $\bar{a}$ to itself and we get a contradiction. By 0.3, we get $\dot{I}(\aleph_1, K) = 2^{\aleph_1}$, contradiction.
2) Easy by 4.6 and part (1). In detail, if $N \leq_{\mathfrak{K}} M_1 \in \mathfrak{K}_{\aleph_0}$ then by the definition of $\Vdash^{\aleph_1}_{\mathfrak{K}}$ and the assumption we can find $(M_2, \bar{a})$ satisfying $M_1 \leq_{\mathfrak{K}} M_2 \in \mathfrak{K}_{\aleph_0}$ and $\bar{a} \in M_2$ such that $(M_2, N) \Vdash^{\aleph_1}_{\mathfrak{K}} \wedge p(\bar{a})$. As L is countable and the definition of $\Vdash^{\aleph_1}_{\mathfrak{K}}$ without loss of generality for every formula $\varphi(\bar{x}) \in L$, $(M_2, N) \Vdash^{\aleph_1}_{\mathfrak{K}} \varphi[\bar{a}]$ or $(M_2, N) \Vdash^{\aleph_1}_{\mathfrak{K}} \neg\varphi[\bar{a}]$. (Why? Simply let $\langle \varphi_n(\bar{x}) : n < \omega \rangle$ list the formulas $\varphi(\bar{x}) \in L$ and choose $M_{2,n} \in \mathfrak{K}_{\aleph_0}$ by induction on n such that $M_{2,0} = M_2$, $M_{2,n} \leq_{\mathfrak{K}} M_{2,n+1}$ such that $(M_{2,n+1}, N) \Vdash^{\aleph_1}_{\mathfrak{K}} \varphi_n(\bar{x})$ or $(M_{2,n+1}, N) \Vdash^{\aleph_1}_{\mathfrak{K}} \neg\varphi_n(\bar{x})$; now replace M_2 by $\cup\{M_{2,n} : n < \omega\}$). Recalling Definition 4.3(4), let $q = \text{gtp}_{L(N)}(\bar{a}, N, M_2)$, it is a complete $(L(N), n)$-type. So clearly

$(M_2, N) \Vdash^{\aleph_1}_{\mathfrak{K}} (\exists \bar{x}) \wedge q(\bar{x})$. Now apply the proof of part (1) to the formula $(\exists \bar{x}) \wedge q(\bar{x})$ so we are done. $\square_{4.10}$

4.11 Claim. *For each countable $L \subseteq \mathbb{L}^0_{\omega_1,\omega}(\tau^{+0})$ and $N \in K_{\aleph_0}$ the number of complete $L(N)$-types p (with no parameters) such that $N \Vdash^{\aleph_1}_{\mathfrak{K}} (\exists \bar{x}) \wedge p(\bar{x})$, is countable.*

Proof. At first glance it seemed that 0.2 will imply this trivially. However, here we need the parameter N as an interpretation of the predicate P and if $2^{\aleph_0} = 2^{\aleph_1}$ there are too many choices. So we shall deal with "every N_α in some presentation". Suppose the conclusion fails. First we choose by induction N_α (for $\alpha < \omega_1$) such that:

- (i) $N_\alpha \in K_{\aleph_0}$ is $\leq_{\mathfrak{K}}$-increasing and $\langle N_\alpha : \alpha < \omega_1 \rangle$ is L-generic
- (ii) for each $\beta < \alpha$, there is $a^\beta_\alpha \in N_{\alpha+1} \backslash N_\alpha$ materializing an $L(N_\beta)$-type not materialized in N_α, (i.e. in (N_α, N_β); see Definition 4.3(2) on materialize), (possible by 4.10 and our assumption toward contradiction)
- (iii) $|N_\alpha| = \omega\alpha$
- (iv) for $\alpha < \beta, N_\beta$ is pseudo $L(N_\alpha)$-generic and $\mathbf{F}(N_{2\beta+1}) \leq_{\mathfrak{K}} N_{2\beta+2}$.

Now let $N = \cup\{N_\alpha : \alpha < \omega_1\}$ and we expand N by all relevant information: the order $<$ on the countable ordinals, $c (c \in N_0)$, enough "set theory", "witness" for $N_\beta \leq_{\mathfrak{K}} N_\alpha$ for $\beta < \alpha$ and the 2-place functions $F, F(\beta, \alpha) = a^\beta_\alpha$ and lastly witnesses of $\mathbf{F}(N_{2\beta+1}) \leq_{\mathfrak{K}} N_{2\beta+2}$ recalling $\mathbf{F}$ is quite definable by Definition 4.8 and names for all formulas in $L(N_\alpha)$ (with α as a parameter), i.e., the relations $R_{\varphi(\bar{x})} = \{\langle \alpha \rangle \hat{\ } \bar{a} : \alpha < \omega_1, \bar{a} \in {}^{\ell g(x)}N$ and for every $\beta < \omega_1$ large enough $(N_\beta, N_\alpha) \Vdash^{\aleph_1}_{\mathfrak{K}}$ "$\varphi(\bar{a})$"$\}$ for $\varphi(\bar{x}) \in L$. Clearly for every $\alpha < \omega_1, \varphi(\bar{x}) \in L(N_\alpha)$ and $\bar{a} \in {}^{\ell g(\bar{x})}N$ we have $(N, N_\alpha) \models \varphi[\bar{a}]$ iff for every $\beta < \omega_1$ large eough we have $(N_\beta, N_\alpha) \Vdash^{\aleph_1}_{\mathfrak{K}}$ "$\varphi[\bar{a}]$". We get a model $\mathfrak{B}$ with countable vocabulary and $\psi \in \mathbb{L}_{\omega_1,\omega}(\mathbf{Q})$ expressing all this. By 0.2(1) applied to the case $\Delta = L$, there are models $\mathfrak{B}_i$ (for $i < 2^{\aleph_1}$) of cardinality $\aleph_1$ (note $N_0 \leq_{\mathfrak{K}} \mathfrak{B} \restriction \tau_{\mathfrak{K}}$), so that the set of $L(N_0)$-types realizes in N^i (the $\tau(K)$-reduct of $\mathfrak{B}_i$) are distinct for distinct i's. So $(N^i, c)_{c \in N_0}$ are pairwise non-isomorphic. If $2^{\aleph_0} < 2^{\aleph_1}$ we finish by 0.3.

So we can assume $2^{\aleph_0} = 2^{\aleph_1}$. In N, uncountably many complete $L(N_0)$-n-types are realized hence by 0.2(2) the set $\{p : p$ a complete $L(N_0) - m$-type is realized in some $N', N_0 \leq_{\mathfrak{K}} N' \in \mathfrak{K}_{\aleph_1}$ for some $m < \omega\}$ has cardinality continuum, hence by 4.10 the set of complete $L(N_0)$-types $p = p(x)$ such that $(N_0, N_0) \Vdash^{\aleph_1}_{\mathfrak{K}} \exists \bar{x} \wedge p(\bar{x})$ has cardinality $2^{\aleph_0}$. So we choose by induction on $\alpha < 2^{\aleph_0}$ a sequence $\langle N^\alpha_i, a^\alpha_i : i < \omega_1\rangle$ such that:

(a) $N^\alpha_i \in \mathfrak{K}_{\aleph_0}$

(b) $N^\alpha_{i_0} \leq_{\mathfrak{K}} N^\alpha_i$ for $i_0 < i < \omega_1$

(c) $a^\alpha_i \in N^\alpha_{i+1} \backslash N^\alpha_i$ materialize a complete $L(N^\alpha_i)$-type p^α_i

(d) if $j < \omega_1$ is a limit ordinal then $N^\alpha_j = \cup\{N^\alpha_i : i < j\}$

(e) $p^\alpha_i \notin \{\text{gtp}(\bar{a}; N^\beta_{j_1}; N^\beta_{j_2}) : j_1 < j_2 < \omega_1, \bar{a} \in {}^{\omega>}(N^\beta_{j_2})$ and $\beta < \alpha\}$ (see Definition 4.3(4))

(f) $\mathbf{F}(N_{2\beta+1}) \leq_{\mathfrak{K}} N_{2\beta+2}$.

As $\aleph_1 < 2^{\aleph_1} = 2^{\aleph_0}$ this is possible, i.e., in clause (e) we should find a type which is not in a set of $\leq \aleph_1 \times |\alpha| < 2^{\aleph_0}$ types, as the number of possibilities is $2^{\aleph_0}$; let $N_\alpha = \cup\{N^\alpha_i : i < \omega_1\}$ for $\alpha < 2^{\aleph_0}$, clearly $N_\alpha \in K_{\aleph_1}$. Now toward contradiction if $\beta < \alpha < 2^{\aleph_0}$ and $N_\alpha \cong N_\beta$ then there is an isomorphism f from N_α onto N_β; necessarily f maps N^α_i onto N^β_i for a club of i. For any such $i, p^\alpha_i \in \text{gtp}_L(f(\bar{a}^\alpha_i); N^\beta_i; N^\beta_j)$ for j large enough, contradiction. $\square_{4.11}$

4.12 Remark. In the proof of 4.11(2), we can fix m and we can combine the two cases, when for $N \in K^{\mathbf{F}}_{\aleph_1}$ represent by $\langle N_\alpha : \alpha < \omega_1\rangle$ we consider $\mathbf{P}_N = \{p : p$ a complete $L - m$-type such that for a club of $\alpha < \omega_1$ for some $\beta \in (\alpha, \omega_1)$ and $\bar{a} \in {}^m(N_\beta)$ materialize p in $(N_\beta, N_\alpha)\}$, can replace "club" by "stationarily many". That is we can prove that $\{\mathbf{P}_N : N \in K^{\mathbf{F}}_{\aleph_1}\}$ has cardinality $2^{\aleph_1}$.

4.13 Lemma. *1) There are countable $L^0_\alpha \subseteq \mathbb{L}^0_{\omega_1,\omega}(\tau^{+0})$ for $\alpha < \omega_1$ increasing continuous in α, closed under finitary operations and subformulas such that, letting $L^0_{<\omega_1} = \cup\{L^0_\alpha : \alpha < \omega_1\}$ we have (some clauses do not metion the L^0_α's):*

(a) *for each* $N \in K_{\aleph_0}$ *and every complete* $L^0_\alpha(N)$*-type* $p(\bar{x})$ *we have* $N \Vdash^{\aleph_1}_{\mathfrak{K}} (\exists \bar{x}) \wedge p(\bar{x}) \Rightarrow \wedge p \in L^0_{\alpha+1}(N)$. *Hence for every* $\mathbb{L}^0_{\omega_1,\omega}(\tau^{+0})$*-formula* $\psi(\bar{x})$ *there are formulas* $\varphi_n(\bar{x}) \in L^0_{<\omega_1}$ *for* $n < \omega$ *such that* $(N, N) \Vdash^{\aleph_1}_{\mathfrak{K}} (\forall \bar{x})[\psi(\bar{x}) \equiv \bigvee_n \varphi_n(\bar{x})]$

(b) *for every* $N_0 \leq_{\mathfrak{K}} N_1 \in K_{\aleph_0}$ *there is* $N_2, N_1 \leq_{\mathfrak{K}} N_2 \in K_{\aleph_0}$, *such that for every* $\bar{a} \in N_2$ *and* $\varphi(\bar{x}) \in \mathbb{L}^0_{\omega_1,\omega}(N_0)$, *of course with* $\ell g(\bar{a}) = \ell g(\bar{x}) < \omega$, *we have* $(N_2, N_0) \Vdash^{\aleph_1}_{\mathfrak{K}} \varphi[\bar{a}]$ *or* $(N_2, N_0) \Vdash^{\aleph_1}_{\mathfrak{K}} \neg\varphi[\bar{a}]$

(c) *If* $N \leq_{\mathfrak{K}} N_\ell \in K_{\aleph_0}(\ell = 1, 2), \bar{a}_\ell \in N_\ell$ *and the* $L^0_{<\omega_1}(N)$*-generic types of* $\bar{a}_\ell$ *in* N_ℓ *are equal (though they are not necessarily complete; i.e., for every* $\varphi(\bar{x}) \in L^0_{<\omega_1}(N)$ *we have* $N_1 \Vdash^{\aleph_1}_{\mathfrak{K}} \varphi(\bar{a}_1)$ *iff* $N_2 \Vdash^{\aleph_1}_{\mathfrak{K}} \varphi[\bar{a}_2]$), *<u>then</u> so are the* $\mathbb{L}^0_{\infty,\omega}(N)$*-generic types. In fact, there is* $M, N \leq_{\mathfrak{K}} M$ *and* $\leq_{\mathfrak{K}}$*-embeddings* $f_\ell : N_\ell \to M$ *such that* f_ℓ *maps* N *onto itself and* $f_1(\bar{a}_1) = f_2(\bar{a}_2)$ *though we do not claim* $f_1 \restriction N = f_2 \restriction N$. *Also if* $N_1 = N_2$ *then there is* $M \in K_{\aleph_0}$ *which* $\leq_{\mathfrak{K}}$*-extends* N_1 *and an automorphism* f *of* M *mapping* N *onto itself and* $\bar{a}_1$ *to* a_2.

(d) *For each* $N \in K_{\aleph_0}$ *and complete* $\mathbb{L}^0_{\omega_1,\omega}(N)$*-type* $p(\bar{x})$, *the class* $K^1 := \{(N, M, \bar{a}) : M \in K_{\aleph_0}, N \leq_{\mathfrak{K}} M$ *and for some* $M', M \leq_{\mathfrak{K}} M' \in K_{\aleph_0}$ *and* $\bar{a}$ *materialize* p *in* $(M; N)\}$ *is a* $PC_{\aleph_0}$*-class.*

(e) *for any complete* $\mathbb{L}^{-1}_{\omega_1,\omega}(N)$*-type* $p(\bar{x})$,
for some complete $\mathbb{L}^0_{\omega_1,\omega}(N)$*-type* q_p, *if* $N \leq_{\mathfrak{K}} M \in K_{\aleph_0}, \bar{a} \in M$ *and* $\bar{a}$ *materialize* p *in* (M, N), *<u>then</u>* $\bar{a}$ *materialize* q_p *in* (M, N); *on* $\mathbb{L}^0, \mathbb{L}^{-1}$ *see Definition 4.2(1),(3)*

(f) *the number of complete* $\mathbb{L}^0_{\omega_1,\omega}(N)$*-types* p *which for some* $\bar{a}, M$ *we have* $\bar{a} \in {}^{\omega>}M, M \in K_{\aleph_0}, N \leq_{\mathfrak{K}} M$ *and* $\bar{a}$ *materialize in* (M, N) *is* $\leq \aleph_1$

(g) *if in clause (f) we get that there are* $\aleph_1$ *such types then* $\dot{I}(\aleph_1, K) \geq \aleph_1$

(h) *let* $L^{-1}_\alpha := L^0_\alpha \cap \mathbb{L}^{-1}_{\omega_1,\omega}(\tau^{+0})$ *then the parallel clauses to (a)-(g) holds.*

2) Clause (e) means that

(*i*) *assume further that* $N_0 \leq_{\mathfrak{K}} N_\ell \in K_{\aleph_0}$ *for* $\ell = 1, 2$ *and* $\bar{a}_\ell \in N_\ell$ *and the* $L^{-1}_{<\omega_1}(N)$*-type which* $\bar{a}_1$ *materializes in* N_1 *is equal to the* $L^{-1}_{<\omega_1}(N)$*-type which* $\bar{a}_2$ *materializes in* N_2. *Then we can find* N_1^+, N_2^+ *such that* $N_\ell \leq_{\mathfrak{K}} N_\ell^+ \in K_{\aleph_0}$ *for* $\ell = 1, 2$ *and isomorphism* f *from* N_1^+ *onto* N_2^+ *mapping* N *onto itself and* $\bar{a}_1$ *to* $\bar{a}_2$.

4.14 Remark. 1) We cannot get rid of the case of $\aleph_1$ types (but see 5.22, 5.27) by the following variant of a well known example of Morley [Mo70] for $\dot{I}(\aleph_0, K) = \aleph_2$. For let $K = \{(A, E, <) : E$ an equivalence relation on A, each E-equivalence class is countable, $x < y \Rightarrow xEy$ and on each E-equivalence class $<$ is a 1-transitive linear order, i.e. $xEy \Rightarrow (x/E, <, x) \cong (y/E, <, y)\}$ and $M \leq_{\mathfrak{K}} N$ if $M \subseteq N$ and $[x \in M \wedge y \in N \wedge xEy \Rightarrow y \in M]$. By the analysis of such countable linear orders, each $(a/E^M, <)$ up to isomorphism is determined by $(\alpha, \ell) \in \omega_1 \times 2$. For appropriate $\mathbf{F}$, if $M = \mathbf{F}(N), a \in N$ and I is an interval of $(a/E^N, <^N)$ which is 1-transitive then for some $b \in M \backslash N, (b/E^M, <^M)$ is isomorphic to $(I, <^N)$. This is enough.
2) In clauses (c),(i) of 4.13 the mapping are not necessarily the identity on N. In clause (i) the assumption is apparently weaker (those by its conclusion the assumption of (c) holds).
3) Note that clause (f) of 4.13 does not follow from clause (a) as there may be $\aleph_1$-Kurepa trees.
4) In clause (c) of 4.13 for the second sentence we can weaken the assumption: if $\varphi(\bar{x}) \in L^0_{<\omega_1}(N)$ and $(N_1; N) \nVdash^{\aleph_1}_{\mathfrak{K}} \varphi(\bar{a}_1)$ then $(N_2, N) \nVdash^{\aleph_1}_{\mathfrak{K}} \varphi(\bar{a}_2)$. This is enough to get the $M_{1,\alpha}, M_{2,\alpha}$ from the proof. (Why? For each $\alpha < \omega_1$, there are $M_{1,\alpha}$ such that $N_1 \leq_{\mathfrak{K}} M_{1,\alpha} \in K_{\aleph_0}$ and a complete $L^0_\alpha - \ell g(\bar{a}_i)$-type $p_*(\bar{x})$ such that $(M_{1,\alpha}, N) \Vdash \wedge p_*(\bar{a}_1)$. But $\neg \wedge p_1(\bar{x}) \in L_{\alpha+1}$ and obviously $(N_1, N) \nVdash \neg \wedge p_*(\bar{a}_1)$ hence $(N_2, N) \nVdash^{\aleph_1}_{\mathfrak{K}} \neg \wedge p_*(\bar{a}_2)$ hence there is $M_{2,\alpha}$ such that $N_2 \leq_{\mathfrak{K}} M_{2,\alpha} \in K_{\aleph_0}$ and $(M_{2,\alpha}; N) \Vdash^{\aleph_1}_{\mathfrak{K}} \wedge p_*[\bar{a}_2]$. Now continue as in the proof below).

Remark. We can prove clause (b) and the last sentence in clause (c) of 4.13 directly not mentioning the L^0_α-s.

Proof. Note that proving clause (e) we say "repeat the proof of clause (a),(b),(c),(d) for $L^{-1}_{\omega,\omega}$".

Clause (a): We choose L^0_α by induction on α using 4.11. The second phrase is proved by induction on the depth of the formula using 4.10.

Clause (b): By iterating ω times, it suffices to prove this for each $\bar{a} \in N_1$, so again by iterating ω times it suffices to prove this for a fix $\bar{a} \in N_1$.

If the conclusion fails we can define by induction on $n < \omega$ for every $\eta \in {}^n 2$, a model M_η and $\varphi_\eta(\bar{x}) \in \mathbb{L}^0_{\omega_1,\omega}(N)$ such that:

(i) $M_{<>} = N_1$

(ii) $M_\eta \leq_{\mathfrak{K}} M_{\eta\char`\^<\ell>} \in K_{\aleph_0}$ for $\ell = 0, 1$

(iii) $(M_\eta, N) \Vdash^{\aleph_1}_{\mathfrak{K}} \varphi_\eta(\bar{a})$

(iv) $\varphi_{\eta\char`\^<1>}(\bar{x}) = \neg\varphi_{\eta\char`\^<0>}(\bar{x})$.

Now for $\eta \in {}^\omega 2$, let $M_\eta = \bigcup\limits_{n<\omega} M_{\eta\restriction n}$. Clearly for $\eta \in {}^\omega 2$ we have $M_\eta \Vdash^{\aleph_1}_{\mathfrak{K}} (\exists \bar{x})[\bigwedge\limits_{n<\omega} \varphi_{\eta\restriction n}(\bar{x})]$ and, after slight work, we get contradiction to 4.11 + 4.10.

Clause (c): In general by clause (a) for each $\alpha < \omega_1$ we can find $M^\alpha_\ell \in K_{\aleph_1}$ for $\ell = 1, 2$ such that $N_\ell \leq_{\mathfrak{K}} M^\alpha_\ell$ and $(M^\alpha_1, \bar{a}_1), (M^\alpha_2, \bar{a}_2)$ are $L^0_\alpha(N)$-equivalent and without loss of generality each of N, N_ℓ, M^α_ℓ have universe an ordinal $< \omega_1$. Let $\mathfrak{A} = (\mathscr{H}(\aleph_2), N, N_1, N_2, \langle M^\alpha_1 : \alpha < \omega_1\rangle, \langle M^\alpha_2 : \alpha < \omega_1\rangle)$ let $\mathfrak{A}_1 \prec \mathfrak{A}$ be countable and recalling 0.4(3) find a non-well ordered countable model $\mathfrak{A}_2$, which is an end extension of $\mathfrak{A}_1$ for $\omega_1^{\mathfrak{A}_1}$, hence $\omega^{\mathfrak{A}_2} = \omega$ so $N^{\mathfrak{A}_2} = N, N^{\mathfrak{A}_2}_\ell = N_\ell$ for $\ell = 1, 2$. For $x \in (\omega_1)^{\mathfrak{A}_2} \backslash \mathfrak{A}_1$ let $M^x_\ell = (M^x_\ell)^{\mathfrak{A}_2}$ so $N_\ell \leq_{\mathfrak{K}} M^x_\ell \in K_{\aleph_0}$. Now there are x_n such that $\mathfrak{A}_2 \models$ "$x_{n+1} < x_n$ are countable ordinals"; so using the hence and forth argument $(M^{x_0}_1, \bar{a}_1, N) \cong (M^{x_0}_2, \bar{a}_2, N)$.
[Why? Let $\mathscr{F}_n = \{(\bar{b}^1, \bar{b}^2) : \bar{b}^\ell \in {}^n(M^{x_0}_\ell)$ and $\mathfrak{A}_2 \models \mathrm{gtp}_{L^0_{x_n}}(\bar{a}^1\char`\^\bar{b}^1, N; M^{x_0}_1) = \mathrm{gtp}_{L^0_{x_n}}(\bar{a}^2\char`\^\bar{b}^2; N; M^{x_0}_2)\}$. Clearly $(<>, <>) \in \mathscr{F}_0$ and if $(\bar{b}^1, \bar{b}^2) \in \mathscr{F}_n, \ell \in \{1, 2\}$ and $b^\ell_n \in M^{x_0}_\ell$ then there is $b^{3-\ell}_n \in M^{x_0}_{3-\ell}$ such that $(\bar{b}^1\char`\^\langle b^1_n\rangle, \bar{b}^2\char`\^\langle b^2_n\rangle) \in \mathscr{F}_{n+1}$. As $M^{x_0}_1, M^{x_0}_2$ are countable we

can find an isomorphism.]
But this is as required in the second phrase of (c).

We still have to prove the first phrase. For this we prove by induction on the ordinal α that

- $\circledast^1_\alpha$ if for $\ell = 1, 2, \bar{a}_\ell \in {}^{\omega>}(N_\ell)$ materialize in (N_ℓ, N_*) a complete $L^0_{<\alpha}$-type $p(\bar{x})$ not depending on ℓ and $\varphi(\bar{x}) \in \mathbb{L}^0_{\infty,\omega}(N_*)$ has quantifier depth $< \alpha$ then: $\ell \in \{1,2\} \Rightarrow (N_\ell, N_*) \Vdash^{\aleph_1}_{\mathfrak{K}} \varphi(\bar{a}_\ell)$ <u>or</u> $\ell \in \{1,2\} \Rightarrow (N_\ell, N_*) \Vdash^{\aleph_1}_{\mathfrak{K}} \neg\varphi(\bar{a}_\ell)$.

For countable $N \leq_{\mathfrak{K}} M$ and $\bar{a} \in {}^{\omega>}N$

- $\odot_1$ let $\mathbf{P}_\alpha(N, M, \bar{a}) = \{\mathrm{gtp}_{L^0_{<\alpha}}(\bar{a}; N; M^\perp) : M \leq_{\mathfrak{K}} M^+ \in K_{\aleph_0}$ and $\mathrm{gtp}_{L^0_\alpha}(\bar{a}; N; M^+)$ is a complete L^0_α-type$\}$.

Now

- $\odot_2$ for $\beta < \alpha < \omega_1$, from $\mathrm{gtp}_{L^0_\alpha}(\bar{a}; N; M)$ we can complete $\mathbf{P}_\beta(N, M, \bar{a})$
- $\odot_3$ for $\alpha < \omega_1$, from $\mathbf{P}_\beta(N, M, \bar{a})$ we can compute $\mathrm{gtp}_{L^0_\alpha}(\bar{a}; N; M)$
- $\odot_4$ assume $N \leq_{\mathfrak{K}} M$ are countable and $\bar{a} \in {}^{\omega>}M$; for $\varphi(\bar{x}) \in L^0_{\omega_1,\omega}(N)$ of quantifier depth $< \alpha$ we have:
  $\varphi(\bar{x}) \in \mathrm{gtp}_{\mathbb{L}^0_{\omega_1,\omega}(N)}(\bar{a}; N; M)$, <u>iff</u> for every $q(\bar{x}) \in \mathbf{P}_\alpha(N, M, \bar{a})$, $\varphi(\bar{x})$ belong to the type computed implicitly in $\circledast_\alpha$, i.e. if $q(\bar{x}) = \mathrm{gtp}_{L^0_{<\alpha}}(\bar{a}'; N'; M')$ then $(N', M') \Vdash^{\aleph_1}_{\mathfrak{K}} \varphi(\bar{x})$.

Those three should be clear and gives the desired conclusion. Also the last sentence is easy.

<u>Clause (d)</u>: Let $N_0 \leq_{\mathfrak{K}} M_0 \in K_{\aleph_0}$ and $\bar{a}_0 \in M_0$ be such that $(M_0, N_0) \Vdash^{\aleph_1}_{\mathfrak{K}} \bigwedge_{\varphi(\bar{x}) \in p} \varphi[\bar{a}_0]$, (if it does not exist, the set of triples is empty). Let $K'' := \{(N, M, \bar{a}) : M \in K_{\aleph_0}, N \in K_{\aleph_0}, N \leq_{\mathfrak{K}} M$, and there are $M'' \in K_{\aleph_0}, M \leq_{\mathfrak{K}} M''$ and $\leq_{\mathfrak{K}}$-embedding $f : M_0 \to M''$, such that $f(N_0) = N, g(\bar{a}_0) = \bar{a}\}$. Clearly it is a $\mathrm{PC}_{\aleph_0}$ class. Also $M_0 \leq_{\mathfrak{K}} M' \in K_{\aleph_0} \Rightarrow \mathrm{gtp}_{\mathbb{L}^0_{\omega_1,\omega}(N_0)}(\bar{a}; N_0; M_0) = \mathrm{gtp}_{\mathbb{L}^0_{\omega_1,\omega}(N_0)}(\bar{a}; N_0, M')$.

Now first if $(N, M, \bar{a}) \in K''$ let (M'', f) witness this so by applying clause (b) of 4.13 $\mathrm{gtp}_{\mathbb{L}^0_{\omega_1,\omega}}(\bar{a}; N; M) \subseteq \mathrm{gtp}_{\mathbb{L}^0_{\omega_1,\omega}}(\bar{a}; N; M'') =$

$\mathrm{gtp}_{\mathbb{L}^0_{\omega_1,\omega}}(\bar{a}; N; f(M_0)) = \mathrm{gtp}_{\mathbb{L}^0_{\omega_1,\omega}}(a_0; N_0; M_0) = p$ so $(N, M, \bar{a}) \in K^1$.

Second, if $(N, M, \bar{a}) \in K^1$ let f_0 be an isomorphism from M_0 onto M_0. Let (M_1, f_1) be such that $N_0 \leq_{\mathfrak{K}} M_1 \in K_{\aleph_0}, f_1 \supseteq f_0$ is an isomorphism from M_1 onto M and $\bar{a}_1 = f_i^{-1}(\bar{a})$ hence $p = \mathrm{gtp}_{\mathbb{L}^0_{\omega_1,\omega}}(\bar{a}_1; N_0; M_1)$ and we apply clause (c) of 4.13 with N_0, M_0, $\bar{a}_0$, M_1, $\bar{a}_1$ here standing for N, M_1, $\bar{a}_1$, M_2, $\bar{a}_2$ there and can finish easily.

Clause (e): We can define $\langle L^{-1}_\alpha : \alpha < \omega_1\rangle$ satisfying the parallel of Clause (a) and repeat the proofs of clauses (b),(c) and we are done.

Clause (f): Suppose this fails.

The proof splits to two cases.

Case A: $2^{\aleph_0} = 2^{\aleph_1}$.

We shall prove $\dot{I}(\aleph_1, K) \geq 2^{\aleph_0}$, thus, (as $2^{\aleph_0} = 2^{\aleph_1}$) contradicting Hypothesis 4.8.

Let p_i (for $i < \omega_2$) be distinct complete $\mathbb{L}^0_{\omega_1,\omega}(\tau^{+0})$-types such that for each i, p_i is materialized in some pair $(M; N)$, so $N \leq_{\mathfrak{K}} M \in K_{\aleph_0}$ (they exist by the assumption that (f) fails). For each $i < \omega_2$ we define $N_{i,\alpha}, \xi_{i,\alpha}$ (for $\alpha < \omega_1$) and $\bar{a}_{i,\alpha}$ such that:

$\boxtimes_1$ (i) $N_{i,\alpha} \in K_{\aleph_0}$ has universe $\omega(1+\alpha), N_{0,0} = N$

(ii) $\langle N_{i,\alpha} : \alpha < \omega_1\rangle$ is $\leq_{\mathfrak{K}}$-increasing continuous

(iii) $\bar{a}_{i,\alpha} \in N_{i,\alpha+1}, \bar{a}_{i,\alpha}$ materialize p_i in $(N_{i,\alpha+1}, N_{i,\alpha})$

(iv) for every $\alpha < \beta < \omega_1$ and $\bar{a} \in {}^{\omega>}(N_{i,\beta})$, the sequence $\bar{a}$ materialize in $(N_{i,\beta}, N_{i,\alpha})$ a complete $\mathbb{L}^0_{\omega_1,\omega}(\tau^{+0})$-type

(v) $\xi_{i,\alpha} < \omega_1$ is strictly increasing continuous in α

(vi) for $\alpha < \beta, N_{i,\beta}$ is pseudo $L^0_\beta(N_{i,\alpha})$-generic, see 4.4(4) and take care of $\mathbf{Q}$, i.e., if $\gamma < \beta, p(y, \bar{x})$ a complete L^0_γ-type and $(N_{i,\beta}, N_{i,\alpha}) \Vdash^{\aleph_1}_{\mathfrak{K}} (\mathbf{Q}y) \wedge p(y, \bar{a})$, then for some $b \in N_{i,\beta+1} \backslash N_{i,\beta}$ we have $(N_{i,\beta+1}, N_{i,\alpha}) \Vdash^{\aleph_1}_{\mathfrak{K}} \wedge p(b, \bar{a})$

(vii) if $\alpha < \beta$ and $\bar{a}, \bar{b} \in N_{\beta-1}$ materialize different $\mathbb{L}^0_{\omega_1,\omega}(N_{i,\alpha})$-types in $N_{i,\beta}$,
then $\bar{a}, \bar{b}$ realize different $(\mathbb{L}_{\omega_1,\omega}(\tau^{+0}) \cap L^{-1}_{\xi_{i,\beta+1}})(N_\alpha)$-types in $N_{i,\beta}$

(viii) $N_i = \cup\{N_{i,\alpha} : \alpha < \omega_1\}$

(ix) if $\alpha_\ell < \beta$ for $\ell = 1,2, \gamma < \beta, n < \omega$ and $\bar a_1 \in {}^n(N_{i,\beta})$ then for some $\bar a_2 \in {}^n(N_{i,\beta})$ we have $\mathrm{gtp}_{L^0_\gamma}(\bar a_1; N_{i,\alpha_1}; N_{i,\beta}) = \mathrm{gtp}_{L^0_\gamma}(\bar a_2; N_{i,\alpha_2}; N_{i,\beta})$

(ix)$^+$ moreover, if $n < \omega, \gamma_1 < \gamma_2 < \beta, \alpha_\ell < \beta, \bar a_\ell \in {}^n(N_{i,\beta})$ for $\ell = 1,2$ and $\mathrm{gtp}_{L^0_{\gamma_2}}(\bar a_1; N_{i,\alpha-1}; N_{i,\beta}) = \mathrm{gtp}_{L^0_{\gamma_2}}(\bar a_2; N_{i,\alpha_2}; N_{i,\beta})$ and $b_1 \in N_{i,\beta}$
<u>then</u> for some $b_2 \in N_{i,\beta}$ we have $\mathrm{gtp}_{L^0_{\gamma_1}}(\bar a_1 {}^\frown \langle b_1\rangle; N_{i,\alpha_1}; N_{i,\beta}) = \mathrm{gtp}_{L^0_{\gamma_1}}(\bar a {}^\frown \langle b_2\rangle; N_{i,\alpha_2}; N_{i,\beta})$.

This is possible by the earlier claims. By clause (e) of 4.13 clearly

$\boxtimes_2$ the pair (N_i, N_0) is $L^{-1}_{<\omega_1}(\tau^{+0})$-homogeneous.

We could below use D_i a set of complete $L^0_{\delta(i)}$-types, the only problem is that the countable $(D_i, \aleph_0)$-homogeneous models have to be redefined using "materialized" instead "realized". As it is we need to use clause (e) to translate the results on $L^0_{\delta(i)}$ to $L^{-1}_{\delta(i)}$.
Let $\tau^* = \{\in, Q_1, Q_2\} \cup \{c_\ell : \ell < 5\}, c_\ell$ an individual constant and $\mathfrak{A}^*_i$ be $(\mathscr{H}(\aleph_2), \in)$ expanded to a τ^*-model, by predicates for $K, \leq_{\mathfrak{K}}$ with $Q_1^{\mathfrak{A}^*_i} = K \cap \mathscr{H}(\aleph_2), Q_2^{\mathfrak{A}^*} = \{(M,N) : M \leq_{\mathfrak{K}} N$ both in $\mathscr{H}(\aleph_2)\}, c_0^{\mathfrak{A}^*_i}, \dots, c_4^{\mathfrak{A}^*_i}$ being $\{\langle N_{i,\alpha} : \alpha < \omega_1\rangle\}, \langle \xi_{i,\alpha} : \alpha < \omega_1\rangle, \{\langle \bar a_{i,\alpha} : \alpha < \omega_1\rangle\}, N_i$ and $\{i\}$ respectively.

Let $\mathfrak{A}_i$ be a countable elementary submodel of $\mathfrak{A}^*_i$ so $|\mathfrak{A}_i| \cap \omega_1$ is an ordinal $\delta(i) < \omega_1$. It is also clear that $c_3^{\mathfrak{A}_i}$ is $N_{i,\delta(i)}$ as $c_3^{\mathfrak{A}^*_i} = N_i$. As $\mathfrak{A}_i$ is defined for $i < \omega_2$, for some unbounded $S \subseteq \omega_2$ and $\delta < \omega_1$, for every $i \in S, \delta(i) = \delta$ and for $i,j \in S$, some sequence from N_j materializes p_i in the pair $(N_j, N_{j,\delta(j)})$ iff $i = j$. For $i \in S$ let $D_i = \{p : p$ is a complete $L^{-1}_{\delta(i)}$-type materialized in $(N_{i,\delta(i)}, N_{i,0})\}$. Because of the $\xi_{i,\alpha}$'s choice and $\boxtimes_2$ the pair $(N_{i,\delta}, N_0)$ is $(D_i, \aleph_0)$-homogeneous and D_i is a countable set of complete L^{-1}_δ-types. Note that by the choice of $S, i \neq j (\in S) \Rightarrow D_i \neq D_j$.

Let $\Gamma = \{D : D$ a countable set of complete L^{-1}_δ-types, such that for some model $\mathfrak{A} = \mathfrak{A}_D$ of $\bigcap\limits_{i\in S} \mathrm{Th}_{\mathbb{L}_{\omega,\omega}}(\mathfrak{A}_i)$, with $\{a : \mathfrak{A}_D \models$ "a countable ordinal$\} = \delta$ (and the usual order) we have $D = \{\{\varphi(\bar x) : \varphi(\bar x) \in L^{-1}_\delta$ and $\mathfrak{A}_D \models$ "$(N; N_0) \Vdash^{\aleph_1}_{\mathfrak{K}} \varphi[\bar a]$"$\} : \bar a \in N$ where $N = c_3^{\mathfrak{A}_D}\}\}$.

So $D_i \in \Gamma$ for $i < \omega_2$, hence Γ is uncountable.

By standard descriptive set theory Γ (is an analytic set hence) has cardinality continuum. So let $D(\zeta) \in \Gamma$ be distinct for $\zeta < 2^{\aleph_0}$. For each ζ, let $\mathfrak{A}^0_{D(\zeta)}$ be as in the definition of Γ. We define by induction on $\alpha < \omega_1, \mathfrak{A}^\alpha_{D(\zeta)}$ such that

(α) $\mathfrak{A}^\alpha_{D(\zeta)}$ is countable

(β) $\alpha < \beta \Rightarrow \mathfrak{A}^\alpha_{D(\zeta)} \prec_{\mathbb{L}_{\omega,\omega}} \mathfrak{A}^\beta_{D(\zeta)}$

(γ) for limit α we have $\mathfrak{A}^\alpha_{D(\zeta)} = \bigcup\limits_{\beta<\alpha} \mathfrak{A}^\beta_{D(\zeta)}$

(δ) if $d \in \mathfrak{A}^{\alpha+1}_{D(\zeta)} \backslash \mathfrak{A}^\alpha_{D(\zeta)}, \mathfrak{A}^{\alpha+1}_{D(\zeta)} \models$ "d a countable ordinal" then for $a \in \mathfrak{A}^\alpha_{D(\zeta)}$ we have $\mathfrak{A}^{\alpha+1}_{D(\zeta)} \models$ "if a is a countable ordinal then $a < d$"

(ε) for $\alpha = 0$ in clause (δ) there is no minimal such d

(ζ) for every α there is $d_{\zeta,\alpha} \in \mathfrak{A}^{\alpha+1}_{D(\zeta)} \backslash \mathfrak{A}^\alpha_{D(\zeta)}$ satisfying $\mathfrak{A}^{\alpha+1}_{D(\zeta)} \models$ "$d_{\zeta,\alpha}$ a countable ordinal" and for $\alpha \neq 0$ it is minimal.

without loss of generality

($*$) $(\mathscr{H}(\aleph_1)^{\mathfrak{A}^0_{D(\zeta)}}, \in^{\mathfrak{A}^0_{D(\zeta)}})$ is equal to its Mostowski collapse (and $\mathbb{L}_{\omega_1,\omega}(N) \subseteq \mathscr{H}(\aleph_1)$).

(We could have fixed also $\text{otp}(\mathfrak{A}_i \cap \omega_2)$, hence ensure that also $(\mathfrak{A}^0_{D(\zeta)}, \in^{\mathfrak{A}^0_{D(\zeta)}})$ is equal to its Mostowski collapse).

Let $M_{\zeta,\alpha}$ be the $d_{\zeta,\alpha}$-th member of the ω_1-sequence of models in $\mathfrak{A}^\beta_{D(\zeta)}$ for $\beta > \alpha$ (remember $c_0^{\mathfrak{A}^*_i} = \langle N_{i,\alpha} : \alpha < \omega_1 \rangle$). Let $M_\zeta = \bigcup\limits_{\alpha<\omega_1} M_{\zeta,\alpha}$. By absoluteness from $\mathfrak{A}^\beta_{D(\zeta)}$ we have $M_{\zeta,\alpha} \leq_{\mathfrak{K}} M_{\zeta,\beta} \in K_{\aleph_0}$. Now

($*$) $0 < \alpha < \beta, (M_{\zeta,\beta}, M_{\zeta,\alpha})$ is $(D(\zeta), \aleph_0)$-homogeneous.

[Why? Assume $\mathfrak{A}^\alpha_{D(\zeta)} \models$ "$d_1 < d_2$ are countable ordinals $> \gamma$" when $\gamma < \delta$.
Now if $\bar{a}, \bar{b} \in {}^{\omega>}(N_{d_2}^{\mathfrak{A}^\alpha_{D(\zeta)}})$ and $[\gamma < \delta \Rightarrow \text{gtp}_{L^0_\gamma}(\bar{a}; N_{d_1}^{\mathfrak{A}^\alpha_{D(\zeta)}}; N_{d_2}^{\mathfrak{A}^\alpha_{D(\zeta)}}) = \text{gtp}_{L^0_\gamma}(\bar{b}; N_{d_1}^{\mathfrak{A}^\alpha_{D(\zeta)}}; N_{d_2}^{\mathfrak{A}^\alpha_{D(\zeta)}})]$ then also $\mathfrak{A}^\alpha_{D(\zeta)}$ satisfies this but "$\mathfrak{A}^\alpha_{D(\zeta)}$

thinks that the countable ordinals are well ordered" hence for some $d, \mathfrak{A}^\alpha_{D(\zeta)} \models$ "d is a countable ordinal $> \gamma$" for each $\gamma < \delta$ and we have $\mathfrak{A}^\alpha_{D(\zeta)} \models$ "$\text{gtp}_{L^0_d}(\bar{a}; N_{d_1}; N_{d_2}) = \text{gtp}_{L^0_d}(\bar{a}; N_{d_1}; N_{d_2})$". Hence if $\mathfrak{A}^\alpha_{D(\zeta)} \models$ "$d' < d$" then for every $a \in N_{d_2}^{\mathfrak{A}^\alpha_{D(\zeta)}}$ for some $b \in N_{d_2}^{\mathfrak{A}^\alpha_{D(\zeta)}}$ we have

$$\mathfrak{A}^\alpha_{D(\zeta)} \models \text{"gtp}_{L^0_d}(\bar{a}\hat{\ }\langle a\rangle; N_{d_1}; N_{d_2}) = \text{gtp}_{L^0_d}(\bar{b}\hat{\ }\langle b\rangle; N_{d_1}; N_{d_2})\text{"}$$

hence $\text{gtp}_{L^0_\gamma}(\bar{a}\hat{\ }\langle a\rangle; N^{\mathfrak{A}^\alpha_{D(\zeta)}}; N_{d_2}^{\mathfrak{A}^\alpha_{D(\zeta)}}) = \text{gtp}(\bar{b}\hat{\ }\langle b\rangle; N_{d_1}^{\mathfrak{A}^\alpha_{D(\zeta)}}; N_{d_2}^{\mathfrak{A}^\alpha_{D(\zeta)}})$.

Also we can replace L^0_δ by L^{-1}_δ. By clause (x) of $\boxtimes_1$ the set $\{\text{gtp}_{L^0_\delta}(\bar{a}; N_{d_1}^{\mathfrak{A}^\alpha_{D(\zeta)}}; N_{d_2}^{\mathfrak{A}^\alpha_{D(\zeta)}}) : a \in {}^{\omega>}(N_{d_2}^{\mathfrak{A}^\alpha_{D(\zeta)}})\}$ is D_i.

So $(N_{d_2}^{\mathfrak{A}^\alpha_{D(\zeta)}}, N_{d_2}^{\mathfrak{A}^\alpha_{D(\zeta)}})$ is $(D_i, \aleph_0)$-homogenous.

So from the isomorphism type of M_ζ we can compute $D(\zeta)$. So $\zeta \neq \xi \Rightarrow M_\zeta \not\cong M_\xi$. As $M_\zeta \in K_{\aleph_1}$ we finish.

Case B: $2^{\aleph_0} < 2^{\aleph_1}$.

By 3.8, $\mathfrak{K}$ has the $\aleph_0$-amalgamation property. So clearly if $N \leq_{\mathfrak{K}} M \in K_{\aleph_0}, \bar{a} \in M$, then $\bar{a}$ materializes in (M, N) a complete $\mathbb{L}^0_{\omega_1,\omega}(\tau^{+0})$-type. We would now like to use descriptive set theory.

We represent a complete $\mathbb{L}^0_{\omega_1,\omega}(\tau^{+0})$-type materialized in some (N, M) by a real, by representing the isomorphism type of some $(N, M, \bar{a}), N \leq_{\mathfrak{K}} M \in K_{\aleph_0}, \bar{a} \in M$. The set of representatives is analytic recalling $\mathfrak{K}$ is $\text{PC}_{\aleph_0}$, and the equivalence relation is Σ^1_1. [As $(N_1, M_1, \bar{a}_1), (N_2, M_2, \bar{a}_2)$ represents the same type if and only if for some $(N, M), N \leq_{\mathfrak{K}} M \in K_{\aleph_0}$, there are $\leq_{\mathfrak{K}}$-embeddings $f_1 : M_1 \to M, f_2 : M_2 \to M$ such that $f_1(N_1) = f_2(N_2) = N$ and $f_1(\bar{a}) = f_2(\bar{a})$.]

By Burgess [Bg] (or see [Sh 202]) as there are $> \aleph_1$ equivalence classes, there is a perfect set of representation, pairwise representing different types.

From this we easily get that without loss of generality that their restriction to some L^0_α are distinct, contradicting part (a).

Clause (g): Easy by the proof of clause (f), Case A above but much simpler as in 4.12.

Clause (h): As in the proof of clause (e).

2) Should by clear by now. $\square_{4.13}$

4.15 Remark. 1) Note that in the proof of Clause (f) of 4.13, in Case (A) we get many types too but it was not clear whether we can make the N_ζ to be generic enough, to get the contradiction we got in Case (B) but this is not crucial here.
2) We may like to replace $\mathbb{L}^0_{\omega_1,\omega}$ by $\mathbb{L}^1_{\omega_1,\omega}$ in 4.10, 4.11 and 4.13 (except that, for our benefit, in 4.13(e), we may retain the definition of $L^1(N)$). We lose the ability to build L-generic models in $K_{\aleph_1}$ (as the number of (even unary) relations on $N \in K_{\aleph_0}$ is $2^{\aleph_0}$, which may be $> \aleph_1$). However, we can say "$\bar{a}$ materializes in $N \in K_{\aleph_0}$ the type $p = p(\bar{x})$ which is a complete type in $\mathbb{L}^1_{\omega_1,\omega}(N_n, N_{n-1}, \dots, N_0)$; where $N_0 \le_{\mathfrak{K}} \dots \le_{\mathfrak{K}} N_n \le_{\mathfrak{K}} N, N_\ell$ countable)".
[Why? Let some $N^1, \bar{a}^1$ be as above, $\bar{a}^1$ materialize p in $(N^1, N_n, \dots, N_0)$ <u>then</u> this holds for $(N, \bar{a})$ <u>iff</u> for some N', f we have $N \le_{\mathfrak{K}} N' \in K_{\aleph_1}$ and f is an isomorphism from N^1 onto N'' mapping $\bar{a}^1$ to $\bar{a}$ and N_ℓ to N_ℓ for $\ell \le n$. If there is no such pair $(N^1, \bar{a}^1)$ this is trivial.]
We can get something on formulas.

This suffices for 4.10.

4.16 Concluding remarks for §4. 0) We can get more information on the case $1 \le \dot{I}(\aleph_1, K) < 2^{\aleph_1}$ (and the case $1 \le \dot{I}(\aleph_1, K^{\mathbf{F}}_{\aleph_1}) < 2^{\aleph_1}$, etc.).
1) As in 3.8, there is no difficulty in getting the results of this section for the class of models of $\psi \in \mathbb{L}_{\omega_1,\omega}(\mathbf{Q})$ because using $(K, \le_{\mathfrak{K}})$ from the proof of 3.18(2) in all constructions we get many non-isomorphic models for appropriate $\mathbf{F}$, as in 4.9(2).
2) For generic enough $N \in K_{\aleph_1}$ with $\le_{\mathfrak{K}}$-representation $\langle N_\alpha : \alpha < \omega_1 \rangle$, we have determined the N_α's (by having that without loss of generality K is categorical in $\aleph_0$). In this section we have shown that for some club E of ω_1, for all $\alpha < \beta$ from E the isomorphism type of (N_β, N_α) essentially[6] is unique. We can continue the analysis, e.g., deal with sequences $N_0 \le_{\mathfrak{K}} N_1 \le_{\mathfrak{K}} \dots \le_{\mathfrak{K}} N_k \in K_{\aleph_0}$ such that $N_{\ell+1}$ is pseudo $L^0_\alpha(N_\ell, N_{\ell-1}, \dots, N_0)$-generic. We can prove by induction on k that for any countable $L \subseteq \mathbb{L}^0_{\omega_1,\omega}(\tau^{+k})$ for some α, any strong L-generic $N \in K_{\aleph_1}$ is L-determined. That is, for

[6] why only essentially? as the number of relevant complete types can be $\aleph_1$; we can get rid of this by shrinking $\mathfrak{K}$

any $\langle N_\alpha : \alpha < \omega_1\rangle, N_\alpha \leq_{\mathfrak{K}} N$ countable $\leq_{\mathfrak{K}}$-increasing continuous with union N, for some club E for all $\alpha_0 < \ldots < \alpha_k$ from N the isomorphic type of $\langle N_{\alpha_k}, N_{\alpha_k}, \ldots, N_{\alpha_0}\rangle$ is the same; i.e., determining for $\mathbb{L}_{\infty,\omega}(aa)$.

3) We can do the same for stronger logics, let us elaborate.

Let us define a logic $\mathscr{L}^*$. It has as variable
variables for elements x_1, x_2...and
variables for filters $\mathscr{Y}_1, \mathscr{Y}_2$...

The atomic formulas are:

- (i) the usual ones
- (ii) $x \in \text{Dom}(\mathscr{Y})$.

The logical operations are:

- (a) $\wedge$ conjunction, $\neg$ negation
- (b) $(\exists x)$ existential quantification where x is individual variable
- (c) the quantifier aa acting on variables $\mathscr{Y}$ so we can form $(aa\,\mathscr{Y})\varphi$
- (d) the quantification $(\exists x \in \text{Dom}(\mathscr{Y}))\varphi$
- (e) the quantification $(\exists^f x \in \text{Dom}(\mathscr{Y}))\varphi$.

It should be clear what are the free variables of a formula φ. The variable $\mathscr{Y}$ vary on pairs (a countable set, a filter on the set). Now in $\exists x[\varphi, \mathscr{Y}], (\exists x \in \text{Dom}(\mathscr{Y}))\varphi, (\exists^f x \in \text{Dom}(\mathscr{Y}))\varphi, x$ is bounded but not $\mathscr{Y}$ and in $aa\mathscr{Y}, \mathscr{Y}$ is bounded. The satisfaction relation is defined as usual plus

- (α) $M \models (\exists x \in \text{Dom}(\mathscr{Y})\varphi(x, \mathscr{Y}, \bar{a})$ if and only if for some b from the domain of $\mathscr{Y}, M \models \varphi[b, \mathscr{Y}, \bar{a}]$
- (β) $M \models \exists^f x \in \text{Dom}(\mathscr{Y})\varphi(x, \mathscr{Y}_{\bar{a}})$ if and only if $\{x \in \text{Dom}(\mathscr{Y}) : \models \varphi(x, \mathscr{Y}, \bar{a})\} \in \mathscr{Y}$
- (γ) $M \models (aa\,\mathscr{Y}, \bar{a})\varphi(\mathscr{Y})$ if and only if there is a function $\mathbf{F}$ from ${}^{\omega>}([M]^{<\aleph_1}) \to [M]^{<\aleph_1}$ such that: if $A_n \subseteq M, |A_n| \leq \aleph_0, A_n \subseteq A_{n+1}$ and $\mathbf{F}(A_0, \ldots, A_n) \subseteq A_{n+1}$ <u>then</u> $M \models \varphi[\mathscr{Y}_{\langle A_n : n<\omega\rangle}, \bar{a}]$ where $\mathscr{Y}_{\langle A_n:n<\omega\rangle}$ is the filter on $\bigcup\limits_{n<\omega} A_n$, generated by $\{\cup\{A_n : n < \omega\} \backslash A_\ell : \ell < \omega\}$.

4) We, of course, can define $\mathscr{L}^*_{\mu,\kappa}$ (extending $\mathbb{L}_{\mu,\ell}$). As we like to analyze models in $\aleph_1$, it is most natural to deal with $\mathscr{L}^*_{\omega_1,\omega}$.

We can prove that (if $1 \leq \dot{I}(\aleph_1, \mathfrak{K}) < 2^{\aleph_1}$) the quantifier $aa\,\mathscr{Y}$ is determined on $K_{\aleph_1}$ (i.e., for almost all $\mathscr{Y}, \varphi(\mathscr{Y})$ iff not for almost all $\mathscr{Y}, \neg\varphi(\mathscr{Y})$.

5) The logic from (3) strengthens the stationary logic $\mathbb{L}(aa)$, see [Sh 43], [BKM78].

Not so strongly: looking at $\mathrm{PC}_{\aleph_0}$ class for $\mathbb{L}_{\omega_1,\omega}(aa)$ (i.e., $\{M \restriction \tau : M$ a model of ψ of cardinal $\aleph_1\}$), we can assume that $\psi \vdash$ " $<$ is an $\aleph_1$-like order". Now we can express $\varphi \in \mathscr{L}^*_{\omega_1,\omega}$, but the determinacy tells us more. Also we can continue to define higher variables $\mathscr{Y}$.

§5 There is a superlimit model in $\aleph_1$

Here we make

5.1 Hypothesis. Like 4.8, but also $2^{\aleph_0} < 2^{\aleph_1}$.

(Note that we can assume that $K_{\aleph_0}$ is the class of atomic models of a first order complete countable theory).

This section is the deepest (of this paper = chapter). The main difficulties are proving the facts which are obvious in the context of [Sh 48]. So while it was easy to show that every $p \in \mathbf{D}^*(N)$ is definable over a finite set ($\mathbf{D}^*(N)$ is defined below), it was not clear to me how to prove that if you extend the type p to $q \in \mathbf{D}^*(M)$ where $N \leq_{\mathfrak{K}} M \in K_{\aleph_0}$, by the same definition, then $q \models p$ (remember p, q are types materialized not realized, and at this point in the paper we still do not have the tools to replace the models by uncountable generic enough models). So we rather have to show that failure is a non-structure property, i.e., implies existence of many models.

Also symmetry of stable amalgamation becomes much more complicated. We prove existence of stable amalgamation by four stages (5.26,5.27(3),5.30,5.32). The symmetry is proved as a consequence of uniqueness of one sided amalgamation, (so it cannot be used in its proof). Originally the intention was the culmination of the section to be the existence of a superlimit models in $\aleph_1$ (5.39). This seems a natural stopping point as it seems reasonable to expect that the next step should be phrasing the induction on n, i.e., dealing with $\aleph_n$ and $\mathscr{P}(n-\ell)$-diagrams of models of power $\aleph_\ell$ as in [Sh 87a], [Sh

87b]; (so this is done in Chapter III). But less is needed in Chapter II.

5.2 Definition. We define functions $\mathbf{D}, \mathbf{D}^*$ with domain $K_{\aleph_0}$.
1) For $N \in K_{\aleph_0}$ let $\mathbf{D}(N) = \{p : p$ is a complete $\mathbb{L}^0_{\omega_1,\omega}(N)$-type over N such that for some $\bar{a} \in M \in K_{\aleph_0}, N \leq_{\mathfrak{K}} M$ and $\bar{a}$ materializes p in $(M, N)\}$, (i.e. the members of p have the form $\varphi(\bar{x}, \bar{a})$, ($\bar{x}$ finite and fixed for p) $\bar{a}$ a finite sequence from N and $\varphi \in \mathbb{L}^0_{\omega_1,\omega}(N)$).
2) For $N \in K_{\aleph_0}$ let $\mathbf{D}^*(N) = \{p : p$ a complete $\mathbb{L}^0_{\omega_1,\omega}(N; N)$-type such that for some $\bar{a} \in M \in K_{\aleph_0}, N \leq_{\mathfrak{K}} M$ and $\bar{a}$ materializes p in $(M, N; N)\}$.
3) For $p(\bar{x}, \bar{y}) \in \mathbf{D}(N)$ let $p(\bar{x}, \bar{y}) \restriction \bar{x} \in \mathbf{D}(N)$ be defined naturally; i.e. if for some $M, N \leq_{\mathfrak{K}} M \in K_{\aleph_0}$ and $\bar{a} \mathbin{\char`\^} \bar{b} \in {}^{\ell g(\bar{x} \char`\^ \bar{y})}M$ materializing $p(\bar{x}, \bar{y})$ such that $\ell g(\bar{x}) = \ell g(\bar{a})$, the sequence $\bar{a}$ materializes $p(\bar{x}, \bar{y}) \restriction x \in \mathbf{D}(N)$. Similarly for permuting the variables.

<u>5.3 Explanation</u>: 0) Recall that any formula in $\mathbb{L}^0_{\omega_1,\omega}(N)$ has finitely many free variables.
1) So for every finite $\bar{b} \in N$ and $\varphi(\bar{x}, \bar{y}) \in \mathbb{L}^0_{\omega_1,\omega}(N)$, if $p \in \mathbf{D}(N)$, then $\varphi(\bar{x}, \bar{b}) \in p$ or $\neg\varphi(\bar{x}, \bar{b}) \in p$.
2) But a formula from $p \in \mathbf{D}^*(N)$ may have all $c \in N$ as parameters whereas a formula from $p \in \mathbf{D}(N)$ can mention only finitely many members of N.

5.4 Lemma. *1) $\mathfrak{K}$ has the $\aleph_0$-amalgamation property.*
2) If $N_* \leq_{\mathfrak{K}} N \in K_{\aleph_0}$, $A_i \subseteq N_*$ for $i \leq n$ <u>then</u> for every sentence $\psi \in \mathbb{L}^1_{\infty,\omega}(N_*, A_n, \ldots, A_1; A_0)$ we have

$$N \Vdash^{\aleph_1}_{\mathfrak{K}} \psi \text{ or } N \Vdash^{\aleph_1}_{\mathfrak{K}} \neg\psi.$$

3) If $N \leq_{\mathfrak{K}} M \in K_{\aleph_0}$, <u>then</u> every $\bar{a} \in M$ materializes in $(M, N; N)$ one and only one type from $\mathbf{D}^*(N)$ and also materializes in (M, N) one and only one type from $\mathbf{D}(N)$. Also for every $N \leq_{\mathfrak{K}} M \in K_{\aleph_0}$ and $q \in \mathbf{D}^*(N)$ for some $M', M \leq_{\mathfrak{K}} M' \in K_{\aleph_0}$ and some $\bar{b} \in M'$ materializes q in $(M; N)$.
4) For every $N \in K_{\aleph_0}$ and countable $L \subseteq \mathbb{L}^0_{\omega_1,\omega}(N; N)$ the number of

complete $L(N;N)$-types p such that $N \Vdash^{\aleph_1}_{\mathfrak{K}}$ "$(\exists \bar{x}) \wedge p$" is countable; note that pedantically $L \subseteq \mathbb{L}_{\omega_1,\omega}(\tau^+ \cup \{c : c \in N\})$ and we restrict ourselves to models M such that $P^M = |N|, c^M = c$.
5) For $N \in K_{\aleph_0}$ there are countable $L^0_\alpha \subseteq \mathbb{L}^0_{\omega_1,\omega}(N;N)$ for $\alpha < \omega_1$ increasing continuous in α, closed under finitary operations (and subformulas) such that:

(∗) *for each complete L^0_α-type p we have*

$$[N \Vdash^{\aleph_1}_{\mathfrak{K}} \exists \bar{x} \wedge p \Rightarrow \wedge p \in L^0_{\alpha+1}].$$

Hence for every $\mathbb{L}^0_{\omega_1,\omega}(N;N)$ formula $\psi(\bar{x})$ for some $\varphi_n(\bar{x}) \in \bigcup_{\alpha<\omega} L^0_\alpha$ for $n < \omega$ for every $N \in K_{\aleph_0}$

$$(N,N) \Vdash^{\aleph_1}_{\mathfrak{K}} (\forall \bar{x})[\psi(\bar{x}) \equiv \bigvee_{n<\omega} \varphi_n(\bar{x})].$$

6) For $N \in K_{\aleph_0}$ we have $|\mathbf{D}^*(N)| \leq \aleph_1$ and $|\mathbf{D}(N)| \leq \aleph_1$.
7) If $p \in \mathbf{D}^*(N)$ then there is q such that: if $N \leq_{\mathfrak{K}} M \in K_\lambda, \bar{a} \in M$ materializes p in $(M;N)$ then the complete $\mathbb{L}^0_{\infty,\omega}(N)$-type which $\bar{a}$ realizes in M over N is q; also q belongs to $\mathbf{D}(N)$ and is unique. Moreover, we can replace q by the complete $\mathbb{L}^{-1}_{\omega_1,\omega}(N)$-type which $\bar{a}$ materializes in M. Similarly for $\mathbf{D}(N), \mathbb{L}^0_{\infty,\omega}(N), \mathbb{L}^{-1}_{\omega_1,\omega}(N)$.
8) If $n < \omega$ and $\bar{b}, \bar{c} \in {}^nN$ realize the same $\mathbb{L}_{\omega_1,\omega}(\tau)$-type in N then they materialize the same $\mathbb{L}^1_{\omega_1,\omega}(\tau^{+0})$-type in (N,N).
9) If f is an isomorphism from $N_1 \in K_{\aleph_0}$ onto $N_2 \in K_{\aleph_0}$ then f induces a one to one function from $\mathbf{D}(N_1)$ onto $\mathbf{D}(N_2)$ and from $\mathbf{D}^*(N_1)$ onto $\mathbf{D}^*(N_2)$.

Proof. 1) By 3.8.
2) By 1).
3) By 2) and 1).
4) Like the proof of 4.11 (just easier).
5) Like the proof of 4.13(a).
6) Like the proof of 4.13(f) (recalling 0.3).
7) Clear as in $p \in \mathbf{D}^*(N)$ we allow more formulas than for $q \in \mathbf{D}(N)$.
8),9) Easy, too. $\square_{5.4}$

We shall use from now on a variant of gtp (in Definition 4.3(4) we define $\mathrm{gtp}_L(\bar{a}; N_*, \bar{A}; A; N)$.

5.5 Definition. 1) If $N_0 \leq_{\mathfrak{K}} N_1 \in K_{\aleph_0}, \bar{a} \in N_1$, $\mathrm{gtp}(\bar{a}, N_0, N_1)$ is the $p \in \mathbf{D}(N_0)$ such that $(N_1, N_0) \Vdash^{\aleph_1}_{\mathfrak{K}} \wedge p[\bar{a}]$. So $\bar{a}$ materializes (but does not necessarily realize) $\mathrm{gtp}(\bar{a}, N_0, N_1)$. We may omit N_1 when clear from context. We define $\mathrm{gtp}^*(\bar{a}, N_0, N_1) \in \mathbf{D}^*(N_0)$ similarly.
2) We say $p = \mathrm{gtp}^*(\bar{b}, N_0, N_1)$ is definable over $\bar{a} \in N_0$ if $\mathrm{gtp}(\bar{b}, N_0, N_1)$ = $p^- := \{\varphi(\bar{x}, \bar{a}) \in p : \varphi(\bar{x}, \bar{y}) \in \mathbb{L}^0_{\omega_1,\omega}(N_0)$ and $\bar{a} \in {}^{\ell g(\bar{y})}(N_0) \subseteq {}^{\omega>}(N_0)\}$ is definable over $\bar{a}$ (see Definition 5.7 below, note that $p \mapsto p^-$ is a one-to-one mapping from $\mathbf{D}^*(N_0)$ onto $\mathbf{D}(N_0)$ by 5.9(1) below). So stationarization is defined for $p \in \mathbf{D}^*(N_0)$, too, after we know 5.9(1).

5.6 Claim. *1) Each $p \in \mathbf{D}(N)$ does not $(\mathbb{L}^0_{\omega_1,\omega}(\tau^{+0}), \mathbb{L}_{\omega_1,\omega}(\tau))$-split (see Definition 5.7 below; also see more below) over some finite subset C of N, hence p is definable over it.*
Moreover, letting $\bar{c}$ list C there is a function g_p satisfying $g_p(\varphi(\bar{x}, \bar{y}))$ is $\psi_{p,\varphi}(\bar{y}, \bar{z}) \in \mathbb{L}_{\omega_1,\omega}(\tau)$ such that for each $\varphi(\bar{x}, \bar{y}) \in \mathbb{L}^0_{\omega_1,\omega}(N)$ and $\bar{a} \in N$ we have $[\varphi(\bar{x}, \bar{a}) \in p \Leftrightarrow N \models \psi_{p,\varphi}(\bar{a}, \bar{c})]$, (in particular, $\mathbf{Q}$ is "not necessary").
2) Every automorphism of N maps $\mathbf{D}(N)$ onto itself and each $p \in \mathbf{D}(N)$ has at most $\aleph_0$ possible images; we may also call them conjugates. So if g is an isomorphism from $N_0 \in K_{\aleph_0}$ onto $N_1 \in K_{\aleph_0}$ then $g(\mathbf{D}(N_0)) = \mathbf{D}(N_1)$.
3) If $N_0 \leq_{\mathfrak{K}} N_1 \leq_{\mathfrak{K}} N_2 \in K_{\aleph_0}$ and $\bar{a} \in N_1$ then $\mathrm{gtp}(\bar{a}, N_0, N_1) = \mathrm{gtp}(\bar{a}, N_0, N_2)$.

Before we prove 5.6:

5.7 Definition. Assume

(a) N is a model

(b) Δ_1 is a set of formulas (possibly in a vocabulary $\nsubseteq \tau_N$) closed under negation

(c) Δ_2 is a set of formulas in the vocabulary $\tau = \tau_N$

(d) p is a (Δ_1, n)-type over N (i.e., each member has the form $\varphi(\bar{x}, \bar{a}), \bar{a}$ from $N, \varphi(\bar{x}, \bar{y})$ from $\Delta_1, \bar{x} = \langle x_\ell : \ell < n\rangle$; no more is required (we may allow other formulas but they are irrelevant)

(e) $A \subseteq N$.

0) We say p is a complete Δ_1-type over B when:

(i) $B \subseteq N$

(ii) $\varphi(\bar{x}, \bar{b}) \in p \Rightarrow \bar{b} \subseteq A \wedge \varphi(\bar{x}, \bar{y}) \in \Delta_1$

(iii) if $\varphi(\bar{x}, \bar{y}) \in \Delta_1$ and $\bar{b} \in {}^{\ell g(\bar{y})}A$ then $\varphi(\bar{x}, \bar{b}) \in p$ or $\neg\varphi(\bar{x}, \bar{b}) \in p$.

The default value here for Δ_1 is $\mathbb{L}_{\omega_1,\omega}(\tau_{\mathfrak{K}})$.
1) We say that p does (Δ_1, Δ_2)-split over A when there are $\varphi(\bar{x}, \bar{y}) \in \Delta_1$ and $\bar{b}, \bar{c} \in {}^{\ell g(\bar{y})}N$ such that

(α) $\varphi(\bar{x}, \bar{b}), \neg\varphi(\bar{x}, \bar{c}) \in p$

(β) $\bar{b}, \bar{c}$ realize the same Δ_2-type over A.

2) We say that p is (Δ_1, Δ_2)-definable over A <u>when</u>: for every formula $\varphi(\bar{x}, \bar{y}) \in \Delta_1$ there is a formula $\psi(\bar{y}, \bar{z}) \in \Delta_2$ and $\bar{c} \in {}^{\ell g(\bar{z})}A$ such that

$$\varphi(\bar{x}, \bar{b}) \in p \Rightarrow N \models \psi[\bar{b}, \bar{c}]$$

$$\neg\varphi(\bar{x}, \bar{b}) \in p \Rightarrow N \models \neg\psi[\bar{b}, \bar{c}]$$

(in the case p is complete over $B, \bar{b} \subseteq B$ we get "iff").
3) Above we may write Δ_2 instead of (Δ_1, Δ_2) when this holds for every Δ_1 (equivalently Δ_1 is $\{\varphi(\bar{x}, \bar{y}) : \varphi(\bar{x}, \bar{a}) \in p\}$).

5.8 Observation. Assume

$(a), (b), (c), (d), (e)$	as in 5.7 and in addition
$(d)^+$	p is a complete (Δ_1, n)-type over N, i.e., if $\varphi(\bar{x}, \bar{y}) \in \Delta_1, \bar{d} \in {}^{\ell g(\bar{y})}N, \bar{x} = \langle x_\ell : \ell < n \rangle$ then $\varphi(\bar{x}, \bar{d}) \in p$ or $\neg\varphi(\bar{x}, \bar{d}) \in d$.

<u>Then</u> the following conditions are equivalent:

(α) p does not (Δ_1, Δ_2)-splits over A

(β) there is a sequence of $\langle g_{\varphi(\bar{x},\bar{y})} : \varphi(\bar{x},\bar{y}) \in \Delta_1 \rangle$ of functions such that:

(i) $g_{\varphi(\bar{x},\bar{y})}$ is a function with domain including $\{\mathrm{tp}_{\Delta_2}(\bar{b}, A, N)$: $\bar{b} \in {}^{\ell g(\bar{y})}N\}$

(ii) the values of $g_{\varphi(\bar{x},\bar{y})}$ are truth values

(iii) if $\varphi(\bar{x},\bar{y}) \in \Delta_1, \bar{b} \in {}^{\ell g(\bar{y})}N$ and $q = \mathrm{tp}_{\Delta_2}(\bar{b}, A, N)$ then:

$$\varphi(\bar{x},\bar{b}) \in p \Rightarrow g_{\varphi(\bar{x},\bar{y})}(q) = \text{true, and}$$
$$\neg\varphi(\bar{x},\bar{b}) \in p \Rightarrow g_{\varphi(\bar{x},\bar{y})}(q) = \text{false.}$$

Proof of 5.8. Reflect on the definitions.

Proof of 5.6. 1) Clearly the second sentence follows from the first, so we shall prove the first. Assume this fails. Let $(M, \bar{a})$ be such that $N \leq_{\mathfrak{K}} M \in K_{\aleph_0}$ the sequence $\bar{a} \in M$ materializes p and clearly for every $\bar{b} \in M$, $(M, N) \Vdash \bigwedge q[\bar{b}]$ for some $q(\bar{x}) \in \mathbf{D}(N)$ and let $\langle b^*_\ell : \ell < \omega \rangle$ list N. We choose by induction on n, $\langle C^0_\eta, C^1_\eta, f_\eta, \bar{a}^0_\eta, \bar{a}^1_\eta : \eta \in {}^n 2 \rangle$ such that

(a) C^ℓ_η is a finite subset of N for $\ell < 2, \eta \in {}^n 2$

(b) f_η is an automorphism of N mapping C^0_η onto C^1_η

(c) $\{b^*_{\ell g(\eta)}\} \cup C^0_\eta \cup C^1_\eta \subseteq C^0_{\eta\char`^<\ell>} \cap C^1_{\eta\char`^<\ell>}$ for $\ell = 0, 1$

(d) $\bar{a}^0_\eta, \bar{a}^1_\eta \in N$ realize in N the same $\mathbb{L}_{\omega_1,\omega}(\tau)$-type over $C^0_\eta \cup C^1_\eta \cup \{b^*_{\ell g(\eta)}\}$ in (M, N) but $\bar{a}\char`^\bar{a}^0_\eta, \bar{a}\char`^\bar{a}^1_\eta$ do not materialize the same $\mathbb{L}^0_{\omega_1,\omega}(\tau^{+0})$ in (M, N) (this exemplifies splitting), so $\varphi_\eta(\bar{x}, \bar{y}_\eta)$ belongs to the first, $\neg\varphi_\eta(\bar{x}, \bar{y}_\eta)$ belongs to the second (where $\ell g(\bar{x}) = \ell g(\bar{a}), \ell g(\bar{y}_\eta) = \ell g(\bar{a}^0_\eta)$)

(e) $f_{\eta\char`^<0>}(\bar{a}^0_\eta) = \bar{a}^1_\eta, f_{\eta\char`^<1>}(\bar{a}^1_\eta) = \bar{a}^1_\eta$

(f) $f_\eta \restriction C^0_\eta \subseteq f_{\eta\char`^<\ell>}$ for $\ell = 0, 1$

(g) $\bar{a}^0_\eta\char`^\bar{a}^1_\eta \subseteq C^0_{\eta\char`^<\ell>} \cap C^1_{\eta\char`^<\ell>}$.

For $n = 0$ let $C^0_\eta, C^1_\eta = \emptyset, f_\eta = \mathrm{id}_N$. Recall that $K_{\aleph_0}$ is categorical in $\aleph_0$ and N is countable, hence if $n < \omega, \bar{b}', \bar{b}'' \in {}^n N$ realize the same

$\mathbb{L}_{\omega_1,\omega}(\tau)$-type over a finite subset B of N, then some automorphism of N over B maps $\bar{b}'$ to $\bar{b}''$ by a theorem of Scott (see [Ke71]). If $(C^0_\eta, C^1_\eta, f_\eta)$ are defined and satisfies clauses (a), (b) we recall that by our assumption toward contradiction as $C^0_\eta \cup C^1_\eta \cup \{b^*_{\ell g(\eta)}\}$ is a finite subset of N, there are $\bar{a}^0_\eta, \bar{a}^1_\eta \in {}^{\omega>}N$ as required in clause (d) again. So clearly there are automorphisms $f_{\eta\hat{}<0>}, f_{\eta\hat{}<1>}$ extending $f_\eta \restriction C^0_\eta$ such that $f_{\eta\hat{}<0>}(\bar{a}^0_\eta) = \bar{a}^1_\eta, f_{\eta\hat{}<1>}(\bar{a}^1_\eta) = \bar{a}^1_\eta$ as required in clause (e), (f).
Lastly, choose $C^0_{\eta\hat{}<\ell>} = C^0_\eta \cup C^1_\eta \cup f^{-1}_{\eta\hat{}<f>}(C^0_\eta) \cup \{b^*_{\ell g(\eta)}, f^{-1}_{\eta\hat{}<\ell>}(b^*_{\ell g(\eta)}), \bar{a}^0_\eta\hat{}\bar{a}^1_\eta, f^{-1}_{\eta\hat{}<\ell>}(\bar{a}^0_\eta\hat{}\bar{a}^1_\eta)\}$ and $C^1_{\eta\hat{}<\ell>} = f_{\eta\hat{}<\ell>}(C^0_{\eta\hat{}<\ell>})$.

Having carried the induction, for every $\eta \in {}^\omega 2$ clearly $f_\eta = \cup\{f_{\eta\restriction n} \restriction C^0_\eta : n < \omega\}$ is an automorphism of N.
[Why? As $\langle f_{\eta\restriction n} \restriction C^0_{\eta\restriction n} : n < \omega\rangle$ is an increasing sequence of functions by clauses (b) + (c) + (f), the union f_η is a partial function; as in addition each f_η is an automorphism of N by clause (b), also f_η is a partial automorphism of N. Recalling $\langle b^*_\ell : \ell < n\rangle$ list N, clearly f_η have domain N by clause (c) and as $f_{\eta\restriction n}(C^0_{\eta\restriction n}) = C^1_{\eta\restriction n}$ the union f_η has range N by clause (c).] Hence for some $M_\eta \in K_{\aleph_0}$ there is an isomorphism f^+_η from M onto M_η extending f. Now for some $p_\eta \in \mathbf{D}(N), f_\eta(\bar{a})$ materialize p_η in (M_η, N). Choose a countable $L \subseteq \mathbb{L}^0_{\omega_1,\omega}(\tau^+)$ which include $\{\varphi_\eta(\bar{x}, \bar{y}_\eta) : \eta \in {}^{\omega>}2\}$. Easily if $\eta\hat{}\langle\ell\rangle \triangleleft \eta_\ell \in {}^\omega 2$ for $\ell = 0, 1$ then $\varphi(\bar{x}, \bar{a}^1_\eta) \in p_0, \neg\varphi(\bar{x}, \bar{a}^1_\eta) \in p_1$. So $\eta \neq \nu \in {}^\omega 2 \Rightarrow p_\eta \cap L \neq p_\nu \cap L$ by clauses (d) + (e), contradiction to 5.4(4) as we can use $\leq \aleph_0$ formulas to distinguish.
2) Follows.
3) Trivial. $\square_{5.6}$

5.9 Claim. *1) Suppose $N_0 \leq_{\mathfrak{K}} N_1 \in K_{\aleph_0}$ and N_1 forces that $\bar{a}, \bar{b}$ (in N_1) realize the same $\mathbb{L}^0_{\omega_1,\omega}(N_0)$-type over N_0, <u>then</u> N_1 forces that they realize the same $\mathbb{L}^0_{\omega_1,\omega}(N_0; N_0)$-type; (the inverse is trivial).*
1A) Suppose $N_0 \subseteq_{\mathfrak{K}} N_\ell \in K_{\aleph_0}$ and $\bar{a}_\ell \in {}^{\omega>}(N_\ell)$ for $\ell = 1, 2$ and $\mathrm{gtp}(\bar{a}_1, N_0, N_1) = \mathrm{gtp}(\bar{a}_2, N_0, N_1)$ *<u>then</u> we can find (N^+_1, N^+_2, f) such that $N_1 \leq_{\mathfrak{K}} N^+_1 \in K_{\aleph_0}, N_2 \leq_{\mathfrak{K}} N^+_2 \in K_{\aleph_0}$ and f is an isomorphism from N^+_1 onto N^+_2 over N_0 mappnig $\bar{a}_1$ to $\bar{a}_2$.*
2) If $N_0 \leq_{\mathfrak{K}} N_1 \leq_{\mathfrak{K}} N_2 \in K_{\aleph_0}$ and $\bar{a}, \bar{b} \in N_2$ (remember N_2 determines the complete $\mathbb{L}^0_{\omega_1,\omega}(N_1)$-generic types of $\bar{a}, \bar{b}$) <u>then</u> from the

$\mathbb{L}^0_{\omega_1,\omega}(N_1)$-generic type of $\bar{a}$ over N_1 we can compute the $\mathbb{L}^0_{\omega_1,\omega}(N_0)$-generic type of $\bar{a}$ over N_0 (hence if the $\mathbb{L}^0_{\omega_1,\omega}(N_1)$-generic types of $\bar{a}, \bar{b}$ over N_1 are equal, then so are the $\mathbb{L}^0_{\omega_1,\omega}(N_0)$-generic types of $\bar{a}, \bar{b}$ over N_0).
3) For every $N_a \in K_{\aleph_0}$ there is a one-to-one function f from $\mathbf{D}(N)$ onto $\mathbf{D}^*(N)$ such that: if $N \subseteq_{\mathfrak{K}} M \in K_{\aleph_0}$ and $\bar{a} \in {}^{\omega>}M$ then $f(\mathrm{gtp}(\bar{a}, N, M)) = \mathrm{gtp}_{\mathbb{L}_{\omega_1,\omega}(N;N)}(\bar{a}; N; N; M)$.

Remark. 1) So there is no essential difference between $\mathbf{D}(N)$ and $\mathbf{D}^*(N)$.
2) Recall that in a formula of $\mathbb{L}^0_{\omega_1,\omega}(N_0; N_0)$ all $c \in N_0$ may appear as individual constants.

Proof. 1) We shall prove there are N_2 such that $N_1 \leq_{\mathfrak{K}} N_2 \in K_{\aleph_0}$ and an automorphism of N_2 over N_0 taking $\bar{a}$ to $\bar{b}$; this clearly suffices; and we prove the existence of such N_2, of course, by hence and forth arguments. We shall use 5.4(2) freely. So by renaming and symmetry, it suffices to prove that

(∗) if $m < \omega$, $N_0 \leq_{\mathfrak{K}} N_0$ and $\bar{a}, \bar{b} \in {}^m(N_1)$ materialize the same $\mathbb{L}^0_{\infty,\omega}(N_0)$-type over N_0 then for every $c \in N_1$, there are N_2 and $d \in N_2$ such that $\bar{a}\hat{}<c>, \bar{b}\hat{}<d>$ materialize the same $\mathbb{L}^0_{\omega_1,\omega}(N_0)$-type over N_0.

However, by the previous claim 5.4, for some $\bar{a}^* \in {}^{\omega>}(N_0)$ the $\mathbb{L}^0_{\omega_1,\omega}(N_0)$-type over N_0 that $\bar{a}\hat{}<c>$ materialize in (N_1, N_0) does not $\mathbb{L}^0_{\omega_1,\omega}(\tau^{+0})$-split over $\bar{a}^*$. Now $\bar{a}, \bar{b}$ materialize in (N_1, N_0) the same $\mathbb{L}^0_{\omega_1,\omega}(N_0)$-type over N_0 hence $\bar{a}^*\hat{}\bar{a}, \bar{a}^*\hat{}\bar{b}$ materialize in (N_1, N_0) the same $\mathbb{L}^0_{\omega_1,\omega}(N_0)$-type. Hence there is $N_2, N_1 \leq_{\mathfrak{K}} N_2 \in K_0$ and an automorphism f of N_2 mapping N_0 onto N_1 and mapping $\bar{a}^*\hat{}\bar{a}$ to $\bar{a}^*\hat{}\bar{b}$ (but possibly $f \restriction N_0 \neq \mathrm{id}_{N_0}$), this holds by the last sentence in 4.13(c). Let $d = f(c)$, hence if $\bar{a}\hat{}<c>, \bar{b}\hat{}<d>$ materialize the same $\mathbb{L}^0_{\omega_1,\omega}(N_0)$-type in (N_2, N_0) then they materialize the same $\mathbb{L}^0_{\omega_1,\omega}(N_0)$-type over N_0 in (N_2, N_0).
1A) Similarly to part (1).
2) Clearly it suffices to prove the "hence " part. By the assumption and proof of 5.9(1) there are N_3 satisfying $N_2 \leq_{\mathfrak{K}} N_3 \in K_{\aleph_0}$ and f

an automorphism of N_3 over N_1 taking $\bar{a}$ to $\bar{b}$. Now the conclusion follows.
3) Should be clear. $\square_{5.9}$

5.10 Definition. 1) We say that $\mathbf{D}_*$ is a $\mathfrak{K}$-diagram function when

- (a) $\mathbf{D}_*$ is a function with domain $K_{\aleph_0}$ (later we shall lift it to K)
- (b) $\mathbf{D}_*(N) \subseteq \mathbf{D}(N)$ and has at least one non-algebraic member for $N \in K_{\aleph_0}$
- (c) if $N_1, N_2 \in K_{\aleph_0}$ and f is an isomorphism from N_1 onto N_2 then f maps $\mathbf{D}_*(N_1)$ onto $\mathbf{D}_*(N_2)$, this applies in particular to an automorphism of $N \in K_{\aleph_0}$.

1A) Such $\mathbf{D}_*$ is called weakly good <u>when</u>:

- (d) (α) $\mathbf{D}_*(N)$ is closed under subtypes, that is: if $p(\bar{x}) \in \mathbf{D}_*(N), \bar{x} = \langle x_\ell : \ell < m\rangle, \pi$ is a function from $\{0, \ldots, m-1\}$ into $\{0, \ldots, n-1\}$
  then some (necessarily unique) $\bar{q}(\langle x_0, \ldots, x_{n-1}\rangle) \in \mathbf{D}_*(N)$ is equal to $\{\varphi(\langle x_0, \ldots, \bar{x}_{n-1}\rangle) : \varphi(x_{\pi(0)}, \ldots, x_{\pi(m-1)}) \in p(\bar{x})\}$

  (β) if $N \leq_{\mathfrak{K}} M \in K_{\aleph_0}, \bar{a}_1, \bar{b}_1 \in {}^{\omega>}N, \bar{a}_2 \in {}^{\ell g(\bar{a}_1)}M$ and $(M, \bar{a}_1) \cong (M, \bar{a}_2)$ and $\mathrm{gtp}_{\mathbb{L}_{\omega_1,\omega}(\tau^+)}(\bar{a}_2; N; M) \in \mathbf{D}(N)$
  <u>then</u> for some $M^+, \bar{b}_2$ we have $M \leq_{\mathfrak{K}} M^+ \in K_{\aleph_0}$, $\bar{b}_2 \in {}^{\ell g(\bar{b}_1)}(M^+)$, $(M^+, \bar{a}_1{}^\frown\bar{b}_1) \cong (M^+, \bar{a}_2{}^\frown\bar{b}_2)$ and $\mathrm{gtp}_{\mathbb{L}_{\omega_1,\omega}(\tau^+)}(\bar{a}_2{}^\frown\bar{b}_2; N; M^+)$

  (γ) if $N \leq_{\mathfrak{K}} M \in K_{\aleph_0}, \bar{a} \in {}^{\omega>}M$ and $\bar{b} \in {}^{\omega>}N$ and $\mathrm{gtp}_{\mathbb{L}_{\omega_1,\omega}(\tau^+)}(\bar{a}; N; M) \in \mathbf{D}(N)$ then $\mathrm{gtp}_{\mathbb{L}_{\omega_1,\omega}(\tau^+)}(\bar{a}{}^\frown\bar{b}; N; M) \in \mathbf{D}(N)$.

2) Such $\mathbf{D}_*$ is called countable if $N \in K_{\aleph_0} \Rightarrow |\mathbf{D}_*(N)| \leq \aleph_0$.
3) Such $\mathbf{D}_*$ is called good when it is weakly good (i.e., clause (d) holds) and

- (e) $\mathbf{D}_*(N)$ has amalgamation (i.e., if $p_0(\bar{x}), p_1(\bar{x}, \bar{y}), p_2(\bar{x}, \bar{z}) \in \mathbf{D}_*(N)$ and $p_0 \subseteq p_1 \cap p_2$ then there is $q(\bar{x}, \bar{y}, \bar{z}) \in \mathbf{D}_*(N)$ which includes $p_1(\bar{x}, \bar{y}) \cup p_2(\bar{x}, \bar{z})$).

4) Such $\mathbf{D}_*$ is called very good if it is good and:

(*f*) $N_0 \leq_{\mathfrak{K}} N_1 \leq_{\mathfrak{K}} N_2 \in K_{\aleph_0}, \bar{a}_0 \subseteq \bar{a}_1 \subseteq \bar{a}_2$ and $\bar{a}_\ell \subseteq N_\ell$ for $\ell = 0, 1, 2$ and $\text{gtp}(\bar{a}_{\ell+1}, N_\ell, N_{\ell+1})$ is definable over $\bar{a}_\ell$ and belongs to $\mathbf{D}_*(N_\ell)$ for $\ell = 0, 1$ then $\text{gtp}(\bar{a}_2, N_0, N_2)$ belongs to $\mathbf{D}_*(N_0)$ and is definable over $\bar{a}_0$.

5.11 Remark. 1) Note that if $\mathbf{D}$ is a weakly good $\mathfrak{K}$-diagram function and $N \in K_{\aleph_0}$ and $p \in \mathbf{D}(N)$ then we can find $(M, \bar{a})$ such that $N \leq_{\mathfrak{K}} M \in K_{\aleph_0}, \bar{a} \in {}^{\omega>}M, p = \text{gtp}_{\mathbb{L}_{\omega_1,\omega}(\tau^+)}(\bar{a}; N; M)$ and for every $\bar{b} \in {}^{\omega>}M$ the type $\text{gtp}_{\mathbb{L}_{\omega_1,\omega}(\tau^+)}(\bar{b}; N; M)$ belongs to $\mathbf{D}(N)$.
2) If moreover, $\mathbf{D}$ is a good $\mathfrak{K}$-diagram function then we can demand above that M is $(\mathbf{D}(N), \aleph_0)^*$-homogeneous, see Definition 5.14(1) below.
3) On very good $\mathbf{D}$ see 5.12(2).
4) The $\mathbf{D}_\alpha$'s in 5.12 below are very good $\mathfrak{K}$-diagrams and for us it sufices to have then the properties mentioned above, so we do not elaborate.

5.12 Fact. 1) There are $\mathbf{D}_\alpha, \mathbf{D}^*_\alpha$ for $\alpha < \omega_1$, functions with domain $K_{\aleph_0}$ such that:

(*a*) for $N \in K_{\aleph_0}, \mathbf{D}_\alpha(N), \mathbf{D}^*_\alpha(N)$ is a countable subset of $\mathbf{D}(N)$, $\mathbf{D}^*(N)$ respectively

(*b*) for each $N \in K_{\aleph_0}, \langle \mathbf{D}_\alpha(N) : \alpha < \omega_1 \rangle$ as well as $\langle \mathbf{D}^*_\alpha(N) : \alpha < \omega_1 \rangle$ are increasing continuous

(*c*) $\mathbf{D}(N) = \bigcup\limits_{\alpha<\omega_1} \mathbf{D}_\alpha(N)$ and $\mathbf{D}^*(N) = \bigcup\limits_{\alpha<\omega_1} \mathbf{D}^*_\alpha(N)$

(*d*) if $N_1, N_2 \in K_{\aleph_0}, f$ is an isomorphism from N_1 onto N_2 then f maps $\mathbf{D}_\alpha(N_1)$ onto $\mathbf{D}_\alpha(N_2)$ and $\mathbf{D}^*_\alpha(N_1)$ onto $\mathbf{D}^*_\alpha(N_2)$ for $\alpha < \omega_1$

(*e*) for every $\alpha < \omega_1$ and $N \in K_{\aleph_0}$ there is a $(\mathbf{D}_\alpha(N), \aleph_0)^*$-homogeneous model (see below Definition 5.14(1)) (obviously it is unique up to isomorphism over N)

(f) if $N_0 \le_{\mathfrak{K}} N_1 \le_{\mathfrak{K}} N_2 \in K_{\aleph_0}$, N_2 is $(\mathbf{D}_\alpha(N_1), \aleph_0)^*$-homogeneous (see Definition 5.14(1) below) and N_1 is $(\mathbf{D}_\alpha(N_0), \aleph_0)^*$-homogeneous or just $(\mathbf{D}_\beta(N_0), \aleph_0)^*$-homogeneous for some $\beta \le \alpha$ or just $\bar{b} \in {}^{\omega >}(N_1) \Rightarrow \text{gtp}_{\mathbb{L}_{\omega_1,\omega}(\tau^+)}(\bar{b}; N_0; N_1) \in \mathbf{D}(N_0)$ <u>then</u> N_2 is $(\mathbf{D}_\alpha(N_0), \aleph_0)^*$-homogeneous

$(f)^+$ if $\langle \alpha_\varepsilon : \varepsilon \le \zeta \rangle$ is increasing continuous sequence of countable ordinals, $\zeta > 0$ and $\langle N_\varepsilon : \varepsilon \le \zeta \rangle$ is $\le_{\mathfrak{K}}$-increasing continuous, $N_\varepsilon \in \mathfrak{K}_{\aleph_0}$, for every $\bar{a} \in N_{\varepsilon+1}$, $\text{gtp}(\bar{a}, N_\varepsilon, N_{\varepsilon+1}) \in \mathbf{D}_\alpha(N_\varepsilon)$ and for every $\xi < \zeta$ for some $\varepsilon \in [\xi, \zeta)$, $N_{\varepsilon+1}$ is $(\mathbf{D}_{\alpha_\varepsilon}(N_\varepsilon), \aleph_0)^*$-homogeneous <u>then</u> N_ζ is $(\mathbf{D}_{\alpha_\zeta}(N_0), \aleph_0)^*$-homogeneous

(g) N_1 is $(\mathbf{D}_\alpha(N_0), \aleph_0)^*$-homogeneous <u>if and only if</u> N_1 is $(\mathbf{D}^*_\alpha(N_0), \aleph_0)^*$-homogeneous where $N_0 \le_{\mathfrak{K}} N_1 \in K_{\aleph_0}$

(h) $\mathbf{D}_\alpha$ is a very good countable $\mathfrak{K}$-diagram function.

2) If $\mathbf{D}$ is very good then clauses (d),(e),(f),$(f)^+$ hold for it (and also (g), defining $\mathbf{D}^*$ as $f''(\mathbf{D})$, f from 5.16(3).

5.13 Remark. 1) We can add

(h) if $\mathfrak{K}, <^*$ are as derived from the $\psi \in \mathbb{L}_{\omega_1,\omega}(\mathbf{Q})$ in the proof of 3.18(2) then we can add: if $N_0 \le_{\mathfrak{K}} N_1 \in K_{\aleph_0}$ and every $p \in \mathbf{D}_0(N_0)$ is materialized in N_1 then $N_0 <^* N_1$.

2) So our results apply to $\psi \in \mathbb{L}_{\omega_1,\omega}(\mathbf{Q})$, too.
3) So it follows that if $\langle N_i : i \le \alpha \rangle$ is $\le_{\mathfrak{K}}$-increasing in $K_{\aleph_0}$, N_{i+1} is $(\mathbf{D}_{\beta_i}(N_0), \aleph_0)^*$-homogeneous and $\langle \beta_i : i < \alpha \rangle$ is non-decreasing with supremum β then N_α is $(\mathbf{D}_\beta, \aleph_0)^*$-homogeneous.
4) So by 5.12(1)(h) each $\mathbf{D}_\alpha$ is very good and countable.

Proof of 5.12. First, $\mathbf{D}$ is a $\mathfrak{K}$-diagram function by Definition 5.2 and 5.4(9). As $\mathbf{D}(N)$ has cardinality $\le \aleph_1$ by 5.4(6) we can find a sequence $\langle \mathbf{D}_\alpha : \alpha < \omega_1 \rangle$ such that

⊛ (a) $\mathbf{D}_\alpha$ is a countable $\mathfrak{K}$-diagram function

(b) for every $N \in K_{\aleph_0}$ the sequence $\langle \mathbf{D}_\alpha(N) : \alpha < \omega_1 \rangle$ is increasing continuous with union $\mathbf{D}(N)$.

Second, $\mathbf{D}$ is very good (clause (f) of 5.10 holds obviously but to prove that it reflects to $\mathbf{D}_\alpha$ for a cub of $\alpha < \omega_1$ we need 5.22 below, no vicious circle; the other - easier).

Third, note that for each of the demands (d),(e),(f) from Definition 5.10, for a club of $\delta < \omega_1, \mathbf{D}_\delta$ satisfies it. So without loss of generality each $\mathbf{D}_\alpha$ is very good.

The parts on $\mathbf{D}^*_\alpha$ follow by 5.9, and see 5.16(1) below which does not rely on 5.12-5.15 (and see proof of 5.18). $\square_{5.12}$

5.14 Definition. Assume $N_0 \leq_{\mathfrak{K}} N_1 \in K_{\aleph_0}$ and $\mathbf{D}_*$ is a $\mathfrak{K}$-diagram.
1) We say that (N_1, N_0) or just N_1 is $(\mathbf{D}_*(N_0), \aleph_0)^*$-homogeneous over N_0 (but we may omit the "over N_0") if:

- (a) every $\bar{a} \in N_1$ materializes in (N_1, N_0) over N_0 some $p \in \mathbf{D}_*(N_0)$ and every $q \in \mathbf{D}_\alpha(N_0)$ is materialized in (N_0, N_1) by some $\bar{b} \in N_1$
- (b) if $\bar{a}, \bar{b} \in N_1, \bar{a}, \bar{b}$ materialize in (N_1, N_0) the same type over N_0 and $c \in N_1$ <u>then</u> for some $d \in N_1$ sequence $\bar{a}\hat{\ } < c >, \bar{b}\hat{\ } < d >$ materialize in (N_1, N_0) the same type from $\mathbf{D}_*(N_0)$.

2) Similarly for $(\mathbf{D}^*_*(N_0), \aleph_0)^*$-homogeneity, pedantically we have to say $(N_1, N_0; N_0)$ is $(\mathbf{D}^*(N), \aleph_0)^*$-homogeneous, but normally say N_1 is.

5.15 Remark. 1) Now this is meaningful only for $N \leq_{\mathfrak{K}} M \in K_{\aleph_0}$, but later it becomes meaningful for any $N \leq_{\mathfrak{K}} M \in K$.
2) Uniqueness for such countable models hold in this context too.

Now by 5.9.

5.16 Conclusion. If (N_1, N_0) is $(\mathbf{D}_\alpha(N_0), \aleph_0)^*$-homogeneous <u>then</u> N_1, i.e. $(N_1, N_0, c)_{c \in N_0}$ is $(\mathbf{D}^*_\alpha(N_0), \aleph_0)^*$-homogeneous.

Proof. This is easy by 5.9(1) and clause (g) of 5.12. $\square_{5.16}$

5.17 Lemma. *There is $N^* \in K_{\aleph_1}$ such that $N^* = \bigcup_{\alpha<\omega_1} N_\alpha$ and $N_\alpha \in K_{\aleph_0}$ is $\leq_{\mathfrak{K}}$-increasing continuous with α and $N_{\alpha+1}$ is $(\mathbf{D}_{\alpha+1}(N_\alpha), \aleph_0)^*$-homogeneous for $\alpha < \omega_1$.*

Proof. Should be clear. $\square_{5.17}$

5.18 Theorem. *The $N^* \in K_{\aleph_1}$ from 5.17 is unique (even not depending on the choice of $\mathbf{D}_\alpha(N)$'s), is universal and is, $(\mathbb{D}(\mathfrak{K}), \aleph_1)$-model-homogeneous hence model-homogeneous (for $\mathfrak{K}$).*

Proof.

Uniqueness: For $\ell = 0, 1$ let $N^\ell_\alpha, \mathbf{D}^\ell_\alpha$ $(\alpha < \omega_1)$ be as in 5.12, 5.17 and we should prove $\bigcup_{\alpha<\omega_1} N^0_\alpha \cong \bigcup_{\alpha<\omega_1} N^1_\alpha$; because of clause (g) of 5.12 it does not matter if we use the $\mathbf{D}$ or $\mathbf{D}^*$ version. As $\mathbf{D}^\ell_\alpha$ $(\alpha < \omega_1)$ is increasing and continuous, $|\mathbf{D}^\ell_\alpha(N)| \leq \aleph_0$ and $\bigcup_{\alpha<\omega_1} \mathbf{D}^\ell_\alpha(N) = \mathbf{D}(N)$ for every $N \in K_{\aleph_0}$ and the $\mathbf{D}^\ell_\alpha$'s commute with isomorphisms, clearly there is a closed unbounded $E \subseteq \omega_1$ consisting of limit ordinals, such that $\alpha \in E \Rightarrow \mathbf{D}^0_\alpha = \mathbf{D}^1_\alpha$. Let $E = \{\alpha(i) : i < \omega_1\}, \alpha(i)$ increasing and continuous. Now we define by induction on $i < \omega_1$, an isomorphism f_i from $N^0_{\alpha(i)}$ on $N^1_{\alpha(i)}$, increasing with i. For $i = 0$ use the $\aleph_0$-categoricity of K and for limit i let $f_i = \bigcup_{j<i} f_j$. Suppose f_i is defined, then by clause (d) of 5.12 the function f_i maps $\mathbf{D}^0_{\alpha(i+1)}(N^0_{\alpha(i)})$ onto $\mathbf{D}^0_{\alpha(i+1)}(N^1_{\alpha(i)})$ and by the choice of $E, \mathbf{D}^0_{\alpha(i+1)} = \mathbf{D}^1_{\alpha(i+1)}$. By the assumption on the N^ℓ_α and clause $(f)^+$ of 5.12, $N^\ell_{\alpha(i+1)}$ is $(\mathbf{D}^\ell_{\alpha(i+1)}(N^\ell_{\alpha(i)}), \aleph_0)^*$-homogeneous. Summing up those facts and 5.12(e) we see that we can extend f_i to an isomorphism f_{i+1} from $N^0_{\alpha(i+1)}$ onto $N^1_{\alpha(i+1)}$.

Now $\bigcup_{i<\omega_1} f_i$ is the required isomorphism.

Universality: Let $M \in K_{\aleph_1}$, so $M = \bigcup_{\alpha<\omega_1} M_\alpha, M_\alpha$ is $\leq_{\mathfrak{K}}$-increasing continuous and $\|M_\alpha\| \leq \aleph_0$. We now define $f_\alpha, N_\alpha, \gamma_\alpha$ by induction on $\alpha < \omega_1$ such that: $\gamma_\alpha \in [\alpha, \omega_1)$ is increasing continuous with α, f_α is a $\leq_{\mathfrak{K}}$-embedding of M_α into $N_\alpha \in K_{\aleph_0}$, N_α is $\leq_{\mathfrak{K}}$-increasing continuous, f_α is increasing and continuous, and for $\beta < \alpha, N_{\beta+1}$ is $(\mathbf{D}_{\gamma_{\beta+1}}(N_\beta), \aleph_0)^*$-homogeneous. For $\alpha = 0$ let $N_\alpha = M_\alpha$ and $f_\alpha = \mathrm{id}_{N_\alpha}$. For α limit use unions. Let $\alpha = \beta + 1$, we use the $\aleph_0$-amalgmation property (which holds by 3.8, 4.8). So there is a pair (f_α, N'_α) such that $N_\beta \leq_{\mathfrak{K}} N'_\alpha \in K_{\aleph_0}$ and f_α is a $\leq_{\mathfrak{K}}$-embedding of M_α into N'_α extending f_β. The set $\{\mathrm{gtp}(\bar{a}, N_\beta, N'_\alpha), \bar{a} \in {}^{\omega>}(N'_\alpha)\}$ is a countable subset of $\mathbf{D}(N_\beta)$ hence is $\subseteq \mathbf{D}_{\gamma_\alpha}(N_\beta)$ for some $\gamma \in (\gamma_\beta, \omega_1)$. By 5.12(1)(c) there is N_α which $\leq_{\mathfrak{K}}$-extends N'_α and is $(\mathbf{D}_{\gamma_\alpha}(N'_\alpha), \aleph_0)^*$-homogeneous; by 5.12(1)(f) we are done. So $f = \cup\{f_\alpha : \alpha < \omega_1\}$ embeds M into $N = \cup\{N_\alpha : \alpha < \omega_1\}$ which is isomorphic to N^* by the uniqueness. So the universality follows from the uniqueness.

<u>$(\mathbb{D}(\mathfrak{K}), \aleph_1)$-Model-homogeneity</u>: So let $\langle N_\alpha : \alpha < \omega_1\rangle, \mathbf{D}_\alpha, N^*$ be as in 5.12, 5.17 and we are given (M_0, M_1, M_0^+, f) such that $M_0 \leq_{\mathfrak{K}} M_0^+ \in K_{\aleph_0}, M_1 \leq_{\mathfrak{K}} N^*, f$ an isomorphism from M_0 onto M_1. For some $\gamma < \omega_1$ we have $M_1 \leq_{\mathfrak{K}} N_\gamma$.

Now $\{\mathrm{gtp}(\bar{a}, M_0, M_0^+) : \bar{a} \in {}^{\omega>}(M_0^+)\}$ is a countable subset of $\mathbf{D}(M_0)$ hence $\subseteq \mathbf{D}_{\gamma_0}(M_0)$ for some $\gamma_0 < \omega_1$; also $\{\mathrm{gtp}(\bar{a}, M_1, N_\gamma) : \bar{a} \in {}^{\omega>}(N_\gamma)\}$ is a countable subset of $\mathbf{D}(M_1)$ hence $\subseteq \mathbf{D}_{\gamma_1}(M_1)$ for some $\gamma_1 < \omega_1$. Let $\beta = \max\{\gamma, \gamma_0, \gamma_1\}$ and let $M_0^* \in K_{\aleph_0}$ be $(\mathbf{D}_\beta(M_0^+), \aleph_0)^*$-homogeneous so $M_0^+ \leq_{\mathfrak{K}} M_0^*$, exists by 5.12(1)(e), hence $M_0^* \in K_{\aleph_0}$ is $(\mathbf{D}_\beta(M_0), \aleph_0)^*$-homogeneous by 5.12(1)(f) because $\beta \geq \gamma_0$. Now N_β is $(\mathbf{D}(N_\gamma), \aleph_0)^*$-homogeneous by 5.12(1), so as $\beta \geq \gamma_1$ is follows that N_β is $(\mathbf{D}_\gamma(M_1), \aleph_0)^*$-homogeneous.

By 5.12(1)(d),(e) we can extend f to an isomorphic g from M_0^* onto N_β, so $g \restriction M_0^+$ is a $\leq_{\mathfrak{K}}$-embedding of M_0^+ into N.

We can deduce "N^* is a model-homogeneous directly; let $M_0, M_1 \leq_{\mathfrak{K}} N^*$ be countable, and f is isomorphic from M_0 onto M_1. Let $\gamma < \omega_1$ be such that $M_0, M_1 \leq_{\mathfrak{K}} N_\gamma$, Let γ_ℓ be such that $\{\mathrm{gtp}(\bar{a}, M_\ell, N_\gamma) : \bar{a} \in {}^{\omega>}(N_\gamma)\} \subseteq \mathbf{D}_{\gamma_\ell}(M_\ell)$ for $\ell = 0, 1$ and let $\beta = \max\{\gamma, \gamma_0, \gamma_1\} + 1$. As above N_β is $(\mathbf{D}_\beta(M_\ell), \aleph_0)^*$-homogeneous, and now we choose an automorphism f_α of N_α increasing with α and extended f for $\alpha \in [\beta, \omega_1)$ by induction. Now $\cup\{f_\alpha : \alpha \in (\beta, \omega_1)\}$ is an automorphism of N^* extending f. $\square_{5.18}$

5.19 Definition. 1) If $N_0 \leq_{\mathfrak{K}} N_1 \in K_{\aleph_0}$ and $p_\ell \in \mathbf{D}(N_\ell)$ for $\ell = 1, 2$ and they are definable in the same way (see Definition 5.7 (and 5.6), so in particular both do not split over the same finite subset of N_0), then we call p_1 the stationarization of p_0 over N_1.
2) For $p_\ell \in \mathbf{D}(N_\ell)$ for $\ell = 0, 1$ let $p_1 \models p_0$ mean that $N_0 \leq_{\mathfrak{K}} N_1$ and if $N_1 \leq_{\mathfrak{K}} N_2 \in K_{\aleph_0}$ and $\bar{a} \in N_2$ materializes p_1 then it materializes p_0.

5.20 Remark. It is easy to justify the uniqueness implied by "the stationarization".

Observe

5.21 Claim. *If* $p_\ell = \text{gtp}(\bar{a}, N_\ell, N_2)$ *for* $\ell = 0, 1$ *and* $N_0 \leq_{\mathfrak{K}} N_1 \leq_{\mathfrak{K}} N_2 \in K_{\aleph_0}$ *then* $p_1 \models p_0$.

Proof. Easy. $\square_{5.21}$

5.22 Claim. *1) Suppose* $N_0 \leq_{\mathfrak{K}} N_1 \leq_{\mathfrak{K}} N_2 \in K_{\aleph_0}, \bar{a}_i \in N_i$, *(for* $i = 0, 1, 2$*),* $\bar{a}_0 \subseteq \bar{a}_1 \subseteq \bar{a}_2$, *i.e. the ranges increase,* $\text{gtp}(\bar{a}_1, N_0, N_1)$ *is definable over* $\bar{a}_0$ *and* $\text{gtp}(\bar{a}_2, N_1, N_2)$ *is definable over* $\bar{a}_1$. *Then* $\text{gtp}(\bar{a}_2, N_0, N_2)$ *is definable over* $\bar{a}_0$. *Moreover, the definition depends only on the definitions mentioned previously.*
2) If $N_0 \leq_{\mathfrak{K}} N_1 \leq_{\mathfrak{K}} N_2$ *and* $p_\ell \in \mathbf{D}(N_\ell)$ *for* $\ell = 0, 1, 2$ *and* $p_{\ell+1}$ *is the stationarization of* p_ℓ *over* $N_{\ell+1}$ *for* $\ell = 0, 1$, *then* p_2 *is the stationarization of* p_0 *over* N_2.

Proof. 1) So we have to prove that $\text{gtp}(\bar{a}_2, N_0, N_2)$ does not split over $\bar{a}_0$. Let $n < \omega$ and $\bar{b}, \bar{c} \in {}^n N_0$ realize the same type in N_0 over $\bar{a}_0$ (in the logic $\mathbb{L}_{\omega_1,\omega}(\tau_{\mathfrak{K}})$, or even first order logic when every $N \in K_{\aleph_0}$ is atomic). Now also $\bar{b}\hat{\ }\bar{a}_1, \bar{c}\hat{\ }\bar{a}_1$ materialize the same $\mathbb{L}_{\omega_1,\omega}(N_0)$-type in N_1 hence they realize the same $\mathbb{L}_{\omega_1,\omega}(\tau_{\mathfrak{K}})$-type (recall 5.4(8)). Hence $\bar{b}, \bar{c}$ realize the same $\mathbb{L}_{\omega_1,\omega}(\tau_{\mathfrak{K}})$-type in N_1 over $\bar{a}_1$ in N_1. But $\text{gtp}(\bar{a}_2, N_0, N_2)$ does not split over $\bar{a}_1$, so by the previous sentence we get that $\bar{b}\hat{\ }\bar{a}_2, \bar{c}\hat{\ }\bar{a}_2$ materializes the same $\mathbb{L}_{\omega_1,\omega}(N_0)$-type in N_2.
2) Easy. The "moreover" is proved similarly. $\square_{5.22}$

5.23 Lemma. *Suppose* $N_0 \leq_{\mathfrak{K}} N_1 \in K_{\aleph_0}, p_\ell \in \mathbf{D}(N_\ell)$ *and* p_1 *is a stationarization of* p_0 *over* N_1, *then* $p_1 \models p_0$, *i.e., every sequence materializing* p_1 *materializes* p_0 *in any* N_2 *such that* $N_1 \leq_{\mathfrak{K}} N_2$.

Remark. 1) In [Sh 48], [Sh 87a], [Sh 87b] and [Sh:c] the parallel proof of the claims were totally trivial, but here we need to invoke $\dot{I}(\aleph_1, K) < 2^{\aleph_1}$.
2) A particular case can be proved in the context of §4.

Proof. So suppose N_0, N_1, p_0, p_1 contradict the claim and let $\bar{a}^* \in N_0$ be such that p_0 is definable over $\bar{a}^*$ so p_1, too. By 5.12(e)+(f) there are $\delta < \omega_1$ and $N_2 \in K_{\aleph_0}$ satisfying $N_1 \leq_{\mathfrak{K}} N_2$ such that N_2 is $(\mathbf{D}^*_\delta(N_\ell), \aleph_0)^*$-homogeneous for $\ell = 0, 1$. We can find $p_2 \in \mathbf{D}(N_2)$ which is the stationarization of p_0, p_1. It is enough to prove that $p_2 \models p_1$.
[Why? First, note that there is an automorphism f of N_2 which maps N_1 onto N_0 and $f(\bar{a}^*) = \bar{a}^*$ hence $f(p_2) = p_2, f(p_1) = p_0$ hence $p_2 \models p_0$. Now assume that $N_1 \leq_{\mathfrak{K}} N_1^+ \in K_{\aleph_0}$ and $\bar{a}_1 \in N_1^+$ materializes p_1 clearly we can find $N_2^+, \bar{a}_2$ such that $N_2 \leq_{\mathfrak{K}} N_2^+ \in K_{\aleph_0}$ and $\bar{a}_2 \in N_2^+$ which materializes p_2, as we are assuming $p_2 \models p_1$ it also materializes p_1 hence there are N_3, f such that $N_1^+ \leq_{\mathfrak{K}} N_3 \in K_{\aleph_0}$ and f is a $\leq_{\mathfrak{K}}$-embedding of N_2^+ into N_3 over N_1 mapping $\bar{a}_2$ to $\bar{a}_1$. But $p_2 \models p_0$ (see above) hence $f(\bar{a}_2) = \bar{a}_1$ materializes p_0 and p_1, too.]

So without loss of generality for some δ

$\circledast$ N_1 is $(\mathbf{D}^*_\delta(N_0), \aleph_0)^*$-homogeneous over N_0.

For $N \in K_{\aleph_0}, N_0 \leq_{\mathfrak{K}} N$, let p_N be the stationarization of p over N; so

$\boxtimes_1$ if $N_0 \leq_{\mathfrak{K}} N \in K_{\aleph_0}$ then p_N is definable over $\bar{a}^*$.

Without loss of generality the universes of N_0, N_1 are $\omega, \omega \times 2$ respectively.

Now we choose by induction on α a model $N_\alpha \in K_{\aleph_0}$ $(\alpha < \omega_1), |N_\alpha| = \omega(1+\alpha), [\beta < \alpha \Rightarrow N_\beta \leq_{\mathfrak{K}} N_\alpha]$; N_0, N_1 are the ones mentioned in the claim and $\bar{a}_\alpha \in N_{\alpha+1}$ materializes the stationarization

$p_\alpha \in \mathbf{D}^*_\delta(N_\alpha)$ of p_0 over N_α and for $\alpha < \beta, N_\beta$ is $(\mathbf{D}^*_\delta(N_\alpha), \aleph_0)$-homogeneous (see 5.12(f),(f)$^+$). Recalling that $\mathfrak{K}$ is categorical in $\aleph_0$ (and uniqueness over N_0 of $(\mathbf{D}_\delta(N_0), \aleph_0)^*$-homogeneous models) we have $\alpha > \beta \Rightarrow (N_\alpha, N_\beta) \cong (N_1, N_0)$ so, recalling $\circledast$ clearly $\bar{a}_\alpha$ does not materialize p_{N_β} (in $N_{\alpha+1}$). Let $N = \cup\{N_\alpha : \alpha < \omega_1\}$. Let $\mathfrak{B}$ be $(\mathscr{H}(\aleph_2), \in)$ expanded by $N, K \cap \mathscr{H}(\aleph_2), \leq_{\mathfrak{K}}\restriction \mathscr{H}(\aleph_2)$ and anything else which is necessary. Let $\mathfrak{B}^-$ be a countable elementary submodel of $\mathfrak{B}$ to which $\langle N_\alpha : \alpha < \omega_1\rangle, N$ belong and let $\delta(*) = \mathfrak{B}^- \cap \omega_1$. For any stationary co-stationary $S \subseteq \omega_1$, let $\mathfrak{B}_S$ be a model which is

- (α) $\mathfrak{B}_S$ an elementary extension of $\mathfrak{B}^-$
- (β) $\mathfrak{B}_S$ is an end extension of $\mathfrak{B}^-$ for ω_1, that is, if $\mathfrak{B}_S \models$ "$s < t$ are countable ordinals" and $t \in \mathfrak{B}^-$ then $s \in \mathfrak{B}^-$
- (γ) among the $\mathfrak{B}_S$-countable ordinals not in $\mathfrak{B}^-$ there is no first one
- (δ) "the set of countable ordinals" of $\mathfrak{B}_S$ is $I_S, I_S = \bigcup\limits_{\alpha<\omega_1} I^S_\alpha$, even I^S_0 is not well ordered, each I_α a countable initial segment of $I_S, \alpha < \beta \Rightarrow I^S_\alpha \subseteq I^S_\beta \wedge I^S_\alpha \neq I^S_\beta$
- (ε) $I_S \backslash I^S_\alpha$ has a first element if and only if $\alpha \in S$ and then we call it $s(\alpha)$.

In particular ω and finite sets are standard in $\mathfrak{B}_S$. For $s \in I_S$, $N_s[\mathfrak{B}_s] := N^{\mathfrak{B}_S}_s$ is defined naturally, and so is $N^S = N^{\mathfrak{B}_S}$; clearly $N^{\mathfrak{B}_S}_s \in K_{\aleph_0}$ is $\leq_{\mathfrak{K}}$-increasing with $s \in I$ as those definitions are Σ^1_1 (as $\mathfrak{K}$ is $\mathrm{PC}_{\aleph_0}$). Let $N^S_\alpha = \bigcup\limits_{s \in I_\alpha} N^{\mathfrak{B}_S}_s$ and let $s+1$ be the successor of s in I_S.

So

$\boxplus$ if $\mathfrak{B}_S \models$ "$s < t$ are countable ordinals", then $(N^{\mathfrak{B}_S}_t, N^{\mathfrak{B}_S}_s)$ is $(\mathbf{D}^*_\delta(N^{\mathfrak{B}_S}_s), \aleph_0)^*$-homogeneous and if $s \in I_\alpha$ then N^S_α is $(\mathbf{D}^*_\delta(N^{\mathfrak{B}_S}_1), \aleph_0)^*$-homogeneous.

If $\alpha \in S$ then clearly the type $p = p_{N^S_\alpha}$ satisfies (using absoluteness from $\mathfrak{B}_S$ because N^S_α is definable in $\mathfrak{B}_S$ as $N^{\mathfrak{B}_S}_{s(\alpha)}$):

- (a) p is materialized in N^S (i.e. in N^S_β for a club of $\beta \in S$)

but by the assumption toward contradiction

(*b*) for a closed unbounded $E \subseteq \omega_1$ for no $\beta \in E \cap S, \beta > \alpha^*$ and $\gamma \in (\beta, \omega_1)$ does a sequence from N^S materialize both $p = p_{N^S_\alpha}$ and its stationarization $p_{N^S_\beta}$ over N^S_β in N^S_γ (again remember $N^S_\alpha = N^{\mathfrak{B}_S}_{s(\alpha)}$ because $\alpha \in S$)

and similarly

(*c*) for a closed unbounded set of $\beta > \alpha, N^S_\beta$ is $(\mathbf{D}^*_\delta(N^S_\alpha), \aleph_0)^*$-homogeneous.

We shall prove that every $\alpha < \omega_1$,

⊡ if $\alpha \notin S$ then α cannot satisfy the statement (c) above.

This is sufficient because if $S(1), S(2) \subseteq \omega_1$ are stationary co-stationary, f is an isomorphism from $N^{S(1)}$ onto $N^{S(2)}$ mapping $\bar{a}^*$ to itself, then for a closed unbounded set $E \subseteq \omega_1$, for each $\alpha < \omega_1$ the function f maps $N^{S(1)}_\alpha$ onto $N^{S(2)}_\alpha$, hence the property above is preserved, hence $S(1) \cap E = S(2) \cap E$. But there is a sequence $\langle S_i : i < 2^{\aleph_1} \rangle$ of subsets of ω_1 such that for $i \neq j$ the set $S_i \backslash S_j$ is stationary. So by 0.3 we have $\dot{I}(\aleph_1, K) = 2^{\aleph_1}$, contradiction.

So suppose $\alpha \in \omega_1 \backslash S, p = p_{N^S_\alpha}$ and clause (c) above hold; but obviously $(c) \Rightarrow (a)$, recalling $p_0 \in \mathbf{D}_\delta(N_0)$ hence $p_{N^S_\alpha} \in \mathbf{D}_\delta(N^S_\alpha)$ so let $\bar{a} \in N^S$ materialize p in N^S and we shall get a contradiction.

There are elements $0 = t(0) < t(1) < \ldots < t(k)$ of I^S and $\bar{a}_0 \in N_0 = N^{\mathfrak{B}_S}_{t(0)}, \bar{a}_{\ell+1} \in N^{\mathfrak{B}_S}_{t(\ell)+1}$ such that $\bar{a} \subseteq \bar{a}_k, \bar{a}^* \subseteq \bar{a}_0, \bar{a}_\ell \subseteq \bar{a}_{\ell+1}$ and $\mathrm{gtp}(\bar{a}_{\ell+1}, N^{\mathfrak{B}_S}_{t(\ell)}, N^{\mathfrak{B}_S}_{t(\ell+1)})$ is definable over $\bar{a}_\ell$ and if $t(\ell+1)$ is a successor (in I_S) then it is the successor of $t(\ell)$ and if limit in I^S then $\bar{a}_\ell = \bar{a}_{\ell+1}$.
[Why do they exist? Because of the sentence saying that for every $\bar{a}$ we can find such $k, t(\ell)(\ell \leq k), \bar{a}_\ell(\ell \leq k)$ as above is satisfied by $\mathfrak{B}$ and involve parameters which belong to $\mathfrak{B}^-$ hence to $\mathfrak{B}_S$, etc., so $\mathfrak{B}_S$ inherits it (and finiteness is absolute from $\mathfrak{B}_S$)]. It follows that $\mathrm{gtp}(\bar{a}, N^{\mathfrak{B}_S}_{t(\ell)}, N^{\mathfrak{B}_S}_{t(k)})$ is definable over $\bar{a}_\ell$ for each $\ell < k$.

Clearly $t(0) = 0 \in I_\alpha$ but $t(k) \notin I_\alpha$ (otherwise $t(k)+1 \in I_\alpha$ hence $\bar{a} \in N^{\mathfrak{B}_S}_{t(k)+1} \leq_{\mathfrak{K}} N^S_\alpha$, impossible as p is a non-algebraic type over

$N_\alpha^{\mathfrak{B}_S}$). Hence for some ℓ we have $t(\ell) \in I_\alpha, t(\ell+1) \notin I_\alpha$. By the construction $t(\ell+1)$ is limit (in I^S) hence $\bar{a}_{\ell+1} = \bar{a}_\ell$. As $\alpha \notin S$ we can choose $t(*) \in I_S \backslash I_\alpha^S, t(*) < t(\ell+1)$. As we are assuming (toward contradiction) that α, p satisfy clause (c), for some $\beta \in S, s(\beta)$ is well defined and $s(\beta) > t(k)$ (on the definition of $s(\gamma)$ for $\gamma \in S$ see clause (ε) above) and N_β^S is $(\mathbf{D}_\delta^*(N_\alpha^S), \aleph_0)^*$-homogeneous. Now $N_{s(\beta)}^{\mathfrak{B}_S} = N_\beta^S, N_{t(\ell+1)}^{\mathfrak{B}_S}$ are isomorphic over $N_{t(*)}$ (being both $(\mathbf{D}_\delta^*(N_{t(*)}^{\mathfrak{B}_S}), \aleph_0)^*$-homogeneous by the choice of $\mathfrak{B}_S$, see $\boxplus$ above).

So as $N_\alpha^S \leq_{\mathfrak{K}} N_{t(\ell+1)}^{\mathfrak{B}_S} \leq_{\mathfrak{K}} N_{s(\beta)}^{\mathfrak{B}_S} = N_\beta^S$ and, as said above, N_β^S is $(\mathbf{D}_\delta^*(N_\alpha^S), \aleph_0)^*$-homogeneous also $N_{t(\ell+1)}^{\mathfrak{B}_S}$ is $(\mathbf{D}_\delta^*(N_\alpha^S), \aleph_0)^*$-homogeneous, too, hence $(N_{t(\ell+1)}^{\mathfrak{B}_S}, N_\alpha^S, \bar{a}^*) \cong (N_1, N_0, \bar{a}^*)$.

As, by $\boxplus$ above, clearly $N_\alpha^S, N_{t(*)}^{\mathfrak{B}_S}$ are $(\mathbf{D}_\delta^*(N_{t(\ell)+1)}^{\mathfrak{B}_S}), \aleph_0)^*$-homogeneous there is an isomorphism f_0 from N_α^S onto $N_{t(*)}^{\mathfrak{B}_S}$ over $N_{t(\ell)+1}^{\mathfrak{B}_S}$. As $N_{t(\ell+1)}^{\mathfrak{B}_S}$ is $(\mathbf{D}_\delta^*(N_{t(*)}^{\mathfrak{B}_S}), \aleph_0)^*$-homogeneous and $(\mathbf{D}_\delta^*(N_\alpha^S), \aleph_0)^*$-homogeneous by the previous paragraph (where we use β) we can extend f_0 to an automorphism f_1 of $N_{t(\ell+1)}^{\mathfrak{B}_S}$. Let $\gamma \in S \cap E$ satisfy $s(\gamma) \geq t(k)+1$. As $\mathrm{gtp}(\bar{a}_k, N_{t(\ell+1)}^{\mathfrak{B}_S}, N_\gamma^S)$ is definable over $\bar{a}_\ell = \bar{a}_{\ell+1}$ and $\bar{a}_\ell = f_0(\bar{a}_\ell) = f_1(\bar{a}_\ell)$ (as $\bar{a}_\ell \in N_{t(\ell)+1}^{\mathfrak{B}_S}$) and $N_{\gamma+1}^S$ is $(\mathbf{D}_\delta^*(N_{t(\ell+1)}^{\mathfrak{B}_S}), \aleph_0)^*$-homogeneous, we can extend f_1 to an automorphism f_2 of N_γ^S satisfying $f_2(\bar{a}_k) = \bar{a}_k$.

Notice that by the choice of $\langle \bar{a}_\ell : \ell \leq k \rangle$ and $\langle t(\ell) : \ell \leq k \rangle$ it follows that for any $m < k$, $\mathrm{gtp}(\bar{a}_k, N_{t(m)}, N_{t(k)+1})$ does not split over $\bar{a}_m$ hence is definable over it by 5.22, and recall that we know that $\bar{a}_\ell = \bar{a}_{\ell+1}$.

So there is in N^S a sequence materializing both $\mathrm{gtp}(\bar{a}, N_\alpha^S, N_\gamma^S) = p_{N_\alpha^S}$ and its stationarization over $N_{t(\ell+1)}^S$: just $\bar{a}(\subseteq \bar{a}_k)$ (so use f_2). This contradicts the assumption as $(N_1, N_0, \bar{a}^*) \cong (N_{t(\ell+1)}^{\mathfrak{B}_S}, N_\alpha^S, \bar{a}^*)$. $\square_{5.23}$

The following claim 5.24(5)-(9) and Definition 5.25 are closely related.

5.24 Claim. *1) If $\bar{a} \in N_0 \leq_{\mathfrak{K}} N_1 \leq_{\mathfrak{K}} N_2 \in K_{\aleph_0}, \bar{b} \in N_2, p_1 = \mathrm{gtp}(\bar{b}, N_1, N_2)$ is definable over $\bar{a} \in N_0$, <u>then</u> $p_0 = \mathrm{gtp}(\bar{b}, N_0, N_2)$ is*

definable in the same way over $\bar{a}$, *hence* $\text{gtp}(\bar{b}, N_1, N_2)$ *is its stationarization.*
2) For a fixed countable $M \in K_{\aleph_0}$ *to have a common stationarization in* $\mathbf{D}(N')$ *for some* N' *satisfying* $M \leq_{\mathfrak{K}} N'$ *or* $N' \leq_{\mathfrak{K}} M$ *is an equivalence relation over* $\{p$: *for some* $N \leq_{\mathfrak{K}} M, p \in \mathbf{D}(N)\}$ *(and we can choose the common stationarization in* $\mathbf{D}(M)$ *as a representative). So if* $N_0 \leq_{\mathfrak{K}} N_1 \leq_{\mathfrak{K}} N_2 \in K_{\aleph_0}, p_\ell \in \mathbf{D}(N_\ell)$ *for* $\ell = 0, 1, 2$ *and* p_1, p_2 *are stationarizations of* p_0 *then* $p_2 \models p_1$.
3) If $N_\alpha \in K_{\aleph_0}$ $(\alpha \leq \omega + 1)$ *is* $\leq_{\mathfrak{K}}$*-increasing and continuous and* $\bar{a} \in N_{\omega+1}$ *then for some* $n < \omega$, *for every* k *we have:* $n < k \leq \alpha \leq \omega$ *implies* $\text{gtp}(\bar{a}, N_\alpha, N_{\omega+1})$ *is the stationarization of* $\text{gtp}(\bar{a}, N_k, N_{\omega+1})$.
4) If $N \leq_{\mathfrak{K}} M \in K, N \in K_{\aleph_0}, \bar{a} \in M$ *then for all* $M' \in K_{\aleph_0}$, *satisfying* $\bar{a} \in M', N \leq_{\mathfrak{K}} M' \leq_{\mathfrak{K}} M, \text{gtp}(\bar{a}, N, M')$ *is the same, we call it* $\text{gtp}(\bar{a}, N, M)$ *(the new point is that* M *is not necessarily countable. This is compatible with Definition 5.25(c) being a special case).*
5) Suppose $N_0 \leq_{\mathfrak{K}} N_1$ *(in* K*),* $\bar{a} \in N_1$, *then there is a countable* $M \leq_{\mathfrak{K}} N_0$, *such that for every countable* M' *satisfying* $M \leq_{\mathfrak{K}} M' \leq_{\mathfrak{K}} N_0$ *we have* $\text{gtp}(\bar{a}, M', N_1)$ *is the stationarization of* $\text{gtp}(\bar{a}, M, N_1)$. *Moreover there is a finite* $A \subseteq N_0$ *such that any countable* $M \leq_{\mathfrak{K}} N_0$ *which includes* A *is O.K. So* $\text{gtp}(\bar{a}, N_0, N_1)$ *from 5.25(c) is well defined and* $\in \mathbf{D}(N_0)$ *and is definable over some finite* $A \subseteq N_0$.
6) The parallel of Part (3) holds for $N_\alpha \in K$, *too, and any limit ordinal instead of* ω. *That is if* $\langle N_\alpha : \alpha \leq \delta + 1 \rangle$ *is* $\leq_{\mathfrak{K}}$*-increasing continuous and* $\bar{a} \in N_{\delta+1}$, *then for some* $\alpha < \delta$ *and countable* $M \leq_{\mathfrak{K}} N_\alpha$ *we have:* $M \leq_{\mathfrak{K}} M' \leq_{\mathfrak{K}} M_\delta \Rightarrow \text{gtp}(\bar{a}, M', M_\delta)$ *is the stationarization of* $\text{gtp}(\bar{a}, M, M_\delta)$; *similarly for every* $p \in \mathbf{D}(N_\delta)$.
7) If $N_0 \leq_{\mathfrak{K}} N_1 \leq_{\mathfrak{K}} N_2 \leq_{\mathfrak{K}} N_3 \leq_{\mathfrak{K}} N_4$ *and* $\bar{a} \in N_4$ *and* $\text{gtp}(\bar{a}, N_3, N_4)$ *is the stationarization of* $\text{gtp}(\bar{a}, N_0, N_4)$ *then* $\text{gtp}(\bar{a}, N_2, N_4)$ *is the stationarization of* $\text{gtp}(\bar{a}, N_1, N_3)$. *Also if* $\bar{b}'$ *satisfies* $\text{Rang}(\bar{b}') \subseteq \text{Rang}(\bar{a})$ *and* $\text{gtp}(\bar{a}, N_2, N_4)$ *is the stationarization of* $\text{gtp}(\bar{a}, N_1, N_4)$ *then this holds also for* $\bar{b}$. *We can replace* $\text{gtp}(\bar{a}, N_3, N_4)$ *by* $p \in \mathbf{D}(N_4)$.
8) If $N_0 \leq_{\mathfrak{K}} N_1 \leq_{\mathfrak{K}} N_2 \in K_{\aleph_0}$ *and* $p_\ell \in \mathbf{D}(N_\ell)$ *for* $\ell = 0, 1, 2$ *and* $p_{\ell+1}$ *is the stationarization of* p_ℓ *for* $\ell = 0, 1$ *then* p_2 *is the stationarization of* p_0.
9) If $\langle M_\alpha : \alpha \leq \delta + 1 \rangle$ *is* $\leq_{\mathfrak{K}}$*-increasing continuous,* δ *a limit ordinal and* $\bar{a} \in {}^{\omega>}(M_{\delta+1})$ *then*

(a) *for some* $\alpha < \delta$ *we have* $\mathrm{gtp}(\bar{a}, M_\beta, M_{\delta+1})$ *is the stationarization of* $\mathrm{gtp}(\bar{a}, M_\alpha, M_{\delta+1})$ *whenever* $\beta \in [\alpha, \delta)$

(b) *if* $\mathrm{gtp}(\bar{a}, M_\alpha, M_{\delta+1})$ *is the stationarization of* $\mathrm{gtp}(\bar{a}, M_0, M_{\delta+1})$ *for every* $\alpha < \delta$ *then this holds for* $\alpha = \delta$, *too.*

10) If $\langle M_\alpha : \alpha \le \delta \rangle$ *is* $\le_{\mathfrak{K}}$*-increasing continuous,* δ *a limit ordinal and* $p_\delta \in \mathbf{D}(M_\delta)$, *<u>then</u> for some* $\alpha < \beta$ *there is* $p_\alpha \in \mathbf{D}(M_\alpha)$ *such that* p_δ *is the stationarization of* p_α.
11) Those definitions in 5.25 are compatible with the ones for countable models.
12) $\mathrm{gtp}(\bar{a}, N, M)$ *(where* $\bar{a} \in M, N \le_{\mathfrak{K}} M$ *are both is* K*) is the stationarization over* N *of* $\mathrm{gtp}(\bar{a}, N', M)$ *for every large enough countable* $N' \le_{\mathfrak{K}} N$, *see 5.24(5).*

Proof. 1) As we can replace N_2 by any N_2' satisfying $N_2 \le_{\mathfrak{K}} N_2' \in K_{\aleph_0}$,. without loss of generality for some α, N_2 is $(\mathbf{D}^*_\alpha(N_0), \aleph_0)^*$-homogeneous and $(\mathbf{D}^*_\alpha(N_1), \aleph_0)^*$-homogeneous. Let $p_2 \in \mathbf{D}(N_2)$ be the stationarization of p_1 over N_2.

So by 5.23 we get $p_2 \models p_1$. On the other hand, clearly there is an isomorphism f_0 from N_0 onto N_1 such that $f_0(\bar{a}) = \bar{a}$; and by the assumption above on N_2, f_0 can be extended to an automorphism f_1 of N_2.

Note that f_1 maps $p_0 = \mathrm{gtp}(\bar{b}, N_0, N_2)$ to $p_0' := \mathrm{gtp}(f_1(\bar{b}), f_1(N_0), N_2)$ and maps p_2 to itself as $f_0(\bar{a}) = \bar{a}$.

Now $p_1 \models p_0$ (by the choices of p_1, p_0) and $p_2 \models p_1$ by 5.9(1), together $p_2 \models p_0$. As $f_1(p_2) = p_2, f_1(p_0) = p_0'$ it follows that $p_2 \models p_0'$. As also $p_2 \models p_1$ and $p_0', p_1 \in \mathbf{D}(N_1)$ it follows that $p_0' = p_1$ hence p_1, p_0' have the same definition over $\bar{a}$, but now also $p_0 \in \mathbf{D}(N_0), p_0' \in \mathbf{D}(N_1)$ have the same definition over $\bar{a}$ (using f_1), together also p_1, p_0 have the same definition over $\bar{a}$, which means that p_1 is the stationarization of p_0 over N_1 and we are done.
2) Trivial.
3) By part (1).
4) Easy.
5) By (3) and (4).
6)-12) Easy by now. $\square_{5.24}$

5.25 Definition. By 5.24(5) the type $\mathrm{gtp}(\bar{a}, M, N)$ can be reasonably defined when $M \leq_{\mathfrak{K}} N, \bar{a} \in {}^{\omega>}N$ and we can define $\mathbf{D}(N)$ and $\mathbf{D}_*(N)$, $\mathrm{gtp}(\bar{a}, N, M)$ and stationarization for not necessarily countable N and $N \leq_{\mathfrak{K}} M \in K$. Everything still holds, except that maybe some p's are not materialized in any $\leq_{\mathfrak{K}}$-extension of N.

More formally

- (a) if $N \leq_{\mathfrak{K}} M$ and $N \in K_{\aleph_0}$ and $p \in \mathbf{D}(N)$ then the stationarization of p over M is $\cup\{q : N_1 \in K_{\aleph_0}$ satisfies $N_1 \leq_{\mathfrak{K}} N_1 \leq_{\mathfrak{K}} M$ and q is the stationarization of $p \in \mathbf{D}(N_1)\}$
- (b) if $M \in \mathfrak{K}$ then $\mathbf{D}(M) = \{q$: for some countable $N \leq_{\mathfrak{K}} M$ and $p \in \mathbf{D}(N)$ the type q is the stationarization of p over $M\}$, similarly for $\mathbf{D}_*$, a $\mathfrak{K}$-diagram
- (c) if $N \leq_{\mathfrak{K}} M$ and $\bar{a} \in {}^{\omega>}M$ then $\mathrm{gtp}(\bar{a}, N, M)$ is defined as $\cup\{\mathrm{gtp}(\bar{a}, N', M') : N_0 \leq_{\mathfrak{K}} N' \leq_{\mathfrak{K}} M' \in K_{\aleph_0}, M' \leq_{\mathfrak{K}} M, N' \leq_{\mathfrak{K}} N\}$ for every countable large enough $N_0 \leq_{\mathfrak{K}} N$; it is well defined and belongs to $\mathbf{D}(N)$ by 5.24(5) and we say $\bar{a}$ materializes $\mathrm{gtp}(\bar{a}, N, M)$ in M
- (d) if $N \in \mathfrak{K}, N \leq_{\mathfrak{K}} M$ and $p \in \mathbf{D}(N)$ is definable over the countable $N_0 \leq_{\mathfrak{K}} N$ equivalently is the stationarization of some $p' \in \mathbf{D}(N_0)$, <u>then</u> the stationarization of p over M is the stationarization of p' over M, see clause (a), equivalently $\cup\{p_{M_0} : N_0 \leq_{\mathfrak{K}} M_0 \leq_{\mathfrak{K}} M, M_0$ is countable$\}$ where p_{M_0} is the stationarization of $p' \in \mathbf{D}(N_0)$ over M_0; it belongs to $\mathbf{D}(N_0)$
- (e) if $p(\bar{x}, \bar{y}) \in \mathbf{D}(M)$ then $p(\bar{x}, \bar{y}) \restriction \bar{x} \in \mathbf{D}(M)$ is naturally defined; 5.2(3) similarly for permuting the variables
- (f) for $N \leq_{\mathfrak{K}} M$ we say that M is $(\mathbf{D}(N), \aleph_0)^*$-homogeneous <u>when</u> for every $p(\bar{x}, \bar{y}) \in \mathbf{D}(N)$ and $\bar{a} \in {}^{\ell g(\bar{x})}(M)$ materializing $p(\bar{x}, \bar{y}) \restriction x$ in M there is $\bar{b} \in {}^{\ell g(\bar{y})}M$ such that $\bar{a}\hat{\ }\bar{b}$ materializes $p(\bar{x}, \bar{y})$ in M.

Remark. Claim 5.26 below strengthens 3.8, it is a step toward non-forking amalgamation.

5.26 Claim. *Suppose* $N_0 \leq_{\mathfrak{K}} N_1 \in K_{\aleph_0}, N_0 \leq_{\mathfrak{K}} N_2 \in K_{\aleph_0}, \bar{a} \in N_1$. *Then we can find* $M, N_0 \leq_{\mathfrak{K}} M \in K_{\aleph_0}$ *and* $\leq_{\mathfrak{K}}$*-embeddings* f_ℓ *of* N_ℓ *into* M *over* N_0 *(for* $\ell = 1, 2$*) such that* $\mathrm{gtp}(f_1(\bar{a}), f_2(N_2), M)$ *is a stationarization of* $p_0 = \mathrm{gtp}(\bar{a}, N_0, N_1)$ *(so* $f_1(\bar{a}) \notin f_2(N_2)$*).*

Proof. Let $p_2 \in \mathbf{D}(N_2)$ be the stationarization of p_0. Clearly we can find an $\alpha < \omega_1$ (in fact, a closed unbounded set of α's) and N'_1, N'_2 from $K_{\aleph_0}$ which are $(D^*_\alpha(N_0), \aleph_0)^*$-homogeneous and $N_\ell \leq_{\mathfrak{K}} N'_\ell$ (for $\ell = 1, 2$) and some $\bar{b} \in N'_2$ materializing p_2. But by 5.23, $\bar{b}$ materializes p_0 hence there is an isomorphism f from N'_1 onto N'_2 over N_0 satisfying $f(\bar{a}) = \bar{b}$, recalling 5.9(1A). Now let $M = N'_2, f_1 = f \restriction N_1, f_2 = \mathrm{id}$. $\square_{5.26}$

5.27 Claim. *1) For any* $N_0 \leq_{\mathfrak{K}} N_1 \in K_{\aleph_1}$ *so* $N_0 \in K_{\leq \aleph_1}$*, there is* N_2 *such that* $N_1 \leq_{\mathfrak{K}} N_2 \in K_{\aleph_1}$ *and* N_2 *is* $(\mathbf{D}(N_0), \aleph_0)^*$*-homogeneous.*
2) Also 5.26 holds for $N_2 \in K_{\aleph_1}$ *(but still* $N_0, N_1 \in K_{\aleph_0}$*).*
3) If $N_0 \leq_{\mathfrak{K}} N_1 \in K_{\aleph_0}$ *and* $N_0 \leq_{\mathfrak{K}} N_2 \in K_{\leq \aleph_1}$ *then we can find* $M \in K_{\leq \aleph_1}$ *and* $\leq_{\mathfrak{K}}$*-embeddings* f_1, f_2 *of* N_1, N_2 *into* M *over* N_0 *respectively such that* $\mathrm{gtp}(f_1(\bar{c}), f_2(N_2), M)$ *is a stationarization of* $\mathrm{gtp}(\bar{c}, N_0, N_1)$ *for every* $\bar{c} \in N_1$*, hence* $f_1(N_1) \cap f_2(N_2) = N_0$*.*
4) $K_{\aleph_2} \neq \emptyset$*.*

Remark. 1) Note that 5.27(3) is another step toward stable amalgamation.
2) Note that 5.27(3) strengthen 5.27(2) hence 5.26.

Proof. 1) As we can iterate $\leq_{\mathfrak{K}}$-increasing N_1 in $K_{\aleph_1}$, it is enough to prove that: if $p(\bar{x}, \bar{y}) \in \mathbf{D}(N_0)$ and $\bar{a} \in N_1$ materializes $p(\bar{x}, \bar{y}) \restriction \bar{x}$ in (N_1, N_0) then for some $N_2 \in K_{\aleph_1}, N_1 \leq_{\mathfrak{K}} N_2$ and for some $\bar{b} \in N_2$ the sequence $\bar{a}\char`^\bar{b}$ materializes $p(\bar{x}, \bar{y})$ in (N_2, N_0). Let $M_0 \leq_{\mathfrak{K}} N_0$ be countable and $q \in \mathbf{D}(M_0)$ be such that $p(\bar{x}, \bar{y})$ a stationarization of q. Without loss of generality if N_0 is countable then $M_0 = N_0$. Note that the case $N_0 = M_0$ is easier. Choose $M_i (0 < i < \omega_1)$ such that $M_i \leq_{\mathfrak{K}} N_1, N_1 = \bigcup_{i<\omega_1} M_i, \langle M_i : i < \omega_1 \rangle$ is $\leq_{\mathfrak{K}}$-increasing continuous sequence of countable models, $M_0 \cup \bar{a} \subseteq M_1$. As $\langle M_i \cap N_0 : i < \omega_1 \rangle$

is an increasing continuous sequence of countable sets with union N_0 clearly for a club of $i < \omega_1, M_i \cap N_0 \leq_{\mathfrak{K}} N_0$ hence $M_i \cap N_0 \leq_{\mathfrak{K}} M_i$. So without loss of generality $i < \omega_1 \Rightarrow M_i \cap N_0 \leq_{\mathfrak{K}} N_0, M_i$. For every $\bar{c} \in N_1$ there is a countable $N_{0,\bar{c}}$ such that $M_0 \leq_{\mathfrak{K}} N_{0,\bar{c}} \leq_{\mathfrak{K}} N_0$ and: if $N_{0,\bar{c}} \leq_{\mathfrak{K}} N' \leq_{\mathfrak{K}} N_0$ and $N' \in K_{\aleph_0}$ then $\mathrm{gtp}(\bar{c}, N', N_1)$ is the stationarization of $\mathrm{gtp}(\bar{c}, N_{0,\bar{c}}, N_1)$. Without loss of generality $\bar{c} \in M_i \Rightarrow N_{0,\bar{c}} \subseteq M_i$ hence

(*) for every $\bar{c} \in M_i$, $\mathrm{gtp}(\bar{c}, N_0, N_1)$ is a stationarization of $\mathrm{gtp}(\bar{c}, N_0 \cap M_i, M_i)$.

We can find $M_1^* \in K_{\aleph_0}$ satisfying $M_1 \leq_{\mathfrak{K}} M_1^*$ and $\bar{b} \in M_1^*$ such that $q = \mathrm{gtp}(\bar{a}\hat{\ }\bar{b}, M_0, M_1^*)$. We can find $\bar{a}_2, \bar{a}_1, \bar{a}_0$ such that $\bar{a}_0 \in M_1 \cap N_0, \bar{a}_1 \in M_1, \bar{a}_2 \in M_1^*, \bar{b} \subseteq \bar{a}_2, \bar{a} \subseteq \bar{a}_1$ and $\bar{a}_0 \trianglelefteq \bar{a}_1 \trianglelefteq \bar{a}_2$ and $\mathrm{gtp}(\bar{a}_2, M_1, M_1^*)$, $\mathrm{gtp}(\bar{a}_1, M_1 \cap N_0, M_1)$ are definable over $\bar{a}_1, \bar{a}_0$, respectively. Now we define $f_j, M_j^*, 1 \leq j < \omega_1$ by induction on i such that:

(*i*) $\langle M_i^* : 1 \leq i \leq j \rangle$ is $\leq_{\mathfrak{K}}$-increasing continuous

(*ii*) M_j^* is countable, M_1^* already given

(*iii*) f_j is a $\leq_{\mathfrak{K}}$-embedding of M_j into M_j^*

(*iv*) f_1 is the identity on M_1

(*v*) f_j is increasing continuous with j

(*vi*) $\mathrm{gtp}(\bar{a}_2, f_j(M_j), M_j^*)$ is the stationarization of $\mathrm{gtp}(\bar{a}_2, M_1, M_1^*)$ (so definable over $\bar{a}_1$).

For $j = 1$ we have it letting $f_j^* = \mathrm{id}_{M_1}$.
For $j > 1$ successor, use 5.26 to define (M_j, f_j) such that $\mathrm{grp}(\bar{a}_2, f_j(M_j), M_j^*)$ is the stationarization of $\mathrm{gtp}(\bar{a}_2, f_{j-1}(M_{j-1}), M_{j-1}^*)$. So clauses (i)-(v) clearly holds. Clause (vi) follows by 5.24(8).
For j limit: let $M_j^* = \bigcup_{1 \leq i < j} M_i^*$ and $f_j = \cup\{f_i : 1 \leq i < j\}$, condition (vi) holds by 5.24(3).

By renaming without loss of generality $f_j = \mathrm{id}_{M_j}$ for $j \in [1, \omega_1)$. By (*) we get that $\mathrm{gtp}(\bar{a}_1, N_0 \cap M_j, M_j^*) = \mathrm{gtp}(\bar{a}_1, N_0 \cap M_j, M_j)$ is definable over $\bar{a}_0$ (as this holds for $j = 1$). Combining this and clause (vi), by 5.22(1) we get for every $j \geq 1$, that $\mathrm{gtp}(\bar{a}_2, N_0 \cap$

$M_j, M_j^*)$ is the stationarization of $\text{gtp}(\bar{a}_2, N_0 \cap M_1, M_1^*)$. Hence by the choice of $\bar{a}_2, \bar{a}_1, a_0$ and 5.24(7), easily $\text{gtp}(\bar{a}\hat{\ }\bar{b}, N_0 \cap M_j, M_j^*)$ is the stationarization of $\text{gtp}(\bar{a}\hat{\ }\bar{b}, N_0 \cap M_1, M_1^*)$ hence of $\text{gtp}(\bar{a}\hat{\ }\bar{b}, M_0, M_1^*)$. Let $N_2 = \cup\{M_j^* : j \in [1, \omega_1)\}$, clearly $N_1 \leq_{\mathfrak{K}} N_2 \in K_{\aleph_1}$.

So by 5.24(9), clause (c) and the first sentence in the proof, we finish.

2) Similar proof[7] (or use the proof of part (3)).

3) Without loss of generality $N_2 \cong N^*$ from 5.17 (as we can replace N_2 by an extension so use 5.18 and 5.24(7)).

Also (by 5.27(1)) there is $M, N_2 \leq_{\mathfrak{K}} M \in K_{\aleph_1}$ such that M is $(\mathbf{D}(N_2), \aleph_0)^*$-homogeneous. As N_1 is countable there is $\alpha < \omega_1$ such that for every $\bar{c} \in N_1$, $\text{gtp}(\bar{c}, N_0, N_1) \in \mathbf{D}_\alpha(N_0)$. Let $M = \bigcup_{i<\omega_1} M_i$ with $M_i \in K_{\aleph_0}$ being $\leq_{\mathfrak{K}}$-increasing continuous. So for some i we have $\alpha < i < \omega_1, M_i \cap N_2 \leq_{\mathfrak{K}} M$ and (recalling 5.24(6)) for every $\bar{c} \in M_i$, $\text{gtp}(\bar{c}, N_2, M)$ is stationarization of $\text{gtp}(\bar{c}, N_2 \cap M_i, M_i)$ and M_i is $(D_i(N_2 \cap M_i), \aleph_0)^*$-homogeneous. Now we can find an isomorphism f_0 from N_0 onto $N_2 \cap M_i$ (as K is $\aleph_0$-categorical) and extend it to an automorphism f_2 of N_2 (by 5.18-model homogeneity). Also there is N_1' such that $N_1 \leq_{\mathfrak{K}} N_1' \in K_{\aleph_0}$ and N_1' is $(\mathbf{D}_i(N_1), \aleph_0)^*$-homogeneous, hence is $(\mathbf{D}_i(N_0), \aleph_0)^*$-homogeneous (by the choice of α as $\alpha < i$ see 5.12(f)), hence there is an isomorphism f_1' from N_1' onto M_i extending f_0. Now $f_0, f_1' \restriction N_1, f_2, M$ show that amalgamation as required exists (we just change names).

4) Immediate, use 1) or 2) or 3) ω_2-times. $\square_{5.27}$

5.28 Definition. For any $\mathbf{D}_* = \mathbf{D}_\alpha$ for some $\alpha < \omega_1$ (or just any very good $\mathfrak{K}$-diagram $\mathbf{D}_*$, see 5.10, i.e., satisfies the demands on each $\mathbf{D}_\alpha$ in 5.12) we define:

1) $M \leq_{\mathbf{D}_*} N$ if $M \leq_{\mathfrak{K}} N$ and for every $\bar{a} \in N$

$$\text{gtp}(\bar{a}, M, N) \in \mathbf{D}_*(M).$$

2) $K_{\mathbf{D}_*}$ is the class of $M \in K$ which are the union of a family of countable submodels, which is directed by $\leq_{\mathbf{D}_*}$.

3) $\mathfrak{K}_{\mathbf{D}_*} = (K_{\mathbf{D}_*} \leq_{\mathbf{D}_*})$, or pedantically $(K_{\mathbf{D}_*}, \leq_{\mathbf{D}_*} \restriction K_{\mathbf{D}_*})$.

---

[7] here $N_1 \in K_{\aleph_1}$ is O.K.; similar to 2.11(1)

5.29 Claim. *Let* $\mathbf{D}_*$ *be countable and as in 5.28.*
1) The pair $(K_{\mathbf{D}_*}, \le_{\mathbf{D}_*})$ *is an* $\aleph_0$*-presentable a.e.c., that is it satisfies all the axioms from 1.2(1) and is* $\mathrm{PC}_{\aleph_0}$.
2) Also for $(K_{\mathbf{D}_*}, \le_{\mathbf{D}_*})$*, we get* $\mathbf{D}(N)$ *countable and equal to* $\mathbf{D}_*(N)$ *for every countable* $N \in K_{\mathbf{D}_*}$.

Proof. 1) Obviously $K_{\mathbf{D}_*}$ is a class of τ-models and $\le_{\mathbf{D}_*}$ is a two-place relation on K_{D_*}; also they are preserved by isomorphisms. About being $\mathrm{PC}_{\aleph_0}$ note that

$\circledast_1$ $M \in K_{\mathbf{D}_*}$ iff $M \in K$ and for some model $\mathfrak{B}$ with universe $|M|$ and countable vocabulary, for every countable $\mathfrak{B}_1 \subseteq \mathfrak{B}_2 \subseteq \mathfrak{B}$ we have $M \restriction \mathfrak{B}_1 \le_{\mathbf{D}_*} M \restriction \mathfrak{B}_2$ iff there is a directed partial order and $\langle M_t : t \in I\rangle$ such that $M_t \in K_{\aleph_0}$ and $s <_I t \Rightarrow M_s \le_{\mathfrak{K}} M_t$ and $\bar{a} \subseteq M_t \Rightarrow \mathrm{gtp}(\bar{a}, M_s, M_t) \in \mathbf{D}_*(M_s)$

$\circledast_2$ similarly for $M \le_{\mathbf{D}_*} N$.

Ax I: If $M \le_{\mathbf{D}_*} N$ then $M \le_{\mathfrak{K}} N$ hence $M \subseteq N$.

Ax II: The transitivity of $\le_{\mathbf{D}_*}$ holds by 5.10(4), 5.22(1) + Definition 5.25 (works as $\mathbf{D}_*$ is closed enough or use clause (f) of 5.12). The demand $M \le_{\mathbf{D}_*} M$ is trivial[8].

Ax III: Assume $\langle M_i : i < \lambda\rangle$ is $\le_{\mathbf{D}_*}$-increasing continuous and $M = \bigcup\{M_i : i < \lambda\}$. As $\mathfrak{K}$ is an a.e.c. clearly $M \in K$ and $i < \lambda \Rightarrow M_i \le_{\mathfrak{K}} M$. Also for each $i < \lambda$ and $\bar{a} \in M$ for some $j \in (i, \lambda)$ we have $\bar{a} \in M_j$ hence $\mathrm{gtp}(\bar{a}, M_i, M_j) \in \mathbf{D}_*(M_i)$ but recalling 5.24(7) it follows that $\mathrm{gtp}(\bar{a}, M_i, M) = \mathrm{gtp}(\bar{a}, M_i, M_j) \in \mathbf{D}_*(M_i)$. So $i < \lambda \Rightarrow M_i \le_{\mathbf{D}_*} M$. By applying $\circledast_1$ to every M_i and coding we can easily show that $M \in K_{\mathbf{D}_*}$ thus finishing.

Ax IV: Assume $\langle M_i : i < \lambda\rangle, M$ are as above and $i < \lambda \Rightarrow M_i \le_{\mathbf{D}_*} N$. To prove $M \le_{\mathbf{D}_*} N$ note that as $\mathfrak{K}$ is an a.e.c., we have $M \le_{\mathfrak{K}} N$ and consider $\bar{a} \in N$. By 5.24(6) for some $i < \lambda$, $\mathrm{gtp}(\bar{a}, M, N)$ is the stationarization of $\mathrm{gtp}(\bar{a}, M_i, N)$ but the latter belongs to $\mathbf{D}_*(M_i)$ hence $\mathrm{gtp}(\bar{a}, M, N) \in \mathbf{D}_*(M)$ as required.

---

[8]recall that $M \restriction \mathfrak{B} = M \restriction \{a \in M : a \in \mathfrak{B}\}$

Ax V: By $\circledast_2$ this is translated to the case $N_0, N_1, M \in K_{\aleph_0}$ but then it holds easily.

Ax VI: By $\circledast_1 + \circledast_2 +$ Ax VI for $\mathfrak{K}$.
2) So we replace $\mathfrak{K}$ by $\mathfrak{K}' = \mathfrak{K}_{\mathbf{D}_*}$ and easily all that we need for $\mathbf{D}$ for $\mathfrak{K}'$ is satisfied by $\mathbf{D}_*$ (actually repeating the works in §5 till now on $\mathfrak{K}'$ we get it) noting that

- $\circledast$ if $M_0 \le_{\mathbf{D}_*} M_\ell \in K_{\aleph_0}$ for $\ell = 1$ and $\mathrm{gtp}(\bar{a}_1, M_0, M_1) = \mathrm{gtp}(\bar{a}_2, M_0, M_2)$ then there is a triple (M_1^+, M_2^+, f) such that $M_\ell \le_{\mathbf{D}_*} M_\ell^+ \in K_{\aleph_0}, M_\ell^+$ is $(\mathbf{D}(M_i), \aleph_0)^*$-homogeneous for $i = 0, \ell$ and f is an isomorphism from M_1^+ onto M_2^+ over M_0 mapping $\bar{a}_1$ to a_2.

This by:

- $\circledast_1$ if $M_0 \le_{\mathbf{D}_*} M_1 \le_{\mathbf{D}_*} M_2$ and $\bar{a} \in M_1$ then $\mathrm{gtp}(\bar{a}, M_0, M_1) = \mathrm{gtp}(\bar{a}, M_0, M_2) \in \mathbf{D}_*(M_0)$
- $\circledast_2$ if $M_0 \in K_{\aleph_0}$ then for some $M_1 \in K_{\aleph_0}$ we have $M_0 \le_{\mathbf{D}_*} M_2$ and M_1 is $(\mathbf{D}_*(M_0), \aleph_0)^*)$-homogeneous
- $\circledast_3$ if $M_0 \le_{\mathbf{D}_*} M_1 \le_{\mathbf{D}_*} M_2$ and M_2 is $(\mathbf{d}_*(M_1), \aleph_0)^*$-homogeneous then M_2 is $(\mathbf{D}_*(M_0), \aleph_0)^*$-homogeneous
- $\circledast_4$ if $M_0 \le_{\mathbf{D}_*} M_\ell \in K_{\aleph_0}$, $\mathrm{gtp}(\bar{a}_1, M_0, M_1) = \mathrm{gtp}(\bar{a}_2, M_0, M_2)$ then there is an isomorphism from M_1 onto M_2 over M_0 mapping $\bar{a}_1$ to $\bar{a}_2$.

$\square_{5.29}$

5.30 Claim. *Suppose $N_0 \le_{\mathfrak{K}} N_\ell \in K_{\aleph_0}$ $(\ell = 1, 2)$ and $\bar{c} \in N_2$, then there is $M, N_0 \le_{\mathfrak{K}} M$ and $\le_{\mathfrak{K}}$-embeddings f_ℓ of N_ℓ into M over N_0 for $\ell = 1, 2$ such that*

- (i) *for every $\bar{a} \in N_1, \mathrm{gtp}(f_1(\bar{a}), f_2(N_2), M)$ is a stationarization of* $\mathrm{gtp}(\bar{a}, N_0, N_1)$
- (ii) $\mathrm{gtp}(f_2(\bar{c}), f_1(N_1), M)$ *is a stationarization of* $\mathrm{gtp}(\bar{c}, N_0, N_2)$.

Remark. This is one more step toward stable amalgamation: in 5.26 we have gotten it for one $\bar{a} \in N_1$, in 5.27(3) for every $\bar{a} \in N_1$, which gives disjoint amalgamation.

Proof. Clearly we can for $\ell = 1, 2$ replace N_ℓ by any $N'_\ell, N_\ell \leq_{\mathfrak{K}} N'_\ell \in K_{\aleph_0}$, and without loss of generality $N_0 = N_1 \cap N_2$. By 5.27(3) there is $N_3 \in K_{\aleph_0}$ such that $N_\ell \leq_{\mathfrak{K}} N_3$ for $\ell < 3$ and $\bar{a} \in {}^{\omega>}(N_1) \Rightarrow$ gtp$(\bar{a}, N_2, N_3)$ is the stationarization of gtp$(\bar{a}, N_0, N_1)$. So we can assume that for some $\mathbf{D}_\alpha$ as in Definition 5.28 we have N_ℓ is $(\mathbf{D}_\alpha(N_0), \aleph_0)^*$-homogeneous for $\ell = 1, 2$. As in the proof of 5.23, we can find a countable linear order I, such that every element $s \in I$ has an immediate successor $s+1, 0$ is first element and I^* has a subset isomorphic to the rationals (follow really) and models $M_s \in K_{\aleph_0}$, (for $s \in I$) such that $s < t \Rightarrow M_s \leq_{\mathfrak{K}} M_t$ and M_t is $(\mathbf{D}_\alpha(M_s), \aleph_0)$-homogeneous when $s <_I t$, etc. So by 5.24(3) for every initial segment J of I and $t \in I$ such that $J < t$, that is, $(\forall s \in J)(s <_I t)$, if J has no last element and $I \backslash J$ has no first element then M_t is $(\mathbf{D}_\alpha(M_J), \aleph_0)^*$-homogeneous, where $M_J = \bigcup_{s \in J} M_s = \bigcap_{t \in I \backslash J} M_t$. We let $N_0^J = M_J, N_1^J = M_I$ and N_2^J be a $(\mathbf{D}_\alpha(N_0^J), \aleph_0)^*$-homogeneous model satisfying $N_0^J \leq_{\mathfrak{K}} N_2^J$ and without loss of generality $N_1^J \cap N_2^J = N_0^J$. Also easily there is $N'_0 <_{\mathfrak{K}} N_0$ such that gtp$(\bar{c}, N_0, N_1)$ is definable over some $\bar{c}_0 \subseteq N'_0$ and N_0 is $(\mathbf{D}_\alpha(N'_0), \aleph_0)$-homogeneous. Clearly the triples $(N_0, N_1, N_2), (N_0^J, N_1^J, N_2^J)$ are isomorphic and let f_0^J, f_1^J, f_2^J be appropriate isomorphisms such that $f_0^J \subseteq f_1^J, f_2^J$ and without loss of generality $f_0^J(N'_0) = M_0$. Now by 5.27(3), there is $M^J \in K_{\aleph_0}$ satisfying $N_\ell^J \leq_{\mathfrak{K}} M^J$ $(\ell = 0, 1, 2)$ such that for every $\bar{a} \in N_1^J$, gtp$(\bar{a}, N_2^J, M^J)$ is the stationarization of gtp$(\bar{a}, N_0^J, N_1^J)$ and there are $N_3 \in K_{\aleph_0}, N_\ell \leq_{\mathfrak{K}} N_3$ for $\ell = 0, 1, 2$ and an isomorphism $f_3^J \supseteq f_1^J \cup f_2^J$ from N_3 onto M^J.

Suppose our conclusion fails, then gtp$(f_2^J(\bar{c}), N_1^J, M^J)$ is not the stationarization of gtp$(f_2^J(\bar{c}), N_0^J, M^J)$. Moreover, as in the proof of 5.23, $t \in I \backslash J \Rightarrow M_I = N_1^J, M_t$ are isomorphic over $N_0^J = M_J$, hence we can replace N_1^J by M_t for any $t \in I \backslash J$ so as we assume that our conclusion fails, $t \in I \backslash J \Rightarrow$ gtp$(f_2^J(\bar{c}), M_t, M^J)$ is not a stationarization of gtp$(f_2^J(\bar{c}), N_0^J, M^J)$ and the latter is the stationarization of gtp$(f_2^J(\bar{c}), M_0, M^J)$. Let $p_J =$ gtp$(f_2^J(\bar{c}), N_1^J, M^J) =$ gtp$(\bar{c}, M_I, M^J)$; all this was done for any appropriate J. So it is easy to check that $J_1 \neq J_2 \Rightarrow p_{J_1} \neq p_{J_2}$, but as $I^* \subseteq I$ & $|I| = \aleph_0$, we have continuum many such J's hence such p_J's. If CH fails, we are done. Otherwise, note that moreover, we can ensure that for $J_1 \neq J_2$ as above there is an automorphism of M_I taking p_{J_1} to p_{J_2}, hence for some $\beta < \omega_1, \{p_J : J \text{ as above}\} \subseteq \mathbf{D}_\beta(M_I)$, i.e., $(f_1^{J_2}) \circ (f_1^{J_1})^{-1}$ maps one to the other, contradiction by clause (d) of 5.12. (Alternatively repeat the proof of 5.23. More elaborately by the way $\mathbf{D}_\alpha$ was chosen, Claim 5.27(3) holds for $\mathfrak{K}_{\mathbf{D}_*}$ hence without loss of generality M^J is $(\mathbf{D}_\alpha(N_1), \aleph_0)$-homogeneous and so without loss of generality for some $t(*) \in I \backslash J, N_1^J = M_{t(*)}), N^J = M_{t(*)+1}$ and we get a contradiction as in the proof of 5.23 (i.e., the choice of $\langle \bar{a}_\ell : \ell \leq \ell(*) \rangle$ there[9]). $\square_{5.30}$

5.31 Definition. 1) $\mathfrak{K}$ has the symmetry property when the following holds: if $N_0 \leq_{\mathfrak{K}} N_\ell \leq_{\mathfrak{K}} N_3$ $(\ell = 1,2)$ and for every $\bar{a} \in N_1$, $\text{gtp}(\bar{a}, N_2, N_3)$ is the stationarization of $\text{gtp}(\bar{a}, N_0, N_3)$, then for every $\bar{b} \in N_2$, $\text{gtp}(\bar{b}, N_1, N_3)$ is the stationarization of $\text{gtp}(\bar{b}, N_0, N_3)$.
2) If $N_0, N_1, N_2 \leq_{\mathfrak{K}} N_3$ satisfies the assumption and conclusion of part (1) we say that N_1, N_2 are in stable amalgamation over N_0 inside N_3 (or in two-sided stable amalgamation over N_0 inside N_3). If only the hypothesis of (1) holds we say they are in a one sided stable amalgamation over N_0 inside N_3 (then the order of (N_1, N_2) is important).
3) We say that $\mathfrak{K}$ has unique [one sided] amalgamation when: if $N_0 \leq_{\mathfrak{K}} N_\ell \in K_{\aleph_0}$ for $\ell = 1,2$ then N_1, N_2 has unique [one sided] stable amalgamation, see part (4).
4) We say N_1, N_2 have a unique [one sided] stable amalgamation over N_0, where for notational simplicity $N_1 \cap N_2 = N_0$, provided that: if $(*)$, i.e. clauses (a)-(d) below hold then $(**)$ below holds, where:

- $(*)$ (a) $N_1 \leq_{\mathfrak{K}} N_3, N_2 \leq_{\mathfrak{K}} N_3$ and (N_1, N_2) in [one sided] stable amalgamation inside N_3 over N_0 and $\|N_3\| \leq \|N_1\| + \|N_2\|$
  - (b) $M_0 \leq_{\mathfrak{K}} M_\ell \leq_{\mathfrak{K}} M_3$ for $\ell = 1,2$ and (M_1, M_2) are in [one sided] stable amalgamation inside M_3 over M_0 (hence $M_1 \cap M_2 = M_0$)
  - (c) f_ℓ is an isomorphism from N_ℓ onto M_ℓ for $\ell = 0,1,2$
  - (d) $f_0 \subseteq f_1$ and $f_0 \subseteq f_2$
- $(**)$ we can find $M_3', M_3 \leq_{\mathfrak{K}} M_3'$ and f_3, a $\leq_{\mathfrak{K}}$-embedding of N_3 into M_3' extending $f_1 \cup f_2$.

[9]A third way is to use forcing and absoluteness to use the case CH fail

We at last get the existence of stable amalgamation (to which we earlier get approximations).

5.32 Claim. *For any $N_0 \leq_{\mathfrak{K}} N_1, N_2$, all from $K_{\aleph_0}$, we can find $M, N_0 \leq_{\mathfrak{K}} M \in K_{\aleph_0}$ and $\leq_{\mathfrak{K}}$-embeddings f_1, f_2 of N_1, N_2 respectively over N_0 into N such that $N_0, f_1(N_1), f_2(N_1)$ are in stable amalgamation.*

Remark. In the proof we could have "inverted the tables" and used $\bar{c}_\zeta$ in the ω_1 direction.

Proof. We define by induction on $\zeta < \omega_1$, $\langle M^\zeta_\alpha : \alpha < \omega_1\rangle$ and $\bar{c}_\zeta$ such that:

(*i*) $\langle M^\zeta_\alpha : \alpha < \omega_1\rangle$ is $\leq_{\mathfrak{K}}$-increasing continuous and $M^\zeta_\alpha \in K_{\aleph_0}$

(*ii*) for $\alpha < \zeta, M^\zeta_\alpha = M^\alpha_\alpha$ and $\xi < \zeta \;\&\; \alpha < \omega_1 \Rightarrow M^\xi_\alpha \leq_{\mathfrak{K}} M^\zeta_\alpha$

(*iii*) for ζ limit, $M^\zeta_\alpha = \bigcup\limits_{\xi<\zeta} M^\xi_\alpha$

(*iv*) for $\zeta \leq \alpha < \omega_1, \zeta$ non-limit $M^\zeta_{\alpha+1}$ is $(\mathbf{D}_{\alpha+1}(M^\zeta_\alpha), \aleph_0)^*$-homogeneous

(*v*) for every $\bar{c} \in M^\zeta_{\alpha+1}$, $\mathrm{gtp}(\bar{c}, M^{\zeta+1}_\alpha, M^{\zeta+1}_{\alpha+1})$ is a stationarization of $\mathrm{gtp}(\bar{c}, M^\zeta_\alpha, M^\zeta_{\alpha+1})$

(*vi*) $\bar{c}_\zeta \in M^{\zeta+1}_{\zeta+1}$ and for $\zeta+1 < \alpha < \omega_1$, $\mathrm{gtp}(\bar{c}_\zeta, M^\zeta_\alpha, M^{\zeta+1}_\alpha)$ is the stationarization of $\mathrm{gtp}(\bar{c}_\zeta, M^\zeta_{\zeta+1}, M^{\zeta+1}_{\zeta+1})$

(*vii*) for every $p \in \mathbf{D}(M^\xi_\alpha)$ for some ζ satisfying $\xi + \alpha < \zeta < \omega_1$ we have $\mathrm{gtp}(\bar{c}_\zeta, M^\zeta_{\zeta+1}, M^{\zeta+1}_{\zeta+1})$ is a stationarization of p.

There is no problem doing this (by 5.30 and as in earlier constructions); in limit stages we use local character 5.24(3) and $\mathbf{D}_\alpha$ being closed under stationarization.
Now easily for a thin enough closed unbounded set of $E \subseteq \omega_1$, for every $\zeta \in E$ we have

$(*)_\zeta(a)$ M^ξ_ζ is $(\mathbf{D}_\zeta(M^0_\zeta), \aleph_0)^*$-homogeneous (*b*) for every $\bar{c} \in M^\zeta_\zeta$, $\text{gtp}(\bar{c}, \bigcup_{\alpha<\omega_1} M^0_\alpha, \bigcup_{\xi<\omega_1} M^\xi_\xi)$ is a stationarization of $\text{gtp}(\bar{c}, M^0_\zeta, M^\zeta_\zeta)$

(*c*) for every $\bar{c} \in M^0_{\zeta+1}$, $\text{gtp}(\bar{c}, M^{\zeta+1}_\zeta, M^{\zeta+1}_{\zeta+1})$ is a stationarization of gtp $(\bar{c}, M^0_\zeta, M^0_{\zeta+1})$.

[Why? Clause (c) holds by clause (v) of the construction (as $\langle M^\zeta_\varepsilon : \varepsilon \le \zeta\rangle$ is $\le_{\mathfrak{K}}$-increasing continuous). Clause (b) holds as E is thin enough, i.e., is proved as in earlier constructions (i.e., see the proof of (*) in the proof of 5.27(1)). As for Clause (a), first note that by clauses (i),(ii),(iii) the sequence $\langle M^\zeta_\varepsilon : \varepsilon \le \zeta\rangle$ is $\le_{\mathfrak{K}}$-increasing continuous. By clause (vi) we have $\varepsilon < \zeta \Rightarrow \text{gtp}(\bar{c}_\varepsilon, M^\varepsilon_\zeta, M^{\varepsilon+1}_\zeta)$ does not fork over M^ε_ζ and clause (vii) of the construction we have: if $p \in \mathbf{D}_\zeta(M^\zeta_\varepsilon), \varepsilon < \zeta$ then for some $\xi \in (\varepsilon, \zeta)$, $\text{gtp}(\bar{c}_\xi, M^\zeta_\xi, M^\zeta_{\xi+1})$ is a non-forking extension of p. As E is thin enough we have $\bar{d} \in M^\zeta_\zeta \Rightarrow \text{gtp}(\bar{d}, M^\zeta_0, M^\zeta_\zeta) \in \mathbf{D}_\zeta(M^\zeta_0)$. Together it is easy to get clause (a), e.g., see 5.41.]

So as in the proof of 5.27(3) we can finish (choose $\zeta \in E, f_0$ an isomorphism from N_0 onto M^0_ζ, $f_1 \supseteq f_0$ is an $\le_{\mathfrak{K}}$-embedding of N_1 into M^ζ_ζ and $f_2 \supseteq f_0$ a $\le_{\mathfrak{K}}$-embedding of N_2 into $M^0_{\zeta+1}$). $\square_{5.32}$

5.33 Remark. Note that in Chapter II we use only the results up to this point.

5.34 Theorem. *1) Suppose in addition to the hypothesis of this section that $2^{\aleph_1} < 2^{\aleph_2}$ and the club ideal on $\aleph_1$ is not $\aleph_2$-saturated and $\dot{I}(\aleph_2, K) < 2^{\aleph_2}$ or just $\dot{I}(\aleph_2, K(\aleph_1\text{-saturated})) < 2^{\aleph_2}$. Then $\mathfrak{K}$ has the symmetry property.*
2) Assume $2^{\aleph_1} < 2^{\aleph_2}$ and $\dot{I}(\aleph_2, K(\aleph_1\text{-saturated})) < \mu_{\text{unif}}(\aleph_2, 2^{\aleph_1})$; this number is always $> 2^{\aleph_1}$, usually $2^{\aleph_2}$, see 0.5. Then $\mathfrak{K}$ has the symmetry property and stable amalgamation in $K_{\aleph_0}$ is unique (we know that it always exists and it follows by (1) + (2) that one sided amalgamation is unique).

5.35 Discussion: 1) This certainly gives a desirable conclusion. However, part (2) is not used so we shall return to it in Chapter VII.

More elaborately, in VII.4.1, in the "lean" version of Chapter VII, see reading plan A in VII§0, assuming the weak diamond ideal is not $\aleph_2$-saturated we prove 5.34(2) hence we also prove a slight weaker version of 5.34(1), replacing "$\dot{I}(\aleph_2, K)(\aleph_1\text{-saturated}) < 2^{\aleph_2}$" by $\dot{I}(\aleph_2, K(\aleph_1\text{-saturated})) < \mu_{\text{unif}}(\aleph_2, 2^{\aleph_1})$.

Better, in VII.4.40 we prove 5.34(2) fully. Still, the proof given below of part (1) is not covered presently by Chapter VII and it gives nicer reasons for non-isomorphisms (essentially diferent natural invariants).

2) As for part (1), we can avoid using it (except in 5.39 below). More fully, in II§3 dealing with $\mathfrak{K}$ as here by II.3.4 for every $\alpha < \omega_1$ we derive a good $\aleph_0$-frame $\mathfrak{s}_\alpha$ with $\mathfrak{K}^{\mathfrak{s}_\alpha} = \mathfrak{K}_{\mathbf{D}_\alpha}$ (if we would have liked to derive a good $\aleph_1$-frame we would need 5.34).

Then in Chapter III if $\mathfrak{s}$ is successful (holds, e.g. if $2^{\aleph_0} < 2^{\aleph_1} < 2^{\aleph_2}$ and $\dot{I}(\aleph_2, \mathfrak{K}^{\mathfrak{s}_\alpha}) < 2^{\aleph_2}$ and $\text{WDmId}_{\aleph_1}$ is not $\aleph_2$-saturated) then we derive the successor $\mathfrak{s}^+_\alpha$, a good $\aleph_1$-frame with $K^{\mathfrak{s}^+_\alpha} \subseteq \{M \in K^{\mathfrak{s}_\alpha}_{\aleph_1} : M$ is $\aleph_1$-saturated for $K^{\mathfrak{s}_\alpha}\}$, and $\mathfrak{s}^+_\alpha$ is even good$^+$ (see Claim III.1.6(2) and Definition III.1.3). This suffices for the main conclusions of II§9 and end of III§12.

3) Still we may wonder is $\leq_{\mathfrak{s}^+_\alpha} = \leq_{\mathfrak{K}} \restriction \mathfrak{K}_{\mathfrak{s}^+_\alpha}$? If $\mathfrak{s}_\alpha$ is good$^+$ then the answer is yes (see III.1.6(1)). That is, the present theorem 5.34 is used in III§1 to prove $\mathfrak{s}$ is "good$^+$", really this is proved in 5.39. In fact part (1) of 5.34 is enough to prove that $\mathfrak{s}_{\mathbf{D}_*}$ is good$^+$, see III.1.5(1A).

3) The proof of 5.34(1) gives that if $\mathfrak{K}$ fails the symmetry property then $\dot{I}(\aleph_2, K) \geq 2^{\aleph_1}$ even if $2^{\aleph_1} = 2^{\aleph_2}$ and do not use $2^{\aleph_0} = 2^{\aleph_1}$ directly (but use earlier results of §5). The case "$\mathscr{D}_{\aleph_1}$ is $\aleph_2$-saturated, $2^{\aleph_0} < 2^{\aleph_1} < 2^{\aleph_2}, \dot{I}(\aleph_2, \aleph_2) < \mu_{\text{unif}}(\aleph_2, 2^{\aleph_2})$" is covered in Chapter VII.

Proof. 1) So in the first part toward contradiction we can assume that $K^4 \neq \emptyset$ where K^4 is the class of quadruple $\bar{N} = (N_0, N_1, N_2, N_3)$ such that N_1, N_2 are one sided stably amalgamated over N_0 inside N_3 but N_2, N_1 are not. Hence there is $\bar{c} \in N_2$ such that $\text{gtp}(\bar{c}, N_1, N_3)$ is not the stationarization of $\text{gtp}(\bar{c}, N_0, N_2) = \text{gtp}(\bar{c}, N_0, N_3)$. We define a two-place relation $\leq$ on K^4 by $\bar{N}^1 \leq \bar{N}^2$ iff $N^1_0 = N^2_0, N^1_\ell \leq_{\mathfrak{K}} N^2_\ell$ for $\ell = 0, 1, 2$ and $\bar{a} \in N^1_1 \Rightarrow \text{gtp}(\bar{a}, N^2_2, N^2_3)$ is definable over some $\bar{b} \in N_0^1$. Easily this is a partial order and K^4 is closed under union of increasing countable sequences. Hence without loss of generality for some $\mathbf{D}_*, \bar{N}^*$

(∗) (a) $\mathbf{D}_* \in \{\mathbf{D}_\alpha : \alpha < \omega_1\}$
(b) $\bar{N}^* \in K^4$
(c) N_ℓ^* is $(\mathbf{D}_*(N_0^*), \aleph_0)^*$-homogeneous over N_0^* for $\ell = 1, 2$
(d) N_3^* is $(\mathbf{D}_*(N_\ell^*), \aleph_0)^*$-homogeneous over N_ℓ^* for $\ell = 1, 2$

So we have proved

5.36 Observation. To prove 5.34, we can assume that $\mathbf{D} = \mathbf{D}_\alpha$ for $\alpha < \omega_1$, i.e., $\mathbf{D}$ is countable.

Continuation of the proof. A problem is that we still have not proven the existence of a superlimit model of K of cardinality $\aleph_1$ though we have a candidate N^* from 5.17. So we use N^*, but to ensure we get it at limit ordinals (in the induction on $\alpha < \aleph_2$), we have to take a stationary $S_0 \subseteq \omega_1$ with $\omega_1 \backslash S_0$ not small, i.e., $\omega_1 \backslash S_0$ does not belong to the ideal $\mathrm{WDmId}_{\aleph_1}$ from Theorem 0.5 and "devote" it to ensure this, using 5.32.

The point of using S_0 is as follows (this is supposed to help to understand the quotation from Chapter VII):

5.37 Definition. 1) Let $K^{\mathrm{qt}} = \{\bar{N} : \bar{N} = \langle N_\alpha : \alpha < \omega_1\rangle$ be $\leq_{\mathfrak{K}}$-increasing continuous, $N_\alpha \in K_{\aleph_0}, N_{\alpha+1}$ is $(\mathbf{D}_\alpha(N_\alpha), \aleph_0)^*$-homogeneous$\}$.
2) On K^{qt} we define a two-place relation $<_S^a$ (for $S \subseteq \omega_1$) as follows: $\bar{N}^1 <_S^a \bar{N}^2$ if and only if for some closed unbounded $E \subseteq \omega_1$

(a) for every $\alpha \in C$ we have $N_\alpha^1 \leq_{\mathfrak{K}} N_\alpha^2$ and $N_{\alpha+1}^1 \leq_{\mathfrak{K}} N_{\alpha+1}^2$
(b) for every $\alpha < \beta$ from E we have $N_\beta^2 \cap \bigcup_{\alpha<\omega_1} N_\alpha^1 = N_\beta^1$ and N_β^1, N_α^2 are in one sided stable amalgamation over N_α^1 inside N_β^2, i.e. if $\bar{a} \in N_\beta^1$ then $\mathrm{gtp}(\bar{a}, N_\alpha^2, N_\beta^2)$ is the stationarization of $\mathrm{gtp}(\bar{a}, N_\alpha^1, N_\beta^1)$
(c) if $\alpha \in S \cap C$ <u>then</u> $N_\alpha^2, N_{\alpha+1}^1$ are in stable amalgamation over N_α^1 inside $N_{\alpha+1}^2$.

5.38 Fact. 0) The two-place relation $<^a_S$ defined in 5.37 are partial orders on $K^{\rm qt}$ for $n < \omega$.
1) If $\bar N^n \le^a_{S_0} \bar N^{n+1}$ and let E_n exemplify this (as in the Definition 5.37) and let $E_\omega = \bigcap_{n<\omega} E_n, E'_\omega = \{\alpha, \alpha+1 : \alpha \in C_\omega\}$ and let $N^\omega_\alpha = \bigcup_{n<\omega} N^n_\beta$ when $\beta = \text{Min}[E'_\omega \backslash \alpha]$. Then $\langle N^\omega_\alpha : \alpha < \omega_1\rangle \in K_{<\aleph_1}$ and $\bar N^n \le^a_{S_0} \langle N^\omega_\alpha : \alpha < \omega_1\rangle$ for $n < \omega$.
2) If $\langle \bar N^\varepsilon : \varepsilon < \omega_1\rangle$ is $<^a_S$-increasing and $N^\varepsilon = \cup\{N^\varepsilon_\alpha : \alpha < \omega_1\} \in K_{\aleph_1}$ is $\le_{\mathfrak{K}}$-increasing continuous, the club $E_{\varepsilon,\zeta}$ witness $\bar N^\varepsilon \le \bar N^\zeta$ for $\varepsilon < \zeta < \aleph_1$ and $\langle N_\alpha : \alpha < \omega_1\rangle$ a $\le_{\mathfrak{K}}$-representation of N, and for a club of $\alpha < \aleph_1, N_\alpha = \cup\{N^\varepsilon_\alpha : \varepsilon < \alpha\}, N_{\alpha+1} = \cup\{N^\varepsilon_{\alpha+1} : \varepsilon < \alpha\}$ then $\varepsilon < \omega_1 \Rightarrow \bar N^\varepsilon \le^a_{S_0} \bar N$.

Proof. Should be easy by now. $\square_{5.38}$

Returning to the proof of 5.34 it is done as follows.

There is $\langle S_\varepsilon : \varepsilon < \omega_1\rangle$ such that $S_\varepsilon \subseteq \omega_1, \zeta < \varepsilon \Rightarrow S_\zeta \cap S_\varepsilon$ countable and $S_0, S_{\varepsilon+1} \backslash S_\varepsilon \in (\mathscr{D}_{\omega_1})^+$, possible by an assumption.

Now for any $u \subseteq \omega_2$ we choose $N^u_\varepsilon, N^u_\varepsilon$ by induction on $\varepsilon < \omega_2$ such that

⊛(a) $\bar N^u_\varepsilon = \langle N^u_{\varepsilon,\alpha} : \alpha < \omega_1\rangle \in K^{\rm qt}$
(b) $N^u_\varepsilon = \cup\{N^u_{\varepsilon,\alpha} : \alpha < \omega_1\} \in K_{\aleph_1}$
(c) for $\zeta < \varepsilon$ we have $\bar N^u_\zeta <^1_{S_\xi} \bar N^u_\varepsilon$ when $\xi \notin [\zeta, \varepsilon) \cap u$ (we can use $S'_{[\zeta,\varepsilon)}$, the complement of the diagonal union of $\{\langle S_\xi : \varepsilon \in [\zeta, \varepsilon)\rangle \cap u\}$
(d) we can demand continuity as defined implicitly in Fact 5.38
(e) for each $\varepsilon \in u$ for a club of $\alpha < \omega_1$ if $\alpha \in S_\varepsilon$ then $N^u_{\varepsilon+1,\alpha}, N^u_{\varepsilon,\alpha+1}$ are not in stable amalgamation over $N^u_{\varepsilon,\alpha}$ inside $N^u_{\varepsilon+1,\alpha+1}$ (though is in one side).

Lastly, let $N^u = \cup\{N^u_\varepsilon : \varepsilon < \omega_1\} \in K_{\aleph_2}$. Now we can prove that if $u, v \subseteq \omega_2$ and $N^u \approx N^v$ then for some club C of $\omega_2, u \cap C = v \cap C$. So we can easily get $\dot I(\aleph_2, \mathfrak{K}) = 2^{\aleph_2}$ and even $\dot I(\aleph_2, \mathfrak{K}(\aleph_1\text{-saturated})) = 2^{\aleph_2}$. $\square_{5.34}$

5.39 Theorem. *Suppose $\mathfrak{K}$ has the symmetry property (holds if the assumption of 5.34(1) hold). Then $\mathfrak{K}$ has a superlimit model in $\aleph_1$.*

Proof. We have a candidate N^* from 5.17. So let $\langle N_i : i < \delta \rangle$ be $\leq_{\mathfrak{K}}$-increasing, $N_i \cong N^*$ and without loss of generality $\delta = \operatorname{cf}(\delta)$. If $\delta = \omega_1$ this is very easy. If $\delta = \omega$, let $N_\omega = \bigcup_{i<\omega} N_i$ and for each $i \leq \omega$ let $\langle N_i^\alpha : \alpha < \omega_1 \rangle$ be $\leq_{\mathfrak{K}}$-increasing continuous with union N_i and $N_i^\alpha \in K_{\aleph_0}$. Now by restricting ourselves to a club E of α's and renaming it $E = \omega_1$, we get: for $i < j \leq \omega, N_i^\alpha = N_i \cap N_j^\alpha$, and

$\circledast_1$ for any $\alpha < \beta < \omega_1, \bar{a} \in N_\omega^\alpha$ and $i < \omega$, the type $\operatorname{gtp}(\bar{a}, N_i^\beta, N_\omega^\beta)$ is a stationarization of $\operatorname{gtp}(\bar{a}, N_i^\alpha, N_\omega^\alpha)$.

To prove $N_\omega \cong N^*$ it is enough to prove:

$\circledast_2$ if $\alpha < \omega_1, p \in \mathbf{D}(N_\omega^\alpha)$ then some $\bar{b} \subseteq N_\omega$ realizes p in N_ω.

By 5.24(3) there is $i < \omega$ such that p is the stationarization of $q = p \restriction N_i^\alpha \in \mathbf{D}(N_i^\alpha)$. As $N_i \cong N^*$, there is $\bar{b} \subseteq N_i$ which realizes q and we can find $\beta \in (\alpha, \omega_1)$ such that $\bar{b} \subseteq N_i^\beta$. By $\circledast_1$ we have $N_\omega^\alpha, N_i^\beta$ is in one sided stable amalgamation over N_i^α inside N_ω^β (see 5.31(2)).

As we assume $\mathfrak{K}$ has the symmetry property, also $N_i^\beta, N_\omega^\alpha$ is in stable amalgamation over N_i^α inside N_ω^β. In particular, as $\bar{b} \subseteq N_i^\beta$, we have $\operatorname{gtp}(\bar{b}, N_\omega^\alpha, N_\omega^\beta)$ is the stationarization of $\operatorname{gtp}(\bar{b}, N_i^\alpha, N_i^\beta)$ but the latter is $p \restriction N_i^\alpha$ so by uniqueness of stationarization, $p = \operatorname{gtp}(\bar{b}, N_\omega^\alpha, N_\omega^\beta)$ which is $\operatorname{gtp}(\bar{b}, N_\omega^\alpha, N_\omega)$, so p is realized in N_ω as required. $\square_{5.39}$

We have implicitly proved

5.40 Claim. *Assume that $N_0 \leq_{\mathfrak{K}} N_1 \in K_{\aleph_0}$ and $\bar{a}_\ell \in {}^{\omega>}(N_1)$ for $\ell = 1, 2$. Then $(*)_1 \Leftrightarrow (*)_2$ where for $\ell = 1, 2$*

$(*)_\ell$ *there are $M_1, M_2, \bar{b}_1, \bar{b}_2$ such that*

(a) $N_0 \leq_{\mathfrak{K}} M_1 \leq_{\mathfrak{K}} M_2 \in K_{\aleph_1}$

(b) $\bar{a}_k \in {}^{\omega>}(M_k)$ *for* $k = 1, 2$

(c) $\operatorname{gtp}(\bar{b}_{3-\ell}, N_0, M_1) = \operatorname{gtp}(\bar{a}_{3-\ell}, N_0, N_1)$

(d) $\text{gtp}(\bar{b}_\ell, M_1, M_2)$ *is the stationarization of* $\text{gtp}(\bar{a}_\ell, N_0, N_1)$ *from* $\mathbf{D}(M_1)$

(e) $\text{gtp}(\bar{b}_1 {}^\frown \bar{b}_2, N_0, M_2) = \text{gtp}(\bar{a}_1 {}^\frown \bar{a}_2, N_0, N_1)$

Proof. We can deduce it from 5.30 (or immitate the proof of 5.23).

In detail by symmetry it is enough to assume $(*)_2$ and prove $(*)_1$. So let $M_1, M_2, \bar{b}_1, \bar{b}_2$ witness $(*)_2$.

By 5.32 we can find M_2', f such that: $M_2 \leq_{\mathfrak{K}} M_2' \in K_{\aleph_0}, f$ is a $\leq_{\mathfrak{K}}$-embedding of M_2 into M_2' over N_0 such that $M_1, f(M_2)$ is in stable amalgamation over N_0 inside M_2'. Now, as $f(M_2), M_1$ are in one sided stable amalgamation over N_0 inside M_2' by the choice of $(M_1, M_2, \bar{b}_1, \bar{b}_2)$ we get $\text{gtp}(f(\bar{b}_2), M_1, M_2') = \text{gtp}(\bar{b}_2, M_1, M_2')$ hence $\text{gtp}(\bar{b}_1 {}^\frown \bar{b}_2, N_0, M_2') = \text{gtp}(\bar{b}_1 {}^\frown f(\bar{b}_2), N_0, M_2')$.

By the choice of M_1^2, f,$\text{gtp}(\bar{b}_1, f(M_2), M_2')$ is the stationarization of $\text{gtp}(\bar{b}_1, N_0, M_2) = \text{gtp}(\bar{a}_1, N_0, N_1)$. Now $(*)_1$ holds as exemplified by $(f(M_2), M_2', f(\bar{b}_2), \bar{b}_1)$. $\square_{5.40}$

5.41 Exercise. Assume $\alpha \leq \omega_1$ and

(a) $\langle M_i : i \leq \delta \rangle$ is $\leq_{\mathfrak{K}}$-increasing continuous, δ a limit ordinal

(b) if $p \in \mathbf{D}(M_i)$ is realized in M_{i+1} then it $\in \mathbf{D}_\alpha(M_i)$ or just $p \restriction M_0 \in \mathbf{D}(M_0)$

(c) if $i < \delta, p \in \mathbf{D}_\alpha(M_i)$ then p is materialized in M_j for some $j \in (i, \delta)$.

Then M_δ is $(\mathbf{D}_\alpha(M_0), \aleph_0)^*$-homogeneous.

Proof. Easy.

5.42 Discussion: 1) Consider $\psi \in \mathbb{L}_{\omega_1,\omega}(\mathbf{Q}), |\tau_\psi| \leq \aleph_0, 1 \leq \dot{I}(\aleph_1, \psi) < 2^{\aleph_0}$. We translate it to $\mathfrak{K}$ and $<^{**}$ as earlier, see 3.18.
2) What if we waive categoricity in $\aleph_0$? The adoption of this was O.K. as we shrink $\mathfrak{K}$ but not too much. But without shrinking probably we still can say something on the models in $\mathfrak{K}^* = \{M \in \mathfrak{K}_{\geq \aleph_0}$: if $N_0 \leq_{\mathfrak{K}} M, N_0 \in K_{\aleph_0}$ then for some $N_1, N_0 <^* N_1 \leq_{\mathfrak{K}} M\}$ as there are good enough approximations.

§6 Counterexamples

In [Sh 48] the statement of Conclusion 3.8 was proved for the first time where K is the class of atomic models of a first order theory assuming Jensen's diamond $\diamondsuit_{\aleph_1}$ (taking $\lambda = \aleph_0$). In [Sh 87a] and [Sh 87b] the same theorem was proved using $2^{\aleph_0} < 2^{\aleph_1}$ only (using 0.5). Let us now concentrate on the case $\lambda = \aleph_0$. We asked whether the assumption $2^{\aleph_0} < 2^{\aleph_1}$ is necessary to get Conclusion 3.8. In this section we construct three classes of models K^1, K^2, K^3, K^4 failing amalgamation, i.e., failing the conclusion of 3.8, K^2, K^3, K^4 are a.e.c. with LS-number $\aleph_0$ while K^1 satisfy all the axioms needed in the proof of Conclusion 3.8 (but it is not an abstract elementary class - fails to satisfy AxIV,AxV).

K^2 is $\mathrm{PC}_{\aleph_0}$ and is axiomatizable in $\mathbb{L}_{\omega_1,\omega}(\mathbf{Q})$.

K^3 is $\mathrm{PC}_{\aleph_0}$ and is axiomatizable in $\mathbb{L}(\mathbf{Q})$.

Now the common phenomena to K^1, K^2, K^3, K^4 are that all of them satisfy the hypothesis of Conclusion 3.8, i.e., for $\ell = 1, 2, 3$ we have $\dot{I}(\aleph_0, K^\ell) = 1$ and the $\aleph_0$-amalgamation property fails in K^ℓ, but assuming $\aleph_1 < 2^{\aleph_0}$ and $\mathrm{MA}_{\aleph_1}$ for $\ell = 1, 2, 3$ we have $\dot{I}(\aleph_1, K^\ell) = 1$.

6.1 Definition. Let Y be an infinite set. A family $\mathscr{P}$ of infinite subsets of Y is called independent if for every $\eta \in {}^{\omega >}2$ and pairwise distinct $X_0, X_1, \dots, X_{\ell g(\eta)-1}$ (notation: for $X \in \mathscr{P}$ denote $X^0 = X$ and $X^1 = Y \backslash X$) the following set $\bigcap\limits_{k<\ell g(\eta)} X_k^{\eta[k]}$ is infinite.

6.2 Definition. 1) The class of models K^0 is defined by

$$\begin{aligned} K^0 = \{M : & M = \langle |M|, P^M, Q^M, R^M\rangle, |M| = P^M \cup Q^M, \\ & P^M \cap Q^M = \emptyset, |P^M| = \aleph_0 \le |Q^M| \text{ and} \\ & R \subseteq P^M \times Q^M\}. \end{aligned}$$

2) For $M \in K^0$, let $A^M_y = \{x \in P^M : xR^My\}$ for every $y \in Q^M$.

3) Let K^1 be the class of $M \in K^0$ such that

(a) the family $\{A^M_y : y \in Q^M\}$ is independent, which means that if $m < n$ and $y_0, \dots, y_{n-1}$ are pairwise distinct members of

Q^M then the set $\{x \in P^M : xR^M y_\ell \equiv \ell < m \text{ for every } \ell < n\}$ is infinite

(b) for every disjoint finite subsets u, w of P^M we have $\|M\| = |A^M_{u,w}|$ where $A^M_{u,w} := \{y \in Q^M : a \in u \Rightarrow (aR^M y)$ and $b \in w \Rightarrow \neg(bR^M y)\}$.

4) The notion of (strict) substructure $\leq_{\mathfrak{K}^1}$ is defined by: for $M_1, M_2 \in K^1$, $M_1 \leq_{\mathfrak{K}^1} M_2$ iff $M_1 \subseteq M_2$, $P^{M_1} = P^{M_2}$ and for any finite disjoint $u, w \subseteq P^{M_2}$ the set $A^{M_2}_{u,w} \backslash M_1$ is infinite when $M_1 \neq M_2$ (equivalently - non-empty).
5) $\mathfrak{K}^1 = (K^1, \leq_{\mathfrak{K}^1})$.

6.3 Lemma. *The class $(K^1, <_{\mathfrak{K}^1})$ satisfies*
0) Ax 0.
1) Ax I.
2) Ax II.
3) Ax III.
4) Ax IV fails even for $\lambda = \aleph_0$; but if $\langle M_\alpha : \alpha \leq \delta\rangle$ is $\leq_{\mathfrak{K}}$-increasing and $\|\bigcup_{\alpha<\delta} M_\alpha\| < \|M_\delta\|$ then $\bigcup_{\alpha<\delta} M_\alpha <_{\mathfrak{K}^1} M_\delta$.
5) Ax V fails for countable models.
6) Ax VI holds with $\mathrm{LS}(\mathfrak{K}^1) = \aleph_0$, in fact it holds for every cardinal.
7) For every $M \in K^1$, $\|M\| \leq 2^{\aleph_0}$.

Proof. 0), 1), 2) follows trivially from the definition.
3) To prove that $M = \bigcup_{i<\lambda} M_i \in K^1$, it is enough to verify that for every finite disjoint $u, w \subseteq P^M, |A^M_{u,w}| = \|M\|$. If $\langle M_i : i < \lambda\rangle$ is eventually constant we are done hence without loss of generality $\langle M_i : i < \lambda\rangle$ is $<_{\mathfrak{K}^1}$-increasing; from the definition of $<_{\mathfrak{K}^1}$ it follows that for each i, M_{i+1} has a new $y = y_i$ as above, i.e., $y_i \in A^{M_{i+1}}_{u,w} \backslash M_i$ for every $i < \lambda$. Also for each i there are at least $\|M_i\|$ many members in $A^{M_i}_{u,w} \subseteq A^M_{u,w}$. Together there are at least $\|M\|$ members in $A^M_{u,w}$.
4) Let $\{M_n : n < \omega\} \subseteq K^1_{\aleph_0}$ be an $<_{\mathfrak{K}^1}$-increasing chain, let $M = \bigcup_{n<\omega} M_n$; by part 3) we have $M \in K^1_{\aleph_0}$. Since $|Q^M| = \aleph_0$ by Claim 6.5(a) below there exists $A \subseteq P^M \backslash \{A^M_y : y \in Q^M\}$ infinite such

that $\{A_y : y \in Q^M\} \cup \{A\}$ is independent. Now define $N \in K^1$ by $P^N = P^M$, let $y_0 \notin M, Q^N = Q^M \cup \{y_0\}$ and finally let $R^N = R^M \cup \{\langle a, y_0\rangle : a \in P^N \ \& \ a \in A\}$. Clearly for every $n < \omega, M_n \leq_{\mathfrak{K}^1} N$ but N is not an $\leq_{\mathfrak{K}^1}$-extension of $M = \bigcup_{n<\omega} M_n$ because the second part in Definition 6.2(4) is violated.
5) Let $N_0 <_{\mathfrak{K}^1} N \in K^1$ be given; as in 4) define $N_1 \subseteq N, |N_0| \subseteq |N_1|$ by adding a single element to Q^{N_0} (from the elements of $Q^N \backslash Q^{N_0}$) it is obvious that $N_0 \leq_{\mathfrak{K}^1} N, N_1 \leq_{\mathfrak{K}^1} N$ but $N_0 \nleq_{\mathfrak{K}^1} N_1$.
6) By closing the set under the second requirement in Definition 6.2(3).
7) Let $y_1 \neq y_2 \in Q^M$, we show that $A^M_{y_1} \neq A^M_{y_2}$; if $A^M_{y_1} \subseteq A^M_{y_2}$ then $A^M_{y_1} \cap (P^M \backslash A^M_{y_2}) = \emptyset$ contradiction to the requirement that $\{A_y : y \in Q\}$ is independent hence $|Q^M| \leq 2^{|P^M|} = 2^{\aleph_0}$ and as $|P^M| = \aleph_0$ we are done. $\square_{6.3}$

6.4 Theorem. $\mathfrak{K}^1 = (K^1, <_{\mathfrak{K}^1})$ *satisfies the hypothesis of Conclusion 3.8. Namely*
1) $\dot{I}(\aleph_0, K^1) = 1$.
2) Every $M \in K^1_{\aleph_0}$ *has a proper* $\leq_{\mathfrak{K}^1}$*-extension in* $K^1_{\aleph_0}$.
3) $\mathfrak{K}^1$ *is closed under chains of length* $\leq \omega_1$.
4) $\mathfrak{K}^1$ *fails the* $\aleph_0$*-amalgamation property.*

Proof. 1) Let $M_1, M_2 \in K^1_{\aleph_0}$, pick the following enumerations $|M_1| = \{a_n : n < \omega\}$ and $|M_2| = \{b_n : n < \omega\}$. It is enough to define an increasing sequence of finite partial isomorphisms $\langle f_n : n < \omega\rangle$ from M_1 to M_2 such that for every $k < \omega$ for some $n(k) < \omega$ satisfy $a_k \in \mathrm{Dom}(f_{n(k)})$ and $b_k \in \mathrm{Range}(f_{n(k)})$, (finally take $f = \bigcup_{n<\omega} f_n$ and this will be an isomorphism from M_1 onto M_2).

Define the sequence $\langle f_n : n < \omega\rangle$ by induction on $n < \omega$: let $f_0 = \emptyset$, if $n = 2m$ denote $k = \min\{k < \omega : a_k \notin \mathrm{Dom}(f_n)\}$. Distinguish between the following two alternatives:

(A) if $a_k \in P^{M_1}$ let $\{a'_0, \ldots, a'_{j-1}\} = Q^{M_1} \cap \mathrm{Dom}(f_n)$. Without loss of generality there exists $i \leq j-1$ such that for all $\ell < i, a_k R^{M_1} a'_\ell$ and for all $i \leq \ell \leq j-1, \neg a_k R a'_\ell$. By 6.2(1),

P^{M_ℓ} is infinite, hence by clause (b) of 6.2(2) also Q^{M_ℓ} is infinite. Hence by clause (a) of 6.2(3) there are infinitely many $y \in P^{M_2}$ such that $yR^{M_2} f_n(a'_\ell)$ for all $\ell < i$ and for all $i \leq \ell < j-1, \neg yR^{M_2} f_n(a'_\ell)$. But $\mathrm{Rang}(f_n)$ is finite. Hence there is such $y \in P^{M_2} \setminus \mathrm{Rang}(f_n)$. Finally let $f_{n+1} = f_n \cup \{\langle a_k, y\rangle\}$

(B) if $a_k \in Q^{M_1}$ let $\{a'_0, \dots, a'_{j-1}\} = P^{M_1} \cap \mathrm{Dom}(f_n)$ and as before we may assume that there exists $i \leq j-1$ such that for all $\ell < i, a'_\ell R^{M_1} a_k$ and for all $i \leq \ell < j-1$ we have $\neg(a'_\ell)R^{M_1} a_k$. By the second requirement in Definition 6.2(3) there exists $y \in Q^{M_2} \setminus \mathrm{Dom}(f_n)$ such that $(\forall \ell < i)[f_n(a'_\ell)R^{M_2} y]$ and $(\forall \ell)[i \leq \ell < j-1 \Rightarrow \neg f_n(a'_\ell)R^{M_2} y]$. Now define $f_{n+1} = f_n \cup \{\langle a_k, y\rangle\}$.

2) First we prove the following.

6.5 Observation.

(a) Let P be a countable set. For every countable family $\mathscr{P}$ of infinite subsets of P if $\mathscr{P}$ is independent then there exists an infinite $A \subseteq P$ such that $\mathscr{P} \cup \{A\}$ is independent and $A \notin \mathscr{P}$, of course

(b) if $A, \mathscr{P}$ are as in (a) then for every infinite $B \subseteq P$ satisfying $|A \Delta B| < \aleph_0$ also $\mathscr{P} \cup \{B\}$ is independent (and $B \notin \mathscr{P}$)

(c) moreover in clause (a) we can require in addition that: for any disjoint finite $u, w \subseteq P$ there exists $A \subseteq P$ as in (a) satisfying $u \subseteq A$ and $A \cap w = \emptyset$.

Proof of Claim 6.5.

Clause (a): Let $\mathscr{P}^* = \{X \subseteq P : (\exists n < \omega)(\exists X_0 \in \mathscr{P}) \dots (\exists X_{n-1} \in \mathscr{P})(\exists k \leq n)$ [X or $P \setminus X$ is equal to $\cap\{X_i : i < k\} \cap \cap\{P \setminus X_i : k \leq i < n\}\}$.

Clearly $|\mathscr{P}^*| = \aleph_0$ hence we can find a sequence $\langle A_n : n < \omega\rangle$ such that $\{A_n : n < \omega\} = \mathscr{P}^*$ and such that for every $k < \omega$ there exists $n > k$ satisfying $A_n = A_k$ hence for some $n > k, A_n = P \setminus A_k$. Let $P = \{a_n : n < \omega\}$ without repetition.

Now define $i(n) < \omega$ by induction on n.
Let $i(0) = 0$.
If $n = k+1$, let $i(n) =$ Min$\{\ell < \omega : i(n-1) < \ell$ and

$$a_\ell \in (A_k \backslash \{a_{i(0)}, \dots, a_{i(n-1)})\}.$$

It is easy to verify that the construction is possible. Directly from the construction it follows that $A = \{a_{i(n)} : n < \omega\}$ is a set as required.

Clause (b): Easy.

Clause (c): Let $u, w \subseteq P$ be finite disjoint and $\mathscr{P}$ a countable family of subsets of P which is independent.

Let $A' \subseteq P$ be as proved in clause (a). According to (b) also $A = (A' \cup u) \backslash w$ satisfies: the family $\mathscr{P} \cup \{A\}$ is independent.

Return to the proof of Theorem 6.4(2). Let $\mathscr{P} = \{A^M_y \subseteq P^M : y \in Q^M\}$. Let $\langle s_n : n < \omega \rangle$ be an enumeration of $[P^M]^{<\aleph_0}$ with repetitions such that for every finite disjoint $u, w \subseteq P^M$ there exists $n < \omega$ such that $s_{2n} = u$, $s_{2n+1} = w$ and for each $k < \omega$, $s_{2k} \cap s_{2k+1} = \emptyset$.

It is enough to define $\{\mathscr{P}_n : n < \omega\}$ increasing chain of countable independent families of subsets of P^M such that $\mathscr{P}_0 = \mathscr{P}$ and for all $k < \omega$ and every finite disjoint $u, w \subseteq P^M, (\exists n < \omega)(\exists A \in \mathscr{P}_n \backslash \mathscr{P}_k)[u \subseteq A \wedge A \cap w = \emptyset]$ because $\bigcup\limits_{n<\omega} \mathscr{P}_n$ enables us to define $N \in K^1_{\aleph_0}$ such that $M \leq_{\mathfrak{K}^1} N$ as required. Assume $\mathscr{P}_n$ is defined; apply Claim 6.5(c) on $P = P^M$ and $\mathscr{P}_n$ when substituting $u = s_{2n}, w = s_{2n+1}$ let $A \subseteq P$ be supplied by the Claim and define $\mathscr{P}_{n+1} = \mathscr{P}_n \cup \{A\}$. It is easy to check that $\{\mathscr{P}_n : n < \omega\}$ satisfies our requirements.

3) This is a special case of Ax III which we checked in Lemma 6.3(3).

4) Let $M \in K^1_{\aleph_0}$ and we shall find $M_\ell \in K^1_{\aleph_0} (\ell = 0, 1), M \leq_{\mathfrak{K}^1} M_\ell$, which cannot be amalgamated over M. By part (2) we can find a model M_1 such that $M <_{\mathfrak{K}^1} M_1 \in K^1_{\aleph_0}$ and choose $y \in Q^{M_1} \backslash Q^M$. Define $M_2 \in K^1_{\aleph_0}$; its universe is $|M_1|, P^{M_2} = P^{M_1}, Q^{M_2} = Q^{M_1}$ and $R^{M_2} = \{(a, b) : aR^{M_1}b \;\; \& \;\; b \neq y$ or $a \in P^M \;\; \& \;\; b = y \;\; \&$

$\neg(aRy)\}$. Clearly M_1, M_2 cannot be amalgamated over M (since the amalgamation must contain a set and its complement). $\square_{6.4}$

6.6 Theorem. *Assume* $\mathrm{MA}_{\aleph_1}$ *(hence* $2^{\aleph_0} > \aleph_1$*). The class* $(K^1, <_{\mathfrak{K}^1})$ *is categorical in* $\aleph_1$.

Proof. Let $M, N \in K^1_{\aleph_1}$ and we shall prove that they are isomorphic. By repeated use of Lemma 6.3(6),(4) for AxVI we get (strictly) $<_{\mathfrak{K}^1}$-increasing continuous chains $\{M_\alpha : \alpha < \omega_1\}, \{N_\alpha : \alpha < \omega_1\} \subseteq K^1_{\aleph_0}$ such that $M = \bigcup_{\alpha<\omega_1} M_\alpha$ and $N = \bigcup_{\alpha<\omega_1} N_\alpha$, so for $\alpha < \beta, M_\alpha <_{\mathfrak{K}^1} M_\beta, N_\alpha <_{\mathfrak{K}^1} N_\beta$.

Now define a forcing notion which supplies an isomorphism $g : M \to N$.

$$\mathbb{P} = \{f : f \text{ is a partial finite isomorphism from } M \text{ into } N \text{ satisfying } (\forall \alpha < \omega_1)(\forall a \in \mathrm{Dom}(f))[a \in M_\alpha \Leftrightarrow f(a) \in N_\alpha]\},$$

the order is inclusion. It is trivial to check that if $G \subseteq \mathbb{P}$ is a directed subset then $g = \cup G$ is a partial isomorphism from M to N, we show that $\mathrm{Dom}(g) = |M|$ if G is generic enough. For every $a \in |M|$ define $\mathscr{I}_a = \{f \in \mathbb{P} : a \in \mathrm{Dom}(f)\}$, and we shall show that for all $a \in |M|$ the set $\mathscr{I}_a$ is dense. For $a \in M$ let $\alpha(a) = \mathrm{Min}\{\alpha < \omega_1 : a \in M_\alpha\}$, clearly it is zero or a successor ordinal. Let $f \in \mathbb{P}$ be a given condition, it is enough to find $h \in \mathscr{I}_a$ such that $f \subseteq h$ and $a \in \mathrm{Dom}(h)$. Let $A = \mathrm{Dom}(f)$, let $B, C \subseteq A$ be disjoint sets such that $B \cup C = A$ and $B = \mathrm{Dom}(f) \cap P^M, C = \mathrm{Dom}(f) \cap Q^M$. Without loss of generality $a \notin B \cup C$. If $a \in P^M$ let $\varphi(x, \bar{c}) = \wedge\{\pm xRc : c \in C \text{ and } M \models \pm aRc\}$. From the definition of K^1 there exists $b \in P^N \setminus \mathrm{Rang}(f)$ such that $N \models \varphi[b, f(\bar{c})]$. If $a \in Q^M$ let $\varphi(x, \bar{b}) = \wedge\{\pm bRx : b \in B, M \models \pm bRa\}$, we can find infinitely many $b \in Q^{N_{\alpha(a)}} \setminus \bigcup_{\beta<\alpha(a)} N_\beta$, satisfying $\varphi(x, f(\bar{b}))$.
Why? This is as $\cup\{N_\beta : \beta < \alpha(a)\} <_{\mathfrak{K}^1} N_{\alpha(a)}$ as C is finite without loss of generality $b \notin f(C)$.
Finally, let $h = f \cup \{\langle a, b\rangle\}$.

The proof that Range$(g) = |N|$ is analogous to the proof that Dom$(g) = |M|$. In order to use MA we just have to show that R has the c.c.c. Let $\{f_\alpha : \alpha < \omega_1\} \subseteq R$ be given. It is enough to find $\alpha, \beta < \omega_1$ such that f_α, f_β have a common extension. Without loss of generality we may assume $|M| \cap |N| = \emptyset$. By the finitary Δ-system lemma there exists $S \subseteq \omega_1, |S| = \aleph_1$ such that $\{\text{Dom}(f_\alpha) \cup \text{Range}(f_\alpha) : \alpha \in S\}$ is a Δ-system with heart A. Let $B \subseteq |M|, C \subseteq |N|$ be such that $A = B \cup C$, now without loss of generality for every $\alpha \in S$, f_α maps B into C.
[Why? If not, $S_1 = \{\alpha \in S$: for some $b = b_\alpha \in B, f_\alpha(b_\alpha) \notin C\}$ is uncountable hence for some $b \in B, S_2 = \{\alpha \in S_1 : b_\alpha = b\}$ is uncountable; so $\langle f_\alpha(b) : \alpha \in S_2\rangle$ is without repetitions hence is uncountable. But $\{f(b) : f \in \mathbb{P}$ and $b \in \text{Dom}(f) \cap B\}$ is countable because $f \in \mathbb{P}$ & $b \in \text{Dom}(f)$ & $\alpha < \omega_1 \Rightarrow [b \in M_\alpha \equiv f(b) \in N_\alpha]$. Similarly, f_α^{-1} maps C into B, so necessarily f_α maps B onto C; but the number of possible functions from B to C is $|C|^{|B|} < \aleph_0$. Hence there exists $S_1 \subseteq S, |S_1| = \aleph_1$ such that for all $\alpha, \beta \in S_1, f_\alpha \restriction B = f_\beta \restriction B$ and $\text{Dom}(f_\alpha) \cap M_0 \subseteq B$, $\text{Rang}(f_\alpha) \cap N_0 \subseteq C$. As $P^{M_\alpha} = P^{M_0} \subseteq M_0, P^{N_\alpha} = P^{N_0} \subseteq N_0$ for every $\alpha \in S_1$ we have $P^M \cap \text{Dom}(f_\alpha) \subseteq B, P^N \cap \text{Range}(f_\alpha) \subseteq C$, therefore for all $\alpha, \beta \in S_1, f_\alpha \cup f_\beta \in \mathbb{P}$ and in particular there exists $\alpha \neq \beta < \omega_1$ such that $f_\alpha \cup f_\beta \in \mathbb{P}$. $\square_{6.6}$

In the terminology of [GrSh 174] Theorems 6.4 and 6.6 give us together:

6.7 Conclusion. Assuming $2^{\aleph_0} > \aleph_1$ and $\text{MA}_{\aleph_1}$, $\mathfrak{K}^1$ is a nice category which has a universal object in $\aleph_1$, moreover it is categorical in $\aleph_1$.

6.8 Definition. 1) K^2 is the class of $M \in K^0$ (see Definition 6.2) satisfying:

(a) $(\forall x \in Q^M)(\forall u \in [P^M]^{<\aleph_0})(\exists y \in Q)[A_x^M \Delta A_y^M = u]$

(b) if $k < \omega$ and $y_0, \ldots, y_{k-1} \in Q$ satisfies $|A_{y_\ell} \Delta A_{y_m}| \geq \aleph_0$ for $\ell < m < k$ <u>then</u> the set $\{A_{y_\ell}^M : \ell < k\}$ is an independent family of subsets of P^M

(c) $Q(y) \wedge Q(z) \wedge (\forall x \in P)[xRy \leftrightarrow xRz] \rightarrow y = z$,

(d) for every $k < \omega$ for some $y_0, \ldots, y_k \in Q^M$ we have
$\bigwedge_{\ell<m\leq k} |A_{y_\ell} \Delta A_{y_m}| \geq \aleph_0$.

2) For $M_1, M_2 \in K^2$

$$M_1 \leq_{\mathfrak{K}^2} M_2 \Leftrightarrow^{\rm df} M_1 \subseteq M_2, P^{M_1} = P^{M_2}.$$

3) $\mathfrak{K}^2 = (K^2, \leq_{\mathfrak{K}^2})$.
4) K^3 is the class of models $M = (|M|, P^M, Q^M, R^M, E^M)$ such that

(a) $(|M|, P^M, Q^M, R^M) \in K^1$
(b) E^M is an equivalence relation on Q^M
(c) E^M has infinitely many equivalence classes
(d) each equivalence class of E^M is countable
(e) if $u, w \subseteq P^M$ are finite disjoint and $y \in Q^M$ then for some $y' \in y/E^M$ we have $a \in u \Rightarrow aR^M y'$ and $b \in w \Rightarrow \neg(bR^M y')$.

5) We define $\leq_{\mathfrak{K}^3}$: $M_1 \leq_{\mathfrak{K}^3} M_2 \Leftrightarrow^{\rm df} M_1 \subseteq M_2$ and $a \in M_1 \Rightarrow a/E^{M_2} = a/E^{M_1}$.
6) $\mathfrak{K}^3 = (K^3, \leq_{\mathfrak{K}^3})$.

If we like to have a class defined by a sentence from $\mathbb{L}_{\omega_1,\omega}$ (rather than $\mathbb{L}_{\omega_1,\omega}(\mathbf{Q})$) we can use:

6.9 Definition. 1) $\mathfrak{K}^4$ is defined as follows:

(A) $\tau(\mathfrak{K}^4) = \{P, Q, R\} \cup \{P_n : n < \omega\}$, R is two-place predicates, P, Q, P_n are unary predicates
(B) $M \in K^4$ <u>iff</u> M is a $\tau(\mathfrak{K}^4)$-model such that $M \restriction \{P, Q, R\} \in K^2$ and
  (a) $\langle P_n^M : n < \omega \rangle$ is a partition of P^M
  (b) P_n^M has exactly 2^n elements
  (c) $(\forall x \in Q)(\forall u \in [P^M]^{<\aleph_0})(\exists y \in Q^M)[A_x^M \Delta A_y^M = u]$
  (d) if $k < \omega$ and $y_0, \ldots, y_{k-1} \in Q$ satisfies $|A_{y_\ell} \Delta A_{y_m}| \geq \aleph_0$ for $\ell < m < k$ <u>then</u> the set $\{A_{y_\ell}^M : \ell < k\}$ is an

independent family of subsets of P^M; moreover for any n large enough for any $\eta \in {}^k2$ the set $P^M_n \cap \bigcap\{A^M_{y_\ell} : \eta(\ell) = 1\} \setminus \bigcup\{A^M_{y_\ell} : \eta(\ell) = 0\}$ has exactly 2^{n-k} elements

(e) $Q^M(y) \wedge Q^M(z) \wedge (\forall x \in P^M)[xR^M y \leftrightarrow xR^M z] \rightarrow y = z$,

(f) for every $k < \omega$ for some $y_0, \ldots, y_k \in Q^M$ we have $\bigwedge_{\ell < m \le k} |A_{y_\ell} \Delta A_{y_m}| \ge \aleph_0$

(C) $M \le_{\mathfrak{K}^4} N$ iff $M, N \in K^4$ and $M \subseteq N$ and $P^M = P^N$.

6.10 Theorem. *1) $(K^2, <_{\mathfrak{K}^2})$ is an $\aleph_0$-presentable abstract elementary class which is categorical in $\aleph_0$.*
2) Also $\mathfrak{K}^3$ and $\mathfrak{K}^4$ are $\aleph_0$-presentable a.e.c. categorical in $\aleph_0$.

Proof. Similar to the proof for $\mathfrak{K}^1$. $\square_{6.10}$

6.11 Theorem. *1) $\mathfrak{K}^1_{\aleph_1}$ has an axiomatization in $\mathbb{L}(\mathbf{Q})$ and $\le_{\mathfrak{K}^1}$ is $<^{**}$ from the proof of 3.18 (this is $<^{**}$ from [Sh 87a] and [Sh 87b]).*
2) $\mathfrak{K}^2$ has an axiomatization in $\mathbb{L}_{\omega_1,\omega}(\mathbf{Q})$ and $\le_{\mathfrak{K}^2}$ is $\le^*$ from the proof of 3.18 (this is $<^*_{\omega_1,\omega}$ from [Sh 87a] and [Sh 87b]).
3) $\mathfrak{K}^3$ has an axiomatization in $\mathbb{L}(\mathbf{Q})$ and $\le_{\mathfrak{K}^3}$ is $<^*$ from [Sh 87a] and [Sh 87b].
4) $\mathfrak{K}^4$ has an axiomatization in $\mathbb{L}_{\omega_1,\omega}$ and $\le_{\mathfrak{K}^4}$ is just being a submodel.

5) $(\forall \ell \in \{1,2,3,4\})[K^\ell$ is $\mathrm{PC}_{\aleph_0}]$.

Proof. Should be clear. $\square_{6.11}$

6.12 Theorem. *If $\mathrm{MA}_{\aleph_1}$ <u>then</u> K^ℓ is categorical in $\aleph_1$ for $\ell = 2, 3$.*

Proof. Easy[10].

6.13 Conclusion. Assuming $\mathrm{MA}_{\aleph_1}$ there exists an abstract elementary class, which is $\mathrm{PC}_{\aleph_0}$, categorical in $\aleph_0, \aleph_1$ but without the $\aleph_0$-amalgamation property.

[10] In the earlier version this was claimed also for $\ell = 4$, but, as Baldwin noted, this was wrong

CATEGORICITY IN ABSTRACT ELEMENTARY CLASSES: GOING UP INDUCTIVELY SH600

§0 Introduction

The paper's main explicit result is proving Theorem 0.1 below. It is done axiomatically, in a "superstable" abstract framework with the set of "axioms" of the frame, verified by applying earlier works, so it suggests this frame as the, or at least a major, non-elementary parallel of superstable.
A major case to which this is applied, is the one from [Sh 576] represented in Chapter VI; we continue this work in several ways but the use of [Sh 576] is only in verifying the basic framework; we refer the reader to the book's introduction or [Sh 576, §0] for background and some further claims but all the definitions and basic properties appear here. Otherwise, the heavy use of earlier works is in proving that our abstract framework applies in those contexts. If $\lambda = \aleph_0$ is O.K. for you, you may use Chapter I or [Sh 48] instead of [Sh 576] as a starting point.

Naturally, our deeper aim is to develop stability theory (actually a parallel of the theory of superstable elementary classes) for non-elementary classes. We use the number of non-isomorphic models as test problem. Our main conclusion is 0.1 below. As a concession to supposedly general opinion, we restrict ourselves here to the λ-good framework and delay dealing with weak relatives (see Chapter VII, Jarden-Shelah [JrSh 875], hopefully [Sh:F888]. Also, we assume that the (normal) weak-diamond ideal on the $\lambda^{+\ell}$ is not saturated (for $\ell = 1, \dots, n-1$). We had intended to rely on [Sh 576, §3], but actually in the end we prefer to rely on the lean version of Chapter

Typeset by $\mathcal{A}\mathcal{M}\mathcal{S}$-TEX

VII, see "reading plan A" in VII§0. Relying on the full version of Chapter VII, we can eliminate this extra assumption "not $\lambda^{+\ell+1}$-saturated[1] (ideal)". On $\mu_{\text{unif}}(\lambda^{+\ell+1}, 2^{\lambda^{+\ell}})$, see, e.g. I.0.5(3)).

0.1 Theorem. *Assume* $2^\lambda < 2^{\lambda^{+1}} < \cdots < 2^{\lambda^{+n+1}}$ *and the (so called weak diamond) normal*[1] *ideal* $\mathrm{WDmId}(\lambda^{+\ell})$ *is not* $\lambda^{+\ell+1}$*-saturated*[2] *for* $\ell = 1, \ldots, n$.
1) Let $\mathfrak{K}$ *be an abstract elementary class (see §1 below) categorical in* λ *and* λ^+ *with* $\mathrm{LS}(\mathfrak{K}) \leq \lambda$ *(e.g. the class of models of* $\psi \in \mathbb{L}_{\lambda^+,\omega}$ *with* $\leq_{\mathfrak{K}}$ *defined naturally). If* $1 \leq \dot{I}(\lambda^{+2}, \mathfrak{K})$ *and* $2 \leq \ell \leq n \Rightarrow \dot{I}(\lambda^{+\ell}, \mathfrak{K}) < \mu_{\text{unif}}(\lambda^{+\ell}, 2^{\lambda^{+\ell-1}})$, *<u>then</u>* $\mathfrak{K}$ *has a model of cardinality* λ^{+n+1}.
2) Assume $\lambda = \aleph_0$, *and* $\psi \in \mathbb{L}_{\omega_1,\omega}(\mathbf{Q})$.
If $1 \leq \dot{I}(\lambda^{+\ell}, \psi) < \mu_{\text{unif}}(\lambda^{+\ell}, 2^{\lambda^{+\ell-1}})$ *for* $\ell = 1, \ldots, n-1$ *<u>then</u>* ψ *has a model in* λ^{+n} *(see [Sh 48]).*

Note that if $n = 3$, then 0.1(1) is already proved in [Sh 576] $\approx$ Chapter VI. If $\mathfrak{K}$ is the class of models of some $\psi \in \mathbb{L}_{\omega_1,\omega}$ this is proved in [Sh 87a], [Sh 87b], but the proof here <u>does not generalize</u> the proofs there. It is a different one (of course, they are related). There, for proving the theorem for n, we have to consider a few statements on $(\aleph_m, \mathscr{P}^-(n-m))$-systems for all $m \leq n$, (going up and down). A major point (there) is that for $n = 0$, as $\lambda = \aleph_0$ we have the omitting type theorem and the types are "classical", that is, are sets of formulas. This helps in proving strong dichotomies; so the analysis of what occurs in $\lambda^{+n} = \aleph_n$ is helped by those dichotomies. Whereas here we deal with $\lambda, \lambda^+, \lambda^{+2}, \lambda^{+3}$ and then "forget" λ and deal with $\lambda^+, \lambda^{+2}, \lambda^{+3}, \lambda^{+4}$, etc. So having started with poor assumptions there is less reason to go back from λ^{+n} to λ. However, there are some further theorems proved in [Sh 87a], [Sh 87b], whose parallels are not proved here, mainly that if for every n, in λ^{+n} we get the "structure" side, then the class has models in every $\mu \geq \lambda$, and theorems about

---

[1] recall that as $2^{\lambda_{\ell-1}} < 2^{\lambda_\ell}$ this ideal is not trivial, i.e., $\lambda^{+\ell}$ is not in the ideal

[2] actually the statement "some normal ideal on μ^+ is μ^{++}-saturated" is "expensive", i.e., of large consistency strength, etc., so it is "hard" for this assumption to fail

categoricity. We shall deal with them in subsequent works, mainly Chapter III. Also in [Sh 48], [Sh 88] = Chapter I we started to deal with $\psi \in \mathbb{L}_{\omega_1,\omega}(\mathbf{Q})$ dealing with $\aleph_1, \aleph_2$. Of course, we integrate them too into our present context. In the axiomatic framework (introduced in §2) we are able to present a lemma, speaking only on 4 cardinals, and which implies the theorem 0.1. (Why? Because in §3 by [Sh 576] $\approx$ Chapter VI we can get a so-called good λ^+-frame $\mathfrak{s}$ with $K^{\mathfrak{s}} \subseteq \mathfrak{K}$, and then we prove a similar theorem on good frames by induction on n, with the induction step done by the lemma mentioned above). For this, parts of the proof are a generalization of the proof of [Sh 576, §8,§9,§10].

A major theme here (and even more so in Chapter III) is:
0.2 Thesis: It is worthwhile to develop model theory (and superstability in particular) in the context of $\mathfrak{K}_\lambda$ or $K_{\lambda^{+\ell}}, \ell \in \{0, \ldots, n\}$, i.e., restrict ourselves to one, few, or an interval of cardinals. We may have good understanding of the class in this context, while in general cardinals we are lost.

As in [Sh:c] for first order classes
0.3 Thesis: It is reasonable first to develop the theory for the class of (quite) saturated enough models as it is smoother and even if you prefer to investigate the non-restricted case, the saturated case will clarify it and you will e able to rely on it. In our case this will mean investigating $\mathfrak{s}^{+n}$ for each n and then $\cap\{\mathfrak{K}^{\mathfrak{s}^{+n}} : n < \omega\}$.

0.4 The Better to be poor Thesis: Better to know what is essential. e.g., you may have better closure properties (here a major point of poverty is having no formulas, this is even more noticeable in Chapter III).

I thank John Baldwin, Alex Usvyatsov, Andres Villaveces and Adi Yarden for many complaints and corrections.

§1 gives a self-contained introduction to a.e.c. (abstract elementary classes), including definitions of types, M_2 is (λ, κ)-brimmed over M_1 and saturativity = universality + model homogeneity. An interesting point is observing that any λ-a.e.c. $\mathfrak{K}_\lambda$ can be lifted to

$\mathfrak{K}_{\geq\lambda}$, uniquely; so it does not matter if we deal with $\mathfrak{K}_\lambda$ or $\mathfrak{K}_{\geq\lambda}$ (unlike the situation for good λ-frames, which if we lift, we in general, lose some essential properties).

The good λ-frames introduced in §2 are a very central notion here. It concentrates on one cardinal λ, in $\mathfrak{K}_\lambda$ we have amalgamation and more, hence types, in the orbital sense, not in the classical sense of set of formulas, for models of cardinality λ can be reasonably defined and "behave" reasonably (we concentrate on so-called basic types) and we axiomatically have a non-forking relation for them.

In §3 we show that starting with classes belonging to reasonably large families, from assumptions on categoricity (or few models), good λ-frames arise. In §4 we deduce some things on good λ-frames; mainly: stability in λ, existence and (full) uniqueness of $(\lambda, *)$-brimmed extensions of $M \in K_\lambda$.

Concerning §5 we know that if $M \in K_\lambda$ and $p \in \mathscr{S}^{\rm bs}(M)$ then there is $(M, N, a) \in K^{3,\rm bs}_\lambda$ such that $\mathbf{tp}(a, M, N) = p$. But can we find a special ("minimal" or "prime") triple in some sense? Note that if $(M_1, N_1, a) \leq_{\rm bs} (M_2, N_2, a)$ then N_2 is an amalgamation of N_1, M_2 over M_1 (restricting ourselves to the case "$\mathbf{tp}(a, M_2, N_2)$ does not fork over M_1") and we may wonder is this amalgamation unique (i.e., allowing to increase or decrease N_2). If this holds for any such (M_2, N_2, a) we say (M_1, N_1, a) has uniqueness (= belongs to $K^{3,\rm uq}_\lambda = K^{3,\rm uq}_{\mathfrak{s}}$). Specifically we ask: is $K^{3,\rm uq}_\lambda$ dense in $(K^{3,\rm bs}_\lambda, \leq_{\rm bs})$? If no, we get a non-structure result; if yes, we shall (assuming categoricity) deduce the "existence for $K^{3,\rm uq}_{\mathfrak{s}}$" and this is used later as a building block for non-forking amalgamation of models.

So our next aim is to find "non-forking" amalgamation of models (in §6). We first note that there is at most one such notion which fulfills our expectations (and "respect" $\mathfrak{s}$). Now if $\bigcup(M_0, M_1, a, M_3)$, $M_0 \leq_{\mathfrak{K}} M_2 \leq_{\mathfrak{K}} M_3$ equivalently $(M_0, M_2, a) \leq_{\rm bs} (M_1, M_3, a)$ and $(M_0, M_2, a) \in K^{3,\rm uq}_\lambda$ by our demands we have to say that M_1, M_2 are in non-forking amalgamation over M_0 inside M_3. Closing this family under the closure demands we expect to arrive to a notion $\mathrm{NF}_\lambda = \mathrm{NF}_{\mathfrak{s}}$ which should be the right one (if a solution exists at all). But then we have to work on proving that it has all the properties it hopefully has.

A major aim in advancing to λ^+ is having a superlimit model in $\mathfrak{K}_{\lambda^+}$. So in §7 we find out who it should be: the saturated model of $\mathfrak{K}_{\lambda^+}$, but is it superlimit? We use our NF_λ to define a "nice" order $\leq^*_{\lambda^+}$ on $\mathfrak{K}_{\lambda^+}$, investigate it and prove the existence of a superlimit model under this partial order. To advance the move to λ^+ we would like to have that the class of λ^+-saturated model with the partial order $\leq^*_{\lambda^+}$ is a λ^+-a.e.c. Well, we do not prove it but rather use it as a dividing line: if it fails we eventually get many models in $\mathfrak{K}_{\lambda^{++}}$ (coding a stationary subset of λ^{++} (really any $S \subseteq \{\delta < \lambda^{++} : \mathrm{cf}(\delta) = \lambda^+\}$)), see §8.

Lastly, we pay our debts: prove the theorems which were the motivation of this work, in §9.

* * *

<u>Reading Plans</u>:

As usual these are instructions on what you can avoid reading.

Note that §3 contains the examples, i.e., it shows how "good λ-frame", our main object of study here, arise in previous works. This, on the one hand, may help the reader to understand what is a good frame and, on the other hand, helps us in the end to draw conclusions continuing those works. However, it is <u>not</u> necessary here otherwise, so you may ignore it.

Note that we treat the subject axiomatically, in a general enough way to treat the cases which exist without trying too much to eliminate axioms as long as the cases are covered (and probably most potential readers will feel they are more than general enough).
We shall assume

$(*)_0$ $2^\lambda < 2^{\lambda^+} < 2^{\lambda^{+2}} < \ldots < 2^{\lambda^{+n}}$ and $n \geq 2$.

In the beginning of §1 there are some basic definitions.

<u>Reading Plan 0</u>: We accept the good frames as interesting per se so ignore §3 (which gives "examples") and: §1 tells you all you need to know on abstract elementary classes; §2 presents frames, etc.

Reading Plan 1: The reader decides to understand why we reprove the main theorem of [Sh 87a], [Sh 87b] so

$(*)_1$ K is the class of models of some $\psi \in \mathbb{L}_{\lambda^+,\omega}$ (with a natural notion of elementary embedding $\prec_{\mathscr{L}}$ for $\mathscr{L}$ a fragment of $\mathbb{L}_{\lambda^+,\omega}$ of cardinality $\leq \lambda$ to which ψ belongs).

So in fact (as we can replace, for this result, K by any class with fewer models still satisfying the assumptions) without loss of generality

$(*)'_1$ if $\lambda = \aleph_0$ then K is the class of atomic models of some complete first order theory, $\leq_{\mathfrak{K}}$ is being elementary submodel.

The theorems we are seeking are of the form

$(*)_2$ if K has few models in $\lambda+\aleph_1, \lambda^+, \ldots, \lambda^{+n}$ then it has a model in λ^{+n+1}.
[Why "$\lambda + \aleph_1$"? If $\lambda > \aleph_0$ this means λ whereas if $\lambda = \aleph_0$ this means that we do not require "few model in $\lambda = \aleph_0$". The reason is that for the class or models of $\psi \in \mathbb{L}_{\omega_1,\omega}$ (or $\in \mathbb{L}_{\omega_1,\omega}(\mathbf{Q})$ or an a.e.c. which is $\mathrm{PC}_{\aleph_0}$, see Definition 3.3) we have considerable knowledge of general methods of building models of cardinality $\aleph_1$, for general λ we are very poor in such knowedge (probably as there is much less).]

But, of course, what we would really like to have are rudiments of stability theory (non-forking amalgamation, superlimit models, etc.). Now reading plan 1 is to follow reading plan 2 below but replacing the use of Claim 3.7 and [Sh 576] by the use of a simplified version of 3.4 and [Sh 87a].

Reading Plan 2: The reader would like to understand the proof of $(*)_2$ for arbitrary $\mathfrak{K}$ and λ. The reader

(a) knows at least the main definitions and results of [Sh 576] $\approx$ Chapter VI,
or just

(b) reads the main definitions of §1 here (in 1.1 - 1.7) and is willing to believe some quotations of results of [Sh 576] $\approx$ Chapter VI.

We start assuming $\mathfrak{K}$ is an abstract elementary class, $\mathrm{LS}(\mathfrak{K}) \leq \lambda$ (or read §1 here until 1.16) and $\mathfrak{K}$ is categorical in λ and λ^+ and $1 \leq \dot{I}(\lambda^{++}, K) < \mu_{\mathrm{unif}}(\lambda^{++}, 2^{\lambda^+})$ and moreover, $1 \leq \dot{I}(\lambda^{++}, K) < \mu_{\mathrm{unif}}(\lambda^{++}, 2^{\lambda^+})$. As an appetizer and to understand types and the definition of types and saturated (in the present context) and brimmed, read from §1 until 1.17.
He should read in §2 Definition 2.1 of λ-good frame, an axiomatic framework and then read the following two Definitions 2.4, 2.5 and Claim 2.6. In §3, 3.7 show how by [Sh 576] $\approx$ Chapter VI the context there gives a λ^+-good frame; of course the reader may just believe instead of reading proofs, and he may remember that our basic types are minimal in this case.
In §4 he should read some consequences of the axioms.
Then in §5 we show some amount of unique amalgamation. Then §6,§7,§8 do a parallel to [Sh 576, §8,§9,§10] in our context; still there are differences, in particular our context is not necessarily uni-dimensional which complicates matters. But if we restrict ourselves to continuing [Sh 576] $\approx$ Chapter VI, our frame is "uni-dimensional", we could have simplified the proofs by using $\mathscr{S}^{\mathrm{bs}}(M)$ as the set of minimal types.

Reading Plan 3: $\psi \in \mathbb{L}_{\omega_1,\omega}(\mathbf{Q})$ so $\lambda = \aleph_0, 1 \leq \dot{I}(\aleph_1, \psi) < 2^{\aleph_1}$ reclling $\mathbf{Q}$ denote the quantifier "there are uncountably many".

For this, [Sh 576] $\approx$ Chapter VI is irrelevant (except if we quote the "black box" use of the combinatorial section §3 of [Sh 576] when using the weak diamond to get many non-isomorphic models in §5, but we prefer to use Chapter VII).

Now reading plan 3 is to follow reading plan 2 but 3.7 is replaced by 3.5 which relies on [Sh 48], i.e., it proves that we get an $\aleph_1$-good frame investigating $\psi \in \mathbb{L}_{\omega_1,\omega}(\mathbf{Q})$.

Note that our class may well be such that $\mathfrak{K}$ is the parallel of "superstable non-multidimension complete first order theory"; e.g., $\psi_1 = (\mathbf{Q}x)(P(x)) \wedge (\mathbf{Q}x)(\neg P(x)), \tau_\psi = \{P\}, P$ a unary predicate; this is categorical in $\aleph_1$ and has no model in $\aleph_0$ and ψ_1 has 3 models in $\aleph_2$. But if we use $\psi_0 = (\forall x)(P(x) \equiv P(x))$ we have $\dot{I}(\aleph_1, \psi_0) = \aleph_0$; however, even starting with ψ_1, the derived a.e.c. $\mathfrak{K}$ has exactly three non-isomorphic models in $\aleph_1$. In general we derived an a.e.c. $\mathfrak{K}$ from

ψ such that: $\mathfrak{K}$ is an a.e.c. with LS number $\aleph_0$, categorical in $\aleph_0$, and the number of somewhat "saturated" models of $\mathfrak{K}$ in λ is $\leq \dot{I}(\lambda, \psi)$ for $\lambda \geq \aleph_1$. The relationship of ψ and $\mathfrak{K}$ is not comfortable; as it means that, for general results to be applied, they have to be somewhat stronger, e.g. "$\mathfrak{K}$ has $2^{\lambda^{++}}$ non-isomorphic λ^+-saturated models of cardinality λ^{++}". The reason is that $\mathrm{LS}(\mathfrak{K}) = \lambda = \aleph_0$; we have to find many somewhat λ^+-saturated models as we have first in a sense eliminate the quantifier $\mathbf{Q} = \exists^{\geq \aleph_1}$, (i.e., the choice of the class of models and of the order guaranteed that what has to be countable is countable, and λ^+-saturation guarantees that what should be uncountable is uncountable). This is the role of $K^{\mathbf{F}}_{\aleph_1}$ in I§3.

Reading Plan 4: $\mathfrak{K}$ an abstract elementary class which is PC_ω (= $\aleph_0$-presentable, see Definition 3.3); see Chapter I or [Mw85a] which includes a friendly presentation of [Sh 88, §1-§3] so of I§1-§3).

Like plan 3 but we have to use 3.4 instead of 3.5 and fortunately the reader is encouraged to read I§4,§5 to understand why we get a λ-good quadruple.

§1 Abstract elementary classes

First we present the basic material on a.e.c. $\mathfrak{K}$, that is types, saturativity and (λ, κ)-brimmness (so most is repeating some things from I§1 and from Chapter V.B).
Second we show that the situation in $\lambda = \mathrm{LS}(\mathfrak{K})$ determine the situation above λ, moreover such lifting always exists; so a λ-a.e.c. can be lifted to a $(\geq \lambda)$-a.e.c. in one and only one way.

1.1 Conventions. Here $\mathfrak{K} = (K, \leq_{\mathfrak{K}})$, where K is a class of τ-models for a fixed vocabulary $\tau = \tau_K = \tau_{\mathfrak{K}}$ and $\leq_{\mathfrak{K}}$ is a two-place relation on the models in K. We do not always strictly distinguish between $\mathfrak{K}$, K and $(K, \leq_{\mathfrak{K}})$. We shall assume that $K, \leq_{\mathfrak{K}}$ are fixed, and $M \leq_{\mathfrak{K}} N \Rightarrow M, N \in K$; and we assume that it is an abstract elementary class, see Definition 1.4 below. When we use $\leq_{\mathfrak{K}}$ in the $\prec$ sense (elementary submodel for first order logic), we write $\prec_{\mathbb{L}}$ as $\mathbb{L}$ is first order logic.

1.2 Definition. For a class of τ_K-models we let $\dot{I}(\lambda, K) = |\{M/\cong: M \in K, \|M\| = \lambda\}|$.

1.3 Definition. 1) We say $\bar{M} = \langle M_i : i < \mu \rangle$ is a representation or filtration of a model M of cardinality μ <u>if</u> $\tau_{M_i} = \tau_M, M_i$ is $\subseteq$-increasing continuous, $\|M_i\| < \|M\|$ and $M = \cup\{M_i : i < \mu\}$ and $\mu = \chi^+ \Rightarrow \|M_i\| = \chi$.
2) We say $\bar{M}$ is a $\leq_{\mathfrak{K}}$-representation or $\leq_{\mathfrak{K}}$-filtration of M <u>if</u> in addition $M_i \leq_{\mathfrak{K}} M$ for $i < \|M\|$ (hence $M_i, M \in K$ and $\langle M_i : i < \mu \rangle$ is $\leq_{\mathfrak{K}}$-increasing continuous, by Av V from Definition 1.4).

1.4 Definition. We say $\mathfrak{K} = (K, \leq_{\mathfrak{K}})$ is an abstract elementary class, a.e.c. in short, if (τ is as in 1.1, $Ax0$ holds and) AxI-VI hold, where:
$Ax0$: The holding of $M \in K, N \leq_{\mathfrak{K}} M$ depends on N, M only up to isomorphism, i.e., $[M \in K, M \cong N \Rightarrow N \in K]$, and [if $N \leq_{\mathfrak{K}} M$ and f is an isomorphism from M onto the τ-model M' mapping N onto N' <u>then</u> $N' \leq_{\mathfrak{K}} M'$], and of course 1.1.

AxI: If $M \leq_{\mathfrak{K}} N$ then $M \subseteq N$ (i.e. M is a submodel of N).

$AxII$: $M_0 \leq_{\mathfrak{K}} M_1 \leq_{\mathfrak{K}} M_2$ implies $M_0 \leq_{\mathfrak{K}} M_2$ and $M \leq_{\mathfrak{K}} M$ for $M \in K$.

$AxIII$: If λ is a regular cardinal, M_i (for $i < \lambda$) is $\leq_{\mathfrak{K}}$-increasing (i.e. $i < j < \lambda$ implies $M_i \leq_{\mathfrak{K}} M_j$) and continuous (i.e. for limit ordinal $\delta < \lambda$ we have
$M_\delta = \bigcup_{i<\delta} M_i$) <u>then</u> $M_0 \leq_{\mathfrak{K}} \bigcup_{i<\lambda} M_i$.

$AxIV$: If λ is a regular cardinal, M_i (for $i < \lambda$) is $\leq_{\mathfrak{K}}$-increasing continuous and $M_i \leq_{\mathfrak{K}} N$ for $i < \lambda$ <u>then</u> $\bigcup_{i<\lambda} M_i \leq_{\mathfrak{K}} N$.

AxV: If $M_0 \subseteq M_1$ and $M_\ell \leq_{\mathfrak{K}} N$ for $\ell = 0, 1$, <u>then</u> $M_0 \leq_{\mathfrak{K}} M_1$.

$AxVI$: LS($\mathfrak{K}$) exists[3], where LS($\mathfrak{K}$) is the minimal cardinal λ such

[3]We normally assume $M \in \mathfrak{K} \Rightarrow \|M\| \geq$ LS($\mathfrak{K}$) so may forget to write $\|M\|$" + LS($\mathfrak{K}$)" instead $\|M\|$, here there is no loss in it. It is also natural to assume $|\tau(\mathfrak{K})| \leq$ LS($\mathfrak{K}$) which means just increasing LS($\mathfrak{K}$), but no real need here; dealing with Hanf numbers it is natural.

that: if $A \subseteq N$ and $|A| \leq \lambda$ then for some $M \leq_{\mathfrak{K}} N$ we have $A \subseteq |M|$ and $\|M\| \leq \lambda$.

1.5 Notation: 1) $K_\lambda = \{M \in K : \|M\| = \lambda\}$ and $K_{<\lambda} = \bigcup_{\mu<\lambda} K_\mu$, etc.

1.6 Definition. 1) The function $f : N \to M$ is $\leq_{\mathfrak{K}}$-embedding when f is an isomorphism from N onto N' where $N' \leq_{\mathfrak{K}} M$, (so $f : N \to N'$ is an isomorphism onto).
2) We say f is a $\leq_{\mathfrak{K}}$-embedding of M_1 into M_2 over M_0 when for some M_1' we have: $M_0 \leq_{\mathfrak{K}} M_1, M_0 \leq_{\mathfrak{K}} M_1' \leq_{\mathfrak{K}} M_2$ and f is an isomorphism from M_1 onto M_1' extending the mapping id_{M_0}.

Recall

1.7 Observation. Let I be a directed set (i.e., I is partially ordered by $\leq = \leq^I$, such that any two elements have a common upper bound).
1) If M_t is defined for $t \in I$, and $t \leq s \in I$ implies $M_t \leq_{\mathfrak{K}} M_s$ then for every $t \in I$ we have $M_t \leq_{\mathfrak{K}} \bigcup_{s \in I} M_s$.
2) If in addition $t \in I$ implies $M_t \leq_{\mathfrak{K}} N$ then $\bigcup_{s \in I} M_s \leq_{\mathfrak{K}} N$.

Proof. Easy or see I.1.6 which does not rely on anything else. $\square_{1.7}$

1.8 Claim. *1) For every $N \in K$ there is a directed partial order I of cardinality $\leq \|N\|$ and sequence $\bar{M} = \langle M_t : t \in I\rangle$ such that $t \in I \Rightarrow M_t \leq_{\mathfrak{K}} N, \|M_t\| \leq \mathrm{LS}(\mathfrak{K}), I \models s < t \Rightarrow M_s \leq_{\mathfrak{K}} M_t$ and $N = \bigcup_{t \in I} M_t$. If $\|N\| \geq \mathrm{LS}(\mathfrak{K})$ we can add $\|M_t\| = \mathrm{LS}(\mathfrak{K})$ for $t \in I$.*
2) For every $N_1 \leq_{\mathfrak{K}} N_2$ we can find $\langle M_t^\ell : t \in I_\ell\rangle$ as in part (1) for $\ell = 1, 2$ such that $I_1 \subseteq I_2$ and $t \in I_1 \Rightarrow M_t^2 = M_t^1$.
3) Any $\lambda \geq \mathrm{LS}(\mathfrak{K})$ satisfies the requirement in the definition of $\mathrm{LS}(\mathfrak{K})$.

Proof. Easy or see I.1.7 which does not require anything else. $\square_{1.8}$

We now (in 1.9) recall the (non-classical) definition of type (note that it is natural to look at types only over models which are amalgamation bases, see part (4) of 1.9 below and consider only extensions of the models of the same cardinality). Note that though the choice of the name indicates that they are supposed to behave like complete types over models as in classical model theory (on which we are not relying), this does not guarantee most of the basic properties. E.g., when $\mathrm{cf}(\delta) = \aleph_0$, uniqueness of $p_\delta \in \mathscr{S}(M_\delta)$ such that $i < \delta \Rightarrow p_\delta \restriction M_i = p_i$ is not guaranteed even if $p_i \in \mathscr{S}(M_i)$, M_i is $\leq_{\mathfrak{K}}$-increasing continuous for $i \leq \delta$ and $i < j < \delta \Rightarrow p_i = p_j \restriction M_i$. Still we have existence: if for $i < \delta, p_i \in \mathscr{S}(M_i)$ increasing with i, then there is $p_\delta \in \mathscr{S}(\cup\{M_i : i < \delta\})$ such that $i < \delta \Rightarrow p_i = p_\delta \restriction M_i$. But when $\mathrm{cf}(\delta) > \aleph_0$ even existence is not guaranteed.

1.9 Definition. 1) For $M \in K_\mu, M \leq_{\mathfrak{K}} N \in K_\mu$ and $a \in N$ let $\mathbf{tp}(a, M, N) = \mathbf{tp}_{\mathfrak{K}}(a, M, N) = (M, N, a)/\mathscr{E}_M$ where $\mathscr{E}_M$ is the transitive closure of $\mathscr{E}^{\mathrm{at}}_M$, and the two-place relation $\mathscr{E}^{\mathrm{at}}_M$ is defined by:

$$
\begin{aligned}
(M, N_1, a_1)\mathscr{E}^{\mathrm{at}}_M(M, N_2, a_2) \ \underline{\text{iff}}\ & M \leq_{\mathfrak{K}} N_\ell,\ a_\ell \in N_\ell,\ \|N_\ell\| = \mu = \|M\| \\
& \text{for } \ell = 1, 2 \\
& \text{and there is } N \in K_\mu \text{ and } \leq_{\mathfrak{K}}\text{-embeddings} \\
& f_\ell : N_\ell \to N \text{ for } \ell = 1, 2 \text{ such that:} \\
& f_1 \restriction M = \mathrm{id}_M = f_2 \restriction M \text{ and } f_1(a_1) = f_2(a_2).
\end{aligned}
$$

We may say $p = \mathbf{tp}(a, M, N)$ is the type which a realizes over M in N. Of course, all those notions depend on $\mathfrak{K}$ so we may write $\mathbf{tp}_{\mathfrak{K}}(a, M, N)$ and $\mathscr{E}_M[\mathfrak{K}], \mathscr{E}^{\mathrm{at}}_M[\mathfrak{K}]$.
(If in Definition 1.4 we do not require $M \in K \Rightarrow \|M\| \geq \mathrm{LS}(\mathfrak{K})$, here we should allow any N such that $\|M\| \leq \|N\| \leq M + \mathrm{LS}(\mathfrak{K})$.) The restriction to $N \in K_\mu$ is essential, and pedantically $(M, N, a)/\mathscr{E}_M$ should be replaced by $((M, N, a)/\mathscr{E}_\mu) \cap \mathscr{H}(\chi_{(M,N,a)})$ where $\chi_{(M,N,a)} = \min\{\chi : ((M, N, a)/\mathscr{E}_M) \cap \mathscr{H}(\chi) \neq \emptyset\}$ so that the equivalence class is a set.
1A) For $M \in \mathfrak{K}_\mu$ let[4] $\mathscr{S}_{\mathfrak{K}}(M) = \{\mathbf{tp}(a, M, N) : M \leq_{\mathfrak{K}} N$ and $N \in K_\mu$ (or just $N \in K_{\leq(\mu+\mathrm{LS}(\mathfrak{K}))}$) and $a \in N\}$ and $\mathscr{S}^{\mathrm{na}}_{\mathfrak{K}}(M) = \{\mathbf{tp}(a, M, N) : M \leq_{\mathfrak{K}} N$ and $N \in K_{\leq(\mu+\mathrm{LS}(\mathfrak{K}))}$ and $a \in N \backslash M\}$, na stands for non-algebraic. We may write $\mathscr{S}^{\mathrm{na}}(M)$ omitting $\mathfrak{K}$ when

[4]if we omit $M \in K \Rightarrow \|M\| \geq \mathrm{LS}(\mathfrak{K})$ in 1.4, still we can insist that $N \in K_\mu$, the difference is not serious

$\mathfrak{K}$ is clear from the context; so omitting na means $a \in N$ rather than $a \in N \backslash M$.
2) Let $M \in K_\mu$ and $M \leq_{\mathfrak{K}} N$. We say "a realizes p in N" and "$p = \mathbf{tp}(a, M, N)$" when: if $a \in N, p \in \mathscr{S}(M)$ and $N' \in K_{\leq(\mu+\mathrm{LS}(\mathfrak{K}))}$ satisfies $M \leq_{\mathfrak{K}} N' \leq_{\mathfrak{K}} N$ and $a \in N'$ then $p = \mathbf{tp}(a, M, N')$ and there is at least one such N'; so $M, N' \in K_\mu$ (or just $M \leq \|N'\| \leq \mu + \mathrm{LS}(\mathfrak{K})$) but possibly $N \notin K_\mu$.
3) We say "a_2 strongly[5] realizes $(M, N^1, a_1)/\mathscr{E}_M^{\mathrm{at}}$ in N" <u>when</u> for some N^2 of cardinality $\leq \|M\| + \mathrm{LS}(\mathfrak{K})$ we have $M \leq_{\mathfrak{K}} N^2 \leq_{\mathfrak{K}} N$ and $a_2 \in N^2$ and $(M, N^1, a_1)\, \mathscr{E}_M^{\mathrm{at}}\, (M, N^2, a_2)$ hence $\mu = \|N^1\|$.
4) We say $M_0 \in K_\lambda$ is an amalgamation base (in $\mathfrak{K}$, but normally $\mathfrak{K}$ is understood from the context) <u>if</u>: for every $M_1, M_2 \in K_\lambda$ and $\leq_{\mathfrak{K}}$-embeddings $f_\ell : M_0 \to M_\ell$ (for $\ell = 1, 2$) there is $M_3 \in K_\lambda$ and $\leq_{\mathfrak{K}}$-embeddings $g_\ell : M_\ell \to M_3$ (for $\ell = 1, 2$) such that $g_1 \circ f_1 = g_2 \circ f_2$. Similarly for $\mathfrak{K}_{\leq\lambda}$.
4A) $\mathfrak{K}$ has amalgamation in λ (or λ-amalgamation or $\mathfrak{K}_\lambda$ has amalgamation) <u>when</u> every $M \in K_\lambda$ is an amalgamation base.
4B) $\mathfrak{K}$ has the λ-JEP or JEP_λ (or $\mathfrak{K}_\lambda$ has the JEP) <u>when</u> any $M_1, M_2 \in K_\lambda$ can be $\leq_{\mathfrak{K}}$-embedded into some $M \in K_\lambda$.
5) We say $\mathfrak{K}$ is stable in λ <u>if</u> ($\mathrm{LS}(\mathfrak{K}) \leq \lambda$ and) $M \in K_\lambda \Rightarrow |\mathscr{S}(M)| \leq \lambda$ and moreover there are no λ^+ pairwise non-$\mathscr{E}_\mu^{\mathrm{at}}$-equivalent triples $(M, N, a), M \leq_{\mathfrak{K}} N \in K_\lambda, a \in N$.
6) We say $p = q \restriction M$ if $p \in \mathscr{S}(M), q \in \mathscr{S}(N), M \leq_{\mathfrak{K}} N$ and for some $N^+, N \leq_{\mathfrak{K}} N^+$ and $a \in N^+$ we have $p = \mathbf{tp}(a, M, N^+)$ and $q = \mathbf{tp}(a, N, N^+)$; see 1.11(1),(2). We may express this also as "q extends p or p is the restriction of q to M".
7) For finite m, for $M \leq_{\mathfrak{K}} N, \bar{a} \in {}^m N$ we can define $\mathbf{tp}(\bar{a}, M, N)$ and $\mathscr{S}_{\mathfrak{K}}^m(M)$ similarly and $\mathscr{S}_{\mathfrak{K}}^{<\omega}(M) = \bigcup\limits_{m<\omega} \mathscr{S}_{\mathfrak{K}}^m(M)$; similarly for $\mathscr{S}^\alpha(M)$ (but we shall not use this in any essential way, so we agree $\mathscr{S}(M) = \mathscr{S}^1(M)$.) Again we may omit $\mathfrak{K}$ when clear from the context.
8) We say that $p \in \mathscr{S}_{\mathfrak{K}}(M)$ is algebraic when some $a \in M$ realizes it.
9) We say that $p \in \mathscr{S}_{\mathfrak{K}}(M)$ is minimal <u>when</u> it is not algebraic and for every $N \in K$ of cardinality $\leq \|M\| + \mathrm{LS}(\mathfrak{K})$ which $\leq_{\mathfrak{K}}$-extend

[5] note that $\mathscr{E}_M^{\mathrm{at}}$ is not necessarily an equivalence relation and hence in general is not $\mathscr{E}_M$

M, the type p has at most one non-algebraic extension in $\mathscr{S}_{\mathfrak{K}}(M)$.

1.10 Remark. 1) Note that here "amalgamation base" means only for extensions of the same cardinality!
2) The notion "minimal type" is important (for categoricity) but not used much in this chapter.

1.11 Observation. 0) Assume $M \in K_\mu$ and $M \le_{\mathfrak{K}} N, a \in N$ <u>then</u> $\mathbf{tp}(a, M, N)$ is well defined and is p <u>if</u> for some $M' \in K_\mu$ we have $M \cup \{a\} \subseteq M' \le_{\mathfrak{K}} N$ and $p = \mathbf{tp}(a, M, M')$.
1) If $M \le_{\mathfrak{K}} N_1 \le_{\mathfrak{K}} N_2, M \in K_\mu$ and $a \in N_1$ <u>then</u> $\mathbf{tp}(a, M, N_1)$ is well defined and equal to $\mathbf{tp}(a, M, N_2)$, (more transparent if $\mathfrak{K}$ has the μ-amalgamation, which is the real case anyhow).
2) If $M \le_{\mathfrak{K}} N$ and $q \in \mathscr{S}(N)$ <u>then</u> for one and only one p we have $p = q \restriction M$.
3) If $M_0 \le_{\mathfrak{K}} M_1 \le_{\mathfrak{K}} M_2$ and $p \in \mathscr{S}(M_2)$ <u>then</u> $p \restriction M_0 = (p \restriction M_1) \restriction M_0$.
4) If $M \in \mathfrak{K}_\mu$ is an amalgamation base <u>then</u> $\mathscr{E}^{\rm at}_M$ is a transitive relation hence is equal to $\mathscr{E}_M$.
5) If $M \le_{\mathfrak{K}} N$ are from $\mathfrak{K}_\lambda, M$ is an amalgamation base and $p \in \mathscr{S}(M)$ <u>then</u> there is $q \in \mathscr{S}(N)$ extending p, so the mapping $q \mapsto q \restriction M$ is a function from $\mathscr{S}(N)$ onto $\mathscr{S}(M)$.

Proof. Easy. $\square_{1.11}$

1.12 Definition. 1) We say <u>N is λ-universal over M</u> when $\lambda \ge \|N\|$ and for every $M', M \le_{\mathfrak{K}} M' \in K_\lambda$, there is a $\le_{\mathfrak{K}}$-embedding of M' into N over M. If we omit λ we mean $\|N\|$; clearly if N is universal over M and both are from K_λ then M is an amalgamation base.
2) $K^{3,\rm na}_\lambda = \{(M, N, a) : M \le_{\mathfrak{K}} N, a \in N\backslash M$ and $M, N \in \mathfrak{K}_\lambda\}$, with the partial order $\le$ defined by $(M, N, a) \le (M', N', a')$ iff $a = a', M \le_{\mathfrak{K}} M'$ and $N \le_{\mathfrak{K}} N'$.
3) We say $(M, N, a) \in K^{3,\rm na}_\lambda$ is minimal <u>when</u>: if $(M, N, a) \le (M', N_\ell, a) \in K^{3,\rm na}_\lambda$ for $\ell = 1, 2$ implies $\mathbf{tp}(a, M', N_1) = \mathbf{tp}(a, M', N_2)$ moreover, $(M', N_1, a)\mathscr{E}^{\rm at}_\lambda(M', N_2, a)$ (this strengthening is not needed if every $M' \in K_\lambda$ is an amalgamation bases).

4) $N \in \mathfrak{K}$ is λ-universal if every $M \in \mathfrak{K}_\lambda$ can be $\leq_{\mathfrak{K}}$-embedded into it.
5) We say $N \in \mathfrak{K}$ is universal for $K' \subseteq \mathfrak{K}$ <u>when</u> every $M \in K'$ can be $\leq_{\mathfrak{K}}$-embedded into N.

Remark. Why do we use $\leq$ on $K^{3,\mathrm{na}}_\lambda$? Because those triples serve us as a representation of types for which direct limit exists.

1.13 Definition. 1) $M^* \in K_\lambda$ is <u>superlimit</u> if: clauses (a) + (b) + (c) below hold, and locally superlimit if clauses $(a)^-$ + (b) + (c) below hold and is pseudo superlimit if clauses (b) + (c) below hold, where:

- (a) it is universal, (i.e. every $M \in K_\lambda$ can be $\leq_{\mathfrak{K}}$-embedded into M^*),
- (b) if $\langle M_i : i \leq \delta \rangle$ is $\leq_{\mathfrak{K}}$-increasing continuous, $\delta < \lambda^+$ and $i < \delta \Rightarrow M_i \cong M^*$ then $M_\delta \cong M^*$
- $(a)^-$ if $M^* \leq_{\mathfrak{K}} M_1 \in K_\lambda$ then there is $M_2 \in K_2$ which $\leq_{\mathfrak{K}}$-extend M_1 and is isomorphic to M^*
- (c) there is M^{**} isomorphic to M^* such that $M^* <_{\mathfrak{K}} M^{**}$.

2) M is λ-saturated above μ <u>when</u> $\|M\| \geq \lambda > \mu \geq \mathrm{LS}(\mathfrak{K})$ and: $N \leq_{\mathfrak{K}} M, \mu \leq \|N\| < \lambda, N \leq_{\mathfrak{K}} N_1, \|N_1\| \leq \|N\| + \mathrm{LS}(\mathfrak{K})$ and $a \in N_1$ then some $b \in M$ strongly realizes $(N, N_1, a)/\mathscr{E}^{\mathrm{at}}_N$ in M, see Definition 1.9(3). Omitting "above μ" means "for some $\mu < \lambda$" hence "M is λ^+-saturated" mean that "M is λ^+-saturated above λ" and $K(\lambda^+\text{-saturated}) = \{M \in K : M \text{ is } \lambda^+\text{-saturated}\}$ and "M is saturated" means "M is $\|M\|$-saturated".

In the following lemma note that amalgamation in $\mathfrak{K}_{<\lambda}$ is not assumed it is even deduced. For variety we allow $K_{<\mathrm{LS}(\mathfrak{K})} \neq \emptyset$.

1.14 The Model-homogeneity = Saturativity Lemma. *Let* $\lambda > \mu + \mathrm{LS}(\mathfrak{K})$ *and* $M \in K$.
1) M *is* λ*-saturated above* μ <u>*iff*</u> M *is* $(\mathbb{D}_{\mathfrak{K}_{\geq\mu}}, \lambda)$*-homogeneous above* μ*, which means: for every* $N_1 \leq_{\mathfrak{K}} N_2 \in K$ *such that* $\mu \leq \|N_1\| \leq$

$\|N_2\| < \lambda$ and $N_1 \leq_{\mathfrak{K}} M$, there is a $\leq_{\mathfrak{K}}$-embedding f of N_2 into M over N_1.
2) If $M_1, M_2 \in K_\lambda$ are λ-saturated above $\mu < \lambda$ and for some $N_1 \leq_{\mathfrak{K}} M_1, N_2 \leq_{\mathfrak{K}} M_2$, both of cardinality $\in [\mu, \lambda)$, we have $N_1 \cong N_2$ then $M_1 \cong M_2$; in fact, any isomorphism f from N_1 onto N_2 can be extended to an isomorphism from M_1 onto M_2.
3) If in (2) we demand only "M_2 is λ-saturated" and $M_1 \in K_{\leq\lambda}$ then f can be extended to a $\leq_{\mathfrak{K}}$-embedding from M_1 into M_2.
4) In part (2) instead of $N_1 \cong N_2$ it suffices to assume that N_1 and N_2 can be $\leq_{\mathfrak{K}}$-embedded into some $N \in K$, which holds if $\mathfrak{K}$ has the JEP *or just θ-*JEP *for some $\theta < \lambda, \theta \geq \mu$. Similarly for part (3).*
*5) If N is λ-universal over $M \in K_\mu$ and $\mathfrak{K}$ has μ-*JEP *then N is λ-universal (where $\lambda \geq$* LS$(\mathfrak{K})$ *for simplicity).*
6) Assume M is λ-saturated above μ. If $N \leq_{\mathfrak{K}} M$ and $\mu \leq \|N\| < \lambda$ then N is an amalgamation base (in $K_{\leq(\|N\|+\mathrm{LS}(\mathfrak{K}))}$ and even in $\mathfrak{K}_{\leq\lambda}$) and $|\mathscr{S}(N)| \leq \|M\|$. So if every $N \in K_\mu$ can be $\leq_{\mathfrak{K}}$-embedded into M then $\mathfrak{K}$ has μ-amalgamation.

Proof. 1) The "if" direction is easy as $\lambda > \mu +$ LS$(\mathfrak{K})$. Let us prove the other direction.

We prove this by induction on $\|N_2\|$. Now first consider the case $\|N_2\| > \|N_1\| + \mathrm{LS}(\mathfrak{K})$ then we can find a $\leq_{\mathfrak{K}}$-increasing continuous sequence $\langle N_{1,\varepsilon} : \varepsilon < \|N_2\|\rangle$ with union N_2 with $N_{1,0} = N_1$ and $\|N_{1,\varepsilon}\| \leq \|N_1\| + |\varepsilon|$. Now we choose f_ε, a $\leq_{\mathfrak{K}}$-embedding of $N_{1,\varepsilon}$ into M, increasing continuous with ε such that $f_0 = \mathrm{id}_{N_1}$. For $\varepsilon = 0$ this is trivial for ε limit take unions and for ε successor use the induction hypothesis. So without loss of generality $\|N_2\| \leq \|N_1\| + \mathrm{LS}(\mathfrak{K})$.

Let $|N_2| = \{a_i : i < \kappa\}$, and we know $\mu \leq \kappa'' := \|N_1\| \leq \kappa := \|N_2\| \leq \kappa' := \|N_1\| + \mathrm{LS}(\mathfrak{K}) < \lambda$; so if, as usual, $\|N_1\| \geq \mathrm{LS}(\mathfrak{K})$ then $\kappa' = \kappa$. We define by induction on $i \leq \kappa$, N_1^i, N_2^i, f_i such that:

(a) $N_1^i \leq_{\mathfrak{K}} N_2^i$ and $\|N_1^i\| \leq \|N_2^i\| \leq \kappa'$
(b) N_1^i is $\leq_{\mathfrak{K}}$-increasing continuous with i
(c) N_2^i is $\leq_{\mathfrak{K}}$-increasing continuous with i
(d) f_i is a $\leq_{\mathfrak{K}}$-embedding of N_1^i into M
(e) f_i is increasing continuous with i
(f) $a_i \in f_i(N_1^{i+1})$

(g) $N_1^0 = N_1, N_2^0 = N_2, f_0 = \mathrm{id}_{N_1}$.

For $i = 0$, clause (g) gives the definition. For i limit let:

$$N_1^i = \bigcup_{j<i} N_1^j, \qquad N_2^i = \bigcup_{j<i} N_2^j, \qquad f_i = \bigcup_{j<i} f_j.$$

Now (a)-(f) continues to hold by continuity (and $\|N_2^i\| \leq \kappa'$ easily).

For i successor we use our assumption; more elaborately, let $M_1^{i-1} \leq_{\mathfrak{K}} M$ be $f_{i-1}(N_1^{i-1})$ and let M_2^{i-1}, g_{i-1} be such that g_{i-1} is an isomorphism from N_2^{i-1}
onto M_2^{i-1} extending f_{i-1}, so $M_1^{i-1} \leq_{\mathfrak{K}} M_2^{i-1}$ (but without loss of generality $M_2^{i-1} \cap M = M_1^{i-1}$). Now apply the saturation assumption, see Definition 1.13(21) with $M, (M_1^{i-1}, M_2^{i-1}), g_{i-1}(a))$ here standing for $M, (N, N_1, a)$ there (note: $a_{i-1} \in N_2 = N_2^0 \subseteq N_2^{i-1}$ and $\lambda > \kappa' \geq \|N_2^{i-1}\| = \|M_2^{i-1}\| \geq \|M_1^{i-1}\| = \|N_1^{i-1}\| \geq \|N_1^0\| = \|N_1\| = \kappa'' \geq \mu$ so the requirements including the requirements on the cardinalities in Definition 1.13(2) holds). So there is $b \in M$ such that $\mathbf{tp}(b, M_1^{i-1}, M) = \mathbf{tp}(g_{i-1}(a_{i-1}), M_1^{i-1}, M_2^{i-1})$. Moreover (if $\mathfrak{K}$ has amalgamation in μ the proof is slightly shorter) remembering the end of the first sentence in 1.13(2) which speaks about "strongly realizes", b strongly realizes $(M_1^{i-1}, M_3^{i-1}, g_{i-1}(a_{i-1}))/\mathscr{E}^{\mathrm{at}}_{M_1^{i-1}}$ in M. This means (see Definition 1.9(3)) that for some $M_1^{i,*}$ we have $b \in M_1^{i,*}$ and $M_1^{i-1} \leq_{\mathfrak{K}} M_1^{i,*} \leq_{\mathfrak{K}} M$ and $(M_1^{i-1}, M_2^{i-1}, g_{i-1}(a_{i-1}))\ \mathscr{E}^{\mathrm{at}}_{M_1^{i-1}}$ $(M_1^{i-1}, M_1^{i,*}, b)$. This means (see Definition 1.9(1)) that $M_1^{i,*}$ too has cardinality $\leq \kappa'$ and there is $M_2^{i,*} \in K_{\leq \kappa'}$ such that $M_1^{i-1} \leq_{\mathfrak{K}} M_2^{i,*}$ and there are $\leq_{\mathfrak{K}}$-embeddings h_2^i, h_1^i of $M_2^{i-1}, M_1^{i,*}$ into $M_2^{i,*}$ over M_1^{i-1} respectively, such that $h_2^i(g_{i-1}(a_{i-1})) = h_1^i(b)$.
Now changing names, without loss of generality h_1^i is the identity. Let N_2^i, h_i be such that $N_2^{i-1} \leq_{\mathfrak{K}} N_2^i$ and h_i an isomorphism from N_2^i onto $M_2^{i,*}$ extending g_{i-1}. Let $N_1^i = h_i^{-1}(M_1^{i,*})$ and $f_i = (h_i \restriction N_1^i)$.

We have carried the induction. Now f_κ is a $\leq_{\mathfrak{K}}$-embedding of N_1^κ into M over N_1, but $|N_2| = \{a_i : i < \kappa\} \subseteq N_1^\kappa$ hence by AxV of Definition 1.4, $N_2 \leq_{\mathfrak{K}} N_1^\kappa$, so $f_\kappa \restriction N_2 : N_2 \to M$ is as required.
2), 3) By the hence and forth argument (or see I.2.4,I.2.5 or see [Sh 300, II,§3] = V.B§3).
4),5),6) Easy, too. $\square_{1.14}$

1.15 Definition.
1) For $\partial = \mathrm{cf}(\partial) \leq \lambda^+$, we say <u>$N$ is (λ, ∂)-brimmed over M</u> if ($M \leq_{\mathfrak{K}} N$ are in K_λ and) we can find a sequence $\langle M_i : i < \partial \rangle$ which is $\leq_{\mathfrak{K}}$-increasing[6], $M_i \in K_\lambda, M_0 = M, M_{i+1}$ is $\leq_{\mathfrak{K}}$-universal[7] over M_i and $\bigcup_{i<\partial} M_i = N$. We say N is (λ, ∂)-brimmed over A if $A \subseteq N \in K_\lambda$ and we can find $\langle M_i : i < \partial \rangle$ as above such that $A \subseteq M_0$ but $M_0 \restriction A \leq_{\mathfrak{K}} M_0 \Rightarrow M_0 = A$; if $A = \emptyset$ we may omit "over A". We say continuously (λ, ∂)-brimmed (over M) <u>when</u> the sequence $\langle M_i : i < \partial \rangle$ is $\leq_{\mathfrak{K}}$-increasing continuous; if $\mathfrak{K}_\lambda$ has amalgamation, the two notions coincide.
2) We say N is $(\lambda, *)$-brimmed over M <u>if</u> for some $\partial \leq \lambda, N$ is (λ, ∂)-brimmed over M. We say N is $(\lambda, *)$-brimmed if for some M, N is $(\lambda, *)$-brimmed over M.
3) If $\alpha < \lambda^+$ let "N is (λ, α)-brimmed over M" mean $M \leq_{\mathfrak{K}} N$ are from K_λ and $\mathrm{cf}(\alpha) \geq \aleph_0 \Rightarrow N$ is $(\lambda, \mathrm{cf}(\alpha))$-brimmed over M.

On the meaning of (λ, ∂)-brimmed for elementary classes, see 3.1(2) below. Recall

1.16 Claim. *Assume $\lambda \geq \mathrm{LS}(\mathfrak{K})$.*
1) If $\mathfrak{K}$ has amalgamation in λ, is stable in λ and $\partial = \mathrm{cf}(\partial) \leq \lambda$, <u>then</u>

- (a) *for every $M \in \mathfrak{K}_\lambda$ there is $N, M \leq_{\mathfrak{K}} N \in K_\lambda$, universal over M*
- (b) *for every $M \in \mathfrak{K}_\lambda$ there is $N \in \mathfrak{K}_\lambda$ which is (λ, ∂)-brimmed over M*
- (c) *if N is (λ, ∂)-brimmed over M <u>then</u> N is universal over M.*

2) If N_ℓ is $(\lambda, \aleph_0)$-brimmed over M for $\ell = 1, 2$, <u>then</u> N_1, N_2 are isomorphic over M.
3) Assume $\partial = \mathrm{cf}(\partial) \leq \lambda^+$, and for every $\aleph_0 \leq \theta = \mathrm{cf}(\theta) < \partial$ any (λ, θ)-brimmed model is an amalgamation base (in $\mathfrak{K}$). <u>Then</u>:

---

[6] we have not asked continuity; because in the direction we are going, it makes no difference if we add "continuous". Then we have in general fewer cases of existence, uniqueness (of being (λ, ∂)-brimmed over $M \in K_\lambda$) does not need extra assumptions and existence is harder

[7] hence M_i is an amalgamation base

(a) *if N_ℓ is (λ, ∂)-brimmed over M for $\ell = 1, 2$ then N_1, N_2 are isomorphic over M*

(b) *if $\mathfrak{K}$ has λ-JEP (i.e., the joint embedding property in λ) and N_1, N_2 are (λ, ∂)-brimmed then N_1, N_2 are isomorphic.*

3A) There is a (λ, ∂)-brimmed model N over $M \in K_\lambda$ when: M is an amalgamation base, and for every $\leq_{\mathfrak{K}_\lambda}$-extension M_1 of M there is a $\leq_{\mathfrak{K}_\lambda}$-extension M_2 of M_1 which is an amalgamation base and there is a λ-universal extension $M_3 \in K_\lambda$ of M_2.
4) Assume $\mathfrak{K}$ has λ-amalgamation and the λ-JEP and $\bar{M} = \langle M_i : i \leq \lambda \rangle$ is $\leq_{\mathfrak{K}}$-increasing continuous and $M_i \in K_\lambda$ for $i \leq \lambda$.

(a) *If λ is regular and for every $i < \lambda, p \in \mathscr{S}(M_i)$ for some $j \in (i, \lambda)$, some $a \in M_j$ realizes p, then M_λ is universal over M_0 and is (λ, λ)-brimmed over M_0*

(b) *if for every $i < \lambda$ every $p \in \mathscr{S}(M_i)$ is realized in M_{i+1} then M_λ is $(\lambda, \mathrm{cf}(\lambda))$-brimmed over M_0.*

5) Assume $\partial = \mathrm{cf}(\partial) \leq \lambda$ and $M \in \mathfrak{K}$ is continuous (λ, ∂)-brimmed. Then M is a locally $(\lambda, \{\partial\})$-strongly limit model in $\mathfrak{K}_\lambda$ (see Definition I.3.3(2),(7), not used).
6) If N is (λ, ∂)-brimmed over M and $A \subseteq N, |A| < \partial$, e.g. $A = \{a\}$ then for some M' we have $M \cup A \subseteq M' <_{\mathfrak{K}} M$ and M is (λ, ∂)-brimmed over M'.

Proof. 1) Clause (c) holds by Definition 1.15.

As for clause (a), for any given $M \in K_\lambda$, easily there is an $\leq_{\mathfrak{K}}$-increasing continuous sequence $\langle M_i : i \leq \lambda \rangle$ of models from $K_\lambda, M_0 = M$ such that $p \in \mathscr{S}(M_i) \Rightarrow p$ is realized in M_{i+1}, this by stability + amalgamation. So $\langle M_i : i \leq \lambda \rangle$ is as in part (4) below hence by clause (b) of part (4) below, we get that M_δ is $\leq_{\mathfrak{K}}$-universal over $M_0 = M$ so we are done. Clause (b) follows by (a).
2) By (3)(a) because the extra assumption in part (3) is empty when $\partial = \aleph_0$.
3) Clause (a) holds by the hence and forth argument, that is assume $\langle N_{\ell,i} : i < \partial \rangle$ is $\leq_{\mathfrak{K}}$-increasing with union $N_{\ell,\partial}, N_{\ell,0} = M, N_{\ell,i+1}$ is universal over $N_{\ell,i}$ and $N_\ell = N_{\ell,\partial}$ so $N_{\ell,i} \in \mathfrak{K}_\lambda$.

Now for each limit $\delta < \partial$ the model $N'_{\ell,\delta} := \cup\{N_{\ell,i} : i < \delta\}$ is an amalgamation base (and is $\leq_{\mathfrak{K}} N_{\ell,\delta+1}$) hence without loss of generality $\langle N_{\ell,i} : i \leq \partial \rangle$ is $\leq_{\mathfrak{K}}$-increasing continuous. We now choose f_i by induction on $i \leq \partial$ such that:

- (i) if i is odd, f_i is a $\leq_{\mathfrak{K}}$-embedding of $N_{1,i}$ into $N_{2,i}$
- (ii) if i is even, f_i^{-1} is a $\leq_{\mathfrak{K}}$-embedding of $N_{2,i}$ into $N_{1,i}$
- (iii) if i is limit then f_i is an isomorphism from $N_{1,i}$ onto $N_{2,i}$
- (iv) f_i is increasing continuous with i
- (v) if $i = 0$ then $f_0 = \mathrm{id}_M$.

For $i = 0$ let $f_0 = \mathrm{id}_M$. If $i = 2j + 2$ use "$N_{1,i}$ is a universal extension of $N_{1,2j+1}$ (in $\mathfrak{K}_\lambda$) and f_{2j+1} is a $\leq_{\mathfrak{K}}$-embedding of $N_{1,2j+1}$ into $N_{2,2j+1}$ (by clause (i) applied to $2j+1$) and $N_{1,2j+1}$ is an amalgamation base". That is, $N_{2,i}$ is a $\leq_{\mathfrak{K}}$-extension of $f_{2j+1}(N_{2j+1})$ which is an amalgamation base so f_{2j+1}^{-1} can be extended to a $\leq_{\mathfrak{K}}$-embedding of f_i^{-1} of $N_{2,i}$ into $N_{1,i}$. For $i = 2j + 1$ use "$N_{2,i}$ is a universal extension (in $\mathfrak{K}_\lambda$) of $N_{2,2j}$ and f_{2j}^{-1} is a $\leq_{\mathfrak{K}}$-embedding of $N_{2,2j}$ into $N_{1,2j}$ and $N_{2,2j}$ is an amalgamation base (in $\mathfrak{K}_\lambda$)".
For i limit let $f_i = \cup\{f_j : j < i\}$. Clearly f_∂ is an isomorphism from $N_1 = N_{1,\partial}$ onto $N_{2,\partial} = N_2$ so we are done, i.e. clause (a) holds.

As for clause (b), for $\ell = 1, 2$ we can assume that $\langle N_{\ell,i} : i \leq \partial \rangle$ exemplifies "N_ℓ is (λ, ∂)-brimmed" so $N_\ell = N_{\ell,\partial}$ and without loss of generality as above $\langle N_{\ell,i} : i \leq \partial \rangle$ is $\leq_{\mathfrak{K}_\lambda}$-increasing continuous. By the λ-JEP there is a pair (g_1, N) such that $N_{1,0} \leq_{\mathfrak{K}} N \in K_\lambda$ and g_1 is a $\leq_{\mathfrak{K}}$-embedding of $N_{2,0}$ into N. As above there is a $\leq_{\mathfrak{K}}$-embedding g_2 of N into $N_{1,1}$ over $N_{1,0}$. Let $f_0 = (g_2 \circ g_1)^{-1}$ and continue as in the proof of clause (a).
3A) Easy, too.
4) We first proved weaker version of (a) and of (b) called (a)$^-$,(b)$^-$ respectively.

<u>Clause $(a)^-$</u>: Like (a) but we conclude only: M_λ is universal over M_0.

So let N satisfy $M_0 \leq_{\mathfrak{K}} N \in K_\lambda$ and we shall prove that N is $\leq_{\mathfrak{K}}$-embeddable into M_λ over M_0. Let $\langle S_i : i < \lambda \rangle$ be a partition of λ such that $|S_i| = \lambda$, $\min(S_i) \geq i$ for $i < \lambda$. We choose a quadruple $(N_i, f_i, \bar{\mathbf{a}}_i, j_i)$ by inductin on $i < \lambda$ such that:

$\circledast$ (a) $N_i \in K_\lambda$ is $\leq_{\mathfrak{K}}$-increasing continuous

(b) $N_0 = N$

(c) $\bar{\mathbf{a}}_i = \langle a_\alpha : \alpha \in S_i \rangle$ list the members of N_i

(d) $j_i < \lambda$ is increasing continuous

(e) f_i is a $\leq_{\mathfrak{K}}$-embedding of M_i into M_i

(f) $f_0 = \mathrm{id}_{M_0}$

(g) f_i is $\subseteq$-increasing continuous

(h) if $i = \alpha + 1$ then $a_\alpha \in \mathrm{Rang}(f_i)$.

There is no problem to carry the definition (below, proving (a) we give more details) and necessarily $f = \cup\{f_i : i < \lambda\}$ is an isomorphism from M_λ onto $N_\lambda := \cup\{N_i : i < \lambda\}$, so $f^{-1} \restriction N$ is a $\leq_{\mathfrak{K}}$-embedding of N into M_λ over M_0 (as $f^{-1} \restriction N \supseteq \mathrm{id}_{M_0}$), so we are done.

<u>Clause $(b)^-$</u>: Like clause (b) but we conclude only: M_λ is universal over M_0.

Similar to the proof of $(a)^-$ except that we demand $j_i = i$.

<u>Clause (a)</u>: Let $M_0 \leq_{\mathfrak{K}} N \in K_\lambda$ and we let $\langle S_i : i < \lambda \rangle$ be a partition of λ to λ sets each with λ members, $i \leq \mathrm{Min}(S_i)$. Let $M_{1,i} = M_i$ for $i \leq \lambda$ and we choose $\langle M_{2,i} : i \leq \delta \rangle$ which is $\leq_{\mathfrak{K}}$-increasing such that $M_{2,i} \in \mathfrak{K}$, $M_{2,0} = M_{1,0}$, $N \leq_{\mathfrak{K}} M_{2,1}$ and $M_{2,i+1} \in K_\lambda$ is $\leq_{\mathfrak{K}}$-universal over $M_{2,i}$, possible as we have already proved clause $(a)^-$ recalling $\mathfrak{K}$ has λ-amalgamation and the λ-JEP.

We shall prove that $M_{1,\lambda}, M_{2,\lambda}$ are isomorphic over $M_0 = M_{1,0}$, this clearly suffices. We choose a quintuple $(j_i, M_{3,i}, f_{1,i}, f_{2,i}, \bar{\mathbf{a}}_i)$ by induction on $i < \lambda$ such that

$\circledast$ (a) $j_i < \lambda$ is increasing continuous

(b) $M_{3,i} \in K_\lambda$ is $\leq_{\mathfrak{K}}$-increasing continuous

(c) $f_{\ell,i}$ is a $\leq_{\mathfrak{K}}$-embedding of M_{ℓ,j_i} into M for $\ell = 1, 2$

(d) $f_{\ell,i}$ is increasing continuous with i for $\ell = 1, 2$

(e) $\bar{\mathbf{a}}_i = \langle a^i_\varepsilon : \varepsilon \in S_i \rangle$ lists the members of $M_{3,i}$

(f) if $\varepsilon \in S_i$ then $a^i_\varepsilon \in \mathrm{Rang}(f_{1,2\varepsilon+1})$ and $a^i_\varepsilon \in \mathrm{Rang}(f_{2,2\varepsilon+2})$.

If we succeed then $f_\ell := \cup\{f_{\ell,i} : i < \lambda\}$ is a $\le_{\mathfrak{K}}$-embedding of $M_{\ell,\lambda}$ into $M_{3,\lambda} := M_3 := \cup\{M_{3,i} : i < \lambda\}$ and this embedding is onto because $a \in M_3 \Rightarrow$ for some $i < \lambda, a \in M_{3,i} \Rightarrow$ for some $i < \lambda$ and $\varepsilon \in S_i, a = a^i_\varepsilon \Rightarrow a = a^i_\varepsilon \in \mathrm{Rang}(f_{\ell,\varepsilon+1}) \Rightarrow a \in \mathrm{Rang}(f_\ell)$. So $f_1^{-1} \circ f_2$ is an isomorphism from $M_{2,\lambda}$ onto $M_{1,\lambda} = M_\lambda$ so as said above we are done.

Carrying the induction; for $i = 0$ use "$\mathfrak{K}$ has the λ-JEP" for $M_{1,0}, M_{2,0}$.

For i limit take unions.

For $i = 2\varepsilon + 1$ let $j_i =$ $\min\{j < \lambda_i : j > j_{2\varepsilon}$ and $(f^1_{2\varepsilon})^{-1}(\mathbf{tp}(a^i_\varepsilon, f^1_{2\varepsilon}(M_{1,i}), M_{3,i})) \in \mathscr{S}_{\mathfrak{K}}(M_{1,i})$ is realized in M_j and continue as in the proof of 1.14(1), so can avoid using "$(f^1_i)^{-2}$ of a type.

For $i = 2\varepsilon + 2$, the proof is similar. So $M_{2,\lambda}$ is $(\lambda, \mathrm{cf}(\lambda))$-brimmed over $M_{2,0} = M_0$ hence also M_λ being isomorphic to $M_{2,\lambda}$ over M_0 is $(\lambda, \mathrm{cf}(\lambda))$-brimmed over M_0, as required.

Clause (b): As in the proof of clause (a) but now $j_i = i$.
5) Easy and not used. (Let $\langle M_i : i \le \partial\rangle$ witness "M is (λ, ∂)-brimmed", so M can be $\le_{\mathfrak{K}}$-embedded into M_i, hence without loss of generality $M_0 \cong M_1$. Now use $\mathbf{F}$ such that $\mathbf{F}(M')$ is a $\le_{\mathfrak{K}_\lambda}$-extension of M' which is $\le_{\mathfrak{K}_\lambda}$-universal over it and is an amalgamation base.)
6) Easy. $\square_{1.16}$

1.17 Claim. *1) Assume that $\mathfrak{K}$ is an* a.e.c., $\mathrm{LS}(\mathfrak{K}) \le \lambda$ *and $\mathfrak{K}$ has λ-amalgamation and is stable in λ and no $M \in K_\lambda$ is $\le_{\mathfrak{K}}$-maximal. Then there is a saturated $N \in K_{\lambda^+}$. Also for every saturated $N \in K_{\lambda^+}$ (in $\mathfrak{K}$, above λ of course) we can find a $\le_{\mathfrak{K}}$-representation $\bar N = \langle N_i : i < \lambda^+\rangle$, with N_{i+1} being $(\lambda, \mathrm{cf}(\lambda))$-brimmed over N_i and N_0 being (λ, λ)-brimmed.*
2) If for $\ell = 1, 2$ we have $\bar N^\ell = \langle N^\ell_i : i < \lambda^+\rangle$ as in part (1), then there is an isomorphism f from N^1 onto N^2 mapping N^1_i onto N^2_i for each $i < \lambda^+$. Moreover, for any $i < \lambda^+$ and isomorphism g from N^1_i onto N^2_i we can find an isomorphism f from N^1 onto N^2 extending g and mappng N^1_j onto N^2_j for each $j \in [i, \lambda^+)$.
3) If $N^0 \le_{\mathfrak{K}} N^1$ are both saturated (above λ) and are in K_{λ^+} (hence $\mathrm{LS}(\mathfrak{K}) \le \lambda$*), then we can find $\le_{\mathfrak{K}}$-representation $\bar N^\ell$ of N^ℓ as in (1)*

for $\ell = 1,2$ *with* $N_i^0 = N^0 \cap N_i^1$, *(so* $N_i^0 \leq_{\mathfrak{K}} N_i^1$*) for* $i < \lambda^+$.
4) If $M \in K_{\lambda^+}$ *and* $\mathfrak{K}$ *has* λ*-amalgamation and is stable in* λ *(and* $\mathrm{LS}(\mathfrak{K}) \leq \lambda$*), then for some* $N \in K_{\lambda^+}$ *saturated (above* λ*) we have* $M \leq_{\mathfrak{K}} N$.

Proof. Easy (for (2),(3) using 1.14(6)), e.g.
4) There is a $\leq_{\mathfrak{K}}$-increasing continuous sequence $\langle M_i : i < \lambda^+ \rangle$ with union M such that $M_i \in K_\lambda$. Now we choose N_i by induction on $i < \lambda$

(∗) (*a*) $N_i \in K_\lambda$ is $\leq_{\mathfrak{K}}$-increasing continuous

(*b*) N_{i+1} is $(\lambda, \mathrm{cf}(\lambda))$-brimmed over N_i

(*c*) $N_0 = M_0$.

This is possible by 1.16(1). Then by induction on $i \leq \lambda^+$ we choose a $\leq_{\mathfrak{K}}$-embedding f_i of M_i into N_i, increasing continuous with i. For $i = 0$ let $f_i = \mathrm{id}_{M_0}$. For i limit use union.

Lastly, for $i = j + 1$ use "$\mathfrak{K}$ has λ-amalgamation" and "N_j is universal over N_i". Now by renaming without loss of generality $f_{\lambda^+} = \mathrm{id}_{N_{\lambda^+}}$ and we are done. (Of course, we hae assumed less). $\square_{1.17}$

You may wonder why in this work we have not restricted ourselves $\mathfrak{K}$ to "abstract elementary class in λ" say in §2 below (or in [Sh 576]); by the following facts (mainly 1.23) this is immaterial.

1.18 Definition. 1) We say that $\mathfrak{K}_\lambda$ is a λ-abstract elementary class or λ-a.e.c. in short, when:

(*a*) $\mathfrak{K}_\lambda = (K_\lambda, \leq_{\mathfrak{K}_\lambda})$,

(*b*) K_λ is a class of τ-models of cardinality λ closed under isomorphism for some vocabulary $\tau = \tau_{\mathfrak{K}_\lambda}$,

(*c*) $\leq_{\mathfrak{K}_\lambda}$ a partial order of K_λ, closed under isomorphisms

(*d*) axioms (0 and) I,II,III,IV,V of abstract elementary classes (see 1.4) hold except that in AxIII we demand $\delta < \lambda^+$ (you can demand this also in AxIV).

2) For an abstract elementary class $\mathfrak{K}$ let $\mathfrak{K}_\lambda = (K_\lambda, \leq_{\mathfrak{K}} \restriction K_\lambda)$ and similarly $\mathfrak{K}_{\geq\lambda}, \mathfrak{K}_{\leq\lambda}, \mathfrak{K}_{[\lambda,\mu]}$ and define $(\leq \lambda)$-a.e.c. and $[\lambda, \mu]$-a.e.c., etc.
3) Definitions 1.9, 1.12, 1.13, 1.15 apply to λ-a.e.c. $\mathfrak{K}_\lambda$.

1.19 Observation. If $\mathfrak{K}^1$ is an a.e.c. with $K^1_\lambda \neq \emptyset$ then

(a) $\mathfrak{K}^1_\lambda$ is a λ-a.e.c.
(b) if $\mathfrak{K}^2_\lambda$ is a λ-a.e.c., and $\mathfrak{K}^1_\lambda = \mathfrak{K}^2_\lambda$ then Definitions 1.9, 1.12, 1.13, 1.15 when applied to $\mathfrak{K}^1$ but restricting ourselves to models of cardinality λ and when applied to $\mathfrak{K}^2_\lambda$ are equivalent.

Proof. Just read the definitions. $\square_{1.19}$

We may wonder
1.20 Problem: Suppose $\mathfrak{K}^1, \mathfrak{K}^2$ are a.e.c. such that for some $\lambda > \mu \geq \mathrm{LS}(\mathfrak{K}^1), \mathrm{LS}(\mathfrak{K}^2)$ and $\mathfrak{K}^1_\lambda = \mathfrak{K}^2_\lambda$. Can we bound the first such λ above μ? (Well, better bound than the Lowenheim number of $\mathbb{L}_{\mu^+,\mu^+}$ (second order)).

1.21 Observation. 1) Let $\mathfrak{K}$ be an a.e.c. with $\lambda = \mathrm{LS}(\mathfrak{K})$ and $\mu \geq \lambda$ and we define $\mathfrak{K}_{\geq\mu}$ by: $M \in \mathfrak{K}_{\geq\mu}$ iff $M \in K$ & $\|M\| \geq \mu$ and $M \leq_{\mathfrak{K}_{\geq\mu}} N$ if $M \leq_{\mathfrak{K}} N$ and $\|M\|, \|N\| \geq \mu$. Then $\mathfrak{K}_{\geq\mu}$ is an a.e.c. with $\mathrm{LS}(\mathfrak{K}_{\geq\mu}) = \mu$.
2) If $\mathfrak{K}_\lambda$ is a λ-a.e.c. then observation 1.7 holds when $|I| \leq \lambda$.
3) Claims 1.11, 1.16 apply to λ-a.e.c.

Proof. Easy. $\square_{1.21}$

1.22 Remark. Recall if $\mathfrak{K}$ is an a.e.c. with Lowenheim-Skolem number λ, then every model of $\mathfrak{K}$ can be written as a direct limit (by $\leq_{\mathfrak{K}}$) of members of $\mathfrak{K}_\lambda$ (see 1.8(1)). Alternating we prove below that given a λ-abstract elementary class $\mathfrak{K}_\lambda$, the class of direct limits of members of $\mathfrak{K}_\lambda$ is an a.e.c. $\mathfrak{K}^{\mathrm{up}}$. We show below $(\mathfrak{K}_\lambda)^{\mathrm{up}} = \mathfrak{K}$, hence $\mathfrak{K}_\lambda$ determines $\mathfrak{K}_{\geq\lambda}$.

1.23 Lemma. *Suppose $\mathfrak{K}_\lambda$ is a λ-abstract elementary class.*
1) The pair $(K', \le_{\mathfrak{K}'})$ is an abstract elementary class with Lowenheim-Skolem number λ which we denote also by $\mathfrak{K}^{\rm up}$ where we define

(a)

$$K' = \Big\{ M : M \text{ is a } \tau_{\mathfrak{K}_\lambda}\text{-model, and for some directed partial order } I \text{ and } \bar{M} = \langle M_s : s \in I \rangle \text{ we have } M = \bigcup_{s \in I} M_s, \quad s \in I \Rightarrow M_s \in K_\lambda, \quad I \models s < t \Rightarrow M_s \le_{\mathfrak{K}_\lambda} M_t \Big\}.$$

We call such $\langle M_s : s \in I\rangle$ a witness for $M \in K'$, we call it reasonable if $|I| \le \|M\|$

(b) $M \le_{\mathfrak{K}'} N$ *<u>iff</u> for some directed partial order J, and directed $I \subseteq J$ and $\langle M_s : s \in J \rangle$ we have*

$$M = \bigcup_{s \in I} M_s, N = \bigcup_{t \in J} M_t, M_s \in K_\lambda \text{ and}$$
$$J \models s < t \Rightarrow M_s \le_{\mathfrak{K}_\lambda} M_t.$$

We call such $I, \langle M_s : s \in J\rangle$ witnesses for $M \le_{\mathfrak{K}'} N$ or say $(I, J, \langle M_s : s \in J\rangle)$ witness $M \le_{\mathfrak{K}'} N$.
2) Moreover, $K'_\lambda = K_\lambda$ and $\le_{\mathfrak{K}'_\lambda}$ (which means $\le_{\mathfrak{K}'} \restriction K'_\lambda$) is equal to $\le_{\mathfrak{K}_\lambda}$ so $(\mathfrak{K}')_\lambda = \mathfrak{K}_\lambda$.
3) If $\mathfrak{K}''$ is an abstract elementary class satisfying (see 1.21) $K''_\lambda = K_\lambda, <_{\mathfrak{K}''} \restriction K_\lambda = \le_{\mathfrak{K}_\lambda}$ and $\mathrm{LS}(\mathfrak{K}'') \le \lambda$ <u>then</u>[8] $\mathfrak{K}''_{\ge\lambda} = \mathfrak{K}'$.
4) If $\mathfrak{K}''$ is an a.e.c., $K_\lambda \subseteq K''_\lambda$ and $\le_{\mathfrak{K}_\lambda} = \le_{\mathfrak{K}''} \restriction K_\lambda$, <u>then</u> $K' \subseteq K''$ and $\le_{\mathfrak{K}'} \subseteq \le_{\mathfrak{K}''} \restriction K'$ and if $\mathrm{LS}(\mathfrak{K}'') \le \lambda$ then equality holds..

---

[8] if we assume in addition that $M \in \mathfrak{K}'' \Rightarrow \|M\| \ge \lambda$ then we can show that equality holds

Proof. The proof of part (2) is straightforward (recalling 1.7) and part (3) follows from 1.8 and part (4) is also straightforward hence we concentrate on part (1). So let us check the axioms one by one.

Ax 0: K' is a class of τ-models, $\leq_{\mathfrak{K}'}$ a two-place relation on K', both closed under isomorphisms.
[Why? Trivially by their definitions.]

Ax I: If $M \leq_{\mathfrak{K}'} N$ then $M \subseteq N$.
[Why? trivial.]

Ax II: $M_0 \leq_{\mathfrak{K}'} M_1 \leq_{\mathfrak{K}'} M_2$ implies $M_0 \leq_{\mathfrak{K}'} M_2$ and $M \in K' \Rightarrow M \leq_{\mathfrak{K}'} M$.
[Why? The second phrase is trivial (as if $\bar{M} = \langle M_t : t \in I \rangle$ witness $M \in K'$ then $(I, I, \bar{M})$ witness $M \leq_{\mathfrak{K}'} M$ above). For the first phrase let for $\ell \in \{1,2\}$ the directed partial orders $I_\ell \subseteq J_\ell$ and $\bar{M}^\ell = \langle M^\ell_s : s \in J_\ell \rangle$ witness $M_{\ell-1} \leq_{\mathfrak{K}'} M_\ell$ and let $\bar{M}^0 = \langle M^0_s : s \in I_0 \rangle$ witness $M_0 \in K'$. Now without loss of generality $\bar{M}^0$ is reasonable, i.e. $|I_0| \leq \|M_0\|$, why? by

- $\boxtimes_1$ every $M \in K'$ has a reasonable witness, in fact, if $\bar{M} = \langle M_t : t \in I \rangle$ is a witness for M then for some $I' \subseteq I$ of cardinality $\leq \|M\|$ we have $\bar{M} \restriction I'$ is a reasonable witness for M.
  [Why? If $\bar{M} = \langle M_t : t \in I \rangle$ is a witness, for each $a \in M$ choose $t_a \in I$ such that $a \in M_{t_a}$ and let $F : [I]^{<\aleph_0} \to I$ be such that $F(\{t_1, \ldots, t_n\})$ is an upper bound of $\{t_1, \ldots, t_n\}$ and let J be the closure of $\{t_a : a \in M\}$ under F; now $\bar{M} \restriction J$ is a reasonable witness of $M \in K'$.]

Similarly

- $\boxtimes_2$ if $(I, J, \langle M_s : s \in J \rangle$ witness $M \leq_{\mathfrak{K}'} N$ then for some directed $I' \subseteq I, |I'| \leq \|M\|$ we have $(I', J, \langle M_s : s \in J \rangle)$ witness $M \leq_{K'} N$
- $\boxtimes_3$ if $I, \bar{M} = \langle M_t : t \in J \rangle$ witness $M \leq_{\mathfrak{K}'} N$ <u>then</u> for some directed $J' \subseteq J$ we have $\|J'\| \leq |I| + \|N\|, I \subseteq J'$ and $I, \bar{M} \restriction J'$ witness $M \leq_{\mathfrak{K}'} N$.

Clearly $\boxtimes_1$ (and $\boxtimes_2, \boxtimes_3$) are cases of the LS-argument. We shall find a witness $(I, J, \langle M_s : s \in J\rangle)$ for $M_0 \le_{\mathfrak{K}'} M_2$ such that $\langle M_s : s \in I\rangle = \langle M^0_s : s \in I_0\rangle$ so $I = I_0$ and $|J| \le \|M_2\|$. This is needed for the proof of Ax III below. Without loss of generality I_1, I_2 has cardinality $\le \|M_0\|, \|M_1\|$ respectively, by $\boxtimes_2$. Also without loss of generality $\bar{M}^1, \bar{M}^1 \restriction I_1, \bar{M}^2, \bar{M}^2 \restriction I_2$ are reasonable as by the same argument we can have $|J_1| \le \|M_1\|, |J_2| \le \|M_2\|$ by $\boxtimes_3$.

As $\langle M^0_s : s \in I_0\rangle$ is reasonable, there is a one-to-one function h from I_0 into M_2 (and even M_0); the function h will be used to get that J defined below is directed. We choose by induction on $m < \omega$, for every $\bar{c} \in {}^m(M_2)$, sets $I_{0,\bar{c}}, I_{1,\bar{c}}, I_{2,\bar{c}}, J_{1,\bar{c}}, J_{2,\bar{c}}$ such that:

$\otimes_1(a)$ $I_{\ell,\bar{c}}$ is a directed subset of I_ℓ of cardinality $\le \lambda$ for $\ell \in \{0,1,2\}$

(b) $J_{\ell,\bar{c}}$ is a directed subset of J_ℓ of cardinality $\le \lambda$ for $\ell \in \{1,2\}$

(c) $\bigcup\limits_{s \in I_{\ell+1,\bar{c}}} M^{\ell+1}_s = (\bigcup\limits_{s \in J_{\ell+1,\bar{c}}} M^{\ell+1}_s) \cap M_\ell$ for $\ell = 0, 1$

(d) $\bigcup\limits_{s \in I_{0,\bar{c}}} M^0_s = (\bigcup\limits_{s \in I_{1,\bar{c}}} M^1_s) \cap M_0$

(e) $\bigcup\limits_{s \in J_{1,\bar{c}}} M^1_s = \bigcup\limits_{s \in I_{2,\bar{c}}} M^2_s$

(f) $\bar{c} \subseteq \bigcup\limits_{s \in J_{2,\bar{c}}} M^2_s$

(g) if $\bar{d}$ is a permutation of $\bar{c}$ (i.e., letting $m = \ell g(\bar{c})$ for some one to one $g : \{0, \ldots, m-1\} \to \{0, \ldots, m-1\}$ we have $d_\ell = c_{g(\ell)}$) <u>then</u> $I_{\ell,\bar{c}} = I_{\ell,\bar{d}}, J_{m,\bar{c}} = J_{m,\bar{d}}$ (for $\ell \in \{0,1,2\}, m \in \{1,2\}$)

(h) if $\bar{d}$ is a subsequence of $\bar{c}$ (equivalently: an initial segment of some permutation of $\bar{c}$) <u>then</u> $I_{\ell,\bar{d}} \subseteq I_{\ell,\bar{c}}, J_{m,\bar{d}} \subseteq J_{m,\bar{c}}$ for $\ell \in \{0,1,2\}, m \in \{1,2\}$

(i) if $h(s) = c$ so $s \in I_0$ then $s \in I_{0,<c>}$.

There is no problem to carry the definition by LS-argument recalling clauses (a) + (b) and $\|M^\ell_s\| = \lambda$ when $\ell = 0 \wedge s \in I_0$ or $\ell = 1 \wedge s \in J_1$ or $\ell = 2 \wedge s \in J_2$. Without loss of generality $I_\ell \cap {}^{\omega>}(M_2) = \emptyset$.

Now let J have as set of elements $I_0 \cup \{\bar{c} : \bar{c}$ a finite sequence from M_2 ordered by: $J \models x \le y$ iff $I_0 \models x \le y$ <u>or</u> $x \in I_0, y \in J \backslash I_0, \exists z \in I_{0,y}[x \le_{I_0} z]$ <u>or</u> $x, y \in J \backslash I_0$ and x is an initial segment of a permutation of y (or you may identify $\bar{c}$ with its set of permutations).

Let $I = I_0$.
Let M_x be M^0_x if $x \in I_0$ and $\bigcup\limits_{s \in J_{2,x}} M^2_s$ if $x \in J \backslash I_0$.

Now

$(*)_1$ J is a partial order
[Clearly $x \le_J y \le_J x \Rightarrow x = y$, hence it is enough to prove transitivity. Assume $x \le_J y \le_J z$; if all three are in I_0 use "I_0 is a partial order", if all three are not in $J \backslash I_0$, use the definition of the order. As $x' \le_J y' \in I_0 \Rightarrow x' \in I_0$ without loss of generality $x \in I_0, z \in J \backslash I_0$. If $y \in I_0$ then (as $y \le_J z$) for some $y', y \le_{I_0} y' \in I_{0,z}$ but $x \le_{I_0} y$ (as $x, y \in I_0, x \le_J y$) hence $x \le_{I_0} y' \in I_{0,z}$ so $x \le_J z$. If $y \notin I_0$ then $I_{0,y} \subseteq I_{0,z}$ (by clause (h)) so we can finish similarly. So we have covered all cases.]

$(*)_2$ J is directed and $I \subseteq J$ is directed
[Let $x, y \in J$ and we shall find a common upper bound. If $x, y \notin I_0$ their concatanation $x\hat{\ }y$ can serve. If $x, y \in I_0$ use "I_0 is directed". If $x \in I_0, y \in J \backslash I_0$, then $\langle h(x) \rangle \in J \backslash I_0$ and $z = y\hat{\ }\langle h(x) \rangle \in J \backslash I_0$ is $<_J$ above y (by the choice of $\le_J$) and is $\le_J$-above x as $x \in I_{0,\langle h(x) \rangle} \subseteq I_{0,z}$ by clause (i) of $\otimes_1$ so we are done. If $x \in J \backslash I_0, y \in J_0$ the dual proof works. Lastly, $I \subseteq J$ as a partial order by the definition of I, J, and I is directed as I_0 is and $I = I_0$.]

$(*)_3$ if $x \in J \backslash I_0$ then $M_x \cap M_\ell \le_{\mathfrak{K}_x} M_x$ for $\ell = 0, 1$
[Why? Clearly $M_x \cap M_0 = (\cup\{M^2_t : t \in J_{1,x}\}) \cap M_0 = ((\cup\{M^2_t : t \in J_{2,x}) \cap M_1) \cap M_0 = (\cup\{M^2_t : t \in I_{2,x}\}) \cap M_0 = (\cup\{M^1_t : t \in J_{1,x}\}) \cap M_0 = \cup\{M^1_t : t \in I_{1,x}\}$ by the choice of M^2_x, as $M_0 \subseteq M_1$, by clause (c) for $\ell = 1$, by clause (e) and by clause (c) for $\ell = 0$, respectively. Similarly $M_x \cap M_1 = \cup\{M^1_t : t \in J_{1,x}\}$. Now the sets $I_{1,x} \subseteq J_{1,x} (\subseteq J_1)$ are directed by $\le_{J_1}$ so by the assumption on $\langle M^1_t : t \in J_1 \rangle$ and Lemma 1.7 we have $M_x \cap M_0 \le_{\mathfrak{K}_\lambda} M_x \cap M_1$. Using J_2

we can similarly prove $M_x \cap M_1 \leq_{\mathfrak{K}_\lambda} M_x \cap M_2$ and trivially $M_x \cap M_2 = M_x$. As $\leq_{\mathfrak{K}_\lambda}$ is transitive we are done.]

$(*)_4$ if $x \leq_J y$ then $M_x \leq_{\mathfrak{K}_\lambda} M_y$
[Why? If $x, y \in I_0$ use the choice of $\langle M^0_s : s \in I_0 \rangle$. If $x, y \in J \backslash I_0$ the proof is similar to that of $(*)_3$ using J_2. If $x \in I_0, y \in J \backslash I_0$ there is $s \in I_{0,y}$ such that $x \leq_{I_0} s$, hence $M_x = M^0_x \leq_{\mathfrak{K}_\lambda} M^0_s$ and as $\langle M^0_t : t \in I_{0,y} \rangle$ is $\leq_{\mathfrak{K}_\lambda}$-directed clearly $M^0_s \leq_{\mathfrak{K}_\lambda} \cup \{M^0_t : t \in I_{0,y}\} = M_y \cap M_0$ and $M_y \cap M_0 \leq_{\mathfrak{K}_\lambda} M_y$ by $(*)_3$. By the transitivity of $\leq_{\mathfrak{K}_\lambda}$ we are done.]

$(*)_5$ $\cup \{M_x : x \in I\} = \cup \{M^0_x : x \in I_0\} = M_0$
[Why? Trivially recalling $I_0 = I$ and $x \in I \Rightarrow M_x = M^0_x$.]

$(*)_6$ $M_2 = \cup \{M_x : x \in J\}$
[Why? Trivially as $\bar{c} \subseteq M^2_{\bar{c}} \subseteq M_2$ for $\bar{c} \in {}^{\omega >}(M_2)$ and $t \in I_0 \Rightarrow M^0_t \subseteq M_0 \subseteq M_1 \subseteq M_2$.]

By $(*)_1 + (*)_2 + (*)_4 + (*)_5 + (*)_6$ we have checked that $I, \langle M_x : x \in J \rangle$ witness $M_0 \leq_{\mathfrak{K}'} M_2$. This completes the proof of AxII, but we also have proved

$\otimes_2$ if $\bar{M} = \langle M_t : t \in I \rangle$ is a reasonable witness to $M \in K'$ and $M \leq_{\mathfrak{K}'} N \in K'$, <u>then</u> there is a witness $I', \bar{M}' = \langle M'_t : t \in J' \rangle$ to $M \leq_{\mathfrak{K}'} N$ such that $I' = I, \bar{M}' \restriction I = \bar{M}$ and $\bar{M}'$ is reasonable and $x \leq_{J'} y \wedge y \in I' \Rightarrow x \in I'$; can add $M = N \Rightarrow I' = I$.]

<u>Ax III</u>: If θ is a regular cardinal, M_i (for $i < \theta$) is $\leq_{\mathfrak{K}'}$-increasing and continuous, <u>then</u> $M_0 \leq_{\mathfrak{K}'} \bigcup_{i<\theta} M_i$ (in particular $\bigcup_{i<\theta} M_i \in \mathfrak{K}'$).

[Why? Let $M_\theta = \bigcup_{i<\theta} M_i$, without loss of generality $\langle M_i : i < \theta \rangle$ is not eventually constant and so without loss of generality $i < \theta \Rightarrow M_i \neq M_{i+1}$ hence $\|M_i\| \geq |i|$; (this helps below to get "reasonable", i.e. $|I_\ell| = \|M_i\|$ for limit i). We choose by induction on $i \leq \theta$, a directed partial order I_i and M_s for $s \in I_i$ such that:

$\otimes_3(a)$ $\langle M_s : s \in I_i \rangle$ witness $M_i \in K'$

(b) for $j < i, I_j \subseteq I_i$ and $(I_j, I_i, \langle M_s : s \in I_i \rangle)$ witness $M_j \leq_{\mathfrak{K}'} M_i$

(c) I_i is of cardinality $\leq \|M_i\|$

(d) if $I_i \models s \leq t$ and $j < i, t \in I_j$ then $s \in I_j$

For $i = 0$ use the definition of $M_0 \in K'$.
For i limit let $I_i := \bigcup_{j<i} I_j$ (and the already defined M_s's) are as required because $M_i = \bigcup_{j<i} M_j$ and the induction hypothesis (and $|I_i| \leq \|M_i\|$ as we have assumed above that $j < i \Rightarrow M_j \neq M_{j+1}$) .
For $i = j+1$ use the proof of Ax.II above with $M_j, M_i, M_i, \langle M_s : s \in I_j\rangle$ here serving as $M_0, M_1, M_2, \langle M^0_j : s \in I_0\rangle$ there, that is, we use $\otimes_2$ from there. Now for $i = \theta, \langle M_s : s \in I_\theta\rangle$ witness $M_\theta \in K'$ and $(I_i, I_\theta, \langle M_s : s \in I_\theta\rangle)$ witness $M_i \leq_{\mathfrak{K}'} M_\theta$ for each $i < \theta$.]

Axiom IV: Assume θ is regular and $\langle M_i : i < \theta\rangle$ is $\leq_{\mathfrak{K}}$-increasingly continuous, $M \in K'$ and $i < \theta \Rightarrow M_i \leq_{\mathfrak{K}'} M$ and $M_\theta = \bigcup_{i<\theta} M_i$ (so $M_\theta \subseteq M$). Then $M_\theta \leq_{\mathfrak{K}'} M$.
[Why? By the proof of Ax.III there are $\langle M_s : s \in I_i\rangle$ for $i < \theta$ satisfying clauses (a),(b),(c) and (d) of $\otimes_3$ there and without loss of generality $I_i \cap \theta = \emptyset$. For each $i < \theta$ as $M_i \leq_{\mathfrak{K}'} M$ there are J_i and M_s for $s \in J_i \backslash I_i$ such that $(I_i, J_i, \langle M_s : s \in J_i\rangle)$ witnesses it; without loss of generality with $\langle \bigcup_{i<\theta} I_i\rangle\hat{}\langle J_i \backslash I_i : i < \theta\rangle$ a sequence of pairwise disjoint sets; exist by $\otimes_2$ above. Let $I := \bigcup_{i<\theta} I_i$, let $\mathbf{i} : I \to \theta$ be $\mathbf{i}(s) = \text{Min}\{i : s \in I_i\}$ and recall $|I| \leq \|M_\theta\|$ hence by clause (d) of $\otimes_3$ we have $s \leq_I t \Rightarrow \mathbf{i}(s) \leq \mathbf{i}(t)$ and let h be a one-to-one function from I into M_θ. Without loss of generality the union below is disjoint and let

$(*)_7$ $J := I \cup \{(A, S) : A \subseteq M$ finite, $S \subseteq I$ finite with max. element$\}$.

ordered by: $J \models x \leq y$ iff $x, y \in I, I \models x \leq y$ or $x \in I, y = (A, S) \in J \backslash I$ and $x \in S$ or $x = (A^1, S^1) \in J \backslash I, y = (A^2, S^2) \in J \backslash I, A^1 \subseteq A^2, S^1 \subseteq S^2$.

We choose N_y for $y \in J$ as follows:

If $y \in I$ we let $N_y = M_y$.

By induction on $n < \omega$, if $y = (A, S) \in J \backslash I$ saisfies $n = |A| + |S|$, we choose the objects $N_y, I_{y,s}, J_{y,s}$ for $s \in S$ such that:

$\otimes_4(a)$ $I_{y,s}$ is a directed subset of $I_{\mathbf{i}(s)}$ of cardinality $\leq \lambda$ and $s \in I_{y,s}$

(b) $J_{y,s}$ is a directed subset of $J_{\mathbf{i}(s)}$ of cardinality $\leq \lambda$

(c) $s \in I_{\mathbf{i}(s)}$ for $s \in S$ (follows from the definition of $\mathbf{i}(s)$)

(d) $I_{y,s} \subseteq J_{y,s}$ for $s \in S$ and for $s <_I t$ from S we have $I_{y,s} \subseteq I_{y,t}$ & $J_{y,s} \subseteq J_{y,t}$

(e) if $y_1 = (A_1, S_1) \in J \backslash I, (A_1, S_1) <_J (A, S)$ and $s \in S_1$ then $I_{y_1,s} \subseteq I_{y,s}, J_{y_1,s} \subseteq J_{y,s}$

(f) $N_y = \bigcup_{t \in J_{y,s}} M_t$ for any $s \in S$

(g) $A \subseteq M_t$ for some $t \in J_{y,s}$ for any $s \in S$, hence $A \subseteq N_y$.

No problem to carry the induction and check that $(I, J, \langle N_y : y \in J \rangle)$ witness $M_\theta \leq_{\mathfrak{K}'} M$.

<u>Axiom V</u>: Assume $N_0 \leq_{\mathfrak{K}'} M$ and $N_1 \leq_{\mathfrak{K}'} M$.
If $N_0 \subseteq N_1$, then $N_0 \leq_{\mathfrak{K}'} N_1$.
[Why? Let $(I_0, J_0, \langle M^0_s : s \in J_0 \rangle)$ witness $N_0 \leq_{\mathfrak{K}'} M$ and without loss of generality $|I_0| \leq \|N_0\|$ and $h_0 : I_0 \to N_0$ be one-to-one. Let $\langle M^1_s : s \in I_1 \rangle$ witness $N_1 \in \mathfrak{K}'$ and without loss of generality I_1 is isomorphic to $([N_1]^{<\aleph_0}, \subseteq)$ and let h_1 be an isomorphism from I_1 onto $([N_1]^{<\aleph_0}, \subseteq)$. Now by induction on n, for $s \in I_1$ satisfying $n = |\{t : t <_{I_1} s\}|$ we choose directed subsets $F_0(s), F_1(s)$ of I_0, I_1 respectively, each of cardinality $\leq \lambda$ such that:

(i) $s \in I_1 \Rightarrow s \in F_1(s)$ and $t <_{I_1} s \Rightarrow F_0(t) \subseteq F_0(s)$ & $F_1(t) \subseteq F_1(s)$

(ii) if $s \in I_1$ then

(α) $\bigcup\{M^0_t : t \in F_0(s)\} = \bigcup\{M^1_t : t \in F_1(s)\} \cap N_0$

(β) $r \in I_0$ & $t \in I_1$ & $h_0(r) \in M^1_t \Rightarrow r \in F_0(s)$.

Now letting $M^2_s = \cup\{M^1_t : t \in F_1(s)\}$ and letting $F = F_0$ we get:

(*iii*) $t \in I_1 \wedge s \in F(t) (\subseteq I_0) \Rightarrow M^0_s \subseteq M^2_t$
(*iv*) F is a function from I_1 to $[I_0]^{\leq \lambda}$
(*v*) for $s \in I_1, F(s)$ is a directed subset of I_0 of cardinality $\leq \lambda$
(*vi*) for $s \in I_1, M^2_s \cap N_0 = \cup\{M^0_t : t \in F(s)\}$
(*vii*) $I_1 \models s \leq t \Rightarrow F(s) \subseteq F(t)$
(*viii*) $\langle M^2_s : s \in I_1\rangle$ witness $N_1 \in K'$.

As $N_1 \leq_{\mathfrak{K}'} M$ by the proof of Ax.II, i.e., by $\otimes_2$ above we can find J_1 extending I_1 and M^2_s for $s \in J_1 \backslash I_1$ such that $(I_1, J_1, \langle M^2_s : s \in J_1\rangle)$ witnesses $N_1 \leq_{\mathfrak{K}'} M$. We now prove

$\boxtimes_4$ if $r \in I_1, s \in I_0$ and $s \in F(r)$ then $M^0_s \leq_{\mathfrak{K}_\lambda} M^2_r$.

[Why? As $\langle M^0_t : t \in J_0\rangle, \langle M^2_t : t \in J_1\rangle$ are both witnesses for $M \in K'$, clearly for $r \in I_1 (\subseteq J_1)$ we can find directed $J'_0(r) \subseteq J_0$ of cardinality $\leq \lambda$ and directed $J'_1(r) \subseteq J_1$ of cardinality $\leq \lambda$ such that $r \in J'_1(r), F(r) \subseteq J'_0(r)$ and $\bigcup_{t \in J'_0(r)} M^0_t = \bigcup_{t \in J'_1(r)} M^2_t$, call it M^*_r. Now $M^*_r \in K'_\lambda = K_\lambda$ (by part (2) and 1.7) and $t \in J'_1(r) \Rightarrow M^2_t \leq_{\mathfrak{K}_\lambda} M^*_r$ (as $\mathfrak{K}_\lambda$ is a λ-abstract elementary class applying the parallel to observation 1.7, i.e., 1.21(2)) and similarly $t \in J'_0(r) \Rightarrow M^0_t \leq_{\mathfrak{K}_\lambda} M^*_r$. Now the s from $\boxtimes_4$ satisfied $s \in F(r) \subseteq J'_0(r)$ hence $M^0_s \subseteq M^1_r$ (why? by clause (iii) above $s \in F(r)$ is as required in $\boxtimes_4$). But above we got $M^0_s \leq_{\mathfrak{K}} M^*_r, M^2_r \leq_{\mathfrak{K}} M^*_r$, so by AxV for $\mathfrak{K}_\lambda$ we have $M^0_s \leq_{\mathfrak{K}} M^1_r$ as required in $\boxtimes_4$.]

Without loss of generality $I_0 \cap I_1 = \emptyset$ and define the partial order J with set of elements $I_0 \cup I_1$ by $J \models x \leq y$ iff $x, y \in I_0, I_0 \models x \leq y$ or $x \in I_0, y \in I_1$ and $x \in F(y)$ or $x, y \in I_1, I_1 \models x \leq y$.

$\boxtimes_5$ J is a partial order and $x \leq_J y in I_0 \Rightarrow x \in I_0$ (hence $x \leq_J y$ & $x \in I_1 \Rightarrow y \in I_1$).

[Why? The second phrase holds by the definition of $\leq_J$. For J being a partial order obviously $x \leq_J y \leq_J x \Rightarrow x = y$, so assume $x \leq_J y \leq_J z$ and we shall prove $x \leq_J z$. If $x \in I_1$ then $y, z \in I_1$ and we use "I_1 is a partial order", and if $z \in I_0$ then $x, y \in I_0$ and we can use "I_0 is a partial order". So assume $x \in I_0, z \in I_1$. If $y \in I_0$ use "$F(z) = F_1(z)$ satisfies clause (i) above. If $y \in I_1$, use clause (vii) above with (y, z) here standing for (s, t) there.]

$\boxtimes_6$ J is directed.

[Why? Note that I_0, I_1 are directed, $x \leq_J y \in I_0 \Rightarrow x \in I_0$ and $(\forall x \in I_0)(\exists y \in I_1)[x \leq_J y]$ because given $r \in I_0, h_0(r) \in N_0$ hence $h_0(r)$ belongs to M_t^1 for some $t \in I_1$, and so by clause (i) we have $t \in F_1(t)$ hence by clause $(ii)(\beta)$ above $r \in F_0(t)$. Together this is easy.]

Define M_s for $s \in J$ as M_s^0 if $s \in I_0$ and as M_s^2 if $s \in I_1$

$\boxtimes_7$ $M_s \in K_\lambda$ for $s \in J$.

[Why? Obvious.]

$\boxtimes_8$ if $x \leq_J y$ then $M_x \leq_{\mathfrak{x}} M_y$.

[Why? If $y \in I_0$ (hence $x \in I_0$) use $\langle M_t^0 : t \in I_0 \rangle$ is a witness for $N_0 \in K'$. If $x \in I_1$ (hence $y \in I_1$) use clazuse (viii) above, i.e. $\langle M_s^2 : s \in I_1 \rangle$ is a witness for $N_1 \in K'$.]

$\boxtimes_9$ $\cup\{M_x : x \in J\} = N_1$.

[Why? As $(\forall x \in I_0)(\exists y \in I_1)(x \leq_J y)$, see the proof of $\boxtimes_6$ recalling $\boxtimes$ we have $\cup\{M_x : x \in J\} = \cup\{M_x : x \in I_1\}$ but the latter is $\cup\{M_x^2 : x \in I_1\}$ which is equal to N_2.]

$\boxtimes_{10}$ $I_0 \subseteq J$ is directed and $\cup\{M_x : x \in J\} = N_1$.

[Why? Obvious.]

Together $(I_0, J, \langle M_s : s \in J \rangle)$ witnesses $N_0 \leq_{\mathfrak{K}'} N_1$ are as required.]

Axiom VI: $\mathrm{LS}(\mathfrak{K}') = \lambda$.

[Why? Let $M \in K', A \subseteq M, |A| + \lambda \leq \mu < \|M\|$ and let $\langle M_s : s \in J \rangle$ witness $M \in K'$. As $\|M\| > \mu$ we can choose a directed $I \subseteq J$ of cardinality $\leq \mu$ such that $A \subseteq M' := \bigcup\limits_{s \in I} M_s$ and so $(I, J, \langle M_s : s \in J \rangle)$ witnesses $M' \leq_{\mathfrak{K}'} M$, so as $A \subseteq M'$ and $\|M'\| \leq |A| + \mu$; this is more than enough.] $\square_{1.23}$

We may like to use $\mathfrak{K}_{\le\lambda}$ instead of $\mathfrak{K}_\lambda$; no need as essentially $\mathfrak{K}$ consists of two parts $\mathfrak{K}_{\le\lambda}$ and $\mathfrak{K}_{\ge\lambda}$ which have just to agree in λ. That is

1.24 Claim. *Assume*

(a) *$\mathfrak{K}^1$ is an abstract elementary class with $\lambda = \mathrm{LS}(\mathfrak{K}^1), K^1 = K^1_{\ge\lambda}$*

(b) *$\mathfrak{K}^2_{\le\lambda}$ is a $(\le\lambda)$-abstract elementary class (defined as in 1.18(1) with the obvious changes so $M \in \mathfrak{K}^2_{\le\lambda} \Rightarrow \|M\| \le \lambda$ and in Axiom III, $\|\bigcup_i M_i\| \le \lambda$ is required)*

(c) *$K^2_\lambda = K^1_\lambda$ and $\le_{\mathfrak{K}^2} \restriction K^2_\lambda = \le_{\mathfrak{K}^1} \restriction K^1_\lambda$*

(d) *we define $\mathfrak{K}$ as follows: $K = K^1 \cup K^2, M \le_{\mathfrak{K}} N$ iff $M \le_{\mathfrak{K}^1} N$ or $M \le_{\mathfrak{K}^2} N$ or for some $M', M \le_{\mathfrak{K}^2} M' \le_{\mathfrak{K}^1} N$.*

Then $\mathfrak{K}$ is an abstract elementary class and $\mathrm{LS}(\mathfrak{K}) = \mathrm{LS}(\mathfrak{K}^2)$ which trivially is $\le \lambda$.

Proof. Straight. E.g.
Axiom V: We shall use freely

$(*)$ $\mathfrak{K}_{\le\lambda} = \mathfrak{K}^2$ and $\mathfrak{K}_{\ge\lambda} = \mathfrak{K}^1$.

So assume $N_0 \le_{\mathfrak{K}} M, N_1 \le_{\mathfrak{K}} M, N_0 \subseteq N_1$.
Now if $\|N_0\| \ge \lambda$ use assumption (a), so we can assume $\|N_0\| < \lambda$. If $\|M\| \le \lambda$ we can use assumption (b) so we can assume $\|M\| > \lambda$ and by the definition of $\le_{\mathfrak{K}}$ there is $M'_0 \in K^1_\lambda = K^2_\lambda$ such that $N_0 \le_{\mathfrak{K}^2} M'_0 \le_{\mathfrak{K}^1} M$. First assume $\|N_1\| \le \lambda$, so we can find $M'_1 \in K^1_\lambda$ such that $N_1 \le_{\mathfrak{K}^2} M'_1 \le_{\mathfrak{K}^1} M$ (why? if $N_1 \in K_{<\lambda}$, by the definition of $\le_{\mathfrak{K}}$ and if $N_1 \in K_\lambda$ just choose $M'_1 = N_1$). Now we can by assumption (a) find $M'' \in K^1_\lambda$ such that $M'_0 \cup M'_1 \subseteq M'' \le_{\mathfrak{K}^1} M$, hence by assumption (a) (i.e. AxV for $\mathfrak{K}^1$) we have $M'_0 \le_{\mathfrak{K}^1} M'', M'_1 \le_{\mathfrak{K}^1} M''$, so by assumption (c) we have $M'_0 \le_{\mathfrak{K}^2} M'', M'_1 \le_{\mathfrak{K}^2} M''$. As $N_0 \le_{\mathfrak{K}^2} M'_0 \le_{\mathfrak{K}^2} M'' \in K_{\le\lambda}$ by assumption (b) we have $N_0 \le_{\mathfrak{K}^2} M''$, and similarly we have $N_1 \le_{\mathfrak{K}^2} M''$. So $N_0 \subseteq N_1, N_0 \le_{\mathfrak{K}^2} M'', N_1 \le_{\mathfrak{K}^2} M'$ so by assumption (b) we have $N_0 \le_{\mathfrak{K}^2} N_1$ hence $N_0 \le_{\mathfrak{K}} N_1$.

We are left with the case $\|N_1\| > \lambda$; by assumption (a) there is $N'_1 \in K_\lambda$ such that $N_0 \subseteq N'_1 \le_{\mathfrak{K}^1} N_1$. By assumption (a) we have $N'_1 \leq_{\mathfrak{K}^1} M$, so by the previous paragraph we get $N_0 \leq_{\mathfrak{K}^2} N'_1$, together with the previous sentence we have $N_0 \leq_{\mathfrak{K}^2} N'_1 \leq_{\mathfrak{K}^1} N_1$ so by the definition of $\leq_{\mathfrak{K}}$ we are done. $\square_{1.24}$

Recall

1.25 Definition. If $M \in K_\lambda$ is locally superlimit or just pseudo superlimit let $K_{[M]} = K^{[M]}_\lambda = \{N \in K_\lambda : N \cong M\}, \mathfrak{K}_{[M]} = \mathfrak{K}^{[M]}_\lambda = (K_{[M]}, \leq_{\mathfrak{K}} \restriction K^{[M]}_\lambda)$ and let $\mathfrak{K}^{[M]}$ be the $\mathfrak{K}'$ we get in 1.23(1) for $\mathfrak{K} = \mathfrak{K}_{[M]} = \mathfrak{K}^{[M]}_\lambda$. We may write $\mathfrak{K}_\lambda[M], \mathfrak{K}[M]$.

Trivially but still important is showing that assuming categoricity in one λ is a not so strong assumption.

1.26 Claim. *1) If $\mathfrak{K}$ is an λ-a.e.c., $M \in K_\lambda$ is locally superlimit or just pseudo superlimit <u>then</u> $\mathfrak{K}_{[M]}$ is a λ-a.e.c. which is categorical (i.e. categorical in λ).*
2) Assume $\mathfrak{K}$ is an a.e.c. and $M \in \mathfrak{K}_\lambda$ is not $\leq_{\mathfrak{K}}$- maximal. M is pseudo superlimit (in $\mathfrak{K}$, i.e., in $\mathfrak{K}_\lambda$) <u>iff</u> $\mathfrak{K}_{[M]}$ is a λ-a.e.c. which is categorical <u>iff</u> $\mathfrak{K}^{[M]}$ is an a.e.c., categorical in λ and $\leq_{\mathfrak{K}[M]} = \leq_{\mathfrak{K}} \restriction K^{[M]}$.
3) In (1) and (2), $\mathrm{LS}(\mathfrak{K}^{[M]}) = \lambda = \mathrm{Min}\{\|N\| : N \in \mathfrak{K}^{[M]}\}$.

Proof. Straightforward. $\square_{1.26}$

<u>1.27 Exercise</u>: Assume $\mathfrak{K}$ is a λ-a.e.c. with amalgamation and stability in λ. <u>Then</u> for every $M_1 \in K_\lambda, p_1 \in \mathscr{S}_{\mathfrak{K}}(M_1)$ we can find $M_2 \in K$ and minimal $p_2 \in \mathscr{S}_{\mathfrak{K}}(M_2)$ such that $M_1 \leq_{\mathfrak{K}} M_2$ and $p_1 = p_2 \restriction M_1$.
[Hint: See VI.2.3(2).]

<u>1.28 Exercise</u>: 1) Any $\leq_{\mathfrak{K}_\lambda}$-embedding f_0 of M^1_0 into M^2_0 can be extended to an isomorphism f from M^1_δ onto M^2_δ such that $f(M^1_{2\alpha}) \leq_{\mathfrak{K}_\lambda} M^2_{2\alpha}, f^{-1}(M^2_{2\alpha+1}) \leq_{\mathfrak{K}_\lambda} M^1_{2\alpha+1}$ for every $\alpha < \delta$, <u>provided that</u>

⊛ (a) $\mathfrak{K}_\lambda$ is a λ-a.e.c. with amalgamation and δ is a limit ordinal $\leq \lambda^+$

(b) $\langle M^\ell_\alpha : \alpha \leq \delta \rangle$ is $\leq_{\mathfrak{K}_\lambda}$-increasing continuous for $\ell = 1, 2$

(c) M^ℓ_α is an amalgamation base in $\mathfrak{K}_\lambda$ (for $\alpha < \delta$ and $\ell = 1, 2$)

(d) $M^\ell_{\alpha+1}$ is $\leq_{\mathfrak{K}_\lambda}$-universal extension of M^ℓ_α for $\alpha < \delta, \ell = 1, 2$.

2) Write the axioms of "a λ-a.e.c." which are used.
3) For $\mathfrak{K}_\lambda, \delta$ as in (a) above, for any $M \in K_\lambda$ there is $N \in K_\lambda$ which is $(\lambda, \mathrm{cf}(\delta))$-brimmed over it.

[Hint: Should be easy; is similar to 1.16 (or 1.17).]

§2 Good Frames

We first present our central definition: good λ-frame (in Definition 2.1). We are given the relation "$p \in \mathscr{S}(N)$ does not fork over $M \leq_{\mathfrak{K}} N$ when p is basic" (by the basic relations and axioms) so it is natural to look at how well we can "lift" the definition of non-forking to models of cardinality λ and later to non-forking of models (and types over them) in cardinalities $> \lambda$. Unlike the lifting of λ-a.e.c. in Lemma 1.23, life is not so easy. We define in 2.4, 2.5, 2.7 and we prove basic properties in 2.6, 2.8, 2.10 and less obvious ones in 2.9, 2.11, 2.12. This should serve as a reasonable exercise in the meaning of good frames; however, the lifting, in general, does not give good μ-frames for $\mu > \lambda$. There may be no $M \in K_\mu$ at all and/or amalgamation may fail. Also the existence and uniqueness of non-forking types is problematic. We do not give up and will return to the lifting problem, under additional assumptions in III§12 and [Sh 842].

In 2.15 (recalling 1.26) we show that the case "$\mathfrak{K}^{\mathfrak{s}}$ categorical in λ" is not so rare among good λ-frames; in fact if there is a superlimit model in λ we can restrict $\mathfrak{K}_\lambda$ to it. So in a sense superstability and categoricity are close, a point which does not appear in first order model theory, <u>but</u> if T is a complete first order superstable theory and $\lambda \geq 2^{|T|}$, <u>then</u> the class $\mathfrak{K} = \mathfrak{K}_{T,\lambda}$ of λ-saturated models of T is in general not an elementary class (though is a PC_λ class) but is an a.e.c. categorical in λ though in general not in λ^+ and for some good λ-frame $\mathfrak{s}, K_{\mathfrak{s}} = \mathfrak{K}_{T,\lambda}$. How justified is our restriction here to something like "the λ-saturated model"? It is O.K. for our test problems but more so it is justified as our approach is to first analyze the quite saturated models.

Last but not least in 2.16 we show that one of the axioms from 2.1, i.e., (E)(i), follows from the rest in our present definition; additional implications are in Claims 2.17, 2.18. Later "Ax(X)(y)" will mean (X)(y) from Definition 2.1.

Recall that good λ-frame is intended to be a parallel to (bare bones) superstable elementary class stable in λ; here we restrict ourselves to models of cardinality λ.

2.1 Definition. We say $\mathfrak{s} = (\mathfrak{K}, \bigcup_{\lambda}, \mathscr{S}^{\mathrm{bs}}_{\lambda}) = (\mathfrak{K}^{\mathfrak{s}}, \bigcup_{\mathfrak{s}}, \mathscr{S}^{\mathrm{bs}}_{\mathfrak{s}})$ is a good frame in λ or a good λ-frame (λ may be omitted when its value is clear, note that $\lambda = \lambda_{\mathfrak{s}} = \lambda(\mathfrak{s})$ is determined by $\mathfrak{s}$ and we may write $\mathscr{S}_{\mathfrak{s}}(M)$ instead of $\mathscr{S}_{\mathfrak{K}^{\mathfrak{s}}}(M)$ and $\mathbf{tp}_{\mathfrak{s}}(a, M, N)$ instead of $\mathbf{tp}_{\mathfrak{K}^{\mathfrak{s}}}(a, M, N)$ when $M \in K^{\mathfrak{s}}_{\lambda}, N \in K^{\mathfrak{s}}$; we may write $\mathbf{tp}(a, M, N)$ for $\mathbf{tp}_{\mathfrak{K}^{\mathfrak{s}}}(a, M, N)$) <u>when</u> the following conditions hold:

(A) $\mathfrak{K} = (K, \leq_{\mathfrak{K}})$ is an abstract elementary class also denoted by $\mathfrak{K}[\mathfrak{s}]$, the Löwenheim Skolem number of $\mathfrak{K}$, being $\leq \lambda$ (see Definition 1.4); there is no harm in assuming $M \in K \Rightarrow \|M\| \geq \lambda$; let $\mathfrak{K}_{\mathfrak{s}} = \mathfrak{K}^{\mathfrak{s}}_{\lambda}$ and $\leq_{\mathfrak{s}} = \leq_{\mathfrak{K}} \restriction K_{\lambda}$, and let $\mathfrak{K}_{\mathfrak{s}} = (K_{\lambda}, \leq_{\mathfrak{s}})$ and $\mathfrak{K}[\mathfrak{s}] = \mathfrak{K}^{\mathfrak{s}}$ so we may write $\mathfrak{s} = (\mathfrak{K}_{\mathfrak{s}}, \bigcup_{\mathfrak{s}}, \mathscr{S}^{\mathrm{bs}}_{\mathfrak{s}})$

(B) $\mathfrak{K}$ has a superlimit model in λ which[9] is not $<_{\mathfrak{K}}$-maximal.

(C) $\mathfrak{K}_{\lambda}$ has the amalgamation property, the JEP (joint embedding property), and has no $\leq_{\mathfrak{K}}$-maximal member.

(D)(a) $\mathscr{S}^{\mathrm{bs}} = \mathscr{S}^{\mathrm{bs}}_{\lambda}$ (the class of basic types for $\mathfrak{K}_{\lambda}$) is included in $\bigcup\{\mathscr{S}(M) : M \in K_{\lambda}\}$ and is closed under isomorphisms including automorphisms; for $M \in K_{\lambda}$ let $\mathscr{S}^{\mathrm{bs}}(M) = \mathscr{S}^{\mathrm{bs}} \cap \mathscr{S}(M)$; no harm in allowing types of finite sequences, i.e., replacing $\mathscr{S}(M)$ by $\mathscr{S}^{<\omega}(M)$, ($\mathscr{S}^{\omega}(M)$) is different as being new (= non-algebraic) is not preserved under increasing unions).

---

[9]in fact, the "is not $<_{\mathfrak{K}}$-maximal" follows by (C)

(*b*) if $p \in \mathscr{S}^{\rm bs}(M)$, <u>then</u> p is non-algebraic (i.e. not realized by any $a \in M$).

(*c*) <u>(density)</u>
if $M \leq_{\mathfrak{K}} N$ are from K_λ and $M \neq N$, <u>then</u> for some $a \in N \backslash M$ we have $\mathbf{tp}(a, M, N) \in \mathscr{S}^{\rm bs}$

[intention: examples are: minimal types in [Sh 576], i.e. Chapter VI,
regular types for superstable first order (= elementary) classes].

(*d*) <u>bs-stability</u>
$\mathscr{S}^{\rm bs}(M)$ has cardinality $\leq \lambda$ for $M \in K_\lambda$.

$(E)(a)$ $\underset{\lambda}{\bigcup}$ denoted also by $\underset{\mathfrak{s}}{\bigcup}$ or just $\bigcup$, is a four place relation[10] called non-forking with $\bigcup(M_0, M_1, a, M_3)$ implying $M_0 \leq_{\mathfrak{K}} M_1 \leq_{\mathfrak{K}} M_3$ are from K_λ, $a \in M_3 \backslash M_1$ and $\mathbf{tp}(a, M_0, M_3) \in \mathscr{S}^{\rm bs}(M_0)$ and
$\mathbf{tp}(a, M_1, M_3) \in \mathscr{S}^{\rm bs}(M_1)$. Also $\bigcup$ is preserved under isomorphisms and we demand: if $M_0 = M_1 \leq_{\mathfrak{K}} M_3$ both in K_λ and $a \in M_3$, then:
$\bigcup(M_0, M_1, a, M_3)$ is equivalent to "$\mathbf{tp}(a, M_0, M_3) \in \mathscr{S}^{\rm bs}(M_0)$".

The assertion $\bigcup(M_0, M_1, a, M_3)$ is also written as $M_1 \overset{M_3}{\underset{M_0}{\bigcup}} a$ and also as "$\mathbf{tp}(a, M_1, M_3)$ does not fork over M_0 (inside M_3)" (this is justified by clause (b) below). So $\mathbf{tp}(a, M_1, M_3)$ forks over M_0 (where $M_0 \leq_{\mathfrak{s}} M_1 \leq_{\mathfrak{s}} M_3, a \in M_3$) is just the negation

[Explanation: The intention is to axiomatize non-forking of types, but we already commit ourselves to dealing with basic types only. Note that in [Sh 576],

---

[10] we tend to forget to write the λ, this is justified by 2.6(2), and see Definition 2.5

i.e. Chapter VI we know something on minimal types but other types are something else.]

(b) <u>(monotonicity)</u>:
if $M_0 \leq_{\mathfrak{K}} M_0' \leq_{\mathfrak{K}} M_1' \leq_{\mathfrak{K}} M_1 \leq_{\mathfrak{K}} M_3 \leq_{\mathfrak{K}} M_3', M_1' \cup \{a\} \subseteq M_3'' \leq_{\mathfrak{K}} M_3'$ all of them in K_λ, <u>then</u> $\bigcup(M_0, M_1, a, M_3) \Rightarrow$

$$\bigcup(M_0', M_1', a, M_3') \text{ and } \bigcup(M_0', M_1', a, M_3') \Rightarrow \bigcup(M_0', M_1', a, M_3''),$$

<u>so</u> it is legitimate to just say "$\mathbf{tp}(a, M_1, M_3)$ does not fork over M_0".

[Explanation: non-forking is preserved by decreasing the type, increasing the basis (= the set over which it does not fork) and increasing or decreasing the model inside which all this occurs, i.e. where the type is computed. The same holds for stable theories only here we restrict ourselves to "legitimate", i.e., basic types. But note that here the "restriction of $\mathbf{tp}(a, M_1, M_3)$ to M_1' is basic" is a worthwhile information.]

(c) <u>(local character)</u>:
if $\langle M_i : i \leq \delta{+}1 \rangle$ is $\leq_{\mathfrak{K}}$-increasing continuous in $\mathfrak{K}_\lambda$, $a \in M_{\delta+1}$ and
$\mathbf{tp}(a, M_\delta, M_{\delta+1}) \in \mathscr{S}^{\mathrm{bs}}(M_\delta)$ <u>then</u> for every $i < \delta$ large enough $\mathbf{tp}(a, M_\delta, M_{\delta+1})$ does not fork over M_i.

[Explanation: This is a replacement for superstability which says that: if $p \in \mathscr{S}(A)$ then there is a finite $B \subseteq A$ such that p does not fork over B.]

(d) <u>(transitivity)</u>:
if $M_0 \leq_{\mathfrak{s}} M_0' \leq_{\mathfrak{s}} M_0'' \leq_{\mathfrak{s}} M_3$ are from K_λ and $a \in M_3$ and $\mathbf{tp}(a, M_0'', M_3)$ does not fork over M_0' and $\mathbf{tp}(a, M_0', M_3)$ does not fork over M_0 (all models are in K_λ, of course, and necessarily the three relevant types are in $\mathscr{S}^{\mathrm{bs}}$), <u>then</u> $\mathbf{tp}(a, M_0'', M_3)$ does not fork over M_0

(e) uniqueness:
if $p, q \in \mathscr{S}^{\mathrm{bs}}(M_1)$ do not fork over $M_0 \leq_{\mathfrak{K}} M_1$ (all in K_λ) and
$p \restriction M_0 = q \restriction M_0$ then $p = q$

(f) symmetry:
if $M_0 \leq_{\mathfrak{K}} M_3$ are in $\mathfrak{K}_\lambda$ and for $\ell = 1, 2$ we have $a_\ell \in M_3$ and $\mathbf{tp}(a_\ell, M_0, M_3) \in \mathscr{S}^{\mathrm{bs}}(M_0)$, then the following are equivalent:

(α) there are M_1, M_3' in K_λ such that $M_0 \leq_{\mathfrak{K}} M_1 \leq_{\mathfrak{K}} M_3'$, $a_1 \in M_1, M_3 \leq_{\mathfrak{K}} M_3'$ and $\mathbf{tp}(a_2, M_1, M_3')$ does not fork over M_0

(β) there are M_2, M_3' in K_λ such that $M_0 \leq_{\mathfrak{K}} M_2 \leq_{\mathfrak{K}} M_3'$, $a_2 \in M_2, M_3 \leq_{\mathfrak{K}} M_3'$ and $\mathbf{tp}(a_1, M_2, M_3')$ does not fork over M_0.

[Explanation: this is a replacement to "$\mathbf{tp}(a_1, M_0 \cup \{a_2\}, M_3)$ forks over M_0 iff $\mathbf{tp}(a_2, M_0 \cup \{a_1\}, M_3)$ forks over M_0" which is not well defined in our context.]

(g) extension existence:
if $M \leq_{\mathfrak{K}} N$ are from K_λ and $p \in \mathscr{S}^{\mathrm{bs}}(M)$ then some $q \in \mathscr{S}^{\mathrm{bs}}(N)$ does not fork over M and extends p

(h) continuity:
if $\langle M_i : i \leq \delta \rangle$ is $\leq_{\mathfrak{K}}$-increasing continuous, all in K_λ (recall δ is always a limit ordinal), $p \in \mathscr{S}(M_\delta)$ and $i < \delta \Rightarrow p \restriction M_i \in \mathscr{S}^{\mathrm{bs}}(M_i)$ does not fork over M_0 then $p \in \mathscr{S}^{\mathrm{bs}}(M_\delta)$ and moreover p does not fork over M_0.

[Explanation: This is a replacement to: for an increasing sequence of types which do not fork over A, the union does not fork over A; equivalently if p forks over A then some finite subtype does.]

(i) non-forking amalgamation:
if for $\ell = 1, 2$, $M_0 \leq_{\mathfrak{K}} M_\ell$ are from $K_\lambda, a_\ell \in M_\ell \backslash M_0$, and $\mathbf{tp}(a_\ell, M_0, M_\ell) \in \mathscr{S}^{bs}(M_0)$,

then we can find f_1, f_2, M_3 satisfying $M_0 \leq_{\mathfrak{K}} M_3 \in K_\lambda$ such that for $\ell = 1, 2$ we have f_ℓ is a $\leq_{\mathfrak{K}}$-embedding of M_ℓ into M_3 over M_0 and $\mathbf{tp}(f_\ell(a_\ell), f_{3-\ell}(M_{3-\ell}), M_3)$ does not fork over M_0 for $\ell = 1, 2$.

[Explanation: This strengthens clause (g), (existence) saying we can do it twice so close to (f), symmetry, but see 2.16.]

* * *

2.2 Discussion: 0) On connections between the axioms see 2.16, 2.17, 2.18.
1) What justifies the choice of the good λ-frame as a parallel to (bare bones) superstability? Mostly starting from assumptions on few models around λ in the a.e.c. $\mathfrak{K}$ and reasonable, "semi ZFC" set theoretic assumptions (e.g. involving categoricity and weak cases of G.C.H., see §3) we can prove that, essentially, for some $\bigcup, \mathscr{S}$ the demands in Definition 2.1 hold. So here we shall get (i.e., applying our general theorem to the case of 3.4) an alternative proof of the main theorem of [Sh 87a], [Sh 87b] in a local version, i.e., dealing with few cardinals rather than having to deal with all the cardinals $\lambda, \lambda^{+1}, \lambda^{+2}, \ldots, \lambda^{+n}$ as in [Sh 87a], [Sh 87b] in an inductive proof. That is, in [Sh 87b], we get dichotomies by the omitting type theorem for countable models (and theories). So problems on $\aleph_n$ are "translated" down to $\aleph_{n-1}$ (increasing the complexity) till we arrive to $\aleph_0$ and then "translated" back. Hence it is important there to deal with $\aleph_0, \ldots, \aleph_n$ together. Here our λ may not have special helpful properties, so if we succeed to prove the relevant claims then they apply to λ^+, too. There are advantages to being poor.
2) Of course, we may just point out that the axioms seem reasonable and that eventually we can say much more.
3) We may consider weakening bs-stability (i.e., Ax$(D)(d)$ in Definition 2.1) to $M \in K_\lambda \Rightarrow |\mathscr{S}^{\mathrm{bs}}(M)| \leq \lambda^+$, we have not looked into

it here; Jarden-Shelah [JrSh 875] will; actually Chapter I deals in a limited way with this in a considerably more restricted framework.
4) On stability in λ and existence of (λ, ∂)-brimmed extensions see 4.2.

From the rest of this section we shall use mainly the defintion of $K_\lambda^{3,\mathrm{bs}}$ in Definition 2.4(3), also 2.20 (restricting ourselves to a superlimit). We sometimes use implications among the axioms (in 2.16 - 2.18). The rest is, for now an exercise to familiarize the reader with λ-frames, in particular (2.3-2.15) to see what occurs to non-forking and basic types in cardinals $> \lambda$. This is easy (but see below). For this we first present the basic definitions.

2.3 Convention. 1) We fix $\mathfrak{s}$, a good λ-frame so $K = K^{\mathfrak{s}}, \mathscr{S}^{\mathrm{bs}} = \mathscr{S}^{\mathrm{bs}}_{\mathfrak{s}}$.
2) By $M \in K$ we mean $M \in K_{\geq\lambda}$ if not said otherwise.

We lift the properties to $\mathfrak{K}_{\geq\lambda}$ by reflecting to the situation in K_λ. But do not be too excited: the good properties do not lift automatically, we shall be working on that later (under additional assumptions). Of course, from the definition below later we shall use mainly $K_{\mathfrak{s}}^{3,\mathrm{bs}} = K_\lambda^{3,\mathrm{bs}}$.

2.4 Definition. 1)

$$K^{3,\mathrm{bs}} = K^{3,\mathrm{bs}}_{\geq \mathfrak{s}} := \Big\{(M,N,a) : M \leq_{\mathfrak{K}} N,\ a \in N \backslash M \text{ and there is } M' \leq_{\mathfrak{K}} M \text{ satisfying } M' \in K_\lambda, \text{ such that for every } M'' \in K_\lambda \text{ we have: } [M' \leq_{\mathfrak{K}} M'' \leq_{\mathfrak{K}} M \Rightarrow \mathbf{tp}(a, M'', N) \in \mathscr{S}^{\mathrm{bs}}(M'') \text{ does not fork over } M']; \text{ equivalently } [M' \leq_{\mathfrak{K}} M'' \leq_{\mathfrak{K}} M \ \&\ M'' \leq_{\mathfrak{K}} N'' \leq_{\mathfrak{K}} N \ \&\ N'' \in K_\lambda \ \&\ a \in N'' \Rightarrow \underset{\lambda}{\bigcup}(M', M'', a, N'')]\Big\}.$$

2) $K^{3,\mathrm{bs}}_{=\mu} = K^{3,\mathrm{bs}}_{\mathfrak{s},\mu} := \{(M,N,a) \in K^{3,\mathrm{bs}}_{\geq \mathfrak{s}} : M, N \in \mathfrak{K}^{\mathfrak{s}}_\mu\}$.
3) $K^{3,\mathrm{bs}}_{\mathfrak{s}} := K^{3,\mathrm{bs}}_{=\lambda,\mathfrak{s}}$; and let $K^{3,\mathrm{bs}}_\mu = K^{3,\mathrm{bs}}_{=\mu}$, used mainly for $\mu = \lambda_{\mathfrak{s}}$ and $K^{3,\mathrm{bs}}_{\mathfrak{s},\geq\mu}$ is defined naturally.

2.5 Definition. We define $\underset{<\infty}{\bigcup}(M_0, M_1, a, M_3)$ (rather than $\underset{\lambda}{\bigcup}$) as follows: it holds <u>iff</u> $M_0 \leq_{\mathfrak{K}} M_1 \leq_{\mathfrak{K}} M_3$ are from K (not necessarily K_λ), $a \in M_3 \backslash M_1$ and there is $M_0' \leq_{\mathfrak{K}} M_0$ which belongs to K_λ satisfying: if $M_0' \leq_{\mathfrak{K}} M_1' \leq_{\mathfrak{K}} M_1, M_1' \in K_\lambda$,
$M_1' \cup \{a\} \subseteq M_3' \leq_{\mathfrak{K}} M_3$ and $M_3' \in K_\lambda$ <u>then</u> $\underset{\lambda}{\bigcup}(M_0', M_1', a, M_3')$.

We now check that $\underset{<\infty}{\bigcup}$ behaves correctly when restricted to K_λ.

2.6 Claim. *1) Assume $M \leq_{\mathfrak{K}} N$ are from K_λ and $a \in N$. <u>Then</u> $(M,N,a) \in K^{3,\mathrm{bs}}_{\mathfrak{s}}$ <u>iff</u> $\mathbf{tp}(a, M, N) \in \mathscr{S}^{\mathrm{bs}}_{\mathfrak{s}}(M)$.*
2) Assume $M_0, M_1, M_3 \in K_\lambda$ and $a \in M_3$. <u>Then</u> $\underset{<\infty}{\bigcup}(M_0, M_1, a, M_3)$

<u>iff</u>

$\bigcup\limits_{\lambda}(M_0, M_1, a, M_3)$.

3) Assume $M \leq_{\mathfrak{K}} N_1 \leq_{\mathfrak{K}} N_2$ *and* $a \in N_1$. *Then*
$(M, N_1, a) \in K^{3,\mathrm{bs}}_{\geq \mathfrak{s}} \Leftrightarrow (M, N_2, a) \in K^{3,\mathrm{bs}}_{\geq \mathfrak{s}}$.
4) Assume $M_0 \leq_{\mathfrak{K}} M_1 \leq_{\mathfrak{K}} M_3 \leq_{\mathfrak{K}} M_3^*$ *and* $a \in M_3$ *then:*
$\bigcup\limits_{<\infty} (M_0, M_1, a, M_3)$ *iff* $\bigcup\limits_{<\infty} (M_0, M_1, a, M_3^*)$.

Proof. 1) First assume $\mathbf{tp}(a, M, N) \in \mathscr{S}^{\mathrm{bs}}_{\mathfrak{s}}(M)$ and check the definition of $(M, N, a) \in K^{3,\mathrm{bs}}$. Clearly $M \leq_{\mathfrak{K}} N, a \in N$ and $a \in N \backslash M$; we have to find M' as required in Definition 2.4(1); we let $M' = M$, so $M' \leq_{\mathfrak{K}} M, M' \in K_\lambda$ and

$$M' \leq_{\mathfrak{K}} M'' \leq_{\mathfrak{K}} M \ \& \ M'' \in K_\lambda \Rightarrow M'' = M$$
$$\Rightarrow \mathbf{tp}_{\mathfrak{K}_\lambda}(a, M'', N) = \mathbf{tp}_{\mathfrak{K}_\lambda}(a, M, N) \in \mathscr{S}^{\mathrm{bs}}_{\mathfrak{s}}(M) = \mathscr{S}^{\mathrm{bs}}_{\mathfrak{s}}(M'')$$

so we are done.

Second assume $(M, N, a) \in K^{3,\mathrm{bs}}$ so there is $M' \leq_{\mathfrak{K}} M$ as asserted in the definition 2.4(1) of $K^{3,\mathrm{bs}}$ so $(\forall M'')[M' \leq_{\mathfrak{K}} M'' \leq_{\mathfrak{K}} M \ \& \ M'' \in K_\lambda \Rightarrow \mathbf{tp}(a, M'', N) \in \mathscr{S}^{\mathrm{bs}}_{\mathfrak{s}}(M'')]$ in particular this holds for $M'' = M$ and we get $\mathbf{tp}(a, M, N) \in \mathscr{S}^{\mathrm{bs}}_{\mathfrak{s}}(M)$ as required.
2) First assume $\bigcup\limits_{<\infty} (M_0, M_1, a, M_3)$.
So there is M_0' as required in Definition 2.5; this means

$$M_0' \in K_\lambda, M_0' \leq_{\mathfrak{K}} M_0 \text{ and}$$

$$(\forall M_1' \in K_\lambda)(\forall M_3' \in K_\lambda)[M_0' \leq_{\mathfrak{K}} M_1' \leq M_1 \ \& \ M_1' \cup \{a\} \subseteq M_3' \leq_{\mathfrak{K}} M_3 \to \bigcup_{\lambda}(M_0', M_1', a, M_3')].$$

In particular, we can choose $M_1' = M_1, M_3' = M_3$ so the antecedent holds hence $\bigcup\limits_{\lambda}(M_0', M_1', a, M_3')$ which means $\bigcup\limits_{\lambda}(M_0', M_1, a, M_3)$ and by clause $(E)(b)$ of Definition 2.1, $\bigcup\limits_{\lambda}(M_0, M_1, a, M_3)$ holds, as required.

Second assume $\bigcup_\lambda(M_0, M_1, a, M_3)$. So in Definition 2.5 the demands $M_0 \leq_{\mathfrak{K}} M_1 \leq_{\mathfrak{K}} M_3, a \in M_3 \backslash M_1$ hold by clause $(E)(a)$ of Definition 2.1; and we choose M_0' as M_0; clearly $M_0' \in K_\lambda$ & $M_0' \leq_{\mathfrak{K}} M_0$. Now suppose $M_0' \leq_{\mathfrak{K}} M_1' \leq_{\mathfrak{K}} M_1$ & $M_1' \in K_\lambda, M_1' \cup \{a\} \leq_{\mathfrak{K}} M_3' \leq M_3$; by clause $(E)(b)$ of Definition 2.1 we have $\bigcup_\lambda(M_0', M_1', a, M_3')$; so M_0' is as required so really $\bigcup_{<\infty}(M_0, M_1, a, M_3)$.

3) We prove something stronger: for any $M' \in \mathfrak{K}_\mathfrak{s}$ which is $\leq_{\mathfrak{K}[\mathfrak{s}]} M, M'$ witnesses $(M, N_1, a) \in K^{3,\mathrm{bs}}$ iff M' witnesses $(M, N_2, a) \in K^{3,\mathrm{bs}}$ (of course, witness means: as required in Definition 2.4). So we have to check the statement there for every $M'' \in K_\lambda$ such that $M' \leq_\mathfrak{s} M'' \leq_{\mathfrak{K}} M$. The equivalence holds because for every $M'' \leq_{\mathfrak{K}} M, M'' \in K_\lambda$ we have $\mathbf{tp}(a, M'', N_1) = \mathbf{tp}(a, M'', N_2)$, by 1.11(2), more transparent as $\mathfrak{K}_\lambda$ has the amalgamation property (by clause (C) of Definition 2.1) and so one is "basic" iff the other is by clause $(E)(b)$ of Definition 2.1.

4) The direction $\Leftarrow$ is because if M_0' witness $\bigcup_{<\infty}(M_0, M_1, a, M_3^*)$ (see Definition 2.5), <u>then</u> it witnesses $\bigcup_{<\infty}(M_0, M_1, a, M_3)$ as there are just fewer pairs (M_1', M_3') to consider. For the direction $\Rightarrow$ the demands $M_0 \leq_{\mathfrak{K}} M_1 \leq_{\mathfrak{K}} M_3, a \in M_3 \backslash M_1$, of course, hold and let M_0' be as required in the definition of $\bigcup_{<\infty}(M_0, M_1, a, M_3)$; let $M_0' \leq_{\mathfrak{K}} M_1' \leq_{\mathfrak{K}} M_1, M_1' \cup \{a\} \subseteq M_3' \leq_{\mathfrak{K}} M_3^*, M_3' \in K_\lambda$. As $\lambda \geq \mathrm{LS}(\mathfrak{K})$ we can find $M_3'' \leq_{\mathfrak{K}} M_3$ such that $M_1' \cup \{a\} \subseteq M_3'' \in K_\lambda$ and then find $M_3''' \leq_\mathfrak{s} M_3^*$ such that $M_3' \cup M_3'' \subseteq M_3''' \in K_\lambda$. So by the choice of M_0' and M_3'' clearly $\bigcup_\lambda(M_0', M_1', a, M_3'')$ and by clause $(E)(b)$ of Definition 2.1 we have

$$\bigcup_\lambda(M_0', M_1', a, M_3'') \Leftrightarrow \bigcup_\lambda(M_0', M_1', a, M_3''') \Leftrightarrow \bigcup_\lambda(M_0', M_1', a, M_3')$$

(note that we know the left statement and need the right statement) so M_1' is as required to complete the checking of $\bigcup_{<\infty}(M_0, M_1, a, M_3^*)$.

$\square_{2.6}$

We extend the definition of $\mathscr{S}^{\mathrm{bs}}_{\mathfrak{s}}(M)$ from $M \in K_\lambda$ to arbitrary $M \in K$.

2.7 Definition. 1) For $M \in K$ we let

$$\mathscr{S}^{\mathrm{bs}}(M) = \mathscr{S}^{\mathrm{bs}}_{\geq \mathfrak{s}}(M) = \Big\{p \in \mathscr{S}(M) : \text{for some } N \text{ and } a, \\ p = \mathbf{tp}(a, M, N) \text{ and } \\ (M, N, a) \in K^{3,\mathrm{bs}}_{\geq \mathfrak{s}}\Big\}$$

(for $M \in K_\lambda$ we get the old definition by 2.6(1); note that as we do not have amalgamation (in general) the meaning of types is more delicate. Not so in $\mathfrak{K}_\lambda$ as in a good λ-frame we have amalgamation in $\mathfrak{K}_\lambda$ but not necessarily in $\mathfrak{K}_{\geq\lambda}$).
2) We say that $p \in \mathscr{S}^{\mathrm{bs}}_{\geq\mathfrak{s}}(M_1)$ does not fork over $M_0 \leq_{\mathfrak{K}} M_1$ <u>if</u> for some M_3, a we have $p = \mathbf{tp}_{\mathfrak{K}[\mathfrak{s}]}(a, M_1, M_3)$ and $\bigcup\limits_{<\infty} (M_0, M_1, a, M_3)$.
(Again, for $M \in K_\lambda$ this is equivalent to the old definition by 2.6).
3) For $M \in K$ let $\mathscr{E}^\lambda_M$ be the following two-place relation on $\mathscr{S}(M)$: $p_1 \mathscr{E}^\lambda_M p_2$ iff $p_1, p_2 \in \mathscr{S}^{\mathrm{bs}}(M)$ and if $p_\ell = \mathbf{tp}(a_\ell, M, M^*), N \leq_{\mathfrak{K}} M, N \in K_\lambda$ then $p_1 \restriction N = p_2 \restriction N$. Let $\mathscr{E}^{\mathfrak{s}}_M = \mathscr{E}^{\lambda(\mathfrak{s})}_M \restriction \mathscr{S}^{\mathrm{bs}}(M)$.
4) $\mathfrak{K}$ is (λ, μ)-local if every $M \in \mathfrak{K}_\mu$ is λ-local which means that $\mathscr{E}^\lambda_M$ is equality; let $(\mathfrak{s}, \mu)$-local means $(\lambda_{\mathfrak{s}}, \mu)$-local.

Though we will prove below some nice things, having the extension property is more problematic. We may define "the extension" in a formal way, for $M \in K_{>\lambda}$ but then it is not clear if it is realized in any $\leq_{\mathfrak{K}}$-extension of M. Similarly for the uniqueness property. That is, assume $M_0 \leq_{\mathfrak{K}} M \leq_{\mathfrak{K}} N_\ell$ and $a_\ell \in N_\ell \backslash M$, and $M_0 \in \mathfrak{K}_{\mathfrak{s}}$ and $\mathbf{tp}(a_\ell, M, N_\ell)$ does not fork over M_0 for $\ell = 1, 2$ and $\mathbf{tp}(a_1, M_0, N_1) = \mathbf{tp}(a_2, M_0, N_1)$. Now does it follow that $\mathbf{tp}(a_1, M, N_1) = \mathbf{tp}(a_2, M, N_2)$? This requires the existence of some form of amalgamation in $\mathfrak{K}$, which we are not justified in assuming. So we may prefer to define $\mathscr{S}^{\mathrm{bs}}(M)$ "formally", the set of stationarization of $p \in \mathscr{S}^{\mathrm{bs}}(M_0), M_0 \in \mathfrak{K}_{\mathfrak{s}}$, see [Sh 842]. We now note that in definition 2.7 "some" can be replaced by "every".

2.8 Fact. 1) For $M \in K$

$$\mathscr{S}^{\mathrm{bs}}_{\geq \mathfrak{s}}(M) = \Big\{ p \in \mathscr{S}_{\mathfrak{K}[\mathfrak{s}]}(M) : \text{for every } N, a \text{ we have if } M \leq_{\mathfrak{K}} N,\ a \in N \backslash M \text{ and } p = \mathbf{tp}_{\mathfrak{K}}(a, M, N) \text{ then } (M, N, a) \in K^{3,\mathrm{bs}}_{\geq \mathfrak{s}} \Big\}.$$

2) The type $p \in \mathscr{S}_{\mathfrak{K}[\mathfrak{s}]}(M_1)$ does not fork over $M_0 \leq_{\mathfrak{K}} M_1$ iff for every a, M_3 satisfying $M_1 \leq_{\mathfrak{K}} M_3 \in K$, $a \in M_3 \backslash M_1$ and $p = \mathbf{tp}_{\mathfrak{K}[\mathfrak{s}]}(a, M_1, M_3)$ we have $\bigcup\limits_{<\infty} (M_0, M_1, a, M_3)$.
3) $(M, N, a) \in K^{3,\mathrm{bs}}_{\geq \mathfrak{s}}$ is preserved by isomorphisms.
4) If $M \leq_{\mathfrak{K}} N_\ell$, $a_\ell \in N_\ell \backslash M$ for $\ell = 1, 2$ and $\mathbf{tp}(a_1, M, N_1)\ \mathscr{E}^{\mathfrak{s}}_M\ \mathbf{tp}(a_2, M, N_2)$ then $(M, N_1, a_1) \in K^{3,\mathrm{bs}}_{\geq \mathfrak{s}} \Leftrightarrow (M, N_2, a_2) \in K^{3,\mathrm{bs}}_{\geq \mathfrak{s}}$.
5) $\mathscr{E}^{\mathfrak{s}}_M$ is an equivalence relation on $\mathscr{S}^{\mathrm{bs}}_{\geq \mathfrak{s}}(M)$ and if $p, q \in \mathscr{S}^{\mathrm{bs}}_{\geq \mathfrak{s}}(M)$ do not fork over $N \in K_\lambda$ so $N \leq_{\mathfrak{K}} M$ then $p \mathscr{E}^{\mathfrak{s}}_M q \Leftrightarrow (p \upharpoonright N = q \upharpoonright N)$.

Proof. 1) By 2.6(3) and the definition of type.
2) By 2.6(4) and the definition of type.
3) Easy.
4) Enough to deal with the case $(M, N_1, a_1) E^{\mathrm{at}}_M, (M, N_2, a_2)$ or (by (3)) even $a_1 = a_2, N_1 \leq_{\mathfrak{K}} N_2$. This is easy.
5) Easy, too. $\square_{2.8}$

We can also get that there are enough basic types, as follows:

2.9 Claim. *If $M \leq_{\mathfrak{K}} N$ and $M \neq N$, then for some $a \in N \backslash M$ we have $\mathbf{tp}_{\mathfrak{K}}(a, M, N) \in \mathscr{S}^{\mathrm{bs}}(M)$.*

Proof. Suppose not, so as we are assuming $K = K_{\geq \lambda}$ by clause $(D)(c)$ of Definition 2.1 necessarily $\|N\| > \lambda$. If $\|M\| = \lambda < \|N\|$ choose N' satisfying $M <_{\mathfrak{K}} N' \leq_{\mathfrak{K}} N, N' \in K_\lambda$ and by clause $(D)(c)$ of Definition 2.1 choose $a^* \in N' \backslash M$ such that $\mathbf{tp}_{\mathfrak{s}}(a^*, M, N') \in \mathscr{S}^{\mathrm{bs}}_{\mathfrak{s}}(M)$. So we can assume $\|M\| > \lambda$; choose $a^* \in N \backslash M$. We choose by induction on $i < \omega, M_i, N_i, M_{i,c}$ (for $c \in N_i \backslash M_i$) such that:

(a) $M_i \leq_{\mathfrak{K}} M$ is $\leq_{\mathfrak{K}}$-increasing

(b) $M_i \in K_\lambda$

(c) $N_i \leq_{\mathfrak{K}} N$ is $\leq_{\mathfrak{K}}$-increasing

(d) $N_i \in K_\lambda$

(e) $a^* \in N_0$

(f) $M_i \leq_{\mathfrak{K}} N_i$

(g) if $c \in N_i \backslash M$, $\mathbf{tp}_{\mathfrak{s}}(c, M_i, N) \in \mathscr{S}^{\mathrm{bs}}_{\mathfrak{s}}(M_i)$ and there is $M' \in K_\lambda$ such that $M_i \leq_{\mathfrak{K}} M' \leq_{\mathfrak{K}} M$ and $\mathbf{tp}_{\mathfrak{s}}(c, M', N)$ forks over M_i then $M_{i,c}$ satisfies this, otherwise $M_{i,c} = M_i$

(h) M_{i+1} includes the set $\bigcup_{c \in N_i \backslash M} M_{i,c} \cup (N_i \cap M)$.

There is no problem to carry the definition; in stage $i+1$ first choose $M_{i,c}$ for $c \in N_i \backslash M$ then choose M_{i+1} and lastly choose N_{i+1}. Let $M^* = \bigcup_{i<\omega} M_i$ and $N^* = \bigcup_{i<\omega} N_i$. It is easy to check that:

(i) $M_i \leq_{\mathfrak{K}} M^* \leq_{\mathfrak{K}} M$ for $i < \omega$
(by clause (a))

(ii) $M^* \in K_\lambda$
(by clause (i) we have $M^* \in K$ and $\|M^*\| = \lambda$ by the choice of M^* and clause (b))

(iii) $N_i \leq_{\mathfrak{K}} N^* \leq_{\mathfrak{K}} N$
(by clause (c))

(iv) $N^* \in K_\lambda$
(by clause (iii) we have $N^* \in K$ and $\|N^*\| = \lambda$ by the choice of N^* and clause (d))

(v) $M_i \leq_{\mathfrak{K}} M^* \leq_{\mathfrak{K}} N^* \leq_{\mathfrak{K}} N$
(by clauses (a) + (f) + (iii) we have $M_i \leq_{\mathfrak{K}} N^*$ hence by clause (a) and the choice of M^* we have $M^* \leq_{\mathfrak{K}} N^*$, and $N^* \leq_{\mathfrak{K}} N$ by clause (iii))

(vi) $M^* = N^* \cap M$
(by clauses (f) + (h) and the choices of M^*, N^*)

(*vii*) $M^* \neq N^*$
(as $a^* \in N \backslash M$ and $a^* \in N_0 \leq_{\mathfrak{K}} N^* \leq_{\mathfrak{K}} N$ and $M^* = N^* \cap M$; they hold by the choice of a^*, clause (e), clause (iii), clause (iii) and clause (vi) respectively)

(*viii*) there is $b^* \in N^* \backslash M^*$ such that $\mathbf{tp}(b^*, M^*, N^*) \in \mathscr{S}^{\mathrm{bs}}(M^*)$
[why? by clause (v) and (viii) recalling Definition 2.1 clause (D)(c) (density)]

(*ix*) for some $i < \omega$ we have $\bigcup(M_i, M^*, b^*, N^*)$, so

$\mathbf{tp}(b^*, M^*, N^*) \in \mathscr{S}^{\mathrm{bs}}_{\mathfrak{s}}(M^*)$ and $\mathbf{tp}_{\mathfrak{s}}(b^*, M_j, N^*) \in \mathscr{S}^{\mathrm{bs}}_{\mathfrak{s}}(M_j)$ for $j \in [i, \omega)$
[why? by Definition 2.1 clause $(E)(c)$ (local character) applied to the sequence $\langle M_n : n < \omega \rangle \hat{\ } \langle M^*, N^* \rangle$ and the element b^*, using of course (E)(a) of Definition 2.1 and clause (viii)]

(*x*) $\bigcup(M_i, M_{i,b^*}, b^*, N^*)$

[why? by clause (ix) and Definition 2.1$(E)(b)$ (monotonicity) as
$M_i \leq_{\mathfrak{K}} M_{i,b^*} \leq_{\mathfrak{K}} M_{i+1} \leq_{\mathfrak{K}} M^*$ by clause (g) in the construction]

(*xi*) if $M_i \leq_{\mathfrak{K}} M' \leq_{\mathfrak{K}} M$ and $M' \cup \{b^*\} \subseteq N' \leq_{\mathfrak{K}} N$ and $M' \in K_\lambda, N' \in K_\lambda$ then $\bigcup(M_i, M', b^*, N')$

[why? by clause (x) and clause (g) in the construction.]

So we are done. $\square_{2.9}$

2.10 Claim. *If $M \leq_{\mathfrak{K}} N, a \in N \backslash M,$ and $\mathbf{tp}(a, M, N) \in \mathscr{S}^{\mathrm{bs}}_{\geq \mathfrak{s}}(M)$ <u>then</u> for some $M_0 \leq_{\mathfrak{K}} M$ we have*

(*a*) $M_0 \in K_\lambda$
(*b*) $\mathbf{tp}(a, M_0, N) \in \mathscr{S}^{\mathrm{bs}}_{\mathfrak{s}}(M_0)$
(*c*) *if $M_0 \leq_{\mathfrak{K}} M' \leq_{\mathfrak{K}} M$, <u>then</u> $\mathbf{tp}(a, M', N) \in \mathscr{S}^{\mathrm{bs}}_{\mathfrak{s}}(M')$ does not fork over M_0.*

Proof. Easy by now. $\square_{2.10}$

2.11 Claim. *1) Assume $M_1 \leq_{\mathfrak{K}} M_2$ and $p \in \mathscr{S}_{\mathfrak{K}}(M_2)$. Then $p \in \mathscr{S}^{\mathrm{bs}}_{\geq \mathfrak{s}}(M_2)$ and p does not fork over M_1 iff for some $N_1 \leq_{\mathfrak{K}} M_1, N_1 \in K_\lambda$ and p does not fork over N_1 iff for some $N_1 \leq_{\mathfrak{K}} M_1, N_1 \in K_\lambda$ and we have $(\forall N)[N_1 \leq_{\mathfrak{K}} N \leq_{\mathfrak{K}} M_2 \;\&\; N \in K_\lambda \Rightarrow p \restriction N \in \mathscr{S}^{\mathrm{bs}}_{\mathfrak{s}}(N) \;\&$ $(p \restriction N$ does not fork over $N_1)]$; we call such N_1 a witness, so every $N'_1 \in K_\lambda, N_1 \leq_{\mathfrak{K}} N'_1 \leq M_1$ is a witness, too.*
2) Assume $M^* \in K$ and $p \in \mathscr{S}_{\mathfrak{K}}(M^*)$.
Then: $p \in \mathscr{S}^{\mathrm{bs}}_{\geq \mathfrak{s}}(M^*)$ iff for some $N^* \leq_{\mathfrak{K}} M^*$ we have $N^* \in K_\lambda, p \restriction N^* \in \mathscr{S}^{\mathrm{bs}}(N^*)$ and $(\forall N \in K_\lambda)(N^* \leq_{\mathfrak{K}} N \leq_{\mathfrak{K}} M^* \Rightarrow p \restriction N \in \mathscr{S}^{\mathrm{bs}}(N)$ and does not fork over $N^*)$ (we say such N^* is a witness, so any $N' \in K_\lambda, N^* \leq_{\mathfrak{K}} N' \leq_{\mathfrak{K}} M$ is a witness, too).
3) (Monotonicity)
If $M_1 \leq_{\mathfrak{K}} M'_1 \leq_{\mathfrak{K}} M'_2 \leq_{\mathfrak{K}} M_2$ and $p \in \mathscr{S}^{\mathrm{bs}}_{\geq \mathfrak{s}}(M_2)$ does not fork over M_1, then
$p \restriction M'_2 \in \mathscr{S}^{\mathrm{bs}}_{\geq \mathfrak{s}}(M'_2)$ and it does not fork over M'_1.
4) (Transitivity)
If $M_0 \leq_{\mathfrak{K}} M_1 \leq_{\mathfrak{K}} M_2$ and $p \in \mathscr{S}^{\mathrm{bs}}_{\geq \mathfrak{s}}(M_2)$ does not fork over M_1 and $p \restriction M_1$ does not fork over M_0, then p does not fork over M_0.
5) (Local character) If $\langle M_i : i \leq \delta + 1\rangle$ is $\leq_{\mathfrak{K}}$-increasing continuous and $a \in M_{\delta+1}$ and $\mathbf{tp}_{\mathfrak{K}}(a, M_\delta, M_{\delta+1}) \in \mathscr{S}^{\mathrm{bs}}_{\geq \mathfrak{s}}(M_\delta)$ then for some $i < \delta$ we have $\mathbf{tp}_{\mathfrak{K}}(a, M_\delta, M_{\delta+1})$ does not fork over M_i.
6) Assume that $\langle M_i : i \leq \delta + 1\rangle$ is $\leq_{\mathfrak{K}}$-increasing and $p \in \mathscr{S}(M_\delta)$ and for every $i < \delta$ we have $p \restriction M_i \in \mathscr{S}^{\mathrm{bs}}_{\geq \mathfrak{s}}(M_i)$ does not fork over M_0. Then $p \in \mathscr{S}^{\mathrm{bs}}_{\geq \mathfrak{s}}(M_\delta)$ and p does not fork over M_0.

Proof. 1), 2) Check the definitions.
3) As $p \in \mathscr{S}^{\mathrm{bs}}_{\geq \mathfrak{s}}(M_2)$ does not fork over M_1, there is $N_1 \in K_\lambda$ which witnesses it.

This same N_1 witnesses that $p \restriction M'_2$ does not fork over M'_1.
4) Let $N_0 \leq_{\mathfrak{K}} M_0$ witness that $p \restriction M_1$ does not fork over M_0 (in particular $N_0 \in K_\lambda$); let $N_1 \leq_{\mathfrak{K}} M_1$ witness that p does not fork over M_1 (so in particular $N_1 \in K_\lambda$). Let us show that N_0 witnesses p does not fork over M_0, so let $N \in K_\lambda$ be such that $N_0 \leq_{\mathfrak{K}} N \leq_{\mathfrak{K}} M_2$ and we should just prove that $p \restriction N$ does not fork over N_0. We can find $N' \leq_{\mathfrak{K}} M_1, N' \in K_\lambda$ such that $N_0 \cup N_1 \subseteq N'$, we can also find $N'' \leq_{\mathfrak{K}} M_2$ satisfying $N'' \in K_\lambda$ such that $N' \cup N \subseteq N''$. As N_1 witnesses that p does not fork over M_1, clearly $p \restriction N'' \in \mathscr{S}^{\mathrm{bs}}_{\mathfrak{s}}(N'')$

does not fork over N_1, hence by monotonicity does not fork over N'. As N_0 witnesses $p \restriction M_1$ does not fork over M_0, clearly $p \restriction N'$ belongs to $\mathscr{S}^{\rm bs}(N')$ and does not fork over N_0, so by transitivity (in $\mathfrak{K}_{\mathfrak{s}}$) we know that $p \restriction N''$ does not fork over N_0; hence by monotonicity $p \restriction N$ does not fork over N_0.
5) Let $p = \mathbf{tp}_{\mathfrak{K}}(a, M_\delta, M_{\delta+1})$ and let $N^* \leq_{\mathfrak{K}} M_\delta$ witness $p \in \mathscr{S}^{\rm bs}(M_\delta)$. Assume toward contradiction that the conclusion fails. Without loss of generality $\mathrm{cf}(\delta) = \delta$.

Case 0: $\|M_\delta\| \leq \lambda (= \lambda_{\mathfrak{s}})$.
Trivial.

Case 1: $\delta < \lambda^+, \|M_\delta\| > \lambda$.
As $\|M_\delta\| > \lambda$, for some i, $\|M_i\| > \lambda$ so without loss of generality $i < \delta \Rightarrow \|M_i\| > \lambda$. We choose by induction on $i < \delta$, models N_i, N'_i such that:

(α) $N_i \in K_\lambda$
(β) $N_i \leq_{\mathfrak{K}} M_i$ (hence $N_i \leq_{\mathfrak{K}} M_j$ for $j \in [i, \delta)$)
(γ) N_i is $\leq_{\mathfrak{K}}$-increasing continuous
(δ) $N'_i \in K_\lambda, N^* \leq_{\mathfrak{K}} N'_0$
(ε) $N_i \leq_{\mathfrak{K}} N'_i \leq_{\mathfrak{K}} M_\delta$,
(ζ) N'_i is $\leq_{\mathfrak{K}}$-increasing continuous
(η) $p \restriction N'_i$ forks over N_i when $i \neq 0$ for simplicity
(θ) $N_i \cup \bigcup_{j \leq i}(N'_j \cap M_{i+1}) \subseteq N_{i+1}$.

No problem to carry the induction, but we give details.
First, if $i = 0$ trivial. Second let i be a limit ordinal.

Let $N_i = \cup\{N_j : j < i\}$, now $N_i \leq_{\mathfrak{K}} M_i$ by clauses $(\beta) + (\gamma)$ and $\mathfrak{K}$ being a.e.c. and $\|N_i\| = \lambda$ by clause (α), as $i \leq \delta < \lambda^+$; so clauses $(\alpha), (\beta), (\gamma)$ hold. Next, let $N'_i = \cup\{N'_j : j < i\}$ and similarly clauses $(\delta), (\varepsilon), (\zeta)$ hold. Lastly, we shall prove clause (η) and assume toward contradiction that it fails; so $p \restriction N'_i$ does not fork over N_i in particular $p \restriction N_i \in \mathscr{S}^{\rm bs}_{\mathfrak{s}}(N_i)$ hence for some $j < i$ the type $p \restriction N'_i$ does not fork over $N_j \leq_{\mathfrak{K}} N_i$, (by (E)(c) of Definition 2.1) hence by transitivity (for $\mathfrak{K}_{\mathfrak{s}}$), $p \restriction N'_i$ does not fork over N_j hence by monotonicity $p \restriction N'_j$ does not fork over N_j (see (E)(b) of Definition 2.1) contradicting the induction hypothesis.

Lastly, clause (θ) is vacuous.

Third assume $i = j+1$, so first choose N_i satisfying clause (θ) (with j, i here standing for $i, i+1$ there), and $(\alpha), (\beta), (\gamma)$; this is possible by the L.S. property. Now N_i cannot witness "p does not fork over M_i" hence for some $N^*_i \in K_\lambda$ we have $N_i \leq_{\mathfrak{K}} N^*_i \leq_{\mathfrak{K}} M_\delta$ and $p \restriction N^*_i$ forks over N_i; again by L.S. choose $N'_i \in K_\lambda$ such that $N'_i \leq_{\mathfrak{K}} M_\delta$ and $N^* \cup N_i \cup N'_j \cup N^*_i \subseteq N'_i$, easily (N_i, N'_i) are as required.

Let $N_\delta = \bigcup\limits_{i<\delta} N_i$, so by clause $(\beta), (\gamma)$ we have $N_\delta \leq_{\mathfrak{K}} M_\delta$ and by clause (α), as $\delta < \lambda^+$ we have $N_\delta \in K_\lambda$ and by clauses $(\delta)+(\theta)$ in the construction we have $i < \delta \Rightarrow N'_i = \cup\{N'_i \cap M_{j+1} : j \in [i, \delta)\} \subseteq N$ so by clause $(\delta), N^* \leq_{\mathfrak{K}} N'_0 \leq_{\mathfrak{K}} N_\delta$. Hence by the choice of $N^*, p \restriction N_\delta \in \mathscr{S}^{\mathrm{bs}}_{\mathfrak{s}}(N_\delta)$ and it does not fork over N^*. Now as $p \restriction N_\delta \in \mathscr{S}^{\mathrm{bs}}_{\mathfrak{s}}(N_\delta)$ by local character, i.e., clause $(E)(c)$ of Definition 2.1, for some $i < \delta, p \restriction N_\delta$ does not fork over N_i (so $p \restriction N_i \in \mathscr{S}^{\mathrm{bs}}_{\mathfrak{s}}(N_i)$). Now $N_i \leq_{\mathfrak{K}} N'_i \leq_{\mathfrak{K}} M_\delta$ and by clause (θ) of the construction $N'_i \subseteq N_\delta$ hence $N_i \leq_{\mathfrak{K}} N'_i \leq_{\mathfrak{K}} N_\delta$ hence by monotonicity of non-forking (i.e. clause (E)(b) of Definition 2.1), $p \restriction N'_i \in \mathscr{S}^{\mathrm{bs}}(N_i)$ does not fork over N_i. But this contradicts the choice of N'_i (i.e., clause (η) of the construction).

Case 2: $\delta = \mathrm{cf}(\delta) > \lambda$.

Recall that $N^* \leq_{\mathfrak{K}} M_\delta, N^*$ is from K_λ and $N^* \leq_{\mathfrak{K}} N \leq_{\mathfrak{K}} M_\delta$ & $N \in K_\lambda \Rightarrow$
$p \restriction N \in \mathscr{S}^{\mathrm{bs}}_{\mathfrak{s}}(N)$. Now as $\delta = \mathrm{cf}(\delta) > \lambda \geq \|N^*\|$ clearly for some $i < \delta$ we have $N^* \subseteq M_i$ hence $N^* \leq_{\mathfrak{K}} M_i$ (hence $i \leq j < \delta \Rightarrow p \restriction M_j \in \mathscr{S}^{\mathrm{bs}}_{\geq \mathfrak{s}}(M_j)$), and N^* witnesses that $p \in \mathscr{S}^{\mathrm{bs}}_{\geq \mathfrak{s}}(M_\delta)$ does not fork over M_i so we are clearly done.

6) Let $N_0 \in K_\lambda, N_0 \leq_{\mathfrak{K}} M_0$ witness $p \restriction M_0 \in \mathscr{S}^{\mathrm{bs}}_{\geq \mathfrak{s}}(M_0)$. By the proof of part (4) clearly $i < \delta$ & $N_0 \leq_{\mathfrak{K}} N \in K_\lambda$ & $N \leq_{\mathfrak{K}} M_i \Rightarrow p \restriction N$ does not fork over N_0. If $\mathrm{cf}(\delta) > \lambda$ we are done, so assume $\mathrm{cf}(\delta) \leq \lambda$. Let $N_0 \leq_{\mathfrak{K}} N^* \in K_\lambda$ & $N^* \leq_{\mathfrak{K}} M_\delta$, and we shall prove that $p \restriction N^*$ does not fork over N_0, this clearly suffices. As in Case 1 in the proof of part (5) we can find $N_i \leq_{\mathfrak{K}} M_i$ for $i \in (0, \delta)$ such

that $\langle N_i : i \leq \delta \rangle$ is $\leq_{\mathfrak{K}}$-increasing with i, each N_i belongs to $\mathfrak{K}_\lambda$ and $N^* \cap M_i \subseteq N_{i+1}$, hence $N^* \subseteq N_\delta := \bigcup_{i<\delta} N_i$. Now $N_\delta \leq_{\mathfrak{K}} M_\delta$ and as said as $i < \delta \Rightarrow p \restriction N_i \in \mathscr{S}^{\text{bs}}_{\geq \mathfrak{s}}(N_i)$ does not fork over N_0 hence $p \restriction N_\delta$ does not fork over N_0 and by monotonicity $p \restriction N^*$ does not fork over N_0, as required. $\square_{2.11}$

2.12 Lemma. *If $\mu = \text{cf}(\mu) > \lambda$ and $M \leq_{\mathfrak{K}} N$ are in K_μ, then we can find $\leq_{\mathfrak{K}}$-representations $\bar{M}, \bar{N}$ of M, N respectively such that:*

(i) *$N_i \cap M = M_i$ for $i < \mu$*

(ii) *if $i < j < \mu$ & $a \in N_i$ then*

(a)
$$\begin{aligned} &\mathbf{tp}(a, M_i, N) \in \mathscr{S}^{\text{bs}}_{\geq \mathfrak{s}}(M_i) \Leftrightarrow \\ &\Leftrightarrow \mathbf{tp}(a, M_j, N) \in \mathscr{S}^{\text{bs}}_{\geq \mathfrak{s}}(M_j) \\ &\Leftrightarrow \mathbf{tp}(a, M, N) \textit{ does not fork over } M_i \\ &\Leftrightarrow \mathbf{tp}(a, M_j, N) \textit{ is a non-forking extension} \\ &\qquad \textit{of } \mathbf{tp}(a, M_i, N) \end{aligned}$$

(b) *$M_i \leq_{\mathfrak{K}} N_i \leq_{\mathfrak{K}} N_j$ and $M_i \leq_{\mathfrak{K}} M_j \leq_{\mathfrak{K}} N_j$*
(and clearly $M_i \leq_{\mathfrak{K}} N_j$ and $M_i \leq_{\mathfrak{K}} M, M_i \leq_{\mathfrak{K}} N, N_i \leq_{\mathfrak{K}} N$).

2.13 Remark. In fact for any representations $\bar{M}, \bar{N}$ of M, N respectively, for some club E of μ the sequences $\bar{M} \restriction E, \bar{N} \restriction E$ are as above.

Proof. Let $\bar{M}$ be a $\leq_{\mathfrak{K}}$-representation of M. For $a \in N$ we define $S_a = \{\alpha < \mu : \mathbf{tp}(a, M_\alpha, N) \in \mathscr{S}^{\text{bs}}_{\geq \mathfrak{s}}(M_\alpha)\}$. Clearly if $\delta \in S_a$ is a limit ordinal then for some $i(a,\delta) < \delta$ we have $i(a,\delta) \leq i < \delta \Rightarrow i \in S_a$ & $(\mathbf{tp}(a, M_i, N)$ does not fork over $M_{i(a,\delta)})$ by 2.11(5). So if S_a is stationary, then for some $i(a) < \mu$ the set $S'_a = \{\delta \in S_a : i(a,\delta) = i(a)\}$ is a stationary subset of λ hence by monotonicity we have $i(a) \leq i \leq \mu \Rightarrow$ $\mathbf{tp}(a, M_i, N)$ does not fork over $M_{i(a)}$. Let E_a be a club of μ such that: if S_a is not stationary (subset of μ) then

$E_a \cap S_a = \emptyset$ and if S_a is not stationary then $S_a \cap E_a = \emptyset$.
Let $\bar{N}$ be a representation of N, and let

$$E^* = \{\delta < \mu : N_\delta \cap M = M_\delta \text{ and } M_\delta \leq_{\mathfrak{K}} M, N_\delta \leq_{\mathfrak{K}} N \\ \text{and for every } a \in N_\delta \text{ we have } \delta \in E_a\}.$$

Clearly it is a club of μ and $\bar{M} \restriction E^*, \bar{N} \restriction E^*$ are as required. $\square_{2.12}$

* * *

We may treat the lifting of $K^{3,\mathrm{bs}}_\lambda$ as a special case of the "lifting" of $\mathfrak{K}_\lambda$ to $\mathfrak{K}_{\geq\lambda} = (\mathfrak{K}_\lambda)^{\mathrm{up}}$ in Claim 1.23; this may be considered a good exercise.

2.14 Claim. *1)* $(K^{3,\mathrm{bs}}_\lambda, \leq_{\mathrm{bs}})$ *is a* λ*-a.e.c.*
2) $(K^{3,\mathrm{bs}}_{\geq\mathfrak{s}}, \leq_{\mathrm{bs}})$ *is* $(K^{3,\mathrm{bs}}_\lambda, \leq_{\mathrm{bs}})^{\mathrm{up}}$.

Remark. What is the class in 2.14(1)? Formally let $\tau^+ = \{R_{[\ell]} : R$ a predicate of $\tau_K, \ell = 1, 2\} \cup \{F_{[\ell]} : F$ a function symbol from τ_K and $\ell = 1, 2\} \cup \{c\}$ where $R_{[\ell]}$ is an n-place predicate when $R \in \tau$ is an n-place predicate and similarly $F_{[\ell]}$ and c is an individual constant. A triple (M, N, a) is identified with the following τ^+-model N^+ defined as follows:

(*a*) its universe is the universe of N

(*b*) $c^{N^+} = a$

(*c*) $R^{N^+}_{[2]} = R^N$

(*d*) $F^{N^+}_{[2]} = F^N$

(*e*) $R^{N^+}_{[1]} = R^M$

(*f*) $F^{N^+}_{[1]} = F^M$

(if you do not like partial functions, extend them to functions with full domain by $F(a_0, \ldots) = a_0$ when not defined if F has arity > 0, if F has arity 0 it is an individual constant, $F^{N^+} = F^N$ so no problem).

Proof. Left to the reader (in particular this means that $K^{3,\mathrm{bs}}_\lambda$ is closed under $\leq_{\mathrm{bs}}$-increasing chains of length $< \lambda^+$). $\square_{2.14}$

Continuing 1.23, 1.26 note that (and see more in 2.20):

2.15 Lemma. *Assume*

(a) $\mathfrak{K}$ *is an abstract elementary class with* $\mathrm{LS}(\mathfrak{K}) \le \mu$

(b) $K'_{\le\mu}$ *is a class of* τ_K*-model,* $K'_{\le\mu} \subseteq K_{\le\mu}$ *is non-empty and closed under* $\le_{\mathfrak{K}}$*-increasing unions of length* $< \mu^+$ *and isomorphisms (e.g. the class of* μ*-superlimit models of* $\mathfrak{K}_\mu$*, if there is one)*

(c) *define* $K' := \{M \in K : M$ *is a* $\le_{\mathfrak{K}}$*-directed union of members of* $K'_\mu\} \cup K'_{\le\mu}$

(d) *let* $\mathfrak{K}' = (K', \le_{\mathfrak{K}} \restriction K')$ *so* $\le_{\mathfrak{K}'}$ *is* $\le_{\mathfrak{K}} \restriction K'$*, so* $\mathfrak{K}'_{\le\mu} := (K'_{\le\mu}, \le_{\mathfrak{K}} \restriction K'_{\le\mu})$*; or* $\le_{\mathfrak{K}}$ *is as in 1.23(1), see 1.23(4).*

<u>*Then*</u>

(A) $\mathfrak{K}'$ *is an abstract elementary class,* $\mathrm{LS}(\mathfrak{K}) \le \mathrm{LS}(\mathfrak{K}') \le \mu$

(B) *If* $\mu \le \lambda$ *and* $(\mathfrak{K}, \bigcup, \mathscr{S}^{\mathrm{bs}})$ *is a good* λ*-frame and* $\mathfrak{K}'_\lambda$ *has amalgamation and JEP and* $M \in \mathfrak{K}'_\lambda \Rightarrow \mathscr{S}_{\mathfrak{K}'}(M) = \mathscr{S}_{\mathfrak{K}}(M)$*,* <u>*then*</u> $(\mathfrak{K}', \bigcup, \mathscr{S}^{\mathrm{bs}})$ *(with* $\bigcup, \mathscr{S}^{\mathrm{bs}}$ *restricted to* $\mathfrak{K}'$*) is a good* λ*-frame*

(C) *in clause (B), instead "*$M \in \mathfrak{K}'_\lambda \Rightarrow \mathscr{S}_{\mathfrak{K}'}(M) = \mathscr{S}_{\mathfrak{K}}(M)$*, it suffices to require: if* $M \in \mathfrak{K}'_\lambda, M \le_{\mathfrak{K}} N \in \mathfrak{K}'_\lambda, p \in \mathscr{S}^{\mathrm{bs}}_{\mathfrak{s}}(N), p$ *does not fork over* M *and* $p \restriction M$ *is realized in some* $M', M \le_{\mathfrak{K}'} M'$ <u>*then*</u> p *is realized in some* $N', N \le_{\mathfrak{K}} N' \in \mathfrak{K}'_\lambda$*.*

Remark. If in 2.15, K'_μ is not closed under $\le_{\mathfrak{K}}$-increasing unions, we can close it but then the "so $\mathfrak{K}'_{\le\mu} = \dots$" in clause (d) may fail.

Proof. <u>Clause (A)</u>: As in 1.23.

<u>Clauses (B),(C)</u>: Check. $\square_{2.15}$

* * *

Next we deal with some implications between the axioms in 2.1.

2.16 Claim. *1) In Definition 2.1 clause (E)(i) is redundant, i.e., follows from the rest, recalling*

(E)(i) *non-forking amalgamation*:
if for $\ell = 1, 2$, $M_0 \leq_{\mathfrak{K}} M_\ell$ *are in* K_λ, $a_\ell \in M_\ell \backslash M_0$, $\mathbf{tp}(a_\ell, M_0, M_\ell) \in \mathscr{S}^{\text{bs}}(M_0)$, *then we can find* f_1, f_2, M_3 *satisfying* $M_0 \leq_{\mathfrak{K}} M_3 \in K_\lambda$ *such that for* $\ell = 1, 2$ *we have* f_ℓ *is a* $\leq_{\mathfrak{K}}$*-embedding of* M_ℓ *into* M_3 *over* M_0 *and* $\mathbf{tp}(f_\ell(a_\ell), f_{3-\ell}(M_{3-\ell}), M_3)$ *does not fork over* M_0.

2) In fact, proving part (1) we use Axioms (A),(C),(E)(b),(d),(f),(g) only.

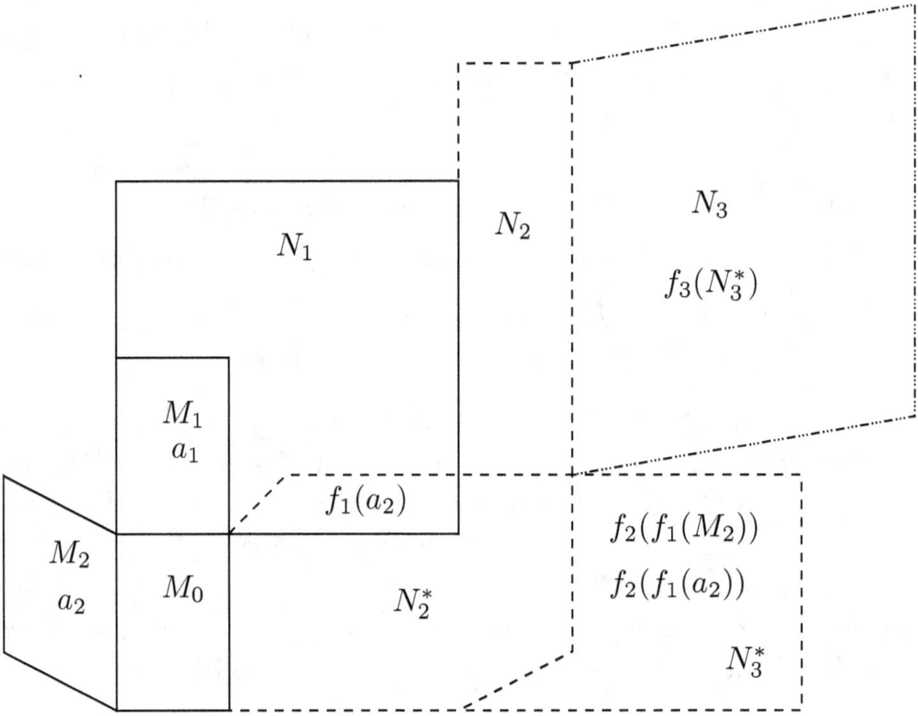

Proof. By Axiom (E)(g) (existence) applied with $\mathbf{tp}(a_2, M_0, M_2)$, M_0, M_1 here standing for p, M, N there; there is q_1 such that:

(*a*) $q_1 \in \mathscr{S}^{\text{bs}}(M_1)$
(*b*) q_1 does not fork over M_0
(*c*) $q_1 \restriction M_0 = \mathbf{tp}(a_2, M_0, M_2)$.

By the definition of types and as $\mathfrak{K}_\lambda$ has amalgamation (by Axiom (C)) there are N_1, f_1 such that

(d) $M_1 \leq_{\mathfrak{K}} N_1 \in K_\lambda$

(e) f_1 is a $\leq_{\mathfrak{K}}$-embedding of M_2 into N_1 over M_0

(f) $f_1(a_2)$ realizes q_1 inside N_1.

Now consider Axiom (E)(f) (symmetry) applied with $M_0, N_1, a_1, f_1(a_2)$ here standing for M_0, M_3, a_1, a_2 there; now as clause (α) of (E)(f) holds (use M_1, N_1 for M_1, M_3') we get that clause (β) of (E)(f) holds which means that there are N_2, N_2^* (standing for M_3', M_2 in clause (β) of (E)(f)) such that:

(g) $N_1 \leq_{\mathfrak{K}} N_2 \in K_\lambda$

(h) $M_0 \cup \{f_1(a_2)\} \subseteq N_2^* \leq_{\mathfrak{K}} N_2$

(i) $\mathbf{tp}(a_1, N_2^*, N_2) \in \mathscr{S}^{\text{bs}}(N_2^*)$ does not fork over M_0.

As $\mathfrak{K}_\lambda$ has amalgamation (see Axiom (C)) and the definition of type and as
$\mathbf{tp}(f_1(a_2), M_0, f_1(M_2)) = \mathbf{tp}(f_1(a_2), M_0, N_2) = \mathbf{tp}(f_1(a_2), M_0, N_2^*)$, we can find N_3^*, f_2 such that

(j) $N_2^* \leq_{\mathfrak{K}} N_3^* \in K_\lambda$

(k) f_2 is a $\leq_{\mathfrak{K}}$-embedding[11] of $f_1(M_2)$ into N_3^* over $M_0 \cup \{f_1(a_2)\}$.

As by clause (i) above $\mathbf{tp}(a_1, N_2^*, N_2) \in \mathscr{S}^{\text{bs}}(N_2^*)$, so by Axiom (E)(g) (extension existence) there are N_3, f_3 such that

(l) $N_2 \leq_{\mathfrak{K}} N_3 \in K_\lambda$

(m) f_3 is a $\leq_{\mathfrak{K}}$-embedding of N_3^* into N_3 over N_2^*

(n) $\mathbf{tp}(a_1, f_3(N_3^*), N_3) \in \mathscr{S}^{\text{bs}}(N_3^*)$ does not fork over N_2^*.

By Axiom (E)(d) (transitivity) using clauses (i) + (n) above we have

(o) $\mathbf{tp}(a_1, f_3(N_3^*), N_3) \in \mathscr{S}^{\text{bs}}(N_3^*)$ does not fork over M_0.

Letting $f = f_3 \circ f_2 \circ f_1$ as $f(M_2) \subseteq f_3(N_3^*)$ by clauses $(e), (k), (m)$ we have

(p) f is a $\leq_{\mathfrak{K}}$-embedding of M_2 into N_3 over M_0.

---

[11] we could have chosen $N_3^* = N_2, f_2 = \text{id}_{f_1(M_2)}$

By (E)(b) (monotonicity) and clause (o) and clause (p)

(*q*) $\mathbf{tp}(a_1, f(M_2), N_3) \in \mathscr{S}^{\mathrm{bs}}(f(M_2))$ does not fork over M_0.

As $\mathbf{tp}(f_1(a_2), M_1, N_3) = \mathbf{tp}(f_1(a_2), M_1, N_1) = q_1$ does not fork over M_0 by clauses (b) + (f), and $f_2(f_1(a_2)) = f_1(a_2)$ by clause (k) and $f_3(f_1(a_2)) = f_1(a_2)$ by clauses (m) + (h), we get

(*r*) $\mathbf{tp}(f(a_2), M_1, N_3) \in \mathscr{S}^{\mathrm{bs}}(M_1)$ does not fork over M_0.

So by clauses (o) and (r) we have $\mathrm{id}_{M_1}, f, N_3$ are as required on f_1, f_2, M_3 in our desired conclusion. $\square_{2.16}$

2.17 Claim. *1) In the local character Axiom (E)(c) of Definition 2.1 if $\mathscr{S}^{\mathrm{bs}}_{\mathfrak{s}} = \mathscr{S}^{\mathrm{na}}_{\mathfrak{K}_\mathfrak{s}}$ recalling $\mathscr{S}^{\mathrm{na}}_{\mathfrak{K}_\mathfrak{s}}(M) = \{\mathbf{tp}(a, M, N) : M \le_\mathfrak{s} N$ and $a \in N \backslash M\}$ then it suffices to restrict ourselves to the case that δ has cofinality $\aleph_0$ (i.e., the general case follows from this special case and the other axioms).*
2) In fact in part (1) we need only Axioms (E)(b),(h) and you may say (A),(D)(a),(E)(a).
3) If $\mathscr{S}^{\mathrm{bs}} = \mathscr{S}^{\mathrm{na}}$ then the continuity Axiom (E)(h) follows from the rest.
4) In (3) actually we need only Axioms (E)(c), (local character) (d), (transitivity) and you may say (A),(D)(a),(E)(a).

Proof. 1), 2) Let $\langle M_i : i \le \delta + 1\rangle$ be $\le_{\mathfrak{K}_\lambda}$-increasing, $a \in M_{\delta+1} \backslash M_\delta$ and without loss of generality $\aleph_0 < \delta = \mathrm{cf}(\delta)$, so for every $\alpha \in S := \{\alpha < \delta : \mathrm{cf}(\alpha) = \aleph_0\}$, $\mathbf{tp}(a, M_\alpha, M_{\delta+1}) \in \mathscr{S}^{\mathrm{bs}}(M_\alpha)$ by the assumption "$S^{\mathrm{bs}}_{\mathfrak{s}} = \mathscr{S}^{\mathrm{na}}_{\mathfrak{K}_\mathfrak{s}}$ hence there is $\beta_\alpha < \alpha$ such that $\mathbf{tp}(a, M_\alpha, M_{\delta+1})$ does not fork over M_{β_α}, so for some $\beta < \delta$ the set $S_1 = \{\alpha \in S : \beta_\alpha = \beta)$ is a stationary subset of δ. By Axiom $(E)(b)$ (monotonicity) it follows that for any $\gamma_1 \le \gamma_2$ from $[\beta, \delta)$ the type $\mathbf{tp}(a, M_{\gamma_2}, M_{\delta+1}) \in \mathscr{S}^{\mathrm{bs}}(M_{\gamma_2})$ does not fork over M_{γ_1}. Now for any $\gamma \in [\beta, \delta)$ the type $\mathbf{tp}(a, M_\delta, M_{\delta+1})$ does not fork over M_γ by applying $(E)(h)$ (continuity) to $\langle M_\alpha : \alpha \in [\gamma, \delta + 1]$ so we have finished.
3),4) So assume $\langle M_i : i \le \delta\rangle$ is $\le_\mathfrak{K}$-increasing continuous, all in K_λ and δ is a limit ordinal, $p \in \mathscr{S}(M_\delta)$ and $p_i := p \restriction M_i \in \mathscr{S}^{\mathrm{bs}}(M_i)$ does not fork over M_0 for each $i < \delta$; we should prove that $p \in \mathscr{S}^{\mathrm{bs}}(M_\delta)$ and p does not fork over M_0.

First, for each $i < \delta, p_i \in \mathscr{S}^{\rm bs}(M_i)$ hence p_i is not realized in M_i. As $M_\delta = \cup\{M_i : i < \delta\}$ clearly p is not realized in M_δ so $p \in \mathscr{S}^{\rm na}(M_\delta) = \mathscr{S}^{\rm bs}(M_\delta)$.

Second, by Ax(E)(c) the type p does not fork over M_j for some $j < \delta$. As $p_j = p \restriction M_j$ does not fork over M_0 (by assumption) by the transitivity Axiom $(E)(d)$, we get that p does not fork over M_0, as required. $\square_{2.17}$

Remark. So in some sense by 2.17 we can omit in 2.1, the local character Axiom $(E)(c)$ or the continuity Axiom $(E)(h)$ but not both. In fact (under reasonable assumptions) they are equivalent.

2.18 Claim. *In Definition 2.1, Clause (E)(d), i.e., transitivity of non-forking follows from (A),(C),(D)(a),(b),(E)(a),(b),(e),(g).*

Proof. As $\mathfrak{K}_\lambda$ is an λ-a.e.c. with amalgamation, types as well as restriction of types are not only well defined but are "rasonable".

So assume $M_0 \leq_{\mathfrak{s}} M_0' \leq_{\mathfrak{s}} M_0'' \leq_{\mathfrak{s}} M_3$, $a \in M_3$ and $p'' := \mathbf{tp}_{\mathfrak{s}}(a, M_0'', M_3)$ does not fork over M_0' and $p' := \mathbf{tp}_{\mathfrak{s}}(a, M_0', M_3)$ does not fork over M_0. Let $p = p' \restriction M_0$. As p' does not fork over M_0, by Axiom $(E)(a)$ we have $p' \in \mathscr{S}^{\rm bs}(M_0')$ and $p = \mathbf{tp}(a, M_0, M_3) = p' \restriction M_0$ belongs to $\mathscr{S}^{\rm bs}(M_0)$. As p'' does not fork over M_0' clearly $p'' \in \mathscr{S}^{\rm bs}(M_0'')$ and recall $p'' \restriction M_0' = p'$. By the existence axiom $(E)(g)$ the type p has an extension $q'' \in \mathscr{S}^{\rm bs}(M_0'')$ which does not fork over M_0. By the monotonicity Axiom (E)(b) the type q'' does not fork over M_0' and $q' = q'' \restriction M_0'$ does not fork over M_0. As $p', q' \in \mathscr{S}^{\rm bs}(M_0')$ do not fork over M_0 and $p' \restriction M_0 = p = q'' \restriction M_0 = q' \restriction M_0$, by the uniqueness Axiom Ax(E)(e), we have $p' = q'$. Similarly $p'' = q''$, but q'' does not fork over M_0 hence p'' does not fork over M_0 as required. $\square_{2.18}$

2.19 Claim. *1) The symmetry axiom (E)(f) is equivalent to (E)(f)′ below if we assume (A),(B),(C),(D)(a),(b),(E)(a),(b),(g) in Definition 2.1*

$(E)(f)'$ *there are no* $M_\ell (\ell \leq 3)$ *and* $a_\ell (\ell \leq 2)$ *such that*

(a) $M_0 \leq_{\mathfrak{s}} M_1 \leq_{\mathfrak{s}} M_2 \leq_{\mathfrak{s}} M_3$

(b) $\mathbf{tp}(a_\ell, M_\ell, M_{\ell+1})$ *does not fork over* M_0 *for* $\ell = 0, 1, 2$

(c) $\mathbf{tp}_{\mathfrak{s}}(a_0, M_0, M_1) = \mathbf{tp}_{\mathfrak{s}}(a_2, M_0, M_3)$

(d) $\mathbf{tp}_{\mathfrak{s}}(\langle a_0, a_1\rangle, M_0, M_1) \neq \mathbf{tp}_{\mathfrak{s}}(\langle a_2, a_1\rangle, M_0, M_1)$.

Proof. Easy.

* * *

A most interesting case of 2.15 is the following. In particular it tells us that the categoricity assumption is not so rare and it will have essential uses here.

2.20 Claim. *If* $\mathfrak{s} = (\mathfrak{K}, \underset{\lambda}{\bigcup}, \mathscr{S}^{\rm bs})$ *is a good* λ*-frame and* $M \in K_\lambda$ *is a superlimit model in* $\mathfrak{K}_\lambda$ *and we define* $\mathfrak{s}' = \mathfrak{s}^{[M]} = \mathfrak{s}[M] = (\mathfrak{K}[\mathfrak{s}^{[M]}], \underset{\lambda}{\bigcup}[\mathfrak{s}^{[M]}], \mathscr{S}^{\rm bs}[\mathfrak{s}^{[M]}])$ *by*

$$\mathfrak{K}[\mathfrak{s}^{[M]}] = \mathfrak{K}^{[M]},\ \textit{see Definition 1.25 so}\ \mathfrak{K}_{\mathfrak{s}[M]} = \mathfrak{K} \restriction \{N : N \cong M\}$$

$$\underset{\lambda}{\bigcup}[\mathfrak{s}^{[M]}] = \{(M_0, M_1, a, M_3) \in \underset{\lambda}{\bigcup} : M_0, M_1, M_3 \in K_\lambda^{[M]}\}$$

$$\mathscr{S}^{\rm bs}[\mathfrak{s}^{[M]}] = \big\{\mathbf{tp}_{\mathfrak{K}[M]}(a, M_0, M_1) : M_0 \leq_{\mathfrak{K}} M_1, M_0 \in K_\lambda^{[M]}, N \in K_\lambda^{[M]}\ \textit{and}\ \mathbf{tp}_{\mathfrak{K}}(a, M_0, M_1) \in \mathscr{S}^{\rm bs}(M_0)\big\}.$$

Then

(a) $\mathfrak{s}'$ *is a good* λ*-frame*

(b) $\mathfrak{K}[\mathfrak{s}'] \subseteq \mathfrak{K}_{\geq\lambda}[\mathfrak{s}]$

(c) $\leq_{\mathfrak{K}[\mathfrak{s}']} = \leq_{\mathfrak{K}} \restriction K[\mathfrak{s}']$

(d) $K_\lambda[\mathfrak{s}']$ *is categorical.*

Proof. Straight by 1.23, 1.26, 2.15. $\square_{2.20}$

§3 EXAMPLES

We show here that the context from §2 occurs in earlier investigation: in [Sh 88] = Chapter I, [Sh 576] that is Chapter VI, [Sh 48] (and [Sh 87a], [Sh 87b]). Of course, also the class K of models of a superstable (first order) theory T (working in $\mathfrak{C}^{eq}$), with $\leq_{\mathfrak{K}}=\prec$ and $\mathscr{S}^{bs}(M)$ being the set of regular types (when we work in $\mathfrak{C}^{eq}$) <u>or</u> just "the set non-algebraic types" works, with $\bigcup(M_0, M_1, a, M_3)$ iff

$M_0 \leq_{\mathfrak{K}} M_1 \leq_{\mathfrak{K}} M_3$ are in K_λ, $a \in M_3$ and $\mathbf{tp}(a, M_1, M_3) \in \mathscr{S}^{bs}(M_1)$ does not fork over M_0, (in the sense of [Sh:c, III], of course). The reader may concentrate on 3.7 (or 3.4) below for easy life.
Note that 3.4 (or 3.5) will be used to continue [Sh 88] = Chapter I and also to give an alternative proof to the theorem of [Sh 87a], [Sh 87b] + (deducing "there is a model in $\aleph_n$" if there are not too many models in $\aleph_\ell$ for $\ell < n$) and note that 3.5 will be used to continue [Sh 48], i.e., on $\psi \in \mathbb{L}_{\omega_1,\omega}(\mathbf{Q})$ and 3.7 will be used to continue [Sh 576]. Many of the axioms from 2.1 are easy.

(A) The superstable prototype.

3.1 Claim. *Assume T is a first order complete theory and λ be a cardinal $\geq |T| + \aleph_0$; let $\mathfrak{K} = \mathfrak{K}_{T,\lambda} = (K_{T,\lambda} \leq_{\mathfrak{K}_{T,\lambda}})$ be defined by:*

(a) *$K_{T,\lambda}$ is the class of models of T of cardinality $\geq \lambda$*

(b) *$\leq_{\mathfrak{K}_{T,\lambda}}$ is "being an elementary submodel".*

0) $\mathfrak{K}$ is an a.e.c. with $\mathrm{LS}(\mathfrak{K}) = \lambda$.
1) If T is superstable, stable in λ, <u>then</u> $\mathfrak{s} = \mathfrak{s}_{T,\lambda}$ is a good λ-frame when $\mathfrak{s} = (\mathfrak{K}_{T,\lambda}\mathscr{S}^{bs}, \bigcup)$ is defined by:

(c) *$p \in \mathscr{S}^{bs}(M)$ <u>iff</u> $p = \mathbf{tp}_{\mathfrak{K}_{t,\lambda}}(a, M, N)$ for some a, N such that $\mathrm{tp}_{\mathbb{L}(\tau_T)}(a, M, N)$, see Definition 3.2 is a non-algebraic complete 1-type over M, so $M \prec N, a \in N \backslash M$*

(d) *$\bigcup(M_0, M_1, a, M_3)$ iff $M_0 \prec M_1 \prec M_3$ are in $K_{T,\lambda}$ and $a \in M_3$ and $\mathrm{tp}_{\mathbb{L}(\tau_T)}(a, M_1, M_3)$ is a type that does not fork over M_0 in the sense of [Sh:c, III].*

2) Let $\kappa = \mathrm{cf}(\kappa) \le \lambda$. *The model* M *is a* (λ,κ)*-brimmed model for* $\mathfrak{K}_{T,\lambda}$ *iff (i)+(ii) or (i)+(iii) where*

- *(i)* T *is stable in* λ
- *(ii)* $\kappa = \mathrm{cf}(\kappa) \ge \kappa(T)$ *and* M *is a saturated model of* T *of cardinality* λ
- *(iii)* $\kappa = \mathrm{cf}(\kappa) < \kappa(T)$ *and there is a* $\prec$*-increasing continuous sequence* $\langle M_i : i \le \kappa\rangle$ *(by* $\prec$*, equivalently by* $\le_{\mathfrak{s}}$*) such that* $M = M_\kappa$ *and* $(M_{i+1}, c)_{c\in M_i}$ *is saturated for* $i < \kappa$.

2A) So there is a (λ,κ)*-brimmed model for* $\mathfrak{K}_{T,\lambda}$ *iff* T *is stable in* λ.
3) M *is* (λ,κ)*-brimmed over* M_0 *in* $\mathfrak{K}_{T,\lambda}$ *iff* $(M,c)_{c\in M_0}$ *is* (λ,κ)*-brimmed.*
4) Assume T *is superstable first order complete theory stable in* λ *and we define* $\mathfrak{s}^{\mathrm{reg}}_{T,\lambda}$ *as above only* $\mathscr{S}^{\mathrm{bs}}(M)$ *is the set of regular types* $p \in \mathscr{S}_{\mathfrak{K}_T}(M)$ *and we work in* T^{eq}. *Then* $\mathfrak{s}^{\mathrm{reg}}_{T,\lambda}$ *is a good* λ*-frame.*
5) For $\kappa \le \lambda$ *or* $\kappa = \aleph_\varepsilon$ *(abusing notation),* $\mathfrak{s}^{\kappa}_{T,\lambda}$ *is defined similarly restricting ourselves to* $\mathbf{F}^a_\kappa$*-saturated models. (Let* $\mathfrak{s}^0_{t,\lambda} = \mathfrak{s}_{T,\lambda}$*.) If* T *is superstable, stable in* λ *then* $\mathfrak{s}^{\kappa}_{T,\lambda}$ *is a good* λ *frame.*

Remark. We can replace (c) of 3.1 by:

- (c)′ $p \in \mathscr{S}^{\mathrm{bs}}(M)$ iff $p = \mathbf{tp}_{\mathfrak{K}_{T,\lambda}}(a,M,N)$ for some a, N such that $\mathrm{tp}_{\mathbb{L}(\tau_T)}(a,M,N)$ is a complete 1-type over M

except that clause (D)(b) of Definition 2.1 fail. In fact the proofs are easier in this case; of course, the two meaning of types essentially agree.

Proof. 0),1),2),2A),3) Obvious (see [Sh:c]).
4) As in (1), except density of regular types which holds by [HuSh 342].
5) Also by [Sh:c]. $\square_{3.1}$

Recall

3.2 Definition. 1) For a logic $\mathscr{L}$ and vocabulary $\tau, \mathscr{L}(\tau)$ is the set of $\mathscr{L}$-formulas in this vocabulary.
2) $\mathbb{L} = \mathbb{L}_{\omega,\omega}$ is first order logic.
3) A theory in $\mathscr{L}(\tau)$ is a set of sentences from $\mathscr{L}(\tau)$ which we assume has a model if not said otherwise. Similarly in a language $L(\subseteq \mathscr{L}(\tau))$

Very central in Chapter I (and Chapter IV) but peripheral here (except when in (parts of) §3 we continue Chapter I in our framework) is:

3.3 Definition. Let T_1 be a theory in $\mathbb{L}(\tau_1), \tau \subseteq \tau_1$ vocabularies, Γ a set of types in $\mathbb{L}(\tau_1)$; (i.e. for some m, a set of formulas $\varphi(x_0, \ldots, x_{m-1}) \in \mathbb{L}(\tau_1)$).
1) $\mathrm{EC}(T_1, \Gamma) = \{M : M$ a τ_1-model of T_1 which omits every $p \in \Gamma\}$. So without loss of generality τ_1 is reconstructible from $T_1, \Gamma)$ and $\mathrm{PC}_\tau(T_1, \Gamma) = \mathrm{PC}(T_1, \Gamma, \tau) = \{M : M$ is a τ-reduct of some $M_1 \in \mathrm{EC}(T_1, \Gamma)\}$.

2) We say that $\mathfrak{K}$ is PC^μ_λ or $\mathrm{PC}_{\lambda,\mu}$ <u>if</u> for some $T_1, T_2, \Gamma_1, \Gamma_2$ and τ_1 and τ_2 we have: (T_ℓ a first order theory in the vocabulary τ_ℓ, Γ_ℓ a set of types in $\mathbb{L}(\tau_\ell)$ and) $K = \mathrm{PC}(T_1, \Gamma_1, \tau_{\mathfrak{K}})$ and $\{(M, N) : M \le_{\mathfrak{K}} N$ and $M, N \in K\} = \mathrm{PC}(T_2, \Gamma_2, \tau')$ where
$\tau' = \tau_{\mathfrak{K}} \cup \{P\}$, ($P$ a new one place predicate and (M, N) means the τ'-model N^+ expanding N where $P^{N^+} = |M|$) and $|T_\ell| \le \lambda, |\Gamma_\ell| \le \mu$ for $\ell = 1, 2$.
3) If $\mu = \lambda$, we may omit μ.

(B) An abstract elementary class which is $\mathrm{PC}_{\aleph_0}$.

3.4 Theorem. *Assume* $2^{\aleph_0} < 2^{\aleph_1}$ *and consider the statements*

(α) $\mathfrak{K}$ *is an abstract elementary class with* $\mathrm{LS}(\mathfrak{K}) = \aleph_0$ *(the last phrase follows by clause* (β)*) and* $\tau = \tau(\mathfrak{K})$ *is countable*

(β) $\mathfrak{K}$ *is* $\mathrm{PC}_{\aleph_0}$, *equivalently for some sentences* $\psi_1, \psi_2 \in \mathbb{L}_{\omega_1,\omega}(\tau_1)$ *where* τ_1 *is a countable vocabulary extending* τ *we have*

$$K = \{M_1 \restriction \tau : M_1 \text{ a model of } \psi_1\}$$
$$\{(N, M) : M \le_{\mathfrak{K}} N\} = \{(N_1 \restriction \tau, M_1 \restriction \tau) : (N_1, M_1) \text{ a model of } \psi_2\}$$

(γ) $1 \leq \dot{I}(\aleph_1, \mathfrak{K}) < 2^{\aleph_1}$

(δ) *$\mathfrak{K}$ is categorical in $\aleph_0$, has the amalgamation property in $\aleph_0$ and is stable in $\aleph_0$*

(δ)$^-$ *like (δ) but "stable in $\aleph_0$" is weakened to: $M \in \mathfrak{K}_{\aleph_0} \Rightarrow |\mathscr{S}(M)| \leq \aleph_1$*

(ε) *all models of $\mathfrak{K}$ are $\mathbb{L}_{\infty,\omega}$-equivalent and $M \leq_{\mathfrak{K}} N \Rightarrow M \prec_{\mathbb{L}_{\infty,\omega}} N$.*

For $M \in \mathfrak{K}_{\aleph_0}$ we define $\mathfrak{K}'_M$ as follows: the class of members is $\{N \in K : N \equiv_{\mathbb{L}_{\infty,\omega}} M\}$ and $N_1 \leq_{\mathfrak{K}'_M} N_2$ iff $N_1 \leq_{\mathfrak{K}} N_2$ & $N_1 \prec_{\mathbb{L}_{\infty,\omega}} N_2$.

1) Assume $(\alpha) + (\beta) + (\gamma)$, <u>then</u> for some $M \in \mathfrak{K}_{\aleph_0}$ the class $\mathfrak{K}'_M$ satisfies $(\alpha) + (\beta) + (\gamma) + (\delta)^- + (\varepsilon)$; in fact any $M \in \mathfrak{K}_{\aleph_0}$ such that $(\mathfrak{K}'_M)_{\aleph_1} \neq \emptyset$ will do and there are such $M \in K_{\aleph_0}$. Moreover, if $\mathfrak{K}$ satisfies (δ) then also $\mathfrak{K}'_M$ satisfies it; also trivially $K'_M \subseteq K$ and $\leq_{\mathfrak{K}'_M} \subseteq \leq_{\mathfrak{K}}$.
1A) Also there is $\mathfrak{K}'$ such that: $\mathfrak{K}'$ satisfies $(\alpha)+(\beta)+(\gamma)+(\delta)+(\varepsilon)$, and for every μ we have $K'_\mu \subseteq K_\mu$. In fact, in the notation of I.5.12 for every $\alpha < \omega_1$ we can choose $\mathfrak{K}' = \mathfrak{K}_{\mathbf{D}_\alpha}$.
2) Assume $(\alpha)+(\beta)+(\gamma)+(\delta)$. <u>Then</u> $(\mathfrak{K}, \bigcup, \mathscr{S}^{\mathrm{bs}})$ is a good $\aleph_0$-frame for some $\bigcup$ and $\mathscr{S}^{\mathrm{bs}}$.

3) In fact, in part (2) we can choose $\mathscr{S}^{\mathrm{bs}}(M) = \{p \in \mathscr{S}(M) : p$ not algebraic$\}$ and $\bigcup$ is defined by I.5.19 (the definable extensions).

Remark. 1) In I.5.34 we use the additional assumption $\dot{I}(\aleph_2, K) < \mu_{\mathrm{unif}}(\aleph_2, 2^{\aleph_1})$. But this Theorem is not used here!
2) Note that $\mathfrak{K}'_M$ is related to $K^{[M]}$ from Definition 1.25 but is different.
3) In the proof we relate the types in the sense of $\mathscr{S}_{\mathfrak{s}}(M)$, and those in I§5. Now in I§5 we have lift types, from $\mathfrak{K}_{\aleph_0}$ to any $\mathfrak{K}_\mu$, i.e., define $\mathbf{D}(N)$ for $N \in \mathfrak{K}_\mu$. In $\mu > \aleph_0$, in general we do not know how to relate them to types $\mathscr{S}_{\mathfrak{K}_{\mathfrak{s}}}(N)$. But when $\mathfrak{s}^+$ is defined (in the "successful" cases, see §8 here and III§1) we can get the parallel claim.

Discussion: 1) What occurs if we do not pass in 3.4 to the case "$\mathbf{D}(N)$ countable for every $N \in K_{\aleph_0}$"? If we still assume "$\mathfrak{K}$ categorical in $\aleph_0$" then as $|\mathbf{D}(N_0)| \le \aleph_1$, if we assume "there is a superlimit model in $\mathfrak{K}_{\aleph_1}$" we can find a good $\aleph_1$-frame $\mathfrak{s}$; this assumption is justified by I.5.34, I.5.39.

Proof. 1) Note that for any $M \in K_{\aleph_0}$, the class $\mathfrak{K}'_M$ satisfies $(\alpha), (\beta), (\varepsilon)$ and it is categorical in $\aleph_0$ and $(K'_M)_\mu \subseteq K_\mu$ hence $\dot{I}(\mu, K'_M) \le \dot{I}(\mu, K)$. By Theorem I.3.10, (note: if you use the original version (i.e., [Sh 88]) by its proof or use it and get a less specified class with the desired properties) for some $M \in K_{\aleph_0}$ we have $(\mathfrak{K}'_M)_{\aleph_1} \neq \emptyset$. By I.3.8 we get that $\mathfrak{K}'_M$ has amalgamation in $\aleph_0$ and by Chapter I almost we get that in $\mathfrak{K}'_M$ the set $\mathscr{S}(M)$ is of small cardinality ($\le \aleph_1$); be careful - the types there are defined differently than here, but by the amalgamation (in $\aleph_0$) and the omitting types theorem in this case they are the same, see more in the proof of part (3) below. So by I.5.2,I.5.4 we have $M \in (\mathfrak{K}'_\mu)_{\aleph_0} \Rightarrow |\mathscr{S}_{\mathfrak{K}'_\mu}(M)| \le \aleph_1$.
Also the second sentence in (1) is easy.
1A) Use I.5.28,I.5.29.
In more detail, (but not much point in reading without some understanding of I§5, however we should not use I.5.34 as long as we do not strengthen our assumptions) by part (1) we can assume that clauses $(\delta)^- + (\varepsilon)$ hold. (Looking at the old version [Sh 88] of Chapter I remember that there $\prec$ means $\le_{\mathfrak{K}}$.) We can find $\mathbf{D}_* = \mathbf{D}^*_\alpha, \alpha < \omega_1$, which is a good countable diagram (see Definition I.5.10 and Fact I.5.12 or I.5.22, I.5.27. So in particular (give the non-maximality of models below) such that for some countable $M_0 <_{\mathfrak{K}} M_1 <_{\mathfrak{K}} M_2$ we have M_m is $(\mathbf{D}^*(M_\ell), \aleph_0)$-homogeneous for $\ell < m \le 2$. In I.5.28 we define $(K_{\mathbf{D}_*}, \le_{\mathbf{D}_*})$. By I.5.29 the pair $(K_{\mathbf{D}_*}, \le_{\mathbf{D}_*})$ is an abstract elementary class (the choice of $\mathbf{D}_*$ a part, e.g. transitivity = Axiom II which holds by the existence of the M_ℓ's above and I.5.22) categorical in $\aleph_0$ and no maximal countable model (by $\le_{\mathbf{D}_*}$, see I.5.12(2). Now $\aleph_0$-stability holds by I.5.29(2) and the equality of the three definitions of types in the proof of parts (2),(3) and $K_{\mathbf{D}_*} \subseteq K$ so we are done by part 3) below.
2),3) The first part of the proof serves also part (1) of the theorem so we assume $(\delta)^-$ instead of (δ). We should be careful: the notion

of type has three relevant meanings here. For $N \in K_{\aleph_0}$ the three definitions for $\mathscr{S}^{<\omega}(N)$ and of $\mathrm{tp}(\bar{a}, N, M)$ when $\bar{a} \in {}^{\omega>}M, N \leq_{\mathfrak{K}} M \in K_{\aleph_0}$ (of course we can use just 1-types) are:

(α) the one we use here (recall 1.9) which uses elementary mappings; for the present proof we call them $\mathscr{S}_0^{<\omega}(M), \mathrm{tp}_0(\bar{a}, M, N)$

(β) $\mathbf{S}_1(N)$ which is (recall: materialzie is close to but different from realize)

$$\begin{aligned}\mathbf{D}(N) = \{p : \ & p \text{ a complete } \mathbb{L}^0_{\aleph_1,\aleph_0}(N)\text{-type over } N \\ & \text{(so in each formula only finitely many parameters} \\ & \text{from } N \text{ appear)} \\ & \text{such that for some } M, \bar{a} \in {}^{\omega>}M, \\ & \bar{a} \text{ materializes } p \text{ in } (M, N)\}\end{aligned}$$

("materializing a type" is defined in I.4.3(2)) so

$$\mathbf{S}_1(N) = \{\mathrm{tp}_1(\bar{a}, N, M) : \bar{a} \in {}^{\omega>}M \text{ and } N \leq_{\mathfrak{K}} M \in K_{\aleph_0}\}$$

where

$$\mathrm{tp}_1(\bar{a}, N, M) = \{\varphi(\bar{x}) \in \mathbb{L}^0_{\aleph_1,\aleph_0}(N) : M \Vdash^{\aleph_1}_{\mathfrak{K}} \varphi(\bar{a})\}$$

(see I.4.3(1) on the meaning of this forcing relation).

(γ) $\mathbf{S}_2(N)$ which is

$$\begin{aligned}\mathbf{D}^*(N) = \{p : \ & p \text{ a complete } \mathbb{L}^0_{\aleph_1,\aleph_0}(N; N)\text{-type over } N \\ & \text{(so in each formula all members of } N \text{ may appear)} \\ & \text{such that for some } M \in K_{\aleph_0} \text{ and} \\ & \bar{a} \in {}^{\omega>}M \text{ satisfying } N \leq_{\mathfrak{K}} M \text{ the sequence} \\ & \bar{a} \text{ materializes } p \text{ in } (M, N)\}\end{aligned}$$

so

$$\mathbf{S}_2(N) = \{\mathrm{tp}_2(\bar{a}, N, M) : \bar{a} \in {}^{\omega>}M \text{ and } N \leq_{\mathfrak{K}} M \in K_{\aleph_0}\}$$

$$\mathrm{tp}_2(\bar{a}, N, M) = \{\varphi(\bar{x}) \in \mathbb{L}^0_{\aleph_1,\aleph_0}(N, N) : M \Vdash^{\aleph_1}_{\mathfrak{K}} \varphi(\bar{a})\}.$$

As we have amalgamation in $K_{\aleph_0}$ it is enough to prove for $\ell, m < 3$ that

$(*)_{\ell,m}$ if $k < \omega, N \leq_{\mathfrak{K}} M \in K_{\aleph_0}$ and $\bar{a}, \bar{b} \in {}^k M$, then
$\mathrm{tp}_\ell(\bar{a}, N, M) = \mathrm{tp}_\ell(\bar{b}, N, M) \Rightarrow \mathrm{tp}_m(\bar{a}, N, M) = \mathrm{tp}_m(\bar{b}, N, M)$.

Now $(*)_{2,1}$ holds trivially (more formulas) and $(*)_{1,2}$ holds by I.5.9. By amalgamation in $\mathfrak{K}_{\aleph_0}$, if $\mathrm{tp}_0(\bar{a}, N, M) = \mathrm{tp}_0(\bar{b}, N, M)$, then for some $M', M \leq_{\mathfrak{K}} M' \in K_{\aleph_0}$ there is an automorphism f of M' over N such that $f(\bar{a}) = \bar{b}$, so trivially $(*)_{0,1}, (*)_{0,2}$ hold (we use the facts that $\mathrm{tp}_\ell(\bar{a}, N, M)$ is preserved by isomorphism and by replacing M by M_1 if $M \leq_{\mathfrak{K}} M_2 \in K_{\aleph_0}$ and $N \cup \bar{a} \subseteq M_1 \leq_{\mathfrak{K}} M_2$). Lastly we prove $(*)_{2,0}$.
So $N \leq_{\mathfrak{K}} M \in K_{\aleph_0}$, hence $\mathrm{tp}_2(\bar{c}, N, M) : \bar{c} \in {}^{\omega>}M\} \subseteq \mathbf{D}^*(N)$ is countable so by I.5.12(b),(c) for some countable $\alpha < \omega_1$ we have $\{\mathrm{tp}_2(\bar{c}, N, M) : \bar{c} \in {}^{\omega>}M\} \subseteq \mathbf{D}^*_\alpha(N)$. Now there is $M' \in K_{\aleph_0}$ such that $M \leq_{\mathfrak{K}} M', M'$ is $(\mathbf{D}^*_\alpha, \aleph_0)^*$-homogeneous (by I.5.12(e) see Definition I.5.14) hence M' is $(\mathbf{D}^*_\alpha(N), \aleph_0)^*$- homogeneous (by I.5.12(f)), and $\mathrm{tp}_2(\bar{a}, N, M') = \mathrm{tp}_2(\bar{b}, N, M')$ by I.5.7(3), (N here means N_0 there, that is increasing the model preserve the type).
Lastly by Definition I.5.14 there is an automorphism f of M' over N mapping $\bar{a}$ to $\bar{b}$, so we have proved $(*)_{2,0}$, so the three definitions of type are equivalent.

Now we define for $M \in K_{\aleph_0}$:

(a) $\mathscr{S}^{\mathrm{bs}}(M) = \{p \in \mathscr{S}_{\mathfrak{K}}(M) : p \text{ not algebraic}\}$

(b) for $M_0, M_1, M_3 \in K_{\aleph_0}$ and an element $a \in M_3$ we define:
$\bigcup(M_0, M_1, a, M_3)$ <u>iff</u> $M_0 \leq_{\mathfrak{K}} M_1 \leq_{\mathfrak{K}} M_3$ and $a \in M_3 \backslash M_1$ and

$\mathrm{tp}_1(a, M_1, M_3) (= \mathrm{gtp}(a, M_1, M_3)$ in Chapter I's notation) is definable over some finite $\bar{b} \in {}^{\omega>}M_0$ (equivalently is preserved by every automorphism of M_1 over $\bar{b}$ (see I.5.19) equivalently $\mathrm{gtp}(a, M_1, M_3)$ is the stationarization of $\mathrm{gtp}(a, M_0, M_3)$.

Now we should check the axioms from Definition 2.1.

<u>Clause (A)</u>: By clause (α) of the assumption.

Clauses (B),(C): By clause (δ) or $(\delta)^-$ of the assumption except "the superlimit $M \in K_{\aleph_0}$ is not $\leq_{\mathfrak{K}}$-maximal" which holds by clause $(\gamma) + (\delta)$ or $(\gamma) + (\delta)^-$.

Clause (D): By the definition (note that about clause (d), bs-stability, that it holds by assumption (δ), and about clause (c), i.e., the density is trivial by the way we have defined $\mathscr{S}^{\rm bs}$).

Subclause (E)(a): By the definition.

Subclause (E)(b)(monotonicity):

Let $M_0 \leq_{\mathfrak{K}} M_0' \leq_{\mathfrak{K}} M_1' \leq_{\mathfrak{K}} M_1 \leq_{\mathfrak{K}} M_3 \leq M_3'$ be all in $\mathfrak{K}_{\aleph_0}$ and assume $\bigcup(M_0, M_1, a, M_3)$. So $M_0' \leq_{\mathfrak{K}} M_1' \leq_{\mathfrak{K}} M_3 \leq M_3'$ and $a \in M_3 \backslash M_1 \subseteq M_3' \backslash M_1'$. Now by the assumption and the definition of $\bigcup$, for some $\bar{b} \in {}^{\omega >}(M_0)$, $\mathrm{gtp}(a, M_1, M_3)$ is definable over $\bar{b}$. So the same holds for $\mathrm{gtp}(a, M_1', M_3)$ by I.5.24, in fact (with the same definition) and hence for $\mathrm{gtp}(a, M_1', M_3') = \mathrm{gtp}(a, M_1', M_3)$ by I.5.7(3), so as $\bar{b} \in {}^{\omega >}(M_0) \subseteq {}^{\omega >}(M_0')$ we have gotten $\bigcup(M_0', M_1', a, M_3')$.

For the additional clause in the monotoncity Axiom, assume in addition $M_1' \cup \{a\} \subseteq M_3'' \leq_{\mathfrak{K}} M_3'$ again by I.5.7(3) clearly $\mathrm{gtp}(a, M_1', M_3'') = \mathrm{gtp}(a, M_1', M_3')$, so (recalling the beginning of the proof) we are done.

Sublcause (E)(c)(local character):

So let $\langle M_i : i \leq \delta + 1 \rangle$ be $\leq_{\mathfrak{K}}$-increasing continuous in $K_{\aleph_0}$ and $a \in M_{\delta+1}$ and $\mathbf{tp}(a, M_\delta, M_{\delta+1}) \in \mathscr{S}^{\rm bs}(M_\delta)$, so $a \notin M_\delta$ and $\mathrm{gtp}(a, M_\delta, M_{\delta+1})$ is definable over some $\bar{b} \in {}^{\omega >}(M_\delta)$ by I.5.6.
As $\bar{b}$ is finite, for some $\alpha < \delta$ we have $\bar{b} \subseteq M_\alpha$, hence we have $(\mathbf{tp}(a, M_\beta, M_{\delta+1}) \in \mathscr{S}^{\rm bs}(M_\beta)$ trivially and) $\mathbf{tp}(a, M_\delta, M_{\delta+1})$ does not fork over M_β.

Sublcause (E)(d)(transitivity):

By I.5.24(2) or even better I.5.22.

Subclause (E)(e)(uniqueness):

Holds by the Definition I.5.19.

Subclause (E)(f)(symmetry):

By I.5.30 + uniqueness we get (E)(f). Actually I.5.30 gives this more directly.

Subclause (E)(g)(extension existence):

By I.5.19 (i.e., by I.5.6 + all $M \in K_{\aleph_0}$ are $\aleph_0$-homogeneous). Alternatively, see I.5.26.

Subclause (E)(h)(continuity):

Suppose $\langle M_\alpha : \alpha \leq \delta \rangle$ is $\leq_{\mathfrak{K}}$- increasingly continuous, $M_\alpha \in K_{\aleph_0}, \delta < \omega_1, p \in \mathscr{S}(M_\delta)$ and $\alpha < \delta \Rightarrow p \restriction M_\alpha$ does not fork over M_0. Now we shall use (E)(c)+(E)(d). As $p \restriction M_\alpha \in \mathscr{S}^{\rm bs}(M_\alpha)$ clearly $p \restriction M_\alpha$ is not realized in M_α hence p is not realized in M_α; as $M_\delta = \bigcup\limits_{\alpha<\delta} M_\alpha$ necessarily p is not realized in M_δ, hence p is not algebraic.

So $p \in \mathscr{S}^{\rm bs}(M_\delta)$. For some finite $\bar{b} \in {}^{\omega>}(M_\delta), p$ is definable over $\bar{b}$, let $\alpha < \delta$ be such that $\bar{b} \in {}^{\omega>}(M_\alpha)$, so as in the proof of (E)(c), (or use it directly) the type p does not fork over M_α. As $p \restriction M_\alpha$ does not fork over M_0, by (E)(d) we get that p does not fork over M_0 as required. Actually we can derive (E)(h) by 2.17.

Subclause (E)(i)(non-forking amalgamation):

One way is by I.5.30; (note that in I.5.34 we get more, but assuming, by our present notation $\dot{I}(\aleph_2, K) < \mu_{\rm wd}(\aleph_2)$); but another way is just to use 2.16.

$\square_{3.4}$

(C) The uncountable cardinality quantifier case, $\mathbb{L}_{\omega_1,\omega}(\mathbf{Q})$.

Now we turn to sentences in $\mathbb{L}_{\omega_1,\omega}(\mathbf{Q})$.

3.5 Conclusion. Assume $\psi \in \mathbb{L}_{\omega_1,\omega}(\mathbf{Q})$ and $1 \leq \dot{I}(\aleph_1, \psi) < 2^{\aleph_1}$ and $2^{\aleph_0} < 2^{\aleph_1}$.

Then for some abstract elementary classes $\mathfrak{K}, \mathfrak{K}^+$ (note $\tau_\psi \subset \tau_{\mathfrak{K}} = \tau_{\mathfrak{K}^+}$) we have:

(*a*) $\mathfrak{K}$ satisfies $(\alpha), (\beta), (\delta), (\varepsilon)$ from 3.4 with $\tau_{\mathfrak{K}} \supseteq \tau_\psi$ countable (for (γ), (*b*) is a replacement)

(*b*) for every $\mu > \aleph_0, \dot{I}(\mu, \mathfrak{K}(\aleph_1\text{-saturated})) \leq \dot{I}(\mu, \psi)$, where[12] "$\aleph_1$-saturated" is well defined as $\mathfrak{K}_{\aleph_0}$ has amalgamation, see 1.14

[12]much less than saturation suffice, like "obeying" $<^{**}$

(c) for some $\bigcup, \mathscr{S}^{\rm bs}$ (and $\lambda = \aleph_0$), the triple $(\mathfrak{K}, \bigcup, S^{\rm bs})$ is as in 3.4(2) so is a good $\aleph_0$-frame

(d) every $\aleph_1$-saturated member of $\mathfrak{K}$ belongs to $\mathfrak{K}^+$ and there is an $\aleph_1$-saturated member of $\mathfrak{K}$ (and naturally it is uncountable, even of cardinality $\aleph_1$)

(e) $\mathfrak{K}^+$ is an a.e.c., has LS number $\aleph_1$ and $\{M \restriction \tau_\psi : M \in \mathfrak{K}^+\} \subseteq \{M : M \models \psi\}$ and every τ-model M of ψ has a unique expansion in $\mathfrak{K}^+$ hence $\mu \geq \aleph_1 \Rightarrow \dot{I}(\mu, \psi) = \dot{I}(\mu, \mathfrak{K}^+)$ and $\mathfrak{K}^+$ is the class of models of some complete $\psi \in \mathbb{L}_{\omega_1,\omega}(\mathbf{Q})$.

Proof. Essentially by [Sh 48] and 3.4.

I feel that upon reading [Sh 48] the proof should not be inherently difficult, much more so having read 3.4, but will give full details. Recall $\mathrm{Mod}(\psi)$ is the class of τ_ψ-models of ψ. We can find a countable fragment $\mathscr{L}$ of $\mathbb{L}_{\omega_1,\omega}(\mathbf{Q})(\tau_\psi)$ to which ψ belongs and a sentence $\psi_1 \in \mathscr{L} \subseteq \mathbb{L}_{\omega_1,\omega}(\mathbf{Q})(\tau_\psi)$ such that ψ_1 is "nice" for [Sh 48, Definition 3.1,3.2], [Sh 48, Lemma 3.1]

$\circledast_1$(a) ψ_1 has uncountable models

(b) $\psi_1 \vdash \psi$, i.e., every model of ψ_1 is a model of ψ

(c) ψ_1 is $\mathbb{L}_{\omega_1,\omega}(\mathbf{Q})$-complete

(d) every model $M \models \psi_1$ realizes just countably many complete $\mathbb{L}_{\omega_1,\omega}(\mathbf{Q})(\tau_\psi)$-types (of any finite arity, over the empty set), each isolated by a formula in $\mathscr{L}$.

The proof of $\circledast_1(d)$ is sketched in Theorem 2.5 of [Sh 48]. The reference to Keisler [Ke71] is to the generalization of theorems 12 and 28 of Keisler's book from $\mathbb{L}_{\omega_1,\omega}$ to $\mathbb{L}_{\omega_1,\omega}(\mathbf{Q})$, see I.0.2.
Let

$\circledast_2$ (*i*) $\mathfrak{K}_0 = (\mathrm{Mod}(\psi), \prec_{\mathscr{L}})$,

(*ii*) $\mathfrak{K}_1 = (\mathrm{Mod}(\psi_1), \prec_{\mathscr{L}})$

$\circledast_3$ $\mathfrak{K}_\ell$ is an a.e.c. with L.S. number $\aleph_1$ for $\ell = 0, 1$.

Toward defining $\mathfrak{K}$, let $\tau_{\mathfrak{K}} = \tau_\psi \cup \{R_{\varphi(\bar{x})} : \varphi(\bar{x}) \in \mathscr{L}\}$, $R_{\varphi(\bar{x})}$ a new $\ell g(\bar{x})$-predicate and let $\psi_2 = \psi_1 \wedge \{(\forall \bar{y})(R_{\varphi(\bar{x})}(\bar{y}) = \varphi(\bar{y}) : \varphi(\bar{x}) \in \mathbb{L}\}$. For every $M \in \mathrm{Mod}(\psi)$ we define M^+ by

$\circledast_4$ M^+ is M expanded to a $\tau_{\mathfrak{K}}$-model by letting $R^{M^+}_{\varphi(\bar{x})} = \{\bar{a} \in {}^{\ell g(\bar{x})}M : M \models \varphi[\bar{a}]\}$

$\circledast_5$ (a) $\mathfrak{K}^+_0 = (\{M^+ : M \in \text{Mod}(\psi)\}, \prec_{\mathbb{L}})$ is an a.e.c. with $\text{LS}(\mathfrak{K}^+_0) = \aleph_1$

(b) $\mathfrak{K}^+_1 = (\{M^+ : M \in \text{Mod}(\psi_1)\}, \prec_{\mathbb{L}})$ is an a.e.c. with $\text{LS}(\mathfrak{K}^+) = \aleph_1$.

Clearly

$\circledast_6$ if $M \models \psi_1$ then M^+ is an atomic model of the complete first-order theory T_{ψ_1} where T_{ψ_1} is the set of first order consequences in $\mathbb{L}(\tau_{\mathfrak{K}})$ of ψ_2.

So it is natural to define $\mathfrak{K}$:

$\circledast_7$(a) $N \in \mathfrak{K}$ iff

(α) N is a $\tau_{\mathfrak{K}}$-model which is an atomic model of T_{ψ_1}

(β) if $\psi_1 \vdash (\forall \bar{x})[\varphi_1(\bar{x}) = (\mathbf{Q}y)\varphi_2(y, \bar{x})]$ and $\varphi_1, \varphi_2 \in \mathscr{L}$ and $N \models \neg R_{\varphi_1(\bar{x})}[\bar{a}]$ then $\{b \in N : N \models R_{\varphi_2(y,\bar{x})}(b, \bar{a})\}$ is countable

(b) $N_1 \leq_{\mathfrak{K}} N_2$ <u>iff</u> $(N_1, N_2 \in K, N_1 \prec_{\mathbb{L}} N_2$ equivalently $N_1 \subseteq N_2$ and) for $\varphi_1(\bar{x}), \varphi_2(y, \bar{x})$ as in subclause (β) of clause (a) above, if $\bar{a} \in {}^{\ell g(\bar{x})}(N_1)$, $N_1 \models \neg R_{\varphi_1(\bar{x})}[\bar{a}]$ and $b \in N_2 \backslash N_1$ then $N_2 \models \neg R_{\varphi_2(y,\bar{x})}[b, \bar{a}]$.

Observe

$\circledast_8$ $N \in \mathfrak{K}$ iff N is an atomic $\tau_{\mathfrak{K}}$-model of the first order $\mathbb{L}(\tau_{\mathfrak{K}})$-consequences ψ_2 (i.e. of ψ and every $\tau_{\mathfrak{K}}$ sentence of the form $\forall \bar{x}[R_\varphi(\bar{x}) \equiv \varphi(\bar{x})]$) and clause ($\beta$) of $\circledast_7(a)$ holds

$\circledast_9$ $\mathfrak{K}$ is an a.e.c. with $\text{LS}(\mathfrak{K}) = \aleph_0$ and is $\text{PC}_{\aleph_0}$, $\mathfrak{K}$ is categorical in $\aleph_0$ (and $\leq_{\mathfrak{K}}$ is called $\leq^*$ in [Sh 48, Definition 3.3]).

Note that $\mathfrak{K}_1, \mathfrak{K}^+_1$ has the same number of models, but $\mathfrak{K}$ has "more models" than $\mathfrak{K}^+_1$, in particular, it has countable members and $\mathfrak{K}_0$ has at least as many models as $\mathfrak{K}_1$. For $N \in \mathfrak{K}$ to be in $\mathfrak{K}^+_1 = \{M^+ : M \in \text{Mod}(\psi_1)\}$ what is missing is the other implications in $\circledast_7(a)(\beta)$.

This is very close to 3.4, but $\mathfrak{K}$ may have many models in $\aleph_1$ (as $\mathbf{Q}$ is not necessarily interpreted as expected). However,

$\circledast_{10}$ constructing $M \in K_{\aleph_1}$ by the union as $\leq_{\mathfrak{K}}$-increasing continuous chain $\langle M_i : i < \omega_1 \rangle$, to make sure $M \in \mathfrak{K}_1^+$ it is enough that for unboundedly many $\alpha < \omega_1, M_\alpha <^{**} M_{\alpha+1}$ and $(\forall M \in \mathfrak{K}_{\aleph_0})(\exists N \in \mathfrak{K}_{\aleph_0})(M <^{**} N)$ where

$\circledast_{11}$ for $M, N \in \mathfrak{K}, M <^{**} N$ iff

(*i*) $M \leq_{\mathfrak{K}} N$

(*ii*) in $\circledast_7(b)$ also the inverse direction holds.

Does $\mathfrak{K}$ have amalgamation in $\aleph_0$? Now [Sh 48, Lemma 3.4], almost says this but it assumed $\diamondsuit_{\aleph_1}$ instead of $2^{\aleph_0} < 2^{\aleph_1}$; and I.3.8 almost says this, but the models are from $\mathfrak{K}_{\aleph_1}$ rather than $\mathfrak{K}_{\aleph_1}^+$ but I.3.16 fully says this using the so called $K_{\aleph_1}^{\mathbf{F}}$, see Definition I.3.14 and using $\mathbf{F}$ such that $M \in \mathfrak{K}_{\aleph_0} \Rightarrow M <^{**} \mathbf{F}(N) \in \mathfrak{K}_{\aleph_0}$; or pedantically $\mathbf{F} = \{(M, N) : M <^{**} N \text{ are from } \mathfrak{K}\}$. So

$\circledast_{12}$ $\mathfrak{K}$ has the amalgamation property in $\aleph_0$.

It should be clear by now that we have proved clauses (a),(b),(d),(e) of 3.5 using $\mathfrak{K}$. We have to prove clause (c); we cannot quote 3.4 as clause (γ) there is only almost true. The proof is similar to (but simpler than) that of 3.4 quoting [Sh 48] instead of Chapter I; a marked difference is that in the present case the number of types over a countable model is countable (in $\mathfrak{K}$) whereas in Chapter I it seemingly could be $\aleph_1$, generally [Sh 48] situation is more similar to the first order logic case.

Recall that all models from $\mathfrak{K}$ are atomic (in the first order sense) and we shall use below $\text{tp}_{\mathbb{L}}$.

As $\mathfrak{K}$ has $\aleph_0$-amalgamation (by $\circledast_{12}$), clearly [Sh 48, §4] applies; now by [Sh 48, Lemma 2.1](B) + Definition 3.5, being $(\aleph_0, 1)$-stable as defined in [Sh 48, Definition 3.5](A) holds. Hence all clauses of [Sh 48, Lemma 4.2] hold, in particular $((D)(\beta)$ there and clause (A), i.e., [Sh 48, Def.3.5](B)), so

⊛$_{13}$ (*i*) if $M \leq_{\mathfrak{K}} N$ and $\bar{a} \in N$ then $\mathrm{tp}_{\mathbb{L}}(\bar{a}, M, N)$ is definable over a finite subset of M

(*ii*) if $M \in \mathfrak{K}_{\aleph_0}$ then $\{\mathrm{tp}_{\mathbb{L}}(\bar{a}, M, N) : \bar{a} \in {}^{\omega>}N$ and $M \leq_{\mathfrak{K}} N\}$ is countable.

By [Sh 48, Lemma 4.4] it follows that

⊛$_{14}$ if $M \leq_{\mathfrak{K}} N$ are countable and $\bar{a} \in M$ then $\mathrm{tp}_{\mathbb{L}}(\bar{a}, M, N)$ determine $\mathrm{tp}_{\mathfrak{K}}(\bar{a}, M, N)$.

Now we define $\mathfrak{s} = (\mathfrak{K}_{\aleph_0}, \mathscr{S}^{\mathrm{bs}}, \bigcup)$ by

⊛$_{15}$ $\mathscr{S}^{\mathrm{bs}}(M) = \{\mathbf{tp}_{\mathfrak{K}}(\bar{a}, M, N) : M \leq_{\mathfrak{K}} N$ are countable and $\bar{a} \in {}^{\omega>}N$ but $\bar{a} \notin {}^{\omega>}M\}$

⊛$_{16}$ $\mathbf{tp}_{\mathfrak{K}}(\bar{a}, M_1, M_3)$ does not fork over M_0 where $M_0 \leq_{\mathfrak{K}} M_1 \leq_{\mathfrak{K}} M_3 \in \mathfrak{K}_{\aleph_0}$ iff $\mathrm{tp}_{\mathbb{L}}(\bar{a}, M_1, M_3)$ is definable over some finite subset of M_0.

Now we check "$\mathfrak{s}$ is a good frame", i.e., all clauses of Definition 2.1.

<u>Clause (A)</u>: By ⊛$_9$ above.

<u>Clause (B)</u>: As $\mathfrak{K}$ is categorical in $\aleph_0$, has an uncountable model and $\mathrm{LS}(\mathfrak{K}) = \aleph_0$ this should be clear.

<u>Clause (C)</u>: $\mathfrak{K}_{\aleph_0}$ has amalgamation by ⊛$_{12}$ and has the JEP by categoricity in $\aleph_0$ and $\mathfrak{K}_{\aleph_0}$ has no maximal model by (categoricity and) having uncountable models (and $\mathrm{LS}(\mathfrak{K}) = \aleph_0$).

<u>Clause (D)</u>: Obvious; stability, i.e., (D)(d) holds by ⊛$_{13}(ii)$ + ⊛$_{14}$.

<u>Subclause (E)(a),(b)</u>: By the definition.

<u>Subclause (E)(c)</u>: (Local character).
If $\langle M_i : i \leq \delta + 1\rangle$ is $\leq_{\mathfrak{K}}$-increasing continuous $M_i \in K_{\aleph_0}$, $\bar{a} \in {}^{\omega>}(M_{\delta+1})$ and $\bar{a} \in {}^{\omega>}(M_\delta)$,
then for some finite $A \subseteq M_\delta$, $\mathrm{tp}_{\mathbb{L}}(\bar{a}, M_\delta, M_{\delta+1})$ is definable over A,

so for some $i < \delta, A \subseteq M_\delta$ hence $j \in [i, \delta) \Rightarrow \mathrm{tp}_{\mathbb{L}}(\bar{a}, M_i, M_{\delta+1})$ is definable over $A \Rightarrow \bigcup(M_i, M_\delta, \bar{a}, M_{\delta+i})$.

Subclause (E)(d): (Transitivity).

As if $M' \leq_{\mathfrak{K}} M'' \in \mathfrak{K}_{\aleph_0}$, two definitions in M' of complete types, which give the same result in M' give the same result in M''.

Sublause (E)(e)(uniqueness): By $\circledast_{14}$ and the justification of transitivity.

Subclause (E)(f)(symmetry): By [Sh 48, Theorem 5.4], we have the symmetry property see [Sh 48, Definition 5.2]. By [Sh 48, 5.5] + the uniqueness proved above we can finish easily.

Subclause (E)(g): Extension existence.

Easy, included in [Sh 48, 5.5].

Subclause (E)(h): Continuity.

As $\mathscr{S}^{\mathrm{bs}}_{\mathfrak{s}}(M)$ is the set of non-algebraic types this follows from "finite character", that is by 2.17(3)(4).

Subclause (E)(i): non-forking amalgamation

By 2.16. $\square_{3.5}$

3.6 Remark. So if $\psi \in \mathbb{L}_{\omega_1,\omega}(\mathbf{Q})$ and $1 \leq \dot{I}(\aleph_1, \psi) < 2^{\aleph_1}$, we essentially can apply Theorem 0.1, exactly see 9.4.

(D) Starting at $\lambda > \aleph_0$.

The next theorem puts the results of [Sh 576] in our context hence rely on it heavily.

(Alternatively, even eliminating "WDmId(λ^+) is λ^{++}-saturated" we can deduce 3.7 by Chapter VI, Chapter VII, i.e. by VI.0.2(2) there is a so called almost good λ-frame $\mathfrak{s}$ and by VII.4.32 it is even a good λ-frame, and by §9 here, also $\mathfrak{s}^+$ is a good λ^+-frame and easily it is the frame described in 3.7(2).)

We use $K^{3,\mathrm{na}}_\lambda$ as in Chapter VI called K^3_λ is [Sh 576]. Note that while the material does not [Sh 576, §1,§2,§4,§7] appears in Chapter

VI, the material in [Sh 576, §8,§9,§10] similar to §6 - §9 here, so we still need some parts of [Sh 576], though as said above we can avoid it.

3.7 Theorem. *Assume* $2^\lambda < 2^{\lambda^+} < 2^{\lambda^{++}}$ *and*

(α) $\mathfrak{K}$ *is an abstract elementary class with* $\mathrm{LS}(\mathfrak{K}) \le \lambda$

(β) $\mathfrak{K}$ *is categorical in* λ *and in* λ^+

(γ) $\mathfrak{K}$ *has a model in* λ^{++}

(δ) $\dot{I}(\lambda^{+2}, K) < \mu_{\mathrm{unif}}(\lambda^{+2}, 2^{\lambda^+})$ *and* $\mathrm{WDmId}(\lambda^+)$ *is not* λ^{++}*-saturated* _or_ *just some consequences: density of minimal types (see by VI.4.13, VI.4.14) and* $\otimes$*, i.e.* $K^{3,\mathrm{uq}}_\lambda \ne \emptyset$ *of [Sh 576, 6.4,pg.99] = VI.6.6 proved by the conclusion of [Sh 576, Th.6.7 (pg.101)] or VI.6.11.*

Then *1) Letting* $\mu = \lambda^+$ *we can choose* $\bigcup\limits_{\mu}, \mathscr{S}^{\mathrm{bs}}$ *such that* $(\mathfrak{K}_{\ge\mu}, \bigcup\limits_{\mu}, \mathscr{S}^{\mathrm{bs}})$ *is a* μ*-good frame.*
2) Moreover, we can let

(a) $\mathscr{S}^{\mathrm{bs}}(M) := \{\mathbf{tp}_{\mathfrak{K}}(a, M, N)$: *for some* M, N, a *we have* $(M, N, a) \in K^{3,\mathrm{na}}_{\lambda^+}$ *and for some* $M' \le_{\mathfrak{K}} M$ *we have* $M' \in K_\lambda$ *and* $\mathbf{tp}_{\mathfrak{K}}(a, M', N) \in \mathscr{S}_{\mathfrak{K}}(M')$ *is minimal*$\}$

(see Definition [Sh 576, 2.3(4),pg.56] and [Sh 576, 2.5(1),(13),pg.57-58] _or_ *(VI.1.6, VI.1.11)*

(b) $\bigcup = \bigcup\limits_{\mu}$ *be defined by:* $\bigcup(M_0, M_1, a, M_3)$ _iff_ $M_0 \le_{\mathfrak{K}} M_1 \le_{\mathfrak{K}} M_3$ *are from* $K_\mu, a \in M_3 \backslash M_1$ *and for some* $N \le_{\mathfrak{K}} M_0$ *of cardinality* λ*, the type* $\mathbf{tp}_{\mathfrak{K}}(a, N, M_3) \in \mathscr{S}_{\mathfrak{K}}(N)$ *is minimal.*

Proof. 1), 2). Note that $\mathfrak{K}$ has amalgamation in λ and in λ^+, see I.3.8. By clause (δ) of the assumption, we can use the "positive" results of [Sh 576] in particular [Sh 576] freely. Now (see Definition 1.12(2))

(∗) if $(M, N, a) \in K^{3,\mathrm{na}}_{\lambda^+}$ and $M' \leq_{\mathfrak{K}} M, M' \in K_\lambda$ and $p = \mathbf{tp}_{\mathfrak{K}}(a, M', N)$ is minimal (see Definition 1.9(0)) <u>then</u>

(a) if $q \in \mathscr{S}_{\mathfrak{K}}(M)$ is not algebraic and $q \restriction M' = p$ then $q = \mathbf{tp}_{\mathfrak{K}}(a, M, N)$

(b) if $\langle M_\alpha : \alpha < \mu\rangle, \langle N_\alpha : \alpha < \mu\rangle$ are $\leq_{\mathfrak{K}}$-representations of M, N respectively <u>then</u> for a club of $\delta < \mu$ we have $\mathbf{tp}_{\mathfrak{K}}(a, M_\delta, N_\delta) \in \mathscr{S}_{\mathfrak{K}}(M_\delta)$ is minimal and reduced

[Why? For clause (b) let $\alpha^* = \mathrm{Min}\{\alpha : M' \leq_{\mathfrak{K}} M_\alpha\}$, so α^* is well defined and as M is saturated (for $\mathfrak{K}$), for a club of $\delta < \mu = \lambda^+$, the model M_δ is $(\lambda, \mathrm{cf}(\delta))$-brimmed over M' hence by [Sh 576, 7.5(2)(pg.106)] we are done.

For clause (a) let $M^0 = M, M^1 = N$ and $a^1 = a$ and $M^2, a^2 = a$ be such that $(M^0, M^2, a^2) \in K^{3,\mathrm{na}}_\mu = K^{3,\mathrm{na}}_{\lambda^+}$ and $q = \mathbf{tp}_{\mathfrak{K}}(a^2, M^0, M)$. Now we repeat the proof of [Sh 576, 9.5(pg.120)] but instead $f(a^2) \notin M^1$ we require $f(a^2) = a^1$; we are using [Sh 576, 10.5(1)(pg.125)] which says $<^*_{\lambda^+} = <_{\mathfrak{K}} \restriction K_{\lambda^+}$.]

In particular we have used

(∗∗) if $M_0 \leq_{\mathfrak{K}_\lambda} M_1, M_1$ is (λ, κ)-brimmed over $M_0, p \in \mathscr{S}_{\mathfrak{K}}(M_1)$ is not algebraic and $p \restriction M_0$ is minimal, <u>then</u> p is minimal and reduced.

<u>Clause (A)</u>:

This is by assumption (α).

<u>Clause (B)</u>:

As K is categorical in $\mu = \lambda^+$, the existence of superlimit $M \in K_\mu$ follows; the superlimit is not maximal as $\mathrm{LS}(\mathfrak{K}) \leq \lambda$ & $K_{\mu^+} = K_{\lambda^{++}} \neq \emptyset$ by assumption (γ).

<u>Clause (C)</u>:

K_{λ^+} has the amalgamation property by I.3.8 or [Sh 576, 1.4(pg.46), 1.6(pg.48)] and $\mathfrak{K}_\lambda$ has the JEP in λ^+ by categoricity in λ^+.

<u>Clause (D)</u>:

<u>Subclause $(D)(a), (b)$</u>:

By the definition of $\mathscr{S}^{\rm bs}(M)$ and of minimal types (in $\mathscr{S}_{\mathfrak{K}}(N), N \in K_\lambda$,
[Sh 576, 2.5(1)+(3)(pg.57), 2.3(4)+(6)(pg.56)], this is clear.

<u>Subclause $(D)(c)$</u>:

Suppose $M \le_{\mathfrak{K}} N$ are from K_μ and $M \ne N$; let $\langle M_i : i < \lambda^+\rangle, \langle N_i : i < \lambda^+\rangle$ be a $\le_{\mathfrak{K}}$-representation of M, N respectively, choose $b \in N\backslash M$ so $E = \{\delta < \lambda^+ : N_\delta \cap M = M_\delta \text{ and } b \in N_\delta\}$ is a club of λ^+. Now for $\delta = \text{Min}(E)$ we have $M_\delta \ne N_\delta, M_\delta \le_{\mathfrak{K}} N_\delta$ and there is a minimal inevitable $p \in \mathscr{S}_{\mathfrak{K}}(M_\delta)$ by [Sh 576, 5.3,pg.94] and categoricity of K in λ; so for some $a \in N_\delta\backslash M_\delta$ we have $p = \mathbf{tp}_{\mathfrak{K}}(a, M_\delta, N_\delta)$. So $\mathbf{tp}_{\mathfrak{K}}(a, M, N)$ is non-algebraic as $a \in M \Rightarrow a \in M \cap N_\delta = M_\delta$, a contradiction, so $\mathbf{tp}_{\mathfrak{K}}(a, M, N) \in \mathscr{S}^{\rm bs}(M)$ as required.

<u>Subclause $(D)(d)$</u>: If $M \in K_\mu$ let $\langle M_i : i < \lambda^+\rangle$ be a $\le_{\mathfrak{K}}$-representation of M, so by $(*)(a)$ above $p \in \mathscr{S}^{\rm bs}(M)$ is determined by $p \restriction M_\alpha$ if $p \restriction M_\alpha$ is minimal and reduced. But for every such p there is such $\alpha(p) < \lambda^+$ by the definition of $\mathscr{S}^{\rm bs}(M)$ and for each $\alpha < \lambda^+$ there are $\le \lambda$ possible such $p \restriction M_\alpha$ as $\mathfrak{K}$ is stable in λ by [Sh 576, 5.7(a)(pg.97)], so the conclusion follows. Alternatively, $M \in K_\mu \Rightarrow |\mathscr{S}^{\rm bs}(M)| \le \mu$ as by [Sh 576, 10.5(pg.125)], we have $\le^*_{\lambda^+} = \le_{\mathfrak{K}} \restriction K_{\lambda^+}$, so we can apply [Sh 576, 9.7(pg.121)]; or use $(*)$ above.

<u>Clause (E)</u>:

<u>Subclause $(E)(a)$</u>:

Follows by the definition.

<u>Subclause $(E)(b)$</u>: (Monotonicity)

Obvious properties of minimal types in $\mathscr{S}(M)$ for $M \in K_\lambda$.

<u>Subclause $(E)(c)$</u>: (Local character)

Let $\delta < \mu^+ = \lambda^{++}$ and $M_i \in K_\mu$ be $\le_{\mathfrak{K}}$-increasing continuous for $i \le \delta$ and $p \in \mathscr{S}^{\rm bs}(M_\delta)$, so for some $N \le_{\mathfrak{K}} M_\delta$ we have $N \in K_\lambda$ and $p \restriction N \in \mathscr{S}_{\mathfrak{K}}(N)$ is minimal. Without loss of generality $\delta = \text{cf}(\delta)$ and if $\delta = \lambda^+$, there is $i < \delta$ such that $N \subseteq M_i$ and easily we are done. So assume $\delta = \text{cf}(\delta) < \lambda^+$.

Let $\langle M^i_\zeta : \zeta < \lambda^+ \rangle$ be a $\leq_{\mathfrak{K}}$-representation of M_i for $i \leq \delta$, hence E is a club of λ^+ where:

$$\begin{aligned} E := \{\zeta < \lambda^+ :& \zeta \text{ a limit ordinal and for } j < i \leq \delta \text{ we have} \\ & M^i_\zeta \cap M_j = M^j_\zeta \text{ and for } \xi < \zeta, i \leq \delta \text{ we have :} \\ & M^i_\zeta \text{ is } (\lambda, \mathrm{cf}(\zeta))\text{-brimmed over } M^i_\xi \text{ and } N \leq_{\mathfrak{K}} M^\delta_\zeta\}. \end{aligned}$$

Let ζ_i be the i-th member of E for $i \leq \delta$, so $\langle \zeta_i : i \leq \delta \rangle$ is increasing continuous, $\langle M^i_{\zeta_i} : i \leq \delta \rangle$ is $\leq_{\mathfrak{K}}$-increasingly continuous in K_λ and $M^{i+1}_{\zeta_{i+1}}$ is $(\lambda, \mathrm{cf}(\zeta_{i+1}))$-brimmed over $M^{i+1}_{\zeta_i}$ hence also over $M^i_{\zeta_i}$. Also $p \restriction M^\delta_{\zeta_\delta}$ is non-algebraic (as p is) and extends $p \restriction N$ (as $N \leq_{\mathfrak{K}} M^\delta_{\zeta_\delta}$ as $\zeta_\delta \in E$) hence $p \restriction M^\delta_{\zeta_\delta}$ is minimal.

Also $M^\delta_{\zeta_\delta}$ is $(\lambda, \mathrm{cf}(\zeta_\delta))$-brimmed over $M^\delta_{\zeta_0}$ hence over N, hence by $(**)$ above we get that $p \restriction M^\delta_{\zeta_\delta}$ is not only minimal but also reduced. Hence by [Sh 576, 7.3(2)(pg.103)] applied to $\langle M^i_{\zeta_i} : i \leq \delta \rangle, p \restriction M^\delta_{\zeta_\delta}$ we know that for some $i < \delta$ the type $p \restriction M^i_{\zeta_i} = (p \restriction M^\delta_{\zeta_\delta}) \restriction M^i_{\zeta_i}$ is minimal and reduced, so it witnesses that $p \restriction M_j \in \mathscr{S}^{\mathrm{bs}}(M_j)$ for every $j \in [i, \delta)$, as required.

Subclause $(E)(d)$: (Transitivity)

Easy by the definition of minimal.

Subclause $(E)(e)$: (Uniqueness)

By $(*)(a)$ above.

Subclause $(E)(f)$: (Symmetry)

By the symmetry in the situation assume $M_0 \leq_{\mathfrak{K}} M_1 \leq_{\mathfrak{K}} M_3$ are from K_μ,
$a_1 \in M_1 \backslash M_0, a_2 \in M_3 \backslash M_1$ and $\mathbf{tp}_{\mathfrak{K}}(a_1, M_0, M_3) \in \mathscr{S}^{\mathrm{bs}}(M_0)$ and $\mathbf{tp}_{\mathfrak{K}}(a_2, M_1, M_3) \in \mathscr{S}^{\mathrm{bs}}(M_1)$ does not fork over M_0; hence for $\ell = 1, 2$ we have $\mathbf{tp}_{\mathfrak{K}}(a_\ell, M_0, M_3) \in \mathscr{S}^{\mathrm{bs}}(M_0)$. By the existence of disjoint amalgamation (by [Sh 576, 9.11 (pg.122), 10.5(1) (pg.125)] there are M_2, M_3', f such that $M_0 \leq_{\mathfrak{K}} M_2 \leq_{\mathfrak{K}} M_3' \in K_\mu$, $M_3 \leq_{\mathfrak{K}} M_3'$, f is an isomorphism from M_3 onto M_2 over M_0, and $M_3 \cap M_2 = M_0$. By $\mathbf{tp}_{\mathfrak{K}}(a_2, M_0, M_3) \in \mathscr{S}^{\mathrm{bs}}(M_1)$ and as $f(a_2) \notin M_1$

being in $M_2 \backslash M_0 = M_2 \backslash M_3$ and $a_2 \notin M_1$ by assumption and as $a_2, f(a_2)$ realize the same type from $\mathscr{S}_{\mathfrak{K}}(M_0)$ clearly by $(*)(a)$ we have $\mathbf{tp}_{\mathfrak{K}}(a_2, M_1, M_3') = \mathbf{tp}_{\mathfrak{K}}(f(a_2), M_1, M_3')$.

Using amalgamation in $\mathfrak{K}_\mu$ (and equality of types) there is M_3'' such that:
$M_3' \leq_{\mathfrak{K}} M_3'' \in K_\mu$, and there is an $\leq_{\mathfrak{K}}$-embedding g of M_3' into M_3'' such that $g \restriction M_1 = \mathrm{id}_{M_1}$ and $g(f(a_2)) = a_2$. Note that as $a_1 \notin g(M_2), M_1 \leq_{\mathfrak{K}} g(M_2) \in K_\mu$ and $\mathbf{tp}_{\mathfrak{K}}(a_1, M_1, M_3'')$ is minimal then necessarily $\mathbf{tp}_{\mathfrak{K}}(a_1, g(M_2), M_3'')$ is its non-forking extension. So $g(M_2), M_3''$ are models as required.

<u>Subclause $(E)(g)$</u>: (Extension existence)
Claims [Sh 576, 9.11(pg.122), 10.5(1)(pg.125)] do even more.

<u>Subclause $(E)(h)$</u>: (Continuity)
Easy.

<u>Subclause $(E)(i)$</u>: (Non-forking amalgamation)
Like $(E)(f)$ or use 2.16. $\square_{3.7}$

<u>3.8 Question</u>: If $\mathfrak{K}$ is categorical in λ and in μ and $\mu > \lambda \geq \mathrm{LS}(\mathfrak{K})$, can we conclude categoricity in $\chi \in (\mu, \lambda)$?

3.9 Fact. In 3.7:
1) If $p \in \mathscr{S}^{\mathrm{bs}}(M)$ and $M \in K_\mu$, <u>then</u> for some $N \leq_{\mathfrak{K}} M, N \in K_\lambda$ and $p \restriction N$ is minimal and reduced.
2) If $M <_{\mathfrak{K}} N, M \in K_\mu$ and $p \in \mathscr{S}^{\mathrm{bs}}(M)$, <u>then</u> some $a \in N \backslash M$ realizes p, (i.e., "a strong version of uni-dimensionality" holds).

Proof. The proof is included in the proof of 3.7.

* * *

<u>(E) An Example</u>:
A trivial example (of an approximation to good λ-frame) is:

3.10 Definition/Claim. 1) Assume that $\mathfrak{K}$ is an a.e.c. and $\lambda \geq \mathrm{LS}(\mathfrak{K})$ or $\mathfrak{K}$ is a λ-a.e.c. We define $\mathfrak{s} = \mathfrak{s}_\lambda[\mathfrak{K}]$ as the triple $\mathfrak{s} = (\mathfrak{K}_\lambda, \mathscr{S}^{\mathrm{na}}, \bigcup\limits_{\mathrm{na}})$ where:

(a) $\mathscr{S}^{\mathrm{na}}(M) = \{\mathbf{tp}_{\mathfrak{K}}(a, M, N), M \leq_{\mathfrak{K}} N \text{ and } a \in N \backslash M\}$

(b) $\bigcup(M_0, M_1, a, M_3)$ iff $M_0 \leq_{\mathfrak{K}_\lambda} M_1 \leq_{\mathfrak{K}_\lambda} M_3$ and $a \in M_3 \backslash M_1$.

2) Then $\mathfrak{s}$ satisfies Definition 2.1 of good λ-frame except possibly: (B), existence of superlimits, (C) amalgamation and JEP, (D)(d) stability and (E)(e),(f),(g),(i) uniqueness, symmetry, extension existence and non-forking amalgamation.

§4 Inside the frame

We investigate good λ-frames. We prove stability in λ (we have assumed in Definition 2.1 only stability for basic types), hence the existence of a (λ, ∂)-brimmed $\leq_{\mathfrak{K}}$-extension in K_λ over $M_0 \in K_\lambda$ (see 4.2), and we give a sufficient condition for "M_δ is $(\lambda, \mathrm{cf}(\delta))$-brimmed over M_0" (in 4.3). We define again $K_\lambda^{3,\mathrm{bs}}$ (like K_λ^3 from 1.12(2) but the type is basic) and the natural order $\leq_{\mathrm{bs}}$ on them as well as "reduced" (Definition 4.5), and indicate their basic properties (4.6). We may like to construct sometimes pairs $N_i \leq_{\mathfrak{K}_\lambda} M_i$ such that M_i, N_i are increasing continuous with i and we would like to guarantee that M_γ is $(\lambda, \mathrm{cf}(\gamma))$-brimmed over N_γ, of course we need to carry more inductive assumptions. Toward this we may give a sufficient condition for building a $(\lambda, \mathrm{cf}(\gamma))$-brimmed extension over N_γ where $\langle N_i : i \leq \gamma \rangle$ is $\leq_{\mathfrak{K}_\lambda}$-increasing continuous, by a triangle of extensions of the N_i's, with non-forking demands of course (see 4.7). We also give conditions on a rectangle of models to get such pairs in both directions (4.11), for this we use nice extensions of chains (4.9, 4.10).

Then we can deduce that if "M_1 is (λ, ∂)-brimmed over M_0" then the isomorphism type of M_1 over M_0 does not depend on ∂ (see 4.8), so the brimmed N over M_0 is unique up to isomorphism (i.e. being (λ, ∂)-brimmed over M_0 does not depend on ∂). We finish giving conclusion about $K_{\lambda^+}, K_{\lambda^{++}}$.

4.1 Hypothesis. $\mathfrak{s} = (\mathfrak{K}, \bigcup, \mathscr{S}^{\rm bs})$ is a good λ-frame.

4.2 Claim. *1) $\mathfrak{K}$ is stable in λ, i.e., $M \in \mathfrak{K}_\lambda \Rightarrow |\mathscr{S}(M)| \le \lambda$.*
2) For every $M_0 \in K_\lambda$ and $\partial \le \lambda$ there is M_1 such that $M_0 \le_{\mathfrak{K}} M_1 \in K_\lambda$ and M_1 is (λ, ∂)-brimmed over M_0 (see Definition 1.15) and it is universal[13] *over M_0.*

Proof. 1) Let $M_0 \in K_\lambda$ and we choose by induction on $\alpha \in [1, \lambda]$, $M_\alpha \in K_\lambda$ such that:

- (i) M_α is $\le_{\mathfrak{K}}$-increasing continuous
- (ii) if $p \in \mathscr{S}^{\rm bs}(M_\alpha)$ then this type is realized in $M_{\alpha+1}$.

No problem to carry this: for clause (i) use Axiom(A), for clause (ii) use Axiom $(D)(d)$ and amalgamation in $\mathfrak{K}_\lambda$, i.e., Axiom (C). If every $q \in \mathscr{S}(M_0)$ is realized in M_λ we are done. So let q be a counterexample, so let $M_0 \le_{\mathfrak{K}} N \in K_\lambda$ be such that q is realized in N. We now try to choose by induction on $\alpha < \lambda$ a triple $(N_\alpha, f_\alpha, \bar{\mathbf{a}}_\alpha)$ such that:

- (A) $N_\alpha \in K_\lambda$ is $\le_{\mathfrak{K}}$-increasingly continuous
- (B) f_α is a $\le_{\mathfrak{K}}$-embedding of M_α into N_α
- (C) f_α is increasing continuous
- (D) $f_0 = \ {\rm id}_{M_0}$ and $N_0 = N$
- (E) $\bar{\mathbf{a}}_\alpha = \langle a_{\alpha,i} : i < \lambda\rangle$ lists the elements of N_α
- (F) if there are $\beta \le \alpha, i < \lambda$ such that $\mathbf{tp}(a_{\beta,i}, f_\alpha(M_\alpha), N_\alpha) \in \mathscr{S}^{\rm bs}(f_\alpha(M_\alpha))$ <u>then</u> for some such pair (β_α, i_α) we have:
  - (i) the pair (β_α, i_α) is minimal in an appropriate sense, that is: if (β, i) is another such pair then $\beta + i > \beta_\alpha + i_\alpha$ or $\beta + i = \beta_\alpha + i_\alpha \ \& \ \beta > \beta_\alpha$ or $\beta + i = \beta_\alpha + i_\alpha \ \& \ \beta = \beta_\alpha \ \& \ i \ge i_\alpha$
  - (ii) $a_{\beta_\alpha, i_\alpha} \in {\rm Rang}(f_{\alpha+1})$.

[13]in fact, this follows

This is easy: for successor α we use the definition of type and let $N_\lambda := \cup\{N_\alpha : \alpha < \lambda\}$. Clearly $f_\lambda := \cup\{f_\alpha : \alpha < \lambda\}$ is a $\leq_{\mathfrak{s}}$-embedding of M_λ into N_λ over M_0.

As in N, the type q is realized and it is not realized in M_λ necessarily $N \nsubseteq f_\lambda(M_\lambda)$ hence $N_\lambda \neq f_\lambda(M_\lambda)$ but easily $f_\lambda(M_\lambda) \leq_{\mathfrak{K}} N_\lambda$. So by Axiom $(D)(c)$ for some $c \in N_\lambda \backslash f_\lambda(M_\lambda)$ we have $p = \mathbf{tp}(c, f_\lambda(M_\lambda), N_\lambda) \in \mathscr{S}^{\mathrm{bs}}(f_\lambda(M_\lambda))$. As $\langle f_\gamma(M_\gamma) : \gamma \leq \lambda\rangle$ is $\leq_{\mathfrak{K}}$-increasing continuous, by Axiom (E)(c) for some $\gamma < \lambda$ we have $\mathbf{tp}(c, f_\lambda(M_\lambda), N_\lambda)$ does not fork over $f_\gamma(M_\gamma)$, also as $c \in N_\lambda = \bigcup\limits_{\beta<\lambda} N_\beta$ clearly $c \in N_\beta$ for some $\beta < \lambda$ and let $i < \lambda$ be such that $c = a_{\beta,i}$. Now if $\alpha \in [\max\{\gamma, \beta\}, \lambda)$ then (β, i) is a legitimate candidate for (β_α, i_α) that is $\mathbf{tp}(a_{\beta,i}, f_\alpha(M_\alpha), N_\alpha) \in \mathscr{S}^{\mathrm{bs}}(f_\alpha(M_\alpha))$ by monotonicity of non-forking, i.e., Axiom (E)(b). So (β_α, i_α) is well defined for any such α and $\beta_\alpha + i_\alpha \leq \beta + i$ by clause (F)(i). But $\alpha_1 < \alpha_2 \Rightarrow a_{\beta_{\alpha_1}, i_{\alpha_1}} \neq a_{\beta_{\alpha_2}, i_{\alpha_2}}$ (as one belongs to $f_{\alpha_1+1}(M_{\alpha_1})$ and the other not), contradiction by cardinality consideration.

2) So $\mathfrak{K}_\lambda$ is stable in λ and has amalgamation, hence (see 1.16) the conclusion holds; alternatively use 4.3 below. $\square_{4.2}$

4.3 Claim. *Assume*

- *(a)* $\delta < \lambda^+$ *is a limit ordinal divisible by* λ
- *(b)* $\bar{M} = \langle M_\alpha : \alpha \leq \delta\rangle$ *is* $\leq_{\mathfrak{K}}$*-increasing continuous sequence in* $\mathfrak{K}_\lambda$
- *(c) if* $i < \delta$ *and* $p \in \mathscr{S}^{\mathrm{bs}}(M_i)$, *then for* λ *ordinals* $j \in (i, \delta)$ *there is* $c' \in M_{j+1}$ *realizing the non-forking extension of* p *in* $\mathscr{S}^{\mathrm{bs}}(M_j)$.

Then M_δ *is* $(\lambda, \mathrm{cf}(\delta))$*-brimmed over* M_0 *and universal over it.*

4.4 Remark. 1) See end of proof of 6.29.

2) Of course, by renaming, M_δ is $(\lambda, \mathrm{cf}(\delta))$-brimmed over M_α for any $\alpha < \delta$.

3) Why in clause (c) of 4.3 we ask for "λ ordinals $j \in (i, \delta)$" rather than "for unboundedly many $j \in (i, \delta)$"? For λ regular there is no difference but for λ singular not so. Think of $\mathfrak{K}$ the class of (A, R), R

an equivalence relation on A; (so it is not categorical) but for some λ-good frames $\mathfrak{s}, \mathfrak{K}_{\mathfrak{s}} = \mathfrak{K}_\lambda$ and exemplifies a problem; some equivalence class of M_δ may be of cardinality $< \lambda$.

Proof. Like 4.2, but we give details.

Let $g : \delta \to \lambda$ be a one to one and choose by induction on $\alpha \leq \delta$ a triple $(N_\alpha, f_\alpha, \bar{\mathbf{a}}_\alpha)$ such that

- (A) $N_\alpha \in K_\lambda$ is $\leq_{\mathfrak{K}}$-increasing continuous
- (B) f_α is a $\leq_{\mathfrak{K}}$-embedding of M_α into N_α
- (C) f_α is increasing continuous
- (D) $f_0 = \mathrm{id}_{M_0}, N_0 = M_0$
- (E) $\bar{\mathbf{a}}_\alpha = \langle a_{\alpha,i} : i < \lambda \rangle$ list the elements of N_α
- (F) $N_{\alpha+1}$ is universal over N_α
- (G) if $\alpha < \delta$ and there is a pair $(\beta, i) = (\beta_\alpha, i_\alpha)$ satisfying the condition $(*)^{\beta,i}_{f_\alpha,N_\alpha}$ stated below and it is minimal in the sense that $(*)^{\beta',i'}_{f_\alpha,N_\alpha} \Rightarrow (**)^{\beta',i',\beta,i}_g$, see below, <u>then</u> $a_{\beta,i} \in \mathrm{Rang}(f_{\alpha+1})$, where

$(*)^{\beta,i}_{f_\alpha,N_\alpha}$ (a) $\beta \leq \alpha$ and $i < \lambda$

(b) $\mathbf{tp}(a_{\beta,i}, f_\alpha(M_\alpha), N_\alpha) \in \mathscr{S}^{\mathrm{bs}}(f_\alpha(M_\alpha))$

(c) some $c \in M_{\alpha+1}$ realizes $f_\alpha^{-1}(\mathbf{tp}(a_{\beta,i}, f_\alpha(M_\alpha), N_\alpha)$, so by clause (b) it follows that $c \in M_{\alpha+1} \backslash M_\alpha$

$(**)^{\beta',i',\beta,i}_g$ $g(\beta) + i < g(\beta') + i'] \vee$
$[g(\beta) + i = g(\beta') + i' \ \& \ g(\beta) < g(\beta')] \vee$
$\vee [g(\beta) + i = g(\beta') + i' \ \& \ g(\beta) = g(\beta') \ \& \ i \leq i']$.

There is no problem to choose f_α, N_α. Now in the end, by clauses (A),(F) clearly N_δ is $(\lambda, \mathrm{cf}(\delta))$-brimmed over N_0, i.e., over M_0, so it suffices to prove that f_δ is onto N_δ. If not, then by Axiom (D)(c), the density, there is $d \in N_\delta \backslash f_\delta(M_\delta)$ such that $p := \mathbf{tp}(d, f_\delta(M_\delta), N_\delta) \in \mathscr{S}^{\mathrm{bs}}(f_\delta(M_\delta))$ hence for some $\beta(*) < \delta$ we have $d \in N_{\beta(*)}$ so for some $i(*) < \lambda, d = a_{\beta(*),i(*)}$. Also by Axiom (E)(c), (the local character) for every $\beta < \delta$ large enough say $\geq \beta_d$ the type p does not fork over $f_\delta(M_\beta)$, without loss of generality $\beta_d = \beta(*)$. Let $q = f_\delta^{-1}(\mathbf{tp}(d, f_\delta(M_\delta), N_\delta)$, so it $\in \mathscr{S}^{\rm bs}(M_\delta)$.

Let $u = \{\alpha : \beta(*) \le \alpha < \delta$ and $q \restriction M_\alpha \in \mathscr{S}^{\rm bs}(M_\alpha)$ (note $\beta(*) \le \alpha$) is realized in $M_{\alpha+1}\}$. By clause (c) of the assumption clearly $|u| = \lambda$. Also by the definition of v for every $\alpha \in u$ the condition $(*)^{\beta(*),i(*)}_{N_\alpha,f_\alpha}$ holds, hence in clause (F) the pair (β_α, i_α) is well defined and is "below" $(\beta(*), i(*))$ in the sense of clause (G). But there are only $\le |g(\beta(*)) \times i(*)| < \lambda$ such pairs hence for some $\alpha_1 < \alpha_2$ in u we have $(\beta_{\alpha_1}, i_{\alpha_1}) = (\beta_{\alpha_2}, i_{\alpha_2})$, a contradiction: $a_{\beta_{\alpha_1}, i_{\alpha_1}} \in \mathrm{Rang}(f_{\alpha_1+1}) \subseteq \mathrm{Rang}(f_{\alpha_2}) = f_{\alpha_2}(M_{\alpha_2})$ hence $\mathbf{tp}(a_{\beta_{\alpha_1}, i_{\alpha_1}}, f_{\alpha_2}(M_{\alpha_2}), N_{\alpha_2}) \notin \mathscr{S}^{\rm bs}(f_{\alpha_2}(M_{\alpha_2}))$, contradiction. So we are done. $\square_{4.3}$

* * *

The following is helpful for constructions so that we can amalgamate disjointly preserving non-forking of a type; we first repeat the definition of $K^{3,\rm bs}_\lambda, <_{\rm bs}$.

4.5 Definition. 1) Let $(M, N, a) \in K^{3,\rm bs}_\lambda$ if $M \le_{\mathfrak{K}} N$ are models from K_λ, $a \in N \backslash M$ and $\mathbf{tp}(a, M, N) \in \mathscr{S}^{\rm bs}(M)$. Let $(M_1, N_1, a) \le_{\rm bs} (M_2, N_2, a)$ or write $\le^{\mathfrak{s}}_{\rm bs}$, when: both triples are in $K^{3,\rm bs}_\lambda$, $M_1 \le_{\mathfrak{K}} M_2$, $N_1 \le_{\mathfrak{K}} N_2$ and $\mathbf{tp}(a, M_2, N_2)$ does not fork over M_1.
2) We say (M, N, a) is bs-reduced <u>when</u> if it belongs to $K^{3,\rm bs}_\lambda$ and $(M, N, a) \le_{\rm bs} (M', N', a) \in K^{3,\rm bs}_\lambda \Rightarrow N \cap M' = M$.
3) We say $p \in \mathscr{S}^{\rm bs}(N)$ is a (really the) stationarization of $q \in \mathscr{S}^{\rm bs}(M)$ if $M \le_{\mathfrak{K}} N$ and p is an extension of q which does not fork over M.

Remark. 1) The definition of $K^{3,\rm bs}_\lambda$ is compatible with the one in 2.4 by 2.6(1).
2) We could have strengthened the definition of bs-reduced (4.5), e.g., add: for no $b \in N' \backslash M'$, do we have $\mathbf{tp}(b, M', N') \in \mathscr{S}^{\rm bs}(M')$ and there are M'', N'' such that $(M', N', a) \le_{\rm bs} (M'', N'', a)$ and $\mathbf{tp}(b, M'', N'')$ forks over M'.

4.6 Claim. *For parts (3),(4),(5) assume $\mathfrak{s}$ is categorical (in λ).*
1) If $\kappa \leq \lambda, (M, N, a) \in K^{3,\mathrm{bs}}_\lambda$, <u>then</u> we can find M', N' such that: $(M, N, a) \leq_{\mathrm{bs}} (M', N', a) \in K^{3,\mathrm{bs}}_\lambda, M'$ is (λ, κ)-brimmed over M, N' is (λ, κ)-brimmed over N and (M', N', a) is bs-*reduced.*
1A) If $(M, N_\ell, a_\ell) \in K^{3,\mathrm{bs}}_\lambda$ for $\ell = 1, 2$, <u>then</u> we can find M^+, f_1, f_2 such that: $M \leq_{\mathfrak{K}} M^+ \in K_\lambda$ and for $\ell \in \{1, 2\}, f_\ell$ is a $\leq_{\mathfrak{K}}$-embedding of N_ℓ into M^+ over M and $(M, f_\ell(N_\ell), f_\ell(a_\ell)) \leq_{\mathrm{bs}} (f_{3-\ell}(N_{3-\ell}), M^+, f_\ell(a_\ell))$, equivalently $\mathbf{tp}(f_\ell(a_\ell), f_{3-\ell}(N_{3-\ell}), M^+)$ does not fork over M.
2) If $(M_\alpha, N_\alpha, a) \in K^{3,\mathrm{bs}}_\lambda$ is $\leq_{\mathrm{bs}}$-increasing for $\alpha < \delta$ and $\delta < \lambda^+$ is a limit ordinal <u>then</u> their union $(\bigcup_{\alpha<\delta} M_\alpha, \bigcup_{\alpha<\delta} N_\alpha, a)$ is a $\leq_{\mathrm{bs}}$-lub. If each (M_α, N_α, a) is bs-reduced then so is their union.
3) Let λ divide $\delta, \delta < \lambda^+$. We can find $\langle N_j, a_i : j \leq \delta, i < \delta\rangle$ such that: $N_j \in K_\lambda$ is $\leq_{\mathfrak{K}}$-increasing continuous, $(N_j, N_{j+1}, a_j) \in K^{3,\mathrm{bs}}_\lambda$ is bs-reduced and if $i < \delta, p \in \mathscr{S}^{\mathrm{bs}}(N_i)$ <u>then</u> for λ ordinals $j \in (i, i + \lambda)$ the type $\mathbf{tp}(a_j, N_j, N_{j+1})$ is a non-forking extension of p; so N_δ is $(\lambda, \mathrm{cf}(\delta))$-brimmed over each $N_i, i < \delta$. We can add "N_0 is brimmed".
4) For any $(M_0, M_1, a) \in K^{3,\mathrm{bs}}_\lambda$ and $M_2 \in K_\lambda$ such that $M_0 \leq_{\mathfrak{K}} M_2$ there are N_0, N_1 such that $(M_0, M_1, a) \leq_{\mathrm{bs}} (N_0, N_1, a), M_0 = M_1 \cap N_0$ and M_2, N_0 are isomorphic over M_0. (In fact, if $(M_0, M_2, b) \in K^{3,\mathrm{bs}}_\lambda$ we can add that for some isomorphism f from M_2 onto N_0 over M_0 we have $(M_0, N_0, f(a)) \leq_{\mathrm{bs}} (M_1, N_1, f(a))$.)
5) If $M_0 \in K_\lambda$ is brimmed and $M_0 \leq_{\mathfrak{s}} M_\ell$ for $\ell = 1, 2$ and there is a disjoint $\leq_{\mathfrak{s}}$-amalgamation of M_1, M_2 over M_0.

Proof. 1) We choose $M_i, N_i, b^\ell_i (\ell = 1, 2), \bar{\mathbf{c}}_i$ by induction on $i < \delta := \lambda$ such that

- (a) $(M_i, N_i, a) \in K^{3,\mathrm{bs}}_{\mathfrak{s}}$ is $\leq_{\mathrm{bs}}$-increasing continuous
- (b) $(M_0, N_0) = (M, N)$
- $(c)_1$ $b^1_i \in M_{i+1} \backslash M_i$ and $\mathbf{tp}(b^1_i, M_i, M_{i+1}) \in \mathscr{S}^{\mathrm{bs}}(M_i)$,
- $(c)_2$ $b^2_i \in N_{i+1} \backslash N_i$ and $\mathbf{tp}(b^2_i, N_i, N_{i+1}) \in \mathscr{S}^{\mathrm{bs}}(N_i)$
- $(d)_1$ if $i < \lambda$ and $p \in \mathscr{S}^{\mathrm{bs}}(M_i)$ <u>then</u> the set $\{j : i \leq j < \lambda$ and $\mathbf{tp}(b^1_j, M_j, M_{j+1})$ is a non-forking extension of $p\}$ has order type λ

$(d)_2$ if $i < \lambda$ and $p \in \mathscr{S}^{\rm bs}(N_i)$ then the set $\{j : i \le j < \lambda$ and $\mathbf{tp}(b_j^2, N_j, N_{j+1})$ is the non-forking extension of $p\}$ has order type λ

(e) $\bar{\mathbf{c}}_i = \langle c_{i,j} : j < \lambda\rangle$ list N_i

(f) if $\alpha < \lambda, i \le \alpha, j < \lambda, c_{i,j} \notin M_\alpha$ but for some (M'', N'') we have $(M_{\alpha+1}, N_{\alpha+1}, a) \le_{\rm bs} (M'', N'', a)$ and $c_{i,j} \in M''$ <u>then</u> for some $i_1, j_1 \le \max\{i,j\}$ we have $c_{i_1,j_1} \in M_{\alpha+1} \backslash M_\alpha$.

Lastly, let $M' = \cup\{M_i : i < \lambda\}, N' = \cup\{N_i : i < \lambda\}$, by 4.3 M' is $(\lambda, {\rm cf}(\lambda))$-brimmed over M (using $(d)_1$), and N' is $(\lambda, {\rm cf}(\lambda))$-brimmed over N (using $(d)_2$).
Lastly, being bs-reduced holds by clauses (e)+(f).
1A) Easy.
2) Recall Ax(E)(h).
3) For proving part (3) use part (1) and the "so" is by using 4.3.
4) For proving part (4), without loss of generality M_2 is $(\lambda, {\rm cf}(\lambda))$-brimmed over M_0, as we can replace M_2 by M_2' if $M_2 \le_{\mathfrak{K}} M_2' \in K_\lambda$. By part (3) there is a sequence $\langle a_i : i < \delta\rangle$ and an $\le_{\mathfrak{K}}$-increasing continuous $\langle N_i : i \le \delta\rangle$ with $N_0 = M_0, N_\delta = M_2$ and $(N_i, N_{i+1}, a_i) \in K_\lambda^{3,\rm bs}$ is reduced. Then use (1A) successively.
5) By part (3) as in the proof of part (4). $\square_{4.6}$

4.7 Claim. *Assume*

(a) $\gamma < \lambda^+$ *is a limit ordinal*

(b) $\delta_i < \lambda^+$ *is divisible by* λ *for* $i \le \gamma, \langle \delta_i : i \le \gamma\rangle$ *is increasing continuous*

(c) $\langle N_i : i < \gamma\rangle$ *is* $\le_{\mathfrak{K}}$*-increasing continuous in* K_λ

(d) $\langle M_i : i < \gamma\rangle$ *is* $\le_{\mathfrak{K}}$*-increasing continuous in* K_λ

(e) $N_i \le_{\mathfrak{K}} M_i$ *for* $i < \gamma$

(f) $\langle M_{i,j} : j \le \delta_i\rangle$ *is* $\le_{\mathfrak{K}}$*-increasing continuous in* K_λ *for each* $i < \gamma$

(g) $M_{i,0} = N_i$, $M_{i,\delta_i} = M_i$, $a_j \in M_{i,j+1} \backslash M_{i,j}$ *and* $\mathbf{tp}(a_j, M_{i,j}, M_{i,j+1}) \in \mathscr{S}^{\rm bs}(M_{i,j})$ *when* $i < \gamma, j < \delta_i$

(h) *if* $j \le \delta_{i(*)}, i(*) < \gamma$ *then* $\langle M_{i,j} : i \in [i(*), \gamma)\rangle$ *is* $\le_{\mathfrak{K}}$*-increasing continuous*

(i) $\mathbf{tp}(a_j, M_{\beta,j}, M_{\beta,j+1})$ *does not fork over* $M_{i,j}$ *when* $i < \gamma, j < \delta_i, i \le \beta < \gamma$

(j) *if* $i < \gamma, j < \delta_i, p \in \mathscr{S}^{\rm bs}(M_{i,j})$ *then for* λ *ordinals* $j_1 \in [j, \delta_i)$ *we have* $\mathbf{tp}(a_{j_1}, M_{i,j_1}, M_{i,j_1+1}) \in \mathscr{S}^{\rm bs}(M_{i,j_1})$ *is a non-forking extension of* p
or we can ask less

$(j)^-$ *if* $i < \gamma, j < \delta_i$ *and* $p \in \mathscr{S}^{\rm bs}(M_{i,j})$ *then for* λ *ordinals* $j_1 \in [j, \delta_\gamma)$ *for some* $i_1 \in [i, \gamma)$ *we have* $\mathbf{tp}(a_{j_1}, M_{i_1,j_1}, M_{i_1,j_1+1}) \in \mathscr{S}^{\rm bs}(M_{i_1,j_1})$ *is a non-forking extension of* p.

Then $M_\gamma := \cup\{M_{i,j} : i < \gamma, j < \delta_i\} = \{M_i : i < \gamma\}$ *is* $(\lambda, \mathrm{cf}(\gamma))$*-brimmed over* $N_\gamma := \cup\{N_i : i < \gamma\}$.

Proof. For $j < \delta_\gamma$ let $M_{\gamma,j} = \cup\{M_{i,j} : i < \gamma\}$, and let $M_{\gamma,\delta_\gamma} = M_\gamma$ be $\cup\{M_{\gamma,j} : j < \delta_\gamma\}$. Easily $\langle M_{\gamma,j} : j \le \delta_\gamma\rangle$ is $\le_{\mathfrak{K}}$-increasing continuous, $M_{\gamma,j} \in K_\lambda$ and $i \le \gamma \wedge j < \delta_i \Rightarrow M_{i,j} \le_{\mathfrak{K}} M_{\gamma,j}$. Also if $i < \gamma, j < \delta_i$ then $\mathbf{tp}(a_j, M_{\gamma,j}, M_{\gamma,j+1}) \in \mathscr{S}^{\rm bs}(M_{\gamma,j})$ does not fork over $M_{i,j}$ by Axiom (E)(h), continuity.
Now if $j < \delta_\gamma$ and $p \in \mathscr{S}^{\rm bs}(M_{\gamma,j})$ then for some $i < \gamma, p$ does not fork over $M_{i,j}$ (by Ax(E)(c)) and without loss of generality $j < \delta_i$.

Hence if clause (j) holds we have $u := \{\varepsilon : j < \varepsilon < \delta_i$ and $\mathbf{tp}(a_\varepsilon, M_{i,\varepsilon}, M_{i,\varepsilon+1})$ is a non-forking extension of $p \restriction M_{i,j}\}$ has λ members. But for $\varepsilon \in u$, $\mathbf{tp}(a_\varepsilon, M_{\gamma,\varepsilon}, M_{\gamma,\varepsilon+1})$ does not fork over $M_{i,\varepsilon}$ (by clause (i) of the assumption) hence does not fork over $M_{i,j}$ and by monotonicity it does not fork over $M_{\gamma,i}$ and by uniqueness it extends p. If clause $(j)^-$ holds the proof is similar. By 4.3 the model M_γ is $(\lambda, \mathrm{cf}(\gamma))$-brimmed over N_γ. $\square_{4.7}$

4.8 Lemma. *1) If* $M \in K_\lambda$ *and the models* $M_1, M_2 \in K_\lambda$ *are* $(\lambda, *)$*-brimmed over* M *(see Definition 1.15(2)), then* M_1, M_2 *are isomorphic over* M.
2) If $M_1, M_2 \in K_\lambda$ *are* $(\lambda, *)$*-brimmed then they are isomorphic.*

We prove some claims before proving 4.8; we will not much use the lemma, but it is of obvious interest and its proof is crucial in one point of §6.

4.9 Claim. *1)*

$(E)(i)^+$ *long non-forking amalgamation for $\alpha < \lambda^+$:*
if $\langle N_i : i \le \alpha \rangle$ is $\le_{\mathfrak{K}}$-increasing continuous sequence of members of K_λ, $a_i \in N_{i+1} \backslash N_i$ for $i < \alpha$, $p_i = \mathbf{tp}(a_i, N_i, N_{i+1}) \in \mathscr{S}^{\rm bs}(N_i)$ and $q \in \mathscr{S}^{\rm bs}(N_0)$, then we can find a $\le_{\mathfrak{K}}$-increasing continuous sequence $\langle N'_i : i \le \alpha \rangle$ of members of K_λ such that: $i \le \alpha \Rightarrow N_i \le_{\mathfrak{K}} N'_i$; some $b \in N'_0 \backslash N_0$ realizes q, $\mathbf{tp}(b, N_\alpha, N'_\alpha)$ does not fork over N_0 and $\mathbf{tp}(a_i, N'_i, N'_{i+1})$ does not fork over N_i for $i < \alpha$.

2) Above assume in addition that there are M, b^* such that $N_0 \le_{\mathfrak{K}} M \in K_\lambda$, $b^* \in M$ and $\mathbf{tp}(b^*, N_0, M) = q$. Then we can add: there is a $\le_{\mathfrak{K}}$-embedding of M into N'_0 over N_0 mapping b^* to b.

Proof. Straight (remembering Axiom (E)(i) on non-forking amalgamation of Definition 2.1). In details
1) Let M_0, b^* be such that $N_0 \le_{\mathfrak{K}[\mathfrak{s}]} M_0$ and $q = \mathbf{tp}(b^*, N_0, M_0)$ and apply part (2).
2) We choose (M_i, f_i) by induction on $i \le \alpha$ such that

⊛ (a) $M_i \in \mathfrak{K}_{\mathfrak{s}}$ is $\le_{\mathfrak{K}}$-increasing continuous
(b) f_i is a $\le_{\mathfrak{K}}$-embedding of N_i into M_i
(c) f_i is increasing continuous with $i \le \alpha$
(d) $M_0 = M$ and $f_0 = \mathrm{id}_{N_0}$
(e) $\mathbf{tp}(b^*, f_i(N_i), M_i)$ does not fork over N_0
(f) $\mathbf{tp}(f_{i+1}(a_i), M_i, M_{i+1})$ does not fork over $f_i(N_i)$.

For $i = 0$ there is nothing to do. For i limit take unions; clause (e) holsd by Ax(E)(h). Lastly, for $i = j + 1$, we can find (M'_i, f'_i) such that $f_j \subseteq f'_i$ and f'_i is an isomorphism from N_i onto M. Hence $f_j(N_j) \le_{\mathfrak{K}[\mathfrak{s}]} N'_i$. Now use Ax(E)(i) for $f_j(N_j), M'_i, N_i, f'_i(a_j), b^*$.

Having carried the induction, we rename to finish. $\square_{4.9}$

In the claim below, we are given a $\le_{\mathfrak{K}_\lambda}$-increasing continuous $\langle M_i : i \le \delta \rangle$ and $u_0, u_1, u_2 \subseteq \delta$ such that: u_0 is where we are already given $a_i \in M_{i+1} \backslash M_i$, $u_1 \subseteq \delta$ is where we shall choose $a_i (\in M'_{i+1} \backslash M'_i)$ and $u_2 \subseteq \delta$ is the place which we "leave for future use"; main case is $u_1 = \delta; u_0 = u_2 = \emptyset$.

4.10 Claim. *1) Assume*

- (a) $\delta < \lambda^+$ *is divisible by* λ
- (b) u_0, u_1, u_2 *are disjoint subsets of* δ
- (c) $\delta = \sup(u_1)$ *and* $\text{otp}(u_1)$ *is divisible by* λ
- (d) $\langle M_i : i \le \delta \rangle$ *is* $\le_{\mathfrak{K}}$*-increasing continuous in* $\mathfrak{K}_\lambda$
- (e) $\bar{\mathbf{a}} = \langle a_i : i \in u_0 \rangle, a_i \in M_{i+1} \backslash M_i, \mathbf{tp}(a_i, M_i, M_{i+1}) \in \mathscr{S}^{\text{bs}}(M_i)$.

Then we can find $\bar{M}' = \langle M'_i : i \le \delta \rangle$ *and* $\bar{\mathbf{a}}' = \langle a_i : i \in u_1 \rangle$ *such that*

- (α) $\bar{M}'$ *is* $\le_{\mathfrak{K}}$*-increasing continuous in* K_λ
- (β) $M_i \le_{\mathfrak{K}} M'_i$
- (γ) *if* $i \in u_0$ *then* $\mathbf{tp}(a_i, M'_i, M'_{i+1})$ *is a non-forking extension of* $\mathbf{tp}(a_i, M_i, M_{i+1})$
- (δ) *if* $i \in u_2$ *then* $M_i = M_{i+1} \Rightarrow M'_i = M'_{i+1}$
- (ε) *if* $i \in u_1$ *then* $\mathbf{tp}(a_i, M'_i, M'_{i+1}) \in \mathscr{S}^{\text{bs}}(M'_i)$
- (ζ) *if* $i < \delta, p \in \mathscr{S}^{\text{bs}}(M'_i)$ *then for* λ *ordinals* $j \in u_1 \cap (i, \delta)$ *the type* $\mathbf{tp}(a_j, M'_j, M'_{j+1})$ *is a non-forking extension of* p.

2) If we add in part (1) the assumption

- (g) $M_0 \le_{\mathfrak{K}} N \in K_\lambda$

then we can add to the conclusion

- (η) *there is an* $\le_{\mathfrak{K}}$*-embedding* f *of* N *into* M'_0 *over* M_0 *and moreover* f *is onto.*

3) If we add in part (1) the assumption

- $(h)^+$ $M_0 \le_{\mathfrak{K}} N \in K_\lambda$ *and* $b \in N \backslash M_0, \mathbf{tp}(b, M_0, N) \in \mathscr{S}^{\text{bs}}(M_0)$

then we can add to the conclusion

- $(\eta)^+$ *as in* (η) *and* $\mathbf{tp}(f(b), M_\delta, M'_\delta)$ *does not fork over* M_0.

4) We can strengthen clause (ζ) *in part (1) to*

- $(\zeta)^+$ *if* $i < \delta$ *and* $p \in \mathscr{S}^{\text{bs}}(M'_i)$ *then for* λ *ordinals* j *we have* $j \in [i, \delta) \cap u_1$ *and* $\mathbf{tp}(a_j, M'_j, M'_{j+1})$ *is a non-forking extension of* p *and* $\text{otp}(u_1 \cap j \backslash i) < \lambda$.

Proof. Straight like 4.9(2). Note that we can find a sequence $\langle u_{1,i,\varepsilon} : i < \delta, \varepsilon < \lambda \rangle$ such that: this is a sequence of pairwise disjoint subsets of u_1 each of cardinality λ satisfying $u_{1,i,\varepsilon} \subseteq \{j : i < j, j \in u_1$ and $|u_1 \cap (i,j)| < \lambda\}$ (or we can demand that $i \leq i_1 < i_2 \leq \delta \wedge |u_1 \cap (i_1, i_2)| = \lambda \Rightarrow |u_{1,i,\varepsilon} \cap (i_1, i_2)| = \lambda$). $\square_{4.10}$

Toward building our rectangles of models with sides of difference lengths (and then we shall use 4.7) we show (to understand the aim of the clauses in the conclusion of 4.11 see the proof of 4.8 below):

4.11 Claim. *Assume*

(a) $\delta_\ell < \lambda^+$ *is divisible by* λ *for* $\ell = 1, 2$
(b) $\bar{M}^\ell = \langle M^\ell_\alpha : \alpha \leq \delta_\ell \rangle$ *is* $\leq_{\mathfrak{K}}$*-increasing continuous for* $\ell = 1, 2$
(c) $u^\ell_0, u^\ell_1, u^\ell_2$ *are disjoint subsets of* $\delta_\ell, \mathrm{otp}(u^\ell_1)$ *is divisible by* λ *and* $\delta_\ell = \sup(u^\ell_1)$ *for* $\ell = 1, 2$
(d) $\bar{\mathbf{a}}^\ell \equiv \langle a^\ell_\alpha : \alpha \in u^\ell_0 \rangle$ *and* $\mathbf{tp}(a^\ell_\alpha, M^\ell_\alpha, M^\ell_{\alpha+1}) \in \mathscr{S}^{\mathrm{bs}}(M^\ell_\alpha)$ *for* $\ell = 1, 2, \alpha \in u^\ell_0$
(e) $M^1_0 = M^2_0$
(f) $\alpha \in u^\ell_1 \cup u^\ell_2 \Rightarrow M^\ell_\alpha = M^\ell_{\alpha+1}$ *for* $\ell = 1, 2$.

<u>Then</u> we can find $\bar{f}^\ell = \langle f^\ell_\alpha : \alpha \leq \delta_\ell \rangle, \bar{\mathbf{b}}^\ell = \langle b^\ell_\alpha : \alpha \in u^\ell_0 \cup u^\ell_1 \rangle$ *for* $\ell = 1, 2$ *and* $\bar{M} = \langle M_{\alpha,\beta} : \alpha \leq \delta_1, \beta \leq \delta_2 \rangle$ *and functions* $\zeta : u^1_1 \to \delta_2$ *and* $\varepsilon : u^2_1 \to \delta_1$ *such that*

$(\alpha)_1$ *for each* $\alpha \leq \delta_1, \langle M_{\alpha,\beta} : \beta \leq \delta_2 \rangle$ *is* $\leq_{\mathfrak{K}}$*-increasing continuous*
$(\alpha)_2$ *for each* $\beta \leq \delta_2, \langle M_{\alpha,\beta} : \alpha \leq \delta_1 \rangle$ *is* $\leq_{\mathfrak{K}}$*-increasing continuous*
$(\beta)_1$ *for* $\alpha \in u^1_0, b^1_\alpha$ *belongs to* $M_{\alpha+1,0}$ *and* $\mathbf{tp}(b^1_\alpha, M_{\alpha,\delta_2}, M_{\alpha+1,\delta_2}) \in \mathscr{S}^{\mathrm{bs}}(M_{\alpha,\delta_2})$ *does not fork over* $M_{\alpha,0}$
$(\beta)_2$ *for* $\beta \in u^2_0, b^2_\beta$ *belongs to* $M_{0,\beta+1}$ *and* $\mathbf{tp}(b^2_\beta, M_{\delta_1,\beta}, M_{\delta_1,\beta+1}) \in \mathscr{S}^{\mathrm{bs}}(M_{\delta_1,\beta})$ *does not fork over* $M_{0,\beta}$
$(\gamma)_1$ *for* $\alpha \in u^1_1, \zeta(\alpha) < \delta_2$ *and we have* $b^1_\alpha \in M_{\alpha+1,\zeta(\alpha)+1}$ *and* $\mathbf{tp}(b^1_\alpha, M_{\alpha,\delta_2}, M_{\alpha+1,\delta_2})$ *does not fork over* $M_{\alpha,\zeta(\alpha)+1}$
$(\gamma)_2$ *for* $\beta \in u^2_1, \varepsilon(\beta) < \delta_1$ *and we have* $b^2_\beta \in M_{\varepsilon(\beta)+1,\beta+1}$ *and* $\mathbf{tp}(b^2_\beta, M_{\delta_1,\beta}, M_{\delta_1,\beta+1})$ *does not fork over* $M_{\varepsilon(\beta)+1,\beta}$

$(\delta)_1$ *if* $\alpha < \delta_1, \beta < \delta_2$ *and* $p \in \mathscr{S}^{\rm bs}(M_{\alpha,\beta})$ *or just* $p \in \mathscr{S}^{\rm bs}(M_{\alpha,\beta+1})$ *then for* λ *ordinals*[14] $\alpha' \in [\alpha, \delta_1) \cap u_1^1$, *the type* $\mathbf{tp}(b^1_{\alpha'}, M_{\alpha',\beta+1}, M_{\alpha+1,\beta+1})$ *is a (well defined) non-forking extension of* p *and* $\beta = \zeta(\alpha')$

$(\delta)_2$ *if* $\alpha < \delta_1, \beta < \delta_2$ *and* $p \in \mathscr{S}^{\rm bs}(M_{\alpha,\beta})$ *or just* $p \in \mathscr{S}^{\rm bs}(M_{\alpha+1,\beta})$ *then for* λ *ordinals*[15] $\beta' \in [\beta, \delta_2) \cap u_1^2$, *the type* $\mathbf{tp}(b^2_{\beta'}, M_{\alpha+1,\beta'}, M_{\alpha+1,\beta'+1})$ *is a non-forking extension of* p *and* $\alpha = \varepsilon(\beta')$

(ε) $M_{0,0} = M^1_0 = M^2_0$

$(\zeta)_1$ f^1_α *is an isomorphism from* M^1_α *onto* $M_{\alpha,0}$ *such that* $\alpha \in u^1_0 \Rightarrow f^1_\alpha(a^1_\alpha) = b^1_\alpha$
$f^1_0 = {\rm id}_{M^1_0}$ *and* f^1_α *is increasing continuous with* α

$(\zeta)_2$ f^2_β *is an isomorphism from* M^2_β *onto* $M_{0,\beta}$ *such that* $\beta \in u^2_0 \Rightarrow f^2_\beta(a^2_\beta) = b^2_\beta$
$f^2_0 = {\rm id}_{M^2_0}$ *and* f^2_α *is increasing continuous with* α

$(\eta)_1$ *if* $\alpha \in u^1_2$ *then* $M_{\alpha,\beta} = M_{\alpha+1,\beta}$ *for every* $\beta \leq \delta_2$

$(\eta)_2$ *if* $\beta \in u^2_2$ *then* $M_{\alpha,\beta} = M_{\alpha,\beta+1}$ *for every* $\alpha \leq \delta_1$.

Proof. Straight, divide u^ℓ_1 to $\delta_{3-\ell}$ subsets large enough), in fact, we can first choose the function $\zeta(-), \varepsilon(-)$. Now choose $\langle M_{\alpha,\beta} : \alpha \leq \delta_1, \beta \leq \beta^* \rangle, \langle f^1_\alpha : \alpha \leq \delta_1 \rangle, \langle f^2_\beta : \beta \leq \beta^* \rangle$ and $\langle b^1_\alpha : \zeta(\alpha) \in \beta^* \rangle, \langle b^2_\beta : \beta < \beta^* \rangle$ by induction on β^* using 4.10. $\square_{4.11}$

Proof of 4.8. By 1.16(3), i.e., uniqueness of the (λ, θ_ℓ)-brimmed model over M, it is enough to show for any regular $\theta_1, \theta_2 \leq \lambda$ that there is a model $N \in K_\lambda$ which is (λ, θ_ℓ)-brimmed over M for $\ell = 1, 2$. Let $\delta_1 = \lambda \times \theta_1, \delta_2 = \lambda \times \theta_2$ (ordinal multiplication, of course), $M^1_\alpha = M^2_\beta = M$ for $\alpha \leq \delta_1, \beta \leq \delta_2, u^1_0 = u^2_0 = \emptyset, u^1_1 = \delta_1, u^2_1 = \delta_2, u^1_2 = u^2_2 = \emptyset$. So there are $\langle M_{\alpha,\beta} : \alpha \leq \delta_1, \beta \leq \delta_2 \rangle, \langle b^1_\alpha : \alpha < \delta_1 \rangle, \langle b^2_\beta : \beta < \delta_2 \rangle$ and $\langle f^1_\alpha : \alpha \leq \delta_1 \rangle, \langle f^2_\beta : \beta \leq \delta_2 \rangle$ as in Claim 4.11. Without loss of generality $f^1_\alpha = f^2_\alpha = {\rm id}_M$. Now

$(*)_1$ $\langle M_{\alpha,\delta_2} : \alpha \leq \delta_1 \rangle$ is $\leq_{\mathfrak{K}}$-increasing continuous in K_λ (by clause $(\alpha)_1$, of 4.11). Also

[14] we can add "and ${\rm otp}(\alpha' \cap u^1_1 \backslash \alpha_2) < \lambda$"
[15] we can add "and ${\rm otp}(\beta' \cap u^2_1 \backslash \beta_2) < \lambda$"

$(*)_2$ if $\alpha < \delta_1$ and $p \in \mathscr{S}(M_{\alpha,\delta_2})$ <u>then</u> for λ ordinals $\alpha' \in (\alpha, \delta_1) \cap u_1^1$ the type $\mathbf{tp}(b^1_{\alpha',\delta_2}, M_{\alpha',\delta_2}, M_{\alpha'+1,\delta_2})$ is a non-forking extension of p.

(Easy, by Axiom (E)(c) for some $\beta < \delta_2, p$ does not fork over $M_{\alpha,\beta+1}$ and use clause $(\delta)_1$ of 4.11).

So by 4.7, M_{δ_1,δ_2} is $(\lambda, \mathrm{cf}(\delta_1))$-brimmed over M_{0,δ_2} which is M.

Similarly M_{δ_1,δ_2} is $(\lambda, \mathrm{cf}(\delta_2))$-brimmed over $M_{\delta_1,0}$ which is M; so together we are done.

$\square_{4.8}$

4.12 Claim. *1) If $M \in K_{\lambda^+}$ and $p \in \mathscr{S}^{\mathrm{bs}}(M_0), M_0 \leq_{\mathfrak{K}} M$ (so $M_0 \in K_\lambda$), <u>then</u> we can find $b, \langle N^0_\alpha : \alpha \leq \lambda^+\rangle$ and $\langle N^1_\alpha : \alpha \leq \lambda^+\rangle$ such that*

- *(a) $\langle N^0_\alpha : \alpha < \lambda^+\rangle$ is a $\leq_{\mathfrak{K}}$-representation of $N^0_{\lambda^+} = M$*
- *(b) $\langle N^1_\alpha : \alpha < \lambda^+\rangle$ is a $\leq_{\mathfrak{K}}$-representation of $N^1_{\lambda^+} \in K_{\lambda^+}$*
- *(c) $N^1_{\alpha+1}$ is (λ, λ)-brimmed over N^1_α (hence $N^1_{\lambda^+}$ is saturated over λ in $\mathfrak{K}$)*
- *(d) $M_0 \leq N^0_0$ and $N^0_\alpha \leq_{\mathfrak{K}} N^1_\alpha$*
- *(e) $\mathbf{tp}_{\mathfrak{s}}(b, N^0_\alpha, N^1_\alpha)$ is a non-forking extension of p for every $\alpha < \lambda^+$.*

2) We can add

- *(f) for $\alpha < \beta < \lambda^+, N^1_\beta$ is $(\lambda, *)$-brimmed over $N^0_\beta \cup N^1_\alpha$.*

Proof. 1) Easy by long non-forking amalgamation 4.9 (see 1.17).
2) Use 4.7. $\square_{4.12}$

4.13 Conclusion. 1) $K_{\lambda^{++}} \neq \emptyset$.
2) $K_{\lambda^+} \neq \emptyset$.
3) No $M \in K_{\lambda^+}$ is $\leq_{\mathfrak{K}}$-maximal.

Proof. 1) By (2) + (3).
2) By (B) of 2.1.
3) By 4.12. $\square_{4.13}$

4.14 Exercise: 1) Let $M \in K_{\mathfrak{s}}$ be superlimit and $\mathfrak{t} = \mathfrak{s}_{[M]}$, so $K_{\mathfrak{t}}$ is categorical. If $(M, N, a) \in K^{\rm bs}_{\mathfrak{t}}$ is reduced for $\mathfrak{t}$, then it is reduced for $\mathfrak{s}$.
2) In 4.6(3),(4),(5), we can omit the assumption "$\mathfrak{s}$ is categorical" if:

- (a) we add in part (3), each N_i is superlimit (equivalently brimmed)
- (b) in parts (4),(5) add the assumption "M_0 is superlimit".

2) Some extra assumption in 4.6(5) is needed.

§5 Non-structure or some unique amalgamation

We shall assuming $2^\lambda < 2^{\lambda^+} < 2^{\lambda^{++}}$ get from essentially $\dot{I}(\lambda^{++}, K) < 2^{\lambda^{++}}$ pedantically $< \mu_{\rm unif}(\lambda^{++}, 2^{\lambda^+})$ or just $\dot{I}(\lambda^{++}, K(\lambda^+\text{-saturated}))$ $< \mu_{\rm unif}(\lambda^{++}, 2^{\lambda^+})$, many cases of uniqueness of amalgamation assuming in addition WDmId(λ^+) is not λ^{++}-saturated, a weak assumption. The proof is similar to [Sh 482], [Sh 576, §3] but now we rely on Chapter VII, the "lean" version; and by the "full version" without we can eliminate the additional assumption.

We define $K^{3,\rm bt}_\lambda$, it is a brimmed relative of $K^{3,\rm bs}_\lambda$ hence the choice of bt; it guarantees much brimness (see Definition 5.2) hence it guarantees some uniqueness, that is, if $(M, N, a) \in K^{3,\rm bt}_\lambda$, M is unique (recalling the uniqueness of the brimmed model) and more crucially, we consider $K^{3,\rm uq}_\lambda$, (the family of members of $K^{3,\rm bs}_\lambda$ for which we have uniqueness in relevant extensions). Having enough such triples is the main conclusion of this section (in 5.9 under "not too many non-isomorphic models" assumptions). In 5.4 we give some properties of $K^{3,\rm bt}_\lambda, K^{3,\rm uq}_\lambda$.

To construct models in λ^{++} we use approximations of cardianlity in λ^+ with "obligation" on the further construction, which are presented as pairs $(\bar{M}, \bar{\mathbf{a}}) \in K^{\rm sq}_\lambda$ ordered by $\le_{\rm ct}$, see Definition 5.5, Claims 5.6, 5.7. We need more: the triples $(\bar{M}, \bar{\mathbf{a}}, \mathbf{f}) \in K^{\rm mqr}_S, K^{\rm nqr}_S$ in Definition 5.12, Claim 5.13. All this enables us to quote results of [Sh 576, §3] or better VII§2, but apart from believing the reader do not need to know non of them.

5.1 Hypothesis.

(a) $\mathfrak{s} = (\mathfrak{K}, \bigcup, \mathscr{S}^{\rm bs})$ is a good λ-frame.

5.2 Definition. 1) Let $K^{3,\rm bt}_\lambda = K^{3,\rm bt}_{\mathfrak{s}}$ be the set of triples (M, N, a) such that for some $\partial = \mathrm{cf}(\partial) \le \lambda$, $M \le_{\mathfrak{K}} N$ are both (λ, ∂)-brimmed members of K_λ, $a \in N \backslash M$ and $\mathbf{tp}(a, M, N) \in \mathscr{S}^{\rm bs}(M)$.
2) For $(M_\ell, N_\ell, a_\ell) \in K^{3,\rm bt}_\lambda$ for $\ell = 1, 2$ let $(M_1, N_1, a_1) <_{\rm bt} (M_2, N_2, a_2)$ mean $a_1 = a_2$, $\mathbf{tp}(a_1, M_2, N_2)$ does not fork over M_1 and for some $\partial_2 = \mathrm{cf}(\partial_2) \le \lambda$, the model M_2 is (λ, ∂_2)-brimmed over M_1 and the model N_2 is (λ, ∂_2)-brimmed over N_1. Finally $(M_1, N_1, a_2) \le_{\rm bt} (M_2, N_2, a_2)$ means $(M_1, N_1, a_1) <_{\rm bt} (M_2, N_2, a_2)$ or $(M_1, N_1, a_1) = (M_2, N_2, a_2)$.

5.3 Definition. 1) Let "$(M_0, M_2, a) \in K^{3,\rm uq}_\lambda$" mean: $(M_0, M_2, a) \in K^{3,\rm bs}_\lambda$ and: for every M_1 satisfying $M_0 \le_{\mathfrak{K}} M_1 \in K_\lambda$, the amalgamation M of M_1, M_2 over M_0, with $\mathbf{tp}(a, M_1, M)$ not forking over M_0, is unique, that is:

- $(*)$ if for $\ell = 1, 2$ we have $M_0 \le_{\mathfrak{K}} M_1 \le_{\mathfrak{K}} M^\ell \in K_\lambda$ and f_ℓ is a $\le_{\mathfrak{K}}$-embedding of M_2 into M^ℓ over M_0 (so $f_1 \restriction M_0 = f_2 \restriction M_0 = \mathrm{id}_{M_0}$) such that $\mathbf{tp}(f_\ell(a), M_1, M^\ell)$ does not fork over M_0, then
  - (a) [uniqueness]:
    for some M', g_1, g_2 we have: $M_1 \le_{\mathfrak{K}} M' \in K_\lambda$ and g_ℓ is a $\le_{\mathfrak{K}}$-embedding of M^ℓ into M' over M_1 for $\ell = 1, 2$ such that $g_1 \circ f_1 \restriction M_2 = g_2 \circ f_2 \restriction M_2$
  - (b) [being reduced] $f_\ell(M_2) \cap M_1 = M_0$
    [this is "for free" in the proofs; and is not really necessary so the decision if to include it is not important but simplify notation, but see 5.4(3)].

2) $K^{3,\rm uq}_\lambda$ is dense (or $\mathfrak{s}$ has density for $K^{3,\rm uq}_\lambda$) when $K^{3,\rm uq}_\lambda$ is dense in $(K^{3,\rm bs}_\lambda, \le_{\rm bs})$, i.e., for every $(M_1, M_2, a) \in K^{3,\rm bs}_\lambda$ there is $(M_1, N_2, a) \in K^{3,\rm uq}_\lambda$ such that $(M_1, M_2, a) \le_{\rm bs} (N_1, N_2, a) \in K^{3,\rm uq}_\lambda$.
3) $K^{3,\rm uq}_\lambda$ has existence or $\mathfrak{s}$ has existence for $K^{3,\rm uq}_\lambda$ when for every

$M_0 \in K_\lambda$ and $p \in \mathscr{S}^{\rm bs}(M_0)$ for some M_1, a we have $(M_0, M_1, a) \in K^{3,{\rm uq}}_\lambda$ and $p = \mathbf{tp}(a, M_0, M_1)$.
4) $K^{3,{\rm uq}}_{\mathfrak{s}} = K^{3,{\rm uq}}_\lambda$.

5.4 Claim. *1) The relation $\le_{\rm bt}$ is a partial order on $K^{3,{\rm bt}}_\lambda$ that is transitive and reflexive (but not necessarily satisfying the parallel of Ax V of* a.e.c. *see Definition 1.4).*
2) If $(M_\alpha, N_\alpha, a) \in K^{3,{\rm bt}}_\lambda$ is $\le_{\rm bt}$-increasing continuous for $\alpha < \delta$ where δ is a limit ordinal $< \lambda^+$ then $(M, N, a) = (\bigcup_{\alpha<\delta} M_\alpha, \bigcup_{\alpha<\delta} N_\alpha, a)$ belongs to $K^{3,{\rm bt}}_\lambda$ and $\alpha < \delta \Rightarrow (M_\alpha, N_\alpha, a) \le_{\rm bt} (M, N, a)$ and so (M, N, a) is a $\le_{\rm bt}$-upper bound of $\langle (M_\alpha, N_\alpha, a) : \alpha < \delta \rangle$.
3) In (*) of 5.3(1), clause (b) follows from (a).

Proof. Easy, e.g. (3) by the uniqueness (i.e., clause (a)) and 4.6(4). $\square_{5.4}$

We now define $K^{\rm sq}_{\lambda^+}$, a family of $\le_{\mathfrak{K}}$-increasing continuous sequences (the reason for sq) in K_λ of length λ^+, will be used to approximate stages in constructing models in $K_{\lambda^{++}}$.

5.5 Definition. 1) Let $K^{\rm sq}_{\lambda^+} = K^{\rm sq}_{\mathfrak{s}}$ be the set of pairs $(\bar M, \bar{\mathbf{a}})$ such that (sq stands for sequence):

- *(a)* $\bar M = \langle M_\alpha : \alpha < \lambda^+ \rangle$ is a $\le_{\mathfrak{K}}$-increasing continuous sequence of models from K_λ
- *(b)* $\bar{\mathbf{a}} = \langle a_\alpha : \alpha \in S \rangle$, where $S \subseteq \lambda^+$ is stationary in λ^+ and $a_\alpha \in M_{\alpha+1} \backslash M_\alpha$
- *(c)* for some club E of λ^+ for every $\alpha \in S \cap E$ we have $\mathbf{tp}(a_\alpha, M_\alpha, M_{\alpha+1}) \in \mathscr{S}^{\rm bs}(M_\alpha)$
- *(d)* if $p \in \mathscr{S}^{\rm bs}(M_\alpha)$ then for stationarily many $\delta \in S$ we have: $\mathbf{tp}(a_\delta, M_\delta, M_{\delta+1}) \in \mathscr{S}^{\rm bs}(M_\delta)$ does not fork over M_α and extends p.

In such cases we let $M = \bigcup_{\alpha<\lambda^+} M_\alpha$.
2) When for $\ell = 1, 2$ we are given $(\bar M^\ell, \bar{\mathbf{a}}^\ell) \in K^{\rm sq}_{\lambda^+}$ we say $(\bar M^1, \bar{\mathbf{a}}^1) \le_{\rm ct}$

$(\bar{M}^2, \bar{\mathbf{a}}^2)$ <u>if</u> for some club E of λ^+, letting $\bar{\mathbf{a}}^\ell = \langle a^\ell_\delta : \delta \in S^\ell \rangle$ for $\ell = 1, 2$, of course, we have

- (a) $S^1 \cap E \subseteq S^2 \cap E$
- (b) if $\delta \in S^1 \cap E$ then
  - (α) $M^1_\delta \leq_{\mathfrak{K}} M^2_\delta$,
  - (β) $M^1_{\delta+1} \leq_{\mathfrak{K}} M^2_{\delta+1}$
  - (γ) $a^2_\delta = a^1_\delta$
  - (δ) $\mathbf{tp}(a^1_\delta, M^2_\delta, M^2_{\delta+1})$ does not fork over M^1_δ, so in particular $a^1_\delta \notin M^2_\delta$.

5.6 *Observation.* 1) If $(\bar{M}, \bar{\mathbf{a}}) \in K^{\mathrm{sq}}_{\lambda^+}$ then $M := \bigcup\limits_{\alpha<\lambda^+} M_\alpha \in K_{\lambda^+}$ is saturated.
2) $K^{\mathrm{sq}}_{\lambda^+}$ is partially ordered by $\leq_{\mathrm{ct}}$. $\square_{5.6}$

5.7 Claim. *Assume $\langle (\bar{M}^\zeta, \bar{\mathbf{a}}^\zeta) : \zeta < \zeta^* \rangle$ is $\leq_{\mathrm{ct}}$-increasing in $K^{\mathrm{sq}}_{\lambda^+}$, and ζ^* is a limit ordinal $< \lambda^{++}$, <u>then</u> the sequence has a $\leq_{\mathrm{ct}}$-lub $(\bar{M}, \bar{\mathbf{a}})$.*

Proof. Let $\bar{\mathbf{a}}^\zeta = \langle a^\zeta_\delta : \delta \in S_\zeta \rangle$ for $\zeta < \zeta^*$ and without loss of generality $\zeta^* = \mathrm{cf}(\zeta^*)$ and for $\zeta < \xi < \zeta^*$ let $E_{\zeta,\xi}$ be a club of λ^+ consisting of limit ordinals witnessing $(\bar{M}^\zeta, \bar{\mathbf{a}}^\zeta) \leq_{\mathrm{ct}} (\bar{M}^\xi, \bar{\mathbf{a}}^\xi)$, i.e. as in 5.5(2).

<u>Case 1</u>: $\zeta^* < \lambda^+$.

Let $E = \cap\{E_{\zeta,\xi} : \zeta < \xi < \zeta^*\}$ and for $\delta \in E$ let $M_\delta = \cup\{M^\zeta_\delta : \zeta < \zeta^*\}$ and $M_{\delta+1} = \cup\{M^\zeta_{\delta+1} : \zeta < \zeta^*\}$ and for any other $\alpha, M_\alpha = M_{\mathrm{Min}(E\setminus\alpha)}$. Let $S = \bigcup\limits_{\zeta<\zeta^*} S_\zeta \cap E$ and for $\delta \in S$ let $a_\delta = a^\zeta_\delta$ for every ζ for which $\delta \in S_\zeta$. Clearly $M_\alpha \in K_\lambda$ is $\leq_{\mathfrak{K}}$-increasing continuous and $\zeta < \zeta^* \wedge \delta \in E \Rightarrow M^\zeta_\delta \leq_{\mathfrak{K}} M_\delta$ & $M^\zeta_{\delta+1} \leq_{\mathfrak{K}} M_{\delta+1}$.

Now if $\delta \in E \cap S_\zeta$ then $\xi \in [\zeta, \zeta^*)$ implies $\mathbf{tp}(a_\delta, M^\xi_\delta, M_{\delta+1}) = \mathbf{tp}(a^\xi_\delta, M^\xi_\delta, M^\xi_{\delta+1})$ does not fork over M^ζ_δ (and $\langle M^\xi_\delta : \xi \in [\zeta, \delta)\rangle$, $\langle M^\xi_{\delta+1} : \xi \in [\zeta, \delta)\rangle$ are $\le_{\mathfrak{K}}$-increasing continuous); hence by Axiom (E)(h) we know that $\mathbf{tp}(a_\delta, M_\delta, M_{\delta+1})$ does not fork over M^ζ_δ and in particular $\in \mathscr{S}^{\mathrm{bs}}(M_\delta)$. Also if $N \le_{\mathfrak{K}} M := \bigcup\limits_{\alpha<\lambda^+} M_\alpha, N \in K_\lambda$ and $p \in \mathscr{S}^{\mathrm{bs}}(N)$ then for some $\delta(*) \in E, N \le_{\mathfrak{K}} M_{\delta(*)}$, let $p_1 \in \mathscr{S}^{\mathrm{bs}}(M_{\delta(*)})$ be a non-forking extension of p, so for some $\zeta < \zeta^*, p$ does not fork over $M^\zeta_{\delta(*)}$ hence for stationarily many $\delta \in S_\zeta, q^0_\delta = \mathbf{tp}(a_\delta, M^\zeta_\delta, M^\zeta_{\delta+1})$ is a non-forking extension of $p_1 \restriction M^\zeta_{\delta(*)}$, hence this holds for stationarily many $\delta \in S \cap E$ and for each such $\delta, q_\delta = \mathbf{tp}(a_\delta, M_\delta, M_{\delta+1})$ is a non-forking extension of $p_1 \restriction M^\zeta_{\delta(*)}$, hence of p_1 hence of p. Looking at the definitions, clearly $(\bar{M}, \bar{\mathbf{a}}) \in K^{\mathrm{sq}}_{\lambda^+}$ and $\zeta < \zeta^* \Rightarrow (\bar{M}^\zeta, \bar{\mathbf{a}}^\zeta) \le_{\mathrm{ct}} (\bar{M}, \bar{\mathbf{a}})$.

Lastly, it is easy to check the $\le_{\mathrm{ct}}$-l.u.b.

<u>Case 2</u>: $\zeta^* = \lambda^+$.

Similarly using diagonal union, i.e., $E = \{\delta < \lambda^+ : \delta$ is a limit ordinal such that $\zeta < \xi < \delta \Rightarrow \delta \in E_{\zeta,\varepsilon}\}$ and we choose $M_\alpha = \cup\{M^\zeta_\alpha : \zeta < \alpha\}$ when $\alpha \in E$ and $M_\alpha = M_{\min(E \setminus (\alpha+1))}$ otherwise. $\square_{5.7}$

5.8 Observation. Assume $K^{3,\mathrm{uq}}_\lambda$ is dense in $K^{3,\mathrm{bs}}_\lambda$, i.e., in $(K^{3,\mathrm{bs}}_\lambda, \le_{\mathrm{bs}})$ and even in $(K^{3,\mathrm{bt}}_\lambda, <_{\mathrm{bt}})$. <u>Then</u>

(*a*) if $M \in K_\lambda$ is superlimit and $p \in \mathscr{S}^{\mathrm{bs}}(M)$ then there are N, a such that $(M, N, a) \in K^{3,\mathrm{uq}}_\lambda$ and $p = \mathbf{tp}(a, M, N)$

(*b*) if in addition $K_{\mathfrak{s}}$ is categorical (in λ) <u>then</u> $\mathfrak{s}$ has existence for $K^{3,\mathrm{uq}}_\lambda$ (recall that this means that for every $M \in K_{\mathfrak{s}}$ and $p \in \mathscr{S}^{\mathrm{bs}}(M)$ for some pair (N, a) we have $(M, N, a) \in K^{3,\mathrm{uq}}_\lambda$ and $p = \mathbf{tp}(a, M, N)$).

Proof. Should be clear. $\square_{5.8}$

Now the assumption of 5.8 are justified by the following theorem (and the categoricity in (b) is justified by Claim 1.26).

5.9 First Main Claim. *Assume that*

(a) *as in 5.1*

(b) $\mathrm{WDmId}(\lambda^+)$ *is not* λ^{++}*-saturated and*[16] $2^\lambda < 2^{\lambda^+} < 2^{\lambda^{++}}$.

If $\dot I(\lambda^{++}, K) < \mu_{\mathrm{unif}}(\lambda^{++}, 2^{\lambda^+}$ *or just* $\dot I(\lambda^{++}, K(\lambda^+\text{-saturated})) < \mu_{\mathrm{unif}}(\lambda^{++}, 2^{\lambda^+})$, *then for every* $(M, N, a) \in K^{3,\mathrm{bs}}_\lambda$ *there is* $(M^*, N^*, a) \in K^{3,\mathrm{bt}}_\lambda$ *such that* $(M, N, a) <_{\mathrm{bt}} (M^*, N^*, a)$ *and* $(M^*, N^*, a) \in K^{3,\mathrm{uq}}_\lambda$.

5.10 Explanation. The reader who agrees to believe in 5.9 can ignore the rest of this section (though it can still serve as a good exercise).

Let $\langle S_\alpha : \alpha < \lambda^{++}\rangle$ be a sequence of subsets of λ^+ such that $\alpha < \beta \Rightarrow |S_\alpha \backslash S_\beta| \le \lambda$ and $S_{\alpha+1} \backslash S_\alpha \ne \emptyset$ mod $\mathrm{WDmId}(\lambda^+)$, exists by assumption.

Why having (M, N, a) failing the conclusion of 5.9 helps us to construct many models in $K_{\lambda^{++}}$? The point is that we can choose $(\bar M^\alpha, \bar{\mathbf{a}}^\alpha) \in K^{\mathrm{sq}}_{\lambda^+}$ with $\mathrm{Dom}(\bar{\mathbf{a}}^\alpha) = S_\alpha$ for $\alpha < \lambda^{++}$, $<_{\mathrm{ct}}$-increasing continuous (see 5.7).

Now for $\alpha = \beta + 1$, having $(\bar M^\beta, \bar{\mathbf{a}}^\beta)$, without loss of generality M^β_{i+1} is brimmed over M^β_i and we shall choose M^α_i by induction on $i < \lambda^+$ (for simplicity we assume $M^\alpha_i \cap \bigcup\{M^\beta_j : j < \lambda^+\} = M^\beta_i$) and $M^\beta_i \le_{\mathfrak{K}} M^\alpha_i \in K_\lambda$ and $\mathbf{tp}(a^\beta_i, M^\alpha_i, M^\alpha_{i+1})$ does not fork over M^β_i and M^α_{i+1} is brimmed over M^α_i).

Given $(\bar M^\beta, \bar{\mathbf{a}}^\beta), \bar M^\beta = \langle M^\beta_i : i < \lambda^+\rangle, \bar{\mathbf{a}}^\beta = \langle a^\beta_i : i \in S_\beta\rangle$ we work toward building $(\bar M^\alpha, \bar{\mathbf{a}}^\alpha), \alpha_{\beta+1}$.

We start with choosing (M^α_0, b) such that no member of $K^{3,\mathrm{bs}}_\lambda$ which is $\le_{\mathrm{bs}}$-above $(M^\beta_0, M^\alpha_0, b) \in K^{3,\mathrm{bs}}_\lambda$ belongs to $K^{3,\mathrm{uq}}_\lambda$ and will choose M^β_i by induction on i such that $(M^\beta_i, M^\alpha_i, b) \in K^{3,\mathrm{bs}}_\lambda$ is $\le_{\mathrm{bs}}$-increasing continuous and even $<_{\mathrm{bt}}$-increasing hence in particular that $\mathbf{tp}(b, M^\beta_i, M^\alpha_i)$ does not fork over M^α_0. Now in each stage $i = j + 1$, as M^β_i is universal over M^β_j, and the choice of M^α_0, b we have some freedom. So it makes sense that we will have many possible outcomes, i.e., models $M = \bigcup\{M^\alpha_i : \alpha < \lambda^{++}, i < \lambda^+\}$ which are in

[16]alternatively the parallel versions for the definitional weak diamond, but not here

$K_{\lambda^{++}}$. The combination of what we have above and §3 better VII§2 gives that $2^\lambda < 2^{\lambda^+} < 2^{\lambda^{++}}$ is enough to materialize this intuition. If in addition $2^\lambda = \lambda^+$ and moreover $\diamondsuit_{\lambda^+}$ it is considerably easier. In the end we still have to define $\bar{\mathbf{a}}^\alpha \restriction (S_\alpha \backslash S_\beta)$ as required in Definition 5.5, [Sh 832]. An alternative is to force a model in λ^{++}. Now below we replace $K^{3,\text{sq}}_{\lambda^+}$ by $K^{\text{mqr}}_{\lambda^+}, K^{\text{nqr}}_S$ but actually $K^{3,\text{sq}}_{\lambda^+}$ is enough. So we need a somewhat more complicated relative as elaborated below which anyhow seems to me more natural.

5.11 Second Main Claim. *Assume $2^\lambda < 2^{\lambda^+} < 2^{\lambda^{++}}$ (or the parallel versions for the definitional weak diamond). If $\dot{I}(\lambda^{++}, K(\lambda^+\text{-saturated})) < \mu_{\text{unif}}(\lambda^{++}, 2^{\lambda^+})$, then for every $(M, N, a) \in K^{3,\text{bt}}_\lambda$ there is $(M^*, N^*, a) \in K^{3,\text{bt}}_\lambda$ such that $(M, N, a) <_{\text{bt}} (M^*, N^*, a)$ and $(M^*, N^*, a) \in K^{3,\text{uq}}_\lambda$.*

We shall not prove here 5.11 and shall not use it, it is proved in the full version of Chapter VII; toward proving 5.9 (by quoting) let

5.12 Definition. Let $S \subseteq \lambda^+$ be a stationary subset of λ^+.
1) Let K^{mqr}_S or $K^{\text{mqr}}_{\lambda^+}[S]$ be the set of triples $(\bar{M}, \bar{\mathbf{a}}, \mathbf{f})$ such that:

(*a*) $\bar{M} = \langle M_\alpha : \alpha < \lambda^+ \rangle$ is $\leq_{\mathfrak{K}}$-increasing continuous, $M_\alpha \in K_\lambda$ (we denote $\bigcup\limits_{\alpha<\lambda^+} M_\alpha$ by M) and demand $M \in K_{\lambda^+}$

(*b*) $\bar{\mathbf{a}} = \langle a_\alpha : \alpha < \lambda \rangle$ with $a_\alpha \in M_{\alpha+1}$

(*c*) $\mathbf{f}$ is a function from λ^+ to λ^+ such that for some club E of λ^+ for every $\delta \in E \cap S$ and ordinal $i < \mathbf{f}(\delta)$ we have $\mathbf{tp}(a_{\delta+i}, M_{\delta+i}, M_{\delta+i+1}) \in \mathscr{S}^{\text{bs}}(M_{\delta+i})$

(*d*) for every $\alpha < \lambda^+$ and $p \in \mathscr{S}^{\text{bs}}(M_\alpha)$, stationarily many $\delta \in S$ satisfies: for some $\varepsilon < \mathbf{f}(\delta)$ we have $\mathbf{tp}(a_{\delta+\varepsilon}, M_{\delta+\varepsilon}, M_{\delta+\varepsilon+1})$ is a non-forking extension of p.

1A) $K^{\text{nqr}}_{\lambda^+}[S] = K^{\text{nqr}}_S$ is the set of triples $(\bar{M}, \bar{\mathbf{a}}, \mathbf{f}) \in K^{\text{mqr}}_S$ such that:

(*e*) for a club of $\delta < \lambda^+$, if $\delta \in S$ then $\mathbf{f}(\delta)$ is divisible by λ

and[17] for every $i < \mathbf{f}(\delta)$ if $q \in \mathscr{S}^{\rm bs}(M_{\delta+i})$ then for λ ordinals $\varepsilon \in [i, \mathbf{f}(\delta))$ the type $\mathbf{tp}(a_{\delta+\varepsilon}, M_{\delta+\varepsilon}, M_{\delta+\varepsilon+1}) \in \mathscr{S}^{\rm bs}(M_{\delta+\varepsilon})$ is a stationarization of q (= non-forking extension of q, see Definition 4.5).

2) Assume $(\bar{M}^\ell, \bar{\mathbf{a}}^\ell, \mathbf{f}^\ell) \in K^{\rm mqr}_S$ for $\ell = 1, 2$; we say $(\bar{M}^1, \bar{\mathbf{a}}^1, \mathbf{f}^1) \le^0_S (\bar{M}^2, \bar{\mathbf{a}}^2, \mathbf{f}^2)$ <u>iff</u> for some club E of λ^+, for every $\delta \in E \cap S$ we have:

(a) $M^1_{\delta+i} \le_{\mathfrak{K}} M^2_{\delta+i}$ for[18] $i \le \mathbf{f}^1(\delta)$

(b) $\mathbf{f}^1(\delta) \le \mathbf{f}^2(\delta)$

(c) for $i < \mathbf{f}^1(\delta)$ we have $a^1_{\delta+i} = a^2_{\delta+i}$ and $\mathbf{tp}(a^1_{\delta+i}, M^2_{\delta+i}, M^2_{\delta+i+1})$ does not fork over $M^1_{\delta+i}$.

3) We define the relation $<^1_S$ on $K^{\rm mqr}_S$ as in part (2) adding

(d) if $\delta \in E$ and $i < \mathbf{f}^1(\delta)$ then $M^2_{\delta+i+1}$ is $(\lambda, *)$-brimmed over $M^1_{\delta+i+1} \cup M^2_{\delta+i}$.

5.13 Claim. *0) If $(\bar{M}, \bar{\mathbf{a}}, \mathbf{f}) \in K^{\rm mqr}_S$ <u>then</u> $\bigcup_{\alpha<\lambda^+} M_\alpha \in K_{\lambda^+}$ is saturated.*

1) The relation $\le^0_S$ is a quasi-order[19] on $K^{\rm mqr}_\lambda$; also $<^1_S$ is.

2) $K^{\rm mqr}_S \supseteq K^{\rm nqr}_S \ne \emptyset$ for any stationary $S \subseteq \lambda^+$.

3) For every $(\bar{M}, \bar{\mathbf{a}}, \mathbf{f}) \in K^{\rm mqr}_\lambda[S]$ for some $(\bar{M}', \bar{\mathbf{a}}, \mathbf{f}') \in K^{\rm nqr}_\lambda[S]$ we have $(\bar{M}, \bar{\mathbf{a}}, \mathbf{f}) <^1_S (\bar{M}', \bar{\mathbf{a}}, \mathbf{f}')$.

4) For every $(\bar{M}^1, \bar{\mathbf{a}}^1, \mathbf{f}^1) \in K^{\rm mqr}_S$ and $q \in \mathscr{S}^{\rm bs}(M^1_\alpha), \alpha < \lambda^+$, <u>there is</u> $(M^2, \bar{\mathbf{a}}^2, \mathbf{f}^2) \in K^{\rm mqr}_S$ such that $(\bar{M}^1, \bar{\mathbf{a}}^1, \mathbf{f}^1) <^1_S (\bar{M}^2, \bar{\mathbf{a}}^2, \mathbf{f}^2) \in K^{\rm nqr}_S$ and $b \in M^2_\alpha$ realizing q such that for every $\beta \in [\alpha, \lambda^+)$ we have $\mathbf{tp}(b, M^1_\beta, M^2_\beta) \in \mathscr{S}^{\rm bs}(M^1_\beta)$ does not fork over M^1_α.

5) If $\langle(\bar{M}^\zeta, \bar{\mathbf{a}}^\zeta, \mathbf{f}^\zeta) : \zeta < \xi(*)\rangle$ is $\le^0_S$-increasing continuous in $K^{\rm mqr}_S$ and $\xi(*) < \lambda^{++}$ a limit ordering, <u>then</u> the sequence has a $\le^0_S$-lub.

---

[17] if we have an a priori bound $\mathbf{f}^* : \lambda^+ \to \lambda^+$ which is a $<_{\mathscr{D}_{\lambda^+}}$-upper bound of the "first" λ^{++} functions in ${}^{\lambda^+}(\lambda^+)/D$, we can use bookkeeping for u_i's as in the proof of 4.10

[18] could have used (systematically) $i < \mathbf{f}^1(\delta)$

[19] quasi order $\le$ is a transitive relation, so we waive $x \le y \le x \Rightarrow x = y$

Proof. 0, 1) Easy.
2) The inclusion $K_S^{\rm mqr} \supseteq K_S^{\rm nqr}$ is obvious, so let us prove $K_S^{\rm nqr} \neq \emptyset$. We choose by induction on $\alpha < \lambda^+, a_\alpha, M_\alpha, p_\alpha$ such that

(a) $M_\alpha \in K_\lambda$ is a super limit model,

(b) M_α is $\leq_{\mathfrak{K}}$-increasingly continuous,

(c) if $\alpha = \beta + 1$, then $a_\beta \in M_\alpha \backslash M_\beta$ realizes $p_\beta \in \mathscr{S}^{\rm bs}(M_\beta)$,

(d) if $p \in \mathscr{S}^{\rm bs}(M_\alpha)$, then for some $i < \lambda$, for every $j \in [i, \lambda)$ for at least one ordinal $\varepsilon \in [j, j+i), p_{\alpha+\varepsilon} \restriction M_\alpha = p$ and $p_{\alpha+\varepsilon}$ does not fork over M_α.

For $\alpha = 0$ choose $M_0 \in K_\lambda$. For α limit, $M_\alpha = \bigcup_{\beta<\alpha} M_\beta$ is as required. Then use Axiom(E)(g) to take care of clause (d) (with careful bookkeeping). Lastly, let $\mathbf{f} : \lambda^+ \to \lambda^+$ be constantly $\lambda, \bar{M} = \langle M_\alpha : \alpha < \lambda\rangle, \bar{\mathbf{a}} = \langle a_\alpha : \alpha < \lambda\rangle$; now for any stationary $S \subseteq \lambda^+$, the triple $(\bar{M}, \bar{\mathbf{a}} \restriction S, \mathbf{f} \restriction S)$ belong to $K_S^{\rm nqr}$.
3) Let E be a club witnessing $(\bar{M}^1, \bar{\mathbf{a}}^1, \mathbf{f}^1) \in K_S^{\rm mqr}$ such that $\delta \in E \Rightarrow \delta + \mathbf{f}^1(\delta) < \mathrm{Min}(E \backslash (\delta+1))$. Choose $\mathbf{f}^2 : \lambda^+ \to \lambda^+$ such that $\alpha < \lambda^+$ implies $\mathbf{f}^1(\alpha) < \mathbf{f}^2(\alpha) < \lambda^+$ and $\mathbf{f}^2(\alpha)$ is divisible by λ. We choose by induction on $\alpha < \lambda^+, f_\alpha, M_\alpha^2, p_\alpha, a_\alpha^2$ such that:

(a), (b), (c) as in the proof of part (2)

(d) f_α is a $\leq_{\mathfrak{K}}$-embedding of M_α^1 into M_α^2

(e) f_α is increasing continuous

(f) if $\delta \in E \cap S$ and $i < \mathbf{f}^1(\delta)$ hence $\mathbf{tp}(a_{\delta+i}^1, M_{\delta+i}^1, M_{\delta+i+1}^1) \in \mathscr{S}^{\rm bs}(M_{\delta+i}^1)$,
<u>then</u> $f_{\delta+i+1}(a_{\delta+i}^1) = a_{\delta+i}^2$ and $p_{\varepsilon+i} = \mathbf{tp}(a_{\delta+i}^2, M_{\delta+i}^2, M_{\delta+i+1}^2) \in \mathscr{S}^{\rm bs}(M_{\delta+i}^2)$ is a stationarization of
$\mathbf{tp}(f_{\delta+i+1}(a_{\delta+i}^1), f_{\delta+i}(M_{\delta+i}^1), f_{\delta+i+1}(M_{\delta+i+1}^1)) =$
$\mathbf{tp}(a_{\delta+i}^2, f_{\delta+i}(M_{\delta+i}^1), M_{\delta+i+1}^2)$

(g) if $\delta \in E$ and $i < \mathbf{f}^2(\delta), q \in \mathscr{S}^{\rm bs}(M_{\delta+i}^2)$ then for some λ ordinals $\varepsilon \in (i, \mathbf{f}^2(\delta))$ the type $p_{\delta+\varepsilon}$ is a stationarization of q

(h) if $\delta \in E, i < \mathbf{f}^2(\delta)$ then $M_{\delta+i+1}$ is $(\lambda, *)$-brimmed over $M_{\delta+i} \cup f_{\delta+i+1}(M_{\delta+i+1}^1)$.

The proof is as in part (2) only the bookkeeping is different. At the end without loss of generality $\bigcup_{\alpha<\lambda^*} f_\alpha$ is the identity and we are done.
4) Similar proof but in some cases we have to use Axiom (E)(i), the non-forking amalgamation of Definition 2.1, in the appropriate cases.
5) Without loss of generality $\mathrm{cf}(\xi(*)) = \xi(*)$. First assume that $\xi(*) \leq \lambda$. For $\varepsilon < \zeta < \xi(*)$ let $E_{\varepsilon,\zeta}$ be a club of λ^+ witnessing $\bar{M}^\varepsilon <^0_{\mathfrak{S}} \bar{M}^\zeta$. Let
$E^* = \bigcap_{\varepsilon<\zeta<\xi(*)} E_{\varepsilon,\zeta} \cap \{\delta < \lambda^+ : \text{for every } \alpha < \delta \text{ we have } \sup_{\varepsilon<\xi(*)} \mathbf{f}^\varepsilon(\alpha) < \delta\}$, it is a club of λ^+. Let $\mathbf{f}^{\xi(*)} : \lambda^+ \to \lambda^+$ be $\mathbf{f}^{\xi(*)}(i) = \sup_{\varepsilon<\xi(*)} \mathbf{f}^\varepsilon(i)$ now define $M_i^{\xi(*)}$ as follows:

Case 1: If $\delta \in E^*$ and $\varepsilon < \xi(*)$ and $i \leq \mathbf{f}^\varepsilon(\delta)$ and $i \geq \bigcup_{\zeta<\varepsilon} \mathbf{f}^\zeta(\delta)$ then

(α) $M^{\xi(*)}_{\delta+i} = \bigcup\{M^\zeta_{\delta+i} : \zeta \in [\varepsilon, \xi(*))\}$

(β) $i < \mathbf{f}^\varepsilon(\delta) \Rightarrow a^{\xi(*)}_{\delta+i} = a^\varepsilon_{\delta+i}$.

(Note: we may define $M^{\xi(*)}_{\delta+i}$ twice if $i = \mathbf{f}^\varepsilon(\delta)$, but the two values are the same).

Case 2: If $\delta \in E^*, i = \mathbf{f}^{\xi(*)}(\delta)$ is a limit ordinal let

$$M^{\xi(*)}_{\delta+i} = \bigcup_{j<i} M^{\xi(*)}_{\delta+i}.$$

Case 3: If $M_i^{\xi(*)}$ has not been defined yet, let it be $M^{\xi(*)}_{\mathrm{Min}(E^*\setminus i)}$.

Case 4: If $a_i^{\xi(*)}$ has not been defined yet, let $a_i^{\xi(*)} \in M^{\xi(*)}_{i+1}$ be arbitrary.

Note that Case 3,4 deal with the "unimportant" cases.
Let $\varepsilon < \xi(*)$, why $(\bar{M}^\varepsilon, \bar{\mathbf{a}}^\varepsilon, \mathbf{f}^\varepsilon) \leq^0_{\mathfrak{S}} (\bar{M}^{\xi(*)}, \bar{\mathbf{a}}^{\xi(*)}, \mathbf{f}^{\xi(*)}) \in K^{\mathrm{mqr}}_{\mathfrak{S}}$? Enough to check that the club E^* witnesses it.

Why $\mathbf{tp}(a_{\delta+i}, M^{\xi(*)}_{\delta+i}, M^{\xi(*)}_{\delta+i+1}) \in \mathscr{S}^{\rm bs}(M^{\xi(*)}_{\delta+i})$ and when $\delta \in E^*, i < \mathbf{f}^{\xi(*)}(i)$, and does not fork over $M^{\xi}_{\delta+i}$ when $i < \mathbf{f}^{\varepsilon}(\delta)$? by Axiom (E)(h) of Definition 2.1.
Why clause (e) of Definition 5.12(1A)? By Axiom (E)(c), local character of non-forking.

The case $\xi(*) = \lambda^+$ is similar using diagonal intersections. $\square_{5.13}$

Remark. If we use weaker versions of "good λ-frames", we should systematically concentrate on successor $i < \mathbf{f}(\delta)$.

Proof of 5.9. We can use VII.2.3 or more explicitly VII.4.20: the older version runs as follows. The use of $\lambda^{++} \notin \mathrm{WDmId}(\lambda^{++})$ is as in the proof of [Sh 576, 3.19(pg.79)]. But now we need to preserve saturation in limit stages $\delta < \lambda^{++}$ of cofinality $< \lambda^+$, we use $<^1_S$, otherwise we act as in [Sh 576, §3]. $\square_{5.9}$

Let us elaborate

5.14 Definition. We define $\mathbf{C} = (\mathfrak{K}^+, \mathbf{S}eq, \leq^*)$ as follows:

(a) $\tau^+ = \tau \cup \{P, <\}, \mathfrak{K}^+$ is the set of $(M, P^M, <^M)$ where $M \in \mathfrak{K}_{<\lambda}, P^M \subseteq M, <^M$ a linear ordering of P^M (but $=^M$ may be as in [Sh 576, 3.1(2)] and $M_1 \leq_{\mathfrak{K}^+} M_2$ iff $(M_1 \restriction \tau) \leq_{\mathfrak{K}} (M_2 \restriction \tau)$ and $M_1 \subseteq M_2$

(b) $\mathbf{S}eq_\alpha = \{\bar{M} : \bar{M} = \langle M_i : i \leq \alpha\rangle$ is an increasing continuous sequence of members of $\mathfrak{K}^+$ and $\langle M_i \restriction \tau : i \leq \alpha\rangle$ is $\leq_{\mathfrak{K}}$-increasing, and for
$i < j < \alpha : P^{M_i}$ is a proper initial segment of $(P^{M_j}, <^{M_j})$ and there is a first element in the difference$\}$
we denote the $<^{M_{i+1}}$-first element of $P^{M_{i+1}} \backslash P^{M_i}$, by $a_i[\bar{M}]$ and we demand $\mathbf{tp}(a_i(\bar{M}), M_i\tau \restriction, M_{i+1} \restriction \tau) \in \mathscr{S}^{\rm bs}(M_i \restriction \tau)$ and if $\alpha = \lambda, M = \cup\{M_i \restriction \tau : i < \lambda^+\}$ is saturated

(c) $\bar{M} <^*_t \bar{N}$ iff
$\bar{M} = \langle M_i : i < \alpha^*\rangle, \bar{N} = \langle N_i : i < \alpha^{**}\rangle$ are from $\mathbf{S}eq, t$ is a set of pairwise disjoint closed intervals of α^* and for any $[\alpha, \beta] \in t$ we have ($\beta < \alpha^*$ and):

$\gamma \in [\alpha, \beta) \Rightarrow M_\gamma \leq_{\mathfrak{K}} N_\gamma$ & $a_\gamma[\bar{M}] \notin N_\gamma$, moreover $a_\gamma[\bar{M}] = a_\gamma[\bar{N}]$ and $\mathbf{tp}(a_j[\bar{M}], N_\gamma \restriction \tau, N_{\gamma+1}, \tau)$ does not fork over $M_\gamma \restriction \tau$.

5.15 Claim. *1)* $\mathbf{C}$ *is a* λ^+*-construction framework (see [Sh 576, 3.3(pg.68)].*
2) $\mathbf{C}$ *is weakly nice (see Definition [Sh 576, 3.14(2)(pg.76)].*
4) $\mathbf{C}$ *has the weakening* λ^+*-coding property.*

Discussion: Is it better to use (see [Sh 576, 3.14(1)(pg.75)]) stronger axiomatization in [Sh 576, §3] to cover this?
But at present this will be the only case.

Proof. Straight. $\square_{5.15}$

Now 5.11 follows by [Sh 576, 3.19(pg.79)].

§6 Non-forking amalgamation in $\mathfrak{K}_\lambda$

We deal in this section only with $\mathfrak{K}_\lambda$.
We would like to, at least, approximate "non-forking amalgamation of models" using as a starting point the conclusion of 5.9, i.e., $K_\lambda^{3,\mathrm{uq}}$ is dense. We use what looks like a stronger hypothesis: the existence for $K_\lambda^{3,\mathrm{uq}}$ (also called "weakly successful"); but in our application we can assume categoricity in λ; the point being that as we have a superlimit $M \in K_\lambda$, this assumption is reasonable when we restrict ourselves to $\mathfrak{K}^{[M]}$, recalling that we believe in first analyzing the saturated enough models; see 5.8. By 4.8, the "$(\lambda, \mathrm{cf}(\delta))$-brimmed over" is the same for all limit ordinals $\delta < \lambda^+$, (but not for $\delta = 1$ or just δ non-limit); nevertheless for possible generalizations we do not use this.

It may help the reader to note, that (assuming 6.8 below, of course), if there is a 4-place relation $\mathrm{NF}_\lambda(M_0, M_1, M_2, M_3)$ on K_λ, satisfying the expected properties of "M_1, M_2 are amalgamated in a non-forking = free way over M_0 inside M_3", i.e., is a $\mathfrak{K}_\lambda$-non-forking relation from Definition 6.1 below then Definition 6.12 below (of NF_λ) gives it (provably!). So we have "a definition" of NF_λ satisfying that: if desirable non-forking relation exists, our definition gives it (assuming the hypothesis 6.8). So during this section we are trying to get better and better approximations to the desirable properties; have the feeling of going up on a spiral, as usual.

For the readers who know on non-forking in stable first order theory we note that in such context $\mathrm{NF}_\lambda(M_0, M_1, M_2, M_3)$ says that $\mathbf{tp}(M_2, M_1, M_3)$, the type of M_2 over M_1 inside M_3, does not fork over M_0. It is natural to say that there are $\langle N_{1,\alpha}, N_{2,\alpha} : \alpha \leq \alpha^* \rangle, N_{\ell,\alpha}$ is increasing continuous. $N_{1,0} = M_0, N_{2,0} = M_2, M_1 \subseteq M_{1,\alpha}, M_3 \subseteq M_3', N_{2,\alpha} \subseteq M_3', N_{\ell,\alpha+2}$ is prime over $N_{\ell,\alpha} + a_\alpha$ for $\ell = 1,2$ and $\mathbf{tp}(a_\alpha, N_{2,\alpha})$ does not fork over $N_{1,\alpha}$ but this is not available. The $K_\lambda^{3,\mathrm{uq}}$ is a substitute.

6.1 Definition. 1) Assume that $\mathfrak{K} = \mathfrak{K}_\lambda$ is a λ-a.e.c. We say NF is a non-forking relation on ${}^4(\mathfrak{K}_\lambda)$ or just a $\mathfrak{K}_\lambda$-non-forking relation <u>when</u>:

$\boxtimes_{\mathrm{NF}}$ (a) NF is a 4-place relation on K_λ and NF is preserved under isomorphisms

(b) $\mathrm{NF}(M_0, M_1, M_2, M_3)$ implies $M_0 \leq_{\mathfrak{K}} M_\ell \leq_{\mathfrak{K}} M_3$ for $\ell = 1,2$

$(c)_1$ (monotonicity): if $\mathrm{NF}(M_0, M_1, M_2, M_3)$ and $M_0 \leq_{\mathfrak{K}} M_\ell' \leq_{\mathfrak{K}} M_\ell$ for $\ell = 1,2$ then $\mathrm{NF}(M_0, M_1', M_2', M_3)$

$(c)_2$ (monotonicity): if $\mathrm{NF}(M_0, M_1, M_2, M_3)$ and $M_3 \leq_{\mathfrak{K}} M_3' \in K_\lambda, M_1 \cup M_2 \subseteq M_3'' \leq_{\mathfrak{K}} M_3'$ <u>then</u> $\mathrm{NF}(M_0, M_1, M_2, M_3'')$

(d) (symmetry) $\mathrm{NF}(M_0, M_1, M_2, M_3)$ iff $\mathrm{NF}(M_0, M_2, M_1, M_3)$

(e) ((long) transitivity) if $\mathrm{NF}(M_i, N_i, M_{i+1}, N_{i+1})$ for $i < \alpha$, $\langle M_i : i \leq \alpha \rangle$ is $\leq_{\mathfrak{K}}$-increasing continuous and $\langle N_i : i \leq \alpha \rangle$ is $\leq_{\mathfrak{K}}$-increasing continuous <u>then</u>
$\mathrm{NF}(M_0, N_0, M_\alpha, N_\alpha)$

(f) (existence) if $M_0 \leq_{\mathfrak{K}} M_\ell$ for $\ell = 1,2$ (all in K_λ) <u>then</u> for some $M_3 \in K_\lambda, f_1, f_2$ we have $M_0 \leq_{\mathfrak{K}} M_3$, f_ℓ is a $\leq_{\mathfrak{K}}$-embedding of M_ℓ into M_3 over M_0 for $\ell = 1,2$ and $\mathrm{NF}(M_0, f_1(M_1), f_2(M_2), M_3)$

(g) (uniqueness) if $\mathrm{NF}(M_0^\ell, M_1^\ell, M_2^\ell, M_3^\ell)$ and for $\ell = 1,2$ and f_i is an isomorphism from M_i^1 onto M_i^2 for $i = 0,1,2$ and

$f_0 \subseteq f_1, f_0 \subseteq f_2$ then $f_1 \cup f_2$ can be extended to an embedding f_3 of M_3^1 into some $M_4^2, M_3^2 \leq_{\mathfrak{K}_\lambda} M_4^2$.

2) We say that NF is a pseudo non-forking relation on ${}^4(K_\lambda)$ or a weak $\mathfrak{K}_\lambda$-non-forking relation if clauses (a)-(f) of $\boxtimes_{\rm NF}$ above holds but not necessarily clause (g).
3) Assume $\mathfrak{s}$ is a good λ-frame and NF is a non-forking relation on $\mathfrak{K}$ or just a weak one. We say that NF respects $\mathfrak{s}$ or NF is an $\mathfrak{s}$-non-forking relation when:

(*h*) if $\mathrm{NF}(M_0, M_1, M_2, M_3)$ and $a \in M_2 \backslash M_0$, $\mathbf{tp}_{\mathfrak{s}}(a, M_0, M_2) \in \mathscr{S}^{\rm bs}(M_0)$ then $\mathbf{tp}_{\mathfrak{s}}(a, M_1, M_3)$ does not fork over M_0 in the sense of $\mathfrak{s}$.

6.2 Observation. Assume $\mathfrak{K}_\lambda$ is a λ-a.e.c. and NF is a non-forking relation on ${}^4(\mathfrak{K}_\lambda)$.
1) Assume $\mathfrak{K}$ is stable in λ. If in clause (g) of 6.1(1) above we assume in addition that M_3^ℓ is (λ, ∂)-brimmed over $M_1^\ell \cup M_2^\ell$, then in the conclusion of (g) we can add $M_3^2 = M_4^2$, i.e., $f_1 \cup f_2$ can be extended to an isomorphism from M_3^1 onto M_3^2. This version of (g) is equivalent to it (assuming stability in λ; note that "$\mathfrak{K}_\lambda$ has amalgamation" follows by clause (f) of Definition 6.1).
2) If $M_0 \leq_{\mathfrak{K}} M_1 \leq_{\mathfrak{K}} M_3$ are from K_λ then $\mathrm{NF}(M_0, M_0, M_1, M_3)$.
3) In Definition 6.1(1), clause (d), symmetry, it is enough to demand "if".

Proof. 1) Chase arrows and the uniqueness from 1.16.
2) By clause (f) of $\boxtimes_{\rm NF}$ of 6.1(1) and clause $(c)_2$, i.e., first apply existence with (M_0, M_0, M_3) here standing for (M_0, M_1, M_2) there, then chase arrows and use the monotonicity as in $(c)_2$.
3) Easy. $\square_{6.2}$

The main point of the following claim shows that there is at most one non-forking relation respecting $\mathfrak{s}$; so it justifies the definition of $\mathrm{NF}_{\mathfrak{s}}$ later. The assumption "NF respects $\mathfrak{s}$" is not so strong by 6.7.

6.3 Claim. *1) If* $\mathfrak{s}$ *is a good* λ*-frame and* NF *is a non-forking relation on* ${}^4(\mathfrak{K}_\mathfrak{s})$ *respecting* $\mathfrak{s}$ *and* $(M_0, N_0, a) \in K^{3,\mathrm{uq}}_\lambda$ *and* $(M_0, N_0, a) \leq_{\mathrm{bs}} (M_1, N_1, a)$ *then* $\mathrm{NF}(M_0, N_0, M_1, N_1)$.
2) If $\mathfrak{s}$ *is a good* λ*-frame, weakly successful (which means* $K^{3,\mathrm{uq}}_\mathfrak{s}$ *has existence in* $K^{3,\mathrm{uq}}_\mathfrak{s}$*, i.e.,* $\mathfrak{s}$ *satisfies hypothesis 6.8 below) and* NF *is a non-forking relation on* ${}^4(\mathfrak{K}_\mathfrak{s})$ *respecting* $\mathfrak{s}$ <u>then</u> *the relation* $\mathrm{NF}_\lambda = \mathrm{NF}_\mathfrak{s}$*, i.e.,* $N_1 \underset{N_0}{\overset{N_3}{\bigcup}} N_2$ *defined in Definition 6.12 below is equivalent to* $\mathrm{NF}(N_0, N_1, N_2, N_3)$. *[Recalling 6.34, but see 6.35(2), 6.36.]*
3) If $\mathfrak{s}$ *is a weakly successful good* λ*-frame and for* $\ell = 1, 2$*, the relation* NF_ℓ *is a non-forking relation on* ${}^4(\mathfrak{K}_\mathfrak{s})$ *respecting* $\mathfrak{s}$*,* <u>then</u> $\mathrm{NF}_1 = \mathrm{NF}_2$.

Proof. Straightforward but we elaborate.
1) We can find (M_1', N_1') such that $\mathrm{NF}(M_0, N_0, M_1', N_1')$ and M_1, M_1' are isomorphic over M_0, say f_1 is such an isomorphism from M_1 onto M_1' over M_0; why such (M_1', N_1', f_1) exists? by clause (f) of $\boxtimes_{\mathrm{NF}}$ of Definition 6.1.

As NF respects $\mathfrak{s}$, see Definition 6.1(2), recalling $\mathbf{tp}(a, M_0, N_0) \in \mathscr{S}^{\mathrm{bs}}(M_0)$ we know that $\mathbf{tp}(a, M_1', N_1')$ does not fork over M_0, so by the definition of $\leq_{\mathrm{bs}}$ we have $(M_0, N_0, a) \leq_{\mathrm{bs}} (M_1', N_1', a)$.

As $(M_0, N_0, a) \in K^{3,\mathrm{uq}}_\lambda$, by the definition of $K^{3,\mathrm{uq}}_\lambda$ (and chasing arrows) we conclude that there are N_2, f_2 such that:

(∗) $N_1 \leq_{\mathfrak{K}[\mathfrak{s}]} N_2 \in K_\lambda$ and f_2 is a $\leq_\mathfrak{K}$-embedding of N_1' into N_2 extending f_1^{-1} and id_{N_0}.

As $\mathrm{NF}(M_0, N_0, M_1', N_1')$ and NF is preserved under isomorphisms (see clause (a) in 6.1(1)) it follows that $\mathrm{NF}(M_0, N_0, M_1, f_2(N_1'))$. By the monotonicity of NF (see clause $(c)_2$ of Definition 6.1) it follows that $\mathrm{NF}(M_0, N_0, M_1, N_2)$. Again by the same monotonicity we have $\mathrm{NF}(M_0, N_0, M_1, N_1)$, as required.
2) First we prove that $\mathrm{NF}_{\lambda,\bar{\delta}}(N_0, N_1, N_2, N_3)$, which is defined in Definition 6.11 below implies $\mathrm{NF}(N_0, N_1, N_2, N_3)$. By definition 6.11, clause (f) there are $\langle (N_{1,i}, N_{2,i} : i \leq \lambda \times \delta_1 \rangle), \langle c_i : i < \lambda \times \delta_1 \rangle$ as there. Now we prove by induction on $j \leq \lambda \times \delta_1$ that $i \leq j \Rightarrow \mathrm{NF}(N_{1,i}, N_{2,i}, N_{1,j}, N_{2,j})$. For $j = 0$ or more generally when $i = j$

this is trivial by 6.2(2). For j a limit ordinal use the induction hypothesis and transitivity of NF (see clause (e) of 6.1(1)).

Lastly, for j successor by the demands in Definition 6.11 we know that $N_{1,j-1} \leq_{\mathfrak{K}} N_{1,j} \leq_{\mathfrak{K}} N_{2,j}, N_{1,j-1} \leq_{\mathfrak{K}} N_{2,j-1} \leq_{\mathfrak{K}} N_{2,j}$ are all in K_λ, $\mathbf{tp}(c_{j-1}, N_{2,j-1}, N_{2,j})$ does not fork over $N_{1,j-1}$ and $(N_{1,j-1}, N_{1,j}, c_{j-1}) \in K^{3,\mathrm{uq}}_\lambda$. By part (1) of this claim we deduce that $\mathrm{NF}(N_{1,j-1}, N_{1,j}, N_{2,j-1}, N_{2,j})$ hence by symmetry (i.e., clause (d) of Definition 6.1(1)) we deduce $\mathrm{NF}(N_{1,j-1}, N_{2,j-1}, N_{1,j}, N_{2,j})$.

So we have gotten $i < j \Rightarrow \mathrm{NF}(N_{1,i}, N_{2,i}, N_{1,j}, N_{2,j})$.
[Why? If $i = j - 1$ by the previous sentence and for $i < j - 1$ note that by the induction hypothesis $\mathrm{NF}(N_{1,i}, N_{2,i}, N_{1,j-1}, N_{1,j-1})$ so by transitivity (clause (e) of 6.1(1) of Definition 6.1) we get $\mathrm{NF}(N_{1,i}, N_{2,i}, N_{1,j}, N_{2,j})$].

We have carried the induction so in particular for $i = 0$, $j = \alpha$ we obtain $\mathrm{NF}(N_{1,0}, N_{2,0}, N_{1,\alpha}, N_{2,\alpha})$ which means $\mathrm{NF}(N_0, N_1, N_2, N_3)$ as promised. So we have proved $\mathrm{NF}_{\lambda,\bar{\delta}}(N_0, N_1, N_2, N_3) \Rightarrow \mathrm{NF}(N_0, N_1, N_2, N_3)$.

Second, if $\mathrm{NF}_\lambda(N_0, N_1, N_2, N_3)$ as defined in Definition 6.12 then there are $M_0, M_1, M_2, M_3 \in K_\lambda$ such that $\mathrm{NF}_{\lambda,\langle\lambda,\lambda\rangle}(M_0, M_1, M_2, M_3)$, $N_\ell \leq_{\mathfrak{K}} M_\ell$ for $\ell < 4$ and $N_0 = M_0$. By what we have proved above we can conclude $\mathrm{NF}(M_0, M_1, M_2, M_3)$. As $N_0 = M_0 \leq_{\mathfrak{K}} N_\ell \leq_{\mathfrak{K}} M_\ell$ for $\ell = 1, 2$ by clause $(c)_1$ of Definition 6.1(1) we get $\mathrm{NF}(M_0, N_1, N_2, M_3)$ and by clause $(c)_2$ of Definition 6.1(1) we get $\mathrm{NF}(N_0, N_1, N_2, N_3)$. So we have proved the implication $\mathrm{NF}_\lambda(N_0, N_1, N_2, N_3) \Rightarrow \mathrm{NF}(N_0, N_1, N_2, N_3)$.

For the other implication assume $\mathrm{NF}(N_0, N_1, N_2, M_3)$. Now as we have existence for NF_λ (as proved below, see 6.21), we can find N'_ℓ for $\ell = 0, 1, 2, 3$ and f_ℓ for $\ell = 0, 1, 2$ such that $\mathrm{NF}_\lambda(N'_0, N'_1, N'_2, N'_3)$, f_ℓ is an isomorphism from N_ℓ onto N'_ℓ for $\ell = 0, 1, 2$ and $f_0 \subseteq f_1, f_0 \subseteq f_2$. But what we have already proved it folows that $\mathrm{NF}(N'_0, N'_1, N'_2, N'_3)$. As we have uniqueness for NF by clause (g) of Definition 6.1 we can find (f_3, N''_3) such that $N'_3 \leq_{\mathfrak{K}_\lambda} N''_3$ and f_3 is a $\leq_{\mathfrak{K}}$-embedding of N_3 into N''_3 extending $f_1 \cup f_2$. As NF_λ satisfies clause $(c)_2$ of 6.1, recalling $\mathrm{NF}_\lambda(N'_0, N'_1, N'_2, N'_3)$ it follows that $\mathrm{NF}_\lambda(N'_0, N'_1, N'_2, f_3(N_3))$ holds. As NF_λ is preserved by isomorphisms, it follows that $\mathrm{NF}_\lambda(N_0, N_1, N_2, N_3)$ holds as required.

3) By the rest of this section, i.e., the main conclusion 6.34, the rela-

tion NF_λ defined in 6.12 is a non-forking relation on ${}^4(K_{\mathfrak{s}})$ respecting $\mathfrak{s}$. Hence by part (2) of the present claim we have $\mathrm{NF}_1 = \mathrm{NF}_\lambda = \mathrm{NF}_2$. $\square_{6.3}$

6.4 Example: Do we need $\mathfrak{s}$ in 6.3(3)? Yes.

Let $\mathfrak{K}$ be the class of graphs and $M \leq_{\mathfrak{K}} N$ iff $M \subseteq N$; so $\mathfrak{K}$ is an a.e.c. with $\mathrm{LS}(\mathfrak{K}) = \aleph_0$. For cardinal λ and $\ell = 1, 2$ we define

- $\mathrm{NF}^\ell = \{(M_0, M_1, M_2, M_3) : M_0 \leq_{\mathfrak{K}} M_1 \leq_{\mathfrak{K}} M_3$ and $M_0 \leq_{\mathfrak{K}} M_2 \leq_{\mathfrak{K}} M_3$ and $M_1 \cap M_2 = M_0$ and if $a \in M_1 \backslash M_0, b \in M_2 \backslash M_0$ then $\{a, b\}$ is an edge of M_3 iff $\ell = 2\}$, and
- $\mathrm{NF}^\ell_\lambda := \{(M_0, M_1, M_2, M_3) \in \mathrm{NF} : M_0, M_1, M_2, M_3 \in K_\lambda\}$.
  Then NF^ℓ_λ is a non-forking relation on ${}^4(\mathfrak{K}_\lambda)$ but $\mathrm{NF}^1_\lambda \neq \mathrm{NF}^2_\lambda$.

6.5 Remark. 1) So the assumption on $\mathfrak{K}_\lambda$ that for some good λ-frame $\mathfrak{s}$ we have $\mathfrak{K}_{\mathfrak{s}} = \mathfrak{K}_\lambda$ is quite a strong demand on $\mathfrak{K}_\lambda$.
2) However, the assumption "respect" essentially is not necessary as it can be deduced when $\mathfrak{s}$ is good enough.
3) Below on "good$^+$" see III§1 in particular Definition III.1.3.

6.6 Exercise: 1) Assume NF_1,NF_2 are non-forking relations on ${}^4(\mathfrak{K}_\lambda)$. If $\mathrm{NF}_1 \subseteq \mathrm{NF}_2$ then $\mathrm{NF}_1 = \mathrm{NF}_2$.
2) In part (1) write down the clauses from 6.1. We need to assume on NF_1, and those we need assume on NF_2.
[Hint: Read the last paragraph of the proof of 6.3(3).]

6.7 Claim. *Assume that $\mathfrak{s}$ is a good$^+$ λ-frame and* NF *is a non-forking relation on ${}^4(\mathfrak{K}_{\mathfrak{s}})$. Then* NF *respects $\mathfrak{s}$.*

Remark. The construction in the proof is similar to the ones in 4.9, 6.14.

Proof. Assume $\mathrm{NF}(M_0, M_1, M_2, M_3)$ and $a \in M_2 \backslash M_0$, $\mathbf{tp}(a, M_0, M_2) \in \mathscr{S}^{\mathrm{bs}}(M_0)$. We define $(N_{0,i}, N_{1,i}, f_i)$ for $i < \lambda^+_{\mathfrak{s}}$ as follows:

$\otimes_1$(a) $N_{0,i}$ is $\leq_{\mathfrak{s}}$-increasing continuous and $N_{0,0} = M_0$

(b) $N_{1,i}$ is $\leq_{\mathfrak{s}}$-increasing continuous and $N_{1,0} = M_1$

(c) $\mathrm{NF}(N_{0,i}, N_{1,i}, N_{0,i+1}, N_{1,i+1})$

(d) f_i is a $\leq_{\mathfrak{K}}$-embedding of M_2 into $N_{0,i+1}$ over $M_0 = N_{0,0}$ such that $\mathbf{tp}(f_i(a), N_{0,i}, N_{0,i+1})$ does not fork over $M_0 = N_{0,0}$.

We shall choose f_i together with $N_{0,i+1}, N_{1,i+1}$.
Why can we define? For $i = 0$ there is nothing to do. For i limit take unions. For $i = j + 1$ choose $f_j, N_{0,i}$ satisfying clause (d) and $N_{0,j} \leq_{\mathfrak{s}} N_{0,i}$, this is possible for $\mathfrak{s}$ as we have the existence of non-forking extensions of $\mathbf{tp}(a, M_0, M_2)$ (and amalgamation).

Lastly, we take care of the rest (mainly clause (c) of $\otimes_1$ by clause (f) of Definition 6.1(1), existence). Now

$\circledast_2$ for $i < j < \lambda^+$ we have $\mathrm{NF}(N_{0,i}, N_{1,i}, N_{0,j}, N_{1,j})$ [why? by transitivity for NF, i.e., clause (e) of Definition 6.1(1), transitivity]

$\circledast_3$ for some i, $\mathbf{tp}(f_i(a), N_{1,i}, N_{1,i+1})$ does not fork over M_0 [why? by the definition of good$^+$].

So for this $i, M_0 \leq_{\mathfrak{s}} f_i(M_2) \leq_{\mathfrak{s}} N_{0,i+1}$ by clause (d) of $\otimes_1$, hence by clause $(c)_1$ of Definition 6.1, monotonicity we have $\mathrm{NF}(M_0, M_1, f_i(M_2), N_{1,i+1})$. Now again by the choice of i, i.e., by $\circledast_3$ we have $\mathbf{tp}(f_i(a), M_1, N_{1,i+1})$ does not fork over M_0. By clause (g) of Definition 6.1(1), i.e., uniqueness of NF (and preservation by isomorphisms) we get $\mathbf{tp}(a, M_1, M_3)$ does not fork over M_0 as required. $\square_{6.7}$

We turn to our main task in this section proving that such NF exist; till 6.34 we assume:

6.8 Hypothesis. 1) $\mathfrak{s} = (\mathfrak{K}, \bigcup, \mathscr{S}^{\mathrm{bs}})$ is a good λ-frame.

2) $\mathfrak{s}$ is weakly successful which just means that it has existence for $K^{3,\mathrm{uq}}_\lambda$: for every $M \in K_\lambda$ and $p \in \mathscr{S}^{\mathrm{bs}}(M)$ there are N, a such that $(M, N, a) \in K^{3,\mathrm{uq}}_\lambda$ (see Definition 5.3) and $p = \mathbf{tp}(a, M, N)$. (This follows by $K^{3,\mathrm{uq}}_{\mathfrak{s}}$ is dense in $K^{3,\mathrm{bs}}_{\mathfrak{s}}$; when $\mathfrak{s}$ is categorical, see 5.8.)

In this section we deal with models from K_λ only.

6.9 Claim. *If $M \in K_\lambda$ and N is (λ, κ)-brimmed over M, then we can find $\bar{M} = \langle M_i : i \le \delta \rangle$, $\le_{\mathfrak{K}}$-increasing continuous, $(M_i, M_{i+1}, c_i) \in K^{3,\mathrm{uq}}_\lambda$, $M_0 = M, M_\delta = N$ and δ any pregiven limit ordinal $< \lambda^+$ of cofinality κ divisible by λ.*

Proof. Let δ be given, e.g., $\delta = \lambda \times \kappa$. By 6.8(2) we can find a $\le_{\mathfrak{K}}$-increasing sequence $\langle M_i : i \le \delta \rangle$ of members of K_λ and $\langle a_i : i < \delta \rangle$ such that $M_0 = M$ and $i < \delta \Rightarrow (M_i, M_{i+1}, a_i) \in K^{3,\mathrm{uq}}_\lambda$ and for every $i < \delta, p \in \mathscr{S}^{\mathrm{bs}}(M_i)$ for λ ordinals $j \in (i, i+\lambda)$ we have $\mathbf{tp}(a_j, M_j, M_{j+1})$ is a non-forking extension of p. So the demands in 4.3 hold hence M_δ is (λ, κ)-brimmed over $M_0 = M$. Now we are done by the uniqueness of N being (λ, κ)-brimmed over M_0, see 1.16(3). $\square_{6.9}$

6.10 Claim. *If $M^\ell_0 \le_{\mathfrak{K}} M^\ell_1 \le_{\mathfrak{K}} M^\ell_3$ and $M^\ell_0 \le_{\mathfrak{K}} M^\ell_2 \le_{\mathfrak{K}} M^\ell_3, c_\ell \in M^\ell_1$ and $(M^\ell_0, M^\ell_1, c_\ell) \in K^{3,\mathrm{uq}}_\lambda$ and $\mathbf{tp}(c_\ell, M^\ell_2, M^\ell_3) \in \mathscr{S}^{\mathrm{bs}}(M^\ell_2)$ does not fork over M^ℓ_0 and M^ℓ_3 is (λ, ∂)-brimmed over $M^\ell_1 \cup M^\ell_2$ all this for $\ell = 1, 2$ and f_i is an isomorphism from M^1_i onto M^2_i for $i = 0, 1, 2$ such that $f_0 \subseteq f_1, f_0 \subseteq f_2$ and $f_1(c_1) = c_2$, then $f_1 \cup f_2$ can be extended to an isomorphism from M^1_3 onto M^2_3.*

Proof. Chase arrows (and recall definition of $K^{3,\mathrm{uq}}_\lambda$), that is by 6.1(1) and Definition 6.2(1) and 1.16(3). $\square_{6.10}$

6.11 Definition. Assume $\bar{\delta} = \langle \delta_1, \delta_2, \delta_3 \rangle, \delta_1, \delta_2, \delta_3$ are ordinals $< \lambda^+$, maybe 1. We say that $\mathrm{NF}_{\lambda, \bar{\delta}}(N_0, N_1, N_2, N_3)$ or, in other words:

N_1, N_2 are brimmedly smoothly amalgamated in N_3 over N_0 for $\bar{\delta}$

when:

- (*a*) $N_\ell \in K_\lambda$ for $\ell \in \{0, 1, 2, 3\}$
- (*b*) $N_0 \le_{\mathfrak{K}} N_\ell \le_{\mathfrak{K}} N_3$ for $\ell = 1, 2$
- (*c*) $N_1 \cap N_2 = N_0$ (i.e. in disjoint amalgamation, actually follows by clause (f))
- (*d*) N_1 is $(\lambda, \mathrm{cf}(\delta_1))$-brimmed over N_0; recall that if $\mathrm{cf}(\delta_1) = 1$ this just means $N_0 \le_{\mathfrak{K}} N_1$

(e) N_2 is (λ,cf(δ_2))-brimmed over N_0; so that if cf$(\delta_2) = 1$ this just means $N_0 \leq_{\mathfrak{K}} N_2$ and

(f) there are $N_{1,i}, N_{2,i}$ for $i \leq \lambda \times \delta_1$ and c_i for $i < \lambda \times \delta_1$ (called witnesses and $\langle N_{1,i}, N_{2,i}, c_j : i \leq \lambda \times \delta_1, j < \lambda \times \delta_1 \rangle$ is called a witness sequence as well as $\langle N_{1,i} : i \leq \lambda \times \delta_1 \rangle, \langle N_{2,i} : i \leq \lambda \times \delta_1 \rangle$) such that:

(α) $N_{1,0} = N_0, N_{1,\lambda \times \delta_1} = N_1$

(β) $N_{2,0} = N_2$

(γ) $\langle N_{\ell,i} : i \leq \lambda \times \delta_1 \rangle$ is a $\leq_{\mathfrak{K}}$-increasing continuous sequence of models for $\ell = 1, 2$

(δ) $(N_{1,i}, N_{1,i+1}, c_i) \in K_\lambda^{3,\mathrm{uq}}$

(ε) $\mathbf{tp}(c_i, N_{2,i}, N_{2,i+1}) \in \mathscr{S}^{\mathrm{bs}}(N_{2,i})$ does not fork over $N_{1,i}$ and $N_{2,i} \cap N_1 = N_{1,i}$, for $i < \lambda \times \delta_1$ (follows by Definition 5.3)

(ζ) N_3 is (λ,cf(δ_3))-brimmed over $N_{2,\lambda \times \delta_1}$; so for cf$(\delta_3) = 1$ this means just $N_{2,\lambda \times \delta_1} \leq_{\mathfrak{K}} N_3$

6.12 Definition. 1) We say $N_1 \overset{N_3}{\underset{N_0}{\bigcup}} N_2$ (or N_1, N_2 are <u>smoothly amalgamated</u> over N_0 inside N_3 or $\mathrm{NF}_\lambda(N_0, N_1, N_2, N_3)$ or $\mathrm{NF}_{\mathfrak{s}}(N_0, N_1, N_2, N_3)$) <u>when</u> we can find $M_\ell \in K_\lambda$ (for $\ell < 4$) such that:

(a) $\mathrm{NF}_{\lambda,\langle \lambda,\lambda,\lambda \rangle}(M_0, M_1, M_2, M_3)$

(b) $N_\ell \leq_{\mathfrak{K}} M_\ell$ for $\ell < 4$

(c) $N_0 = M_0$

(d) M_1, M_2 are $(\lambda, \mathrm{cf}(\lambda))$-brimmed over N_0 (follows by (a) see clauses (d), (e) of 6.11).

2) We call (M, N, a) <u>strongly bs-reduced</u> if $(M, N, a) \in K_\lambda^{3,\mathrm{bs}}$ and $(M, N, a) \leq_{\mathrm{bs}} (M', N', a) \in K_\lambda^{3,\mathrm{bs}} \Rightarrow \mathrm{NF}_\lambda(M, N, M', N')$; not used.

Clearly we expect "strongly bs-reduced" to be equivalent to "$\in K_\lambda^{3,\mathrm{uq}}$", e.g. as this occurs in the first order case. We start by proving existence for $\mathrm{NF}_{\lambda,\bar{\delta}}$ from Definition 6.11.

6.13 Claim. *1) Assume* $\bar\delta = \langle \delta_1, \delta_2, \delta_3\rangle, \delta_\ell$ *an ordinal* $< \lambda^+$ *and* $N_\ell \in K_\lambda$ *for* $\ell < 3$ *and* N_1 *is* $(\lambda, \mathrm{cf}(\delta_1))$*-brimmed over* N_0 *and* N_2 *is* $(\lambda, \mathrm{cf}(\delta_2))$*-brimmed over* N_0 *and* $N_0 \le_{\mathfrak{K}} N_1$ *and* $N_0 \le_{\mathfrak{K}} N_2$ *and for simplicity* $N_1 \cap N_2 = N_0$. *Then we can find* N_3 *such that* $\mathrm{NF}_{\lambda,\bar\delta}(N_0, N_1, N_2, N_3)$.
2) Moreover, we can choose any $\langle N_{1,i} : i \le \lambda \times \delta_1\rangle, \langle c_i : i < \lambda \times \delta_1\rangle$ *as in 6.11 subclauses (f)*$(\alpha), (\gamma), (\delta)$ *as part of the witness.*
3) If $\mathrm{NF}_\lambda(N_0, N_1, N_2, N_3)$ *then* $N_1 \cap N_2 = N_0$.

Proof. 1) We can find $\langle N_{1,i} : i \le \lambda \times \delta_1\rangle$ and $\langle c_i : i < \lambda \times \delta_1\rangle$ as required in part (2) by Claim 6.9, the $(\lambda, \mathrm{cf}(\lambda \times \delta_1))$-brimness holds by 4.3 and apply part (2).
2) We choose the $N_{2,i}$ (by induction on i) by 4.9 preserving $N_{2,i} \cap N_{1,\lambda\times\delta_2} = N_{1,i}$; in the successor case use Definition 5.3 + Claim 5.4(3). We then choose N_3 using 4.2(2).
3) By the definitions of NF_λ, $\mathrm{NF}_{\lambda,\bar\delta}$. $\square_{6.13}$

The following claim tells us that if we have "$(\lambda, \mathrm{cf}(\delta_3))$-brimmed" in the end, then we can have it in all successor stages.

6.14 Claim. *In Definition 6.11, if* δ_3 *is a limit ordinal and* $\kappa = \mathrm{cf}(\kappa) \ge \aleph_0$, *then without loss of generality (even without changing* $\langle N_{1,i} : i \le \lambda \times \delta_1\rangle, \langle c_i : i < \lambda \times \delta_1\rangle$*)*

- (g) $N_{2,i+1}$ *is* (λ, κ)*-brimmed over* $N_{1,i+1} \cup N_{2,i}$ *(which means that it is* (λ, κ)*-brimmed over some* N*, where* $N_{1,i+1} \cup N_{2,i} \subseteq N \le_{\mathfrak{K}} N_{2,i+1}$*).*

Proof. So assume $\mathrm{NF}_{\lambda,\bar\delta}(N_0, N_1, N_2, N_3)$ holds as being witnessed by $\langle N_{\ell,i} : i \le \lambda \times \delta_1\rangle, \langle c_i : i < \lambda \times \delta_1\rangle$ for $\ell = 1, 2$. Now we choose by induction on $i \le \lambda \times \delta_1$ a model $M_{2,i} \in K_\lambda$ and f_i such that:

- (i) f_i is a $\le_{\mathfrak{K}}$-embedding of $N_{2,i}$ into $M_{2,i}$
- (ii) $M_{2,0} = f_i(N_2)$
- (iii) $M_{2,i}$ is $\le_{\mathfrak{K}}$-increasing continuous and also f_i is increasing continuous
- (iv) $M_{2,j} \cap f_i(N_{1,i}) = f_i(N_{1,j})$ for $j \le i$

(*v*) $M_{2,i+1}$ is (λ, κ)-brimmed over $M_{2,i} \cup f_i(N_{2,i+1})$

(*vi*) $\mathbf{tp}(f_{i+1}(c_i), M_{2,i}, M_{2,i+1}) \in \mathscr{S}^{\rm bs}(M_{2,i})$ does not fork over $f_i(N_{1,i})$.

There is no problem to carry the induction. Using in the successor case $i = j+1$ the existence Axiom (E)(g) of Definition 2.1 there is a model $M'_{2,i} \in K_{\mathfrak{s}}$ such that $M_{2,j} \leq_{\mathfrak{K}} M'_{2,i}$ and $f_i \supseteq f_j$ as required in clauses (i), (iv), (vi) and then use Claim 4.2 to find a model $M_{2,i} \in K_\lambda$ which is (λ, κ)-brimmed over $M_{2,j} \cup f_i(N_{2,i})$.

Having carried the induction, without loss of generality $f_i = \mathrm{id}_{N_{2,i}}$. Let M_3 be such that $M_{2,\lambda\times\delta_1} \leq_{\mathfrak{K}} M_3 \in K_\lambda$ and M_3 is $(\lambda,\mathrm{cf}(\delta_3))$-brimmed over $M_{2,\lambda\times\delta_1}$, it exists by 4.2(2) but $N_{2,\lambda\times\delta_1} \leq_{\mathfrak{K}} M_{2,\lambda\times\delta_1}$, hence it follows that M_3 is (λ, κ)-brimmed over $N_{1,\lambda\times\delta_1}$. So both M_3 and N_3 are $(\lambda,\mathrm{cf}(\delta_3))$-brimmed over $N_{2,\lambda\times\delta_1}$, hence they are isomorphic over $N_{2,\lambda\times\delta_1}$ (by 1.16(1)) so let f be an isomorphism from M_3 onto N_3 which is the identity over $N_{2,\lambda\times\delta_1}$.
Clearly $\langle N_{1,i} : i \leq \lambda \times \delta_1\rangle, \langle f(M_{2,i}) : i \leq \lambda \times \delta_1\rangle$ are also witnesses for
$\mathrm{NF}_{\lambda,\bar\delta}(N_0, N_1, N_2, N_3)$ satisfying the extra demand (g) from 6.14. $\square_{6.14}$

The point of the following claim is that having uniqueness in every atomic step we have uniqueness in the end (using the same "ladder" $N_{1,i}$ for now).

6.15 Claim. *(Weak Uniqueness).*

Assume that for $x \in \{a, b\}$, *we have* $\mathrm{NF}_{\lambda,\bar\delta^x}(N^x_0, N^x_1, N^x_2, N^x_3)$ *holds as witnessed by* $\langle N^x_{1,i} : i \leq \lambda \times \delta^x_1\rangle, \langle c^x_i : i < \lambda \times \delta^x_1\rangle, \langle N^x_{2,i} : i \leq \lambda \times \delta^x_1\rangle$ *and* $\delta_1 := \delta^a_1 = \delta^b_1, \mathrm{cf}(\delta^a_2) = \mathrm{cf}(\delta^b_2)$ *and* $\mathrm{cf}(\delta^a_3) = \mathrm{cf}(\delta^b_3) \geq \aleph_0$. *(Note that* $\mathrm{cf}(\lambda \times \delta^a_1) \geq \aleph_0$ *by the definition of* NF*).*

Suppose further that f_ℓ *is an isomorphism from* N^a_ℓ *onto* N^b_ℓ *for* $\ell = 0, 1, 2$, *moreover:* $f_0 \subseteq f_1, f_0 \subseteq f_2$ *and* $f_1(N^a_{1,i}) = N^b_{1,i}, f_1(c^a_i) = c^b_i$.

<u>*Then*</u> *we can find an isomorphism* f *from* N^a_3 *onto* N^b_3 *extending* $f_1 \cup f_2$.

Proof. Without loss of generality for each $i < \lambda \times \delta_1$, the model $N^x_{2,i+1}$ is (λ, λ)-brimmed over $N^x_{1,i+1} \cup N^x_{2,i}$ (by 6.14, note there the

statement "without changing the $N_{1,i}$'s"). Now we choose by induction on $i \le \lambda \times \delta_1$ an isomorphism g_i from $N^a_{2,i}$ onto $N^b_{2,i}$ such that: g_i is increasing with i and g_i extends $(f_1 \restriction N^a_{1,i}) \cup f_2$.
For $i = 0$ choose $g_0 = f_2$ and for i limit let g_i be $\bigcup_{j<i} g_j$ and for $i = j+1$ it exists by 6.10, whose assumptions hold by $(N^x_{1,i}, N^x_{1,i+1}, c^x_i) \in K^{3,\mathrm{uq}}_\lambda$ (see 6.11, clause (f)(δ)) and the extra brimness clause from 6.14. Now by 1.16(3) we can extend $g_{\lambda\times\delta_1}$ to an isomorphism from N^a_3 onto N^b_3 as N^x_3 is $(\lambda, \mathrm{cf}(\delta_3))$-brimmed over $N^x_{2,\lambda\times\delta_1}$ (for $x \in \{a, b\}$). $\square_{6.15}$

Note that even knowing 6.15 the choice of $\langle N_{1,i} : i \le \lambda \times \delta_1\rangle, \langle c_i : i < \lambda \times \delta_1\rangle$ still possibly matters. Now we prove an "inverted" uniqueness, using our ability to construct a "rectangle" of models which is a witness for $\mathrm{NF}_{\lambda,\bar\delta}$ in two ways.

6.16 Claim. *Suppose that*

(a) *for* $x \in \{a, b\}$ *we have* $\mathrm{NF}_{\lambda,\bar\delta^x}(N^x_0, N^x_1, N^x_2, N^x_3)$

(b) $\bar\delta^x = \langle \delta^x_1, \delta^x_2, \delta^x_3\rangle, \delta^a_1 = \delta^b_2,\ \delta^a_2 = \delta^b_1, \mathrm{cf}(\delta^a_3) = \mathrm{cf}(\delta^b_3)$, *all limit ordinals*

(c) f_0 *is an isomorphism from* N^a_0 *onto* N^b_0

(d) f_1 *is an isomorphism from* N^a_1 *onto* N^b_2

(e) f_2 *is an isomorphism from* N^a_2 *onto* N^b_1

(f) $f_0 \subseteq f_1$ *and* $f_0 \subseteq f_2$.

Then there is an isomorphism from N^a_3 *onto* N^b_3 *extending* $f_1 \cup f_2$.

Before proving we shall construct a third "rectangle" of models such that we shall be able to construct appropriate isomorphisms each of N^a_3, N^b_3

6.17 Subclaim. *Assume*

(a) $\delta^a_1, \delta^a_2, \delta^a_3 < \lambda^+$ *are limit ordinals*

$(b)_1$ $\bar M^1 = \langle M^1_\alpha : \alpha \le \lambda \times \delta^a_1\rangle$ *is* $\le_{\mathfrak{K}}$*-increasing continuous in* K_λ *and* $(M^1_\alpha, M^1_{\alpha+1}, c_\alpha) \in K^{3,\mathrm{bs}}_\lambda$

$(b)_2$ $\bar{M}^2 = \langle M^2_\alpha : \alpha \le \lambda \times \delta^a_2 \rangle$ *is* $\le_{\mathfrak{K}}$*-increasing continuous in* K_λ *and* $(M^2_\alpha, M^2_{\alpha+1}, d_\alpha) \in K^{3,\mathrm{bs}}_\lambda$

(c) $M^1_0 = M^2_0$ *we call it* M *and* $M^1_\alpha \cap M^2_\beta = M$ *for* $\alpha \le \lambda \times \delta^a_1, \beta \le \lambda \times \delta^a_2$.

Then we can find $M_{i,j}$ *(for* $i \le \lambda \times \delta^a_1$ *and* $j \le \lambda \times \delta^a_2$*) and* M_3 *such that:*

(A) $M_{i,j} \in K_\lambda$ *and* $M_{0,0} = M$ *and* $M_{i,0} = M^1_i, M_{0,j} = M^2_j$

(B) $i_1 \le i_2$ & $j_1 \le j_2 \Rightarrow M_{i_1,j_1} \le_{\mathfrak{K}} M_{i_2,j_2}$

(C) *if* $i \le \lambda \times \delta^a_1$ *is a limit ordinal and* $j \le \lambda \times \delta^a_2$ *then* $M_{i,j} = \bigcup_{\zeta < i} M_{\zeta,j}$

(D) *if* $i \le \lambda \times \delta^a_1$ *and* $j \le \lambda \times \delta^a_2$ *is a limit ordinal then* $M_{i,j} = \bigcup_{\xi < j} M_{i,\xi}$

(E) $M_{\lambda \times \delta^a_1, j+1}$ *is* $(\lambda, \mathrm{cf}(\delta^a_1))$*-brimmed over* $M^a_{\lambda \times \delta^a_1, j}$ *for* $j < \lambda \times \delta^a_2$

(F) $M_{i+1, \lambda \times \delta^a_2}$ *is* $(\lambda, \mathrm{cf}(\delta^a_2))$*-brimmed over* $M_{i, \lambda \times \delta^a_2}$ *for* $i < \lambda \times \delta^a_1$

(G) $M_{\lambda \times \delta^a_1, \lambda \times \delta^a_2} \le_{\mathfrak{K}} M_3 \in K_\lambda$ *moreover* M_3 *is* $(\lambda, \mathrm{cf}(\delta^a_3))$*-brimmed over* $M_{\lambda \times \delta^a_1, \lambda \times \delta^a_2}$

(H) *for* $i < \lambda \times \delta^a_1, j \le \lambda \times \delta^a_2$ *we have* $\mathbf{tp}(c_i, M_{i,j}, M_{i+1,j})$ *does not fork over* $M_{i,0}$

(I) *for* $j < \lambda \times \delta^a_2, i \le \lambda \times \delta^a_1$ *we have* $\mathbf{tp}(d_j, M_{i,j}, M_{i,j+1})$ *does not fork over* $M_{0,j}$.

We can add

(J) *for* $i < \lambda \times \delta^a_1, j < \lambda \times \delta^b_2$ *the model* $M_{i+1,j+1}$ *is* $(\lambda, *)$*-brimmed over* $M_{i,j+1} \cup M_{i+1,j}$.

Remark. 1) We can replace in 6.17 the ordinals $\lambda \times \delta^a_\ell$ $(\ell = 1, 2, 3)$ by any ordinal $\alpha^a_\ell < \lambda^+$ (for $\ell = 1, 2, 3$) we use the present notation just to conform with its use in the proof of 6.16.
2) Why do we need u^ℓ_1 in the proof below? This is used to get the brimmness demands in 6.17.

Proof. We first change our towers, repeating models to give space for bookkeeping. That is we define ${}^*M^1_\alpha$ for $\alpha \le \lambda \times \lambda \times \delta^a_1$ as follows:

if $\lambda \times \beta < \alpha \le \lambda \times \beta + \lambda$ and $\beta < \lambda \times \delta^a_1$ then ${}^*M^1_\alpha = M^1_{\beta+1}$

if $\alpha = \lambda \times \beta$, then ${}^*M^1_\alpha = M^1_\beta$.

Let $u^1_0 = \{\lambda\beta : \beta < \delta^a_1\}, u^1_1 = \lambda \times \lambda \times \delta^a_1 \backslash u^1_0, u^1_2 = \emptyset$ and for $\alpha = \lambda\beta \in u^1_0$ let $a^1_\alpha = c_\beta$.

Similarly let us define ${}^*M^2_\alpha$ (for $\alpha \le \lambda \times \lambda \times \delta^a_2$),$u^2_0, u^2_1, u^2_2$ and $\langle a^2_\alpha : \alpha \in u^2_0 \rangle$.

Now apply 4.11 (check) and get ${}^*M_{i,j}, (i \le \lambda \times \lambda \times \delta^a_1, j \le \lambda \times \lambda \times \delta^a_2)$. Lastly, for $i \le \delta^a_1, j \le \delta^a_2$ let $M_{i,j} = {}^*M_{\lambda \times i, \lambda \times j}$. By 4.3 clearly ${}^*M_{\lambda \times i + \lambda, \lambda \times j + \lambda}$ is $(\lambda, \mathrm{cf}(\lambda))$-brimmed over ${}^*M_{\lambda \times i+1, \lambda \times j+1}$ hence $M_{i+1,j+1}$ is $(\lambda, \mathrm{cf}(\lambda))$-brimmed over $M_{i+1,j} \cup M_{i,j+1}$. And, by 4.2(1) choose $M_3 \in K_\lambda$ which is $(\lambda, \mathrm{cf}(\delta^a_3))$-brimmed over $M_{\lambda \times \delta^a_1, \lambda \times \delta^a_2}$. $\square_{6.17}$

Proof of 6.16. We shall let $M_{i,j}, M_3$ be as in 6.17 for $\bar{\delta}^a$ and $\bar{M}^1, \bar{M}^2$ determined below. For $x \in \{a, b\}$ as $\mathrm{NF}_{\lambda, \bar{\delta}^x}(N^x_0, N^x_1, N^x_2, N^x_3)$, we know that there are witnesses $\langle N^x_{1,i} : i \le \lambda \times \delta^x_1 \rangle, \langle c^x_i : i < \lambda \times \delta^x_1 \rangle, \langle N^x_{2,i} : i \le \lambda \times \delta^x_1 \rangle$ for this. So $\langle N^x_{1,i} : i \le \lambda \times \delta^x_1 \rangle$ is $\le_{\mathfrak{K}}$-increasing continuous and $(N^x_{1,i}, N^x_{1,i+1}, c^x_i) \in K^{3,\mathrm{uq}}_\lambda$ for $i < \lambda \times \delta^x_1$. Hence by the freedom we have in choosing $\bar{M}^1$ and $\langle c_i : i < \lambda \times \delta_1 \rangle$ without loss of generality there is an isomorphism g_1 from $N^a_{1,\lambda \times \delta^a_1}$ onto $M_{\lambda \times \delta^a_1}$ mapping $N^a_{1,i}$ onto $M^1_i = M_{i,0}$ and c^a_i to c_i; remember that $N^a_{1,\lambda \times \delta^a_1} = N^a_1$. Let $g_0 = g_1 \restriction N^a_0 = g_1 \restriction N^a_{1,0}$ so $g_0 \circ f_0^{-1}$ is an isomorphism from N^b_0 onto $M_{0,0}$.

Similarly as $\delta^b_1 = \delta^a_2$, and using the freedom we have in choosing $\bar{M}^2$ and $\langle d_i : i < \lambda \times \delta^b_1 \rangle$ without loss of generality there is an isomorphism g_2 from $N^b_{1,\lambda \times \delta^a_2}$ onto $M^2_j = M_{0,\lambda \times \delta^a_2}$ mapping $N^b_{1,j}$ onto $M_{0,j}$ (for $j \le \lambda \times \delta^a_2$) and mapping c^b_i to d_i and g_2 extends $g_0 \circ f_0^{-1}$.
Now would like to use the weak uniqueness 6.15 and for this note:

(α) $\mathrm{NF}_{\lambda, \bar{\delta}^a}(N^a_0, N^a_1, N^a_2, N^a_3)$ is witnessed by the sequences $\langle N^a_{1,i} : i \le \lambda \times \delta^a_1 \rangle$, and $\langle N^a_{2,i} : i \le \lambda \times \delta^a_1 \rangle$
[why? an assumption]

(β) $\mathrm{NF}_{\lambda,\bar{\delta}^a}(M_{0,0}, M_{\lambda\times\delta_1^a,0}, M_{0,\lambda\times\delta_2^a}, M_3)$ is witnessed by the sequences $\langle M_{i,0} : i \leq \lambda \times \delta_1^a\rangle, \langle M_{i,\lambda\times\delta_2^a} : i \leq \lambda \times \delta_1^a\rangle$
[why? check]

(γ) g_0 is an isomorphism from N_0^a onto $M_{0,0}$
[why? see its choice]

(δ) g_1 is an isomorphism from N_1^a onto $M_{\lambda\times\delta_1^a,0}$ mapping $N_{1,i}^a$ onto $M_{i,0}$ for $i < \lambda \times \delta_1^a$ and c_i^a to c_i for $i < \lambda \times \delta_1^a$ and extending g_0
[why? see the choice of g_1 and of g_0]

(ε) $g_2 \circ f_2$ is an isomorphism from N_2^a onto $M_{0,\lambda\times\delta_2^a}$ extending g_0
[why? f_2 is an isomorphism from N_2^a onto N_1^b and g_2 is an isomorphism from N_1^b onto $M_{0,\lambda\times\delta_1^a}$ extending $g_0 \circ f_0^{-1}$ and $f_0 \subseteq f_2$].

So there is by 6.15 an isomorphism g_3^a from N_3^a onto M_3 extending both g_1 and $g_2 \circ f_2$.

We next would like to apply 6.15 to the N_i^b's; so note:

(α)′ $\mathrm{NF}_{\lambda,\bar{\delta}^b}(N_0^b, N_1^b, N_2^b, N_3^b)$ is witnessed by the sequences $\langle N_{1,i}^b :$ $i \leq \lambda \times \delta_2^a\rangle$,
$\langle N_{2,i}^b : i \leq \lambda \times \delta_2^a\rangle$

(β)′ $\mathrm{NF}_{\lambda,\bar{\delta}^b}(M_{0,0}, M_{0,\lambda\times\delta_2^a}, M_{\lambda\times\delta_1^a,0}, M_3)$ is witnessed by the sequences $\langle M_{0,j} : j \leq \lambda \times \delta_2^a\rangle, \langle M_{\lambda\times\delta_1^a,j} : j \leq \lambda \times \delta_2^a\rangle$

(γ)′ $g_0 \circ (f_0)^{-1}$ is an isomorphism from N_0^b onto $M_{0,0}$
[why? Check.]

(δ)′ g_2 is an isomorphism from N_1^b onto $M_{0,\lambda\times\delta_2^a}$ mapping $N_{1,j}^b$ onto $M_{0,j}$ and c_j^a to d_j for $j \leq \lambda\times\delta_2^a$ and extending $g_0 \circ (f_2)^{-1}$
[why? see the choice of g_2: it maps $N_{1,j}^b$ onto $M_{0,j}$]

(ε)′ $g_1 \circ (f_1)^{-1}$ is an isomorphism from N_2^b onto $M_{\lambda\times\delta_0^a}$ extending g_0
[why? remember f_1 is an isomorphism from N_1^a onto N_2^b extending f_0 and the choice of g_1: it maps N_1^a onto $M_{\lambda\times\delta_1^a,0}$].

So there is an isomorphism g_3^b form N_3^b onto M_3 extending g_2 and $g_1 \circ (f_1)^{-1}$.

Lastly $(g_3^b)^{-1} \circ g_3^a$ is an isomorphism from N_3^a onto N_3^b (chase arrows). Also

$$\begin{aligned}((g_3^b)^{-1} \circ g_3^a) \restriction N_1^a &= (g_3^b)^{-1}(g_3^a \restriction N_1^a)\\ &= (g_3^b)^{-1} g_1 = ((g_3^b)^{-1} \restriction M_{\lambda \times \delta_1^a, 0}) \circ g_1\\ &= (g_3^b \restriction N_2^b)^{-1} \circ g_1 = ((g_1 \circ (f_1)^{-1})^{-1}) \circ g_1\\ &= (f_1 \circ (g_1)^{-1}) \circ g_1 = f_1.\end{aligned}$$

Similarly $((g_3^b)^{-1} \circ g_3^a) \restriction N_2^a = f_2$.
So we have finished. $\square_{6.16}$

But if we invert twice we get straight; so

6.18 Claim. *[Uniqueness]. Assume for $x \in \{a, b\}$ we have $\mathrm{NF}_{\lambda, \bar{\delta}^x}(N_0^x, N_1^x, N_2^x, N_3^x)$ and $\mathrm{cf}(\delta_1^a) = \mathrm{cf}(\delta_1^b), \mathrm{cf}(\delta_2^a) = \mathrm{cf}(\delta_2^b), \mathrm{cf}(\delta_3^a) = \mathrm{cf}(\delta_3^b)$, all δ_ℓ^x limit ordinals $< \lambda^+$.*

If f_ℓ is an isomorphism from N_ℓ^a onto N_ℓ^b for $\ell < 3$ and $f_0 \subseteq f_1, f_0 \subseteq f_2$ then there is an isomorphism f from N_3^a onto N_3^b extending f_1, f_2.

Proof. Let $\bar{\delta}^c = \langle \delta_1^c, \delta_2^c, \delta_3^c \rangle = \langle \delta_2^a, \delta_1^a, \delta_3^a \rangle$; by 6.13(1) there are N_ℓ^c (for $\ell \leq 3$) such that $\mathrm{NF}_{\lambda, \bar{\delta}^c}(N_0^c, N_1^c, N_2^c, N_3^c)$ and $N_0^c \cong N_0^a$. There is for $x \in \{a, b\}$ an isomorphism g_0^x from N_0^x onto N_0^c and without loss of generality $g_0^a = g_0^b \circ f_0$. Similarly for $x \in \{a, b\}$ there is an isomorphism g_1^x from N_1^x onto N_2^c extending g_0^x (as N_1^x is $(\lambda, \mathrm{cf}(\delta_1^x))$-brimmed over N_0^x and also N_2^c is $(\lambda, \mathrm{cf}(\delta_2^c))$-brimmed over N_0^c and $\mathrm{cf}(\delta_2^c) = \mathrm{cf}(\delta_1^a) = \mathrm{cf}(\delta_1^x)$) and without loss of generality $g_1^b = g_1^a \circ f_1$. Similarly for $x \in \{a, b\}$ there is an isomorphism g_2^x from N_2^x onto N_1^c extending g_0^x (as N_2^x is $(\lambda, \mathrm{cf}(\delta_2^x))$-brimmed over N_0^x and also N_1^c is $(\lambda, \mathrm{cf}(\delta_1^c))$-brimmed over N_0^c and $\mathrm{cf}(\delta_1^c) = \mathrm{cf}(\delta_2^a) = \mathrm{cf}(\delta_2^x)$) and without loss of generality $g_2^a = g_2^b \circ f_2$.
So by 6.16 for $x \in \{a, b\}$ there is an isomorphism g_3^x from N_3^x onto N_3^c extending g_1^x and g_2^x. Now $(g_3^b)^{-1} \circ g_3^a$ is an isomorphism from N_3^a onto N_3^b extending f_1, f_2 as required. $\square_{6.18}$

So we have proved the uniqueness for $\mathrm{NF}_{\lambda, \bar{\delta}}$ when all δ_ℓ are limit ordinals; this means that the arbitrary choice of $\langle N_{1,i} : i \leq \lambda \times \delta_1 \rangle$

and $\langle c_i : i < \lambda \times \delta_1 \rangle$ is immaterial; it figures in the definition and, e.g. existence proof but does not influence the net result. The power of this result is illustrated in the following conclusion.

6.19 Conclusion. [Symmetry].

If $\mathrm{NF}_{\lambda,\langle \delta_1,\delta_2,\delta_3 \rangle}(N_0, N_1, N_2, N_3)$ where $\delta_1, \delta_2, \delta_3$ are limit ordinals $< \lambda^+$ then $\mathrm{NF}_{\lambda,\langle \delta_2,\delta_1,\delta_3 \rangle}(N_0, N_2, N_1, N_3)$.

Proof. By 6.17 we can find $N'_\ell(\ell \leq 3)$ such that: $N'_0 = N_0, N'_1$ is $(\lambda, \mathrm{cf}(\delta_1))$-brimmed over N'_0, N'_2 is $(\lambda, \mathrm{cf}(\delta_2))$-brimmed over N'_0 and N'_3 is $(\lambda, \mathrm{cf}(\delta_3))$-brimmed over $N'_1 \cup N'_2$ and $\mathrm{NF}_{\lambda,\langle \delta_1,\delta_2,\delta_3 \rangle}(N'_0, N'_1, N'_2, N'_3)$ and $\mathrm{NF}_{\lambda,\langle \delta_2,\delta_1,\delta_3 \rangle}(N'_0, N'_2, N'_1, N'_3)$. Let f_1, f_2 be an isomorphism from N_1, N_2 onto N'_1, N'_2 over N_0, respectively. By 6.18 (or 6.16) there is an isomorphism f'_3 form N_3 onto N'_3 extending $f_1 \cup f_2$. As isomorphisms preserve NF we are done. $\square_{6.19}$

Now we turn to smooth amalgamation (not necessarily brimmed, see Definition 6.12). If we use Lemma 4.8, of course, we do not really need 6.20.

6.20 Claim. *1) If* $\mathrm{NF}_{\lambda,\bar\delta}(N_0, N_1, N_2, N_3)$ *and* $\delta_1, \delta_2, \delta_3$ *are limit ordinals, then* $\mathrm{NF}_\lambda(N_0, N_1, N_2, N_3)$ *(see Definition 6.12).*
2) In Definition 6.12(1) we can add:

- $(d)^+$ *M_ℓ is $(\lambda, \mathrm{cf}(\lambda))$-brimmed over N_0 and moreover over N_ℓ,*
- (e) *M_3 is $(\lambda, \mathrm{cf}(\lambda))$-brimmed over $M_1 \cup M_2$ (actually this is given by clause $(f)(\zeta)$ of Definition 6.11).*

3) If $N_0 \leq_{\mathfrak{K}} N_\ell$ for $\ell = 1, 2$ and $N_1 \cap N_2 = N_0$, then we can find N_3 such that $\mathrm{NF}_\lambda(N_0, N_1, N_2, N_3)$.

Proof. 1) Note that even if every δ_ℓ is limit and we waive the "moreover" in clause $(d)^+$, the problem is in the case that e.g. $(\mathrm{cf}(\delta^a), \mathrm{cf}(\delta^b), \mathrm{cf}(\delta^c)) \neq (\mathrm{cf}(\lambda), \mathrm{cf}(\lambda), \mathrm{cf}(\lambda))$. For $\ell = 1, 2$ we can find $\bar M^\ell = \langle M^\ell_i : i \leq \lambda \times (\delta_\ell + \lambda) \rangle$ and $\langle c^\ell_i : i < \lambda \times (\delta_i + \lambda) \rangle$ such that $M^\ell_0 = N_0, \bar M^1$ is $\leq_{\mathfrak{K}}$-increasing continuous $(M^\ell_i, M^\ell_{i+1}, c_i) \in K^{3,\mathrm{uq}}_{\mathfrak{s}}$ and if $p \in \mathscr{S}^{\mathrm{bs}}(M^\ell_i)$ and $i < \lambda \times (\delta_\ell + \lambda)$ then for λ ordinals $j < \lambda$, $\mathbf{tp}(c_i, M^\ell_{i+j}, M^\ell_{i+j+1})$ is a non-forking extension of p. So $M^\ell_{\lambda \times \delta_\ell}$

is $(\lambda, \mathrm{cf}(\delta_\ell))$-brimmed over $M^\ell_0 = N_0$ and $M^\ell_{\lambda\times(\delta_\ell+\lambda)}$ is $(\lambda, \mathrm{cf}(\lambda))$-brimmed over $M^\ell_{\lambda\times\delta_\ell}$; so without loss of generality $M^\ell_{\lambda\times\delta_\ell} = N_\ell$ for $\ell = 1, 2$.
By 6.17 we can find $M_{i,j}$ for $i \le \lambda \times (\delta_1 + \lambda), j \le \lambda \times (\delta_2 + \lambda)$ for $\bar\delta' := \langle \delta_1 + \lambda, \delta_2 + \lambda, \delta_3\rangle$ such that they are as in 6.17 for $\bar M^1, \bar M^2$ so $M_{0,0} = N_0$; then choose $M'_3 \in K_\lambda$ which is $(\lambda, \mathrm{cf}(\delta_3))$-brimmed over $M_{\lambda\times\delta_1,\lambda\times\delta_2}$. So $\mathrm{NF}_{\lambda,\bar\delta}(M_{0,0}, M_{\lambda\times\delta_1,0}, M_{0,\lambda\times\delta_2}, M'_3)$, hence by 6.18 without loss of generality $M_{0,0} = N_0, M_{\lambda\times\delta_1,0} = N_1, M_{0,\lambda\times\delta_2} = N_2$, and $N_3 = M'_3$. Lastly, let M_3 be $(\lambda, \mathrm{cf}(\lambda))$-brimmed over M'_3. Now clearly also
$\mathrm{NF}_{\lambda,\langle\delta_1+\lambda,\delta_2+\lambda,\delta_3+\lambda\rangle}(M_{0,0}, M_{\lambda\times(\delta_1+\lambda),0}, M_{0,\lambda\times(\delta_2+\lambda)}, M_3)$ and
$N_0 = M_{0,0}, N_1 = M_{\lambda\times\delta_2,0} \le_{\mathfrak{K}} M_{\lambda\times(\delta_2+\lambda),0}, N_2 = M_{0,\lambda\times\delta_2} \le_{\mathfrak{K}} M_{0,\lambda\times(\delta_2+\lambda)}$
and $M_{\lambda\times(\delta_1+\lambda),0}$ is $(\lambda, \mathrm{cf}(\lambda))$-brimmed over $M_{\lambda\times\delta_1,0}$ and $M_{0,\lambda\times(\delta_2+\lambda)}$ is
$(\lambda, \mathrm{cf}(\lambda))$-brimmed over $M_{0,\lambda\times\delta_2}$ and $N_3 = M'_3 \le_{\mathfrak{K}} M_3$. So we get all the requirements for $\mathrm{NF}_\lambda(N_0, N_1, N_2, N_3)$ (as witnessed by $\langle M_{0,0}, M_{\lambda\times(\delta_1+\lambda),0}, M_{0,\lambda\times(\delta_2+\lambda)}, M_3\rangle$). 2) Similar proof.
3) By 6.13 and the proof above. $\square_{6.20}$

Now we turn to NF_λ; existence is easy.

6.21 Claim. NF_λ *has existence, i.e., clause (f) of 6.1(1).*

Proof. By 6.20(3). $\square_{6.21}$

Next we deal with real uniqueness

6.22 Claim. *[Uniqueness of smooth amalgamation]:*
1) If $\mathrm{NF}_\lambda(N^x_0, N^x_1, N^x_2, N^x_3)$ for $x \in \{a, b\}$, f_ℓ an isomorphism from N^a_ℓ onto N^b_ℓ for $\ell < 3$ and $f_0 \subseteq f_1, f_0 \subseteq f_2$ then $f_1 \cup f_2$ can be extended to a $\le_{\mathfrak{K}}$-embedding of N^a_3 into some $\le_{\mathfrak{K}}$-extension of N^b_3.
2) So if above N^x_3 is (λ, κ)-brimmed over $N^x_1 \cup N^x_2$ for $x = a, b$, we can extend $f_1 \cup f_2$ to an isomorphism from N^a_3 onto N^b_3.

Proof. 1) For $x \in \{a, b\}$ let the sequence $\langle M^x_\ell : \ell < 4\rangle$ be a witness to
$\mathrm{NF}_\lambda(N^x_0, N^x_1, N^x_2, N^x_3)$ as in 6.12, 6.20(2), so in particular

$\mathrm{NF}_{\lambda,\langle\lambda,\lambda,\lambda\rangle}(M_0^x, M_1^x, M_2^x, M_3^x)$. By chasing arrows (disjointness) and uniqueness, i.e. 6.18 without loss of generality $M_\ell^a = M_\ell^b$ for $\ell < 4$ and $f_0 = \mathrm{id}_{N_0^a}$. As M_1^a is $(\lambda, \mathrm{cf}(\lambda))$-brimmed over N_1^a and also over N_1^b (by clause $(d)^+$ of 6.20(2)) and f_1 is an isomorphism from N_1^a onto N_1^b, clearly by 1.16 there is an automorphism g_1 of M_1^a such that $f_1 \subseteq g_1$, hence also $\mathrm{id}_{N_0^a} = f_0 \subseteq f_1 \subseteq g_1$. Similarly there is an automorphism g_2 of M_2^a extending f_2 hence f_0. So $g_\ell \in \mathrm{AUT}(M_\ell^a)$ for $\ell = 1, 2$ and $g_1 \restriction M_0^a = f_0 = g_2 \restriction M_0^a$. By the uniqueness of $\mathrm{NF}_{\lambda,\langle\lambda,\lambda,\lambda\rangle}$ (i.e. Claim 6.18) there is an automorphism g_3 of M_3^a extending $g_1 \cup g_2$. This proves the desired conclusion.
2) Should be clear. $\square_{6.22}$

We now show that in the cases the two notions of non-forking amalgamations are meaningful then they coincide, one implication already is a case of 6.20.

6.23 Claim. *Assume*

(a) $\bar\delta = \langle \delta_1, \delta_2, \delta_3 \rangle, \delta_\ell < \lambda^+$ is a limit ordinal for $\ell = 1, 2, 3$;
$N_0 \le_{\mathfrak{K}} N_\ell \le_{\mathfrak{K}} N_3$ are in K_λ for $\ell = 1, 2$

(b) N_ℓ is $(\lambda, \mathrm{cf}(\delta_\ell))$-brimmed over N_0 for $\ell = 1, 2$

(c) N_3 is $\mathrm{cf}(\delta_3)$-brimmed over $N_1 \cup N_2$.

Then $\mathrm{NF}_\lambda(N_0, N_1, N_2, N_3)$ *iff* $\mathrm{NF}_{\lambda,\bar\delta}(N_0, N_1, N_2, N_3)$.

Proof. The "if" direction holds by 6.20(1). As for the "only if" direction, basically it follows from the existence for $\mathrm{NF}_{\lambda,\bar\delta}$ and uniqueness for NF_λ; in details by the proof of 6.20(1) (and Definition 6.11, 6.12) we can find $M_\ell (\ell \le 3)$ such that $M_0 = N_0$ and $\mathrm{NF}_{\lambda,\bar\delta}(M_0, M_1, M_2, M_3)$ and clauses (b), (c), (d) of Definition 6.12 and $(d)^+$ of 6.20(2) hold so by 6.20 also $\mathrm{NF}_\lambda(M_0, M_1, M_2, M_3)$. Easily there are for $\ell < 3$, isomorphisms f_ℓ from M_ℓ onto N_ℓ such that $f_0 = f_\ell \restriction M_\ell$ where $f_0 = \mathrm{id}_{N_0}$. By the uniqueness of smooth amalgamations (i.e., 6.22(2)) we can find an isomorphism f_3 from M_3 onto N_3 extending $f_1 \cup f_2$. So as $\mathrm{NF}_{\lambda,\bar\delta}(M_0, M_1, M_2, M_3)$ holds also $\mathrm{NF}_{\lambda,\bar\delta}, (f_0(M_0), f_3(M_1), f_3(M_2), f_3(M_3))$; that is $\mathrm{NF}_{\lambda,\bar\delta}(N_0, N_1, N_2, N_3)$ is as required. $\square_{6.23}$

6.24 Claim. *[Monotonicity]: If* $\mathrm{NF}_\lambda(N_0, N_1, N_2, N_3)$ *and* $N_0 \le_{\mathfrak{K}} N_1' \le_{\mathfrak{K}} N_1$ *and* $N_0 \le_{\mathfrak{K}} N_2' \le_{\mathfrak{K}} N_2$ *and* $N_1' \cup N_2' \subseteq N_3' \le_{\mathfrak{K}} N_3'', N_3 \le_{\mathfrak{K}} N_3''$ *then* $\mathrm{NF}_\lambda(N_0, N_1', N_2', N_3')$.

Proof. Read Definition 6.12(1). $\square_{6.24}$

6.25 Claim. *[Symmetry]:* $\mathrm{NF}_\lambda(N_0, N_1, N_2, N_3)$ *holds if and only if* $\mathrm{NF}_\lambda(N_0, N_2, N_1, N_3)$ *holds.*

Proof. By Claim 6.19 (and Definition 6.12). $\square_{6.25}$

We observe

6.26 Conclusion. If $\mathrm{NF}_\lambda(N_0, N_1, N_2, N_3)$, N_3 is (λ, ∂)-brimmed over $N_1 \cup N_2$ and $\lambda \ge \partial, \kappa \ge \aleph_0$, then there is N_2^+ such that

(a) $\mathrm{NF}_\lambda(N_0, N_1, N_2^+, N_3)$
(b) $N_2 \le_{\mathfrak{K}} N_2^+$
(c) N_2^+ is (λ, κ)-brimmed over N_0 and even over N_2
(d) N_3 is (λ, ∂)-brimmed over $N_1 \cup N_2^+$.

Proof. Let N_2^+ be (λ, κ)-brimmed over N_2 be such that $N_2^+ \cap N_3 = N_2$. So by existence 6.21 there is N_3^+ such that $\mathrm{NF}_\lambda(N_0, N_1, N_2^+, N_3^+)$ and N_3^+ is (λ, ∂)-brimmed over $N_1 \cup N_2^+$. By monotonicity 6.24 we have $\mathrm{NF}_\lambda(N_0, N_1, N_2, N_3^+)$. So by uniqueness (i.e., 6.22(2)) without loss of generality $N_3 = N_3^+$, so we are done. $\square_{6.26}$

The following claim is a step toward proving transitivity for NF_λ; so we first deal with $\mathrm{NF}_{\lambda,\bar{\delta}}$. Note below: if we ignore N_i^c we have problem showing $\mathrm{NF}_{\lambda,\bar{\delta}}(N_0^a, N_\alpha^a, N_0^b, N_\alpha^b)$. Note that it is not clear at this stage whether, e.g. N_ω^b is even universal over N_ω^a, but N_ω^c is; note that the N_i^c are $\le_{\mathfrak{K}}$-increasing with i but not necessarily continuous. However once we finish proving that NF_λ is a non-forking relation on $\mathfrak{K}_{\mathfrak{s}}$ respecting $\mathfrak{s}$ this claim will lose its relevance.

6.27 Claim. *Assume $\alpha < \lambda^+$ is an ordinal and for $x \in \{a,b,c\}$ the sequence $\bar N^x = \langle N^x_i : i \le \alpha\rangle$ is a $\le_{\mathfrak{K}}$-increasing sequence of members of K_λ, and for $x = a,b$ the sequence $\bar N^x$ is $\le_{\mathfrak{K}}$-increasing continuous, $N^b_i \cap N^a_\alpha = N^a_i, N^c_i \cap N^a_\alpha = N^a_i, N^a_i \le_{\mathfrak{K}} N^b_i \le_{\mathfrak{K}} N^c_i$ and N^b_0 is (λ,δ_2)-brimmed over N^a_0 and $\mathrm{NF}_{\lambda,\bar\delta^i}(N^a_i, N^a_{i+1}, N^c_i, N^b_{i+1})$ (so necessarily $i < \alpha \Rightarrow N^c_i \le_{\mathfrak{K}} N^b_{i+1}$) where $\bar\delta^i = \langle \delta^i_1, \delta^i_2, \delta^i_3\rangle$ with $\delta^i_1, \delta^i_2, \delta^i_3$ are ordinals $< \lambda^+$ and $\delta_3 < \lambda^+$ is limit, N^c_α is $(\lambda, \mathrm{cf}(\delta_3))$-brimmed over $N^b_\alpha, \delta_1 = \sum\limits_{\beta<\alpha} \delta^\beta_1$ and $\delta_3 = \delta^\alpha_3$ and $\delta_2 = \delta^0_2, \bar\delta = \langle\delta_1,\delta_2,\delta_3\rangle$.*

<u>Then</u> $\mathrm{NF}_{\lambda,\bar\delta}(N^a_0, N^a_\alpha, N^b_0, N^c_\alpha)$.

Proof. For $i < \alpha$ let $\langle N^i_{1,\varepsilon}, N^i_{2,\varepsilon}, d^i_\zeta : \varepsilon \le \lambda\times\delta^i_1, \zeta < \lambda\times\delta^i_1\rangle$ be a witness to $\mathrm{NF}_{\lambda,\bar\delta^i}(N^a_i, N^a_{i+1}, N^c_i, N^b_{i+1})$. Now we define a sequence $\langle N_{1,\varepsilon}, N_{2,\varepsilon}, d^i_\zeta : \varepsilon \le \lambda\times\delta_1$ and $\zeta < \lambda\times\delta_1\rangle$ where

(a) $N_{1,0} = N^a_0, N_{2,0} = N^b_0$ and

(b) if $\lambda\times(\sum\limits_{j<i}\delta^j_1) < \zeta \le \lambda\times(\sum\limits_{j\le i}\delta^j_1)$ then we let $N_{1,\zeta} = N^i_{1,\varepsilon_\zeta}, N_{2,\zeta} = N^i_{2,\varepsilon_\zeta}$ where $\varepsilon_\zeta = \zeta - \lambda\times(\sum\limits_{j<i}\delta^j_1)$ and

(c) if $0 < \zeta = \lambda\times\sum\limits_{j<\alpha}\delta^j_1$ we let $N_{1,\zeta} = N^a_i, N_{2,\zeta} = N^b_i = \alpha$ (if i is non-limit we should note that this is compatible with clause (b), note that by this if $i = \alpha$ then $N_{1,\zeta} = N^a_\alpha, N_{2,\zeta} = \cup\{N^i_{2,\lambda\times\delta_1} : i < \alpha\}$

(d) if $\lambda\times(\sum\limits_{j<i}\delta^j_1) \le \zeta < \lambda\times(\sum\limits_{j\le i}\delta^j_1)$ then we let $d_\zeta = d^i_{\varepsilon_\zeta}$ where $\varepsilon_\zeta = \zeta - \lambda\times(\sum\limits_{j<i}\delta^j_j) = \cup\{N^*_{2,\zeta} : \zeta < \lambda\times(\sum\limits_{j<\alpha}\delta^j_1)$.

Clearly $\langle N_{1,\zeta} : \zeta \le \lambda\times\delta_1\rangle$ is $\le_{\mathfrak{K}}$-increasing continuous, and also $\langle N_{2,\zeta} : \zeta \le \lambda\times\delta_1\rangle$ is. Obviously $(N_{1,\zeta}, N_{1,\zeta+1}, d_\zeta) \in K^{3,\mathrm{uq}}_\lambda$ as this just means $(N^i_{1,\varepsilon_\zeta}, N^i_{1,\varepsilon_\zeta+1}, d^i_\zeta) \in K^{3,\mathrm{uq}}_\lambda$ when $\lambda\times\sum\limits_{j<i}\delta^j_1 : j \le \zeta <$

$\lambda \times \sum_{j \le i} \delta_1^j$ and ε_ζ as above.

Why $\mathbf{tp}(d_\zeta, N_{2,\zeta}, N_{2,\zeta+1})$ does not fork over $N_{1,\zeta}$ for ζ, i such that $\lambda \times (\sum_{j<i} \delta_1^j)\zeta < \lambda \times (\sum_{j \le i} \delta_j^j)$? If $\lambda \times \sum_{j<i} \delta_1^j < \zeta$ this holds as it means $\mathbf{tp}(d^i_{\varepsilon_\zeta}, N^i_{2,\varepsilon_\zeta}, N^i_{2,\varepsilon_\zeta+1})$ does not fork over $N^i_{1,\zeta}$. If $\lambda \times \sum_{j<i} \delta_1^j = \zeta$ this is not the case but $N^i_{1,0} = N_{1,\zeta} \le_{\mathfrak{K}} N_{2,\zeta} \le_{\mathfrak{K}} N^c_i = N^i_{2,0}$ and we know that $\mathbf{tp}(d_\zeta, N^i_{2,0}, N^i_{2,1})$ does not fork over $N^i_{1,0} = N_{1,\zeta}$ hence by monotonicity of non-forking $\mathbf{tp}(d_\zeta, N_{2,\zeta}, N_{2,\zeta+1})$ does not fork over $N_{1,\zeta}$ is as required.
Note that we have not demanded or used "$\bar{N}^c$ continuous"; the N^c_i is really needed for i limit as we do not know that N^b_i is brimmed over N^a_i. $\square_{6.27}$

6.28 Claim. *[transitivity] 1) Assume that $\alpha < \lambda^+$ and for $x \in \{a, b\}$ we have $\langle N^x_i : i \le \alpha\rangle$ is a $\le_{\mathfrak{K}}$-increasing continuous sequence of members of K_λ.
If $\mathrm{NF}_\lambda(N^a_i, N^a_{i+1}, N^b_i, N^b_{i+1})$ for each $i < \alpha$ <u>then</u> $\mathrm{NF}_\lambda(N^a_0, N^a_\alpha, N^b_0, N^b_\alpha)$.*

2) Assume that $\alpha_1 < \lambda^+, \alpha_2 < \lambda^+$ and $M_{i,j} \in K_\lambda$ (for $i \le \alpha_1, j \le \alpha_2$) satisfy clauses (B), (C), (D), from 6.17, and for each $i < \alpha_1, j < \alpha_2$ we have:

$$M_{i,j+1} \underset{M_{i,j}}{\overset{M_{i+1,j+1}}{\bigcup}} M_{i+1,j}.$$

<u>Then</u> $$M_{i,0} \underset{M_{0,0}}{\overset{M_{\alpha_1,\alpha_2}}{\bigcup}} M_{0,j} \text{ for } i \le \alpha_1, j \le \alpha_2.$$

Proof. 1) We first prove special cases and use them to prove more general cases.

<u>Case A</u>: N^a_{i+1} is (λ, κ_i)-brimmed over N^a_i and N^b_{i+1} is (λ, ∂_i)-brimmed over $N^a_{i+1} \cup N^b_i$ for $i < \alpha$ (∂_i infinite, of course).

In essence the problem is that we do not know "N_i^b is brimmed over N_i^a" (i limit) so we shall use 6.27; for this we introduce appropriate N_i^c.

Let $\delta_1^i = \kappa_i, \delta_2^i = \kappa_i, \delta_3^i = \partial_i$ where we stipulate $\partial_\alpha = \lambda$. For $i \le \alpha$ we can choose $N_i^c \in K_\lambda$ such that

(a) $N_i^b \le_{\mathfrak{K}} N_i^c \le_{\mathfrak{K}} N_{i+1}^b$, N_i^c is (λ, κ_i)-brimmed over N_i^b, and $\mathrm{NF}_{\lambda,\langle\delta_1^i,\delta_2^i,\delta_3^i\rangle}(N_i^a, N_{i+1}^a, N_i^c, N_{i+1}^b)$

(b) $N_\alpha^c \in K_\lambda$ is $(\lambda, \delta_3^\alpha)$-brimmed over N_α^b

(c) $\langle N_i^c : i < \alpha\rangle$ is $\le_{\mathfrak{K}}$-increasing (in fact follows)

(Possible by 6.26). Now we can use 6.27.

<u>Case B</u>: For each $i < \alpha$ we have: N_{i+1}^a is (λ, κ_i)-brimmed over N_i^a.

In essence our problem is that we do not know anything about brimmness of the N_i^b, so we shall "correct it".

Let $\bar\delta^i = (\kappa_i, \lambda, \lambda)$.

We can find a $\le_{\mathfrak{K}}$-increasing sequence $\langle M_i^x : i \le \alpha\rangle$ of models in K_λ for $x \in \{a, b, c\}$, continuous for $x = a, b$ such that $i < \alpha \Rightarrow M_i^a \le_{\mathfrak{K}} M_i^b \le_{\mathfrak{K}} M_i^c \le_{\mathfrak{K}} M_{i+1}^b$ and $M_\alpha^b \le_{\mathfrak{K}} M_\alpha^c$ and M_i^c is (λ, κ_i)-brimmed over M_i^b (hence over M_i^a) and $\mathrm{NF}_{\lambda,\bar\delta^i}(M_i^a, M_{i+1}^a, M_i^c, M_{i+1}^b)$ by choosing M_i^a, M_i^b, M_i^c by induction on i, $M_0^a = N_0^a$ and M_0^b is universal over M_0^a recalling that the $\mathrm{NF}_{\lambda,\bar\delta^i}$ implies some brimness condition, e.g. M_{i+1}^b is $(\lambda, \mathrm{cf}(\delta_3^i))$-brimmed over $M_{i+1}^a \cup M_i^b$. By Case A we know that $\mathrm{NF}_\lambda(M_0^a, M_\alpha^a, M_0^b, M_\alpha^c)$ holds.

We can now choose an isomorphism f_0^a from N_0^a onto M_0^a, as the identity (exists as $M_0^a = N_0^a$) and then a $\le_{\mathfrak{K}}$-embedding f_0^b of N_0^b into M_0^b extending f_0^a. Next we choose by induction on $i \le \alpha$, f_i^a an isomorphism from N_i^a onto M_i^a such that: $j < i \Rightarrow f_j^a \subseteq f_i^a$, possible by "uniqueness of the (λ, κ_i)-brimmed model over M_i^a" so here we are using the assumption of this case.

Now we choose by induction on $i \le \alpha$, a $\le_{\mathfrak{K}}$-embedding f_i^b of N_i^b into M_i^b extending f_i^a and f_j^b for $j < i$. For $i = 0$ we have done it, for i limit use $\bigcup\limits_{j<i} f_j^b$, lastly for i a successor ordinal let $i = j + 1$, now we have

$(*)_2$ $\mathrm{NF}_\lambda(M^a_j, M^a_{j+1}, f^b_j(N^b_j), M^b_{j+1})$
[why? because $\mathrm{NF}_{\lambda,\bar{\delta}^j}(M^a_j, M^a_{j+1}, M^c_j, M^b_{j+1})$ by the choice of the
M^x_ζ's hence by 6.23 we have $\mathrm{NF}_\lambda(M^a_j, M^a_{j+1}, M^c_j, M^b_{j+1})$ and as
$M^a_j = f^a_j(N^a_j) \leq_{\mathfrak{K}} f^b_j(N^b_j) \leq M^b_j \leq_{\mathfrak{K}} M^c_j$ by 6.24 we get $(*)_2$.]

By $(*)_2$ and the uniqueness of smooth amalgamation 6.22 and as M^b_{j+1} is $(\lambda, \mathrm{cf}(\delta^3_j))$-brimmed over $M^a_{j+1} \cup M^b_j$ hence over $M^a_{j+1} \cup f^b_j(N^b_j)$ clearly there is f^b_i as required.
So without loss of generality f^a_α is the identity, so we have $N^a_0 = M^a_0$, $N^a_\alpha = M^a_\alpha$, $N^b_0 \leq_{\mathfrak{K}} M^b_0$, $N^b_\alpha \leq_{\mathfrak{K}} M^b_\alpha$; also as said above, $\mathrm{NF}_\lambda(M^a_0, M^a_\alpha, M^b_0, M^b_\alpha)$ holds (using Case A) so by monotonicity, i.e., 6.24 we get $\mathrm{NF}_\lambda(N^a_0, N^a_\alpha, N^b_0, N^b_\alpha)$ as required.

Case C: General case.
We can find M^ℓ_i for $\ell < 3, i \leq \alpha$ such that (note that $M^1_0 = M^0_0$):

(a) $M^\ell_i \in K_\lambda$
(b) for each $\ell < 3$, M^ℓ_i is $\leq_{\mathfrak{K}}$-increasing in i (but for $\ell = 1, 2$ they are not required to be continuous)
(c) $M^0_i = N^a_i$
(d) $M^{\ell+1}_{i+1}$ is (λ, λ)-brimmed over $M^\ell_{i+1} \cup M^{\ell+1}_i$ for $\ell < 2, i < \alpha$
(e) $\mathrm{NF}_\lambda(M^\ell_i, M^\ell_{i+1}, M^{\ell+1}_i, M^{\ell+1}_{i+1})$ for $\ell < 2, i < \alpha$
(f) $M^1_0 = M^0_0$ and M^2_0 is $(\lambda, \mathrm{cf}(\lambda))$-brimmed over M^1_0
(g) for $\ell < 2$ and $i < \alpha$ limit we have

$$M^{\ell+1}_i \text{ is } (\lambda, \lambda)\text{-brimmed over } \bigcup_{j<i} M^{\ell+1}_j \cup M^\ell_i$$

(h) for $i < \alpha$ limit we have

$$\mathrm{NF}_\lambda(\bigcup_{j<i} M^1_j, M^1_i, \bigcup_{j<i} M^2_j, M^2_i).$$

[How? As in the proof of 6.17 or just do by hand.]

Now note:

$(*)_3$ $M_i^{\ell+1}$ is $(\lambda, \mathrm{cf}(\lambda \times (1+i)))$-brimmed over M_i^ℓ if $\ell = 1 \vee i \neq 0$ [why? If $i = 0$ by clause (f), if i a successor ordinal by clause (d) and if i is a limit ordinal then by clause (g)]

$(*)_4$ for $i < \alpha, \mathrm{NF}_\lambda(M_i^0, M_{i+1}^0, M_i^2, M_{i+1}^2)$.
[Why? If $i = 0$ by clause (e) for $\ell = 1, i = 0$ we get $\mathrm{NF}_\lambda(M_0^1, M_1^1, M_0^2, M_1^2)$ so by clause (f) (i.e., $M_0^1 = M_0^0$) and monotonicity (i.e., Claim 6.24) we have $\mathrm{NF}_\lambda(M_0^0, M_0^1, M_0^2, M_1^2)$ as required. If $i > 0$ we use Case B for $\alpha = 2$ with M_i^0, M_{i+1}^0, M_i^1, M_{i+1}^1, M_i^2, M_{i+1}^2 here standing for N_0^a, N_0^b, N_1^a, N_1^b, N_2^a, N_2^b there (and symmetry).]

Let us define N_i^ℓ for $\ell < 3$, $i \leq \alpha$ by: N_i^ℓ is M_i^ℓ if i is non-limit and $N_i^\ell = \cup\{N_j^\ell : j < i\}$ if i is limit.

$(*)_5(i)$ $\langle N_i^\ell : i \leq \alpha \rangle$ is $\leq_{\mathfrak{K}}$-increasing continuous, $N_i^0 = N_i^a$ and $N_i^\ell \leq_{\mathfrak{K}} M_i^\ell$

(ii) for $i < \alpha$, $\mathrm{NF}_\lambda(N_i^0, N_{i+1}^0, N_i^2, N_{i+1}^2)$
[why? by $(*)_4+$ monotonicity of NF_λ]

(iii) for $i < \alpha$, N_{i+1}^2 is $(\lambda, \mathrm{cf}(\lambda))$-brimmed over $N_{i+1}^0 \cup N_i^2$ and even over $N_{i+1}^1 \cup N_i^2$
[why? by clause (d)]

$(*)_6$ $\mathrm{NF}_{\lambda,\langle\lambda,\lambda,1\rangle}(N_0^1, N_\alpha^1, N_0^2, N_\alpha^2)$.
[Why? As we have proved case A (or, if you prefer, by 6.27; easily the assumption there holds).]

Choose $f_i^a = \mathrm{id}_{N_i^a}$ for $i \leq \alpha$ and let f_0^b be a $\leq_{\mathfrak{K}}$-embedding of N_0^b into N_0^2.

Now we continue as in Case B defining by induction on i a $\leq_{\mathfrak{K}}$-embedding f_i^b of N_i^b into N_i^2, the successor case is possible by $(*)_5(ii)+$ $(*)_5(iii)$. In the end by $(*)_6$ and monotonicity of NF_λ (i.e., Claim 6.24) we are done.

2) Apply for each $i < \alpha_2$ part (1) to the sequences $\langle M_{\beta,i} : \beta \leq \alpha_1\rangle, \langle M_{\beta,i+1} : \beta \leq \alpha_1\rangle$ so we get $M_{\alpha_1,i} \underset{M_{0,i}}{\overset{M_{\alpha_1,i+1}}{\bigcup}} M_{0,i+1}$ hence by

symmetry (i.e., 6.22) we have $M_{0,i+1} \underset{M_{0,i}}{\overset{M_{\alpha_1,i+1}}{\bigcup}} M_{\alpha_1,i}$.

Applying part (1) to the sequences $\langle M_{0,j} : j \le \alpha_2\rangle, \langle M_{\alpha_1,j} : j \le \alpha_2\rangle$ we get $M_{0,\alpha_2} \underset{M_{0,0}}{\overset{M_{\alpha_1,\alpha_2}}{\bigcup}} M_{\alpha_1,0}$ hence by symmetry (i.e. 6.22) we have

$M_{\alpha_1,0} \underset{M_{0,0}}{\overset{M_{\alpha_1,\alpha_2}}{\bigcup}} M_{0,\alpha_2}$; so we get the desired conclusion. $\square_{6.28}$

6.29 Claim. *Assume $\alpha < \lambda^+, \langle N^\ell_i : i \le \alpha\rangle$ is $\le_{\mathfrak{K}}$-increasing continuous sequence of models for $\ell = 0,1$ where $N^\ell_i \in K_\lambda$ and N^1_{i+1} is (λ,κ_i)-brimmed over $N^0_{i+1} \cup N^1_i$ and $\mathrm{NF}_\lambda(N^0_i, N^1_i, N^0_{i+1}, N^1_{i+1})$.*
<u>*Then*</u> *N^1_α is $(\lambda, \mathrm{cf}(\sum_{i<\alpha} \kappa_i))$-brimmed over $N^0_\alpha \cup N^1_0$.*

6.30 Remark. 1) If our framework is uni-dimensional (see III§2; as for example when it comes from [Sh 576]) we can simplify the proof.
2) Assuming only "N^1_{i+1} is universal over $N^0_{i+1} \cup N^1_i$" suffices when α is a limit ordinal, i.e., we get N^1_α is $(\lambda, \mathrm{cf}(\alpha))$-brimmed over N^0_α. Why? We choose N^2_j for $j \le i$ such that $N^2_j = N^1_j$ if $j = 0$ or j a limit ordinal and N^2_j is a model $\le_{\mathfrak{s}} N^1_j$ and (λ,κ_1)-brimmed over $N^0_j \cup N^1_i$ when $j = i+1$. Now $\langle N^2_j : j \le \alpha\rangle$ satisfies all the requirements in $\langle N^1_j : j \le \alpha\rangle$ in 6.29.
3) We could have proved this earlier and used it, e.g. in 6.28.

Proof. The case α not a limit ordinal is trivial so assume α is a limit ordinal. We choose by induction on $i \le \alpha$, an ordinal $\varepsilon(i)$ and a sequence $\langle M_{i,\varepsilon} : \varepsilon \le \varepsilon(i)\rangle$ and $\langle c_\varepsilon : \varepsilon < \varepsilon(i)$ non-limit$\rangle$ such that:

- (*a*) $\langle M_{i,\varepsilon} : \varepsilon \le \varepsilon(i)\rangle$ is (strictly) $<_{\mathfrak{K}}$-increasing continuous in K_λ
- (*b*) $N^0_i \le_{\mathfrak{K}} M_{i,\varepsilon} \le_{\mathfrak{K}} N^1_i$
- (*c*) $N^0_i = M_{i,0}$ and $N^1_i = M_{i,\varepsilon(i)}$
- (*d*) $\varepsilon(i)$ is (strictly) increasing continuous in i and $\varepsilon(i)$ is divisible by λ

(e) $j < i$ & $\varepsilon \le \varepsilon(j) \Rightarrow M_{i,\varepsilon} \cap N^1_j = M_{j,\varepsilon}$

(f) for $j < i$ and $\varepsilon \le \varepsilon(j+1)$, the sequence $\langle M_{\beta,\varepsilon} : \beta \in (j,i]\rangle$ is $\le_{\mathfrak{K}}$-increasing continuous

(g) for $j < i, \varepsilon < \varepsilon(j)$ non-limit; the type $\mathbf{tp}(c_\varepsilon, M_{i,\varepsilon}, M_{i,\varepsilon+1}) \in \mathscr{S}^{\text{bs}}(M_{i,\varepsilon})$ does not fork over $M_{j,\varepsilon}$ (actually, here allowing all ε is O.K., too)

(h) $M_{i+1,\varepsilon+1}$ is $(\lambda, \text{cf}(\lambda))$-brimmed over $M_{i+1,\varepsilon} \cup M_{i,\varepsilon+1}$

(i) if $\varepsilon < \varepsilon(i)$ and $p \in \mathscr{S}^{\text{bs}}(M_{i,\varepsilon})$ then for λ successor ordinals $\xi \in [\varepsilon, \varepsilon(i))$ the type $\mathbf{tp}(c_\xi, M_{i,\xi}, M_{i,\xi+1})$ is a non-forking extension of p.

If we succeed, then $\langle M_{\alpha,\varepsilon} : \varepsilon \le \varepsilon(\alpha)\rangle$ is a (strictly) $<_{\mathfrak{K}}$-increasing continuous sequence of models from K_λ, $M_{\alpha,0} = N^0_\alpha$, and $M_{\alpha,\varepsilon(\alpha)} = N^1_\alpha$. We can apply 4.3 and we conclude that $N^1_\alpha = M_{\alpha,\varepsilon(\alpha)}$ is $(\lambda, \text{cf}(\alpha))$-brimmed over $M_{\alpha,\varepsilon(j)}$ hence over $N^0_\alpha \cup N^1_0$ (both $\le_{\mathfrak{K}} M_{\alpha,1}$).

Carrying the induction is easy. For $i = 0$, there is not much to do. For i successor we use "N^j_{i+1} is brimmed over $N^0_{i+1} \cup N^1_i$" the existence of non-forking amalgamations and 4.2, bookkeeping and the extension property $(E)(g)$. For i limit we have no problem. $\square_{6.29}$

6.31 Conclusion. 1) If $\text{NF}_\lambda(N_0, N_1, N_2, N_3)$ and $\langle M_{0,\varepsilon} : \varepsilon \le \varepsilon(*)\rangle$ is an $\le_{\mathfrak{K}}$-increasing continuous sequence of models from K_λ, $N_0 \le_{\mathfrak{K}} M_{0,\varepsilon} \le_{\mathfrak{K}} N_2$ <u>then</u> we can find $\langle M_{1,\varepsilon} : \varepsilon \le \varepsilon(*)\rangle$ and N'_3 such that:

(a) $N_3 \le_{\mathfrak{K}} N'_3 \in K_\lambda$

(b) $\langle M_{1,\varepsilon} : \varepsilon \le \varepsilon(*)\rangle$ is $\le_{\mathfrak{K}}$-increasing continuous

(c) $M_{1,\varepsilon} \cap N_2 = M_{0,\varepsilon}$

(d) $N_1 \le_{\mathfrak{K}} M_{1,\varepsilon} \le_{\mathfrak{K}} N'_3$

(e) if $M_{0,0} = N_0$ then $M_{1,0} = N_1$

(f) $\text{NF}_\lambda(M_{0,\varepsilon}, M_{1,\varepsilon}, N_2, N'_3)$, for every $\varepsilon \le \varepsilon(*)$.

2) If N_3 is universal over $N_1 \cup N_2$, then without loss of generality $N'_3 = N_3$.
3) In part (1) we can add

(g) $M_{1,\varepsilon+1}$ is brimmed over $M_{0,\varepsilon+1} \cup M_{1,\varepsilon}$.

Proof. 1) Define $M'_{0,i}$ for $i \le \varepsilon^* := 1 + \varepsilon(*) + 1$ by $M'_{0,0} = N_0, M'_{0,1+\varepsilon} = M_{0,\varepsilon}$ for $\varepsilon \le \varepsilon(*)$ and $M'_{0,1+\varepsilon(*)+1} = N_2$. By existence (6.21) we can find an $\le_{\mathfrak{K}}$-increasing continuous sequence $\langle M'_{1,\varepsilon} : \varepsilon \le \varepsilon^* \rangle$ with $M'_{1,0} = N_1$ and $\le_{\mathfrak{K}}$-embedding f of N_2 into M'_{1,ε^*} such that $\varepsilon < \varepsilon^* \Rightarrow \mathrm{NF}_\lambda(f(M'_{0,\varepsilon}), M'_{1,0}, f(M'_{0,\varepsilon+1}), M'_{1,\varepsilon+1})$. By transitivity we have $\mathrm{NF}_\lambda(f(M'_{0,0}), M'_{1,0}, f(M'_{0,\varepsilon^*}), M'_{1,\varepsilon^*})$. By disjointness (i.e., $f(M'_{0,\varepsilon^*}) \cap M'_{1,0} = M'_{0,0}$, see 6.13(3)) without loss of generality f is the identity. By uniqueness for NF there are $N'_3, N_3 \le_{\mathfrak{K}} N'_3 \in K_\lambda$ and $\le_{\mathfrak{K}}$-embedding of M'_{1,ε^*} onto N'_3 over $N_1 \cup N_2 = M'_{0,\varepsilon^*} \cup M'_{1,0}$ so we are done.
2) Follows by (1).
3) Similar to (1). $\square_{6.31}$

6.32 Claim. NF_λ *respects* $\mathfrak{s}$.
That is, assume $\mathrm{NF}_\lambda(M_0, M_1, M_2, M_3)$ *and* $a \in M_1 \backslash M_0$ *satisfies* $\mathbf{tp}(a, M_0, M_3) \in \mathscr{S}^{\mathrm{bs}}(M_0)$, *then* $\mathbf{tp}(a, M_2, M_3) \in \mathscr{S}^{\mathrm{bs}}(M_2)$ *does not fork over* M_0.

Proof. Without loss of generality M_1 is $(\lambda, *)$-brimmed over M_0. [Why? By the existence we can find M_1^+ which is a $(\lambda, *)$-brimmed extension of M_1. By the existence for NF_λ without loss of generality we can find M_3^+ such that $\mathrm{NF}_\lambda(M_1, M_1^+, M_3, M_3^+)$, hence by transitivity for NF_λ we have $\mathrm{NF}_\lambda(M_0, M_1^+, M_2, M_3^+)$.] By the hypothesis of the section there are M'_1, a' such that $M_0 \cup \{a'\} \subseteq M'_1$ and $\mathbf{tp}(a', M_0, M'_1) = \mathbf{tp}(a, M_0, M_1)$ and $(M_0, M'_1, a) \in K_\lambda^{3,\mathrm{uq}}$; as M_1^+ is $(\lambda, *)$-brimmed over M_0 without loss of generality $M' \le_{\mathfrak{K}} M_1^+$ and $a' = a$ and M_1 is $(\lambda, *)$-brimmed over M'_1. We can apply 6.9 to M'_1, M_1^+ getting $\langle M_i^*, a_i : i \le \delta < \lambda^+ \rangle$ as there. Let M'_i be: M_0 if $i = 0, M_j^*$ if $1 + j = i$ so $M'_1 = M_0^* = M'_1$ and let a_i be a if $i = 0, a_j$ if $1 + j = i$. So we can find M'_3 and f such that $M_2 \le_{\mathfrak{K}} M'_3, f$ is a $\le_{\mathfrak{K}}$-embedding of M_1^+ into M'_3 extending id_{M_0} such that $\mathrm{NF}_{\lambda,\langle \delta,\lambda,\lambda \rangle}(M_0, f(M_1^+), M_2, M'_3)$ and M'_3, this is witnessed by $\langle f(M'_i) : i \le \delta \rangle, \langle M''_i : i \le \delta \rangle, \langle f(a_i) : i < \delta \rangle$ and $M''_0 = M_2$; this is possible by 6.13(2). Hence $\mathrm{NF}_\lambda(M_0, f(M_1^+), M_2, N) = \mathrm{NF}_\lambda(f(M'_0), f(M'_\delta), M''_0, N)$ hence by the uniqueness for NF_λ without loss of generality $f = \mathrm{id}_{M_1^+}$ and $M_3 \le_{\mathfrak{K}} N$. By the choice of f, N we have $\mathbf{tp}(a, M_2, M_3) = \mathbf{tp}(a_0, M_2, N) = \mathbf{tp}(a_0, M''_0, M'_1) \in$

$\mathscr{S}^{\rm bs}(M''_0) = \mathscr{S}^{\rm bs}(M_2)$ does not fork over $M'_0 = M_0$ as required. $\square_{6.32}$

6.33 Conclusion. If $M_0 \le_{\mathfrak{K}} M_\ell \le_{\mathfrak{K}} M_3$ for $\ell = 1,2$ and $(M_0, M_1, a) \in K^{3,\rm uq}_\lambda$ and $\mathbf{tp}(a, M_2, M_3) \in \mathscr{S}^{\rm bs}(M_2)$ does not fork over M_0 then $\mathrm{NF}(M_0, M_1, M_2, M_3)$.

Proof. By the definition of $K^{3,\rm uq}_\lambda$ and existence for NF_λ and 6.32 (or use 6.3 + 6.34. $\square_{6.33}$

We can sum up our work by

6.34 Main Conclusion. NF_λ is a non-forking relation on ${}^4(\mathfrak{K}_\lambda)$ which respects $\mathfrak{s}$.

Proof. We have to check clauses (a)-(g)+(h) from 6.1. Clauses (a),(b) hold by the Definition 6.12 of NF_λ. Clauses $(c)_1, (c)_2$, i.e., monotonicity hold by 6.24. Clause (d), i.e., symmetry holds by 6.25. Clause (e), i.e., transitivity holds by 6.28. Clause (f), i.e., existence hold by 6.21. Clause (g), i.e., uniqueness holds by 6.22.

Lastly, clause (h), i.e., NF_λ respecting $\mathfrak{s}$ by 6.32. $\square_{6.34}$

The following definition is not needed for now but is natural (of course, we can omit "there is superlimit" from the assumption and the conclusion). For the rest of the section we stop assuming Hypothesis 6.8.

6.35 Definition. 1) A good λ-frame $\mathfrak{s}$ is type-full when for $M \in \mathfrak{K}_{\mathfrak{s}}, \mathscr{S}^{\rm bs}(M) = \mathscr{S}^{\rm na}_{\mathfrak{K}_\lambda}(M)$.
2) Assume $\mathfrak{K}_\lambda$ is a λ-a.e.c. and NF is a 4-place relation on K_λ. We define $\mathfrak{t} = \mathfrak{t}_{\mathfrak{K}_\lambda,\mathrm{NF}} = (K_{\mathfrak{t}}, \bigcup\limits_{\mathfrak{t}}, \mathscr{S}^{\rm bs}_{\mathfrak{t}})$ as follows:

- (a) $\mathfrak{K}_{\mathfrak{t}}$ is the λ-a.e.c. $\mathfrak{K}_\lambda$
- (b) $\mathscr{S}^{\rm bs}_{\mathfrak{t}}(M)$ is $\mathscr{S}^{\rm na}_{\mathfrak{K}_\lambda}(M)$ for $M \in \mathfrak{K}_\lambda$
- (c) $\bigcup\limits_{\mathfrak{t}}$ is defined by: $(M_0, M_1, a, M_3) \in \bigcup\limits_{\mathfrak{t}}$ when we can find M_2, M'_3 such that $M_0 \le_{\mathfrak{K}_\lambda} M_2 \le_{\mathfrak{K}_\lambda} M'_3, M_3 \le_{\mathfrak{K}_\lambda} M'_3, a \in M_2 \backslash M_0$ and $\mathrm{NF}(M_0, M_1, M_2, M'_3)$.

6.36 Claim. *1) Assume that*

(*a*) $\mathfrak{K}_\lambda$ *is a* λ*-a.e.c. with amalgamation (actually follows by (c)) and a superlimit model*

(*b*) $\mathfrak{K}_\lambda$ *is stable*

(*c*) NF *is a* $\mathfrak{K}_\lambda$*-non-forking relation, see Definition 6.1(1).*

Then $\mathfrak{t} = \mathfrak{t}_{\mathfrak{K}_\lambda,\mathrm{NF}}$ *is a type-full good* λ*-frame.*
2) Assume that $\mathfrak{s}$ *is a good* λ*-frame which has existence for* $K^{3,\mathrm{uq}}_\lambda$ *(see 6.8(2)) and* $\mathrm{NF} = \mathrm{NF}_\lambda$*. Then* $\mathfrak{t}$ *is very close to* $\mathfrak{s}$*, i.e.:*

(*a*) $\mathfrak{K}_\mathfrak{s} = \mathfrak{K}_\mathfrak{t}$

(*b*) *if* $p \in \mathscr{S}^{\mathrm{bs}}_\mathfrak{s}(M_1)$ *and* $M_0 \le_{\mathfrak{K}_\lambda} M_1$ *then* $p \in \mathscr{S}^{\mathrm{bs}}_\mathfrak{t}(M_1)$ *and* p *forks over* M_0 *for* $\mathfrak{s}$ *iff* p *forks over* M_0 *for* $\mathfrak{t}$*.*

Proof. For the time being, left to the reader (but before it is really used it is proved in III.9.6).

Remark. Note that this actually says that from now on we could have used type-full $\mathfrak{s}$, but it is not necessary for a long time.

6.37 Definition. 1) Let $\mathfrak{s}$ be a good λ-frame. We say that NF is a weak $\mathfrak{s}$-non-forking relation when

(*a*) NF is a pseudo $\mathfrak{K}_\mathfrak{s}$-non-forking relation, see Definition 6.1(2), i.e., uniqueness is omitted

(*b*) NF respects $\mathfrak{s}$, see Definition 6.1(3)

(*c*) NF satisfies 6.31, (NF-lifting of an $\le_\mathfrak{K}$-increasing sequence).

1A) If in part (1) we replace "$\mathfrak{s}$-non-forking" by "non-forking", we mean that we omit clause (c).
1B) In part (1) we omit "weak" when we omit the "pseudo" in clause (a), so clause (c) becomes redundant.
2) We say $\mathfrak{s}$ is pseudo-successful if some NF is a weak $\mathfrak{s}$-non-forking relation witnesses it.

6.38 Observation. 1) If $\mathfrak{s}$ is a good λ-frame which is weakly successful (i.e., has existence for $K^{3,\mathrm{uq}}_\lambda$, i.e., 6.8) then $\mathrm{NF}_\lambda = \mathrm{NF}_\mathfrak{s}$ is a $\mathfrak{s}$-non-forking relation.
2) If $\mathfrak{s}$ is a good λ-frame and NF is a weak $\mathfrak{s}$-non-forking relation then 6.33 holds.
3) If $\mathfrak{s}$ is a good λ-frame and NF is an $\mathfrak{s}$-non-forking relation then NF is a weak $\mathfrak{s}$-non-forking relation which implies NF is a pseudo non-forking relation.

Proof. Straight.
1) Follows by 6.34, NF_λ satisfies clauses (a)+(b) and by 6.31 it satisfies also clause (c) of Definition 6.1(1).
2) Also easy.
3) We have just to check the proof of 6.31 still works.

6.39 Remark. 1) In Chapter III ,§1 -§11 we can use "$\mathfrak{s}$ is pseudo successful as witnessed by NF" so has lifting of decompositions instead of "$\mathfrak{s}$ is weakly successful". We shall return to this elsewhere, see Chapter VII, [Sh 842].

§7 Nice extensions in K_{λ^+}

7.1 Hypothesis. Assume the hypothesis 6.8.

So by §6 we have reasonable control on smooth amalgamation in K_λ. We use this to define "nice" extensions in K_{λ^+} and prove some basic properties. This will be treated again in §8.

7.2 Definition. 1) $K^{\mathrm{nice}}_{\lambda^+}$ is the class of saturated $M \in K_{\lambda^+}$.
2) Let $M_0 \le^*_{\lambda^+} M_1$ mean:

> $M_0 \le_\mathfrak{K} M_1$ and they are from K_{λ^+} and we can find $\bar M^\ell = \langle M^\ell_i : i < \lambda^+\rangle$, a $\le_\mathfrak{K}$-representation of M_ℓ for $\ell = 0, 1$ such that:
> $\mathrm{NF}_\lambda(M^0_i, M^0_{i+1}, M^1_i, M^1_{i+1})$ for $i < \lambda^+$.

3) Let $M_0 <^+_{\lambda^+,\kappa} M_1$ mean[20] that ($M_0, M_1 \in K_{\lambda^+}$ and) $M_0 \le^*_{\lambda^+} M_1$ by some witnesses M^ℓ_i (for $i < \lambda^+, \ell < 2$) such that $\mathrm{NF}_{\lambda,\langle 1,1,\kappa\rangle}(M^0_i, M^0_{i+1}, M^1_i, M^1_{i+1})$ for $i < \lambda^+$; of course $M_0 \le_{\mathfrak{K}} M_1$ in this case. Let $M_0 \le^+_{\lambda^+,\kappa} M_1$ mean $(M_0 = M_1 \in K_{\lambda^+}) \vee (M_0 <^+_{\lambda^+,\kappa} M_1)$. If $\kappa = \lambda$, we may omit it.
4) Let $K^{3,\mathrm{bs}}_{\lambda^+} = \{(M, N, a) : M \le^*_{\lambda^+} N$ are from K_{λ^+} and $a \in N \backslash M$ and for some $M_0 \le_{\mathfrak{K}} M, M_0 \in K_\lambda$ we have $[M_0 \le_{\mathfrak{K}} M_1 \le_{\mathfrak{K}} M$ & $M_1 \in K_\lambda$ implies $\mathbf{tp}(a, M_1, N) \in \mathscr{S}^{\mathrm{bs}}(M_1)$ and does not fork over $M_0]\}$. We call M_0 or $\mathbf{tp}(a, M_0, N)$ a witness for $(M, N, a) \in K^{3,\mathrm{bs}}_{\lambda^+}$. (In fact this definition on $K^{3,\mathrm{bs}}_{\lambda^+}$ is compatible with the definition in §2 for triples such that $M \le^*_{\lambda^+} N$ but we do not know now whether even $(K^{\mathrm{nice}}_{\lambda^+}, \le^*_{\lambda^+})$ is a λ^+-a.e.c..)

7.3 Claim. *0) $K^{\mathrm{nice}}_{\lambda^+}$ has one and only one model up to isomorphism and $M \in K^{\mathrm{nice}}_{\lambda^+}$ implies $M \le^*_{\lambda^+} M$ and $M \le^+_{\lambda^+} M$; moreover, $M \in K_{\lambda^+} \Rightarrow M \le^*_{\lambda^+} M$. Also $\le^*_{\lambda^+}$ is a partial order and if $M_\ell \in K_{\lambda^+}$ for $\ell = 0, 1, 2$ and $M_0 \le_{\mathfrak{K}} M_1 \le_{\mathfrak{K}} M_2$ and $M_0 \le^*_{\lambda^+} M_2$ then $M_0 \le^*_{\lambda^+} M_1$.*
1) If $M_0 \le^*_{\lambda^+} M_1$ and $\bar{M}^\ell = \langle M^\ell_i : i < \lambda^+\rangle$ is a representation of M_ℓ for $\ell = 0, 1$ <u>then</u>

- (∗) *for some club E of λ^+,*
  - (a) *for every $\alpha < \beta$ from E we have $\mathrm{NF}_\lambda(M^0_\alpha, M^0_\beta, M^1_\alpha, M^1_\beta)$*
  - (b) *if $\ell < 2$ and $M_\ell \in K^{\mathrm{nice}}_{\lambda^+}$ then for $\alpha < \beta$ from E the model M^ℓ_β is $(\lambda, *)$-brimmed over M^ℓ_α.*

2) Similarly for $<^+_{\lambda^+,\kappa}$: if $M_0 <^+_{\lambda^+,\kappa} M_1, \bar{M}^\ell = \langle \bar{M}^\ell_i : i < \lambda^+\rangle$ a representation of M_ℓ for $\ell = 0, 1$ <u>then</u> for some club E of λ^+ for every $\alpha < \beta$ from E we have $\mathrm{NF}_{\lambda,\langle 1,1,\kappa\rangle}(M^0_\alpha, M^0_\beta, M^1_\alpha, M^1_\beta)$, moreover $NF_{\lambda,\langle 1,\mathrm{cf}(\lambda\times(1+\beta)),\kappa\rangle}(M^0_\alpha, M^0_\beta, M^1_\alpha, M^1_\beta)$ and if $(M_\alpha, \bar{M}^0_\beta, M^1_\alpha, M^1_\beta)$, $M_0 \in K^{\mathrm{nice}}_{\lambda^+}$ then we can add $NF_{\lambda,\langle\lambda,\mathrm{cf}(\lambda\times(1+\beta)),\kappa\rangle}(M^0_\alpha, M^0_\beta, M'_\alpha, M'_\beta)$.
3) The κ in Definition 7.2(3) does not matter.

---

[20]Note that $M_0 <^+_{\lambda^+,\kappa} M_1$ implies $M_1 \in K^{\mathrm{nice}}_{\lambda^+}$ but in general $M_0 \in K^{\mathrm{nice}}_{\lambda^+}$ does not follow.

4) If $M_0 <^+_{\lambda^+,\kappa} M_1$, *then* $M_1 \in K^{\rm nice}_{\lambda^+}$.
5) If $M \in K_{\lambda^+}$ *is saturated, equivalently* $M \in K^{\rm nice}_{\lambda^+}$ *then* M *has a* $\le_{\mathfrak{K}}$*-representation* $\bar M = \langle M_\alpha : \alpha < \lambda^+\rangle$ *such that* M_{i+1} *is* (λ,λ)*-brimmed over* M_i *for* $i < \lambda^+$ *and also the inverse is true.*
6) If $M \le^*_{\lambda^+} N$ *and* $N_0 \le_{\mathfrak{K}} N, N_0 \in K_\lambda$ *then we can find* $M_1 \le_{\mathfrak{K}} N_1$ *from* K_λ *such that* $M_1 \le_{\mathfrak{K}} M, N_0 \le_{\mathfrak{K}} N_1 \le_{\mathfrak{K}} N$ *and: for every* $M_2 \in K_\lambda$ *satisfying* $M_1 \le_{\mathfrak{K}} M_2 \le_{\mathfrak{K}} M$ *there is* $N_2 \le_{\mathfrak{K}} N$ *such that* $\mathrm{NF}_{\mathfrak{s}}(M_1, M_2, N_1, N_2)$.

Proof. 0) Obvious by now (for the second sentence use part (1) and $\mathrm{NF}_{\mathfrak{s}}$ being a non-forking relation on $\mathfrak{K}_{\mathfrak{s}}$) in particular transitivity and monotonicity.
1) Straight by 6.28 as any two representations agree on a club.
2) Up to "moreover" quite straight. For the "moreover" use 6.29 to show that M^1_β is $(\lambda, \mathrm{cf}(\beta))$-brimmed over M^0_β. Lastly, for the "we can add" just use part (5), choosing thin enough club E of λ^+ then use $\{\alpha \in E : \mathrm{otp}(\alpha \cap E)$ is divisible by $\lambda\}$.
3) By 6.29.
4) By 6.29.
5) Trivial.
6) Easy. $\square_{7.3}$

7.4 Claim. *0) For every* $M_0 \in K_{\lambda^+}$ *for some* $M_1 \in K^{\rm nice}_{\lambda^+}$ *we have* $M_0 \le_{\mathfrak{K}} M_1$.
1) For every $M_0 \in K_{\lambda^+}$ *and* $\kappa = \mathrm{cf}(\kappa) \le \lambda$ *for some* $M_1 \in K_{\lambda^+}$ *we have* $M_0 <^+_{\lambda^+,\kappa} M_1$ *so* $M_1 \in K^{\rm nice}_{\lambda^+}$.
1A) Moreover, if $N_0 \le_{\mathfrak{K}} M_0 \in K_{\lambda^+}, N_0 \in K_\lambda, p \in \mathscr{S}^{\rm bs}(N_0)$ *then in (1) we can add that for some* $a, (M_0, M_1, a) \in K^{3,\rm bs}_\lambda$ *as witnessed by* p.
2) $\le^*_{\lambda^+}$ *and* $<^+_{\lambda^+,\kappa}$ *are transitive.*
3) If $M_0 \le_{\mathfrak{K}} M_1 \le_{\mathfrak{K}} M_2$ *are in* K_{λ^+} *and* $M_0 \le^*_{\lambda^+} M_2$, *then* $M_0 \le^*_{\lambda^+} M_1$.
4) If $M_1 <^+_{\lambda^+,\kappa} M_2$, *then* $M_1 <^*_{\lambda^+} M_2$.
5) If $M_0 <^*_{\lambda^+} M_1 <^+_{\lambda,\kappa} M_2$ *then* $M_0 <^+_{\lambda,\kappa} M_2$.

Proof. 0) Easy and follows by the proof of part (1) below.
1), 1A) Let $\langle M^0_i : i < \lambda^+\rangle$ be a $\le_{\mathfrak{K}}$-representation of M_0 with M^0_i

brimmed and brimmed over M^0_j for $j < i$ and for part (1A) we have $M^0_0 = N_0$, and for part (1) let p be any member of $\mathscr{S}^{\rm bs}(M^0_0)$. We choose by induction on i a model $M^1_i \in K_\lambda$ and $a \in M^1_0$ such that M^1_i is $(\lambda, {\rm cf}(\lambda \times (1+i)))$-brimmed over M^0_i, $\langle M^1_i : i < \lambda^+\rangle$ is $<_{\mathfrak{K}}$-increasing continuous, $M^1_i \cap M_0 = M^0_i$ and $\mathbf{tp}(a, M^0_0, M^1_0) = p$ and M^1_{i+1} is (λ, κ)-brimmed over $M^0_{i+1} \cup M^1_i$ and ${\rm NF}_{\lambda,\langle 1,{\rm cf}(\lambda\times(1+i)),\kappa\rangle}(M^0_i, M^0_{i+1}, M^1_i, M^1_{i+1})$ for $i < \lambda^+$. Note that for limit i, by 6.29, M^1_i is $(\lambda, {\rm cf}(i))$-brimmed over $M^0_i \cup M^1_j$ for any $j < i$.

Note that for $i < \lambda^+$, the type $\mathbf{tp}(a, M^0_i, M^1_i)$ does not fork over $M^0_0 = N_0$ and extends p by 6.32 (saying ${\rm NF}_\lambda$ respects $\mathfrak{s}$) 6.25 (symmetry) and 6.23. So clearly we are done.

2) Concerning $<^+_{\lambda^+,\kappa}$ use 7.3 and 6.28 (i.e. transitivity for smooth amalgamations). The proof for $<^*_{\lambda^+}$ is the same.

3) By monotonicity for smooth amalgamations in $\mathfrak{K}_\lambda$; i.e., 6.24.

4), 5) Check. $\square_{7.4}$

7.5 Claim. *1) If $(M_0, M_1, a) \in K^{3,\rm bs}_{\lambda^+}$ and $M_1 \le^*_{\lambda^+} M_2 \in K_{\lambda^+}$ then $(M_0, M_2, a) \in K^{3,\rm bs}_{\lambda^+}$.*
2) If $M_0 <^*_{\lambda^+} M_1$, then for some $a, (M_0, M_1, a) \in K^{3,\rm bs}_{\lambda^+}$.

Proof. 1) By the transitivity of $\le^*_{\lambda^+}$ which holds by 7.4(2).
2) As in the proof of 2.9, in fact it follows from it. $\square_{7.5}$

Remark. Note that the parallel to 7.4(1A) is problematic in §2 as, .e.g. locality may fail, i.e., $(M, N_i, a_i) \in K^{3,\rm bs}_{\lambda^+}$ and $M' \le_{\mathfrak{K}} M \wedge M' \in K_\lambda \Rightarrow \mathbf{tp}_{\mathfrak{s}}(a_1, M', N_1) = \mathbf{tp}_{\mathfrak{s}}(a_2, M', N_2)$ but $\mathbf{tp}_{K^{\mathfrak{s}}_{\lambda^+}}(a_1, M, N_1) \neq \mathbf{tp}_{K^{\mathfrak{s}}_{\lambda^+}}(\bar a_2, M, N_2)$. $\square_{7.5}$

7.6 Claim. *1) [Amalgamation of $\le^*_{\lambda^+}$ and toward extending types] If $M_0 \le^*_{\lambda^+} M_\ell$ for $\ell = 1, 2, \kappa = {\rm cf}(\kappa) \le \lambda$ and $a \in M_2 \backslash M_0$ is such that $(M_0, M_2, a) \in K^{3,\rm bs}_{\lambda^+}$ is witnessed by p, then for some M_3 and f we have: $M_1 <^+_{\lambda^+,\kappa} M_3$ and f is an $\le_{\mathfrak{K}}$-embedding of M_2 into M_3 over M_0 with $f(a) \notin M_1$, moreover, $f(M_2) \le^*_{\lambda^+} M_3$ and $(M_1, M_3, f(a)) \in K^{3,\rm bs}_{\lambda^+}$ is witnessed by p.*
2) [uniqueness] Assume $M_0 <^+_{\lambda^+,\kappa} M_\ell$ for $\ell = 1, 2$ then there is an

isomorphism f from M_1 onto M_2 over M_0.
3) [locality] Moreover[21]*, in (2) if $a_\ell \in M_\ell \backslash M_0$ for $\ell = 1,2$ and $[N \leq_{\mathfrak{K}} M_0 \ \& \ N \in K_\lambda \Rightarrow \mathbf{tp}(a_1, N, M_1) = \mathbf{tp}(a_2, N, M_2)]$, then we can demand $f(a_1) = a_2$ (so in particular $\mathbf{tp}(a_1, M_0, M_1) = \mathbf{tp}(a_2, M_0, M_2)$ where the types are as defined in $\mathfrak{K}_{\lambda^+}$ and even in $(K_{\lambda^+}, \leq^*_{\lambda^+})$.*
4) Moreover in (2), assume further that for $\ell = 1,2$, the following hold: $N_0 \leq_{\mathfrak{K}} N_\ell \leq_{\mathfrak{K}} M_\ell, N_0 \in K_\lambda, N_0 \leq_{\mathfrak{K}} N_\ell, N_\ell \in K_\lambda$ and $(\forall N \in K_\lambda)[N_0 \leq_{\mathfrak{K}} N \leq_{\mathfrak{K}} M_0 \rightarrow (\exists N' \in K_\lambda)(N \cup N_\ell \subseteq N' \leq_{\mathfrak{K}} M_\ell \wedge \mathrm{NF}_\lambda(N_0, N_\ell, N, N')]$. If f_0 is an isomorphism from N_1 onto N_2 over N_0 then we can add $f \supseteq f_0$.

Proof. We first prove part (2).
2) By 7.3(1) + (2) there are representations $\bar{M}^\ell = \langle M^\ell_i : i < \lambda^+ \rangle$ of M_ℓ for $\ell < 3$ such that for $\ell = 1,2$ we have: $M^\ell_i \cap M_0 = M^\ell_0$ and $\mathrm{NF}_{\lambda,\langle 1,1,\kappa \rangle}(M^0_i, M^0_{i+1}, M^\ell_i, M^\ell_{i+1})$ and without loss of generality M^ℓ_0 is (λ,κ)-brimmed over M^0_0 for $\ell = 1,2$.

Now we choose by induction on $i < \lambda^+$ an isomorphism f_i from M^1_i onto M^2_i, increasing with i and being the identity over M^0_i. For $i = 0$ use "M^ℓ_0 is (λ,κ)-brimmed over M^0_0 for $\ell = 1,2$" which we assume above. For i limit take unions, for i successor ordinal use uniqueness (Claim 6.18).

Proof of part (1). By 7.4(1) there are for $\ell = 1,2$ models $N^*_\ell \in K_{\lambda^+}$ such that $M_\ell <^+_{\lambda^+,\kappa} N^*_\ell$. Now let $\bar{M}^\ell = \langle M^\ell_i : i < \lambda^+ \rangle$ be a representation of M_ℓ for $\ell = 0,1,2$ and let $\bar{N}^\ell = \langle N^\ell_i : i < \lambda^+ \rangle$ be a representation of N^*_ℓ for $\ell = 1,2$. By 7.4(4) and 7.3(2) without loss of generality N^ℓ_0 is (λ,κ)-brimmed over M^ℓ_0 and $\mathrm{NF}_\lambda(M^0_i, M^0_{i+1}, M^\ell_i, M^\ell_{i+1})$ and $\mathrm{NF}_{\lambda,\langle 1,1,\kappa \rangle}(M^\ell_i, M^\ell_{i+1}, N^\ell_i, N^\ell_{i+1})$ respectively for $i < \lambda^+, \ell = 1,2$. Let M^*_0 be such that $p \in \mathscr{S}^{\mathrm{bs}}(M^*_0), M^*_0 \in K_\lambda, M^*_0 \leq_{\mathfrak{K}} M_0$; without loss of generality $M^*_0 \leq_{\mathfrak{K}} M^0_0$ and $a \in M^2_0 \leq_{\mathfrak{K}} N^2_0$. Now N^ℓ_0 is (λ,κ)-brimmed over M^ℓ_0 hence over M^0_0 (for $\ell = 1,2$) so there is an isomorphism f_0 from N^2_0 onto N^1_0 extending $\mathrm{id}_{M^0_0}$. There is $a' \in N^1_0$ such that $\mathbf{tp}(a', M^1_0, N^1_0)$ is a non-forking extension of p and without

[21]the meaning of this will be that types over $M \in K^{\mathrm{nice}}_{\lambda^+}$ for $(K^{\mathrm{nice}}_{\lambda^+}, \leq^*_{\lambda^+})$ can be reduced to basic types over a model in K_λ, i.e., locality

loss of generality $f_0(a) = a'$ hence $\mathbf{tp}(f_0(a), M_0^1, N_0^1) \in \mathscr{S}^{\rm bs}(M_0^1)$ does not fork over M_0^0.

We continue as in the proof of part (2). In the end $f = \bigcup_{i<\lambda^+} f_i$ is an isomorphism of N_2^* onto N_1^* over M_0 and as $f_0(a)$ is well defined and in $N_0^1 \backslash M_0^1$ clearly $\mathbf{tp}(f(a), M_i^1, N_i^1)$ does not fork over M_0^1 and extends p hence the pair $(N_1^*, f \restriction M_2)$ is as required.

Proof of part (3), (4). Like part (2). $\square_{7.6}$

7.7 Claim. *1) If δ is a limit ordinal $< \lambda^{+2}$ and $\langle M_i : i < \delta\rangle$ is a $\leq^*_{\lambda^+}$-increasing continuous (in K_{λ^+}) and $M_\delta = \bigcup_{i<\delta} M_i$ (so $M_\delta \in K_{\lambda^+}$), <u>then</u> $M_i \leq^*_{\lambda^+} M_\delta$ for each $i < \delta$.*
2) If δ is a limit ordinal $< \lambda^{+2}$ and $\langle M_i : i < \delta\rangle$ is a $\leq^*_{\lambda^+}$-increasing sequence, each M_i is in $K^{\rm nice}_{\lambda^+}$, <u>then</u> $\bigcup_{i<\delta} M_i$ is in $K^{\rm nice}_{\lambda^+}$.
3) If δ is a limit ordinal $< \lambda^{+2}$ and $\langle M_i : i < \delta\rangle$ is a $<^+_{\lambda^+}$-increasing continuous (or just $<^*_{\lambda^+}$-increasing continuous, and $M_{2i+1} <^+_{\lambda^+} M_{2i+2}$ for $i < \delta$), <u>then</u> $i < \delta \Rightarrow M_i <^+_{\lambda^+} \bigcup_{j<\delta} M_j$.

Proof. 1) We prove it by induction on δ. Now if C is a club of δ, (as $\leq^*_{\lambda^+}$ is transitive) then we can replace $\langle M_j : j < \delta\rangle$ by $\langle M_j : j \in C\rangle$ so without loss of generality $\delta = \mathrm{cf}(\delta)$, so $\delta \leq \lambda^+$; similarly it is enough to prove $M_0 \leq^*_{\lambda^+} M_\delta := \bigcup_{j<\delta} M_j$. For each $i \leq \delta$ let $\langle M^i_\zeta : \zeta < \lambda^+\rangle$ be a $<^*_{\mathfrak{K}}$-representation of M_i.

<u>Case A</u>: $\delta < \lambda^+$.
Without loss of generality (see 7.3(1)) for every $i < j < \delta$ and $\zeta < \lambda^+$ we have:
$M^j_\zeta \cap M_i = M^i_\zeta$ and $\mathrm{NF}_\lambda(M^i_\zeta, M^i_{\zeta+1}, M^j_\zeta, M^j_{\zeta+1})$. Let $M^\delta_\zeta = \bigcup_{i<\delta} M^i_\zeta$, so
$\langle M^\delta_\zeta : \zeta < \lambda^+\rangle$ is $\leq_{\mathfrak{K}}$-increasing continuous sequence of members

of K_λ with limit M_δ, and for $i < \delta, M^\delta_\zeta \cap M_i = M^i_\zeta$. By symmetry (see 6.25) we have $\mathrm{NF}_\lambda(M^i_\zeta, M^{i+1}_\zeta, M^i_{\zeta+1}, M^{i+1}_{\zeta+1})$ so as $\langle M^i_\zeta : i \le \delta\rangle$,$\langle M^i_{\zeta+1} : i \le \delta\rangle$ are $\le_{\mathfrak{K}}$-increasing continuous, by 6.28, the transitivity of $\mathrm{NF}_{\mathfrak{s}}$, we know $\mathrm{NF}_\lambda(M^0_\zeta, M^\delta_\zeta, M^0_{\zeta+1}, M^\delta_{\zeta+1})$ hence by symmetry (6.25) we have $\mathrm{NF}_\lambda(M^0_\zeta, M^0_{\zeta+1}, M^\delta_\zeta, M^\delta_{\zeta+1})$.
So $\langle M^0_\zeta : \zeta < \lambda^+\rangle, \langle M^\delta_\zeta : \zeta < \lambda^+\rangle$ are witnesses to $M_0 \le^*_{\lambda^+} M_\delta$.

Case B: $\delta = \lambda^+$.

By 7.3(1) (using normality of the club filter, restricting to a club of λ^+ and renaming), without loss of generality for $i < j \le 1+\zeta < 1+\xi < \lambda^+$ we have $M^j_\zeta \cap M_i = M^i_\zeta$, and $\mathrm{NF}_\lambda(M^i_\zeta, M^i_\xi, M^j_\zeta, M^j_\xi)$. Let us define $M^{\lambda^+}_\zeta = \bigcup\limits_{j<1+\zeta} M^j_\zeta$. So $\langle M^{\lambda^+}_\zeta : \zeta < \lambda^+\rangle$ is a $<_{\mathfrak{K}}$-representation of $M_{\lambda^+} = M_\delta$ and continue as before.
2) Again without loss of generality $\delta = \mathrm{cf}(\delta)$ call it κ. Let $\langle M^i_\zeta : \zeta < \lambda^+\rangle$ be a $<_{\mathfrak{K}}$-representation of M_i for $i < \delta$.

Case A: $\delta = \kappa < \lambda^+$.

Easy by now, yet we give details, noting 7.8. So without loss of generality (see 7.3(1)) for every $i < j < \delta$ and $\zeta < \xi < \lambda^+$ we have: $M^j_\zeta \cap M_i = M^i_\zeta$, $\mathrm{NF}_\lambda(M^i_\zeta, M^i_\xi, M^j_\zeta, M^j_\xi)$ and $M^i_{\zeta+1}$ is (λ,λ)-brimmed over M^i_ζ. Let $M^\delta_\zeta = \bigcup\limits_{\beta<\delta} M^\beta_\zeta$. Let $\xi < \lambda^+$. Now if $p \in \mathscr{S}^{\mathrm{bs}}(M^\delta_\xi)$ then by the local character Axiom (E)(c) + the uniqueness Axiom (E)(e), for some $i < \delta, p$ does not fork over M^i_ξ. As M_i is λ^+-saturated above λ, the type $p \restriction M^i_\xi$ is realized in M_i. So let $b \in M_i$ realize $p \restriction M^i_\xi$ and by Axiom $(E)(h)$, continuity, it suffices to prove that for every $j \in (i,\delta), b$ realizes $p \restriction M^j_\xi$ in M_j which holds by 6.32 (note that $b \in M_i \le_{\mathfrak{K}} M_j$ as $j \in [i,\delta)$). So p is realized in $M_\delta = \bigcup\limits_{i<\delta} M_i$.
As this holds for every $\xi < \lambda^+$ and $p \in \mathscr{S}^{\mathrm{bs}}(M^\delta_\xi)$, the model M_δ is saturated.

Case B: $\mathrm{cf}(\delta) = \lambda^+$.

Straight, in fact true for $\mathfrak{K}$ a.e.c. with the λ-amalgamation property.

3) Similar. $\square_{7.7}$

7.8 Remark. Note that in Ax(E)(c),Ax(E)(h) the continuity of the sequences is not required.

7.9 Claim. *1) If $M_0 \in K_{\lambda^+}$ <u>then</u> there is M_1 such that $M_0 <^+_{\lambda^+} M_1 \in K^{\mathrm{nice}}_{\lambda^+}$, and any such M_1 is universal over M_0 in $(K_{\lambda^+}, \le^*_{\lambda^+})$.*
2) Assume $\boxtimes_{\bar N_1,\bar N_2,M_1,M_2}$ below holds. <u>Then</u> $M_1 <^+_{\lambda^+} M_2$ <u>iff</u> for every $\alpha < \lambda^+$ for stationarily many $\beta < \lambda^+$ there is N such that $N^1_\beta \cup N^2_\alpha \subseteq N \le_{\mathfrak{K}} N^2_\beta$ and N^2_β is $(\lambda, *)$-brimmed over N where

$_{,\bar N_2,M_1,M_2}$ *$M_1 \le^*_{\lambda^+} M_2$ is being witnessed by $\bar N_1, \bar N_2$ that is $\bar N_\ell = \langle N^\ell_\alpha : \alpha < \lambda^+\rangle$ is a $\le_{\mathfrak{K}}$-representation of M_ℓ for $\ell = 1,2$ and $\alpha < \lambda^+ \Rightarrow \mathrm{NF}_\lambda(N^1_\alpha, N^1_{\alpha+1}, N^2_\alpha, N^2_{\alpha+1})$ (hence $\alpha \le \beta < \lambda^+ \Rightarrow \mathrm{NF}_\lambda(N^1_\alpha, N^1_\beta, N^2_\alpha, N^2_\beta)$).*

Proof. 1) The existence by 7.4(1). Why "any such $M_1, \ldots$?" if $M_0 \le^*_{\lambda^+} M_2$ then for some $M_2^+ \in K^{\mathrm{nice}}_{\lambda^+}$ we have $M_2 <^+_\lambda M_2^+ \in K^{\mathrm{nice}}_{\lambda^+}$ so $M_0 \le^*_{\lambda^+} M_1 <^+_{\lambda^+} M_2^+$ hence by 7.4(5) we have $M_0 <^+_\lambda M_2^+$; so by 7.6(2) the models M_2^+, M_1 are isomorphic over M_0, so M_2 can be $\le^*_{\lambda^+}$-embedded into M_1 over M_0, so we are done.
2) Not hard. $\square_{7.9}$

§8 Is $K^{\mathrm{nice}}_{\lambda^+}$ with $\le^*_{\lambda^+}$ an a.e.c.?

8.1 Hypothesis. The hypothesis 6.8.

An important issue is whether $(K^{\mathrm{nice}}_{\lambda^+}, \le^*_{\lambda^+})$ satisfies Ax IV of a.e.c. So a model $M \in K_{\lambda^{++}}$ may be the union of a $\le^*_{\lambda^+}$-increasing chain of length λ^{++}, but we still do not know if there is a continuous such sequence.

E.g. let $\langle M_\alpha : \alpha < \lambda^{++}\rangle$ be $\le^*_{\lambda^+}$-increasing with union $M \in K_{\lambda^{++}}$ let $M'_n = M_n, M'_{\omega+\alpha+1} = M_{\omega+\alpha}$ and $M'_\delta = \cup\{M_\beta : \beta < \delta\}$ for δ limit. So $\langle M'_\alpha : \alpha < \lambda^{++}\rangle$ is $\le_{\mathfrak{K}}$-increasing continuous, $\langle M'_{\alpha+1} : \alpha < \lambda^{++}\rangle$ is $\le^*_{\lambda^+}$-increasing, but we do not know whether $M'_\delta \le^*_{\lambda^+} M'_{\delta+1}$ for limit $\delta < \lambda^{++}$.

8.2 Definition. Let $M \in \mathfrak{K}_{\lambda^{++}}$ be the union of an $\le_{\mathfrak{K}}$-increasing continuous chain from $(K^{\rm nice}_{\lambda^+}, \le^*_{\lambda^+})$ or just $(K_{\lambda^+}, \le^*_{\lambda^+}), \bar M = \langle M_i : i < \lambda^{++}\rangle$ such that $\langle M_i : i < \lambda^{++}$ non-limit$\rangle$ is $\le^*_{\lambda^+}$-increasing.
1) Let $S(\bar M) = \{\delta : M_\delta \nleq^*_{\lambda^+} M_{\delta+1}$ (see 8.3(3) below)$\}$, so $S(\bar M) \subseteq \lambda^{++}$.
2) For such M let $S(M)$ be $S(\bar M)/\mathscr{D}_{\lambda^{++}}$ where $\bar M$ is a $\le_{\mathfrak{K}}$-representation of M and $\mathscr{D}_{\lambda^{++}}$ is the club filter on λ^{++}; it is well defined by 8.3 below.
3) We say $\langle M_i : i < \delta\rangle$ is non-limit $<^*_{\lambda^+}$-increasing <u>if</u> for non-limit $i < j < \delta$ we have $M_i \le^*_{\lambda^+} M_j$.

8.3 Claim. *1) If $\bar M^\ell = \langle M^\ell_i : i < \lambda^{++}\rangle$ for $\ell \in \{1,2\}$ is $\le_{\mathfrak{K}}$-increasing continuous and $i < j < \lambda^{++} \Rightarrow M_0 \le^*_{\lambda^+} M_{i+1} \le^*_{\lambda^+} M_{j+1}$ and $M = \bigcup\limits_{i<\lambda^{++}} M^1_i = \bigcup\limits_{i<\lambda^{++}} M^2_i$ has cardinality λ^{++} <u>then</u> $S(\bar M^1) = S(\bar M^2)$ mod $\mathscr{D}_{\lambda^{++}}$.*
2) If $M, \bar M$ are as in 8.2 hence $M = \bigcup\limits_{i<\lambda^{++}} M_i$ <u>then</u> $S(\bar M)/\mathscr{D}_{\lambda^{++}}$ depends just on $M/\cong$.
3) If $\bar M$ is as in 8.2 or, equivalently as in part (1), and $i < j < \lambda^{++}$, <u>then</u> $M_i \le^*_{\lambda^+} M_{i+1} \Leftrightarrow M_i \le^*_{\lambda^+} M_j$.
4) If $M \in \mathfrak{K}_{\lambda^{++}}$ is the union of a $\le^*_{\lambda^+}$-increasing chain from $(K^{\rm nice}_{\lambda^+}, \le^*_{\lambda^+}$), not necessarily continuous, <u>then</u> there is $\bar M$ as in Defintion 8.2, that is $\bar M = \langle M_i : i < \lambda^{++}\rangle$, a $\le_{\mathfrak{K}}$-representation of M with $M_i \le^*_{\lambda^+} M_j$ for non-limit $i < j$.

Proof. 1) We can find a club E of λ^{++} consisting of limit ordinals such that $i \in E \Rightarrow M^1_i = M^2_i$. Now if $\delta_1 < \delta_2$ are from E then $\delta_1 \in S(\bar M^1) \Leftrightarrow M^1_{\delta_1} \le^*_{\lambda^+} M^1_{\delta_1+1} \Leftrightarrow M^1_{\delta_1} \le^*_{\lambda^+} M^1_{\delta_2} \Leftrightarrow M^2_{\delta_1} \le^*_{\lambda^+} M^2_{\delta_2} \Leftrightarrow M^2_{\delta_1} \le^*_{\lambda^+} M^2_{\delta_1+1} \Leftrightarrow \delta_1 \in S(M^2)$.
[Why? By the definition of $S(\bar M^1)$, by part (3), by "$\delta_1, \delta_2 \in E$", by part (3), by the definition of $S(\bar M^2)$, respectively.] So we are done.
2) Follows by parts (1) and (3).
3) The implication $\Leftarrow$ is by 7.4(3); for the implication $\Rightarrow$, note that assuming $M_i <^*_{\lambda^+} M_{i+1}$, as $\le^*_{\lambda^+}$ is a partial order, noting that by the assumption on $\bar M$ we have $M_{i+1} \le^*_{\lambda^+} M_{j+1}$, and by 7.4(3) we

are done.
4) Trivial. $\square_{8.3}$

8.4 Claim. *If* $(*)$ *below holds* *<u>then</u>* *for every stationary* $S \subseteq S^{\lambda^{++}}_{\lambda^+}$ $(= \{\delta < \lambda^{++} : {\rm cf}(\delta) = \lambda^+\})$ *for some* λ^+*-saturated* $M \in K_{\lambda^{++}}$ *we have* $S(M)$ *is well defined and equal to* $S/\mathscr{D}_{\lambda^{++}}$, *where*

- $(*)$ *we can find* $\langle M_i : i \le \lambda^+ + 1\rangle$ *which is* $<_{\mathfrak{K}}$*-increasing continuous sequence of members of* $K^{\rm nice}_{\lambda^+}$ *such that* $i < j \le \lambda^+ + 1$ & $(i,j) \ne (\lambda^+, \lambda^+ + 1) \Rightarrow M_i <^+_{\lambda^+} M_j$ *but* $\neg(M_{\lambda^+} \le^*_{\lambda^+} M_{\lambda^+ + 1})$.

Proof. Fix $S \subseteq S^{\lambda^{++}}_{\lambda^+}$ and $\langle M_i : i \le \lambda^+ + 1\rangle$ as in $(*)$.
Without loss of generality $|M_{\lambda^+ + 1} \backslash M_{\lambda^+}| = \lambda^+$.

We choose by induction on $\alpha < \lambda^{+2}$ a model M^S_α such that:

- (a) $M^S_\alpha \in K^{\rm nice}_{\lambda^+}$ has universe an ordinal $< \lambda^{++}$
- (b) for $\beta < \alpha$ we have $M^S_\beta \le_{\mathfrak{K}} M^S_\alpha$
- (c) if $\alpha = \beta + 1$, $\beta \notin S$ then $M^S_\beta <^+_{\lambda^+} M^S_\alpha$
- (d) if $\alpha = \beta + 1, \beta \in S$ then $(M^S_\beta, M^S_\alpha) \cong (M_{\lambda^+}, M_{\lambda^+ + 1})$
- (e) if $\beta < \alpha, \beta \notin S$ then $M^S_\beta \le^+_{\lambda^+} M^S_\alpha$
- (f) if α is a limit ordinal, then $M_\alpha = \cup\{M_\beta : \beta < \alpha\}$.

We use freely the transitivity and continuity of $\le^*_\lambda$ and of $<^+_\lambda$.

For $\alpha = 0$ no problem.

For α limit no problem; choose an increasing continuous sequence $\langle \gamma_i : i < {\rm cf}(\alpha)\rangle$ of ordinals with limit α each of cofinality $< \lambda, \gamma_i \notin S$, and use 7.7(3) for clause (e).

For $\alpha = \beta + 1, \beta \notin S$ no problem.

For $\alpha = \beta + 1, \beta \in S$ so ${\rm cf}(\beta) = \lambda^+$, let $\langle \gamma_i : i < \lambda^+\rangle$ be increasing continuous with limit β and ${\rm cf}(\gamma_i) \le \lambda$, hence $\gamma_i \notin S$ and each γ_{i+1} a successor ordinal. By clause (e) above and 7.4(5) we have $M^S_{\gamma_i} <^+_{\lambda^+} M^S_{\gamma_{i+1}}$, hence $\langle M_{\gamma_i} : i < \lambda^+\rangle$ is $<^+_{\lambda^+}$-increasing continuous. Now there is an isomorphism f_β from M_{λ^+} onto M^S_β mapping M_i

onto $M^S_{\gamma_i}$ for $i < \lambda$ (why? choose $f_\beta \restriction M_i$ by induction on i, for $i = 0$ by 7.3(0), for i successor $M^S_{\gamma_i} <^+_\lambda M^S_{\gamma_{i+1}}$ by 7.4(3) as $M^S_{\gamma_i} <^*_{\lambda^+} M^S_{\gamma_{i+1}} <^+_{\lambda^+} M^S_{\gamma_{i+1}}$ so we can use 7.6(2)). So we can choose a one-to-one function f_α from M_{λ^++1} onto some ordinal $< \lambda^{++}$ extending f_β and let $M_\alpha = f_\alpha(M_{\lambda^++1})$.

Finally having carried the induction, let $M_S = \bigcup\limits_{\alpha<\lambda^{+2}} M^S_\alpha$, it is easy to check that $M_S \in K_{\lambda^{++}}$ is λ^+-saturated and $\bar{M} = \langle M^S_\alpha : \alpha < \lambda^{++}\rangle$ witnesses that $S(M_S)/\mathscr{D}_{\lambda^{++}}$ is well defined and $S(M_S)/\mathscr{D}_{\lambda^{++}} = S(\langle M^S_\alpha : \alpha < \lambda^{++}\rangle)/\mathscr{D}_{\lambda^{++}} = S/\mathscr{D}_{\lambda^{++}}$ as required. $\square_{8.4}$

Below we prove that some versions of non-smoothness are equivalent.

8.5 Claim. *1) We have $(**)_{M^*_1,M^*_2} \Rightarrow (***)$ (see below).*
2) If $(*)$ then $(**)_{M^*_1,M^*_2}$ for some M^*_1, M^*_2 and trivially $(***) \Rightarrow (*)$.
3) In part (1) we get $\langle M_i : i \le \lambda^+ + 1\rangle$ as in $(***)$, see below, such that $M_{\lambda^+} = M^*_1, M_{\lambda^++1} = M^*_2$ if we waive $i < \lambda^+ \Rightarrow M_i <^+_\lambda M_{\lambda+1}$ or assume $M^*_1 <_{\mathfrak{K}} M^* <^+_\lambda M^*_2$ for some M^*.
4) If $M^*_1 \le^*_{\lambda^+} M^*_2$ and $M^*_1 \in K^{\rm nice}_{\lambda^+}$ and $N_1 \le_{\mathfrak{K}} N_2 \in K_\lambda, N_\ell \le M^*_\ell$ for $\ell = 1,2$ and $p \in \mathscr{S}^{\rm bs}(N_2)$ does not fork over N_1 <u>then</u> some $c \in M^*_1$ realizes p
<u>where</u>

- $(*)$ *there are limit $\delta < \lambda^{++}, N$ and $\bar{M} = \langle M_i : i \le \delta\rangle$ a $\le^*_{\lambda^+}$-increasing continuous sequence with $M_i, N \in K^{\rm nice}_{\lambda^+}$ such that: $M_i \le^*_{\lambda^+} N \Leftrightarrow i < \delta$*
- $(**)_{M^*_1,M^*_2}$
    - *(i) $M^*_1 \in K^{\rm nice}_{\lambda^+}, M^*_2 \in K^{\rm nice}_{\lambda^+}$*
    - *(ii) $M^*_1 \le_{\mathfrak{K}} M^*_2$*
    - *(iii) $M^*_1 \not\le^*_{\lambda^+} M^*_2$*
    - *(iv) if $N_1 \le_{\mathfrak{K}} N_2$ are from K_λ, $N_\ell \le_{\mathfrak{K}} M^*_\ell$ for $\ell = 1,2$ and $p \in \mathscr{S}^{\rm bs}(N_2)$ does not fork over N_1, <u>then</u> some $a \in M^*_1$ realizes p in M^*_2*
- $(***)$ *there is $\bar{M} = \langle M_i : i \le \lambda^+ + 1\rangle, \le_{\mathfrak{K}}$-increasing continuous, every $M_i \in K^{\rm nice}_{\lambda^+}$ and $M_{\lambda^+} \not\le^*_{\lambda^+} M_{\lambda^++1}$ but $i < j \le \lambda^+ + 1$ &*

$i \ne \lambda^+ \Rightarrow M_i <^+_{\lambda^+} M_j$;
note that this is $(*)$ of 8.4.

Proof. 1),3) Let $\langle a^\ell_i : i < \lambda^+\rangle$ list the elements of M^*_ℓ for $\ell = 1, 2$. Let $\langle N^*_{2,i} : i < \lambda^+\rangle$ be a $\le_{\mathfrak{K}}$-representation of M^*_2.
Let $\langle (p_\zeta, N^*_\zeta, \gamma_\zeta) : \zeta < \lambda^+\rangle$ list the triples (p, N, γ) such that $\gamma < \lambda^+, p \in \mathscr{S}^{\rm bs}(N), N \in \{N^*_{2,i} : i < \lambda^+\}$ with each such triple appearing λ^+ times. By induction on $\alpha < \lambda^+$ we choose $\langle N^\alpha_i : i \le \alpha\rangle, N_\alpha$ such that:

(a) $N^\alpha_i \in K_\lambda$ and $N^\alpha_i \le_{\mathfrak{K}} M^*_1$

(b) $N_\alpha \le_{\mathfrak{K}} M^*_2$ and $N_\alpha \in K_\lambda$
(c) $\langle N^\alpha_i : i \le \alpha\rangle$ is $\le_{\mathfrak{K}}$-increasing continuous
(d) $N^\alpha_\alpha \le_{\mathfrak{K}} N_\alpha, N_\alpha \cap M^*_1 = N^\alpha_\alpha$
(e) if $i \le \alpha$ then $\langle N^\beta_i : \beta \in [i, \alpha]\rangle$ is $\le_{\mathfrak{K}}$-increasing continuous
(f) $\langle N_\beta : \beta \le \alpha\rangle$ is $\le_{\mathfrak{K}}$-increasing continuous
(g) if $\alpha = \beta + 1, i \le \beta$ then $\mathrm{NF}_\lambda(N^\beta_i, N_\beta, N^\alpha_i, N_\alpha)$
(h) if $\alpha = 2\beta + 1$ then $a^2_\beta \in N_{\alpha+1}$
(i) if $\alpha = 2\beta + 2$ and $i < \alpha$ then N^α_{i+1} is brimmed over $N^\alpha_i \cup N^{2\beta+1}_{i+1}$ and N^α_0 is brimmed over $N^{2\beta}_0$.

<u>Why is this enough?</u>

We let $M_{\lambda^+} = M^*_1, M_{\lambda^++1} = M^*_2$ and let $M'_{\lambda^++1} \in K^{\rm nice}_{\lambda^+}$ be such that $M_{\lambda^++1} <^+_{\lambda^+} M'_{\lambda^++1}$ and for $i < \lambda^+$ we let $M_i = \cup\{N^\alpha_i : \alpha \in [i, \lambda^+)\}$; now

(α) $M^*_1 = \bigcup\limits_{\alpha<\lambda^+} N^\alpha_\alpha = \bigcup\limits_{i<\lambda^+} M_i$ and $M^*_2 = \bigcup\limits_{\alpha<\lambda^+} N_\alpha$
[why? the second by clause (h) (and (b) of course), the first as $N_\alpha \cap M^*_1 = N^\alpha_\alpha$].

Now:

(β) $\langle M_i : i \le \lambda^+ + 1\rangle$ is $\le_{\mathfrak{K}}$-increasing continuous
[trivial by clauses (c) + (e) if $i < \lambda^+$ and (d) if $i = \lambda^+$]

(γ) for $i < \lambda^+$, M_i is saturated, i.e., $\in K^{\rm nice}_{\lambda^+}$.
[Why? Clearly $\langle N^\alpha_i : \alpha \in (i, \lambda^+)\rangle$ is a $\le_{\mathfrak{K}}$-representation of M_i by clause (e) and the choice of M_i. If $i = 0$ this follows by clauses (i) + (e). If $i = j+1$ this follows by clauses (e) + (i). If i is a limit ordinal use 7.7(2) and clause (g)]

(δ) for $i < \lambda^+, i < j \le \lambda^+ + 1$ we have $M_i \le^*_{\lambda^+} M_j$.
[Why? Let $N^\alpha_{\lambda^+} := N^\alpha_\alpha$, $N^\alpha_{\lambda^++1} = N_\alpha$ for $\alpha < \lambda^+$ and let γ be i if $j = \lambda^+, \lambda^+ + 1$ and be j if $j < \lambda^+$; so in any case $\gamma < \lambda^+$. Now as $\langle N^\alpha_i : \alpha \in [\gamma, \lambda^+)\rangle$ is a $\le_{\mathfrak{K}}$-representation of M_i and $\langle N^\alpha_j : \alpha \in [\gamma, \lambda^+)\rangle$ is a $\le_{\mathfrak{K}}$-representation of M_j and if $\gamma \le \beta < \lambda^+$ then by clause (g) we have $\mathrm{NF}_\lambda(N^\beta_i, N_\beta, N^{\beta+1}_i, N_{\beta+1})$ hence by symmetry $\mathrm{NF}_\lambda(N^\beta_i, N^{\beta+1}_i, N_\beta, N_{\beta+1})$, hence by monotonicity $\mathrm{NF}_\lambda(N^\beta_i, N^{\beta+1}_i, N^\beta_j, N^{\beta+1}_j)$; this suffices]

(ε) if $i < j \le \lambda^+$ then $M_i <^+_{\lambda^+} M_j$
[why? by 7.7(3) it suffices to prove this in the cases $j = i+1$. Now claim 7.9(2), clause (i) guaranteed this.]

Clearly $\langle M_i : i \le \lambda^+ + 1\rangle$ is as required for part (1) and for part (3) for first possibility (with waiving) obviously. For the second possibility in part (2), easily $\langle M_i : i \le \lambda^+\rangle\hat{}\langle M'_{\lambda^++1}\rangle$ is as required but $M^*_2, M^1_{\lambda+1}$ are isomorphic over M^*, so also $\langle M_i : i \le \lambda^+ + 1\rangle$ is O.K.
So we are done.

So let us carry the construction.

For $\alpha = 0$ trivially.

For α limit: straightforward.

For $\alpha = 2\beta + 1$ we let $N^\alpha_i = N^{2\beta}_i$ for $i \le 2\beta$ and $N_\alpha \in K_\lambda$ is chosen such that $N_{2\beta} \cup \{a^2_\beta\} \subseteq N_\alpha \le_{\mathfrak{K}} M^*_2$ and $N_\alpha \restriction M^*_1 \le_{\mathfrak{K}} M^*_1$, easy by the properties of abstract elementary class and we let $N^\alpha_{2\beta+1} = N_\alpha \restriction M^*_1$.

For $\alpha = 2\beta+2$ we choose by induction on $\varepsilon < \lambda^2$, a triple $(N^\oplus_{\alpha,\varepsilon}, N^\otimes_{\alpha,\varepsilon}, a_{\alpha,\varepsilon})$ such that:

(A) $N^\otimes_{\alpha,\varepsilon} \le_{\mathfrak{K}} M^*_2$ belongs to K_λ and is $\le_{\mathfrak{K}}$-increasing continuous with ε

(B) $N^\otimes_{\alpha,0} = N_{2\beta+1}$ and $N^\otimes_{\alpha,\varepsilon} \restriction M^*_1 \le^*_{\mathfrak{K}} M^*_1$

(C) $N^\oplus_{\alpha,\varepsilon} \le_{\mathfrak{K}} M^*_1$ belongs to K_λ and is $\le_{\mathfrak{K}}$-increasing continuous with ε

(D) $N^\oplus_{\alpha,0} = N^{2\beta+1}_{2\beta+1}$

(E) $(N^\oplus_{\alpha,\varepsilon}, N^\oplus_{\alpha,\varepsilon+1}, a_{\alpha,\varepsilon}) \in K^{3,\rm uq}_\lambda$

(F) $\mathbf{tp}(a_{\alpha,\varepsilon}, N^\otimes_{\alpha,\varepsilon}, M^*_2)$ does not fork over $N^\oplus_{\alpha,\varepsilon}$

(G) $N^\oplus_{\alpha,\varepsilon} \le_{\mathfrak{K}} N^\otimes_{\alpha,\varepsilon}$

(H) for every $p \in \mathscr{S}^{\rm bs}(N^\oplus_{\alpha,\varepsilon})$ for some odd $\zeta \in [\varepsilon, \varepsilon+\lambda)$ the type $\mathbf{tp}(a_{\alpha,\zeta}, N^\otimes_{\alpha,\zeta}, N^\otimes_{\alpha,\zeta+1})$ is a non-forking extension of p.

No problem to carry this. [Why? For $\varepsilon = 0$ and ε limit there are no problems. In stage $\varepsilon+1$ by bookkeeping gives you a type $p_\varepsilon \in \mathscr{S}^{\rm bs}(N^\oplus_{\alpha,\varepsilon})$ and let $q_\varepsilon \in \mathscr{S}^{\rm bs}(N^\otimes_{\alpha,\varepsilon})$ be a non-forking extension of p_ε. By assumption (iv) of $(**)_{M^*_1,M^*_2}$ there is an element $a_{\alpha,\varepsilon} \in M^*_1$ realizing q_ε. Now M^*_1 is saturated hence there is a model $N^\oplus_{\alpha,\varepsilon+1} \in K_\lambda$ such that $N^\oplus_{\alpha,\varepsilon+1} \le_{\mathfrak{K}} M^*_1$ and $(N^\oplus_{\alpha,\varepsilon}, N^\oplus_{\alpha,\varepsilon+1}, a_{\alpha,\varepsilon}) \in K^{3,\rm uq}_\lambda$.

Lastly, choose $N^\otimes_{\alpha,\varepsilon+1}$ satisfying clauses (A),(B),(G) so we have carried the induction on ε.]

Note that $\mathrm{NF}_\lambda(N^\oplus_{\alpha,\varepsilon}, N^\otimes_{\alpha,\varepsilon}, N^\oplus_{\alpha,\varepsilon+1}, N^\otimes_{\alpha,\varepsilon+1})$ for each $\varepsilon < \lambda^2$ by clauses (E),(F) and 6.33, hence $\mathrm{NF}(N^{2\beta+1}_{2\beta+1}, N_{2\beta+1}, \cup\{N^\oplus_{\alpha,\varepsilon} : \varepsilon < \lambda^2\}, \cup\{N^\otimes_{\alpha,\varepsilon} : \varepsilon < \lambda^2\})$ by 6.28 as $(N^\oplus_{\alpha,0}, N^\otimes_{\alpha,0}) = (N^{2\beta+1}_{2\beta+1}, N_{2\beta+1})$ and the sequence $\langle N^\oplus_{\alpha,\varepsilon} : \varepsilon < \lambda^+\rangle, \langle N^\otimes_{\alpha,\varepsilon} : \varepsilon < \lambda^+\rangle$ are increasing continuous.

Now let $N_\alpha = \bigcup\{N^\otimes_{\alpha,\varepsilon} : \varepsilon < \lambda^2\}$, $N^\alpha_\alpha = N_\alpha \cap M^*_1$ recalling clauses (A)+(B).

Now $\cup\{N^\oplus_{\alpha,\varepsilon} : \varepsilon < \lambda^2\} \le_{\mathfrak{K}} M^*_1$ is $(\lambda,*)$-brimmed over $N^{2\beta+1}_{2\beta+1}$ by 4.3 (and clause (H) above). Hence there is no problem to choose $N^\alpha_i \le_{\mathfrak{K}} N^\alpha_\alpha$ for $i \le 2\beta+1$ as required, that is $N^{2\beta+1}_i \le_{\mathfrak{K}} N^\alpha_i$, $\langle N^\alpha_i : i \le 2\beta+1\rangle$ is $\le_{\mathfrak{K}}$-increasing continuous, $\mathrm{NF}_\lambda(N^{2\beta+1}_i, N^{2\beta+1}_{i+1}, N^\alpha_i, N^\alpha_{i+1})$ and N^α_{i+1} is $(\lambda,*)$-brimmed over $N^{2\beta+1}_{i+1} \cup N^\alpha_i$ and N^α_0 is $(\lambda,*)$-brimmed over $N^{2\beta+1}_0$. So we have finished the induction step on $\alpha = 2\beta+2$.

Having carried the induction we are done.

2) So assume $(*)$ and let $M_{\delta+1} := N$ from $(*)$. It is enough to

prove that $(**)_{M_\delta,M_{\delta+1}}$ holds. Clearly clauses (i), (ii), (iii) hold, so we should prove (iv). Without loss of generality $\delta = \mathrm{cf}(\delta)$ so $\delta = \lambda^+$ or $\delta \le \lambda$. For $i \le \delta+1$ let $\langle M_{i,\alpha} : \alpha < \lambda^+\rangle$ be a $\le_{\mathfrak{K}}$-representation of M_i and for $i < \delta, j \in (i,\delta+1]$ let $E_{i,j}$ be a club of λ^+ witnessing $M_i \le^*_{\lambda^+} M_j$ for $\bar{M}^i, \bar{M}^j$. First assume $\delta \le \lambda$. Let $E = \cap\{E_{i,j} : i < \delta, j \in (i,\delta+1]\}$, it is a club of λ^+. So assume $N_2 \le_{\mathfrak{K}} M_{\delta+1}, N_1 \le_{\mathfrak{K}} N_2, N_1 \le_{\mathfrak{K}} M_\delta$ and $N_1, N_2 \in K_\lambda$ and $p \in \mathscr{S}^{\mathrm{bs}}(N_2)$ does not fork over N_1. We can choose $\zeta \in E$ such that $N_2 \subseteq M_{\delta+1,\zeta}$, let $p_1 \in \mathscr{S}^{\mathrm{bs}}(M_{\delta+1,\zeta})$ be a non-forking extension of p, so p_1 does not fork over N_1 hence (by monotonicity) over $M_{\delta,\zeta}$ so $p_2 := p_1 \restriction M_{\delta,\zeta} \in \mathscr{S}^{\mathrm{bs}}(M_{\delta,\zeta})$. By Axiom $(E)(c)$ for some $\alpha < \delta, p_2$ does not fork over $M_{\alpha,\zeta}$ hence $p_2 \restriction M_{\alpha,\zeta} \in \mathscr{S}^{\mathrm{bs}}(M_{\alpha,\zeta})$. As $M_\alpha \in K^{\mathrm{nice}}_{\lambda^+}$, i.e., M_α is λ^+-saturated (above λ), clearly for some $\xi \in (\zeta,\lambda^+) \cap E$ some $c \in M_{\alpha,\xi}$ realizes $p_2 \restriction M_{\alpha,\zeta}$ but $\mathrm{NF}_\lambda(M_{\alpha,\zeta}, M_{\delta+1,\zeta}, M_{\alpha,\xi}, M_{\delta+1,\xi})$ hence by 6.32 we know that $\mathbf{tp}(c, M_{\delta+1,\zeta}, M_{\delta+1,\xi})$ belongs to $\mathscr{S}^{\mathrm{bs}}(M_{\delta+1,\zeta})$ and does not fork over $M_{\alpha,\zeta}$ hence c realizes p_2 and even p_1 hence p and we are done.

Second, assume $\delta = \lambda^+$, then for some $\delta^* < \delta$ we have $N_1 \le_{\mathfrak{K}} M_{\delta^*}$, and use the proof above for $\langle M_i : i \le \delta^*\rangle, M_{\delta+1}$ (or use $M_{\delta^*} \le^*_{\lambda^+} M_{\delta+1}$).

4) Straight, in fact included the proof of 7.7(2). $\square_{8.5}$

The definition below has affinity to "blowing $\mathfrak{K}_\lambda$ to $\mathfrak{K}^{\mathrm{up}}_\lambda$" in §1.

8.6 Definition. 0) $K^{3,\mathrm{cs}}_{\lambda^+} = \{(M,N,a) \in K^{3,\mathrm{bs}}_{\lambda^+} : M, N$ are from $K^{\mathrm{nice}}_{\lambda^+}\}$; we say $N' \in K_\lambda$ (or p') witness $(M,N,a) \in K^{3,\mathrm{cs}}_{\lambda^+}$ if it witnesses $(M,N,a) \in K^{3,\mathrm{bs}}_\lambda$.

1) $\mathscr{S}^{\mathrm{cs}}_{\lambda^+} := \{\mathbf{tp}(a,M,N) : M \le^*_{\lambda^+} N$ are in $K^{\mathrm{nice}}_{\lambda^+}, a \in N$ and $(M,N,a) \in K^{3,\mathrm{cs}}_{\lambda^+}\}$, the type being for $\mathfrak{K}^{\mathrm{nice}}_{\lambda^+} = (K^{\mathrm{nice}}_{\lambda^+}, \le^*_{\lambda^+})$, see below[22] so the notation is justified by 8.7(1).

2) We define $\mathfrak{K}^\otimes = (K^\otimes, \le^\otimes)$ as follows

(a) $K^\otimes = \mathfrak{K} \restriction \{M \in K : M = \cup\{M_s : s \in I\}$ where $M_s \in K^{\mathrm{nice}}_{\lambda^+}$, I is a directed partial order and $s <_I t \Rightarrow M_s \le^*_{\lambda^+} M_t\}$

[22] actually to define $\mathbf{tp}_{\mathfrak{K}_\lambda}(a,M,N)$ where $M \le_{\mathfrak{K}_\lambda} N, \bar{a} \in N$ we need less that "$\mathfrak{K}_\lambda$ is a λ-a.e.c.", and we know on $(K^{\mathrm{nice}}_{\lambda^+}, \le^*_{\lambda^+})$ more than enough

(b) Let $M_1 \le^\otimes M_2$ if $M_1, M_2 \in K^\otimes, M_1 \le_{\mathfrak{K}} M_2$ and:

$(*)_{M_1,M_2}$ if $N_\ell \in K_\lambda, N_\ell \le_{\mathfrak{K}} M_\ell$, for $\ell = 1,2, p \in \mathscr{S}^{\rm bs}(N_2)$ does not fork over N_1 and $N_1 \le_{\mathfrak{K}} N_2$ <u>then</u> some $a \in M_1$ realizes p in M_2

(c) let $\le^\otimes_{\lambda^+} = \le^\otimes \restriction K^\otimes_{\lambda^+}$.

3) $\bigcup\limits_{\lambda^+} = \{(M_0, M_1, a, M_3) : M_0 \le^*_{\lambda^+} M_1 \le^*_{\lambda^+} M_3$ are in $K^{\rm nice}_{\lambda^+}$ and $(M_1, M_3, a) \in K^{3,\rm cs}_{\lambda^+}$ as witnessed by some $N \le_{\mathfrak{K}} M_0$ from $K_\lambda\}$.
4) $\mathfrak{K}^{\rm nice}_{\lambda^+} = (K^{\rm nice}_{\lambda^+}, \le^*_{\lambda^+})$, that is $(K^{\rm nice}_{\lambda^+}, \le^*_{\lambda^+} \restriction K^{\rm nice}_{\lambda^+})$.
5) We say that M' or p' witness $p = \mathbf{tp}_{\mathfrak{K}^{\rm nice}_{\lambda^+}}(a, M, N)$ when $M' \le_{\mathfrak{K}} M, M' \in K_\lambda$ and $[M' \le_{\mathfrak{K}_\lambda} M'' \le_{\mathfrak{K}} M \Rightarrow \mathbf{tp}_{\mathfrak{s}}(a, M'', N)$ does not fork over M' and $p' = \mathbf{tp}_{\mathfrak{s}}(a, M', N)$.

8.7 Conclusion. Assume[23] (recalling 8.4):

> $\boxtimes$ not for every $S \subseteq S^{\lambda^{++}}_{\lambda^+}$ is there λ^+-saturated $M \in K_{\lambda^{++}}$ such that $S(M) = S/\mathscr{D}_{\lambda^{++}}$.

0) On $K^{\rm nice}_{\lambda^+}$, the relations $\le^*_{\lambda^+}, \le^\otimes$ agree.
1) $\mathfrak{K}^{\rm nice}_{\lambda^+} = (K^{\rm nice}_{\lambda^+}, \le^*_{\lambda^+})$ is a λ^+-abstract elementary class and is categorical in λ^+ and has no maximal member and has amalgamation.
2) $K^\otimes$ is included in the class of λ^+-saturated models in $\mathfrak{K}$ and $K^\otimes_{\lambda^+} = K^{\rm nice}_{\lambda^+}$.
3) $\mathfrak{K}^\otimes$ is an a.e.c. with $\mathrm{LS}(K^\otimes) = \lambda^+$ and is the lifting of $\mathfrak{K}^{\rm nice}_{\lambda^+}$.
4) On $K^{\rm nice}_{\lambda^+}, (\mathscr{S}^{\rm cs}_{\lambda^+}, \bigcup\limits_{\lambda^+})$ are equal to $(\mathscr{S}^{\rm bs} \restriction K^{\rm nice}_{\lambda^+}, \bigcup\limits_{<\infty} \restriction K^{\rm nice}_{\lambda^+})$ where they are defined in 2.4, 2.5.
5) $(\mathfrak{K}^{\rm nice}_{\lambda^+}, \mathscr{S}^{\rm cs}_{\lambda^+}, \bigcup\limits_{\lambda^+})$ is a good λ^+-frame.
6) For $M_1 \le^*_{\lambda^+} M_2$ from $K^\otimes_{\lambda^+}$ and $a \in M_2 \backslash M_1$, the type $\mathbf{tp}_{K^\otimes}(a, M_1, M_2)$ is determined by $\mathbf{tp}_{\mathfrak{K}_\lambda}(a, N_1, M_2)$ for all $N_1 \le_{\mathfrak{K}} M_1, N_1 \in K_\lambda$.

Proof. 0) By 8.4 and our assumption $\boxtimes$, we have $M_1, M_2 \in K^{\rm nice}_{\lambda^+}$ & $M_1 \le^\otimes M_2 \Rightarrow M_1 \le^*_{\lambda^+} M_2$ (otherwise $(**)_{M_1,M_2}$ of 8.5 holds hence

[23]this is like $(**)_{M_1,M_2}$ from 8.5, particularly see clause (iv) there

$(***)$ of 8.5 holds and by 8.4 we get $\neg\boxtimes$, contradiction). The other direction is easier just see 8.5(4).
1) We check the axioms for being a λ^+-a.e.c.:

Ax 0: (Preservation under isomorphisms) Obviously.

Ax I: Trivially.

Ax II: By 7.4(2).

Ax III: By 7.7(2) the union belongs to $K^{\rm nice}_{\lambda^+}$ and it $\le^*_{\lambda^+}$-extends each member of the union by 7.7(1).

Ax IV: Otherwise $(*)$ of 8.5 holds, hence by 8.5 also $(***)$ of 8.5 holds. So by 8.4 our assumption $\boxtimes$ fail, contradiction; this is the only place we use $\boxtimes$ in the proof of (1).

Ax V: By 7.4(3) and Ax V for $\mathfrak{K}$.

Also $\mathfrak{K}^{\rm nice}_{\lambda^+}$ is categorical by the uniqueness of the saturated model in λ^+ for $\mathfrak{K}$ has no maximal model by 7.4(1). $\mathfrak{K}^{\rm nice}_{\lambda^+}$ has amalgamation by 7.6(1).
2) Every member of $K^{\otimes}$ is λ^+-saturated in $\mathfrak{K}$ by 7.7(2) (prove by induction on the cardinality of the directed family in Definition 8.6(2), i.e. by the LS-argument it is enough to deal with the index family of $\le \lambda^+$ models each of cardinality λ^+, which holds by part (0) + (1)). If $M \in K_{\lambda^+}$ is λ^+-saturated, clearly $\in K^{\rm nice}_{\lambda^+}$.
3),4) Easy by now (or see §1).
5) We have to check all the clauses in Definition 2.1. We shall use parts (0)-(3) freely.

Axiom (A): By part (3) (of 8.7).

Axiom (B):

There is a superlimit model in $K^{\otimes}_{\lambda^+} = K^{\rm nice}_{\lambda^+}$ by part (1) and uniqueness of the saturated model.

Axiom (C):

By part (1), i.e., 7.6(1) we have amalgamation; JEP holds as $K^{\rm nice}_{\lambda^+}$ is categorical in λ^+. "No maximal member in $\mathfrak{K}^{\otimes}_{\lambda^+}$" holds by 7.4(1).

Axiom (D)(a),(b): By the definition 8.6(1).

<u>Axiom (D)(c)</u>:

By 2.9 (and Definition 8.6(1)). Clearly $K^{3,\rm cs}_{\lambda^+} = K^{3,\rm bs} \restriction K^{\rm nice}_{\lambda^+}$.

<u>Axiom (D)(d)</u>:

For $M \in \mathfrak{K}^{\otimes}_{\lambda^+}$ let $\bar M = \langle M_i : i < \lambda^+\rangle$ $\le_{\mathfrak{K}}$-represent M, so if $M \le^{\otimes} N \in K^{\otimes}_{\lambda^+}$, (hence $M \le^*_{\lambda^+} N \in K^{\otimes}_{\lambda^+} = K^{\rm nice}_{\lambda^+}$) and $a \in N$, $\mathbf{tp}_{\mathfrak{K}^{\rm nice}_{\lambda^+}}(a, M, N) \in \mathscr{S}^{\rm cs}_{\lambda^+}(M)$, we let $\alpha(a, N, \bar M) = \ \mathrm{Min}\{\alpha : \mathbf{tp}(a, M_\alpha, N) \in \mathscr{S}^{\rm bs}(M_\alpha)$ and for every $\beta \in (\alpha, \lambda^+)$, $\mathbf{tp}(a, M_\beta, N) \in \mathscr{S}^{\rm bs}(M_\beta)$ is a non-forking extension of $\mathbf{tp}(a, M_\alpha, N)\}$.

Now

- (a) $\alpha(a, N, \bar M)$ is well defined for a, N as above
  [Why? By Defintion 2.7 + 8.6(1)]
- (b) if a_ℓ, N_ℓ are above for $\ell = 1, 2$ and $\alpha(a_1, N_1, \bar M) = \alpha(a_2, N_2, \bar M)$ call it α and $\mathbf{tp}_{\mathfrak{s}}(a_1, M_\alpha, N) = \ \mathbf{tp}_{\mathfrak{s}}(a_2, M_\alpha, N_2)$ then
  - $(*)$ for $\beta < \lambda^+$ we have $\mathbf{tp}_{\mathfrak{s}}(a_1, M_\beta, N_1) = \ \mathbf{tp}_{\mathfrak{s}}(a_1, M_\beta, N_2) \in \mathscr{S}^{\rm bs}(M_\beta)$
    [Why? By the non-forking uniqueness (Ax(E)(e)) when $\beta \ge \alpha$ by monotonicity if $\beta \le \alpha$]
- (c) if a_ℓ, N_ℓ are as above for $\ell = 1, 2$ and $(*)$ above holds then
  - $(**)$ $\mathbf{tp}_{\mathfrak{K}^{\otimes}_{\lambda^+}}(a_1, M, N_1) = \mathbf{tp}_{\mathfrak{K}^{\otimes}_{\lambda^+}}(a_2, M, N_2)$
    [Why? Use 7.6(3) or by part (6) below].

As $\alpha < \lambda \Rightarrow |\mathscr{S}^{\rm bs}_{\mathfrak{s}}(M_\alpha)| \le \lambda$ (by the stability Axiom (D)(d) for $\mathfrak{s}$), clearly $|\mathscr{S}^{\rm cs}_{\lambda^+}(M)| \le \sum\limits_{\alpha<\lambda^+} |\mathscr{S}^{\rm bs}(M_\alpha)| \le \lambda^+ = \|M\|$ as required.

The reader may ask why do we not just quote the parallel result from §2: The answer is that the equality of types there is "a formal, not the true one". The crux of the matter is that we prove locality (in clause (c) above).

<u>Axiom (E)(a)</u>: By 2.4 - 2.7.

<u>Axiom (E)(b); monotonicity</u>:

Follows by Axiom (E)(b) for $\mathfrak{s}$ and the definition.

Axiom (E)(c); local character:
By 2.11(5) or directly by translating it to the $\mathfrak{s}$-case.

Axiom (E)(d); (transitivity): By 2.11(4).

Axiom (E)(e); uniqueness: By 7.6(3) or by part (6) below.

Axiom (E)(f); symmetry:
So assume $M_0 \leq^*_{\lambda^+} M_1 \leq^*_{\lambda^+} M_2$ are from $K^{\otimes}_{\lambda^+}$ and for $\ell = 1, 2$ we have $a_\ell \in M_\ell$, $\mathbf{tp}_{\mathfrak{K}^{\text{nice}}_{\lambda^+}}(a_\ell, M_0, M_\ell) \in \mathscr{S}^{\text{cs}}_{\lambda^+}(M_0)$ as witnessed by $p_\ell \in \mathscr{S}^{\text{bs}}_{\mathfrak{s}}(N^*_\ell), N^*_\ell \in K_\lambda, N^*_\ell \leq_{\mathfrak{K}} M_0$ and $\mathbf{tp}_{\mathfrak{K}^{\otimes}_{\lambda^+}}(a_2, M_1, M_2)$ does not fork (in the sense of $\underset{\lambda^+}{\bigcup}$) over M_0 (note that M_0, M_1, M_2 here stand for M_0, M_1, M'_3 in clause (i) of Ax(E)(f) from Definition 2.1). As we know by monotonicity without loss of generality $M_1 <^+_{\lambda^+} M_2$. We can finish by 7.6(4) (and Axiom (E)(e) for $\mathfrak{s}$).

In more details, we can find N_0, N_1, N_2 such that: $N_\ell \leq_{\mathfrak{K}} M_\ell$ and $N_\ell \in K_\lambda$ for $\ell = 0, 1, 2$ and $N^*_1 \cup N^*_2 \subseteq N_0 \leq_{\mathfrak{K}} N_1 \leq_{\mathfrak{K}} N_2$ and $a_1 \in N_1, a_2 \in N_2$ and N_2 is $(\lambda, *)$-brimmed over N_1 hence over N_0, and $(\forall N \in K_\lambda)[N_0 \leq_{\mathfrak{K}} N \leq_{\mathfrak{K}} M_0 \rightarrow (\exists M \in K_\lambda)(M \leq_{\mathfrak{K}} M_2 \;\&\; \mathrm{NF}_\lambda(N_0, N, N_2, M))]$.
By Axiom (E)(f) for $\mathfrak{s} = (\mathfrak{K}, \mathscr{S}^{\text{bs}}, \underset{\lambda}{\bigcup})$ we can find N' such that $N_0 \leq_{\mathfrak{K}} N' \leq_{\mathfrak{K}} N_2$ such that $a_2 \in N'$ and $\mathbf{tp}_{\mathfrak{s}}(a_1, N', N_2)$ does not fork over N_0. Now we can find f'_0, M'_1 such that $M_0 \leq^+_{\lambda^+} M'_1, f'_0$ is a $\leq_{\mathfrak{K}}$-embedding of N' into M'_1 and $(\forall N \in K_\lambda)[N_0 \leq_{\mathfrak{K}} N \leq_{\mathfrak{K}} M_0 \rightarrow (\exists M \in K_\lambda)(M \leq_{\mathfrak{K}} M'_1 \;\&\; \mathrm{NF}_\lambda(N_0, N, f'_0(N'), M))]$. Next we can find f''_0, M'_2 such that $M'_1 <^+_{\lambda^+} M'_2, f''_0 \supseteq f'_0$ and f''_0 is a $\leq_{\mathfrak{K}}$-embedding of N_2 into M'_2 and $(\forall N \in K_\lambda)[N_0 \leq_{\mathfrak{K}} N \leq_{\mathfrak{K}} M_0 \rightarrow (\exists M \in K_\lambda)(M \leq_{\mathfrak{K}} M'_2 \;\&\; \mathrm{NF}_\lambda(N_0, N, f''_0(N_2), M))]$.
Lastly, by 7.6(4) there is an isomorphism f from M_2 onto M'_2 over M_0 extending f''_0. Now $f^{-1}(M'_1)$ is a model as required.

Axiom (E)(g); extension existence:
Assume $M_0 \leq^*_{\lambda^+} M_1$ are from $K^{\text{nice}}_{\lambda^+}, p \in \mathscr{S}^{\text{cs}}_{\lambda^+}(M_0)$, hence there is $N_0 \leq_{\mathfrak{K}} M_0, N_0 \in K_\lambda$ such that $(\forall N \in K_\lambda)(N_0 \leq_{\mathfrak{K}} N <_{\mathfrak{K}} M_0 \rightarrow$

$p \restriction N$ does not fork over N_0). By 7.4(1A) there are $M_2 \in K^{\otimes}_{\lambda^+}$ and $a \in M_2$ such that $M_1 \leq^*_{\lambda^+} M_2$ and $\mathbf{tp}_{\mathfrak{K}^{\text{nice}}_{\lambda^+}}(a, M_1, M_2) \in \mathscr{S}^{\text{cs}}_{\lambda^+}(M_1)$ is witnessed by $p \restriction N_0$ and by part (6) we have $\mathbf{tp}_{\mathfrak{K}^{\text{nice}}_{\lambda^+}}(a, M_0, M_2) = p$. Checking the definition of does not fork, i.e., $\bigcup\limits_{\lambda^+}$ we are done.

Axiom (E)(h), (continuity): By 2.11(6).

Axiom (E)(i):
It follows from the rest by 2.16.

6) So assume $M \leq^*_{\lambda^+} M_\ell, a_\ell \in M_\ell \backslash M$ for $\ell = 1, 2$ and $N \leq_{\mathfrak{K}} M \wedge N \in K_\lambda \Rightarrow \mathbf{tp}_{\mathfrak{K}}(a_1, N, M_1) = \mathbf{tp}_{\mathfrak{K}}(a_2, N, M_2)$. By 7.4(1) there are $M_1^+, M_2^+ \in K^{\text{nice}}_{\lambda^+}$ such that $M_\ell <^+_{\lambda^+} M_\ell^+$ for $\ell = 1, 2$. By 7.6(2),(3) there is an isomorphism f from M_1^+ onto M_2^+ over M which maps a_1 to a_2. This clearly suffices. $\square_{8.7}$

§9 Final conclusions

We now show that we have actually solved our specific test questions about categoricity and few models. First we deal with good λ-frames.

9.1 Main Lemma. *1) Assume*

(a) (α) $2^\lambda < 2^{\lambda^+} < 2^{\lambda^{++}} < \ldots < 2^{\lambda^{+n}}$, *and* $n \geq 2$
(β) *and* $\text{WDmId}(\lambda^{+\ell})$ *is not* $\lambda^{+\ell+1}$*-saturated (normal ideal on* $\lambda^{+\ell}$*) for* $\ell = 1, \ldots, n-1$

(b) $\mathfrak{s} = (\mathfrak{K}, \mathscr{S}^{\text{bs}}, \bigcup)$ *is a good* λ*-frame*

(c) $\dot{I}(\lambda^{+\ell}, \mathfrak{K}(\lambda^+\text{-saturated})) < \mu_{\text{unif}}(\lambda^{+\ell}, 2^{\lambda^{\ell-1}})$ *for* $\ell = 2, \ldots, n$.

Then

(α) *K has a member of cardinality* λ^{+n+1}

(β) *for* $\ell < n$ *there is a good* $\lambda^{+\ell}$*-frame* $\mathfrak{s}_\ell = (\mathfrak{K}^\ell, \mathscr{S}^{\text{bs}}_{\mathfrak{s}_\ell}, \bigcup\limits_{\mathfrak{s}_\ell})$ *such that* $K^\ell_{\lambda^{+\ell}} \subseteq K_{\lambda^{+\ell}}$ *and* $\leq_{\mathfrak{K}^\ell} \subseteq \leq_{\mathfrak{K}}$

(γ) $\mathfrak{s}_0 = \mathfrak{s}$ *and if* $\ell < m < n$ *then* $K^\ell_{\lambda^{+m}} \supseteq K^m_{\lambda^{+m}}$ & $\leq_{\mathfrak{K}^\ell} \restriction K^m \supseteq \leq_{\mathfrak{K}^m}$.

2) Like part (1) omitting (β) of clause (a).

Proof. 1) We prove this by induction on n.

For $n = m+1 \geq 2$, by the induction hypothesis for $\ell = 0, \ldots, m-1$, there is a frame $\mathfrak{s}_\ell = (\mathfrak{K}^\ell, \bigcup_{\mathfrak{s}_\ell}, \mathscr{S}^{\text{bs}}_{\mathfrak{s}_\ell})$ which is $\lambda^{+\ell}$-good and $K_{\mathfrak{s}_\ell} \subseteq K^{\mathfrak{s}}_{\lambda^{+\ell}}$ and $\leq_{\mathfrak{K}^\ell} \subseteq \leq_{\mathfrak{K}} \restriction \mathfrak{K}^\ell$. By 5.9 and clause (c) of the assumption we know that $\mathfrak{s}$ has density for $K^{3,\text{uq}}_{\mathfrak{s}}$. Now without loss of generality K^{m-1} is categorical in $\lambda^{+(m-1)}$ (by 2.20 really necessary only for $\ell = 0$) and by Observation 5.8 we get the assumption 6.8 of §6 hence the results of §6, §7, §8 apply. Now apply 8.7 to $(\mathfrak{K}^{m-1}, \mathscr{S}^{\text{bs}}_{\mathfrak{s}_{m-1}}, \bigcup_{\mathfrak{s}_{m-1}})$ and get a λ^{+m}-frame $\mathfrak{s}_m$ as required in clause (β). By 4.13 we have $K^m_{\lambda^{+m+1}} \neq \emptyset$ which is clause (α) in the conclusion. Clause (β) has already been proved and clause (γ) should be clear.
2) Similarly but we use 5.11 instead of 5.9, i.e. we use the full version. $\square_{9.1}$

Second (this fulfills the aim of [Sh 576] equivalently Chapter VI).

9.2 Theorem. *1) Assume $2^{\lambda^{+\ell}} < 2^{\lambda^{+(\ell+1)}}$ for $\ell = 0, \ldots, n-1$ and the normal ideal* WDmId$(\lambda^{+\ell})$ *is not $\lambda^{+\ell+1}$-saturated for $\ell = 1, \ldots, n-1$.*

If $\mathfrak{K}$ is an abstract elementary class with LS$(\mathfrak{K}) \leq \lambda$ *which is categorical in λ, λ^+ and $1 \leq \dot{I}(\lambda^{+2}, K)$ and $\dot{I}(\lambda^{+m}, \mathfrak{K}) < \mu_{\text{unif}}(\lambda^{+m}, 2^{\lambda^{+(m-1)}})$, see I.0.5(3). For $m \in [2, n)$ (or just $\dot{I}(\lambda^{+m}, \mathfrak{K}(\lambda^+\text{-saturated}))$ $< \mu_{\text{unif}}(\lambda^{+m}, 2^{\lambda^{+(m-1)}})$), <u>then</u> $\mathfrak{K}_{\lambda^{+n}} \neq \emptyset$ (and there are $\mathfrak{s}_\ell (\ell < n)$ as in (γ) of 9.1).*
2) We can omit the assumption "not $\lambda^{+\ell+1}$-saturated".

Proof. 1) By 3.7 and 9.1(1).
2) By 3.7 and 9.1(2), i.e. using the full version of Chapter VII. $\square_{9.2}$

Next we fulfill an aim of Chapter I.

9.3 Theorem. *1) Assume $2^{\aleph_\ell} < 2^{\aleph_{(\ell+1)}}$ for $\ell = 0, \ldots, n-1$ and $n \geq 2$ and* WDmId$(\lambda^{+\ell})$ *is not $\lambda^{+\ell+1}$-saturated for $\ell = 1, \ldots, n-1$.*

If $\mathfrak{K}$ is an abstract elementary class which is PC$_{\aleph_0}$ *and $1 \leq \dot{I}(\aleph_1, \mathfrak{K}) < 2^{\aleph_1}$ and $\dot{I}(\aleph_\ell, \mathfrak{K}) < \mu_{\text{unif}}(\aleph_\ell, 2^{\aleph_{\ell-1}})$, for $\ell = 2, \ldots, n$, <u>then</u> $\mathfrak{K}$ has a*

model of cardinality $\aleph_{n+1}$ *(and there are* $\mathfrak{s}_\ell(\ell < n)$ *as in 9.2.*
2) We can omit the assumption "not $\lambda^{+\ell+1}$*-saturated".*

Remark. Compared with Theorem 9.2 our gains are no assumption on $\dot{I}(\lambda, K)$ and weaker assumption on $\dot{I}(\lambda^+, K)$, i.e., $< 2^{\aleph_1}$ (and ≥ 1) rather than $= 1$. The price is $\lambda = \aleph_0^+$ and being $\mathrm{PC}_{\aleph_0}$.

Proof. 1) By 3.4 and 9.1(1).
2) By 3.4 and 9.1(2), i.e. using the full version of Chapter VII. $\square_{9.3}$

Lastly, we fulfill an aim of [Sh 48].

9.4 Theorem. *1) Assume* $2^{\aleph_\ell} < 2^{\aleph_{\ell+1}}$ *for* $\ell \leq n-1$ *and* $\mathrm{WDmId}(\lambda^{+\ell})$ *is not* $\lambda^{+\ell+1}$*-saturated for* $\ell = 1, \ldots, n-1$, $\psi \in \mathbb{L}_{\omega_1,\omega}(\mathbf{Q})$, $\dot{I}(\aleph_1, \psi) \geq 1$ *and* $\dot{I}(\aleph_\ell, \psi) < \mu_{\mathrm{unif}}(\aleph_\ell, 2^{\aleph_{\ell-1}})$*, for* $\ell = 1, \ldots, n$. *Then* ψ *has a model in* $\aleph_{n+1}$ *and there are* $\mathfrak{s}_1, \ldots, \mathfrak{s}_{n-1}$ *as in 9.3 with* $K_{\mathfrak{s}_\ell} \subseteq \mathrm{Mod}_\psi$ *and appropriate* $\leq_{\mathfrak{K}}$.
2) We can omit the assumption "not $\lambda^{+\ell+1}$*-saturated".*

Proof. 1) By 3.5 mainly clauses (c)-(d) and 9.1(1). Note that this time in 9.1 we use the $\dot{I}(\lambda^{+\ell}, \mathfrak{K}(\lambda^+\text{-saturated})) < \mu_{\mathrm{unif}}(\aleph_\ell, 2^{\aleph_{\ell-1}})$.
2) As in part (1) using 9.1(2). $\square_{9.4}$

TOWARD CLASSIFICATION THEORY OF GOOD λ-FRAMES AND ABSTRACT ELEMENTARY CLASSES SH705

§0 Introduction

For us the family of good λ-frames is a good family of (enriched) classes of models for which to study generalizations of superstability theory to a.e.c.. A priori our main line is to start with a good$^+$ λ-frame $\mathfrak{s}$, categorical in λ, m-successful for $m \le n$, where n is large enough and try to have parallel of superstability theory for $\mathfrak{K}_{\mathfrak{s}(+\ell)}$ for $\ell < n$ not too large. Characteristically from time to time we have to increase n relative to ℓ to get our desirable properties; considering our intentions a priori we do not critically mind the exact n, so you can think of an ω-successful $\mathfrak{s}$. Usually each claim or definition is for a fixed $\mathfrak{s}$, assumed to be successful enough. So using assumptions on λ^{+2} rather than λ^{+3} is not so crucial now.

But a postriori we are interested in the model theory of such classes $\mathfrak{K}_{\mathfrak{s}}$ per-se so use small n, however eventually we mainly were interested in finishing so delay sorting out what is needed to [Sh:F735]. The original aim which we see as a test for this theory, is that in the ω-successful case we can understand also models in higher cardinals, e.g., prove that $\mathfrak{K}^{\mathfrak{s}}_\mu \neq \emptyset$ for every $\mu \ge \lambda$. Recall there are reasonable λ-frames which are not n-successful but still we can say a lot on models in $\mathfrak{K}_{\mathfrak{s}(+\ell)}$ for $\ell < n$ so we have worked to reduce the assumption.

Moving from λ to λ^+ we would have preferred not to restrict ourselves to saturated models but at present we do not know how to eliminate this. However, in the ω-successful case we can prove that $\mathfrak{s}^{+n}$ is n-beautiful (see §12) and using this we shall be able to understand essentially the class of $\lambda^{+\omega}$-saturated models in $\mathfrak{K}^{\mathfrak{s}}$, in all cardinals, i.e., more exactly the class $\mathfrak{K}^{\mathfrak{s}(+\omega)}$. Recall that $K_{\mathfrak{s}}^{+(n+1)}$ is the class of models in $\mathfrak{K}^{\mathfrak{s}(+n)}$ which are saturated for $\mathfrak{K}^{\mathfrak{s}(+n)}$, but we do not know if $K_{\mathfrak{s}(+(n+2))}$ is the class of models from $K^{\mathfrak{s}(+n)}$ of cardinality λ^{+n+2} which are saturated for $\mathfrak{K}^{\mathfrak{s}(+n)}$ as we do not know that $\mathfrak{K}^{\mathfrak{s}(+n)}$ has amalgamation in $\lambda^{+(n+1)}$. (Actually here we draw conclusions on the existence of models in every $\mu \geq \lambda$ and on the categoricity spectrum and the full consequences are delayed to subsequent work). This fits well the thesis that it is reasonable to first analyze the quite saturated case which guides [Sh:c].

Why are we interested in $\mathfrak{K}_{\mathfrak{s}}$ a $\lambda_{\mathfrak{s}}$-a.e.c. rather than $K^{\mathfrak{s}}$, an a.e.c.? (see 0.2(1)). We can "blow a $\lambda_{\mathfrak{s}}$-.a.e.c., e.g. $\mathfrak{K}_{\mathfrak{s}}$, up to all cardinals $\geq \lambda$" by II.1.23, what we get is an a.e.c. but for being good frames this is not necessarily preserved.

Note that for our main purpose it is reasonable to assume always that $\mathfrak{s}$ is a successful good λ-frame, to assume that it is good$^+$ from 1.9 on and to assume that $\mathfrak{s}$ has primes after 4.9. Also we can assume all the time that $\mathfrak{s}$ is type-full (that is $\mathscr{S}^{\rm bs}_{\mathfrak{s}}(M) = \mathscr{S}^{\rm na}_{\mathfrak{s}}(M)$ for $M \in K_{\mathfrak{s}}$), see II.6.36 and 9.6, note that the assumptions are "weakly successful". Also we may assume categoricity in $\lambda_{\mathfrak{s}}$, hence $\underset{\rm wk}{\perp} = \perp$, etc., after 6.10(5) or 6.11. On the other hand on weakening the assumptions see [Sh:F735].

Concerning the framework note that the uni-dimensional (or just non-multi-dimensional) case is easier. In the characteristic uni-dimensional case, each $p \in \mathscr{S}^{\rm bs}_{\mathfrak{s}}(M)$ is (regular and morever) minimal and any $p, q \in \mathscr{S}^{\rm bs}_{\mathfrak{s}}(M)$ are not orthogonal. In the characteristic non-multi-dimensional case for any $M \in K_\lambda, \mathscr{S}^{\rm bs}_{\mathfrak{s}}(M)$ contains up to non-orthogonality every $p \in \mathscr{S}^{\rm bs}_{\mathfrak{s}}(N), M \leq_{\mathfrak{s}} N \in K_\lambda$.
Generally the uni-dimensional case is easiest and is enough to continue [Sh 576] = Chapter VI, and to deal with categoricity.

A drawback in II§5 is that we need to assume that the normal ideal WDmId(λ^+) is not λ^{++}-saturated. This will be essentially eliminated in Chapter VII; it is easier to do it when we have the theory developed here. Let me stress again most work here is in one cardinal, $\lambda_{\mathfrak{s}}$.

We sometimes give first a proof from stronger assumptions, which as explained above suffice for our purposes.

We thank John Baldwin, Adi Yarden for helpful remarks and Alex Usvyatsov for doing much to improve this work.
0.1 Notation: Let $\mathfrak{s}$ denote a good λ-frame and rarely just a pre-λ-frame, but we may omit λ, that is

0.2 Definition. 1) We say $\mathfrak{s}$ is a pre-λ-frame if $\mathfrak{s} = (\mathfrak{K}_{\mathfrak{s}}, \mathscr{S}^{\rm bs}_{\mathfrak{s}}, \bigcup\limits_{\mathfrak{s}})$ with $\mathfrak{K}_{\mathfrak{s}} = \mathfrak{K}(\mathfrak{s})$ a $\lambda_{\mathfrak{s}}$-a.e.c., $\mathscr{S}^{\rm bs}[\mathfrak{s}] = \mathscr{S}^{\rm bs}_{\mathfrak{s}}, \bigcup[\mathfrak{s}] = \bigcup\limits_{\mathfrak{s}}, \le_{\mathfrak{s}} = \le_{\mathfrak{K}_{\mathfrak{s}}}$ and they satisfy axioms (A), (D)(a),(b), (E)(a)(b) from II.2.1.
1A) We say $\mathfrak{s}$ is a weak frame if it satisfies axioms (A), (B), (C), (D)(a),(b), (E)(a),(b) from II§2 and $\mathfrak{s}$ is a frame if it satisfies also (D)(c),(E)(d),(e),(f),(g),(i). Recall that $\mathfrak{s}$ is a good frame if it satisfies all the axioms there.
2) For a pre-λ-frame $\mathfrak{s}$ let $\mathfrak{K}^{\mathfrak{s}} = \mathfrak{K}[\mathfrak{s}]$ be the a.e.c. derived from $\mathfrak{K}_{\mathfrak{s}}$ and $\mathfrak{K}^{\mathfrak{s}}_{\mu} = (\mathfrak{K}[\mathfrak{s}])_{\mu}$ so $\mathfrak{K}_{\mathfrak{s}} = \mathfrak{K}^{\mathfrak{s}}_{\lambda(\mathfrak{s})}$ and let $\mathfrak{K}(\mathfrak{s}) = \mathfrak{K}_{\mathfrak{s}}$. Recall that if $\mathfrak{K}$ is a λ-a.e.c., then the a.e.c.-derived from it, $\mathfrak{K}^{\rm up}$ is the unique a.e.c. $\mathfrak{K}'$ with $\tau(\mathfrak{K}') = \tau(\mathfrak{K}), {\rm LS}(\mathfrak{K}') = \lambda, \mathfrak{K}'_{\lambda} = \mathfrak{K}_{\mathfrak{s}}$, see II.1.23.
3) For a frame $\mathfrak{s}$ let $\le^{\mathfrak{s}} = \le_{\mathfrak{K}^{\mathfrak{s}}}$ be $\le_{\mathfrak{K}[\mathfrak{s}]}$ (and $\le_{\mathfrak{s}} = \le_{\mathfrak{K}_{\mathfrak{s}}} = \le_{\mathfrak{K}(\mathfrak{s})} = \le_{\mathfrak{K}[\mathfrak{s}]} \restriction K_{\mathfrak{s}}$).
4) Let $<^+_{\mathfrak{s}}$ be the following two place relation on $K_{\mathfrak{s}} : M <^+_{\mathfrak{s}} N$ iff $M \le_{\mathfrak{s}} N$ and N is $\mathfrak{K}_{\mathfrak{s}}$-universal over M.

0.3 Convention: For notational simplicity we (sometimes) assume $\mathfrak{K}$ is such that if $\bar{a} \in {}^{\omega>}M, M \in K$ then $\bar{a}$ can be considered an element of M. This can be trivially justified.

0.4 Definition. Let $\mathfrak{s}$ be a good frame and $\mu \ge \lambda_{\mathfrak{s}}$.
1) Let $\mathfrak{s}\langle\mu\rangle = (\mathfrak{K}^{\mathfrak{s}}_{\mu}, \mathscr{S}^{\rm bs}_{\mathfrak{s},\mu}, \bigcup\limits_{\mathfrak{s},\mu})$ with $\mathscr{S}^{\rm bs}_{\mathfrak{s},\mu}, \bigcup\limits_{\mathfrak{s},\mu}$ as defined in II§2; also $\mathscr{S}^{\rm bs}_{\mathfrak{s},<\mu}, \mathscr{S}^{\rm bs}_{\mathfrak{s},<\infty}, \bigcup\limits_{\mathfrak{s},<\mu}$ and $\bigcup\limits_{\mathfrak{s},<\infty}$ are from there.
2) Let $\mathfrak{s}[\mu] := (\mathfrak{K}_{\mathfrak{s}[\mu]}, \mathscr{S}^{\rm bs}_{\mathfrak{s},\mu}, \bigcup\limits_{\mathfrak{s}(\mu)})$ where $K_{\mathfrak{s}[\mu]} := \{M \in K^{\mathfrak{s}}_{\mu} : M$ is superlimit in $\mathfrak{K}^{\mathfrak{s}}_{\mu}\}, \le_{K_{\mathfrak{s}[\mu]}} = \le^{\mathfrak{s}}_{\mathfrak{K}} \restriction K_{\mathfrak{s}[\mu]}$, and of course $\mathscr{S}^{\rm bs}_{\mathfrak{s}[\mu]} = \mathscr{S}^{\rm bs}_{\mathfrak{s},\mu} \restriction K_{\mathfrak{s}[\mu]}, \bigcup\limits_{\mathfrak{s},[\mu]} = \bigcup\limits_{\mathfrak{s}} \restriction K^{\mathfrak{s},s}_{\mu}$; of course on the one hand $K_{\mathfrak{s}[\mu]}$ may be empty and on the other hand maybe $K_{\mathfrak{s}[\mu]} \ne \emptyset$ but still $\mathfrak{s}[\mu]$ is not a good frame.

3) For $M \leq_{\mathfrak{s}} N$ let $\mathbf{I}_{M,N} = \{a \in N : \mathbf{tp}_{\mathfrak{s}}(a, M, N) \in \mathscr{S}^{\rm bs}_{\mathfrak{s}}(M)\}$.
4) If $\mathfrak{s}$ is ω-successful let $\mathfrak{s}^{+\omega} = \mathfrak{s}(+\omega)$ be $\mathfrak{s}(\lambda^{+\omega}_{\mathfrak{s}})$; used only in §12.

Remark. Note that Definition 0.2(1), 0.4(1),(2) are, in this Chapter , really peripheral.

0.5 Remark. On $\mathfrak{s}^{+n} = \mathfrak{s}(+n)$ and, in particular, $\mathfrak{s}^+ = \mathfrak{s}(+)$, see 1.7.

0.6 Definition. For a good λ-frame $\mathfrak{s}$ and $M \in K_{\mathfrak{s}}$ let

$$\mathscr{S}_{\mathfrak{s}}(M) = \mathscr{S}^{\rm all}_{\mathfrak{s}}(M) = \{\mathbf{tp}_{\mathfrak{s}}(a, M, N) : a \in N \text{ and } M \leq_{\mathfrak{s}} N\}$$

$$\mathscr{S}^{\rm na}_{\mathfrak{s}}(M) = \{\mathbf{tp}_{\mathfrak{s}}(b, M, N) : b \in N \backslash M \text{ and } M \leq_{\mathfrak{s}} N\}.$$

§1 Good$^+$ frames

In II.8.7 there was what may look like a minor drawback: moving from λ to λ^+ the derived abstract elementary class not only have fewer models of cardinality $\geq \lambda^+$ but also the notion of being a submodel changes; this is fine there, and, it seemed, unavoidable in some circumstances. More specifically, for proving the main theorem there, it was enough to move from $\mathfrak{s}$ to a good frame $\mathfrak{t}$ satisfying $\lambda_{\mathfrak{t}} = \lambda^+_{\mathfrak{s}}, \lambda_{\mathfrak{s}} < \mu < \lambda^{+\omega} \Rightarrow \dot{I}(\mu, K^{\mathfrak{t}}) \leq \dot{I}(\mu, K^{\mathfrak{s}})$ and forget $\mathfrak{s}$. But for us now this is undesirable (as arriving to $\lambda^{+\omega}$ we have forgotten everything!) and toward this we consider a (quite mild) strengthening of good.

We prove that we do not lose much: the examples of good λ-frames from II§3 are all good$^+$ and when $\mathfrak{s}^+$ is well defined and $\mathfrak{s}$ is good$^+$ then it is good$^+$, see 1.5, 1.6(2); moreover, by 1.9, even if $\mathfrak{s}$ is just good and successful then $\mathfrak{s}^+$ is good$^+$.

We say some things on [weakly] n-successful, this will become important in §12. Also we present basic properties of $\mathrm{NF}_{\mathfrak{s}}$, which is well defined when $\mathfrak{s}$ is weakly successful and are widely used.

Recall

1.1 Definition. 1) Let $\mathfrak{s}$ be a good λ-frame. We say $\mathfrak{s}$ is successful <u>if</u> the conclusions of Chapter II under "no non-structure assumptions in λ^{++}" hold, that is:

$(*)(a)$ it has existence for $K^{3,\text{uq}}_\lambda$; i.e., for every $M \in K_\lambda$ and $p \in \mathscr{S}^{\text{bs}}_{\mathfrak{s}}(M)$ there is $(M,N,a) \in K^{3,\text{uq}}_\lambda$ such that $\mathbf{tp}_{\mathfrak{s}}(a,M,N) = p$ (see II§5, follows from density of $K^{3,\text{uq}}_{\mathfrak{s}}$ if $K_{\mathfrak{s}}$ is categorical)

(b) if $\langle N_i : i \le \delta\rangle$ is $\le^*_{\lambda^+}[\mathfrak{s}]$-increasing continuous in $K^{\text{nice}}_{\lambda^+}[\mathfrak{s}]$ and $i < \delta \Rightarrow N_i \le^*_{\lambda^+} N \in K^{\text{nice}}_{\lambda^+}$ <u>then</u> $N_\delta \le^*_{\lambda^+} N$ (see II.8.7(1), recall $K^{\text{nice}}_{\lambda^+}[\mathfrak{s}]$ consists of the saturated $M \in K^{\mathfrak{s}}_{\lambda^+}$).

2) We say (the good λ-frame) $\mathfrak{s}$ is weakly successful <u>if</u> clause (a) of $(*)$ above holds.

Usually at least "$\mathfrak{s}$ is weakly successful" is used, but sometimes less suffices (this is helpful though not crucial).

1.2 Remark. For successful $\mathfrak{s}$ we define a successor, $\mathfrak{s}^+ = \mathfrak{s}(+)$, a good λ^+-frame (see 1.7 below), but not with the most desirable $\le_{\mathfrak{K}_{\mathfrak{s}(+)}}$, for rectifying this we consider below good$^+$ frames. Together with locality of types for models in $\mathfrak{K}^{\mathfrak{s}(+)}_{\lambda^+_{\mathfrak{s}}}$, see 1.10 or II.7.6(3) this seems to be in the right direction. Less central, still worthwhile, is that $\mathfrak{s}^+$ has a strong property we call saturative such that: for a good λ^+-frame $\mathfrak{t}$, being saturative can be used in several cases as an alternative assumption to "$\mathfrak{t}$ has the form $\mathfrak{s}^+$ with $\mathfrak{s}$ a successful good$^+$ frame". Usually we do not adopt it and we feel it is really too restrictive.

1.3 Definition. 1) We say that $\mathfrak{s} = (\mathfrak{K}_\lambda, \mathscr{S}^{\text{bs}}, \bigcup_\lambda) = (K_\lambda, \le_{\mathfrak{K}_\lambda}, \mathscr{S}^{\text{bs}}, \bigcup_{\mathfrak{s}})$ is a good$^+$ λ-frame <u>when</u>:

(a) $\mathfrak{s}$ is a good λ-frame

(b) the following is impossible

$(*)$ $\langle M_i : i < \lambda^+\rangle$ is $\le_{\mathfrak{s}}$-increasing continuous (so each M_i is from K_λ) and $\langle N_i : i < \lambda^+\rangle$ is $\le_{\mathfrak{s}}$-increasing continuous (so each N_i is from K_λ) and $i < \lambda^+ \Rightarrow M_i \le_{\mathfrak{s}}$

$N_i, \cup\{M_i : i < \lambda^+\} \in K^{\mathfrak{s}}_{\lambda^+}$ is saturated, $p^* \in \mathscr{S}^{\rm bs}_{\mathfrak{s}}(M_0)$ and for each $i < \lambda^+$ we have:
$a_{i+1} \in M_{i+2}, \mathbf{tp}_{\mathfrak{s}}(a_{i+1}, M_{i+1}, M_{i+2})$ is a non-forking extension of p^* but $\mathbf{tp}(a_{i+1}, N_0, N_{i+2})$ is not.
We then say $\langle M_i, N_i, a_i : i < \lambda^+\rangle$ is a counterexample (well, actually a_i being defined only for successor i; we could for non-successor i let $a_i \in M_{i+2}$ and $\mathbf{tp}_{\mathfrak{s}}(a_i, M_i, M_{i+2})$ does not fork over M as above, this follows by monotonicity of non-forking or requiring $M_\delta = M_{\delta+1}$).

2) We say a good λ-frame $\mathfrak{s}$ is saturative <u>if</u>:

(*) if $M_0 \le_{\mathfrak{s}} M_1 \le_{\mathfrak{s}} M_2$ and M_1 is $(\lambda, *)$-brimmed over M_0 <u>then</u> M_2 is $(\lambda, *)$-brimmed over M_0.

1.4 Remark. 1) The "$\mathfrak{s}$ is saturative" is a relative of "non-multi-dimensional". But be careful, see clause (iii) of 1.5(3) below, so in the first order case, we may really look at the saturated models in λ of a superstable class.
2) Well, do we lose much by adopting the good$^+$ version? First, are the old cases covered? Yes, by the following claim (and 1.6(2)).

1.5 Claim. *1) In II§3 essentially all the cases where we prove "good λ-frame" we actually get "good$^+$ λ-frames; more fully:*
1A) In II.3.4(2) we get good$^+$ $\aleph_0$-frame <u>when</u> $\mathfrak{K}$ has the symmetry property (this is defined in I.5.31(1), actually deal with countable models and is proved in I.5.34(1) when we add $2^{\aleph_1} < 2^{\aleph_2}$ and e.g. $\dot{I}(\aleph_2, \mathfrak{K}) < 2^{\aleph_2}$ and, presently "$\mathscr{D}_{\aleph_1}$ is not $\aleph_2$-saturated", see more in VII.4.40).
1B) Similarly for II.3.5 because I.5.34(1) speaks on many non-isomorphic models in $K_{\aleph_2}$ which are $\aleph_1$-saturated.
1C) Similarly for II.3.7 which rely on [Sh 576].
2) In fact the frames from II.3.7, II.3.4 are also saturative.
3) If T is a complete superstable first order theory stable in λ and $\kappa \le \lambda$ (so $\kappa \ge \aleph_0$) or $\kappa = \aleph_\varepsilon$ (in an abuse of notation stipulating $0 < \varepsilon < 1, \aleph_0 < \aleph_\varepsilon < \aleph_1$) or $\kappa = 0, \lambda \ge |T|$ and $[\kappa > 0 \Rightarrow$

T stable in λ] and $\mathfrak{s} = \mathfrak{s}^{\kappa}_{T,\lambda}$ (so $K_{\mathfrak{s}} = \{M : M \models T, \|M\| = \lambda$ and M is κ-saturated$\}, \prec\restriction K_{\mathfrak{s}}$), that is the $\mathfrak{s}$ which is defined in II.3.1(5), then

(i) *$\mathfrak{s}$ is a good$^+$ λ-frame*

(ii) *assume $\aleph_\varepsilon \le \kappa < \lambda$, then: $\mathfrak{s}$ is saturative iff T is non-multi-dimensional (see 2.2(5))*

(iii) *if $\mathfrak{s}' = \mathfrak{s}^{\kappa}_{T,\lambda}[M]$; see Definition II.2.20 of $\mathfrak{s}[M]$; where $M \in \mathfrak{K}_\lambda$ is the superlimit model (i.e., the saturated one), then $\mathfrak{s}'$ is saturative*

(iv) *if $\kappa = \lambda, \mathfrak{s}$ is saturative.*

Proof. 1), 2).

Case 1: Concerning Claim II.3.7.

So $2^\lambda < 2^{\lambda^+} < 2^{\lambda^{++}}$, $\mathfrak{K}$ is an abstract elementary class categorical in λ, λ^+ and $1 \le \dot{I}(\lambda^{++}, K) < 2^{\lambda^{++}}$, with $\mathrm{LS}(\mathfrak{K}) \le \lambda$, $\mathrm{WDmId}(\lambda^+)$ is not λ^{++}-saturated (or $\dot{I}(\lambda^{++}, K) < \mu_{\mathrm{unif}}(\lambda^{++}, 2^{\lambda^+})$, see II.3.7 or just a model theoretic consequence). Recall that we have defined $\mathfrak{s}$ as follows: we let $\lambda_{\mathfrak{s}} = \lambda^+, \mathfrak{K}_{\mathfrak{s}} = \mathfrak{K}_{\lambda^+}$ and for $M \in K_{\mathfrak{s}}$ we let $\mathscr{S}^{\mathrm{bs}}_{\mathfrak{s}}(M) = \{p \in \mathscr{S}_{\mathfrak{K}}(M) : p$ is not algebraic and for some $M_0 \le_{\mathfrak{K}} M, M_0 \in K_\lambda$ and $p \restriction M_0$ is minimal$\}$ and $\bigcup_{\mathfrak{s}}(M_0, M_1, a, M_3)$ iff $M_0 \le_{\mathfrak{K}} M_1 \le_{\mathfrak{K}} M_3$ are in $K_{\mathfrak{s}}$ and $a \in M_3 \backslash M_1$ and for some $M_0' \le_{\mathfrak{K}} M_0$ from K_λ the type $\mathbf{tp}_{\mathfrak{K}}(a, M_0', M_3)$ is minimal. So we know that $\mathfrak{s}$ is a λ^+-good frame, etc.

To prove "$\mathfrak{s}$ is good$^+$" assume toward contradiction that $\langle (M_i, N_i, a_i) : i < \lambda^+_{\mathfrak{s}} \rangle$ is as in $(*)$ of clause (b) of Definition 1.3. So for successor $i < \lambda^+_{\mathfrak{s}}, \mathbf{tp}_{\mathfrak{s}}(a_i, M_0, N_{i+1}) \in \mathscr{S}^{\mathrm{bs}}_{\mathfrak{s}}(M_0)$ hence $a_i \in N_{i+1} \backslash N_i$ and for some $M_0' \le_{\mathfrak{K}} M_0, M_0' \in K_\lambda$ and $M_0' \le_{\mathfrak{K}_\lambda} M_0'' \le_{\mathfrak{K}} M_i \Rightarrow \mathbf{tp}_{\mathfrak{K}}(a_i, M_0'', N_i)$ is minimal while $\mathbf{tp}_{\mathfrak{s}}(a_i, N_0, N_{i+1})$ is not its non-forking extension, hence necessarily $i < \lambda^+_{\mathfrak{s}} \Rightarrow a_{i+1} \in N_0 \backslash M_0$. But recall that $\langle a_{i+1} : i < \lambda^+_{\mathfrak{s}} \rangle$ is a sequence with no repetitions of members from $N_0 \backslash M_0$ while by the last sentence $\{a_{i+1} : i < \lambda^+_{\mathfrak{s}}\} \subseteq N_0$ and $N_0 \in K_{\lambda_{\mathfrak{s}}}$ so $\|N_0\| = \lambda_{\mathfrak{s}} < \lambda^+_{\mathfrak{s}}$, contradiction. Also saturativity should be clear.

Case 2: Claim II.3.4; actually from Chapter I.

Let $\mathfrak{s}$ be defined as there so $\lambda_\mathfrak{s} = \aleph_0$. Toward contradiction let $\langle (M_i, N_i, a_i) : i < \omega_1 \rangle$ be as in clause (b) of 1.3. Recall that $\mathbf{tp}_\mathfrak{s}(a_i, M_i, M_{i+1})$ does not fork over M_0 (i a successor ordinal) hence there is a finite $A_0 \subseteq M_0$ such that $\text{gtp}(a_{i+1}, M_{i+1}, M_{i+2})$ is definable over A_0 (see I.5.19), but $\text{gtp}(a_{i+1}, N_0, N_{i+2})$ does not have the same definition hence it splits over A_0 hence $\alpha \leq i \Rightarrow \text{gtp}(a_{i+1}, N_\alpha, N_{i+2})$ is not the non-forking extension of $\text{gtp}(a_{i+1}, M_\alpha, N_{i+1})$. By I.5.24(5) for some club of E of ω_1 we have

$\boxtimes$ $\delta \in E \;\;\&\;\; \delta < \alpha < \omega_1 \;\;\&\;\; \bar{a} \in N_\delta \Rightarrow \text{gtp}(\bar{a}, M_\alpha, N_\alpha)$ is definable over some finite $B_{\bar{a}} \subseteq M_\delta$.

We get a contradiction to "$\mathfrak{K}$ has the symmetry property", which is defined in I.5.31(1), and is proved in I.5.34(1). (Note that this is not equivalent to the symmetry axiom Ax(E)(f) of good λ-frame proved inside the proof of II.3.4). Also saturatively should be clear.

Case 3: Claim II.3.5; actually fro m [Sh 48].

Similar to case 2 (note that saturativity is unreasonable here).

3) Naturally this proof assumes knowledge of first order (superstable) classes and use types as in [Sh:c] and as in II.3.1, we can replace $\mathbf{tp}(a, M, N)$ by $\text{tp}(a, M, N)$. We leave to the reader the proof. By II.3.5 we know clause (a) of 1.3(1), that is "$\mathfrak{s}$ is a good λ-frame" and clause (iii) of 1.5(3), in fact it is like clause (iv). But we prove clause (b) of 1.3(1), i.e. we prove $\mathfrak{s}$ is a good$^+$ frame; so assume toward contradiction that $\langle M_i : i < \lambda^+ \rangle, \langle N_i : i < \lambda^+ \rangle, p^*$ and a_{i+1} for $i < \lambda^+$ are as in $(*)$ of 1.3(2) clause (b). Let $M = \cup\{M_i : i < \lambda^+\}$ and $N = \cup\{N_i : i < \lambda^+\}$. Now for every finite sequence $\bar{c}$ from N_0, there is $i_c < \lambda^+$ such that $\text{tp}_\mathfrak{s}(\bar{c}, M, N)$ does not fork over $M_{i_{\bar{c}}}$ (in the first order sense!), and let $i^* = \sup\{i_{\bar{c}} : \bar{c} \in {}^{\omega>}(N_0)\}$ so $i^* < \lambda^+$ and easily $i \in [i^*, \lambda^+) \;\;\&\;\; \bar{c} \in {}^{\omega>}(N_0) \Rightarrow \text{tp}(\bar{c}, M_{i+2}, N)$ does not fork over M_{i+1}, hence by symmetry and finite character ([Sh:c, III,§0]) we have $\text{tp}(a_{i+1}, M_{i+1} \cup N_0, N)$ does not fork over M_{i+1} hence (transitivity) over M_0, contradiction. So clause (i) holds.

As for saturativeness, we have two cases. One is clause (iv), for it notes that for $M, N \in K_\mathfrak{s}$, "N is $(\lambda, *)$-brimmed over M iff $(N, c)_{c \in M}$ is a saturated model". So we have to show that the model $(M_2, c)_{c \in M_0}$ is saturated, for this it is enough to show that for every $A \subseteq M_2, |A| < \aleph_0$ and regular $p \in \mathbf{S}^1(A \cup M_0)$, we have $\dim(p, M_2) = \lambda$. Why this holds? If $p \pm M_0$ then we can find a regular $q \in \mathbf{S}(M_0), q \pm p$ and as M_1 is $(\lambda, *)$-brimmed over M_0, $\dim(q, M_2) \geq \dim(q, M_1) = \lambda$ and easily $\dim(p, M_2) = \dim(q, M_2)$. If $p \perp M_0$ then see [Sh 225a].

The other case concerning saturativeness is clause (ii). The proof of clause (ii) is easy too; if T is multi-dimensional then by [Sh 429] in a model $M_0 \in K_{\mathfrak{s}}$ we can find $\langle \mathbf{I}_\alpha : \alpha < \lambda \rangle, \mathbf{I}_\alpha$ an infinite indiscernible set, $\alpha \neq \beta \Rightarrow \mathbf{I}_\alpha \perp \mathbf{I}_\beta, \mathbf{I}_\alpha \supseteq \{\bar{a}_{\alpha,n} : n < \omega\}$ and $\langle \bar{a}_{\alpha,0} \hat{\ } \bar{a}_{\alpha,1} \hat{\ } \ldots : \alpha < \lambda \rangle$ is an indiscernible sequence.

Now we can find $M_1 \in K_{\mathfrak{s}}$ brimmed over M_0 and $M_2 \in K_{\mathfrak{s}}$ which $\leq_{\mathfrak{s}}$-extends M_1 and $\langle \bar{a}_{\lambda,n} : n < \omega \rangle$ such that:

$(*)_1$ $\langle \bar{a}_{\alpha,} \hat{\ } \bar{a}_{\alpha,1} \ldots : \alpha \leq \lambda \rangle$ is an indiscernible sequence

$(*)_2$ $\mathbf{I}_\lambda = \{\bar{a}_{\alpha,n} : n < \omega\}$ is orthogonal to M_1 and is included in M_2

$(*)_3$ $\mathrm{Av}(\mathbf{I}_\lambda, \cup \mathbf{I}_\lambda)$ is omitted by M_2.

The other direction follows, too. (The reader may wonder about the case $\kappa = 0$ when for some M, $\mathrm{Th}(M, c)_{c \in M}$ is categorical in λ^+, see [Sh:c] and properties as in [ShHM 158] and the analysis of Laskowski ([Las88]) of models of T in $\lambda = |T|$ when T is categorical in λ^+). $\square_{1.5}$

Also in the main result of Chapter II we can get good$^+$, see more below in 1.14.

1.6 Goodness Plus Claim. *1) Assume that* $\mathfrak{s} = (\mathfrak{K}_\lambda, \mathscr{S}^{\mathrm{bs}}, \bigcup_\lambda)$ *is a weakly successful good*$^+$ *λ-frame. Then:*

(a) *if* $M_1^* \leq_{\mathfrak{K}} M_2^*$ *are from* $K^{\mathrm{nice}}_{\lambda^+}$ *and* $M_1^* \not\leq^*_{\lambda^+} M_2^*$ *then* $(**)_{M_1^*, M_2^*}$ *from II.8.5 holds*

(b) *if* $\boxtimes$ *from II.8.7 (which holds if* $\dot{I}(\lambda^{++}, K) < 2^{\lambda^{++}}$*) then*

(α) $\leq^*_{\lambda^+}, \leq_{\mathfrak{K}}$ *agree on* $K^{\mathrm{nice}}_{\lambda^+}$

(β) $(K^{\mathrm{nice}}_{\lambda^+}, \leq^*_{\lambda^+}, \mathscr{S}^{\mathrm{bs}}_{\lambda^+}, \bigcup_{\lambda^+})$ *(as defined there, called* $\mathfrak{s}^+$ *below) is a good*$^+$ *λ^+-frame.*

2) If $\mathfrak{s} = (\mathfrak{K}_\lambda, \mathscr{S}^{\rm bs}, \underset{\lambda}{\bigcup})$ is a successful good$^+$ λ-frame, then $\mathfrak{s}^+$ (defined in 1.7 below) is a good$^+$ λ^+-frame.

Remark. Recall that in Chapter II we get a weak version of (α) of (b), that is $\leq^*_\lambda, \leq^\otimes_{\mathfrak{K}}$ agree on $K^{\rm nice}_{\lambda^+}$.

Before proving 1.6 we see a conclusion. Recall

1.7 Definition. 1) For a good λ-frame $\mathfrak{s} = (\mathfrak{K}, \mathscr{S}^{\rm bs}, \bigcup)$ define $\mathfrak{s}^+ = \mathfrak{s}(+)$, a λ^+-frame, as follows (so $\lambda(\mathfrak{s}^+) = \lambda^+$):

- (a) $K(\mathfrak{s}^+) = K_{\lambda^+}[\mathfrak{s}^+]$ = the class of λ^+-saturated models from $K^{\mathfrak{s}}$ of cardinality λ^+ (also called $K^{\rm nice}_{\lambda^+}[\mathfrak{s}]$)
- (b) $\leq_{\mathfrak{K}(\mathfrak{s}^+)} = \leq^*_{\lambda^+} \restriction K_{\lambda^+}[\mathfrak{s}^+]$
- (c) $\mathscr{S}^{\rm bs}[\mathfrak{s}^+] = \mathscr{S}^{\rm bs}_{\mathfrak{s}(+)} = \{\mathbf{tp}_{\mathfrak{s}(+)}(a, M_1, M_2) : M_1 \leq_{\mathfrak{s}(+)} M_2$ are from $K_{\mathfrak{s}(+)} = K_{\lambda^+}[\mathfrak{s}^+], a \in M_2 \backslash M_1$ and there is $N_1 \leq_{\mathfrak{K}[\mathfrak{s}]} M_1$ called a witness, $N_1 \in K_\lambda$, such that $N_1 \leq_{\mathfrak{K}} N \leq_{\mathfrak{K}} M_1 \ \& \ N \in K_{\mathfrak{s}} \Rightarrow \mathbf{tp}_{\mathfrak{s}}(a, N, M_2) \in \mathscr{S}^{\rm bs}_{\mathfrak{s}}(N)$ does not fork over N_1 (in $\mathfrak{s}$'s sense)$\}$; recalling $\mathbf{tp}_{\mathfrak{s}(+)}(a, M_1, M_2) = \mathbf{tp}_{\mathfrak{K}_{\mathfrak{s}(+)}}(a, M_1, M_2)$ and note that for $p = \mathbf{tp}_{\mathfrak{s}(+)}(a, M_1, M_2) \in \mathscr{S}_{\mathfrak{s}(+)}(M_1)$, $M_2 \in K_{\mathfrak{s}(+)}$ and $M \leq_{\mathfrak{K}[\mathfrak{s}]} M_1, M \in K_\lambda$ the type $p \restriction M \in \mathscr{S}^{\rm bs}_{\mathfrak{s}}(M)$ is well defined as $\mathbf{tp}_{\mathfrak{K}_{\mathfrak{s}}}(a, M, M_2)$ is well defined
- (d) $\bigcup = \{(M_0, M_1, a, M_2) : M_0 \leq_{\mathfrak{s}(+)} M_1 \leq_{\mathfrak{s}(+)} M_2$ so of cardinality λ^+, $a \in M_2 \backslash M_1$ and $\mathbf{tp}_{\mathfrak{s}(+)}(a, M_1, M_2) \in \mathscr{S}^{\rm bs}_{\mathfrak{s}(+)}(M_1)$ has a witness $N_1 \leq_{\mathfrak{K}} M_0\}$.

2) If a, M_1, M_2, N_1 are as in clause (c) then we call N_1 or $\mathbf{tp}_{\mathfrak{s}}(a, N_1, M_2)$, a witness for $\mathbf{tp}_{\mathfrak{s}(+)}(a, M, N)$; we may abuse our notation and say that $\mathbf{tp}_{\mathfrak{s}(+)}(a, M_1, M_2)$ does not fork over N_1. Similarly for stationarization (= non-forking extension).

1.8 Conclusion. Assume $\mathfrak{s}$ is a good λ-frame and is successful (see Definition 1.1). If $\mathfrak{s}$ is good$^+$, then $\leq_{\mathfrak{s}(+)} = \leq_{\mathfrak{K}[\mathfrak{s}]} \restriction K_{\mathfrak{s}(+)}$ and $\mathfrak{s}^+$ is a

good$^+$ λ^+-frame, so: if $\langle M^\ell_\alpha : \alpha < \lambda^+\rangle$ is an $\le_{\mathfrak{K}}$-representation of a saturated $M_\ell \in K^{\mathfrak{s}}_{\lambda^+}$ for $\ell = 1, 2$ and $M_1 \le_{\mathfrak{K}[\mathfrak{s}]} M_2$ then for some club E of λ^+ for every $\alpha < \beta$ from E we have $\mathrm{NF}_{\mathfrak{s}}(M^1_\alpha, M^2_\alpha, M^1_\beta, M^2_\beta)$.
2) For $M_1, M_2 \in K^{\mathrm{nice}}_{\lambda^+}[\mathfrak{s}]$ we have $M_1 \le^+_{\lambda^+} M_2$ for $\mathfrak{s}$ (see Definition II.7.2(3)) iff M_2 is $(\lambda^+, *)$-brimmed over M_1 for $\mathfrak{s}^+$.
3) $\mathfrak{K}^{\mathfrak{s}(+)}$ is the class of λ^+-saturated models from $\mathfrak{K}^{\mathfrak{s}}$ and if $\mathfrak{s}$ is good$^+$ then $\le_{\mathfrak{K}^{\mathfrak{s}(+)}} = \le_{\mathfrak{K}[\mathfrak{s}]} \restriction K^{\mathfrak{s}(+)}$.

Proof. 1) By clause (b) of 1.6(1) we know that $\le_{\mathfrak{s}(+)} = \le_{\mathfrak{K}_{\mathfrak{s}}} \restriction K_{\mathfrak{s}(+)}$. By this and II.8.7 clearly $\mathfrak{s}^+$ is a good λ^+-frame and by Definition II.7.2(2) the equality $\le_{\mathfrak{s}(+)} = \le_{\mathfrak{K}[\mathfrak{s}]} \restriction K_{\mathfrak{s}(+)}$ it follows that $\mathfrak{s}^+$ is good$^+$. The last phrase holds by Definition II.7.2(2) and the first sentence.
2) By the proof of II.7.6(2).
3) Easy. $\square_{1.8}$

We shall use this conclusion freely.

Proof of 1.6(1).

<u>Clause (a)</u>:

Assume that $(**)_{M^*_1, M^*_2}$ fails, <u>then</u> by the assumptions of clause (a), from the clauses of $(**)_{M^*_1, M^*_2}$ all except possibly clause (iv) there follows, hence clause (iv) there has to fail. So we can find $N^*_1 \le_{\mathfrak{s}} N^*_2$ from K_λ satisfying $N^*_\ell \le_{\mathfrak{K}[\mathfrak{s}]} M^*_\ell$ for $\ell = 1, 2$ and $p \in \mathscr{S}^{\mathrm{bs}}_{\mathfrak{s}}(N^*_2)$ which does not fork over N^*_1 such that no $a \in M^*_1$ realizes p in M^*_2. Let $\langle M^\ell_\alpha : \alpha < \lambda^+\rangle$ be a $\le_{\mathfrak{s}}$-representation of M^*_ℓ for $\ell = 1, 2$. Without loss of generality $N^*_2 \le_{\mathfrak{s}} M^2_0$ and $M^2_\alpha \cap M^*_1 = M^1_\alpha$ for $\alpha < \lambda^+$ and as $M^*_1 \in K^{\mathrm{nice}}_{\lambda^+}$ also $M^1_{\alpha+1}$ is $(\lambda, *)$-brimmed over M^1_α for $\alpha < \lambda^+$. For each $\alpha < \lambda^+$ the type $p \restriction N^*_1 \in \mathscr{S}^{\mathrm{bs}}_{\mathfrak{s}}(N^*_1)$ has a non-forking extension $p_\alpha \in \mathscr{S}^{\mathrm{bs}}_{\mathfrak{s}}(M^1_\alpha)$ which is equal to $p \restriction M^1_\alpha$. As $M^1_{\alpha+1}$ is $(\lambda, *)$-brimmed over M^1_α clearly for every $\alpha < \lambda^+$ there is $a_\alpha \in M^1_{\alpha+1} \backslash M^1_\alpha$ realizing p_α.

Let $M_\alpha := M^1_\alpha, N_\alpha := M^2_\alpha, p^* = p_0$; note that $N^*_1 \le_{\mathfrak{s}} M_0, N^*_2 \le_{\mathfrak{s}} N_0$, so all the demands in $(*)$ of clause (b) of Definition 1.3(1) hold, in particular $a_{i+1} \in M_{i+2} \backslash M_{i+1}, \mathbf{tp}_{\mathfrak{s}}(a_{i+1}, M_{i+1}, M_{i+2}) = p_{i+1} \in \mathscr{S}^{\mathrm{bs}}_{\mathfrak{s}}(M_{i+1})$ is a non-forking extension of $p \restriction N^*_1$ hence of $p^* := p_0$ but if $\mathbf{tp}_{\mathfrak{s}}(a_{i+1}, N_0, N_{i+1})$ does not fork over M_0 then it is also a non-forking extension of p (recall $N^*_2 \le_{\mathfrak{s}} N_0$) impossible by the choice of

p, N_1^*, N_2^*. So we have gotten a counterexample to "$\mathfrak{s}$ is good$^+$", i.e., clause (b) of Definition 1.3. In other words, as we assume that $\mathfrak{s}$ is good$^+$, some a_i realizes p so actually clause (iv) of $(**)$ of II.8.5 holds.

Clause (b): Subclause (α).

Note that II.8.5 show the equivalence of $(**)_{M_1^*,M_2^*}$ to some relatives. Now II.8.4 proves that if one of those relatives holds then every stationary set $S \subseteq \{\delta < \lambda^{++}\colon \mathrm{cf}(\delta) = \lambda^+\}$ can (modulo $\mathscr{D}_{\lambda^{++}}$) be coded by the isomorphic type of a model $M_S \in K_{\lambda^+}$, i.e., the failure of $\boxtimes$ from the assumption of II.8.7 which we are assuming (in (b) of 1.6(1)). So we have proved that $(**)_{M_1,M_2}$ fails and now we have gotten a contradiction by clause (a).

Clause (b): Subclause (β).

Recalling II.8.7, the only new point is the + of the good$^+$ (for $\mathfrak{s}^+$).

So assume that $\langle (M_i, N_i, a_i) : i < \lambda^{++}\rangle$ is a counterexample to the "$\mathfrak{s}^+$ is a good$^+$ λ^+-frame". So in particular $M_i, N_i \in K_{\mathfrak{s}(+)}$ and $p \in \mathscr{S}^{\mathrm{bs}}_{\mathfrak{s}(+)}(M_0)$ and $p_i = \mathbf{tp}_{\mathfrak{s}(+)}(a_{i+1}, M_{i+1}, M_{i+2})$ the type, for $\mathfrak{K}_{\mathfrak{s}(+)} = \mathfrak{K}^{\mathrm{nice}}_{\lambda^+}$ of course, which by subclause $(b)(\alpha)$ is $(K^{\mathrm{nice}}_{\lambda^+}, \leq_{\mathfrak{K}} \restriction K^{\mathrm{nice}}_{\lambda^+})$. As $p := p_i \restriction M_0 \in \mathscr{S}^{\mathrm{bs}}_{\mathfrak{s}(+)}(M_0)$ there are $M' \leq_{\mathfrak{K}[\mathfrak{s}]} M_0, M' \in K_\lambda$ and $q \in \mathscr{S}^{\mathrm{bs}}_{\mathfrak{s}}(M')$ which witness $p \in \mathscr{S}^{\mathrm{bs}}_{\mathfrak{s}(+)}(M_0)$ (see Definition 1.7). Let $\langle N'_\varepsilon : \varepsilon < \lambda^+\rangle$ be a sequence which $\leq_{\mathfrak{K}[\mathfrak{s}]}$-represents N_0. For each $i < \lambda^{++}$, as $p'_i = \mathbf{tp}_{\mathfrak{s}(+)}(a_{i+1}, N_0, N_{i+2})$ is not a non-forking extension of p necessarily there is $\varepsilon = \varepsilon_i < \lambda^+$ such that $M' \leq_{\mathfrak{K}[\mathfrak{s}]} N'_\varepsilon$ and $p'_i \restriction N'_\varepsilon = \mathbf{tp}_{\mathfrak{s}}(a_{i+1}, N'_\varepsilon, N_{i+2})$ is not a non-forking extension of q. So for some $\varepsilon < \lambda^+$ the set $S_\varepsilon = \{i < \lambda^{++} : \varepsilon_i = \varepsilon\}$ is unbounded in λ^{++}. We now choose by induction on $\zeta < \lambda^+$ a triple $(i_\zeta, M^*_\zeta, N^*_\zeta)$ such that:

- (a) $i_\zeta < \lambda^{++}$ is increasing continuous
- (b) $\zeta = \xi + 1 \Rightarrow i_\zeta \in S_\varepsilon$
- (c) $M^*_\zeta \leq_{\mathfrak{K}[\mathfrak{s}]} M_{i_\zeta}$
- (d) $M^*_\zeta \in K_{\mathfrak{s}}$ is $\leq_{\mathfrak{s}}$-increasing continuous
- (e) $N^*_\zeta \leq_{\mathfrak{K}[\mathfrak{s}]} N_{i_\zeta}$
- (f) $N^*_\zeta \in K_{\mathfrak{s}}$ is $\leq_{\mathfrak{s}}$-increasing continuous

(g) $M^*_\zeta = N^*_\zeta \cap M_{i_\zeta}$

(h) $\zeta = \xi + 1 \Rightarrow a_{i_\xi+1} \in M^*_{\zeta+1}$

(i) $M^*_0 = M'$ and $N^*_0 = N'_\varepsilon$ and $i_0 =$ Min(S_ε).

There is no problem to do this and letting $a^*_\zeta = a_{i_\zeta}$ clearly $\langle (M^*_\zeta, N^*_\zeta, a^*_\zeta) : \zeta < \lambda^+ \rangle$ contradict "$\mathfrak{s}$ is good$^+$".
2) By II.8.7, $\mathfrak{s}^+$ is a good λ- frame. The good$^+$ holds by clause $(b)(\beta)$ of part (1) above. $\square_{1.6}$

1.9 Claim. : *If $\mathfrak{s}$ is a successful good λ-frame <u>then</u> $\mathfrak{s}^+$ is a good$^+$ λ^+-frame.*

Remark. 1) This is a strong justification for assuming good$^+$ here.
2) So by this we can improve 4.16.

Proof.: By II.8.4, II.8.7 it is a good λ^+-frame, so it is enough to prove that it is good$^+$.

Let $\langle (M_i, N_i, a_i) : i < \lambda^{++} \rangle, p_i, M', q, \langle N'_\varepsilon : \varepsilon < \lambda^+ \rangle$ and $\langle \varepsilon_i : i < \lambda^{++} \rangle$ be as in the proof of clause (b), subclause (β) of 1.6. So for some $\varepsilon(*)$ the set $S_{\varepsilon(*)} := \{i < \lambda^{++} : \varepsilon_i = \varepsilon(*)\}$ is unbounded in λ^{++}, and let i_ζ be the ζ-th member of $S_{\varepsilon(*)}$ for $\zeta < \lambda^+$ and let $i(*) = \cup\{i_\zeta : \zeta < \lambda^+\}$. Let $\langle M_{i(*),\zeta} : \zeta < \lambda^+ \rangle, \langle N_{i(*),\zeta} : \zeta < \lambda^+ \rangle$ be a $\leq_{\mathfrak{K}}$-representation of $M_{i(*)}, N_{i(*)}$ respectively.

As $M_{i(*)} \leq_{\mathfrak{s}(+)} N_{i(*)}$ by the definition of λ^+ for some club E of λ^+ we have:
$(*)$ if $\zeta_1 < \zeta_1$ are from E then $\mathrm{NF}_{\mathfrak{s}}(M_{i(*),\zeta_1}, N_{i(*),\zeta_1}, M_{i(*),\zeta_2}, N_{i(*),\zeta_2})$.

Recalling that as $\mathfrak{s}$ is weakly successful, $\mathrm{NF}_{\mathfrak{s}}$ is a non-forking relation on $\mathfrak{K}_{\mathfrak{s}}$ respecting $\mathfrak{s}$, see II.6.1, for some limit ordinal $\zeta(*)$ we have:

⊛ (a) $\zeta(*)$ is from E

(b) $M_{i(*),\zeta(*)} \leq_{\mathfrak{s}} N_{i(*),\zeta(*)}$

(c) $M' \leq_{\mathfrak{s}} M_{i(*),\zeta(*)}$

(d) $N'_{\varepsilon(*)} \leq_{\mathfrak{s}} N_{i(*),\zeta(*)}$

(e) $M_{i(*),\zeta(*)} \leq_{\mathfrak{K}[\mathfrak{s}]} M_{i_{\zeta(*)}}$.

As $M' \le_{\mathfrak{s}} M_{i(*),\zeta(*)} \le_{\mathfrak{K}[\mathfrak{s}]} M_{i_{\zeta(*)}}$ holds, and as $\varepsilon_{i_{\zeta(*)}} = \varepsilon(*)$ clearly the type $\mathbf{tp}_{\mathfrak{s}}(a_{i_{\zeta(*)}+1}, M_{i(*),\zeta(*)}, M_{i_{\zeta(*)}+2})$ is a non-forking extension of $\mathbf{tp}_{\mathfrak{s}}(a_{i_{\zeta(*)}+1}, M', M_{i_{\zeta(*)}+2})$ which is equal to q.
We can find $\xi(*) \in E$ which is $> \zeta(*) + 2$ such that $a_{i_{\zeta(*)}+1} \in M_{i(*),\xi(*)}$, hence $\mathbf{tp}_{\mathfrak{s}}(a_{i_{\zeta(*)}+1}, M_{i(*),\zeta(*)}, M_{i(*),\xi(*)})$ is a non-forking extension of q.
By $(*)$ above we have $\mathrm{NF}_{\mathfrak{s}}(M_{i(*),\zeta(*)}, N_{i(*),\zeta(*)}, M_{i(*),\xi(*)}, N_{i(*),\xi(*)})$ hence, as $\mathrm{NF}_{\mathfrak{s}}$ respects $\mathfrak{s}$ by $(*)$ we know that

$$\mathbf{tp}_{\mathfrak{s}}(a_{i_{\zeta(*)}+1}, N_{i(*),\zeta(*)}, N_{i(*),\xi(*)})$$

is a non-forking extension of $\mathbf{tp}_{\mathfrak{s}}(a_{i_{\zeta(*)}+1}, M_{i(*),\zeta(*)}, M_{i(*),\xi(*)})$ hence of q. As $q \in \mathscr{S}^{\mathrm{bs}}_{\mathfrak{s}}(M')$ and $M' \le_{\mathfrak{s}} N'_{\varepsilon(*)} \le_{\mathfrak{s}} M_{i(*),\zeta(*)}$ recalling clause (c) of $\circledast$ we conclude that

$$\mathbf{tp}_{\mathfrak{s}}(a_{i_{\zeta(*)}+1}, N'_{\varepsilon(*)}, M_{i(*),\xi(*)}) = \mathbf{tp}_{\mathfrak{s}}(a_{i_{\zeta(*)}+1}, N'_{\varepsilon(*)}, M_{i(*)}) =$$

$$= \mathbf{tp}_{\mathfrak{s}}(a_{i_{\zeta(*)}+1}, N'_{\varepsilon(*)}, M_{i_{\varepsilon(*)}+1})$$

is a non-forking extension of q. But this contradicts the choice of $\varepsilon_{i_{\zeta(*)}} = \varepsilon(*)$. $\square_{1.9}$

The following claim sums up the "localness" of the basic types for $\mathfrak{s}(+)$, i.e., how to translate their properties to ones in $\mathfrak{s}$.

1.10 Claim. *[$\mathfrak{s}$ is a successful good λ-frame].*
Assume $\langle M_\alpha : \alpha < \lambda^+ \rangle$ is a $\le_{\mathfrak{s}}$-representation of $M \in K_{\mathfrak{s}(+)}$.
1) If $p_1, p_2 \in \mathscr{S}^{\mathrm{bs}}_{\mathfrak{s}(+)}(M)$ <u>then</u> $p_1 = p_2 \Leftrightarrow \bigwedge_{\alpha<\lambda^+} p_1 \restriction M_\alpha = p_2 \restriction M_\alpha \Leftrightarrow$ $(\exists^{\lambda^+} \alpha)(p_1 \restriction M_\alpha = p_2 \restriction M_\alpha) \Leftrightarrow (\exists \beta < \lambda^+)[p_1 \restriction M_\beta = p_2 \restriction M_\beta$ & $(\forall \alpha)(\forall \ell \in \{1,2\})(\beta \le \alpha < \lambda^+ \rightarrow p_\ell \restriction M_\alpha \in \mathscr{S}^{\mathrm{bs}}_{\mathfrak{s}}(M_\alpha)$ does not fork over $M_\beta)] \Leftrightarrow (\exists N \in K_{\mathfrak{s}})[(N$ is a witness for $p_1, p_2) \wedge (p_1 \restriction N = p_2 \restriction N)]$; see Definition 1.7.
2) If $S \subseteq \lambda^+ = \sup(S), \alpha_* \le \min(S)$ and for $\alpha \in S$ the type $p_\alpha \in \mathscr{S}^{\mathrm{bs}}_{\mathfrak{s}}(M_\alpha)$ does not fork over M_{α_*} and $p_\alpha \restriction M_{\alpha_*} = p_*$, <u>then</u>

(a) there is $p \in \mathscr{S}^{\mathrm{bs}}_{\mathfrak{s}(+)}(M)$ satisfying $\alpha \in S \Rightarrow p \restriction M_\alpha = p_\alpha$

(b) there is no $p' \in \mathscr{S}^{\mathrm{bs}}_{\mathfrak{s}(+)}(M) \backslash \{p\}$ satisfying this, i.e., p is unique

3) If $p = \mathbf{tp}_{\mathfrak{s}(+)}(a, M, N) \in \mathscr{S}_{\mathfrak{s}(+)}(M)$ *so* $M \leq_{\mathfrak{s}(+)} N$ *and* $a \in N$, *then*
$p \in \mathscr{S}^{\mathrm{bs}}_{\mathfrak{s}(+)}(M) \Leftrightarrow$
[for every $\alpha < \lambda^+$ *large enough,* $\mathbf{tp}_{\mathfrak{s}}(a, M_\alpha, N) \in \mathscr{S}^{\mathrm{bs}}_{\mathfrak{s}}(M_\alpha)] \Leftrightarrow$
for stationarily many $\alpha < \lambda^+$, $\mathbf{tp}_{\mathfrak{s}}(a, M_\alpha, N) \in \mathscr{S}^{\mathrm{bs}}_{\mathfrak{s}}(M_\alpha)]$.
4) Assume $M_1 \leq_{\mathfrak{s}(+)} M_2 \leq_{\mathfrak{s}(+)} M_3$ *and* $\langle M^\ell_\alpha : \alpha < \lambda^+\rangle$ *is a* $\leq_{\mathfrak{s}}$*-representation of* M_ℓ *for* $\ell = 1, 2, 3$ *and assume* $a \in M_3$.
Then $\mathbf{tp}_{\mathfrak{s}(+)}(a, M_2, M_3)$ *belongs to* $\mathscr{S}^{\mathrm{bs}}_{\mathfrak{s}(+)}(M_2)$ *and does not fork over* M_1 *(for* $\mathfrak{s}^+$*) iff for some club* E *of* λ^+*, for every* $\alpha < \beta$ *from* $E, \mathbf{tp}_{\mathfrak{s}}(a, M^2_\beta, M^3_\beta) \in \mathscr{S}^{\mathrm{bs}}_{\mathfrak{s}}(M^2_\beta)$ *does not fork over* M^1_α *(for* $\mathfrak{s}$*) iff for some stationary subset* $S \subseteq \lambda^+$ *for every* δ *from* $S, \mathbf{tp}_{\mathfrak{s}}(a, M^2_\delta, M^3_\delta) \in \mathscr{S}^{\mathrm{bs}}_{\mathfrak{s}}(M^2_\delta)$ *does not fork over* M^1_δ.

Proof. 1) Among the four statements which we have to prove equivalent, the first implies the second trivially, the second implies the third trivially, the third implies the second easily by monotonicity for equality of types and it implies the fourth by our assumption "$p_1, p_2 \in \mathscr{S}^{\mathrm{bs}}_{\mathfrak{s}(+)}(M)$" and the definition of $\mathscr{S}^{\mathrm{bs}}_{\mathfrak{s}(+)}(M)$. Also the first implies the fifth by the definition of $\mathscr{S}^{\mathrm{bs}}_{\mathfrak{s}(+)}(M)$ and the fifth implies the third by the closure properties of "a type does not fork over M" for $\mathfrak{s}$. To finish we shall prove that the fourth implies the first.

Let $M^0 = M, M^0_\alpha = M_\alpha$, let M^ℓ, a_ℓ for $\ell = 1, 2$ be such that $M^0 \leq_{\mathfrak{s}(+)} M^\ell, a_\ell \in M^\ell$ and $\mathbf{tp}_{\mathfrak{s}(+)}(a_\ell, M^0, M^\ell) = p_\ell$ and let $\langle M^\ell_\alpha : \alpha < \lambda^+\rangle$ be a $\leq_{\mathfrak{K}}$-representation of M^ℓ for $\ell = 1, 2$. As we can replace M^1 by some M' satisfying $M^1 <^+_{\lambda^+} M'$ (where $<^+_\lambda$ is defined in II.7.2), and similarly for M^2, recalling the definition of $<^+_{\lambda^+}$ without loss of generality $\alpha < \beta \Rightarrow \mathrm{NF}_{\mathfrak{s}}(M^0_\alpha, M^\ell_\alpha, M^0_\beta, M^\ell_\beta), M^\ell_{\alpha+1}$ is $(\lambda, *)$-brimmed over $M^0_{\alpha+1} \cup M^\ell_\alpha$ and $a_\ell \in M^\ell_0$. Clearly $\mathbf{tp}_{\mathfrak{s}}(a_1, M^0_0, M^1_0) = \mathbf{tp}_{\mathfrak{s}}(a_2, M^0_0, M^2_0)$ and we can build an isomorphism f from M_1 onto M_2 over M_0 mapping a_1 to a_2 by choosing $f \restriction M^1_\alpha : M^1_\alpha \xrightarrow[\text{onto}]{} M^2_\alpha$ by induction on $\alpha < \lambda^+$.
2) For proving clause (a), we choose $M^1_{\alpha_*} \in \mathfrak{K}_{\mathfrak{s}}$ such that $M_{\alpha_*} \leq_{\mathfrak{s}} M^1_{\alpha_*}$ and $M^1_{\alpha_*} \cap M = M_{\alpha_*}$ and $a \in M^1_{\alpha_*}$ realizes p_*. Now choose M^1_α by induction on $\alpha \in [\alpha_*, \lambda^+)$ satisfying $M^1_\alpha \cap M = M_\alpha$ and M'_α is $\leq_{\mathfrak{s}}$-increasing continuous with α such that $\alpha_* \leq \beta < \alpha \Rightarrow \mathrm{NF}_{\mathfrak{s}}(M_\beta, M^1_\beta, M_\alpha, M^1_\alpha)$, this should be clear.

Lastly, let $M^1 = \cup\{M^1_\alpha : \alpha < \lambda^+, \alpha \geq \alpha_*\}$ and let $p = \mathbf{tp}_{\mathfrak{s}(+)}(a, M, M^1)$. Now for $\alpha \in S$ the type $p \restriction M_\alpha = \mathbf{tp}_{\mathfrak{K}[\mathfrak{s}]}(a, M_\alpha, M^1) = \mathbf{tp}_{\mathfrak{s}}(a, M_\alpha, M^1_\alpha)$ is a non-forking extension of $\mathbf{tp}_{\mathfrak{K}[\mathfrak{s}]}(a, M_{\alpha_*}, M^1_{\alpha_*}) = p_*$ but also p_α satisfies this hence $p_\alpha = \mathbf{tp}_{\mathfrak{K}_\mathfrak{s}}(a, M_\alpha, M^1) = p \restriction M_\alpha$ so we are done by part (1). We can prove clause (b) of part (2) as in the proof of part (1).
3), 4) By the definition of $\mathfrak{s}^+$ and properties of $\mathrm{NF}_\mathfrak{s}$. $\square_{1.10}$

We may like in 1.10(1) to replace basic types by any types (later this is needed and more is done), note that if $\mathfrak{s}$ is type-full (Definition II.6.35) this is not needed. Also we may like in 1.10(3),(4) to replace "stationary" by "unbounded", i.e. add it as another equivalent clause. The answer is positive by the following.

1.11 Claim. *[$\mathfrak{s}$ is a successful good λ-frame]. (λ^+-locality)*
Assume $\langle M_\alpha : \alpha < \lambda^+ \rangle$ is a $\leq_\mathfrak{K}$-representation of $M \in K_{\mathfrak{s}(+)}$.
1) For any $p_1, p_2 \in \mathscr{S}_{\mathfrak{s}(+)}(M)$, <u>then</u> $p_1 = p_2 \Leftrightarrow (\forall \alpha)(p_1 \restriction M_\alpha = p_2 \restriction M_\alpha) \Leftrightarrow (\exists^{\lambda^+} \alpha)(p_1 \restriction M_\alpha = p_2 \restriction M_\alpha)$.
2) In 1.10(2),(3) we can replace stationary by unboundedly.

Proof. 1) The first condition implies the second by the basic properties of types, see II§1 mainly II.1.11. The second conditions implies the third condition trivially and the third condition implies the second by the basic properties of types. Lastly, the second condition implies the first by the proof of 1.10(1).
2) Note that by 1.16 below and Fodor lemma for any $p \in \mathscr{S}_{\mathfrak{K}[\mathfrak{s}]}(M)$, $M \in K^\mathfrak{s}_{\lambda^+}$ if $S := \{\alpha < \lambda^+ : p \restriction M_\alpha \in \mathscr{S}^{\mathrm{bs}}_\mathfrak{s}(M_\alpha)\}$ is unbounded in λ^+ then it contains an end-segment of λ^+. $\square_{1.11}$

1.12 Definition. 1) We shall define by induction on n:

(*a*) $\mathfrak{s}$ is n-successful

(*b*) $\mathfrak{s}^{+m} = \mathfrak{s}(+m)$ for $m \leq n$.

<u>For $n = 0$</u>: We say $\mathfrak{s}$ is 0-successful if it is a good $\lambda_\mathfrak{s}$-frame.
Let $\mathfrak{s}^{+0} = \mathfrak{s}$.

<u>For $n = 1$</u>: We say $\mathfrak{s}$ is 1-successful <u>if</u> it is good ($\lambda_\mathfrak{s}$-frame) and successful; let $\mathfrak{s}^{+1} = \mathfrak{s}^+$.

For $n = m + 1 \geq 2$: We say $\mathfrak{s}$ is n-successful if it is m-successful and $\mathfrak{s}^{+m}$ is 1-successful.

We let $\mathfrak{s}^{+n} = (\mathfrak{s}^{+m})^+$ (so $\mathfrak{s}^{+m}$ is well defined iff $\mathfrak{s}$ is m-successful).

2) We say $\mathfrak{s}$ is $(n + \frac{1}{2})$-successful or say is weakly $(n + 1)$-successful if it is n-successful and $\mathfrak{s}^{+n}$ satisfies clause (a) of 1.1.

3) We say $\mathfrak{s}$ is ω-successful if it is n-successful for every n.

4) If $\mathfrak{s}^{+n}$ is well defined let $\mathfrak{B}_n = \mathfrak{B}_n^{\mathfrak{s}} = \mathfrak{B}_n^{\mathfrak{s}(+n)} = \mathfrak{B}(\mathfrak{s}^{+n})$ be a superlimit model in $\mathfrak{K}[\mathfrak{s}^{+n}]$; it is well defined, i.e., is unique only up to isomorphism.

1.13 Claim. *Assume $\mathfrak{s}$ is an n-successful good frame.*

1) $\mathrm{NF}_{\mathfrak{s}[+n]} = \mathrm{NF}[\mathfrak{s}^{+n}]$ is well defined if $\mathfrak{s}$ is $(n + \frac{1}{2})$-successful.

2) There is $\mathfrak{B}_n^{\mathfrak{s}} \in K_{\lambda^{+n}}$, that is a $\mathfrak{K}[\mathfrak{s}^{+n}]$-superlimit is well defined.

3) $\mathfrak{K}_{\mathfrak{s}(+n)} = \mathfrak{K}_{\mathfrak{s}(+n)}[\mathfrak{B}_n^{\mathfrak{s}}]$ if $n > 0$.

4) If $k+m = n$ then $\mathfrak{s}^{+k}$ is m-successful good frame and $((\mathfrak{s}^{+k})^{+m}) = \mathfrak{s}^{+n}$.

5) $\mathfrak{s}^k$ is m-successful iff $\mathfrak{s}$ is $(k+m)$-successful; and if this holds then $(\mathfrak{s}^{+k})^{+m} = \mathfrak{s}^{+(k+m)}$. Also $\mathfrak{s}^k$ is $(m+\frac{1}{2})$-successful iff $\mathfrak{s}$ is $(k+m+\frac{1}{2})$-successful.

6) If $\mathfrak{s}$ is a good$^+$ λ-frame and $0 < n < \omega$ then

- *(a) $\leq_{\mathfrak{s}(+n)} = \leq_{\mathfrak{K}[\mathfrak{s}]} \restriction K_{\mathfrak{s}(+n)}$,*
- *(b) $\mathfrak{B}_n^{\mathfrak{s}}$ is pseudo superlimit also in $\mathfrak{K}_{\lambda^{+n}}^{\mathfrak{s}}$, i.e.: it has a $<_{\mathfrak{K}[\mathfrak{s}]}$-extension isomorphic to itself and if $\langle M_i : i \leq \delta \rangle$ is $\leq_{\mathfrak{K}[\mathfrak{s}]}$-increasing continuous, $\delta < \lambda^{+(n+1)}$ a limit ordinal of course, and $i < \delta \Rightarrow M_i \cong \mathfrak{B}_n^{\mathfrak{s}}$ then $M_\delta \cong \mathfrak{B}_n^{\mathfrak{s}}$, but the universality of $\mathfrak{B}_n^{\mathfrak{s}}$ in $\mathfrak{K}_{\lambda^{+n}}^{\mathfrak{s}}$ is not clear*
- *(c) $\mathfrak{K}_{\mathfrak{s}(+n)} = \mathfrak{K}_{\lambda^{+n}}^{\mathfrak{s}}[\mathfrak{B}_n^{\mathfrak{s}}]$.*

Proof. Easy, by induction on n (and for (5) on $k + m$). For (6)(a) use 1.8 hence $(b)(\alpha)$ of 1.6 holds. $\square_{1.13}$

1.14 Conclusion. In the main lemma II.9.1, if we strengthen the assumptions (which includes $\mathfrak{s}$ is successful) by "$\mathfrak{s} = (\mathfrak{K}_\lambda, \mathscr{S}^{\mathrm{bs}}, \bigcup_\lambda)$ is a good$^+$ λ-frame", $\mathfrak{K} = \mathfrak{K}^{\rm up}_\lambda$, then we can strengthen the conclusion to:

(α) $\mathfrak{s}_\ell = (\mathfrak{K}_{\mathfrak{s}_\ell}, \mathscr{S}^{\rm bs}_{\mathfrak{s}_\ell}, \bigcup_{\mathfrak{s}_\ell})$ is a good$^+$ $\lambda^{+\ell}$-frame

(β) $\mathfrak{s}_\ell = \mathfrak{s}^{+\ell}$, hence for $\mathfrak{B}^{\mathfrak{s}}_\ell \in K_{\lambda^{+\ell}}$ we have:

- (i) $\mathfrak{B}^{\mathfrak{s}}_\ell$ is superlimit (in $\mathfrak{K}_{\mathfrak{s}_\ell} = \mathfrak{K}^{\mathfrak{s}(+\ell)}_{\lambda^{+\ell}}$ and if $\ell = m+1$, even in $\mathfrak{K}^{\mathfrak{s}(+m)}_{\lambda^{+\ell}}$)
- (ii) $\mathfrak{K}_{\mathfrak{s}_\ell}$ is $\mathfrak{K}^{[\mathfrak{B}^{\mathfrak{s}}_\ell]}$, see Definition II.1.25, and $\leq_{\mathfrak{K}^{\mathfrak{s}}_{\lambda^{+\ell}}} = \leq_{\mathfrak{K}} \restriction K^{\mathfrak{s}_\ell}_{\lambda^{+\ell}}$ (this equality is the new point)
- (iii) $\mathscr{S}^{\rm bs}_{\mathfrak{s}_\ell}, \bigcup_{\mathfrak{s}_\ell}$ are defined as in II§2 but restricted to $\mathfrak{K}_{\mathfrak{s}_\ell}$ of course

(γ) if $\mathfrak{B}^{\mathfrak{s}}_\ell$ is the unique superlimit in $\mathfrak{K}^{\mathfrak{s}}_{\lambda^{+\ell}}$ then $\mathfrak{s}_\ell = \mathfrak{s}[\lambda^{+\ell}]$, see 0.4(2); note that $\mathfrak{B}^{\mathfrak{s}}_\ell$ may be just locally superlimit in $\mathfrak{K}^{\mathfrak{s}}_{\lambda^{+\ell}}$.

Proof. Should be clear (or combine II.9.1 with 1.6). $\square_{1.14}$

Note that in particular

1.15 Claim. *Assume $\mathfrak{s}$ is successful good$^+$ frame. Then $\mathfrak{s}^+ = \mathfrak{s}[\lambda^+_{\mathfrak{s}}]$ where on $\mathfrak{s}[\lambda^+_{\mathfrak{s}}]$ see Definition 0.4(2).*

Proof. Easy because $\mathfrak{B}^{\mathfrak{s}}_1$ is $\leq_{\mathfrak{K}^{\mathfrak{s}}_{\lambda^+}}$-universal. $\square_{1.15}$

1.16 Claim. *[Assume $\mathfrak{s}$ is a weakly successful good λ-frame.] Let $\delta < \lambda^+$, be a limit ordinal and $\langle M_i : i \leq \delta + 1\rangle$ be $\leq_{\mathfrak{s}}$-increasing continuous.*
If $b \in M_{\delta+1}$ satisfies $\mathbf{tp}_{\mathfrak{s}}(b, M_i, M_{\delta+1}) \in \mathscr{S}^{\rm bs}_{\mathfrak{s}}(M_i)$ for arbitrarily large $i < \delta$, then $\mathbf{tp}_{\mathfrak{s}}(b, M_\delta, M_{\delta+1}) \in \mathscr{S}^{\rm bs}_{\mathfrak{s}}(M_\delta)$ hence $\mathbf{tp}_{\mathfrak{s}}(b, M_\delta, M_{\delta+1})$ does not fork over M_i for every $i < \delta$ large enough.

Remark. If $\mathbf{tp}_{\mathfrak{s}}(b, M_\delta, M_{\delta+1}) \in \mathscr{S}^{\rm bs}_{\mathfrak{s}}(M_\delta)$, this is an axiom of good frames.

Proof. Let $\langle N_i : i \le \delta \rangle$ be as in Claim 1.17(1) below, so in particular N_δ is $(\lambda, *)$-brimmed over M_δ hence $\le_{\mathfrak{s}}$-universal over N_δ, so without loss of generality $M_{\delta+1} \le_{\mathfrak{s}} N_\delta$. But $N_\delta = \cup\{N_i : i < \delta\}$, so for some $i < \delta$ we have $b \in N_i$, so without loss of generality $\mathbf{tp}_{\mathfrak{s}}(b, M_i, M_{\delta+1}) \in \mathscr{S}^{\rm bs}_{\mathfrak{s}}(M_i)$ (by the assumptions on b), now as $\mathrm{NF}_{\mathfrak{s}}(M_i, M_\delta, N_i, N_\delta)$ holds by 1.17(2), we can by 1.18 below deduce that have $\mathbf{tp}_{\mathfrak{s}}(b, M_\delta, N_\delta) = \mathbf{tp}_{\mathfrak{s}}(b, M_\delta, M_{\delta+1})$ is a non-forking extension of $\mathbf{tp}_{\mathfrak{s}}(b, M_i, N_i)$, and so we are done. $\square_{1.16}$

* * *

Recall from Chapter II some claims on non-forking which we shall use.

1.17 Claim. *[Assume $\mathfrak{s}$ is a weakly successful good λ-frame.]*
1) If $\langle M_i : i \le \delta \rangle$ is $\le_{\mathfrak{s}}$-increasing continuous, then we can find an $\le_{\mathfrak{s}}$-increasing continuous sequence $\langle N_i : i \le \delta \rangle$ such that $M_i \le_{\mathfrak{s}} N_i, i < \delta \Rightarrow \mathrm{NF}_{\mathfrak{s}}(M_i, N_i, M_{i+1}, N_{i+1})$ and N_{i+1} is universal over $N_i \cup M_{i+1}$.
2) If $\langle M_i, N_i : i \le \delta \rangle$ are as in part (1), then $i \le j \le \delta \Rightarrow \mathrm{NF}_{\mathfrak{s}}(M_i, N_i, M_j, N_j)$ and N_δ is $(\lambda, *)$-brimmed over M_δ and even over $M_\delta \cup M_i$ for $i < \delta$.
3) In part (1) we can add "N_i is brimmed over M_i for every i".
4) If $\alpha < \lambda^+$, $\langle M_i : i \le \alpha \rangle$ is $\le_{\mathfrak{s}}$-increasing continuous and $\mathrm{NF}_{\mathfrak{s}}(M_0, N_0, M_\alpha, N')$ then we can find N'' and $\langle N_i : i \le \alpha, i \ne 0 \rangle$ such that $\langle (M_i, N_i) : i \le \alpha \rangle$ is as in (1),(3) above, $N_\alpha \le_{\mathfrak{s}} N'', N' \le_{\mathfrak{s}} N''$ and even $N' \le_{\mathfrak{s}} N_\alpha$.
5) In parts (1),(3) we can allow $M_\delta \in K^{\mathfrak{s}}_{\lambda^+}$ though $i < \delta \Rightarrow M_i \in K_{\mathfrak{s}}$ (so $\delta = \lambda^+$).

Proof. 1), 2), 3) By II.6.29, II.6.30(2) and the properties of NF.
4) Let $\langle N'_i : i \le \alpha \rangle$ be as in part (1) + (3), let f_0 be a $\le_{\mathfrak{s}}$-embedding of N_0 into N'_0 over M_α. By the uniqueness for $\mathrm{NF}_{\mathfrak{s}}$ we can find N''_α, f' such that $N'_\alpha \le_{\mathfrak{s}} N''$ and $f' \supseteq f \cup \mathrm{id}_{M_\alpha}$ is a $\le_{\mathfrak{s}}$-embedding of N' into N'', (in fact without loss of generality $N'' = N'_\alpha$ by part (2)).
5) Should be clear. $\square_{1.17}$

1.18 Claim. *[Assume $\mathfrak{s}$ is a weakly successful good λ-frame.] If* $\mathrm{NF}_{\mathfrak{s}}(M_0, M_1, M_2, M_3)$ *and* $(M_0, M_2, a) \in K_{\mathfrak{s}}^{3,\mathrm{bs}}$ *<u>then</u>* $\mathbf{tp}_{\mathfrak{s}}(a, M_1, M_3) \in \mathscr{S}_{\mathfrak{s}}^{\mathrm{bs}}(M_1)$ *does not fork over* M_0.

Remark. This holds as "$\mathrm{NF}_{\mathfrak{s}}$ respects $\mathfrak{s}$" which is defined in II.6.1 and holds by II.6.34.

Proof. See II.6.32.

1.19 Claim. *Assume $\mathfrak{s}$ is a weakly successful good λ-frame. If* $M_0 \le_{\mathfrak{s}} M_\ell \le_{\mathfrak{s}} M_3$ *for* $\ell = 1, 2$ *and* (M_0, M_1, a) *belongs to* $K_{\mathfrak{s}}^{3,\mathrm{uq}}$ *and* $\mathbf{tp}_{\mathfrak{s}}(a, M_2, M_3)$ *does not fork over* M_0 *(e.g.* $\mathbf{tp}_{\mathfrak{s}}(a, M_0, M_1)$ *has a unique extension in* $\mathscr{S}_{\mathfrak{s}}(M_2)$*) <u>then</u>* $\mathrm{NF}_{\mathfrak{s}}(M_0, M_1, M_2, M_3)$.

Remark. This is close to the definition of $K_{\mathfrak{s}}^{3,\mathrm{uq}}$, which says that there is a unique such amalgamation up to embeddings.

Proof. By II.6.33. $\square_{1.19}$

1.20 Claim. *[$\mathfrak{s}$ is a weakly successful good λ-frame.] Assume* $M_0 \le_{\mathfrak{s}} M_1 \le_{\mathfrak{s}} M_2, a \in M_2$ *and* $\mathbf{tp}_{\mathfrak{s}}(a, M_1, M_2)$ *does not fork over* M_0 *and* $b \in M_1, \mathbf{tp}_{\mathfrak{s}}(b, M_0, M_1) \in \mathscr{S}_{\mathfrak{s}}^{\mathrm{bs}}(M_0)$*. <u>Then</u> there are* M_1^*, M_2^* *such that* $M_2 \le_{\mathfrak{s}} M_2^*, M_0 \le_{\mathfrak{s}} M_1^* \le_{\mathfrak{s}} M_2^*, (M_0, M_1^*, a) \in K_{\mathfrak{s}}^{3,\mathrm{uq}}$ *and* $\mathbf{tp}_{\mathfrak{s}}(b, M_1^*, M_2^*)$ *does not fork over* M_0.

Proof. By NF calculus (i.e. using II.6.34 and Definition II.6.1). That is, by $\mathfrak{s}$ being weakly successful we can find (M_1^*, a^*) such that $(M_0, M_1^*, a^*) \in K_{\mathfrak{s}}^{3,\mathrm{uq}}$ and $\mathbf{tp}_{\mathfrak{s}}(a^*, M_0, M_1^*) = \mathbf{tp}_{\mathfrak{s}}(a, M_0, M_2)$. By the amalgamation property for $\mathfrak{K}_{\mathfrak{s}}$ and the definition of $\mathbf{tp}_{\mathfrak{s}}$ without loss of generality for some M_2^* we have $M_2 \le_{\mathfrak{s}} M_2^*, M_1^* \le_{\mathfrak{s}} M_2^*$ and $a^* = a$. As $\mathbf{tp}_{\mathfrak{s}}(a, M_1, M_2)$ does not fork over M_0 and by 1.19 above we have $\mathrm{NF}_{\mathfrak{s}}(M_0, M_1^*, M_2, M_2^*)$ and by symmetry for $\mathrm{NF}_{\mathfrak{s}}$ we have $\mathrm{NF}_{\mathfrak{s}}(M_0, M_2, M_1^*, M_2^*)$. As $\mathbf{tp}_{\mathfrak{s}}(b, M_0, M_1) \in \mathscr{S}^{\mathrm{bs}}(M_0)$ and $M_0 \le_{\mathfrak{s}} M_1 \le_{\mathfrak{s}} M_2$ clearly $(M_0, M_2, b) \in K_{\mathfrak{s}}^{3,\mathrm{bs}}$ hence by 1.18 we get $\mathbf{tp}_{\mathfrak{s}}(b, M_1^*, M_2^*)$ does not fork over M_0 as required. $\square_{1.20}$

We could have mentioned in Chapter II:

1.21 Claim. *[$\mathfrak{s}$ is a good λ-frame].*
Assume that $M_1 \le_{\mathfrak{s}} M_2$ are superlimit in $K_{\mathfrak{s}}$ and $p_i \in \mathscr{S}^{bs}_{\mathfrak{s}}(M_2)$ does not fork over M_1 for $i < \alpha < \lambda_{\mathfrak{s}}$. Then there is an isomorphism f from M_1 onto M_2 such that $i < \alpha \Rightarrow f(p_i \restriction M_1) = p_i$.

Proof. First assume that M_2 is $(\lambda, *)$-brimmed over M_1. Clearly we can find a regular cardinal θ such that $\alpha < \theta \le \lambda$. Now we can find a sequence $\langle N_\beta : \beta < \theta \rangle$ which is $\le_{\mathfrak{s}}$-increasing continuous, $N_{\beta+1}$ being universal over N_β (of course, we are using II§4). Clearly $\bigcup_{\beta<\theta} N_\beta \in K_{\mathfrak{s}}$ is $(\lambda, *)$-brimmed over N_0, so without loss of generality is equal to M_1.

So for each $i < \alpha$ for some $\beta(i) < \theta$ the type $p_i \restriction M_1$ which belongs to $\mathscr{S}^{bs}_{\mathfrak{s}}(M_1)$ does not fork over $N_{\beta(i)}$, so $\beta = \sup\{\beta(i) : i < \alpha\} < \theta$, hence by transitivity and monotonicity of non-forking $i < \alpha \Rightarrow p_i$ does not fork over N_β. Clearly M_2 is $(\lambda, *)$-brimmed over N_β and by the choice of $\langle N_\gamma : \gamma < \theta \rangle$ also M_1 is $(\lambda, *)$-brimmed over N_β hence there is an isomorphism f from M_2 onto M_1 over N_β. Now for $i < \alpha$ the types $p_i \restriction M_1$ and $f(p_i)$ are members of $\mathscr{S}^{bs}_{\mathfrak{s}}(M_1)$ which do not fork over N_β and have the same restriction to N_β hence are equal. So f^{-1} is as required.

Second without the assumption "M_2 is $(\lambda, *)$-brimmed over M_1" we can find $M_3 \in K_{\mathfrak{s}}$ which is $(\lambda, *)$-brimmed over M_2 hence also over M_1 and let $q_i \in \mathscr{S}^{bs}_{\mathfrak{s}}(M_3)$ be a non-forking extension of p_i for $i < \alpha$.

Applying what we have already proved to the pair (M_1, M_3) there is an isomorphism f_1 from M_1 onto M_3 mapping $p_i \restriction M_1 = q_i \restriction M_1$ to q_i for $i < \alpha$. Applying what we have already proved to the pair (M_2, M_3), there is an isomorphism f_2 from M_2 onto M_3 mapping p_i to q_i for $i < \alpha$. Now $f_2^{-1} \circ f_1$ is as required. $\square_{1.21}$

1.22 Claim. *Let $\mathfrak{s}$ be a good λ-frame.*
1) Assume $\bar{M} = \langle M_i : i \le \delta+1 \rangle$ is $\le_{\mathfrak{s}}$-increasing continuous. If $\mathbf{P} \subseteq \mathscr{S}^{bs}_{\mathfrak{s}}(M_{\delta+1})$, $|\mathbf{P}| < \mathrm{cf}(\delta)$ and $M_\delta \le_{\mathfrak{s}} N$ and $N, M_{\delta+1}$ are brimmed over N_i for $i < \delta$ and $p \in \mathbf{P} \Rightarrow p$ does not fork over M_δ. Then for every large enough $i < \delta$ there is an isomorphism f from N onto $M_{\delta+1}$ over M_i such that $p \in \mathbf{P} \Rightarrow f^{-1}(p) \restriction M_\delta = p \restriction M_\delta$.

2) Instead $|\mathbf{P}| < \text{cf}(\delta)$ *it is enough to demand: for some* $i < \delta$ *we have* $p \in \mathbf{P} \Rightarrow p$ *does not fork over* M_i.

Proof. 1) For $p \in \mathbf{P}$, choose $i(p) < \delta$ such that $p \in \mathbf{P} \Rightarrow p \restriction M_\delta$ does not fork over $M_{i(p)}$ and let $i(*) = \sup\{i(p) : p \in \mathbf{P}\}$ it is $< \delta$ or $|\Gamma| <$ cf(δ). Continue as in 1.20.
2) Similar. $\square_{1.22}$

1.23 Exercise: Assume $\mathfrak{s}$ is a good λ-frame.

If M_0 is $(\lambda, *)$-brimmed, M_1 is $(\lambda, *)$-brimmed over M_0, f an automorphism of M_1 and $p_i, q_i \in \mathscr{S}(M_1)$ do not fork over M_0 for $i < \alpha, f(p_i) = q_i$ and $\alpha < \lambda$ then for some automorphism g of $M_0, i < \alpha \Rightarrow g(p_i \restriction M_0) = q_i \restriction M_0$.

1.24 Exercise: Assume $\mathfrak{s}$ is a good λ-frame.

If $N \in \mathfrak{K}_{\mathfrak{s}}$ is $(\lambda, *)$-brimmed and $\mathbf{P} \subseteq \mathscr{S}^{\text{bs}}_{\mathfrak{s}}(N)$ has cardinality $< \lambda_{\mathfrak{s}}$ then for some M:

(a) $M \in \mathfrak{K}_{\mathfrak{s}}$ is $(\lambda, *)$-brimmed

(b) N is $(\lambda, *)$-brimmed over M

(c) p does not fork over M for every $p \in \mathbf{P}$.

1.25 Exercise: If $\mathfrak{s}$ is successful and $M <^+_{\lambda,\kappa} N$, see Definition II.7.2(2), then N is (λ, κ)-brimmed over M for $\mathfrak{s}^+$.

[Hint: Obvious by now. E.g. for every $M' \in K_{\mathfrak{s}(+)}$ there is $M'' \in K_{\mathfrak{s}(+)}$ such that $M' <_{\lambda^+,\kappa} M''$, by II.7.4(1). Hence there is $\leq_{\mathfrak{s}(+)}$-increasing continuous sequence $\langle M_\alpha : \alpha \leq \kappa \rangle$ such that $M_{2\alpha+1} <^+_{\lambda,\kappa} M_{2\alpha+2}$ for $\alpha < \kappa$ and $M_0 = M$. Now $M_{2\alpha+2}$ is $\leq_{\mathfrak{s})(+)}$-universal over $M_{2\alpha+1}$ for $\alpha < \kappa$ by II.7.9(1) hence M_κ is (λ^*, κ)-brimmed over M_0.

By II.7.7(3) it follows that $M = M_0 <^+_{\lambda^+,\kappa} M_\kappa$ and uniqueness, II.7.6(2), without loss of generality $N = M_\kappa$.]

§2 UNI-DIMENSIONALITY AND NON-SPLITTING

Dealing all the time with good frames, we may wonder what occurs to the question on the spectrum of categoricity. As in the first order case it is closely related to being uni-dimensional. So we may wonder how to define "uni-dimensional" and whether: if $\mathfrak{s}$ is categorical in λ and is uni-dimensional and $\mathfrak{s}$ is n-successful (see 1.12), <u>then</u> $K^{\mathfrak{s}}_{\lambda^{+n}}$ is categorical. By 2.11 below the answer is yes. There are several variants of uni-dimensional but when $\mathfrak{K}_{\mathfrak{s}}$ is categorical (in λ) we have: $\mathfrak{K}^{\mathfrak{s}}$ is catgorical in λ^+ iff $\mathfrak{s}$ is weakly uni-dimensional iff all models from $K^{\mathfrak{s}}_{\lambda^+}$ are saturated and it implies that $\mathfrak{s}^+$ is weakly uni-dimensional; concerning the other variants see 2.15(3). By this in the end of §12 we shall be able to derive results on the categoricity spectrum.

We may consider a more restricted framework, $\mathfrak{s}$ is saturative, it is close to categoricity. Note that "saturative" is closed to non-multi-dimensional but see 1.4. Note that if we know more (e.g. $\mathfrak{s}$ has primes, see §4 (and is categorical in λ)) we can invert the implications of 2.10.

2.1 Hypothesis. $\mathfrak{s}$ is a good λ-frame.

2.2 Definition. 1) We say $\mathfrak{s}$ is semi$^{\rm bs}$-uni-dimensional <u>when</u> for any model $M \in K_{\mathfrak{s}}$, if $M <_{\mathfrak{s}} N_k \in K_{\mathfrak{s}}$ for $k = 1, 2$ <u>then</u> some $p \in \mathscr{S}^{\rm bs}_{\mathfrak{s}}(M)$ is realized in N_1 and in N_2. Let "$\mathfrak{s}$ is semi$^{\rm na}$-uni-dimensional" be defined similarly but we allow $p \in \mathscr{S}^{\rm na}_{\mathfrak{s}}(M)$ and let "$\mathfrak{s}$ is semi$^{\rm bs}$-uni-dimensional" be called "$\mathfrak{s}$ is semi-uni-dimensional". Instead of na,bs as superscripts to semi we may write 0, 1 respectively.
2) We say $\mathfrak{s}$ is almost uni-dimensional <u>if</u> for any model $M \in K_{\mathfrak{s}}$, there is an unavoidable $p \in \mathscr{S}^{\rm bs}_{\mathfrak{s}}(M)$ (see below).
3) For $M \in K_{\mathfrak{s}}$ we say $p \in \mathscr{S}^{\rm all}_{\mathfrak{s}}(M)$ is unavoidable, <u>if</u> for every N satisfying $M <_{\mathfrak{s}} N \in K_{\mathfrak{s}}$, some $a \in N \backslash M$ realizes p. This is well defined for any λ-a.e.c. $\mathfrak{K}_\lambda$.
4) We say $\mathfrak{s}$ is explicitly uni-dimensional <u>if</u> every $p \in \mathscr{S}^{\rm bs}_{\mathfrak{s}}(M)$ where $M \in K_{\mathfrak{s}}$, is unavoidable.
5) We call $\mathfrak{s}$ non-multi-dimensional <u>if</u> for every $M_0 \in K_\lambda$ whenever $M_0 \le_{\mathfrak{s}} M_1 <_{\mathfrak{s}} M_2$, there is $p \in \mathscr{S}^{\rm bs}_{\mathfrak{s}}(M_1)$ which does not fork over M_0 and is realized in M_2.
6) We say $\mathfrak{s}$ is weakly uni-dimensional <u>when</u>: if for every $M <_{\mathfrak{s}} M_\ell$ for $\ell = 1, 2$, <u>then</u> there is $c \in M_2 \backslash M$ such that $\mathbf{tp}_{\mathfrak{s}}(c, M, M_2)$ belongs to $\mathscr{S}^{\rm bs}_{\mathfrak{s}}(M)$ and has more than one extension in $\mathscr{S}^{\rm all}_{\mathfrak{s}}(M_1)$.

7) We say a λ-a.e.c. $\mathfrak{K}$ with amalgamation is weakly [or is semi] uni-dimensional <u>when</u>: if $M <_{\mathfrak{K}} N_\ell$ for $\ell = 1,2$ then there is $p \in \mathscr{S}^{\rm na}_{\mathfrak{K}}(M)$ realized in N_1 and having at least two extensions in $\mathscr{S}^{\rm all}_{\mathfrak{K}}(M)$ or realized in N_1 and in N_2.
8) We say a λ-a.e.c. $\mathfrak{K}$ with amalgamation is almost uni-dimensional <u>when</u> for every $M \in K$, there an unavoidable $p \in \mathscr{S}_{\mathfrak{K}}(M)$

On the meaning in the first order case see 2.6(5) below; the definition in 2.2(5) fits the first order one for superstable T.
We naturally first look at the natural implications. Note that being "semi$^{\rm bs}$/almost/ explicitly/weakly uni-dimensional" may be influenced by the choice of the basic types (compare 2.6(1) with 2.6(2)) as well as non-multi-dimensional but not so semi$^{\rm na}$-uni-dimensional.

2.3 Claim. *1) If $\mathfrak{s}$ is explicitly uni-dimensional, <u>then</u> $\mathfrak{s}$ is almost uni-dimensional.*
2) If $\mathfrak{s}$ is almost uni-dimensional, <u>then</u> $\mathfrak{s}$ is semi^x-uni-dimensional for $x \in \{\rm na,bs\}$; if $\mathfrak{s}$ is semi$^{\rm bs}$-uni-dimensional <u>then</u> $\mathfrak{s}$ is semi$^{\rm na}$-uni-dimensional.
3) If $\mathfrak{s}$ is semi$^{\rm na}$-uni-dimensional, <u>then</u> $K^{\mathfrak{s}}$ is categorical in $\lambda^+_{\mathfrak{s}}$.
4) If $\mathfrak{s}$ is weakly uni-dimensional, <u>then</u> $K^{\mathfrak{s}}$ is categorical in $\lambda^+_{\mathfrak{s}}$.
5) If $\mathfrak{s}$ is semi$^{\rm bs}$-uni-dimensional <u>then</u> $\mathfrak{s}$ is weakly uni-dimensional.
6) If $\mathfrak{s}$ is weakly uni-dimensional <u>then</u> $\mathfrak{K}_{\mathfrak{s}}$ is weakly uni-dimensional.
7) If $\mathfrak{s}$ is a type-full <u>then</u> $\mathfrak{s}$ is weakly uni-dimensional iff $\mathfrak{K}_{\mathfrak{s}}$ is weakly uni-dimensional.
8) If a λ-a.e.c. $\mathfrak{K}_\lambda$ with amalgamation and the JEP is almost uni-dimensional (or semi-uni-dimensional) <u>then</u> $\mathfrak{K}_{\lambda^+} = (\mathfrak{K}^{\rm up}_\lambda)_{\lambda^+}$ is cateogorical in λ^+.

2.4 Remark. 1) Concerning non-multi-dimensionality and minimal $\leq_{\mathfrak{s}}$-extension, see 3.8.
2) See more in 2.10.
3) If $\mathfrak{s}$ has primes for $\mathscr{S}^{\rm na}_{\mathfrak{s}}$ (see later 3.2(5)) and $\mathfrak{s}$ is semi$^{\rm na}$-uni-dimensional <u>then</u> $\mathfrak{s}$ has semi$^{\rm bs}$-uni-dimensional; this helps for $\mathfrak{s}^+$ because $\mathfrak{s}^+$ has primes by 4.9, and even for $p \in \mathscr{S}^{\rm na}_{\mathfrak{s}}(M)$ by II.6.34,II.6.36.
4) See more in 2.15(3).
5) Assume $\mathfrak{K}_\lambda$ is a λ-a.e.c. with amalgamation and the JEP and $\mathfrak{K}_\lambda$ is stable in λ, (i.e. if $M \in \mathfrak{K}_\lambda$ then $|\mathscr{S}_{\mathfrak{K}_\lambda}(M)| \leq \lambda$). If $\mathfrak{K}_\lambda$ is weakly

uni-dimensional then $\mathfrak{K}_{\lambda^+} = (K^{\rm up}_\lambda)_{\lambda^+}$ is categorical in λ. See more in [Sh:F735].

Proof. 1) By the definitions and as $\mathscr{S}^{\rm bs}_{\mathfrak{s}}(M) \neq \emptyset$ for every $M \in K_{\mathfrak{s}}$. (Why? Because there is $N, M <_{\mathfrak{s}} N$ and Ax(D)(c) of good λ-frames (density of basic types)).
2) Check the definitions.
3) Let $M_0, M_1 \in K^{\mathfrak{s}}_{\lambda^+}$ and we shall prove that they are isomorphic. Let $\langle M^\ell_\alpha : \alpha < \lambda^+\rangle$ be $<_{\mathfrak{s}}$-representation of M_ℓ such that $\alpha < \lambda^+ \Rightarrow M^\ell_\alpha \neq M^\ell_{\alpha+1}$ for $\ell = 0, 1$. Let $\langle a^\ell_i : i < \lambda^+\rangle$ list the elements of M_ℓ. We choose by induction on $\varepsilon < \lambda^+$ a tuple $(N_\varepsilon, \alpha^0_\varepsilon, f^0_\varepsilon, \alpha^1_\varepsilon, f^1_\varepsilon)$ such that:

- (a) $N_\varepsilon \in K_{\mathfrak{s}}$ is $\leq_{\mathfrak{s}}$-increasing continuous
- $(b)_\ell$ $\alpha^\ell_\varepsilon < \lambda^+$ is increasing continuous
- $(c)_\ell$ f^ℓ_ε is a $\leq_{\mathfrak{s}}$-embedding of $M^\ell_{\alpha^\ell_\varepsilon}$ into N_ε
- $(d)_\ell$ f^ℓ_ε is increasing continuous with ε
- $(e)_\ell$ if $\varepsilon = 4\zeta + \ell, \ell \in \{0,1\}$ and $j_\varepsilon = \text{Min}\{i : f^\ell_\varepsilon(\mathbf{tp}(a^\ell_i, M^\ell_{\alpha^\ell_\varepsilon}, M_\ell))$ is realized by some $d \in N_{4\zeta} \backslash \text{Rang}(f^\ell_\varepsilon)\}$ is well defined <u>then</u> $a^\ell_{j_\varepsilon} \in \text{Dom}(f^\ell_{\varepsilon+1})$ and $f^\ell_{\varepsilon+1}(a^\ell_{j_\varepsilon}) \in N_{4\zeta}\}$; also in any case $f^{1-\ell}_{\varepsilon+1} = f^{1-\ell}_\varepsilon$
- $(f)_\ell$ if $\varepsilon = 4\zeta + 2 + \ell, \ell \in \{0,1\}$ <u>then</u> $\alpha^\ell_{\varepsilon+1} > \alpha^\ell_\varepsilon$.

First, assume that we succeed, then for some club E of λ^+ for every $\ell < 2, \alpha < \lambda^+$ and $\delta \in E$ we have $a^\ell_\alpha \in M^\ell_\delta \Leftrightarrow \alpha < \delta$ and $\delta \in E \Rightarrow N_\delta \cap \bigcup\limits_{\varepsilon<\lambda^+} \text{Rang}(f^\ell_\varepsilon) = \text{Rang}(f^\ell_\delta)$. Let $\delta \in E$ and $\ell \in \{0,1\}$ and note that $f^\ell_{\delta+\ell} = f^\ell_\delta$.
[Why? If $\ell = 0$ trivially and if $\ell = 1$ by the demand $f^{1-0}_{\varepsilon+1} = f^{1-0}_\varepsilon$ in $(e)_0$.] Now if $\text{Rang}(f^\ell_\delta) \neq N_\delta$ then by the assumption ("$\mathfrak{s}$ is semi$^{\rm na}$-uni-dimensional") see Definition 2.2(1), for some $c \in M^\ell_{\alpha^\ell_\delta+1} \backslash M^\ell_{\alpha^\ell_\delta}$ and $d \in N_\delta \backslash \text{Rang}(f^\ell_\delta)$ we have $\mathbf{tp}_{\mathfrak{s}}(d, \text{Rang}(f^\ell_\delta), N_\delta) = f^\ell_\delta(\mathbf{tp}_{\mathfrak{s}}(c, M^\ell_{\alpha_\delta}, M_\ell))$ and recall $f^\ell_{\delta+\ell} = f^\ell_\delta$ for $\ell = 0, 1$, so by $(e)_\ell$ we have $\text{Rang}(f^\ell_{\delta+2}) \cap N_\delta \backslash \text{Rang}(f^\ell_\delta) \neq \emptyset$ contradiction. So $\delta \in E \wedge \ell \in$

$\{0,1\} \Rightarrow \text{Rang}(f^\ell_\delta) = N_\delta$ hence $f_\ell := \bigcup_{\delta \in E} f^\ell_\delta$ is an isomorphism from M_ℓ onto $N := \bigcup_{\delta \in E} N_\delta$, so $M_1 \cong N \cong M_2$ and we are done.

So we have just to carry the induction, which is straight as $K_{\mathfrak{s}}$ is a $\lambda_{\mathfrak{s}}$-a.e.c. with amalgamation and the hypothesis[1].

4) The proof is similar to that of part (3) but we replace clause $(e)_\ell$ by

$(e)^*_\ell$ if $\varepsilon = 4\zeta + \ell, \ell \in \{0,1\}$ and for some $c \in N_{4\zeta} \setminus \text{Rang}(f^\ell_\varepsilon)$ we have $\mathbf{tp}_{\mathfrak{s}}(c, f^\ell_\varepsilon(M^\ell_{\alpha^\ell_\varepsilon}), N_\varepsilon) \in \mathscr{S}^{\text{bs}}_{\mathfrak{s}}(f^\ell_\varepsilon(M^\ell_{\alpha^\ell_\varepsilon}))$ and $(f^\ell_\varepsilon)^{-1}(\mathbf{tp}_{\mathfrak{s}}(c, f^\ell_\varepsilon(M^\ell_{\alpha^\ell_\varepsilon}), N_\varepsilon))$ has at least two extensions in $\mathscr{S}^{\text{all}}_{\mathfrak{s}}(M^\ell_\beta)$ for some $\beta \in (\alpha^\ell_\varepsilon, \lambda^+)$ <u>then</u>:[2]
for some $c' \in N_{4\zeta} \setminus \text{Rang}(f^\ell_\varepsilon)$ the type $\mathbf{tp}_{\mathfrak{s}}(c', f^\ell_\varepsilon(M^\ell_{\alpha^\ell_\varepsilon}), N_\varepsilon)$ belongs to $\mathscr{S}^{\text{bs}}_{\mathfrak{s}}(f^\ell_\varepsilon(M^\ell_{\alpha_\varepsilon}))$, and $\mathbf{tp}_{\mathfrak{s}}(c', f^\ell_{\varepsilon+1}(M^\ell_{\alpha^\ell_{\varepsilon+1}}), N_{\varepsilon+1})$ is not the non-forking extension of $\mathbf{tp}_{\mathfrak{s}}(c, f^\ell_\varepsilon(M^\ell_{\alpha^\ell_\varepsilon}), N_\varepsilon)$ in $\mathscr{S}^{\text{bs}}_{\mathfrak{s}}(f^\ell_{\varepsilon+1}(M^\ell_{\alpha^\ell_{\varepsilon+1}}))$; also $f^{1-\ell}_{\varepsilon+1} = f^{1-\ell}_\varepsilon$.

Again there is no problem to carry the definition. So it is enough to prove for $\ell = 0,1$ that $\cup\{\text{Rang}(f^\ell_\varepsilon) : \varepsilon < \lambda^+\} = N_{\lambda^+}$, and for this it suffices to prove that $S_\ell = \{\delta < \lambda^+ : \delta$ is a limit ordinal such that $\varepsilon < \delta \Rightarrow \alpha^\ell_\varepsilon < \delta$ and $N_\delta \neq \text{Rang}(f^\ell_\delta)\}$ is not stationary. Toward contradiction assume S_ℓ is stationary; for every $\delta \in S_\ell$ by the assumption "$\mathfrak{s}$ is weakly uni-dimensional" we know that the assumption of $(e)^*_\ell$ holds hence there is $c = c^\ell_\delta$ as there. By Fodor lemma for some c_ℓ the set $S'_\ell = \{\delta \in S_\ell : c^\ell_\delta = c_\ell\}$ is stationary. Choose ordinals $\delta(1) < \delta(2)$ from S'_ℓ so $a^\ell_{j_{\delta(1)}} = c^\ell_{\delta(1)} = c^\ell_{\delta(2)} = a^\ell_{j_{\delta(2)}}$, easy contradiction.

5) So assume $\mathfrak{s}$ is semi^{bs}-uni-dimensional. To prove that $\mathfrak{s}$ is weakly uni-dimensional, let us have $M <_{\mathfrak{s}} M_\ell$ for $\ell = 1,2$ and we should find $c \in M_2 \setminus M_1$ as required in 2.2(6). As we are assuming "$\mathfrak{s}$ is semi^{bs}-uni-dimensional", by Definition 2.2(2) there is $p \in \mathscr{S}^{\text{bs}}_{\mathfrak{s}}(M)$ realized in M_1 and M_2. So there are $c_1 \in M_1, c_2 \in M_2$ such that

---

[1] actually the "semi^{na}-uni-dimensional" can be weakened - in Definition 2.2(1) we may ask $N_1 \in K^{\mathfrak{s}}_{\lambda^+}$

[2] by the proof here also in the proof of part (3) we can avoid using $\langle a^\ell_\varepsilon : \varepsilon < \lambda^+ \rangle$

$\mathbf{tp}_{\mathfrak{s}}(c_1, M, M_1) = p = \mathbf{tp}_{\mathfrak{s}}(c_2, M, M_2)$. Let $c = c_2$ so as c_2 realizes $p \in \mathscr{S}^{\rm bs}_{\mathfrak{s}}(M)$ in M_2 necessarily $c_2 \in M_2 \backslash M$ and $\mathbf{tp}_{\mathfrak{s}}(c_2, M, M_2) = p \in \mathscr{S}^{\rm bs}_{\mathfrak{s}}(M)$.

Now p has at least two extensions in $\mathscr{S}^{\rm all}_{\mathfrak{s}}(M_1) : \mathbf{tp}_{\mathfrak{s}}(c_1, M_1, M_1) \in \mathscr{S}^{\rm all}_{\mathfrak{s}}(M_1) \backslash \mathscr{S}^{\rm na}_{\mathfrak{s}}(M_1)$ and the non-forking extension of p in $\mathscr{S}^{\rm bs}_{\mathfrak{s}}(M_1)$.
6),7) Easy.
8) Similar to the proof of parts (3),(4). $\square_{2.3}$

A conclusion is (see Definitions 0.4, 1.7):
2.5 Conclusion: [$\mathfrak{s}$ is successful good$^+$ λ-frame.]
1) If $\mathfrak{s}$ is semi$^{\rm na}$-uni-dimensional then $\mathfrak{s}\langle\lambda^+\rangle = \mathfrak{s}^+$ so $\mathfrak{K}^{\mathfrak{s}}_{\lambda^+} = \mathfrak{K}_{\mathfrak{s}(+)}$; see Definition 0.4.
2) Similarly if $K^{\mathfrak{s}}$ is categorical in $\lambda^+_{\mathfrak{s}}$.

Proof. 1) Clearly $K_{\mathfrak{s}(+)} \subseteq K^{\mathfrak{s}}_{\lambda^+}$ and by 1.8 we know $\leq_{\mathfrak{K}^{\mathfrak{s}}} \restriction K_{\mathfrak{s}(+)} = \leq_{\mathfrak{s}(+)}$. But by 2.3(3), $K^{\mathfrak{s}}$ is categorical in $\lambda_{\mathfrak{s}(+)} = \lambda^+$. Hence $K_{\mathfrak{s}(+)} = K^{\mathfrak{s}}_{\lambda^+}$ and even $\mathfrak{K}_{\mathfrak{s}(+)} = \mathfrak{K}^{\mathfrak{s}}_{\lambda^+}$ and check similarly for $\bigcup$ and $\mathscr{S}^{\rm bs}$.

2) So again $K_{\mathfrak{s}(+)} = K^{\mathfrak{s}}_{\lambda^+}$ and just check. $\square_{2.5}$

Remark. 1) But we may need "if $\mathfrak{s}$ is non-multi-dimensional then so is $\mathfrak{s}^+$", similarly for uni-dimensional. On this see 2.10 by later results.
2) We also note that the cases we have dealt with in II§3 using categoricity hypothesis, give not just good frames but even uni-dimensional ones.

2.6 Claim. *1) In II.3.7 (= 1.5(1) Case 1 above) we can add: the $\mathfrak{s}$ obtained there is explicitly uni-dimensional.*
2) In II.3.4, if $\mathfrak{K}$ is categorical in $\aleph_1$ (see above 1.5(1), Case 2) then the $\mathfrak{s}$ obtained there is almost uni-dimensional.
3) In II.3.5, if ψ is categorical in $\aleph_2$ (see above 2.6(1), Case 3) then the $\mathfrak{s}$ obtained there is almost uni-dimensional.
4) Assume that $\mathfrak{s}$ is a successful good λ-frame (not necessarily good$^+$), if $\mathfrak{s}$ is almost uni-dimensional, then also the good λ^+-frame $\mathfrak{s}^+$ obtained in II.8.7 is almost uni-dimensional.

5) If T is complete superstable first order and $\mathfrak{s} = \mathfrak{s}^{\kappa}_{T,\lambda}$ (see 1.5(3)) and $\lambda \geq |T| + \kappa^+$ (and $\kappa \neq 0 \Rightarrow T$ stable in λ) <u>then</u>:

(i) *$\mathfrak{s}$ is saturative iff T is non-multi-dimensional (see [Sh:c]; this is (ii) of 1.5(3))*

(ii) *$\mathfrak{K}^{\mathfrak{s}}$ is categorical in λ^+ <u>iff</u> T is uni-dimensional <u>iff</u> $\mathfrak{s}$ is almost uni-dimensional.*

Proof. 1) As for $\mathfrak{s}$, every minimal type $\in \mathscr{S}_{\mathfrak{s}}(M)$ is unavoidable (see the proof of II.3.7).
2) We can easily show that for $M \in K_{\aleph_0}$ there is a minimal $p \in \mathscr{S}^{\rm bs}_{\mathfrak{s}}(M)$, and every minimal $p \in \mathscr{S}^{\rm bs}_{\mathfrak{s}}(M)$ is unavoidable (see the proof of II.3.4), using the categoricity in $\aleph_1$ of course.
3) Similarly.
4) By the analysis of the types for $\mathfrak{s}^+$ by those for $\mathfrak{s}$.
5) Look at 1.5(3), left to the reader. $\square_{2.6}$

2.7 Claim. *Assume that $\mathfrak{s}$ is a weakly successful good λ-frame and $\mathfrak{t}$ is the full good λ-frame constructed from it in Definition II.6.35, Claim II.6.36, (the type-full one).*
1) If $\mathfrak{s}$ is weakly uni-dimensional <u>then</u> $\mathfrak{t}$ is.
2) If $\mathfrak{s}$ is categorical in λ also the inverse holds.

Proof. 1) By the definition.
2) By 2.3(4) and 2.9 below we have: $\mathfrak{s}$ is weakly uni-dimensional iff $K^{\mathfrak{s}}$ is categorical in $\lambda^+_{\mathfrak{s}}$ iff $K^{\mathfrak{t}}$ is categorical in $\lambda^+_{\mathfrak{t}}$ iff $\mathfrak{t}$ is weakly uni-dimensional. $\square_{2.7}$

<u>2.8 Exercise</u>: Assuming $\mathfrak{s}, \mathfrak{t}$ are as in 2.7 and sort out the easy implications for the properties defined in 2.2.

The following is an inverse to 2.3(4).

2.9 Claim. *$[K_{\mathfrak{s}}$ categorical in $\lambda_{\mathfrak{s}}]$. If $K^{\mathfrak{s}}_{\lambda^+}$ is categorical in λ^+ <u>then</u> $\mathfrak{s}$ is weakly uni-dimensional.*

Proof. Assume toward contradiction that $\mathfrak{s}$ is not weakly uni-dimensional, hence we can find $M_0 <_{\mathfrak{s}} M_\ell$ for $\ell = 1, 2$ such that:

if $c \in M_2 \backslash M_0$ and $\mathbf{tp}_{\mathfrak{s}}(c, M_0, M_2) \in \mathscr{S}^{\rm bs}_{\mathfrak{s}}(M_0)$ then it has a unique extension in $\mathscr{S}_{\mathfrak{s}}(M_1)$. By Axiom (D)(d) of good λ-frames (existence) we can choose $c \in M_2 \backslash M_0$ such that $p = \mathbf{tp}_{\mathfrak{s}}(c, M_0, M_2) \in \mathscr{S}^{\rm bs}_{\mathfrak{s}}(M_0)$. Now we choose by induction on $\alpha < \lambda^+$ a model $N_\alpha \in K_{\mathfrak{s}}, \leq_{\mathfrak{s}}$-increasing continuous, $N_\alpha \neq N_{\alpha+1}, N_0 = M_0$ and p has a unique extension in $\mathscr{S}_{\mathfrak{s}}(N_\alpha)$, call it p_α and by Axiom (E)(g) (extension) we know that $p_\alpha \in \mathscr{S}^{\rm bs}_{\mathfrak{s}}(N_\alpha)$ does not fork over N_0. For $\alpha = 0$ this is trivial, for $\alpha = \beta+1$ by 1.21 (noting that every $M \in K_{\mathfrak{s}}$ is isomorphic to M_0 and is $(\lambda, *)$-brimmed as $K_{\mathfrak{s}}$ is categorical in λ) there is an isomorphism f_β from $N_0 = M_0$ onto N_β such that $f_\beta(p_0) = p_\beta$, so we can find $N_\alpha = N_{\beta+1}$ and isomorphism g_β from M_1 onto N_α extending f_β. Hence $f(p) = p_\beta$ and so p_β has a unique extension in $\mathscr{S}_{\mathfrak{s}}(g_\beta(M_1)) = \mathscr{S}_{\mathfrak{s}}(N_\alpha)$ as required. For β limit use Axiom (E)(h), continuity.

Now $N := \bigcup\limits_{\alpha<\lambda^+} N_\alpha \in K^{\mathfrak{s}}_{\lambda^+}$ (recall $N_\alpha \neq N_{\alpha+1}$), and p_β is not realized in N_β for $\beta < \lambda^+$ hence $p = p_0$ is not realized in N, so $N \in K^{\mathfrak{s}}_{\lambda^+}$ is not saturated contradicting categoricity in λ^+. $\square_{2.9}$

Now we consider those properties and how they are related in $\mathfrak{s}$ and $\mathfrak{s}^+$.

2.10 Claim. *Assume $\mathfrak{s}$ is a successful good λ-frame.*
1) If $\mathfrak{s}$ is semi^x-uni-dimensional where $x \in \{$na,bs$\}$ <u>then</u> $\mathfrak{s}^+$ is semi^x-uni-dimensional.
2) If $\mathfrak{s}$ is weakly uni-dimensional, <u>then</u> $\mathfrak{s}^+$ is weakly uni-dimensional.
3) If $\mathfrak{s}$ is almost uni-dimensional, <u>then</u> $\mathfrak{s}^+$ is almost uni-dimensional.
4) If $\mathfrak{s}$ is explicitly uni-dimensional, <u>then</u> $\mathfrak{s}^+$ is explicitly uni-dimensional.
5) If $\mathfrak{s}$ is non-multi-dimensional, <u>then</u> so is $\mathfrak{s}^+$.

Proof. 1) First let $x =$ na. Assume toward contradiction, that $\mathfrak{s}^+$ is not semi$^{\rm na}$-uni-dimensional, so we can find $M_0 <_{\mathfrak{s}(+)} M_\ell$ for $\ell = 1, 2$ such that $c_1 \in M_1 \backslash M_0 \ \ \& \ \ c_2 \in M_2 \backslash M_0 \Rightarrow \mathbf{tp}_{\mathfrak{s}}(c_1, M_0, M_1) \neq \mathbf{tp}_{\mathfrak{s}}(c_2, M_0, M_2)$. Let $\langle M^\ell_\alpha : \alpha < \lambda^+ \rangle$ be a $\leq_{\mathfrak{s}}$-representation of M_ℓ for $\ell < 3$, hence by 1.11 and the definition of $\leq_{\mathfrak{s}(+)}$ for some club E of λ^+ the following holds:

(∗) for $\ell \in \{1,2\}$ and $\alpha < \beta$ in E, we have $\mathrm{NF}_{\mathfrak{s}}(M^0_\alpha, M^\ell_\alpha, M^0_\beta, M^\ell_\beta)$ (hence $M^\ell_\alpha \cap M_0 = M^0_\alpha$) and $M^0_\alpha \neq M^\ell_\alpha$ and for any c_1, c_2 we have:
if $c_1 \in M^1_\alpha \backslash M^0_\alpha$ & $c_2 \in M^2_\alpha \backslash M^0_\alpha$
& $(\exists \gamma < \lambda^+)[\mathbf{tp}_{\mathfrak{s}}(c_1, M^0_\gamma, M^1_\gamma) \neq \mathbf{tp}_{\mathfrak{s}}(c_2, M^0_\gamma, M^2_\gamma)]$
then $\mathbf{tp}_{\mathfrak{s}}(c_1, M^0_\alpha, M^1_\alpha) \neq \mathbf{tp}_{\mathfrak{s}}(c_2, M^0_\alpha, M^2_\alpha)$.

Let $\delta =$ Min(E) and apply the assumption on $\mathfrak{s}$ so there are $c_1 \in M^1_\delta \backslash M^0_\delta, c_2 \in M^2_\delta \backslash M^0_\delta$ satisfying $\mathbf{tp}_{\mathfrak{s}}(c_1, M^0_\delta, M^1_\delta) = \mathbf{tp}_{\mathfrak{s}}(c_2, M^0_\delta, M^2_\delta)$. By the choice of E we have $\beta < \lambda^+ \Rightarrow \mathbf{tp}_{\mathfrak{s}}(c_1, M^0_\beta, M_1) = \mathbf{tp}_{\mathfrak{s}}(c_2, M^0_\beta, M_2)$ and use 1.11(1).
If $x =$ bs the proof is similar using basic types (and 1.10(3), 1.18).
2) So assume $M_0 <_{\mathfrak{s}(+)} M_\ell$ for $\ell = 1, 2$. Let $\langle M^\ell_\alpha : \alpha < \lambda^+ \rangle$ be a $\leq_{\mathfrak{s}}$-representation of M_ℓ and E a thin enough club of λ^+.

For each $\delta \in E$ we have $M^0_\delta <_{\mathfrak{s}} M^\ell_\delta$ for $\ell = 1, 2$ but $\mathfrak{s}$ is weakly uni-dimensional hence for some $c_\delta \in M^2_\delta \backslash M^0_\delta$ the type $p_\delta = \mathbf{tp}_{\mathfrak{s}}(c_\delta, M^0_\delta, M^2_\delta)$ belongs to $\mathscr{S}^{\mathrm{bs}}_{\mathfrak{s}}(M^0_\delta)$ and has more than one extension in $\mathscr{S}_{\mathfrak{s}}(M^1_\delta)$. Clearly there is $\alpha_\delta < \delta$ such that p_δ does not fork for $\mathfrak{s}$ over $M^0_{\alpha_\delta}$ and trivially $p_\delta \restriction M^0_{\alpha_\delta} \in \mathscr{S}^{\mathrm{bs}}_{\mathfrak{s}}(M^0_{\alpha_\delta})$, a set which has cardinality $\leq \lambda$. We also demand that if $\delta = \sup(E \cap \delta)$ then $\alpha_\delta \in E$. By Fodor lemma for some $S \subseteq E$, a stationary subset of λ^+ we have $\delta \in S \Rightarrow \alpha_\delta = \alpha_*$ & $c_\delta = c_*$ & $p_\delta \restriction M^0_{\alpha_*} = p_*$ so $\alpha_* \in E$. For $\alpha \in [\alpha_*, \lambda^+)$ let $p_\alpha = p_\delta \restriction M^0_\alpha$ for every (equivalent for some) $\delta \in S \backslash \alpha$, so $\alpha_* \leq \alpha < \beta < \lambda^+ \Rightarrow p_\alpha = p_\beta \restriction M^0_\alpha$ and p_α does not fork over $p_\alpha \restriction M^0_{\alpha_*} = p_*$. Now c_* realizes p_α for $\alpha \in [\alpha_*, \lambda^+)$ hence by 1.10 the type $p := \mathbf{tp}_{\mathfrak{s}(+)}(c_*, M_0, M_2)$ belongs to $\mathscr{S}^{\mathrm{bs}}_{\mathfrak{s}(+)}(M_0)$. Also by the choice of c_* there are $q^1_{\alpha_*} \neq q^2_{\alpha_*} \in \mathscr{S}_{\mathfrak{s}}(M^1_{\alpha_*})$ extending $p_* \in \mathscr{S}^{\mathrm{bs}}(M^0_{\alpha_*})$.

By 1.17 we can find $M_3, \langle M^3_\alpha : \alpha < \lambda^+ \rangle$ such that $M_1 \leq_{\mathfrak{s}(+)} M_3, \langle M^3_\alpha : \alpha < \lambda^+ \rangle$ is $\leq_{\mathfrak{s}}$-representation of M_3 and $\alpha < \beta < \lambda^+ \Rightarrow \mathrm{NF}_{\mathfrak{s}}(M^1_\alpha, M^3_\alpha, M^1_\beta, M^3_\beta)$ and M^3_α is brimmed over M^1_α (and $M^3_{\alpha+1}$ is brimmed over $M^3_{\alpha+1} \cup M^1_\alpha$; help in building the M^3_α's). Hence we can find $a_1, a_2 \in M^3_{\alpha_*}$ realizing $q^1_{\alpha_*}, q^2_{\alpha_*}$ respectively.

Now if $\beta \in (\alpha_*, \lambda^+) \cap S$ then $\mathrm{NF}_{\mathfrak{s}}(M^1_{\alpha_*}, M^3_{\alpha_*}, M^1_\beta, M^3_\beta)$ and also $\mathrm{NF}_{\mathfrak{s}}(M^0_{\alpha_*}, M^1_{\alpha_*}, M^0_\beta, M^1_\beta)$ by the choice of E (recall $M_0 \leq_{\mathfrak{s}(+)} M_1$), so as $\mathrm{NF}_{\mathfrak{s}}$ satisfies transitivity we have $\mathrm{NF}_{\mathfrak{s}}(M^0_{\alpha_*}, M^3_{\alpha_*}, M^0_\beta, M^3_\beta)$, so together with the previous sentence $\mathbf{tp}_{\mathfrak{s}}(a_\ell, M^0_\beta, M^3_\beta)$ extends p_* and does not fork over $M^0_{\alpha_*}$ hence is equal to p_β which is $\mathbf{tp}_{\mathfrak{s}(+)}(c_*, M^0_\beta, M^1_\beta)$. As this holds for every $\beta \in (\delta, \lambda^+) \cap S$ by 1.10 it follows that $\mathbf{tp}_{\mathfrak{s}(+)}(a_\ell, M_0, M_3) = p = \mathbf{tp}_{\mathfrak{s}(+)}(c_*, M_0, M_3)$.

So for $\ell = 1, 2$ the type $q_\ell := \mathbf{tp}_{\mathfrak{s}}(a_\ell, M_1, M_3) \in \mathscr{S}_{\mathfrak{s}(+)}(M_0)$ extend $\mathbf{tp}_{\mathfrak{s}(+)}(c_*, M_0, M_3)$. But $q_1 = \mathbf{tp}_{\mathfrak{s}}(a_1, M_1, M_3) \neq \mathbf{tp}_{\mathfrak{s}}(a_2, M_1, M_3) = q_2$ because $q_1 \restriction M^1_{\alpha_*} \neq q_2 \restriction M^1_{\alpha_*}$. So q_1, q_2 are as required in the definition of "$\mathfrak{s}^+$ is weakly uni-dimensional".

3) Similar. So let $M_0 \in K_{\mathfrak{s}(+)}$ and let $\langle M^0_\alpha : \alpha < \lambda^+ \rangle$ be a $<_{\mathfrak{K}[\mathfrak{s}]}$-representation of M_0. For limit $\delta < \lambda^+$ there is an unavoidable $p_\delta \in \mathscr{S}^{\mathrm{bs}}_{\mathfrak{s}}(M_\delta)$ and there is an ordinal $\alpha_\delta < \delta$ such that p_δ does not fork over $M^0_{\alpha_\delta}$. For some stationary $S \subseteq \lambda^+$, we have $[\delta \in S \Rightarrow \alpha_\delta = \alpha_* \wedge p_\delta \restriction M^0_{\alpha_\delta} = p_*]$. For $\alpha \in [\alpha_*, \lambda^+)$ define $p_\alpha \in \mathscr{S}^{\mathrm{bs}}_{\mathfrak{s}}(M_\alpha)$ such that $p_\alpha = p_\delta \restriction M^0_\alpha$ for every $\delta \in S \backslash \alpha$. So for some $p \in \mathscr{S}^{\mathrm{bs}}_{\mathfrak{s}(+)}(M_0)$ we have $\alpha_* \leq \alpha < \lambda^+ \Rightarrow p \restriction M^0_\alpha = p_\alpha$ hence $\{\alpha < \lambda^+ : p \restriction M^0_\alpha \in \mathscr{S}^{\mathrm{bs}}_{\mathfrak{s}}(M^0_\alpha)$ is unavoidable$\}$ is stationary. It is easy to check that p is unavoidable (for $\mathfrak{K}_{\mathfrak{s}(+)}$).

4) Essentially the same proof as of part (3).

5) So suppose "$\mathfrak{s}$ is non-multi-dimensional". To prove that $\mathfrak{s}^+$ is non-multi-dimensional, (see Definition 2.2(5)) let $M_0 \leq_{\mathfrak{s}(+)} M_1 <_{\mathfrak{s}(+)} M_2$ be given. We have to find $p \in \mathscr{S}^{\mathrm{bs}}_{\mathfrak{s}(+)}(M_1)$ which is realized in M_2 and does not fork over M_0.

Let $\langle M^\ell_\alpha : \alpha < \lambda^+ \rangle$ be a $<_{\mathfrak{K}[\mathfrak{s}]}$-representation of M_ℓ for $\ell = 0, 1, 2$. So for some club E of λ^+

$\circledast_1$ if $\ell(1) < \ell(2) \leq 2$ and $\alpha < \beta$ are from E then $\mathrm{NF}_{\mathfrak{s}}(M^{\ell(1)}_\alpha, M^{\ell(2)}_\alpha, M^{\ell(1)}_\beta, M^{\ell(2)}_\beta)$.

Also without loss of generality

$\circledast_2$ for $\alpha \in E, M^1_\alpha \neq M^2_\alpha$.

Choosing $\alpha \in E$, by the assumption on $\mathfrak{s}$, for some $a \in M^2_\alpha \backslash M^1_\alpha$ we have

$\circledast_3$ $\mathbf{tp}_{\mathfrak{s}}(a, M^1_\alpha, M^2_\alpha) \in \mathscr{S}^{\mathrm{bs}}_{\mathfrak{s}}(M^1_\alpha)$ does not fork over M^0_α.

By $\circledast_1$ and 1.18 + 1.10, we have: if $\beta \in E \backslash \alpha \Rightarrow \mathbf{tp}_{\mathfrak{s}}(a, M^1_\beta, M^2_\beta)$ does not fork over M^1_α hence (transitivity) over M^0_α. Hence we can deduce that $p := \mathbf{tp}_{\mathfrak{s}(+)}(a, M_1, M_2) \in \mathscr{S}^{\mathrm{bs}}_{\mathfrak{s}(+)}(M_1)$ does not fork over M_0, as required. $\square_{2.10}$

Together we can "close the circle"; to continuing "up" we shall get more (see more in 4.18).

2.11 Conclusion: Assume $\mathfrak{s}$ is a good λ-frame categorical in λ.
1) The frame $\mathfrak{s}$ is weakly uni-dimensional iff $K^{\mathfrak{s}}_{\lambda^+}$ is categorical in λ^+.
2) If in addition $\mathfrak{s}$ is successful and good$^+$ we can add: iff $\mathfrak{s}^+$ is weakly uni-dimensional and $\mathfrak{K}^{\mathfrak{s}}_{\lambda^+} = \mathfrak{K}_{\mathfrak{s}(+)}$.

Proof. 1) The second condition implies the first by 2.9, the first condition implies the second by 2.3(4).
2) The first implies the third as by 2.10(2) we have $\mathfrak{s}^+$ is weakly uni-dimensional and $\mathfrak{K}^{\mathfrak{s}}_{\lambda^+} = \mathfrak{K}_{\mathfrak{s}(+)}$ by 2.3(4) + 2.5. The third condition implies the second as $\mathfrak{K}_{\mathfrak{s}(+)}$ is categorical in λ^+ by the definition of $\mathfrak{K}_{\mathfrak{s}(+)}$. $\square_{2.11}$

A strengthening of 2.11 is

2.12 Claim. *Assume $\mathfrak{s}$ is categorical (in λ) and successful good$^+$ (λ-frame). Then $\mathfrak{s}$ is weakly uni-dimensional iff $\mathfrak{s}^+$ is.*

Remark. We use in the proof that: if $M \in K_{\mathfrak{s}}$ is brimmed and there are so-called "weakly orthogonal" $p_1, p_2 \in \mathscr{S}^{\mathrm{bs}}_{\mathfrak{s}}(M)$, (but this is defined later) then $\mathfrak{s}^+$ is not weakly uni-dimensional. We also use a special case of independence (see 3.9 and §5) in the proof.

Proof. The "only if" direction holds by 2.10(2). For the other direction assume that $\mathfrak{s}$ is not weakly uni-dimensional and we shall prove this for $\mathfrak{s}^+$, so let $M <_{\mathfrak{s}} M_\ell$ for $\ell = 1, 2$ be such that

> $(*)_1$ if $c \in M_2 \backslash M$ and $p := \mathbf{tp}_{\mathfrak{s}}(c, M, M_2) \in \mathscr{S}^{\mathrm{bs}}_{\mathfrak{s}}(M)$ then p has a unique extension in $\mathscr{S}^{\mathrm{all}}_{\mathfrak{s}}(M_1)$.

For $\ell = 1, 2$ let $a_\ell \in M_\ell \backslash M$ be such that $p_\ell := \mathbf{tp}_{\mathfrak{s}}(a_\ell, M, M_1) \in \mathscr{S}^{\mathrm{bs}}_{\mathfrak{s}}(M)$.

By the categoricity of $K_{\mathfrak{s}}$, every $M' \in K_{\mathfrak{s}}$ is brimmed hence by 1.21 it follows that

$(*)_2$ if $M <_{\mathfrak{s}} N$ and $b_1, b_2 \in N$ realizes p_1, p_2 respectively <u>then</u> we can find N_1, N_2, N_3 such that $M \cup \{b_{3-\ell}\} \subseteq N_\ell <_{\mathfrak{s}} N_3$ and $\mathbf{tp}_{\mathfrak{s}}(b_\ell, N_\ell, N_3)$ does not fork over M for $\ell = 1, 2$ and $N \leq_{\mathfrak{s}} N_3$.

As $\mathfrak{s}$ is weakly successful and categorical in λ we can find $\langle M^0_\alpha : \alpha < \lambda^+\rangle$ which is $<_{\mathfrak{s}}$-increasing continuous, $M^0_{\alpha+1}$ brimmed over M^0_α and $M^0_\alpha = M$. Also for $\ell = 1, 2$ we can find $\langle M^\ell_\alpha : \alpha < \lambda^+\rangle$ which is $<_{\mathfrak{s}}$-increasing continuous, $M^\ell_{\alpha+1}$ brimmed over M^ℓ_α and $M^0_\alpha <_{\mathfrak{s}} M^\ell_\alpha$ and $c_\ell \in M^\ell_0$ and $\mathbf{tp}_{\mathfrak{s}}(c_\ell, M^\ell_\alpha, M^\ell_\alpha)$ does not fork over $M^0_0 = M$ and $\alpha < \lambda^+ \Rightarrow M^\ell_\alpha \leq_{\mathfrak{s}} M_\ell, (M^0_\alpha, M^\ell_\alpha, c_\ell) \in K^{3,\mathrm{uq}}_{\mathfrak{s}}$ and $p^\ell_\alpha := \mathbf{tp}_{\mathfrak{s}}(c_\ell, M^0_\alpha, M^\ell_\alpha)$ is a non-forking extension of p_ℓ (why? see later in 4.3, we shall not use this claim till then and note that Hypothesis 4.1 holds by the assumption of our claim 2.12).
So $M'_\ell = \cup\{M^\ell_\alpha : \alpha < \lambda^+\}$ satisfies $M_0 \cup \{c_\ell\} \subseteq M'_\ell \leq_{\mathfrak{K}[\mathfrak{s}]} M_\ell$, hence without loss of generality $M_\ell = M'_\ell$.

Now for $\alpha < \lambda^+$ by $(*)_2$ and 1.21 we have

$(*)^3_\alpha$ if N is a $<_{\mathfrak{s}}$-extension of M^0_α and $b_1, b_2 \in N$ realizes p^1_α, p^2_α respectively then we can find N_1, N_2, N_3 such that $N \leq_{\mathfrak{s}} N_\ell \leq_{\mathfrak{s}} N_3, b_\ell \in N_{3-\ell}, \mathbf{tp}_{\mathfrak{s}}(b_\ell, N_\ell, N_3)$ does not fork over M^0_α for $\ell = 1, 2$.

So if in $(*)^3_\alpha$ also $N'_\ell \leq_{\mathfrak{s}} N_3$ and $(M^0_\alpha, N'_{3-\ell}, a_\ell) \in K^{3,\mathrm{uq}}_{\mathfrak{s}}$, then $\mathrm{NF}_{\mathfrak{s}}(M^0_\alpha, N_{3-\ell}, N'_\ell, N_3)$, i.e. we can replace N_ℓ by N'_ℓ hence as $(M^0_\alpha, M^1_\alpha, c_1) \in K^{3,\mathrm{uq}}_{\mathfrak{s}}$

$(*)^4_\alpha$ for $\alpha < \lambda^+, p^2_\alpha$ has a unique extension in $\mathscr{S}^{\mathrm{bs}}_{\mathfrak{s}}(M^1_\alpha)$.

Let $M_\ell := \cup\{M^\ell_\alpha : \alpha < \lambda^+\}$ so $M_\ell \in K_{\mathfrak{s}(+)}, M_0 \leq_{\mathfrak{s}(+)} M_\ell$ for $\ell = 1, 2$. By $(*)^4_\alpha$ and 1.10(2) we have

$(*)_5$ $\mathbf{tp}_{\mathfrak{s}(+)}(c_2, M_0, M_2)$ has a unique extension in $\mathscr{S}^{\mathrm{all}}_{\mathfrak{s}(+)}(M_1)$.

Still we have to show more. Let $b_2 \in M_2 \backslash M_0$ be such that $\mathbf{tp}_{\mathfrak{s}}(b_2, M_0, M_2) \in \mathscr{S}^{\mathrm{bs}}_{\mathfrak{s}(+)}(M_0)$ and let $\alpha < \lambda^+$ be large enough such that $b_2 \in M^2_\alpha$. For $\beta \in (\alpha, \lambda^+)$ by 1.18 clearly $p^+_\beta = \mathbf{tp}_{\mathfrak{s}}(b_2, M^0_\beta, M^2_\beta)$ belongs to

$\mathscr{S}^{\rm bs}_{\mathfrak{s}}(M^0_\beta)$ and it is a witness for $\mathbf{tp}_{\mathfrak{s}(+)}(b_2, M_0, M_2)$. Now if p^+_β has at least two extensions in $\mathscr{S}^{\rm bs}_{\mathfrak{s}}(M^1_\beta)$ then as above we easily get contradiction to $(M^0_\beta, M^2_\beta, c_2) \in K^{3,{\rm uq}}_{\mathfrak{s}}$.

As this holds for every $\beta \in [\alpha, \lambda^+)$, we get that $\mathbf{tp}_{\mathfrak{s}(+)}(b_2, M_0, M_2)$ has a unique extension in $\mathscr{S}^{\rm bs}_{\mathfrak{s}(+)}(M_1)$. So we are done. $\square_{2.12}$

* * *

Earlier (say in [Sh 576]) minimal type were central, so let us mention them:

2.13 Definition. 1) We say $\mathfrak{s}$ is (a good λ-frame) of minimals when the following holds: $p \in \mathscr{S}^{\rm bs}_{\mathfrak{s}}(M_0)$ implies p is $\mathfrak{s}$-minimal which means: if $M_0 \le_{\mathfrak{s}} M_1 \le_{\mathfrak{s}} N_1, M_0 \le_{\mathfrak{s}} N_0 \le_{\mathfrak{s}} N_1, a \in M_1, p = \mathbf{tp}_{\mathfrak{s}}(a, M_0, M_1)$ and $a \notin N_0$ <u>then</u> $\mathbf{tp}_{\mathfrak{s}}(a, N_0, N_1)$ is a non-forking extension of p (so $p \in \mathscr{S}^{\rm bs}_{\mathfrak{s}}(M_0)$).
Also the triple (M_0, M_1, a) is called $\mathfrak{s}$-minimal when $\mathbf{tp}_{\mathfrak{s}}(a, M_0, M_1)$ is $\mathfrak{s}$-minimal.
2) For an λ-a.e.c. $\mathfrak{K}$ we say that $p \in \mathscr{S}^{\rm na}_{\mathfrak{K}}(M)$ is minimal or $\mathfrak{K}$-minimal or $\le_{\mathfrak{K}}$-minimal <u>when</u> for every $N, M \le_{\mathfrak{K}} N \in K_\lambda, p$ has at most one extension in $\mathscr{S}^{\rm na}_{\mathfrak{K}}(N)$.
3) For an λ-a.e.c. $\mathfrak{K}_\lambda$ let ${\rm frame}^{\rm min}(\mathfrak{K}_\lambda) = {\rm frame}^{\rm min}(\mathfrak{K}, \lambda)$ be defined as in II.3.7 so $\mathscr{S}^{\rm bs}_{\mathfrak{s}}(M) = \{p \in \mathscr{S}_{\mathfrak{K}}(M) : p \text{ is minimal}\}$.

<u>2.14 Exercise</u>: Assume $\mathfrak{s}$ is a good λ-frame. We have $\mathfrak{s}$ is of minimals <u>iff</u> for every $M \in K_{\mathfrak{s}}$ every $p \in \mathscr{S}^{\rm bs}_{\mathfrak{s}}(M)$ is $\mathfrak{K}$-minimal.

How is weakly uni-dimensional preserved in passing to minimal and in II§6? First

2.15 Definition/Claim. *Let $\mathfrak{s}$ be a good λ-frame and $\mathfrak{t} =$ ${\rm frame}^{\rm min}(\mathfrak{K}_{\mathfrak{s}})$.*
1) $\mathfrak{t}$ satisfies all the demands of being a "good λ-frame" from Definition II.2.1 except possibly axiom(D)(c) "density of basic types".
2) $\mathfrak{t}$ is a good λ-frame <u>iff</u> for every $M <_{\mathfrak{s}} N$ for some $c \in N \backslash M$ the type $\mathbf{tp}_{\mathfrak{s}}(c, M, N)$ is minimal.
3) If $\mathfrak{s}$ is categorical (in λ) and is weakly uni-dimensional <u>then</u> $\mathfrak{t}$ is

a good λ-frame almost and even explicitly uni-dimensional; also $\mathfrak{s}$ is almost uni-dimensional.

4) Assume $\mathfrak{s}$ is type-full categorical in λ. If $\mathfrak{t}$ is a good λ-frame, then $\mathfrak{s}$ is weakly uni-dimensional iff $\mathfrak{t}$ is weakly uni-dimensional.

Proof. 1) As in the proof of II.3.7 or of II.6.36, except the extension axioms (E)(c),(E)(g). The latter holds because

- (*a*) minimal types has at most one non-algebraic extension which necessarily is minimal
- (*b*) the existence of disjoint amalgamation proved in II.4.6(5).

We prove Ax(E)(c). So assume that $\langle M_\alpha : \alpha \leq \delta+1 \rangle$ is $\leq_{\mathfrak{s}}$-increasing continuous, $c \in M_{\delta+1} \backslash M_\delta$ and $\mathbf{tp}_{\mathfrak{K}_{\mathfrak{s}}}(c, M_\alpha, M_{\delta+1})$ is not minimal for $\alpha < \delta$. By 1.17 applied to $\mathfrak{s}$ we can find M^ℓ_α for $\alpha \leq \delta, \ell \leq 2$ such that

⊛ (*a*) $M^0_\alpha = M_\alpha$
(*b*) $\mathrm{NF}_{\mathfrak{s}}(M^\ell_\alpha, M^{\ell+1}_\alpha, M^\ell_\beta, M^{\ell+1}_\beta)$ when $\ell \in \{0,1\}$ and $\alpha < \beta < \delta$
(*c*) $M^{\ell+1}_{\alpha+1}$ is brimmed over $M^\ell_{\alpha+1} \cup M^{\ell+1}_\alpha$ for $\alpha \leq \delta$
(*d*) $M^{\ell+1}_\alpha$ is brimmed over M^ℓ_α for $\alpha \leq \delta$ (actually follows).

So there is an $\leq_{\mathfrak{s}}$-embedding f of $M_{\delta+1}$ into M^1_δ over $M_\delta = M_\delta$, so for some $\alpha < \delta$ we have $f(c) \in M^1_{\alpha+1}$. As $\mathbf{tp}_{\mathfrak{s}}(c, M_{\alpha+1}, M_{\delta+1}) = \mathbf{tp}_{\mathfrak{s}}(f(c), M^0_{\alpha+1}, M^1_{\alpha+1})$ is not a minimal type there are distinct $p_1, p_2 \in \mathscr{S}^{\mathrm{na}}_{\mathfrak{K}_{\mathfrak{s}}}(M^1_{\alpha+1})$ extending $\mathbf{tp}(c, M_\alpha, M_{\delta+1})$ and let $c_1, c_2 \in M^2_{\alpha+1}$ realizes p_1, p_2 respectively.
As $\mathrm{NF}_{\mathfrak{s}}(M^0_{\alpha+1}, M^2_{\alpha+1}, M^0_\delta, M^2_\delta)$ and $\mathbf{tp}_{\mathfrak{s}}(c_1, M^0_{\alpha+1}, M^2_{\alpha+1}) = \mathbf{tp}_{\mathfrak{s}}(c, M_{\alpha+1}, M_{\delta+1}) = \mathbf{tp}_{\mathfrak{s}}(c_2, M^0_{\alpha+1}, M^2_{\alpha+1})$ and the uniqueness of NF we have $\mathbf{tp}_{\mathfrak{s}}(c_1, M_\delta, M^2_\delta) = \mathbf{tp}_{\mathfrak{s}}(c, M_\delta, M_{\delta+1}) = \mathbf{tp}_{\mathfrak{s}}(c_2, M_\delta, M^2_\delta)$, but $M^1_{\alpha+1} \leq_{\mathfrak{s}} M^1_\delta$ and $p_1 \neq p_2$ hence
$\mathbf{tp}_{\mathfrak{s}}(c_1, M^1_\delta, M^2_{\delta+1}) \neq \mathbf{tp}_{\mathfrak{s}}(c_2, M^1_\delta, M^2_\delta)$.
This shows that also $\mathbf{tp}_{\mathfrak{s}}(c, M_\delta, M_{\alpha+1})$ is not minimal, as required.
2) By part (1) we have to just check Ax(D)(c) which obviously holds.
3) We shall use part (1) freely.

We can show that for some $M <_{\mathfrak{s}} N$ and $c \in N \backslash M$ the type $\mathbf{tp}_{\mathfrak{s}}(c, M, N)$ is minimal (this is proved as in [Sh 576], essentially as otherwise we contradict $M \in K_{\mathfrak{s}} \Rightarrow |\mathscr{S}_{\mathfrak{s}}(M)| \leq \lambda$ proved in II.4.2).

Assume toward contradiction that $p \in \mathscr{S}^{\rm na}_{\mathfrak{K}[\mathfrak{s}]}(M)$ is minimal and not unavoidable. Then there is N such that $M <_{\mathfrak{s}} N$ and N does not realize p. Now if $M \le_{\mathfrak{s}} M'$ then by part (1) the type p has an extension $p' \in \mathscr{S}^{\rm na}_{\mathfrak{K}_{\mathfrak{s}}}(M')$. So M, M' are superlimit (as $K_{\mathfrak{s}}$ is categorical). By the proof of 1.21 there is an isomorphism f from M onto M' mapping p to p', hence there is N' such that $M' <_{\mathfrak{s}} N'$ and p' is not realized in N'; so by the definition of minimal also p is not realized in N'. Hence we can choose $M_i \in K_{\mathfrak{s}}$ which is $<_{\mathfrak{s}}$-increasing continuous in $i, M_0 = M, M_1 = N$ and M_i omits p. So $M^* = \cup\{M_i : i < \lambda^+\} \in \mathfrak{K}^{\mathfrak{s}}_{\lambda^+}$ is not saturated. But $\mathfrak{K}^{\mathfrak{s}}$ has amalgamation and JEP in λ and is stable in λ hence there is a saturated $M \in K^{\mathfrak{s}}_{\lambda^+}$. Together we get a contradiction by 2.3(4) to "$\mathfrak{s}$ is weakly uni-dimensional". This proves that $M <_{\mathfrak{K}_{\mathfrak{t}}} N \Rightarrow$ every $p \in \mathscr{S}^{\rm bs}_{\mathfrak{t}}(M)$ is realized in N; hence $\mathfrak{t}$ satisfies the density of basic type axiom (D)(c).

So by part (1), $\mathfrak{t}$ is a good λ-frame. Also (by categoricity in λ) for every $M \in K_{\mathfrak{s}} = K_{\mathfrak{t}}$ every $p \in \mathscr{S}^{\rm bs}_{\mathfrak{t}}(M)$ is unavoidable. This means that $\mathfrak{t}$ is explictly uni-dimensinal hence is weakly uni-dimensional and almost uni-dimensional by 2.3(1),(2),(5)).

4) First assume that $\mathfrak{t}$ is weakly uni-dimensional, note that $\mathfrak{K}_{\mathfrak{t}} = \mathfrak{K}_{\mathfrak{s}}$ and as $\mathfrak{s}$ is type-full $M \in K_{\mathfrak{s}} \Rightarrow \mathscr{S}^{\rm bs}_{\mathfrak{t}}(M) \subseteq \mathscr{S}^{\rm bs}_{\mathfrak{s}}(M)$, so by the definitions $\mathfrak{s}$ is weakly uni-dimensional. The other direction holds by (3). $\square_{2.15}$

2.16 Remark. 1) In 2.15(3),(4), if we know and assume enough about orthogonality, primes, etc., (which follows by reasonable assumption, e.g. (relying on later sections) $\mathfrak{s}$ is type-full and $\perp = \underset{\rm wk}{\perp}$ and rk:$\cup\mathscr{S}^{\rm bs}_{\mathfrak{s}}(M) \to$ Ord a "reasonable" rank function, see on it later in this section), then we can omit the assumption of categoricity in λ. [Why? It is enough to prove 2.15(3). We can assume $\mathfrak{s} = \mathfrak{s}^{\rm full}$ is a good λ-frame see 9.6 and $\perp = \underset{\rm wk}{\perp}$. Toward contradiction assume $p \in \mathscr{S}^{\rm bs}_{\mathfrak{s}}(M)$ is with minimal rk(p), for this M but not unavoidable so there is N such that $M \le_{\mathfrak{s}} N$ and N omits p. So there is $b \in N \backslash M$ such that $q := \mathbf{tp}_{\mathfrak{s}}(b, M, N) \in \mathscr{S}^{\rm bs}_{\mathfrak{s}}(M)$. Now p has a unique extension in $\mathscr{S}^{\rm na}_{\mathfrak{K}_{\mathfrak{s}}}(M)$ hence $p \underset{\rm wk}{\perp} q$ (in any reasonable definition of $\underset{\rm wk}{\perp}$) hence by a hypothesis $p \perp q$. So we can choose M_α, a_α

by induction on $\alpha < \lambda^+$ such that M_α is $\le_{\mathfrak{s}}$-increasing continuous and $\mathbf{tp}_{\mathfrak{s}}(a_\alpha, M_\alpha, M_{\alpha+1})$ is a non-forking extension of q and p has a unique extension in $\mathscr{S}^{\rm na}_{\mathfrak{K}_{\mathfrak{s}}}(M_\alpha)$. Then continue as in the proof of (3).]

2.17 Claim. *1) In II.3.7 (= above in 1.5(1),2.6(1)) we can add: $\mathfrak{s}$ is a good λ-frame of minimals.*
2) Similarly[3] *in 2.6(2),(3).*
3) In Definition 1.7, if $\mathfrak{s}$ is a frame of minimals and is successful then so is $\mathfrak{s}^+$.
4) If (M_0, M_1, a) is $\mathfrak{s}$-minimal (i.e., see end of Definition 2.13(1)) then:

- *(i) $p = \mathbf{tp}_{\mathfrak{K}_{\mathfrak{s}}}(a, M_0, M_1)$ is minimal for $\mathfrak{K}_{\mathfrak{s}}$, (and $\in \mathscr{S}^{\rm bs}_{\mathfrak{s}}(M_0)$)*
- *(ii) if $M_0 \le_{\mathfrak{s}} M_2$ and $q \in \mathscr{S}_{\mathfrak{s}}(M_2)$ extends p and is not algebraic then q does not fork over M_0; hence, in particular, $q \in \mathscr{S}^{\rm bs}_{\mathfrak{s}}(M_0)$.*

5) If $M_0 \le_{\mathfrak{s}} M_1$ and $p = \mathbf{tp}_{\mathfrak{s}}(b, M_0, M_1)$ satisfies clauses (i) and (ii) from part (4) (or just clause (i)) then (M_0, M_1, b) is $\mathfrak{s}$-minimal.

Proof. Straight (for (2) see II.3.4).

* * *

We now deal with splitting.

2.18 Definition. Let $\mathfrak{K}$ be a λ-a.e.c. (so $\mathfrak{K} = \mathfrak{K}_\lambda$) with amalgamation (in λ) for simplicity.
1) We say that $p \in \mathscr{S}^\beta_{\mathfrak{K}}(M_1)$ does α-splits or $(\alpha, \mathfrak{K})$-split over $A \subseteq M_1$ if there are $\bar a_1, \bar a_2 \in {}^\alpha(M_1)$ such that:

- (α) $\bar a_1, \bar a_2$ realize the same type over A inside M_1 that is,
  - $(*)$ for some M_2, f we have:
    - (i) $M_1 \le_{\mathfrak{K}} M_2$ (so necessarily $\|M_1\| = \|M_2\| = \lambda$)
    - (ii) f is an automorphism of M_2 over A mapping $\bar a_1$ to $\bar a$

[3] this is not just under the assumptions of Chapter I as in 2.6(2), 2.6(3) we are assuming categoricity in $\aleph_1, \aleph_2$ respectively

(β) if $M_1 \leq_{\mathfrak{K}} M_2$ and $\bar{c} \in {}^{\beta}(M_2)$ realizes p inside M_2 then $\bar{a}_1, \bar{a}_2$ do not realize the same type over $A \cup \bar{c}$ inside M_2, that is for no M_3, f do we have $M_2 \leq_{\mathfrak{K}} M_3$ and f is an automorphism of M_3 over A mapping $\bar{a}_1 \hat{} \bar{c}$ to $\bar{a}_2 \hat{} \bar{c}$.

3) We may write $\bar{a}$ instead of $A = \text{Rang}(\bar{a})$ and M_0 instead of $A = |M_0|$. If we omit α (and write split or $\mathfrak{K}$-split) we mean "for some $\alpha < \lambda^+$"; in fact for now always $\alpha < \lambda_{\mathfrak{s}}^+$.
4) We say $\mathfrak{K}$ has χ-non-splitting if for every $M \in K_\lambda$ and $p \in \mathscr{S}_{\mathfrak{K}}(M)$ there is $A \subseteq M, |A| \leq \chi$ such that p does not split over A (in $\mathfrak{K}$).
5) We say $\mathfrak{s}$ has χ-non-splitting if $\mathfrak{K}_{\mathfrak{s}}$ has basically χ-non-splitting which means that this holds for $p \in \mathscr{S}_{\mathfrak{s}}^{\text{bs}}(M)$.
6) In part (1), (2), (3) though not (4) writing $\mathfrak{s}$ instead of $\mathfrak{K}$ means $\mathfrak{K}_{\mathfrak{s}}$, and $(< \kappa)$-non-splitting has the natural meaning.

Remark. 1) In part (4),(5) we may instead consider $\leq_{\mathfrak{s}}$-increasing chains $\langle M_\alpha : \alpha \leq \delta \rangle$, when $\text{cf}(\delta) \geq \chi$ (or $\text{cf}(\delta) = \chi$).
2) If in 2.21(1) we add "M_α is $(\lambda, *)$-brimmed over M_α for $\alpha < \delta$", then we do not need 2.19, similarly 2.21(1A).

For the rest of this section (though not always needed)

2.19 Hypothesis. The good λ-frame $\mathfrak{s}$ is weakly successful, so $\text{NF}_{\mathfrak{s}}$ is well defined.

2.20 Claim. *1) If* $\text{NF}_{\mathfrak{s}}(M_0, M_1, M_2, M_3)$ *and* $c \in M_2$ *then* $\mathbf{tp}_{\mathfrak{s}}(c, M_1, M_3)$ *does not split over* M_0.
2) Similarly for $\bar{c} \in {}^{\alpha}(M_2)$ *where* $\alpha < \lambda_{\mathfrak{s}}^+$ *if not said otherwise.*

Proof. Straightforward (by uniqueness of $\text{NF}_{\mathfrak{s}}$). $\square_{2.20}$

We could have noted earlier (actually this holds for a λ-a.e.c. $\mathfrak{K} = \mathfrak{K}_\lambda$ which has amalgmation and is stable in λ):

2.21 Claim. *1) Assume*

(a) $\delta < \lambda_{\mathfrak{s}}^+$ *is a limit ordinal*

(b) $\langle M_\alpha : \alpha \leq \delta \rangle$ *is* $\leq_{\mathfrak{s}}$*-increasing continuous*

(c) $p \in \mathscr{S}_{\mathfrak{s}}(M_\delta)$.

Then for some $i < \delta$ *the type* p *does not* λ*-split over* M_i *for* $\mathfrak{K}_{\mathfrak{s}}$.
1A) If $p \in \mathscr{S}^{\mathrm{bs}}_{\mathfrak{s}}(N)$ *does not fork oever* $M \leq_{\mathfrak{s}} N$, *then* p *does not split over* M.
1B) Assume $\lambda^{\mathrm{cf}(\delta)} > \lambda, \mathfrak{K}_\lambda$ *is a* λ*-a.e.c. with amalgamation,* $\delta < \lambda^+$ *a limit ordinal* $\langle M_\alpha : \alpha \leq \delta \rangle$ *is* $\leq_{\mathfrak{K}}$*-increasing continuous, and* $p \in \mathscr{S}_{\mathfrak{K}}(M_\delta)$. *Then for some* $i < \delta$ *for every* $j \in [i, \delta)$ *the type* $p \restriction M_j$ *does not split over* M_i.
2) Assume $\mathfrak{K}$ *is an a.e.c.,* $\mathrm{LS}(\mathfrak{K}) \leq \lambda < \mu, \mathfrak{K}_\lambda$ *has amalgamation, is stable in* λ *and* $M \in K_\mu$. *If* $p = \mathbf{tp}(a, M, N) \in \mathscr{S}_{\mathfrak{K}}(M)$, *then for some* $M_0 \leq_{\mathfrak{K}} M$ *of cardinality* λ, *the type* p *does not* λ*-split over* M_δ.

Proof. 1) By 1.17(1) we can find a $<_{\mathfrak{s}}$-increasing continuous sequence $\langle N_\alpha : \alpha \leq \delta \rangle$ such that $M_\alpha \leq_{\mathfrak{s}} N_\alpha$ and $N_{\alpha+1}$ is $(\lambda, *)$-brimmed over $M_{\alpha+1} \cup N_\alpha$ and $\mathrm{NF}_{\mathfrak{s}}(M_\alpha, N_\alpha, M_{\alpha+1}, N_{\alpha+1})$ for every $\alpha < \delta$. We know (see 1.17(2)) that N_δ is $(\lambda, *)$-brimmed over M_δ, hence some $c \in N_\delta$ realizes p, so for some $i < \delta, c \in N_i$. Easily this i is as required by 2.20.
1A) Similar to (1).
1B) By building a tree of types as in [Sh 31], [Sh 576].
2) Recalling $\lambda < \lambda^{\mathrm{cf}(\lambda)}$ this follows by part (1A). $\square_{2.21}$

We define rank (we can define it for any λ-a.e.c. $\mathfrak{K}$)

2.22 Definition. $\mathrm{rk} = \mathrm{rk}_{\mathfrak{s}}$ is defined as follows:

(a) $\mathrm{rk}_{\mathfrak{s}}(p)$ is defined if $p \in \mathscr{S}_{\mathfrak{s}}(M)$ for some $M \in K_{\mathfrak{s}}$

(b) it is an ordinal or ∞

(c) $\mathrm{rk}_{\mathfrak{s}}(p) \geq \alpha$ iff for every $\beta < \alpha$ we can find (M_1, p_1) such that

(α) $M \leq_{\mathfrak{s}} M_1$

(β) $p_1 \in \mathscr{S}_{\mathfrak{s}}(M_1)$ is an extension of p which splits over M and

(γ) $\mathrm{rk}_{\mathfrak{s}}(p_1) \geq \beta$.

Lastly,

(*d*) $\mathrm{rk}_{\mathfrak{s}}(p) = \alpha$ iff $\mathrm{rk}_{\mathfrak{s}}(p) \geq \alpha$ and $\mathrm{rk}_{\mathfrak{s}}(p) \ngeq \alpha + 1$.

A basic properties of $\mathrm{rk}_{\mathfrak{s}}$ is

2.23 Claim. *If $M \in K_{\mathfrak{s}}$ and $p \in \mathscr{S}_{\mathfrak{s}}(M)$, <u>then</u> $\mathrm{rk}_{\mathfrak{s}}(p) < \infty$.*

Remark. So if $\mathfrak{s}$ is successful, then by 2.27 below, the claim 2.27 below applies to $\mathfrak{s}^+$, too, in fact by 2.26 we have $p \in \mathscr{S}^{\mathrm{bs}}_{\mathfrak{s}(+)}(M) \Rightarrow \mathrm{rk}_{\mathfrak{s}(+)}(p) < \alpha^*$ for some $\alpha^* < \lambda^+_{\mathfrak{s}}$.

Before proving we note that trivially (part (3) see proof of 2.26).

2.24 Exercise. 1) $\mathrm{rk}_{\mathfrak{s}}(p)$ is a well defined ordinal or ∞ <u>when</u> $p \in \mathscr{S}_{\mathfrak{s}}(M), M \in K_{\mathfrak{s}}$.
2) rk is preserved by isomorphisms.
3) For some ordinal $\alpha^* < (2^\lambda)^+$ the set $\{\mathrm{rk}_{\mathfrak{s}}(p) : p \in \mathscr{S}(M), M \in K_{\mathfrak{s}}\}$ is α^* or is $\alpha^* \cup \{\infty\}$.

Proof of 2.23. Assume $\mathrm{rk}_{\mathfrak{s}}(p) = \infty$ hence we can choose by induction on n a triple (M_n, N_n, a), $M_n \leq_{\mathfrak{s}} N_n, a \in N_n$, $\mathrm{rk}_{\mathfrak{s}}(\mathbf{tp}_{\mathfrak{s}}(a, M_n, N_n)) = \infty$ and $M_n \leq_{\mathfrak{s}} M_{n+1}, N_n \leq_{\mathfrak{s}} N_{n+1}$ and $\mathbf{tp}_{\mathfrak{s}}(a, M_{n+1}, N_{n+1})$ does λ-split over M_n and $p = \mathbf{tp}_{\mathfrak{s}}(a, M_0, N_0)$ so $M_0 = M$.
(Why? For $n = 0$ use $M_0 = M$ and N, a such that $\mathbf{tp}_{\mathfrak{s}}(a, M_0, N_0) = p$. Let the ordinal α^* be as 2.24(3); if (M_n, N_n, a) has been chosen then $p_n = \mathbf{tp}_{\mathfrak{s}}(a, M_n, N_n)$ satisfies $\mathrm{rk}_{\mathfrak{s}}(p_n) = \infty > \alpha^*$ hence by the definition, there is a pair (M_{n+1}, p_{n+1}) such that $M_n \leq_{\mathfrak{s}} M_{n+1}$ and $p_{n+1} \in \mathscr{S}_{\mathfrak{s}}(M_{n+1})$ extends p_n, splits over M_n and $\mathrm{rk}_{\mathfrak{s}}(p_{n+1}) \geq \alpha^*$. By the choice of α^* in 2.24(3) we have $\mathrm{rk}_{\mathfrak{s}}(p_{n+1}) = \infty$, and without loss of generality for some N_{n+1} we have: $M_{n+1} \leq_{\mathfrak{s}} N_{n+1}, N_n \leq_{\mathfrak{s}} N_{n+1}$ and a realizes p_{n+1} (we use $\mathfrak{K}_{\mathfrak{s}}$ has amalgamation and the definition of $\mathbf{tp}_{\mathfrak{s}}$).

Clearly we can find $\langle N^+_n : n < \omega\rangle$ such that $M_n \leq_{\mathfrak{s}} N^+_n$ and $\mathrm{NF}_{\mathfrak{s}}(M_n, M_{n+1}, N^+_n, N^+_{n+1})$, N^+_{n+1} is $(\lambda, *)$-brimmed over $M_{n+1} \cup N^+_n$ for $n < \omega$. By 1.17 we know that $N^+_\omega = \cup\{N^+_n : n < \omega\}$ is $(\lambda, *)$-brimmed over $M_\omega = \cup\{M_n : n < \omega\}$, hence we can $\leq_{\mathfrak{s}}$-embed $N_\omega =$

$\cup\{N_n : n < \omega\}$ into N^+_ω over M_ω so without loss of generality $n < \omega \Rightarrow N_n \le_{\mathfrak{s}} N^+_\omega$. So for some $n < \omega$ we have $a \in N^+_n$, and by long transitivity for $\mathrm{NF}_{\mathfrak{s}}$ we have $\mathrm{NF}_{\mathfrak{s}}(M_n, N^+_n, M_\omega, N^+_\omega)$. We get easy contradiction by 2.20 to $\mathbf{tp}_{\mathfrak{s}}(a, M_{n+1}, N^+_{n+1}) = \mathbf{tp}_{\mathfrak{s}}(a, M_{n+1}, N^+_\omega) = \mathbf{tp}_{\mathfrak{s}}(a, M_{n+1}, N_{n+1})$ does λ-split over M_n. $\square_{2.23}$

2.25 Remark. An important point is that for any $\langle M_i : i \le \delta\rangle$ which is $\le_{\mathfrak{s}}$-increasing continuous and $p_i \in \mathscr{S}_{\mathfrak{s}}(M_i)$ for $i < \delta$ such that $i < j \Rightarrow p_i = p_j \restriction M_i$ in general there is no $p \in \mathscr{S}_{\mathfrak{s}}(\cup\{M_i : i < \delta\})$ such that $i < \delta \Rightarrow p_i = p \restriction M_i$, but for $\delta = \omega$ there is (essentially the proof is included in 2.23).

2.26 Claim. *1) If* $\mathrm{rk}_{\mathfrak{s}}(p)$ *is well defined* <u>*then*</u> *it is* $< \lambda^+_{\mathfrak{s}}$ *even* $< \alpha_{\mathfrak{s}}$ *for some* $\alpha_{\mathfrak{s}} < \lambda^+_{\mathfrak{s}}$.
2) If $M <_{\mathfrak{s}} N$ *and* $p \in \mathscr{S}_{\mathfrak{s}}(N)$ *splits over* M, <u>*then*</u> $\mathrm{rk}_{\mathfrak{s}}(p) < \mathrm{rk}_{\mathfrak{s}}(p \restriction M)$.
3) If $\mathrm{NF}_{\mathfrak{s}}(M_0, M_1, M_2, M_3)$ *and* $a \in M_2$, <u>*then*</u> $\mathrm{rk}_{\mathfrak{s}}(\mathbf{tp}_{\mathfrak{s}}(a, M_0, M_3)) = \mathrm{rk}_{\mathfrak{s}}(\mathbf{tp}_{\mathfrak{s}}(a, M_1, M_3))$.
4) If $M \le_{\mathfrak{s}} N$ *and* $p \in \mathscr{S}^{\mathrm{bs}}_{\mathfrak{s}}(N)$ *does not fork over* M <u>*then*</u> *p does not split over* M *and* $\mathrm{rk}_{\mathfrak{s}}(p) = \mathrm{rk}_{\mathfrak{s}}(p \restriction M)$.
5) If M_1 *is brimmed over* $M_0, p \in \mathscr{S}_{\mathfrak{s}}(M_1), p \restriction M_0 \in \mathscr{S}^{\mathrm{bs}}_{\mathfrak{s}}(M_0)$ *and p is not the non-forking extension of* $p \restriction M_0$ *in* $\mathscr{S}_{\mathfrak{s}}(M_1)$ <u>*then*</u> *p does* $\lambda_{\mathfrak{s}}$*-splits over* M_0.
6) Assume $p \in \mathscr{S}_{\mathfrak{s}}(N), M \le_{\mathfrak{s}} N$ *and* $p \restriction M \in \mathscr{S}^{\mathrm{bs}}_{\mathfrak{s}}(M)$. <u>*Then*</u> *p does not fork over* M *iff* $\mathrm{rk}_{\mathfrak{s}}(p) = \mathrm{rk}_{\mathfrak{s}}(p \restriction M)$.

Proof. 1) The set $S = \{\mathrm{rk}_{\mathfrak{s}}(p) : p \in \mathscr{S}_{\mathfrak{s}}(M), M \in \mathfrak{K}_{\mathfrak{s}}\}$ does not contain ∞ by 2.23 and has cardinality $\le 2^\lambda$ by preservation by automorphisms and is downward closed by its definition (and specifically by 2.24(3)) hence is an ordinal (as $|S| \le 2^\lambda$ it cannot be the class of ordinals). By part (3), the set S is equal to $\{\mathrm{rk}_{\mathfrak{s}}(p) : p \in \mathscr{S}(M), M$ is superlimit in $\mathfrak{K}_{\mathfrak{s}}\}$, and, of course, if $p_\ell \in \mathscr{S}_{\mathfrak{s}}(M_\ell), \ell = 1, 2, M_\ell \in \mathfrak{K}_{\mathfrak{s}}, \pi$ an isomorphism from M_1 to M_2 then it induces a mapping $\hat{\pi}$ from $\mathscr{S}_{\mathfrak{s}}(M_1)$ onto $\mathscr{S}_{\mathfrak{s}}(M_2)$ which preserves $\mathrm{rk}_{\mathfrak{s}}$, see 2.24(2). So $S = \{\mathrm{rk}_{\mathfrak{s}}(p) : p \in \mathscr{S}_{\mathfrak{s}}(M)\}$, for any M a superlimit model of $\mathfrak{K}_{\mathfrak{s}}$, but $|\mathscr{S}_{\mathfrak{s}}(M)| \le \lambda_{\mathfrak{s}}$. As S is an ordinal we are done.

2) By the definition and part (1) or 2.23.
3) Prove by induction on the ordinal α that

$$rk_{\mathfrak{s}}(\mathbf{tp}_{\mathfrak{s}}(a, M_0, M_3)) \geq \alpha \Leftrightarrow \text{rk}_{\mathfrak{s}}(\mathbf{tp}_{\mathfrak{s}}(a, M_2, M_3)) \geq \alpha$$

using the $\text{NF}_{\mathfrak{s}}$-calculus.
4) By (3) or recall 2.20(1) or recall 2.21(1A).
5) We can find a $\leq_{\mathfrak{s}}$-increasing continuous sequence $\langle M_{1,\alpha} : \alpha \leq \omega \rangle$ such that $M_{1,0} = M_0, M_{1,\omega} = M_1$ and $M_{1,n+1}$ is $(\lambda_{\mathfrak{s}}, *)$-brimmed over $M_{1,n}$. We can also find $M_{0,n}$ which is $(\lambda_{\mathfrak{s}}, *)$-brimmed over M_0 and $\text{NF}_{\mathfrak{s}}(M_0, M_{1,n}, M_{0,n}, M_{1,n+1})$; note that $M_{0,n}$ does not increase with n. We can also find an $\leq_{\mathfrak{s}}$-increasing continuous $\langle M_{2,\alpha} : \alpha \leq \omega \rangle$ such that $M_{1,n} \leq_{\mathfrak{s}} M_{2,n}$ and $M_{2,n+1}$ is $(\lambda, *)$-brimmed over $M_{1,n+1} \cup M_{2,n}$ and $\text{NF}_{\mathfrak{s}}(M_{1,n}, M_{2,n}, M_{1,n+1}, M_{2,n+1})$ for $n < \omega$. So by 1.17 the model $M_{2,\omega}$ is $(\lambda_{\mathfrak{s}}, *)$-brimmed over $M_{1,\omega} = M_1$ hence some $a \in M_{2,\omega}$ realizes p hence for some $n, a \in M_{2,n}$. So necessarily $\mathbf{tp}_{\mathfrak{s}}(a, M_{1,n}, M_{2,\omega})$ is not a non-forking extension of $p \restriction M_0$ because $\mathbf{tp}_{\mathfrak{s}}(a, M_1, M_{2,\omega})$ forks over M_0 and $M_1 = M_{1,\omega}$ and $\text{NF}_{\mathfrak{s}}(M_{1,n}, M_{2,n}, M_{1,\omega}, M_{2,\omega})$. However, $\mathbf{tp}_{\mathfrak{s}}(a, M_{0,n}, M_{2,\omega})$ is a non-forking extension of $p \restriction M_0$ because $\text{NF}_{\mathfrak{s}}(M_0, M_{0,n}, M_{2,n}, M_{2,n+1})$ follows by transitivity of $\text{NF}_{\mathfrak{s}}$. Also as $M_{1,n}, M_{0,n}$ both are $(\lambda, *)$-brimmed over M_0 there is an isomorphism f_0 from $M_{1,n}$ onto $M_{0,n}$ over M_0. As $M_{1,\omega} = M_1$ is $(\lambda, *)$-brimmed over $M_{1,n+1} \supseteq M_{0,n} \cup M_{1,n}$ and $M_{0,n} \leq_{\mathfrak{s}} M_{0,n+1}, M_{1,n} \leq_{\mathfrak{s}} M_{1,n+1}$ we can extend f_0 to an automorphism of M_1 over M_0. Let $\bar{a}_0$ list the members of M_1 (or $M_{1,n}$), $\bar{a}_1 = f_1(\bar{a}_0)$, so $\bar{a}, \bar{a}_1$ exemplifies the splitting.
6) Should be clear, e.g. we can find $N_\ell\, (\ell \leq 3)$ such that $N_0 = N$, $\text{NF}_{\mathfrak{s}}(N_0, N_1, N_2, N_3)$, $a \in N_2$ realizes p and N_1 is $(\lambda, *)$-brimmed over $N_0 = N$; by part (3) we have $\text{rk}_{\mathfrak{s}}(\mathbf{tp}(a, N_1, N_3)) = \text{rk}_{\mathfrak{s}}(\text{tp}(a, N_0, N_3)) = p$ and we apply part (5) to M, N_1, $\mathbf{tp}(a, N_1, N_3))$. $\square_{2.26}$

We may like to translate ranks between $\mathfrak{s}$ and $\mathfrak{s}^+$.

2.27 Claim. *[$\mathfrak{s}$ is a successful good λ-frame]*

Assume $N_1 <_{\mathfrak{K}[\mathfrak{s}]} M_1, N_1 \in K_{\mathfrak{s}}, M_1 \in K_{\mathfrak{s}(+)}$.

1) If $p \in \mathscr{S}_{\mathfrak{s}(+)}(M_1)$ does not λ-split over N_1 for $\mathfrak{K}^{\mathfrak{s}}$, then $\text{rk}_{\mathfrak{s}(+)}(p) = \text{rk}_{\mathfrak{s}}(p \restriction N_1)$.

2) Also the inverse holds.

3) If $p \in \mathscr{S}^{\rm bs}_{\mathfrak{s}(+)}(M_1)$ and $N_1 \in K_{\mathfrak{s}}$ witnesses it <u>then</u> p does not λ-split over N_1 and moreover does not split over N_1.
4) If $p \in \mathscr{S}_{\mathfrak{s}(+)}(M_1)$ <u>then</u> for some $N_0 <_{\mathfrak{K}[\mathfrak{s}]} M_1$ of cardinality λ, the type p does not λ-split over N_0 and even does not split over N_0.
5) If $p \in \mathscr{S}^{\rm bs}_{\mathfrak{s}(+)}(M_1)$ <u>then</u> N_1 is a witness for $p \in \mathscr{S}^{\rm bs}_{\mathfrak{s}(+)}(M_1)$ iff p does not λ-split over N_1. ("N_1 is a witness for p" is defined in Definition 1.7(1C)).
6) If $p \in \mathscr{S}^{\rm bs}_{\mathfrak{s}(+)}(M)$ is witnessed by $N \leq_{\mathfrak{K}[\mathfrak{s}]} M$ and $N \in K_{\mathfrak{s}}$ <u>then</u> p does not split over N (for $\mathfrak{K}_{\mathfrak{s}(+)}$).

Before proving

2.28 Remark. 1) No real harm in assuming "$\mathfrak{s}$ is type-full" (see Definition II.6.34 and Claim II.6.36 or Definition 9.2). Also $\text{rk}_{\mathfrak{s}}(p)$ is defined in $\mathfrak{K}_{\mathfrak{s}}$.
2) Note that 2.27(1),(2) are similar to 2.26(2),(6) when we replace "$N_1 \in K_{\mathfrak{s}}$" by $N_1 \in K_{\mathfrak{s}(+)}$ and replace λ-split by split.

Proof of 2.27. 1) We prove by induction α that

$\circledast_\alpha$ for any such (N_1, M_1, p) we have
$\text{rk}_{\mathfrak{s}(+)}(p) \geq \alpha \Leftrightarrow \text{rk}_{\mathfrak{s}}(p \restriction N_1) \geq \alpha$

This clearly suffices.
For $\alpha = 0$ and α limit there are no problems. So assume $\alpha = \beta + 1$. First assume $\text{rk}_{\mathfrak{s}(+)}(p) \geq \alpha$ hence by the definition of $\text{rk}_{\mathfrak{s}(+)}$ we can find q, M_2 such that $M_1 \leq_{\mathfrak{s}(+)} M_2, q \in \mathscr{S}_{\mathfrak{s}(+)}(M_2), q \restriction M_1 = p$ and q does λ^+-split over M_1 and $\text{rk}_{\mathfrak{s}(+)}(q) \geq \beta$. Clearly if $\mathfrak{s}(+)$ is weakly successful then by 2.23 we have $\text{rk}_{\mathfrak{s}(+)}(q) < \text{rk}_{\mathfrak{s}(+)}(p)$, hence there is no isomorphism from M_1 onto M_2 mapping p to q; if $\mathfrak{s}^+$ is not necessarily weakly successful still there is no such isomorphism by part (6) proved below. Also N_1 is a witness to p by part (5) and $p \restriction N_1 = q \restriction N_1$ and there is an isomorphism from M_1 onto M_2 extending id_{N_1}. Hence q is not witnessed by N_1 hence by part (5) the type q does λ-split over N_1 hence for some $N_2 \in K_{\mathfrak{s}}$ we have $N_1 \leq_{\mathfrak{s}} N_2 \leq_{\mathfrak{K}[\mathfrak{s}]} M_2$ and $q \restriction N_2$ does λ-split over N_1 for $\mathfrak{K}_{\mathfrak{s}}$ and without loss of generality N_2 is a witness for q. So by the induction hypothesis $\text{rk}_{\mathfrak{s}(+)}(q) \geq \beta \Leftrightarrow \text{rk}_{\mathfrak{s}}(q \restriction N_2) \geq \beta$.

But by the choice of q, $\mathrm{rk}_{\mathfrak{s}(+)}(q) \geq \beta$ hence $\mathrm{rk}_{\mathfrak{s}}(q \restriction N_2) \geq \beta$. By the definition of $\mathrm{rk}_{\mathfrak{s}}$, as $q \restriction N_2$ does λ-split over N_1 for $\mathfrak{s}$, we get $\mathrm{rk}_{\mathfrak{s}}(p \restriction N_1) > \mathrm{rk}_{\mathfrak{s}}(q \restriction N_2) \geq \beta$ so $\mathrm{rk}_{\mathfrak{s}}(p \restriction N_1) \geq \beta + 1 = \alpha$ as required.

Second assume $\mathrm{rk}_{\mathfrak{s}}(p \restriction N_1) \geq \alpha$ so we can find N_2, N_3, a such that $N_1 \leq_{\mathfrak{s}} N_2 \leq_{\mathfrak{s}} N_3, a \in N_3, q^- = \mathbf{tp}_{\mathfrak{s}}(a, N_2, N_3)$ is a λ-splitting (for $\mathfrak{s}$) extension of $p^- = p \restriction N_1$. We use $\mathrm{NF}_{\mathfrak{s}}$ amalgamation to lift this to M_2, q.

2) It is enough to prove $\mathrm{rk}_{\mathfrak{s}(+)}(p) < \mathrm{rk}_{\mathfrak{s}}(p \restriction N_1)$ assuming that p does λ-split over N_1. Now we can find $N_2 \in K_{\mathfrak{s}}$ such that $N_1 \leq_{\mathfrak{s}} N_2 \leq_{\mathfrak{K}[\mathfrak{s}]} M_1$ and N_2 is a witness for p hence by part (5) the type p does not λ-split over N_2 but $p \restriction N_2$ does λ-split over N_1. So by part (1) we have $\mathrm{rk}_{\mathfrak{s}(+)}(p) = \mathrm{rk}_{\mathfrak{s}}(p \restriction N_2)$, and by the definition of $\mathrm{rk}_{\mathfrak{s}}$ we know that $\mathrm{rk}_{\mathfrak{s}}(p \restriction N_2) < \mathrm{rk}_{\mathfrak{s}}(p \restriction N_1)$. Together we are done.

3),4) Left to the reader.

5) Using 1.11 similarly to the proof of 2.26(5).

6) So assume $\gamma < \lambda^{++}$ and $\bar{b}, \bar{c} \in {}^{\gamma}M$, realize the same type over N inside M which means here that we can find (M_1, f_1) such that $M \leq_{\mathfrak{s}(*)} M_1$ and f_1 is an automorphism of M_1 over N mapping $\bar{b}$ to $\bar{c}$. Let M_2 be $(\lambda^+, *)$-brimmed over M_1 for $\mathfrak{K}_{\mathfrak{s}(+)}$ (which is a good λ^+-frame). There is isomorphism g_2 from M onto M_2 over N and let $p_2 := g(p)$, also $p_2 (\in \mathscr{S}^{\mathrm{bs}}_{\mathfrak{s}(+)}(M_2))$ is witnessed by N, (and this implies that p_2 does not λ-split over N). Also easily $N' \leq_{\mathfrak{K}[\mathfrak{s}]} M \wedge N' \in K_{\mathfrak{s}} \Rightarrow p_2 \restriction N' = p \restriction N'$ hence by 1.11 we have $p = p_2 \restriction M$. Also f_1 can be extended to an automorphism f_2 of M_2.

Again by 1.11 we know $p_2 = f_2(p_2)$ hence $\bar{b}, \bar{c} = f_2(\bar{b})$ cannot witness that p_2 split over N hence this holds for $p_2 \restriction M$, as required. $\square_{2.27}$

§3 PRIMES TRIPLES

3.1 Hypothesis. $\mathfrak{s} = (\mathfrak{K}, \bigcup, \mathscr{S}^{\mathrm{bs}})$ is a good λ-frame.

3.2 Definition. 1) Let $K^{3,\mathrm{pr}}_{\lambda} = K^{3,\mathrm{pr}}_{\mathfrak{s}}$ be the family (pr stands for prime) of triples $(M, N, a) \in K^{3,\mathrm{bs}}_{\lambda} = K^{3,\mathrm{bs}}_{\mathfrak{s}}$ such that: if $(M, N', a') \in$

$K^{3,\mathrm{bs}}_\lambda$ and $\mathbf{tp}_{\mathfrak{s}}(a, M, N) = \mathbf{tp}_{\mathfrak{s}}(a', M, N')$ then there is a $\le_{\mathfrak{s}}$-embedding $f : N \to N'$ over M satisfying $f(a) = a'$. Such triples are called prime.
2) We say that $\mathfrak{s} = (\mathfrak{K}, \bigcup, \mathscr{S}^{\mathrm{bs}})$ has primes if ($\mathfrak{s}$ is a good λ-frame and)

(a) if $M \in K_\lambda$ and $p \in \mathscr{S}^{\mathrm{bs}}_{\mathfrak{s}}(M)$ then for some N, a we have $(M, N, a) \in K^{3,\mathrm{pr}}_\lambda$ and $p = \mathbf{tp}_{\mathfrak{s}}(a, M, N)$.

3) We say that (M, N, a) is model-minimal if it belongs to $K^{3,\mathrm{bs}}_\lambda$ and there is no N' such that $M <_{\mathfrak{s}} N' <_{\mathfrak{s}} N$ and $a \in N'$ (this notion is close to "$\mathbf{tp}_{\mathfrak{s}}(a, M, N)$ is of depth zero, N prime over $M \cup \{a\}$" in the context of [Sh:c]).
4) We say $\mathfrak{s}$ has model-minimality if for every $M \in K_{\mathfrak{s}}$ and $p \in \mathscr{S}^{\mathrm{bs}}_{\mathfrak{s}}(M)$ there is $(M, N, a) \in K^{3,\mathrm{bs}}_\lambda$ in which a realizes p and (M, N, a) is model-minimal (compare with Definition 2.13).
5) We say that $\mathfrak{s}$ has primes for $\mathscr{S}'$ where $M \in K_{\mathfrak{s}} \Rightarrow \mathscr{S}'(M) \subseteq \mathscr{S}^{\mathrm{all}}_{\mathfrak{s}}(M)$ when for every $p \in \mathscr{S}'(M), M \in K_{\mathfrak{s}}$ we can find N, a such that $M \le_{\mathfrak{s}} N, a \in N, p = \mathbf{tp}_{\mathfrak{s}}(a, M, N)$ and (M, N, a) belongs to $K^{3,\mathrm{pr}(*)}_{\mathfrak{s}}$, which is defined as in part (1) but the triple is not necessarily in $K^{3,\mathrm{bs}}_{\mathfrak{s}}$.

3.3 Definition. 1) We say $\langle M_i, a_j : i \le \alpha, j < \alpha\rangle$ is a pr-decomposition of N over M or of (M, N) if: M_i is $\le_{\mathfrak{s}}$-increasing continuous, $(M_i, M_{i+1}, a_i) \in K^{3,\mathrm{pr}}_{\mathfrak{s}}, M_0 = M$ and $M_\alpha = N$; we may allow $N \in K^{\mathfrak{s}}_{\lambda^+}$ but $i < \alpha \Rightarrow M_{i+1} \in K_{\mathfrak{s}}$ and $M_0 \in K_{\mathfrak{s}}$. If we demand just $M_\alpha \le_{\mathfrak{s}} N$ (instead $M_\alpha = N$) we say "inside N" instead of "of N". If we also allow $M \le_{\mathfrak{s}} M_0, M_\alpha \le_{\mathfrak{s}} N$ we say in (M, N). Instead "over M" we can say M-based. We call α the length of the decomposition.
2) Similarly for uq ($K^{3,\mathrm{uq}}_{\mathfrak{s}}$ is from Definition II.5.3) and we define uq-decomposition and similarly bs-decomposition. We may write just decomposition (or $\mathfrak{s}$-decomposition) instead pr-decomposition.

Existence of uq-decomposition was used extensively in II§6, (see II.6.9).

3.4 Observation. Assume $\mathfrak{s}$ is weakly successful. If $N_1 \in K_{\mathfrak{s}}$ is $(\lambda, *)$-brimmed over N_0 <u>then</u> we can find a uq-decomposition $\langle M_i, a_j : i \leq \lambda, j < \lambda\rangle$ of N_1 over N_0. For any $a \in N_1$ such that $\mathbf{tp}_{\mathfrak{s}}(a, N_0, N_1) \in \mathscr{S}^{\rm bs}_{\mathfrak{s}}(N_0)$ we can add $a_0 = a$.

Proof. Should be clear. $\square_{3.4}$

3.5 Claim. *1) If $(M, N, a) \in K^{3,\rm pr}_{\mathfrak{s}}$ and $M \cup \{a\} \subseteq N' \leq_{\mathfrak{s}} N$ <u>then</u> $(M, N', a) \in K^{3,\rm pr}_{\lambda}$.*
2) Similarly for $K^{3,\rm uq}_{\mathfrak{s}}$.
3) If $(M, N_1, a_1) \in K^{3,\rm bs}_{\mathfrak{s}}$ is model-minimal and $(M, N_2, a_2) \in K^{3,\rm pr}_{\mathfrak{s}}$ and $p = \mathbf{tp}_{\mathfrak{s}}(a_1, M, N_1) = \mathbf{tp}_{\mathfrak{s}}(a_2, M, N_2)$ <u>then</u> there is an isomorphism from N_1 onto N_2 over M, mapping a_1 to a_2 (so both triples are model-minimal and prime and so if (M, N', a') is prime or is model minimal with $\mathbf{tp}_{\mathfrak{s}}(a', M, N') = \mathbf{tp}_{\mathfrak{s}}(a_\ell, M, N_\ell)$ for some $\ell = 1, 2$ then for $\ell = 1, 2$ there is an isomorphism f_ℓ from N' onto N_ℓ mapping a' to a_ℓ and being the identity on M).
4) If $M_0 \leq_{\mathfrak{s}} M_1 \leq_{\mathfrak{s}} M_2, a_\ell \in M_{\ell+1}$ and $\mathbf{tp}_{\mathfrak{s}}(a_\ell, M_\ell, M_{\ell+1}) \in \mathscr{S}^{\rm bs}_{\mathfrak{s}}(M_\ell)$ does not fork over M_0 for $\ell = 0, 1$ <u>then</u> $(M_0, M_2, a_0) \notin K^{3,\rm uq}_{\mathfrak{s}}$.

Proof. Easy, e.g.
4) By symmetry, (Ax(E)(f) of II.2.1) there are $M_0^+, M_2^+ \in K_{\mathfrak{s}}$ such that $M_0 \leq_{\mathfrak{s}} M_0^+ \leq_{\mathfrak{s}} M_2^+, M_2 \leq_{\mathfrak{s}} M_2^+$ and $a_1 \in M_0^+$ and $\mathbf{tp}_{\mathfrak{s}}(a_0, M_0^+, M_2^+)$ does not fork over M_0. Now $(M_0, M_2, a_0) \leq_{\rm bs} (M_0^+, M_2^+, a_0)$ and a_1 exemplifies $M_0^+ \cap M_2 \neq M_0$. If $\mathfrak{s}$ is weakly successful we recall that non-forking amalgamation is disjoint hence $\neg\, {\rm NF}_{\mathfrak{s}}(M_0, M_2, M_0^+, M_2^+)$. Now $(M_0, M_2, a_0) \in K^{3,\rm uq}_{\mathfrak{s}} \wedge (M_0, M_2, a_0) \leq_{\rm bs} (M_0^+, M_2^+, a_0) \Rightarrow {\rm NF}_{\mathfrak{s}}(M_0, M_2, M_0^+, M_2^+)$ by 1.19, contradiction.

In general we can find (M_0', M_2', f) such that $(M_0, M_2, a_0) \leq_{\rm bs} (M_0', M_2', a_0)$ and f is an isomorphism from M_0^+ onto M_0' over M_0 and $M_0' \cap M_2 = M_0$.
(Why? By the existence of disjoint amalgamation; see II.4.6). Together f, (M_0^+, M_2^+, a_0), (M_0', M_2', a_0) exemplify that $(M_0, M_2, a) \notin K^{3,\rm uq}_{\mathfrak{s}}$. $\square_{3.5}$

<u>3.6 Exercise</u>: 1) In 3.5(4) also $(M_0, M_2, a_1) \notin K^{3,\rm uq}_{\mathfrak{s}}$.
2) Another proof of 3.5(4) is to find $M_0^+ \leq_{\mathfrak{s}} M_1^+ \leq_{\mathfrak{s}} M_2^+$ such that

$M_\ell \le_{\mathfrak{s}} M_\ell^+$ for $\ell = 0,1,2$ and $(M_0, M_1, a_0) \le_{\rm bs} (M_0^+, M_1^+, a_0)$ and $(M_1, M_2, a_1) \le_{\rm bs} (M_1^+, M_2^+, a_1)$.

[Hint: 1) First, find (M_0', M_1', f') such that $(M_0, M_2, a_0) \le_{\rm bs} (M_0', M_1', a_0)$ and f' is an isomorphism from M_1 onto M_0' and $\mathbf{tp}_{\mathfrak{s}}(f'(a_0), M_1, M_1')$ does not fork over M_0, exists as $\mathbf{tp}(a_0, M_0, M_2)$ has a non-forking extension p_0' in $\mathscr{S}^{\rm bs}_{\mathfrak{s}}(M_1)$ and the definitions and use symmetry.

Second, find (M_1'', M_2'', f'') such that $(M_1, M_2, a_1) \le_{\rm bs} (M_1'', M_2'', a_1)$ and f'' is an isomorphism from M_1' onto M_1'' over M_1 such that $\mathbf{tp}_{\mathfrak{s}}(f''(f'(a_0)), M_2, M_2'')$ does not fork over M_1, exists for the same reason. Let $M_0'' = f''(M_0')$ so we have: $(M_0, M_2, a_1) \le_{\rm bs} (M_0'', M_2'', a_1)$ and $f'' \circ f'$ is an isomorphism from M_1 onto M_0'' and $a_0 \in M_1 \le_{\mathfrak{s}} M_2, f'' \circ f'(a_0) \notin M_2$ but trivially $(M_0, M_2, a_1) \le_{\rm bs} (M_1, M_2, a_1)$ easy contradiction to $(M_0, M_2, a_1) \in K^{3,\rm uq}_{\mathfrak{s}}$.]

3.7 Claim. *1) Assume that $\mathfrak{s}$ has primes; if $(M, N, a) \in K^{3,\rm uq}_{\mathfrak{s}}$ <u>then</u> for some N' we have $M \cup \{a\} \subseteq N' \le_{\mathfrak{K}} N$ and $(M, N', a) \in K^{3,\rm pr}_{\mathfrak{s}} \cap K^{3,\rm uq}_{\mathfrak{s}}$; also $K^{3,\rm pr}_{\mathfrak{s}} \subseteq K^{3,\rm uq}_{\mathfrak{s}}$.*
2) If $(M, N, a) \in K^{3,\rm pr}_{\mathfrak{s}}$ and $\mathfrak{s}$ has existence for $K^{3,\rm uq}_{\mathfrak{s}}$ (i.e., $\mathfrak{s}$ is weakly successful) or just for some $N', a', (M, N', a') \in K^{3,\rm uq}_{\mathfrak{s}}$ and $\mathbf{tp}_{\mathfrak{s}}(a', M, N') = \mathbf{tp}_{\mathfrak{s}}(a, M, N)$ <u>then</u> $(M, N, a) \in K^{3,\rm uq}_{\mathfrak{s}}$.

Proof. Immediate:
1) By the definition and monotonicity of $K^{3,\rm uq}_{\mathfrak{s}}$.
2) Easy. $\square_{3.7}$

3.8 Claim. *1) If $\mathfrak{s}$ is non-multi-dimensional and is weakly successful, <u>then</u> all $(M, N, a) \in K^{3,\rm pr}_{\mathfrak{s}}$ are model minimal.*
2) If in addition $\mathfrak{s}$ has primes, <u>then</u> $\mathfrak{s}$ has model-minimality.

Proof. For part (2), let $M \in K_{\mathfrak{s}}$ and $p \in \mathscr{S}^{\rm bs}_{\mathfrak{s}}(M)$. We know (by Definition 3.2(2)(a)) that there is $(M, N_2, a) \in K^{3,\rm bs}_{\mathfrak{s}}$ which is prime and $p = \mathbf{tp}_{\mathfrak{s}}(a, M, N_2)$ and if we know part (1) we are done. For proving part (1) assume (M, N_2, a) is prime. If (M, N_2, a) is model-minimal we are done, otherwise there is N_1 satisfying $M \cup \{a\} \subseteq N_1 <_{\mathfrak{s}} N_2$. So by non-multi-dimensionality there is $b \in N_2 \backslash M$ such that $\mathbf{tp}_{\mathfrak{s}}(b, N_1, N_2) \in \mathscr{S}^{\rm bs}_{\mathfrak{s}}(N_1)$ does not fork over M hence by 3.5(4) we have $(M, N_2, a) \notin K_{\mathfrak{s}}^{3,\mathrm{uq}}$ (where M, N_1, N_2, a, b here correspond to M_0, M_1, M_2, a_0, a_1 there). This easily contradicts "$(M, N_2, a) \in K_{\mathfrak{s}}^{3,\mathrm{pr}} \subseteq K_{\mathfrak{s}}^{3,\mathrm{uq}}$" which holds by 3.7(2).

$\square_{3.8}$

3.9 Claim. *[$\mathfrak{s}$ is a weakly successful (good λ-frame)].*
1) Assume $M_0 \leq_{\mathfrak{s}} M_\ell \leq_{\mathfrak{s}} M_3, a_\ell \in M_\ell$ for $\ell = 1, 2$ and $(M_0, M_\ell, a_\ell) \in K_{\mathfrak{s}}^{3,\mathrm{uq}}$ for $\ell = 1, 2$.

Then $\mathbf{tp}_{\mathfrak{s}}(a_2, M_1, M_3)$ does not fork over M_0 iff $\mathbf{tp}(a_1, M_2, M_3)$ does not fork over M_0.
2) Assume $M_0 \leq_{\mathfrak{s}} M_\ell \leq_{\mathfrak{s}} M_3$ and $a_\ell \in M_\ell$ for $\ell = 1, 2$ and $(M_0, M_1, a_1) \in K_{\mathfrak{s}}^{3,\mathrm{uq}}$ and $\mathbf{tp}_{\mathfrak{s}}(a_2, M_0, M_2) \in \mathscr{S}_{\mathfrak{s}}^{\mathrm{bs}}(M_0)$.

If $\mathbf{tp}_{\mathfrak{s}}(a_1, M_2, M_3)$ does not fork over M_0 then $\mathbf{tp}_{\mathfrak{s}}(a_2, M_1, M_3)$ does not fork over M_0.

Proof. 1) By the symmetry in the claim it is enough to prove the if part, so assume that $\mathbf{tp}_{\mathfrak{s}}(a_1, M_2, M_3)$ does not fork over M_0. As $(M_0, M_1, a_1) \in K_{\mathfrak{s}}^{3,\mathrm{uq}}$ by 1.19 it follows that $\mathrm{NF}_{\mathfrak{s}}(M_0, M_1, M_2, M_3)$, hence by symmetry of $\mathrm{NF}_{\mathfrak{s}}$ (see II.6.25) we have $\mathrm{NF}_{\mathfrak{s}}(M_0, M_2, M_1, M_2)$ which implies that $\mathbf{tp}(a_2, M_1, M_3)$ does not fork over M_0 by 1.18.
2) The proof is included in the proof of part (1). $\square_{3.9}$

3.10 Exercise: In 3.9 we can omit the assumption "$\mathfrak{s}$ is weakly successful".

[Hint: It is enough to prove part (2). By axiom (E)(i) of the definition of good λ-frame (II.2.1) we can find (M_3', f_1, f_2) such that:

$\circledast$ (a) $M_0 \leq_{\mathfrak{s}} M_3'$
(b) f_ℓ is a $\leq_{\mathfrak{s}}$-embedding of M_ℓ into M_3' over M_0 for $\ell = 1, 2$
(c) $\mathbf{tp}(f_\ell(a_\ell), f_{3-\ell}(M_{3-\ell}), M_3')$ does not fork over M_0 for $\ell = 1, 2$.

Now use the definition of $K_{\mathfrak{s}}^{3,\mathrm{uq}}$.]

3.11 Claim. *Assume $\mathfrak{s}$ has primes.*

(1) *If $M \leq_{\mathfrak{s}} N$ then there is a decomposition of N over M (see Definition 3.3(1),(2)). Moreover, if $(M, N, a) \in K_{\mathfrak{s}}^{3,\mathrm{bs}}$ then without loss of generality $a_0 = a$.*

(2) *If* $M \leq_{\mathfrak{K}[\mathfrak{s}]} N, M \in K_{\mathfrak{s}}, N \in K^{\mathfrak{s}}_{\leq \lambda^+}$, *then there is a decomposition of* N *over* M

(3) *In part (2), if* $N \in K^{\mathfrak{s}}_{\lambda^+}$ *the length of the decomposition is* λ^+

(4) *In part (1) there is a decomposition of* N *over* M *of length* $\leq \lambda$

(5) *In part (1) if* N *is* $(\lambda, *)$*-brimmed over* M *or just* $\leq_{\mathfrak{s}}$*-universal over* M, *then there is a decomposition of* N *over* M *of length exactly* λ.

Proof. 1) Without loss of generality $M \neq N$.

We try to choose a_i, M_i by induction $i < \lambda^+$ such that $M_i \leq_{\mathfrak{s}} N$ is $\leq_{\mathfrak{s}}$-increasing continuous, $M_0 = M$ and $i = j+1 \Rightarrow (M_j, M_{j+1}, a_j) \in K^{3,\mathrm{pr}}_{\mathfrak{s}}$. Arriving to i, if $i = 0$ let $M_i = M$. If i is limit let $M_i = \cup\{M_j : j < i\}$. If $M_i = N$ we are done, if not then for some $a_i \in N \backslash M_i$ we have $\mathbf{tp}_{\mathfrak{s}}(a_i, M_i, N) \in \mathscr{S}^{\mathrm{bs}}_{\mathfrak{s}}(M_i)$ and if $i = 0$, and a is given we choose $a_i = a$. If $i = j+1$ then M_j, a_j are already well defined and we know (as $\mathfrak{s}$ has primes and the definition of a prime triple) that there is $M_{j+1} \leq_{\mathfrak{s}} N$ such that $(M_j, M_{j+1}, a_j) \in K^{3,\mathrm{pr}}_{\mathfrak{s}}$; again if $M_i = N$ we are done and otherwise we can choose $a_i \in N \backslash M_i$ such that $\mathbf{tp}(a_i, M_i, N) \in \mathscr{S}^{\mathrm{bs}}_{\mathfrak{s}}(M_i)$. So by cardinality consideration at some point we are stuck, i.e., $M_i = N$ so we are done.
2) By part (1) without loss of generality $N \in K_{\lambda^+}$. Let $\langle b_\varepsilon : \varepsilon < \lambda^+ \rangle$ list the elements of N. Repeating the proof of part (1), now in choosing a_i when $i > 0$ we can choose any $a \in \mathbf{I}_i = \{a \in N \backslash M_i : \mathbf{tp}_{\mathfrak{s}}(a, M_i, N) \in \mathscr{S}^{\mathrm{bs}}_{\mathfrak{s}}(M_i)\}$ so we can demand that $a_i = b_{\varepsilon_1}$ & $(\forall \varepsilon_2 < \lambda^+)[b_{\varepsilon_2} \in \mathbf{I}_i \Rightarrow \varepsilon_1 \leq \varepsilon_2]$. Let $M_{\lambda^+} = \cup\{M_i : i < \lambda^+\}$, obviously $M_{\lambda^+} \leq_{\mathfrak{K}[\mathfrak{s}]} N$. If $N = M_{\lambda^+}$ we are done, otherwise let $\langle N_i : i < \lambda^+ \rangle$ be a $\leq_{\mathfrak{K}}$-representation of N. Clearly for some club E of λ^+ we have:

(∗) if $\delta \in E$ then $N_\delta \cap M_{\lambda^+} = M_\delta$ and $M_\delta \leq_{\mathfrak{s}} N_\delta$ and $N_\delta \nsubseteq M$ (and δ is a limit ordinal).

For each such δ for some $a_\delta \in N_\delta \backslash M_\delta$ and $i_\delta < \delta$ we have $\mathbf{tp}_{\mathfrak{s}}(a_\delta, M_\delta, N_\delta) \in \mathscr{S}^{\mathrm{bs}}_{\mathfrak{s}}(M_\delta)$ does not fork over M_{i_δ}. So by Fodor lemma we can find $a \in N \backslash M_{\lambda^+}$ such that $S = \{i : \mathbf{tp}_{\mathfrak{s}}(a, M_i, N) \in$

$\mathscr{S}^{\rm bs}_{\mathfrak{s}}(M_i)\}$ is stationary, so $i \in S \Rightarrow a \in \mathbf{I}_i$, hence (by the "we can demand" above) we have: if $a = b_{\varepsilon(*)}$ then $i \in S \Rightarrow a_i \in \{b_\varepsilon : \varepsilon < \varepsilon(*)\}$, so we have a 1-to-1 function from S into $[0, \varepsilon(*))$, contradiction. Actually we can get $S = [i(*), \lambda^+)$.
3)-5) Left to the reader. $\square_{3.11}$

3.12 Claim. *1) [$\mathfrak{s}$ is a weakly successful (good λ-frame) with primes].*

If $\mathfrak{C} \in K^{\mathfrak{s}}_{\lambda^+}$ is λ^+-saturated (above λ of course), $M \in K_{\mathfrak{s}}, M \le_{\mathfrak{K}[\mathfrak{s}]} \mathfrak{C}$ and $a_1, a_2 \in \mathfrak{C}$ satisfy $\mathbf{tp}_{\mathfrak{s}}(a_\ell, M, \mathfrak{C}) \in \mathscr{S}^{\rm bs}_{\mathfrak{s}}(M)$ for $\ell = 1, 2$, then the following are equivalent:

- (a) *there are M_1, M_2 from $K_{\mathfrak{s}}$ such that $\mathrm{NF}_{\mathfrak{s}}(M, M_1, M_2, \mathfrak{C})$ and $a_1 \in M_1, a_2 \in M_2$ (the meaning of $\mathrm{NF}_{\mathfrak{s}}$ above is that for some $M_3 \le_{\mathfrak{K}[\mathfrak{s}]} \mathfrak{C}$ from $K_{\mathfrak{s}}$ we have $\mathrm{NF}_{\mathfrak{s}}(M, M_1, M_2, M_3)$)*
- $(b)_\ell$ *there is $M_\ell \le_{\mathfrak{K}[\mathfrak{s}]} \mathfrak{C}$ from $K_{\mathfrak{s}}$ satisfying $M \le_{\mathfrak{s}} M_\ell \le_{\mathfrak{K}[\mathfrak{s}]} \mathfrak{C}$ such that $a_\ell \in M_\ell$ and $\mathbf{tp}_{\mathfrak{s}}(a_{3-\ell}, M_\ell, \mathfrak{C})$ does not fork over M*
- $(c)_\ell$ *if $(M, M_\ell, a_\ell) \in K^{3,\rm uq}_{\mathfrak{s}}$ and $M_\ell \le_{\mathfrak{K}[\mathfrak{s}]} \mathfrak{C}$ then $\mathbf{tp}_{\mathfrak{s}}(a_{3-\ell}, M_\ell, \mathfrak{C})$ does not fork over M*
- $(d)_\ell$ *if $(M, M_\ell, a_\ell) \in K^{3,\rm pr}_{\mathfrak{s}}$ and $M_\ell \le_{\mathfrak{K}[\mathfrak{s}]} \mathfrak{C}$ then $\mathbf{tp}_{\mathfrak{s}}(a_{3-\ell}, M_\ell, \mathfrak{C})$ does not fork over M.*

2) [$\mathfrak{s}$ is a (good) weakly successful λ-frame.] Above $(a) \Leftrightarrow (b)_\ell \Leftrightarrow (c)_\ell \Rightarrow (d)_\ell$.

Proof. 1)

$(a) \Rightarrow (b)_\ell$ by 1.18 (and the symmetry of NF).

$(b)_\ell \Rightarrow (a) + (c)_{3-\ell}$. To prove $(c)_{3-\ell}$ assume $(M, M_{3-\ell}, a_{3-\ell}) \in K^{3,\rm uq}_{\mathfrak{s}}$ so only $M_{3-\ell}$ (and a_1, a_2) are defined.

As we assume $(b)_\ell$ for some $M_\ell \le_{\mathfrak{K}} \mathfrak{C}$ in $K_{\mathfrak{s}}$, we have $\mathbf{tp}_{\mathfrak{s}}(a_{3-\ell}, M_\ell, \mathfrak{C})$ does not fork over M and $a_\ell \in M_\ell$, so $M \le_{\mathfrak{s}} M_\ell$. As $\mathbf{tp}_{\mathfrak{s}}(a_\ell, M, M_\ell) \in \mathscr{S}^{\rm bs}_{\mathfrak{s}}(M)$ clearly $(M, M_\ell, a_\ell) \in K^{3,\rm bs}_{\mathfrak{s}}$.
By 1.19 we have $\mathrm{NF}_{\mathfrak{s}}(M, M_\ell, M_{3-\ell}, \mathfrak{C})$ hence by 1.18 the desired conclusion of $(c)_{3-\ell}$ holds. This proves also clause (a) if we note that, as $\mathfrak{s}$ is weakly successful, for some $(M, N, b) \in K^{3,\rm uq}_{\mathfrak{s}}$, $\mathbf{tp}_{\mathfrak{s}}(b, M, N) =$

$\mathbf{tp}_{\mathfrak{s}}(a_{3-\ell}, M, \mathfrak{C})$, so as $\mathfrak{C}$ is λ^+-saturated without loss of generality $a_{3-\ell} = b, N \leq_{\mathfrak{K}[\mathfrak{s}]} \mathfrak{C}$ and we let $M_\ell := N$.

<u>$(c)_{3-\ell} \Rightarrow (d)_{3-\ell}$</u>: To prove $(d)_{3-\ell}$ assume that $(M, M_{3-\ell}, a_{3-\ell}) \in K^{3,\mathrm{pr}}_{\mathfrak{s}}$ and $M_{3-\ell} \leq_{\mathfrak{K}[\mathfrak{s}]} \mathfrak{C}$; now as "$\mathfrak{s}$ is weakly successful" Claim 3.7(2) implies that $(M, M_{3-\ell}, a_{3-\ell}) \in K^{3,\mathrm{uq}}_{\mathfrak{s}}$, and we can apply clause $(c)_{3-\ell}$ to get the desired conclusion of $(d)_{3-\ell}$.

<u>$(d)_\ell \Rightarrow (b)_\ell$</u>: As $\mathfrak{s}$ has primes there is M_ℓ such that $(M, M_\ell, a_\ell) \in K^{3,\mathrm{pr}}_{\mathfrak{s}}$ and $M_\ell \leq_{\mathfrak{K}[\mathfrak{s}]} \mathfrak{C}$ so by clause $(d)_\ell$ we have $\mathbf{tp}_{\mathfrak{s}}(a_{3-\ell}, M_\ell, \mathfrak{C})$ does not fork over M. So M_ℓ is as required in clause $(b)_\ell$.
Clearly those implications are enough.
2) The proof is included in the proof of part (1) except <u>$(c)_\ell \Rightarrow (b)_\ell$</u> which is proved like $(d)_\ell \Rightarrow (b)_{3-\ell}$ using "weakly successful". $\square_{3.12}$

§4 Prime existence

We give some easy properties of primes for $\mathfrak{s}^+$. A major point is 4.9: existence of primes. We also note how various properties reflect from $K_{\mathfrak{s}^+}$ to $K_{\mathfrak{s}}$. How much the "$\mathfrak{s}$ being good$^+$" rather than just "$\mathfrak{s}$ being good" is necessary? It plays a role, e.g., in the end of proof of 4.3 (and similarly 4.7). If $K^{3,\mathrm{uq}}_{\mathfrak{s}}$ is closed under union of $<^*_{\mathrm{bs}}$-sequences (or less), we could avoid it (in nice cases we shall show it), see 4.10 below.

4.1 Hypothesis. $\mathfrak{s} = (\mathfrak{K}_{\mathfrak{s}}, \bigcup, \mathscr{S}^{\mathrm{bs}})$ is a successful good$^+$ λ-frame, $\mathfrak{K} = \mathfrak{K}[\mathfrak{s}]$ as usual.

Recall (Definition II.4.5) and add

4.2 Definition. 1) We let $\leq_{\mathrm{bs}}$ be the following two-place relation (really quasi order) on $K^{3,\mathrm{bs}}_{\mathfrak{s}}$: $(M, N, a) \leq_{\mathrm{bs}} (M', N', a)$ if both are in $K^{3,\mathrm{bs}}_{\mathfrak{s}}, M \leq_{\mathfrak{s}} M', N \leq_{\mathfrak{s}} N'$ and $\mathbf{tp}_{\mathfrak{s}}(a, M', N')$ does not fork over M.
2) $\leq^*_{\mathrm{bs}}$ is the following quasi order on $K^{3,\mathrm{bs}}_{\mathfrak{s}}$: $(M, N, a) \leq^*_{\mathrm{bs}} (M', N', a)$ if (they are in $K^{3,\mathrm{bs}}_{\mathfrak{s}}$ and) $(M, N, a) \leq_{\mathrm{bs}} (M', N', a)$ and if they are

not equal then M', N' is $\leq_{\mathfrak{s}}$-universal over M, N respectively (and $<^*_{\rm bs}$ has the obvious meaning).
3) $\leq^{**}_{\rm bs}$ is defined similarly with brimmed instead of universal.
4) We may write $\leq^{\mathfrak{s}}_{\rm bs}, \leq^{*,\mathfrak{s}}_{\rm bs}, \leq^{**,\mathfrak{s}}_{\rm bs}$ to recall the frame.

4.3 Claim. *Assume $M_0 \in K_{\mathfrak{s}(+)}$ and $p \in \mathscr{S}^{\rm bs}_{\mathfrak{s}(+)}(M_0)$. Then we can find $a, \bar{M}_0, M_1, \bar{M}_1$ such that:*

(i) $M_0 \leq_{\mathfrak{s}(+)} M_1$
(ii) $a \in M_1$
(iii) $p = \mathbf{tp}_{\mathfrak{s}(+)}(a, M_0, M_1)$
(iv) $\bar{M}_\ell = \langle M_{\ell,\alpha} : \alpha < \lambda^+ \rangle$ *is a* $\leq_{\mathfrak{K}[\mathfrak{s}]}$*-representation of* M_ℓ *for* $\ell = 0, 1$
(v) $a \in M_{1,0}$
(vi) $(M_{0,\alpha}, M_{1,\alpha}, a) \in K^{3,\rm uq}_{\mathfrak{s}}$ *for every* $\alpha < \lambda^+$
(vii) $M_{\ell,i+1}$ *is* $(\lambda, *)$*-brimmed over* $M_{\ell,i}$ *for* $i < \lambda^+, \ell < 2$
(viii) $(M_{0,\alpha}, M_{1,\alpha}, a)$ *is* $<^{\mathfrak{s}}_{\rm bs}$*-increasing and even* $<^{**}_{\rm bs}$*-increasing with* α.

4.4 Definition. We say (M_0, M_1, a) is canonically $\mathfrak{s}^+$-prime (over $\mathfrak{s}$) <u>if</u> there are $\bar{M}^0, \bar{M}^1$ as in claim 4.3 above (see 4.9 below, of course, this depends on $\mathfrak{s}$, but our $\mathfrak{s}$ is constant).

Proof. Let $M_{0,0} \leq_{\mathfrak{K}[\mathfrak{s}]} M_0, M_{0,0} \in K_{\mathfrak{s}}$ be such that $M_{0,0}$ is a witness for p.

We choose by induction on $\alpha < \lambda^+$, a pair $(M_{0,\alpha}, M_{1,\alpha})$ and a such that:

(a) $(M_{0,\alpha}, M_{1,\alpha}, a) \in K^{3,\rm bs}_{\mathfrak{s}}$ and $\mathbf{tp}_{\mathfrak{s}}(a, M_{0,0}, M_{1,0})$ is $p \restriction M_{0,0}$
(b) $(M_{0,\beta}, M_{1,\beta}, a) \leq_{\rm bs} (M_{0,\alpha}, M_{1,\alpha}, a)$ for $\beta < \alpha$
(c) if α is a limit ordinal then $M_{\ell,\alpha} = \bigcup\limits_{\beta<\alpha} M_{\ell,\beta}$ for $\ell = 0, 1$
(d) for every even α, if $(M_{0,\alpha}, M_{1,\alpha}, a) \notin K^{3,\rm uq}_{\mathfrak{s}}$ then $\neg\, {\rm NF}_{\mathfrak{s}}(M_{0,\alpha}, M_{1,\alpha}, M_{0,\alpha+1}, M_{1,\alpha+1})$
(e) for odd α, $M_{\ell,\alpha+1}$ is brimmed over $M_{\ell,\alpha}$ for $\mathfrak{s}$, for $\ell = 1, 2$.

There is no problem to carry the definition (concerning clause (d), it follows that we can do it by 4.5 below).

Before we continue recall

4.5 Claim. *1) If $(M,N,a) \in K^{3,\mathrm{bs}}_{\mathfrak{s}}$ <u>then</u>: $(M,N,a) \notin K^{3,\mathrm{uq}}_{\mathfrak{s}}$ <u>iff</u> for some $(M',N',a) \in K^{3,\mathrm{bs}}_{\mathfrak{s}}$ we have $(M,N,a) \le_{\mathrm{bs}} (M',N',a)$ and $\neg \mathrm{NF}_{\mathfrak{s}}(M,N,M',N')$.*
2) If $(M_\ell,N_\ell,a) <^*_{\mathrm{bs}} (M_{\ell+1},N_{\ell+1},a)$ for $\ell = 0,1$ and $(M_\ell,N_\ell,a) \in K^{3,\mathrm{uq}}_{\mathfrak{s}}$ for $\ell = 0$ <u>then</u> (M_2,N_2,a) is universal over (M_0,N_0,a) for $\le_{\mathrm{bs}}$.
3) In part (1) we can add $(M,N,a) <^{**}_{\mathrm{bs}} (M',N',a)$.

Proof. 1) The implication $\Leftarrow$ (if) holds by Claim 1.19. The other direction holds by the definition of $K^{3,\mathrm{uq}}_{\mathfrak{s}}$ and the uniqueness of $\mathrm{NF}_{\mathfrak{s}}$-amalgamation by II§6.
2) Easy (by the definition of $\le^*_{\mathrm{bs}}$ and $\le_{\mathrm{bs}}$ being transitive).
Assume $(M_0,N_0,a) \le_{\mathrm{bs}} (M',N',a)$. Now M_1 is $\le_{\mathfrak{s}}$-universal over M_0 (as $(M_0,N_0,a) <^*_{\mathrm{bs}} (M_1,N_1,a)$) hence there is a $\le_{\mathfrak{s}}$-embedding f_0 of M' into M_1 over M_0. By uniqueness of $\mathrm{NF}_{\mathfrak{s}}$ as $(M_0,N_0,a) \in K^{3,\mathrm{uq}}_{\mathfrak{s}}$, there is a pair (N^+_1, f_1) such that: $N_1 \le_{\mathfrak{s}} N^+_1, f_1$ is a $\le_{\mathfrak{s}}$-embedding of N' into N^+_1 extending $f_0 \cup \mathrm{id}_{N_0}$. As $(M_1,N_1,a) <^*_{\mathrm{bs}} (M_1,N_2,a)$ we know that N_2 is $\le_{\mathfrak{s}}$-universal over N_1 so as $N_1 \le_{\mathfrak{s}} N^+$ there is a $\le_{\mathfrak{s}}$-embedding f_2 of N^+ into N_2 over N. Now $f_2 \circ f_1$ is a $\le_{\mathfrak{s}}$-embedding as required.
3) Should be clear. $\square_{4.5}$

<u>Continuation of the proof of 4.3</u>

By clause (e), necessarily $M'_\ell := \bigcup\limits_{\alpha<\lambda^+} M_{\ell,\alpha} \in K^{\mathfrak{s}}_{\lambda^+}$ are saturated for $\ell = 0,1$. Also $M'_0 \le_{\mathfrak{K}[\mathfrak{s}]} M'_1$ and by clause (b) we have $\mathbf{tp}_{\mathfrak{s}(+)}(a, M'_0, M'_1) \in \mathscr{S}^{\mathrm{bs}}_{\mathfrak{s}(+)}(M'_0)$ is a stationarization of $p \restriction M_{0,0}$, i.e., is witnessed by it. Now $M'_0, M_0 \in K_{\mathfrak{s}(+)}$ are saturated and $<_{\mathfrak{K}[\mathfrak{s}]}$-extend $M_{0,0}$ which $\in K_{\mathfrak{s}}$ hence clearly there is an isomorphism h from M_0 onto M'_0 over $M_{0,0}$. So $h(p) \in \mathscr{S}^{\mathrm{bs}}_{\mathfrak{s}(+)}(M'_0)$ does not fork over $M_{0,0}$ and $h(p) \restriction M_{0,0} = p \restriction M_{0,0} = \mathbf{tp}_{\mathfrak{s}}(a, M_{0,0}, M_{1,0})$. Now also $\mathbf{tp}_{\mathfrak{s}}(a, M'_0, M'_1)$ has those properties hence by 1.10 we have

$h(p) = \mathbf{tp}_{\mathfrak{s}}(a, M_0', M_0'')$. So by renaming without loss of generality $M_0' = M_0$ and $\mathbf{tp}_{\mathfrak{s}}(a, M_0', M_1') = p$ and by 1.6(1), clause (b) we have $M_0 \le^*_{\lambda^+} M_1$, so by its definition (see II.7.2) for some club E of λ^+ we have $\alpha \in E \;\&\; \alpha < \beta \in E \Rightarrow \mathrm{NF}_{\mathfrak{s}}(M_{0,\alpha}, M_{1,\alpha}, M_{0,\beta}, M_{1,\beta})$, hence by monotonicity of $\mathrm{NF}_{\mathfrak{s}}$ and clause (d) of the construction we have $\alpha \in E \Rightarrow (M_{0,\alpha}, M_{1,\alpha}, a) \in K^{3,\mathrm{uq}}_\lambda$. By renaming we get the conclusion. $\square_{4.3}$

<u>4.6 Conclusion</u>: 1) $K^{3,\mathrm{uq}}_{\mathfrak{s}}$ is $<^*_{\mathrm{bs}}$-dense and $<^{**}_{\mathrm{bs}}$-dense in $K^{3,\mathrm{bs}}_{\mathfrak{s}}$.
2) If $(M_1, N_1, a) <^*_{\mathrm{bs}} (M_2, N_2, a) \le_{\mathrm{bs}} (M_3, N_3, a)$ <u>then</u> $(M_1, N_1, a) <^*_{\mathrm{bs}} (M_3, N_3, a)$.
3) If $(M_1, N_1, a) \le_{\mathrm{bs}} (M_2, N_2, a) <^*_{\mathrm{bs}} (M_3, N_3, a)$ <u>then</u> $(M_1, N_1, a) <^*_{\mathrm{bs}} (M_3, N_3, a)$.
4) If $(M_1, N_1, a) \le_{\mathrm{bs}} (M_2, N_2, a) <^{**}_{\mathrm{bs}} (M_3, N_3, a)$ <u>then</u> $(M_1, N_1, a) <^{**}_{\mathrm{bs}} (M_3, N_3, a)$.
5) If $(M_1, N_1, a) <^{**}_{\mathrm{bs}} (M_2, N_2, a)$ <u>then</u> $(M_1, N_1, a) <^*_{\mathrm{bs}} (M_2, N_2, a)$. If $(M_1, N_1, a) <^*_{\mathrm{bs}} (M_2, N_2, a)$ <u>then</u> $(M_1, N_1, a) <_{\mathrm{bs}} (M_1, N_1, a)$.

Proof. 1) By the proof of 4.3.
2),3),4),5) Immediate. $\square_{4.6}$

The point of the following claim is in clause (iv).

4.7 Claim. *1) Assume $\beta < \lambda^+$, $\langle M_i : i \le \beta\rangle$ is $\le_{\mathfrak{s}}$-increasing continuous and $(M_i, M_{i+1}, a_i) \in K^{3,\mathrm{bs}}_{\mathfrak{s}}$ for $i < \beta$ and $M_0 \le_{\mathfrak{s}} M^+$. <u>Then</u> we can find $\langle N_i : i \le \beta\rangle$ such that:*

- *(i)* $M_i \le_{\mathfrak{s}} N_i$
- *(ii)* *N_i is $\le_{\mathfrak{s}}$-increasing continuous*
- *(iii)* *$\mathbf{tp}_{\mathfrak{s}}(a_i, N_i, N_{i+1})$ does not fork over M_i*
- *(iv)* $(N_i, N_{i+1}, a_i) \in K^{3,\mathrm{uq}}_{\mathfrak{s}}$
- *(v)* *M^+ can be $\le_{\mathfrak{s}}$-embedded into N_0 over M_0*
- *(vi)* *N_i is $(\lambda, *)$-brimmed over M_i for $i \le \beta$*

2) Assume further $\mathrm{NF}_{\mathfrak{s}}(M_0, M^+, M_\beta, M^*)$, so in particular $M^+ = M_0, M_\beta \le_{\mathfrak{s}} M^*$, <u>then</u> we can replace (v) by

$(v)^+$ $M^+ \le_{\mathfrak{s}} N_0$ and $M^* \le_{\mathfrak{s}} N_\beta$.

3) If in part (1) we have $(M_i, M_{i+1}, a_i) \in K^{3,\mathrm{uq}}_{\mathfrak{s}}$ for $i < \beta$ then $i < j \le \beta \Rightarrow \mathrm{NF}_{\mathfrak{s}}(M_i, N_i, M_j, N_j)$ so in particular $\mathrm{NF}_{\mathfrak{s}}(M_0, N_0, M_\beta, N_\beta)$ so $\mathrm{NF}_{\mathfrak{s}}(M_0, M^+, M_\beta, N_\beta)$.

Proof. 1) We try to choose by induction on $\zeta < \lambda^+$ a sequence $\bar{M}^\zeta = \langle M^\zeta_i : i \le \beta\rangle$ such that

$\circledast$ (a) $\bar{M}^\zeta$ is $\le_{\mathfrak{s}}$-increasing continuous

(b) $\bar{M}^0 = \langle M_i : i \le \beta\rangle$

(c) for each $i \le \beta$ the sequence $\langle M^\varepsilon_i : \varepsilon \le \zeta\rangle$ is $\le_{\mathfrak{s}}$-increasing continuous

(d) $\mathbf{tp}_{\mathfrak{s}}(a_i, M^\zeta_i, M^\zeta_{i+1})$ belongs to $\mathscr{S}^{\mathrm{bs}}_{\mathfrak{s}}(M^\zeta_i)$ and does not fork over M^0_i

(e) if $\zeta = 1$ then M^+ can be $\le_{\mathfrak{s}}$-embedded into M^ζ_0 over $M_0 = M^0_0$

(f) if $\zeta = \varepsilon + 1$ and ε limit, then for some $i < \beta$ we have $\neg\, \mathrm{NF}_{\mathfrak{s}}(M^\varepsilon_i, M^\varepsilon_{i+1}, M^\zeta_i, M^\zeta_{i+1})$

(g) if $\zeta = \varepsilon + 2$ then M^ζ_i is $(\lambda, *)$-brimmed over $M^{\varepsilon+1}_i$ for $i \le \beta$

(h) if $\zeta = \varepsilon + 2$ and $i = j + 1 \le \beta$ then M^ζ_i is $(\lambda, *)$-brimmed over $M^\zeta_j \cup M^{\varepsilon+1}_i$.

There is no problem to define for $\zeta = 0, \zeta = 1$ and ζ limit. For $\zeta = \varepsilon + 1, \varepsilon$ not limit straightforward: we choose M^ζ_i by induction on $i \le \beta$ such that $\mathrm{NF}_{\mathfrak{s}}(M^\varepsilon_i, M^\varepsilon_{i+1}, M^\zeta_i, M^\zeta_{i+1})$ and $M^\zeta_{i+1} \cap M^\varepsilon_\beta = M^\varepsilon_{i+1}$ for $i < \beta$ and M^ζ_{i+1} is brimmed over $M^\varepsilon_{i+1} \cup M^\zeta_i$, recalling 1.17. So assume ε is a limit ordinal, if $i < \beta \Rightarrow (M^\varepsilon_i, M^\varepsilon_{i+1}, a_i) \in K^{3,\mathrm{uq}}_{\mathfrak{s}}$ then we are done: let $\langle N_i : i \le \beta\rangle = \langle M^\zeta_i : i \le \beta\rangle$, obviously clause (iv) holds and clause (vi) holds by clause (g).

So assume $i < \beta$ and $(M^\varepsilon_i, M^\varepsilon_{i+1}, a_i) \notin K^{3,\mathrm{uq}}_{\mathfrak{s}}$, but of course $(M^\varepsilon_j, M^\varepsilon_j, a_i) \in K^{3,\mathrm{bs}}_{\mathfrak{s}}$ for $j \le \beta$. Now we choose $\langle M^\zeta_j : j \le i\rangle$ as in the case of $\zeta = \varepsilon' + 2$ and $M^\zeta_i \cap M^\varepsilon_\beta = M^\varepsilon_i$. As above, M^ζ_i is $(\lambda, *)$-brimmed over M^ε_i. By 4.5(3) there is (M', N') such that $(M^\varepsilon_i, M^\varepsilon_{i+1}, a_i) <^{**}_{\mathrm{bs}} (M', N', a_i)$ and $\neg\mathrm{NF}(M^\varepsilon_i, M^\varepsilon_{i+1}, M', N')$. But

M', M^ζ_i are both $(\lambda, *)$-brimmed over M^ε_i, so without loss of generality $M' = M^\zeta_i$ and let $M^\zeta_{i+1} := N'$ and without loss of generality $M^\zeta_{i+1} \cap M^\varepsilon_\beta = M^\varepsilon_{i+1}$.

Lastly, we choose $\langle M^\zeta_j : j \in (i+1, \beta]\rangle$ as in the case $\zeta = \varepsilon' + 2$.

Next assume that we succeed to carry the induction for all $\zeta < \lambda^+$. As $\mathfrak{s}$ is successful and good$^+$ by 1.8(1), for each $i < \beta$ for some club E_i of λ^+, for every $\varepsilon < \zeta$ from E_i we have $\mathrm{NF}_\mathfrak{s}(M^\varepsilon_i, M^\varepsilon_{i+1}, M^\zeta_i, M^\zeta_{i+1})$. Let $\varepsilon < \zeta$ be successive members of $E = \cap\{E_i : i < \beta\}$, so by monotonicity of non-forking we have $i < \beta \Rightarrow \mathrm{NF}_\mathfrak{s}(M^\varepsilon_i, M^\varepsilon_{i+1}, M^\zeta_i, M^\zeta_{i+1})$, contradiction to the construction.

2) Repeat the construction in (1) but in clause (f) this is demanded only if possible. So obviously we can carry the construction. Let f be a $\leq_\mathfrak{s}$-embedding of M^+ into $M^0_{0,1}$, this guarantees clause (e).

Now $M_i := \cup\{M^\zeta_i : \zeta < \lambda^+\}$ is saturated for $i \leq \beta$ so we can find a $\leq_\mathfrak{s}$-embedding f of M^* into N_β over M_β and by uniqueness of NF. Without loss of generality g extends f hence f is into $M^{\zeta(1)}_\beta$ for some $\zeta(1) < \lambda^+$. In the end of the proof of (1) demand $\zeta > \zeta(1)$. (Alternatively use clause (g) for $i = \beta$).

3) Easy. $\square_{4.7}$

<u>4.8 Exercise</u>: Even if in 4.7 we omit the assumption $\langle M_i : i < \beta\rangle$ is continuous, we can still get the result.

4.9 Claim. *(Prime Existence) 1) If $M \in K_{\mathfrak{s}(+)}, p \in \mathscr{S}^{\mathrm{bs}}_{\mathfrak{s}(+)}(M)$, <u>then</u> there are $N \in K_{\mathfrak{s}(+)}$ and an element a satisfying $(M, N, a) \in K^{3,\mathrm{pr}}_{\mathfrak{s}(+)}$ and $p = \mathbf{tp}_{\mathfrak{s}(+)}(a, M, N)$. This means that if $M \leq_\mathfrak{K} N' \in K_{\mathfrak{s}(+)}$ and $a' \in N'$ realizes p <u>then</u> there is a $\leq_\mathfrak{K}$-embedding f of N into N' such that $f \restriction M = \mathrm{id}_M, f(a) = a'$.*
2) In fact if (M, N, a) is like (M_0, M_1, a) of 4.3 <u>then</u> this holds, i.e., if (M, N, a) is canonically $\mathfrak{s}^+$-prime <u>then</u> it is prime for $\mathfrak{s}^+$.

Proof of 4.9. Let $M_0 = M$ and let $\bar{M}_0, M_1, \bar{M}_1, a$ be as in 4.3 and let $N = M_1$ and we shall prove that N, a are as required. So let $M \leq_{\mathfrak{s}(+)} M' \in K_{\mathfrak{s}(+)}$ and $a' \in M'$ be such that $\mathbf{tp}_{\mathfrak{s}(+)}(a', M, M') = p$. We choose by induction on $\alpha < \lambda^+$ a $\leq_\mathfrak{K}$-embedding f_α of $M_{1,\alpha}$ into M', such that $f_\alpha(a) = a'$, f_α is increasing continuous and $f_\alpha \restriction$

$M_{0,\alpha} \equiv \mathrm{id}_{M_{0,\alpha}}$. For $\alpha = 0$, as M' is saturated in $\mathfrak{K}^{\mathfrak{s}}_{\lambda^+}$ (above λ) and a' realizes in M' the type $p \restriction M_{0,0}$ this should be clear. For α limit take the unions. For $\alpha = \beta + 1$, choose a model $N_\alpha \leq_{\mathfrak{K}[\mathfrak{s}]} M'$ from $K_{\mathfrak{s}}$ which includes $f_\beta(M_{1,\beta}) \cup M_{0,\beta+1}$ and is $(\lambda, *)$-brimmed for $\mathfrak{s}$ over this set; there is such N_α as M' is saturated above λ. So as $(M_{0,\beta}, M_{1,\beta}, a) \in K^{3,\mathrm{uq}}_\lambda$ and $\mathbf{tp}_{\mathfrak{s}}(a, M_{0,\beta+1}, M')$ does not fork over $M_{0,\beta}$, we have $\mathrm{NF}_{\mathfrak{s}}(M_{0,\beta}, f_\beta(M_{1,\beta}), M_{0,\beta+1}, N_\alpha)$ hence by the uniqueness of $\mathrm{NF}_{\mathfrak{s}}$ (i.e., the Definition of $K^{3,\mathrm{uq}}_{\mathfrak{s}}$) we can extend $f_\beta \cup \mathrm{id}_{M_{0,\alpha}}$ to a $\leq_{\mathfrak{K}}$-embedding f_α of $M_{1,\alpha}$ into N_α.

So having carried the induction, $f = \cup\{f_\alpha : \alpha < \lambda^+\}$ is a $\leq_{\mathfrak{K}[\mathfrak{s}]}$-embedding of $M_1 = \bigcup_{\alpha<\lambda^+} M_{1,\alpha}$ into M' over $M = M_0$ mapping a to a', so we are done. $\square_{4.9}$

Closely related to claim 4.9 is:

4.10 Claim. *$(K^{3,\mathrm{uq}}_{\mathfrak{s}}, \leq^{\mathfrak{s}}_{\mathrm{bs}})$ is λ^+-strategically closed inside $(K^{3,\mathrm{bs}}_{\mathfrak{s}}, \leq^{\mathfrak{s}}_{\mathrm{bs}})$. Moreover, $(K^{3,\mathrm{uq}}_{\mathfrak{s}}, <^*_{\mathrm{bs}})$ is λ^+-closed, i.e., if $\delta < \lambda^+$ is a limit ordinal, $\langle (N_{0,i}, N_{1,i}, a) : i < \delta\rangle$ is $\leq_{\mathrm{bs}}$-increasing continuous in $K^{3,\mathrm{uq}}_{\mathfrak{s}}$ and $N_{0,i+1}, N_{1,i+1}$ is $(\lambda, *)$-brimmed or just universal over $N_{0,i}, N_{1,i}$ respectively for each $i < \delta$ (equivalently, the sequence is $\leq^*_{\mathrm{bs}}$-increasing continuous),*

<u>then</u> $(\bigcup_{i<\delta} N_{0,i} \bigcup_{i<\delta} N_{1,i}, a) \in K^{3,\mathrm{uq}}_{\mathfrak{s}}$ is $<^{**}_{\mathrm{bs}}$-above $(N_{0,j}, N_{1,j}, a)$ for every $j < \delta$.

Recalling

4.11 Definition. Let I, J be partial orders and $I \subseteq J$. We say I is δ-strategically closed inside J if in the following game the COM player has a winning strategy. A play last δ moves, for $\alpha < \delta$ the COM player chooses $s_\alpha \in I$ such that $\beta < \alpha \Rightarrow t_\beta \leq_J s_\alpha$ and if α is a limit ordinal, s_α is the $\leq_I$-lub of $\langle s_\beta : \beta < \alpha\rangle$ in I and then the player INC (for incomplete) chooses t_α such that $s_\alpha \leq_J t_\alpha$. The player COM wins the play if for every $\alpha < \delta$ he has a legal move; otherwise, the player INC wins the play.

Proof of 4.10. It suffices to prove the second sentence by 4.6(1). Let $\langle (M_{0,i}, M_{1,i}, a) : i < \lambda^+ \rangle$ be as constructed in 4.3 such that $i = 0 \Rightarrow (M_{0,i}, M_{1,i}, a) = (N_{0,i}, N_{1,i}, a)$. Recall that $\langle (N_{0,i}, N_{1,i}, a) : i < \delta \rangle$ is $<^*_{\mathrm{bs}}$-increasing.

We now by induction on i choose f_i, α_i such that:

(*a*) f_i is an $\leq_{\mathfrak{s}}$-embedding of $N_{1,i}$ into M_{1,α_i}

(*b*) f_i increasing continuous in i

(*c*) f_i maps $N_{0,i}$ into M_{0,α_i}

(*d*) α_i is increasing continuous

(*e*) $\alpha_0 = 0$ and f_0 is the identity

(*f*) $f_i(N_{1,i}) \cap \bigcup\limits_{\gamma<\lambda^+} M_{0,\gamma} = f_i(N_{0,i})$

(*g*) $N_{0,\alpha_i} \leq_{\mathfrak{s}} f_{i+1}(M_{0,i+1})$ and $N_{1,\alpha_i} \leq_{\mathfrak{s}} f_{i+1}(M_{1,i+1})$.

We may add

(*h*) $\alpha_i < i + \omega$ and $\alpha_i = i$ for i limit.

Note that clause (f) follows automatically as $(N_{0,i}, N_{1,i}, a) \in K^{3,\mathrm{uq}}_\lambda$ implies $\alpha_i \leq j \Rightarrow \mathrm{NF}(M_{0,i}, f_i(N_{j,i}), M_{0,j}, M_{1,j})$ but non-forking amalgamation of models is disjoint which in our case means that $i < j \Rightarrow M_{0,j} \cap f_i(N_{1,i}) = M_{0,i}$. Also for limit i, clearly $f_i(N_{0,i}) = M_{0,\alpha_i}$ and $f_i(N_{1,i}) = M_{1,\alpha_i}$ and $\alpha_i = i$.

During the induction the case $i = 0$ is trivial, the case i limit is easy and the case $i = j + 1$ is done using 4.5(2).

Lastly f_i maps $(\bigcup\limits_{i<\delta} N_{0,i}, \bigcup\limits_{i<\delta} N_{1,\delta}, a)$ isomorphically into $(M_{0,\alpha_\delta}, M_{1,\alpha_\delta}, a)$ and maps $\bigcup\limits_{i<\delta} N_{\ell,i}$ onto M_{ℓ,α_δ} (see clause (g), i.e., a previous paragraph). The latter belongs to $K^{3,\mathrm{uq}}_\lambda$, so (actually only $f_\varepsilon(N_{1,\delta}) \subseteq M_{1,\alpha_\delta}$ is needed by 1.19 monotonicity for $K^{3,\mathrm{uq}}_{\mathfrak{s}}$) we are done. $\square_{4.10}$

4.12 Claim. *Assume $(*)_{\bar M}$ holds (see below) and $p \in \mathscr{S}^{\mathrm{bs}}_{\mathfrak{s}}(M_0)$ <u>then</u> we can find $\bar N$ and a such that $(*)_{\bar N, \bar M}$ holds, where:*

$(*)_{\bar N, \bar M, a}$ *$\ell g(\bar N) = \ell g(\bar M)$, $M_i \leq_{\mathfrak{s}} N_i$ are from $K_{\mathfrak{s}}$, $\bar N$ is $<_{\mathfrak{s}}$-increasing continuous, $a \in N_0$, $\mathbf{tp}_{\mathfrak{s}}(a, M_i, N_i)$ is a non-forking extension* *of p and $(M_i, N_i, a) \in K^{3,\mathrm{uq}}_{\mathfrak{s}}$ for every $i < \ell g(\bar M)$ and N_{i+1} is $(\lambda, *)$-brimmed over N_i when $i + 1 < \ell g(\bar M)$ and N_0 is $(\lambda, *)$-brimmed; (we can even add N_0 is $(\lambda, *)$-brimmed over M_0 and N_{i+1} is $(\lambda, *)$-brimmed over $M_{i+1} \cup N_i$)*

$(*)_{\bar M}$ *$\bar M = \langle M_i : i < \alpha \rangle$ is $\le_{\mathfrak{s}}$-increasing continuous, $\alpha \le \lambda^+$, M_0 is $(\lambda, *)$-brimmed and M_{i+1} is $(\lambda, *)$-brimmed over M_i for any i such that $i + 1 < \alpha$ (hence $[i < j < \alpha \Rightarrow M_j$ is $(\lambda, *)$-brimmed over $M_i]$ and if $\alpha < \lambda^+$ is limit, then $[i < \alpha \Rightarrow \cup\{M_j : j < \alpha\}$ is $(\lambda, *)$-brimmed over $M_i])$.*

Proof. By 4.10. $\square_{4.12}$

In 4.13(1) below we prove the inverse of 4.9.
In 4.13(2) below we show how relevant situations in $\mathfrak{s}^+$ reflects to $\mathfrak{s}$ (some of its clauses repeat 1.10).

4.13 Claim. *1) If $(M_0, M_1, a) \in K^{3,\mathrm{pr}}_{\mathfrak{s}(+)}$ and $\bar M^\ell = \langle M^\ell_\alpha : \alpha < \lambda \rangle$ is a $\le_{\mathfrak{s}}$-representation of M_ℓ for $\ell = 0, 1$ <u>then</u> for some club E of λ^+ we have*

$(*)$ *if $\alpha \in E$ then $(M^0_\alpha, M^1_\alpha, a) \in K^{3,\mathrm{uq}}_{\mathfrak{s}}$.*

2) Assume $M_0 \le_{\mathfrak{s}(+)} M_\ell \le_{\mathfrak{s}(+)} M_3$ and $a_\ell \in M_\ell$ and $p_\ell = \mathbf{tp}_{\mathfrak{s}(+)}(a_\ell, M_0, M_\ell)$ for $\ell = 1, 2$ and $\langle M^\ell_\alpha : \alpha < \lambda^+ \rangle$ be a $\le_{\mathfrak{s}}$-representation of M_ℓ for $\ell \le 3$. <u>Then</u> for some club E of λ^+ for every $\delta \in E$ we have

- *(i) for $\ell = 1, 2$ if $p_\ell \in \mathscr{S}^{\mathrm{bs}}_{\mathfrak{s}(+)}(M_0)$ then $p_{\ell,\delta} = \mathbf{tp}_{\mathfrak{s}}(a_\ell, M^0_\delta, M_{\ell,\delta}) \in \mathscr{S}^{\mathrm{bs}}_{\mathfrak{s}}(M^0_\delta)$ and M^ℓ_δ (and $p_{\ell,\delta}$) are witnesses for p_ℓ*
- *(ii) if (M_0, M_ℓ, a_ℓ) is canonically prime and $\mathbf{tp}_{\mathfrak{s}(+)}(a_\ell, M_{3-\ell}, M_3)$ is an $\mathfrak{s}^+$-non-forking extension of p_ℓ for $\ell \in \{1, 2\}$ <u>then</u> $\mathbf{tp}_{\mathfrak{s}}(a_\ell, M^{3-\ell}_\delta, M^3_\delta)$ is an $\mathfrak{s}$-non-forking extension of $\mathbf{tp}_{\mathfrak{s}}(a_\ell, M^0_\delta, M^\ell_\delta)$ (hence of $\mathbf{tp}_{\mathfrak{s}}(a_\ell, M^0_{\min(E)}, M^\ell_{\min(E)})$)*
- *(iii) if $\alpha < \delta \in E$ then M^ℓ_δ is $(\lambda, *)$-brimmed over M^ℓ_α for $\ell < 4$.*

Remark. In 4.13(2), if we know enough on $\mathfrak{s}$ we can add

- *(iv)* if $(M_0, M_\ell, a_\ell) \in K^{3,\mathrm{uq}}_{\mathfrak{s}(+)}$ <u>then</u> $(M^0_\delta, M^\ell_\delta, a_\ell) \in K^{3,\mathrm{uq}}_{\mathfrak{s}}$.

This can be proved after we analyze $K_{\mathfrak{s}}^{3,\mathrm{uq}}$, see later.

Proof. 1) We can find M_2, a_2 such that $(M_0, M_2, a_2) \in K^{3,\mathrm{pr}}_{\mathfrak{s}(+)}$ and this triple is canonically $\mathfrak{s}^+$-prime, i.e., is as in 4.3 (with M_0, M_2, a_2 here standing for M_0, M_1, a there, of course) and $\mathbf{tp}_{\mathfrak{s}(+)}(a_2, M_0, M_2) = \mathbf{tp}_{\mathfrak{s}(+)}(a, M_0, M_1)$. As (M_0, M_1, a) is a prime triple there is a $\leq_{\mathfrak{s}(+)}$-embedding of M_1 into M_2 over M_0 mapping a to a_2 so without loss of generality $a_2 = a$ & $M_1 \leq_{\mathfrak{s}(+)} M_2$. Let $\bar{M}^\ell = \langle M^\ell_\alpha : \alpha < \lambda^+\rangle$ be a $\leq_{\mathfrak{s}}$-representation of M_ℓ, for $\ell < 3$ and E a thin enough club of λ^+. As $M_\ell \leq_{\mathfrak{s}(+)} M_{\ell+1}$, for any $\alpha < \beta$ from E we have $\mathrm{NF}_{\mathfrak{s}}(M^\ell_\alpha, M^{\ell+1}_\alpha, M^\ell_\beta, M^{\ell+1}_\beta)$ for $\ell = 0, 1$ and by the choice of M_2 for any α from E we have $(M^0_\alpha, M^2_\alpha, a) \in K^{3,\mathrm{uq}}_{\mathfrak{s}}$. By monotonicity of $K^{3,\mathrm{uq}}_{\mathfrak{s}}$, i.e., 3.5(2) as $\alpha \in E \Rightarrow M^0_\alpha \cup \{a\} \subseteq M^1_\alpha \leq_{\mathfrak{K}[\mathfrak{s}]} M^2_\alpha$ we get $\alpha \in E \Rightarrow (M^0_\alpha, M^1_\alpha, a) \in K^{3,\mathrm{uq}}_{\mathfrak{s}}$ as required.
2) Straightforward. (For (iii) recall that $M_\ell \in K^{\mathfrak{s}}_{\lambda^+}$ is saturated (above λ) for $\mathfrak{s}$).

$\square_{4.13}$

4.14 Claim. *1) $(M, N, a) \in K^{3,\mathrm{pr}}_{\mathfrak{s}(+)}$ <u>iff</u> the triple (M, N, a) is canonically $\mathfrak{s}^+$-prime.*
2) Uniqueness: if $(M, N_\ell, a_\ell) \in K^{3,\mathrm{pr}}_{\mathfrak{s}(+)}$ and $\mathbf{tp}_{\mathfrak{s}(+)}(a_1, M, N_1) = \mathbf{tp}_{\mathfrak{s}(+)}(a_2, M, N_2)$ <u>then</u> there is an isomorphism f from N_1 onto N_2 over M satisfying $f(a_1) = a_2$.

Proof. 1) The "if" direction by 4.9 and the "only if" direction by 4.13(1) and Definition 4.4.
2) By part (1), we know that (M, N_ℓ, a_ℓ) is canonically $\mathfrak{s}^+$-prime. Now we build the isomorphism by hence and forth as in the proof of 4.9. $\square_{4.1}$

It is good to know that also $\mathrm{NF}_{\mathfrak{s}(+)}$ reflects down (when we have it).

4.15 Claim. *Assume that also $\mathfrak{s}^+$ is weakly successful so $\mathrm{NF}_{\mathfrak{s}(+)}$ is well defined. If $M_\ell \in K_{\mathfrak{s}(+)}$ for $\ell = 0, 1, 2, 3$ and $\mathrm{NF}_{\mathfrak{s}(+)}(M_0, M_1, M_2, M_3)$ and $\bar{M}_\ell = \langle M_{\ell,\alpha} : \alpha < \lambda^+\rangle$ does $\leq_{\mathfrak{s}}$-represent M_ℓ for $\ell < 4$, <u>then</u> for a club of $\delta < \lambda^+$ we have $\mathrm{NF}_{\mathfrak{s}}(M_{0,\delta}, M_{1,\delta}, M_{2,\delta}, M_{3,\delta})$.*

Proof. Without loss of generality M_ℓ is $(\lambda^+,*)$-brimmed over M_0 for $\mathfrak{s}^+$ for $\ell = 1,2$ and M_3 is $(\lambda^+,*)$-brimmed over $M_1 \cup M_2$ (by density of $(\lambda^+,*)$-brimmed extension recalling $\mathfrak{s}^+$ is a weakly successful good λ^+-frame, and the existence property of $\mathrm{NF}_{\mathfrak{s}(+)}$ and the monotonicity of $\mathrm{NF}_{\mathfrak{s}(+)}$ and $\mathrm{NF}_{\mathfrak{s}}$).

Let $\langle N_{1,\alpha}, a^1_\alpha : \alpha < \lambda^+\rangle$ be as in 3.11 applied to $\mathfrak{s}(+)$ for (M_0, M_1) (the length being λ^+ is somewhat more transparent and is allowed as M_ℓ is $(\lambda^+,*)$-brimmed over M_0 for $\mathfrak{s}^+$ by 3.11(5)). As M_3 is $(\lambda^+,*)$-brimmed over $M_1 \cup M_2$, chasing arrows (by the definition of primes and properties of $\mathrm{NF}_{\mathfrak{s}(+)}$) without loss of generality there is a sequence $\langle N_{3,\alpha} : \alpha \le \lambda^+\rangle$ which is $\le_{\mathfrak{s}(+)}$-increasing continuous, $N_{3,0} = M_2, N_{3,\lambda^+} \le_{\mathfrak{s}(+)} M_3, N_{3,\alpha} \cap M_1 = N_{1,\alpha}$ and $(N_{1,\alpha}, N_{1,\alpha+1}, a^1_\alpha) \le^{\mathfrak{s}(+)}_{\mathrm{bs}} (N_{3,\alpha}, N_{3,\alpha+1}, a^1_\alpha)$; see 1.17(3). For each $\alpha \le \lambda^+, \ell = 1,3$ let $\langle N_{\ell,\alpha,i} : i < \lambda^+\rangle$ be a sequence which $\le_{\mathfrak{s}}$-represent $N_{\ell,\alpha}$. For $\ell = 1,3$ and $\alpha < \lambda^+$ let $E_{\ell,\alpha}$ be a club of λ^+ such that $i \in E_{\ell,\alpha}$ implies $(N_{\ell,\alpha,i}, N_{\ell,\alpha+1,i}, a^1_\alpha) \in K^{3,\mathrm{uq}}_{\mathfrak{s}}$ and $\mathbf{tp}_{\mathfrak{s}}(a_\alpha, N_{\ell,\alpha,i}, N_{1,\alpha+1,i})$ does not fork over $N_{\ell,\mathrm{Min}(E_{1,\alpha})}$; note that $E_{\ell,\alpha}$ exists by 4.13(1).

Let $E = \{\delta < \lambda^+ : \delta \text{ limit}, \delta \in E_{\ell,\alpha}$ for $\alpha < \delta, \ell = 1,3$ and $M_{m,\delta} \cap M_\ell = M_{\ell,\delta}$ for $\ell < m \le 3, (\ell,m) \ne (1,2)$ and $M_{1,\delta} = \bigcup_{\alpha<\delta} N_{1,\alpha,\delta} = \bigcup_{\alpha<\delta} N_{1,\alpha,\alpha}, M_{3,\delta} \cap N_{3,\lambda^+} = N_{3,\lambda^+,\delta} = \bigcup_{\alpha<\delta} N_{3,\alpha,\delta} = \bigcup_{\alpha<\delta} N_{3,\alpha,\alpha}, M_{0,\delta} = N_{1,0,\delta}$ and $M_{2,\delta} = N_{3,0,\delta}\}$. Now let $\delta \in E$. The sequences $\langle N_{1,\alpha,\delta} : \alpha < \delta\rangle, \langle N_{3,\alpha,\delta} : \alpha < \delta\rangle$ are $\le_{\mathfrak{s}}$-increasing continuous, also for $\alpha < \delta$, $(N_{1,\alpha}, N_{1,\alpha+1,\delta}, a_\alpha) \le^{\mathfrak{s}}_{\mathrm{bs}} (N_{3,\alpha,\delta}, N_{3,\alpha+1,\delta}, a_\alpha)$ are both in $K^{3,\mathrm{bs}}_{\mathfrak{s}}$ and $(N_{1,\alpha,\delta}, N_{1,\alpha+1,\delta}, a_\alpha)$ belongs to $K^{3,\mathrm{uq}}_{\mathfrak{s}}$ hence $\mathrm{NF}_{\mathfrak{s}}(N_{1,\alpha,\delta}, N_{3,\alpha,\delta}, N_{1,\alpha+1,\delta}, N_{3,\alpha+1,\delta})$. Without loss of generality, for $\delta \in E$ we have $N_{1,\delta,\delta} = \cup\{N_{1,\alpha,\delta} : \alpha < \delta\}$ and $N_{3,\delta,\delta} = \cup\{N_{3,\alpha,\delta} : \alpha < \delta\}$ and $N_{3,\lambda^+,\delta} = N_{3,\delta,\delta}$ so $N_{1,\delta} = M_{1,\delta}$ and $N_{3,\delta,\delta} = N_{3,\lambda^+,\delta} \le_{\mathfrak{s}} M_{3,\delta}$.

By long transivity for $\mathrm{NF}_{\mathfrak{s}}$ we get $\mathrm{NF}_{\mathfrak{s}}(N_{1,0,\delta}, N_{3,0,\delta}, N_{1,\delta,\delta}, N_{3,\delta,\delta})$ which means $\mathrm{NF}_{\mathfrak{s}}(M_{0,\delta}, M_{2,\delta}, M_{1,\delta}, N_{3,\lambda^+,\delta})$ hence by monotonicity $\mathrm{NF}_{\mathfrak{s}}(M_{0,\delta}, M_{2,\delta}, M_{1,\delta}, M_{3,\delta})$.

By symmetry of $\mathrm{NF}_{\mathfrak{s}}$ we are done. $\square_{4.15}$

We can summarize what we can say so far of the n-successors of $\mathfrak{s}$.

4.16 Claim. *Assume $\mathfrak{s}$ is good$^+$ and n-successful and $n > 0$.*
1) $\mathfrak{s}^{+n}$ is a good$^+$ λ^{+n}-frame.
2) $\mathscr{S}^{\rm bs}_{\mathfrak{s}(+n)} = \mathscr{S}^{\rm bs}_{\mathfrak{s}<\lambda^{+n}>} \restriction K^{\mathfrak{s}^{+n}}_{\lambda^{+n}}$ (see 0.4, the $\mathscr{S}^{\rm bs}_{\mathfrak{s}<\lambda^{+n}>}$ is from II§2 and is $\{p \in \mathscr{S}^{\rm bs}_{\mathfrak{s}}(M) : M \in K^{\mathfrak{s}^{+n}}_{\lambda^{+n}}\}$).
3) If $M \in K^{\mathfrak{s}^{+n}}_{\lambda^{+n}}$ and $p \in \mathscr{S}^{\rm bs}_{\mathfrak{s}(+n)}(M)$ then for some $(M,N,a) \in K^{3,\rm pr}_{\lambda^{+n}}[\mathfrak{s}^{+n}]$ we have $\mathbf{tp}_{\mathfrak{s}}(a,M,N) = p$.
4) If $(M,N,a) \in K^{3,\rm pr}_{\lambda^{+n}}[\mathfrak{s}^{+n}]$ then $(M,N,a) \in K^{3,\rm uq}_{\lambda^{+n}}[\mathfrak{s}^{+n}]$ provided that $\mathfrak{s}^{+n}$ is weakly successful.
5) If $M \le_{\mathfrak{K}} N^1 \le_{\mathfrak{K}} N^2$ are in $K^{\mathfrak{s}^{+n}}_{\lambda^{+n}}$ and $a \in N^1$ then $(M,N^2,a) \in K^{3,\rm uq}_{\lambda^{+n}}[\mathfrak{s}^{+n}] \Rightarrow (M,N^1,a) \in K^{3,\rm uq}_{\lambda^{+n}}[\mathfrak{s}^{+n}]$ and $(M,N^2,a) \in K^{3,\rm pr}_{\lambda^{+n}}[\mathfrak{s}^{+n}] \Rightarrow (M,N^1,a) \in K^{3,\rm pr}_{\lambda^{+n}}[\mathfrak{s}^{+n}]$.
6) Assume $n = m+1$, and $(M_0,M_1,a) \in K^{3,\rm bs}_{\lambda^{+n}}[\mathfrak{s}^{+n}]$ and $\bar{M}_\ell = \langle M_{\ell,\alpha} : \alpha < \lambda^{+n}\rangle$ a $\le_{\mathfrak{K}}$-representation of M_ℓ for $\ell = 1,2$. Then:

> (∗) *$(M_0,M_1,a) \in K^{3,\rm pr}_{\lambda^{+n}}[\mathfrak{s}^{+n}]$ iff for some club E of λ^+ we have: for $\alpha < \beta$ in E, $(M_{0,\alpha},M_{1,\alpha},a) \in K^{3,\rm uq}_{\mathfrak{s}}[\mathfrak{s}^{+m}]$ and $(M_{0,\alpha},M_{1,\alpha},a) <^{*,\mathfrak{s}^{+m}}_{\rm bs} (M_{0,\beta},M_{1,\beta},a)$, see 4.2.*

Proof. Straight; all by induction on n; part (3) by 4.9 + 4.3, part (4) by 3.7, part (5) by 3.5(1),3.5(2), part (6) by 4.13(1) + the proof of 4.9, see also 4.17(2). $\square_{4.16}$

4.17 Remark. 1) If we assume $\mathfrak{s}_0$ is uni-dimensional (see §2), life is easier: $(M,N,a) \in K^{3,\rm pr}_{\lambda^{+n}}$ implies model-minimality, see 3.8. On categoricity see 4.18 below.
2) For 4.16(6), note that M_1 is saturated (in $\mathfrak{K}^{\mathfrak{s}^{+(n-1)}}$ above $\lambda^{+(n-1)}$). Now if $(M_0,M_2,b) \in K^{3,\rm bs}_{\lambda^{+n}}[\mathfrak{s}^{+n}]$ and $\mathbf{tp}_{\mathfrak{s}(+n)}(b,M_0,M_2) = \mathbf{tp}_{\mathfrak{s}(+n)}(a,M_0,M_1)$ then by induction on $\alpha \in E$ we can choose an $\le_{K[\mathfrak{s}]}$-embedding f_α of $M_{1,\alpha}$ into $M_{2,\alpha}$, increasing continuous with α, mapping a to b. For $\alpha = \min(E)$ use the saturation, for $\alpha \in E$ the successor in E of β use $(M_{1,\alpha},M_{2,\alpha},a) \in K^{3,\rm uq}_{\mathfrak{s}(+(n-1))}+$ saturation.

4.18 Claim. *If $\mathfrak{s}$ is an n-successful good$^+$ λ-frame and weakly uni-dimensional and categorical in λ, <u>then</u>*

(i) $\mathfrak{K}^{\mathfrak{s}(+n)} = \mathfrak{K}^{\mathfrak{s}}_{\ge \lambda^{+n}}$

(ii) *$\mathfrak{s}(+n)$ is weakly uni-dimensional, and categorical in λ^{+n}.*

Proof. By 4.16 and 2.10 and 2.3(4). $\square_{4.18}$

* * *

<u>4.19 Discussion</u>: By the above in the cases we construct good λ-frames $\mathfrak{s}$ in II.3.7, $\mathfrak{s}$ is essentially $\mathfrak{t}^+, \mathfrak{t}$ is a λ-frame which is very closed to be a good λ-frame, also by [Sh 576] we know that $\mathfrak{t}$ is successful. Now $\mathfrak{t}$ is good$^+$ by the definition of minimal types hence we can deduce that $\mathfrak{s}$ has primes; we cannot apply 4.9 itself but $\mathfrak{t}$ is close enough to being good$^+$ and successful that $\mathfrak{s}$ itself has primes. In another one of the cases of 1.5(1) there are primes for different reason: $\aleph_0$ is easier. Alternatively, see Chapter VII.

4.20 Claim. *1) If $\mathfrak{s}$ is as in II.3.7, <u>then</u> $\mathfrak{s}$ has primes.*
2) If $\mathfrak{s}$ is as in II.3.5, i.e. Case 3 of 1.5(1), <u>then</u> $\mathfrak{s}$ has primes.

Proof. 1) Like the proof of 4.9.
2) Using stability in $\aleph_0$ we can construct primes directly. $\square_{4.20}$

Concerning 4.20:
<u>4.21 Example</u>: We define

(a) let $\langle V_n : n < \omega \rangle$ be a (strictly) decreasing sequence of vector spaces over the rational field, each of dimension $\aleph_0$, with intersection $\{0\}$

(b) let $\tau = \{R\} \cup \{P_t : t \in V_0\}$ where R and each P_t is a two-place relation

(c) K is the class of τ-models M such that

 (α) R^M is an equivalence relation

 (β) $P_t^M \subseteq R^M$

(γ) for each R^M-equivalence class A there are $n < \omega$ and one to one function h from V_n onto A such that for every $t \in V_0$

$$P_t^M \cap (A \times A) = \{(h(s_1), h(s_2)) : s_1, s_2 \in V_n,\ V_n \models \text{“}t = s_1 - s_2\text{”}\}$$

(d) $\leq_{\mathfrak{K}}$ is the partial order of $\subseteq$ on K.

A) Prove that $\mathfrak{K} := (K, \leq_{\mathfrak{K}})$ is an a.e.c.
B) $\mathfrak{K}$ is $\mathrm{PC}_{\aleph_0}$, is stable in $\aleph_0$, has amalgamation.
C) $\mathfrak{K}$ is not categorical in $\aleph_0$.
D) If $M \leq_{\mathfrak{K}} N_\ell, a_\ell \in N_\ell$ is not R^{N_ℓ}-equivalent to any $b \in M$ then $\mathbf{tp}(a_1, M, N_1) = \mathbf{tp}(a_2, M, N)$, call this type p_M.
E) If $M \leq_{\mathfrak{K}} N$ and $P_M = \mathbf{tp}(a, M, N)$ then $(M, N, a) \in K^{3,\mathrm{pr}}$.
F) Change to get categoricity.

4.22 Example: Assume $\mathfrak{K}_{\mathfrak{s}}$ is the class of structures of the form $(A, \mathscr{E})$ such that $|A| = \lambda_{\mathfrak{s}}, \mathscr{E}$ an equivalence relation on $A, \leq_{\mathfrak{s}} = \subseteq \restriction K_{\mathfrak{s}}, \mathscr{S}_{\mathfrak{s}}^{\mathrm{bs}} = \mathscr{S}_{\mathfrak{s}}^{\mathrm{na}}$ and $\mathbf{tp}_{\mathfrak{s}}(a, M_1, N)$ does not fork over $M_0 \leq_{\mathfrak{s}} M_1$ iff $a \in N \backslash M_1$ and $M_1 \cap (a/\mathscr{E}^N) \neq \emptyset \Rightarrow M_0 \cap (a/\mathscr{E}^N) \neq \emptyset$. Assume $M \subseteq N$ are from $\mathfrak{K}_{\mathfrak{s}}, a \in N \backslash M$ and $N \backslash M = a/\mathscr{E}^N$ so $a/\mathscr{E}^N$ is disjoint to M, then

(a) $(M, N, a) \in K_{\mathfrak{s}}^{3,\mathrm{bs}}$,

(b) if $a/\mathscr{E}^N$ is not a singleton then $(M, N, a) \notin K_{\mathfrak{s}}^{3,\mathrm{pr}}$.

If we like an elementary class we let $\lambda_{\mathfrak{s}} = \aleph_0$ and restrict ourselves to $K^{\mathfrak{s}\prime} = \{M \in K^{\mathfrak{s}} : M/\mathscr{E}^M$ is infinite and $a \in M \Rightarrow |a/\mathscr{E}^M| \geq \aleph_0\}$, then above $(M, N, a) \in K_{\mathfrak{s}}^{3,\mathrm{pr}} \Leftrightarrow |a/\mathscr{E}^N| = \aleph_0$.

§5 Independence

Here we make a real step forward: independence (of set of elements realizing basic types) can be defined and proved to be as required. In an earlier version we have used existence of primes but eventually eliminate it. Note that good$^+$ is used in proving 5.5(2)(6); we can weaken 5.1, see [Sh:F735].

5.1 Hypothesis.

(a) $\mathfrak{s}$ is a successful good$^+$ λ-frame

or at least

$(a)^-$ $\mathfrak{s}$ is a weakly successful good λ-frame and 4.7 holds.

5.2 Definition. Let $M \leq_{\mathfrak{s}} N$ (hence from $K_{\mathfrak{s}} = K^{\mathfrak{s}}_\lambda$).
1) Let $\mathbf{I}_{M,N} = \{a \in N : \mathbf{tp}_{\mathfrak{s}}(a, M, N) \in \mathscr{S}^{\rm bs}_{\mathfrak{s}}(M)\}$.
2) We say that $\mathbf{J}$ is independent in (M, A, N) if $(*)$ below holds; when $A = N'$ we may write N' instead of A; if N is understood from the context we may write "over (M, A)"; if $A = M$ we may omit it and then we say "in (M, N)" or "for (M, N)" or "over M"; where:

$(*)$ $\mathbf{J} \subseteq \mathbf{I}_{M,N}, M \leq_{\mathfrak{s}} N, M \subseteq A \subseteq N$ and we can find a witness $\langle M_i, a_j : i \leq \alpha, j < \alpha\rangle$ and N^+ which means:

- (a) $\langle M_i : i \leq \alpha\rangle$ is $\leq_{\mathfrak{K}}$-increasing continuous[4]
- (b) $M \cup A \subseteq M_i \leq_{\mathfrak{s}} N^+$, (usually $M \subseteq A$) and $N \leq_{\mathfrak{s}} N^+$
- (c) $a_i \in M_{i+1} \backslash M_i$, (if we forget to mention M_α we may stipulate $M_\alpha = N^+$)
- (d) $\mathbf{tp}_{\mathfrak{s}}(a_i, M_i, M_{i+1})$ does not fork over M,
- (e) $\mathbf{J} = \{a_i : i < \alpha\}$.

The notion independent indicates we expect various properties, like finite character, so we start to prove them.

5.3 Claim. *Assume that* $\mathrm{NF}_{\mathfrak{s}}(M_0, M_1, M_2, M_3)$ *and* $\mathbf{J} \subseteq M_2$. *Then* $\mathbf{J}$ *is independent in* (M_0, M_2) *iff* $\mathbf{J}$ *is independent in* (M_0, M_1, M_3).

Proof. The "if" implication is trivial. The "only if" implication is easy by 1.17(4).

$\square_{6.16}$

---

[4] note that omitting "continuous" makes no difference in the present context

5.4 Theorem. *Assume $M \leq_{\mathfrak{s}} N$ so are from $K_{\mathfrak{s}}$ and $\mathbf{J} \subseteq \mathbf{I}_{M,N}$.*
1) The following are equivalent:

$(*)_0$ *every finite $\mathbf{J}' \subseteq \mathbf{J}$ is independent in (M,N)*

$(*)_1$ *$\mathbf{J}$ is independent in (M,N)*

$(*)_2$ *like $(*)$ of Definition 5.2 adding*

- (f) *M_{i+1} is $(\lambda,*)$-brimmed over $M_i \cup \{a_i\}$ for $i < \alpha$*

$(*)_3$ *for every ordinal β satisfying $|\beta| = |\mathbf{J}|$ and a list $\langle a_i : i < \beta\rangle$ with no repetitions of $\mathbf{J}$ there are N^+, M_i (for $i \leq \beta$) such that $\langle M_j, a_i : j \leq \beta, i < \beta\rangle$ and N^+ which satisfy:*

- (a) *M_i is $\leq_{\mathfrak{s}}$-increasing continuous,*
- (b) *$M \leq_{\mathfrak{s}} M_i \leq_{\mathfrak{s}} N^+$ and $N \leq_{\mathfrak{s}} N^+$*
- (c) *$a_i \in M_{i+1} \backslash M_i$*
- (d) *$\mathbf{tp}_{\mathfrak{s}}(a_i, M_i, M_{i+1})$ does not fork over M*
- (e) *$\mathbf{J} = \{a_i : i < \beta\}$*
- $(f)_3$ *M_{i+1} is $(\lambda,*)$-brimmed[5] over $M_i \cup \{a_i\}$ for $i < \alpha$*

$(*)_4$ *like $(*)_3$ replacing $(f)_3$ by*

- $(f)_4$ *$(M_i, M_{i+1}, a_i) \in K^{3,\mathrm{uq}}_{\mathfrak{s}}$ and $M_0 = M$*

$(*)_5$ *like $(*)_4$ adding*

- (g) *if there is $M'_{i+1} \leq_{\mathfrak{s}} M_{i+1}$ such that $(M_i, M'_{i+1}, a_i) \in K^{3,\mathrm{pr}}_{\mathfrak{s}}$ then $(M_i, M_{i+1}, a_i) \in K^{3,\mathrm{pr}}_{\mathfrak{s}}$*

$(*)_6$ *like $(*)_5$ adding*

- (h) *if $(M_i, M_{i+1}, a_i) \in K^{3,\mathrm{pr}}_{\mathfrak{s}}$ for each $i < \beta$ (holds e.g. if $\mathfrak{s}$ has primes) then $M_\beta \leq_{\mathfrak{s}} N$.*

(so in $(*)_6$ the sequence $\langle (M_j, a_i) : j \leq \beta, i < \beta\rangle$ is a witness for "$\mathbf{J}$ independent for (M,N)" and it is an M-based pr*-decomposition*

[5] omitting $\{a_i\}$ give an equivalent condition

inside (M, N) *see Definition 3.3).*
2) If $M \cup \mathbf{J} \subseteq N^- \leq_{\mathfrak{s}} N$ *and* $N \leq_{\mathfrak{s}} N^+ \in K_{\mathfrak{s}}$, *then:* $\mathbf{J}$ *is independent in* (M, N^+) *iff* $\mathbf{J}$ *is independent in* (M, N) *iff* $\mathbf{J}$ *is independent in* (M, N^-).
3) If $\mathbf{J}$ *is independent in* (M, N) *and* $a_i \in \mathbf{J}$ *for* $i < \beta$ *are with no repetitions,* $M_0 = M$ *and* $\langle M_i : i \leq \beta\rangle, \langle a_i : i < \beta\rangle$ *are as in* $(*)_4$ *clauses (a),(c),(d),(f)*$_4$ *from part (1) and* $M_\beta \leq_{\mathfrak{s}} N$, *then* $\mathbf{J}\backslash\{a_i : i < \beta\}$ *is independent in* (M, M_β, N).
4) If $M^- \leq_{\mathfrak{s}} M, \mathbf{J}$ *is independent in* (M, N) *and* $\mathbf{J}' \subseteq \mathbf{J}$ *and* $[a \in \mathbf{J}' \Rightarrow \mathbf{tp}_{\mathfrak{s}}(a, M, N)$ *does not fork over* $M^-]$, *then* $\mathbf{J}'$ *is independent in* (M^-, N); *moreover if* $M^- \subseteq A \subseteq M$ *then* $\mathbf{J}'$ *is independent in* (M^-, A, N).
5) If $\mathbf{J}$ *is independent in* (M, M_0, N) *and* $M \leq_{\mathfrak{s}} M' \leq_{\mathfrak{s}} M_0 \leq_{\mathfrak{s}} N$, *then* J *is independent in* (M', N), *e.g. in* (M_0, N).

Proof. First note that part (2) is immediate by the amalgamation property. Also parts (4),(5) are straightforward so it is enough to prove parts (1) + (3). Clearly

$\boxtimes_0$ $(*)_4 \Rightarrow (*)_3$
[Why? We choose (f_i, M_i') by induction on $i \leq \beta$ such that M_i' is $\leq_{\mathfrak{s}}$-increasing continuous, f_i is a $\leq_{\mathfrak{s}}$-embedding of M_i into M_i', f_i is increasing continuous, $M_0' = M_0, f_0 = \mathrm{id}_{M_0}, \mathbf{tp}_{\mathfrak{s}}(f_{i+1}(a_i), M_i', M_{i+1}')$ does not fork over M, M_{i+1}' is $(\lambda, *)$-brimmed over $M_i' \cup \{f_{i+1}(a_i)\}$. By amalgamation without loss of generality there are (g, N') such that $M_\beta' \leq_{\mathfrak{s}} N', g$ is a $\leq_{\mathfrak{s}}$-embedding of N^+ into N' extending f_β. Renaming, without loss of generality $g = \mathrm{id}_{N^+}$ so clearly we are done (note that this is similar to the proof of $\boxtimes_5$ below but use less).]

$\boxtimes_1$ $(*)_6 \Rightarrow (*)_5 \Rightarrow (*)_4 \Rightarrow (*)_3 \Rightarrow (*)_2 \Rightarrow (*)_1 \Rightarrow (*)_0$
[Why? The implication $(*)_4 \Rightarrow (*)_3$ holds by $\boxtimes_0$, $(*)_5 \Rightarrow (*)_4$ holds trivially (and note that $K^{3,\mathrm{pr}} \subseteq K^{3,\mathrm{uq}}$ by 3.7(2) as we assume 5.1, i.e., $\mathfrak{s}$ is weakly successful). For the others, just read them.]

$\boxtimes_2$ if $\langle M_i' : i \leq \alpha'\rangle, \langle a_i : i < \alpha'\rangle$ and N^+ are as in $(*)_3$, so in particular witnessing "$\mathbf{J}$ is independent in (M, N)" then we can find $M_i'' \leq_{\mathfrak{s}} M_i'$ for $i \leq \alpha'$ such that $\langle M_i'' : i \leq \alpha'\rangle, N^+$

is as required in $(*)_4$ clauses (a)-(e),$(f)_4$ of part (1).
[Why? Choose M''_i by induction on i such that M''_i is $\leq_{\mathfrak{s}}$-increasing continuous, $M''_0 = M \leq_{\mathfrak{s}} M'_0$ and $(M''_i, M''_{i+1}, a_i) \in K^{3,\mathrm{uq}}_{\mathfrak{s}}$ and $i \leq \alpha \Rightarrow M''_i \leq_{\mathfrak{K}} M'_i$, using the hypothesis "$\mathfrak{s}$ is weakly successful" and "M'_{i+1} is $(\lambda, *)$-brimmed over $M'_i + a_i$"].

Hence

$\boxtimes_3$ $(*)_3 \Rightarrow (*)_4$

and similarly (recalling $K^{3,\mathrm{pr}}_{\mathfrak{s}} \subseteq K^{3,\mathrm{uq}}_{\mathfrak{s}}$ and the definition of $(M, N, a) \in K^{3,\mathrm{pr}}_{\mathfrak{s}}$)

$\boxtimes_4$ $(*)_4 \Rightarrow (*)_5$.

Also if $\langle M_i : i \leq \alpha\rangle, N^+, \langle a_i : i < \alpha\rangle$ are as in $(*)_1$, i.e., satisfy clauses (a)-(e) of $(*)_3$ with α instead of β then we can choose (M^+_i, f_i) by induction on $i \leq \alpha$ such that M^+_i is $\leq_{\mathfrak{s}}$-increasing continuous, f_i is a $\leq_{\mathfrak{s}}$-embedding of M_i into M^+_i, f_i is increasing continuous, $\mathrm{NF}_{\mathfrak{s}}(f_i(M_i), M^+_i, f_{i+1}(M_{i+1}), M^+_{i+1})$ and M^+_i is $(\lambda, *)$-brimmed over $M^+_i \cup \{a_i\}$. By renaming without loss of generality $f_i = \mathrm{id}_{M_i}$ and by amalgamation without loss of generality $M^+_i \leq_{\mathfrak{s}} N^+$ (actually we could have used 1.16(1)). So

$\boxtimes_5$ $(*)_1 \Rightarrow (*)_2$.

Now we prove parts (1) + (3) of the Lemma by induction on $|\mathbf{J}|$.

Case 1: $|\mathbf{J}| \leq 1$.
Trivial.

Case 2: $n = |\mathbf{J}|$ finite > 1.
As $\mathbf{J}$ is finite (and monotonicity of independence) clearly

$\otimes_1$ $(*)_0 \Leftrightarrow (*)_1$.

We first show

$\otimes_2$ $(*)_2 \Rightarrow (*)_3$.

Note that in $(*)_3$, as $\mathbf{J}$ is finite, necessarily $\beta = |\mathbf{J}| = n$. As the permutations exchanging $m, m+1$ generate all permutations of $\{0, \ldots, n-1\}$, it is enough to show

$\otimes_3$ if $\langle M_k, a_\ell : k \le n, \ell < n\rangle$ and N^+ are as in $(*)_2$ and $m < n-1$ and a'_ℓ is a_ℓ if $\ell < n$ & $\ell \ne m$ & $\ell \ne m+1$, is a_{m+1} if $\ell = m$ and is a_m if $\ell = m+1$, <u>then</u> for some M'_ℓ for $\ell < n$ we have $\langle M'_k, a'_\ell : k \le n, \ell < n\rangle$ and N^+ are as in $(*)_3$.

Why does $\otimes_3$ hold? Let M'_ℓ be M_ℓ if $\ell \le m \vee \ell \ge m+2$.
As $\mathfrak{s}$ is a good frame and $\mathbf{tp}_{\mathfrak{s}}(a_{m+1}, M_{m+1}, M_{m+2}) \in \mathscr{S}^{\text{bs}}_{\mathfrak{s}}(M_{m+1})$ does not fork over M and $M \le_{\mathfrak{s}} M_m \le_{\mathfrak{s}} M_{m+1}$,
clearly $\mathbf{tp}_{\mathfrak{s}}(a_{m+1}, M_{m+1}, M_{m+2}) \in \mathscr{S}^{\text{bs}}_{\mathfrak{s}}(M_{m+1})$ does not fork over M_m and similarly $\mathbf{tp}_{\mathfrak{s}}(a_m, M_m, M_{m+1}) \in \mathscr{S}^{\text{bs}}_{\mathfrak{s}}(M_m)$. Hence there are by symmetry (see Ax(E)(f) from Definition II.2.1) M', M'' such that $M_{m+2} \le_{\mathfrak{s}} M'', M_m \le_{\mathfrak{s}} M' \le_{\mathfrak{s}} M'', a_{m+1} \in M'$ and $\mathbf{tp}_{\mathfrak{s}}(a_m, M', M'')$ does not fork over M_m and without loss of generality M' is $(\lambda, *)$-brimmed over $M_m \cup \{a_{m+1}\}$ and M'' is $(\lambda, *)$-brimmed over $M' \cup \{a_{m+1}\}$ which include $M_m \cup \{a_m, a_{m+1}\}$. As M_{m+2} is $(\lambda, *)$-brimmed over $M_{m+1} \cup \{a_{m+1}\}$ which include $M_m \cup \{a_m, a_{m+1}\}$, by the previous sentence there is an isomorphism f from M'' onto M_{m+2} over $M_m \cup \{a_m, a_{m+1}\}$, (in fact, even over $M_{m+1} \cup \{a_{m+1}\}$), so by renaming without loss of generality $M'' = M_{m+2}$. Let $M'_{m+1} = M'$ so we have finished proving $\otimes_3$ hence $\otimes_2$ holds (in the present case!).
So (in the present case, the $\Rightarrow$ by $\boxtimes_1$ and the $\Leftarrow$ by $\otimes_1, \boxtimes_5, \otimes_2, \boxtimes_3, \boxtimes_4$) we have

$\otimes_4$ $(*)_0 \Leftrightarrow (*)_1 \Leftrightarrow (*)_2 \Leftrightarrow (*)_3 \Leftrightarrow (*)_4 \Leftrightarrow (*)_5$.

Next

$\otimes_5$ part (3) holds.

Why? By the induction hypothesis and parts (5),(4), it is enough to deal with the case $\beta = 1$, i.e., to prove that $\mathbf{J}\backslash\{a_0\}$ is independent in (M_1, N) assuming $(M, M_1, a_0) \in K^{3,\text{uq}}_{\mathfrak{s}}, a_0 \in \mathbf{J}$, also without loss of generality $\mathbf{J}\backslash\{a_0\} \ne \emptyset$ (otherwise the conclusion is trivial) hence $n \ge 2$. Choose $b_0, \ldots, b_{n-2}$ such that they list $\mathbf{J}\backslash\{a_0\}$ and we can let $b_{n-1} = a_0$ and possibly increasing N let $\langle M'_\ell : \ell \le n\rangle$ be such that $\langle b_\ell : \ell < n\rangle, \langle M'_\ell : \ell \le n\rangle$ are as in $(*)_4$ clauses (a)-(e),$(f)_4$, they exist as we have already proved most of part (1) in $\otimes_4$ above in the present case. As $a_0 = b_{n-1}$ in particular we have $\mathbf{tp}_{\mathfrak{s}}(a_0, M'_{n-1}, M'_n) = \mathbf{tp}_{\mathfrak{s}}(b_{n-1}, M'_{n-1}, N)$ does not fork over M so as $(M, M_1, a_0) \in K^{3,\mathrm{uq}}_{\mathfrak{s}}$ we can deduce that $\mathrm{NF}_{\mathfrak{s}}(M, M'_{n-1}, M_1, N)$ by 1.19. Hence easily (by 1.17(4)) by the $\mathrm{NF}_{\mathfrak{s}}$-calculus for some N^+ satisfying $N \leq_{\mathfrak{K}} N^+ \in K_{\mathfrak{s}}$ we can find $M_2, \ldots, M_n$ such that $M_1 \leq_{\mathfrak{s}} M_2 \leq_{\mathfrak{s}} \ldots \leq_{\mathfrak{s}} M_n \leq_{\mathfrak{K}} N^+, \ell \in \{1, 2, \ldots, n-1\} \Rightarrow \mathrm{NF}_{\mathfrak{s}}(M'_\ell, M_{\ell+1}, M'_{\ell+1}, M_{\ell+2})$. By 1.18 the type $\mathbf{tp}_{\mathfrak{s}}(b_\ell, M_{\ell+1}, M_{\ell+2})$ does not fork over M'_ℓ hence by transitivity of non-forking for $\ell < n-1$ the type $\mathbf{tp}_{\mathfrak{s}}(b_\ell, M_{\ell+1}, M_{\ell+2})$ does not fork over M. So $\langle M_{1+\ell} : \ell \leq n-1 \rangle$ witness that $\langle b_0, \ldots, b_{n-2} \rangle$ is independent in (M_1, N^+). So by part (2), i.e., for $n-1$, clearly $\langle b_0, \ldots, b_{n-2} \rangle$ is independent in (M_1, N), hence by part (4) we have shown part (3), i.e., $\otimes_5$.

To complete the proof in the present case we need

$\otimes_6$ $(*)_5 \Rightarrow (*)_6$.

We do more: we prove this in the general case provided that part (3) has already been proved.

So let $\langle a_i : i < \beta \rangle$ list $\mathbf{J}$ with no repetitions and let $N^+, \langle M_i : i \leq \beta \rangle$ be as in $(*)_5$. The only non-trivial case is when $i < \beta \Rightarrow (M_i, M_{i+1}, a_i) \in K^{3,\mathrm{pr}}_{\mathfrak{s}}$. We now choose by induction on $i \leq \beta$ a $\leq_{\mathfrak{s}}$-embedding f_i of M_i into N such that $f_0 = \mathrm{id}_{M_0}$ and $f_{i+1}(a_i) = a_i$ and $i < j \Rightarrow f_i \subseteq f_j$.

Now f_0 is defined and in limit stages we take the union.

Lastly, if f_i is defined, then by part (3) which we have already proved for this case we know that $\mathbf{J} \backslash \{a_j : j < i\}$ is independent in $(M, f_i(M_i), N^+)$ so in particular $\mathbf{tp}_{\mathfrak{s}}(a_i, f_i(M_i), N^+)$ does not fork over M hence $f_i(\mathbf{tp}_{\mathfrak{s}}(a_i, M_i, M_{i+1})) = \mathbf{tp}_{\mathfrak{s}}(a_i, f_i(M_i), N^+)$. Hence f_{i+1} exists as $(M_i, M_{i+1}, a_i) \in K^{3,\mathrm{pr}}_{\mathfrak{s}}$ and the definition of $K^{3,\mathrm{pr}}_{\mathfrak{s}}$. Lastly $\langle f_\ell(M_\ell) : \ell \leq n \rangle$ witnesses that $(*)_6$ holds.

<u>Case 3</u>: $|\mathbf{J}| = \mu \geq \aleph_0$.

We first prove what we now call $(3)^-$, a weaker variant of part (3), which is: replacing in the conclusion "independent" by "every finite subset is independent", this will be subsequently used to prove the other parts, and part (3) itself. We prove $(3)^-$ by induction on the ordinal β (for all possibilities) and for a fixed β by induction on $|\mathbf{J}\backslash\{a_i : i < \beta\}|$.

First, for $\beta = 0$ it is trivial.

Second, assume $\beta = \gamma + 1$ and let $b_0, \ldots, b_{n-1} \in \mathbf{J}\backslash\{a_i : i < \beta\}$ be pairwise distinct, so $M_0 = M, \langle M_i : i \leq \beta\rangle, \langle a_i : i < \beta\rangle$ are as in (a),(c),(d),(f)$_4$ of $(*)_4, M_\beta \leq_{\mathfrak{s}} N$ and we should prove that $\{b_0, \ldots, b_{n-1}\}$ is independent in (M, M_β, N). Now by the induction hypothesis on β applied to $\langle M_j, a_i : j \leq \gamma, i < \gamma\rangle$ and $\{b_0, \ldots, b_{n-1}, a_\gamma\}$ we deduce that $\{b_0, \ldots, b_{n-1}, a_\gamma\}$ is independent in (M, M_γ, N) hence in (M_γ, N).

Now by the case with $\mathbf{J}$ finite, $\{b_0, \ldots, b_{n-1}\}$ is independent as (M_β, N). So by part (4), $\{b_0, \ldots, b_{n-1}\}$ is indepedent in (M, M_β, N) as required.

Lastly, assume β is a limit ordinal and let $n < \omega, b_0, \ldots, b_{n-1} \in \mathbf{J}' := \mathbf{J}\backslash\{a_i : i < \beta\}$ be pairwise distinct. We should prove that $\{b_0, \ldots, b_{n-1}\}$ is independent in (M, M_β, N); the case $n = 0$ is trivial. By the induction hypothesis on β we have $\varepsilon < \beta$ implies $\mathbf{tp}(b_0, M_\varepsilon, N)$ does not fork over M, hence by the continuity axiom (E)(h) the type $\mathbf{tp}_{\mathfrak{s}}(b_0, M_\beta, N)$ does not fork over M. So if $n = 1$ we are done hence without loss of generality $n \geq 2$. By Claim 5.5(2) (if $\mathfrak{s}$ has primes we can use 5.5(1)) we can find $\bar{M}', N^+$ such that

- (α) $\bar{M}' = \langle M'_\varepsilon : \varepsilon \leq \beta\rangle$
- (β) $\bar{M}'$ is $\leq_{\mathfrak{s}}$-increasing continuous
- (γ) $N \leq_{\mathfrak{s}} N^+ \in K_{\mathfrak{s}}$
- (δ) $\mathrm{NF}_{\mathfrak{s}}(M_\varepsilon, M_\zeta, M'_{1+\varepsilon}, M'_{1+\zeta})$ for $\varepsilon < \zeta \leq \beta$
- (ε) $M'_\beta \leq_{\mathfrak{s}} N^+, M_0 \leq_{\mathfrak{s}} M'_0$
- (ζ) $\mathbf{tp}_{\mathfrak{s}}(a_\varepsilon, M'_{1+\varepsilon}, M'_{1+\varepsilon+1})$ does not fork over M for $\varepsilon < \beta$
- (η) $\mathbf{tp}_{\mathfrak{s}}(b_0, M'_0, N^+)$ does not fork over M
- (ι) letting $a'_0 = b_0, a'_{1+\varepsilon} = a_\varepsilon$ we have $(M'_\varepsilon, M'_{\varepsilon+1}, a'_\varepsilon) \in K^{3,\mathrm{uq}}_{\mathfrak{s}}$; note that $1 + \beta = \beta$ as β is a limit ordinal.

Now noting $M_\varepsilon \leq_{\mathfrak{s}} M'_{1+\varepsilon}$ for $\varepsilon < \beta$

$\otimes_7$ $\langle M'_\varepsilon, a'_\zeta : \varepsilon \leq \beta, \zeta < \beta\rangle$ is as in $(*)_4$ for (M'_0, N^+).

By clauses $(\varepsilon) + (\zeta)$ necessarily $\mathbf{J}$ is independent in (M, M'_0, N^+) hence in (M'_0, N^+).
Hence by the induction hypothesis on n, $\{b_1, \ldots, b_{n-1}\}$ is independent in (M'_0, M'_β, N^+) so by part (4) also in (M, M'_β, N^+) and recall $\mathbf{tp}_{\mathfrak{s}}(b_0, M_\beta, N^+)$ does not fork over M while $b_0 \in M'_\beta, M \leq_{\mathfrak{s}} M_\beta \leq_{\mathfrak{s}} M'_\beta$, so easily $\{b_0, \ldots, b_{n-1}\}$ is independent in (M, M_β, N^+), i.e. by 5.6(1) below, hence by part (2) for the case of finite $\mathbf{J}$ the set $\{b_0, \ldots, b_{n-1}\}$ is independent in (M, M_β, N) as required so we have proved $(3)^-$

Next we prove $(*)_0 \Rightarrow (*)_4$ in part (1).

For proving $(*)_4$ let $\langle a_i : i < \beta\rangle$ be a given list of $\mathbf{J}$ and we will find $N^+, \langle M_i : i \leq \beta\rangle$ as required; we do it by induction on β, i.e. we prove $(*)_{4,\beta}$. We now choose by induction on i a pair of models $M_i \leq_{\mathfrak{s}} N_i$ such that

> $\boxdot$ N_i is $\leq_{\mathfrak{s}}$-increasing continuous, $N_0 = N, M_i$ is $\leq_{\mathfrak{s}}$-increasing continuous, $M_0 = M$ and $i = j+1$ implies $(M_j, M_i, a_j) \in K^{3,\mathrm{uq}}_\lambda$ and every finite subset of $\mathbf{J}\backslash\{a_j : j < i\}$ is independent in (M, M_i, N_i).

<u>Subcase a</u>: For $i = 0$ there is no problem.

<u>Subcase b</u>: For i limit let $M_i = \bigcup\limits_{j<i} M_j, N_i = \bigcup\limits_{j<i} N_j$, the least trivial part is the clause in $\boxdot$ on independence. Now for $i = \beta$ this clause is trivial, in fact $(*)_{4,\beta}$ is already proved. We can assume $i < \beta$ and let $\{a_{i_0}, \ldots, a_{i_{n-1}}\} \subseteq \mathbf{J}\backslash\{a_j : j < i\}$ be with no repetitions. Now if $i < \mu$ then by renaming without loss of generality $\max[\{i_\ell : \ell < n\} \cup \{i\}] < \mu$ so we can use our induction hypothesis on μ. So we can assume $\beta > \mu \vee (i = \beta = \mu)$ and the case $i = \beta = \mu$ is trivial, so as we are inducting on all listings of subsets of $\mathbf{J}$ of a given length we have actually proved $(*)_0 \Rightarrow (*)_{4,\beta}$ for $\beta = \mu$, so $\mathbf{J}$ in independent in (M, N) (as witnessed by some list of length μ!). So we can apply $(3)^-$ and get that $\{a_{i_0}, \ldots, a_{i_{n-1}}\}$ is independent in (M, M_i, N_i) as required.

<u>Subcase c</u>: For $i = j+1$, as for $\mathfrak{s}$ we know that $K^{3,\mathrm{uq}}_{\mathfrak{s}}$ satisfies existence (as $\mathfrak{s}$ is weakly successful and good) we can find N_i, M_i as

required and the induction assumption on i holds as we have proved the claims for finite $\mathbf{J}$.
So we have finished the induction on i, thus proving $(*)_0 \Rightarrow (*)_4$ for $\mathbf{J}$ of cardinality $\leq \mu$. Hence we have part (1) for μ because $(*)_5 \Rightarrow (*)_4 \Rightarrow (*)_3 \Rightarrow (*)_2 \Rightarrow (*)_1 \Rightarrow (*)_0$ by $\boxtimes_1$ and $(*)_0 \Rightarrow (*)_4$ was just proved, $(*)_4 \Rightarrow (*)_5$ by $\boxtimes_4$ above and $(*)_5 \Rightarrow (*)_6$ was proved inside the proof of the finite case. Hence we have proved part (1). Now part (3) for μ follows from (1) + $(3)^-$. So we have finished the induction step for μ also in the infinite case (case 3) so have finished the proof. $\square_{5.4}$

Still to finish the proof of 5.4 we have to show 5.5(2), 5.6(1) below.

5.5 Claim. *1) [Assume $\mathfrak{s}$ has primes.] If $\langle N_i, a_j : i \leq \alpha, j < \alpha \rangle$ is an M-based* pr*-decomposition for $\mathfrak{s}$ inside N (so $N_0 = M$ and $N_\alpha \leq_{\mathfrak{s}} N$, see Definition 3.3), $b \in N, \mathbf{tp}_{\mathfrak{s}}(b, N_\alpha, N)$ is a non-forking extension of $p \in \mathscr{S}^{\mathrm{bs}}_{\mathfrak{s}}(N_0)$, <u>then</u> we can find $\langle N'_i, a'_j : i \leq 1+\alpha, j < 1+\alpha \rangle$, an M-based* pr*-decomposition for $\mathfrak{s}$ inside N^+, such that $N_i \leq_{\mathfrak{s}} N'_{1+i}, N_0 = N'_0, b = a'_0, a_i = a'_{1+i}, \mathbf{tp}_{\mathfrak{s}}(a_i, N'_{1+i}, N'_{1+i+1})$ does not fork over $N_i, N'_\alpha \leq_{\mathfrak{s}} N^+$ and $N \leq_{\mathfrak{K}} N^+$; note that $\mathrm{NF}_{\mathfrak{s}}(N_i, N'_{1+i}, N_{i+1}, N'_{1+i+1})$ for $i < \alpha$ follows.*
2) Similarly for uq*-decomposition except that we require only $N_0 \leq_{\mathfrak{s}} N'_0$ (instead equality) but still require $\mathbf{tp}_{\mathfrak{s}}(b, N'_0, N^+)$ does not fork over M so necessary $\mathrm{NF}_{\mathfrak{s}}(N_i, N'_{1+i}, N_{i+1}, N'_{1+i+1})$ holds for $i < \alpha$.*
3) In part (1) if $N' \leq_{\mathfrak{s}} N$ and $(M, N', b) \in K^{3,\mathrm{pr}}_{\mathfrak{s}}$ <u>then</u> we can add $N'_1 = N'$.
4) In part (2) if $M \leq_{\mathfrak{s}} M'$ we can demand that M' is $\leq_{\mathfrak{s}}$-embeddable into N'_0 over M and if M' is $(\lambda, *)$-brimmed over M we can add M' is isomorphic to N'_0 over M.

Proof. 1) Chasing arrows: first ignore $b = a'_0$, demand just $\mathbf{tp}_{\mathfrak{s}}(a'_0, N_0, N'_1) = \mathbf{tp}_{\mathfrak{s}}(b, N_0, N)$ and ignore $N \leq_{\mathfrak{s}} N^+$. After proving this we can use equality of types.

In details, we choose by induction on $i \leq \alpha$ a pair (N^*_i, f_i) and b^*, a^*_i (if $i < \alpha$) such that:

(a) f_i is a $\leq_{\mathfrak{s}}$-embedding of N_i into N^*_i

(b) N^*_i is $\leq_{\mathfrak{s}}$-increasing continuous

(c) N_0^* satisfies $(N_0, N_0^*, b^*) \in K_{\mathfrak{s}}^{3,\mathrm{pr}}$ and $\mathbf{tp}_{\mathfrak{s}}(b^*, N_0, N_0^*) = \mathbf{tp}_{\mathfrak{s}}(b, N_0, N)$

(d) $f_0 = \mathrm{id}_{N_0}$

(e) if $i = j+1$ then $(N_j^*, N_i^*, a_j^*) \in K_{\mathfrak{s}}^{3,\mathrm{pr}}$ and $f_i(a_j) = a_j^*$

(f) $\mathbf{tp}_{\mathfrak{s}}(b^*, f_i(N_i), N_i^*)$ does not fork over N_0

(g) $i < j \Rightarrow f_i \subseteq f_j$.

For $i = 0$ just use "$\mathfrak{s}$ has primes".

For $i = j+1$ first choose $p_j \in \mathscr{S}_{\mathfrak{s}}^{\mathrm{bs}}(N_j^*)$, a non-forking extension of $p_j^- = f_j(\mathbf{tp}_{\mathfrak{s}}(a_j, N_j, N_{j+1}))$ and second choose N_i^*, a_j^* such that $(N_j^*, N_i^*, a_j^*) \in K_{\mathfrak{s}}^{3,\mathrm{pr}}$ and $\mathbf{tp}_{\mathfrak{s}}(a_j^*, N_j^*, N_i^*) = p_j$; now clause (f) is satisfied (using "$\mathfrak{s}$ has primes"), lastly choose a $\le_{\mathfrak{s}}$-embedding $f_i \supseteq f_j$ of N_i into N_i^* mapping a_j to a_j^* using the assumption $(N_j, N_i, a_j) \in K_{\mathfrak{s}}^{3,\mathrm{pr}}$ and a_j^*'s realizing p_j.

For i limit take union.

Having finished the induction without loss of generality each f_i is the identity on N_i hence $j < \alpha \Rightarrow a_j^* = a_j$. So $N_\alpha \le_{\mathfrak{s}} N_\alpha^*$ and $N_\alpha \le_{\mathfrak{s}} N$ and $\mathbf{tp}_{\mathfrak{s}}(b^*, N_\alpha, N_\alpha^*)$ does not fork over N_0 (by clause (f)) and extends $\mathbf{tp}_{\mathfrak{s}}(b, N_0, N)$; also $\mathbf{tp}_{\mathfrak{s}}(b, N_\alpha, N)$ satisfies this so as $K_{\mathfrak{s}}$ has amalgamation without loss of generality $b = b^*$ and for some $N^+ \in K_{\mathfrak{s}}$ we have $N \le_{\mathfrak{s}} N^+$ & $N_\alpha^* \le_{\mathfrak{s}} N^+$.

Letting $N_0' = N', N_{1+i}' = N_i^*$ we are done.

2) Using 4.7.

3) Similarly.

4) Similarly. $\square_{5.5}$

Some trivial properties are:

5.6 Claim. *1) If $\langle M_i : i \le \alpha\rangle$ is $\le_{\mathfrak{s}}$-increasing continuous, and $\mathbf{J}_i$ is independent in (M_0, M_i, M_{i+1}) for each $i < \alpha$ <u>then</u> $\cup\{\mathbf{J}_i : i < \alpha\}$ is independent in (M_0, M_α).*

2) If $\mathrm{NF}_{\mathfrak{s}}(M_0, M_1, M_2, M_3)$ and $\mathbf{J}$ is independent in (M_0, M_1) <u>then</u> $\mathbf{J}$ is independent in (M_2, M_3) and even in (M_0, M_1, M_3).

If $\mathrm{NF}_{\mathfrak{s}}(M_0, M_1, M_2, M_3)$, $M_0 \le_{\mathfrak{s}} M_1^- \le_{\mathfrak{s}} M_1$, $\mathbf{J}_1$ independent in (M_1^-, M_1) and $\mathbf{J}_2$ independent in (M_0, M_2) <u>then</u> $\mathbf{J}_1 \cup \mathbf{J}_2$ is independent in (M_1^-, M_3).

3) [Monotonicity] If $\mathbf{J}$ *is independent in* (M, A, N) *and* $\mathbf{I} \subseteq \mathbf{J}$, *then* $\mathbf{I}$ *is independent in* (M, A, N).
4) If $\mathbf{J}$ *is independent in* (M_1, N) *and* $M_0 \leq_{\mathfrak{s}} M_1 \leq_{\mathfrak{s}} N$ *and* $c \in \mathbf{J} \Rightarrow \mathbf{tp}_{\mathfrak{s}}(c, M_1, N)$ *does not fork over* M_0, *then* $\mathbf{J}$ *is independent in* (M_0, M_1, N).
5) [$\mathfrak{s}$ has primes]. Assume that $\mathrm{NF}_{\mathfrak{s}}(M_0, M_1, M_2, M_3)$ *and* $\langle M_{0,i}, a_j : i \leq \alpha, j < \alpha \rangle$ *is a decomposition of* M_2 *over* M_0. *Then we can find* M_3^+ *satisfying* $M_3 \leq_{\mathfrak{s}} M_3^+$ *and* $\langle M_{1,i} : i \leq \alpha \rangle$ *such that* $M_{1,\alpha} \leq_{\mathfrak{s}} M_3^+$, $\langle M_{1,i}, a_j : i \leq \alpha, j < \alpha \rangle$ *is a decomposition of* $M_{1,\alpha}$ *over* M_1 *and* $M_{0,i} \leq_{\mathfrak{s}} M_{1,i}$ *and* $\mathbf{tp}_{\mathfrak{s}}(a_i, M_{1,i}, M_3^+)$ *does not fork over* $M_{0,i}$ *for* $i < \alpha$.
6) Similarly for uq*-decompositions except that* $M_1 \leq_{\mathfrak{s}} M_{1,0}$ *(not necessarily equal); we may add* $M_{1,i}$ *is brimmed over* $M_{0,i}$.
7) The set $\{a\}$ *is* $\mathfrak{s}$*-independent in* (M, N) *iff* $(M, N, a) \in K_{\mathfrak{s}}^{3,\mathrm{bs}}$.

Proof. 1) Should be clear (e.g., as in the proof of $(*)_1 \Rightarrow (*)_2$ inside the proof of 5.4(1) without loss of generality M_{i+1} is $(\lambda, *)$-brimmed over $M_i \cup \mathbf{J}_i$).
2) The first phrase is satisfied by 5.3.

For the second phrase by symmetry $\mathrm{NF}_{\mathfrak{s}}(M_0, M_2, M_1, M_3)$.

The first phrase of part (2) applied with $M_0, M_2, M_1, M_3, \mathbf{J}_2$ here standing for $M_0, M_1, M_2, M_3, \mathbf{J}$ there we get $\mathbf{J}_2$ is independent in (M_0, M_1, M_3) hence in (M_1^-, M_1, M_3). By part (1) applied to $M_1^- \leq_{\mathfrak{s}} M_3^- \leq_{\mathfrak{s}} M_3^+$ and $\mathbf{J}_2, \mathbf{J}_1$ we can deduce that $\mathbf{J}_1 \cup \mathbf{J}_2$ is independent in (M_1^-, M_3^-), hence in (M_1^-, M_3) (and even in (M_0, M_1^-, M_3)) as required.
3) Trivial.
4) Easy by the non-forking calculus.
5) As in the proof of the previous claim 5.5 there is an $\leq_{\mathfrak{s}}$-increasing continuous sequence $\langle M_{1,i} : i \leq \alpha \rangle$ and a $\leq_{\mathfrak{s}}$-embedding f_α of $M_2 = M_{0,\alpha}$ into $M_{1,\alpha}$ such that $M_{1,0} = M_1, f \restriction M_0 = \mathrm{id}_{M_0}$ and $f_i(M_{0,i}) \leq_{\mathfrak{s}} M_{1,i}$ and $\mathbf{tp}_{\mathfrak{s}}(f(a_i), M_{1,i}, M_{1,i+1})$ does not fork over $f(M_{0,i})$ and $(M_{1,i}, M_{1,i+1}, f_{i+1}(a_i)) \in K_{\mathfrak{s}}^{3,\mathrm{pr}}$ (e.g., just choose $M_{1,i}, f_i = f \restriction M_{0,i}$ by induction on $i \leq \alpha$).

As $(M_{0,i}, M_{0,i+1}, a_i) \in K_{\mathfrak{s}}^{3,\mathrm{pr}}$ also $(f(M_{0,i}), f(M_{0,i+1}), f(a_i))$ belongs to $K_{\mathfrak{s}}^{3,\mathrm{pr}}$ hence to $K_{\mathfrak{s}}^{3,\mathrm{uq}}$ hence $\mathrm{NF}_{\mathfrak{s}}(f(M_{0,i}), f(M_{0,i+1}), M_{1,i}, M_{1,i+1})$ hence by long transitivity $\mathrm{NF}(f(M_{0,0}), f(M_{0,\alpha}), M_{1,0}, M_{1,i})$.

By the uniqueness of NF, without loss of generality f is the identity and for some M_3^+ we have $M_{1,\alpha} \leq_{\mathfrak{s}} M_3^+$ and $M_3 \leq_{\mathfrak{s}} M_3^+$.
6) Using 4.7, see details inside the proof of 6.16.
7) Trivial by the definitions. $\square_{5.6}$

* * *

5.7 Definition. 1) We say N is prime over $M \cup \mathbf{J}$ (for $\mathfrak{s}$), or $(M, N, \mathbf{J}) \in K_{\mathfrak{s}}^{3,\text{qr}}$ if:

- (a) $M \leq_{\mathfrak{s}} N$ in K_λ
- (b) $\mathbf{J} \subseteq \mathbf{I}_{M,N}$ and $\mathbf{J}$ is independent in (M, N), actually the second statement follows from clause (c)
- (c) if $M \leq_{\mathfrak{s}} N', \mathbf{J}' \subseteq \mathbf{I}_{M,N'}$ and $\mathbf{J}'$ is independent in (M, N') and h is a one to one mapping from $\mathbf{J}$ onto $\mathbf{J}'$ such that $\mathbf{tp}_{\mathfrak{s}}(a, M, N) = \mathbf{tp}_{\mathfrak{s}}(h(a), M, N')$ for every $a \in \mathbf{J}$, then there is a $\leq_{\mathfrak{s}}$-embedding of N into N' over M extending h.

2) Let $(M, N, \mathbf{J}) \in K_{\mathfrak{s}}^{3,\text{bs}}$ mean that $\mathbf{J}$ is independent in the pair (M, N).
3) We define $\leq_{\text{bs}}^*$ as in Definition 4.2(2):

$(M_1, N_1, \mathbf{J}_1) \leq_{\text{bs}}^* (M_2, N_2, \mathbf{J}_2)$

when both are from $K_{\mathfrak{s}}^{3,\text{bs}}$ and either they are equal or $\mathbf{J}_1 \subseteq \mathbf{J}_2$ is independent in (M_1, M_2, N_2) and M_2, N_2 is $\leq_{\mathfrak{s}}$ -universal over M_1, N_2 respectively.

4) We defined $\leq_{\text{bs}}^{**}$ as in Definition 4.2(3):
$(M_1, N_1, \mathbf{J}_1) \leq_{\text{bs}}^{**} (M_2, N_2, \mathbf{J}_2)$

when both are from $K_{\mathfrak{s}}^{3,\text{bs}}$ and either they are equal or $\mathbf{J}_1 \subseteq \mathbf{J}_2$ is independent in (M_1, M_2, N_2) and M_2, N_2 is $\leq_{\mathfrak{s}}$ -brimmed over M_1, N_1 respectively.

Some basic properties are

5.8 Claim. *1) [Assume $\mathfrak{s}$ has primes.] If $M \le_{\mathfrak{s}} N$ (in $K_{\mathfrak{s}}$) and $\mathbf{J} \subseteq \mathbf{I}_{M,N}$ is independent in (M,N), <u>then</u> there is $N' \le_{\mathfrak{s}} N$ which is prime over $M \cup \mathbf{J}$.*
2) If $\mathbf{J}$ is independent in (M,N) and $\langle M_i, a_j : i \le \alpha, j < \alpha\rangle$ is an M-based pr-decomposition of (M_0,N) (see Definition 3.3(1)) and $\mathbf{J} = \{a_j : j < i\}$, <u>then</u> M_α is prime over $M_0 \cup \mathbf{J}$.
3) $(M,N,\{a\}) \in K^{3,\mathrm{qr}}_{\mathfrak{s}}$ <u>iff</u> $(M,N,a) \in K^{3,\mathrm{pr}}_{\mathfrak{s}}$ and also $(M,N,\{a\}) \in K^{3,\mathrm{bs}}_{\mathfrak{s}}$ iff $(M,N,a) \in K^{3,\mathrm{bs}}_{\mathfrak{s}}$.
4) If $(M,N,\mathbf{J}) \in K^{3,\mathrm{qr}}_{\mathfrak{s}}$ and $M \cup \mathbf{J} \subseteq N^- \le_{\mathfrak{s}} N$ <u>then</u> $(M,N^-,\mathbf{J}) \in K^{3,\mathrm{qr}}_{\mathfrak{s}}$.
5) Assume $\delta < \lambda^+_{\mathfrak{s}}$ is a limit ordinal, $\langle M_i : i \le \delta\rangle$ is $\le_{\mathfrak{s}}$-increasing continuous; $\langle N_i : i \le \delta\rangle$ is $\le_{\mathfrak{s}}$-increasing continuous, $\mathbf{J}_i$ is $\subseteq$-increasing continuous and $i < \delta \Rightarrow (M_i,N_i,\mathbf{J}_i) \in K^{3,\mathrm{bs}}_{\mathfrak{s}}$, <u>Then</u> $(M_\delta,N_\delta,\mathbf{J}_\delta) \in K^{3,\mathrm{bs}}_{\mathfrak{s}}$; note: $i < \delta \Rightarrow N_i = N_\delta$ is O.K. Moreover, we can weaken the assumption to $i < \delta \Rightarrow (M_{i+1},N_{i+1},\mathbf{J}_{i+1}) \in K^{3,\mathrm{bs}}_{\mathfrak{s}}$.
6) If $p_i \in \mathscr{S}^{\mathrm{bs}}_{\mathfrak{s}}(M)$ for $i < \alpha$ <u>then</u> for some N and $\mathbf{J} = \{a_i : i < \alpha\}$ with no repetitions we have $(M,N,\mathbf{J}) \in K^{3,\mathrm{bs}}_{\mathfrak{s}}$ and $\mathbf{tp}_{\mathfrak{s}}(a_i,M,N) = p_i$ for $i < \alpha$.

Proof. 1) By 5.4(1), $(*)_1 \Leftrightarrow (*)_6$, letting $\langle a_i : i < \alpha\rangle$ list $\mathbf{J}$ we can find $M_i \le_{\mathfrak{s}} N$ for $i < \alpha$ such that $\langle M_i, a_j : i \le \alpha, j < \alpha\rangle$ as in $(*)_6$ of 5.4. Now we can use part (2).
2) Let N^* satisfying $M = M_0 \le_{\mathfrak{s}} N^*$ and a one-to-one function $h : \mathbf{J} \to \mathbf{J}' \subseteq N^*$ satisfying $c \in \mathbf{J} \Rightarrow \mathbf{tp}_{\mathfrak{s}}(h(c),M_0,N^*) = \mathbf{tp}_{\mathfrak{s}}(c,M_0,M_\alpha)$ and $\mathbf{J}'$ independent in (M_0,N^*) be given. Let $h(a_j) = c_j$. We now choose by induction on $i \le \alpha$ a $\le_{\mathfrak{s}}$-embedding f_i of M_i into N^*, increasing continuous with i and mapping a_j to c_j. For $i = 0$ this is given, for i limit take union. For $i = j+1$, we know that $\mathbf{tp}_{\mathfrak{s}}(c_j,f_j(M_j),N^*)$ does not fork over $f_0(M_0)$ by 5.4(3) (because $(M_j,M_{j+1},a_j) \in K^{3,\mathrm{uq}}_{\mathfrak{s}}$ by 3.7(2)) and so as $\mathbf{tp}_{\mathfrak{s}}(a_j,M_j,M_i)$ does not fork over M_0 and $f_j[\mathbf{tp}_{\mathfrak{s}}(a_j,M_0,M_1)] = \mathbf{tp}_{\mathfrak{s}}(c_j,f_0(M_0),N^*)$ clearly $f_j[\mathbf{tp}_{\mathfrak{s}}(a_j,M_j,M_\alpha)] = \mathbf{tp}_{\mathfrak{s}}(c_j,f_j(M_j),N^*)$. But $(M_j,M_{j+1},a_j) \in K^{3,\mathrm{pr}}_{\mathfrak{s}}$ so we can find $f_i \supseteq f_j$ as required. So f_α is as required in Definition 5.7.
3) By the definitions.
4) Easy, like in 3.5(1).
5) By 5.4(1) it suffices to prove that for any finite non-empty $\mathbf{I} \subseteq \mathbf{J}_\delta$,

we have $(M_\delta, N_\delta, \mathbf{I}) \in K^{3,\mathrm{bs}}_{\mathfrak{s}}$. By 1.16 for each $c \in \mathbf{I}$ for some $i(c) < \delta$ we have $c \in \mathbf{J}_{i(c)}$ and $\mathbf{tp}_{\mathfrak{s}}(c, M_\delta, N_\delta)$ does not fork over $M_{i(c)}$. Let $i(*) = \max\{i(c) : c \in \mathbf{I}\}$, so $i(*) < \delta$ and $i \in [i(*), \delta) \Rightarrow (M_{i+1}, N_{i+1}, \mathbf{I}) \in K^{3,\mathrm{bs}}_{\mathfrak{s}}$. By renaming $i(*) = 0$ so $\mathbf{I} \subseteq N_0$ and we shall prove the statement by induction on $n = |\mathbf{I}|$. If $n = 1$, this is said above (or use 5.6(7) and Ax(E)(h)). So assume $n = m + 1$, choose $c \in \mathbf{I}$, let $\mathbf{I}' = \mathbf{I} \backslash \{c\}$.

We can choose (f_i, M'_i, N'_i) for $i \leq \delta$ such that

⊛ (a) $(M'_i, N'_i, c) \in K^{3,\mathrm{bs}}_{\mathfrak{s}}$ is $\leq_{\mathfrak{s}}$-increasing continuous

(b) $M'_0 = M_0, N'_0 = N_0$

(c) f_i is a $\leq_{\mathfrak{s}}$-embedding of M_i into M'_i, increasing with i

(d) if $i = j + 1$ then $(M'_i, N'_i, c) \in K^{3,\mathrm{uq}}_{\mathfrak{s}}$

(e) $\mathrm{NF}_{\mathfrak{s}}(f_i(M_i), M'_i, f_{i+1}(M_{i+1}), M'_{i+1})$.

Without loss of generality $f_i = \mathrm{id}_{M_i}$ for $i \leq \delta$. By the existence of non-forking amalgamation and existence of amalgamation without loss of generality for some $N^+ \in K_\lambda$ we have $N'_\delta \leq_{\mathfrak{s}} N^+$ and $N_\delta \leq_{\mathfrak{s}} N^+$ and $\mathrm{NF}_{\mathfrak{s}}(M_\delta, M'_\delta, N_\delta, N^+)$. Now by long transitivity for any successor $i < \delta$ by clauses (e) and (a) we have $\mathrm{NF}_{\mathfrak{s}}(M_i, M'_i, M_\delta, M'_\delta)$ hence (by transitivity for $\mathrm{NF}_{\mathfrak{s}}$) we have $\mathrm{NF}_{\mathfrak{s}}(M_i, M'_i, N_\delta, N^+)$ hence by monotonicity $\mathrm{NF}_{\mathfrak{s}}(M_i, M'_i, N_i, N^+)$.

Recall that for successor $i < \delta$, the set $\mathbf{I}$ is independent in (M_i, N_i) hence it follows by 5.3 that $\mathbf{I}$ is independent in (M_i, M'_i, N^+) and even (M_0, M'_i, N^+). By 5.4(3) for every successor $i < \delta$ the set $\mathbf{I}' = \mathbf{I} \backslash \{c\}$ is independent in (M_0, N'_i, N^+) hence by monotonicity, 5.4(4) this holds for every $i < \delta$. By the induction hypothesis $\mathbf{I}'$ is independent in (N'_δ, N^+), hence in (M_0, N'_δ, N^+) hence by 5.6(1) the set $\mathbf{I} = \mathbf{I}' \cup \{c\}$ is independent in (M_0, M'_δ, N^+) but $M_0 \leq_{\mathfrak{s}} M_\delta \subseteq_{\mathfrak{s}} M'_\delta$ hence $\mathbf{I}$ is independent in (M_δ, N^+) but $\mathbf{I} \cup M_\delta \subseteq N_\delta \subseteq_{\mathfrak{s}} N^+$ hence $\mathbf{I}$ is independent in (M_δ, N_δ) as required.

6) Easy. $\square_{5.8}$

* * *

5.9 Claim. *1) If $(M_0, M_1, \mathbf{J}) \in K^{3,\mathrm{qr}}_{\mathfrak{s}}$ (hence $\mathbf{J}$ is independent in (M_0, M_1)) and $N_0 \leq_{\mathfrak{s}} N_1, f_0$ is an isomorphism from M_0 onto N_0 and $\{c_a : a \in \mathbf{J}\}$ is an independent set in (N_0, N_1) satisfying $\mathbf{tp}_{\mathfrak{s}}(c_a, N_0, N_1) = f_0[\mathbf{tp}_{\mathfrak{s}}(a, M_0, M_1)]$ for $a \in \mathbf{J}$ (and of course $\langle c_a : a \in \mathbf{J}\rangle$ is with no repetitions) <u>then</u> there is a $\leq_{\mathfrak{s}}$-embedding f of M_1 into N_1 extending f_0 and mapping each $b \in \mathbf{J}$ to c_b.*
2) [Assume $\mathfrak{s}$ has primes]. Assume $(M_0, M_1, \mathbf{J}) \in K^{3,\mathrm{qr}}_{\mathfrak{s}}$ and $M_0 \leq_{\mathfrak{s}} M_2 \leq_{\mathfrak{s}} M_3, M_0 \leq_{\mathfrak{s}} M_1 \leq_{\mathfrak{s}} M_3$ and $\mathbf{J}$ is independent in (M_0, M_2, M_3). <u>Then</u> $\mathrm{NF}_{\mathfrak{s}}(M_0, M_1, M_2, M_3)$.
3) [Assume $\mathfrak{s}$ has primes] If $(M, N, \mathbf{J}) \in K^{3,\mathrm{qr}}_{\mathfrak{s}}$ and $\{a_\alpha : \alpha < \alpha^*\}$ list $\mathbf{J}$ with no repetitions of course <u>then</u> we can find $\bar{M} = \langle M_\alpha : \alpha \leq \alpha(*)\rangle$ such that: $\bar{M}$ is $\leq_{\mathfrak{s}}$-increasing continuous[6], $M = M_0, N \leq_{\mathfrak{s}} M_{\alpha(*)}, (M_\alpha, M_{\alpha+1}, a_\alpha) \in K^{3,\mathrm{pr}}_{\mathfrak{s}}$ (hence $\mathbf{tp}_{\mathfrak{s}}(a_\alpha, M_\alpha, M_{\alpha+1})$ does not fork over M).
4) If $M_0 \leq_{\mathfrak{s}} M_1 \leq_{\mathfrak{s}} M_2$ and $(M_0, M_1, \mathbf{J}_0) \in K^{3,\mathrm{qr}}_{\mathfrak{s}}$ and $(M_0, M_2, \mathbf{J}_1) \in K^{3,\mathrm{bs}}_{\mathfrak{s}}$ and $\mathbf{J}_0 \subseteq \mathbf{J}_1$ <u>then</u> $\mathbf{J}_1 \backslash \mathbf{J}_0$ is independent in (M_0, M_1, M_2).

Proof. 1) This just rephrases Definition 5.7.
2) We are allowed to increase M_1, M_3, i.e. if $M_3 \leq_{\mathfrak{s}} M'_3, M'_1 \leq_{\mathfrak{s}} M'_3, M_1 \leq_{\mathfrak{s}} M'_1$ but still $(M_0, M'_1, \mathbf{J}) \in K^{3,\mathrm{qr}}_{\mathfrak{s}}$ then by the monotonicity of the relation NF we can replace M_1, M_3 by M'_1, M'_3. By 5.4(1), specifically $(*)_5$, we can find M' and $\langle M^0_i, a_j : i \leq \alpha, j < \alpha\rangle$ such that $M^0_0 = M_0, M^0_i$ is $\leq_{\mathfrak{s}}$-increasing continuous, $(M^0_i, M^0_{i+1}, a_i) \in K^{3,\mathrm{pr}}_{\mathfrak{s}} \subseteq K^{3,\mathrm{uq}}_{\mathfrak{s}}$ for $i < \alpha$ and $\langle a_i : i < \alpha\rangle$ list $\mathbf{J}$ with no repetitions and $M^0_\alpha \leq_{\mathfrak{s}} M'_3$ and $M_3 \leq_{\mathfrak{s}} M'_3$. As "$M_1$ is prime over $M_0 \cup \mathbf{J}$" by assumption, there is an $\leq_{\mathfrak{s}}$-embedding f from M_1 into M^0_α over M_0, and by amalgamation we can extend f^{-1} to a $\leq_{\mathfrak{s}}$-embedding f^+ of M^0_α into some M''_3 where $M'_3 \leq_{\mathfrak{s}} M''_3$. So without loss of generality f^+ is the identity hence $M_1 \leq_{\mathfrak{s}} M^0_\alpha$ and so by the first sentence in this proof without loss of generality $M''_3 = M_3, M^0_\alpha = M_1$. We now choose by induction on $i \leq \alpha, M^2_i, M^3_i$ such that:

(α) M^2_i is $\leq_{\mathfrak{s}}$-increasing continuous

(β) M^3_i is $\leq_{\mathfrak{s}}$-increasing continuous

(γ) $M^2_0 = M_2, M^3_0 = M_3$

[6] of course, we can have $M_{\alpha(*)} \leq_{\mathfrak{s}} N$, but having equality is harder, see later

(δ) $M_i^0 \le_{\mathfrak{s}} M_i^2 \le_{\mathfrak{s}} M_i^3$

(ε) $(M_i^2, M_{i+1}^2, a_i) \in K_{\mathfrak{s}}^{3,\mathrm{pr}}$

(ζ) $\mathbf{tp}_{\mathfrak{s}}(a_i, M_i^2, M_i^3)$ does not fork over M_i^0

(η) $M_i^0 \le_{\mathfrak{s}} M_i^3$.

Why is this enough? For each i we have $(M_i^0, M_{i+1}^0, a_i) \in K_{\mathfrak{s}}^{3,\mathrm{uq}}$ (by the choice of M_i^0, M_{i+1}^0, a_i we have $(M_i^0, M_{i+1}^0, a_i) \in K_{\mathfrak{s}}^{3,\mathrm{pr}}$ and use claim 3.7(2)). By this, 1.19 and clauses (δ) and (ζ) we have $\mathrm{NF}_{\mathfrak{s}}(M_i^0, M_{i+1}^0, M_i^2, M_{i+1}^2)$. By the symmetry property of $\mathrm{NF}_{\mathfrak{s}}$ we have $\mathrm{NF}_{\mathfrak{s}}(M_i^0, M_i^2, M_{i+1}^0 M_{i+1}^2)$. As this holds for every $i < \alpha$ and clauses $(\alpha)+(\beta)$ by the long transitivity property of $\mathrm{NF}_{\mathfrak{s}}$ (see II.6.28) we get $\mathrm{NF}_{\mathfrak{s}}(M_0^0, M_0^2, M_\alpha^0, M_\alpha^2)$, which means $\mathrm{NF}_{\mathfrak{s}}(M_0, M_2, M_1, M_\alpha^2)$. Now by monotonicity we can replace M_α^2 first by M_α^3 then by M_3 so we got $\mathrm{NF}_{\mathfrak{s}}(M_0, M_2, M_1, M_3)$ as required.

Why is it possible to carry the induction? Having arrived to i we can find models $M_{i+1}^{2,*} \le_{\mathfrak{s}} M_{i+1}^3$ such that $M_i^3 \le_{\mathfrak{s}} M_{i+1}^3$ and $(M_i^2, M_{i+1}^{2,*}, a_i) \in K_{\mathfrak{s}}^{3,\mathrm{pr}}$. By 5.4(3) the type $\mathbf{tp}_{\mathfrak{s}}(a_i, M_i^{2,*}, M_i^3)$ does not fork over M_i^0.

Now by the definition of prime, there is a $\le_{\mathfrak{s}}$-embedding f_i of M_{i+1}^0 into M_{i+1}^3 over M_i^0 satisfying $f_i(a_i) = a_i$. As $(M_i^0, M_{i+1}^0, a_i) \in K_{\mathfrak{s}}^{3,\mathrm{pr}} \subseteq K_{\mathfrak{s}}^{3,\mathrm{uq}}$, possibly replacing M_{i+1}^3 by a $\le_{\mathfrak{s}}$-extension we can extend f_i^{-1} to an $\le_{\mathfrak{s}}$-embedding g_i of $M_{i+1}^{2,*}$ into M_{i+1}^3 extending $\mathrm{id}_{M_i^2}$; lastly let $M_{i+1}^2 := g_i(M_i^{2,*})$.

3) Easy and included in the proof of part (2).

4) By part (3) and 5.4(3). $\square_{5.9}$

5.10 Claim. *Assume $\langle M_i : i \le \delta + 1\rangle$ is $\le_{\mathfrak{s}}$-increasing continuous and $\mathbf{J} \subseteq \mathbf{I}_{M_\delta, M_{\delta+1}}$.*

1) If $|\mathbf{J}| < \mathrm{cf}(\delta)$ and $\mathbf{J}$ is independent in $(M_\delta, M_{\delta+1})$ then for every $i < \delta$ large enough, $\mathbf{J}$ is independent in $(M_i, M_\delta, M_{\delta+1})$.

2) If $\mathbf{J} \subseteq \mathbf{I}_{M_\delta, M_{\delta+1}}$ is independent in $(M_i, M_{\delta+1})$ for every $i < \delta$, then $\mathbf{J}$ is independent in $(M_\delta, M_{\delta+1})$. If $\mathbf{J}$ is independent in $(M_0, M_i, M_{\delta+1})$ for every $i < \delta$ then $\mathbf{J}$ is independent in $(M_i, M_\delta, M_{\delta+1})$ for every $i < \delta$.

3) Assume $\langle M_i : i \le \delta\rangle, \langle N_i : i \le \delta\rangle$ are $\le_{\mathfrak{s}}$-increasing continuous,

$\langle \mathbf{J}_i : i < \delta \rangle$ is $\subseteq$-increasing continuous and $(M_i, N_i, \mathbf{J}_i) \in K^{3,\mathrm{bs}}_{\mathfrak{s}}$ for $i < \delta$. Then $(M_\delta, N_\delta, \mathbf{J}_\delta) \in K^{3,\mathrm{bs}}_{\mathfrak{s}}$ for $i < \alpha$.

Discussion: 1) At this point, if $\langle M_i : i \leq \alpha \rangle$ is $\leq_{\mathfrak{s}}$-increasing continuous $M_\alpha \leq_{\mathfrak{s}} N, a \in N$ and $\mathbf{tp}_{\mathfrak{s}}(a, M_\alpha, N) \in \mathscr{S}^{\mathrm{bs}}_{\mathfrak{s}}(M_\alpha)$ does not fork over M_0 we do not know if there is $\langle N_i : i \leq \alpha \rangle$ which is $\leq_{\mathfrak{s}}$-increasing continuous $a \in N_0, M_i \leq_{\mathfrak{s}} N_i$ and $(M_i, N_i, a) \in K^{3,\mathrm{pr}}_{\mathfrak{s}}$. So we go around this. The claim 5.10 is used in 6.16.
2) Note that 5.8(5) implies 5.10(2),(3) and also the inverse is easy, but we give different proofs.

Proof. 1) For each $c \in \mathbf{J}$ for some $i_c \in \delta, \mathbf{tp}_{\mathfrak{s}}(c, M_\delta, M_{\delta+1})$ does not fork over M_{i_c} let $i(*) = \sup\{i_c : c \in \mathbf{J}\}$ and use 5.4(4).
2) By 5.4(1) it suffices to deal with finite $\mathbf{J}$, say $\mathbf{J} = \{b_\ell : \ell < n\}$ with no repetitions.
By the $\mathrm{NF}_{\mathfrak{s}}$-calculus, i.e. by 1.17 there is a $\leq_{\mathfrak{s}}$-increasing continuous sequence $\langle M^+_i : i \leq \delta + 1 \rangle$ such that $\mathrm{NF}_{\mathfrak{s}}(M_i, M^+_i, M_j, M^+_j)$ for any $i < j \leq \delta + 1$ and M^+_{i+1} is $(\lambda, *)$-brimmed over $M_{i+1} \cup M^+_i$ for $i < \delta$, hence M^+_δ is $(\lambda, *)$-brimmed over M_δ, see 1.17(1). Hence there is a $\leq_{\mathfrak{s}}$-embedding h of $M_{\delta+1}$ into M^+_δ over M_δ, so without loss of generality $M_{\delta+1} \leq_{\mathfrak{s}} M^+_\delta$. As $\mathbf{J}$ is finite and M^+_δ is the union of the $\leq_{\mathfrak{s}}$-increasing sequence $\langle M^+_i : i < \delta \rangle$ clearly for some $i < \delta$ we have $\mathbf{J} \subseteq M^+_i$ hence "$\mathbf{J}$ is independent in (M_i, M^+_i). But $\mathrm{NF}_{\mathfrak{s}}(M_i, M^+_i, M_\delta, M^+_\delta)$ hence by Claim 5.3 we deduce "$\mathbf{J}$ is independent in $(M_i, M_\delta, M^+_\delta)$ hence in $(M_i, M_\delta, M_{\delta+1})$ as required.
3) For every finite $\mathbf{J} \subseteq \mathbf{J}_\delta$, for some $\alpha < \delta$ we have $\mathbf{J} \subseteq \mathbf{J}_\alpha$ hence by monotonicity, $\beta \in (\alpha, \delta) \Rightarrow (M_\beta, N_\delta, \mathbf{J}) \in K^{3,\mathrm{bs}}_{\mathfrak{s}}$, hence by part (2) we know that $(M_\delta, N_\delta, \mathbf{J}) \in K^{3,\mathrm{bs}}_{\mathfrak{s}}$. By 5.4(1) we deduce $(M_\delta, N_\delta, \mathbf{J}_\delta) \in K^{3,\mathrm{bs}}_{\mathfrak{s}}$. $\square_{5.10}$

5.11 Claim. *Assume $\mathfrak{s} = \mathfrak{t}^+, \mathfrak{t}$ a successful good$^+$ $\lambda_{\mathfrak{t}}$-frame so $\lambda = \lambda_{\mathfrak{s}} = \lambda^+_{\mathfrak{t}}$. Assume $M_\ell \in K_{\mathfrak{s}}$ and $\langle M^\ell_\alpha : \alpha < \lambda \rangle$ is a $\leq_{\mathfrak{K}[\mathfrak{t}]}$-representation of M_ℓ for $\ell = 0, 1$.*
If $M_0 \leq_{\mathfrak{s}} M_1$ and $\mathbf{J} \subseteq \mathbf{I}_{M_0,M_1}$ then: $\mathbf{J}$ is independent (for $\mathfrak{s}$) in (M_0, M_1) iff for a club of $\delta < \lambda$ the set $\mathbf{J} \cap M_{1,\delta}$ is independent (for

$\mathfrak{t}$) *in* $(M_{0,\delta}, M_{1,\delta})$ *iff for stationarily many* $\delta < \lambda, \mathbf{J} \cap M_{1,\delta}$ *is independent (for* $\mathfrak{t}$) *in* $(M_{0,\delta}, M_{1,\delta})$ *iff for unboundedly many* $\alpha < \lambda, \mathbf{J} \cap M_{1,\delta}$ *is independent (for* $\mathfrak{t}$) *in* $(M_{0,\alpha}, M_{1,\alpha})$.

Proof of 5.11. By 5.4(1), applied to $\mathfrak{s}$ and to $\mathfrak{t}$ without loss of generality $\mathbf{J}$ is finite. As $\mathfrak{t}$ satisfies the Hypothesis 4.1, clearly the results of §4 apply. Note that $\mathfrak{s}$ has primes by 4.9 applied to $\mathfrak{t}$, also $\mathfrak{s}$ is good$^+$ by 1.6(2) and moreover by 1.9 and is weakly successful by 5.1.

Using 5.8(5), the "moreover" and see last sentence, the third clause implies the second clause and trivially second implies third; similarly for the fourth. So assume the failure of the first and we show the failure of the third. Let $\mathbf{J} = \{a_\ell : \ell < n\}$ without repetitions. We can try to choose by induction on ℓ a model $M'_\ell \leq_{\mathfrak{s}} M_1$ such that $M'_0 = M_0, (M'_\ell, M'_{\ell+1}, a_\ell) \in K^{3,\mathrm{pr}}_{\mathfrak{s}}$, moreover is as constructed in 4.9 + 4.3 and $\mathbf{tp}_{\mathfrak{s}}(a_\ell, M'_\ell, M_1) \in \mathscr{S}^{\mathrm{bs}}_{\mathfrak{s}}(M'_\ell)$ does not fork over M_0. We cannot succeed so for some $m < n$ we have $M'_0, \ldots, M'_m$ as above but $\mathbf{tp}_{\mathfrak{s}}(a_m, M'_m, M_1)$ forks over M_0. Rename M_1 as M'_{m+1} and let $\langle M'_{\ell,\alpha} : \alpha \leq \lambda\rangle$ be a $\leq_{\mathfrak{t}}$-representation of M'_ℓ for $\ell \leq m+1$ and $M'_{0,\alpha} = M^0_\alpha, M'_{m+1,\alpha} = M^1_\alpha$. Now by 4.13(1) for some club E of λ, if δ is from E and $\ell < m$ then $(M'_{\ell,\delta}, M'_{\ell+1,\delta}, a_\ell) \in K^{3,\mathrm{uq}}_{\mathfrak{s}}$ and $a_m \in M'_{m+1,\delta}$ and $\mathbf{tp}_{\mathfrak{s}}(a_m, M_{m,\delta}, M_{m+1,\delta})$ forks over $M_{0,\delta}$ while $M_{m,\delta} \leq_{\mathfrak{t}} M_{m+1,\delta}$. By 5.4(3) for $\mathfrak{t}$ we get $\{a_\ell : \ell \leq m\} \subseteq \mathbf{I}_{M_{0,\delta}, M_{m,\delta}}$ is not independent. So we have gotten the failure of the third clause.

The proof that the first clause implies the third one is similar. $\square_{5.11}$

We can deal with dimension as in [Sh:c, Ch.III].

5.12 Definition. Assume that $M \leq_{\mathfrak{s}} N$ and $p \in \mathscr{S}^{\mathrm{bs}}_{\mathfrak{s}}(M)$, then we let

$$\begin{aligned} \dim(p, N) = \ & \mathrm{Min}\{|\mathbf{J}| : \mathbf{J} \text{ satisfies} \\ & (i) \ \ \mathbf{J} \text{ is a subset of } \{c \in N : \mathbf{tp}_{\mathfrak{s}}(c, M, N) \text{ is equal to } p\}, \\ & (ii) \ \ \text{the triple } (M, N, \mathbf{J}) \text{ belongs to } K^{3,\mathrm{bs}}_{\mathfrak{s}} \text{ and} \\ & (iii) \ \ \mathbf{J} \text{ is maximal under those restrictions}\}. \end{aligned}$$

We shall say more on dim after we understand regular types (see §10, mainly 10.15).

5.13 Claim. *Assume $M \in K_{\mathfrak{s}}$ and $\mathbf{J}$ is independent in (M, N^*).*
1) If $\mathbf{tp}_{\mathfrak{s}}(a, M, N^*) \in \mathscr{S}^{\mathrm{bs}}_{\mathfrak{s}}(M)$, then for some finite $\mathbf{J}' \subseteq \mathbf{J}$ the set $(\mathbf{J} \backslash \mathbf{J}') \cup \{a\}$ is independent in (M, N^*) and $a \notin \mathbf{J} \backslash \mathbf{J}'$, of course.
2) [$\mathfrak{s}$ has primes] If $a \in N^*$ then for some finite $\mathbf{J}' \subseteq \mathbf{J}$ and M' we have: $M \cup \{a\} \subseteq M' \leq_{\mathfrak{s}} N$ and $\mathbf{J} \backslash \mathbf{J}'$ is independent in (M, M', N^*).
3) If $a \in N^*$ ubthen for some finite $\mathbf{J}' \subseteq \mathbf{J}$ and M', N' we have: $M \cup \{a\} \subseteq M' \leq_{\mathfrak{s}} N', N \leq_{\mathfrak{s}} N'$ and $\mathbf{J} \backslash \mathbf{J}'$ is independent in (M, M', N').

Proof. 1) Let $\mathbf{J} = \{a_i : i < \alpha\}$, we prove the statement by induction on α. For $\alpha = 0, \alpha$ successor this is trivial. For α limit $< \lambda^+$ by the definition there are $N^+ \in K_{\mathfrak{s}}$ and a $\leq_{\mathfrak{s}}$-increasing continuous sequence $\langle M_i : i \leq \alpha \rangle$ such that $M_0 = M, M_i \leq_{\mathfrak{s}} N^+, N^* \leq_{\mathfrak{s}} N^+$ and $\mathbf{tp}_{\mathfrak{s}}(a_i, M_i, M_{i+1}) \in \mathscr{S}^{\mathrm{bs}}_{\mathfrak{s}}(M_i)$ does not fork over M for each $i < \alpha$. By 1.17, as in the proof of 5.10(2) we can find a $\leq_{\mathfrak{s}}$-increasing continuous $\langle N_i : i \leq \alpha \rangle$ such that for each $i < \alpha, M_i \leq_{\mathfrak{s}} N_i$ and $\mathrm{NF}_{\mathfrak{s}}(M_i, N_i, M_{i+1}, N_{i+1})$ hence $\mathbf{tp}_{\mathfrak{s}}(a_i, N_i, N_{i+1})$ does not fork over M_i and N_α is $(\lambda, *)$-brimmed over M_α. Hence we can $\leq_{\mathfrak{s}}$-embed N^+ into N_α over M_α so by renaming without loss of generality $N^+ = N_\alpha$.

Lastly, replace the M_i by N_i so without loss of generality $M_\alpha = N^*$.

Now for some $\beta < \alpha, a \in M_\beta$ and by the induction hypothesis on α for some finite $u \subseteq \beta$ the set $\{a_i : i \in \beta \backslash u\} \cup \{a\}$ is independent in (M, M_β). Clearly by the Definition 5.2 the set $\{a_i : i \in \alpha \backslash \beta\}$ is independent in (M, M_β, M_α). By 5.6(1) and the last two sentences $(\{a_i : i \in \beta \backslash u\} \cup \{a\}) \cup (\{a_i : i \in \alpha \backslash \beta\})$ is independent in (M, M_α) hence in (M, N^+) hence in (M, N^*) by 5.4(2). But the set is $\{a_i : i \in \alpha \backslash u\} \cup \{a\}$ so we are done.
2),3) Similar. $\square_{5.13}$

5.14 Conclusion. *Assume that $M \leq_{\mathfrak{s}} N$ and $p \in \mathscr{S}^{\mathrm{bs}}_{\mathfrak{s}}(M)$. Then any two sets $\mathbf{J}$ satisfying the demands (i) + (ii) + (iii) from Definition 5.12 have the same cardinality or are both finite.* $\square_{5.14}$

Proof. By 5.13(3) and 5.8(5). $\square_{5.14}$

As $K_{\mathfrak{s}}^{3,\mathrm{pr}}$ was generalized to $K_{\mathfrak{s}}^{3,\mathrm{qr}}$ above so now we generalize $K_{\mathfrak{s}}^{3,\mathrm{uq}}$ to $K_{\mathfrak{s}}^{3,\mathrm{vq}}$.

5.15 Definition. 1) Let $(M, N, \mathbf{J}) \in K_{\mathfrak{s}}^{3,\mathrm{vq}}$ mean:

- (a) $M \leq_{\mathfrak{s}} N$
- (b) $\mathbf{J}$ is independent in (M, N)
- (c) if $N \leq_{\mathfrak{s}} M_3, M \leq_{\mathfrak{s}} M_2 \leq_{\mathfrak{s}} M_3$ and $\mathbf{J}$ is independent in (M, M_2, M_3) <u>then</u> $\mathrm{NF}_{\mathfrak{s}}(M, M_2, N, M_3)$.

2) We say $(M, N, \mathbf{J}) \in K_{\mathfrak{s}}^{3,\mathrm{bs}}$ is thick <u>when</u> for every $p \in \mathscr{S}_{\mathfrak{s}}^{\mathrm{bs}}(M)$ the set $\{c \in \mathbf{J} : \mathbf{tp}_{\mathfrak{s}}(c, M, N) = p\}$ has cardinality λ.
3) We say that $(M, N, \mathbf{J}) \in K_{\mathfrak{s}}^{3,\mathrm{bs}}$ is $\mathscr{S}^*$-thick when $\mathscr{S}^* \subseteq \mathscr{S}_{\mathfrak{s}}^{\mathrm{bs}}$ (or $\mathscr{S}^* \subseteq \mathscr{S}_{\mathfrak{s}}^{\mathrm{bs}}(M)$) and for every $p \in \mathscr{S}^*(M)$ (or $p \in \mathscr{S}_{\mathfrak{s}}^*$) the set $\{c \in \mathbf{J} : \mathbf{tp}_{\mathfrak{s}}(c, M, N) = p\}$ has cardinality λ (we can below replace $\mathscr{S}_{\mathfrak{s}}^{\mathrm{bs}}$ by any dense $\mathscr{S}^* \subseteq \mathscr{S}_{\mathfrak{s}}^{\mathrm{bs}}$ where dense means that Ax(D)(c) of Definition II.2.1 holds).
4) We say that $(M, N, \mathbf{J}) \in K_{\mathfrak{s}}^{3,\mathrm{bs}}$ is weakly $\mathscr{S}^*$-thick[7] <u>when</u> for every $p \in \mathscr{S}^*$ (or $p \in \mathscr{S}^*(M)$) the set $\{c \in \mathbf{J} : \mathbf{tp}_{\mathfrak{s}}(c, M, N) \pm p\}$, see 6.2 below has cardinality λ (where $\mathscr{S}^* \subseteq \mathscr{S}_{\mathfrak{s}}^{\mathrm{bs}}$ or $\mathscr{S}^* \subseteq \mathscr{S}_{\mathfrak{s}}^{\mathrm{bs}}(M)$). If $\mathscr{S}^* = \mathscr{S}_{\mathfrak{s}}^{\mathrm{bs}}(M)$ we may omit $\mathscr{S}^*$.

5.16 Claim. *Assume $\mathfrak{s}$ is successful good$^+$ in (2)(b), (2)(c), (3), (4), (8) hence for them 4.4 holds hence all §4 apply.*
1) [$\mathfrak{s}$ has primes] If $(M, N, \mathbf{J}) \in K_{\mathfrak{s}}^{3,\mathrm{qr}}$ <u>then</u> $(M, N, \mathbf{J}) \in K_{\mathfrak{s}}^{3,\mathrm{vq}}$.
2) If $(M, N, \mathbf{J}) \in K_{\mathfrak{s}}^{3,\mathrm{bs}}$ <u>then</u>

- (a) *for some thick $(M', N', \mathbf{J}') \in K_{\mathfrak{s}}^{3,\mathrm{bs}}$ we have $(M, N, \mathbf{J}) <_{\mathrm{bs}}^{**} (M', N', \mathbf{J}')$*
  moreover $(M', N', \mathbf{J}' \backslash \mathbf{J})$ is thick
- (b) *for some $(M', N', \mathbf{J}) \in K_{\mathfrak{s}}^{3,\mathrm{vq}}$ we have $(M, N, \mathbf{J}) <_{\mathrm{bs}}^{**} (M', N', \mathbf{J})$; hence if $(M, N, \mathbf{J}) \leq_{\mathrm{bs}} (M^*, N^*, \mathbf{J})$ and M^* is $(\lambda, *)$-brimmed over M then without loss of generality $M' = M^*$ and $N' \leq_{\mathfrak{s}} N'', N^* \leq_{\mathfrak{s}} N''$ for some N''*

[7] This notion is not really used now. It becomes convenient when one restricts himself to regular types, see §10.

(c) *for some thick* $(M', N', \mathbf{J}') \in K^{3,\mathrm{vq}}_{\mathfrak{s}}$ *we have* $(M, N, \mathbf{J}) <^{**}_{\mathrm{bs}} (M', N', \mathbf{J}')$ *moreover* $(M', N', \mathbf{J}' \backslash \mathbf{J})$ *is thick.*

3) If $\langle M_i : i \leq \alpha \rangle$ *is* $\leq_{\mathfrak{s}}$*-increasing continuous and* $i < \alpha \Rightarrow (M_i, M_{i+1}, a_i) \in K^{3,\mathrm{uq}}_{\mathfrak{s}}$ *and* $\mathbf{tp}_{\mathfrak{s}}(a_i, M_i, M_{i+1})$ *does not fork over* M_0, *then* $(M_0, M_\alpha, \{a_i : i < \alpha\})$ *belongs to* $K^{3,\mathrm{vq}}_{\mathfrak{s}}$.
4) If $(M_\ell, M_{\ell+1}, \mathbf{J}_\ell) \in K^{3,\mathrm{vq}}_{\mathfrak{s}}$ *for* $\ell = 0, 1$ *and* $\mathbf{J}_1$ *is* (M_0, M_1, M_2)*-independent then* $(M_0, M_2, \mathbf{J}_0 \cup \mathbf{J}_1)$ *belongs to* $K^{3,\mathrm{vq}}_{\mathfrak{s}}$. *Moreover if* $\langle M_i : i \leq \alpha \rangle$ *is* $\leq_{\mathfrak{s}}$*-increasing continuous,* $\mathbf{J}_i$ *is independent in* (M_0, M_i, M_{i+1}) *for* $i < \alpha$ *and* $(M_i, M_{i+1}, \mathbf{J}_i) \in K^{3,\mathrm{vq}}_{\mathfrak{s}}$
then $(M_0, M_\alpha, \cup\{\mathbf{J}_i : i < \alpha\}) \in K^{3,\mathrm{vq}}_{\mathfrak{s}}$.
5) If $(M_0, M_1, \mathbf{J}_0) \in K^{3,\mathrm{vq}}_{\mathfrak{s}}, \mathbf{J}_0 \cap \mathbf{J}_1 = \emptyset$ *and* $\mathbf{J}_0 \cup \mathbf{J}_1$ *is independent in* (M_0, M_2) *and* $M_1 \leq_{\mathfrak{s}} M_2$ *then* $\mathbf{J}_1$ *is independent in* (M_0, M_1, M_2).
6) If $\langle M_i : i \leq \alpha + 1 \rangle$ *is* $\leq_{\mathfrak{s}}$*-increasing continuous,* $\mathbf{J}_i$ *is independent in* (M_0, M_i, M_{i+1}) *for* $i < \alpha, (M_i, M_{i+1}, \mathbf{J}_i) \in K^{3,\mathrm{vq}}_{\mathfrak{s}}$ *for* $i < \alpha, \mathbf{J}_\alpha \subseteq M \backslash M_0 \backslash \cup \{\mathbf{J}_i : i < \alpha\}$ *and* $\cup\{\mathbf{J}_i : i < \alpha + 1\}$ *is independent in* $(M_0, M_{\alpha+1})$ *then* $\mathbf{J}_\alpha$ *is independent in* $(M_0, M_\alpha, M_{\alpha+1})$.
7) Assume that $M_0 \leq_{\mathfrak{s}} M_1 \leq_{\mathfrak{s}} M_2, (M_0, M_2, \mathbf{J}_2) \in K^{3,\mathrm{vq}}_{\mathfrak{s}}$ *and* $\mathbf{J}_0 = M_1 \cap \mathbf{J}_2$ *and* $\mathbf{J}_2 \backslash \mathbf{J}_0$ *is independent in* (M_0, M_1, M_2).
Then $(M_0, M_1, \mathbf{J}_0) \in K^{3,\mathrm{vq}}_{\mathfrak{s}}$.
8) If $\langle M_i : i \leq \alpha \rangle$ *is* $\leq_{\mathfrak{s}}$*-increasing continuous,* $\alpha < \lambda^+$ *and* $\mathbf{J}_i$ *is independent in* (M_0, M_i, M_{i+1}) *for* $i < \alpha$ *then we can find a* $\leq_{\mathfrak{s}}$*-increasing continuous sequence* $\langle N_i : i \leq \alpha \rangle$ *such that* $M_i \leq_{\mathfrak{s}} N_i, N_i$ *is* $(\lambda, *)$*-brimmed over* M_i *and* $(M_i, M_{i+1}, \mathbf{J}_i) \leq_{\mathrm{bs}} (N_i, N_{i+1}, \mathbf{J}_i) \in K^{3,\mathrm{vq}}_{\mathfrak{s}}$ *for* $i < \alpha$.

Remark. See 4.2(3) and 5.7(4) for the definition of $<^{**}_{\mathrm{bs}}$.

Note also:

5.17 Observation. Assume $\mathfrak{s}$ is a successful good$^+$ frame.
If $(M_i, N_i, \mathbf{J}_i) \in K^{3,\mathrm{vq}}_{\mathfrak{s}}$ for $i < \delta, \delta$ a limit ordinal $< \lambda^+_{\mathfrak{s}}, (M_i, N_i, \mathbf{J}_i)$ is $<^{**}_{\mathrm{bs}}$-increasing for $i < \delta$ (hence M_i is $\leq_{\mathfrak{s}}$-increasing, N_i is $\leq_{\mathfrak{s}}$-increasing, we have not demanded continuity) then

(a) $(M_\delta, N_\delta, \mathbf{J}_\delta) \in K^{3,\mathrm{vq}}_{\mathfrak{s}}$ when we let $M_\delta = \cup\{M_i : i < \delta\}, N_\delta = \cup\{N_i : i < \delta\}$ and $\mathbf{J}_\delta = \cup\{\mathbf{J}_i : i < \delta\}$

(b) for $i < j \leq \delta$ we have $\mathrm{NF}_{\mathfrak{s}}(M_i, N_i, M_j, N_j)$

(c) if $\alpha \leq j \leq \delta$ and α is a limit ordinal then
$\mathrm{NF}_{\mathfrak{s}}(\bigcup_{i<\alpha} M_i, \bigcup_{i<\alpha} N_i, M_j, N_j)$.

Remark. Note that in the proof of 5.16 + 5.17 actually only the following order is required:

(a) 5.16(2),(5),(7) before 5.17,
(b) 5.16(3),(5),(8) before 5.16(4) and
(c) 5.16(5) before 5.16(6).

Proof of 5.17. By 5.8(5) for non-zero $\alpha \leq \delta$ we have $(\cup\{M_i : i < \alpha\}, \cup\{N_i : i < \alpha\}, \cup\{\mathbf{J}_i : i < \alpha\})$ belongs to $K^{3,\mathrm{bs}}_{\mathfrak{s}}$. We get that $(M_\delta, N_\delta, \mathbf{J}_\delta)$ belongs to $K^{3,\mathrm{bs}}_{\mathfrak{s}}$; this partially proves clause (a).

We prove the observation by induction on δ, so by the induction hypothesis without loss of generality $\langle (M_i, N_i, \mathbf{J}_i) : i < \delta \rangle$ is $\leq_{\mathrm{bs}}$-increasing continuous as we can for limit i redefine $(M_i, N_i, \mathbf{J}_i)$ as $(\cup\{M_j : j < i\}, \cup\{N_j : j < i\}, \cup\{\mathbf{J}_j : j < i\})$. Hence it belongs to $K^{3,\mathrm{vq}}_{\mathfrak{s}}$ by the induction hypothesis and $<^{**}_{\mathfrak{s}}$ continues to hold.

Now (using clause (a) of 5.16(2)) we choose an $<^{**}_{\mathfrak{s}}$-increasing continuous sequence $\langle (M'_i, N'_i, \mathbf{J}'_i) : i < \lambda^+ \rangle$ such that

(*) (a) $(M'_i, N'_i, \mathbf{J}'_i) \in K^{3,\mathrm{bs}}_{\mathfrak{s}}$ is thick
(b) $M'_0 = M_0, N_0 \leq_{\mathfrak{s}} N'_0$ and $\mathbf{J}_0 \subseteq \mathbf{J}'_0$
(c) if $(M'_i, N'_i, \mathbf{J}'_i) \notin K^{3,\mathrm{vq}}_{\mathfrak{s}}$ then $\neg\, \mathrm{NF}_{\mathfrak{s}}(M'_i, N'_i, M'_{i+1}, N'_{i+1})$
(d) $(M'_i, N'_i, \mathbf{J}'_i \backslash \mathbf{J}_j)$ is thick when $i = j + 1$.

As $\mathfrak{s}$ is successfull good$^+$ the set $\{\delta < \lambda^+ \colon \mathrm{NF}_{\mathfrak{s}}(M_\delta, N_\delta, M_{\delta+1}, N_{\delta+1})\}$ contains a club of λ^+. Let $\langle \alpha_i : i < \lambda^+ \rangle$ list such club in increasing order so clearly without loss of generality $M'_{\alpha_{i+1}}, N'_{\alpha_{i+1}}$ is brimmed over $M'_{\alpha_i+1}, N'_{\alpha_i+1}$ respectively.

Now we choose f_i by induction on $i \leq \delta$ such that

⊡ (a) f_i is a $\leq_{\mathfrak{s}}$-embedding of N_i into N'_{α_i+2}
(b) f_i maps M_i onto M'_{α_i}

(c) f_i maps $\mathbf{J}_i$ into $\mathbf{J}'_{\alpha_i}$

(d) f_i is increasing continuous in i

(e) for $j < i$ we have $f_i(\mathbf{J}_i) \cap \mathbf{J}'_{\alpha_j} = f_i(\mathbf{J}_j)$.

For $i = 0$ this holds for $f_i = \mathrm{id}_{N_0}$ by $(*)(b)$.

For i limit take unions. So let $i = j + 1$. As M'_i is brimmed over M'_j we can find an isomorphism f_i^1 from M_i onto M'_i extending $f_j \restriction M_j$; this will ensure clause (b) of $\boxdot$.

Let N_i^+, N_i^- be such that $N_i \leq_{\mathfrak{s}} N_i^+$ and $N_i^- \leq_{\mathfrak{s}} N_i^+$ and $(M_j, N_j, \mathbf{J}_j) \leq_{\mathrm{bs}} (M_i, N_i^-, \mathbf{J}_j) \in K_{\mathfrak{s}}^{3,\mathrm{vq}}$ and N_i^+ is brimmed over N_j (possibly by 5.16(2b), the "hence..."). We know that $(M'_{\alpha_j}, N'_{\alpha_j}, \mathbf{J}'_{\alpha_j}) \leq_{\mathrm{bs}} (M'_{\alpha_i}, N'_{\alpha_i}, \mathbf{J}'_{\alpha_i})$ and $f_j(\mathbf{J}_j) \subseteq \mathbf{J}'_{\alpha_j}$ hence $f_j(\mathbf{J}^+_{\alpha_j})$ is independent in $(M'_{\alpha_j}, M'_{\alpha_i}, N'_{\alpha_i})$ and, of course, $M'_{\alpha_i} \leq_{\mathfrak{s}} N'_{\alpha_i} \wedge f_j(N_j) \leq_{\mathfrak{s}} N'_{\alpha_j+1} \leq_{\mathfrak{s}} N'_{\alpha_i}$. Hence as $(M_j, N_j, \mathbf{J}_j) \in K_{\mathfrak{s}}^{3,\mathrm{vq}}$ we know that $\mathrm{NF}_{\mathfrak{s}}(M'_{\alpha_j}, f_j(N_j), M'_{\alpha_i}, N'_{\alpha_i})$ and of course $\mathrm{NF}_{\mathfrak{s}}(M_j, N_j, M_i, N_i^+)$.

So as N'_{α_i+1} is brimmed over N'_{α_i}, there is a $\leq_{\mathfrak{s}}$-embedding f_i^2 of N_i^+ into N'_{α_i+1} extending $f_i^1 \cup f_j$. By 5.16(5) from $(M'_{\alpha_i}, N'_{\alpha_i+1}, \mathbf{J}'_{\alpha_i}) \in K_{\mathfrak{s}}^{3,\mathrm{bs}}$ we can deduce that $\mathbf{J}_{\alpha_i} \backslash f_j(\mathbf{J}_j)$ is independent in $(M'_{\alpha_i}, f_i^2(N_i^-), N'_{\alpha_{i+1}})$.

So we can find a one to one $g_i : \mathbf{J}_i \backslash \mathbf{J}_j \to \mathbf{J}'_{\alpha_i} \backslash f_j(\mathbf{J}_j)$ such that $a \in \mathbf{J}_i \backslash \mathbf{J}_j \Rightarrow \mathbf{tp}_{\mathfrak{s}}(g_i(a), f_i^2(N_i^-), N_{\alpha_{i+1}}) = f_i^2(\mathbf{tp}_{\mathfrak{s}}(a, N_i^-, N_i^+))$. Next, we can choose $f_i^3 \supseteq (f_i^2 \restriction N_i^-) \cup g_i$ which is an $\leq_{\mathfrak{s}}$-embedding of N_i^+ into N'_{α_i+2}. The $f_i^3 \supseteq g_i$ will ensure clause (c) of $\boxdot$ and $f_i^3 \supseteq f_i^2 \supseteq f_j$ will ensure (d) of $\boxdot$.

Lastly, let $f_i = f_i^3 \restriction N_i$ so the relevant cases of clauses (a)-(d) are satisfied.

So we have carried the inductive definition of $\bar{f} = \langle f_i : i \leq \delta \rangle$ as required in $\boxdot$. Clearly $f_i(N_i) \leq_{\mathfrak{s}} N_{\alpha_i}$ for every limit $i \leq \delta$ and

$\odot$(a) $f_\delta(M_\delta) = M'_{\alpha_\delta} \leq_{\mathfrak{s}} f_\delta(N_\delta) \leq_{\mathfrak{s}} N'_\delta$

(b) for every $i < \delta, \mathbf{J}'_{\alpha_i} \backslash f_i(\mathbf{J}_i)$ is independent in $(M'_{\alpha_i}, f_i(N_i), N'_{\alpha_i+1})$

(c) $\langle \mathbf{J}'_{\alpha_i} \backslash f_i(\mathbf{J}_i) : i \leq \delta \rangle$ is $\subseteq$-increasing continuous

(d) $\mathbf{J}'_\delta \backslash f_\delta(\mathbf{J}_\delta)$ is independent in $(M'_{\alpha_\delta}, f_\delta(N_\delta), N'_{\alpha_\delta})$.

[Why? Clause (a) of $\odot$ holds by the choice of $\bar{f}$. As for clause (b) note that for every $i < \delta$ we have $(M'_{\alpha_i}, N'_{\alpha_i}, \mathbf{J}'_{\alpha_i}) \in K^{3,\text{bs}}_{\mathfrak{s}}$ and $M'_{\alpha_i} \leq_{\mathfrak{s}} f_i(N_i) \leq_{\mathfrak{s}} N'_{\alpha_i+1}$ and $(M_i, N_i, \mathbf{J}_i) \in K^{3,\text{vq}}_{\mathfrak{s}}$ by an assumption of the observation 5.17 which we are proving hence $(M'_{\alpha_i}, f_i(N_i), f_i(\mathbf{J}_i)) = (f_i(M_i), f_i(N_i), f_i(\mathbf{J}_i)) \in K^{3,\text{vq}}_{\mathfrak{s}}$ hence by 5.16(5) really clause (b) of $\odot$ holds.

Now, clause (c) of $\odot$ holds by $\boxdot(d),(e)$. By $\odot(a)$ and 5.8(5) we get also that $\mathbf{J}'_{\alpha_\delta} \backslash f_\delta(\mathbf{J}_\delta) = \cup\{\mathbf{J}'_i \backslash f_i(\mathbf{J}_i) : i < \delta\}$ is independent in $(M'_{\alpha_\delta}, f_\delta(N_\delta), N'_{\alpha_\delta})$. So clause (d) of $\odot$ holds.]

Now by 5.16(7) and $\odot$ as $(M'_{\alpha_\delta}, f_\delta(N_\delta)), f_\delta(\mathbf{J}_\delta)) = (f_\delta(M_\delta), f_\delta(N_\delta), f_\delta(\mathbf{J}_\delta))$ belongs to $K^{3,\text{vq}}_{\mathfrak{s}}$ we get that $(M_\delta, N_\delta, \mathbf{J}_\delta) \in K^{3,\text{vq}}_{\mathfrak{s}}$ as required in clause (a) of the conclusion of 5.17.

As for clause (b) there note that obviously $(M_i, N_i, \mathbf{J}_i) \in K^{3,\text{uq}}_{\mathfrak{s}}$ is $\leq_{\text{bs}} (M_\delta, N_\delta, \mathbf{J}_\delta)$ and use the definition of $K^{3,\text{vq}}_{\mathfrak{s}}$ and the definition of $<_{\text{bs}}$, i.e., 4.2(1), which is included in the definition of $<^{**}_{\text{bs}}$ and clause (c) there holds as presently it is a case of clause (b). $\square_{5.17}$

Proof of 5.16. 1) By 5.9(2) and the definition of $K^{3,\text{vq}}_{\mathfrak{s}}$.
2) <u>Clause (a)</u>:

Let M' be a $<_{\mathfrak{s}}$-extension of M which is brimmed over M and without loss of generality $M' \cap N = M$. So for some N' we have $\text{NF}_{\mathfrak{s}}(M, N, M', N')$. Let N'' be a $<_{\mathfrak{s}}$-extension of N' which is brimmed over N' and moreover there is a $\leq_{\mathfrak{s}}$-increasing continuous sequence $\langle N'_i : i \leq \lambda \rangle$ such that $N'_0 = N', N'_\lambda = N'', N'_{i+1}$ is brimmed over $N'_i, a_i \in N'_{i+1}, \mathbf{tp}_{\mathfrak{s}}(a_i, N'_i, N'_{i+1})$ does not fork over M' and for each $p \in \mathscr{S}^{\text{bs}}_{\mathfrak{s}}(M')$ for λ ordinals $i < \lambda$ the element a_i realizes p. Now choose $\mathbf{J}' = \mathbf{J} \cup \{a_i : i < \lambda\}$ and so $(M', N'', \mathbf{J}')$ exemplify the desired conclusion. In fact also $(M', N'', \mathbf{J}' \backslash \mathbf{J})$ is thick.

<u>Clause (b),(c)</u>: [We use $\mathfrak{s}$ successful good$^+$]

We give the details for clause (c), for clause (b) below in $\boxtimes$ we demand $\mathbf{J}_\alpha = \mathbf{J}$ and omit the thickness demands or just see 5.24.

We choose $(M_\alpha, N_\alpha, \mathbf{J}_\alpha)$ by induction on $\alpha < \lambda^+$ such that

$\boxtimes(\alpha)$ $(M_\alpha, N_\alpha, \mathbf{J}_\alpha) \in K^{3,\text{bs}}_{\mathfrak{s}}$ is $\leq_{\text{bs}}$-increasing continuous

(β) $(M_0, N_0, \mathbf{J}_0) = (M, N, \mathbf{J})$

(γ) if α is non-limit then $(M_{\alpha+1}, N_{\alpha+1}, \mathbf{J}_{\alpha+1})$ is thick; moreover $(M_{\alpha+1}, N_{\alpha+1}, \mathbf{J}_{\alpha+1} \backslash \mathbf{J}_\alpha)$ is thick and $(M_\alpha, N_\alpha, \mathbf{J}_\alpha) <^{**}_{\rm bs} (M_{\alpha+1}, N_{\alpha+1}, \mathbf{J}_{\alpha+1})$

(δ) if α is a limit ordinal then, if possible under clause (α) then $\neg$ $\mathrm{NF}_{\mathfrak{s}}(M_\alpha, N_\alpha, M_{\alpha+1}, N_{\alpha+1})$.

By 5.8(5) for every limit $\alpha < \lambda^+$ and $\beta < \alpha$ we have $(M_\beta, N_\beta, \mathbf{J}_\beta) <^{**}_{\rm bs} (M_\alpha, N_\alpha, \mathbf{J}_\alpha)$.
Also by the local character axiom (E)(c), also for limit α, $(M_\alpha, N_\alpha, \mathbf{J}_\alpha)$ is thick. Lastly, by $\mathfrak{s}$ being good$^+$ we know that $\le_{\mathfrak{K}} \restriction K_{\mathfrak{s}(+)} = \le^*_{\lambda^+} \restriction K_{\mathfrak{s}(+)}$ hence by clause (δ) for a club of $\delta < \lambda^+$, $(M_\delta, N_\delta, \mathbf{J}_\delta) \in K^{3,\rm vq}_{\mathfrak{s}}$. So we are done.
3) Let $\mathbf{J} = \{a_i : i < \alpha\}$. We prove this by induction on α. So assume $(M_0, M_\alpha, \mathbf{J}) \le_{\rm bs} (M'_0, M'_\alpha, \mathbf{J})$ and we have to prove that $\mathrm{NF}_{\mathfrak{s}}(M_0, M'_0, M_\alpha, M'_\alpha)$.
We can find (M''_0, M''_α) such that $(M'_0, M'_\alpha, \mathbf{J}) <_{\rm bs} (M''_0, M''_\alpha, \mathbf{J})$, M''_0 is brimmed over M'_0 and M''_α is brimmed over $M''_0 \cup M'_\alpha$. So it is enough to prove that $\mathrm{NF}_{\mathfrak{s}}(M_0, M_\alpha, M''_0, M''_\alpha)$.

By 4.7 we can find a $\le_{\mathfrak{s}}$-increasing continuous sequence $\langle N_i : i \le \alpha\rangle$ such that

$\circledast_1$ (a) $M_i \le_{\mathfrak{s}} N_i$ for $i \le \alpha$

(b) N_0 is brimmed over M_0

(c) if $i < \alpha$ then $\mathbf{tp}_{\mathfrak{s}}(a_i, N_i, N_{i+1})$ does not fork over M_i hence over M_0

(d) $(N_i, N_{i+1}, a_i) \in K^{3,\rm uq}_{\mathfrak{s}}$ for $i < \alpha$.

Now we choose (N^*_i, f_i) by induction on $i \le \alpha$ such that

$\circledast_2$ (a) N^*_i is $\le_{\mathfrak{s}}$-increasing continuous

(b) $N^*_0 = M''_\alpha$

(c) f_i is a $\le_{\mathfrak{s}}$-embedding of N_i into N^*_i

(d) $f_i \restriction M_i = \mathrm{id}_{M_i}$

(e) f_0 maps N_0 into M''_0 and $M'_0 \le_{\mathfrak{s}} f_0(N_0)$

(f) $f_i(a_j) = a_j$ if $j < i$.

For $i = 0$ this is possible by $N_0^* = M''_\alpha$ and N_0 being $\leq_{\mathfrak{s}}$-universal over M_0 and $M_0 \leq M'_0$ and M''_α being $\leq_{\mathfrak{s}}$-universal over M_0. For i limit take union. So let $i = j + 1$.

Now note that:

$(*)_1$ $\mathbf{J}$ is independent in (M_0, M''_0, M''_α) hence in (M_0, M''_0, N_j^*).

[Why? As $(M_0, M_\alpha, \mathbf{J}) \leq_{\rm bs} (M'_0, M'_\alpha, \mathbf{J}) \leq_{\rm bs} (M''_0, M''_\alpha, \mathbf{J})$ and $M''_\alpha = N_0^* \leq_{\mathfrak{s}} N_j^*$.]

$(*)_2$ $\mathbf{J}\backslash\{a_i : i < j\}$ is independent in $(M''_0, f_j(N_j), N_j^*)$.

[Why? By 5.4(3) recalling $M_0 \leq_{\mathfrak{s}} f_j(N_0) \subseteq M''_0$.]

$(*)_3$ $\mathbf{tp}_{\mathfrak{s}}(a_j, f_j(N_j), N_j^*)$ does not fork over M''_0 hence over M_0.

[Why? By $(*)_2$ and the definition of independent.]

$(*)_4$ $\mathrm{NF}_{\mathfrak{s}}(M_j, M_{j+1}, f_j(N_j), N_j^*)$.

[Why? By $(*)_3$ as $(M_j, M_{j+1}, a_j) \in K_{\mathfrak{s}}^{3,\rm uq}$.]

$(*)_5$ $\mathrm{NF}_{\mathfrak{s}}(M_j, M_{j+1}, N_j, N_{j+1})$.

[Why? By $\circledast_1$ as $(M_j, M_{j+1}, a_j) \in K_{\mathfrak{s}}^{3,\rm uq}$ by an assumption.]

By $(*)_4 + (*)_5$ we can find (N_i^*, f_i^*) as required.

Having carried the induction we have $\mathrm{NF}_{\mathfrak{s}}(M_i, M_{i+1}, f_i(N_i), f_{i+1}(N_{i+1}))$ for every $i < \alpha$ and the sequences $\langle M_i : i \leq \alpha\rangle, \langle f_i(N_i) : i \leq \alpha\rangle$ are increasing continuous. So by long transitivity for $\mathrm{NF}_{\mathfrak{s}}$ we have $\mathrm{NF}_{\mathfrak{s}}(M_0, M_\alpha, f_0(N_0), f(N_\alpha))$. But $f_0(M_0) = M_0 \leq_{\mathfrak{s}} M'_0 \leq_{\mathfrak{s}} f_0(N_0) \leq_{\mathfrak{s}} M''_0, M_0 \leq_{\mathfrak{s}} f_0(N_0) \leq_{\mathfrak{s}} M''_0$ and $f_\alpha(N_\alpha) \leq_{\mathfrak{s}} N_\alpha^*$ and $M_\alpha \leq_{\mathfrak{s}} M'_\alpha \leq_{\mathfrak{s}} M''_\alpha \leq_{\mathfrak{s}} N_\alpha^*$ so together we get $\mathrm{NF}_{\mathfrak{s}}(M_0, M_\alpha, M'_0, M'_\alpha)$ as required.

4) We prove this by induction on α. We repeat the proof of part (3) replacing a_i by $\mathbf{J}_i, K_{\mathfrak{s}}^{3,\rm uq}$ by $K_{\mathfrak{s}}^{3,\rm vq}$. However, in the parallel to $\circledast_1$ instead of 4.7 we use part (8); then in the proof of $\circledast_2$, in $(*)_2$ we use part (5) instead 5.4(3) and in $(*)_3$ now say $\mathbf{J}_j$ is independent in $(M_0, f_j(N_j), N_\gamma^*)$.

5) By the assumption $\mathbf{J}_1$ is independent in (M_0, M_2). Let $\mathbf{J}_1 = \{a_i : i < \alpha\}$ without repetitions of course and so we can find (M_i^0, M_i^2) for $i \leq \alpha$ such that

(∗) (a) $(M_i^0, M_i^2) = (M_0, M_2)$ for $i = 0$

(b) M_i^ℓ $(i \le \alpha)$ is $\le_{\mathfrak{s}}$-increasing continuous for $\ell = 0, 2$

(c) $(M_i^0, M_{i+1}^0, a_i) \in K_{\mathfrak{s}}^{3,\mathrm{uq}}$

(d) $\mathbf{tp}_{\mathfrak{s}}(a_i, M_i^0, M_{i+1}^0)$ does not fork over M_0.

Easy to construct by 5.4(1).

Now we can prove by induction on $i \le \alpha$ that $\mathbf{I}_i = \mathbf{J}_0 \cup \{a_\varepsilon : \varepsilon \in [i, \alpha)\}$ is independent in (M_0, M_i^0, M_i^2).

[Why? For $i = 0$ this is assumed. For $i = j + 1$ this follows by $\mathbf{I}_j = \mathbf{I}_i \cup \{a_j\}$ is independent in (M_0, M_j^0, M_j^2) hence in (M_j^0, M_j^2) hence by monotonicity in (M_j^0, M_i^2) hence by 5.4(3) the set $\mathbf{I}_i$ is independent in (M_j^0, M_i^0, M_i^2) and in (M_i^0, M_i^2) hence by 5.4(4) is independent in (M_0, M_i^0, M_i^2). For i limit use monotonicity and 5.10(2).]

So we have proved that $\mathbf{J}_0 = \mathbf{I}_\alpha$ is independent in $(M_0, M_\alpha^0, M_\alpha^2)$.

As $(M_0, M_1, \mathbf{J}_0) \in K_{\mathfrak{s}}^{3,\mathrm{vq}}$ by the definition of $K_{\mathfrak{s}}^{3,\mathrm{vq}}$ it follows that $\mathrm{NF}_{\mathfrak{s}}(M_0, M_1, M_\alpha^0, M_\alpha^2)$ and this implies by 5.3 that $\mathbf{J}_1$ is independent in (M_0, M_1, M_α^2) hence in (M_0, M_1, M_2) as required.

6) For $i \le \delta$ let $\mathbf{I}_i = \cup\{\mathbf{J}_j : i \le j < \delta + 1\}$ and we prove that $\mathbf{I}_i$ is independent in $(M_0, M_i, M_{\delta+1})$ for $i \le \delta$ by induction on i. For $i = 0$ this is given, for i successor use part (5) and for i limit use 5.8(5).

7) Toward contradiction assume that this fails. Then we can find N_0, N_1 such that $M_0 \le_{\mathfrak{s}} N_0 \le_{\mathfrak{s}} N_1$ and $M_1 \le_{\mathfrak{s}} N_1$ and $\mathbf{J}_0$ is independent in (M_0, N_0, M_2) but $\neg\, \mathrm{NF}_{\mathfrak{s}}(M_0, M_1, N_0, N_1)$ hence by symmetry of NF we have $\neg \mathrm{NF}_{\mathfrak{s}}(M_0, N_0, M_1, N_1)$. For simplicity without loss of generality $N_1 \cap M_2 = M_1$. By the existence of non-forking amalgamation (which is necessarily disjoint) we can find N_2 such that $\mathrm{NF}_{\mathfrak{s}}(M_1, N_1, M_2, N_2)$. From this and the monotonicity of $\mathrm{NF}_{\mathfrak{s}}$ we conclude that $\neg\, \mathrm{NF}_{\mathfrak{s}}(M_0, N_0, M_2, N_2)$. As $\mathbf{J}_2 \backslash \mathbf{J}_0$ is independent in (M_1, M_2) by an assumption and $\mathrm{NF}_{\mathfrak{s}}(M_1, N_1, M_2, N_2)$ we deduce that $\mathbf{J}_2 \backslash \mathbf{J}_0$ is independent in (M_1, N_1, N_2) hence in (M_0, N_1, N_2). Also we have $\mathbf{J}_0$ is independent in (M_0, N_0, N_1) so we deduce that $\mathbf{J} = \mathbf{J}_0 \cup (\mathbf{J}_2 \backslash \mathbf{J}_0)$ is independent in (M_0, N_0, N_2). As $(M_0, M_2, \mathbf{J}) \in K_{\mathfrak{s}}^{3,\mathrm{vq}}$ we get a contradiction.

8) As in the proof of 4.7 and 5.24 below. $\square_{5.16}$

5.18 Conclusion. 1) Assume $\mathfrak{s}$ is good$^+$ and successful and

(a) $\langle M_i : i \leq \delta \rangle$ is $\leq_{\mathfrak{s}}$-increasing continuous
(b) M_{i+1} is $(\lambda, *)$-brimmed over M_i
(c) $(M_0, N_0, \mathbf{J}) \in K_{\mathfrak{s}}^{3,\mathrm{vq}}$
(d) $N_0 \cap M_\delta = M_0$.

Then we can find N_i for $i \in (0, \delta]$ such that

(α) $\bar{N} = \langle N_i : i \leq \delta \rangle$ is $\leq_{\mathfrak{s}}$-increasing continuous
(β) $(M_i, N_i, \mathbf{J}) \in K_{\mathfrak{s}}^{3,\mathrm{vq}}$ for $i \leq \delta$.

Proof. We choose (M_i', N_i') by induction on $i \leq \delta$ such that

(a) M_i' is $\leq_{\mathfrak{s}}$-increasing continuous
(b) N_i' is $\leq_{\mathfrak{s}}$-increasing continuous
(c) $(M_i', N_i', \mathbf{J}) \in K_{\mathfrak{s}}^{3,\mathrm{vq}}$
(d) $(M_0', N_0') = (M_0, N_0)$
(e) $(M_j', N_j', \mathbf{J}) <_{\mathfrak{s}}^{**} (M_i', N_i', \mathbf{J})$ when $i = j + 1 < \delta$.

For $i = 0$ use clause (d). For $i = j + 1$ use 5.16(2)(b).

For i limit take unions and use 5.17.

Having carried the definition we can choose an isomorphism f_i from M_i onto M_i' increasing continuous with i such that $f_0 = \mathrm{id}_{M_0}$. This is possible for $i = 0$ as $M_0' = M_0$, for i limit by taking union and by $i = j + 1$ by the uniqueness of a $(\lambda, *)$-brimmed extension in $\mathfrak{K}_{\mathfrak{s}}$. So by renaming $M_i' = M_i$ for $i \leq \delta$ so we are done.

$\square_{5.18}$

Recall (Definition 4.4)

5.19 Definition. Assume $\mathfrak{s} = \mathfrak{t}^+, \mathfrak{t}$ is a successful good$^+$ $\lambda_{\mathfrak{t}}$-frame, so $\lambda = \lambda_{\mathfrak{s}} = \lambda_{\mathfrak{t}}^+$. We say that $(M, N, \mathbf{J})$ is canonically prime for $\mathfrak{s}$ (over $\mathfrak{t}$) when there are $\leq_{\mathfrak{K}[\mathfrak{t}]}$-representations $\langle M_\alpha : \alpha < \lambda \rangle, \langle N_\alpha : \alpha < \lambda \rangle$ of M, N respectively such that $(M_\alpha, N_\alpha, \mathbf{J} \cap N_\alpha) \in K_{\mathfrak{t}}^{3,\mathrm{vq}}$ for every $\alpha < \lambda$.

5.20 Claim. *[$\mathfrak{s} = \mathfrak{t}^+, \mathfrak{t}$ a successful good$^+$ λ-frame.]*
If $(M_0, M_2, \mathbf{J})$ is canonically prime for $\mathfrak{s}$ (see Definition 5.19) and $M_0 \cup \mathbf{J} \subseteq M_1 \leq_{\mathfrak{s}} M_2$ then $(M_0, M_1, \mathbf{J})$ is canonically prime for $\mathfrak{s}$.

5.21 Remark. In 5.20 it is enough to assume $\mathfrak{t}$ is a successful good λ-frame.

Proof. Let $\langle M_{\ell,\alpha} : \alpha < \lambda \rangle$ be a $\leq_{\mathfrak{K}[\mathfrak{t}]}$-representation of M_ℓ for $\ell = , 0, 1, 2$. Clearly $M_0 \leq_{\mathfrak{s}} M_1 \leq_{\mathfrak{s}} M_2$ hence for some club E for every $\alpha < \beta$ from E and $\ell \leq 1$ we have $\mathrm{NF}_{\mathfrak{t}}(M_{\ell,\alpha}, M_{\ell+1,\alpha}, M_{\ell,\beta}, M_{\ell+1,\beta})$. Also without loss of generality $\alpha \in E \Rightarrow (M_{0,\alpha}, M_{2,\alpha}, \mathbf{J}_\alpha) \in K_{\mathfrak{t}}^{3,\mathrm{vq}}$ where $\mathbf{J}_\alpha := \mathbf{J} \cap M_{2,\alpha}$. Now by 5.16(7) for $\mathfrak{t}$ applied with $M_{0,\alpha}, M_{1,\alpha}, M_{2,\alpha}, \mathbf{J}_\alpha$ here standing for $M_0, M_1, M_2, \mathbf{J}_2$ there (but in this case $\mathbf{J}_0 = \mathbf{J}_2$, a simpler case), we have $\alpha \in E \Rightarrow (M_0, M_1, \mathbf{J}_\delta) \in K_{\mathfrak{t}}^{3,\mathrm{vq}}$, so we are done.
Note that 5.16(7) does not need "good$^+$". $\square_{5.20}$

5.22 Claim. *[$\mathfrak{s} = \mathfrak{t}^+, \mathfrak{t}$ a successful good$^+$ λ_t-frame, so $\lambda = \lambda_{\mathfrak{t}}^+$, and recall $\mathfrak{s}$ is successful (and good$^+$, of course by 1.9).]*
Assume

- (a) *$\langle M_\alpha^\ell : \alpha < \lambda_{\mathfrak{s}} \rangle$ is a $\leq_{\mathfrak{t}}$-representation of $M_\ell \in K_{\mathfrak{s}}$ for $\ell = 1, 2$*
- (b) *$M_1 \leq_{\mathfrak{s}} M_2$*
- (c) *$\mathbf{J} \subseteq \mathbf{I}_{M_1,M_2}$ is independent in (M_1, M_2) and let $\mathbf{J}_\alpha = \mathbf{J} \cap M_\alpha^2$.*

<u>Then</u> the following conditions are equivalent:

- (α) *$(M_1, M_2, \mathbf{J}) \in K_{\mathfrak{s}}^{3,\mathrm{qr}}$*
- (β) *for a club of $\delta < \lambda_{\mathfrak{s}}$ the triple $(M_\alpha^1, M_\alpha^2, \mathbf{J}_\alpha)$ belongs to $K_{\mathfrak{t}}^{3,\mathrm{vq}}$ (i.e. $(M_1, M_2, \mathbf{J})$ is canonically prime)*
- (γ) *for stationarily many $\delta < \lambda_{\mathfrak{s}}$ the triple $(M_\alpha^1, M_\alpha^2, \mathbf{J}_\alpha)$ belongs to $K_{\mathfrak{t}}^{3,\mathrm{vq}}$*
- (δ) *for arbitrarily large $\delta \in E$ the triple $(M_\alpha^1, M_\alpha^2, \mathbf{J}_\alpha)$ belongs to $K_{\mathfrak{t}}^{3,\mathrm{vq}}$ wherever E is a club of λ satisfying: if $\alpha < \beta$ are from E then $NF_{\mathfrak{t}}(M_\alpha^1, M_\alpha^2, M_\beta^1, M_\beta^2)$ and $(M_\alpha^1, M_\alpha^2, \mathbf{J}_\alpha) <_{\mathrm{bs}}^{**} (M_\beta^1, M_\beta^2, \mathbf{J}_\alpha)$.*

Proof. For clauses $(\beta),(\gamma)$ as we can restrict ourselves to a club (for (δ) - the given E) so without loss of generality $M^\ell_{\alpha+1}$ is brimmed over M^ℓ_α (for $\alpha<\lambda^+,\ell=1,2$) and $\alpha<\beta\Rightarrow \mathrm{NF}_{\mathfrak{t}}(M^1_\alpha,M^2_\alpha,M^1_\beta,M^2_\beta)$ and recalling 5.11 also $(M^1_\alpha,M^2_\alpha,\mathbf{J}_\alpha)\in K^{3,\mathrm{bs}}_{\mathfrak{s}}$ for $\alpha<\lambda_{\mathfrak{s}}$ and $\alpha<\beta<\lambda_{\mathfrak{s}}\wedge c\in\mathbf{J}_\alpha\Rightarrow \mathbf{tp}_{\mathfrak{t}}(c,M^1_\beta,M^2_\beta)$ does not fork over M^1_α. For clause (δ) without loss of generality $E=\lambda$.

Recalling Definition 5.19 by repeating the proof of 4.9 we can prove "clause (β) is equivalent to clause (α)". Clause (β) implies clause (γ) which implies clause (δ) trivially. Now clause (δ) implies clause (β) by 5.17 as for every $\alpha<\beta<\lambda^+_{\mathfrak{t}}$ we have $(M^1_\alpha,M^2_\alpha,\mathbf{J}_\alpha)<^{**}_{\mathrm{bs}}(M^1_\beta,M^2_\beta,\mathbf{J}_\beta)$. $\square_{5.22}$

5.23 Conclusion. [$\mathfrak{t},\mathfrak{s}$ are successful good$^+$ and $\mathfrak{s}=\mathfrak{t}^+$]
1) Uniqueness for $K^{3,\mathrm{qr}}_{\mathfrak{s}}$. If $(M,N_\ell,\{a^\ell_t:t\in I\})\in K^{3,\mathrm{qr}}_{\mathfrak{s}}$ for $\ell=1,2$ and $\mathbf{tp}_{\mathfrak{s}}(a^1_t,M,N_1)=\mathbf{tp}_{\mathfrak{s}}(a^2_t,M,N_2)$ for $t\in I$ then there is an isomorphism f from N_1 onto N_2 over M mapping a^1_t to a^2_t for $t\in I$.
2) [Existence] If $(M,N,\mathbf{J})\in K^{3,\mathrm{bs}}_{\mathfrak{s}}$ then some $N'\le_{\mathfrak{s}}N$ we have $(M,N',\mathbf{J})\in K^{3,\mathrm{qr}}_{\mathfrak{s}}$.

Remark. Of course, we have existence as $\mathfrak{s}=\mathfrak{t}^+$.

Proof. 1) As in the proof for $K^{3,\mathrm{pr}}_{\mathfrak{s}}$.
2) By 4.9 we know that $\mathfrak{s}$ has primes (i.e. existence for $K^{3,\mathrm{pr}}_{\mathfrak{s}}$) hence by 5.8(1) we are done. $\square_{5.23}$

Close to 5.16(2)(b) and 5.16(8) is:

5.24 Claim. *[$\mathfrak{s}$ is successful good$^+$.]*

If $\mathbf{J}$ is independent in (M_0,N_0), then we can find (M_1,N_1) such that:

(a) $M_0\le_{\mathfrak{s}}M_1\le_{\mathfrak{s}}N_1$ *and* $M_0\le_{\mathfrak{s}}N_0\le_{\mathfrak{s}}N_1$
(b) M_1 *is* $(\lambda,*)$*-brimmed over* M_0
(c) N_1 *is* $(\lambda,*)$*-brimmed over* N_0
(d) $(M_1,N_1,\mathbf{J})\in K^{3,\mathrm{vq}}_{\mathfrak{s}}$

(e) $\mathbf{tp}_{\mathfrak{s}}(c, M_1, N_1)$ *does not fork over* M_0 *for every* $c \in \mathbf{J}$ *so* $\mathbf{J}$ *is independent in* (M_0, M_1, N_1).

Proof. We try to choose by induction on $\zeta < \lambda^+_{\mathfrak{s}}$, a pair (M'_ζ, N'_ζ) such that:

$(*)_1$ $(M'_0, N'_0) = (M_0, N_0)$

$(*)_2$ M'_ζ is $\leq_{\mathfrak{s}}$-increasing continuous,

$(*)_3$ N'_ζ is $\leq_{\mathfrak{s}}$-increasing continuous

$(*)_4$ $M'_\zeta \leq_{\mathfrak{s}} N'_\zeta, \mathbf{J}$ is independent in $(M_0, M'_\zeta, N'_\zeta)$

$(*)_5$ $\neg\, \mathrm{NF}_{\mathfrak{s}}(M'_\zeta, N'_\zeta, M'_{\zeta+1}, N'_{\zeta+1})$ for ζ even

$(*)_6$ $M'_{\zeta+1}$ is brimmed over M'_ζ and $N'_{\zeta+1}$ is brimmed over N'_ζ if ζ is odd.

For $\zeta = 0$ this is trivial. For $\zeta = \xi + 1$, if $(M'_\zeta, N'_\zeta, \mathbf{J}) \notin K^{3,\mathrm{vq}}_{\mathfrak{s}}$ there are no problems and otherwise (M'_ζ, N'_ζ) satisfies the demands on (M_1, N_1) in the claim so we are done and for limit stages use 5.10(2).

We necessarily (as $\leq_{\mathfrak{s}(+)} = <^*_{\lambda^+} [\mathfrak{s}]$ by 1.6) get stuck for some ζ and (M'_ζ, N'_ζ) can serve as (M_1, N_1).

$\square_{5.24}$

5.25 Exercise: 1) Assume $M_0 \leq_{\mathfrak{s}} M_\ell \leq_{\mathfrak{s}} M_3$ for $\ell = 1, 2$ and $(M_0, M_2, \mathbf{J}) \in K^{3,\mathrm{qr}}_{\mathfrak{s}}$ and $\mathbf{J}$ is independent in (M_0, M_1, M_3). Then we can find M^+_3, M^-_3 such that $M_{\mathfrak{s}} \leq_{\mathfrak{s}} M^+_{\mathfrak{s}}, M_1 \cup M_2 \subseteq M^-_{\mathfrak{s}} \leq_{\mathfrak{s}} M^+_3$ and $(M_1, M^-_3, \mathbf{J}) \in K^{3,\mathrm{qr}}_{\mathfrak{s}}$.

[Hint: Clearly $\mathrm{NF}_{\mathfrak{s}}(M_0, M_1, M_2, M_3)$. We can find M^+_3 which is $(\lambda, *)$-brimmed over M_3. As $\mathfrak{s}$ has primes there is $M'_{\mathfrak{s}} \leq_{\mathfrak{s}} M^+_3$ such that $(M_1, M'_{\mathfrak{s}}, \mathbf{J}) \in K^{3,\mathrm{qr}}_{\mathfrak{s}}$. As $(M_0, M_2, \mathbf{J}) \in K^{3,\mathrm{qr}}_{\mathfrak{s}}$ there is a $\leq_{\mathfrak{s}}$-embedding f of M_2 into M'_3 over $M_0 \cup \mathbf{J}$. As above $\mathrm{NF}_{\mathfrak{s}}(M, M_1, f(M_2), M^+_3)$ and by uniqueness of NF possibly increasing M^+_3, there is an automorphism g of M^+_3 extending f. Let $M^-_{\mathfrak{s}} := g^{-1}(M'_3)$. Used in 8.13. Actually just $\mathfrak{s}$ is weakly successful with primes.]

2) First we choose N^+_i and if $i = j + 1$ also $M^+_j, \mathbf{J}_j$ such that

⊛ (a) $\langle N^+_j : j \leq i \rangle$ is $\leq_{\mathfrak{s}}$-increasing continuous

(b) $N_i^+ = M$ if $i = 0$

(c) if $i = j + 1$ then $M_j \leq_{\mathfrak{s}} M_j^+$ and $(M, M_j^+, \mathbf{J}_j)$ belongs to $K_{\mathfrak{s}}^{3,\mathrm{vq}}$ and $\mathrm{NF}_{\mathfrak{s}}(M_j, N_j^+, M_j^+, N_d^+)$.

This is similar to the proof of part (1). Second, we prove by induction on n that if u_1, u_2 are disjoint finite subsets of α and $n \geq |u_2|$ then $\{M_i : i \in u_1\} \cup \{M_i^+ : i \in u_2\}$ is independent inside (M, N^+). This as in part (1) we use part (3).

§6 Orthogonality

Note that presently the case "orthogonality = weak orthogonality" is the main one for us. In the latter part of the section "$\mathfrak{s}$ has primes" is usually used and we shall later weaken this, but this is not a serious flaw here. We can weaken the hypothesis see [Sh:F735].

6.1 Hypothesis. As in 5.1:

(a) $\mathfrak{s}$ is a good$^+$ λ-frame, successful or at least

(b) $\mathfrak{s}$ is a weakly successful good λ-frame and 4.7 holds.

6.2 Definition. 1) For $p, q \in \mathscr{S}_{\mathfrak{s}}^{\mathrm{bs}}(M)$ we say that they are weakly orthogonal, $p \underset{\mathrm{wk}}{\perp} q$ when: if $(M, N, b) \in K_{\mathfrak{s}}^{3,\mathrm{uq}}$ and $\mathbf{tp}_{\mathfrak{s}}(b, M, N) = q$ <u>then</u> p has a unique extension in $\mathscr{S}_{\mathfrak{s}}(N)$; equivalently, every extension of p in $\mathscr{S}_{\mathfrak{s}}(N)$ does not fork over M; note: the order of p, q is seemingly important. (In the first order case the symmetry is essentially by the definition and here it will be proved).
2) For $p, q \in \mathscr{S}_{\mathfrak{s}}^{\mathrm{bs}}(M)$ we say that they are orthogonal, $p \underset{\mathrm{st}}{\perp} q$ or $p \perp q$ <u>if</u> p_1, q_1 are weakly orthogonal whenever $M \leq_{\mathfrak{s}} M_1$ and $p_1, q_1 \in \mathscr{S}_{\mathfrak{s}}^{\mathrm{bs}}(M_1)$ are non-forking extensions of p, q respectively.
3) If $p \in \mathscr{S}_{\mathfrak{s}}^{\mathrm{bs}}(M_1), q \in \mathscr{S}_{\mathfrak{s}}^{\mathrm{bs}}(M_2)$ and $M_\ell \leq_{\mathfrak{s}} N$ for $\ell = 1, 2$ <u>then</u> orthogonality of p and q means that p', q' are orthogonal where $p' \in \mathscr{S}_{\mathfrak{s}}^{\mathrm{bs}}(N)$ is the unique non-forking extension of p and $q' \in \mathscr{S}_{\mathfrak{s}}^{\mathrm{bs}}(N)$ is the unique non-forking extension of q (this is justified by 6.8(1)).
4) If $p \in \mathscr{S}_{\mathfrak{s}}^{\mathrm{bs}}(M_1), q \in \mathscr{S}_{\mathfrak{s}}^{\mathrm{bs}}(M_2)$ and $M_\ell \leq_{\mathfrak{s}} M$ for $\ell = 1, 2$ <u>then</u> p, q being weakly orthogonal means that for some M', M'' we have $M \leq_{\mathfrak{s}} M'', M' \leq_{\mathfrak{s}} M'', M_1 \cup M_2 \subseteq M'$ and letting $p' \in$

$\mathscr{S}^{\rm bs}_{\mathfrak{s}}(M''), q' \in \mathscr{S}^{\rm bs}_{\mathfrak{s}}(M'')$ be non-forking extensions of p, q respectively, we have $(p' \restriction M') \underset{\rm wk}{\perp} (q' \restriction M')$, (see 6.7(7)).

Naturally we now show that the definition is equivalent to some variants (e.g. for some such pair (N, b) rather than all such (N, b); so if $\mathfrak{s}$ has primes and they are unique this is trivial).

6.3 Claim. *1) Assume that $p, q \in \mathscr{S}^{\rm bs}_{\mathfrak{s}}(M)$ and $(M, N, b) \in K^{3,\rm uq}_{\mathfrak{s}}$ and $q = \mathbf{tp}_{\mathfrak{s}}(b, M, N)$. Then $p \underset{\rm wk}{\perp} q$ iff p has a unique extension in $\mathscr{S}_{\mathfrak{s}}(N)$.*
2) Assume $(M, N, b) \in K^{3,\rm uq}_{\mathfrak{s}}, (M, N_2, b) \in K^{3,\rm uq}_{\mathfrak{s}}$ and $M \le_{\mathfrak{s}} N_1 \le_{\mathfrak{s}} N^+$ and $M \le_{\mathfrak{s}} N \le_{\mathfrak{s}} N^+, M \le_{\mathfrak{s}} N_2 \le_{\mathfrak{s}} N^+$.
Then $\mathrm{NF}_{\mathfrak{s}}(M, N, N_1, N^+) \equiv \mathrm{NF}_{\mathfrak{s}}(M, N_2, N_1, N^+) \equiv (\mathbf{tp}_{\mathfrak{s}}(b, N_1, N^+)$ does not fork over $M)$.

Proof. 1) The implication $\Rightarrow$ holds by the "every" in the definition. So assume p has a unique extension in $\mathscr{S}^{\rm bs}_{\mathfrak{s}}(N)$ and we shall prove $p \underset{\rm wk}{\perp} q$.
So assume $(M, N_2, b_2) \in K^{3,\rm uq}_{\mathfrak{s}}$ and $\mathbf{tp}_{\mathfrak{s}}(b_2, M, N_2) = q$ and let $p_2 \in \mathscr{S}_{\mathfrak{s}}(N_2)$ extend p. So there are $N^+ \in K_{\mathfrak{s}}$ and $a \in N^+$ such that $N_2 \le_{\mathfrak{s}} N^+$ and a realizes p_2 in N^+. As a realizes p_2 in N^+ it also realized $p_2 \restriction M$ which is p, so $\mathbf{tp}_{\mathfrak{s}}(a, M, N^+) \in \mathscr{S}^{\rm bs}_{\mathfrak{s}}(M)$. As $\mathfrak{s}$ is weakly successful, possibly replacing N^+ by a $\le_{\mathfrak{s}}$-extension, there is $N_1 \le_{\mathfrak{s}} N^+$ such that $(M, N_1, a) \in K^{3,\rm uq}_{\mathfrak{s}}$. Also without loss of generality $N \le_{\mathfrak{s}} N^+$ and $b_2 = b$ (as N, N^+ are $\le_{\mathfrak{s}}$-extensions of M and $b \in N \le_{\mathfrak{s}} N^+, b_2 \in N_2 \le_{\mathfrak{s}} N^+$ realizes the same type so we can amalgamate).

Now as a realizes p_2 in N^+ it also realizes $p_2 \restriction M$ which is p so $\mathbf{tp}_{\mathfrak{s}}(a, N, N^+)$ is an extension of p in $\mathscr{S}_{\mathfrak{s}}(N)$ hence by our present assumption it does not fork over M. As the triple $(M, N_1, a) \in K^{3,\rm uq}_{\mathfrak{s}}$, we can conclude by Claim 1.19 that $\mathrm{NF}_{\mathfrak{s}}(M, N, N_1, N^+)$, but $b \in N, \mathbf{tp}_{\mathfrak{s}}(b, M, N) \in \mathscr{S}^{\rm bs}_{\mathfrak{s}}(M)$ hence by 1.18 the type $\mathbf{tp}_{\mathfrak{s}}(b, N_1, N^+)$ does not fork over M. But we have $(M, N_2, b) \in K^{3,\rm uq}_{\mathfrak{s}}$ hence $\mathrm{NF}_{\mathfrak{s}}(M, N_2, N_1, N^+)$ from which (as $a \in N_1, \mathbf{tp}_{\mathfrak{s}}(a, M, N^+) \in \mathscr{S}^{\rm bs}_{\mathfrak{s}}(M)$) we deduce $\mathbf{tp}_{\mathfrak{s}}(a, N_2, N^+)$ does not fork over M, but this last type is p_2 so we are done.

2) By the same proof. Actually 2) is the idea of the proof of 1). $\square_{6.3}$

6.4 Claim. *1) Assume $(M,N,b) \in K_{\mathfrak{s}}^{3,\mathrm{pr}}$ or just $(M,N,b) \in K_{\mathfrak{s}}^{3,\mathrm{uq}}$ and $q = \mathbf{tp}_{\mathfrak{s}}(b,M,N)$ and $p \in \mathscr{S}_{\mathfrak{s}}^{\mathrm{bs}}(M)$.*
If p is realized in N, then p,q are not orthogonal and even not weakly orthogonal.
2) Let $p,q \in \mathscr{S}_{\mathfrak{s}}^{\mathrm{bs}}(M)$. Then $p \underset{\mathrm{wk}}{\pm} q$ iff for some a,b,N' we have $M \le_{\mathfrak{s}} N', a \in N'$ realizes $p, b \in N'$ realizes q and the (indexed) set $\{a,b\}$ is not independent in (M,N'), e.g. $a=b$.

Proof. 1) The type p has at least two extensions in $\mathscr{S}_{\mathfrak{s}}(N)$: one algebraic, is $\mathbf{tp}_{\mathfrak{s}}(a,N,N)$ where $a \in N$ realizes p and the second is in $\mathscr{S}_{\mathfrak{s}}^{\mathrm{bs}}(N)$, hence non-algebraic, in fact a non-forking extension of p. So by the definition we have $p \underset{\mathrm{wk}}{\pm} q$.
2) Let $(M,N,b) \in K_{\mathfrak{s}}^{3,\mathrm{uq}}$ be such that $q = \mathbf{tp}_{\mathfrak{s}}(b,M,N)$. If $p \underset{\mathrm{wk}}{\pm} q$ then by 6.3 there is $p_1 \in \mathscr{S}_{\mathfrak{s}}(N)$ extending p forking over M, and let a,N' be such that $N \le_{\mathfrak{s}} N'$, and $p_1 = \mathbf{tp}_{\mathfrak{s}}(a,N,N')$; now by 5.4(3) we get $\{a,b\}$ is not independent in (M,N').
[Why? Use it with $\mathbf{J},M,N,\beta,\langle a_i : i<\beta\rangle, \mathbf{J}\backslash\{a_i : i<\beta\},(M,M_\beta,N)$ there standing for $\{a,b\},M,N',1,\langle a\rangle,\{b\},(M,N,N')$ here so we get that $\{b\}$ is independent in (M,N,N'), contradicting the choice of (N',b).]

So we have proved that the first phrase implies the second. If $p \underset{\mathrm{wk}}{\perp} q$ then we shall show that the second phrase fails. So assume that $M \le_{\mathfrak{s}} N'$ and $a,b \in N$ realize p,q respectively. Then we can find N_1,N_2 such that $N' \le_{\mathfrak{s}} N_2, M \le_{\mathfrak{s}} N_1 \le_{\mathfrak{s}} N_2$ and $(M,N_1,b) \in K_{\mathfrak{s}}^{3,\mathrm{uq}}$. By Definition 6.2(1) clearly $\mathbf{tp}_{\mathfrak{s}}(a,M_1,N_2)$ does not fork over M. Now $\{a,b\}$ is independent over M inside N_2 hence in N' by the Definition 5.2 and 6.2. $\square_{6.4}$

6.5 Definition. Fixing $\mathfrak{C} \in K^{\mathfrak{s}}$, if $p_\ell \in \mathscr{S}_{\mathfrak{s}}^{\mathrm{bs}}(M_\ell)$ and $M_\ell \le_{\mathfrak{K}[\mathfrak{s}]} \mathfrak{C}$ for $\ell=1,2$, then let $p_1 \| p_2$, in words p_1,p_2 are parallel inside $\mathfrak{C}$, mean that for some M,p we have $M_1 \cup M_2 \subseteq M <_{\mathfrak{K}[\mathfrak{s}]} \mathfrak{C}$ and $M \in K_{\mathfrak{s}}, p \in \mathscr{S}_{\mathfrak{s}}^{\mathrm{bs}}(M)$ does not fork over M_ℓ and extends p_ℓ for $\ell=1,2$.

Remark. 1) As we shall note, e.g. (see 6.7(4), 6.8(2)), "p orthogonal to $-$" is a property of p up to parallelism (as being parallel is an equivalence relation). For this we have to use the extended definition of being orthogonal from 6.2(3).

Similarly for $p \perp M$ defined in 6.9 below.

Obvious properties of parallelism are

6.6 Claim. *1) Parallelism inside $\mathfrak{C} \in K^{\mathfrak{s}}$ is an equivalence relation.*
2) If $M \in K_{\mathfrak{s}}$ and $M \leq_{\mathfrak{K}[\mathfrak{s}]} \mathfrak{C}$, <u>then</u> on $\mathscr{S}^{\rm bs}_{\mathfrak{s}}(M)$, parallelism inside $\mathfrak{C}$ is equality.

Proof. Easy.

6.7 Claim. *1) If $p, q \in \mathscr{S}^{\rm bs}_{\mathfrak{s}}(M)$ and f is an isomorphism from M onto N <u>then</u> $p \underset{\rm wk}{\perp} q \Leftrightarrow f(p) \underset{\rm wk}{\perp} f(q)$. Similarly for $\perp$.*
2) [symmetry] If $p, q \in \mathscr{S}^{\rm bs}_{\mathfrak{s}}(M)$ <u>then</u> $p \underset{\rm wk}{\perp} q \Leftrightarrow q \underset{\rm wk}{\perp} p$. Similarly for $\perp$.
3) Assume that $M, N \in K_{\mathfrak{s}}$ are brimmed (e.g. $K_{\mathfrak{s}}$ categorical). If $M \leq_{\mathfrak{s}} N$, and $p, q \in \mathscr{S}^{\rm bs}_{\mathfrak{s}}(N)$ do not fork over M, <u>then</u> $p \underset{\rm wk}{\perp} q \Leftrightarrow (p \restriction M) \perp_{\rm wk} (q \restriction M)$.
4) Assume $M, N \in K_{\mathfrak{s}}$ are brimmed. If $p_1, p_2 \in \mathscr{S}^{\rm bs}_{\mathfrak{s}}(M)$ and $q_1, q_2 \in \mathscr{S}^{\rm bs}_{\mathfrak{s}}(N)$ and $M \leq_{\mathfrak{K}[\mathfrak{s}]} \mathfrak{C}, N \leq_{\mathfrak{K}[\mathfrak{s}]} \mathfrak{C}$ and $p_1 \| q_1, p_2 \| q_2$ inside $\mathfrak{C}$ <u>then</u> $p_1 \underset{\rm wk}{\perp} p_2 \Leftrightarrow q_1 \underset{\rm wk}{\perp} q_2$.
5) If $\langle M_i, a_j : i \leq \alpha, j < \alpha \rangle$ is an M_0-based pr-*decomposition or just* uq-*decomposition of (M_0, M_α) and $p \in \mathscr{S}^{\rm bs}_{\mathfrak{s}}(M_0)$ is weakly orthogonal to $\mathbf{tp}_{\mathfrak{s}}(a_j, M_j, M_{j+1})$ for every $j < \alpha$ <u>then</u> p has a unique extension in $\mathscr{S}_{\mathfrak{s}}(M_\alpha)$.*
6) Assume $M_0 \leq_{\mathfrak{s}} M_1$ and $p, q \in \mathscr{S}^{\rm bs}_{\mathfrak{s}}(M_1)$ do not fork over M_0. If $p \underset{\rm wk}{\perp} q$ <u>then</u> $(p \restriction M_0) \underset{\rm wk}{\perp} (q \restriction M_0)$.
7) Assume that $M_1 \leq_{\mathfrak{s}} M_2$ and $p, q \in \mathscr{S}^{\rm bs}_{\mathfrak{s}}(M_2)$. If p, q does not fork over M_1 <u>then</u> $p \underset{\rm wk}{\perp} q \Leftrightarrow (p \restriction M_1) \underset{\rm wk}{\perp} q \Leftrightarrow p \underset{\rm wk}{\perp} (q \restriction M_2)$. Also if only p does not fork over M_1, then still the first $\Leftrightarrow$ holds.

Proof. 1) Immediate.
2) By 3.9 alternatively this follows from 6.4(2).
3) By 1.21 there is an isomorphism f from M onto N such that

$f(p \restriction M) = p, f(q \restriction M) = q$ so the result holds by part (1).
4) We can find a model $\mathfrak{B} \le_{\mathfrak{K}[\mathfrak{s}]} \mathfrak{C}$ of cardinality λ which includes $M \cup N$ hence $M \le_{\mathfrak{s}} \mathfrak{B}$ and $N \le_{\mathfrak{s}} \mathfrak{B}$. We can find $\mathfrak{B}^+$ such that $\mathfrak{B} \le_{\mathfrak{s}} \mathfrak{B}^+ \in K_{\mathfrak{s}}$ and $\mathfrak{B}^+$ is brimmed over $\mathfrak{B}$ hence over M and over N. Let $r_\ell \in \mathscr{S}^{\rm bs}_{\mathfrak{s}}(\mathfrak{B}^+)$ be a non-forking extension of $p_\ell \in \mathscr{S}^{\rm bs}_{\mathfrak{s}}(M)$ for $\ell = 1, 2$. As $p_\ell \| q_\ell, M \le_{\mathfrak{s}} \mathfrak{B} \le_{\mathfrak{s}} \mathfrak{B}^+, \mathfrak{C}, N \le_{\mathfrak{s}} \mathfrak{B} \le_{\mathfrak{s}} \mathfrak{B}^+$ and $\mathfrak{B} \le_{\mathfrak{K}[\mathfrak{s}]} \mathfrak{C}$ it follows that r_ℓ is a non-forking extension of q_ℓ for $\ell = 1, 2$.

Now we apply part (3) with $r_1, r_2, M, \mathfrak{B}^+$ here standing for p, q, M, N there, (recalling that $M \in K_{\mathfrak{s}}$ is brimmed by the assumption of part (4)), so we get its conclusion $r_1 \underset{\rm wk}{\perp} r_2 \Leftrightarrow (r_1 \restriction M) \underset{\rm wk}{\perp} (r_2 \restriction M)$. But $r_1 \restriction M = p_1, r_2 \restriction M = p_2$ so $r_1 \underset{\rm wk}{\perp} r_2 \Leftrightarrow p_1 \underset{\rm wk}{\perp} p_2$.

Similarly $r_1 \underset{\rm wk}{\perp} r_2 \Leftrightarrow q_1 \underset{\rm wk}{\perp} q_2$, so as $\Leftrightarrow$ is transitive we can deduce $p_1 \underset{\rm wk}{\perp} p_2 \Leftrightarrow q_1 \underset{\rm wk}{\perp} q_2$, which is the first desired conclusion.
5) Let $M_\alpha \le_{\mathfrak{s}} N$ and $c \in N$ realizes p. We prove by induction on $\beta \le \alpha$ that $\mathbf{tp}_{\mathfrak{s}}(c, M_\beta, N) \in \mathscr{S}^{\rm bs}_{\mathfrak{s}}(M_\beta)$ does not fork over M_0. For $\beta = 0$ this is trivial, and for β successor use part (7) and for β a limit ordinal use the continuity axiom (E)(h) of good frames. Lastly, for $\beta = \alpha$ we get the desired result.
6) Note that here we do not have the brimness assumption. Assume toward contradiction $(p \restriction M_0) \underset{\rm wk}{\pm} (q \restriction M_0)$. So there are N_1, N_2, b such that $M_0 \le_{\mathfrak{s}} N_1 \le_{\mathfrak{s}} N_2, b \in N_1$ realizes $q \restriction M_0, (M_0, N_1, b) \in K^{3,\rm uq}_{\mathfrak{s}}$ and $a \in N_2$ realizes $p \restriction M_0$ but $\mathbf{tp}_{\mathfrak{s}}(a, N_1, N_2)$ forks over M_0. By our knowledge on $\mathrm{NF}_{\mathfrak{s}}$ without loss of generality for some N_3 we have $\mathrm{NF}_{\mathfrak{s}}(M_0, M_1, N_2, N_3)$ hence a, b realize p, q in N_3 respectively (see 1.18). By 5.3, if $\{a, b\}$ is independent in (M_1, N_3) then $\{a, b\}$ is independent in (M_0, N_3) which fails. So N_1, N_3, a, b witness $p \underset{\rm wk}{\pm} q$ by 6.4(2).
7) By the definition 6.2(4) and part (6); (note that in 6.2(4), our case $M_1 \le_{\mathfrak{s}} M_2$, by part (6) without loss of generality $M' = M_2$). $\square_{6.7}$

6.8 Claim. *1) If $p, q \in \mathscr{S}^{\rm bs}_{\mathfrak{s}}(M_1)$ does not fork over M_0 where $M_0 \le_{\mathfrak{s}} M_1$ <u>then</u> $(p \perp q) \Leftrightarrow (p \restriction M_0) \perp (q \restriction M_0)$.*
1A) Assume $M_\ell \le_{\mathfrak{s}} N, p_\ell \in \mathscr{S}^{\rm bs}_{\mathfrak{s}}(M_\ell), q_\ell \in \mathscr{S}^{\rm bs}_{\mathfrak{s}}(M_\ell)$ for $\ell = 1, 2$. If

p_1, p_2 are parallel and q_1, q_2 are parallel <u>*then*</u> *$p_1 \perp q_2 \Leftrightarrow p_2 \perp q_2$.*
2) If $p \perp q$ <u>*then*</u> *$p \underset{\rm wk}{\perp} q$.*
3) Assume that $\langle M_\alpha : \alpha \le \delta \rangle$ is $\le_{\mathfrak{s}}$-increasing continuous, $\delta < \lambda^+$ limit ordinal and $p, q \in \mathscr{S}^{\rm bs}_{\mathfrak{s}}(M_\delta)$. <u>*Then*</u> *$p \underset{\rm wk}{\perp} q$ iff for every $\alpha < \delta$ large enough $(p \restriction M_\alpha) \underset{\rm wk}{\perp} (q \restriction M_\alpha)$. Similarly for $\perp$.*
4) If $M \in K_{\mathfrak{s}}$ is brimmed and $p, q \in \mathscr{S}^{\rm bs}_{\mathfrak{s}}(M)$ <u>*then*</u> *$p \underset{\rm wk}{\perp} q \Leftrightarrow p \perp q$.*
5) If $K_{\mathfrak{s}}$ is categorical <u>*then*</u> *$\perp, \underset{\rm wk}{\perp}$ are equal.*
6) If $M <_{\mathfrak{s}} N, N$ is universal over $M, p, q \in \mathscr{S}^{\rm bs}_{\mathfrak{s}}(N)$ do not fork over M and $p \underset{\rm wk}{\perp} q$ <u>*then*</u> *$p \perp q$ (hence $(p \restriction M) \perp q \restriction M$).*

Proof. 1) The implication $\Leftarrow$ is by the definition 6.2(2). For the other direction assume $p \perp q$ and $M_0 \le_{\mathfrak{s}} M_2$ and $p_2, q_2 \in \mathscr{S}^{\rm bs}_{\mathfrak{s}}(M_2)$ are non-forking extensions of $p \restriction M_0, q \restriction M_0$ respectively. Without loss of generality for some M_3 we have $M_2 \le_{\mathfrak{s}} M_3, M_1 \le_{\mathfrak{s}} M_3$ and let $p_3, q_3 \in \mathscr{S}^{\rm bs}_{\mathfrak{s}}(M_3)$ be non-forking extensions of p_2, q_2 respectively hence of $p \restriction M_0, q \restriction M_0$ respectively. As $p \perp q$ we have $p_3 \underset{\rm wk}{\perp} q_3$ and by 6.7(6) also $(p_3 \restriction M_2) \underset{\rm wk}{\perp} (q_3 \restriction M_2)$ which means $p_2 \underset{\rm wk}{\perp} q_2$, as required.
1A) Follows by part (1).
2) Read the definitions.
3) By axiom (E)(c) of good λ-frame for some $\alpha_* < \delta$ the types p, q does not fork over M_{α_*}.

If $\alpha \in [\alpha_*, \delta)$ and $p \restriction M_\alpha, q \restriction M_\alpha$ are not weakly orthogonal then by 6.7(6) the types p, q are not weakly orthogonal.

If p, q are not weakly orthogonal then by 6.4(2) there are a $\le_{\mathfrak{s}}$-extension N of M_δ and $a, b \in N$ realizing p, q respectively such that the (indexed) set $\{a, b\}$ is not independent in (M_δ, N). By 5.10(2) applied to the sequence $\langle M_\alpha : \alpha \in [\alpha_*, \delta) \rangle$ for some $\alpha \in [\alpha_*, \delta)$, the (indexed) set $\{a, b\}$ is not independent in (M_α, N). As a, b realizes $p \restriction M_\alpha, q \restriction M_\alpha$ respectively by 6.4(2) the types $p \restriction M_\alpha, q \restriction M_\alpha$ are not weakly orthogonal hence by 6.7(6) for every $\beta \in [\alpha, \delta)$ the types $p \restriction M_\beta, q \restriction M_\alpha$ are not weakly orthogonal. The last two paragraphs prove the desirable equivalence concerning $\underset{\rm wk}{\perp}$. As for $\perp$ this follows by part (1).
4) First assume $\neg(p \underset{\rm wk}{\perp} q)$ then by the definitions $\neg(p \perp q)$, the coun-

terexample being M itself (or use part (2)).
Second, assume $p \underset{\text{wk}}{\perp} q$ and let N_1 be such that $M \le_{\mathfrak{s}} N_1$ and $p_1, q_1 \in \mathscr{S}^{\text{bs}}_{\mathfrak{s}}(N_1)$ be non-forking extensions of p, q respectively; we shall prove $p_1 \underset{\text{wk}}{\perp} q_1$, this suffices for $p \perp q$ hence finishes the proof. Now there are N_2, p_2, q_2 such that $N_1 \le_s N_2, N_2$ is $(\lambda, *)$-brimmed and $p_2, q_2 \in \mathscr{S}^{\text{bs}}_{\mathfrak{s}}(N_2)$ are non-forking extensions of p, q respectively hence $p_2 \restriction N_1 = p_1, q_2 \restriction N_1 = q_1$. By 6.7(4) we have $(p \underset{\text{wk}}{\perp} q) \equiv (p_2 \underset{\text{wk}}{\perp} q_2)$ but our present assumption is $p \underset{\text{wk}}{\perp} q$ so necessarily $p_2 \underset{\text{wk}}{\perp} q_2$. Now by 6.7(6) we have $p_1 \underset{\text{wk}}{\perp} q_1$ so we are done.
[If we know part (6) first then we can say that: if $p \underset{\text{wk}}{\perp} q$ by 1.24 there is M such that $M_0 \le_{\mathfrak{s}} M$ and M is brimmed over M_0 and p, q does not fork over M_0. Now by (6) we get $p \perp q$ as required.]
5) By part (4).
6) As N is universal over M, clearly there is $M_1 \le_{\mathfrak{s}} N$ which is brimmed over M. By monotonicity, $p, q \in \mathscr{S}^{\text{bs}}_{\mathfrak{s}}(N)$ does not fork over M_1 and as we assume that $p \underset{\text{wk}}{\perp} q$, by 6.7(6) it follows that $(p \restriction M_1) \underset{\text{wk}}{\perp} (q \restriction M_1)$. But M_1 is brimmed so by part (4) we get $(p \restriction M_1) \perp (q \restriction M_1)$, hence by part (1) we get $p \perp q$ as required.

$\square_{6.8}$

6.9 Definition. 1) Assuming $M \le_{\mathfrak{s}} N$ and $p \in \mathscr{S}^{\text{bs}}_{\mathfrak{s}}(N)$, we let $p \perp M$ (p is orthogonal to M) mean that: for any q, if $q \in \mathscr{S}^{\text{bs}}_{\mathfrak{s}}(N)$ does not fork over M <u>then</u> $p \perp q$ (but see 6.10(1) below). Similarly for $p \underset{\text{wk}}{\perp} M, p$ weakly orthogonal to M.
2) Assuming $M \le_{\mathfrak{s}} N$ and $p \in \mathscr{S}^{\text{bs}}_{\mathfrak{s}}(N)$, we say that p is super-orthogonal to M and write $p \underset{\text{su}}{\perp} M$ <u>when</u>: if $\text{NF}_{\mathfrak{s}}(M, N, M', N')$ and $q \in \mathscr{S}^{\text{bs}}_{\mathfrak{s}}(M')$ then $p \perp q$.

6.10 Claim. *0) Automorphism of any $\mathfrak{C} \in K^{\mathfrak{s}}_{\ge\lambda}$ preserves $p \| q$, $p \underset{x}{\perp} q$, $p \underset{x}{\perp} M$ for $x \in \{\text{wk,st,su}\}$ where $p \in \mathscr{S}^{\text{bs}}_{\mathfrak{s}}(M), q \in \mathscr{S}^{\text{bs}}_{\mathfrak{s}}(N)$, and $M, N \in K_{\mathfrak{s}}$ are $\le_{\mathfrak{K}[\mathfrak{s}]} \mathfrak{C}$.*
1) If $M \le_{\mathfrak{s}} N_\ell \, (\le_{\mathfrak{K}[\mathfrak{s}]} \mathfrak{C}), p_\ell \in \mathscr{S}^{\text{bs}}_{\mathfrak{s}}(N_\ell)$ for $\ell = 1, 2$ and $p_1 \| p_2$ <u>then</u> $p_1 \perp M \Leftrightarrow p_2 \perp M$. Similarly for $\underset{\text{su}}{\perp}$ (so for $x =$ st,su we can write $p \underset{x}{\perp} N$

if for some $p' \in \mathscr{S}^{\rm bs}_{\mathfrak{s}}(N')$ *parallel to* p *we have* $M \le_{\mathfrak{s}} N'$ & $p' \underset{x}{\perp} N$*).*
2) Assume $\langle M_\alpha : \alpha \le \delta\rangle$ *is* $\le_{\mathfrak{s}}$*-increasing continuous,* $p \in \mathscr{S}^{\rm bs}_{\mathfrak{s}}(N)$ *where* $N \le_{\mathfrak{K}[\mathfrak{s}]} \mathfrak{C}, M_\delta \le_{\mathfrak{K}[\mathfrak{s}]} \mathfrak{C}$ *(so* $N \in K_{\mathfrak{s}}$*). If* $\alpha < \delta \Rightarrow p \perp M_\alpha$ *then* $p \perp M_\delta$*. If* $\alpha < \delta \Rightarrow p \underset{\rm su}{\perp} M_\alpha$ *then* $p \underset{\rm su}{\perp} M_\delta$*.*
3) If M *is brimmed,* $M \le_{\mathfrak{s}} N_\ell \le_{\mathfrak{s}} N$ *and* $p_\ell \in \mathscr{S}^{\rm bs}_{\mathfrak{s}}(N_\ell)$ *for* $\ell = 1, 2$ *and* $\mathrm{NF}_{\mathfrak{s}}(M, N_1, N_2, N)$ *then* $p_2 \perp M \Rightarrow p_2 \perp p_1$ *(hence* $p_2 \perp M \Rightarrow p_2 \perp N_1$*, i.e.,* $p_2 \perp M \Rightarrow p_2 \underset{\rm su}{\perp} M$*).*
4) (monotonicity) If $p \in \mathscr{S}^{\rm bs}_{\mathfrak{s}}(M_3), M_0 \le_{\mathfrak{s}} M_\ell \le_{\mathfrak{s}} M_3$ *for* $\ell = 1, 2$*, the type* p *does not fork over* M_2 *and* $p \underset{x}{\perp} M_1$ *then* $p \restriction M_2 \underset{x}{\perp} M_0$ *when* $x \in \{\mathrm{st,wk,su}\}$*.*
5) If $\mathfrak{K}_{\mathfrak{s}}$ *is categorical,* $M \le_{\mathfrak{s}} N$ *and* $p \in \mathscr{S}^{\rm bs}_{\mathfrak{s}}(M)$*, then* $p \perp M \Leftrightarrow p \underset{\rm wk}{\perp} M \Leftrightarrow p \underset{\rm su}{\perp} M$*.*

Proof. 0) Trivial.
1) Let $N_3 \le_{\mathfrak{K}[\mathfrak{s}]} \mathfrak{C}$ be such that $N_1 \cup N_2 \subseteq N_3 \in \mathfrak{K}_{\mathfrak{s}}$ and let $p_3 \in \mathscr{S}^{\rm bs}_{\mathfrak{s}}(N_3)$ be a non-forking extension of p_1 (and of p_2 (see 6.6(1)). Toward proving the first sentence assume $q \in \mathscr{S}^{\rm bs}_{\mathfrak{s}}(M)$ and $q_\ell \in \mathscr{S}^{\rm bs}_{\mathfrak{s}}(N_\ell)$ is a non-forking extension of q for $\ell = 1, 2$ and for the first sentence it is enough to show that $q_1 \perp p_1 \Leftrightarrow q_2 \perp p_2$ and let $q_3 \in \mathscr{S}^{\rm bs}_{\mathfrak{s}}(N_3)$ be the non-forking extension of q (hence of q_1 and of q_2). By 6.8(1) we have $p_1 \perp q_1 \Leftrightarrow p_3 \perp q_3 \Leftrightarrow p_2 \perp q_2$ so we are done.

For the second sentence, by symmetry it is enough to assume that $p_1 \underset{\rm su}{\perp} M$ and to prove $p_2 \underset{\rm su}{\perp} M$. So assume $\mathrm{NF}_{\mathfrak{s}}(M, N_2, M^+, N_2^+)$ and $p_2^+ \in \mathscr{S}^{\rm bs}_{\mathfrak{s}}(N_2^+)$ be a non-forking extension of p_2; and we should prove that ($p_2 \perp M^+$, equivalently) $p_2^+ \perp M^+$. By part (0) without loss of generality for some $N_3^+ \le_{\mathfrak{K}[\mathfrak{s}]} \mathfrak{C}$ of cardinality $\lambda_{\mathfrak{s}}$ we have $N_3 \le_{\mathfrak{s}} N_3^+, N_2^+ \le_{\mathfrak{s}} N_3^+$ and $\mathrm{NF}_{\mathfrak{s}}(N_2, N_2^+, N_3, N_3^+)$ and let $p_3^+ \in \mathscr{S}^{\rm bs}_{\mathfrak{s}}(N_3^+)$ be a non-forking extension of p_2^+ so of p_2 and of p_1.

By symmetry and transitivity of $\mathrm{NF}_{\mathfrak{s}}$ we get $\mathrm{NF}_{\mathfrak{s}}(M, M^+, N_3, N_3^+)$ hence by monotonicity $\mathrm{NF}_{\mathfrak{s}}(M, M^+, N_1, N_3^+)$. As we are assuming $p_1 \underset{\rm su}{\perp} M$ we deduce $p_1 \perp M^+$, but p_3^+ is the non-forking extension of p in $\mathscr{S}^{\rm bs}_{\mathfrak{s}}(N_3^+)$ hence $p_3^+ \perp M^+$ but $p_2^+ \| p_3^+$ hence by the first sentence of part (1) we get $p_2^+ \perp M^+$ as required for proving $p_2 \underset{\rm su}{\perp} M$.
2) The first implication is easy by the local character (i.e., Axiom

(E)(c) of good frames) and 6.8(3).

For the second implication by part (1) for $\underset{\text{su}}{\perp}$, without loss of generality $M_\delta \leq_{\mathfrak{s}} N$; so assume that $\mathrm{NF}_{\mathfrak{s}}(M_\delta, N, M'_\delta, N')$ and $q \in \mathscr{S}^{\mathrm{bs}}_{\mathfrak{s}}(M'_\delta)$ and we should prove that $p \perp q$. We know that there is a $\leq_{\mathfrak{s}}$-increasing continuous sequence $\langle M^*_\alpha : \alpha \leq \delta \rangle$ such that $\alpha < \beta \leq \delta \Rightarrow \mathrm{NF}_{\mathfrak{s}}(M_\alpha, M^*_\alpha, M_\beta, M^*_\beta)$ and M^*_δ is $(\lambda, *)$-brimmed over M_δ. So without loss of generality $M'_\delta \leq_{\mathfrak{s}} M^*_\delta$. Also without loss of generality for some N^* we have $\mathrm{NF}_{\mathfrak{s}}(M'_\delta, N', M^*_\delta, N^*)$. By the $\mathrm{NF}_{\mathfrak{s}}$ calculus, $\mathrm{NF}_{\mathfrak{s}}(M_\alpha, M^*_\alpha, N, N^*)$. Let $q^* \in \mathscr{S}^{\mathrm{bs}}_{\mathfrak{s}}(M^*_\delta)$ be a non-forking extension of q, hence for some $\alpha < \delta$, the type q^* does not fork over M^*_α so by the last sentence as $p \perp M_\alpha$ by an assumption of the clause we get $p \perp (q^* \restriction M_\alpha)$ hence $p \perp q$.

3) By part (1) without loss of generality

$(*)_1$ N_ℓ is $(\lambda, *)$-brimmed over M for $\ell = 1, 2$.
[Why? We can find N^+_1, N^+_2, N^+ such that $N_\ell \leq_{\mathfrak{s}} N^+_\ell \leq_{\mathfrak{s}} N^+$ and N^+_ℓ is $(\lambda, *)$-brimmed over N_ℓ for $\ell = 1, 2$ and $N \leq_{\mathfrak{s}} N^+$ and $\mathrm{NF}_{\mathfrak{s}}(M, N^+_1, N^+_2, N^+)$. Let $p^+_\ell \in \mathscr{S}^{\mathrm{bs}}(N^+_\ell)$ be a non-forking extension of p_ℓ for $\ell = 1, 2$. Clearly $p^+_2 \perp M$ and $p^+_2 \perp p^+_1 \Rightarrow p_2 \perp p_1$; so we can replace $(M, N_1, N_2, N, p_1, p_2)$ by $(M, N^+_2, N^+_2, N^+, p^+_1, p^+_2)$ and for them $(*)_1$ holds.]

Also we can find $\langle M_n, N_{2,n} : n < \omega \rangle$ such that: $\mathrm{NF}_{\mathfrak{s}}(M_n, N_{2,n}, M_{n+1}, N_{2,n+1})$, M_{n+1} is $(\lambda, *)$-brimmed over M_n, $N_{2,n+1}$ is $(\lambda, *)$-brimmed over $M_{n+1} \cup N_{2,n}$ (by NF calculus). So by 1.17 we know $\bigcup_{n<\omega} N_{2,n}$ is $(\lambda, *)$-brimmed over $\bigcup_{n<\omega} M_n$ which is $(\lambda, *)$-brimmed so by $(*)_1+$ "M is brimmed" without loss of generality $\bigcup_{n<\omega} N_{2,n} = N_2$ and $\bigcup_{n<\omega} M_n = M$. So for some $k < \omega$ the type p_2 does not fork over $N_{2,k}$. By the $\mathrm{NF}_{\mathfrak{s}}$ calculus we have $\mathrm{NF}_{\mathfrak{s}}(M_k, N_{2,k}, N_1, N)$. Let $\mathfrak{C} \in \mathfrak{K}_{\mathfrak{s}}$ be $(\lambda, *)$-brimmed over N, so we can find an automorphism f of $\mathfrak{C}$ such that $f \restriction N_{2,k} = \mathrm{id}_{N_{2,k}}, f(N_1) \subseteq M_{k+1} \subseteq M$ recalling that M_{k+1} and N_1 are $(\lambda, *)$-brimmed over M_k and $\mathrm{NF}_{\mathfrak{s}}(M_k, N_{2,k}, N_1, N)$. Let $p'_1 = f(p_1) \in \mathscr{S}^{\mathrm{bs}}_{\mathfrak{s}}(f(N_1))$ and let $p''_1 \in \mathscr{S}^{\mathrm{bs}}_{\mathfrak{s}}(M)$ be a non-forking extension of p'_1 recalling that $f(N_1) \leq_{\mathfrak{s}} M$. Now $p''_1 \perp p_2$ as $p_2 \perp M$,

hence $p_1'\perp(p_2 \restriction M_k)$ by 6.8(1A). By part (0) we have $p_1\perp(p_2 \restriction M_k)$ and lastly $p_1\perp p_2$ by 6.8(1A).
4) Let $q_0 \in \mathscr{S}^{\rm bs}_{\mathfrak{s}}(M_0)$ and let $q_3 \in \mathscr{S}^{\rm bs}_{\mathfrak{s}}(M_3)$ be its non-forking extension. If $p\underset{\rm wk}{\perp} M_1$ then by the definition $p\underset{\rm wk}{\perp}(q_3 \restriction M_1)$ which by 6.7(7) means $p\underset{\rm wk}{\perp} q_3$ hence by 6.7(6) we get $(p \restriction M_2)\underset{\rm wk}{\perp}(q_3 \restriction M_2)$ which by the definitions mean $p \restriction M_2\underset{\rm wk}{\perp} q_0$; this is enough for $x =$ wk. The case $x =$ st is even easier. For the case $x =$ su, use the second sentence of part (1) of this claim so it is enough to prove $p\underset{\rm su}{\perp} M_0$. Assume that $\mathrm{NF}_{\mathfrak{s}}(M_0, M_3, M_0', M_3')$ and we should prove that $p\perp M_0'$. By uniqueness of NF for some M_0'', M_1'', M_3'' we have $\mathrm{NF}_{\mathfrak{s}}(M_0, M_0'', M_1, M_1'')$ and $\mathrm{NF}_{\mathfrak{s}}(M_1, M_1'', M_3, M_3'')$ and $M_0' \leq_{\mathfrak{s}} M_0'', M_{\mathfrak{s}}' \leq_{\mathfrak{s}} M_3''$. As $p \in \mathscr{S}^{\rm bs}_{\mathfrak{s}}(M_3)$ and $p\underset{\rm su}{\perp} M_1$ clearly $p\perp M_1''$ hence $p\perp M_0''$ as required.
5) By part (3), $p\perp M \Leftrightarrow p\underset{\rm su}{\perp} M$ and by 6.8(4) we get $p\underset{\rm wk}{\perp} M \Leftrightarrow p\perp M$.
$\square_{6.10}$

Naturally, we would like to reduce orthogonality for $\mathfrak{s} = \mathfrak{t}^+$, to orthogonality for $\mathfrak{t}$.

6.11 Claim. *Assume $\mathfrak{s} = \mathfrak{t}^+, \mathfrak{t}$ a successful good$^+$ $\lambda_{\mathfrak{t}}$-frame, so $\lambda = \lambda_{\mathfrak{s}} = \lambda_{\mathfrak{t}}^+$.*
Below if $M_\ell \in K_{\mathfrak{s}}$ then $\langle M^\ell_\alpha : \alpha < \lambda\rangle$ will be some $\leq_{\mathfrak{t}}$-representation of M_ℓ and $M_\ell \leq_{\mathfrak{s}} \mathfrak{C}_{\mathfrak{s}}$ for $\ell = 0, 1, 2$.
0) For $\mathfrak{s}$ we have $\perp = \underset{\rm wk}{\perp} = \underset{\rm su}{\perp}$.
1) If $p_1, p_2 \in \mathscr{S}^{\rm bs}_{\mathfrak{s}}(M_0)$ <u>then</u>:
$p_1\perp_{\mathfrak{s}} p_2$ <u>iff</u> for unboundedly many $\alpha < \lambda$ we have $(p_1 \restriction M^0_\alpha)\underset{\rm wk}{\perp}_{\mathfrak{t}}(p_2 \restriction M^0_\alpha)$ <u>iff</u> for every large enough $\alpha < \lambda$, we have $(p_1 \restriction M^0_\alpha)\perp_{\mathfrak{t}}(p_2 \restriction M^0_\alpha)$.
2) If $M_0 \leq_{\mathfrak{s}} M_1 \leq_{\mathfrak{s}} M_2$ and $a \in M_2\backslash M_1$, <u>then</u>: $\mathbf{tp}_{\mathfrak{s}}(a, M_1, M_2) \in \mathscr{S}^{\rm bs}_{\mathfrak{s}}(M_1)$ and is orthogonal (for $\mathfrak{s}$) to M_0 <u>iff</u> for a club of ordinals $\delta < \lambda$ we have $\mathbf{tp}_{\mathfrak{t}}(a, M^1_\delta, M^2_\delta) \in \mathscr{S}^{\rm bs}_{\mathfrak{t}}(M^1_\delta)$ and is orthogonal (for $\mathfrak{t}$) to M^0_δ <u>iff</u> for a stationary set of ordinals $\delta < \lambda$ we have $\mathbf{tp}_{\mathfrak{t}}(a, M^1_\delta, M^2_\delta) \in \mathscr{S}^{\rm bs}_{\mathfrak{t}}(M^1_\delta)$ and is $\perp_{\mathfrak{t}} M^0_\delta$.
3) In part (2), "for all but boundedly many $\delta < \lambda$", "for unboundedly many $\delta < \lambda$" can replace "club of $\delta < \lambda$", "stationarily many $\delta < \lambda$" respectively.

Proof. 0) By 6.10(5) or 6.8(5) and 6.10(3) and the definition of $\mathfrak{t}^+$, i.e. as $\mathfrak{s} = \mathfrak{t}^+$ implies $K_{\mathfrak{s}}$ is categorical. .
1) We can find a club E of λ such that for every $\alpha \in E$ and $\ell = 1, 2$ we have M^0_α is $\mathfrak{t}$-brimmed and $p_\ell \restriction M^0_\alpha \in \mathscr{S}^{\rm bs}_{\mathfrak{t}}(M^0_\alpha)$ does not fork over $M^0_{\min(E)}$ (and so $p_\ell \restriction M^0_\alpha$ is a witness for p_ℓ) and without loss of generality $0 \in E, 1 \in E$. Hence by 6.7(6) + 6.8(4) + 6.8(1) for every $\alpha < \lambda$ we have $(p_1 \restriction M^0_\alpha) \underset{\rm wk}{\perp}_{\mathfrak{t}} (p_2 \restriction M^0_\alpha) \Leftrightarrow (p_1 \restriction M^0_\alpha) \perp_{\mathfrak{t}} (p_2 \restriction M^0_\alpha) \Leftrightarrow (p_1 \restriction M^0_0) \perp_{\mathfrak{t}} (p_2 \restriction M^0_0)$ and by transitivity of equivalence $(p_1 \restriction M^0_\alpha) \underset{\rm wk}{\perp}_{\mathfrak{t}} (p_2 \restriction M^0_\alpha) \Leftrightarrow (p_1 \restriction M^0_\beta) \perp_{\mathfrak{t}} (p_2 \restriction M^0_\beta)$ for $\alpha, \beta < \lambda$.

Case 1: Assume that $(p_1 \restriction M^0_0) \perp_{\mathfrak{t}} (p_2 \restriction M^0_0)$.

Assume that M_2, a_1, a_2 satisfy $M_0 \le_{\mathfrak{s}} M_2$ and $a_1, a_2 \in M_2$ and $p_\ell = \mathbf{tp}_{\mathfrak{s}}(a_\ell, M_0, M_2)$ for $\ell = 1, 2$, and it suffices to prove that $\{a_1, a_2\}$ is independent in (M_0, M_2).

Now there are M_1, a'_1 such that $M_0 \le_{\mathfrak{s}} M_1, a'_1 \in M_1$ and $p_1 = \mathbf{tp}_{\mathfrak{s}}(a'_1, M_0, M_1)$ and (M_0, M_1, a'_1) is canonically prime. As $\mathfrak{s}$ has primes (by 4.9(1)), 4.3 so without loss of generality $M_1 \le_{\mathfrak{s}} M_2$ and $a'_1 = a_1$. Clearly it suffices to prove that $\mathbf{tp}_{\mathfrak{s}}(a_2, M_1, M_2)$ is a non-forking extension of p_2. So for some club $E' \subseteq E$ of λ, for every $\delta \in E'$ we have:

(∗) $(M^0_\delta, M^1_\delta, a_1) \in K^{3,\rm uq}_{\mathfrak{t}}$ and $M^1_\delta \le_{\mathfrak{t}} M^2_\delta$, $a_2 \in M^2_\delta$, $\mathbf{tp}_{\mathfrak{t}}(a_2, M^0_\delta, M^2_\delta)$ is a non-forking extension of $p_2 \restriction M^0_0$ and $\mathbf{tp}_{\mathfrak{t}}(a_1, M^0_\delta, M^1_\delta)$ is a non-forking extension of $p_1 \restriction M^0_0$.

We are assuming $(p_1 \restriction M^0_0) \perp_{\mathfrak{t}} (p_2 \restriction M^0_0)$ hence $(p_1 \restriction M^0_\delta) \perp_{\mathfrak{t}} (p_2 \restriction M^0_\delta)$, so we get by (∗) and the definition of orthogonality that $\mathbf{tp}_{\mathfrak{t}}(a_2, M^1_\delta, M^2_\delta)$ is a non-forking extension of $\mathbf{tp}_{\mathfrak{t}}(a_2, M^0_\delta, M^2_\delta)$ hence it does not fork over M^0_0. As this holds for every $\delta \in E$ clearly M^0_0 witnesses that $\mathbf{tp}_{\mathfrak{s}}(a_2, M_1, M_2)$ does not fork over M_0 as required in this case.

Case 2: Assume that $(p_1 \restriction M^0_0) \pm_{\mathfrak{t}} (p_2 \restriction M^0_0)$.
We shall prove that $p_1 \pm p_2$, this suffices. We can assume that $M_0 <_{\mathfrak{s}} M_1, a_1 \in M_1, (M_0, M_1, a_1) \in K^{3,\rm pr}_{\mathfrak{s}}$ (recall that $\mathfrak{s}$ has primes being $\mathfrak{t}^+$) and $p_1 = \mathbf{tp}_{\mathfrak{s}}(a_1, M_0, M_1)$ and so as $\mathfrak{t}$ is successful (i.e. the definition of $\mathfrak{t}^+$), without loss of generality $\alpha < \beta < \lambda = \lambda^+_{\mathfrak{t}} \Rightarrow \mathrm{NF}_{\mathfrak{s}}(M^0_\alpha, M^1_\alpha, M^0_\beta, M^1_\beta)$ and (as $\mathfrak{t}$ is good$^+$) we have $\alpha <$

$\lambda_{\mathfrak{t}}^+ \Rightarrow (M_0^0, M_\alpha^1, a_1) \in K_{\mathfrak{t}}^{3,\mathrm{uq}}$. As p_ℓ is witnessed by $p_\ell \restriction M_0^0$ and M_0^0 is brimmed and without loss of generality also M_0^1 is brimmed and there is $q_2 \in \mathscr{S}_{\mathfrak{t}}(M_0^1)$ extending $p_2 \restriction M_0^0$ which forks over M_0^0. We can choose M_0^2, a_2 such that $M_0^1 \le_{\mathfrak{t}} M_0^2, a_2 \in M_0^2$ and $q_2 = \mathbf{tp}_{\mathfrak{t}}(a_2, M_0^1, M_0^2)$. Now we can choose inductively f_α, M_α^2 such that M_α^2 is $\le_{\mathfrak{t}}$-increasing continuous, f_α is a $\le_{\mathfrak{t}}$-embedding of M_α^1 into M_α^2, increasing continuous with α, $f_0 = \mathrm{id}_{M_0^1}$ and $\alpha = \beta+1 \Rightarrow \mathrm{NF}_{\mathfrak{t}}(M_\beta^1, M_\beta^2, M_\alpha^1, M_\alpha^2)$ and M_α^2 is $(\lambda, *)$-brimmed over M_β^2. No problem to do it and at the end without loss of generality $\bigcup_{\alpha<\lambda_{\mathfrak{s}}} f_\alpha = \mathrm{id}_{M_1}$ and let $M_2 := \cup\{M_\alpha^2 : \alpha < \lambda\}$. Clearly $M_2 \in K^{\mathfrak{t}}$ is saturated above $\lambda_{\mathfrak{t}}$ hence $M_2 \in \mathfrak{K}_{\mathfrak{s}}$. Clearly for $\alpha < \beta$ from E we have by symmetry $\mathrm{NF}_{\mathfrak{s}}(M_\alpha^0, M_\beta^0, M_\alpha^1, M_\beta^1)$ and $\mathrm{NF}_{\mathfrak{s}}(M_\alpha^1, M_\beta^1, M_\alpha^2, M_\beta^2)$ hence by transitivity we get $\mathrm{NF}_{\mathfrak{s}}(M_\alpha^0, M_\beta^0, M_\alpha^2, M_\beta^2)$ hence by symmetry $\mathrm{NF}_{\mathfrak{s}}(M_\alpha^0, M_\alpha^2, M_\beta^0, M_\beta^2)$. Easily M_1, a_1, M_2, a_2 exemplify $p_1 \pm p_2$; that is by 1.18 for every $\alpha < \lambda$ the type $\mathbf{tp}_{\mathfrak{t}}(a_2, M_\alpha^0, M_\alpha^2)$ is a non-forking extension of $\mathbf{tp}_{\mathfrak{t}}(a_2, M_0^0, M_0^2) = p_2 \restriction M_0^0$, hence $\mathbf{tp}_{\mathfrak{s}}(a_2, M_0, M_2) = p_2$. We conclude $\mathbf{tp}_{\mathfrak{s}}(a_2, M_1, M_2)$ extends p_2 but is not its non-forking extension in $\mathscr{S}_{\mathfrak{s}}(M_1)$ as required for proving $p_1 \pm_{\mathfrak{s}} p_2$.

2) Let E be a club of $\lambda_{\mathfrak{s}}$ such that for every $\alpha < \beta$ from E, M_β^ℓ is brimmed over M_α^ℓ, M_α^ℓ is brimmed (for $\ell = 0,1,2$) and $\mathrm{NF}_{\mathfrak{t}}(M_\alpha^\ell, M_\alpha^{\ell+1}, M_\beta^\ell, M_\beta^{\ell+1})$ for $\ell = 0,1$ and $a \in M_\alpha^2$ and M_α^1 is a witness for $p = \mathbf{tp}_{\mathfrak{s}}(b, M_1, M_2)$ if $b \in M_\alpha^2 \backslash M_\alpha^1$ and $\mathbf{tp}_{\mathfrak{t}}(b, M_\alpha^1, M_\alpha^2) \in \mathscr{S}_{\mathfrak{s}}^{\mathrm{bs}}(M_1)$. Now each of the statements which we should prove are equivalent implies that $\mathbf{tp}_{\mathfrak{s}}(a, M_1, M_2) \in \mathscr{S}_{\mathfrak{s}}^{\mathrm{bs}}(M_1)$ (by 1.10(4) and 1.16) so we can assume this. Without loss of generality $E = \lambda$ and $a \in M_0^2$. We shall use part (1) freely. Clearly $\mathbf{tp}_{\mathfrak{s}}(a, M_1, M_2) \perp_{\mathfrak{s}} M_0$ iff for every $q \in \mathscr{S}_{\mathfrak{s}}^{\mathrm{bs}}(M_0)$ we have $\mathbf{tp}_{\mathfrak{s}}(a, M_1, M_2) \perp_{\mathfrak{s}} q$ iff for each $\alpha < \lambda$ for every $q \in \mathscr{S}_{\mathfrak{s}}^{\mathrm{bs}}(M_0)$ which does not fork over M_α^0 we have $\mathbf{tp}_{\mathfrak{s}}(a, M_1, M_2) \perp_{\mathfrak{s}} q$ <u>iff</u> for each $\alpha < \lambda$ for every $q \in \mathscr{S}_{\mathfrak{t}}^{\mathrm{bs}}(M_\alpha^0)$, for some $\beta \in [\alpha, \lambda)$, the type $p \restriction M_\beta^1$ is $\mathfrak{t}$-orthogonal to the non-forking extension of q in $\mathscr{S}_{\mathfrak{t}}^{\mathrm{bs}}(M_\beta^0)$ <u>iff</u> for each $\gamma < \lambda$ we have $(p \restriction \mathscr{S}_{\mathfrak{t}}^{\mathrm{bs}}(M_\gamma^1)) \perp_{\mathfrak{t}} M_\gamma^0$. Thus we finish.

3) By monotonicity, i.e. 6.10(4) and 1.16. $\square_{6.11}$

6.12 Claim. *If* $\mathbf{J}$ *is independent in* (M, N) *and* $\mathbf{tp}_{\mathfrak{s}}(a, M, N) \in$

$\mathscr{S}^{\rm bs}_{\mathfrak{s}}(M)$ is orthogonal to $\mathbf{tp}_{\mathfrak{s}}(c, M, N)$ for every $c \in \mathbf{J}, M \le_{\mathfrak{s}} N_1 \le_{\mathfrak{s}} N$ and $(M, N_1, a) \in K^{3,\rm uq}_{\mathfrak{s}}$ then $\mathbf{J}$ is independent in (M, N_1, N).

Remark. 1) We shall below replace $K^{3,\rm uq}_{\mathfrak{s}}$ by $K^{3,\rm vq}_{\mathfrak{s}}$.
2) In 6.12, weak orthogonality suffice.

Proof. Let $\mathbf{J} = \{c_i : i < \alpha\}$.

By 5.4(1) there are N^+ and $\le_{\mathfrak{s}}$-increasing continuous sequence $\langle M_i : i \le \alpha\rangle$ such that $M_0 = M, M_\alpha \le_{\mathfrak{s}} N^+, N \le_{\mathfrak{s}} N^+$ and $(M_i, M_{i+1}, c_i) \in K^{3,\rm uq}_{\mathfrak{s}}$ for every $i < \alpha$. Now we prove by induction on $i \le \alpha$ that $\mathbf{tp}_{\mathfrak{s}}(a, M_i, N^+)$ does not fork over M_0. For $i = 0$ this is trivial. For i limit this holds by the continuity Ax$(E)(h)$. For $i = j+1$, we know that $\mathbf{tp}_{\mathfrak{s}}(a, M_j, N^+), \mathbf{tp}_{\mathfrak{s}}(c_i, M_j, N^+)$ are weakly orthogonal and $(M_j, M_{j+1}, c_i) \in K^{3,\rm uq}_{\mathfrak{s}}$ hence by the definition of orthogonality $\mathbf{tp}_{\mathfrak{s}}(a, M_{j+1}, N^+) = \mathbf{tp}_{\mathfrak{s}}(a, M_i, N^+)$ does not fork over M_j hence by transitivity of non-forking this type does not fork over M_0. Having carried the induction, we can conclude that $\mathbf{J} \cup \{a\}$ is independent in (M_0, N^+). As $(M_0, N_1, a) \in K^{3,\rm uq}_{\mathfrak{s}}$, by 5.4(3) we get that $\mathbf{J}$ is independent in (M_0, N_1, N^+) hence by monotonicity in (M_0, N_1, N), as required.

$\square_{6.12}$

6.13 Claim. : *We can in 6.12 replace $(M, N_1, a) \in K^{3,\rm uq}_{\mathfrak{s}}$ by $(M, N_1, \mathbf{I}) \in K^{3,\rm vq}_{\mathfrak{s}}$, which means: if $M \le_{\mathfrak{s}} N_1 \le_{\mathfrak{s}} N, \mathbf{J}$ is independent in (M, N) and $(M, N_1, \mathbf{I}) \in K^{3,\rm vq}_{\mathfrak{s}}$ and $a \in \mathbf{I} \wedge c \in \mathbf{J} \Rightarrow \mathbf{tp}_{\mathfrak{s}}(a, M, N) \perp \mathbf{tp}_{\mathfrak{s}}(c, M, N_1)$ then $\mathbf{J}$ is independent in (M, N_1, M).*

Proof. Let $\langle c_i : i < \alpha\rangle, N^+, \langle M_i : i \le \alpha\rangle$ be as in the proof of 6.12. We now prove by induction on $\beta \le \alpha$ that $\mathbf{I}$ is independent in (M, M_β, N^+). For $\beta = 0$ this is given. For β limit this holds by 5.8(5). For $\beta = \gamma + 1$, use 6.12. For $\beta = \alpha$ we get that $\mathbf{I}$ is independent in (M, M_α, N^+). As $M \le_{\mathfrak{s}} N_1 \le_{\mathfrak{s}} N \le_{\mathfrak{s}} N^+$ and $(M, N_1, \mathbf{I}) \in K^{3,\rm vq}_{\mathfrak{s}}$, by the definition of $K^{3,\rm vq}_{\mathfrak{s}}$ we get that $\mathrm{NF}_{\mathfrak{s}}(M, N_1, M_\alpha, N^+)$. By symmetry for $\mathrm{NF}_{\mathfrak{s}}$ we deduce that $\mathrm{NF}_{\mathfrak{s}}(M, M_\alpha, N_1, N^+)$, so as clearly $(M, M_\alpha, \mathbf{J}) \in K^{3,\rm bs}_{\mathfrak{s}}$ by 5.3 we get that $\mathbf{J}$ is independent in (M, N_1, N^+), so by monotonicity also in (M, N_1, N) as required. $\square_{6.13}$

For understanding $K^{3,\mathrm{uq}}_{\mathfrak{s}}$ the following claim is crucial.

6.14 Claim. *1) Assume $(M,N,a) \in K^{3,\mathrm{uq}}_{\mathfrak{s}}$.*
If $M \cup \{a\} \subseteq N' <_{\mathfrak{s}} N, b \in N \backslash N'$ and $q = \mathbf{tp}_{\mathfrak{s}}(b,N',N) \in \mathscr{S}^{\mathrm{bs}}_{\mathfrak{s}}(N')$ then q is weakly orthogonal to M.
2) [$\mathfrak{s}$ has primes.] Assume $(M,N,a) \in K^{3,\mathrm{bs}}_{\mathfrak{s}}$. We can find $\langle M_i, a_j : i \le \alpha, j < \alpha\rangle$ for some $\alpha < \lambda^+$, which is a pr-decomposition of N over M with $a_0 = a$ (so we stipulate $M_\alpha := N$), i.e., such that:

- (a) $a_0 = a$
- (b) $M_0 = M$.
- (c) $M_i \le_{\mathfrak{s}} N$ *is* $\le_{\mathfrak{s}}$*-increasing continuous for* $i \le \alpha$
- (d) $\mathbf{tp}_{\mathfrak{s}}(a_i, M_i, N) \in \mathscr{S}^{\mathrm{bs}}_{\mathfrak{s}}(M_i)$
- (e) $(M_i, M_{i+1}, a_i) \in K^{3,\mathrm{pr}}_{\mathfrak{s}}$ *for* $i < \alpha$.

3) [$\mathfrak{s}$ has primes]. In part (2) if also $(M,N,a) \in K^{3,\mathrm{uq}}_{\mathfrak{s}}$ and $\langle M_i, a_j : i \le \alpha, j < \alpha\rangle$ is as there then we can add, in fact necessarily have

- (f) *if* $i > 0$ *then* $\mathbf{tp}_{\mathfrak{s}}(a_i, M_i, N)$ *is weakly orthogonal to* M.

4) If $(M,N,\mathbf{J}) \in K^{3,\mathrm{vq}}_{\mathfrak{s}}$ and $M \cup \mathbf{J} \subseteq N' \le_{\mathfrak{s}} N$ and $b \in N \backslash N'$ and $q = \mathbf{tp}_{\mathfrak{s}}(b,N',N) \in \mathscr{S}^{\mathrm{bs}}_{\mathfrak{s}}(N')$ then q is weakly orthogonal to M.

Proof. (1) If $q \underset{\mathrm{wk}}{\pm} M$ then for some c, N^+, r we have $N \le_{\mathfrak{s}} N^+, c \in N^+, r = \mathbf{tp}_{\mathfrak{s}}(c, N', N^+) \in \mathscr{S}^{\mathrm{bs}}_{\mathfrak{s}}(N')$ does not fork over M but the set $\{b,c\}$ is not independent in (N', N^+) (or $b = c$). Possibly $\le_{\mathfrak{s}}$-increasing N^+, as $\mathbf{tp}_{\mathfrak{s}}(c, N', N^+)$ does not fork over $M \le_{\mathfrak{s}} N'$, clearly there is M' such that $M \cup \{c\} \subseteq M'$ and $\mathrm{NF}_{\mathfrak{s}}(M, M', N', N^+)$. As $a \in N'$ and $\mathbf{tp}_{\mathfrak{s}}(a, M, N') \in \mathscr{S}^{\mathrm{bs}}_{\mathfrak{s}}(M)$ this implies that $\mathbf{tp}_{\mathfrak{s}}(a, M', N^+) \in \mathscr{S}^{\mathrm{bs}}_{\mathfrak{s}}(M')$ does not fork over M. As $(M,N,a) \in K^{3,\mathrm{uq}}_{\mathfrak{s}}$ it follows that $\mathrm{NF}_{\mathfrak{s}}(M, N, M', N^+)$, and this implies that $\{b,c\}$ is independent in (N', N^+) by 5.6(2), second sentence contradicting the choice of c.
2) This is 3.11(1).
3) Follows by (part (2) and) part (1).
4) Like part (1). $\square_{6.14}$

6.15 Claim. *1) [$\mathfrak{s}$ has primes]. Assume $(M, N, a) \in K_{\mathfrak{s}}^{3,\mathrm{uq}}$. Then we can find $\langle M_i, a_j : i \leq \alpha, j < \alpha \rangle$ such that:*

$(a) - (e)$ *as in 6.14(2)*

(f) *as in 6.14(3)*

(g) $\alpha \leq \lambda$.

2) If in addition $\mathfrak{s} = \mathfrak{t}^+, \mathfrak{t}$ is a successful good$^+$ $\lambda_{\mathfrak{t}}$-frame (so $\lambda = \lambda_{\mathfrak{t}}^+$) then we can add

(h) *for each $i < \alpha, (M_i, M_{i+1}, a_i)$ is canonically prime, that is for any $<_{\mathfrak{t}}$-representations $\langle M_\varepsilon^i : \varepsilon < \lambda_{\mathfrak{t}}^+ \rangle, \langle M_\varepsilon^{i+1} : \varepsilon < \lambda^+ \rangle$ of M^i, M^{i+1} respectively, for a club of ordinals $\delta < \lambda_{\mathfrak{t}}^+$ we have $(M_\delta^i, M_\delta^{i+1}, a_i) \in K_{\mathfrak{t}}^{3,\mathrm{uq}}$.*

Proof. 1) Exactly as in 3.11(5), i.e., in the proof of 3.11(1) use a bookkeeping in order to get clause (g).
2) By 4.13(1). $\square_{6.15}$

6.16 Claim. *Assume*

(a) $\mathrm{NF}_{\mathfrak{s}}(M_0, M_0^+, M_1, M_3)$

(b) $\mathbf{J}$ *is independent in* (M_1, M_3)

(c) $\mathbf{tp}_{\mathfrak{s}}(c, M_1, M_3)$ *is super-orthogonal to M_0 for every $c \in \mathbf{J}$, see 6.9(2).*

Then we can find M_1^+, M_3^+ such that:

(α) $M_3 \leq_{\mathfrak{s}} M_3^+$

(β) $M_1 \cup M_0^+ \subseteq M_1^+ \leq_{\mathfrak{s}} M_3^+$

(γ) $\mathbf{tp}_{\mathfrak{s}}(c, M_1^+, M_3^+)$ *does not fork over M_1 for $c \in \mathbf{J}$*

(δ) $\mathbf{J}$ *is independent*[8] *in* (M_1^+, M_3^+).

Remark. 1) A related claim is 6.20.
2) No great harm at present if in 6.16 - 6.20 we assume that "$\mathfrak{s}$ has

[8]so $(\gamma) + (\delta)$ says that $\mathbf{J}$ is independent in (M_1, M_1^+, M_3^+)

primes".
3) Before we prove 6.16 note that:

6.17 Conclusion. If to the assumptions of 6.16 we add

(*d*) $(M_1, M_2, \mathbf{J}) \in K_{\mathfrak{s}}^{3,\mathrm{qr}}$ or just $(M_1, M_2, \mathbf{J}) \in K_{\mathfrak{s}}^{3,\mathrm{vq}}$ and $M_2 \leq_{\mathfrak{s}} M_3$,

then we can add to the conclusion (in fact follows from it):

(ε) $\mathrm{NF}_{\mathfrak{s}}(M_1, M_2, M_1^+, M_3^+)$.

Proof. If $(M_1, M_2, \mathbf{J}) \in K_{\mathfrak{s}}^{3,\mathrm{qr}}$, by 6.16 and 5.9(2). In general by the definition of $K_{\mathfrak{s}}^{3,\mathrm{vq}}$ (see 5.15) and clause (δ) of the conclusion of 6.16. $\square_{6.17}$

Proof of 6.16. First assume that $\mathfrak{s}$ has primes. By 6.14(2) we can find $\langle M_i^0, a_j : i \leq \alpha, j < \alpha\rangle$ which is a decomposition of M_0^+ over M_0. We can now choose by induction on $i \leq \alpha$ a pair (M_i^1, f_i) such that $M_i^1 \in K_{\mathfrak{s}}$ is $\leq_{\mathfrak{s}}$-increasing continuous, $M_0^1 = M_1, f_0 = \mathrm{id}_{M_0}, f_i$ is an $\leq_{\mathfrak{s}}$-embedding of M_i^0 into M_i^1, increasing continuous with i and $(M_i^1, M_{i+1}^1, f_{i+1}(a_i)) \in K_{\mathfrak{s}}^{3,\mathrm{pr}}$ and $\mathbf{tp}_{\mathfrak{s}}(f_{i+1}(a_i), M_i^1, M_{i+1}^1)$ does not fork over $f_i(M_i^0)$. There is no problem to do this, (as in stage $i = j + 1$ first choose $p_i = f_i(\mathbf{tp}_{\mathfrak{s}}(a_i, M_i^0, M_{i+1}^0))$ and then M_{i+1}^1 such that some $b_i \in M_{i+1}$ realizes p_i and as $(M_i^0, M_{i+1}^0, a_i) \in K_{\mathfrak{s}}^{3,\mathrm{pr}}$ we can choose a $\leq_{\mathfrak{K}}$-embedding f_i of M_{i+1}^0 into M_{i+1}^1 extending f_j and mapping a_i to b_i). As $K_{\mathfrak{s}}^{3,\mathrm{pr}} \subseteq K_{\mathfrak{s}}^{3,\mathrm{uq}}$ and the definition of $K_{\mathfrak{s}}^{3,\mathrm{uq}}$ easily $\mathrm{NF}_{\mathfrak{s}}(f_i(M_i^0),\ f_{i+1}(M_{i+1}^0), M_i^1, M_{i+1}^1)$ hence by $\mathrm{NF}_{\mathfrak{s}}$-symmetry $\mathrm{NF}_{\mathfrak{s}}(f_i(M_i^0),\ M_i^1,\ f_{i+1}(M_{i+1}^0),\ M_{i+1}^1)$ for every i hence by long transitivity we have $\mathrm{NF}_{\mathfrak{s}}(f_0(M_0), M_0^1, f_\alpha(M_\alpha^0), M_\alpha^1)$, and recalling $f_0 = \mathrm{id}_{M_0}, M_0^0 = M_0, M_0^1 = M_1, M_\alpha^0 = M_0^+$ this means $\mathrm{NF}_{\mathfrak{s}}(M_0, M_1, f_\alpha(M_0^+), M_\alpha^1)$. But also we assume $\mathrm{NF}_{\mathfrak{s}}(M_0, M_1, M_0^+, M_3)$, hence by $\mathrm{NF}_{\mathfrak{s}}$-uniqueness without loss of generality for some $M_3^+, M_3 \leq_{\mathfrak{s}} M_3^+, f_i = \mathrm{id}_{M_i^0}$ and $M_\alpha^1 \leq_{\mathfrak{s}} M_3^+$. This actually repeats the proof of 5.6(5).

For $i < \alpha$ for each $c \in \mathbf{J}$, note that $\mathbf{tp}_{\mathfrak{s}}(c, M_1, M_3)$ is super-orthogonal to M_0 (by a hypothesis) hence it is orthogonal to M_i^0 for $i \leq \alpha$. We prove by induction on $i \leq \alpha$ that $\mathbf{J}$ is independent over (M_1, M_i^1) inside M_3^+ and for every $c \in \mathbf{J}, \mathbf{tp}_{\mathfrak{s}}(c, M_i^1, M_3^+)$ (does not fork over $M_0^1 = M_1$ and) is orthogonal to M_i^0. For $i = 0$ this is trivial. For i limit easy by 5.10(2) for independence and by 6.8(3) for orthogonality. For $i+1$, as $\mathbf{tp}_{\mathfrak{s}}(a_i, M_i^1, M_3^+)$ does not fork over M_i^0, it is orthogonal to $\mathbf{tp}_{\mathfrak{s}}(c, M_i^1, M_3^+)$ for $c \in \mathbf{J}$ hence by 6.12 we know that $\mathbf{J} \cup \{a_i\}$ is independent over M_i^1. As $(M_i^1, M_{i+1}^1, a_i) \in K_{\mathfrak{s}}^{3,\mathrm{pr}} \subseteq K_{\mathfrak{s}}^{3,\mathrm{uq}}$ we get $\mathbf{tp}_{\mathfrak{s}}(c, M_{i+1}^1, M_3^+)$ does not fork over M_i^1 hence over M_1 for $c \in \mathbf{J}$ and $\mathbf{J}$ is independent in (M_1, M_i^1, M_3^+). Let M_1^+ be chosen as M_α^1 so we are done.

Let us turn to the general case where $\mathfrak{s}$ does not necessarily have primes. First

$(*)_1$ without loss of generality M_0^+ is $(\lambda, *)$-brimmed over M_1.

[Why? By assumption (a) and symmetry we have $\mathrm{NF}_{\mathfrak{s}}(M_0, M_1, M_0^+, M_3)$. We can find M_0^*, M_3^* such that $\mathrm{NF}_{\mathfrak{s}}(M_0^+, M_3, M_0^*, M_3^*)$ and M_0^* is $(\lambda, *)$-brimmed over M_0^+ hence over M_0; by transitivity for $\mathrm{NF}_{\mathfrak{s}}$ and the previous sentence we know that $\mathrm{NF}_{\mathfrak{s}}(M_0, M_1, M_0^*, M_3^*)$ hence by symmetry $\mathrm{NF}_{\mathfrak{s}}(M_0, M_0^*, M_1, M_3^*)$. Now by monotonicity $\mathbf{J}$ is independent in (M_1, M_3^*) and $c \in \mathbf{J} \Rightarrow \mathbf{tp}_{\mathfrak{s}}(c, M_1, M_3^*) = \mathbf{tp}_{\mathfrak{s}}(c, M_1, M_3) \underset{\mathrm{su}}{\perp} M_0$ so the assumptions of the claim holds for $(M_0, M_1, M_0^*, M_3^*, \mathbf{J})$.

Lastly, the conclusion for this quintuple implies the desired conclusion, so we can replace M_0^+, M_3 by M_0^*, M_3^* so really without loss of generality M_0^+ is $(\lambda, *)$-brimmed over M_0.]

$(*)_2$ without loss of generality M_1 is $(\lambda, *)$-brimmed over M_0 and M_3 is $(\lambda, *)$-brimmed over $M_0^+ \cup M_1$.

[Why? Similar to the proof of $(*)_1$.]

Hence there is a uq-decomposition $\langle M_{0,i}, a_j : i \leq \alpha, j < \alpha \rangle$ of M_0^+ over M_0 even with $\alpha = \lambda$ and $M_0^+ = M_{0,\alpha} := \cup\{M_{0,i} : i < \alpha\}$. By 4.7 we can find a $\leq_{\mathfrak{s}}$-increasing continuous sequence $\langle M_{1,i} : i \leq \alpha \rangle$ such that $M_{0,i} \leq_{\mathfrak{s}} M_{1,i}$, $\mathbf{tp}_{\mathfrak{s}}(a_i, M_{1,i}, M_{1,i+1})$ does not fork over $M_{0,i}$ and $(M_{1,i}, M_{1,i+1}, a_i) \in K_{\mathfrak{s}}^{3,\mathrm{uq}}$ and $M_{1,0}$ is brimmed over

$M_{0,0} = M_0$. Hence $\mathrm{NF}_{\mathfrak{s}}(M_{0,i}, M_{1,i}, M_{0,i+1}, M_{1,i+1})$ for each $i < \alpha$ so by long transitivity $\mathrm{NF}_{\mathfrak{s}}(M_{0,0}, M_{1,0}, M_{0,\alpha}, M_{1,\alpha})$ so by symmetry $\mathrm{NF}_{\mathfrak{s}}(M_{0,0}, M_{0,\alpha}, M_{1,0}, M_{1,\alpha})$.

Now $\mathrm{NF}_{\mathfrak{s}}(M_0, M_0^+, M_1, M_3)$ and $\mathrm{NF}_{\mathfrak{s}}(M_0, M_{0,\alpha}, M_{1,0}, M_{1,\alpha}), M_{0,\alpha} = M_0^+$, so by the uniqueness of $\mathrm{NF}_{\mathfrak{s}}$ and $M_{1,0}, M_1$ being brimmed over $M_{0,0} = M_0$ and M_3 being brimmed over $M_0^+ \cup M_3$ without loss of generality $M_{1,0} = M_1$ and $M_{1,\alpha} \leq_{\mathfrak{s}} M_3$ (we could have quoted 5.6(6) but here we give a detailed proof). Let $M_3^+ = M_3$ and $M_1^+ = M_{1,\alpha}$, so clauses $(\alpha), (\beta)$ of the desired conclusion obviously holds. By Definition 6.9(2) clearly $c \in \mathbf{J} \Rightarrow \mathbf{tp}_{\mathfrak{s}}(c, M_1, M_3) \perp \mathbf{tp}_{\mathfrak{s}}(a_i, M_{0,i}, M_{0,i+1})$; hence $c \in \mathbf{J} \Rightarrow \mathbf{tp}_{\mathfrak{s}}(c, M_1, M_3) \perp \mathbf{tp}_{\mathfrak{s}}(a_i, M_{1,i}, M_{1,i+1})$ so by 6.18 below this shows that clause (γ) of the desired conclusion. Lastly, for clause (δ) of the desired conclusion, we apply 6.18 below. $\square_{6.16}$

6.18 Claim. *Assume $x \in \{\mathrm{pr}, \mathrm{uq}\}$ and*

- (a) $M_0 \leq_{\mathfrak{s}} M_\ell \leq_{\mathfrak{s}} M_3$ *for* $\ell = 1, 2$
- (b) $\langle M_{0,i}, a_i : i < \alpha \rangle$ *is a x-decomposition of M_2 over M_0 (so $M_{0,\alpha} := M_2$)*
- (c) $\mathbf{J}$ *is independent in* (M_0, M_1)
- (d) $\mathbf{tp}_{\mathfrak{s}}(c, M_0, M_1) \perp \mathbf{tp}_{\mathfrak{s}}(a_i, M_{0,i}, M_2)$ *for $i < \alpha$ and $c \in \mathbf{J}$.*

Then $\mathbf{J}$ *is independent in (M_2, M_3) moreover in (M_0, M_2, M_3).*

6.19 Remark. We can replace clauses (b),(d) by:

- (b)′ $\langle M_{0,i} : i \leq \alpha \rangle$ is $\leq_{\mathfrak{s}}$-increasing continuous, $M_{0,0} = M_0, M_{0,\alpha} = M_2$ and $(M_i, M_{i+1}, \mathbf{J}_i) \in K_{\mathfrak{s}}^{3,\mathrm{vq}}$
- (d)′ $\mathbf{tp}_{\mathfrak{s}}(c, M_0, M_1) \perp \mathbf{tp}_{\mathfrak{s}}(a, M_{0,i}, M_2)$ for $i < \alpha, c \in \mathbf{J}$ and $a \in \mathbf{J}_i$.

See the after 6.13.

Proof. We prove by induction on $i \leq \alpha$ that $\mathbf{J}$ is independent in $(M_0, M_{0,i}, M_3)$. For $i = 0$ this holds by assumption (c) as $M_{0,i} = M_0$ and $M_1 \leq_{\mathfrak{s}} M_3$. For i a limit ordinal use 5.8(5). For $i = j + 1$ by the induction hypothesis $\mathbf{J}$ is independent in $(M_0, M_{0,j}, M_3)$ hence is independent in $(M_{0,j}, M_3)$. By clause (d) we have $c \in \mathbf{J}$ implies $\mathbf{tp}(c, M_{0,j}, M_3) \perp \mathbf{tp}_{\mathfrak{s}}(a_i, M_{0,j}, M_i)$ and by the induction hypothesis $\mathbf{tp}(c, M_0, M_1) \| \mathbf{tp}(c, M_{0,j}, M_3)$ hence clearly Claim 6.12 implies that $\mathbf{J} \cup \{a_i\}$ is independent in $(M_{0,j}, M_3)$. As $(M_{0,j}, M_{0,j+i}, a_i) \in K^{3,\mathrm{uq}}_{\mathfrak{s}}$ it follows by 5.4(3) that $\mathbf{J}$ is independent in $(M_{0,j}, M_{0,j+i}, M_3)$. As $c \in \mathbf{J} \Rightarrow \mathbf{tp}_{\mathfrak{s}}(c, M_{0,j+1}, M_3)$ does not fork over M_0 it follows that $\mathbf{J}$ is independent in $(M_0, M_{0,j+1}, M_3) = (M_0, M_{0,i}, M_3)$ so we have carried the induction.

For $i = \alpha$ we get the desired conclusion. $\square_{6.18}$

Below the restriction $\gamma \leq \omega$ may seem quite undesirable but it will be used as a stepping stone for better things. Note that in the proof of 6.20(1), clause (γ) in the induction hypothesis on $\langle M^n_i : i \leq \alpha \rangle$, primeness is not proved to hold for $\langle M^\omega_i : i \leq \alpha \rangle$, though enough is proved to finish the proof, this is why the proof does not naturally work for $\gamma > \omega$. It will be used in the proof of 7.7(3).

6.20 Claim. *1) Assume*

(a) $\langle M_\beta : \beta \leq \gamma \rangle$ *is* $\leq_{\mathfrak{s}}$*-increasing continuous with* $\gamma \leq \omega$
(b) $M_0 \leq_{\mathfrak{s}} M^+_0 \leq_{\mathfrak{s}} M^+_3$ *and* $M_\gamma \leq_{\mathfrak{s}} M^+_3$
(c) $\mathrm{NF}_{\mathfrak{s}}(M_0, M_1, M^+_0, M^+_3)$
(d) *if* $0 < \beta < \gamma$ *then* $(M_\beta, M_{\beta+1}, \mathbf{J}_\beta) \in K^{3,\mathrm{qr}}_{\mathfrak{s}}$ *or at least* $\in K^{3,\mathrm{vq}}_{\mathfrak{s}}$ *(so* $\mathbf{J}_\beta$ *is independent in* $(M_\beta, M_{\beta+1})$*)*
(e) *for every* $\beta \in (0, \gamma)$ *and* $a \in \mathbf{J}_\beta$ *the type* $\mathbf{tp}_{\mathfrak{s}}(a, M_\beta, M_{\beta+1})$ *is super-orthogonal to* M_0.

Then $\mathrm{NF}_{\mathfrak{s}}(M_0, M_\gamma, M^+_0, M^+_3)$.
2) If $\langle M_\beta : \beta \leq \gamma \rangle, \langle \mathbf{J}_\beta : 0 < \beta < \gamma \rangle$ *satisfy clauses (a), (d), (e) above and* $(M_0, M_1, \mathbf{J}) \in K^{3,\mathrm{vq}}_{\mathfrak{s}}$ *then* $(M_0, M_\gamma, \mathbf{J}) \in K^{3,\mathrm{vq}}_{\mathfrak{s}}$.

Proof. 1) If $\mathfrak{s}$ has primes we choose $\langle M^*_i, a_j : i \leq \alpha, j < \alpha \rangle$, a decomposition of M^+_0 over M_0 so $M^*_\alpha = M^+_0$ (as in the proof of 6.16). If $\mathfrak{s}$ does not necessarily have primes, as the proof of 6.16 without loss of generality M^+_0 is $(\lambda, *)$-brimmed over M_0, and so there is a uq-decomposition $\langle M^*_i, a_j : i \leq \alpha, j < \alpha \rangle$ of M^+_0 over M_0. Now by induction on $n \leq \gamma, n < \omega$ we choose $N^3_n, \bar{M}^n = \langle M^n_i : i \leq \alpha \rangle$ such that:

$\boxtimes(\alpha)$ $N^3_0 = M^+_3$ and $\bar{M}^0 = \langle M^*_i : i \le \alpha \rangle$ so, i.e. $M^0_i = M^*_i$

(β) $N^3_n \le_{\mathfrak{s}} N^3_{n+1}$

(γ) $\langle M^n_i, a_j : i \le \alpha, j < \alpha \rangle$ is a uq-decomposition inside N^3_n over M^n_0

(δ) $M^n_i \le_{\mathfrak{s}} M^{n+1}_i$

(ε) $\mathbf{tp}_{\mathfrak{s}}(a_i, M^n_i, N^3_n)$ does not fork over M^0_i

(ζ) (i) if $\mathfrak{s}$ has primes $M_n = M^n_0$ for $n < \omega$; in general
(ii) $M_n \le_{\mathfrak{s}} M^n_0$ and $n = 0 \Rightarrow M^n_0 = M_0$ and $\mathrm{NF}_{\mathfrak{s}}(M_n, M_\gamma, M^n_0, N^3_n)$
(this holds trivially when $n = 0$ by clause (c) and also holds in subclause (i) trivially).

For $n = 0$ this is done. The step from n to $n + 1$ is by the first paragraph of the proof of 6.16 or 5.6(5) when $\mathfrak{s}$ has primes, by 5.6(6) in general, but for this we need to know that

$\circledast$ $\mathrm{NF}_{\mathfrak{s}}(M_n, M_{n+1}, M^n_\alpha, N^3_n)$.
[Why does $\circledast$ hold? First if $n = 0$ this holds by clause (c) of the assumption as $M^0_\alpha = M^*_\alpha = M^+_0$. Second if $n > 0$ then by clause (ζ) we have $\mathrm{NF}_{\mathfrak{s}}(M_n, M_\gamma, M^n_0, N^3_n)$ hence by monotonicity $\mathrm{NF}_{\mathfrak{s}}(M_n, M_{n+1}, M^n_0, N^3_n)$. Now we can prove by induction on i that $\mathbf{J}_n$ is independent in (M_n, M^n_i, N^3_n) and note that as $(M_n, M_{n+1}, \mathbf{J}_n) \in K^{3,\mathrm{vq}}_{\mathfrak{s}}$ this implies that $\mathrm{NF}_{\mathfrak{s}}(M_n, M_{n+1}, M^n_i, N^3_n)$. For $i = 0$ this was proved in the sentence before last and for i limit by continuity of independence by 5.10(3). For $i = j + 1$ this holds as by clause (ε) of $\boxtimes$ the type $\mathbf{tp}_{\mathfrak{s}}(a_i, M^n_i, M^n_{i+1})$ does not fork over $M^0_i = M_{0,i}$ hence over M_0 whereas for every $c \in \mathbf{J}_n$ by the induction hypothesis for j the type $\mathbf{tp}_{\mathfrak{s}}(c, M^n_i, M^*_n)$ is parallel to $\mathbf{tp}_{\mathfrak{s}}(c, M_n, M_{n+1})$ which is super-orthogonal to M_0, so they are orthogonal. As $(M^n_j, M^n_{j+1}, a_j) \in K^{3,\mathrm{uq}}_{\mathfrak{s}}$ by clause (γ) of $\boxtimes$ clearly by claim 6.12 the statement holds for i. So we have carried the induction so we are done.]

If $\gamma < \omega$ we are done. So assume $\gamma = \omega$, and for $i < \alpha$ let $M^\omega_i := \bigcup\limits_{n<\omega} M^n_i$. Now for each $i < \alpha$, and $n < \omega$ we have (by clauses

$(\gamma) + (\delta) + (\varepsilon)$ and the Definition of $K_{\mathfrak{s}}^{3,\mathrm{uq}}$ or see the proof of 6.16), $\mathrm{NF}_{\mathfrak{s}}(M_i^n, M_{i+1}^n, M_i^{n+1}, M_{i+1}^{n+1})$, hence by long transitivity of $\mathrm{NF}_{\mathfrak{s}}$ (see II.6.28) we have $\mathrm{NF}_{\mathfrak{s}}(M_i^0, M_{i+1}^0, M_i^\omega, M_{i+1}^\omega)$. By symmetry we get $\mathrm{NF}_{\mathfrak{s}}(M_i^0, M_i^\omega, M_{i+1}^0, M_{i+1}^\omega)$ for $i < \omega$. As $\langle M_i^0 : i \le \alpha\rangle, \langle M_i^\omega : i \le \alpha\rangle$ are $\le_{\mathfrak{s}}$-increasing continuous, by long transitivity of $\mathrm{NF}_{\mathfrak{s}}$ we get $\mathrm{NF}_{\mathfrak{s}}(M_0^0, M_0^\omega, M_\alpha^0, M_\alpha^\omega)$ which means $\mathrm{NF}_{\mathfrak{s}}(M_0, M_0^\omega, M_0^+, M_\alpha^\omega)$ so by using monotonicity twice we get
$\mathrm{NF}_{\mathfrak{s}}(M_0, M_\omega, M_0^+, M_3^+)$ as required.
2) By definition 5.15 we are given (M_0^+, M_3^+) such that $M_0 \le_{\mathfrak{s}} M_0^+ \le_{\mathfrak{s}} M_3^+, M_\gamma \le_{\mathfrak{s}} M_3^+$ and $\mathbf{J}$ is independent in (M_0, M_0^+, M_3^+); we should prove $\mathrm{NF}_{\mathfrak{s}}(M_0, M_\gamma, M_0^+, M_3^+)$. As we are assuming $(M_0, M_1, \mathbf{J}) \in K_{\mathfrak{s}}^{3,\mathrm{vq}}$ we can deduce that $\mathrm{NF}_{\mathfrak{s}}(M_0, M_1, M_0^+, M_3^+)$, i.e., clause (c) of the assumption of part (1). Now clause (b) of the assumption of part (1) holds trivially, so the assumption of part (1) hence its conclusion, i.e., $\mathrm{NF}_{\mathfrak{s}}(M_0, M_\gamma, M_0^+, M_3^+)$ is as required. $\square_{6.20}$

We could have noted earlier

6.21 Claim. *Assume $p_i = \mathbf{tp}_{\mathfrak{s}}(a_i, M, N) \in \mathscr{S}_{\mathfrak{s}}^{\mathrm{bs}}(M)$ for $i < \alpha$ are pairwise orthogonal. Then $\{a_i : i < \alpha\}$ is independent in (M, N).*

Proof. By $(*)_0 \Rightarrow (*)_1$ from 5.4(1) and renaming it is enough to deal with finite α, (not really needed).
We now choose a pair (M_ℓ, N_ℓ) by induction on $\ell \le \alpha$ such that

- $\circledast(i)$ $M_\ell \le_{\mathfrak{s}} N_\ell$
- (ii) $M_0 = M, N_0 = N$
- (iii) if $m < \ell$ then $M_m \le_{\mathfrak{s}} M_\ell$ and $N_m \le_{\mathfrak{s}} N_\ell$
- (iv) if $\ell = m + 1$ then $(M_m, M_\ell, a_m) \in K_{\mathfrak{s}}^{3,\mathrm{uq}}$
- (v) $\mathbf{tp}_{\mathfrak{s}}(a_m, M_m, M_{m+1})$ does not fork over M_0.

For $\ell = 0$ this is trivial. For $\ell = m + 1$, first we prove by induction on $k \le m$ that $p_m^k = \mathbf{tp}_{\mathfrak{s}}(a_m, M_k, N_m)$ is the non-forking extension of p_m in $\mathscr{S}_{\mathfrak{s}}(M_k)$. Now for $k = 0$ this is trivial by the choice of p_m. For $k + 1 \le m$ by the induction hypothesis on k, $\mathbf{tp}_{\mathfrak{s}}(a_m, M_k, N_m)$ is a non-forking extension of p_m. Now $p_k \perp p_m$ and by clause (v) for k, $\mathbf{tp}_{\mathfrak{s}}(a_k, M_k, M_{k+1})$ is a non-forking extension of p_k. So $\mathbf{tp}_{\mathfrak{s}}(a_m, M_k, N_m)$ is orthogonal to $\mathbf{tp}_{\mathfrak{s}}(a_k, M_k, M_{k+1})$.

As $(M_k, M_{k+1}, a_k) \in K^{3,\mathrm{uq}}_{\mathfrak{s}}$ we get that $\mathbf{tp}_{\mathfrak{s}}(a_m, M_{k+1}, M_{m+1})$ does not fork over M_k. Together with the induction hypothesis by transitivity of non-forking of types $\mathbf{tp}_{\mathfrak{s}}(a_m, M_{k+1}, M_{m+1})$ does not fork over M_0. So we have carried the induction on $k \leq m$.
Second, as $\mathfrak{s}$ is weakly successful there are b_m, M^*_ℓ such that $(M_m, M^*_\ell, b_m) \in K^{3,\mathrm{uq}}_{\mathfrak{s}}$ and $\mathbf{tp}_{\mathfrak{s}}(b_m, M_m, M^*_\ell) = p^m_m$. By the definition of types and as $K_{\mathfrak{s}}$ has amalgamation by renaming there is N_ℓ such that $M^*_\ell \leq_{\mathfrak{s}} N_\ell, N_m \leq_{\mathfrak{s}} N_\ell$ and $b_m = a_m$ and let $M_\ell = M^*_\ell$. So we can define (M_ℓ, N_ℓ) for $\ell \leq n$ as in $\circledast$. By the definition of independence we are done. Alternatively use 6.12.

$\square_{6.21}$

6.22 Claim. *If $p_i \in \mathscr{S}^{\mathrm{bs}}_{\mathfrak{s}}(M)$ for $i < \alpha$ are pairwise orthogonal and $q \pm p_i$ for $i < \alpha$ <u>then</u> $\alpha < \omega$.*

Proof. By 5.13 and 6.21. That is assume $\alpha \geq \omega$, let $q = \mathbf{tp}_{\mathfrak{s}}(b, M, N_0)$ and we can find N_n for $n < \omega$ such that (N_0 is as above and) $N_n \leq_{\mathfrak{K}} N_{n+1}$ and $a_n \in N_{n+1}$ realizing p_n such that $\{b, a_n\}$ is not independent. By 6.21, the set $\{a_n : n < \omega\}$ is independent in $N_\omega = \cup\{N_n : n < \omega\}$ and so by 5.13, we get a contradiction.

$\square_{6.22}$

<u>6.23 Exercise</u>: Assume that

- (a) $\mathbf{J}_i$ is independent in (M, N) for $i < \alpha$
- (b) if $i_1 \neq i_2$ are $< \alpha$ and $c_1 \in \mathbf{J}_{i_1}, c_2 \in \mathbf{J}_{i_2}$ then $\mathbf{tp}_{\mathfrak{s}}(c_1, M, N) \perp \mathbf{tp}_{\mathfrak{s}}(c_2, M, N)$.

<u>Then</u> $\cup\{\mathbf{J}_i : i < \alpha\}$ is independent in (M, N) and the $\mathbf{J}_i$'s are pairwise disjoint.

<u>6.24 Exercise</u>: 1) Assume that

- (a) $\mathscr{T} \subseteq {}^{\omega>}\mathrm{Ord}$
- (b) $\bar{\mathbf{J}} = \langle \mathbf{J}_\eta : \eta \in \mathscr{T} \rangle$ is a sequence of pairwise disjoint subsets of $\mathbf{I}_{M,N}$
- (c) for $\eta \in \mathscr{T}$ the set $\cup\{\mathbf{I}_\nu : \nu \trianglelefteq \eta\}$ is independent in (M, N)

(*d*) if $\eta_1, \eta_2 \in \mathscr{T}$ are $\trianglelefteq$-incomparable and $c_\ell \in \mathbf{J}_{\eta_\ell}$ for $\ell = 1, 2$, then

$$\mathbf{tp}_{\mathfrak{s}}(c_1, M, N) \perp \mathbf{tp}_{\mathfrak{s}}(c_2, M, N).$$

Then $\cup\{\mathbf{J}_\eta : \eta \in \mathscr{T}\}$ is independent in (M, N).
2) Similar for any partial order $\mathscr{T}$ which is a tree.

[Hint: first reduce to finite $\mathscr{T}$, then prove by induction on $|\mathscr{T}|$. If $\mathscr{T}$ has no root use the induction hypothesis and 6.23.

If η_0 is the root, without loss of generality it is $<>$ and we can find N' such that $N \leq_{\mathfrak{s}} N'$ and $\leq_{\mathfrak{s}}$-increasing continuous $\langle M_i : i \leq \alpha\rangle, M_0 = M, M_\alpha \leq_{\mathfrak{s}} N, (M_i, M_{i+1}, a_i) \in K^{3,\mathrm{uq}}_{\mathfrak{s}}$ for $i < \alpha$ where $\langle a_i : i < \alpha\rangle$ list $\mathbf{J}_{\eta_0}$ with no repetitions. Let $u = \{\eta(0) : \eta \in \mathscr{T}\backslash\{<>\}\}$ and for each $\alpha \in u$ let $\mathscr{T}'_\alpha = \{\nu : \langle\alpha\rangle\char`^\nu \in \mathscr{T}\}$. Now first, by the induction hypothesis for each $\alpha \in u$ the set $\cup\{\mathbf{J}_{\langle\alpha\rangle\char`^\nu} : \nu \in \mathscr{T}'_\alpha\}$ is independent in (M, M_α, N') and then by 6.23, also $\cup\{\mathbf{J}_{\langle\alpha\rangle\char`^\nu} : \nu \in \mathscr{T}'_\alpha$ and $\alpha \in u\}$ is independent in (M_α, N^*) hence in (M, M_α, N'). As $\mathbf{J}_{<>}$ is independent in (M, M_α) we are easily done.]

6.25 Exercise: Assume $M_* \leq_{\mathfrak{s}} M \leq_{\mathfrak{s}} N, (M, N, a_1) \in K^{3,\mathrm{uq}}_{\mathfrak{s}}$ and $a_2 \in N$ and $p_\ell = \mathbf{tp}_{\mathfrak{s}}(a_\ell, M, N) \in \mathscr{S}^{\mathrm{bs}}_{\mathfrak{s}}(M)$ for $\ell = 1, 2$. Then $p_1 \perp M_* \Rightarrow p_2 \perp M_*$ and $p_1 \underset{\mathrm{su}}{\perp} M_* \Rightarrow p_2 \underset{\mathrm{su}}{\perp} M_*$.

[Hint: Use 6.26 below, reducing to the case of brimmed M over M_* as in the proof of 6.16.]

6.26 Exercise: 1) Assume $M_0 \leq_{\mathfrak{s}} M_1 \leq_{\mathfrak{s}} M_2$ and $(M_1, M_2, a) \in K^{3,\mathrm{uq}}_{\mathfrak{s}}$ and $(M_1, M_2, b) \in K^{3,\mathrm{bs}}_{\mathfrak{s}}$. If $\mathbf{tp}_{\mathfrak{s}}(a, M_1, M_2) \underset{\mathrm{wk}}{\perp} M_0$ then $\mathbf{tp}_{\mathfrak{s}}(b, M_1, M_2) \underset{\mathrm{wk}}{\perp} M_0$, recalling Definition 6.9(1).
2) If $M_0 \leq_{\mathfrak{s}} M_1 \leq_{\mathfrak{s}} M_2$ and $(M_1, M_2, \mathbf{J}) \in K^{3,\mathrm{vq}}_{\mathfrak{s}}$ and $a \in \mathbf{J} \Rightarrow \mathbf{tp}_{\mathfrak{s}}(a, M_1, M_2) \underset{\mathrm{wk}}{\perp} M$ and $q \in \mathscr{S}^{\mathrm{bs}}_{\mathfrak{s}}(M_1)$ is realized in M_2 then $q \underset{\mathrm{wk}}{\perp} M_0$.

[Hint: 1) If not, then there is $q \in \mathscr{S}^{\mathrm{bs}}_{\mathfrak{s}}(M_1)$ which does not fork over M_0 such that $q \underset{\mathrm{wk}}{\pm} \mathbf{tp}_{\mathfrak{s}}(b, M_1, M_2)$. We can find a pair (M_3, c) such that $M_2 \leq_{\mathfrak{s}} M_3$ and $c \in M_3$ realizes q but $\{c, b\}$ is not independent in (M_1, M_3). As $\mathbf{tp}_{\mathfrak{s}}(a, M_1, M_2) \underset{\mathrm{wk}}{\perp} M_0$, necessarily $\mathbf{tp}_{\mathfrak{s}}(a, M_1, M_2) \underset{\mathrm{wk}}{\perp} q$ hence $\{a, c\}$ is independent in (M_1, M'_3). As $(M_1, M_2, a) \in K^{3,\mathrm{uq}}_{\mathfrak{s}}$ necesarily $\mathbf{tp}_{\mathfrak{s}}(c, M_2, M_3)$ does not fork over M_1. This implies $\{c, b\}$

is independent in (M_1, M_3), contradiction.
2) Similar to part (1) using 6.23.]

6.27 Exercise [$\mathfrak{s}$ has primes] Assume:

⊛ (a) $\langle N^\ell_\alpha : \alpha \le \delta \rangle$ is $\le_{\mathfrak{s}}$-increasing continuous for $\ell = 1,2$ and $\lambda | \delta$

(b) $\mathbf{P}_\ell \subseteq \mathscr{S}^{\rm bs}_{\mathfrak{s}}(N^\ell_0)$

(c) if $N^\ell_0 \le_{\mathfrak{s}} M <_{\mathfrak{s}} N^\ell_\delta$ then for some $c \in N^\ell_\delta \backslash M$ the type $p = \mathbf{tp}_{\mathfrak{s}}(c, M, N^\ell_\delta)$ belongs to $\mathscr{S}^{\rm bs}_{\mathfrak{s}}(M)$ and is orthogonal to $\mathbf{P}_\ell$, i.e., p to q for every $q \in \mathbf{P}_\ell$

(d) if $p \in \mathscr{S}^{\rm bs}_{\mathfrak{s}}(N^\ell_\alpha), \alpha < \delta, \ell \in \{1,2\}$ and p is orthogonal to $\mathbf{P}_\ell$, then for λ ordinals $\beta \in (\alpha, \delta)$ there is $c \in N^\ell_{\beta+1}$ such that $\mathbf{tp}_{\mathfrak{s}}(c, N^\ell_\beta, N^\ell_{\beta+1})$ is a non-forking extension of p

(e) f_0 is an isomorphism from N^1_0 onto N^2_0 mapping $\mathbf{P}_1$ onto $\mathbf{P}_2$.

Then there is an isomorphism from N^1_δ onto N^2_δ extending f_0.

[Hint: Hence and forth, as usual (using the existence of primes).]

6.28 Exercise 1) If $\circledast_{\bar N, \mathbf{P}}$ then $\boxtimes_{\bar N, \bar{\mathbf{P}}}$ where:

$\circledast_{\bar N, \mathbf{P}}$ (a) $\bar N = \langle N_0, N_1 \rangle, N_0 \le_{\mathfrak{s}} N_1$

(b) $\mathbf{P} \subseteq \mathscr{S}^{\rm bs}_{\mathfrak{s}}(N_0)$

(c) if $N_0 \le_{\mathfrak{s}} M <_{\mathfrak{s}} N_1$ then for some $c \in N_1, M$ the type $\mathbf{tp}(c, M, N_1)$ belongs to $\mathscr{S}^{\rm bs}_{\mathfrak{s}}(M)$ and is orthogonal to $\mathbf{P}$, i.e., $q \in \mathbf{P} \Rightarrow q \perp \mathbf{tp}_{\mathfrak{s}}(c, M, N_1)$

$\circledast_{\bar N, P}$ there is $\le_{\mathfrak{s}}$-increasing continuous sequence $\langle M_\alpha : \alpha \le \alpha(*) \rangle$ such that $N_0 = M_0, N_1 \le_{\mathfrak{s}} M_{\alpha(*)}$ and for each $\alpha < \alpha(*)$ for some a_α we have $(M_\alpha, M_{\alpha+1}, a_\alpha) \in K^{3,\rm uq}_{\mathfrak{s}}$ and $\mathbf{tp}_{\mathfrak{s}}(a_\alpha, M_\alpha, M_{\alpha+1}) \perp \mathbf{P}$.

[Hint: See 8.3.]

§7 Understanding $K^{3,\mathrm{vq}}_{\mathfrak{s}}$

We would like to show that $K^{3,\mathrm{vq}}_{\mathfrak{s}} = K^{3,\mathrm{qr}}_{\mathfrak{s}}$ and $K^{3,\mathrm{pr}}_{\mathfrak{s}} = K^{3,\mathrm{uq}}_{\mathfrak{s}}$ and more when we assume that $\mathfrak{s}$ is categorical (in λ).
The hypothesis below holds if $\mathfrak{t}$ is a good$^+$ successful frame, $\mathfrak{s} = \mathfrak{t}^+$ is successful[9].

7.1 Hypothesis.

(a) $\mathfrak{s}$ is a good$^+$ λ-frame

(b) $\mathfrak{s}$ is successful

(c) $\mathfrak{s}$ has primes

(d) $\perp = \underset{\mathrm{wk}}{\perp}$

(e) $\underset{\mathrm{su}}{\perp} = \perp$, i.e. $p \perp M \Leftrightarrow p \underset{\mathrm{su}}{\perp} M$ when $M \le_{\mathfrak{s}} N$ and $p \in \mathscr{S}^{\mathrm{bs}}_{\mathfrak{s}}(M)$.

In the definition below note that our aim is to analyze $(M, N, \mathbf{J}_0) \in K^{3,\mathrm{bs}}_{\mathfrak{s}}$ so $\mathbf{J}_0$ has a special role.

7.2 Definition. 1) $\mathscr{W}_\alpha = \{(N, \bar{M}, \bar{\mathbf{J}}) : \bar{M} = \langle M_i : i < \alpha\rangle$ is $\le_{\mathfrak{s}}$-increasing continuous, $M_i \le_{\mathfrak{s}} N$ for $i < \alpha$ and $\bar{\mathbf{J}} = \langle \mathbf{J}_i : i < \alpha\rangle$ and $\mathbf{J}_i$ is independent in (M_i, M_{i+1}) for $i < \alpha$ stipulating $M_\alpha = N\}$ and we let

$$\mathscr{W} = \bigcup_{\alpha < \lambda^+} \mathscr{W}_\alpha$$

Let $(\langle M_i : i \le \alpha\rangle, \langle \mathbf{J}_i : i < \alpha\rangle)$ mean $(M_\alpha, \langle M_i : i < \alpha\rangle, \langle \mathbf{J}_i : i < \alpha\rangle)$.
2) $\le_{\mathscr{W}} = \le_{\mathscr{W}[\mathfrak{s}]}$ is the following two place relation on $\mathscr{W}$:

$(N^1, \bar{M}^1, \bar{\mathbf{J}}^1) \le_{\mathscr{W}} (N^2, \bar{M}^2, \bar{\mathbf{J}}^2)$ <u>iff</u> both are from $\mathscr{W}$ and $(a)+(b)$ where

(a) $N^1 \le_{\mathfrak{s}} N^2$, $\ell g(\bar{M}^1) \le \ell g(\bar{M}^2)$, $i < \ell g(\bar{M}^1) \Rightarrow M^1_i \le_{\mathfrak{s}} M^2_i$ & $\mathbf{J}^1_i \subseteq \mathbf{J}^2_i$

(b) $a \in \mathbf{J}^1_i \Rightarrow$ $\mathbf{tp}_{\mathfrak{s}}(a, M^2_i, M^2_{i+1})$ does not fork over M^1_i

[9] where do we use weakly successful good$^+$ rather than weakly successful good? E.g. in 7.3(3).

3) $\leq^{\rm fx}_{\mathscr{W}}$ is defined like $\leq_{\mathscr{W}}$ but also $\bar{\mathbf{J}}^1 = \bar{\mathbf{J}}^2$ (so $\ell g(\bar{M}^1) = \ell g(\bar{M}^2)$ in particular).
4)We say that $(N, \bar{M}, \bar{\mathbf{J}}) \in \mathscr{W}$ is prime <u>if</u> $(M_n, M_{n+1}, \mathbf{J}_n) \in K^{3,\rm qr}_{\mathfrak{s}}$ for $n < \ell g(\bar{M})$.

7.3 Claim. *1) $\leq_{\mathscr{W}}$ is a partial order.*
2) If $\delta < \lambda^+_{\mathfrak{s}}$ is a limit ordinal and $\langle (N^\alpha, \bar{M}^\alpha, \bar{\mathbf{J}}^\alpha) : \alpha < \delta \rangle$ is $\leq_{\mathscr{W}}$-increasing, <u>then</u> this sequence has a $\leq_{\mathscr{W}}$-lub $(N, \bar{M}, \mathbf{J})$, with $\ell g(\bar{M}) = \sup\{\ell g(\bar{M}^\alpha) : \alpha < \delta\}, N = \cup\{N^\alpha : \alpha < \delta\}, M_i = \cup\{M^\alpha_i : \alpha < \delta$ satisfies that $i < \ell g(\bar{M}^\alpha)\}, \mathbf{J}_i = \cup\{\mathbf{J}^\alpha_i : \alpha < \delta$ satisfies that $i < \ell g(\bar{M}^\alpha)\}$. We call this $\leq_{\mathscr{W}}$-lub the union of the chain.
3) If $(N^1, \bar{M}^1, \bar{\mathbf{J}}^1) \in \mathscr{W}_\alpha$ <u>then</u> for some $(N^2, \bar{M}^2)$ we have

- (α) $(N^1, \bar{M}^1, \bar{\mathbf{J}}^1) \leq^{\rm fx}_{\mathscr{W}} (N^2, \bar{M}^2, \bar{\mathbf{J}}^1) \in \mathscr{W}_\alpha$
- (β) $(M^2_i, M^2_{i+1}, \mathbf{J}^1_i) \in K^{3,\rm vq}_{\mathfrak{s}}$ *for each* $i < \ell g(\bar{M})$
- (γ) $N^2 = \cup\{M^2_i : i < \ell g(\bar{M}^2)\}$.

Proof. Straight: part (1) is trivial, part (2) holds by 5.10(3), and part (3) is proved repeating, e.g. the proof of 5.24 but using part (2) here. $\square_{7.3}$

We are interested in "nice" such sequences; we define several variants.

7.4 Definition. 1) $K^{\rm or}_{\mathfrak{s}} = \{(N, \bar{M}, \bar{\mathbf{J}}) \in \mathscr{W}_\omega$: if ($n < \omega$ and) $a \in \mathbf{J}_{n+1}$ <u>then</u> $\mathbf{tp}_{\mathfrak{s}}(a, M_{n+1}, M_{n+2})$ is orthogonal to $M_0\}$, if we omit N we mean $N = \cup\{M_n : n < \omega\}$.
2) $K^{\rm ar}_{\mathfrak{s}} = \{(N, \bar{M}, \bar{\mathbf{J}}) \in \mathscr{W}_\omega$: if $a \in \mathbf{J}_{n+1}$ <u>then</u> $\mathbf{tp}_{\mathfrak{s}}(a, M_{n+1}, M_{n+2})$ is orthogonal to $M_n\}$.
3) $K^{\rm br}_{\mathfrak{s}} = \{(N, \bar{M}, \bar{\mathbf{J}}) \in \mathscr{W}_\omega$: if $b \in \mathbf{J}_{n+1}$ then for some $m = m(b) \leq n$ we have $\mathbf{tp}_{\mathfrak{s}}(b, M_{n+1}, M_{n+2})$ does not fork over M_{m+1} and is orthogonal to $M_m\}$.
4) We say that $(N, \bar{M}, \bar{\mathbf{J}})$ is $K^{\rm or}_{\mathfrak{s}}$-fat <u>if</u> it belongs to $K^{\rm or}_{\mathfrak{s}}$ and[10] $p \underset{\rm wk}{\perp} M_0$ & $p \in \mathscr{S}^{\rm bs}_{\mathfrak{s}}(M_{n+1}) \Rightarrow \lambda_{\mathfrak{s}} = |\{c \in \mathbf{J}_{n+1} : p = \mathbf{tp}_{\mathfrak{s}}(c, M_{n+1}, M_{n+2})\}|$ and $N = \cup\{M_n : n < \omega\}$.

---

[10]Note that on $\mathbf{J}_0$ there are no demands; we can write $\perp$ here or "weakly orthogonal" above by 7.1(d).

5) We say that $(N, \bar{M}, \bar{\mathbf{J}})$ is $K^{\mathrm{ar}}_{\mathfrak{s}}$-fat if it $\in K^{\mathrm{ar}}_{\mathfrak{s}}$ and $p \in \mathscr{S}^{\mathrm{bs}}_{\mathfrak{s}}(M_{n+1})$ & $p \underset{\mathrm{wk}}{\perp} M_n \Rightarrow \lambda_{\mathfrak{s}} = |\{c \in \mathbf{J}_{n+1} : p = \mathbf{tp}_{\mathfrak{s}}(c, M_{n+1}, M_{n+2})\}|$ and $N = \cup\{M_n : n < \omega\}$.
6) We say that $(N, \bar{M}, \bar{\mathbf{J}})$ is $K^{\mathrm{br}}_{\mathfrak{s}}$-fat if it $\in K^{\mathrm{br}}_{\mathfrak{s}}$ and for every $m \leq n < \omega$ we have $p \in \mathscr{S}^{\mathrm{bs}}_{\mathfrak{s}}(M_{n+1})$ & p does not fork over M_{m+1} & $p \underset{\mathrm{wk}}{\perp} M_m \Rightarrow \lambda_{\mathfrak{s}} = |\{c \in \mathbf{J}_{n+1} : c \text{ realizes } p\}|$ and $N = \cup\{M_n : n < \omega\}$.
7) $\leq_{\mathrm{or}} = \leq^{\mathfrak{s}}_{\mathrm{or}}$ is the following two place relation over $K^{\mathrm{or}}_{\mathfrak{s}}$:

$$(N^1, \bar{M}^1, \bar{\mathbf{J}}^1) \leq_{\mathrm{or}} (N^2, \bar{M}^2, \bar{\mathbf{J}}^2) \text{ \underline{iff} } (N^1, \bar{M}^1, \bar{\mathbf{J}}^1) \leq_{\mathscr{W}} (N^2, \bar{M}^2, \bar{\mathbf{J}}^2) \text{ and } \mathbf{J}^1_0 = \mathbf{J}^2_0.$$

7.5 Definition. We say $\mathfrak{s}$ weakly has regulars when: if $\bar{M} = \langle M_\alpha : \alpha \leq \beta + 1\rangle$ is $\leq_{\mathfrak{s}}$-increasing continuous and $M_{\beta+1} \neq M_\beta$, <u>then</u> there are $\alpha < \beta$ and $c \in M_{\beta+1}$ such that $\mathbf{tp}_{\mathfrak{s}}(c, M_\beta, M_{\beta+1}) \in \mathscr{S}^{\mathrm{bs}}_{\mathfrak{s}}(M_\beta)$ does not fork over M_α and $\alpha = 0$ or $\alpha = \gamma + 1$ & $p \perp M_\gamma$ for some γ; this definition is meaningful for any good λ-frame.

Remark. 1) The name will be justified in claim 10.9(2), see also Definition 7.18.
2) If we are dealing with $\mathfrak{s}^{\kappa}_{T,\lambda}$, see 1.5(3), (so T is superstable first order complete theory) then (using in the proof regular types) this property holds.
3) Note that "fat" is closely related to "thick" from Definition 5.15. We do not use the same word as then in Definition 7.11(2) we get a contradiction.

7.6 Claim. *1) $K^{\mathrm{ar}}_{\mathfrak{s}} \subseteq K^{\mathrm{br}}_{\mathfrak{s}} \subseteq K^{\mathrm{or}}_{\mathfrak{s}}$.*
2) $\leq_{\mathrm{or}}$ is a partial order on $K^{\mathrm{or}}_{\mathfrak{s}}$; for an $\leq_{\mathrm{or}}$-increasing sequence of length $\delta < \lambda^+_{\mathfrak{s}}$, it has a $\leq_{\mathrm{or}}$-lub.
3) If $(N^\alpha, \bar{M}^\alpha, \bar{\mathbf{J}}^\alpha) \in K^{\mathrm{or}}_{\mathfrak{s}}$ for $\alpha < \delta < \lambda^+$ is $\leq_{\mathrm{or}}$-increasing, <u>then</u> its $\leq_{\mathscr{W}}$-lub (see 7.2) is its $\leq_{\mathrm{or}}$-lub (so it belongs to $K^{\mathrm{or}}_{\mathfrak{s}}$).
4) In part (2), if $(N^\alpha, \bar{M}^\alpha, \bar{\mathbf{J}}^\alpha) \in K^{\mathrm{ar}}_{\mathfrak{s}}$ for $\alpha < \delta$ is $\leq_{\mathrm{or}}$-increasing <u>then</u> the $\leq_{\mathrm{or}}$-lub belongs to $K^{\mathrm{ar}}_{\mathfrak{s}}$.

5) In part (2), if $(N^\alpha, \bar{M}^\alpha, \bar{\mathbf{J}}^\alpha) \in K^{\rm br}_{\mathfrak{s}}$ for $\alpha < \delta$ is $\le_{\rm or}$-increasing, then the $\le_{\rm or}$-lub (of this sequence) belongs to $K^{\rm br}_{\mathfrak{s}}$.
6) If $(N^1, \bar{M}^1, \bar{\mathbf{J}}^1) \in K^{\rm or}_{\mathfrak{s}}$ then there is a $K^{\rm or}_{\mathfrak{s}}$-fat $(N^2, \bar{M}^2, \bar{\mathbf{J}}^2)$ such that $(N^1, \bar{M}^1, \bar{\mathbf{J}}^1) \le_{\rm or} (N^2, \bar{M}^2, \bar{\mathbf{J}}^2)$.
7) If $(N^1, \bar{M}^1, \bar{\mathbf{J}}^1) \in K^{\rm ar}_{\mathfrak{s}}$ then there is a $K^{\rm ar}_{\mathfrak{s}}$-fat $(N^2, \bar{M}^2, \bar{\mathbf{J}}^2)$ such that $(N^1, \bar{M}^1, \bar{\mathbf{J}}^1) \le_{\rm or} (N^2, \bar{M}^2, \bar{\mathbf{J}}^2)$.
8) If $(N^1, \bar{M}^1, \bar{\mathbf{J}}^1) \in K^{\rm br}_{\mathfrak{s}}$ then there is a $K^{\rm br}_{\mathfrak{s}}$-fat $(N^2, \bar{M}^2, \bar{\mathbf{J}}^2)$ such that $(N^1, \bar{M}^1, \bar{\mathbf{J}}^1) \le_{\rm or} (N^2, \bar{M}^2, \bar{\mathbf{J}}^2)$.
9) Like parts (3), (4), (5) for $K^{\rm or}_{\mathfrak{s}}$-fat, $K^{\rm ar}_{\mathfrak{s}}$-fat, $K^{\rm br}_{\mathfrak{s}}$-fat triples.

Proof. Straight, e.g.
1) By the definition and monotonicity, i.e. Claim 6.10(4).
2) Concerning "$\le_{\rm or}$ is a partion order" read the definition. The second phrase follows from part 3).
3),4),5) The independence holds by 5.10 or use 7.3(2), the orthogonality holds by 6.10(2).
6),7),8) As in the proof of 5.16(8), so 4.7, recalling part (2) and its proof.
9) Easy, too. $\square_{7.6}$

7.7 Claim. *1) Assume that $(M, N, \mathbf{J}) \in K^{3,\rm vq}_{\mathfrak{s}}$ or at least $(*)_{(M,N,\mathbf{J})}$ below. Then we can find $(\bar{M}, \bar{\mathbf{J}})$ such that $(**)_{(M,N,\mathbf{J}),\bar{M},\bar{\mathbf{J}}}$ below holds, where*

$(*)_{(M,N,\mathbf{J})}$ *$(M, N, \mathbf{J}) \in K^{3,\rm bs}_{\mathfrak{s}}$ and for no N', b do we have $\mathbf{J} \subseteq N', M \le_{\mathfrak{s}} N' \le_{\mathfrak{s}} N, b \in N \backslash N'$ and $\mathbf{tp}(b, N', N) \underset{\rm wk}{\pm} M$*

$(**)_{(M,N,\mathbf{J}),\bar{M},\bar{\mathbf{J}}}$
- *(a)* $(M, N, \mathbf{J}) \in K^{3,\rm bs}_{\mathfrak{s}}$
- *(b)* $\bar{M} = \langle M_n : n < \omega \rangle, M_n \le_{\mathfrak{s}} M_{n+1}$
- *(c)* $M_0 = M$ *and* $\cup\{M_n : n < \omega\} = N$
- *(d)* $(\bar{M}, \bar{\mathbf{J}}) \in K^{\rm or}_{\mathfrak{s}}$
- *(e)* $(M_n, M_{n+1}, \mathbf{J}_n) \in K^{3,\rm qr}_{\mathfrak{s}}$
- *(f)* $\mathbf{J}_0 = \mathbf{J}$
- *(g)* *if $n < \omega$ and $b \in \mathbf{J}_{n+1}$ then* $\mathbf{tp}_{\mathfrak{s}}(b, M_{n+1}, M_{n+2}) \underset{\rm wk}{\perp} M_0$ *(follows by clause (d)).*

2) Assume $\mathfrak{s}$ weakly has regulars. If $(M, N, \mathbf{J}) \in K^{3,\rm bs}_{\mathfrak{s}}$ then we can find $\bar{M}, \bar{\mathbf{J}}$ such that (a)-(e),(g) above hold and

$(d)^+$ $(\bar{M}, \bar{\mathbf{J}}) \in K^{3,\mathrm{ar}}_{\mathfrak{s}}$

$(f)'$ $\mathbf{J} \subseteq \mathbf{J}_0$, *so: if* $\mathbf{J}$ *is maximal s.t.* $(M, N, \mathbf{J}) \in K^{3,\mathrm{bs}}_{\mathfrak{s}}$ *then they are equal*

$(g)'$ *if* $n < \omega, b \in \mathbf{J}_{n+1}$ *then* $\mathbf{tp}_{\mathfrak{s}}(b, M_{n+1}, M_{n+2}) \underset{\mathrm{wk}}{\perp} M_n$, *(follows by clause* $(d)^+$*).*

3) If (in part (2)) $\mathbf{J}$ *is maximal such that* $(M, N, \mathbf{J}) \in K^{3,\mathrm{bs}}_{\mathfrak{s}}$, *then* $(M, N, \mathbf{J}) \in K^{3,\mathrm{vq}}_{\mathfrak{s}}$.

Proof. 1) By 6.14(4), we know that $(*)_{(M,N,\mathbf{J})}$ holds in both cases.

We shall choose $M_n, \mathbf{J}_n$ by induction on n satisfying the relevant clauses in $(**)$. Let $M_0 = M$, let $\mathbf{J}_0 = \mathbf{J}$ and let $M_1 \leq_{\mathfrak{s}} N$ be such that $(M_0, M_1, \mathbf{J}) \in K^{3,\mathrm{qr}}_{\mathfrak{s}}$, exists by 5.8(1). If $M_n \leq_{\mathfrak{s}} N$ is well defined, $n \geq 1$ let $\mathbf{J}_n$ be a maximal subset of $\mathbf{I}_{M_n,N}$ independent in (M_n, N) such that $b \in \mathbf{J}_n \Rightarrow \mathbf{tp}_{\mathfrak{s}}(b, M_n, N) \perp M_0$.

Lastly, let $M_{n+1} \leq_{\mathfrak{s}} N$ be such that $(M_n, M_{n+1}, \mathbf{J}_n) \in K^{3,\mathrm{qr}}_{\mathfrak{s}}$, exists by 5.8(1). To finish we need to prove that $M_\omega := \bigcup_{n<\omega} M_n$ is equal to N. Clearly $M_\omega \leq_{\mathfrak{s}} N$, if $M_\omega \neq N$ then for some $b \in N \backslash M_\omega$ we have $\mathbf{tp}_{\mathfrak{s}}(b, M_\omega, N) \in \mathscr{S}^{\mathrm{bs}}_{\mathfrak{s}}(M_\omega)$. Now by $(*)_{(M,N,\mathbf{J})}$ clearly $\mathbf{tp}_{\mathfrak{s}}(b, M_\omega, N) \underset{\mathrm{wk}}{\perp} M_0$ and clearly for some $n < \omega$, $\mathbf{tp}_{\mathfrak{s}}(b, M_\omega, N)$ does not fork over M_n (and necessarily $n \geq 1$), and similarly (by 7.1(d)) we have $\mathbf{tp}_{\mathfrak{s}}(b, M_n, N) \underset{\mathrm{wk}}{\perp} M_0$ so b contradicts the choice of $\mathbf{J}_n$ (as "maximal such that ...", see 5.6(1)). So we are done.

2) Let $\mathbf{J}_0 \subseteq \mathbf{I}_{M,N}$ be maximal such that $(M, N, \mathbf{J}_0) \in K^{3,\mathrm{bs}}_{\mathfrak{s}}$ and $\mathbf{J} \subseteq \mathbf{J}_0$.

We repeat the proof of part (1) (except requiring $(g)'$ instead of (g)), till the proof that $M_\omega = N$. If $M_\omega \neq N$ stipulate $M_{\omega+1} = N$ and apply Definition 7.5 for $\beta = \omega$ and so there are $n = \alpha < \omega$ and c as there. If $n = 0$ we get contradiction to the choice of $\mathbf{J}_0$ and if $n > 0$ we get contradiction to the choice of $\mathbf{J}_n$.

3) By part (1) and 6.20(2). $\square_{7.7}$

7.8 Claim. *[$\mathfrak{s}$ weakly has regulars, see Definition 7.5].*

1) In 7.7(1) we can get

$(**)^+_{(M,N,\mathbf{J}),\bar{M},\bar{\mathbf{J}}}$ *(a)-(f) as in 7.7* *(d)$^+$ $(\bar M, \bar{\mathbf{J}}) \in K^{\rm ar}_{\mathfrak s}$ (i.e. we strengthen clause (d)).*

2) In 7.9 below we can add

$(B)^+$ like (B) there adding $(N, \bar M, \bar{\mathbf{J}}) \in K^{\rm ar}_{\mathfrak s}$.

Proof. 1) Similar to the proof of 7.7(2) noting that $\mathbf{J}$ is maximal such that $(M, N, \mathbf{J}) \in K^{3,\rm bs}_{\mathfrak s}$.
2) In 7.9 note that $(C) \Rightarrow (B)^+$ by 7.8(1) and $(B)^+ \Rightarrow (B)$ trivially. $\square_{7.8}$

Now we arrive to "understanding $K^{3,\rm vq}_{\mathfrak s}$"; this is reformulated in 12.6.

7.9 Theorem. *For every triple $(M, N, \mathbf{J})$, the following conditions are equivalent:*

- (A) $(M, N, \mathbf{J}) \in K^{3,\rm vq}_{\mathfrak s}$
- (B) *We can find $\bar M, \bar{\mathbf{J}}$ such that $(N, \bar M, \bar{\mathbf{J}}) \in K^{\rm or}_{\mathfrak s}, \mathbf{J}_0 = \mathbf{J}, M_0 = M, N = \bigcup_{n<\omega} M_n$ and $(M_n, M_{n+1}, \mathbf{J}_n) \in K^{3,\rm qr}_{\mathfrak s}$ (so in particular $(N, \bar M, \bar{\mathbf{J}})$ is prime, see Definition 7.2(4)), recall that by the definition of $K^{\rm or}_{\mathfrak s}$, $\mathbf{tp}_{\mathfrak s}(b, M_{n+1}, M_{n+2})$ is orthogonal to M_0 for every $n < \omega, b \in \mathbf{J}_{n+1}$*
- (C)
  - (a) *$M \le_{\mathfrak s} N$*
  - (b) *$\mathbf{J}$ is independent in (M, N)*
  - (c) *if $M \cup \mathbf{J} \subseteq N' \le_{\mathfrak s} N, b \in N \backslash N', \mathbf{tp}_{\mathfrak s}(b, N', N) \in \mathscr{S}^{\rm bs}_{\mathfrak s}(N')$ <u>then</u> $\mathbf{tp}_{\mathfrak s}(b, N', N) \underset{\rm wk}{\perp} M$*
- (D)
  - (a), (b) *as above*
  - (c) *if $N \le_{\mathfrak s} N^+, b \in N^+ \backslash M \backslash \mathbf{J}$ and $\mathbf{J} \cup \{b\}$ is independent in (M, N^+) <u>then</u> $\mathbf{tp}_{\mathfrak s}(b, N, N^+) \in \mathscr{S}^{\rm bs}_{\mathfrak s}(N)$ does not fork over M, (so in particular $b \notin N$)*
- (E) *there is a uq-decomposition $\langle M_i, a_j : i \le \alpha, j < \alpha\rangle$ inside (M_0, N) such that $(M, M_0, \mathbf{J}) \in K^{3,\rm vq}_{\mathfrak s}, M_\alpha = N$ and each $\mathbf{tp}_{\mathfrak s}(a_j, M_j, M_{j+1})$ is weakly orthogonal to M.*

From this we shall deduce (after the proof of 7.9):

7.10 Conclusion. If $\delta < \lambda_{\mathfrak{s}}^+$ and $\langle M_i : i \leq \delta \rangle$ is $\leq_{\mathfrak{s}}$-increasing continuous, and $\langle \mathbf{J}_i : i \leq \delta \rangle$ is $\subseteq$-increasing continuous and $(M, M_i, \mathbf{J}_i)$ belongs to $K_{\mathfrak{s}}^{3,\mathrm{vq}}$ for $i < \delta$, then $(M, M_\delta, \mathbf{J}_\delta)$ belongs to $K_{\mathfrak{s}}^{3,\mathrm{vq}}$.

Remark. See also 7.15 for more (changing the basis) and also 7.16.

Proof of 7.9. The following implications clearly suffice.

(A) ⇒ (E): Let $\alpha = 0, M_0 = N$.

(E) ⇒ (D): Clauses (a), (b) are obvious, so let us turn to (c). Assume b, N^+ are as in clause (c) of (D), so by Claim 5.16(5) we know that $\mathbf{tp}_{\mathfrak{s}}(b, M_0, N^+)$ does not fork over M; now we prove by induction on $i \leq \alpha$ that $\mathbf{tp}_{\mathfrak{s}}(b, M_i, N^+)$ does not fork over M. For $i = 0$ see above, for i limit use Axiom (E)(h), for i successor by the definition of orthogonality. For $i = \alpha$, $M_\alpha = N$, so we are done.

(C) ⇒ (B): By 7.7(1).

(B) ⇒ (A): By 6.20(2); here we use the hypothesis "super-orthogonality is equal to orthogonality", i.e. 7.1(e).

(A) ⇒ (D): (Actually not used). Clauses (a), (b) are obvious. For clause (c), as $\mathbf{tp}_{\mathfrak{s}}(b, M, N^+) \in \mathscr{S}_{\mathfrak{s}}^{\mathrm{bs}}(M)$, there is $M' \leq_{\mathfrak{s}} N^+$ such that $(M, M', b) \in K_{\mathfrak{s}}^{3,\mathrm{pr}}$ (recalling $\mathfrak{s}$ has primes). By 5.4(3) we know $\mathbf{J}$ is independent over (M, M', N^+) hence by Definition 5.15, we have $\mathrm{NF}_{\mathfrak{s}}(M, M', N, N^+)$ hence by 1.18 we get that $\mathbf{tp}_{\mathfrak{s}}(b, N, N^+) \in \mathscr{S}_{\mathfrak{s}}^{\mathrm{bs}}(N)$ does not fork over M as required.

(D) ⇒ (C):

Again the problem is to prove clause (c) of (C) so toward contradiction assume that $M \cup \mathbf{J} \subseteq N' <_{\mathfrak{s}} N$ and $b \in N \backslash N'$ and $p = \mathbf{tp}_{\mathfrak{s}}(b, N', N) \in \mathscr{S}_{\mathfrak{s}}^{\mathrm{bs}}(N')$ is not weakly orthogonal to M. So for some $q \in \mathscr{S}_{\mathfrak{s}}^{\mathrm{bs}}(M)$ we have $p \underset{\mathrm{wk}}{\pm} q$, and let $q_1 \in \mathscr{S}_{\mathfrak{s}}^{\mathrm{bs}}(N')$ be a non-forking extension of q. We can find N_2 such that $N' \cup \{b\} \subseteq N_2 \leq_{\mathfrak{s}} N$ and[11] $(N', N_2, b) \in K_{\mathfrak{s}}^{3,\mathrm{pr}}$. So (see 6.3) the type q_1 has some extension

[11] if we like to avoid using "$\mathfrak{s}$ has primes" here we find N'', N_2 such that $N \leq_{\mathfrak{s}} N'', N_2 \leq_{\mathfrak{s}} N''$ and $(N', N_2, b) \in K_{\mathfrak{s}}^{3,\mathrm{uq}}$.

$q_2 \in \mathscr{S}_{\mathfrak{s}}(N_2)$ which is not a non-forking extension of q, and so we can find N_4 and c such that $N \leq_{\mathfrak{s}} N_4$ and $q_2 = \mathbf{tp}_{\mathfrak{s}}(c, N_2, N_4)$. Now as c realizes q_1 clearly $\mathbf{tp}_{\mathfrak{s}}(c, N', N_4)$ does not fork over M hence $\mathbf{J} \cup \{c\}$ is independent in (M, N_4). By the choice of $q_2 \in \mathscr{S}_{\mathfrak{s}}(N_2)$, as c realizes q_2 clearly $\mathbf{tp}_{\mathfrak{s}}(c, N_2, N_4)$ forks over M, hence as $N_2 \leq_{\mathfrak{s}} N \leq_{\mathfrak{s}} N_4$ also $\mathbf{tp}_{\mathfrak{s}}(c, N, N_4)$ forks over M. So we have gotten a contradiction to clause (c) of (D) thus finishing. $\square_{7.9}$

Proof of 7.10. It is enough to check clause (D) of 7.9, now clause (a) is trivial, clause (b) holds by 5.10(3). For proving clause (c) we assume $M_\delta \leq_{\mathfrak{s}} N^+, b \in N^+ \backslash \mathbf{J} \backslash M$ and $\mathbf{J} \cup \{b\}$ is independent in (M, N^+), and we should prove that "$\mathbf{tp}_{\mathfrak{s}}(b, M_\delta, N^+)$ belongs to $\mathscr{S}^{\mathrm{bs}}_{\mathfrak{s}}(M_\delta)$ and does not fork over M". Now clearly for each $i < \delta$ the set $\mathbf{J}_i \cup \{b\}$ is independent in (M, N^+) by monotonicity of independence. Hence by 7.9 $(A) \Rightarrow (D)$ as we are assuming $(M, M_i, \mathbf{J}_i) \in K^{3,\mathrm{vq}}_{\mathfrak{s}}$ we can conclude that $\mathbf{tp}_{\mathfrak{s}}(b, M_i, N) \in \mathscr{S}^{\mathrm{bs}}_{\mathfrak{s}}(M_i)$ does not fork over M; so this holds for every $i < \delta$. Now $\mathbf{tp}_{\mathfrak{s}}(b, M_\delta, N^+)$ does not fork over M by Axiom (E)(h).

$\square_{7.10}$

* * *

We now try to show that there is a parallel to "N brimmed over M" among $\{(M, N, a) \in K^{3,\mathrm{uq}}_{\mathfrak{s}} : \mathbf{tp}_{\mathfrak{s}}(a, M, N) = p\}$

7.11 Definition. 1) If $M^* \in K_{\mathfrak{s}}, p^* \in \mathscr{S}^{\mathrm{bs}}_{\mathfrak{s}}(M^*)$ <u>then</u> let $K^{3,\mathrm{uq}}_{\mathfrak{s},p^*} = \{(M, N, a) \in K^{3,\mathrm{uq}}_{\mathfrak{s}} : M = M^*$ and $p^* = \mathbf{tp}_{\mathfrak{s}}(a, M, N)\}$.
1A) If $M^* \in K_{\mathfrak{s}}$ and $\bar{p} = \langle p_t : t \in I\rangle$ is a sequence of members of $\mathscr{S}^{\mathrm{bs}}_{\mathfrak{s}}(M^*)$ and $|I| \leq \lambda$, <u>then</u> $K^{3,\mathrm{vq}}_{\mathfrak{s},\bar{p}} = \{(M, N, \mathbf{J}) \in K^{3,\mathrm{vq}}_{\mathfrak{s}} : M = M^*$ and $\mathbf{J} = \{a_t : t \in I\}$ with no repetitions such that $t \in I \Rightarrow p_t = \mathbf{tp}_{\mathfrak{s}}(a_t, M, N)\}$.
2) We say $(M, N, \mathbf{J}) \in K^{3,\mathrm{vq}}_{\mathfrak{s}}$ is fat or is $K^{3,\mathrm{vq}}_{\mathfrak{s}}$-fat <u>if</u> there is a fat $(N, \bar{M}, \bar{\mathbf{J}}) \in K^{3,\mathrm{or}}_{\mathfrak{s}}$ satisfying $\mathbf{J}_0 = \mathbf{J}, M_0 = M$ and $\cup\{M_n : n < \omega\} = N$. If $\mathbf{J} = \{a\}$ we may write a instead of $\mathbf{J}$ and say (M, N, a) is $K^{3,\mathrm{uq}}_{\mathfrak{s}}$-fat and if $p = \mathbf{tp}_{\mathfrak{s}}(a, M, N)$ we may say (M, N, a) is $K^{3,\mathrm{uq}}_{\mathfrak{s},p}$-fat. We define "$K^{3,\mathrm{vq}}_{\mathfrak{s},\bar{p}}$-fat" similarly.

7.12 Universality/Uniqueness Claim.
1) If $(M,N,a) \in K^{3,\mathrm{uq}}_{\mathfrak{s},p^*}$ and $(N^1,\bar{M}^1,\bar{\mathbf{J}}^1)$ is $K^{\mathrm{or}}_{\mathfrak{s}}$-fat, see Definition 7.4(4), $\mathbf{J}^1_0 = \{a^*\}, (M^1_0, M^1_1, a^*) \in K^{3,\mathrm{uq}}_{\mathfrak{s},p^*}$ and $M = M^1_0$ then there is a $\leq_{\mathfrak{s}}$-embedding f of N into $N^1 = \cup\{M^1_n : n < \omega\}$ over M mapping a to a^*.
2) If $(N^\ell, \bar{M}^\ell, \bar{\mathbf{J}}^\ell)$ is $K^{\mathrm{or}}_{\mathfrak{s}}$-fat, $\mathbf{J}^\ell_0 = \{a_\ell\}, (M^\ell_0, N^\ell, \mathbf{J}^\ell_0) \in K^{3,\mathrm{vq}}_{\mathfrak{s}}$, $\mathbf{tp}_{\mathfrak{s}}(a_\ell, M^\ell_0, M^\ell_1) = p$ for $\ell = 1,2$ (so $M^\ell_0 = \mathrm{Dom}(p)$ does not depend on ℓ). Then there is an isomorphism f from N^1 onto N^2 over $\mathrm{Dom}(p)$ which maps a_1 to a_2.
3) Similar to (2) with $\mathbf{J}^\ell_0 = \mathbf{J}, M^\ell_0 = M_0$ and $\mathbf{tp}_{\mathfrak{s}}(c, M^1_0, M^1_1) = \mathbf{tp}_{\mathfrak{s}}(c, M^2_0, M^2_1)$ for $c \in \mathbf{J}$ for $\ell = 1,2$.
4) If $(M,N,\mathbf{J}) \in K^{3,\mathrm{vq}}_{\mathfrak{s}}$ and $(N^1,\bar{M}^1,\bar{\mathbf{J}}^1)$ is $K^{\mathrm{or}}_{\mathfrak{s}}$-fat and $M = M^1_0, \mathbf{J} = \mathbf{J}^1_0$ and $c \in \mathbf{J} \Rightarrow \mathbf{tp}_{\mathfrak{s}}(c,M,N) = \mathbf{tp}_{\mathfrak{s}}(c, M^1_0, N^1)$ then there is an embedding of N into $\cup\{M^1_n : n < \omega\}$ which is the identity on $M \cup \mathbf{J}$.

Remark. 1) In 7.12 we can make stronger demands on f. In part (1) if $M \cup \{a\} \subseteq N' \leq_{\mathfrak{s}} N, f'$ a $\leq_{\mathfrak{s}}$-embedding of N' into M^1_n for some $n < \omega, f' \subseteq \mathrm{id}_M, f'(a) = a^*$, then we can require $f' \subseteq f$.
2) In part (2) of 7.12, if $M^\ell_0 \cup \{a_\ell\} \subseteq M'_\ell \leq M^\ell_n$ for $\ell = 1,2, f'$ an isomorphism from M'_1 onto M'_2 extending $\mathrm{id}_{\mathrm{Dom}(p)} \cup \{\langle a_1, a_2\rangle\}$ then we can require $f' \subseteq f$.
3) We can in 7.12 use $K^{\mathrm{ar}}_{\mathfrak{s}}, K^{\mathrm{br}}_{\mathfrak{s}}$ instead of $K^{\mathrm{or}}_{\mathfrak{s}}$ but then parts (1),(2) of the remark may fail.

Proof. 1) By 7.9 $(A) \Rightarrow (B)$, we can find $(N,\bar{M},\bar{\mathbf{J}}) \in K^{\mathrm{or}}_{\mathfrak{s}}$ with $M_0 = M, \mathbf{J}_0 = \{a\}$, as in 7.9, clause (B) so $N = \cup\{M_n : n < \omega\}$ and $(M_n, M_{n+1}, \mathbf{J}_n) \in K^{3,\mathrm{qr}}_{\mathfrak{s}}$ for $n < \omega$. Now we choose by induction on $n < \omega$ a $\leq_{\mathfrak{s}}$-embedding f_n of M_n into M^1_n increasing with $n, f_0 = \mathrm{id}_{M_0}, f_1(a) = a^*$. For $n = 0$ this is trivial, for $n = 1$ note that $\mathbf{tp}_{\mathfrak{s}}(a, M, M_1) = \mathbf{tp}_{\mathfrak{s}}(a^*, M, M^1_1)$ and recall the definition of $(M, M_1, a) \in K^{3,\mathrm{pr}}_{\mathfrak{s}}$. For $n = m+1 > 1$, for every $b \in \mathbf{J}_m$, by the definition of $K^{\mathrm{or}}_{\mathfrak{s}}, p := \mathbf{tp}_{\mathfrak{s}}(b, M_m, M_{m+1}) \perp M_0$. So the non-forking extension of $f_m(p)$ to M^1_m is orthogonal to M_0. So by the definition of "fat" in 7.4(4) we can find a one-to-one mapping h_m from $\mathbf{J}_m$ into $\mathbf{J}^1_m$ such that

(i) $b \in \mathbf{J}_m \Rightarrow \mathbf{tp}_{\mathfrak{s}}(h_m(b), M^1_m, M^1_{m+1})$ does not fork over $\mathrm{Rang}(f_m)$

(ii) $b \in \mathbf{J}_m \Rightarrow f_m(\mathbf{tp}_{\mathfrak{s}}(b, M_m, M_{m+1})) = \mathbf{tp}_{\mathfrak{s}}(h_m(b), \mathrm{Rang}(f_m), M^1_{m+1})$.

Then choose a $\leq_{\mathfrak{s}}$-embedding f_n of M_n into M^1_n satisfying $f_n \supseteq f_m, f_m(c) = h_m(c)$ for $c \in \mathbf{J}_m$, this is possible by the definition of $(M_m, M_{m+1}, \mathbf{J}_m) \in K^{3,\mathrm{qr}}_{\mathfrak{s}}$. Having carried the induction, $f := \cup\{f_n : n < \omega\}$ is as required.

2) We choose by induction on n a tuple $(N^1_n, N^2_n, f_n, \mathbf{I}^1_n)$ such that (with $\ell \in \{1, 2\}$):

(a) $N^\ell_n \leq_{\mathfrak{s}} M^\ell_n$ for $\ell = 1, 2$

(b) f_n is an isomorphism from N^1_n onto N^2_n

(c) $N^\ell_n \leq_{\mathfrak{s}} N^\ell_{n+1}$ and $f_n \subseteq f_{n+1}$ for $\ell = 1, 2$

(d) $N^\ell_0 = M^\ell_0$ and f_0 is the identity

(e) $(N^1_0, N^1_1, \mathbf{J}^1_0) \in K^{3,\mathrm{qr}}_{\mathfrak{s}}$ and $f_1(a_1) = a_2$

(f) if $n = \ell$ mod 2 (where $\ell \in \{1, 2\}$) and $n \geq 1$ then

- (α) $\mathbf{I}_n$ is a maximal subset of $\{b \in \mathbf{I}_{N^\ell_n, M^\ell_{n+1}} : \mathbf{tp}_{\mathfrak{s}}(b, N^\ell_n, M^\ell_{n+1}) \perp M_0\}$ which is independent in (N^ℓ_n, M^ℓ_{n+1})
- (β) $(N^\ell_n, N^\ell_{n+1}, \mathbf{I}_n) \in K^{3,\mathrm{qr}}_{\mathfrak{s}}$
- $(\gamma)_1$ if $\ell = 1$, $f_{n+1} \restriction \mathbf{I}_n$ is a one-to-one mapping from $\mathbf{I}_n$ into $\mathbf{J}^2_{n+1}$
- $(\gamma)_2$ if $\ell = 2$, $f^{-1}_{n+1} \restriction \mathbf{I}_n$ is a one-to-one mapping from $\mathbf{I}_n$ into $\mathbf{J}^1_{n+1}$.

There is no problem to carry the induction. Let $N_\ell := \cup\{N^\ell_n : n < \omega\}$ so $N_\ell \leq_{\mathfrak{s}} N^\ell$ recalling $N^\ell = \cup\{M^\ell_n : n < \omega\}$ by Definition 7.4(4) and $f := \bigcup_{n<\omega} f_n$ is an isomorphism from N_1 onto N_2. We shall show that they are as required as in the proof of 7.7. That is, we prove that $N_\ell = N^\ell$, if not we have $N_\ell := \cup\{N^\ell_n : n < \omega\} <_{\mathfrak{s}} N^\ell$ so we can find $b \in N^\ell \backslash N_\ell$ such that $\mathbf{tp}_{\mathfrak{s}}(b, N_\ell, N^\ell) \in \mathscr{S}^{\mathrm{bs}}_{\mathfrak{s}}(N_\ell)$, hence for some $n_1 < \omega$, this type does not fork over N^ℓ_n and for some

$n_2 < \omega, b \in M^\ell_{n_2}$. By 7.7 clearly $(M, N^0, \mathbf{J}^\ell_0) \in K^{3,\mathrm{vq}}_{\mathfrak{s}}$, hence by 6.14(1) clearly $\mathbf{tp}_{\mathfrak{s}}(b, N_\ell, N^\ell) \perp M = M_0$. Choose $n < \omega, n = \ell \mod 2, n > n_1, n > n_2$ and we could have added b to $\mathbf{I}_n$, contradiction.
3), 4) Similarly. $\square_{7.12}$

7.13 Conclusion. 1) If $(M, N_\ell, a) \in K^{3,\mathrm{uq}}_{\mathfrak{s},p}$ is fat for $\ell = 1, 2$ then N_1, N_2 are isomorphic over $M \cup \{a\}$.
2) Similarly for $K^{3,\mathrm{vq}}_{\mathfrak{s},\bar{p}}$.
3) If $(M, N, \mathbf{J}) \in K^{3,\mathrm{bs}}_{\mathfrak{s}}$ then for some $N' \leq_{\mathfrak{s}} N''$ we have $\mathbf{J} \subseteq N'$ and $N \leq_{\mathfrak{s}} N''$ and $(M, N', \mathbf{J})$ is $K^{3,\mathrm{vq}}_{\mathfrak{s}}$-fat.

Proof. 1),2) By the proof of 7.12(2) (note that under a stronger assumption in 7.14 below (i.e. $\mathfrak{s} = \mathfrak{t}^+$) we get uniqueness even without assuming fatness).
3) Easy. $\square_{7.13}$

7.14 Claim. *[$\mathfrak{s} = \mathfrak{t}^+$, $\mathfrak{t}$ is a good$^+$ and successful frame.]*
1) $K^{3,\mathrm{uq}}_{\mathfrak{s}} = K^{3,\mathrm{pr}}_{\mathfrak{s}}$; so together with 4.14(2) we get uniqueness for $K^{3,\mathrm{uq}}_{\mathfrak{s}}$.
2) $(M, N, \mathbf{J}) \in K^{3,\mathrm{qr}}_{\mathfrak{s}}$ iff $(M, N, \mathbf{J}) \in K^{3,\mathrm{vq}}_{\mathfrak{s}}$ so together with 5.10(5)(2) we get uniqueness for $K^{3,\mathrm{vq}}_{\mathfrak{s}}$.
3) $(M, N, \mathbf{J}) \in K^{3,\mathrm{qr}}_{\mathfrak{s}}$ iff $(M, N, \mathbf{J})$ belongs to $K^{3,\mathrm{vq}}_{\mathfrak{s}}$ and is fat.

Question: Can we assume less than $\mathfrak{s} = \mathfrak{t}^+$?

Proof of 7.14. 1) This is a special case of (2).
2) Note that by changing $\mathfrak{t}$ without loss of generality $K_{\mathfrak{t}}$ is categorical hence it satisfies 7.1. The "only if" implication we already proved in 3.7(2), more exactly 5.16(1). For the other direction assume $(M, N, \mathbf{J}) \in K^{3,\mathrm{vq}}_{\mathfrak{s}}$ and by 7.7(1) applied to $(M, N, \mathbf{J})$ we get $(\bar{M}, \bar{\mathbf{J}})$ satisfying $(**)_{(M,N,\mathbf{J}),\bar{M},\bar{\mathbf{J}}}$ of 7.7(1) which means that it is as in clause (B) of 7.9 in particular $\mathbf{J}_0 = \mathbf{J}$ and $(M_n, M_{n+1}, \mathbf{J}_n) \in K^{3,\mathrm{qr}}_{\mathfrak{s}}$ for $n < \omega$ and $b \in \mathbf{J}_{n+1} \Rightarrow \mathbf{tp}_{\mathfrak{s}}(b, M_{n+1}, M_{n+2}) \underset{\mathrm{wk}}{\perp} M_0$ hence by 7.1 we get $b \in \mathbf{J}_{n+1} \Rightarrow \mathbf{tp}_{\mathfrak{s}}(b, M_{n+1})$. Let $\langle M^\beta_i : i < \lambda_{\mathfrak{s}} \rangle$ be $\leq_{\mathfrak{t}}$-representations

of M_β for $\beta \le \omega$. Now by 5.22 there is a club E of $\lambda_{\mathfrak{s}} = \lambda_{\mathfrak{t}}^+$ such that ($\mathrm{NF}_{\mathfrak{t}}(M^n_\alpha, M^{n+1}_\alpha, M^n_\beta, M^{n+1}_\beta)$ for every $n < \omega$ and $\alpha < \beta$ from E and) for $\delta \in E$ the triple $(M^n_\delta, M^{n+1}_\delta, \mathbf{J}_n \cap M^{n+1}_\delta)$ belongs to $K^{3,\mathrm{vq}}_{\mathfrak{t}}$ for each $n < \omega$.

Let $\delta \in E$ and $b \in \mathbf{J}_{n+1} \cap M^{n+2}_\delta$. By 6.11(2) (reflecting non-super-orthogonality) for some $\varepsilon \in [\delta, \lambda_{\mathfrak{s}})$ we have $\mathbf{tp}_{\mathfrak{t}}(b, M^{n+1}_\varepsilon, M^{n+2}_\varepsilon) \underset{\mathrm{su}}{\perp} M^0_\varepsilon$. But $\mathbf{tp}_{\mathfrak{t}}(b, M^{n+1}_\varepsilon, M^{n+2}_\varepsilon)$ does not fork over M^{n+1}_δ, so by monotonicity (6.10(4)) we have $\mathbf{tp}_{\mathfrak{t}}(b, M^{n+1}_\delta, M^{n+2}) \underset{\mathrm{su}}{\perp} M$.

So by 6.20(2) we can deduce that for $\delta \in E$ we have $(M^0_\delta, M^\omega_\delta, \mathbf{J} \cap M^\omega_\delta) \in K^{3,\mathrm{vq}}_{\mathfrak{t}}$.

Hence (see 4.9 as $M_0, M_\omega \in K_{\mathfrak{s}}$, more exactly by 4.13(1) if $\mathbf{J}$ is a singleton, by 5.22 in general) the triple $(M_0, M_\omega, \mathbf{J})$ belongs to $K^{3,\mathrm{qr}}_{\mathfrak{s}}$, as required.

3) The "if" direction is trivial by part (2).

For the "only if" part we can use the uniqueness from part (2), but we give a direct proof. Assume $(M, N, \mathbf{J}) \in K^{3,\mathrm{qr}}_{\mathfrak{s}}$. By part (2) we have $(M, N, \mathbf{J}) \in K^{3,\mathrm{vq}}_{\mathfrak{s}}$ and again by claim 5.23(2) (uniqueness of $K^{3,\mathrm{qr}}_{\mathfrak{s}}$) it is enough to find N' such that

⊛ (a) $(M, N', \mathbf{J}) \in K^{3,\mathrm{vq}}_{\mathfrak{s}}$

(b) $\mathbf{tp}_{\mathfrak{s}}(c, M, N') = \mathbf{tp}_{\mathfrak{s}}(c, M, N)$ for $c \in \mathbf{J}$

(c) $(M, N', \mathbf{J})$ is fat.

By 7.13(3) this holds so we are done. $\square_{7.14}$

7.15 Claim. *Assume $(M_i, N_i, \mathbf{J}_i) \in K^{3,\mathrm{vq}}_{\mathfrak{s}}$ for $i < \delta$ where $\delta < \lambda^+_{\mathfrak{s}}$. Assume further that $\langle M_i : i < \delta \rangle$ is $\le_{\mathfrak{s}}$-increasing continuous, $\langle N_i : i < \delta \rangle$ is $\le_{\mathfrak{s}}$-increasing continuous and $\langle \mathbf{J}_i : i < \delta \rangle$ is $\subseteq$-increasing continuous and $i < j < \delta$ & $c \in \mathbf{J}_i \Rightarrow \mathbf{tp}_{\mathfrak{s}}(c, M_j, N_j)$ does not fork over M_i. Let $M_\delta = \cup\{M_i : i < \delta\}, N_\delta = \cup\{N_i : i < \delta\}$ and $\mathbf{J}_\delta = \cup\{\mathbf{J}_i : i < \delta\}$. Then $(M_\delta, N_\delta, \mathbf{J}_\delta) \in K^{3,\mathrm{vq}}_{\mathfrak{s}}$.*

Remark. 1) We should compare this claim to 5.17. Here we assume less on $\langle (M, M_i, \mathbf{J}_i) : i \le \delta \rangle$, as M_i being $<^*_{\mathfrak{s}}$-increasing is not demanded here. However, we assume more on $\mathfrak{s}$ as the hypothesis of

this section is stronger.
2) Compare also with 8.21, there we do not assume that $\mathfrak{s}$ has primes.

Proof. 1) Note that $\mathbf{J}_\delta$ is independent in (M_δ, N_δ) by 5.10(3).

We shall use Claim 7.9, our desired conclusion is clause (A) for $(M_\delta, N_\delta, \mathbf{J}_\delta)$ so it is enough to check clause (D). So let $M_\delta \leq_{\mathfrak{s}} N^+_\delta, b \in N^+_\delta \backslash \mathbf{J}_\delta \backslash M_\delta$ and assume that $\mathbf{J}_\delta \cup \{b\}$ is independent in (M_δ, N^+_δ). So $\mathbf{tp}_{\mathfrak{s}}(b, M_\delta, N^+_\delta) \in \mathscr{S}^{\mathrm{bs}}_{\mathfrak{s}}(M_\delta)$ hence for some $i(*) < \delta$ the type $\mathbf{tp}_{\mathfrak{s}}(b, M_\delta, N^+_\delta)$ does not fork over $M_{i(*)}$. It is enough to prove that for every $i \in [i(*), \delta)$, the type $\mathbf{tp}_{\mathfrak{s}}(b, N_i, N^+_\delta) \in \mathscr{S}^{\mathrm{bs}}_{\mathfrak{s}}(N_i)$ does not fork over M_i. (Why? Recalling $\mathbf{tp}_{\mathfrak{s}}(b, M_i, N^+_i)$ does not fork over $M_{i(*)}$, by transitivity of non-forking $\mathbf{tp}_{\mathfrak{s}}(b, N_i, N^+_\delta)$ does not fork over $M_{i(*)}$; as this is true for every $i \in [i(*), \delta)$ and $\langle N_i : i \in [t(*), \delta] \rangle$ is $\leq_{\mathfrak{s}}$-increasing continuous, clearly then also $\mathbf{tp}_{\mathfrak{s}}(b, N_\delta, N^+_\delta)$ does not fork over $M_{i(*)}$ hence over M_δ as required). Let i be any ordinal $\in (i(*), \delta)$, so by monotonicity $\mathbf{J}_i \cup \{b\}$ is independent in (M_δ, N^+); as $\mathbf{J}_i \cup \{b\} \subseteq \mathbf{J}_\delta \cup \{b\}$. As $c \in \mathbf{J}_i \cup \{b\} \Rightarrow \mathbf{tp}_{\mathfrak{s}}(c, M_\delta, N^+_\delta)$ does not fork over M_i it follows that $\mathbf{J}_i \cup \{b\}$ is independent in $(M_i, M_\delta, N^+_\delta)$ hence in (M_i, N^+_δ). As $(M_i, N_i, \mathbf{J}_i) \in K^{3,\mathrm{vq}}_{\mathfrak{s}}$ by claim 7.9 clearly $\mathbf{tp}_{\mathfrak{s}}(b, N_i, N^+_\delta)$ does not fork over M_i, and as said earlier this suffices. $\square_{7.15}$

7.16 Claim. *Assume $\langle M_i : i \leq \alpha \rangle$ is $\leq_{\mathfrak{s}}$-increasing continuous and $(M_i, M_{i+1}, \mathbf{J}_i) \in K^{3,\mathrm{vq}}_{\mathfrak{s}}$ and $\mathbf{J}_i$ is independent in (M_0, M_i, M_{i+1}) for $i < \alpha$. Then $(M_0, M_\alpha, \cup\{\mathbf{J}_i : i < \alpha\}) \in K^{3,\mathrm{vq}}_{\mathfrak{s}}$.*

Remark. Compare with 7.10.

Proof. We prove the statement by induction on α. Let $\mathbf{J} = \cup\{\mathbf{J}_i : i < \alpha\}$. First note that $\mathbf{J}$ is independent in (M_0, M_α) by 5.10(3). To prove that $(M_0, M_\alpha, \mathbf{J}) \in K^{3,\mathrm{vq}}_{\mathfrak{s}}$ by 7.9 it suffices to prove clause (D) there, so assume toward contradiction that $M_\alpha \leq_{\mathfrak{s}} N$ and $b \in N \backslash \mathbf{J} \backslash M_0$ and $\mathbf{J} \cup \{b\}$ is independent in (M_0, N). If $\alpha = 0$ this is trivial. For α limit, note that each $i < \alpha$, clearly $\{\mathbf{J}_j : j < i\} \cup \{b\}$ is independent in (M_0, N^+) hence by the induction hypothesis and 7.9 we know that $\mathbf{tp}_{\mathfrak{s}}(b, M_i, N)$ does not fork over M_i; as this holds for

every $i < \alpha$, we can deduce that $\mathbf{tp}_{\mathfrak{s}}(b, M_\alpha, N)$ does not fork over M_0 as required.
So we are left with the case $\alpha = \beta + 1$. So $\mathbf{J} \cup \{b\}$ is independent in (M_0, N) hence so is $\bigcup\limits_{i<\beta} \mathbf{J}_i \cup \{b\}$ and by the induction hypothesis for β the type $\mathbf{tp}_{\mathfrak{s}}(b, M_\beta, N)$ does not fork over M_0 hence $b \notin M_\beta$. Hence clearly $(\mathbf{J} \cup \{b\}) \cap M_\beta = \cup\{\mathbf{J}_i : i < \beta\}$, so by 5.16(6) we know that $\mathbf{J} \cup \{b\} \backslash M_\beta = \mathbf{J}_\beta \cup \{b\}$ is independent in (M_0, M_β, N). As $(M_\beta, M_\alpha, \mathbf{J}_\beta) \in K^{3,\mathrm{vq}}_{\mathfrak{s}}$ by 5.16(5), $\mathbf{tp}_{\mathfrak{s}}(b, M_\alpha, N)$ does not fork over M_β hence over M_0. $\square_{7.16}$

The following claim will be used in §12; this is a strengthening of "$\mathfrak{s}$ weakly has regulars" replacing a chain by a finite partial order. Note that the conclusion of 7.17 is given a name in 7.18.

7.17 Claim. *Assume $\mathfrak{s}$ weakly has regulars; (see Definition 7.5, enough for sequence of length 4), I any set.*
1) Assume

- *(a) $u_\ell \subseteq I$ is finite, $M_\ell \leq_{\mathfrak{s}} M \leq_{\mathfrak{s}} N$ for $\ell < n$ and $u_\ell \subset u_m \Rightarrow M_\ell \leq_{\mathfrak{s}} M_m$*
- *(b) $\mathbf{J}$ satisfies*
  - $(\alpha)_{\mathbf{J}}$ *$(M, N, \mathbf{J}) \in K^{3,\mathrm{bs}}_{\mathfrak{s}}$*
  - $(\beta)_{\mathbf{J}}$ *if $a \in \mathbf{J}, \ell < n$ and $\mathbf{tp}_{\mathfrak{s}}(a, M, N) \underset{\mathrm{wk}}{\pm} M_\ell$ then for some $k < n$ the type $\mathbf{tp}_{\mathfrak{s}}(a, M, N)$ does not fork over M_k and is $\underset{\mathrm{wk}}{\perp} M_m$ whenever $u_m \subset u_k$.*

<u>Then</u> there is $\mathbf{J}' \supseteq \mathbf{J}$ such that $(\alpha)_{\mathbf{J}'}, (\beta)_{\mathbf{J}'}$ and $\mathbf{J}'$ is maximal under those conditions.
1A) In part (1) if $\mathbf{J}'$ is any set satisfying its conclusions <u>then</u>

$(\gamma)_{\mathbf{J}'}$ $(M, N, \mathbf{J}') \in K^{3,\mathrm{vq}}_{\mathfrak{s}}$.

2) Assume $M_\ell \leq_{\mathfrak{s}} M \leq_{\mathfrak{s}} M' <_{\mathfrak{s}} N, u_\ell \subseteq I$ finite for $\ell < n$ and $u_\ell \subset u_m \Rightarrow M_\ell \leq_{\mathfrak{s}} M_m$ and for some $a \in \mathbf{I}_{M',N}$, $\mathbf{tp}_{\mathfrak{s}}(a, M', N)$ is not (weakly) orthogonal to M (and $\in \mathscr{S}^{\mathrm{bs}}_{\mathfrak{s}}(M')$). <u>Then</u> for some

$a \in \mathbf{I}_{M',N}$ the type $\mathbf{tp}_{\mathfrak{s}}(a, M', N)$ does not fork over M and is weakly orthogonal to M_ℓ for $\ell < n$ or for some $k < n$ does not fork over M_k and either is weakly orthogonal to M_ℓ whenever $u_\ell \subset u_k$.
3) If $\langle M_i : i \leq \alpha \rangle$ is $\leq_{\mathfrak{s}}$-increasing continuous, $\beta < \alpha$ and $M_\alpha <_{\mathfrak{s}} N$ and for some $a \in N \backslash M_\alpha$ the type $\mathbf{tp}_{\mathfrak{s}}(a, M_\alpha, N) \in \mathscr{S}^{\mathrm{bs}}_{\mathfrak{s}}(M_\alpha)$ is not (weakly) orthogonal to M_β <u>then</u> for some non-limit $\gamma \leq \beta$ and $b \in N \backslash M_\alpha$ we have:

(a) $\mathbf{tp}_{\mathfrak{s}}(b, M_\gamma, N)$ does not fork over M_γ

(b) $\mathbf{tp}_{\mathfrak{s}}(b, M_\gamma, N)$ is (weakly) orthogonal to $M_{\gamma'}$ for every $\gamma' < \gamma$.

Remark. In parts (1),(2) we can allow $\langle M_i : i < \beta \rangle, \langle u_i : i < \beta \rangle, \beta, \alpha$ not necessarily finite if we can find $n < \omega, S_\ell \subseteq \beta$ for $\ell < n$ such that $\beta = \cup \{S_\ell : \ell < n\}$ and $\langle u_i : i \in S_\ell \rangle$ is increasing (see in 12.18).

Proof. 1) The conclusion is obvious by 5.4(1), i.e. $(*)_0 \Leftrightarrow (*)_1$ there.
1A) If not, then by $(C) \Leftrightarrow (A)$ from claim 7.9 there are N', b such that $M \cup \mathbf{J} \subseteq N' <_{\mathfrak{s}} N$ and $b \in N \backslash \mathbf{J} \backslash N'$ such that $\mathbf{tp}_{\mathfrak{s}}(b, N', N) \in \mathscr{S}^{\mathrm{bs}}_{\mathfrak{s}}(N')$ is $\pm M$. Hence by part (2) of the present claim there is $a \in N \backslash N'$ such that one of the following cases holds.

<u>Case 1</u>: $\mathbf{tp}_{\mathfrak{s}}(a, N', N)$ does not fork over M and is $\perp M_\ell$ for $\ell < n$.

<u>Case 2</u>: For some $k < n$, the type $\mathbf{tp}_{\mathfrak{s}}(a, N', N)$ does not fork over M_k but $\ell < n \wedge u_\ell \subset u_k \Rightarrow \mathbf{tp}_{\mathfrak{s}}(a, N', N) \perp M_\ell$.

Now both cases contradict the assumption that "$\mathbf{J}'$ is a maximal set such that ..." as exemplified by $\mathbf{J}' \cup \{a\}$.
2) We prove this by induction on n. We can easily choose an ordinal $\alpha < \lambda^+$ and $\langle N'_i : i \leq \alpha \rangle, \langle b_i : i < \alpha \rangle$ such that

⊛ (a) $N'_i \leq_{\mathfrak{s}} N$ is $\leq_{\mathfrak{s}}$-increasing continuous

(b) $N'_0 = M'$

(c) $(N'_i, N'_{i+1}, b_i) \in K^{3,\mathrm{pr}}_{\mathfrak{s}}$

(d) $\mathbf{tp}_{\mathfrak{s}}(b_i, N'_i, N'_{i+1})$ is orthogonal to M

(e) for no $a \in N \backslash N'_\alpha$ is $\mathbf{tp}_{\mathfrak{s}}(a, N'_\alpha, N) \in \mathscr{S}^{\mathrm{bs}}_{\mathfrak{s}}(N'_\alpha)$ orthogonal to M.

If $q \in \mathscr{S}^{\rm bs}_{\mathfrak{s}}(M') = \mathscr{S}^{\rm bs}_{\mathfrak{s}}(N'_0)$ does not fork over M then we can prove by induction on $i \le \alpha$ that q has a unique extension in $\mathscr{S}_{\mathfrak{s}}(N'_i)$. Hence $a \in \mathbf{I}_{M',N'_\alpha} \Rightarrow \mathbf{tp}_{\mathfrak{s}}(a, M', N'_\alpha) \perp M$ so if $N'_\alpha = N$ it follows that the assumption never holds, so assume $N \ne N'_\alpha$. Without loss of generality $u_\ell \subseteq u_k \Rightarrow \ell \le k$. Now we prove by induction on $m \le n$ that for some $a_m \in N \backslash N'_\alpha$

(∗) the type $\mathbf{tp}_{\mathfrak{s}}(a_m, N'_\alpha, N) \in \mathscr{S}^{\rm bs}_{\mathfrak{s}}(N'_\alpha)$ does not fork over M and is orthogonal to $M_0, \dots, M_{m-1}$.

For $m = 0$ note that $M \le_{\mathfrak{s}} N'_\alpha <_{\mathfrak{s}} N$, so by the assumption "$\mathfrak{s}$ weakly has regulars" there is $a_0 \in \mathbf{I}_{N'_\alpha,N}$ such that $\mathbf{tp}_{\mathfrak{s}}(a_0, N'_\alpha, N)$ either does not fork over M or is orthogonal to M. The latter is impossible by clause (e) of ⊛, so a_0 is as required. Next assume that a_m is well defined and $m < n$. Let $N'' \le_{\mathfrak{s}} N$ be such that $(N'_\alpha, N'', a_m) \in K^{3,\rm pr}_{\mathfrak{s}}$. Now $M_m \le_{\mathfrak{s}} M \le_{\mathfrak{s}} N'_\alpha <_{\mathfrak{s}} N''$ and apply "$\mathfrak{s}$ weakly has regulars" to this sequence. So there is $a_{m+1} \in \mathbf{I}_{N'_\alpha,N''}$ such that $p_{m+1} = \mathbf{tp}_{\mathfrak{s}}(a_{m+1}, N'_\alpha, N'')$ satisfies one of the following cases:

(*i*) p_{m+1} does not fork over M_m

(*ii*) p_{m+1} does not fork over M and is orthogonal to M_m

(*iii*) p_{m+1} is orthogonal to M.

Now the case (*iii*) is impossible by clause (e) of ⊛.

Also if case (*i*) holds, by (∗) above we have obtained the second possibility in the conclusion, hence without loss of generality case (*ii*) holds. Now for each $\ell < m, p_{m+1} \perp M_\ell$ by (∗) above and Example 6.25. So a_{m+1} is as required so we have carried the induction and a_n is as required in the second possibility in the conclusion so we are done.

3) Similar; without loss of generality β is minimal, i.e. $\gamma < \beta \wedge b \in N \backslash M_\alpha \wedge \mathbf{tp}_{\mathfrak{s}}(b, M_\alpha, N) \in \mathscr{S}^{\rm bs}_{\mathfrak{s}}(M_\alpha) \Rightarrow \mathbf{tp}_{\mathfrak{s}}(b, M_\gamma, N) \perp M_\gamma$. Let $\gamma \le \beta$ be minimal such that $\beta \le \gamma + 1$ so if β is successor then $\beta = \gamma + 1$ and if not, (i.e. β is zero or limit) then $\beta = \gamma$. Now the sequence $\langle M_\gamma, M_\beta, M_\alpha, N \rangle$ is $\le_{\mathfrak{s}}$-increasing and by the choice of β, for some $b \in N \backslash M_\alpha$ the type $\mathbf{tp}_{\mathfrak{s}}(b, M_\alpha, N)$ belongs to $\mathscr{S}^{\rm bs}_{\mathfrak{s}}(M_\alpha)$ and is $\pm M_\beta$. As $\mathfrak{s}$ weakly has regulars, there is $c \in N \backslash M_\alpha$ such that either $\mathbf{tp}_{\mathfrak{s}}(c, M_\alpha, N)$ does not fork over M_β and is $\perp M_\gamma$ or

$\mathbf{tp}_{\mathfrak{s}}(c, M_\alpha, N)$ does not fork over M_γ. If $\beta = \gamma + 1$ and the first case holds, we are done. If $\beta = \gamma + 1$ and the second case occurs we get a contradiction to "β is minimal such that ..."

If $\beta = \gamma$ the first case is impossible, so the second case holds; if $\beta = \gamma = 0$ we are done; otherwise $\beta = \gamma$ is a limit ordinal hence by Ax(E)(c) for some $\varepsilon < \beta$ the type $\mathbf{tp}_{\mathfrak{s}}(c, M_\alpha, N)$ does not fork over M_ε and we get a contradiction to the choice of β. So in all cases we are done. $\square_{7.17}$

7.18 Definition. We say that $\mathfrak{s}$ almost has regulars when the conclusions of 7.17 holds (even if $\mathfrak{s}$ is just a good λ-frame, e.g. does not have primes).

7.19 Claim. *1) In the Definition 7.5 of "weakly has regulars" without loss of generality the sequence $\langle M_\alpha : \alpha \le \beta + 1\rangle$ is finite, i.e. $\beta < 4$.*
[Actually "$\mathfrak{s}$ is a good λ-frame" suffice instead 7.1.]
2) If $\mathfrak{s}$ is weakly has regulars, then $\mathfrak{s}^+$ weakly has regulars and even almost has regulars.

Proof. 1) By the proof of 7.17(3).
2) By the translations between $\mathfrak{s}^+$ and $\mathfrak{s}$. $\square_{7.19}$

Explanation: In §10 regular types are defined and investigated, they are useful, but in the main theorem in §12 it is enough to assume weaker properties than actual denseness of regulars. Also we sometimes work with sets $\mathbf{P} \subseteq \mathscr{S}^{\mathrm{bs}}_{\mathfrak{s}}(M)$ which have some of the properties of the set of regular types which we call auto-dense $\mathbf{P}$.

7.20 Definition. 1) We say that $\mathbf{P} \subseteq \mathscr{S}^{\mathrm{bs}}_{\mathfrak{s}}(M)$ is auto-dense when: if $M \le_{\mathfrak{s}} N_1 <_{\mathfrak{s}} N_2$ and for some $c \in N_2 \backslash N_1$, the type $\mathbf{tp}_{\mathfrak{s}}(c, N_1, N_2)$ is not orthogonal to some $p \in \mathbf{P}$ then for some $c \in N_2 \backslash N_1$ the type $\mathbf{tp}_{\mathfrak{s}}(c, N_1, N_2)$ is a non-forking extesnion of some $p \in \mathbf{P}$.
2) For $\mathbf{P} \subseteq \mathscr{S}^{\mathrm{bs}}_{\mathfrak{s}}(M)$ let $\mathbf{P}^\perp = \{q \in \mathscr{S}^{\mathrm{bs}}_{\mathfrak{s}}(M) : p \perp q$ for every $p \in \mathbf{P}\}$. If $\mathbf{P} = \emptyset$ this is $\mathscr{S}^{\mathrm{bs}}_{\mathfrak{s}}(M)$, but pedantically we have to say who is M.

7.21 Definition. We say that $\mathbf{P}$ is a type base of $(M, \bar{M})$ when

(a) $\bar{M} = \langle M_\ell : \ell < n \rangle$
(b) $M_\ell \le_{\mathfrak{s}} M$
(c) $\mathbf{P} \subseteq \mathscr{S}^{\rm bs}_{\mathfrak{s}}(M)$
(d) if $p \in \mathbf{P}$ and $\ell < n$ then $p \perp M_\ell$
(e) if $M \le_{\mathfrak{s}} N' <_{\mathfrak{s}} N''$ and $a \in N'' \backslash N'$, $\mathbf{tp}_{\mathfrak{s}}(a, N', N'')$ does not fork over M and is orthogonal to M_ℓ for $\ell < n$ then for some $b \in N'' \backslash N'$, $\mathbf{tp}_{\mathfrak{s}}(b, N', N'')$ is the non-forking extension of some $p \in \mathbf{P}$.

The following notion is used in the proof 12.30(10); used implicitly in what is quoted there, in particularly in "base".

7.22 Definition. 1) Assume $M_\ell \le_{\mathfrak{s}} N$ for $\ell = 1, 2$. We say that $q \in \mathscr{S}^{\rm bs}_{\mathfrak{s}}(M_1)$ is dominated by $p \in \mathscr{S}^{\rm bs}_{\mathfrak{s}}(M_2)$ when: if $M \le_{\mathfrak{s}} N', N \le_{\mathfrak{s}} N'$ and $p', q' \in \mathscr{S}^{\rm bs}_{\mathfrak{s}}(M)$ are parallel to p, q respectively and $(M, M^+, a) \in K^{3,\rm bs}_{\mathfrak{s}}$ and $\mathbf{tp}_{\mathfrak{s}}(a, M, M^+) = p'$ then for some $b \in M^+$ realizes q'.
2) We say above "essentially dominated" when we require that M is brimmed (equivalently for some brimmed $M' \le_{\mathfrak{s}} M$ the types p', q' does not fork over M'). We say weakly if we demand $M = N = M_2$.

7.23 Claim. *1) The property "p weakly dominates q" depends on p and on q only up to parallelism. Also in Definition 7.22 without loss of generality $(M, M^+, a) \in K^{3,\rm pr}_{\mathfrak{s}}$.*
2) If p, r are orthogonal and q is dominated by p then q, r are orthogonal.

3) [$\mathfrak{s}$ weakly has regulars] If $M \le_{\mathfrak{s}} N$ and $p \in \mathscr{S}^{\rm bs}_{\mathfrak{s}}(N)$ is not orthogonal to M, then some $q \in \mathscr{S}^{\rm bs}_{\mathfrak{s}}(M)$ is essentially dominated by p (hence $M' <_{\mathfrak{s}} N \wedge p \perp M' \Rightarrow q \perp M'$).
4) If $p, q \in \mathscr{S}^{\rm bs}_{\mathfrak{s}}(N)$ do not fork over M and $p \underset{\rm wk}{\perp} q$ and q dominates $q_1 \in \mathscr{S}^{\rm bs}_{\mathfrak{s}}(N)$ then $p \underset{\rm wk}{\perp} q_1$.
5) If p dominates q then it essentially dominates q and weakly dominates p; if $\mathfrak{s}$ is categorical in λ all three are equivalent.

Proof. 1) Obvious.
2) By 6.25.
3) Let $(N, N_1, a) \in K_{\mathfrak{s}}^{3,\mathrm{pr}}$ be such that $\mathbf{tp}_{\mathfrak{s}}(a, N, N_1) = p$. As "$\mathfrak{s}$ weakly has regulars", for some $b \in N_1 \backslash N$, we have $\mathbf{tp}_{\mathfrak{s}}(b, N, N_1) \in \mathscr{S}_{\mathfrak{s}}^{\mathrm{bs}}(N)$ does not fork over M (we can let $N_2 \leq_{\mathfrak{s}} N$ we have $(N, N_2, b) \in K_{\mathfrak{s}}^{3,\mathrm{pr}}$). Now use part (2).
4) Use 6.26.
5) Left to the reader. $\square_{7.23}$

Remark. We may replace "$\mathfrak{s}$ has prime" by the weaker demands.

Important but not presently used in

7.24 Definition. $\mathfrak{s}$ has super density for $K_{\mathfrak{s}}^{3,\mathrm{uq}}$ when:

If $(M, N, a) \in K_{\mathfrak{s}}^{3,\mathrm{bs}}$ then for some N' we have $N \cup \{a\} \subseteq N' \subseteq_{\mathfrak{s}} N$ and $(M, N', a) \in K_{\mathfrak{s}}^{3,\mathrm{uq}}$.

§8 Tries to decompose and independence of sequences of models

We try to find smooth or otherwise good decompositions; at present only 8.3 works. We shall get really what we want after using type-fullness hence regular types. On this assumption, having $\mathscr{S}_{\mathfrak{s}}^{\mathrm{na}}$ equal to $\mathscr{S}_{\mathfrak{s}}^{\mathrm{bs}}$, i.e., fullness being "soft", see §9.

On $\langle M_\eta : \eta \in I \rangle, I \subseteq {}^{\omega>}\lambda$ and the needed assumptions, see [Sh:F735] and [Sh 842]. We put sometimes as an exercise an easy case of weakening the hypothesis. Also we try to understand $\mathrm{NF}_{\mathfrak{s}}$ better - it is closed under union of increasing chains.

8.1 Hypothesis. 1) (*a*) $\mathfrak{s}$ is a successful good$^+$ frame.
2)

(*b*) $\mathfrak{s}$ has primes,

(*c*) $K_{\mathfrak{s}}^{3,\mathrm{uq}} = K_{\mathfrak{s}}^{3,\mathrm{pr}}$, moreover $K_{\mathfrak{s}}^{3,\mathrm{vq}} = K_{\mathfrak{s}}^{\mathrm{qr}}$

(*d*) $\perp = \underset{\mathrm{wk}}{\perp}$ as a relation between two types

(*e*) $\perp = \underset{\mathrm{su}}{\perp}$, as a relation between a type and a model.

Remark. Hypothesis 8.1(2)(d),(e) are reasonable as if $\mathfrak{s}$ is categorical by 6.10(5) and 6.8(5) they hold, but in some examples it holds without going to the successor, so we use them as an hypothesis. Hypothesis 8.1(a),(b) are justified by 4.9, 7.14.

8.2 Definition. We say that $\langle M_i, a_j : i \le \alpha, j < \alpha \rangle$ is a smooth x-decomposition inside N over M (where $x \in \{\text{pr,uq}\}$, if $x =$ pr we may omit it); <u>when</u>:

(a) it is a decomposition inside N over M (see Definition 3.3), which means
$\langle M_i : i \le \alpha \rangle$ is $\le_{\mathfrak{s}}$-increasing continuous
$M_\alpha \le_{\mathfrak{s}} N$
$M = M_0$
$\mathbf{tp}_{\mathfrak{s}}(a_i, M_i, M_{i+1}) \in \mathscr{S}^{\text{bs}}_{\mathfrak{s}}(M_i)$
$(M_i, M_{i+1}, a_i) \in K^{3,\text{x}}_{\mathfrak{s}}$

(b) for every $i < \beta$ there is $j \le i$ such that $\mathbf{tp}_{\mathfrak{s}}(a_i, M_i, M_{i+1})$ does not fork over M_j and if $j > \varepsilon$, then the type is weakly orthogonal to M_ε; so by 6.26 necessarily j is non-limit and by 8.1(d) the type is orthogonal to M_ε.

In the claim below, if N is $(\lambda, *)$-brimmed over M then we can choose $M_\alpha = N$, see 8.6(2) similarly in other cases.

8.3 Claim. *1) If $M \le_{\mathfrak{s}} N$ <u>then</u> we can find $\langle M_i, a_j : i \le \alpha, j < \alpha \rangle$ such that*

$\boxtimes_{M,N,\bar{M},\bar{a}}$ (a) *$M_0 = M$ and*[12] *$N \le_{\mathfrak{K}} M_\alpha$*

(b) *M_i is $\le_{\mathfrak{K}}$-increasing continuous*

(c) *$(M_i, M_{i+1}, a_i) \in K^{3,\text{uq}}_{\mathfrak{s}}$ moreover $\in K^{3,\text{pr}}_{\mathfrak{s}}$ as $\mathfrak{s}$ has primes, i.e., 8.1(b)*

(d) *for each $j < \alpha$ either $\mathbf{tp}_{\mathfrak{s}}(a_j, M_j, M_{j+1})$ does not fork over M_0 <u>or</u> it is weakly orthogonal to M_0.*

[12] recalling $\mathfrak{s}$ has primes, can we add $N = M_\alpha$? Of course, under additional assumptions,e.g. on regular types

2) If $(M, N, \mathbf{J}) \in K^{3,\mathrm{bs}}_{\mathfrak{s}}$ then we can demand in (1) that $\mathbf{J} = \{a_i : i < \alpha'\}$ for some $\alpha' \le \alpha$.
3) In (1) we can add $\langle M_j, a_i : j \le \alpha\rangle$ is smooth (i.e. we strengthen clause (d)).

Remark. Proving 8.3 we do not use the hypothesis 8.1(2).

Proof. 1) We try to choose by induction on $i < \lambda^+$ a pair (M_i, N_i) and if $i = j+1$ also a_j such that: $M_i \le_{\mathfrak{s}} N_i, M_i$ is $\le_{\mathfrak{s}}$-increasing continuous, N_i is $\le_{\mathfrak{s}}$-increasing continuous, and the M_i, a_j satisfy the relevant cases of clauses (b), (c), (d) and $M_0 = M, N_0 = N$ and $i = j+1 \Rightarrow \neg\mathrm{NF}_{\mathfrak{s}}(M_j, N_j, M_i, N_i)$. We cannot succeed (as $\mathfrak{s}$ is good$^+$ and successful, see §1) and we can define for $i = 0$ and i limit. Hence for some i we have (M_i, N_i) but cannot choose M_{i+1}, N_{i+1}, a_i. If $M_i = N_i$ then we are done. If $M_i \ne N_i$ there is $b_i \in N_i$ such that $\mathbf{tp}_{\mathfrak{s}}(b, M_i, N_i) \in \mathscr{S}^{\mathrm{bs}}_{\mathfrak{s}}(M_i)$. Trivially, one of the following cases occurs and in each case we get a contradiction.

<u>Case 1</u>: $\mathbf{tp}_{\mathfrak{s}}(b_i, M_i, N_i)$ is weakly orthogonal to M_0.
Then we let $a_i = b_i$ and we can find a pair $M_{i+1} \le_{\mathfrak{s}} N_{i+1}$ such that $N_i \le_{\mathfrak{s}} N_{i+1}$ and $(M_i, M_{i+1}, a_i) \in K^{3,\mathrm{uq}}_{\mathfrak{s}}$ and if $\mathfrak{s}$ has primes then $N_{i+1} = N_i$ and $(M_i, M_{i+1}, a_i) \in K^{3,\mathrm{pr}}_{\mathfrak{s}}$. All the induction demands hold and $\mathbf{tp}_{\mathfrak{s}}(a_i, M_{i+1}, N_i)$ is algebraic so does not belong to $\mathscr{S}^{\mathrm{bs}}_{\mathfrak{s}}(M_{i+1})$ hence this type forks over M_i. As $\mathrm{NF}_{\mathfrak{s}}$ respects $\mathfrak{s}$ (see II.6.1(3)), it follows that $\neg\mathrm{NF}_{\mathfrak{s}}(M_i, N_i, M_{i+1}, N_{i+1})$.

<u>Case 2</u>: $\mathbf{tp}_{\mathfrak{s}}(b_i, M_i, N_i)$ is not weakly orthogonal to M_0.
So there is $p_i \in \mathscr{S}^{\mathrm{bs}}_{\mathfrak{s}}(M_i)$ which does not fork over M_0 such that p_i, $\mathbf{tp}_{\mathfrak{s}}(b_i, M_i, N_i)$ are not weakly orthogonal. Hence we can find $N_{i+1} \in K_{\mathfrak{s}}$ such that $N_i \le_{\mathfrak{s}} N_{i+1}$ and some $a_i \in N_{i+1}$ realizing p_i in N_{i+1} and $\mathbf{tp}_{\mathfrak{s}}(a_i, N_i, N_{i+1})$ is not the non-forking extension of p_i in $\mathscr{S}^{\mathrm{bs}}_{\mathfrak{s}}(N_i)$. As we can increase N_{i+1} without loss of generality there is $M_{i+1} \le_{\mathfrak{s}} N_{i+1}$ such that (M_i, M_{i+1}, a_i) is in $K^{3,\mathrm{pr}}_{\mathfrak{s}}$ (if not assuming 8.1(b), then just in $K^{3,\mathrm{uq}}_{\mathfrak{s}}$), so clearly $\neg\mathrm{NF}_{\mathfrak{s}}(M_i, N_i, M_{i+1}, N_{i+1})$ again, by "$\mathrm{NF}_{\mathfrak{s}}$ respects $\mathfrak{s}$", see 1.18. So all the demands hold.
So if $M_i \ne N_i$ then we can continue the induction, contradiction, hence $M_i = N_i$ and so $\alpha = i, \langle M_j : j \le \alpha\rangle, \langle a_j : j < \alpha\rangle$ are as

required.
2) First use $(*)_2 \Leftrightarrow (*)_3$ from 5.4(1) and then continue as above.
3) In the proof of part (1) when $M_i \neq N_i$ are defined after we choose b, let $j = j_i \leq i$ be minimal such that $\mathbf{tp}_{\mathfrak{s}}(b, M_i, N_i) \underset{\text{wk}}{\pm} M_j$, well defined as $j = i$ is O.K. Now continue as above. $\square_{8.3}$

8.4 Claim. *If $\langle M_i : i \leq \alpha \rangle$ is $\leq_{\mathfrak{s}}$-increasing continuous and $p \in \mathscr{S}^{\text{bs}}_{\mathfrak{s}}(M_\alpha)$ does not fork over M_0, <u>then</u> we can find an $\leq_{\mathfrak{s}}$-increasing continuous sequence $\langle N_i : i \leq \alpha \rangle$ and a such that $M_i \leq_{\mathfrak{s}} N_i, a \in N_0$, $\mathbf{tp}_{\mathfrak{s}}(a, M_\alpha, N_\alpha) = p$ and (M_i, N_i, a) is prime for $i \leq \alpha$.*

Proof. We choose by induction on $i \leq \alpha$ a pair (N_i, f_i) and a such that N_i is $\leq_{\mathfrak{s}}$-increasing, f_i is an $\leq_{\mathfrak{s}}$-embedding of M_i into N_i, f_i is increasing continuous, $f_0 = \text{id}_{M_0}$, $\mathbf{tp}_{\mathfrak{s}}(a, M_0, N_0) = p \restriction M_0, (f_i(M_i), N_i, a) \in K^{3,\text{pr}}_{\mathfrak{s}}$ and $\mathbf{tp}_{\mathfrak{s}}(a, f_i(M_i), N_i)$ does not fork over $f_0(M_0) = M_0$. For $i = 0$ use existence of primes.

For i limit use 7.15 and the hypothesis (c) of 8.1, and for $i = j+1$ use the definition of a prime triple, chasing arrows. In the end, renaming, without loss of generality $f_i = \text{id}_{M_i}$ for $i \leq \alpha$. $\square_{8.4}$

8.5 Claim. *If $\boxtimes_{M,N,\bar{M},\bar{a}}$ from Claim 8.3 holds and $\mathbf{J} = \{a_i : \mathbf{tp}_{\mathfrak{s}}(a_i, M_i, M_{i+1})$ does not fork over $M = M_0\}$ <u>then</u> $(M, N, \mathbf{J})$ belongs to $K^{3,\text{vq}}_{\mathfrak{s}}$.*

Proof. We shall use Claim 7.9, this is O.K. as hypothesis 7.1 holds. Now the desired conclusion is clause (A) there, so it suffices to prove clause (D) there. Now subclauses (a), (b) are obvious so let us prove subclause (c). For this we prove by induction on $\beta \leq \alpha = \ell g(\bar{a})$ that letting $\mathbf{J}_\beta = \{a_i : i < \beta$ and $\mathbf{tp}_{\mathfrak{s}}(a_i, M_i, M_{i+1})$ does not fork over $M_0\}$, we have:

> $\circledast_\beta$ if $M_\beta \leq_{\mathfrak{s}} N^+, n < \omega$ and $b_\ell \in N^+$ for $\ell < n$ and $\mathbf{J}_\beta \cup \{b_0, \ldots, b_{n-1}\}$ is independent in (M_0, N^+) (and $b_\ell \notin \mathbf{J}_\beta$ and $\ell \neq k \Rightarrow b_\ell \neq b_k$ of course), <u>then</u> $\{b_0, \ldots, b_{n-1}\}$ is independent in (M, M_β, N^+).

For $\beta = 0$ this is trivial. For β limit this holds by 5.10(2) as $(M_0, M_\gamma, \{b_\ell : \ell < n\})$ is independent by the induction hypothesis for each $\gamma < \beta$. Lastly, let $\beta = \gamma + 1$; then by 7.9, as we have proved $\circledast_\gamma$, we have $(M_0, M_\gamma, \mathbf{J}_\gamma) \in K^{3,\text{vq}}_{\mathfrak{s}}$. First assume $a_\gamma \in \mathbf{J}_\beta$. So

- $(*)_1$ we are given $n, b_0, \dots, b_{n-1}$ we let $b_n = a_\gamma$ and we are assuming $\mathbf{J}_\beta \cup \{b_0, \dots, b_{n-1}\}$ is independent in (M_0, N^+). Hence trivially
- $(*)_2$ $\mathbf{J}_\gamma \cup \{b_0, \dots, b_n\}$ is independent in (M_0, N^+).

Now apply $\circledast_\gamma$ for $n+1, b_0, \dots, b_{n-1}, b_n$, so $\{b_0, \dots, b_n\}$ is independent in (M, M_γ, N^+). As $(M_\gamma, M_\beta, a_{\gamma+1}) \in K^{3,\text{uq}}_{\mathfrak{s}}$ by 5.16(5) we deduce that $\{b_0, \dots, b_{n-1}\}$ is independent in (M_0, M_β, N^+), which gives the desired conclusion.
Second, assume $a_\gamma \notin \mathbf{J}_\beta$ and $n, b_0, \dots, b_{n-1}$ are given as in $\circledast_\beta$. By the induction hypothesis $\{b_0, \dots, b_{n-1}\}$ is independent in (M, M_γ, N^+). But $\mathbf{tp}_{\mathfrak{s}}(a_\gamma, M_\gamma, M_\beta) = \mathbf{tp}_{\mathfrak{s}}(a_\gamma, M_\gamma, N^+)$ is orthogonal to M_0, $\mathbf{tp}_{\mathfrak{s}}(b_\ell, M_\gamma, N^+)$ does not fork over $M = M_0$ and $(M_\gamma, M_\beta, a_\gamma) \in K^{3,\text{uq}}_{\mathfrak{s}}$ so by 6.18, (with $\alpha = 1$ there) necessarily $\{b_0, \dots, b_{n-1}\}$ is independent in $(M, M_\beta, N^+), M_0$, as required.

Having carried the induction we got $\circledast_\alpha$ which for $n = 1$ is the statement (D) of 7.9 hence gives the desired conclusion. $\square_{8.5}$

Now we can show that any type in a sense is below a non-forking combination of basic ones.

8.6 Conclusion. 1) If $M \leq_{\mathfrak{s}} N$ <u>then</u> for some pair $(N', \mathbf{J})$ we have $N \leq_{\mathfrak{s}} N'$ and $(M, N', \mathbf{J}) \in K^{3,\text{vq}}_{\mathfrak{s}}$.
2) If N is brimmed over M, we can demand $N' = N$. Also, if $(M, N, \mathbf{J}) \in K^{3,\text{bs}}_{\mathfrak{s}}$ <u>then</u> for some $(N', \mathbf{J}')$ we have $N \leq_{\mathfrak{s}} N'$ and $\mathbf{J} \subseteq \mathbf{J}'$ and $(M, N', \mathbf{J}') \in K^{3,\text{vq}}_{\mathfrak{s}}$.

Proof. 1) By 8.3 + 8.5.
2) Easy. $\square_{8.6}$

8.7 Claim. *If $p \in \mathscr{S}_{\mathfrak{s}}(M)$ <u>then</u> for some $n, M_\ell (\ell \leq n), a_k (k < n)$ and b we have:*

(a) $M_0 = M$

(b) $(M_\ell, M_{\ell+1}, a_\ell) \in K^{3,\mathrm{uq}}_{\mathfrak{s}}$ *and as* $\mathfrak{s}$ *has primes (i.e. 8.1(b)) even* $\in K^{3,\mathrm{pr}}_{\mathfrak{s}}$

(c) $\mathbf{tp}_{\mathfrak{s}}(a_\ell, M_\ell, M_{\ell+1})$ *does not fork over* M_0, *so*

(d) $\{a_\ell : \ell < n\}$ *is independent in* (M_0, M_n) *and* $(M_0, M_n, \{a_\ell : \ell < n\}) \in K^{3,\mathrm{vq}}_{\mathfrak{s}}$ *and as* $\mathfrak{s}$ *has primes even* $\in K^{3,\mathrm{qr}}_{\mathfrak{s}}$

(e) $b \in M_n$ *and* $p = \mathbf{tp}_{\mathfrak{s}}(b, M, M_n)$.

Proof. We can find N, b such that $M \leq_{\mathfrak{s}} N$ and $b \in N$ and $\mathbf{tp}_{\mathfrak{s}}(b, M, N) = p$. By 8.3, without loss of generality, possibly increasing N, we have $\boxtimes_{M,N,\bar{M},\bar{a}}$ for some $\bar{M}, \bar{a}$ as there. By 8.5 as we have clauses (c),(d) of the Hypothesis 8.1 we have $(M, N, \mathbf{J}) \in K^{3,\mathrm{vq}}_{\mathfrak{s}}$ for some $\mathbf{J}$. Among all such triples $(N, \mathbf{J}, b)$ choose one $(N^*, \mathbf{J}^*, b^*)$ with the cardinality of $\mathbf{J}^*$ being minimal. Let $\mathbf{J}^* = \{a_i : i < \theta\}$; by hypothesis 8.1(b) we know $(M, N^*, \mathbf{J}^*) \in K^{3,\mathrm{qr}}_{\mathfrak{s}}$ so we can find an M-based pr-decomposition $\langle M_i, b_j : i \leq \theta, j < \theta\rangle$ over M, i.e., $M_0 = M$ such that $\mathbf{tp}_{\mathfrak{s}}(b_j, M, M_{j+1}) = \mathbf{tp}_{\mathfrak{s}}(a_i, M, N^*)$ and $\mathbf{tp}_{\mathfrak{s}}(b_j, M_j, M_{j+1})$ does not fork over M, of course. So there is a $\leq_{\mathfrak{s}}$-embedding of N^* into M_θ over M mapping b_i to a_i for $i < \theta$ so without loss of generality $N^* \leq_{\mathfrak{s}} M_\theta$ and $i < \theta \Rightarrow a_i = b_i$. Now if $\theta < \aleph_0$ we have gotten the desired conclusion, otherwise $b \in N^* \subseteq M_\theta = \bigcup\limits_{i<\theta} M_i$ so for some $\beta < \theta$ we have $b \in M_\beta$ and has clearly $(M_0, M_\beta, \{a_i : i < \beta\}) \in K^{3,\mathrm{qr}}_{\mathfrak{s}}$ by 5.8(2) so we have gotten a contradiction to the minimality of $|\mathbf{J}^*|$. $\square_{8.7}$

<u>Exercise</u>: Prove 8.7 for $\mathfrak{s}$ a weakly successful full a good λ-frame so demanding $(M_\ell, M_{\ell+1}, a_\ell) \in K^{3,\mathrm{uq}}_{\mathfrak{s}}$ only.

[Hint: E.g. do it for full $\mathfrak{s}$, justified by §9.]

8.8 Definition. 1) For $\alpha < \lambda^+$ we say that $\langle M_i : i < \alpha\rangle$ is $\mathfrak{s}$-independent over M inside N or inside (M, N) <u>when</u> $M \leq_{\mathfrak{s}} M_i \leq_{\mathfrak{s}} N$ and we can find a $\leq_{\mathfrak{s}}$-increasing continuous sequence $\bar{N} = \langle N_i : i \leq \alpha\rangle$ such that $N_0 = M, N \leq_{\mathfrak{s}} N_\alpha$ and $\mathrm{NF}_{\mathfrak{s}}(M, N_i, M_i, N_{i+1})$ for every $i < \alpha$. We call such $\langle N_i : i \leq \alpha\rangle$ a witness.
2) For $\alpha = \lambda^+$ we define similarly so $i < \alpha \Rightarrow N_i \in K_{\mathfrak{s}}$, but $N_\alpha \in K^{\mathfrak{s}}$.

8.9 Observation. 1) Assume $\langle N_i : i \le \alpha \rangle$ is $\le_{\mathfrak{s}}$-increasing but not necessarily continuous, and $\bar{N}, M, \langle M_i : i < \alpha \rangle$ satisfies the other requirements in 8.8, so $N \le_{\mathfrak{s}} N_\alpha$. Then the sequence $\langle N'_i : i \le \alpha' \rangle$ is a witness for $\langle M'_i : i < \alpha \rangle$ being $\mathfrak{s}$-independent over M inside N when:

- (a) let $\alpha' = \alpha$
- (b) N'_i is N_i if $i \le \alpha$ is non-limit, is $\cup\{N_j : j < i\}$ if $i \le \alpha$ is limit
- (c) M'_i is M_i if $i < \alpha$.

2) Moreover, there is a sequence $\langle N'_i : i \le \alpha \rangle$ witnessing $\bar{M}$ is independent in (M, N) such that $N_\alpha \le_{\mathfrak{s}} N'_\alpha$.

Proof. 1) Straight.
2) We can choose a $\le_{\mathfrak{s}}$-increasing continuous sequence $\langle N''_i : i \le \alpha \rangle$ such that $i < \alpha \Rightarrow N''_i \cap N'_\alpha = N'_i$ and $i < j \le \alpha \Rightarrow \mathrm{NF}_{\mathfrak{s}}(N'_i, N''_i, N'_j, N''_j)$ and N''_α is brimmed over N'_α. Hence there is a $\le_{\mathfrak{s}}$-embedding of $N'_{\alpha+1}$ into N''_α over N'_α, so without loss of generality $N'_{\alpha+1} \le_{\mathfrak{s}} N''_\alpha$. So $\langle N''_i : i \le \alpha \rangle$ is as required. $\square_{8.9}$

8.10 Weak Uniqueness Claim. *Assume*

- (a) *for* $\ell = 1, 2, \langle M^\ell_i : i < \alpha \rangle$ *is* $\mathfrak{s}$*-independent over* M_ℓ *inside* N_ℓ
- (b) f *is an isomorphism from* M_1 *onto* M_2
- (c) *for* $i < \alpha, f_i$ *is an isomorphism from* M^1_i *onto* M^2_i *extending* f.

Then there is N_3 *such that* $N_2 \le_{\mathfrak{s}} N_3$ *and a* $\le_{\mathfrak{s}}$*-embedding* f^* *of* N_1 *into* N_3 *extending every* f_i.

Proof. Let $\langle N^\ell_i : i \le \alpha \rangle$ be a witness to $\langle M^\ell_i : i < \alpha \rangle$ being $\mathfrak{s}$-independent over M_ℓ inside N_ℓ for $\ell = 1, 2$. Without loss of generality $M_1 = M_2$ call it M and f is the identity on M, so $N^\ell_0 = M$.

We choose by induction on $i \le \alpha$ the tuple (N^3_i, g^1_i, g^2_i) such that

- (α) $N^3_0 = M, g^\ell_0 = \mathrm{id}_M$

(β) N^3_i is $\le_{\mathfrak{s}}$-increasing continuous

(γ) g^ℓ_i is a $\le_{\mathfrak{s}}$-embedding of N^ℓ_i into N^3_i for $\ell = 1, 2$

(δ) g^ℓ_i is increasing continuous for i

(ε) $(\forall a \in M^1_i)(g^2_{i+1}(f_i(a)) = g^1_{i+1}(a))$.

For $i = 0, i$ limit this is obvious. For $i = j + 1$ use the uniqueness of $\mathrm{NF}_{\mathfrak{s}}$-amalgamation. Having carried the induction, by renaming we get the conclusion.

$\square_{8.10}$

8.11 Claim. *Assume that $\langle M_i : i < \alpha\rangle$ is $\mathfrak{s}$-independent over M, inside N with $\bar N = \langle N_i : i \le \alpha\rangle$ a witness.*
1) If $M \le_{\mathfrak{s}} M'_i \le_{\mathfrak{s}} M_i$ for $i < \alpha$, then $\langle M'_i : i < \alpha\rangle$ is $\mathfrak{s}$-independent over M inside N.
2) In part (1), $\bar N$ is also a witness for $\langle M'_i : i < \alpha\rangle$ being $\mathfrak{s}$-independent over M inside N.
3) If $M_i \le_{\mathfrak{s}} M^+_i$ for $i < \alpha$ then we can find $N^+, \langle N^+_i : i \le \alpha\rangle, \langle f_i : i < \alpha\rangle$ such that:

(a) *$N_i \le_{\mathfrak{s}} N^+_i \le_{\mathfrak{s}} N^+$ and $N \le_{\mathfrak{s}} N^+$*

(b) *$\langle N^+_i : i \le \alpha\rangle$ is $\le_{\mathfrak{s}}$-increasing continuous*

(c) *f_i is a $\le_{\mathfrak{s}}$-embedding of M^+_i into N_{i+1} over M_i*

(d) *$\langle N^+_i : i \le \alpha\rangle$ witness $\langle f_i(M^+_i) : i < \alpha\rangle$ is $\mathfrak{s}$-independent over M inside N^+_1.*

4) There are $\langle M^*_i : i < \alpha\rangle, \langle N^+_i : i \le \alpha\rangle, N^+, \langle \mathbf{J}_i : i < \alpha\rangle$ such that:

(a) *$M_i \le_{\mathfrak{s}} M^*_i \le_{\mathfrak{s}} N^+$ for $i < \alpha, N \le_{\mathfrak{s}} N^+$ and $N_i \le_{\mathfrak{s}} N^+_i$ for $i \le \alpha$*

(b) *$\langle N^+_i : i \le \alpha\rangle$ witness that $\langle M^*_i : i < \alpha\rangle$ is $\mathfrak{s}$-independent over M inside N^+*

(c) *$(M, M^*_i, \mathbf{J}_i) \in K^{3,\mathrm{vq}}_{\mathfrak{s}}$ for $i < \alpha$.*

5) If $N \le_{\mathfrak{s}} N^+$ or $\cup\{M_i : i < \alpha\} \subseteq N^+ \le_{\mathfrak{s}} N$ then $\langle M_i : i < \alpha\rangle$ is $\mathfrak{s}$-independent over M inside N^+.
6) If $\langle M_i : i < \alpha\rangle$ is independent inside (M, N^+) as witnessed by $\langle N_i : i \le \alpha\rangle$ and $\alpha \le \beta, M_i = M \wedge N_i = N_\alpha$ for $i \in [\alpha, \beta)$ then $\langle M_i :$

$i < \beta\rangle$ is independent inside (M, N^+) as witnessed by $\langle N_i : i \le \beta\rangle$. If $\alpha = 0$, then $\langle M_i : i < \alpha\rangle$ is independent inside (M, M).

Proof. 1), 2) Straightforward.
3) By induction on $i \le \alpha$ we choose N'_i, N^+_i, f_i and if $i = j+1 \le \alpha$ also g_j such that

$\circledast$ (a) N^+_i is $\le_{\mathfrak{s}}$-increasing continuous
(b) f_i is a $\le_{\mathfrak{s}}$-embedding of N_i into N^+_i
(c) f_i is increasing continuous with i
(d) if $i = j+1$ then $N'_j \le_{\mathfrak{s}} N^+_j \le_{\mathfrak{s}} N'_i$
(e) g_j is a $\le_{\mathfrak{s}}$-embedding of M^+_i into N^+_i when $i = j+1 \le \alpha$
(f) $g_j \restriction M_i = f_i \restriction M_i \supseteq f_j \restriction M$ when $i = j+1 \le \alpha$
(g) if $i = j+1 \le \alpha$ then $\mathrm{NF}_{\mathfrak{s}}(f_j(N_j), f_i(N_i), N^+_j, N'_i)$
(h) if $i = j+1 \le \alpha$ then $\mathrm{NF}_{\mathfrak{s}}(f_i(M_j), N'_i, g_j(M^+_j), N^+_i)$.

There is no problem to carry the induction by the basic properties of $\mathrm{NF}_{\mathfrak{s}}$. That is, for $i = 0$ we let $N^+_i = M, f_i = \mathrm{id}_M$. For i limit let $N^+_i = \cup\{N^+_j : j < i\}$ and $f_i = \cup\{f_j : j < i\}$. Lastly, for $i = j+1$ we apply the "existence property for $\mathrm{NF}_{\mathfrak{s}}$" twice to have clauses (g) and then (h).

Having carried the induction clearly $\langle N^+_i : i \le \alpha\rangle$ is $\le_{\mathfrak{s}}$-increasing continuous. Let $i = j+1 \le \alpha$, so $\mathrm{NF}_{\mathfrak{s}}(M, f_i(M_j), f_j(N_j), f_i(N_i))$ as it means $\mathrm{NF}_{\mathfrak{s}}(M, M_j, N_j, N_i)$ which is assumed. But by clause (g) of $\circledast$ we have $\mathrm{NF}_{\mathfrak{s}}(f_j(N_j), f_i(N_i), g_j(M^+_j), N'_i)$ hence by transitivity of $\mathrm{NF}_{\mathfrak{s}}$ we get $\mathrm{NF}_{\mathfrak{s}}(M, f_i(M_j), N^+_j, N'_i)$.
By symmetry we have $\mathrm{NF}_{\mathfrak{s}}(M, N^+_j, f_i(M_j), N'_i)$, but by clause (h) of $\circledast$ we have $\mathrm{NF}_{\mathfrak{s}}(f_i(M_j), N'_i, g_j(M^+_j), N^+_i)$ so again by transitivity we get $\mathrm{NF}_{\mathfrak{s}}(M, N^+_j, g_j(M^+_j), N^+_i)$, so renaming $i < \alpha \Rightarrow f_i = \mathrm{id}_{N_i}$ and so $\langle g_i : i < \alpha\rangle$ satisfies the requirements on $\langle f_i : i < \alpha\rangle$.
4) For each $i < \alpha$ choose $M^+_i \in K_{\mathfrak{s}}$ which is brimmed over M_i hence in particular $M_i \le_{\mathfrak{s}} M^*_i$. Now we apply part (3) and get $N^+, \langle N^+_i : i \le \alpha\rangle, \langle f_i : i < \alpha\rangle$ as there and let $M^*_i := f_i(M^+_i)$. Clearly $N^+, \langle N^+_i : i < \alpha\rangle, \langle M^*_i : i < \alpha\rangle$ satisfies clauses (a),(b) of part (4). As for clause (c), recalling $M \le_{\mathfrak{s}} M_i \le_{\mathfrak{s}} M^*_i$ by 8.6(1),(2)

for $i < \alpha$ there is $\mathbf{J}_i$ such that $(M, M_i^*, \mathbf{J}_i)$ belongs to $K_{\mathfrak{s}}^{3,\mathrm{vq}}$, so clause (c) holds and we are done.
5), 6) Trivial. $\square_{8.11}$

8.12 Conclusion. If $M \le_{\mathfrak{s}} M_i'$ for $i < \alpha < \lambda^+$ <u>then</u> we can find N and $\bar{M} = \langle M_i : i < \alpha \rangle$ such that $\bar{M}$ is independent inside (M, N) and M_i, M_i' are isomorphic over M for $i < \alpha$.

Proof. We apply 8.11(3) with $M, \langle M : i < \alpha \rangle, \langle M : i \le \alpha \rangle, \langle M_i : i < \alpha \rangle$ here standing for $M, \langle M_i : i < \alpha \rangle, \langle N_i : i \le \alpha \rangle, \langle M_i^+ : i < \alpha \rangle$ there. Its assumption holds by 8.11(6). Its conclusion gives the desired conclusion. $\square_{8.12}$

8.13 Claim. *Assume $M \le_{\mathfrak{s}} M_i \le_{\mathfrak{s}} N$ for $i < \alpha$.*
1) For any $\bar{M}' = \langle M_i' : i < \alpha' \rangle$, a permutation of $\bar{M} = \langle M_i : i < \alpha \rangle$ (that is for some one to one function π from α onto α' we have $i < \alpha \Rightarrow M_i = M_{\pi(i)}'$) we have: $\bar{M}$ is $\mathfrak{s}$-independent over M inside N <u>iff</u> $\bar{M}'$ is $\mathfrak{s}$-independent over M inside N.
2) $\bar{M} = \langle M_i : i < \alpha \rangle$ is $\mathfrak{s}$-independent over M inside N <u>iff</u> every finite subsequence $\bar{M}'$ of $\bar{M}$ is $\mathfrak{s}$-independent over M inside N.
3) Assume $(M, M_i, \mathbf{J}_i) \in K_{\mathfrak{s}}^{3,\mathrm{vq}}$ for $i < \alpha$. <u>Then</u>: $\langle M_i : i < \alpha \rangle$ is $\mathfrak{s}$-independent over M inside N <u>iff</u> $\cup\{\mathbf{J}_i : i < \alpha\}$ is independent in (M, N) and, of course, the $\mathbf{J}_i$ are pairwise disjoint.

Proof. 1) By the symmetry assume $\langle M_i : i < \alpha \rangle$ is $\mathfrak{s}$-independent over M inside N. By 8.11(4) there are $\langle M_i^* : i < \alpha \rangle, \langle N_i^+ : i < \alpha \rangle, N^+, \langle \mathbf{J}_i : i < \alpha \rangle$ as there, in particular $\langle M_i^* : i < \alpha \rangle$ is independent inside (M, N^+) as witnessed by $\langle N_i^+ : i \le \alpha \rangle$ and $(M, M_i^*, \mathbf{J}_i)$ belongs to $K_{\mathfrak{s}}^{3,\mathrm{vq}}$ and $M_i \le_{\mathfrak{s}} M_i^*$ for each $i < \alpha$. By 8.11(1),(5) it suffices to prove that $\langle M_{\pi^{-1}(i)}^* : i < \alpha' \rangle$ is independent inside (M, N^+). Together without loss of generality, for each $i < \alpha$ for some $\mathbf{J}_i$ we have $(M, M_i, \mathbf{J}_i) \in K_{\mathfrak{s}}^{3,\mathrm{vq}}$. Now using part (3) which is proved below, part (1) is translated to parts of 5.4.
2) Similarly.
3) First assume that $\langle M_i : i < \alpha \rangle$ is $\mathfrak{s}$-independent over M inside N,

let $\langle N_i : i \leq \alpha \rangle$ witness this; of course, $i \neq j \Rightarrow \mathbf{J}_i \cap \mathbf{J}_j = 0$ because $i \neq j \Rightarrow M_i \cap M_j = M$ (by properties of $\mathrm{NF}_{\mathfrak{s}}$).

We prove by induction on $\beta \leq \alpha$ that $\cup\{\mathbf{J}_i : i < \beta\}$ is independent in $(M, N_\beta) = (N_0, N_\beta)$; of course, we can increase N_β (see 5.4(2)). For $\beta = 0$ this is trivial, for β limit use by e.g. 5.10(3), for $\beta = \gamma + 1$, by 5.6(2) we know that $\mathbf{J}_\gamma$ is independent in (M, N_γ, N_β) and so by 5.6(1) we deduce that $(\cup\{\mathbf{J}_i : i < \gamma\}) \cup \mathbf{J}_\gamma = \cup\{\mathbf{J}_i : i < \beta\}$ is independent in (M, N_β). For $\beta = \alpha$ we get that $\cup\{\mathbf{J}_i : i < \alpha\}$ is independent in (M, N) as required.

Second assume that the $\mathbf{J}_i$-s are pairwise disjoint and $\cup\{\mathbf{J}_i : i < \alpha\}$ is independent in (M, N). Let $\mathbf{J}_{<\beta} = \cup\{\mathbf{J}_i : i < \beta\}$, so $\langle \mathbf{J}_{<\beta} : \beta \leq \alpha \rangle$ is $\subseteq$-increasing continuous.

We now choose by induction on $\beta \leq \alpha$, the tuple[13] $(M^*_\beta, N^*_\beta, \mathbf{J}^*_\beta)$ such that:

(a) M^*_β is $\leq_{\mathfrak{s}}$-increasing continuous

(b) N^*_β is $\leq_{\mathfrak{s}}$-increasing continuous

(c) $M^*_0 = M, N^*_0 = N$

(d) $M^*_\beta \leq_{\mathfrak{s}} N^*_\beta$

(e) $M_i \leq_{\mathfrak{s}} M^*_\beta$ for $i < \beta$

(f) $\mathbf{J}^*_\beta \subseteq N^*_\beta \backslash M \backslash \mathbf{J}_{<\alpha}$

(g) $\mathbf{J}^*_\beta$ is $\subseteq$-increasing continuous

(h) $(M, M^*_\beta, \mathbf{J}^*_\beta \cup \mathbf{J}_{<\beta})$ belongs to $K^{3,\mathrm{vq}}_{\mathfrak{s}}$

(i) $\mathbf{J}^*_\beta \cup \mathbf{J}_{<\alpha}$ is independent in (M, N^*_β)

(j) $\mathrm{NF}_{\mathfrak{s}}(M, M^*_\gamma, M_\gamma, N^*_\beta)$ for $\beta = \gamma + 1$.

Note that by clauses (a),(b),(e),(j) this is enough to prove that $\langle M^*_\beta : \beta \leq \alpha \rangle$ witness that $\langle M_i : i < \alpha \rangle$ is independent in (M, N), (see Definition 8.8) as required.
For $\beta = 0$ let $M^*_\beta = M, N^*_\beta = N$ and $\mathbf{J}^*_\beta = \emptyset$; easy to check.

For β a limit ordinal let $M^*_\beta = \cup\{M^*_\gamma : \gamma < \beta\}, N^*_\beta = \cup\{N^*_\gamma : \gamma < \beta\}$ and $\mathbf{J}^*_\beta = \cup\{\mathbf{J}^*_\gamma : \gamma < \beta\}$; the least obvious points are clause (h)

[13] the sequence $\langle M^*_\beta : \beta \leq \alpha \rangle$ will witness that $\langle M_i : i < \alpha \rangle$ is independent in (M, N)

which holds by 7.10 and clause (i) which holds by the local character of independence (by 5.4(1) and, of course, 5.4(2)).

Lastly, for $\beta = \gamma + 1$ as $(M, M^*_\gamma, \mathbf{J}^*_\gamma \cup \mathbf{J}_{<\gamma}) \in K^{3,\mathrm{vq}}_{\mathfrak{s}}$ by clause (h), and $\mathbf{J}^*_\gamma \cup \mathbf{J}_{<\gamma} \subseteq \mathbf{J}^*_\gamma \cup \mathbf{J}_{<\beta} \subseteq \mathbf{J}^*_\gamma \cup \mathbf{J}_{<\alpha}$ obviously as the last one, $\mathbf{J}^*_\gamma \cup \mathbf{J}_{<\alpha}$ is independent in (M, N^*_γ) by clause (i) clearly $\mathbf{J}^*_\gamma \cup \mathbf{J}_{<\beta}$ is independent in (M, N'_γ). Together by 5.16(5) we deduce that $\mathbf{J}_\gamma = \mathbf{J}^*_\gamma \cup \mathbf{J}_{<\beta} \backslash \mathbf{J}^*_\gamma \cup \mathbf{J}_{<\gamma}$ is independent in $(M, M^*_\gamma, N^*_\gamma)$. So as $(M, M_\gamma, \mathbf{J}_\gamma) \in K^{3,\mathrm{vq}}_{\mathfrak{s}}$, by the definition of $K^{3,\mathrm{vq}}_{\mathfrak{s}}$ (see Definition 5.15) we get $\mathrm{NF}_{\mathfrak{s}}(M, M^*_\gamma, M_\gamma, N^*_\gamma)$, i.e., clause (j) holds.

By Example 5.25 we can find N^*_β which $\leq_{\mathfrak{s}}$-extends N^*_γ and M^*_β which $\leq_{\mathfrak{s}}$-extends M_γ. Hence easily clauses (h),(i) holds. So we have carried the induction on $\beta \leq \alpha$ hence $\langle M^*_\beta : \beta \leq \alpha \rangle$ witness that $\langle M_\beta : \beta < \alpha \rangle$ is $\mathfrak{s}$-independent inside (M, N) so we are done. $\square_{8.13}$

Remark. Alternatively, by 5.24 we can find (M'_γ, N'_γ) such that: $M^*_\gamma \leq_{\mathfrak{s}} M'_\gamma \leq_{\mathfrak{s}} N'_\gamma, N^*_\gamma \leq_{\mathfrak{s}} N'_\gamma, M'_\gamma$ is $(\lambda, *)$-brimmed over M^*_γ, N'_γ is $(\lambda, *)$-brimmed over N^*_γ and $(M'_\gamma, N'_\gamma, \mathbf{J}_\gamma) \in K^{3,\mathrm{vq}}_{\mathfrak{s}}$ and $\mathbf{J}_\gamma$ is independent in $(M, M'_\gamma, N'_\gamma)$. By Definition 5.15 (of $K^{3,\mathrm{vq}}_{\mathfrak{s}}$) this implies $\mathrm{NF}_{\mathfrak{s}}(M, M_\gamma, M'_\gamma, N'_\gamma)$. There are also (f_β, N''_γ) such that $N'_\gamma \leq_{\mathfrak{s}} N''_\gamma, f_\beta$ is a $\leq_{\mathfrak{s}}$-embedding of M'_γ into N''_γ over M^*_γ and $\mathrm{NF}_{\mathfrak{s}}(M^*_\gamma, N^*_\gamma, f_\beta(M'_\gamma), N''_\gamma)$, simply by the existence of $\mathrm{NF}_{\mathfrak{s}}$-amalgamation.]

As we also have $\mathrm{NF}_{\mathfrak{s}}(M, M_\gamma, M^*_\gamma, N^*_\gamma)$ (see above), by transitivity for $\mathrm{NF}_{\mathfrak{s}}$ we have $\mathrm{NF}_{\mathfrak{s}}(M, M_\gamma, f_\beta(M'_\gamma), N''_\gamma)$. As we have $\mathrm{NF}_{\mathfrak{s}}(M, M_\gamma, M'_\gamma, N'_\gamma)$ by the uniqueness of $\mathrm{NF}_{\mathfrak{s}}$-amalgamation, possibly increasing N''_γ, we can extend f_β to f'_β, a $\leq_{\mathfrak{s}}$-embedding of N'_γ into N''_γ such that $\mathrm{id}_{M_\gamma} \subseteq f'_\beta$. Let $N^*_\beta = N''_\gamma, M^*_\beta = f'_\beta(N'_\gamma)$, note that $f'_\beta(N'_\gamma)$ is $(\lambda, *)$-brimmed over $f'_\beta(M^*_\gamma) = M^*_\gamma$ and $\mathbf{J}^*_\gamma \cup \mathbf{J}_{<\gamma}$ is independent in $(M, f'_\beta(M'_\gamma))$ and $\mathbf{J}^*_\gamma \cup \mathbf{J}_{<\gamma} \subseteq f'_\beta(M^*_\gamma) = M^*_\gamma$, hence we can find $\mathbf{J}'_\gamma \subseteq M^*_\beta \backslash (\mathbf{J}^*_\gamma \cup \mathbf{J}_{<\gamma}) \backslash M$ such that: $\mathbf{J}'_\gamma \cup \mathbf{J}^*_\gamma \cup \mathbf{J}_{<\gamma}$ is independent in $(M, f'_\beta(M'_\gamma))$ and $(M, f'_\beta(M'_\gamma), \mathbf{J}'_\gamma \cup \mathbf{J}^*_\gamma \cup \mathbf{J}_{<\gamma}) \in K^{3,\mathrm{vq}}_{\mathfrak{s}}$ by 8.5 + 8.3. As $M \leq_{\mathfrak{s}} f'_\beta(M'_\gamma) \leq_{\mathfrak{s}} f'_\beta(N'_\gamma) = M^*_\beta$ and $(M, f'_\beta(M'_\gamma), \mathbf{J}'_\gamma \cup \mathbf{J}^*_\gamma \cup \mathbf{J}_{<\gamma}) \in K^{3,\mathrm{vq}}_{\mathfrak{s}}$ and $(f'_\beta(M'_\gamma), f'_\beta(N'_\gamma), \mathbf{J}_\gamma) = (f'_\beta(M'_\gamma), f'_\beta(N'_\gamma), f'_\beta(\mathbf{J}_\gamma)) \in K^{3,\mathrm{vq}}_{\mathfrak{s}}$ we get by 7.16 that $(M, f'_\beta(N'_\gamma),$

$\mathbf{J}'_\gamma \cup \mathbf{J}^*_\gamma \cup \mathbf{J}_{<\gamma} \cup \mathbf{J}_\gamma) = (M, M^*_\beta, (\mathbf{J}'_\gamma \cup \mathbf{J}^*_\gamma) \cup \mathbf{J}_{<\beta}) \in K^{3,\mathrm{vq}}_{\mathfrak{s}}$.

Let $\mathbf{J}^*_\beta = \mathbf{J}'_\gamma \cup \mathbf{J}^*_\gamma$, so we have almost finished proving the induction step, we still need: $\mathbf{J}^*_\beta$ disjoint to $\mathbf{J}^*_{<\alpha}$ and $\mathbf{J}^*_\beta \cup \mathbf{J}_{<\alpha}$ is independent in (M, N^*_β); for this we recall that $\mathbf{J}^*_\beta \cup \mathbf{J}_{<\gamma}$ is independent in $(M, f'_\beta(M'_\gamma))$ and $M \leq_{\mathfrak{s}} f'_\beta(M'_\gamma) <_{\mathfrak{s}} N^*_\beta$ and $(\mathbf{J}^*_\beta \cap \mathbf{J}_{<\alpha}) \backslash f'_\beta(M'_\gamma) = (\mathbf{J}_{<\alpha} \backslash \mathbf{J}_{<\gamma})$ hence it suffices to prove that $\mathbf{J}_{<\alpha} \backslash \mathbf{J}_{<\gamma}$ is independent in $(M, f_\beta(M'_\gamma), N^*_\beta)$. But $\mathrm{NF}_{\mathfrak{s}}(M^*_\gamma, f_\beta(M'_\gamma), N^*_\gamma, N''_\gamma)$ and $\mathbf{J}_{<\alpha} \backslash \mathbf{J}_{<\gamma} \subseteq N^*_\gamma$ is independent in $(M, M^*_\gamma, N^*_\gamma)$ as stated above, so by 5.6(2) we are done.

8.14 Conclusion. 1) If $\langle M_i : i < \alpha \rangle$ is $\mathfrak{s}$-independent over M inside $N, \alpha \geq 2, f_i$ is an isomorphism from M_0 onto M_i over M for $i < \alpha$ and π is a permutation of α and N^+ is $(\lambda, *)$-brimmed over N, then for some automorphism f of N^+ over M we have $\pi(i) = j \Rightarrow f_j \circ f_i^{-1} \subseteq f$. So $\langle f_i : i < \alpha \rangle$ is $(< \lambda^+)$-indiscernible over M inside N, see Definition 8.15 below.
2) [uniqueness] Assume that $\langle M^\ell_i : i < \alpha \rangle$ is $\mathfrak{s}$-independent over M_ℓ inside N_ℓ for $\ell = 1, 2$ and $f_i \supseteq f$ is an isomorphism from M^1_i onto M^2_i for $i < \delta, f$ an isomorphism from M_1 onto M_2 and N_ℓ is $(\lambda, *)$-brimmed over $\cup\{M^\ell_i : i < \alpha\}$. Then there is an isomorphism from N_1 onto N_2 extending $\cup\{f_i : i < \alpha\}$.

Proof. 1) By (2).
2) By 8.10 and uniqueness of the $(\lambda, *)$-brimmed model over a model in $K_{\mathfrak{s}}$.

$\square_{8.14}$

8.15 Definition. 1) We say that $\bar{f} = \langle f_i : i < \alpha \rangle$ is $(< \theta)$-indiscernible over M inside N when: for some sequence $\bar{\mathbf{a}} = \langle a_\varepsilon : \varepsilon < \zeta \rangle$ (possibly infinite) we have $\mathrm{Dom}(f_i) = \{a_\varepsilon : \varepsilon < \zeta\}$ for every $i < \alpha, M \leq_{\mathfrak{s}} N$ and $\cup\{\mathrm{Rang}(f_i) : i < \alpha\} \subseteq N$ and for every partial one to one function π such that $\mathrm{Dom}(\pi) \cup \mathrm{Rang}(\pi) \subseteq \alpha$ and $|\mathrm{Dom}(\pi)| < \theta$ there are N^+, g such that $N \leq_{\mathfrak{s}} N^+, g$ is an automorphism of N^+ over M and for every $i \in \mathrm{Dom}(\pi)$ the function g maps $f_i(\bar{\mathbf{a}})$ to $f_{\pi(i)}(\bar{\mathbf{a}})$.

8.16 Claim. *If $\langle M_i : i < \alpha \rangle$ is $\mathfrak{s}$-independent over M inside N <u>then</u> we can find $\langle M_i^+ : i < \alpha \rangle, M^+, N^+$ such that:*

- (a) $M \le_{\mathfrak{s}} M^+ \le_{\mathfrak{s}} M_i^+ \le_{\mathfrak{s}} N^+$
- (b) $N \le_{\mathfrak{s}} N^+$
- (c) $\langle M_i^+ : i < \alpha \rangle$ *is independent over* M^+ *inside* N^+
- (d) $\mathrm{NF}_{\mathfrak{s}}(M, M^+, N, N^+)$ *hence* $\mathrm{NF}_{\mathfrak{s}}(M, M^+, M_i, M_i^+)$ *for* $i < \alpha$
- (e) M^+ *is brimmed over* M
- (f) M_i^+ *is brimmed over* M^+ *and even over* $M_i \cup M^+$
- (g) N^+ *is brimmed over* $\cup\{M_i^+ : i < \alpha\} \cup M^+$
- (h) *if* $p \in \mathscr{S}^{\mathrm{bs}}_{\mathfrak{s}}(M_i^+)$ *does not fork over* M_i *then* $p \perp M^+ \Leftrightarrow (p \restriction M_i) \perp M$.

Proof. Should be easy by now. By the assumption we can find $N', \langle M_i' : i < \alpha \rangle, \langle \mathbf{J}_i' : i < \alpha \rangle$ such that $N \le_{\mathfrak{s}} N', M_i \le_{\mathfrak{s}} M_i', \langle M_i' : i < \alpha \rangle$ is independent over M inside N' and $(M, M_i', \mathbf{J}_i) \in K^{3,\mathrm{vq}}_{\mathfrak{s}}$ for $i < \alpha$.
We can find $M^+, \langle M_i^+ : i < \alpha \rangle, N^+$ such that $\mathrm{NF}_{\mathfrak{s}}(M, N', M^+, N^+)$ and clauses (c),(e),(f) (without "and even") and (g).

We can find a $\le_{\mathfrak{s}}$-embedding f_i of M_i' into M_i^+ over M such that $\mathrm{NF}_{\mathfrak{s}}(M, M^+, f_i(M_i), M_i^+)$ and M_i^+ is brimmed over $M^+ \cup f_i(M_i)$. Now $\cup\{f_i : i < \alpha\}$ can be extended to an $\le_{\mathfrak{s}}$-embedding of N' into N^+. Renaming f is the identity and lastly clause (h) follows by 8.1(e), see Definition 6.9(2). $\square_{8.16}$

8.17 Claim. *Assume $\alpha \ge 2$ and $\langle M_i : i < \alpha \rangle$ is $\mathfrak{s}$-independent over M inside N, f_i $(i < \alpha)$ is an isomorphism from M_0 onto M_i over M for $i < \alpha$, and $p \in \mathscr{S}^{\mathrm{bs}}_{\mathfrak{s}}(M_0)$. <u>Then</u> the following are equivalent:*

- (A) $p \perp M$
- (B) $p \perp f_1(p)$
- (C) *for some* $i < j < \alpha$ *we have* $f_i(p) \perp f_j(p)$

Proof. By 8.16 without loss of generality M is $(\lambda, *)$-brimmed, M_i is brimmed over M and by 8.11(6), 8.16 without loss of generality α is infinite.

$\underline{(B) \Leftrightarrow (C)}$: by the indiscernibility (i.e., by 8.14(1)).

$\underline{\neg(C) \Rightarrow \neg(A)}$: So we have $i < j < \alpha \Rightarrow f_i(p) \pm f_j(p)$.
First we can find $\langle M_n^* : n < \omega \rangle$ which is $\leq_{\mathfrak{s}}$-increasing and M_{n+1}^* is $(\lambda, *)$-brimmed over M_n^* such that $\cup\{M_n^* : n < \omega\} = M$. Second, we can also find $\langle M_{0,n}^* : n < \omega \rangle$ which is $\leq_{\mathfrak{s}}$-increasing $M_{0,n+1}$ brimmed over $M_{0,n} \cup M_{n+1}$ and $\mathrm{NF}_{\mathfrak{s}}(M_n^*, M_{0,n}^*, M_{n+1}^*, M_{0,n+1}^*)$ for $n < \omega$ and $\cup\{M_{0,n}^* : n < \omega\} = M_0$, see 1.17 and uniqueness of being brimmed over M. Third, for some n the type p does not fork over $M_{0,n}^*$ so without loss of generality $n = 0$. Hence we can consider $M_0^*, \langle f_i(M_{0,0}^*) : i < \alpha \rangle, \langle f_i' = f_i \restriction M_{0,0}^* : i < \alpha \rangle, \langle p_i' = f_i(p \restriction M_{0,0}^*) : i < \alpha \rangle$ and can choose one more copy f_α' inside M, that is, f_α' has domain $M_{0,0}^*, f_\alpha', (M_{0,0}^*) \leq_{\mathfrak{s}} M$ such that $\langle f_i' \restriction M_{0,0}^* : i \leq \alpha \rangle$ is indiscernible over M_0^*, there is such f_α' as M is brimmed over M_0^*. By the indiscernibility clearly $p_0' \pm p_\alpha'$ but $p_0' \| p_0$ and there is $q \in \mathscr{S}_{\mathfrak{s}}^{\mathrm{bs}}(M), q \| p_\alpha'$, so we are done. [This is similar to the proof of 6.10].

$\underline{\neg(A) \Rightarrow \neg(C)}$:

Assume $q \in \mathscr{S}_{\mathfrak{s}}^{bs}(M), q \pm p$, so $q \pm p_i$ for $i < \alpha$ where $p_i = f_i(p)$. Let N^+, b be such that $N \leq_{\mathfrak{s}} N^+, b \in N^+$ and $\mathbf{tp}_{\mathfrak{s}}(b, N, N^+)$ is a non-forking extension of q. So by 6.4(2), possibly increasing N^+, for each $i < \alpha$ there is $a_i \in N^+$ such that $\mathbf{tp}_{\mathfrak{s}}(a_i, N, N^+)$ is a non-forking extension of p_i and $\{b, a_i\}$ is not independent in (N, N^+). But if clause (C) holds then by 6.21(1) the set $\{a_i : i < \alpha\}$ is independent in (N, N^+), contradicting 5.13(1) as α is infinite. Alternatively use 6.22. $\square_{8.17}$

A conclusion of 8.11 is

8.18 Claim. *1) If $\langle M_i : i < \alpha \rangle$ is $\mathfrak{s}$-independent over M inside N and $a_i \in M_i$, $\mathbf{tp}_{\mathfrak{s}}(a_i, M, M_i) \in \mathscr{S}_{\mathfrak{s}}^{\mathrm{bs}}(M)$ for $i < \alpha$ <u>then</u> $\{a_i : i < \alpha\}$ is independent over M inside N.*
2) If above $(M, M_i, \mathbf{J}_i) \in K_{\mathfrak{s}}^{3,\mathrm{bs}}$, <u>then</u> $\cup\{\mathbf{J}_i : i < \alpha\}$ is independent in (M, N) and the $\mathbf{J}_i (i < \alpha)$ are pairwise disjoint.

Proof. Easy.

* * *

We now return to investigating $\mathrm{NF}_{\mathfrak{s}}$.

8.19 Claim. *1) Assume that $\langle M_i : i \le \delta\rangle$ is $\le_{\mathfrak{s}}$-increasing continuous sequence, $\delta < \lambda_{\mathfrak{s}}^+$ and $\mathrm{NF}_{\mathfrak{s}}(M_0, N_0, M_i, N^+)$ for every $i < \delta$. Then $\mathrm{NF}_{\mathfrak{s}}(M_0, N_0, M_\delta, N^+)$.*
2) If $\langle M_i : i \le \delta+1\rangle$ is $\le_{\mathfrak{s}}$-increasing continuous and $\langle N_i : i \le \delta+1\rangle$ is increasing continuous and $\mathrm{NF}_{\mathfrak{s}}(M_i, N_i, M_{\delta+1}, N_{\delta+1})$ for each $i < \delta$ then $\mathrm{NF}_{\mathfrak{s}}(M_\delta, , N_\delta, M_{\delta+1}, N_{\delta+1})$.
3) Assume that $\langle M_i^\ell : i \le \delta\rangle$ is $\le_{\mathfrak{s}}$-increasing for $\ell \le 3$.
If $\mathrm{NF}_{\mathfrak{s}}(M_i^0, M_i^1, M_i^2, M_i^3)$ for each $i < \delta$ then this holds for $i = \delta$, too.

Proof of 8.19. In all cases, of course δ is a limit ordinal and without loss of generality $\delta < \lambda_{\mathfrak{s}}^+$.
1) Note that

$(*)_1$ without loss of generality N_0 is $(\lambda, *)$-brimmed over M_0 and N^+ is $(\lambda, *)$-brimmed over $N_0 \cup M_\delta$.

[Why? We can find N' such that $N_0 \le_{\mathfrak{s}} N', N'$ is $(\lambda, *)$-brimmed over N_0 and $N' \cap N^+ = N_0$. Also we can find N'' such that $\mathrm{NF}_{\mathfrak{s}}(N_0, N', N^+, N'')$ hence $N' <_{\mathfrak{s}} N'', N^+ <_{\mathfrak{s}} N''$ and such that N'' is $(\lambda, *)$-brimmed over $N' \cup N^+$ hence over $N' \cup M_\delta$. Now for $i < \delta$ clearly $\mathrm{NF}_{\mathfrak{s}}(M_0, M_i, N_0, N^+)$ and $\mathrm{NF}_{\mathfrak{s}}(N_0, N^+, N', N'')$ hence by transtivity $\mathrm{NF}_{\mathfrak{s}}(M_0, M_i, N', N'')$. Also if we prove that $\mathrm{NF}_{\mathfrak{s}}(M_0, M_\delta, N', N'')$ then by monotonicity we shall get $\mathrm{NF}_{\mathfrak{s}}(M_0, M_\delta, N_0, N^+)$ and we are done. So we can replace N_0, N^+ by N', N'' and they are as required in $(*)_1$.]

$(*)_2$ without loss of generality M_{i+1} is $(\lambda, *)$-brimmed over M_i for each $i < \delta$.

[Why? We can choose M_i' by induction on $i \le \delta$ such that: M_i' is $\le_{\mathfrak{s}}$-increasing continuous, $M_i' \cap N^+ = M_i, M_0' = M_0, M_i \le_{\mathfrak{s}} M_i'$ for $i \le \delta$ and $\mathrm{NF}_{\mathfrak{s}}(M_i, M_i', M_{i+1}, M_{i+1}')$ and M_{i+1}' is $(\lambda, *)$-brimmed over $M_i' \cup M_{i+1}$ for $i < \delta$. Then let N^* be such that $\mathrm{NF}_{\mathfrak{s}}(M_\delta, N^+, M_\delta', N^*)$ and N^* is $(\lambda, *)$-brimmed over $M_\delta' \cup N^+$. Now by long transitivity for each $i \le j \le \delta$ we have $\mathrm{NF}_{\mathfrak{s}}(M_i, M_i', M_j, M_j')$.

Let $i < \delta$, in particular $\mathrm{NF}_{\mathfrak{s}}(M_i, M'_i, M_\delta, M'_\delta)$, also $\mathrm{NF}_{\mathfrak{s}}(M_\delta, N^+, M'_\delta, N^*)$ hence by symmetry $\mathrm{NF}_{\mathfrak{s}}(M_\delta, M'_\delta, N^+, N^*)$ so by transitivity we have $\mathrm{NF}_{\mathfrak{s}}(M_i, M'_i, N^+, N^*)$.

By an assumption we have $\mathrm{NF}_{\mathfrak{s}}(M_0, N_0, M_i, N^+)$ and by the previous sentence and symmetry $\mathrm{NF}_{\mathfrak{s}}(M_i, N^+, M'_i, N^*)$ hence by transitivity $\mathrm{NF}_{\mathfrak{s}}(M_0, N_0, M'_i, N^*)$. As this holds for every $i < \delta$ and $M'_0 = M_0$ clearly the sequence $\langle M'_i : i \leq \delta \rangle$, with N_0, N^* satisfies the assumptions (so far) on $\langle M_i : i \leq \delta \rangle, N_0, N^+$.

Lastly, if we shall prove that $\mathrm{NF}_{\mathfrak{s}}(M_0, N_0, M'_\delta, N^*)$ then by monotonicity we get $\mathrm{NF}_{\mathfrak{s}}(M_0, N_0, M_\delta, N^+)$ so we can replace $\langle M_i : i \leq \delta \rangle, N_0, N^+$ by $\langle M'_i : i \leq \delta \rangle, N_0, N^*$, so $(*)_2$ is O.K.]

$(*)_3$ there is $\mathbf{J}_i \subseteq \mathbf{I}_{M_i, M_{i+1}}$ such that $(M_i, M_{i+1}, \mathbf{J}_i) \in K^{3,\mathrm{vq}}_{\mathfrak{s}}$.

[Why? By 8.6(2) as M_{i+1} is $(\lambda, *)$-brimmed over M_i.]

$(*)_4$ we can find $\bar{N}', \mathbf{I}$ such that

- (a) $\bar{N}' = \langle N'_i : i \leq \delta \rangle$ is $\leq_{\mathfrak{s}}$-increasing continuous
- (b) $M_i \leq_{\mathfrak{s}} N'_i$
- (c) $\mathrm{NF}_{\mathfrak{s}}(M_i, N'_i, M_j, N'_j)$ for $i < j \leq \delta$
- (d) $(M_i, N'_i, \mathbf{I}) \in K^{3,\mathrm{vq}}_{\mathfrak{s}}$ for $i \leq \delta$
- (e) N'_0 is $(\lambda, *)$-brimmed over M_0
- (f) $N'_i \cap N^+ = M'_i$.

[Why? First choose $N'_0, \mathbf{I}$ by 8.6(2) satisfying (b),(d),(e) such that $N'_0 \cap M_\delta = M_0$. Second, first ignoring clause (f), by $(*)_2$ we can choose N'_i for $i \in (0, \delta]$, by 5.18 (as $\mathfrak{s}$ is good$^+$ and successful). But for clause (f), by clause (c) clearly $N'_i \cap M_\delta = M_i$ so by renaming we get it.]

$(*)_5$ we can choose (f_i, N^+_i) by induction on $i \leq \delta$ such that

- (a) N^+_i is $\leq_{\mathfrak{s}}$-increasing continuous
- (b) f_i is a $\leq_{\mathfrak{s}}$-embedding of N'_i into N^+_i
- (c) f_i is increasing continuous
- (d) f_i is the identity on M_i

(e) f_0 is onto N_0

(f) $N_0^+ = N^+$.

[Why? For $i = 0$ note that both N'_0 and N_0 are $\leq_{\mathfrak{s}}$-extensions of M_0 which are $(\lambda, *)$-brimmed over it.
For i limit take unions. For $i = j+1$, note that ($i < \delta$ and)

(α) $f_0(\mathbf{I}) = f_j(\mathbf{I})$ is independent in $(M_0, f_0(N'_0)) = (M_0, f_j(N'_0)) = (M_0, N_0)$.
[Why? By $(*)_4(d)$ we have $\mathbf{I}$ is independent in (M_0, N'_0).]

(β) $\mathrm{NF}_{\mathfrak{s}}(M_0, N_0, M_i, N_j^+)$ hence $f_0(\mathbf{I})$ is independent in $(M_0, M_i, N^+$

[Why? By an assumption of the claim $\mathrm{NF}_{\mathfrak{s}}(M_0, N_0, M_i, N^+)$ and by the induction hypothesis $N^+ = N_0^+ \leq_{\mathfrak{s}} N_j$ so by monotonicity $\mathrm{NF}_{\mathfrak{s}}(M_0, N_0, M_i, N_j^+)$. The hence is by 5.3.]

(γ) $M_0 \leq_{\mathfrak{s}} M_j \leq_{\mathfrak{s}} M_i$ and $(M_j, f_j(N'_j), f_0(\mathbf{I})) \in K_{\mathfrak{s}}^{3,\mathrm{vq}}$.
[Why? Recall $(*)_4(d)$.]

(δ) $\mathrm{NF}_{\mathfrak{s}}(M_j, f_j(N'_j), M_i, N_j^+)$.
[Why? By $(\beta) + (\gamma)$ and the definition of $K_{\mathfrak{s}}^{3,\mathrm{vq}}$.]

and recall from $(*)_4(c)$

(ε) $\mathrm{NF}_{\mathfrak{s}}(M_j, N'_j, M_i, N'_i)$.

Now by the uniqueness of $\mathrm{NF}_{\mathfrak{s}}$ and (δ)+(ε), recalling by the induction hypothesis, $(*)_5(b) + (c)$ for j we can extend $f_j \cup \mathrm{id}_{M_i}$ to a $\leq_{\mathfrak{s}}$-embedding of N'_i into some $\leq_{\mathfrak{s}}$-extension of N_j^+ which we call N_i^+. So we have finished carrying the induction hence proving $(*)_5$.]

$(*)_6$ $\mathrm{NF}_{\mathfrak{s}}(M_0, N_0, M_\delta, f_\delta(N'_\delta))$.

[Why? As by $(*)_4(c)$ we have $\mathrm{NF}_{\mathfrak{s}}(M_0, N'_0, M_\delta, N'_\delta)$ and f_δ maps N'_0 onto N_δ and is the identity on M_δ.]
As $f(N'_\delta) \leq_{\mathfrak{s}} N_\delta^+, N_0 \leq N_\delta^+$ by monotonicity of NF we are done.
2) We should prove that $\mathrm{NF}_{\mathfrak{s}}(M_\delta, N_\delta, M_{\delta+1}, N_{\delta+1})$ when $\langle M_i : i \leq \delta + 1\rangle$ is $\leq_{\mathfrak{s}}$-increasing continuous, $\langle N_i : i \leq \delta + 1\rangle$ is $\leq_{\mathfrak{s}}$-increasing continuous and for all $i < \delta$ $\mathrm{NF}_{\mathfrak{s}}(M_i, N_i, M_{\delta+1}, N_{\delta+1})$ holds. Now

$(*)_1$ without loss of generality there is $\mathbf{J}$ such that $(M_\delta, M_{\delta+1}, \mathbf{J}) \in K^{3,\mathrm{vq}}_{\mathfrak{s}}$.

[Why? By 8.6(1) there are M', $\mathbf{J}$ such that $M_{\delta+1} \le_{\mathfrak{s}} M'$ and $(M_\delta, M', \mathbf{J}) \in K^{3,\mathrm{vq}}_{\mathfrak{s}}$ and without loss of generality $M' \cap N_{\delta+1} = M_\delta$. We can find N' such that $\mathrm{NF}_{\mathfrak{s}}(M_{\delta+1}, N_{\delta+1}, M', N')$ hence by transitivity $i < \delta \Rightarrow \mathrm{NF}_{\mathfrak{s}}(M_i, N_i, M', N')$. Now if we shall prove $\mathrm{NF}_\delta(M_\delta, N_\delta, M', N')$ then by monotonicity it follows that $\mathrm{NF}_{\mathfrak{s}}(M_\delta, N_\delta, M_{\delta+1}, N_{\delta+1})$ so we can replace $(M_{\delta+1}, N_{\delta+1})$ by (M', N') and $\mathbf{J}$ is as required.]

$(*)_2$ it suffices to prove that $\mathbf{J}$ is independent in $(M_\delta, N_\delta, N_{\delta+1})$.

[Why? By a basic property of $K^{3,\mathrm{vq}}_{\mathfrak{s}}$.]

Let $\mathbf{J}_i = \{c \in \mathbf{J}\colon \mathbf{tp}_{\mathfrak{s}}(c, M_\delta, M_{\delta+1})$ does not fork over $M_i\}$

$(*)_3$ $\langle \mathbf{J}_i : i < \delta\rangle$ is increasing with union $\mathbf{J}$.

[Why? By local character, i.e., Ax(E)(c) of good λ-frames.]

$(*)_4$ it suffices to prove that $\mathbf{J}_i$ is independent in $(M_\delta, N_\delta, N_{\delta+1})$ for each $i < \delta$.

[Why? By the finite character of being independent, i.e., 5.4(1) it suffices to prove that $\mathbf{J}_i$ is independent in $(M_\delta, N_\delta, N_{\delta+1})$.]

$(*)_5$ $\mathbf{J}_i$ is independent in $(M_i, M_\delta, M_{\delta+1})$ hence in $(M_i, M_j, N_{\delta+1})$ for $j \in [i, \delta]$.

[Why? As $\mathbf{J}_i$ is independent in $(M_\delta, M_{\delta+1})$ and $c \in \mathbf{J}_i \Rightarrow \mathbf{tp}_{\mathfrak{s}}(c, M_\delta, M_{\delta+1})$ does not fork over M_i, using 5.6(4).]

$(*)_6$ $\mathbf{J}_i$ is independent in $(M_i, N_j, N_{\delta+1})$ when $i \le j < \delta$.

[Why? As $\mathbf{J}_i$ is independent in $(M_i, M_j, N_{\delta+1})$ and $\mathrm{NF}_{\mathfrak{s}}(M_j, N_j, M_{\delta+1}, N_{\delta+1})$ and $\mathbf{J}_i \subseteq M_{\delta+1}$ by 5.3.]

$(*)_7$ $\mathbf{J}_i$ is independent in $(M_i, N_\delta, N_{\delta+1})$ for $i < \delta$.

[Why? By $(*)_6$ and using 5.10(2) recalling the definition of $\mathbf{J}_i$.]
So by $(*)_4 + (*)_7$ we are done.
3) We prove by induction on δ hence without loss of generality $\langle M^\ell_i :$

$i \le \delta\rangle$ is $\le_{\mathfrak{s}}$-increasing continuous for each $\ell \le 3$. We would like to show that without loss of generality M^2_δ is $(\lambda, *)$-brimmed over M^0_δ, this will be done in $(*)_2$ below. Toward this we shall first prove

$(*)_1$ we can choose (M^4_i, M^5_i) by induction on $i \le \delta$ such that

- (a) M^5_i is $\le_{\mathfrak{s}}$-increasing continuous
- (b) $M^3_i \le_{\mathfrak{s}} M^5_i$ and $M^5_i \cap M^3_\delta = M^3_i$
- (c) $\mathrm{NF}_{\mathfrak{s}}(M^3_j, M^5_j, M^3_i, M^5_i)$ when $i = j+1 < \delta$
- (d) M^4_i is $\le_{\mathfrak{s}}$-increasing continuous
- (e) $M^2_i \le_{\mathfrak{s}} M^4_i \le_{\mathfrak{s}} M^5_i$
- $(f)_1$ M^4_{i+1} is $(\lambda, *)$-brimmed over $M^4_i \cup M^2_{i+1}$
- $(f)_2$ M^5_{i+1} is $(\lambda, *)$-brimmed over $M^5_i \cup M^3_{i+1}$
- (g) $\mathrm{NF}_{\mathfrak{s}}(M^2_j, M^4_j, M^2_i, M^5_i)$ when $i = j+1 < \delta$
- (h) $\mathrm{NF}_{\mathfrak{s}}(M^2_i, M^4_i, M^3_i, M^5_i)$ if $i < \delta$.

Why?

Case 1: $i = 0$.

Easy.

Case 2: i a limit ordinal.

Let $M^5_i = \cup\{M^5_j : j < i\}$, $M^4_i = \cup\{M^4_j : j < i\}$. Clauses (a)-(e) holds trivially. Clauses $(f)_1, (f)_2, (g)$ are irrelevant and lastly clause (h) holds by the induction hypothesis on δ if $i < \delta$ and is empty otherwise.

Case 3: $i = j+1$.

First we can find M^5_i such that $\mathrm{NF}_{\mathfrak{s}}(M^3_j, M^5_j, M^3_i, M^5_i)$. Moreover, we can choose M^5_i such that

$\odot_1$ $M^5_i \cap M^3_\delta = M^3_i$ and M^5_i is $(\lambda, *)$-brimmed over $M^5_j \cup M^3_i$.

Notice that:

(i) $M^2_j \le_{\mathfrak{s}} M^3_j \le M^3_i$.
[Why? By the assumptions of the claim.]

(ii) $M^4_j \le_{\mathfrak{s}} M^5_j \le M^5_i$.
[Why? By clause (e) for j and the choice of M^5_i.]

(iii) $M^2_j \le_{\mathfrak{s}} M^4_j, M^3_j \le_{\mathfrak{s}} M^5_j$ and $M^3_i \le M^5_i$.
[Why? By clause (e) for j, by clause (b) for j and by the choice of M^5_i.]

(iv) $\mathrm{NF}_{\mathfrak{s}}(M^2_j, M^4_j, M^3_j, M^5_j)$.
[Why? By clause (h) in the present induction.]

(v) $\mathrm{NF}_{\mathfrak{s}}(M^3_j, M^5_j, M^3_i, M^5_i)$.
[Why? By the choice of M^5_i.]

(vi) $\mathrm{NF}_{\mathfrak{s}}(M^2_j, M^4_j, M^3_i, M^5_i)$.
[Why? By (iv) + (v) and transitivity of $\mathrm{NF}_{\mathfrak{s}}$.]

But $M^2_j \le_{\mathfrak{s}} M^2_i \le_{\mathfrak{s}} M^3_i$ hence by ((vi) and) monotonicity for $\mathrm{NF}_{\mathfrak{s}}$ we have

(vii) $\mathrm{NF}_{\mathfrak{s}}(M^2_j, M^4_j, M^2_i, M^5_i)$.

Recall

(viii) $M^2_j \le_{\mathfrak{s}} M^2_i \le_{\mathfrak{s}} M^3_i$ and $M^2_j \le_{\mathfrak{s}} M^4_j$

and by (vi) we have $\mathrm{NF}_{\mathfrak{s}}(M^2_j, M^4_j, M^3_i, M^5_i)$, hence

(ix) $M^4_j \cap M^3_i = M^2_j$.

Next by (viii) we have $M^2_j \le_{\mathfrak{s}} M^4_j, M^2_j \le_{\mathfrak{s}} M^2_i$ and by (viii) first clause + (ix) we have $M^4_j \cap M^2_i = M^2_j$ hence we can find N^4_i such that

(x) $\mathrm{NF}_{\mathfrak{s}}(M^2_j, M^4_j, M^2_i, N^4_i)$, and N^4_i is $(\lambda, *)$-brimmed over $M^4_j \cup M^2_i$ and N^4_i is disjoint to $M^3_i \backslash M^2_i$.

So $M^2_i \le_{\mathfrak{s}} M^3_i$ (by (viii)) and $M^2_i \le_{\mathfrak{s}} N^4_i$ (by (x)) and $M^3_i \cap N^4_i = M^2_i$ (by the last phrase in (x)), hence there is N^5_i such that

(xi) $\mathrm{NF}_{\mathfrak{s}}(M^2_i, N^4_i, M^3_i, N^5_i)$

so by (x) + (xi) and transitivity of $\mathrm{NF}_{\mathfrak{s}}$

(xii) $\mathrm{NF}_{\mathfrak{s}}(M^2_j, M^4_j, M^3_i, N^5_i)$.

By (xii) + (vi) and uniqueness of $\mathrm{NF}_{\mathfrak{s}}$ recalling M^5_i is $(\lambda,*)$-brimmed over $M^5_j \cup M^3_i$

(*xiii*) there is a $\le_{\mathfrak{s}}$-embedding $f = f_i$ of N^5_i into M^5_i which is the identity on $M^3_i \cup M^4_j$.

Lastly, we choose

$\odot_2$ $M^4_i = f(N^4_i)$,

well defined as $N^4_i \le_{\mathfrak{s}} N^5_i = \mathrm{Dom}(f)$ so we have chosen (M^4_i, M^5_i) and just have to check that clauses (a)-(h) of $(*)_1$ are satisfied.
Clauses (a),(b) and (c) hold by the choice of M^5_i in $\odot_1$.
Clause (d), i.e. $M^4_j \le_{\mathfrak{s}} M^4_i$ holds as $M^4_j \le_{\mathfrak{s}} N^4_i, M^4_i = f(N^4_i)$ and f is the identity on M^4_j by (x),$\odot_2$,(xiii) respectively.
Clause (e) holds, i.e. $M^2_i \le_{\mathfrak{s}} M^4_i \le_{\mathfrak{s}} M^5_i$ as $M^2_i \le_{\mathfrak{s}} N^4_i$ by (x) and $f \restriction N^4_i$ is a $\le_{\mathfrak{s}}$-embedding of N^4_i into M^5_i over M^2_i by (xiii) as $M^2_i \le_{\mathfrak{s}} M^3_i$ by the assumptions of 8.19(3).
Clause $(f)_1$ with i, j here standing for $i+1, i$ there holds as N^4_i is $(\lambda,*)$-brimmed over $M^4_j \cup M^2_i$ (see (x)) and f preserves this as it is the identity on $M^4_j \cup M^2_i \subseteq M^4_j \cup M^3_i$ by (xiii).
Clause $(f)_2$ holds by the choice of M^5_i in $\odot_1$.
Clause (g) holds by (vii) above.
Clause (h) holds because $\mathrm{NF}_{\mathfrak{s}}(M^2_i, N^4_i, M^3_i, N^5_i)$ holds by clause (xi), but f is a $\le_{\mathfrak{s}}$-embedding of N^5_i into M^5_i so we also have $\mathrm{NF}_{\mathfrak{s}}(f(M^2_i), f(N^4_i), f(M^3_i), f(N^5_i))$, but $f(M^2_i) = M^2_i$ by (xiii) + $M^2_i \le_{\mathfrak{s}} M^3_i$ and $f(N^4_i) = M^4_i$ by $\odot_2$ and $f(M^3_i) = M^3_i$ by (xiii) and $f(N^5_i) \le_{\mathfrak{s}} M^5_i$ by (xiii), hence $\mathrm{NF}_{\mathfrak{s}}(M^2_i, M^4_i, M^3_i, M^5_i)$ as required.
So we have finished proving $(*)_1$.

$(*)_2$ without loss of generality M^2_δ is $(\lambda,*)$-brimmed over M^0_δ.

Why? Note that:

$(i)_2$ $\langle M^\ell_i : i \le \delta\rangle$ is $\le_{\mathfrak{s}}$-increasing continuous for $\ell \le 5$.
[Why? By an assumption of 8.19(3) for $\ell \le 3$ and by clauses (a),(d) of $(*)_1$, for $\ell = 4, 5$ respectively.]

$(ii)_2$ $M^0_i \le_{\mathfrak{s}} M^2_i \le_{\mathfrak{s}} M^4_i$.
[Why? By the assumptions of 8.19(3) and $(*)_1(e)$.]

$(iii)_2$ $M_i^1 \leq_{\mathfrak{s}} M_i^3 \leq_{\mathfrak{s}} M_i^5$.
[Why? By the assumptions of 8.19(3) and $(*)_1(b)$.]

$(iv)_2$ $M_i^\ell \leq_{\mathfrak{s}} M_i^{\ell+1}$ for $\ell = 0, 2, 4$.
[Why? By an assumption of 8.19(3) for $\ell = 0, 2$ and $(*)_1(e)$ for $\ell = 4$.]

$(v)_2$ $\mathrm{NF}_{\mathfrak{s}}(M_i^0, M_i^1, M_i^2, M_i^3)$ for $i < \delta$.
[Why? By an assumption of the claim 8.19(3).]

$(vi)_2$ $\mathrm{NF}_{\mathfrak{s}}(M_i^2, M_i^3, M_i^4, M_i^5)$ for $i < \delta$.
[Why? By clause (h) of $(*)_1$ and the symmetry property of $\mathrm{NF}_{\mathfrak{s}}$.]

$(vii)_2$ $\mathrm{NF}_{\mathfrak{s}}(M_i^0, M_i^1, M_i^4, M_i^5)$ for $i < \delta$.
[Why? By $(v)_2 + (vi)_2+$ transitivity of $\mathrm{NF}_{\mathfrak{s}}$.]

$(viii)_2$ if $\mathrm{NF}_{\mathfrak{s}}(M_\delta^0, M_\delta^1, M_\delta^4, M_\delta^5)$ then $\mathrm{NF}_{\mathfrak{s}}(M_\delta^0, M_\delta^1, M_\delta^2, M_\delta^3)$.
[Why? By monotonicity of $\mathrm{NF}_{\mathfrak{s}}$ as (using $(i)_2$ + smoothness from $M_\delta^0 \leq_{\mathfrak{s}} M_\delta^4$ by $(ii)_2$, $M_\delta^1 \leq_{\mathfrak{s}} M_\delta^5$ by $(iii)_2$ and $M_\delta^1 \leq_{\mathfrak{s}} M_\delta^2$, e.g. by $(iv)_2$.]

So by $(i)_2$ for $\ell = 0, 1, 4, 5$ and $(ii)_2$ and $(iii)_2$ and $(iv)_2$ for $\ell = 0, 4$ and $(viii)_2$ if we replace $\langle M_i^2 : i \leq \delta\rangle, \langle M_i^3 : i \leq \delta\rangle$ by $\langle M_i^4 : i \leq \delta\rangle, \langle M_i^5 : i \leq \delta\rangle$ respectively, the assumptions still holds for the new case; and also, by $(viii)_2$, the conclusion for the new case implies the conclusion for the original case. However, M_δ^4 is $(\lambda, *)$-brimmed over M_δ^0 by clauses (d),(e),$(f)_1$ of $(*)_1$, so we have proved $(*)_2$.

$(*)_3$ There is $\mathbf{J}_2$ such that $(M_\delta^0, M_\delta^2, \mathbf{J}_2) \in K_{\mathfrak{s}}^{3,\mathrm{vq}}$.
[Why? By $(*)_2$ and 8.6(2).]

$(*)_4$ without loss of generality there is $\mathbf{J}_1$ such that $(M_\delta^0, M_\delta^1, \mathbf{J}_1) \in K_{\mathfrak{s}}^{3,\mathrm{vq}}$.
[Why? The claim is symmetric for $\langle M_i^1 : i \leq \delta\rangle, \langle M_i^2 : i \leq \delta\rangle$ as $\mathrm{NF}_{\mathfrak{s}}$ is symmetric, and in $(*)_1 + (*)_2$, $\langle M_i^1 : i \leq \delta\rangle$ was not changed as well as $\langle M_i^0 : i \leq \delta\rangle$.]

For $\ell = 1, 2$ define a function $\mathbf{i}_\ell : \mathbf{J}_\ell \to \delta$ by $\mathbf{i}_\ell(c) = \mathrm{Min}\{i : c \in M_i^\ell$ and $\mathbf{tp}_{\mathfrak{s}}(c, M_\delta^0, M_\delta^\ell)$ does not fork over $M_i^0\}$. Now

$(*)_5$ $\mathbf{i}_\ell(i) < \delta$ is well defined for $c \in \mathbf{J}_c, \ell = 1, 2$.

[Why? If $c \in \mathbf{J}_\ell$ then $c \in M^\ell_\delta$, but $M^\ell_\delta = \cup\{M^\ell_i : i < \delta\}$ hence for some $i < \delta, c \in M^\ell_i$. Also $\mathbf{tp}_{\mathfrak{s}}(c, M^0_\delta, M^1_\delta) \in \mathscr{S}^{\rm bs}_{\mathfrak{s}}(M^0_\delta)$ as $(M^0_\delta, M^\ell_\delta, \mathbf{J}_\ell) \in K^{3,\rm vq}_{\mathfrak{s}}$ by $(*)_3 + (*)_4$ hence by Ax(E)(c) (local character) of good λ-frames, for some $j < \delta$, $\mathbf{tp}_{\mathfrak{s}}(c, M^0_\delta, M^\ell_\delta)$ does not fork over M^0_j. So $\max\{i, j\}$ show that $\mathbf{i}_\ell(c)$ is well defined and $< \delta$.]

Let $\mathbf{J}^\ell_i = \{c \in \mathbf{J}_\ell : \mathbf{i}_\ell(c) \le i\}$, so clearly

$(*)_6$ $\langle \mathbf{J}^\ell_i : i < \delta\rangle$ is $\subseteq$-increasing with union $\mathbf{J}_\ell$

$(*)_7$ $\mathbf{J}_1 \cap \mathbf{J}_2 = \emptyset$.
[Why? Easy as for $i < \delta$, $\mathrm{NF}_{\mathfrak{s}}(M^0_0, M^1_i, M^2_i, M^3_i)$ hence $\mathrm{M}^1_i \cap M^2_i = M^0_i$, but $\langle M^\ell_i : i \le \delta\rangle$ is $\le_{\mathfrak{s}}$-increasing continuous, so $M^1_\delta \cap M^2_\delta = M^0_\delta$ but $\mathbf{J}_\ell \subseteq M^\ell_\delta \backslash M^0_\delta$ so we are done.]

$(*)_8$ $\mathbf{J}^1_i \cup \mathbf{J}^2_i$ is independent in $(M^0_i, M^0_j, M^3_\delta)$ when $i \le j < \delta$.
[Why? By the definition of and the choice of $\mathbf{J}^\ell_i, \mathbf{i}_\ell(c)$ for $c \in \mathbf{J}^1_i \cup \mathbf{J}^2_i$ clearly $\mathbf{tp}_{\mathfrak{s}}(c, M^0_j, M^3_j)$ does not fork over M^0_i. Also by $(*)_3 + (*)_4$ we know that $\mathbf{J}^\ell_i$ is independent in $(M^0_\delta, M^\ell_\delta)$ hence by the previous sentence and 5.6(4) it is independent in $(M^0_i, M^0_\delta, M^\ell_\delta)$ but $i \le j < \delta$ hence by monotonicity, e.g. 5.4(4) + 5.6(4) the set $\mathbf{J}^\ell_i$ is independent in $(M^0_i, M^0_j, M^3_\delta)$ but $\mathbf{J}^\ell_i \subseteq M^\ell_i \subseteq M^\ell_j$ hence in (M^0_i, M^0_j, M^3_j). As $M^0_i \le_{\mathfrak{s}} M^0_j$ by an assumption and $\mathrm{NF}_{\mathfrak{s}}(M^0_j, M^1_j, M^2_j, M^3_j)$ by an assumption, clearlya by 5.6(2) it follows that $\mathbf{J}^1_i \cup \mathbf{J}^2_i$ is independent in (M^0_i, M^0_j, M^3_j).]

$(*)_9$ $\mathbf{J}^1_i \cup \mathbf{J}^2_i$ is independent in $(M^0_i, M^0_\delta, M^3_\delta)$.
[Why? By $(*)_8$ this holds for every $j \in [i, \delta)$ hence it holds for $j = \delta$ by 5.10(2) as required.]

$(*)_{10}$ $\mathbf{J}_1 \cup \mathbf{J}_2$ is independent in (M^0_δ, M^3_δ).
[Why? By 5.4(1) as the sequence $\langle \mathbf{J}^1_i \cup \mathbf{J}^2_i : i < \delta\rangle$ is $\subseteq$-increasing with union $\mathbf{J}_1 \cup \mathbf{J}_2$.]

So $M^0_\delta \le_{\mathfrak{s}} M^\ell_\delta \le_{\mathfrak{s}} M^3_\delta$ for $\ell = 1, 2$ (by the assumptions of the claim) and $(M^0_\delta, M^\ell_\delta, \mathbf{J}_\ell) \in K^{3,\rm vq}_{\mathfrak{s}}$ by $(*)_3 + (*)_4$ and $\mathbf{J}_1 \cup \mathbf{J}_2$ is independent in (M^0_δ, M^3_δ) (by $(*)_{10}$) and $\mathbf{J}_1 \cap \mathbf{J}_2 = \emptyset$ by $(*)_7$ hence by claim 8.13(2) (so again we use $\mathfrak{s}$ is good$^+$, successful, $\perp = \underset{\rm wk}{\perp}$) we have $\mathrm{NF}_{\mathfrak{s}}(M^0_\delta, M^1_\delta, M^2_\delta, M^3_\delta)$ as desired in the claim. $\square_{8.19}$

8.20 Exercise: 1) Assume that $\beta < \lambda^+$ and their sequences $\langle M_i^2 : i \le \beta\rangle, \langle M_i^3 : i \le \beta\rangle$ are $\le_{\mathfrak{s}}$-increasing continuous and $M_i^2 \le_{\mathfrak{s}} M_i^3$ for $i \le \beta$.

Then we can choose $\langle M_i^4 : i \le \beta\rangle, \langle M_i^5 : i \le \beta\rangle$ which satisfy clauses (a)-(h) of $(*)_1$ from the proof of 8.19(3).
2) Similarly for $\langle M_i^2 : i < \lambda^+\rangle, \langle M_i^3 : i < \lambda^+\rangle$.

Remark. 1) On 8.19 see more in [Sh 842].
2) Compare the following conclusion with 7.15, but we give a different proof.

8.21 Conclusion. Assume that $\delta < \lambda^+$ is a limit ordinal, the sequences $\langle M_i : i < \delta\rangle$ and $\langle N_i : i < \delta\rangle$ are $\le_{\mathfrak{s}}$-increasing, $\mathbf{J}_i$ $(i \le \delta)$ is $\subseteq$-increasing and $(M_i, N_i, \mathbf{J}_i) \in K_{\mathfrak{s}}^{3,\mathrm{vq}}$ for $i < \delta$ and $c \in \mathbf{J}_i \wedge i < j < \delta \Rightarrow \mathbf{tp}_{\mathfrak{s}}(c, M_j, N_j)$ does not fork over M_i.

Then $(M_\delta, N_\delta, \mathbf{J}_\delta) \in K_{\mathfrak{s}}^{3,\mathrm{vq}}$ when we let $M_\delta = \cup\{M_i : i < \delta\}, N_\delta = \cup\{N_i : i < \delta\}, \mathbf{J}_\delta \equiv \cup\{\mathbf{J}_i : i < \delta\}$.

Proof. We shall use 8.19(2). We prove this by induction on δ hence without loss of generality $\langle M_i : i < \delta\rangle, \langle N_i : i < \delta\rangle$ are $\le_{\mathfrak{s}}$-increasing continuous and $\langle \mathbf{J}_i : i < \delta\rangle$ is $\subseteq$-increasing continuous. Now $i < \delta \Rightarrow (M_i, N_i, \mathbf{J}_i) \in K_{\mathfrak{s}}^{3,\mathrm{bs}}$ as $i < \delta \Rightarrow (M_i, N_i, \mathbf{J}_i) \in K_{\mathfrak{s}}^{3,\mathrm{vq}}$. Hence by 5.10(3) we have $(M_\delta, N_\delta, \mathbf{J}_\delta) \in K_{\mathfrak{s}}^{3,\mathrm{bs}}$. For proving the desired conclusion assume that $M_\delta \le_{\mathfrak{s}} M_{\delta+1} \le_{\mathfrak{s}} N_{\delta+1}$ and $N_\delta \le_{\mathfrak{s}} N_{\delta+1}$ and $\mathbf{J}_\delta$ is independent in $(M_\delta, M_{\delta+1}, N_{\delta+1})$ and we should prove that $\mathrm{NF}_{\mathfrak{s}}(M_\delta, N_\delta, M_{\delta+1}, N_{\delta+1})$.

Now for each $i < \delta, (M_i, N_i, \mathbf{J}_i) \in K_{\mathfrak{s}}^{3,\mathrm{bs}}$. As $c \in \mathbf{J}_i \wedge j \in (i, \delta) \Rightarrow \mathbf{tp}_{\mathfrak{s}}(c, M_j, N_j)$ does not fork over M_j, clearly $\mathbf{J}_i$ is independent in (M_i, M_j, N_j) hence by 5.10(2) we also have $\mathbf{J}_i$ is independent in $(M_i, M_\delta, N_\delta)$.

As $\mathbf{J}_i \subseteq \mathbf{J}_\delta$ and $\mathbf{J}_\delta$ is independent in $(M_\delta, M_{\delta+1}, N_{\delta+1})$ we get that $\mathbf{J}_i$ is independent in $(M_i, M_{\delta+1}, N_{\delta+1})$. As $(M_i, N_i, \mathbf{J}_i) \in K_{\mathfrak{s}}^{\mathrm{vq}}$ we can deduce $\mathrm{NF}_{\mathfrak{s}}(M_i, N_i, M_{\delta+1}, N_{\delta+1})$. As this holds for every $i < \delta$ by 8.19(2) we get $\mathrm{NF}_{\mathfrak{s}}(M_\delta, N_0, M_{\delta+1}, N_{\delta+1})$ as required.

$\square_{8.21}$

8.22 Claim. *1) [$\mathfrak{s}$ weakly has regulars] Assume that* $(M, N, \mathbf{J}) \in K^{3,\mathrm{vq}}_{\mathfrak{s}}$ *is fat and is thick (or just* $\mathscr{S}^*$*-thick for some dense* $\mathscr{S}^* \subseteq \mathscr{S}^{\mathrm{bs}}_{\mathfrak{s}}$*; see Definition 7.11(2) and 5.15(2)),* *__then__* N *is* $(\lambda, *)$*-brimmed over* M*.*
2) [$\mathfrak{s}$ weakly has regulars.] If $(M, N, \mathbf{J}) \in K^{3,\mathrm{bs}}_{\mathfrak{s}}$ *is thick,* *__then__* N *is* $(\lambda, *)$*-brimmed over* M*.*

Remark. 1) Note 8.22(1) is used later in the proof of 10.20 and 8.22(2) and is used in 12.5.
2) In the proof of 8.22 we try to minimize the use of 8.1(b),(c).

Proof. 1) Let N' be $(\lambda, *)$-brimmed over M. Now we choose by induction on $n < \omega$ first M_n and then $\mathbf{J}_n$ such that

$\circledast$ (*a*) $M_n \leq_{\mathfrak{s}} N'$
(*b*) $M_0 = M$
(*c*) $n = m + 1 \Rightarrow M_m <_{\mathfrak{s}} M_n$
(*d*) $\mathbf{J}_n \subseteq \{c \in N' : \mathbf{tp}_{\mathfrak{s}}(c, M_n, N') \in \mathscr{S}^{\mathrm{bs}}(M_n)$ and if $n = m + 1$ then $\mathbf{tp}_{\mathfrak{s}}(c, M_n, N')$ is orthogonal to $M_m\}$
(*e*) $\mathbf{J}_n$ is independent in (M_n, N')
(*f*) under (d)+(e), $\mathbf{J}_n$ is maximal
(*g*) if $n = m + 1$ then $(M_m, M_n, \mathbf{J}_m) \in K^{3,\mathrm{qr}}_{\mathfrak{s}}$
(*h*) there is a one to one mapping h from $\mathbf{J}$ onto $\mathbf{J}_0$ such that $\mathbf{tp}_{\mathfrak{s}}(c, M, N) = \mathbf{tp}_{\mathfrak{s}}(h(c), M, N')$ for $c \in \mathbf{J}$.

There is no problem to carry the definition; (for $n = 0$ use $(M, N, \mathbf{J})$ is thick"). As "$\mathfrak{s}$ weakly has regulars" see 7.5 it follows that $N' = \cup\{M_n : n < \omega\}$. By 7.9[$(B) \Rightarrow (A)$] we know that $(M, N', \mathbf{J}_0) \in K^{3,\mathrm{vq}}_{\mathfrak{s}}$, and we can prove that it is fat see 7.11(2) (or use part (2)). Using h from clause (h) and the uniqueness of fat triples there is an isomorphism f from N onto N' over N extending h. As N' is brimmed over M also N is brimmed over M.
2) Let $\mathbf{J}^+$ be a maximal subset of $\mathbf{I}_{M,N}$ which is independent in (M, N) and extend $\mathbf{J}$, exists by the local character of independence (see 5.4(3)). As $\mathfrak{s}$ weakly has regulars by 7.7(3) we know that $(M, N, \mathbf{J}^+) \in K^{3,\mathrm{vq}}_{\mathfrak{s}}$. As $\mathfrak{s}$ has uniqueness for $K^{3,\mathrm{vq}}_{\mathfrak{s}}$ it follows that

$(M, N, \mathbf{J})$ is fat (by the existence theorem for fat members of $K^{3,\mathrm{vq}}_{\mathfrak{s}}$). Now by part (1) the conclusion follows. $\square_{8.22}$

8.23 Exercise: Prove 8.6 replacing "Hypothesis 8.1" by: $\mathfrak{s}$ is a good λ-frame, which is categorical and (close to II§5)

⊛ we cannot find $M_{i,j} \in K_{\mathfrak{s}}$ for $i < \lambda^+, j < \lambda \times (1+i)$ such that $M_{i,j}$ is $\leq_{\mathfrak{s}}$-increasing continuous with i and with j and $M_{i+1,j+1}$ is $\leq_{\mathfrak{s}}$-universal over $M_{i+1,j} \cup M_{i,j+1}$ and for every $i < \lambda^+$ for some $j < \lambda \times (1+i)$ we have $\neg \mathrm{NF}_{\mathfrak{s}}(M_{i,j}, M_{i,j+1}, M_{i+1,j}, M_{i+1,j+1})$.

§9 Between cardinals, Non-splitting and getting fullness

Our major aim is to get type-full good λ-frames. Fullness seems naturally desirable (being closer to superstability) and will help in proving the existence of enough regular types. This fulfills a promise from the end of II§6. We also deal with "type-closed" and $\mathfrak{s}^{\mathrm{nsp}}$ but they will not be used.

9.1 Hypothesis. $\mathfrak{s}$ is a good λ-frame.

Below note that $a \underset{M}{\overset{N}{\bigcup}} b$ means "$a \neq b$ and $\{a, b\}$ is independent in (M, N)" but in §5 we assume $\mathfrak{s}$ is weakly successful.

9.2 Definition. 1) We say that a pre-frame $\mathfrak{t}$ is type-full (or just full) if $\mathscr{S}^{\mathrm{bs}}_{\mathfrak{t}} = \mathscr{S}^{\mathrm{na}}_{\mathfrak{t}}$.
2) Let "$\{a_0, \ldots, a_{n-1}\}$ be independent in (M, N)" mean that $\mathbf{tp}_{\mathfrak{s}}(a_\ell, M, N) \in \mathscr{S}^{\mathrm{bs}}_{\mathfrak{s}}(M)$ and for some M_ℓ (for $\ell \leq n$) we have $M_0 = M, N \leq_{\mathfrak{s}} M_n$ for $\ell \leq n, M_\ell \leq_{\mathfrak{s}} M_{\ell+1}, a_\ell \in M_{\ell+1}$ and $\mathbf{tp}_{\mathfrak{s}}(a_\ell, M_\ell, M_{\ell+1})$ does not fork over M_0 for $\ell < n$. So necessarily $\langle a_\ell : \ell < n \rangle$ is without repetitions.
3) We may allow not to distinguish types of elements and of finite tuples, so we use $\mathscr{S}_{\mathfrak{s}}(M) = \cup\{\mathscr{S}^m_{\mathfrak{s}}(M) : m < \omega\}$, we say such $\mathfrak{s}$ deals

with $(< \omega)$-types, then let ab be the concatanation (this makes no real difference).
4) We say that $\mathfrak{s}$ is type-closed <u>if</u> $\mathfrak{s}$ deals with $(< \omega)$-types and: $\mathbf{tp}_{\mathfrak{s}}(a_\ell, M, N) \in \mathscr{S}^{\mathrm{bs}}_{\mathfrak{s}}(M)$ for $\ell = 1, 2$ and $\{a_1, a_2\}$ is independent in (M, N) implies $\mathbf{tp}_{\mathfrak{s}}(a_1 a_2, M, N) \in \mathscr{S}^{\mathrm{bs}}_{\mathfrak{s}}(M)$.

9.3 Definition. For a good λ-frame $\mathfrak{s}$ we define a frame $\mathfrak{t} = \mathfrak{s}^{\mathrm{tc}} = \mathfrak{s}[\mathrm{tc}]$:

$$\lambda_{\mathfrak{t}} = \lambda_{\mathfrak{s}}, \quad \mathfrak{K}_{\mathfrak{t}} = \mathfrak{K}_{\mathfrak{s}}$$

$$\begin{aligned}\mathscr{S}^{\mathrm{bs}}_{\mathfrak{t}} = \big\{\mathbf{tp}_{\mathfrak{s}}(\langle a_0 \ldots a_{n-1}\rangle, M, N) : \; & M \le_{\mathfrak{s}} N, \, n < \omega, \\ & a_\ell \in N \backslash M \text{ for } \ell < n \\ & \text{and } \{a_\ell : \ell < n\} \text{ is independent} \\ & \text{in } (M, N)\big\}\end{aligned}$$

$$\begin{aligned}\underset{\mathfrak{t}}{\bigcup} = \big\{(M_0, M_1, \bar{a}, M_3) : \; & M_0 \le_{\mathfrak{s}} M_1 \le_{\mathfrak{s}} M_3, \\ & \text{for some } n \text{ and } \langle M^*_\ell : \ell \le n\rangle \text{ we have} \\ & \bar{a} \in {}^n(M_3), \bar{a} = \langle a_0 \ldots a_{n-1}\rangle, M_3 \le_{\mathfrak{s}} M^*_n, \\ & M_1 = M^*_0 \le_{\mathfrak{s}} M^*_1 \le_{\mathfrak{s}} \ldots \le_{\mathfrak{s}} M^*_n \\ & \text{and } \mathbf{tp}_{\mathfrak{s}}(a_\ell, M^*_\ell, M^*_{\ell+1}) \in \mathscr{S}^{\mathrm{bs}}_{\mathfrak{s}}(M_\ell) \\ & \text{does not fork over } M_0 \text{ for } \ell < n\big\}.\end{aligned}$$

<u>9.4 Exercise</u> Assume $\mathfrak{s}$ is a good λ-frame.
1) $\mathfrak{t} := \mathfrak{s}^{\mathrm{tc}}$ is a good λ-frame and deals with $(< \omega)$-type and is type closed.
2) If $\mathfrak{s}$ is a good$^+$ λ-frame, <u>then</u> $\mathfrak{t}$ is a good$^+$ λ-frame.
3) In (1), if $\mathfrak{s}$ has primes <u>then</u> $\mathfrak{t}$ has primes.
4) If $\mathfrak{s}$ is (weakly) successful, <u>then</u> $\mathfrak{t}$ is (weakly) successful.
5) If $\bar{a}_\alpha = \langle a_{\alpha,\ell} : \ell < n_\alpha\rangle$ for $\alpha < \alpha^*$ and $\mathbf{tp}_{\mathfrak{t}}(\bar{a}_\alpha, M, N) \in \mathscr{S}_{\mathfrak{t}}(M)$ for $\alpha < \alpha^*$ and there are no repetitions in $\langle a_{\alpha,n} : n < n_\alpha, \alpha < \alpha^*\rangle$ <u>then</u>

(α) $(M, N, \{\bar{a}_\alpha : \alpha < \alpha^*\}) \in K^{3,\mathrm{bs}}_{\mathfrak{t}} \Leftrightarrow (M, N, \{a_{\alpha,n} : \alpha < \alpha^*, n < n_\alpha\}) \in K^{3,\mathrm{bs}}_{\mathfrak{s}}$

(β) $(M, N, \{\bar{a}_\alpha : \alpha < \alpha^*\}) \in K^{3,\mathrm{vq}}_{\mathfrak{t}} \Leftrightarrow (M, N, \{a_{\alpha,n} : n < n_\alpha, \alpha < \alpha^*\}) \in K^{3,\mathrm{vq}}_{\mathfrak{s}}$

(γ) $(M, N, \{\bar{a}_\alpha : \alpha < \alpha^*\}) \in K^{3,\mathrm{qr}}_{\mathfrak{t}} \Leftrightarrow (M, N, \{a_{\alpha,n} : n < n_\alpha, \alpha < \alpha^*\}) \in K^{3,\mathrm{qr}}_{\mathfrak{s}}$.

[Hint: See [Sh:F735], Chapter VII.]

5) [$\mathfrak{s}$ is successful and a type-full or type-closed good λ-frame].

If $M_0 \le_{\mathfrak{s}} M_1 \le_{\mathfrak{s}} M_2$ and $(M_\ell, M_{\ell+1}, a_\ell) \in K^{3,\mathrm{pr}}_\lambda$ for $\ell = 1, 2$ and $\mathbf{tp}_{\mathfrak{s}}(a_1, M_1, M_2)$ does not fork over M_0, <u>then</u> $(M_0, M_2, a_0 a_1) \in K^{3,\mathrm{pr}}_\lambda$.

[Hint: Use 5.8(2).

* * *

Now we return to trying to deal with all types in $\mathscr{S}_{\mathfrak{s}}(M)$, i.e. fullness keeping a promise from II.6.36.

9.5 Definition. 1) [$\mathfrak{s}$ is a weakly successful good λ-frame].
Let $\mathfrak{s}^{\mathrm{nf}} = \mathfrak{s}(\mathrm{nf})$ be the following $\lambda_{\mathfrak{s}}$-frame (see below)

(a) $\mathfrak{K}_{\mathfrak{s}(\mathrm{nf})} = \mathfrak{K}_{\mathfrak{s}}$

(b) $\mathscr{S}^{\mathrm{bs}}_{\mathfrak{s}(\mathrm{nf})}(M) = \mathscr{S}^{\mathrm{na}}_{\mathfrak{s}}(M)$

(c) $\bigcup\limits_{\mathfrak{s}(\mathrm{nf})} (M_0, M_1, a, M_3)$ holds <u>iff</u> $M_0 \le_{\mathfrak{s}} M_1 \le_{\mathfrak{s}} M_3$ and $a \in M_3 \backslash M_1$ and there are M_3', M_2 such that $M_0 \le_{\mathfrak{s}} M_2 \le_{\mathfrak{s}} M_3', M_3 \le_{\mathfrak{s}} M_3'$ and $a \in M_2$ and $\mathrm{NF}_{\mathfrak{s}}(M_0, M_1, M_2, M_3')$.

2) [$\mathfrak{s}$ a successful good λ-frame]
Let $\mathfrak{s}^{+\mathrm{nsp}}$ be the $\lambda^+_{\mathfrak{s}}$-frame which we also denote by $\mathfrak{s}(*)$ or $\mathfrak{s}(+\mathrm{nsp})$, defined by

(a) $\mathfrak{K}_{\mathfrak{s}(*)} = \mathfrak{K}_{\mathfrak{s}(+)}$

(b) $\mathscr{S}^{\mathrm{bs}}_{\mathfrak{s}(*)}(M) = \mathscr{S}^{\mathrm{nsp}}_{\mathfrak{s}(+)}(M) := \{p \in \mathscr{S}^{\mathrm{na}}_{\mathfrak{s}(*)}(M) :$ for some $M_0 \le_{\mathfrak{K}[\mathfrak{s}]} M$ from $K^{\mathfrak{s}}_\lambda, p$ does not $\lambda_{\mathfrak{s}}$-split over $M_0\}$, (see Definition 2.18(1); note that in our case as M is $K^{\mathfrak{s}}$-saturated above λ, this means that every automorphism g of M over M_0 maps p

to itself); note as the types are not necessarily basic, we use splitting rather than non-forking

(c) $\underset{\mathfrak{s}(*)}{\bigcup}(M_0, M_1, a, M_3)$ <u>if</u> $M_0 \le_{\mathfrak{s}} M_1 \le_{\mathfrak{s}} M_3, a \in M_3 \backslash M_1$ and $\mathbf{tp}_{\mathfrak{s}(*)}(a, M_1, M_3)$ does not $\lambda_{\mathfrak{s}}$-split over some $N_0 \le_{\mathfrak{K}[\mathfrak{s}]} M_0$, $N_0 \in K_{\mathfrak{s}}$.

3) If $\mathfrak{K}_\lambda$ has amalgamation and JEP, and NF is a non-forking relation on ${}^4(K_\lambda)$ (see Definition II.6.1) then let $\mathfrak{s}_{\mathfrak{K}_\lambda,\mathrm{NF}_\lambda}$ be the λ-frame defined as in part (1).

9.6 Claim. *1) ($\mathfrak{s}^+$ is local over $\mathfrak{s}$): Assume $\mathfrak{s}$ is a successful good λ-frame. If $M \in K_{\mathfrak{s}(+)}$ and (the not necessarily basic types) $p, q \in \mathscr{S}_{\mathfrak{s}(+)}(M)$ and $[N \le_{\mathfrak{K}[\mathfrak{s}]} M \;\&\; N \in K_{\mathfrak{s}} \Rightarrow (p \restriction N) = (q \restriction N)]$ <u>then</u> $p = q$ and for every $M \le_{\mathfrak{s}(+)} N$ and $p \in \mathscr{S}^{\mathrm{bs}}_{\mathfrak{s}(+)}(N)$ we have p does not fork over M in the sense of $\mathfrak{s}^{+\mathrm{nsp}}$ iff p does not fork over M in the sense of $\mathfrak{s}^+$.*
2) If $\mathfrak{s}$ is a weakly successful good λ-frame <u>then</u> $t := \mathfrak{s}^{\mathrm{nf}}$ is a type-full good λ-frame.
2A) If $\mathfrak{s}$ is a weakly successful good λ-frame then $K^{3,\mathrm{uq}}_{\mathfrak{s}} \subseteq K^{3,\mathrm{uq}}_{\mathfrak{t}}$.
2B) If $\mathfrak{s}$ is a categorical successful good$^+$ λ-frame <u>then</u> $\mathfrak{t} = \mathfrak{s}^{\mathrm{nf}}$ is a successful good$^+$ λ-frame.
3) In part (2), $\mathrm{NF}_{\mathfrak{s}}$ is a non-forking relation on $K_{\mathfrak{s}} = K_{\mathfrak{t}}$ respecting $\mathfrak{t}$, so $\mathrm{NF}_{\mathfrak{t}} = \mathrm{NF}_{\mathfrak{s}}$.
4) In part (2), if $\mathbf{J}$ is independent in (M_0, M, N) for $\mathfrak{s}$ <u>then</u> it is independent in (M_0, M, N) for $\mathfrak{s}^{\mathrm{nf}}$. Also the inverse holds.
5) In part (2),

(a) $(M, N, a) \in K^{3,\mathrm{uq}}_{\mathfrak{s}}$ <u>then</u> $(M, N, a) \in K^{3,\mathrm{uq}}_{\mathfrak{t}}$

(a)$^+$ if $(M, N, a) \in K^{3,\mathrm{bs}}_{\mathfrak{s}}$ <u>then</u> $(M, N, a) \in K^{3,\mathrm{uq}}_{\mathfrak{s}}$ iff $(M, N, a) \in K^{3,\mathrm{uq}}_{\mathfrak{t}}$

(b) if $p, q \in \mathscr{S}^{\mathrm{bs}}_{\mathfrak{s}}(M)$ <u>then</u> $p \underset{\mathrm{wk}}{\perp} q$ for $\mathfrak{s}$ iff $p \underset{\mathrm{wk}}{\perp} q$ for $\mathfrak{t}$

(c) like clause (c) for $\perp$

(d) if $M \le_{\mathfrak{s}} N$ and $p \in \mathscr{S}^{\mathrm{bs}}_{\mathfrak{s}}(N)$ <u>then</u> $p \perp M$ for $\mathfrak{s}$ iff $p \perp M$ for $\mathfrak{t}$

(e) if $\mathfrak{s}$ weakly has regulars (see Definition 7.5) <u>then</u> $\mathfrak{t}$ weakly has regulars.

6) If NF *is a non-forking relation on* $\mathfrak{K}_{\mathfrak{s}}$ *respecting* $\mathfrak{s}$, *then* $\mathfrak{s}_{\mathfrak{K}_{\mathfrak{s}},\mathrm{NF}}$ *is a full good* λ*-frame.*
6A) If $\mathfrak{t}$ *is a weakly successful good* λ*-frame,* *then* $\mathfrak{t}(\mathrm{nf}) = \mathfrak{s}_{\mathfrak{K}_{\mathfrak{t}},\mathrm{NF}_{\mathfrak{t}}}$.
$\square_{9.6}$

Proof. 1) By 1.11(1) or similarly to 2.21.
2) This is promised in II.6.36.

Axioms (A),(B),(C):
As $\mathfrak{K}_{\mathfrak{t}} = \mathfrak{K}_{\mathfrak{s}}$.

Axiom (D)(a),(b),(c):
By the definition of $\mathscr{S}^{\mathrm{bs}}_{\mathfrak{t}}(M)$.

Axiom (D)(d):
We have proved $M \in \mathfrak{K}_{\mathfrak{s}} \Rightarrow |\mathscr{S}_{\mathfrak{s}}(M)| \le \lambda$ in II.4.2(1), this implies that $M \in \mathfrak{K}_{\mathfrak{t}} \Rightarrow |\mathscr{S}^{\mathrm{bs}}_{\mathfrak{t}}(M)| = |\mathscr{S}^{\mathrm{na}}_{\mathfrak{s}}(M)| \le |\mathscr{S}_{\mathfrak{s}}(M)| \le \lambda$.

Axiom (E)(a):
By the definitions.

Axiom (E)(b)(monotonicity):
So assume $p \in \mathscr{S}_{\mathfrak{s}(\mathrm{nf})}(M_3)$ does not fork over M_0 and $M_0 \le_{\mathfrak{K}_{\mathfrak{s}}} M_1 \le_{\mathfrak{K}_{\mathfrak{s}}} M_2 \le_{\mathfrak{K}_{\mathfrak{s}}} M_3$.

As $p \in \mathscr{S}^{\mathrm{bs}}_{\mathfrak{s}(\mathrm{nf})}(M_3)$ does not fork over M_0 for $\mathfrak{s}(\mathrm{nf})$ we can find N_0, N_3 such that $M_0 \le_{\mathfrak{K}_{\mathfrak{s}}} N_0 \le_{\mathfrak{K}_{\mathfrak{s}}} N_3, M_3 \le_{\mathfrak{K}_{\mathfrak{s}}} N_3$ and $a \in N_0$ such that $\mathrm{NF}_{\mathfrak{s}}(M_0, N_0, M_3, N_3)$ and $p = \mathbf{tp}_{\mathfrak{K}_{\mathfrak{s}}}(a, M_3, N_3)$. Also by $\mathrm{NF}_{\mathfrak{s}}$-existence we can find N'_ℓ for $\ell \le 3$ such that $\mathrm{NF}_{\mathfrak{s}}(M_\ell, N'_\ell, M_{\ell+1}, N'_{\ell+1})$ for $\ell = 0, 1, 2$ and N'_0 is isomorphic to N_0 over M_0. By transitivity for $\mathrm{NF}_{\mathfrak{s}}$ we have $\mathrm{NF}_{\mathfrak{s}}(M_0, N'_0, M_3, N'_3)$. By uniqueness for $\mathrm{NF}_{\mathfrak{s}}$ without loss of generality $N'_0 = N_0, N'_3 \le_{\mathfrak{s}} N''_3, N_3 \le_{\mathfrak{s}} N''_3$. So $\mathrm{NF}_{\mathfrak{s}}(M_1, N'_1, M_2, N''_3)$ and $a \in N_0 = N'_0 \le_{\mathfrak{K}_{\mathfrak{s}}} N'_1$ realizes $\mathbf{tp}_{\mathfrak{s}}(a, M_2, N'_3)$ and (by the definition of $\mathfrak{s}(\mathrm{nf})$ this type does not fork for $\mathfrak{s}(\mathrm{nf})$ over M_1, but this type is $p \restriction M_2$. So we are done.

Axiom (E)(c)(local character):
So let $\langle M_i : i \le \delta{+}1\rangle$ be $\le_{\mathfrak{K}[\mathfrak{s}]}$-increasing and $p = \mathbf{tp}_{\mathfrak{s}}(a, M_\delta, M_{\delta+1}) \in \mathscr{S}^{\mathrm{bs}}_{\mathfrak{t}}(M) = \mathscr{S}^{\mathrm{na}}_{\mathfrak{s}}(M)$.

By 1.17 we can find $\langle N_i : i \le \delta\rangle$ which is $\le_{\mathfrak{s}}$-increasing continuous, $i < j \le \delta \Rightarrow \mathrm{NF}_{\mathfrak{s}}(M_i, N_i, M_j, N_j)$ and N_δ is $(\lambda, *)$-brimmed over M_δ. Hence without loss of generality $M_{\delta+1} \leq_{\mathfrak{s}} N_\delta$, so $a \in M_{\delta+1} \leq_{\mathfrak{s}} N_\delta = \cup\{N_i : i < \delta\}$ hence for some $i < \delta, a \in N_i$. By $\mathrm{NF}_{\mathfrak{s}}(M_i, N_i, M_\delta, N_\delta)$ we deduce $\mathbf{tp}_{\mathfrak{s}}(a, M_\delta, M_{\delta+1})$ does not fork over M_i for $\mathfrak{s}^{\mathrm{nf}}$ as required.

Axiom (E)(d)(transitivity):

By II.2.18 it follows from Axioms which we have proved and (E)(e),(E)(g) proved below.

Axiom (E)(e)(uniqueness):

By uniqueness for $\mathrm{NF}_{\mathfrak{s}}$ and the definition of (orbital) type.

Axiom (E)(f)(symmetry):

So assume that $M_0 \leq_{\mathfrak{K}(\mathfrak{s})} M_1 \leq_{\mathfrak{K}(\mathfrak{s})} M_3$ and $a_2 \in M_3 \backslash M_1$ and $a_1 \in M_1 \backslash M_0$ and $\mathbf{tp}_{\mathfrak{K}(\mathfrak{s})}(a_2, M_1, M_3)$ does not fork over M_0 for $\mathfrak{t}$. By the definition of non-forking for $\mathfrak{t}$ we can find N_0, N_3 such that $\mathrm{NF}_{\mathfrak{s}}(M_0, M_1, N_0, N_3)$ and $M_3 \leq_{\mathfrak{s}} N_3$ and $a_2 \in N_0$. By the symmetry for $\mathrm{NF}_{\mathfrak{s}}$ we have $\mathbf{tp}_{\mathfrak{K}(\mathfrak{s})}(a_1, N_0, N_3)$ does not fork over M_0 for $\mathfrak{t}$ so as $a_2 \in N_0$ clearly we are done.

Axiom (E)(g)(extension existence):

By existence for $\mathrm{NF}_{\mathfrak{s}}$.

Axiom (E)(h):

By II.2.17(3),(4) it follows from axioms which we prove.

Axiom (E)(i):

By II.2.16.

So we have proved that $\mathfrak{t} = \mathfrak{s}^{\mathrm{nf}}$ is a good λ-frame and trivially it is a type-full good λ-frame as required.

2A) Trivial.

2B) Why $\mathfrak{t}$ has density for $K^{3,\mathrm{uq}}_{\mathfrak{t}}$? Let $(M, N, a) \in K^{3,\mathrm{bs}}_{\mathfrak{t}}$ and we try to choose $(M_\alpha, N_\alpha, a) \in K^{3,\mathrm{bs}}_{\mathfrak{s}}$ such that

- ⊛ $(M_0, N_0, a) = (M, N, a)$
- (b) (M_α, N_α, a) is $\leq^{\mathfrak{t}}_{\mathrm{bs}}$-increasing continuous
- (c) if $\alpha = \beta + 1, \beta$ even and $(M_\beta, N_\beta, a) \notin K^{3,\mathrm{uq}}_{\mathfrak{t}}$ then $\neg\mathrm{NF}_{\mathfrak{t}}(M_\beta, N_\beta, M_\alpha, N_\alpha)$
- (d) if $\alpha = \beta + 1, \beta$ odd then M_α, N_β is brimmed (for $\mathfrak{K}_{\mathfrak{s}}$) over M_β, N_β respectively.

For $\alpha = 0$ and α limit there are no problems. So let $\alpha = \beta + 1$; if β is odd this should be clear, so assume β is even. <u>If</u> $(M_\beta, N_\beta, a) \notin K^{3,\mathrm{uq}}_{\mathfrak{t}}$ then we can find $(M_\alpha, N^1_\alpha, N^2_\alpha)$ such that $(M_\beta, N_\beta, a) \leq^{\mathfrak{t}}_{\mathrm{bs}} (M_\alpha, N^\ell_\alpha, a)$ for $\ell = 1, 2$ and

> $(*)$ we cannot find $N^3_\alpha \in \mathfrak{K}_{\mathfrak{t}}$ which $\leq_{\mathfrak{t}}$-extends N^2_α and a $\leq_{\mathfrak{t}}$-embedding f of N^1_α into N^3_α over $M_\alpha \cup N_\beta$.

As $\mathfrak{K}_{\mathfrak{s}} = \mathfrak{K}_{\mathfrak{t}}$ clearly $(*)$ holds for $\mathfrak{K}_{\mathfrak{s}}$ too so for some $\ell \in \{1, 2\}$ we have $\neg\mathrm{NF}_{\mathfrak{s}}(M_\beta, N_\beta, M_\alpha, N^\ell_\alpha)$ so we can choose $N_\alpha = N^\ell_\alpha$.

If we are stuck we have proved the given case of density of $K^{3,\mathrm{uq}}_{\mathfrak{t}}$. If we have carried the induction <u>then</u> $M^* := \cup\{M_\alpha : \alpha < \lambda^+\}$ is $\leq_{\mathfrak{K}[\mathfrak{s}]} N := \cup\{N_\alpha : \alpha < \lambda^+\}$ are saturated (recalling $\circledast(d)$) but we know that $\leq_{\mathfrak{K}[\mathfrak{s}]} \restriction K_{\lambda^+} = \leq_{\mathfrak{s}(+)}$, hence for some club E of λ^+, for every $\alpha < \beta$ from E we have $\mathrm{NF}_{\mathfrak{s}}(M_\alpha, N_\alpha, M_\beta, N_\beta)$, contradiction.

We can deduce "$\mathfrak{s}^{\mathrm{nf}}$ is weakly successful" by "$\mathfrak{s}$, i.e. $\mathfrak{K}_{\mathfrak{s}}$ is categorical". Hence $\mathrm{NF}_{\mathfrak{t}}$ is well defined and it is a non-forking relation on ${}^4(K_{\mathfrak{t}}) = {}^4(K_{\mathfrak{s}})$ respecting $\mathfrak{t}$. But by part (3), also $\mathrm{NF}_{\mathfrak{s}}$ is a non-forking relation on ${}^4(\mathfrak{K}_{\mathfrak{t}}) = {}^4(K_{\mathfrak{s}}))$ respecting $\mathfrak{t}$.

So by the uniqueness of such relations, (see II.6.3(4)), $\mathrm{NF}_{\mathfrak{s}} = \mathrm{NF}_{\mathfrak{t}}$.

Now the proof of the other half of "$\mathfrak{t}$ is successful" is similar to the proof of the density of $K^{3,\mathrm{uq}}_{\mathfrak{s}}$ or just use the $\mathrm{NF}_{\mathfrak{s}} = \mathrm{NF}_{\mathfrak{t}}$. The "good$^+$" holds as $\leq_{\mathfrak{s}(+)} = \leq_{\mathfrak{t}(+)}$ because $\mathfrak{K}_{\mathfrak{t}} = \mathfrak{K}_{\mathfrak{s}} \wedge \mathrm{NF}_{\mathfrak{t}} = \mathrm{NF}_{\mathfrak{s}}$.

3) Trivially $\mathrm{NF}_{\mathfrak{s}}$ is a non-forking four-place relation on $K_{\mathfrak{s}} = K_{\mathfrak{t}}$, i.e., on ${}^4(K_{\mathfrak{s}})$. Does it respect $\mathfrak{t}$? Of course, it does by the definition of $\mathfrak{t}$.

4) First assume $M_0 \leq_{\mathfrak{s}} M_1 \leq_{\mathfrak{s}} M_3$ and $a \in M_3$ satisfies "$\mathbf{tp}_{\mathfrak{s}}(a, M_1, M_3)$ does not fork over M_0 for $\mathfrak{s}$". By existence for $\mathrm{NF}_{\mathfrak{s}}$, we can find N_0, N_1, f such that $\mathrm{NF}_{\mathfrak{s}}(M_0, N_0, M_3, N_3)$ and f is an isomorphism from N_0 onto M_0 over M_0. Now as $\mathrm{NF}_{\mathfrak{s}}$ respects $\mathfrak{s}$, "$\mathbf{tp}_{\mathfrak{t}}(f(a), M_1, N_3)$ does not fork over M_0 for $\mathfrak{s}$ and extends $\mathbf{tp}_{\mathfrak{s}}(a, M_0, M_3)$ hence it is equal to $\mathbf{tp}_{\mathfrak{s}}(a, M_1, N_3)$. So we get $\mathbf{tp}_{\mathfrak{K}_{\mathfrak{t}}}(a, M_1, M_3) = \mathbf{tp}_{\mathfrak{K}_{\mathfrak{t}}}(f(a), M_1, N_0) = \mathbf{tp}_{\mathfrak{K}_{\mathfrak{t}}}(f(a), M_1, N_3)$ and the latter does not fork over M_0 for $\mathfrak{t}$ by the definition of $\mathfrak{t}$.

From this by the definition of independence we can deduce part (4). The inverse is easy, too.

5) <u>Clause (a)</u>:

By part (4), if $M \leq_{\mathfrak{K}_{\mathfrak{s}}} M' <_{\mathfrak{K}_{\mathfrak{s}}} N', N \leq_{\mathfrak{K}_{\mathfrak{s}}} N'$ and $\mathbf{tp}_{\mathfrak{s}}(a, M, N) \in \mathscr{S}^{\mathrm{bs}}_{\mathfrak{s}}(M)$ then $\mathbf{tp}_{\mathfrak{K}_{\mathfrak{s}}}(a, M', N')$ does not fork over M for $\mathfrak{s}$ iff it does not fork over M for $\mathfrak{t}$. By the definition of $K^{3,\mathrm{uq}}_{\mathfrak{s}}$ this is enough.

Clause (a)$^+$:

Similarly.

Clause (b):

Follows by clause (a) and the definition of $\underset{\mathrm{wk}}{\perp}$.

Clauses (c),(d):

Should be clear.

Clause (e):

By Definition 7.5(1) we are given a $\leq_{\mathfrak{s}}$-increasing sequence $\langle M_\alpha : \alpha \leq \beta + 1\rangle$ and $M_\beta \neq M_{\beta+1}$. As $\mathfrak{s}$ weakly has regulars, for some $c \in M_{\beta+1} \backslash M_\beta$ and non-limit $\alpha \leq \beta$ we have $\mathbf{tp}_{\mathfrak{K}_{\mathfrak{s}}}(c, M_\beta, M_{\beta+1})$ does not fork over M_α is $\perp M_{\alpha-1}$ if $\alpha > 0$. Now $\mathbf{tp}_{\mathfrak{K}_{\mathfrak{t}}}(c, M_\beta, M_{\beta+1})$ does not fork over M_α by part (4), and is $\perp M_{\alpha-1}$ if $\alpha > 0$ by (part (5)), clause (c).

6) Left to the reader (and not used).

6A) Easy. $\square_{9.6}$

9.7 Lemma. *1) Assume that $\mathfrak{s}$ is a successful good$^+$ λ-frame. Then the frame $\mathfrak{s}(+\mathrm{nsp}) = \mathfrak{s}(*)$ is a full good$^+$ $\lambda^+_{\mathfrak{s}}$-frame and $\mathscr{S}^{\mathrm{bs}}_{\mathfrak{s}(*)}(M) = \mathscr{S}^{\mathrm{na}}_{\mathfrak{s}(*)}(M)$.*

2) If in addition $\mathfrak{s}(*)$ is weakly successful then weak orthogonality is equivalent to orthogonality and to super-orthogonality for $\mathfrak{s}(*)$.

Exercise: In 9.7, $\mathfrak{s}(*)$ has primes and $\mathfrak{K}_{\mathfrak{s}(*)}$ is categorical and is equal to $\mathfrak{K}_{(\mathfrak{s}^{\mathrm{nf}})^+}$.

Remark. 1) We can actually omit the assumption "$\mathfrak{s}^+$ is weakly successful" in 9.7(2) but for this we have to define those notions.

2) In some sense we do not really need both 9.7 and 9.6, so we make both proofs self contained.

Proof. Let $\lambda = \lambda_{\mathfrak{s}}$. Recall that $\mathfrak{s}^+$ is a good$^+$ $\lambda^+_{\mathfrak{s}}$-frame. We have to check the axioms there.

Axioms: (A),(B),(C).

As $\mathfrak{K}_{\mathfrak{s}(*)} = \mathfrak{K}_{\mathfrak{s}(+)}$ this follows from 1.8.

<u>Axiom</u>: (D),(a),(b) by the Definition of $\mathscr{S}^{\rm bs}_{\mathfrak{s}(*)}$.

<u>Axiom (D)(c)</u>: If $M <_{\mathfrak{s}(*)} N$ then any $a \in N \backslash M$ is O.K. by 2.27(4), i.e. $\mathbf{tp}_{\mathfrak{K}_{\mathfrak{s}(*)}}(a, M, N) \in \mathscr{S}^{\rm bs}_{\mathfrak{s}(*)}(M)$. Hence we get also $\mathscr{S}^{\rm bs}_{\mathfrak{s}(*)}(M) = \mathscr{S}^{\rm na}_{\mathfrak{s}(*)}(M)$, i.e. fullness.

<u>Axiom (D)(d)</u>: This holds as $\mathfrak{K}_{\mathfrak{s}(+)}$ is stable in $\lambda^+_{\mathfrak{s}}$ by 1.8(1) and II.4.2 but $\mathfrak{K}_{\mathfrak{s}(*)} = \mathfrak{K}_{\mathfrak{s}(+)}$, alternately use II.7.6(3). In more detail note below;

$\circledast_1$ $p_1 = p_2$ when:

(a) $N_0 \leq_{\mathfrak{s}} N_1 \leq_{\mathfrak{K}[\mathfrak{s}]} M_0$

(b) N_1 is $\leq_{\mathfrak{s}}$-universal over N_0

$(c)_\ell$ $M_0 \leq_{\mathfrak{s}(*)} M_\ell$

$(d)_\ell$ $p_\ell = \mathbf{tp}_{\mathfrak{s}(*)}(a_\ell, M_0, M_\ell)$

$(e)_\ell$ p_ℓ does not $\lambda_{\mathfrak{s}}$-split over N_0

(f) $p_1 \restriction N_1 = p_2 \restriction N_1$, i.e., $\mathbf{tp}_{\mathfrak{K}[\mathfrak{s}]}(a_1, N_1, M_1) = \mathbf{tp}_{\mathfrak{K}[\mathfrak{s}]}(a_2, N_1, M_2)$.

[Why? First note that $N_1 \leq_{\mathfrak{s}} N_2 \leq_{\mathfrak{K}[\mathfrak{s}]} M_0 \Rightarrow \mathbf{tp}_{\mathfrak{K}[\mathfrak{s}]}(a_1, N_2, M_1) = \mathbf{tp}_{\mathfrak{K}[\mathfrak{s}]}(a_2, N_2, M_1)$.
Second, use II.7.6(3).]

$\circledast_2$ for every $M_0 \in \mathfrak{K}_{\mathfrak{s}(*)}$ and $\leq_{\mathfrak{K}[\mathfrak{s}]}$-representation $\langle M_{0,\alpha} : \alpha < \lambda^+_{\mathfrak{s}}\rangle$ we have: for every $p \in \mathscr{S}^{\rm na}_{\mathfrak{s}(*)}(M_0)$ for some $\alpha < \beta < \lambda^+_{\mathfrak{s}}$, the quadruple $(p, M_0, M_{0,\alpha}, M_{0,\beta})$ satisfies the demands (p_1, M_0, N_0, N_1) satisfies in $\circledast_1$, i.e. clause (a),(b),$(e)_1$ and $p_1 \in \mathscr{S}_{\mathfrak{s}(*)}(M_0)$.

<u>Axiom (E)(a)</u>: By the definitions.

<u>Axiom (E)(b)</u>: [monotonicity].

So assume $M_0 \leq_{\mathfrak{s}(*)} M'_0 \leq_{\mathfrak{s}(*)} M'_1 \leq_{\mathfrak{s}(*)} M_1 \leq_{\mathfrak{s}(*)} M_3 \leq_{\mathfrak{s}(*)} M'_3$ and $\bigcup\limits_{\mathfrak{s}(*)} (M_0, M_1, a, M_3)$ so it is witnessed by some $N_0 \leq_{\mathfrak{K}[\mathfrak{s}]} M_0$ with $N_0 \in K_{\mathfrak{s}}$.

Now the same N_0 witnesses also $\bigcup\limits_{\mathfrak{s}(*)} (M'_0, M'_1, a, M_3)$. The other statement

$$\Big(\bigcup_{\mathfrak{s}(*)} (M_0, M_1, a, M_3) \Leftrightarrow \bigcup_{\mathfrak{s}(*)} (M_0, M_1, a, M'_3)\Big)$$

is immediate by $\mathbf{tp}_{\mathfrak{s}(*)}(a, M_1', M_3') = \mathbf{tp}_{\mathfrak{s}(*)}(a, M_1', M_3)$.

Axiom (E)(c): (local character).

So assume that $\langle M_i : i \le \delta + 1\rangle$ is $\mathfrak{s}(*)$-increasing continuous, $\delta < (\lambda_{\mathfrak{s}(*)})^+ = \lambda^{++}, c \in M_{\delta+1} \backslash M_\delta$, and assume toward contradiction that $\mathbf{tp}_{\mathfrak{s}(*)}(c, M_\delta, M_{\delta+1}) \in \mathscr{S}^{\rm bs}_{\mathfrak{s}(*)}(M_\delta)$ is a counterexample. Without loss of generality $\delta = \text{cf}(\delta)$, so $\delta \le \lambda^+_{\mathfrak{s}}$. Let $\bar M^i = \langle M^i_\alpha : \alpha < \lambda^+_{\mathfrak{s}}\rangle$ be a $\le_{\mathfrak{s}}$-representation of M_i, E a thin enough club of λ^+, so e.g.

(a) $\alpha \in E \Rightarrow c \in M^{\delta+1}_\alpha$

(b) $\alpha \in E$ & $i < j \le \delta + 1$ & $[(j < \alpha) \vee (i < (\delta \cap \alpha) \wedge j \ge \delta) \vee (i = \delta \wedge j = \delta + 1)] \Rightarrow M^i_\alpha \le_{\mathfrak{s}} M^j_\alpha$

(c) $\alpha \in E$ & $\alpha < \beta \in E$ & $i \le \delta$ & $i < \alpha \Rightarrow$ $\mathbf{tp}_{\mathfrak{s}}(c, M^i_\beta, M^{\delta+1}_\alpha)$ does not λ-split over M^i_α

(d) $\alpha < \beta \in E$ & $i < \delta \Rightarrow$ $\mathbf{tp}_{\mathfrak{s}}(c, M^\delta_\beta, M^{\delta+1}_\beta)$ does $\lambda_{\mathfrak{s}}$-split over M^i_α

(e) $\alpha < \beta \in E$ & $i \le \delta + 1 (i < \beta \vee i \ge \delta) \Rightarrow M^i_\beta$ is $(\lambda_{\mathfrak{s}}, *)$-brimmed over M^i_α.

Choose $\varepsilon_i \in E$ for $i \le \delta$, increasing continuous, so $\langle M^i_{\varepsilon_i} : i \le \delta\rangle$ is $<_{\mathfrak{s}}$-increasing continuous, each $M^i_{\varepsilon_i}$ is $(\lambda_{\mathfrak{s}}, *)$-brimmed for $\mathfrak{s}$ and $i < j \le \delta \Rightarrow M^j_{\varepsilon_j}$ is $(\lambda_{\mathfrak{s}}, *)$-brimmed over $M^i_{\varepsilon_i}$ for $\mathfrak{s}$. If $\delta < \lambda^+$, by Subclaim 2.21, for some $i < \delta$, $\mathbf{tp}_{\mathfrak{s}}(c, M^\delta_{\varepsilon_\delta}, M^{\delta+1}_{\varepsilon_{\delta+1}})$ does not $\lambda_{\mathfrak{s}}$-split over $M^i_{\varepsilon_i}$ for $\mathfrak{K}_{\mathfrak{s}}$, contradiction to the choice of E above (and obvious monotonicity of non-splitting). If $\delta = \lambda^+_{\mathfrak{s}}$, use what we proved for every limit $\delta' < \delta$ and Fodor's lemma.

Axiom (E)(d): [transitivity]

Assume

(α) $M_1 \le_{\mathfrak{s}(*)} M_2 \le_{\mathfrak{s}(*)} M_3 \le_{\mathfrak{s}(*)} M_4$

(β) $a \in M_4 \backslash M_3$

(γ) $\mathbf{tp}_{\mathfrak{s}(*)}(a, M_2, M_4)$ does not $\mathfrak{s}(*)$-fork over M_1 and

(δ) $\mathbf{tp}_{\mathfrak{s}(*)}(a, M_3, M_4)$ does not $\mathfrak{s}(*)$-fork over M_2.

Let $\bar M^\ell = \langle M^\ell_\zeta : \zeta < \lambda^+\rangle$, for $\ell = 1, 2, 3, 4$ be a $\le_{\mathfrak{s}}$-representation of M_ℓ such that $a \in M^4_0$ and without loss of generality $\alpha < \beta < \lambda^+$ & $1 \le \ell < m \le 4 \Rightarrow$ $\text{NF}_{\mathfrak{s}}(M^\ell_\alpha, M^m_\alpha, M^\ell_\beta, M^m_\beta)$ and for $\ell = 1, 2$

$\boxtimes_\ell$ M_0^ℓ witnesses that $\mathbf{tp}_{\mathfrak{s}(*)}(a, M_{\ell+1}, M_4)$ does not $\mathfrak{s}(*)$-fork over M_ℓ.

Let $\bar{\mathbf{a}}_\ell$ list M_0^ℓ so $\bar{\mathbf{a}}_\ell \in {}^{\lambda_\mathfrak{s}}(M_0^\ell)$. Now assume $\bar{\mathbf{b}}, \bar{\mathbf{c}} \in {}^{\lambda_\mathfrak{s}}(M_3)$ are such that

(ε) $\mathbf{tp}_{\mathfrak{K}[\mathfrak{s}]}(\bar{\mathbf{b}}, M_0^1, M_4) = \mathbf{tp}_{\mathfrak{K}[\mathfrak{s}]}(\bar{\mathbf{c}}, M_0^1, M_4)$.

As M_2 is $\mathfrak{K}_\mathfrak{s}$-saturated above $\lambda_\mathfrak{s}$ we can find $\bar{\mathbf{b}}' \in {}^{\lambda_\mathfrak{s}}(M_2)$ such that

(ζ) $\mathbf{tp}_{\mathfrak{K}[\mathfrak{s}]}(\bar{\mathbf{b}}', M_0^2, M_4) = \mathbf{tp}(\bar{\mathbf{b}}, M_0^2, M_4)$

similarly we can find $\bar{\mathbf{c}}' \in {}^{\lambda}(M_2)$ such that

(η) $\mathbf{tp}_{\mathfrak{K}[\mathfrak{s}]}(\bar{\mathbf{c}}', M_0^2, M_4) = \mathbf{tp}(\bar{\mathbf{c}}, M_0^2, M_4)$.

Chasing equalities $(\varepsilon) + (\zeta) + (\eta)$, as $M_0^1 \subseteq M_0^2$, clearly $\mathbf{tp}_{\mathfrak{K}[\mathfrak{s}]}(\bar{\mathbf{b}}', M_0^1, M_4) = \mathbf{tp}_{\mathfrak{K}[\mathfrak{s}]}(\bar{\mathbf{c}}', M_0^1, M_4)$, hence by clause (γ) more exactly by $\boxtimes_1$ we have

(θ) $\mathbf{tp}_{\mathfrak{K}[\mathfrak{s}]}(\langle a \rangle \hat{\ } \bar{\mathbf{b}}', M_0^1, M_4) = \mathbf{tp}_{\mathfrak{K}[\mathfrak{s}]}(\langle a \rangle \hat{\ } \bar{\mathbf{c}}', M_0^1, M_4)$.

By clause (δ), i.e., by $\boxtimes_2$ and the statement (η) we have

(ι) $\mathbf{tp}_{\mathfrak{K}[\mathfrak{s}]}(\langle a \rangle \hat{\ } \bar{\mathbf{c}}, M_0^2, M_4) = \mathbf{tp}_{\mathfrak{K}[\mathfrak{s}]}(\langle a \rangle \hat{\ } \bar{\mathbf{c}}', M_0^2, M_4)$

and similarly by $\boxtimes_2$ and (ζ)

(κ) $\mathbf{tp}_{\mathfrak{K}[\mathfrak{s}]}(\langle a \rangle \hat{\ } \bar{\mathbf{b}}, M_0^2, M_4) = \mathbf{tp}_{\mathfrak{K}[\mathfrak{s}]}(\langle a \rangle \hat{\ } \bar{\mathbf{b}}', M_0^2, M_4)$.

By chasing the equalities $(\theta)+(\iota)+(\kappa)$ we get $\mathbf{tp}_{\mathfrak{K}[\mathfrak{s}]}(\langle a \rangle \hat{\ } \bar{\mathbf{b}}, M_0^1, M_4) = \mathbf{tp}_{\mathfrak{K}[\mathfrak{s}]}(\langle a \rangle \hat{\ } \bar{\mathbf{c}}, M_0^1, M_4)$ as required.

Alternatively use II.2.18.

Axiom (E)(e): [Unique non-forking extension].

So let $M_0 <_{\mathfrak{s}(*)} M_1$ and $p, q \in \mathscr{S}^{\mathrm{bs}}_{\mathfrak{s}(*)}(M_1)$ do not fork over M_0 and $p \restriction M_0 = q \restriction M_0$. Let $M_1 <_{\mathfrak{s}(*)} M_2$ and $a_1, a_2 \in M_2$ be such that $\mathbf{tp}_{\mathfrak{s}(*)}(a_1, M_1, M_2) = p$ and $\mathbf{tp}_{\mathfrak{s}(*)}(a_2, M_1, M_2) = q$ and without loss of generality $M_1 <^+_{\lambda^+} M_2$. Let $\langle M_{\ell,\zeta} : \zeta < \lambda^+ \rangle$ be a $\leq_\mathfrak{s}$-representation of M_ℓ for $\ell = 0, 1, 2$ with $a_1, a_2 \in M_{2,0}$. By the assumption and the definition of $\mathfrak{s}(*)$ for a club E of $\lambda_\mathfrak{s}^+$ we have, for $\zeta \in E \Rightarrow p \restriction M_{0,\zeta} = q \restriction M_{0,\zeta} \in \mathscr{S}_\mathfrak{s}(M_{0,\zeta})$ call it r_ζ and

$p \restriction M_{1,\zeta}, q \restriction M_{1,\zeta}$ belong to $\mathscr{S}_{\mathfrak{s}}(M_{1,\zeta})$ and without loss of generality do not $\lambda_{\mathfrak{s}}$-split over $M_{0,0}$ and they extend r_ζ; also for $\zeta < \xi$ in E and $\ell < 2$ we have $\mathrm{NF}_{\mathfrak{s}}(M_{\ell,\zeta}, M_{\ell+1,\zeta}, M_{\ell,\xi}, M_{\ell+1,\xi})$ and for $\zeta < \xi \in E, M_{\ell,\xi}$ is $(\lambda_{\mathfrak{s}}, *)$-brimmed over $M_{\ell,\zeta}$ for $\mathfrak{s}$. Also without loss of generality $\zeta \in E \Rightarrow M_{2,\zeta}$ is $(\lambda_{\mathfrak{s}}, *)$-brimmed over $M_{1,\zeta}$. Hence for $\xi \in E, \mathbf{tp}_{\mathfrak{s}}(\bar{a}_1, M_{1,\xi}, M_{2,\xi}) = \mathbf{tp}_{\mathfrak{s}}(\bar{a}_2, M_{1,\xi}, M_{2,\xi})$ because their restriction to $M_{0,\xi}$ are equal and for $\ell = 1, 2$, $\mathbf{tp}_{\mathfrak{s}}(\bar{a}_\ell, M_{1,\xi}, M_{2,\xi})$ does not $\lambda_{\mathfrak{s}}$-split over $M_{0,0} \le_{\mathfrak{s}} M_{1,\zeta}$ (by the hypothesis). So clearly $p \restriction M_{1,\xi} = q \restriction M_{1,\xi}$. By $\circledast_1$ above we get $p = q$ (recall that $\mathfrak{K}_{\mathfrak{s}(+)} = \mathfrak{K}_{\mathfrak{s}(*)}$).

Axiom (E)(f): [Symmetry].

So assume that $M_0 \le_{\mathfrak{s}(*)} M_1 \le_{\mathfrak{s}(*)} M_3, a_1 \in M_2, a_2 \in M_2$ and $\mathbf{tp}_{\mathfrak{s}(*)}(a_2, M_1, M_3)$ does not $\mathfrak{s}(*)$-fork over M_0.

For $\ell = 0, 1, 3$ there is a $\le_{\mathfrak{K}[\mathfrak{s}]}$-representation $\langle M_{\ell,\alpha} : \alpha < \lambda_{\mathfrak{s}}^+\rangle$ of M_ℓ, without loss of generality each $M_{\ell,\alpha}$ is brimmed for $\mathfrak{s}$ and $M_{\ell,\beta}$ is brimmed over $M_{\ell,\alpha}$ for $\mathfrak{s}$ when $\alpha < \beta < \lambda_{\mathfrak{s}}^+$ and without loss of generality $\mathrm{NF}_{\mathfrak{s}}(M_{\ell,\alpha}, M_{m,\alpha}, M_{\ell,\beta}, M_{m,\beta})$ when $\alpha < \beta < \lambda_{\mathfrak{s}}^+$ and $\ell < m, \{\ell, m\} \subseteq \{0, 1, 3\}$. Also without loss of generality $a_1 \in M_{1,0}, a_2 \in M_{3,0}$ and $\mathbf{tp}_{\mathfrak{s}(*)}(a_{\ell+1}, M_\ell, M_3)$ does not $\lambda_{\mathfrak{s}}$-split over $M_{0,0}$ for $\ell = 0, 1$.

By 4.9 we can find a club E of $\lambda_{\mathfrak{s}}^+$ and $\le_{\mathfrak{s}}$-increasing continuous sequence $\langle M_{2,\alpha} : \alpha \in E\rangle$ such that $M_{2,\alpha} \le_{\mathfrak{s}} M_{3,\alpha}$ and $(M_{0,\alpha}, M_{2,\alpha}, a_2) \in K_{\mathfrak{s}}^{3,\mathrm{uq}}$, by renaming $E = \lambda_{\mathfrak{s}}^+$. By 5.4 for every $\alpha < \lambda_{\mathfrak{s}}^+$ the type $\mathbf{tp}_{\mathfrak{s}}(a_1, M_{2,\alpha}, M_{3,\alpha})$ does not fork over $M_{0,0}$, hence it does not $\lambda_{\mathfrak{s}}$-split over $M_{0,0}$. Letting $M_2 = \cup\{M_{2,\alpha} : \alpha < \lambda_{\mathfrak{s}}^+\}$ we are easily done.

Axiom(E)(g): [extension existence]

So let $M_0 \le_{\mathfrak{s}(*)} M_1$ and $p \in \mathscr{S}^{\mathrm{bs}}_{\mathfrak{s}(*)}(M_0)$ so for some $N_0, N_1 \in \mathfrak{K}_{\mathfrak{s}}, N_0 \le_{\mathfrak{s}} N_1 \le_{\mathfrak{K}[\mathfrak{s}]} M_0, N_1$ is $\le_{\mathfrak{s}}$-universal over N_0 and p does not $\lambda_{\mathfrak{s}}$-split over N_0. So M_0, M_1 are saturated models in $\lambda_{\mathfrak{s}}^+$ for $\mathfrak{K}^{\mathfrak{s}}$ above $\lambda_{\mathfrak{s}}$ hence there is an isomorphism f from M_0 onto M_1 over N_1 and $f(p) \in \mathscr{S}^{\mathrm{bs}}_{\mathfrak{s}(*)}(M_1)$ is witnessed by N_0 and extends $p \restriction N_0$ hence it extends p by the uniqueness proved above.

Axiom (E)(h): By claim II.2.17(3),(4).

Axiom (E)(i): By II.2.16.

Lastly
$\mathfrak{s}(*)$ is good$^+$: when $\mathfrak{s}$ is good$^+$

So assume $\bar{M}^\ell = \langle M^\ell_\alpha : \alpha < \lambda^{++}\rangle$ is $\leq_{\mathfrak{s}(*)}$-increasing continuous for $\ell = 0, 1$ and $M^0_\alpha \leq_{\mathfrak{s}} M^1_\alpha$, $a_\alpha \in M^0_{\alpha+1}$, $\mathbf{tp}_{\mathfrak{s}(*)}(a_{\alpha+1}, M^0_{\alpha+1}, M^0_{\alpha+2}) \in \mathscr{S}^{\rm bs}_{\mathfrak{s}(*)}(M^0_{\alpha+1})$ is an $\mathfrak{s}(*)$-non-forking extension of $p^* \in \mathscr{S}^{\rm bs}_{\mathfrak{s}(*)}(M^0_0)$ but $\mathbf{tp}_{\mathfrak{s}(*)}(a_{\alpha+1}, M^1_0, M^1_{\alpha+2})$ does $\mathfrak{s}(*)$-fork over M^0_0 and we shall get a contradiction.

As $p^* \in \mathscr{S}^{\rm bs}_{\mathfrak{s}(*)}(M^0_0)$ clearly for some $N^* \in K_{\mathfrak{s}}$ we have $N^* \leq_{\mathfrak{K}[\mathfrak{s}]} M^0_0$ and p^* does not $\lambda_{\mathfrak{s}}$-split over N^* hence (by 9.8(2) below) also $\mathbf{tp}_{\mathfrak{s}(*)}(a_{\alpha+2}, M^0_{\alpha+1}, M^0_{\alpha+2})$ does not $\lambda_{\mathfrak{s}}$-split over N^*. Let $\langle N_\varepsilon : \varepsilon < \lambda^+\rangle$ be a $\leq_{\mathfrak{s}}$-representation of M^1_0, and without loss of generality $N^* \leq_{\mathfrak{s}} N_0$.

Now for each $\alpha < \lambda^{++}_{\mathfrak{s}}$ the type $\mathbf{tp}_{\mathfrak{s}(*)}(a_{\alpha+1}, M^1_0, M^1_{\alpha+2})$ does $\mathfrak{s}(*)$-fork over M^0_0 hence it does $\lambda_{\mathfrak{s}}$-split over N^*, but clearly for some $\zeta_\alpha < \lambda^+ - \mathfrak{s}$ it does not λ-split over N_{ζ_α}. So for some $\zeta^* < \lambda^+_{\mathfrak{s}}$ the set $S = \{\alpha < \lambda^{++}_{\mathfrak{s}} : \zeta_\alpha = \xi\}$ is unbounded in $\lambda^{++}_{\mathfrak{s}}$. Now choose by induction on $\varepsilon < \lambda$ a triple $(\alpha_\varepsilon, M_{0,\varepsilon}, M_{1,\varepsilon})$ such that:

(*a*) $\alpha_\varepsilon \in S$ is increasing

(*b*) $M_{0,\varepsilon} \leq_{\mathfrak{K}[\mathfrak{s}]} M^0_{\alpha_\varepsilon}$ is $\leq_{\mathfrak{s}}$-increasing continuous

(*c*) $M_{1,\varepsilon} \leq_{\mathfrak{K}[\mathfrak{s}]} M^1_{\alpha_\varepsilon}$ is $\leq_{\mathfrak{s}}$-increasing continuous

(*d*) $M_{0,\varepsilon} \leq_{\mathfrak{s}} M_{1,\varepsilon}$

(*e*) $a \in M_{0,\varepsilon+1}$

(*f*) $N^* \subseteq M_{0,\varepsilon}, N_{\zeta^*} \subseteq M_{1,\varepsilon}$.

There is no problem to carry the definition and $\langle (M_{0,\varepsilon}, M_{1,\varepsilon}; a_\varepsilon) : \varepsilon < \lambda^+_{\mathfrak{s}} a\rangle$ provide a counterexample to "$\mathfrak{s}$ is good$^+$".

2) Equivalence of the three versions of orthogonality

Follows by "$\mathfrak{s}(*)$ is categorical" by claim 6.10(5).

$\square_{9.7}$

9.8 Observation. In clause (c) of Definition 9.5(2), an equivalent condition is

(∗) if $N_0 \leq_{\mathfrak{K}[\mathfrak{s}]} M_0, N_0 \in K_{\mathfrak{s}}$ and $\mathbf{tp}_{\mathfrak{s}(*)}(a, M_0, M_3)$ does not λ-split over N_0 then also $\mathbf{tp}_{\mathfrak{s}(*)}(a, M_1, M_3)$ does not λ-split over it.

Proof. Easy. $\square_{9.8}$

Of course

9.9 Claim. *1) If $\mathfrak{s}$ is a successful type-full λ-good-frame, then $\mathfrak{s}^+$ is a type-full λ^+-good frame and $\mathfrak{s}^* = \mathfrak{s}^+$.*
2) Similarly for type-closed.

Proof. Easy.

Exercise: Clarify on $\mathfrak{s}^{\mathrm{nf}}$ for $\mathfrak{s}$ saturative.

§10 Regular types

10.1 Hypothesis. $\mathfrak{s}$ is a λ-good$^+$ successful frame with primes such that $\mathfrak{s}$ is type-full.

Remark. So the earlier Hypothesis 2.1, 2.18, 3.1, 4.1, 5.1, 6.1, 9.1 hold.

10.2 Definition. 1) We say that $p \in \mathscr{S}^{\mathrm{bs}}_{\mathfrak{s}}(M)$ is regular if there are M_0, M_1, a, M_2 such that:

- (a) M_1 is $(\lambda, *)$-brimmed over M_0
- (b) $M \leq_{\mathfrak{s}} M_1$ and $M_0 \leq_{\mathfrak{s}} M_1 \leq_{\mathfrak{s}} M_2$ and $a \in M_2$
- (c) $p' = \mathbf{tp}_{\mathfrak{s}}(a, M_1, M_2)$ is parallel to p
- (d) p' does not fork over M_0
- (e) if $c \in M_2 \backslash M_1$ realizes $p' \restriction M_0$ then c realizes p' (in other words for every $c \in M_2 \backslash M_1$ realizing $p' \restriction M_0$ the type $\mathbf{tp}_{\mathfrak{s}}(c, M_1, M_2)$ does not fork over M_0).

2) We say that $p \in \mathscr{S}^{\mathrm{bs}}_{\mathfrak{s}}(M)$ is regular$^+$ if there are M_1, M_2, a such that clauses (a)-(d) above holds and (see Definition 2.22 and the rest of §2)

$(e)'$ if $c \in M_2 \backslash M_1$, then $\mathrm{rk}_{\mathfrak{s}}(\mathbf{tp}(c, M, M_2)) \geq \mathrm{rk}_{\mathfrak{s}}(p)$.

3) We say $p \in \mathscr{S}^{\text{bs}}_{\mathfrak{s}}(M)$ is "directly regular" (or "directly regular$^+$") <u>when</u> in part (1) (or in part (2)) we add

(*f*) $M_1 = M$.

4) We say that $\mathfrak{s}$ has regulars or truely has regulars <u>when</u> every $p \in \mathscr{S}^{\text{bs}}_{\mathfrak{s}}(M)$ is not orthogonal to some $q \in \mathscr{S}^{\text{bs}}_{\mathfrak{s}}(M)$.

10.3 Remark. 1) Note that regular $\neq$ regular$^+$. For example let T be the first order theory of $M = (\lambda \times \lambda \cup \lambda, P^M, Q^M, F^M)$ where $P^M = \lambda \times \lambda, Q^M = \lambda, F^M((\alpha,\beta)) = \alpha, F^M(\alpha) = \alpha$. Lastly, choose the $p(x) \in \mathbf{S}(M)$ which contains $\{P(x) \wedge x \neq a \wedge F(x) \neq b : a \in P^M$ and $b \in Q^M\}$.

Now let

$$M_0 = M \restriction (\{(\alpha,\beta) : \alpha,\beta < \lambda \text{ are odd}\} \cup \{\alpha : \alpha < \lambda \text{ is odd}\})$$

$$M_1 = M \restriction (\{(\alpha,\beta) : \alpha,\beta < \lambda \text{ and } \alpha \neq 0\} \cup \{\alpha : \alpha < \lambda, \alpha \neq 0\})$$

$$M_2 = M$$

$a = (0,0) \in M_2 \backslash M_1$. Now easily $M_0 \prec M_1 \prec M_2$ and T is superstable (even $\aleph_0$-stable).

Now $p = \mathbf{tp}(a, M_1, M_2)$ is regular in the sense of Definition 10.2(1) as M_0, M_1, M_2 witness but p is not regular$^+$ as $\text{rk}_{\mathfrak{s}}(p) = 2$ and if $M_1 \prec M_2', p$ realized in M_2 then $Q^{M_2'} \neq Q^{M_1}$ and $b \in Q^{M_2'} \backslash Q^{M_1} \Rightarrow \text{rk}_{\mathfrak{s}}(\mathbf{tp}_{\mathfrak{s}}(b, M_1, M_2')) = 1$.

2) But for our purposes every regular type is "equivalent" to a regular$^+$ type so those suffice, i.e., no loss in using them.

3) Naturally "$\mathfrak{s}$ has regulars" $\Rightarrow$ "$\mathfrak{s}$ almost has regular" $\Rightarrow$ "$\mathfrak{s}$ weakly has regulars" but we shall not use this here so this is delayed.

10.4 Claim. *1)*

(a) If $p_1 \| p_2$ then p_1 is regular iff p_2 is regular

(b) if $M \in K_{\mathfrak{s}}$ is $(\lambda, *)$-brimmed (trivially holds if $\mathfrak{K}_{\mathfrak{s}}$ is categorical) and $p \in \mathscr{S}^{\rm bs}_{\mathfrak{s}}(M)$, then p is regular iff it is directly regular.

2) Assume $p \in \mathscr{S}^{\rm bs}_{\mathfrak{s}}(M), M$ is $(\lambda, *)$-brimmed over M_0, p does not fork over M_0 and $(M, M_2, a) \in K^{3,\rm pr}_{\mathfrak{s}}$, $\mathbf{tp}_{\mathfrak{s}}(a, M, M_2) = p$ and we let $M_1 = M$. Then p is regular iff clause (e) of 10.2(1) holds.
3) The parallel of parts (1),(2) holds for regular$^+$.
4) If $(M, N, a) \in K^{3,\rm pr}_{\mathfrak{s}}$ and $p = \mathbf{tp}_{\mathfrak{s}}(a, M, N)$ is regular$^+$, then $c \in N \backslash M \Rightarrow \mathrm{rk}_{\mathfrak{s}}(\mathbf{tp}_{\mathfrak{s}}(c, M, N)) \geq \mathrm{rk}_{\mathfrak{s}}(p)$.
5) If p is regular$^+$ then p is regular.
6) If M_0, M_1, M_2 satisfies clauses (a),(b) of 10.2(1), $a \in M_2 \backslash M_1$ realizes $p \in \mathscr{S}^{\rm bs}_{\mathfrak{s}}(M_1)$ which does not fork over M_0 but for every $b \in M_2 \backslash M_1$ realizing $p \restriction M_0$ we have $\mathrm{rk}_{\mathfrak{s}}(\mathbf{tp}_{\mathfrak{s}}(b, M_1, M_2)) \geq \mathrm{rk}_{\mathfrak{s}}(\mathbf{tp}_{\mathfrak{s}}(a, M_1, M_2)$ then $\mathbf{tp}_{\mathfrak{s}}(a, M_1, M_2)$ is regular.
7) If $M_0 \leq_{\mathfrak{s}} M_1, (M_1, M_2, a) \in K^{3,\rm pr}_{\mathfrak{s}}$ and $p = \mathbf{tp}_{\mathfrak{s}}(a, M_1, M_2)$ does not fork over M_0 and is regular and $c \in M_2 \backslash M_1$ realizes $p \restriction M_0$ then c realizes p.
8) If $(M_1, M_2, a) \in K^{3,\rm pr}_{\mathfrak{s}}$ and M_1 is $(\lambda, *)$-brimmed over M_0 and $p = \mathbf{tp}_{\mathfrak{s}}(a, M_1, M_2)$ does not fork over M_0 then

(a) p is regular iff $(\forall c)[c \in M_2 \backslash M_1$ realizes $p \restriction M_0 \Rightarrow c$ realizes $p]$

(b) p is regular$^+$ iff $(\forall c)[c \in M_2 \backslash M_1 \Rightarrow \mathrm{rk}_{\mathfrak{s}}(\mathbf{tp}(c, M_1, M_2)) \geq \mathrm{rk}_{\mathfrak{s}}(\mathbf{tp}(a, M_2, M_2)]$.

Proof. 1) Clause (a): So assume that $M' \leq_{\mathfrak{s}} M$ and $M'' \leq_{\mathfrak{s}} M$ and $p' \in \mathscr{S}^{\rm bs}(M'), p'' \in \mathscr{S}^{\rm bs}_{\mathfrak{s}}(M'')$ are parallel, that is some $p \in \mathscr{S}^{\rm bs}(M)$ does not fork over M' and over M'' and $p \restriction M' = p', p \restriction M'' = p''$ and we should prove that p' is regular iff p'' is regular. By the symmetry it suffices to show that p' is regular iff p is regular. Now the "if" direction is trivial (the same witnesses M_0, M_1, M_2, a work). For the "only if" direction, let (M'_0, M'_1, M'_2, a) witness p' is regular.

As $\mathfrak{K}_{\mathfrak{s}}$ has amalgamation and $M' \leq_{\mathfrak{s}} M, M' \leq_{\mathfrak{s}} M'_1$ without loss of generality for some M_1 we have $M'_1 \leq_{\mathfrak{s}} M_1$ and $M \leq_{\mathfrak{s}} M_1$ and without loss of generality M_1 is $(\lambda, *)$-brimmed over $M'_1 \cup M$. There is an isomorphism f from M_1' onto M_1 over M_0' as both are $(\lambda, *)$-brimmed over it, and we can find f^*, M_2, a^* such that $M_1 \leq_{\mathfrak{s}} M_2, f^* \supseteq f, f^*$ an isomorphism from M_2' onto M_2 and $f^*(a) = a^*$.

Now $M_0 := f^*(M_0'), M_1 = f^*(M_1'), M_2 = f^*(M_2')$ and a^* witness the regularity of p.

Clause (b) of (1): The if direction is obvious (same witnesses).

For the other direction assume that M_0, M_1, M_2, a witness that $p \in \mathscr{S}^{\rm bs}_{\mathfrak{s}}(M)$ is regular; so $p^+ := \mathbf{tp}_{\mathfrak{s}}(a, M_1, M_2)$ does not fork over M_0 and over M and extends p.

As $M \leq_{\mathfrak{s}} M_1$ are $(\lambda, *)$-brimmed and $p^+ \in \mathscr{S}^{\rm bs}_{\mathfrak{s}}(M_1)$ does not fork over M and extend p, by 1.21 there is an isomorphism h from M_1 onto M mapping p^+ to p. There is a pair (h^+, M_2') such that $M \leq_{\mathfrak{s}} M_2$ and h^+ is an isomorphism from M_2 onto M_2' extending h. Now $(h(M_0), M, M_2', h^+(a))$ witness the desired conclusion.

2) If no $b \in M_2 \backslash M_1$ realizes $p \restriction M_0$ then (M_0, M_1, M_2) witness that p is regular. If p is regular recalling M is $(\lambda, *)$-brimmed by 10.4(1)(b) clearly $p \in \mathscr{S}^{\rm bs}_{\mathfrak{s}}(M)$ is directly regular, so let it be witnessed by M_0', M_2', a', but as $(M, N, a) \in K^{3,\rm pr}_{\mathfrak{s}}$ without loss of generality $a' = a, N \leq_{\mathfrak{s}} M_2'$, hence every $b \in N \backslash M$ realizing $p \restriction M_0$ in N realzies it in M_2 hence realizes p in M_2 hence in N. So clause (e) of 10.2(1) holds.

3) Similarly.

4) If M is $(\lambda, *)$-brimmed the proof is similar to the proof of part (2). Otherwise we can find M^+ such that $M \leq_{\mathfrak{s}} M^+$ and M^+ is $(\lambda, *)$-brimmed over M and we let $p^+ \in \mathscr{S}^{\rm bs}_{\mathfrak{s}}(M^+)$ be the non-forking extension of p. Now by part (3), the parallel to part (1), $p^+ \in \mathscr{S}^{\rm bs}_{\mathfrak{s}}(M^+)$ is regular$^+$.

So there are M_0, M_1, M_2, a' witnessing it. By part (3), the parallel to part (2), without loss of generality $M_1 = M^+$, and let $N^+ = M_2$ hence, $p^+ = \mathbf{tp}_{\mathfrak{s}}(a', M^+, N^+)$. So a' realizes $p = p^+ \restriction M$ inside N^+ hence, recalling $(M, N, a) \in K^{3,\rm pr}_{\mathfrak{s}}$, there is a $\leq_{\mathfrak{s}}$-embedding f of N into N^+ mapping a to a'. So without loss of generality $f = \mathrm{id}_N$. By the choice of $M_1 = M^+, N^+ = M_2$, recalling clause $(e)'$ of 10.2(2) we have:

$$(*)\quad c \in N^+ \backslash M^+ \Rightarrow \mathrm{rk}_{\mathfrak{s}}(\mathbf{tp}_{\mathfrak{s}}(c, M^+, N^+) \geq \mathrm{rk}_{\mathfrak{s}}(p^+).$$

But $\mathbf{tp}_{\mathfrak{s}}(a, M^+, N^+)$ is p^+ hence does not fork over M and $(M, N, a) \in$

$K^{3,\mathrm{pr}}_{\mathfrak{s}} \subseteq K^{3,\mathrm{uq}}_{\mathfrak{s}}$ hence by 1.19 we have $\mathrm{NF}_{\mathfrak{s}}(M, N, M^+, N^+)$. So $c \in N \backslash M \Rightarrow \mathbf{tp}_{\mathfrak{s}}(c, M^+, N^+)$ does not fork over M hence if $c \in N \backslash M$ then $c \in N^+ \backslash M^+$ and $\mathrm{rk}_{\mathfrak{s}}(\mathbf{tp}_{\mathfrak{s}}(c, M, N)) = \mathrm{rk}_{\mathfrak{s}}(\mathbf{tp}(c, M^+, N^+)) \geq \mathrm{rk}_{\mathfrak{s}}(p^+) = \mathrm{rk}_{\mathfrak{s}}(p) = \mathrm{rk}_{\mathfrak{s}}(\mathbf{tp}(a, M, N))$ by 2.26(3), by $(*)$ above, by 2.26(4), and by an assumption respectively. So we are done.
5) Because if M_1 is $(\lambda, *)$-brimmed over M_0 and $p \in \mathscr{S}^{\mathrm{bs}}_{\mathfrak{s}}(M_1)$ does not fork over M_0 and $q \in \mathscr{S}_{\mathfrak{s}}(M_1), q \neq p, q \restriction M_0 = p \restriction M_0$ then $\mathrm{rk}_{\mathfrak{s}}(p) = \mathrm{rk}_{\mathfrak{s}}(p \restriction M_0) > \mathrm{rk}_{\mathfrak{s}}(q)$, see 2.26(6), i.e., the inequality holds as q forks over M_0.
6) Easy, follows by the definition as in the proof of part (5).
7) Similar to the proof of part (4).
8) Easy by part (2), (and see Exercise 1.24). $\square_{10.4}$

10.5 Claim. *1) If $M <_{\mathfrak{s}} N$ and M is $(\lambda, *)$-brimmed, __then__ for some $c \in N \backslash M$ the type $\mathbf{tp}_{\mathfrak{s}}(c, M, N)$ is regular$^+$ (hence regular).*
2) If $M' <_{\mathfrak{s}} M <_{\mathfrak{s}} N, M$ is $(\lambda, *)$-brimmed and $p \in \mathscr{S}^{\mathrm{bs}}_{\mathfrak{s}}(M')$ is realized by some member of $N \backslash M$ and moreover M is $(\lambda, *)$-brimmed over M', __then__ for some $c \in N \backslash M$ realizing p we have $\mathbf{tp}_{\mathfrak{s}}(c, M, N)$ is regular. Note that possibly $\mathbf{tp}_{\mathfrak{s}}(c, M, N)$ forks over M_0.

Proof. 1) Choose $c \in N \backslash M$ such that $\mathrm{rk}_{\mathfrak{s}}(\mathbf{tp}_{\mathfrak{s}}(c, M, N))$ is minimal. Then choose $(\lambda, *)$-brimmed $M_0 <_{\mathfrak{s}} M$ such that M is $(\lambda, *)$-brimmed over M_0 and $\mathbf{tp}_{\mathfrak{s}}(c, M, N)$ does not fork over M_0, exists by Exercise 1.24. Let $M_1 := M, M_2 := N, a := c$ so by Claim 10.4(3), the parallel to 10.4(2) the type $\mathbf{tp}_{\mathfrak{s}}(c, M, N) \in \mathscr{S}^{\mathrm{bs}}_{\mathfrak{s}}(M)$ is regular$^+$ (hence regular by 10.4(5)).
2) Choose $c \in N \backslash M$ realizing p with $\mathrm{rk}_{\mathfrak{s}}(\mathbf{tp}_{\mathfrak{s}}(c, M, N))$ minimal. Let $M_0 <_{\mathfrak{s}} M$ be such that $M' \leq_{\mathfrak{s}} M_0$ and M is $(\lambda, *)$-brimmed over M_0 and M_0 is $(\lambda, *)$-brimmed over M'; it follows that $\mathbf{tp}_{\mathfrak{s}}(c, M, N)$ does not fork over M_0.
Now we are done by 10.4(6). $\square_{10.5}$

10.6 Claim. *[$\mathfrak{s} = \mathfrak{t}^+, \mathfrak{t}$ is a good$^+$ $\lambda_{\mathfrak{s}}$-frame, successful with primes and $K^{3,\mathrm{uq}}_{\mathfrak{t}} = K^{3,\mathrm{pr}}_{\mathfrak{t}}$.]*

Assume $(M, N, a) \in K^{3,\mathrm{bs}}_{\mathfrak{s}}, p = \mathbf{tp}_{\mathfrak{s}}(a, M, N)$ and $q \in \mathscr{S}^{\mathrm{bs}}_{\mathfrak{s}}(M)$ and $M_0 \in K_{\mathfrak{t}}, M_0 \leq_{\mathfrak{K}[\mathfrak{t}]} M$ (so M is $(\lambda, *)$-brimmed as $\mathfrak{s} = \mathfrak{t}^+$).
1) If p does not fork over M_0 (i.e. M_0 is a witness for p) __then__

(a) *p is regular (for* $\mathfrak{s}$*) iff* $p \restriction M_0$ *is regular (for* $\mathfrak{t}$*)*

(b) *Similarly for regular*$^+$

(c) *if* $(M, N, a) \in K_{\mathfrak{s}}^{3,\mathrm{pr}}$ *and* $c \in N \backslash M$ *realizes* $p \restriction M_0$ *and p is regular (for* $\mathfrak{s}$*) then c realizes p.*

2) There is a regular$^+$ *type* $p_1 \in \mathscr{S}_{\mathfrak{s}}^{\mathrm{bs}}(M)$ *not orthogonal to p, and realized in N such that* $\mathrm{rk}_{\mathfrak{s}}(p_1) \le \mathrm{rk}_{\mathfrak{s}}(p)$ *and* $\mathrm{rk}_{\mathfrak{s}}(r) < \mathrm{rk}_{\mathfrak{s}}(p_1) \Rightarrow r \perp p$ *for every* $r \in \mathscr{S}_{\mathfrak{s}}^{\mathrm{bs}}(M)$ *or even*[14] *every* $r \in \mathscr{S}_{\mathfrak{s}}(M'), M \le_{\mathfrak{s}} M'$*. In fact also* $\mathrm{rk}_{\mathfrak{s}}(r) < \mathrm{rk}_{\mathfrak{s}}(p_1) \Rightarrow r \perp p_1$ *for every r as above.*
3) If p is regular$^+$ *not orthogonal to q, then* $\mathrm{rk}_{\mathfrak{s}}(q) \ge \mathrm{rk}_{\mathfrak{s}}(p)$*.*
4) If $M^* \le_{\mathfrak{s}} M, p \restriction M^* = q \restriction M^*, p \ne q, p$ *does not fork over* M^* *and p is regular then* $p \perp q$*.*
5) If p is regular$^+$ *and* $\mathrm{rk}_{\mathfrak{s}}(q) < \mathrm{rk}_{\mathfrak{s}}(p)$ *then* $p \perp q$*.*
6) Let $p_1 \in \mathscr{S}_{\mathfrak{s}}^{\mathrm{bs}}(M)$ *be not orthogonal to p with minimal rank. Then*

(α) p_1 *is realized in N and is regular*$^+$

(β) *if p is regular and* $(M, N^1, a^1) \in K_{\mathfrak{s}}^{3,\mathrm{pr}}$ *and* $\mathbf{tp}_{\mathfrak{s}}(a^1, M, N^1) = p_1$ *then p is realized in* N^1

(γ) *if p is regular then* $p \perp q \Leftrightarrow p_1 \perp q$ *(recall q is any member of* $\mathscr{S}_{\mathfrak{s}}^{\mathrm{bs}}(M)$*))*

(δ) *if* $p \perp q$ *then* $p_1 \perp q$*.*

7) Assume $(M, N, a) \in K_{\mathfrak{s}}^{3,\mathrm{pr}}$*. If* $a_1 \in N \backslash M$ *and (recalling* $q \in \mathscr{S}_{\mathfrak{s}}^{\mathrm{bs}}(M)$*) we have* $p \perp q$ *and* $\mathbf{tp}_{\mathfrak{s}}(a_1, M, N) \in \mathscr{S}_{\mathfrak{s}}^{\mathrm{bs}}(M)$ *then* $\mathbf{tp}_{\mathfrak{s}}(a_1, M, N) \perp q$ *and* $M' <_{\mathfrak{s}} M$ *&* $p \perp M' \Rightarrow \mathbf{tp}_{\mathfrak{s}}(a_1, M, N) \perp M'$*.*

8) If q is regular and p, q are not orthogonal (or just q has at least two extensions in $\mathscr{S}_{\mathfrak{s}}(N)$*) then q is realized in N.*
9) If q is regular and $(M, N, a) \in K_{\mathfrak{s}}^{3,\mathrm{pr}}$ *then* $p \pm q$ *iff q is realized in N.*

Remark. 1) Note that by part (6) we can "replace" a regular p by a regular$^+$ one which is advantageous and helps, e.g. in part (8) in this claim 10.6.
2) Note that in 10.6 the proof of (6)(γ) depends on part (7).

[14] We may wonder, is the "existence of $\mathfrak{t}$" is necessary? But anyhow we usually have arrived to good λ-frames with primes by deriving it from such $\mathfrak{t}$.

Proof. Clearly $\mathfrak{t}$ is as reuried in 10.2 so it satisfies what we hae proved so far; in particular $\mathfrak{t}$ is full.

As $\mathfrak{s} = \mathfrak{t}^+$ let $\langle M_\alpha : \alpha < \lambda_\mathfrak{t}^+ \rangle, \langle N_\alpha : \alpha < \lambda_\mathfrak{t}^+ \rangle$ be $\le_\mathfrak{t}$-representations of M, N respectively. Checking all parts of the claim, clearly without loss of generality $(M, N, a) \in K_\mathfrak{s}^{3,\mathrm{pr}}$ and even is canonically prime, see 4.9, recalling that $\mathfrak{t}$ being successful good$^+$ λ-frame satisfies Hypothesis 4.1. Hence without loss of generality $\alpha < \lambda_\mathfrak{t}^+ \Rightarrow (M_\alpha, N_\alpha, a) \in K_\mathfrak{t}^{3,\mathrm{uq}}$ and $\alpha < \beta \Rightarrow \mathrm{NF}_\mathfrak{t}(M_\alpha, N_\alpha, M_\beta, N_\beta)$ and $\alpha < \beta \Rightarrow N_\beta, M_\beta$ is $(\lambda, *) - \mathfrak{t}$-brimmed over N_α, M_α respectively and $\alpha \Rightarrow \lambda_\mathfrak{t}^+ \Rightarrow M_\alpha, N_\alpha$ are $(\lambda, *) - \mathfrak{t}$-brimmed.
1) Note that assumption $K_\mathfrak{t}^{3,\mathrm{uq}} = K_\mathfrak{t}^{3,\mathrm{pr}}$ helps. (The proof is similar to the proof of 2.27 which deals with rk).

Clause (a):

First, assume that $p \restriction M_0 \in \mathscr{S}_\mathfrak{t}^{\mathrm{bs}}(M_0)$ is not regular; then $p \restriction M_1$ is not regular, M_1 is $(\lambda_\mathfrak{s}, *)$-brimmed over $M_0, a \in N_0 \subseteq N_1$ realizes $p \restriction M_1$ hence by 10.4(2) some $c \in N_1 \backslash M_1$ realizes $p \restriction M_0$ but not $p \restriction M_1$. Now we can choose $M'_\alpha \in K_\mathfrak{t}$ for $\alpha < \lambda_\mathfrak{t}^+, \le_\mathfrak{t}$-increasing continuous such that $M'_0 = M_0, M'_1 = M_0, M'_\alpha \le_\mathfrak{s} M_\alpha$ and $\beta < \alpha \Rightarrow \mathrm{NF}_\mathfrak{t}(M'_\beta, M_\beta, M'_\alpha, M_\alpha)$ and if $\alpha = \beta + 1 \ge 2$ then M'_α is $(\lambda_\mathfrak{t}, *)$-brimmed over M'_β for $\mathfrak{t}$ and M_α is $(\lambda_\mathfrak{t}, *)$-brimmed over $M'_\alpha \cup M_\beta$. Let $M' = \cup\{M'_\alpha : \alpha < \lambda_\mathfrak{t}^+\}$. Easily $M' \in K_\mathfrak{s}$ and M is $(\lambda_\mathfrak{t}, *)$-brimmed over M'. Also by symmetry for $\mathrm{NF}_\mathfrak{t}$ recalling $M'_0 = M'_1$ for every $\alpha \ge 2$, $\mathrm{NF}_\mathfrak{t}(M'_0, M'_\alpha, M_1, M_\alpha)$ but also $\mathrm{NF}_\mathfrak{t}(M_1, M_\alpha, N_1, N_\alpha)$ hence by transitivity for $\mathrm{NF}_\mathfrak{t}$ we have $\mathrm{NF}_\mathfrak{t}(M'_0, M'_\alpha, N_1, N_\alpha)$. As $\mathbf{tp}_\mathfrak{t}(b, M_0, N_1) = p \restriction M_0$ it follows that $\mathbf{tp}_\mathfrak{t}(b, M'_\alpha, N_\alpha)$ hence is a non-forking extension of $p \restriction M_0$ hence is $p \restriction M'_\alpha$.

We can conclude by 1.11 that $\mathbf{tp}_\mathfrak{s}(b, M', N) = p \restriction M'$, but trivially $\mathbf{tp}_\mathfrak{s}(b, M, N) \neq p$ and as $M_0 \le_{\mathfrak{K}[\mathfrak{t}]} M$ we know that p does not fork over M' for $\mathfrak{s}$. So (M', M, M) witness that p is not regular for $\mathfrak{s}$. This gives one implication of clause (a).

Second, asume that $p \restriction M_0$ is regular (for $\mathfrak{t}$). For every $\alpha < \lambda_\mathfrak{s}^+$ clearly $(M_\alpha, N_\alpha, a) \in K_\mathfrak{s}^{3,\mathrm{uq}} = K_\mathfrak{s}^{3,\mathrm{pr}}$, hence by 10.4(2) every $b \in N_\alpha \backslash M_\alpha$ realizing $p \restriction M_0$ realizes $p \restriction N_\alpha$. As this holds for every $\alpha < \lambda_\mathfrak{s}$, it follows by 1.10(1) that every $b \in N \backslash M$ realizing $p \restriction M_0$ realizes p. This is more than eough to show that p is regular.

Clause (b):

Similar using 2.27 with $\mathfrak{s}$ there standing for $\mathfrak{t}$ here.

Clause (c):

By the proof of clause (a).

2) Choose $p_1 \in \mathscr{S}^{\rm bs}_{\mathfrak{s}}(M)$ realized by some $c_1 \in N \backslash M$ with $\mathrm{rk}_{\mathfrak{s}}(p_1)$ minimal and let $N_1 \le_{\mathfrak{s}} N$ be such that $(M, N_1, c_1) \in K^{3,\rm pr}_{\mathfrak{s}}$. As in the proof above we can find M' such that $M_0 \le_{\mathfrak{K}[\mathfrak{t}]} M' \le_{\mathfrak{s}} M$ and M is $(\lambda_{\mathfrak{s}}, *)$-brimmed over M' and p_1 does not fork over M'. Now p_1 is regular$^+$ by 10.4(2),(3), is not orthogonal to p and it is realized in N as exemplified by c_1. Also $\mathrm{rk}_{\mathfrak{s}}(p_1) \le \mathrm{rk}_{\mathfrak{s}}(p)$ by the minimality of $\mathrm{rk}_{\mathfrak{s}}(p_1)$.

Lastly, assume $r \in \mathscr{S}^{\rm bs}_{\mathfrak{s}}(M''), r \pm p, M \le_{\mathfrak{s}} M''$ and we should prove that $\mathrm{rk}_{\mathfrak{s}}(r) \ge \mathrm{rk}_{\mathfrak{s}}(p_1)$. Without loss of generality $M = M''$. [Why? Because without loss of generality, in fact by the assumption on $\mathfrak{s}$ the model M'' is $(\lambda_{\mathfrak{s}}, *)$-brimmed and by 1.21 there is an isomorphism f from M onto M'' which maps p to the extension p' of p which does not fork over M. Now $p \perp r \Leftrightarrow p' \perp r \Leftrightarrow p \perp f^{-1}(r)$ and $\mathrm{rk}_{\mathfrak{s}}(r) = \mathrm{rk}_{\mathfrak{s}}(f^{-1}(r))$, so we can replace (r, M') by $(f^{-1}(r), M)$.]

Without loss of generality p, p_1, r do not fork over M_0, so as N is $\lambda^+_{\mathfrak{t}}$-saturated (above $\lambda_{\mathfrak{t}}$ for $\mathfrak{K}^{\mathfrak{t}}$) and $p \pm r$ there is $c_2 \in N$ realizing $r \restriction M_0$ such that $\{a, c_2\}$ is not independent over M_0 inside N for $\mathfrak{t}$. Now this implies $c_2 \notin M$ [as for every $c' \in M \backslash M_0$, the pair $\{a, c'\}$ is independent over M_0 inside N for $\mathfrak{t}$ as a realizes $p \in \mathscr{S}^{\rm bs}_{\mathfrak{s}}(M)$ which does not fork over M_0] hence $\mathrm{rk}_{\mathfrak{s}}(\mathbf{tp}_{\mathfrak{s}}(c_2, M, N)) \ge \mathrm{rk}_{\mathfrak{s}}(p_1)$ by the choice of p_1 so necessarily using 2.27(1) we have $\mathrm{rk}_{\mathfrak{s}}(r) = \mathrm{rk}_{\mathfrak{t}}(r \restriction M_0) = \mathrm{rk}_{\mathfrak{t}}(\mathbf{tp}_{\mathfrak{t}}(c_2, M_0, N)) \ge \mathrm{rk}_{\mathfrak{s}}(\mathbf{tp}_{\mathfrak{s}}(c_2, M, N) \ge \mathrm{rk}_{\mathfrak{s}}(p_1)$ as required. The proof of the "in fact" is similar.

3) Without loss of generality p, q do not fork over $M_0 \in K_{\mathfrak{t}}$, so $q \restriction M_0$ is regular$^+$ not orthogonal to $p \restriction M_0$ (for $\mathfrak{t}$, by part (1) and by 6.11 respectively). As $a \in N$ realizes $p \restriction M_0$ and N is $\lambda^+_{\mathfrak{t}}$-saturated and $P \perp q$ clearly there is $c \in N$ realizing $q \restriction M_0$ such that $\{a, c\}$ is not independent over M_0 inside N for $\mathfrak{t}$. Hence $c \notin M$.

As p is regular$^+$ and $(M, N, a) \in K^{3,\rm pr}_{\mathfrak{s},p}$, by 10.4(4) we know that $\mathrm{rk}_{\mathfrak{s}}(\mathbf{tp}_{\mathfrak{s}}(c, M, N)) \ge \mathrm{rk}_{\mathfrak{s}}(p)$. As $\mathbf{tp}_{\mathfrak{s}}(c, M, N)$ extends $q \restriction M_0$ it has rank $\le \mathrm{rk}_{\mathfrak{s}}(q \restriction M_0) = \mathrm{rk}_{\mathfrak{s}}(q)$ so by the previous sentence $\mathrm{rk}_{\mathfrak{s}}(p) \le \mathrm{rk}_{\mathfrak{s}}(q)$.

4) Without loss of generality $M^*_0 := M^* \cap M_0 <_{\mathfrak{t}} M_0$ and p does not fork over it.

Without loss of generality also q does not fork over (i.e. is witnessed by) M_0 and again using 6.11 we have $p \perp q \Leftrightarrow (p \restriction M_0 \perp q \restriction M_0)$. Assume toward contradition $p \pm q$ hence $(p \restriction M_0) \pm (q \restriction M_0)$ so for some $c \in N$ realizing $q \restriction M_0, \{a, c\}$ is not independent over M_0 for $\mathfrak{t}$ inside N, hence $c \in N \backslash M$.

Now choose a non-zero $\alpha < \lambda_{\mathfrak{t}}^+$ such that $c \in N_\alpha$ and recall (M_α, N_α, a) belongs to $K_{\mathfrak{t}}^{3,\mathrm{uq}}$ hence by an assumption of the claim it belongs to $K_{\mathfrak{t}}^{3,\mathrm{pr}}$, and N_α is $(\lambda_{\mathfrak{t}}, *)$ – $\mathfrak{t}$-brimmed over M_0.
So use $M_0, M_\alpha, N_\alpha, a, p \restriction M_\alpha, c$ to apply 10.4(7) and we get that $\mathbf{tp}_{\mathfrak{t}}(c, M_\alpha, N_\alpha)$ does not fork over M_0; hence is equal to $p \restriction M_\alpha$. As $\alpha < \beta < \lambda_{\mathfrak{t}}^+ \Rightarrow \mathrm{NF}_{\mathfrak{t}}(M_\alpha, N_\alpha, M_\eta, N_\beta) \Rightarrow \mathbf{tp}_{\mathfrak{t}}(c, M_\beta, N_\beta)$ does not fork over $M_\alpha \Rightarrow \mathbf{tp}_{\mathfrak{t}}(c, M_\beta, N_\beta) = p \restriction M_\beta$ we get that c realizes p inside N for $\mathfrak{s}$. Hence indeed $p \perp q$.
5) Proof similar to (4).
6) <u>Clause (α)</u>: As in the proof of part (2), starting with "lastly".

<u>Proof of Clause (β)</u>:

By clause (α) we know that p_1 is realized in N, so without loss of generality $N^1 \leq_{\mathfrak{s}} N$ hence $a^1 \in N$. Without loss of generality both p and p_1 do not fork over M_0 and so as in earlier cases there is $c \in N^1$ realizing $p \restriction M_0$ such that $\{c, a^1\}$ is not independent in N over M_0. This implies $c \in N^1 \backslash M$ hence by clause (c) of part (1) we know that c realizes p, as required.

<u>Proof of Clause (γ)</u>:

First assume $p \perp q$. Let $(M, N', b) \in K_{\mathfrak{s}}^{3,\mathrm{pr}}$ where $\mathbf{tp}_{\mathfrak{s}}(b, M, N') = q$ and without loss of generality $N \leq_{\mathfrak{s}} N^+, N' \leq_{\mathfrak{s}} N^+$. By clause (α) some $a_1 \in N$ realizes p_1 by part (7) below with M, N', b, a_1, q here by standing for M, N, a, a_1, q there we conclude that $\mathbf{tp}_{\mathfrak{s}}(a_1, M, N') \perp q$ which means $p_1 \perp q$.

For the other direction let $(M, N_1, b_1) \in K_{\mathfrak{s}}^{3,\mathrm{pr}}$ be such that $\mathbf{tp}(b_1, M, N_1) = p_1$, by clause (β) some $a_1 \in N'$ realizes p and continue as above (interchanging p and p_1).

<u>Clause (δ)</u>:

By the first paragraph in the proof of claue (γ).
7) Easy (and do not use part (6) and less than 10.1) but we elaborate.

Let $N_1 \leq_{\mathfrak{s}} N$ be such that $(M, N_1, a_1) \in K^{3,\mathrm{pr}}_{\mathfrak{s}}$. Now as $q \perp p$ clearly $p \underset{\mathrm{wk}}{\perp} q$ so by 6.3 it follows that q has a unique extension in $\mathscr{S}^{\mathrm{bs}}_{\mathfrak{s}}(N)$ hence it has a unique extension in $\mathscr{S}^{\mathrm{bs}}_{\mathfrak{s}}(N_1)$ which by 6.3 implies that $\mathbf{tp}_{\mathfrak{s}}(a, M, N_1) \underset{\mathrm{wk}}{\perp} p$; but for $\mathfrak{s}, \perp = \underset{\mathrm{wk}}{\perp}$ by 6.8(5) as its assumption, "categoricity in $\lambda_{\mathfrak{s}}$" holds. The second phrase (or $\perp M'$) follows.

Clause (β):

Without loss of generality p and q does not fork over M_0. Clearly if $p \perp q$ then q has at least two extensions in $\mathscr{S}_{\mathfrak{s}}(M)$, so we can assume the latter. We try by induction on $\alpha < \lambda^+_{\mathfrak{s}}$ to choose M_α and if $\alpha = \beta + 1$ also a_β such that:

- $(*)$ (a) M_α is $\leq_{\mathfrak{s}}$-increasing continuous
  - (b) $M_0 = M$
  - (c) $M_\alpha \leq_{\mathfrak{s}} N$
  - (d) if $\alpha = \beta + 1$ then $(M_\beta, M_\alpha, a_\beta) \in K^{3,\mathrm{pr}}_{\mathfrak{s}}$
  - (e) $\mathbf{tp}_{\mathfrak{s}}(a_\beta, M_\beta, N)$ is orthogonal to p, q.

Necessarily for some $\alpha_* < \lambda^+_{\mathfrak{s}}, M_\alpha$ is well defined iff $\alpha < \alpha_*$. As for $\alpha = 0$ and α limit there are no problems, necessarily α_* has from $\beta_* + 1$. Now we can prove by induction on $\alpha \leq \beta_*$ that q has a unique extension in $\mathscr{S}_{\mathfrak{s}}(M_\alpha)$. But $M_{\beta_*} \leq_{\mathfrak{s}} N$ and q has at least two extensions in $\mathscr{S}^{\mathrm{bs}}_{\mathfrak{s}}(N)$ hence $M_{\beta_*} \neq N$. So we can choose $a_{\beta_*} \in N \backslash M_{\beta_*}$ as $\mathfrak{s}$ is full and has primes, necessarily $\mathbf{tp}_{\mathfrak{s}}(a_{\beta_*}, M_{\beta_*}, N) \perp q$. 8) Let $q_{\beta_*} \in \mathscr{S}_{\mathfrak{s}}(M_{\beta_*})$ be the unique extension of q in $\mathscr{S}_{\mathfrak{s}}(M_{\beta_*})$. So there is a $M_{\beta_*,0} \leq_{\mathfrak{K}[\mathfrak{t}]} M_{\beta_*}$ from $K_{\mathfrak{t}}$, brimmed for $\mathfrak{t}$ such that $\mathbf{tp}_{\mathfrak{s}}(a_{\beta_*}, M_{\beta_*}, N)$ and q^+ does not fork oever $M_{\beta_*,0}$ hence $q^+ \restriction M_{\beta_*,0}$ is regular for $\mathfrak{t}$ and is not orthogonal to $\mathbf{tp}_{\mathfrak{t}}(a_{\beta_*}, M_{\beta_*,0}, N)$ for $\mathfrak{t}$.

As before we can choose $c \in N \backslash M_{\beta_*}$ realizing $q^+ \restriction M_{\beta_*,0}$ such that $\{c, a_{\beta_*}\}$ is not independent in $(M_{\beta_*,0}, N)$ for $\mathfrak{t}$ hence) $c \in N \backslash M_{\beta_*}$. By the choice of β_* the type $\mathbf{tp}_{\mathfrak{s}}(c, M_{\beta_*}, N)$ is not orthogonal to q, so for some $M_{\beta_*,1}$ we have $M_{\beta_*,0} \leq_{\mathfrak{t}} M_{\beta_*,1} \leq_{\mathfrak{K}[\mathfrak{t}]} M_{\beta_*}$, the type $\mathbf{tp}_{\mathfrak{s}}(c, M_{\beta_*}, N)$ does not fork over $M_{\beta_*,1}$ and $M_{\beta_*,1}$ is brimmed over $M_{\beta_*,0}$ for $\mathfrak{t}$. So necessarily $\mathbf{tp}_{\mathfrak{t}}(c, M_{\beta_*,1}, N)$ forks over $M_{\beta_*,0}$ for $\mathfrak{t}$ hence is orthogonal to $q^+ \restriction M_{\beta_*,1}$. This implies then $\mathbf{tp}_{\mathfrak{s}}(c, M_{\beta_*}, N)$

and q^+ are orthogonal contradiction to the choice of β_*.
9) If q is realized in N then by 6.4 we know $p \pm q$. So assume that $p \pm q$ and then use part (8). $\square_{10.6}$

10.7 Hypothesis. Assume, in addition to 10.1

(a) $\mathfrak{s} = \mathfrak{t}^+, \mathfrak{t}$ is $\lambda_\mathfrak{t}$-good$^+$ successful with primes and $K^{3,\mathrm{qr}}_\mathfrak{t} = K^{3,\mathrm{vq}}_\mathfrak{t}$.

Remark. So $\mathfrak{t}$ satisfies Hypotheis 10.1; the "$\mathfrak{t}$ type-full" require a short argument.

10.8 Conclusion. 1) Non-orthogonality among regular types is an equivalence relation.
2) For non-orthogonal regular $p, q \in \mathscr{S}^{\mathrm{bs}}_\mathfrak{s}(N)$ and $M \leq_\mathfrak{s} N$, we have $p \perp M \Leftrightarrow q \perp M$.
3) If $p, q \in \mathscr{S}^{\mathrm{bs}}_\mathfrak{s}(M), p \pm q$ the type q is regular, $(M, N, a) \in K^{3,\mathrm{pr}}_\mathfrak{s}$ or just $(M, N, a) \in K^{3,\mathrm{bs}}_\mathfrak{s}$ and $\mathbf{tp}_\mathfrak{s}(a, M, N) = p$ <u>then</u> q is realizes in N.
4) For $p, q, r \in \mathscr{S}^{\mathrm{bs}}_\mathfrak{s}(M), q$ regular, $p \pm q, q \pm r$ we have $p \pm r$.
5) For regular $p, q \in \mathscr{S}^{\mathrm{bs}}_\mathfrak{s}(M)$ and $r \in \mathscr{S}^{\mathrm{bs}}_\mathfrak{s}(M)$ we have $p \pm q, q \pm r \Rightarrow p \pm r$.

Remark. Alternative proof of 10.8(3),(4),(5) appear after the proof of 10.16.

Proof. 1) Assume $M \in K_\mathfrak{s}$ and regular p_1, p_2, p_3 belongs to $\mathscr{S}^{\mathrm{bs}}_\mathfrak{s}(M)$ and $p_1 \pm p_2, p_2 \pm p_3$. Let a_1, N_1 be such that $(M, N_1, a_1) \in K^{3,\mathrm{pr}}_{\mathfrak{s},p_1}$, so by 10.6(8) for some $a_2 \in N_2$ and $N_2 \leq_\mathfrak{s} N_1$ we have $(M, N_2, a_2) \in K^{3,\mathrm{pr}}_{\mathfrak{s},p_2}$. Similarly there is $a_3 \in N_2 \leq_\mathfrak{s} N_1$ which realizes p_3, so by 6.4(1) easily $p_1 \pm p_3$. This proves that non-orthogonality is transitive, but symmetry was proved in 6.7(2) and reflexivity is obvious so we are done.
2) By symmetry we assume $p \pm M$ and we shall prove $q \pm M$, this suffices. By 10.9(1) below there is a $r \in \mathscr{S}^{\mathrm{bs}}_\mathfrak{s}(M)$ not orthogonal to

p and regular. Now use part (1).
3) If $(M, N, a) \in K_{\mathfrak{s}}^{3,\mathrm{bs}}$ then for some $N' \leq_{\mathfrak{s}} N$ we have $a \in N'$ and $(M, N', a) \in K_{\mathfrak{s}}^{3,\mathrm{pr}}$ so without loss of generality $(M, N, a) \in K_{\mathfrak{s}}^{3,\mathrm{pr}}$. Now apply 10.6(9), noting that the assumption of 10.6 holds by Hypothesis 10.7.
4) Let $(M, N, a_\ell) \in K_{\mathfrak{s}}^{3,\mathrm{pr}}$ and $p_\ell = \mathbf{tp}_{\mathfrak{s}}(a_\ell, M, N_\ell)$ for $\ell = 1, 2, 3$ be such that $p_1 = p, p_2 = q, p_3 = r$. Now as q is regular and $p \pm q$, by part (3) the type q is realized in N_1 say by b_1. So $b_1 \in N_1 \backslash M$. Similarly q is realized by some $b_3 \in N_3 \backslash M$. Now N_1, N_3 can be amalgamated in two incompatible ways over M: identifying b_3 with b_1 or not, which gives $p = p_1 \pm p_3 = r$.
5) Follows by part (3). $\square_{10.8}$

10.9 Claim. *1) If $M \leq_{\mathfrak{s}} N$ and $p \in \mathscr{S}_{\mathfrak{s}}^{\mathrm{bs}}(N)$ is not orthogonal to M <u>then</u> there is $q \in \mathscr{S}_{\mathfrak{s}}^{\mathrm{bs}}(N)$ not orthogonal to p, conjugate to p (i.e., $f(p) = q$ for some $f \in \mathrm{Aut}(M)$) and q does not fork over M.*
2) If $\langle M_i : i \leq \delta + 1\rangle$ is $\leq_{\mathfrak{s}}$-increasing continuous and $M_\delta \neq M_{\delta+1}$, <u>then</u> for some $c \in M_{\delta+1} \backslash M_\delta$ and non-limit $i < \delta$, we have $\mathbf{tp}_{\mathfrak{s}}(c, M_\delta, M_{\delta+1})$ is regular, does not fork over M_i, and is orthogonal to M_{i-1} if $i > 0$ so $\mathfrak{s}$ weakly has regulars (see Definition 7.5).
3) If in part (2), $q \in \mathscr{S}_{\mathfrak{s}}^{\mathrm{bs}}(M_\delta)$ is regular realized by some member of $M_{\delta+1}$, <u>then</u> we can demand $\mathbf{tp}_{\mathfrak{s}}(c, M_i, M_{\delta+1})$ is conjugate to q hence regular.
4) In part (1), if for some M_0 the model M is $(\lambda, *)$-brimmed over M_0, <u>then</u> we can get q conjugate to p over M_0.

Remark. Part (1) of 10.9 could have appeared earlier.

Proof. 1) Without loss of generality N is $(\lambda, *)$-brimmed over M (using 10.7(a)). Let $r \in \mathscr{S}_{\mathfrak{s}}^{\mathrm{bs}}(M)$ be not orthogonal to p. Let $\langle M_\alpha : \alpha \leq \omega\rangle, \langle N_\alpha : \alpha \leq \omega\rangle$ be as in the proof of 8.17, i.e., $M = M_\omega = \cup\{M_n : n < \omega\}, N = N_\omega = \cup\{N_n : n < \omega\}, \mathrm{NF}_{\mathfrak{s}}(M_n, N_n, M_{n+1}, M_{n+1}, N_{n+1})$ and M_{n+1}, N_{n+1} are $(\lambda, *)$-brimmed over M_n, N_n respectively for $n < \omega$; without loss of generality p does not fork over N_0 and r does not fork over M_0. We can find $\langle f_i : i < \lambda\rangle$ such that f_{1+i} is a $\leq_{\mathfrak{s}}$-embedding of N_0 into M over $M_0, f_0 = \mathrm{id}_{N_0}$, such that $\langle f_i(N_0) : i < \omega\rangle$ is independent over M_0 (see 8.17) and clearly

$f_i(p \restriction N_0) \pm r \restriction M_0$ for $i < \lambda$ hence $f_i(p \restriction N_0) \pm M_0$. By 8.17 clearly $p \restriction N_0 \pm f_1(p \restriction N_0)$ and let $q \in \mathscr{S}^{\rm bs}_{\mathfrak{s}}(N)$ be a non-forking extension of $f_1(p \restriction N_0)$.
2) By 10.5(1) for some $d \in M_{\delta+1} \backslash M_\delta$, the type $\mathbf{tp}_{\mathfrak{s}}(d, M_\delta, M_{\delta+1})$ is regular, and apply part (3).
3) Let $j = \text{Min}\{i \le \delta : q \pm M_i\}$, as $q \pm M_\delta$ clearly j is well defined. By 6.10(2), j is a non-limit ordinal and by part (1) there is $r \in \mathscr{S}^{\rm bs}_{\mathfrak{s}}(M_\delta)$ not forking over M_j not orthogonal to q and conjugate to q hence r is regular and by 10.8(2) is orthogonal to M_{j_1} for $j_1 < j$ but not orthogonal to p.

By 10.6(8) some $c \in M_{\delta+1} \backslash M_\delta$ realizes r.
4) By a similar proof. $\square_{10.9}$

10.10 Claim. *If* $\text{NF}_{\mathfrak{s}}(M_0, M_1, M_2, M_3)$ *and* $p \in \mathscr{S}^{\rm bs}_{\mathfrak{s}}(M_3)$ *is regular and* $p \pm M_1, p \pm M_2$ *<u>then</u>* $p \pm M_0$.

Proof. As $p \pm M_1$, by 10.9(1) there is $q \in \mathscr{S}^{\rm bs}_{\mathfrak{s}}(M_1)$ conjugate to p and not orthogonal to p. As q is conjugate to p it is regular. As $p \pm q$ by 10.8(2) it is enough to prove $q \pm M_0$ and we know that $q \pm M_2$. Now if q is orthogonal to M_0, by 6.10(5) (recalling Definition 6.9(2)), as $\mathfrak{s}$ is categorical, (see Hypothesis 10.7), it is super-orthogonal to M_0, which implies that it is orthogonal to M_2, contradiction so we are done.

$\square_{10.10}$

10.11 Definition. We call $(\bar{M}, \bar{\mathbf{J}}) \in \mathscr{W}$ (from Definition 7.2) regular if $c \in \mathbf{J}_i \Rightarrow \mathbf{tp}_{\mathfrak{s}}(c, M_i, M_{i+1})$ is regular; we say "regular except $\mathbf{J}$" if the $c \in \mathbf{J}$ are excluded.

10.12 Claim. *1) Assume* $M \le_{\mathfrak{s}} N$ *and* $\mathbf{J} \subseteq \mathbf{I}_{M,N}$ *is independent in* (M, N). *<u>Then</u> we can find a prime* $(\bar{M}, \bar{\mathbf{J}}) \in K^{3,\rm ar}_{\mathfrak{s}}$, *see Definition 7.4(2),(8) so* $\bar{M} = \langle M_n : n < \omega \rangle$ *with* $\mathbf{J} \subseteq \mathbf{J}_0, M_0 = M, N = \cup\{M_n : n < \omega\}$ *and* $(\bar{M}, \bar{\mathbf{J}})$ *is regular except (possibly)* $\mathbf{J}$.
2) If $(M, N, a) \in K^{3,\rm uq}_{\mathfrak{s}}$ *<u>then</u> we can find prime* $(\bar{M}, \bar{\mathbf{J}}) \in K^{3,\rm ar}_{\mathfrak{s}}$ *with* $\mathbf{J}_0 = \{a\}, M_0 = M, N = \cup\{M_n : n < \omega\}$ *and* $(\bar{M}, \bar{\mathbf{J}})$ *is regular except possibly* $\mathbf{J}_0$. *Of course, we can replace* $\{a\}$ *by* $\mathbf{J} \subseteq \mathbf{I}_{M,N}$ *if* $(M, N, \mathbf{J}) \in K^{3,\rm vq}_{\mathfrak{s}}$ *so* $\mathbf{J}_0 = \mathbf{J}$.

3) If $N_0 \leq_{\mathfrak{s}} N_1 \leq_{\mathfrak{s}} N_2, c \in N_2 \backslash N_1$ and $\mathbf{tp}_{\mathfrak{s}}(c, N_1, N_2) \pm N_0$ and $\mathbf{tp}_{\mathfrak{s}}(c, N_1, N_2) \in \mathscr{S}^{\mathrm{bs}}_{\mathfrak{s}}(N_1)$ (actualy follows) then for some $b \in N_2 \backslash N_1$ the type $\mathbf{tp}_{\mathfrak{s}}(b, N_1, N_2)$ does not fork over N_0 and is regular.

Remark. Can be viewed as changing $\mathscr{S}^{\mathrm{bs}}_{\mathfrak{s}}$.

Proof. 1) Clearly $(M, N, \mathbf{J}) \in K^{3,\mathrm{bs}}_{\mathfrak{s}}$ so the assumption of 7.7(2) holds by 10.9(2). By 7.7(2) the desired conclusion almost holds, i.e. holds except "regular except $\mathbf{J}$". So repeat the proof of 7.7(2) such that $c \in \mathbf{J}_{n+1} \Rightarrow \mathbf{tp}_{\mathfrak{s}}(c, M_{n+1}, M_{n+1})$ is regular and $c \in \mathbf{J}_0 \backslash \mathbf{J} \Rightarrow \mathbf{tp}_{\mathfrak{s}}(c, M, N)$ is regular using 10.9(2); alternatively use 10.19.
2) The same proof using 6.14(1) and 7.9(A) $\Rightarrow$ (C) for $K^{3,\mathrm{vq}}_{\mathfrak{s}}$.
3) Apply part (1) with $(N_1, N_2, \emptyset)$ here standing for $(M, N, \mathbf{J})$ there and get $(\bar{M}, \bar{\mathbf{J}})$ as there. If for some $b \in \mathbf{J}_0$, $\mathbf{tp}_{\mathfrak{s}}(b, N_1, N_2) \pm N_0$, then we get the desired conclusion. (Why? By 10.9(1) there is $q \in \mathscr{S}^{\mathrm{bs}}_{\mathfrak{s}}(N_1)$ not orthogonal to $\mathbf{tp}_{\mathfrak{s}}(b, N_1, N_2)$ conjugate to it, and not forking over N_0; now by 10.6(8) the type q is realized by some $b' \in N_2 \backslash N_1$ and it is as requried.)

Otherwise, now 7.7(3) is applicable as $\mathfrak{s}$ weakly has regulars by 10.9(2), so we have $(M, N, \mathbf{J}_0) \in K^{3,\mathrm{vq}}_{\mathfrak{s}}$. Now if $q \in \mathscr{S}^{\mathrm{bs}}_{\mathfrak{s}}(N_1)$ does not fork over N_0 then $c \in \mathbf{J}_0 \Rightarrow \mathbf{tp}_{\mathfrak{s}}(c, N_1, N_2) \perp q$ so by Claim 6.13 the type q has a unique extension in $\mathscr{S}^{\mathrm{bs}}_{\mathfrak{s}}(N_2)$. Hence: if $N_2 \leq_{\mathfrak{s}} N'$ and $b \in N'$ realizes q then $\{b, c\}$ is independent in (N_1, N') (recall $\mathfrak{s}$ is type-full) so $q \underset{\mathrm{wk}}{\perp} \mathbf{tp}_{\mathfrak{s}}(c, N_1, N_2)$. As q was any member of $\mathscr{S}^{\mathrm{bs}}_{\mathfrak{s}}(N_1)$ which does not fork over N_0, we get $\mathbf{tp}_{\mathfrak{s}}(c, N_1, N_2) \underset{\mathrm{wk}}{\perp} N_0$, but $\mathfrak{s}$ is categorical hence $\mathbf{tp}_{\mathfrak{s}}(c, N_1, N_2) \perp N_0$, contradicting an assumption. $\square_{10.12}$

10.13 Claim. *1) If $(M_0, M_\ell, a_\ell) \in K^{3,\mathrm{uq}}_{\mathfrak{s}}$ for $\ell = 1, 2$ and $M_1 \cap M_2 = M_0$ then we can find M_3 such that $(M_\ell, M_3, a_{3-\ell}) \in K^{3,\mathrm{uq}}_{\mathfrak{s}}$ for $\ell = 1, 2$.*
2) Similarly for $(M_0, M_\ell, \mathbf{J}_\ell) \in K^{3,\mathrm{vq}}_{\mathfrak{s}}$.

Proof. 1) Let $p_1' \in \mathscr{S}^{\mathrm{bs}}_{\mathfrak{s}}(M_2)$ be a non-forking extension of $p_1 := \mathbf{tp}_{\mathfrak{s}}(a_1, M_0, M_1)$. Let (M_3, a_1') be such that $(M_2, M_3, a_1') \in K^{3,\mathrm{uq}}_{\mathfrak{s}}$

and $p'_1 = \mathbf{tp}_{\mathfrak{s}}(a'_1, M_2, M_3)$; we can choose M_3, a'_1 because $\mathfrak{s}$ has existence for $K^{3,\mathrm{pr}}_{\mathfrak{s}}$.

Similarly we can find $M'_1 \le_{\mathfrak{s}} M_3$ such that $(M_0, M'_1, a'_1) \in K^{3,\mathrm{pr}}_{\mathfrak{s}}$. As $\{M''_1 : M''_1 \le_{\mathfrak{s}} M_3$ and $(M_0, M'_1, a'_1) \in K^{3,\mathrm{uq}}_{\mathfrak{s}}\}$ is non-empty (as M'_1 belongs to it) and is closed under $\le_{\mathfrak{s}}$-increasing unions (by 7.10), necessarily it has a $\le_{\mathfrak{s}}$-maximal member, call it M^*_1. By uniqueness for $K^{3,\mathrm{uq}}_{\mathfrak{s}} = K^{3,\mathrm{pr}}_{\mathfrak{s}}$, see 7.14, there is an isormophism f from M_1 onto M^*_1 over M_0 mapping a_1 to a'_1.

As $M_0 \le_{\mathfrak{s}} M_2 \le_{\mathfrak{s}} M_3$ and $\mathbf{tp}(a'_1, M_2, M_3)$ does not fork over M_0 and $M_0 \le_{\mathfrak{s}} M^*_1 \le_{\mathfrak{s}} M_3$ and $(M_0, M^*_1, a'_1) \in K^{3,\mathrm{uq}}_{\mathfrak{s}}$ it follows that $\mathrm{NF}_{\mathfrak{s}}(M_0, M^*_1, M_2, M_3)$ hence $M^*_1 \cap M_2 = M_0$. So without loss of generality $f = \mathrm{id}_{M_1}$ so $a'_1 = a_1, M^*_1 = M_1$.

Do we have $(M^*_1, M_3, a_2) \in K^{3,\mathrm{uq}}_{\mathfrak{s}}$? If not, then as above, let $M^-_3 \le_{\mathfrak{s}} M_3$ be $\le_{\mathfrak{s}}$-maximal such that $(M^*_1, M^-_3, a_2) \in K^{3,\mathrm{uq}}_{\mathfrak{s}}$. Let $b \in M_3 \backslash M^-_3$ be such that $\mathbf{tp}_{\mathfrak{s}}(b, M^-_3, M_3) \in \mathscr{S}^{\mathrm{bs}}_{\mathfrak{s}}(M^-_3)$. By a case of 10.9(2), without loss of generality $\mathbf{tp}_{\mathfrak{s}}(b, M^-_3, M_3)$ is regular and is $\perp M^*_1$ or does not fork over M^*_1 (not used) and is $\perp M_0$ or does not fork over M_0.

Now first if $\mathbf{tp}_{\mathfrak{s}}(b, M^-_3, M_3) \perp M_0$ we can find $M'_3 \le_{\mathfrak{s}} M_3$ such that $(M^-_3, M'_3, b) \in K^{3,\mathrm{pr}}_{\mathfrak{s}}$, so M'_3 contradict the maximality of M^*_1.

Second, if $\mathbf{tp}_{\mathfrak{s}}(b, M^-_3, M_3) \pm M_0$ then it does not fork over M_0 so as $\{a_1, a_2\}$ is independent in (M_0, M^-_3) also $\{a_1, a_2, b\}$ is independent in (M_0, M_3). However, recall $(M_0, M_2, a_2), (M_2, M_3, a_1) \in K^{3,\mathrm{uq}}_{\mathfrak{s}}$, which by 5.4(3) gives $\mathbf{tp}_{\mathfrak{s}}(b, M_3, M_3)$ does not fork over M_0, contradiction.

2) Similar (and not used). $\square_{10.13}$

10.14 Definition. 1) For $M \le_{\mathfrak{s}} N$ let $\mathbf{I}^{\mathrm{reg}}_{M,N} = \{c \in N : \mathbf{tp}_{\mathfrak{s}}(c, M, N)$ is regular$\}$.

2) For $M <_{\mathfrak{s}} N$, we define on $\mathbf{I}^{\mathrm{reg}}_{M,N}$ a dependence relation called the regular (M,N)-dependence relation by:

- (a) $\mathbf{J} \subseteq \mathbf{I}^{\mathrm{reg}}_{M,N}$ is (M,N)-independent or regularly independent in (M,N) <u>if</u> if it is independent
- (b) $c \in \mathbf{J}^{\mathrm{reg}}_{M,N}$ is (M,N)-dependent on $\mathbf{J} \subseteq \mathbf{J}^{\mathrm{reg}}_{M,N}$ or c is regularly dependent on $\mathbf{J} \subseteq \mathbf{J}^{\mathrm{reg}}_{M,N}$ <u>if</u> there is an independent $\mathbf{J}' \subseteq \mathbf{J}$ such that $c \in \mathbf{J}'$ or $\mathbf{J} \cup \{c\}$ is not independent.

We may omit (M, N) if clear and may omit the "regular" if clear.

Remark. We can use only regular$^+$ types; somewhat simplify.

10.15 Claim. *1) Assume* $\mathbf{J}_i \subseteq \mathbf{I}^{\rm reg}_{M,N}$ *for* $i < i^*$ *and* $i \neq j$ & $a \in \mathbf{J}_i$ & $b \in \mathbf{J}_j \Rightarrow \mathbf{tp}_{\mathfrak{s}}(a, M, N) \perp \mathbf{tp}_{\mathfrak{s}}(b, M, N)$.
Then

- (α) $i \neq j \Rightarrow \mathbf{J}_i \cap \mathbf{J}_j = \emptyset$
- (β) $\cup\{\mathbf{J}_i : i < i^*\}$ *is independent in* (M, N) *iff for each* $i, \mathbf{J}_i$ *is independent in* (M, N).

2) Assume $\mathbf{J} \subseteq \mathbf{I}^{\rm reg}_{M,N}$ *and* $\mathscr{E}$ *is the following equivalence relation on* $\mathbf{I}^{\rm reg}_{M,N} : a\mathscr{E}b \Leftrightarrow \mathbf{tp}_{\mathfrak{s}}(a, M, N) \pm \mathbf{tp}_{\mathfrak{s}}(b, M, N)$. *Then* $\mathbf{J}$ *is independent in* (M, N) *iff for every* $a \in \mathbf{I}_{M,N}$ *the set* $\mathbf{J} \cap (a/\mathscr{E})$ *is independent in* (M, N) *iff for every* $a \in \mathbf{J}$ *the set* $\mathbf{J} \cap$ $(a/\mathscr{E})$ *is independent in* (M, N).

Proof. Easy. $\square_{10.15}$

10.16 Claim. *Assume* $M \leq_{\mathfrak{s}} N$.
1) The relations in 10.14 and their negations are preserved if we replace N *by a* $\leq_{\mathfrak{s}}$*-extension.*
2) If $\mathbf{J}_1, \mathbf{J}_2 \subseteq \mathbf{I}^{\rm reg}_{M,N}$ *are* (M, N)*-independent, every* $b \in \mathbf{J}_2$ *does* (M, N)*-depend on* $\mathbf{J}_1$ *and* $c \in \mathbf{I}^{\rm reg}_{M,N}$ *depend on* $\mathbf{J}_2$, *then* c *does* (M, N)*-depend on* $\mathbf{J}_1$.
3) The regular (M, N)*-dependence relation satisfies the axioms of dependence relation, so dimension is well defined. Also if* $p, q \in \mathscr{S}^{\rm bs}_{\mathfrak{s}}(M)$ *are regular not orthogonal then* $\dim(p, N) = \dim(q, M)$.
4) $\mathbf{J} \subseteq \mathbf{I}^{\rm reg}_{M,N}$ *is a maximal* (M, N)*-independent subset of* $\mathbf{I}^{\rm reg}_{M,N}$ *iff* $(M, N, \mathbf{J}) \in K^{3,\rm vq}_{\mathfrak{s}}$.
5) If $\mathbf{P} \subseteq \{p \in \mathscr{S}^{\rm bs}_{\mathfrak{s}}(M) : p$ *regular*$\}$ *is a maximal set of pairwise orthogonal types and* $\mathbf{J} \subseteq \mathbf{I}^{\rm reg}_{M,N}$ *is independent, then we can find* $\mathbf{J}', h$ *such that:*

- (a) $\mathbf{J}' \subseteq \mathbf{I}^{\rm reg}_{M,N}$ *is* (M, N)*-independent*

(b) *h is a function from $\mathbf{J}$ onto $\mathbf{J}'$ such that $h(c)$ does (M,N)-depend on $\{c\}$*

(c) $c \in \mathbf{J}' \Rightarrow \mathbf{tp}_{\mathfrak{s}}(c,M,N) \in \mathbf{P}$.

6) Assume $\mathbf{P} \subseteq \cup\{\mathscr{S}^{\rm bs}_{\mathfrak{s}}(M') : M' \leq_{\mathfrak{s}} M\}$ and every $q \in \mathscr{S}^{\rm bs}_{\mathfrak{s}}(M)$ is non-orthogonal to some $p \in \mathbf{P}$

(a) *if $M \leq_{\mathfrak{s}} N$ and $q \in \mathscr{S}^{\rm bs}_{\mathfrak{s}}(N)$ is not orthogonal to M then q is not orthogonal to some $p \in \mathbf{P}$*

(b) *if $q \in M$ then we can find $n < \omega$ and a $\leq_{\mathfrak{s}}$-increasing sequence $\langle M_\ell : \ell \leq n\rangle$ with $M_0 = M, a_\ell \in M_{\ell+1}$ such that $(M_\ell, M_{\ell+1}, a_\ell) \in K^{3,\rm pr}_{\mathfrak{s}}$ for $\ell < n$ such that $\mathbf{tp}_{\mathfrak{s}}(a_\ell, M_{\ell+1})$ is parallel to some $p \in \mathbf{P}$ and q is realized in M_n, i.e. as in 8.7.*

Proof. 1) Trivial.
2) If $c \in \mathbf{J}_2$ then conclusion is trivial, so we can assume that $c \notin \mathbf{J}_2$ and even more obviously if $c \in \mathbf{J}_1$ the conclusion holds so we can assume that $c \notin \mathbf{J}_1$.

Let

$$\mathbf{J}'_2 = \{b \in \mathbf{J}_2 : \mathbf{tp}_{\mathfrak{s}}(b,M,N) \text{ is not orthogonal to } \mathbf{tp}_{\mathfrak{s}}(c,M,N)\}$$

and let

$$\mathbf{J}'_1 = \{a \in \mathbf{J}_1 : \mathbf{tp}_{\mathfrak{s}}(a,M,N) \text{ is not orthogonal to } \mathbf{tp}_{\mathfrak{s}}(c,M,N)\}.$$

We know that $\mathbf{J}_2 \cup \{c\}$ is not independent, hence by 10.15(2) necessarily $\mathbf{J}'_2 \cup \{c\}$ is not independent. Similarly for every $b \in \mathbf{J}'_2$ we have $b \in \mathbf{J}'_1$ or $\mathbf{J}'_1 \cup \{b\}$ is not independent (trivially $b \notin \mathbf{J}_1 \backslash \mathbf{J}'_1$).

Toward contradiction assume that "$\mathbf{J}'_1 \cup \{c\}$ is independent". Let $\langle a_i : i < \alpha_1\rangle$ list $\mathbf{J}'_1$, without repetition and let $\langle a_i : i \in [\alpha_1,\alpha_2)\rangle$ list $\mathbf{J}'_2$ without repetitions and let $a_{\alpha_2} = c$. We can find a pr-decomposition $\langle M_i, a_i : i \leq \alpha_2\rangle$ over M inside N (recall that $\mathfrak{s}$ is type-full; also uq-decomposition inside $N', N \leq_{\mathfrak{s}} N'$ is O.K.; possibly $a_i \in M_i$ and the $M_{i+1} = M_i$).

Now by 5.4(3) the type $\mathbf{tp}_{\mathfrak{s}}(c, M_{\alpha_1}, N_1)$ does not fork over M. Also for each $i \in [\alpha_1, \alpha_2)$ the type $\mathbf{tp}_{\mathfrak{s}}(a_i, M_{\alpha_1}, N)$ forks over M (as a_i depends on $\mathbf{J}'_1$ in (M, N)) hence by Claim 10.6(4) the type $\mathbf{tp}_{\mathfrak{s}}(a_i, M_{\alpha_1}, N)$ is orthogonal to $\mathbf{tp}_{\mathfrak{s}}(a_i, M, N)$ hence by 10.8(1) also to $\mathbf{tp}_{\mathfrak{s}}(c, M, N)$ recalling that $a_i \in \mathbf{J}'_2$, hence to $\mathbf{tp}_{\mathfrak{s}}(c, M_{\alpha_1}, N)$. So we can prove by induction on $i \in [\alpha_1, \alpha_2]$ that $\mathbf{tp}_{\mathfrak{s}}(c, M_i, N)$ does not fork over M. For $i = \alpha_2$ we get a contradiction, the reason being the assumption "$\mathbf{J}'_2 \cup \{c\}$ is not independent in (M, N)", so we are done.

3) The finite character holds by 5.4, transitivity holds by part (2), monotonicity is trivial by the definitions and also the exchange principle. The "also" follows by this and 10.6(8).

4) First assume $(M, N, \mathbf{J}) \in K^{3,\mathrm{vq}}_{\mathfrak{s}}$, but $\mathbf{J}$ is not a maximal independent set, so for some $c \in N \backslash M \backslash \mathbf{J}$ the set $\mathbf{J} \cup \{c\}$ is independent. Let $N_1 \leq_{\mathfrak{s}} N$ be such that $(M, N_1, \mathbf{J}) \in K^{3,\mathrm{qr}}_{\mathfrak{s}}$, so by 5.16(5) the type $\mathbf{tp}_{\mathfrak{s}}(c, N_1, N)$ does not fork over M, but by 7.9(A) $\Rightarrow$ (C), the type $\mathbf{tp}_{\mathfrak{s}}(c, N_1, N)$ is orthogonal to M, contradiction; so $\mathbf{J}$ is maximal. That is the second implies the first.

Second assume $\mathbf{J}$ is maximal and let $N_0 \leq_{\mathfrak{s}} N$ be such that $(M, N_0, \mathbf{J}) \in K^{3,\mathrm{qr}}_{\mathfrak{s}}$. Now by 5.4, 5.8(2), 5.11(2) and 10.6(2) we can find a pr-decomposition $\langle (M_i, a_i) : i < \alpha \rangle$ of N over N_0, with each $\mathbf{tp}_{\mathfrak{s}}(a_i, N_i, N)$ regular.

If for some $i < \alpha$, $\mathbf{tp}_{\mathfrak{s}}(a_i, N_i, N) \pm M$ then by 10.9(1) there is a regular $q \in \mathscr{S}^{\mathrm{bs}}_{\mathfrak{s}}(N_i)$ not orthogonal to $\mathbf{tp}_{\mathfrak{s}}(a_i, N_i, N)$ which does not fork over M. Now by 10.6(8) some $b \in N_{i+1} \backslash N_i$ realizes q, but this contradicts $\mathbf{J}$'s maximality.

Hence $i < \alpha \Rightarrow \mathbf{tp}_{\mathfrak{s}}(a_i, N_i, N) \perp M$ then by 6.20(2) the triple $(M, N_\alpha, \mathbf{J})$ belongs to $K^{3,\mathrm{vq}}_{\mathfrak{s}}$ as required.

5) For every regular $p \in \mathscr{S}^{\mathrm{bs}}_{\mathfrak{s}}(M)$ let r_p be the unique $r \in \mathbf{P}$ not orthogonal to p, (exists by the maximality of $\mathbf{P}$, unique as $\pm$ is an equivalence relation on the family of regular types, see 10.8(2)). Now for each $a \in \mathbf{J}$ let $r_a \in \mathbf{P}$ be $r_{\mathbf{tp}_{\mathfrak{s}}(a,M,N)}$, by 10.6(8) there is $b_a \in N$ realizing r_a such that $\{a, b_a\}$ is not independent (why? choose $N_a \leq_{\mathfrak{s}} N$ such that $(M, N_a, a) \in K^{3,\mathrm{pr}}_{\mathfrak{s}}$ and then choose $b_a \in N$ realizing r_a by 10.6(8) they are not independent by 6.3).

Now by part (3) we can finish easily.
6) Let $\mathbf{P}' = \{r \in \mathscr{S}^{\rm bs}_{\mathfrak{s}}(M) : p$ is regular not orthogonal to some $p \in \mathbf{P}\}$, the rest should be clear. $\square_{10.16}$

Alternative Proof of 10.8(3),(4),(5):
(3) Let $\mathbf{J} \subseteq \mathbf{I}^{\rm reg}_{M,N}$ be a maximal set indepndent in (M,N). So $(M,N,\mathbf{J}) \in K^{3,\rm vq}_{\mathfrak{s}}$ by 10.12(2). Now if $c \in \mathbf{J} \Rightarrow \mathbf{tp}_{\mathfrak{s}}(c,M,N) \perp q$ then by 6.13 we can deduce that q has unique extension in $\mathscr{S}^{\rm bs}_{\mathfrak{s}}(N)$. As $(M,N,a) \in K^{3,\rm pr}_{\mathfrak{s}}$ it follows that $q \perp \mathbf{tp}_{\mathfrak{s}}(a,M,N)$ but the latter is p, contradicting an assumption.

So we can choose $c \in \mathbf{J}$ such that $q_1 = \mathbf{tp}_{\mathfrak{s}}(c,M,N)$ is $\pm q$, hence by 10.6(8) (as q, q_1 are regular) some $c' \in N$ realizes q so we are done.
(4) We can find $(M,N_1,a_1) \in K^{3,\rm pr}_{\mathfrak{s}}$ and $(M,N_2,a_2) \in K^{3,\rm pr}_{\mathfrak{s}}$ such that $p = \mathbf{tp}_{\mathfrak{s}}(a_1,M,N_1)$ and $r = \mathbf{tp}_{\mathfrak{s}}(a_2,M,N_2)$.

For $\ell = 1,2$ by part (3) there is $c_\ell \in N_\ell$ realizing r (in N_ℓ). So without loss of generality for some N we have $N_1 \le_{\mathfrak{s}} N, N_2 \le_{\mathfrak{s}} N$ and $c_1 = c_2$. Now if $\{a_1,a_2\}$ is independent in (M,N) then it follows that $\mathrm{NF}_{\mathfrak{s}}(M,N_1,N_2,N)$, easy contradiction, so $\{a_1,a_2\}$ is not independent in (M,N) hence $p = \mathbf{tp}_{\mathfrak{s}}(a_1,M,N) \pm \mathbf{tp}_{\mathfrak{s}}(a_2,M,N) = r$, as required.
(5) Let $(M,N,c) \in K^{3,\rm pr}_{\mathfrak{s}}$ be such that $\mathbf{tp}_{\mathfrak{s}}(c,M,N) = r$. As $q \pm r$ by part (3) clause (a) there is $b \in N$ realizing p. As $p \pm q$ by part (3) or 10.6(8) there is $a \in N$ realizing p. This gives $p \pm r$ by 6.4(1). $\square_{10.8}$

10.17 Remark.: 1) On weight of types and $\mathbf{P}$-simple types (parallel to [Sh:c, V,§5]) see [Sh 839].
2) If $M_0 \le_{\mathfrak{s}} M_1 \le_{\mathfrak{s}} M_2$ and there is no $c \in M_2 \backslash M_1$ such that $\mathbf{tp}_{\mathfrak{s}}(c,M_1,M_2)$ does not fork over M_0 and $(M_0,M_1,\mathbf{J}) \in K^{3,\rm vq}_{\mathfrak{s}}$ then $(M_0,M_2,\mathbf{J}) \in K^{3,\rm vq}_{\mathfrak{s}}$.
[Why? As above by 10.12(2) it is enough to show that $\mathbf{J}$ is a maximal subset of $\mathbf{I}_{M_0,M_2}$ which is independent in (M,N). If not, then for some $c \in M_2 \backslash M_0 \backslash \mathbf{J}$ the set $\mathbf{J} \cup \{c\}$ is independent in (M_0,M_2). But then by 6.13 the set $\{c\}$ is independent in (M_0,M_1,M_2), contradiction.]

10.18 Definition. We define a pre-frame $\mathfrak{t} := \mathfrak{s}^{\text{reg}}$ ($= \mathfrak{s}[\text{reg}]$) as follows:

(a) $\lambda_{\mathfrak{t}} = \lambda_{\mathfrak{s}}$

(b) $\mathfrak{K}_{\mathfrak{t}} = \mathfrak{K}_{\mathfrak{s}}$

(c) $\mathscr{S}^{\text{bs}}_{\mathfrak{t}}(M) = \{p \in \mathscr{S}^{\text{bs}}_{\mathfrak{s}}(M) : p \text{ is regular (for } \mathfrak{s})\}$

(d) $p = \mathbf{tp}_{\mathfrak{K}[\mathfrak{t}]}(a, M_1, M_2)$ does not fork over M_0 (for $\mathfrak{t}$) <u>when</u> $M_0 \leq_{\mathfrak{s}} M_1 \leq_{\mathfrak{s}} M_2, a \in M_2 \backslash M_1, p$ is regular and does not fork over M_0 for $\mathfrak{s}$.

10.19 Claim. *1)* $\mathfrak{t} = \mathfrak{s}^{\text{reg}} = \mathfrak{s}[\text{reg}]$ *is a good* λ*-frame.*
2) Moreover, $\mathfrak{s}^{\text{reg}}$ *is a successful good*$^+$ λ*-frame with primes (so as in 10.1 except the type-full).*
3) If $(M, N, \mathbf{J}) \in K^{3,\text{bs}}_{\mathfrak{s}[\text{reg}]}$ *then*

(a) $(M, N, \mathbf{J}) \in K^{3,\text{bs}}_{\mathfrak{s}}$

(b) $\leq^{\mathfrak{s}[\text{reg}]}_{\text{bs}} = \leq^{\mathfrak{s}}_{\text{bs}} \restriction K^{3,\text{bs}}_{\mathfrak{s}[\text{reg}]}$

(c) $(M, N, \mathbf{J}) \in K^{3,\text{qr}}_{\mathfrak{s}} \Leftrightarrow (M, N, \mathbf{J}) \in K^{3,\text{qr}}_{\mathfrak{s}[\text{reg}]}$

(d) *N is $(\lambda, *)$-brimmed over M for $\mathfrak{s}$ iff for $\mathfrak{s}^{\text{reg}}$.*

Proof. 1) As $\mathscr{S}^{\text{bs}}_{\mathfrak{t}}(M) \subseteq \mathscr{S}^{\text{bs}}_{\mathfrak{s}}(M)$ for every $M \in \mathfrak{K}_{\mathfrak{s}} = \mathfrak{K}_{\mathfrak{t}}$ and the definition of "does not fork for $\mathfrak{t}$" the main point we should check is density, i.e., Ax(D)(d), i.e., if $M <_{\mathfrak{s}} N$, then for some $c \in N \backslash M$ the type $\mathbf{tp}_{\mathfrak{s}}(c, M, N)$ belongs to $\mathscr{S}^{\text{bs}}_{\mathfrak{t}}(M)$, i.e., is regular for $\mathfrak{s}$. But this holds by 10.5(1).

A minor point we should notice is that if $M_1 \leq_{\mathfrak{K}[\mathfrak{s}]} M_2, p \in \mathscr{S}^{\text{bs}}_{\mathfrak{s}}(M_2)$ does not fork over M_1 for $\mathfrak{s}$, then $p \in \mathscr{S}^{\text{bs}}_{\mathfrak{t}}(M_2) \Leftrightarrow p \restriction M_1 \in \mathscr{S}^{\text{bs}}_{\mathfrak{t}}(M)$ as this just means that regularity is preserved by parallelism which holds by 10.4(1)(a).
2) Easy, too. Follows from "$\mathfrak{s}$ has primes".
3) <u>Clause (a)</u>: Any sequence $\langle N_i : i \leq \alpha \rangle$ witnessing $\mathbf{J}$ is independent for $\mathfrak{s}^{\text{reg}}$ do it for $\mathfrak{s}$, recalling $\mathscr{S}^{\text{bs}}_{\mathfrak{s}[\text{reg}]}(M) \subseteq \mathscr{S}^{\text{bs}}_{\mathfrak{s}}(M)$ for $M \in K_{\mathfrak{s}} = K_{\mathfrak{s}[\text{reg}]}$.

<u>Clause (b)</u>: Similarly easy as regularity is preserved by parallelism.

Clause (c): Check the definition.

Clause (d): Follows from $\mathfrak{K}_{\mathfrak{s}[\mathrm{reg}]} = K_{\mathfrak{s}}$. $\square_{10.19}$

10.20 Claim. *Assume*

(a) $M \le_{\mathfrak{s}} N$ *are* $(\lambda, *)$*-brimmed*

(b) *if* $p \in \mathscr{S}^{\mathrm{bs}}_{\mathfrak{s}}(M)$ *is regular then for some regular* $q \in \mathscr{S}^{\mathrm{bs}}_{\mathfrak{s}}(M)$ *we have* $\dim(q, N) = \lambda_{\mathfrak{s}}$ *and* $p \pm q$, *(see Definition 5.12).*

Then N *is* $(\lambda, *)$*-brimmed over* M*.*

Remark. Used in 12.36.

Proof. By 10.19 we can work in $\mathfrak{s}^{\mathrm{reg}}$. By the claim on dimension 10.16(3), we have $p \in \mathscr{S}^{\mathrm{bs}]}_{\mathfrak{s}[\mathrm{reg}]}(M) \Rightarrow \dim(p, N) = \lambda_{\mathfrak{s}}$.

Hence we can find $\mathbf{J} \subseteq \mathbf{I}^{\mathrm{reg}}_{M,N}$ independent in (M, N) such that for every $p \in \mathscr{S}^{\mathrm{bs}}_{\mathfrak{s}[\mathrm{reg}]}(M) \Rightarrow$ for λ elements $c \in \mathbf{J}$, c realizes p. Without loss of generality $\mathbf{J} \subseteq \mathbf{I}^{\mathrm{reg}}_{M,N}$ is maximal such that $\mathbf{J}$ is independent in (M, N).

So $(M, N, \mathbf{J}) \in K^{3,\mathrm{vq}}_{\mathfrak{s}}$ by 10.16(4) and by the definition it is thick for $\mathfrak{s}^{\mathrm{reg}}$, see Definition 5.15.
By 7.14(3) we know that $(M, N, \mathbf{J})$ is fat noting that the assumption of 7.14 holds by 10.7(a). Hence by 8.22 we know that N is $(\lambda, *)$-brimmed over M as required. $\square_{10.20}$

§11 DOP

We start to look at the parallel of the DOP/NDOP dichotomy in our context. We note some equivalent forms and then assuming DOP prove a non-structure result: build many complicated models in λ^{++}. This section is not needed for continuing to read in this work, and we do not deal with the complimentary side: proving a decomposition theorem assuming NDOP, we shall return to it elsewhere. Note that decomposition is meaningful even if $\mathfrak{s}$ is categorical and we look only at models from $K_{\mathfrak{s}}$, as we can decompose N over M when $M \le_{\mathfrak{s}}$

N. Note that if $\mathfrak{s}$ is n-beautiful[1] (see end of §12) we shall get for $M \in K^{\mathfrak{s}}$ of cardinality $\leq \lambda^{+n}$, a decomposition by a tree of models of cardinality λ, but not here. Note that this decomposition theorem is meaningful for non-beautiful frames.

11.1 Hypothesis.

(a) $\mathfrak{s}$ is a good$^+$ λ-frame which is successful,

(b) $\mathfrak{s}$ has primes

(c) $\underset{\text{wk}}{\perp} = \perp$ and $\underset{\text{su}}{\perp} = \perp$

(d) Hypothesis 10.7 or at least the conclusion of 10.13 (used 11.2(C), 10.12(3)) (used in 11.5, 10.16).

Remark. So Hypothesis 5.1, 7.1 hold.

Convention: Let $\mathfrak{C} \in K^{\mathfrak{s}}_{\lambda^+}$ be saturated above λ; clearly exists.

11.2 Definition. 1) We say $\mathfrak{s}$ has DOP when: we can find M_ℓ (for $\ell < 4$) and a_ℓ (for $\ell = 1, 2$) and q which exemplifies it, which means

⊛ (a) $\text{NF}_{\mathfrak{s}}(M_0, M_1, M_2, M_3)$

(b) $(M_0, M_\ell, a_\ell) \in K^{3,\text{uq}}_\lambda$ for $\ell = 1, 2$

(c) $(M_\ell, M_3, a_{3-\ell}) \in K^{3,\text{uq}}_\lambda$ for $\ell = 1, 2$

(d) no $q \in \mathscr{S}^{\text{bs}}_{\mathfrak{s}}(M_3)$ is orthogonal to M_1 and to M_2.

2) We say that (p_1, p_2) has the DOP when there are M_ℓ ($\ell < 4$), a_ℓ ($\ell = 1, 2$) exemplifying it which means satisfying clauses (a)-(d) from part (1) and $M_3 \leq_{\mathfrak{K}[\mathfrak{s}]} \mathfrak{C}$ and $\mathbf{tp}_{\mathfrak{s}}(a_\ell, M_0, M_\ell) \| p_\ell$ for $\ell = 1, 2$; we say (p_1, p_2) has the explicit DOP if $\mathbf{tp}_{\mathfrak{s}}(a_\ell, M_0, M_\ell) = p_\ell$ for $\ell = 1, 2$.
3) We say $\mathfrak{s}$ has NDOP if it fails to have DOP; similarly for (p_1, p_2).

We shall use freely

11.3 Observation. In 11.2(1) it follows that $(M_0, M_3, \{a_1, a_2\}) \in K^{3,\text{vq}}_{\mathfrak{s}}$.

Proof. By 5.16(3). $\square_{11.3}$

11.4 Claim. *Assume* $\mathbf{P} \subseteq \cup\{\mathscr{S}^{\rm bs}_{\mathfrak{s}}(M') : M' \le_{\mathfrak{s}} M\}$ *and for every* $q \in \mathscr{S}^{\rm bs}_{\mathfrak{s}}(M)$ *and* $q \pm p$ *for some* $p \in \mathbf{P}$. *If* $M \le_{\mathfrak{s}} N$ *and* $q \in \mathscr{S}^{\rm bs}_{\mathfrak{s}}(N)$ *is not orthogonal to* M, *then it is not orthogonal to some* $p \in \mathbf{P}$ *used in 11.8.*

Proof. I.e. by Hypothesis 10.1, 10.7, by 10.16. $\square_{11.4}$

11.5 Claim. *[*$\mathfrak{s}$ *has NDOP] If* $(M_0, M_\ell, \mathbf{J}_\ell) \in K^{3,\rm vq}_{\mathfrak{s}}$ *for* $\ell = 1, 2$ *and* $M_1 \cap M_2 = M_0$, *then we can find* M_3 *such that*

(a) $\mathrm{NF}_{\mathfrak{s}}(M_0, M_1, M_2, M_3)$

(b) $(M_\ell, M_3, \mathbf{J}_{3-\ell}) \in K^{3,\rm vq}_{\mathfrak{s}}$ *for* $\ell = 1, 2$

(c) *no* $q \in \mathscr{S}^{\rm bs}_{\mathfrak{s}}(M_3)$ *is orthogonal to* M_1 *and to* M_2 *but not to* M_0.

Proof. Case 1: $\mathbf{J}_1$ is a singleton say $\{b\}$.

We can find $\langle M_{2,i} : i \le \alpha_2\rangle, \langle a_{2,i} : i < \alpha_2\rangle$ and $\alpha'_2 \le \alpha_2$

$\circledast_1$ (a) $M_{2,i}$ is $\le_{\mathfrak{s}}$-increasing continuous

(b) $M_{2,i} = M_0, M_{\alpha_2} = M_2$

(c) $(M_{2,i}, M_{2,i+1}, a_{2,i}) \in K^{3,\rm uq}_{\mathfrak{s}}$

(d) $\mathbf{J}_2 = \{a_{2,i} : i < \alpha'_2\}$

(e) if $i \in (\alpha'_2, \alpha_2)$ then $\mathbf{tp}_{\mathfrak{s}}(a_{2,i}, M_{2,i}, M_{2,i+1})$ is orthogonal to M_0

(f) if $i < \alpha'_2$ then $\mathbf{tp}_{\mathfrak{s}}(a_{2,i}, M_{2,i}, M_{2,i+1})$ does not fork over M_0.

[Why? As $\mathfrak{s}$ weakly has regulars.]

Next by 5.6(5) we can find $\langle N_{2,i} : i \le \alpha_2\rangle$ such that

$\circledast_2$ (a) $N_{2,i}$ is $\le_{\mathfrak{s}}$-increasing continuous

(b) $M_{2,i} \le_{\mathfrak{s}} N_{2,i}$

(c) $N_{2,0} = M_1$

(d) $\mathbf{tp}_{\mathfrak{s}}(a_{2,i}, N_{2,i}, N_{2,i+1})$ does not fork over $M_{2,i}$

(e) $(N_{2,}, N_{2,i+1}, a_{2,i}) \in K^{3,\rm pr}_{\mathfrak{s}}$

(f) $\mathrm{NF}_{\mathfrak{s}}(M_{2,j}, N_{2,j}, M_{2,i}, N_{2,i})$ for $j \le i$

(g) $(M_{2,i}, N_{2,i}, b) \in K^{3,\rm uq}_{\mathfrak{s}}$.

[Why is this possible? We choose $N_{2,i}$ by induction on i such that the relevant clauses $\circledast_2$ holds.

For $i = 0$ this is given. For $i = j + 1$ note that $(M_{2,i}, M_{2,i+1}, a_{2,i})$ and $(N_{2,i}, N_{2,i+1}, a_{2,i})$ belongs to $K_{\mathfrak{s}}^{3,\mathrm{uq}}$ (and $(M_{2,i}, N_{2,i}, b) \in K_{\mathfrak{s}}^{3,\mathrm{uq}}$ by the induction hypothesis, hence by 10.13 such $N_{2,i}$ exists. Clause (f) holds by long transitivity.

For i limit clause (g) holds by 7.15 so we have carried the induction.]

Lastly, we choose $M_3 = N_{2,\alpha_2}$. Now we have to prove that M_3 is as required.

Toward this, we prove by induction on $i \leq \alpha_2$ that

$(*)_i^1$ no $p \in \mathscr{S}_{\mathfrak{s}}^{\mathrm{bs}}(N_{2i})$ is orthogonal to M_1 and to $M_{2,i}$.

Subcase 1a: For $i = 0$.

This is trivial.

Subcase 1b: For i limit.

If $p \in \mathscr{S}_{\mathfrak{s}}^{\mathrm{bs}}(N_{2,i})$ then p does not fork over some $N_{2,j}$ for some $j < i$ and so $p \restriction N_{2,j}$ is $\pm M_1$ or $\pm M_{2,j}$ by the induction hypothesis, but the second possibility, $p \pm M_{2,j}$ implies $p \pm M_2$, so we are done.

Subcase 1c: $i = j + 1$.

Let $q \in \mathscr{S}_{\mathfrak{s}}^{\mathrm{bs}}(N_{2,i})$. As "$\mathfrak{s}$ has NDOP" it follows that $q \pm M_{2,i}$ or $q \pm N_{2,j}$. In the first case we are done. In the second case use 10.12(3) and the induction hypothesis. So $(*)_i^1$ holds for every $i \leq \alpha_2$.

By $\circledast_2(f)$ we have $\mathrm{NF}_{\mathfrak{s}}(M_0, M_1, M_2, M_3)$ and by $\circledast_2(g)$ for $i = \alpha_2$ we get

$(*)_2$ $(M_2, M_3, \mathbf{J}_1) \in K_{\mathfrak{s}}^{3,\mathrm{vq}}$.

Lastly $(M_1, M_3, \mathbf{J}_2) \in K_{\mathfrak{s}}^{3,\mathrm{vq}}$ by 7.10.

Case 2: General.

Similar proof using Case 1 in the successor case of the induction $i = j + 1$. $\square_{11.5}$

11.6 Definition. Assume $M_1, M_2 \in K_{\mathfrak{s}}$ are $\leq_{\mathfrak{s}}$-extensions of M and $\mathbf{P} \subseteq \mathbf{P}[M_*] := \cup\{\mathscr{S}_{\mathfrak{s}}(N) : N \leq_{\mathfrak{K}[\mathfrak{s}]} M$ and $N \in K_{\mathfrak{s}}\}$. Then $M_1 \leq_{\mathfrak{s},\mathbf{P}} M_2$ means that $M_1 \leq_{\mathfrak{s}} M_2$ and if $p \in \mathbf{P}, p_\ell$ the non-forking extension of p in $\mathscr{S}_{\mathfrak{s}}(M_\ell)$ for $\ell = 1, 2$, then p_2 is the unique extension of p_1 in $\mathscr{S}_{\mathfrak{s}}(M_2)$.

11.7 Claim. *Let $M_* \in K_{\mathfrak{s}}$ and $\mathbf{P}$ be as in 11.6.*
1) $\leq_{\mathfrak{s},\mathbf{P}}$ is a partial order on $\{M' : M \leq_{\mathfrak{K}} M' \in K_{\mathfrak{s}}\}$ and if $M_1 \leq_{\mathfrak{s},\mathbf{P}} M_2$ and $M_1 \leq_{\mathfrak{s}} M_1' \leq_{\mathfrak{s}} M_2' \leq_{\mathfrak{s}} M_2$ then $M_1' \leq_{\mathfrak{s},\mathbf{P}} M_2'$.
2) If $\langle M_i : i < \delta\rangle$ is $\leq_{\mathfrak{s},\mathbf{P}}$-increasing continuous, $\delta < \lambda^+$ and $M_\delta = \bigcup_{i<\delta} M_i$ then $i < \delta \Rightarrow M_i \leq_{\mathfrak{s},\mathbf{P}} M_\delta$.
3) If $M_* \leq_{\mathfrak{s}} M$ and $r \in \mathscr{S}^{\mathrm{bs}}_{\mathfrak{s}}(M)$ is (weakly) orthogonal to every $p \in \mathbf{P}$ and $(M, N, a) \in K^{3,\mathrm{uq}}_{\mathfrak{s}}$,
$\mathbf{tp}_{\mathfrak{s}}(a, M, N) = r$ then $M \leq_{\mathfrak{s},\mathbf{P}} N$.
4) $M \leq_{\mathfrak{s},\mathbf{P}} N$ iff there is a pr*-decomposition $\langle M_i, a_i : i < \alpha\rangle$ of N over M (so letting $M_\alpha := N, M_i$ is $\leq_{\mathfrak{s}}$-increasing continuous, $M_0 = M, (M_i, M_{i+1}, a_i) \in K^{3,\mathrm{pr}}_{\mathfrak{s}}$) such that $\mathbf{tp}_{\mathfrak{s}}(a_i, M_i, N) \perp \mathbf{P}$ for every $i < \alpha$ (where $q \perp \mathbf{P}$ means $q \in \mathbf{P} \Rightarrow q \perp p$).*

Proof. Straight. E.g.

4) The "if" direction:

So assume $\langle M_i : i \leq \alpha\rangle, \langle a_i : i < \alpha\rangle$ are as in the claim, so in particular $M_0 = M, M_\alpha = N$. We prove that $(\forall j \leq i)(M_j \leq_{\mathfrak{s},\mathbf{P}} M_i)$ by induction on $i \leq \alpha$.

For $i = 0$ this is trivial (and included in part (1) of the claim).

For i limit use part (2) of the claim.

For $i = j + 1$ note that $M_j \leq_{\mathfrak{s},\mathbf{P}} M_i$ holds by part (3), hence $j' \leq j \Rightarrow M_{j'} \leq_{\mathfrak{s}} M_i$ holds by part (1) as $M_{j'} \leq_{\mathfrak{s},\mathbf{P}} M_j$ by the induction hypothesis. So in particular $M = M_0 \leq_{\mathfrak{s},\mathbf{P}}, M_\alpha = N$ as required.

The "only if" direction:

So assume $M \leq_{\mathfrak{s},\mathbf{P}} N$. We now try by induction on $i < \lambda^+$ to choose M_i and if $i = j + 1$ also a_j such that

⊛ (a) $\langle M_j : j \leq i\rangle$ is $\leq_{\mathfrak{s}}$-increasing continuous

(b) $M_0 = M$

(c) $M_j \leq_{\mathfrak{s}} N$ for $j \leq i$

(d) $(M_j, M_{j+1}, a_j) \in K^{3,\text{pr}}_{\mathfrak{s}}$ for $j < i$

(e) $\mathbf{tp}_{\mathfrak{s}}(a_j, M_j, M_{j+1}) \perp \mathbf{P}$ for $j < i$.

For $i = 0$ let $M_0 = M$.

For i limit let $M_i = \cup\{M_j : j < i\}$.

For $i = j + 1$ if $M_j = N$ we are done. Otherwise, for some $a_j \in N \backslash M_j$ we have $p_j = \mathbf{tp}_{\mathfrak{s}}(a_j, M_j, N) \in \mathscr{S}^{\text{bs}}_{\mathfrak{s}}(M_i)$.

If $p_j \underset{\text{wk}}{\pm} \mathbf{P}$, let $q \in \mathbf{P}$ be not orthogonal to p_j and let $q_j \in \mathscr{S}^{\text{bs}}_{\mathfrak{s}}(M_j)$ be the non-forking extension of q in $\mathscr{S}^{\text{bs}}_{\mathfrak{s}}(M_j)$ hence q_j has an extension $q^* \in \mathscr{S}_{\mathfrak{s}}(N)$ which forks over M_j hence over M. This easily contradicts $M \leq_{\mathfrak{s},\mathbf{P}} N$. So $p_j \underset{\text{wk}}{\perp} \mathbf{P}$ and as $\mathfrak{s}$ has primes we can find $M_i \leq_{\mathfrak{s}} N$ so $(M_j, M_i, a_j) \in K^{3,\text{pr}}_{\mathfrak{s}}$. So we have carried the induction hence finished the proof. $\square_{11.7}$

11.8 Claim. *1) Assume $p \in \mathscr{S}_{\mathfrak{s}}(M)$ and*

(∗) $\mathbf{P}$ *is a type base for M which means:*

(a) $\mathbf{P} \subseteq \mathbf{P}[M] = \cup\{\mathscr{S}^{\text{bs}}_{\mathfrak{s}}(N) : N \leq_{\mathfrak{s}} M,$ *(so $N \in K_{\mathfrak{s}}$)*$\}$

(b) *for every $q \in \mathscr{S}^{\text{bs}}_{\mathfrak{s}}(M)$ there is $r \in \mathbf{P}$ not orthogonal to it; (moreover, if $M \leq_{\mathfrak{s}} N, q \in \mathscr{S}^{\text{bs}}_{\mathfrak{s}}(N)$ is $\pm M$ then $q \pm \mathbf{P}$, follows by 11.4).*

Then we can find a decomposition $\langle M_i : \ell \leq n \rangle, \langle a_\ell : \ell < n \rangle$ such that

(i) $M_0 = M$,

(ii) *p is realized in M_n*

(iii) *for each $\ell < n$, the triple $(M_\ell, M_{\ell+1}, a_\ell)$ belongs to $K^{3,\text{pr}}_{\mathfrak{s}}$*

(iv) *for each $\ell < n$, either $\mathbf{tp}_{\mathfrak{s}}(a_\ell, M_\ell, M_{\ell+1})$ is a non-forking extension of some $q \in \mathbf{P}$ or $\mathbf{tp}_{\mathfrak{s}}(a_\ell, M_\ell, M_{\ell+1})$ is orthogonal to M (the second possibility can be waived if $K^{3,\text{uq}}_{\mathfrak{s}} = K^{3,\text{pr}}_{\mathfrak{s}}$).*

2) Assume that $\mathfrak{s}$ *has* NDOP *and* $\mathrm{NF}_{\mathfrak{s}}(M_0, M_1, M_2, M_3)$ *and* $(M_0, M_\ell, a_\ell) \in K^{3,\mathrm{pr}}_{\mathfrak{s}}$ *and* $(M_\ell, M_3, a_{3-\ell}) \in K^{3,\mathrm{uq}}_{\mathfrak{s}}$ *for* $\ell = 1, 2$. *Then* $\mathscr{S}^{\mathrm{bs}}_{\mathfrak{s}}(M_1) \cup \mathscr{S}^{\mathrm{bs}}_{\mathfrak{s}}(M_2)$ *is a type base for* M_3.

Proof. Easy.
1) By 10.16.
2) By the definitions. $\square_{11.8}$

The following claim will be helpful when starting with (M_0, M_1, M_2, M_3) q as in 11.12 and creating many copies of M_1, M_2 over M_0 and need the orthogonality of the copies of q.

11.9 Claim. *Assume*

(a) $\langle M^*_\ell : \ell < 4\rangle, \langle a_\ell : \ell = 1, 2\rangle, q$ *are as in* ⊛ *of 11.2(1)*

(b) $a^k_\ell \in \mathfrak{C}$ *realizes* $\mathbf{tp}_{\mathfrak{s}}(a_\ell, M^*_0, M_\ell)$ *for* $\ell, k = 1, 2$ *and* $\langle a^k_\ell : \ell = 1, 2$ *and* $k = 1, 2\rangle$ *is independent over* M_0

(c) $M^k_\ell <_{\mathfrak{K}[\mathfrak{s}]} \mathfrak{C}$ *and* f^k_ℓ *is an isomorphism from* M_ℓ *onto* M^k_ℓ *over* M^*_0 *for* $\ell = 1, 2$, $k = 1, 2$ *and* $f^k_\ell(a_\ell) = a^k_\ell$

(d) f^{k_1,k_2} *is an isomorphism from* M_3 *onto* $M^{k_1,k_2} <_{\mathfrak{K}} \mathfrak{C}$ *extending* $f^{k_1}_1 \cup f^{k_2}_2$

(e) $q^{k_1,k_2} = f^{k_1,k_2}(q)$.

Then the types $q^{1,1}, q^{1,2}, q^{2,1}, q^{2,2}$ *are pairwise orthogonal and each of them orthogonal to* M^k_ℓ *for* $k_1 = 1, 2, k_2 = 1, 2$.

Proof. Straightforward.

We shall use $\underset{\mathrm{su}}{\perp} = \perp$ which holds by Hypothesis 11.1(c). Clearly $q^{k_1,k_2} \perp M^1_{k_1}$ and $q^{k_1,k_2} \perp M^2_{k_2}$. By the symmetry in the situation and as $\perp = \underset{\mathrm{su}}{\perp}$ it is enought to note

$(*)_1$ $\{M^{1,1}, M^{2,2}\}$ is independent over M^*_0 (hence $q^{1,1} \perp q^{2,2}$).
[Why? As $\{a^k_\ell : \ell = 1, 2$ and $k = 1, 2\}$ is independent over M^*_0 and $(M^*_0, M^{1,1}, \{a^1_1, a^2_1\}) \in K^{3,\mathrm{vq}}_{\mathfrak{s}}$ and $(M^*_0, M^{2,2}, \{a^1_2, a^2_2\}) \in K^{3,\mathrm{vq}}_{\mathfrak{s}}$ the statement in $(*)_2$ follows.]

$(*)_2$ $\{M^{1,1}, M^{1,2}\}$ is independent over M^1_1 (hence $q^{1,1} \perp q^{1,2}$).

$\square_{11.9}$

11.10 Claim. *[$\mathfrak{s} = \mathfrak{t}^+, \mathfrak{t}$ as in 11.1], $\mathfrak{s}$ has* DOP *iff $\mathfrak{t}^+$ has* DOP.

Proof. First assume that $\mathfrak{s}$ has DOP. Let $\langle M^*_\ell : \ell \le 3\rangle, \langle a_1, a_2\rangle, q$ exemplify it. Let $\langle M^*_{\ell,\alpha} : \alpha < \lambda_{\mathfrak{s}}\rangle$ be a $\le_{\mathfrak{K}[\mathfrak{t}]}$-representation of M^*_ℓ for $\ell \le 3$. Let E be a thin enough club of $\lambda_{\mathfrak{s}}$.

Now for every $\delta \in E$ and $\ell = 1, 2$ by 4.13, 7.14 we have $(M^*_{0,\delta}, M^*_{\ell,\delta}, a_\ell) \in K^{3,\mathrm{uq}}_{\mathfrak{t}}$ and $(M^*_{\ell,\delta}, M^*_{3,\delta}, a_{3-\ell}) \in K^{3,\mathrm{uq}}_{\mathfrak{t}}$.

Also q is witnessed by $M^*_{3,\delta}$ and $q \restriction M^*_{3,\delta} \perp M^*_{\ell,\delta}$ for $\ell = 1, 2$ by 6.11(2) so we are done.

Second, assume $\mathfrak{t}$ has the DOP as exemplified by $\langle M^*_\ell : \ell \le 3\rangle, \langle a_1, a_2\rangle, q$, i.e. they satisfy ⊛ of 11.2(1). Let $(M^*_3, M^{**}_3, b) \in K^{3,\mathrm{bs}}_{\mathfrak{t}}$ be such that $q = \mathbf{tp}_{\mathfrak{s}}(b, M^*_3, M^{**}_3)$. Let $\langle M^*_{0,\alpha} : \alpha < \lambda^+_{\mathfrak{t}}\rangle$ be $\le_{\mathfrak{t}}$-increasing continuous, $M^*_{0,\alpha+1}$ brimmed over $M^*_{0,\alpha}$ for $\mathfrak{t}$, $M^*_{0,0} = M_0$. Let $N^*_0 = \cup\{M^*_{0,\alpha} : \alpha < \lambda^+_{\mathfrak{t}}\} \in K_{\mathfrak{s}}$ and let $p^+_\ell \in \mathscr{S}^{\mathrm{bs}}_{\mathfrak{s}}(N^*_0)$ be witnessed by $p_\ell = \mathbf{tp}_{\mathfrak{t}}(a_\ell, M^*_0, M^*_\ell)$ for $\ell = 1, 2$. By 10.13 we can find $N^*_\ell (\ell = 1, 2, 3)$ and a^+_1, a^+_2 such that $\mathrm{NF}_{\mathfrak{s}}(N^*_0, N^*_1, N^*_2, N^*_3)$ and $(N^*_0, N^*_\ell, a^+_\ell) \in K^{3,\mathrm{uq}}_{\mathfrak{s}}, (N^*_\ell, N^*_3, a^+_{3-\ell}) \in K^{3,\mathrm{uq}}_{\mathfrak{s}}$ for $\ell = 1, 2$. Now easily there is $\le_{\mathfrak{K}[\mathfrak{t}]}$-embedding f of M^*_3 into N^*_3 over M^*_0 mapping a_1, a_2 to a^+_1, a^+_2 respectively and let $q^* \in \mathscr{S}^{\mathrm{bs}}_{\mathfrak{s}}(N^*_3)$ be witnessed by $f(q)$. Now check. $\square_{11.10}$

<u>Discussion</u>: Our aim is to get strong non-structure in λ^{++} when $\mathfrak{s}$ has DOP. [Why in λ^{++}? We have quite strong independence but it speaks on λ-tuples, hence it is hard to get many models in λ^+, and if we deal with $K^{\mathfrak{s}}_{\lambda^{++}}$, why not ask λ^+-saturation. See more [Sh 839].]

For simplicity

11.11 Hypothesis. $\mathfrak{s}$ and $\mathfrak{s}^+$ are as in 11.1.

11.12 Definition. 1) We call $\mathfrak{a}$ an approximation or an $\mathfrak{s}$-approximation (in symbols $\mathfrak{a} \in \mathfrak{A} = \mathfrak{A}_{\mathfrak{s}}$) <u>if</u> $\mathfrak{a}$ consists of the following objects, satisfying the following demands

(a) $I^{\mathfrak{a}}_1, I^{\mathfrak{a}}_2$ disjoint index sets of cardinality $\le \lambda^+$

(b) $R_\mathfrak{a} \subseteq I^\mathfrak{a}_1 \times I^\mathfrak{a}_2$, we write $sR_\mathfrak{a}t$ for $(s,t) \in R_\mathfrak{a}, \neg sR_\mathfrak{a}t$ for $s \in I^\mathfrak{a}_1$, $t \in I^\mathfrak{a}_2$ such that $(s,t) \notin R_\mathfrak{a}$

(c) $M^\mathfrak{a}_\ell$ for $\ell < 4, a^\mathfrak{a}_\ell$ for $\ell = 1,2$ and $q_\mathfrak{a}$ exemplifying DOP

(d) $M^\mathfrak{a} \in K^\mathfrak{s}_{\lambda^+}$ saturated (so $\in K_{\mathfrak{s}(+)}$) such that $M^\mathfrak{a}_0 \leq_\mathfrak{K} M^\mathfrak{a}$

(e) $f^\mathfrak{a}_{\ell,t}$ an $\leq_\mathfrak{K}$-embeddiing of $M^\mathfrak{a}_\ell$ into $M^\mathfrak{a}$ over $M^\mathfrak{a}_0$ for $\ell = 1,2, t \in I^\mathfrak{a}_\ell$ and we let $M^\mathfrak{a}_{\ell,t} = f^\mathfrak{a}_{\ell,t}(M^\mathfrak{a}_\ell), a^\mathfrak{a}_{\ell,t} = f^\mathfrak{a}_{\ell,t}(a^\mathfrak{a}_\ell)$

(f) $\{a^\mathfrak{a}_{\ell,t} : \ell = 1,2$ and $t \in I^\mathfrak{a}_\ell\} \subseteq \mathbf{I}_{M^\mathfrak{a}_0,M^\mathfrak{a}}$ is independent over $M^\mathfrak{a}_0$; hence
$\langle M^\mathfrak{a}_{\ell,t} : \ell = 1,2, t \in I^\mathfrak{a}_\ell\}$ is independent by, see 8.8, 8.13

(g) if $sR_\mathfrak{a}t$ then $f^\mathfrak{a}_{s,t}$ is a $\leq_\mathfrak{K}$-embedding of $M^\mathfrak{a}_3$ into $M^\mathfrak{a}$ extending $f^\mathfrak{a}_{1,s} \cup f^\mathfrak{a}_{2,t}$; we let $M^\mathfrak{a}_{s,t} = f^\mathfrak{a}_{s,t}(M_3)$ and $q^\mathfrak{a}_{s,t} = f^\mathfrak{a}_{s,t}(q^\mathfrak{a})$.

2) For an approximation $\mathfrak{a}$ let $\mathbf{P}^+_\mathfrak{a} = \{q^\mathfrak{a}_{s,t} : sR_\mathfrak{a}t$ hence $s \in I^\mathfrak{a}_1$ and $t \in I^\mathfrak{a}_2\}$ and $\mathbf{P}^-_\mathfrak{a} = \{f(q_\mathfrak{a})$: for some f and $(s,t) \in I^\mathfrak{a}_1 \times I^\mathfrak{a}_2$ we have $\neg sR_\mathfrak{a}t$ and f is a $\leq_\mathfrak{K}$-embedding of $M^\mathfrak{a}_3$ into M extending $f^\mathfrak{a}_{1,s} \cup f^\mathfrak{a}_{2,t}\}$.
3) Let $\mathfrak{A} = \mathfrak{A}_\mathfrak{s}$ be the class of $\mathfrak{s}$-approximations.
4) We call $\mathfrak{a}^-$ a DOP witness if it consists of just $M^{\mathfrak{a}^-}_\ell$ $(\ell < 4), a^{\mathfrak{a}^-}_\ell$ $(\ell = 1,2), q^{\mathfrak{a}^-}$ which are as above. If $\mathfrak{b}$ is an approximation let $\mathfrak{b}^-$ be defined naturally.

Remark. We can weaken the demands on $\mathfrak{s}$ so that $\mathfrak{s}^+$ is not necessarily well defined.

11.13 Definition. 1) If $\mathfrak{a}, \mathfrak{b}$ are approximations, let $\mathfrak{a} \leq \mathfrak{b}$ means:

(α) $M^\mathfrak{a}_\ell = M^\mathfrak{b}_\ell$ for $\ell < 4, a^\mathfrak{a}_\ell = a^\mathfrak{b}_\ell$ for $\ell = 1,2$ and $q^\mathfrak{a} = q^\mathfrak{b}$

(β) $I^\mathfrak{a}_\ell \subseteq I^\mathfrak{a}_\mathfrak{b}$ for $\ell = 1,2$ and $R_\mathfrak{a} = R_\mathfrak{b} \cap (I^\mathfrak{a}_1 \times I^\mathfrak{a}_2)$

(γ) for $\ell = 1,2, t \in I^\mathfrak{a}_1$ we have $f^\mathfrak{a}_{\ell,t} = f^\mathfrak{b}_{\ell,t}$

(δ) for $(s,t) \in R^\mathfrak{a}$ we have $f^\mathfrak{a}_{s,t} = f^\mathfrak{b}_{s,t}$

(ε) $M^\mathfrak{a} \leq_{\mathfrak{K}[\mathfrak{s}]} M^\mathfrak{b}$, moreover $M^\mathfrak{a} \leq_{\mathfrak{K},\mathbf{P}^-_\mathfrak{a}} M^\mathfrak{b}$.

2) If $\langle \mathfrak{a}_\zeta : \zeta < \delta\rangle$ is $\leq$-increasing in $\mathfrak{A}$ and $\delta < \lambda^{++}$ let their union $\mathfrak{a} = \bigcup\limits_{\zeta<\delta} \mathfrak{a}_\zeta$ be defined by $I^\mathfrak{a}_\ell = \bigcup\limits_{\zeta<\delta} I^{\mathfrak{a}_\ell}, R_\mathfrak{a} = \bigcup\limits_{\zeta<\delta} R_{\mathfrak{a}_\zeta}, f^\mathfrak{a}_{\ell,t} = f^{\mathfrak{a}_\zeta}_{\ell,t}$ for

$\zeta < \delta$ large enough, $f^{\mathfrak{a}}_{s,t} = f^{\mathfrak{a}_\zeta}_{s,t}$ for $\zeta < \delta$ large enough when $sR_{\mathfrak{a}}t$ and $M^{\mathfrak{a}} = \cup\{M^{\mathfrak{a}_\zeta} : \zeta < \delta\}$.

Below we restrict ourselves to $K_{\mathfrak{s}(+)}$ for the application we have in mind but K_{λ^+} would be also O.K.

11.14 Claim. *1)* $(\mathfrak{s}, \leq)$ *is a partial order.*
2) If $\langle \mathfrak{a}_\zeta : \zeta < \delta \rangle$ *is increasing in* $\mathfrak{A}$ *and* $\delta < \lambda^{++}$ *then* $\mathfrak{a} = \bigcup_{\zeta<\delta} \mathfrak{a}_\zeta$ *belong to* $\mathfrak{A}$ *is the* lub *of the sequence.*

Proof. Straight.

11.15 Claim. *1) If* $\mathfrak{a} \in \mathfrak{A}_{\mathfrak{s}}, p \in \mathscr{S}^{\text{bs}}_{\mathfrak{s}}(M^{\mathfrak{a}})$ *is orthogonal to every* $q \in \mathbf{P}^-_{\mathfrak{a}}$ *and* $(M^{\mathfrak{a}}, N, a) \in K^{3,\text{uq}}_{\mathfrak{s}(+)}$ *and* $p = \mathbf{tp}_{\mathfrak{s}}(a, M^{\mathfrak{a}}, N)$, *then for some* $\mathfrak{b} \in \mathfrak{A}$ *we have* $\mathfrak{a} \leq \mathfrak{b}$ *and* $M^{\mathfrak{b}} = N$.
2) Assume $\mathfrak{a} \in \mathfrak{A}, \ell(*) \in \{1,2\}, Y \subseteq I^{\mathfrak{a}}_{3-\ell(*)}, t^* \notin I^{\mathfrak{a}}_{\ell(*)}$ *and* $\langle M_\alpha : \alpha < \lambda^+ \rangle$ *is a* $\leq_{\mathfrak{s}}$*-representation of* $M^{\mathfrak{a}}$ *such that* $M_0 = M^{\mathfrak{a}}_0$ *(and of course* $M_{\alpha+1}$ *is* $(\lambda, *)$*-brimmed over* M_α *in* $\mathfrak{K}_\lambda$*). Then we can find* $\mathfrak{b}$ *and* $\langle N_\alpha : \alpha < \lambda^+ \rangle$ *such that:*

(A)(a) N_α *is* $\leq_{\mathfrak{K}[\mathfrak{s}]}$*-increasing continuous in* $K_{\mathfrak{s}}, N_{\alpha+1}$ *is* $(\lambda, *)$*-brimmed over* N_α
(b) $\text{NF}_{\mathfrak{s}}(M_\alpha, N_\alpha, M_{\alpha+1}, N_{\alpha+1})$
(c) $(M_\alpha, N_\alpha, a) \in K^{3,\text{uq}}_{\mathfrak{s}}$
(d) N_0 *is isomorphic to* $M^{\mathfrak{a}}_{\ell(*)}$ *over* $M^{\mathfrak{a}}_0$
(e) $(\bigcup_\alpha M_\alpha, \bigcup_\alpha N_\alpha, a) \in K^{3,\text{pr}}_{\mathfrak{s}(+)}$

(B)(a) $\mathfrak{b} \in \mathfrak{A}$ *and* $\mathfrak{a} \leq \mathfrak{b}$
(b) $I^{\mathfrak{b}}_{\ell(*)} = I^{\mathfrak{a}}_{\ell(*)} \cup \{t^*\}$ *and* $I^{\mathfrak{b}}_{3-\ell(*)} = I^{\mathfrak{a}}_{3-\ell(*)}$
(c) $R^{\mathfrak{b}}$ *is* $R^{\mathfrak{a}} \cup \{\langle t^*, s \rangle : s \in Y\}$ *if* $\ell(*) = 1$ *and is* $R^{\mathfrak{a}} \cup \{\langle s, t^* \rangle : s \in y\}$ *if* $\ell(*) = 2$
(d) $f^{\mathfrak{a}}_{\ell(*),t^*}$ *is an isomorphism from* $M^{\mathfrak{a}}_{\ell(*)}$ *onto* N_0 *mapping* $a^{\mathfrak{a}}_{\ell(*)}$ *to* a.

Proof. 1) Easy (as we can choose $I^{\mathfrak{b}}_\ell = I^{\mathfrak{a}}_\ell, f^{\mathfrak{b}}_{\ell,t} = M^{\mathfrak{a}}_{\ell,t}, f^{\mathfrak{b}}_{s,t} = f^{\mathfrak{a}}_{s,t}$).
2) First choose a, N_α to satisfy (A). Then the choice of $\mathfrak{b}$ is actually described in (B); the orthogonality hold by 11.7. $\square_{11.15}$

11.16 Claim. *Let $\mathfrak{a}$ be a* DOP*-witness and $R \subseteq \lambda^{++} \times \lambda^{++}$ be given. For $\alpha < \lambda^{++}$ let $I^\alpha_1 = \{i : 3i+1 \le \alpha\}, I^\alpha_2 = \{i : 3i+2 \le \alpha\}, R^*_\alpha = R \cap (I^\alpha_1 \times I^\alpha_2)$. We can find $\langle \mathfrak{a}^\alpha : \alpha < \lambda^{++}\rangle$ such that*

(a) *$\mathfrak{a}^\alpha \in \mathfrak{A}_{\mathfrak{s}}$ is increasing continuous and $\mathfrak{a}^-_\alpha = \mathfrak{a}$*
(b) *$(I^{\mathfrak{a}^\alpha}_1, I^{\mathfrak{a}^\alpha}_2, R_{\mathfrak{a}^\alpha}) = (I^\alpha_1, I^\alpha_2, R_\alpha)$,*
(c) *for $(s,t) \in R_\alpha$ for arbitrarily large $\beta \in (\alpha, \lambda^{++})$ (by some bookkeeping), some $b \in M^{\mathfrak{a}^{3\beta+3}} \backslash M^{\mathfrak{a}^{3\beta+2}}$ the type $\mathbf{tp}_{\mathfrak{s}}(b, M^{\mathfrak{a}^{3\beta+2}}, M^{\mathfrak{a}^{3\beta+3}})$ is a non-forking extension of $q^{\mathfrak{a}_\alpha}_{s,t}$*
(d) *$\mathfrak{a}^\alpha$ depends just on $(I^\alpha_1, I^\alpha_2, R_\alpha)$*
(e) *the universe of $M^{\mathfrak{a}^\alpha}$ is $\gamma_\alpha < \lambda^{++}$ (really $\gamma_\alpha = \lambda^* \times (1+\alpha)$ is O.K. for non-trivial cases.*

Proof. We choose $\mathfrak{a}^\alpha$ by induction on α. For $\alpha = 0$ this is trivial, for α limit by 11.14(2), for $\alpha = 3\beta+1$ by 11.15(2) for $\ell(*) = 1$, for $\alpha = 3\beta+2$ by 11.15(2) for $\ell(*) = 2$ for $\alpha = 3\beta+3$ bookkeeping gives as a pair (s_α, t_α) and we use 11.15(1).

$\square_{11.16}$

11.17 Claim. *In 11.16 we can add: letting $M^* = \{M^{\mathfrak{a}^\alpha} : \alpha < \lambda^{++}\}$*

(*) *for $(s,t) \in \lambda^{++} \times \lambda^{++}$, the following are equivalent*
 (α) $(s,t) \in R$
 (β) $\dim(q^{\mathfrak{a}^\alpha}_{s,t}, M^*) = \lambda^{++}$ *when $(s,t) \in \mathscr{T}^{\mathfrak{a}^\alpha}$, that is, there is a sequence $\langle b_\gamma : \gamma < \lambda^{++}\rangle$ independent in $(M^{\mathfrak{a}^\alpha}, M^*)$ of elements realizing $q^{\mathfrak{a}^\alpha}_{s,t}$, i.e. if $(s,t) \in R_{\mathfrak{a}^\alpha}, \alpha < \beta < \lambda$ then $\{b_\gamma : \gamma < \lambda^{++}$ and $b_\gamma \in M^{\mathfrak{a}^\beta}\}$ is independent in $(M^{\mathfrak{a}^\alpha}, M^{\mathfrak{a}^\beta})$ and each member realizes the non-forking extension of $q^{\mathfrak{a}^\alpha}_{s,1}$ in $\mathscr{S}^{\mathrm{bs}}_{\mathfrak{s}}(M^{\mathfrak{a}^\alpha})$*

(γ) *if* $(s,t) \in R^{\mathfrak{a}^\alpha}, \alpha < \lambda^{++}$ *then there is a* $\le_{\mathfrak{K}}$*-embedding* f *of* $M_3^{\mathfrak{a}^\alpha}$ *into* M^**, extending* $f_{1,s}^{\mathfrak{a}^\alpha} \cup f_{2,t}^{\mathfrak{a}^\alpha}$ *such that* $\dim(f(q^{\mathfrak{a}^\alpha}), M^*) = \lambda^{++}$ *interpreted as in* (β)

(δ) *for no* $\alpha < \lambda^{++}$ *and* f *as in clause* (γ)*, we have:* $q^* \in \mathscr{S}_{\mathfrak{s}}^{\mathrm{bs}}(M^{\mathfrak{a}^\alpha})$*, the non-forking extension of* $f(q_{\mathfrak{b}})$ *in* $\mathscr{S}_{\mathfrak{s}}^{\mathrm{bs}}(M^{\mathfrak{a}^\alpha})$ *satisfies: for every* $\beta \in (\alpha, \lambda^{++}), q^*$ *has a unique extension in* $\mathscr{S}_{\mathfrak{s}}^{\mathrm{bs}}(M^{\mathfrak{a}_\beta})$.

Proof. Easy.

11.18 Claim. $[2^{\lambda^+} < 2^{\lambda^{++}}]$. *If* $\mathfrak{s}$ *has* DOP *then* $\dot{I}(\lambda^{++}, K^{\mathfrak{s}(+)}) = 2^{\lambda^{++}}$.

11.19 Remark. 1) It is a strong non-structure (i.e., neither like for deepness, no even like unsuperstable.
2) We can in 11.16, 11.17 restrict more the types realized.
3) Alternatively, we may use [Sh 300, III] or [Sh:e, III] for proving 11.18 even without the assumption $2^{\lambda^+} < 2^{\lambda^{++}}$.

Proof. We use the construction above in the framework of [Sh 576, §3] or better Chapter VII. $\square_{11.18}$

§12 Brimmed Systems

This section generalizes [Sh 87b], [Sh:c, XII,§4,§5]. Here every system is in the context of some good frame $\mathfrak{s}$ and usually we look at models of cardinality $\lambda = \lambda_{\mathfrak{s}}$ (in this section), but we vary $\mathfrak{s}$. The problem is that unlike [Sh:c, XII], the type $\mathrm{tp}(A, B, M)$ does not make much sense for $A, B \subseteq M \in \mathfrak{K}$ and unlike [Sh 87b], we cannot restrict ourselves to finite A and suitable $B = (\cup\{M_u : u \in \mathscr{P}^-(n)\} \cup \bar{a}$, with $\langle M_u : u \in \mathscr{P}^-(n)\rangle$ a so-called stable system and $\bar{a}$ a finite sequence. This has ramifications which further complicate our task, anyhow we do not rely on [Sh 87b], [Sh:c, XII,§4,§5].

The adoption of "$\mathfrak{K}_\mathfrak{s}$ categorical in λ" (in 12.3) is very helpful here but there is a price: when we shall work on "all $\lambda^{+\omega}$-saturated models in $K^\mathfrak{s}$", more exactly on $\cap\{K^{\mathfrak{s}(+n)} : n < \omega\}$ we cannot just quote the results. Note that this restriction fits well the thesis that the main road is first to understand a class of models is first to analyze the quite saturated models.

Note that there is no real harm in assuming $\mathfrak{s}$ is type-full or just that the regular types are dense and even that $\mathfrak{s}$ is as in 12.2.

<u>12.1 Convention</u>: 1) Without loss of generality always $u \cap \mathscr{P}(u) = \emptyset$ for the index sets $u =$ Dom(I) which we shall use.
2) From 12.3 we use freely (a)-(d). In the cases we assume clauses (e)* and or (f)* of 12.3 we add $*$ (e.g., 12.6(2)), and we add in brackets (f)* or (f)** when used.

Justifying the Hypothesis 12.3 below is

12.2 Claim. *If $\mathfrak{s}$ is a 3-successful good λ-frame and $\mathfrak{s}' = (\mathfrak{s}^{\rm nf})^{+3}$, see Definition 9.5, Claim 9.6 <u>then</u> $\mathfrak{s}'$ satisfies Hypothesis 12.3(1) +(2) + (3) as well as 2.1, 2.19, 3.1, 4.1, 5.1, 6.1, 7.1, 8.1, 9.1, 10.1, 10.7, 11.1, 11.11.*

Proof. Note that what is proved below for $\mathfrak{s}^{+\ell}$ holds for $\mathfrak{s}^{+m}$, when $m \in [\ell, 3]$. Clearly $\mathfrak{s}^{+0}$ is a successful full good $\lambda_\mathfrak{s}$-frame hence satisfies Hypothesis 2.1, 3.1, 9.1.

Just collect the relevant results: first $\mathfrak{s}^+$ is a good$^+$ λ-frame by 1.6 hence satisfies by 1.6. Second, $\mathfrak{s}^{+2}$ has primes (i.e. clause (c)) by 4.9, as $\mathfrak{s}^+$ satisfies Hypothesis 4.1, i.e. is a good$^+$ $\lambda_{\mathfrak{s}(+)}$-frame. Third, $\mathfrak{K}_{\mathfrak{s}^+}$ is categorical (i.e. clause (e)*) by its definition hence $\mathfrak{s}^{+2}$ satisfies $\perp = \underset{\rm wk}{\perp}$ and $\underset{\rm su}{\perp} = \perp$ by 6.8(5), 6.10(5) respectively hence $K^{3,{\rm vq}}_{\mathfrak{s}(+2)} = K^{3,{\rm qr}}_{\mathfrak{s}(2)}$ by 7.14 hence Hypothesis 8.1 holds. Fourth, $\mathfrak{s}^{+2}$ satisfies Hypothesis 10.1 hence $\mathfrak{s}^{+3}$ satisfies Hypothesis 10.7, hence by 10.9(2) the frame $\mathfrak{s}^{+3}$ weakly has regulars and even almost has regulars by 7.19(2) and satisfies Hypothesis 11.1.

Lastly, $\mathfrak{s}^{+3}$ satisfies $K^{3,{\rm vq}}_{\mathfrak{s}(+)} = K^{3,{\rm qr}}_{\mathfrak{s}(+)}$, i.e. clause (f)** by 8.22 hence also (f)* holds by part (1). $\square_{12.2}$

12.3 Hypothesis. 1)

(*a*) $\mathfrak{s}$ is a successful good$^+$ λ-frame (hence 4.1 and 4.7 hold)

(*b*) $\bot = \underset{\text{wk}}{\bot}$ is well defined and $\underset{\text{su}}{\bot} = \bot$; this follows from (e)* by 6.8(5), 6.10(5)

(*c*) $\mathfrak{s}$ has primes (in the sense of $K_{\mathfrak{s}}^{3,\text{qr}}$) and they are unique

(*d*) (*i*) $\mathfrak{s}$ "weakly has regulars" (see Definition 7.5 and Claim 7.6, 10.9; if $\mathfrak{s}$ is type-full and (e)* then this follows from 10.9(2) noting that the Hypothesis 10.1 is satisfied by part (1)); moreover

(*ii*) $\mathfrak{s}$ almost has regulars (see Definition 7.18 and Claim 7.17).

2) Possibly, in addition

(*e*)* $\mathfrak{K}_{\mathfrak{s}}$ is categorical in $\lambda = \lambda_{\mathfrak{s}}$

(*f*)* if $(M, N, \mathbf{J}) \in K_{\mathfrak{s}}^{3,\text{bs}}$ is thick and N is brimmed then N is brimmed over M (see Definition 5.15(2) and see 12.6 below)

$(f)^{**}$ $K_{\mathfrak{s}}^{3,\text{vq}} = K_{\mathfrak{s}}^{3,\text{qr}}$ (justified by 7.14).

3) When we mention regular types we assume

(*g*) $\mathfrak{s}$ has regulars and satisfies the Hypothesis of §10, i.e. 10.1, 10.7, note that on part (1) this adds "type-full"; (or just the conclusions in §10).

12.4 Remark. The use of regulars is just for simplicity; note that you can ignore all mentioning of regular types.

12.5 Observation. 0) The following Hypothesis holds: those from §2,§7, i.e. (2.1, 2.19, 3.1, 4.1, 5.1, 6.1, 7.1) and 9.1; if $(f)^{**}$ then also 8.1 and if $\mathfrak{s}$ is type-full then also 10.1.
1) In 12.3, clause $(f)^{**}$ implies $(f)^{*}$, even without "N is brimmed".
2) If $\mathfrak{s}^+$ is successful then $\mathfrak{s}^+$ satisfies the hypothesis 12.3(1) and also 12.1(2). Moreover we can weaken the Hypothesis on $\mathfrak{s}$ to $\mathfrak{s}$ satisfies just 12.1(1) and if $\mathfrak{s}$ satisfies 12.1(3) then $\mathfrak{s}^+$ satisfies it too.
3) Hypothesis 12.3(3) together with categoricity in $\lambda_{\mathfrak{s}}$ implies 12.3(2).

Proof. 0) Check.
1) Let $(M, N, \mathbf{J}) \in K^{3,\mathrm{bs}}_{\mathfrak{s}}$ be thick. Let $\mathbf{J}^+$ be maximal such that $(M, N, \mathbf{J}^+) \in K^{3,\mathrm{bs}}_{\mathfrak{s}}$ and $\mathbf{J} \subseteq \mathbf{J}^+$, exist by the local character of independence. By the definition inside 12.6(1) below this means that $(M, N, \mathbf{J}^+) \in K^{3,\mathrm{mx}}_{\mathfrak{s}}$, and by the claim itself implies that $(M, N, \mathbf{J}^+) \in K^{3,\mathrm{vq}}_{\mathfrak{s}}$.

As $(M, N, \mathbf{J})$ is thick (see Definition 5.15(2)), trivially also $(M, N, \mathbf{J}^+)$ is thick. Now we can apply claim 8.22(2), noting the clause $(f)^{**}$ of 12.3 is one of the assumptions of §8, we deduce N is brimmed over M.
2) Just collect the relevant results: first $\mathfrak{s}^+$ is a good$^+$ λ-frame (i.e. clause (a) of 12.3(1)), by 1.6. Second, $\mathfrak{s}$ has primes (i.e. clause (c)) by 4.9, as $\mathfrak{s}^+$ satisfies 4.1, i.e. is a good$^+$ $\lambda_{\mathfrak{s}(+)}$-frame. Third, $\mathfrak{K}_{\mathfrak{s}}$ is categorical (i.e. clause (e)*) by its definition hence $\perp = \underset{\mathrm{wk}}{\perp}$ and $\underset{\mathrm{su}}{\perp} = \perp$, i.e. claues (b) of 12.3(1) holds. Fourth, $\mathfrak{s}$ weakly has regulars and even almost has regulars by 7.19(2) recalling that Hypothesis 7.1 of §7 holds by (a),(b),(c) and "$\mathfrak{s}$ weakly has regulars" assumed in 7.19(2) holds by (d). Lastly, $\mathfrak{s}^+$ satisfies $K^{3,\mathrm{vq}}_{\mathfrak{s}(+)} = K^{3,\mathrm{qr}}_{\mathfrak{s}(+)}$, i.e. clause (f)** by 8.22 hence also (f)* holds by part (1). Lastly, 12.1(3) holds for $\mathfrak{s}^+$ when it holds for $\mathfrak{s}$.
3) It suffices to prove $(f)^{**}$ which is proved by 7.14.

Recall (by 7.9)

12.6 Claim. *1) The following*[15] *conditions on* $(M, N, \mathbf{J})$ *are equivalent*

- $\circledast_1$ $(M, N, \mathbf{J}) \in K^{3,\mathrm{mx}}_{\mathfrak{s}}$ *which we define to mean that* $(M, N, \mathbf{J}) \in K^{3,\mathrm{bs}}_{\mathfrak{s}}$, *and* $\mathbf{J}$ *is maximal*
- $\circledast_2$ $(M, N, \mathbf{J}) \in K^{3,\mathrm{bs}}_{\mathfrak{s}}$ *and if* $M \cup \mathbf{J} \subseteq N' \leq_{\mathfrak{s}} N, b \in N \backslash N'$ *and* $\mathbf{tp}_{\mathfrak{s}}(b, N', N) \in \mathscr{S}^{\mathrm{bs}}_{\mathfrak{s}}(N')$ *<u>then</u>* $\mathbf{tp}_{\mathfrak{s}}(b, N', N) \perp M$,
- $\circledast_3$ $(M, N, \mathbf{J}) \in K^{3,\mathrm{vq}}_{\mathfrak{s}}$.

2)* [12.3$(f)^{**}$] We have $K^{3,\mathrm{mx}}_{\mathfrak{s}} = K^{3,\mathrm{qr}}_{\mathfrak{s}}$.
3) $K^{3,\mathrm{qr}}_{\mathfrak{s}} \subseteq K^{3,\mathrm{mx}}_{\mathfrak{s}}$.

---

[15]we may be interested in the case we replace $K^{3,\mathrm{mx}}_{\mathfrak{s}}$ by $K^{3,\mathrm{qr}}_{\mathfrak{s}}$, but we have troubles enough

4) For every $M \in K_{\mathfrak{s}}, i^* < \lambda_{\mathfrak{s}}^+$ and $p_i \in \mathscr{S}_{\mathfrak{s}}^{\rm bs}(M)$ for $i < i^*$ we can find N and a_i $(i < i^*)$ such that: $(M, N, \{a_i : i < i^*\}) \in K_{\mathfrak{s}}^{3,\rm qr} \subseteq K_{\mathfrak{s}}^{3,\rm mx} \subseteq K_{\mathfrak{s}}^{3,\rm bs}$ and $p_i = \mathbf{tp}_{\mathfrak{s}}(a_i, M, N)$ and, of course, $\langle a_i : i < i^* \rangle$ is without repetition.
5) If $M \le_{\mathfrak{s}} N_1 \le_{\mathfrak{s}} N_2$ and $(M, N_1, \mathbf{J}_1) \in K_{\mathfrak{s}}^{3,\rm mx}, (M, N_2, \mathbf{J}_2) \in K_{\mathfrak{s}}^{3,\rm bs}$ and $\mathbf{J}_1 \subseteq \mathbf{J}_2$ then $(N_1, N_2, \mathbf{J}_2 \backslash \mathbf{J}_1) \in K_{\mathfrak{s}}^{3,\rm bs}$.

Proof. 1) If $\circledast_1$ holds, i.e. $(M, N, \mathbf{J}) \in K_{\mathfrak{s}}^{3,\rm mx}$ than by 7.7(3) we get $\circledast_3$, i.e. $(M, N, \mathbf{J}) \in K_{\mathfrak{s}}^{3,\rm vq}$. But by 7.7(3), we have $\circledast_3 \Rightarrow \circledast_1$.

By 7.9 several conditions on $(M, N, \mathbf{J})$ are equivalent, now (A) there is $\circledast_3$ here, and (C) there is $\circledast_2$ here, hence $\circledast_3 \Leftrightarrow \circledast_2$ so by the previous paragraph we are done.
2) We have $K_{\mathfrak{s}}^{3,\rm ms} = K_{\mathfrak{s}}^{3,\rm vq}$ by part (1) and $K_{\mathfrak{s}}^{3,\rm vq} = K_{\mathfrak{s}}^{3,\rm qr}$ by clause (f)** of 12.3.
3) Similar to (2) as $K_{\mathfrak{s}}^{3,\rm qr} \subseteq K_{\mathfrak{s}}^{3,\rm vq}$ by 5.16.
4) Holds by 5.8(6) + 5.8(1).
5) Holds by 5.16(5) recalling part (1). $\square_{12.6}$

12.7 Definition. Let u_* be a set (usually finite) and I a family of finite subsets of u_* (so for u_* finite this is automatic) satisfying $I \subseteq \mathscr{P}(u_*)$ is downward closed; I, J will denote such sets in this section; let $\mathrm{Dom}(I) = \cup\{u : u \in I\}$.
1) We say $\mathbf{m}$ is an I-system or $(I, \mathfrak{s})$-system or I-system for $\mathfrak{s}$ when:

- (a) $\mathbf{m}$ consists of M_u (for $u \in I$), and $h_{v,u}$ (for $u \subseteq v \in I$) (mappings, with $h_{u,u} = \mathrm{id}_{M_u}$ so we may ignore $\langle h_{u,u} : u \in I \rangle$ when defining $\mathbf{m}$)
- (b) $M_u \in K_{\mathfrak{s}}$ for $u \in I$
- (c) if $u \subseteq v \in I$ then $h_{v,u}$ is a $\le_{\mathfrak{s}}$-embedding of M_u into M_v, and the diagram of the $h_{u,v}$'s commutes and we let $M_{v,u} := h_{v,u}(M_u)$ and recall $h_{u,u} = \mathrm{id}_{M_u}$; so if $u \subseteq v \subseteq w \in I$ we have $M_{w,u} \le_{\mathfrak{s}} M_{w,v}$ and $M_{u,u} = M_u$.

2) We say $\mathbf{m}$ is a (μ, I)-system or $(\mu, I, \mathfrak{s})$-system if we replace (b) by

$(b)^+$ $M_u \in K_{\mu}^{\mathfrak{s}}$.

Similarly for $(\geq \mu, I, \mathfrak{s})$, etc.
3) We shall write $u^{\mathbf{m}}_*$ for $\mathrm{Dom}(I)$ and $h^{\mathbf{m}}_{v,u}, I^{\mathbf{m}}, M^{\mathbf{m}}_u$, for $h_{u,v}, I, M_u$, respectively. If $h_{v,u} = \mathrm{id}_{M_u}$ for $u \subseteq v \in I^{\mathbf{m}}$ and $M^s_u \cap M^s_v = M^{\mathbf{m}}_{v \cap u}$ for $u, v \in I$ we call $\mathbf{m}$ normal.
4) We say $\bar{g}$ is an isomorphism from the I-system $\mathbf{m}^1$ onto the I-system $\mathbf{m}^2$ if $\bar{g} = \langle g_u : u \in I\rangle, g_u$ is an isomorphism from $M^{\mathbf{m}^1}_u$ onto $M^{\mathbf{m}^2}_u$ such that $u \subseteq v \in I \Rightarrow h^{\mathbf{m}^2}_{v,u} \circ g_u = g_v \circ h^{\mathbf{m}_1}_{v,u}$. If $\mathbf{m}^1, \mathbf{m}^2$ are normal we may say $g = \bigcup\limits_u g_u$ is an isomorphism from $\mathbf{m}^1$ onto $\mathbf{m}^2$. Similarly $\bar{g}$ is a $\leq_{\mathfrak{s}}$-embedding of $\mathbf{m}^1$ into $\mathbf{m}^2$ when $g_u(M^{\mathbf{m}^1}_u) \leq_{\mathfrak{s}} M^{\mathbf{m}^2}_u$ or[16] $g_u(M^{\mathbf{m}^1}_u) \leq_{\mathfrak{K}[\mathfrak{s}]} M^{\mathbf{m}^2}_u$ for $u \in I$ and $u \subset v \in I \Rightarrow h^{\mathbf{m}^2}_{v,u} \circ g_u = g_v \circ h^{\mathbf{m}^1}_{v,u}$. Let $\mathbf{m}_1 \leq \mathbf{m}_2$ or $\mathbf{m}_1 \leq_{\mathfrak{K}_{\mathfrak{s}}} \mathbf{m}_2$ when $\langle \mathrm{id}_{M^{\mathbf{m}_1}_u} : u \in I\rangle$ is a $\leq_{\mathfrak{s}}$-embedding of $M^{\mathbf{m}_1}_u$ into $M^{\mathbf{m}_2}_u$ for $u \in I$, similarly $\leq_{\mathfrak{K}[\mathfrak{s}]}$.
5) We say $\bar{f}$ is an $\leq_{\mathfrak{s}}$-embedding of an I-system $\mathbf{m}$ to a model M if $\bar{f} = \langle f_u : u \in I\rangle, f_u$ is a $\leq_{\mathfrak{s}}$-embedding of M_u into M and $u \subseteq v \in I \Rightarrow f_u = f_v \circ g_{v,u}$.
6) Similarly for $(\geq \mu, I)$-systems, $(\geq \mu, I, \mathfrak{s})$-systems, so we use $\leq_{\mathfrak{K}[\mathfrak{s}]}$-embeddings. We may omit the "$\leq_{\mathfrak{s}}$ "-before embedding when $\mathfrak{s}$ is clear from the context.
7) If $I_1 \subseteq I_2$ and $\mathbf{m}_2$ is an I_2-system, let $\mathbf{m}_2 \restriction I_1 = \langle M^{\mathbf{m}_2}_u, h^{\mathbf{m}_2}_{v,u} : u \subseteq v \in I_1\rangle$.

We now define the class of systems which we are really interested in here: the stables ones.
For $\mathbf{m}$ to be stable, there should be witnesses and a system expanded by such witnesses is called an expanded stable system. We shall prove that "all witnesses look alike", the point is their existence.

12.8 Definition. 1) We say that $\mathbf{d}$ is an expanded stable $(I, \mathfrak{s})$-system or I-system or (λ, I)-system or $(\lambda, I, \mathfrak{s})$-system if it consists of $\mathbf{m}$ and $\mathbf{J}_{v,u}$ for $u \subseteq v \in I$ such that:

(a) $\mathbf{m} = \langle M_u, h_{v,u} : u \subseteq v \in I\rangle$ is an $(\lambda, I, \mathfrak{s})$-system

(b) $\mathbf{J}_{v,u} \subseteq \mathbf{I}_{M_{v,u}, M_v} \backslash \cup \{M_{v,w} : w \subset v\}$ for $u \subseteq v \in I$ so $\mathbf{J}_{u,u} = \emptyset$.

---

[16]the difference is meaningful only if $M^{\mathbf{m}^2}_u \in K^{\mathfrak{s}} \backslash K_{\mathfrak{s}}$

such that

(c) if $u_0 \subset u_1 \subset u_2 \in I$ and $c \in \mathbf{J}_{u_2,u_1}$ then $\mathbf{tp}_{\mathfrak{s}}(c, M_{u_2,u_1}, M_{u_2})$ is orthogonal to M_{u_2,u_0}, recalling that $M_{u_2,u_1} = h_{u_2,u_1}(M_{u_1})$

(d) $\mathbf{J}_{v,u}$ is a maximal subset of $\{c \in M_v : c \notin \cup\{h_{v,w}(\mathbf{J}_{w,u}) : w$ satisfies $u \subset w \subset v\}$ and $\mathbf{tp}_{\mathfrak{s}}(c, M_{v,u}, M_v)$ belongs to $\mathscr{S}^{\rm bs}_{\mathfrak{s}}(M_{v,u})$ and is orthogonal to $M_{v,w}$ for every $w \subset u\}$ such that $\mathbf{J}^*_{v,u} := \mathbf{J}_{v,u} \cup \bigcup\{\mathbf{J}_{v,w,u} : w$ satisfies $u \subset w \subset v\}$ is independent in $(M_{v,u}, M_v)$, see (1A)(α) below

(e) if $u_\ell \subset w_\ell \subseteq v \in I$ for $\ell = 1, 2$ and $(u_1, w_1) \neq (u_2, w_2)$ then $a_1 \in \mathbf{J}_{w_1,u_1} \wedge a_2 \in \mathbf{J}_{w_2,u_2} \Rightarrow h_{v,w_1}(a_1) \neq h_{v,w_2}(a_2)$.

If we omit clauses (d),(e) we say "an expanded $(I, \mathfrak{s})$-system".
1A) For $u_0 \subseteq u_1 \subseteq u_2 \in I$ we define (if $u_2 = u_1$ we may omit it, this catches the "main action", so $\mathbf{J}^0_{u_1,u_1,u_0} = \mathbf{J}^0_{u_1,u_0} = \mathbf{J}_{u_1,u_0}$)

(α) $\mathbf{J}^0_{u_2,u_1,u_0} = \{h_{u_2,u_1}(c) : c \in \mathbf{J}_{u_1,u_0}\}$ and call it also $\mathbf{J}_{u_2,u_1,u_0}$

(β) $\mathbf{J}^1_{u_2,u_1,u_0} = \cup\{\mathbf{J}^0_{u_2,u_1,u} : u \subseteq u_0\}$

(γ) $\mathbf{J}^2_{u_2,u_1,u_0} = \cup\{\mathbf{J}^0_{u_2,w_1,w_0} : w_0 \subseteq w_1 \subseteq u_1$ and $w_0 \subseteq u_0$ and $w_1 \nsubseteq u_0\}$

(δ) $\mathbf{J}^*_{u_2,u_1,u_0} := \{h_{u_2,u_1}(c) : c \in \mathbf{J}^*_{u_1,u_0}\}$.

2) We say $\mathbf{d}$ is normal if each $h_{v,u}$ is the identity (on M_u) and $M_u \cap M_v = M_{u \cap v}$, that is $\mathbf{s}$ is normal. We say that $\mathbf{d}$ is reduced in $u \in I$ when $w \subset u \Rightarrow \mathbf{J}^{\mathbf{d}}_{u,w} = \emptyset$.
2A) Above we let $\mathbf{m}^{\mathbf{d}} = \mathbf{m}[\mathbf{d}], M^{\mathbf{d}}_u = M^{\mathbf{m}}_u, h^{\mathbf{d}}_{v,u} = h^{\mathbf{m}}_{v,v}, \mathbf{J}^{\mathbf{d}}_{v,u} = \mathbf{J}_{v,u}, \mathbf{J}^{\ell,\mathbf{d}}_{u_2,u_1,u_0} = \mathbf{J}^{\ell}_{u_2,u_1,u_0}$ but we do not write the superscript $\mathbf{d}$ when clear from the context.
3) For an expanded stable (λ, I)-system $\mathbf{d}$ and $M \in K_{\mathfrak{s}}$ we say that $\bar{f}$ is an embedding (or $\leq_{\mathfrak{s}}$-embedding) of $\mathbf{d}$ into M when:

(A) $\bar{f}$ embeds $\mathbf{m}^{\mathbf{d}}$ into M, i.e.

 (a) $\bar{f} = \langle f_u : u \in I\rangle$

 (b) f_u is a $\leq_{\mathfrak{s}}$-embedding of $M^{\mathbf{d}}_u$ into M

 (c) if $u \subset v \in I$ then $f_u = f_v \circ h^{\mathbf{d}}_{v,u}$

(B) if $u \in I$ then $\cup\{f_u(\mathbf{J}^{2,\mathbf{d}}_{v,u}) : v$ satisfies $u \subset v \in I\}$, is an independent set in $(f_u(M^{\mathbf{d}}_u), M)$ and, of course, $u \subseteq v_1 \in I \wedge u \subseteq v_2 \in I \cap v_1 \neq v_2 \Rightarrow f_u(\mathbf{J}^{\mathbf{d}}_{v_1,u}) \cap f_u(\mathbf{J}^{\mathbf{d}}_{v_2,u}) = \emptyset$.

4) We say that a normal expanded stable (λ, I)-system $\mathbf{d}$ is $\leq_{\mathfrak{s}}$-embedded into $M \in K_{\mathfrak{s}}$ if ($\mathbf{d}$ is normal and) $\bar{f} = \langle f_u : u \in I\rangle$ is an embedding of $\mathbf{d}$ into M when we choose $f_u = \mathrm{id}_{M^{\mathbf{d}}_u}$.
5) We say $\mathbf{d}$ an expanded stable I-system is explicitly regular if:

(a) if $u \subset v \in I$ and $c \in \mathbf{J}^{\mathbf{d}}_{v,u}$ then $\mathbf{tp}_{\mathfrak{s}}(c, M^{\mathbf{d}}_{v,u}, M^{\mathbf{d}}_v)$ is regular (see 12.3(3))

(b) if $c_1 \neq c_2 \in \mathbf{J}^{\mathbf{d}}_{v,u}$ and $u \subset v \in I$ then $\mathbf{tp}_{\mathfrak{s}}(c_1, M^{\mathbf{d}}_{v,u}, M^{\mathbf{d}}_v)$, $\mathbf{tp}_{\mathfrak{s}}(c_2, M^{\mathbf{d}}_{v,u}, M^{\mathbf{d}}_v)$ are equal or orthogonal

(c) if $u \subseteq v_\ell \in I$ and $c_\ell \in \mathbf{J}^{\mathbf{d}}_{v_\ell,u}$ for $\ell = 1,2$ and $v_1 \neq v_2$ then $h^{-1}_{v_1,u}(\mathbf{tp}_{\mathfrak{s}}(c_1, M^{\mathbf{d}}_{v_1,u}, M^{\mathbf{d}}_{v_1}))$ and $h^{-1}_{v_2,u}(\mathbf{tp}_{\mathfrak{s}}(c_2, M^{\mathbf{d}}_{v_2,u}, M^{\mathbf{d}}_{v_2}))$ are equal or orthogonal.

6) An expanded stable (λ, I)-system $\mathbf{d}$ is called regular if

(a) if $u \subset v \in I, c \in \mathbf{J}^{\mathbf{d}}_{v,u}$ then $\mathbf{tp}_{\mathfrak{s}}(c, M_{v,u}, M_v)$ is regular.

7) If $\mathbf{d}_2$ is an expanded stable I_2-system and $I_1 \subseteq I_2$ then $\mathbf{d}_1 = \mathbf{d}_2 \restriction I_1$ is defined by $\mathbf{m}^{\mathbf{d}_1} = \mathbf{m}^{\mathbf{d}_2} \restriction I_1$ and $\mathbf{J}^{\mathbf{d}_1}_{v,u} = \mathbf{J}^{\mathbf{d}_2}_{v,u}$ for $u \subseteq v \in I_1$.

12.9 Definition. 1) We say $\bar{f}$ is an isomorphism from the expanded stable I-system $\mathbf{d}^1$ onto the expanded stable system $\mathbf{d}^2$ if $\bar{f}$ is an isomorphism from $\mathbf{m}^{\mathbf{d}^1}$ onto $\mathbf{m}^{\mathbf{d}^2}$ and f_v maps $\mathbf{J}^{\mathbf{d}^1}_{v,u}$ onto $\mathbf{J}^{\mathbf{d}^2}_{v,u}$ for $u \subseteq v \in I$.
2) For expanded stable (λ, I)-systems $\mathbf{d}_1, \mathbf{d}_2$, we say $\bar{f}$ is an $\leq_{\mathfrak{s}}$-embedding of $\mathbf{d}_1$ into $\mathbf{d}_2$ if

(α) $\bar{f}$ is an $\leq_{\mathfrak{s}}$-embedding of $\mathbf{m}_1$ into $\mathbf{m}_2$

(β) for $u \subset v \in I$, f_v maps $\mathbf{J}^{\mathbf{d}_1}_{v,u}$ into $\mathbf{J}^{\mathbf{d}_2}_{v,u}$

(β)$^+$ moreover, if $u \subset v \in I$ and $c \in \mathbf{J}^{\mathbf{d}_1}_{v,u}$ then $\mathbf{tp}_{\mathfrak{s}}(f_{v,u}(c), M^{\mathbf{d}_2}_{v,u}, M^{\mathbf{d}_2}_v)$ does not fork over $f_v(M^{\mathbf{d}_1}_{v,u})$.

3) A system $\mathbf{m}$ is called stable if for some expanded stable system $\mathbf{d}$ we have $\mathbf{m} = \mathbf{m}^{\mathbf{d}}$ ($\mathbf{d}$ is called an expansion of $\mathbf{m}$), we call the $\mathbf{J}^{\mathbf{d}}_{v,u}$'s witnesses.
Similarly for other properties, e.g. we say that $\bar{f} = \langle f_u : u \in I \rangle$ is a stable embedding of $\mathbf{m}$ into N where $\mathbf{m}$ is a stable $\mathscr{P}(n)$-system when for some expanded stable I-system $\mathbf{d}, \mathbf{m}_{\mathbf{d}} = \mathbf{m}$ and $\bar{f}$ is a stable embedding of $\mathbf{d}$ into M.
4) A (λ, I)-system $\mathbf{m}$ (or an expanded stable system $\mathbf{d}$ with $\mathbf{m}^{\mathbf{d}} = \mathbf{m}$) is called very brimmed if:

(g) for every $v \in I, \mathbf{m}$ is very brimmed in v which means

$(g)^0_v$ $M^{\mathbf{m}}_v$ is $(\lambda_{\mathfrak{s}}, *)$-brimmed

$(g)^1_v$ $M^{\mathbf{m}}_v$ is $(\lambda_{\mathfrak{s}}, *)$-brimmed over $\cup\{M^{\mathbf{m}}_{v,u} : u \subset v\}$ when $v \neq \emptyset$.

5) A stable I-system $\mathbf{m}$ (or an expanded stable system $\mathbf{d}$ with $\mathbf{m}^{\mathbf{d}} = \mathbf{m}$) is weakly brimmed if for every $v \in I$ it is weakly brimmed at v which means

$(h)^-_v$ $M^{\mathbf{m}}_v$ is brimmed[17].

6) An expanded stable system $\mathbf{d}$ is called brimmed if for every $v \in I_s$ it is brimmed in v which means

$(h)^0_v$ $M^{\mathbf{d}}_v$ is brimmed

$(h)^1_v$ if $u \subset v \in I$ and $p \in \mathscr{S}^{\rm bs}_{\mathfrak{s}}(M^{\mathbf{d}}_{v,u})$ is orthogonal to $M^{\mathbf{d}}_{v,w}$ for every $w \subset u$, then the set $\{c \in \mathbf{J}_{v,u} : \mathbf{tp}_{\mathfrak{s}}(c, M^{\mathbf{d}}_{v,u}, M^{\mathbf{d}}_v) \pm p\}$ has cardinality $\|M_v\| = \lambda_{\mathfrak{s}}$.

7) An I-system is [very][weakly] brimmed if there is a [very][weakly] brimmed expanded stable system $\mathbf{d}$ expanding it (so the system is stable). Similarly for "in u".

12.10 Definition. 1) We say $\mathbf{m}$ is a $(I, \mathfrak{s})^\ell$-system or a brimmed$^\ell$ I-system or ℓ-brimmed I-system when it is a I-system and: if $\ell = 0$ no additional demands; if $\ell = 1$, it is stable; if $\ell = 2$ it is a stable system which is weakly brimmed, that is each $M^{\mathbf{m}}_u$ is brimmed; if

[17]note that if $K_{\mathfrak{s}}$ is categorical (in $\lambda_{\mathfrak{s}}$) then this follows

$\ell = 3$, it is stable and brimmed; if $\ell = 4$, it is a very brimmed stable $(I, \mathfrak{s})$-system.
2) Similarly for $\mathbf{d}$, an expanded stable $(I, \mathfrak{s})$-system (so $\ell = 0, \ell = 1$ become equivalent).
3) We say M is brimmed$^\ell$ if letting $I = \{\emptyset\}, M_\emptyset = M$, we get that $\langle M_t : t \in I \rangle$ is a brimmed$^\ell$ system.
4) For $\ell = 1, 2, 3, 4$ we say that an expanded stable I-system $\mathbf{d}$ is brimmed$^\ell$ in $u \in I$ when the demand in Definition 12.9 holds for u (so for $\ell = 1$: no demand).

Remark. The central case here will be $\ell = 3$. A posteori we would like to have $\ell = 1$ (e.g. for analyzing $K^{\mathfrak{s}(+\omega)}$).

Notation: Let $\mathscr{P}^-(u_*) = \{v \subseteq u_* : v \neq u_*\}$.

12.11 Claim. *Let $\mathbf{d}$ be an expanded stable I-system (so we may omit the superscript $\mathbf{d}$ here).*
0) If $u \subset v \in I$ then $(M_{v,u}, M_v, \mathbf{J}^2_{v,u})$ belongs to $K^{3,\mathrm{mx}}_{\mathfrak{s}}$; equivalently (see 12.6(1)) it belongs to $K^{3,\mathrm{vq}}_{\mathfrak{s}}$, recalling $\mathbf{J}^2_{v,v,u} = \mathbf{J}^2_{v,u}$.
0A) In Definition 12.8(1), we can weaken clause (b) to $(b)^-$ (and get equivalent definition) where:

> $(b)^-$ $\mathbf{J}_{v,u} \subseteq \mathbf{I}_{M_{v,u},M_v} \setminus \cup \{\mathbf{J}_{v,w,u} : w$ *satisfies* $u \subset w \subset v\}$, *recalling that* $\mathbf{J}_{v,w,u} = h_{v,w}(\mathbf{J}_{w,u})$.

1) If $u_1 \subseteq u \in I, u_2 \subseteq u$ and $u_0 = u_1 \cap u_2$, then $\mathrm{NF}_{\mathfrak{s}}(M_{u,u_0}, M_{u,u_1}, M_{u,u_2}, M_u)$ hence $M_{u,u_1} \cap M_{u,u_2} = M_u$.
1A) If $u_0, u_1, u_2 \subseteq v \in I, u_0 \subseteq u_1, c \in \mathbf{J}^{\mathbf{d}}_{v,u_1,u_0}$ so $u_0 \subset u_1$ and $u_0 \nsubseteq u_2$ then $\mathbf{tp}_{\mathfrak{s}}(c, M_{v,u_0}, M_{v,u_1})$ is orthogonal to M_{v,u_2}. Moreover, if $u_0, u_2 \subseteq v \in I, u_0 \subseteq u_1 \in I, c \in \mathbf{J}^{\mathbf{d}}_{u_1,u_0}$ so $u_0 \subseteq u_1$ and $u_0 \nsubseteq u_2$ then $h^{-1}_{v,u_1}(\mathbf{tp}_{\mathfrak{s}}(c, M_{u_1,u_0}, M_{u_1}))$ is orthogonal to M_{v,u_2}.
2) If $p \in \mathscr{S}^{\mathrm{bs}}_{\mathfrak{s}}(M_{v,u})$ is regular, then there is a unique set $w \subseteq u$ such that p is not orthogonal to $M_{v,w}$ but p is orthogonal to $M_{v,w'}$ whenever $w' \subseteq u$ & $w \nsubseteq w'$ hence even when $w' \subseteq v$ & $w \nsubseteq w'$.
3) $\mathbf{d}$ is isomorphic to some normal $\mathbf{d}'$ (see Definition 12.7(3)).
4) If $\ell \in \{1, 2, 3\}$ and the expanded stable I-system $\mathbf{d}$ is brimmed$^{\ell+1}$ then it is brimmed$^\ell$. If the I-system $\mathbf{m}$ is brimmed$^{\ell+1}$ and $\ell =$

$0, 1, 2, 3$ <u>then</u> $\mathbf{m}$ *is brimmed*$^\ell$.

5) If $u_0 \subset u_2 \in I$ <u>then</u> $\mathbf{J}^{\mathbf{d}}_{u_2,u_0}$ *is a maximal set* $\mathbf{J}$ *such that*

- (α) $\mathbf{J} \subseteq M_{u_2}$ *is disjoint to* $\cup\{M_{u_2,w} : w \subset u_2\}$
- (β) *for each* $c \in \mathbf{J}$ *we have* $\mathbf{tp}_{\mathfrak{s}}(c, M_{u_2,u_0}, M_{u_2}) \in \mathscr{S}^{\rm bs}_{\mathfrak{s}}(M_{u_2,u_0})$ *and it is orthogonal to* $M_{u_2,w}$ *whenever* $w \subset u_0$
- (γ) $\mathbf{J} \cup \bigcup\{\mathbf{J}_{u_2,w_1,w_0} : w_1 \subseteq u_2, \neg(w_1 = u_2 \wedge w_0 = u_0)$ *and* $w_1 \nsubseteq u_0, w_0 \subseteq u_0 \cap w_1, w_0 \neq w_1\}$ *is independent in* (M_{u_2,u_0}, M_{u_2}); *note that necessarily there are no repetitions in the union.*

6) Assume that in addition to $\mathbf{d}$ *being an expanded stable* I*-system we have*

- (a) *for* $u \subset v \in I, \mathbf{J}'_{v,u}$ *is a maximal subset of* $\{c \in \mathbf{I}_{M^{\mathbf{d}}_{v,u}, M^{\mathbf{d}}_v} :$ $\mathbf{tp}_{\mathfrak{s}}(c, M^{\mathbf{d}}_{v,u}, M^{\mathbf{d}}_v) \perp M^{\mathbf{d}}_w$ *for* $w \subset u$ *and* $c \notin M^{\mathbf{d}}_{v,w}$ *for* $w \subset v\}$ *such that the set* $\mathbf{J}'_{v,u} \cup \bigcup\{\mathbf{J}_{v,w_1,u} : u \subset w \subset w\}$ *is independent in* $(M^{\mathbf{d}}_{v,u}, M^{\mathbf{d}}_v)$; *for the last phrase alternatively:* $\mathbf{J}'_{v,u} \cup \bigcup\{h_{v,w_1}(\mathbf{J}'_{w_1,w_0}) : w_1 \subseteq v, \neg(w_1 = v \wedge w_0 = u_0), w_1 \nsubseteq u_0, w_0 \subseteq u_0 \cap w_1, w_0 \neq w_1\}$ *is independent in* $(M^{\mathbf{d}}_{v,u}, M^{\mathbf{d}}_v)$
- (b) $\mathbf{d}'$ *is defined by* $\mathbf{m}^{\mathbf{d}'} = \mathbf{m}^{\mathbf{d}}, \mathbf{J}^{\mathbf{d}'}_{v,u} = \mathbf{J}'_{v,u}$ *for* $u \subset v \in I$.

<u>*Then*</u> $\mathbf{d}'$ *is an expanded stable* I*-system.*

7) If $\mathbf{d}$ *is a brimmed*^{3} I*-system and* $u \in I$ <u>then</u> $M^{\mathbf{d}}_u \backslash \cup \{M^{\mathbf{d}}_{u,w} : w \subset u\}$ *has cardinality* λ; *moreover, for* $w_1 \subset u, \mathbf{J}_{u,w_1}$ *is a subset of* $M^{\mathbf{d}}_u \backslash \cup \{M^{\mathbf{d}}_{u,w_2} : w_2 \subset u\}$ *of cardinality* λ.

8) In part (5) we can replace clause (α) *by*

- (α)$'$ $\mathbf{J} \subseteq M_{u_2}$ *is disjoint to* $\cup\{\mathbf{J}_{u_2,u_1,u_0} : u_0, u_1$ *satisfies* $u_0 \subset u_1 \subset u_2\}$.

9) In part (6) in clause (a) *we can change the demand "*$\mathbf{J}'_{v,u}$ *is a maximal subset of ... such that" to "*$\mathbf{J}'_{v,u}$ *is a maximal subset of* $\{c \in \mathbf{I}_{M^{\mathbf{d}}_{v,u}, M^{\mathbf{d}}_v} : \mathbf{tp}_{\mathfrak{s}}(c, M^{\mathbf{d}}_{v,u}, M^{\mathbf{d}}_v) \perp M^{\mathbf{d}}_w$ *for* $w \subset u$ *and* $c \notin \mathbf{J}^{\mathbf{d}}_{v,w,u}$ *when* $u \subset w \subset v\}$.

*10) [(f)** or just (f)* of 12.3]*

If $\mathbf{d}$ *is a brimmed*^{3} I*-system and* $u \subset v \in I$ <u>then</u> $M^{\mathbf{d}}_v$ *is brimmed over* $M^{\mathbf{d}}_{v,u}$.

11) If $I_1 \subseteq I$ *then* $\mathbf{d} \restriction I_1$ *is an expanded stable* I_1*-system; if* $\mathbf{d}$ *is* *brimmed*$^\ell$ *then so is* $\mathbf{d}_1$; *if* $\mathbf{d}$ *is normal then so is* $\mathbf{d}_1$. *If* $\mathbf{m}$ *is an* I*-system [a stable* I*-system]* <u>*then*</u> $\mathbf{m} \restriction I_1$ *is an* I_1*-system [a stable* I*-system].*

Proof. We prove (0),(0A),(1),(1A) together. More specifically, we prove by induction on $n < \omega$ and then on $m \le n$ that (0),(0A) hold when $|v| \le n, |u| \le m$ and (1) holds when $|u| \le n$ and (1A) holds when $|v| \le n$.

0) Note that the case $(v \backslash u)$ is a singleton is easy as then $\mathbf{J}^2_{v,u} = \mathbf{J}_{v,u}$, also note that for $v = u$ we actually have $\mathbf{J}^2_{v,u} = \emptyset$ and so the statement is trivial hence without loss of generality $|u|+1 < |v| \le n$.

Let $u_1 \subseteq u$ then by clause (d) of Definition 12.8(1) the set $\mathbf{J}^*_{v,u_1}$ is a maximal subset of $\{c \in \mathbf{I}_{M_{v,u_1}} : \mathbf{tp}_{\mathfrak{s}}(c, M_{v,u_1}, M_v)$ belongs to $\mathscr{S}^{\rm bs}_{\mathfrak{s}}(M_{v,u_1})$ and is $\perp M_{v,u_0}$ for $u_0 \subset u_1\}$ independent in (M_{v,u_1}, M_v). By the induction hypothesis $(M_{u,u_1}, M_{u_1}, \mathbf{J}^2_{u,u_1})$ belongs to $K^{3,\rm mx}_{\mathfrak{s}}$ hence, by preservation by isomorphisms, $(M_{v,u_1}, M_{v,u}, \mathbf{J}^2_{v,u,u_1}) \in K^{3,\rm mx}_{\mathfrak{s}} = K^{3,\rm vq}_{\mathfrak{s}}$.

If $u_1 \subset u$ then, as $|u_1| < |u| \le m$ and $|v| \le n$ by the induction hypothesis we know that $(M_{v,u_1}, M_v, \mathbf{J}^2_{v,u_1}) \in K^{3,\rm bs}_{\mathfrak{s}}$.

But $M_{v,u_1} \le_{\mathfrak{s}} M_{v,u} \le_{\mathfrak{s}} M_v$ and $\mathbf{J}^2_{v,u,u_1} \subseteq \mathbf{J}^2_{v,u_1}$, hence by the last two sentences recalling 12.6(5)

$(*)_1$ (a) $(M_v, M_{v,u}, \mathbf{J}^2_{v,u_1} \backslash \mathbf{J}^2_{v,u,u_1}) \in K^{3,\rm bs}_{\mathfrak{s}}$

(b) if $c \in \mathbf{J}^2_{v,u_1} \backslash \mathbf{J}^2_{v,u,u_1}$ then $\mathbf{tp}_{\mathfrak{s}}(c, M_{v,u}, M_v)$ does not fork over M_{v,u_1}.

But by clause (e) of Definition 12.8(1) easily $\mathbf{J}^*_{v,u_1} \backslash \mathbf{J}^*_{v,u,u_1} \subseteq \mathbf{J}^2_{v,u_1} \backslash \mathbf{J}^2_{v,u,u_1}$ hence

$(*)_2$ (a) $(M_v, M_{v,u}, \mathbf{J}^*_{v,u_1} \backslash \mathbf{J}^*_{v,u,u_1}) \in K^{3,\rm bs}_{\mathfrak{s}}$

(b) if $c \in \mathbf{J}^*_{v,u_1} \backslash \mathbf{J}^*_{v,u,u_1}$ then $\mathbf{tp}_{\mathfrak{s}}(c, M_{v,u}, M_v)$ does not fork over M_{v,u_1}.

Now

$(*)_3$ if $u_1 \ne u_2$ are subsets of u and $c_\ell \in \mathbf{J}^*_{v,u_\ell} \backslash \mathbf{J}^*_{v,u,u_1}$ for $\ell = 1, 2$ then $\mathbf{tp}_{\mathfrak{s}}(c_1, M_{v,u}, M_v) \perp \mathbf{tp}_{\mathfrak{s}}(c_2, M_{v,u}, M_v)$.

[Why? Let $u_0 = u_1 \cap u_2$, by applying part (1), i.e. the induction hypothesis, we know that $\mathrm{NF}_{\mathfrak{s}}(M_{v,u_0}, M_{v,u_1}, M_{v,u_2}, M_{v,u})$. Now for $\ell = 1,2$ the type $\mathbf{tp}_{\mathfrak{s}}(c_\ell, M_{v,u}, M_v)$ does not fork over M_{v,u_ℓ} by $(*)_2(b)$ and is orthogonal $M_{v,w}$ if $w \subset u_\ell$. So if $u_1 \subset u_2$ or $u_2 \subset u_1$ we are easily done. So without loss of generality $u_0 \subset u_1, u_0 \subset u_2$ hence as $\underset{\mathrm{su}}{\perp} = \perp$ we know that $\mathbf{tp}_{\mathfrak{s}}(c_\ell, M_{v,u}, M_v)$ is orthogonal to $M_{3-\ell}$. Together we get the desired orthogonality.]

Hence by 6.24, clearly $\cup\{\mathbf{J}^*_{v,w} \backslash \mathbf{J}^*_{v,u,w} : w \subseteq u\}$ is independent in $(M_{v,u}, M_v)$, but checking the definitions (see 12.3(1),(1A)) this union is $\mathbf{J}^2_{v,u}$, hence $\mathbf{J}^2_{v,u}$ is independent in $(M_{v,u}, M_v)$.

If it is maximal we are done. Othewise as $\mathfrak{s}$ almost has regulars (see 12.3(1)(d)(ii)) there is a pair (c, u_1) satisfying $c \in M_v \backslash M_{v,u} \backslash \mathbf{J}^2_{v,u}$ such that $\mathbf{J}^2_{v,u} \cup \{c\}$ is independent in $(M_{v,u}, M_v)$ and $\mathbf{tp}_{\mathfrak{s}}(c, M_{v,u}, M_v)$ does not fork over M_{v,u_1} and is $\perp M_{v,u_0}$ whenever $u_0 \subset u_1$. But this contradicts the demand (d) in 12.8(1).

0A) By part (0).

1) We have

- (i) $M_{u,u_1 \cap u_2} \leq_{\mathfrak{s}} M_{u,u_\ell} \leq_{\mathfrak{s}} M_u$ for $\ell = 1,2$ when $u_1 \cup u_2 \subseteq u \in I$
  [Why? Each case by a different instance of clause (c) of 12.7(1)]
- $(ii)_\ell$ $(M_{u,u_1 \cap u_2}, M_{u,u_\ell}, \mathbf{J}^2_{u,u_\ell,u_1 \cap u_2}) \in K^{3,\mathrm{mx}}_{\mathfrak{s}}$ hence $\in K^{3,\mathrm{vq}}_{\mathfrak{s}}$
  [Why? Without loss of generality $u_1 \neq u_2$ so $u_1 \cap u_2 \subset u_1$ and $u_1 \cap u_2 \subset u_2$. By part (0), we know that $(M_{u_\ell,u_1 \cap u_2}, M_{u_\ell}, \mathbf{J}^2_{u_\ell,u_1 \cap u_2}) \in K^{3,\mathrm{mx}} = K^{3,\mathrm{vq}}_{\mathfrak{s}}$ and by the definition of $\mathbf{J}^2_{u,u_\ell,u}$ we know that h_{u,u_ℓ} maps this triple to the one mentioned in clause $(ii)_\ell$]
- (iii) $\mathbf{J}^2_{u,u_1,u_1 \cap u_2}, \mathbf{J}^2_{u,u_2,u_1 \cap u_2}$ are disjoint
  [Why? By clause (e) of Definition 12.8(1).]
- (iv) $\mathbf{J}^2_{u,u_1,u_1 \cap u_2} \cup \mathbf{J}^2_{u,u_2,u_1 \cap u_2} \subseteq \mathbf{J}^2_{u,u_1 \cap u_2}$
  [Why? By their definitions]
- (v) $\mathbf{J}^2_{u,u_1,u_1 \cap u_2} \cup \mathbf{J}^2_{u_1,u_2,u_1 \cap u_2}$ is independent in $(M_{u,u_1 \cap u_2}, M_u)$ hence also in $(M_{u,u_1 \cap u_2}, M_{u,u_1 \cup u_2})$
  [Why? By monotonicity properties.]

Now by 8.13(3), 8.8 we are done. The "hence $M_{v,u_1} \cap M_{v,u_2} = M_{v,u_1 \cap u_2}$" follows by II.6.11 as $\mathrm{NF}_{\mathfrak{s}}$ is a non-forking relation on $\mathfrak{K}_{\mathfrak{s}}$.

1A) As $u_0 \cap u_2 \subset u_0$, by 12.8(1)(c) the type $p := \mathbf{tp}_{\mathfrak{s}}(c, M_{v,u_0}, M_{v,u_1})$ is orthogonal to $M_{v,u_0 \cap u_2}$ (and belongs to $\mathscr{S}^{\rm bs}_{\mathfrak{s}}(M_{v,u_0})$). By part (1) we know that $\mathrm{NF}_{\mathfrak{s}}(M_{v,u_0 \cap u_2}, M_{v,u_0}, M_{v,u_2}, M_v)$ holds hence by the definition of $\underset{\rm su}{\perp}$ (and clause (b) of the Hypothesis 12.3(1)) we get the conclusion. The "moreover" is proved similarly.
2) As u is finite, there is $w \subset u$ such that p is not orthogonal to M_w but is orthogonal to $M_{v,w'}$ if $w' \subset w$. The "hence" follows by part (1) and 10.10 (quoting 10.10 is O.K. by Hypothesis 12.3(3)). Note that by 10.10 if $M_0 \le_{\mathfrak{s}} M_\ell \le_{\mathfrak{s}} M_3$ for $\ell = 1, 2$ and $p \in \mathscr{S}^{\rm bs}_{\mathfrak{s}}(M_3)$ is regular orthogonal to M_0 but not to M_1 then p is orthogonal to M_2 and this is what we use.
3) By renaming; possible as $M_{v,u_1} \cap M_{v,u_2} = M_{v,u_1 \cap u_2}$ whenever $u_1, u_2 \subseteq v \in I$ by part (1) and properties of $\mathrm{NF}_{\mathfrak{s}}$.
4) The least easy part is $\ell = 3$. So we have to check clauses $(h)^0_v$, $(h)^1_v$ of 12.9(6). For the first, clearly every $M^{\mathbf{d}}_u (u \in I)$ is brimmed, because clause $(g)^0_v$ in 12.9(4) holds so only the second clause $(h)^1_v$ there may fail.
Assume toward contradiction that $u \subset v \in I, p \in \mathscr{S}^{\rm bs}_{\mathfrak{s}}(M^{\mathbf{d}}_{v,u})$ is orthogonal to $M^{\mathbf{d}}_{v,w}$ for every $w \subset u$ and the set $\mathbf{J} := \{c \in \mathbf{J}^{\mathbf{d}}_{v,u} : \mathbf{tp}_{\mathfrak{s}}(c, M^{\mathbf{d}}_{v,u}, M^{\mathbf{d}}_v) \pm p\}$ has cardinality $< \lambda_{\mathfrak{s}}$.

By the assumption (i.e. $\mathbf{d}$ is brimmed[4]) there is N such that $\cup\{M^{\mathbf{d}}_{v,w} : w \subset v\} \subseteq N \le_{\mathfrak{s}} M^{\mathbf{d}}_v$ and $M^{\mathbf{d}}_v$ is $(\lambda, *)$-brimmed over N. Hence there is $\mathbf{I}$ of cardinality $\lambda_{\mathfrak{s}}$ independent in $(M^{\mathbf{d}}_{v,u}, N, M^{\mathbf{d}}_v)$ such that $\mathbf{tp}_{\mathfrak{s}}(c, N, M^{\mathbf{d}}_v)$ is a non-forking extension of $p \in \mathscr{S}^{\rm bs}_{\mathfrak{s}}(M^{\mathbf{d}}_{v,u})$ for every $c \in \mathbf{I}$.

Clearly $\mathbf{J}_{v,u} \subseteq \mathbf{J}^*_{v,u}$ and $\mathbf{J}^*_{v,u} \backslash \mathbf{J}_{v,u} \subseteq \cup\{M_{v,w} : w \subset v\} \subseteq N$, hence $\mathbf{I}$ is disjoint to $(\mathbf{J}^*_{v,u} \backslash \mathbf{J}_{v,u})$ and their union is independent in $(M_{v,u}, M_v)$. Also $\mathbf{J}_{v,u}$ hence also $\mathbf{J}$ is disjoint to $\mathbf{J}^*_{v,u} \backslash \mathbf{J}_{v,u}$ and their union is independent in $(M_{v,u}, M_v)$.

As $|\mathbf{J}| < |\mathbf{I}| = \lambda$ it follows that for some $c \in \mathbf{I} \backslash \mathbf{J}$ also the set $(\mathbf{J}^*_{v,u} \backslash \mathbf{J}_{v,u}) \cup \mathbf{J} \cup \{c\}$ is independent in $(M_{v,u}, M_v)$ (using models from $K^{3,\rm vq}_{\mathfrak{s}}$).

By orthogonality consideration, i.e. 6.24 also $c \notin \mathbf{J}^{2,\mathbf{d}}_{v,u}$ and $\mathbf{J}^{2,\mathbf{d}}_{v,u} \cup \{c\}$ is independent in $(M^{\mathbf{d}}_{v,u}, M^{\mathbf{d}}_v)$, but by clause (d) of Definition 12.8(1) we get contradiction to $\mathbf{d}$ being an expanded stable I-system.
5) Left to the reader using 12.6 and 12.3(1)(d)(i),(ii); note that the

case of regular systems is more transparent.
6) Also easy first assume that this is a unique pair (u, v) such that $\mathbf{J}^{\mathbf{d}}_{v,u} \neq \mathbf{J}'_{v,u}$; second prove by induction on $|\{(u,v) : u \subset v \in I, \mathbf{J}^{\mathbf{d}}_{v,u} \neq \mathbf{J}'_{v,u}\}|$.
7) The first conclusion follows from the second which holds by clause (d) of Definition 12.8(1) and the definition of brimmed[3] in 12.9(6).
8), 9) Left to the reader.
10) First note that the triple $(M^{\mathbf{d}}_{v,u}, M^{\mathbf{d}}_{v_d}, \mathbf{J}^2_{v,v,u})$ is thick.
[Why? If $p \in \mathscr{S}^{\mathrm{bs}}_{\mathfrak{s}}(M^{\mathbf{d}}_{v,u})$ then for some $w \subset u, p \pm M^{\mathbf{d}}_{u,w}$ and $w_1 \subset w \Rightarrow p \perp M^{\mathbf{d}}_{u,w_1}$, hence there is $q \in \mathscr{S}^{\mathrm{bs}}_{\mathfrak{s}}(M^{\mathbf{d}}_{u,w})$ orthogonal to M_{u,w_1} for every $w_1 \subset w$ such that q is dominated by p (see Definition 7.22); (we assume that $\mathfrak{s}$ is as in §10, we can use q is regular and simplify). Now $\{c \in \mathbf{J}^{\mathbf{d}}_{v,w} : \mathbf{tp}_{\mathfrak{s}}(c, M^{\mathbf{d}}_{v,w}, M^{\mathbf{d}}_v) \pm q\}$ has cardinality λ and for each c in this set, by the choice of q, $\mathbf{tp}_{\mathfrak{s}}(c, M^{\mathbf{d}}_{v,w}, M^{\mathbf{d}}_v) \pm p$.]
Also $M^{\mathbf{d}}_{v,u}$ is brimmed. Now clause $(f)^*$ of 12.3 holds by 12.5, and it gives the required result.
11) Trivial. $\square_{12.11}$

12.12 Conclusion. 1) [Density of explicitly regular expanded stable systems, see Definition 12.8(5) so 12.3(3)]. Assume $\mathfrak{s}$ has regulars.
Assume

(a) $\mathbf{m}$ is a stable I-system
(b) for each $u \in I, \mathbf{P}_u \subseteq \{p \in \mathscr{S}^{\mathrm{bs}}_{\mathfrak{s}}(M_u) : p$ is regular orthogonal to $M^{\mathbf{d}}_{u,w}$ whenever $w \subset u\}$ is a maximal subset of pairwise orthogonal types.

<u>Then</u> there is an expanded stable I-system $\mathbf{d}^*$ such that:

(α) $\mathbf{m}^{\mathbf{d}^*} = \mathbf{m}$
(β) $\mathbf{d}^*$ is regular, moreover, $\mathbf{d}$ obeys $\bar{\mathbf{P}} = \langle \mathbf{P}_u : u \in I^{\mathbf{d}} \rangle$ which means:
   ⊠ if $u \subset v \in I^{\mathbf{d}}$ and $c \in \mathbf{J}^{\mathbf{d}}_{v,u}$ then for some $q \in \mathbf{P}_u$ we have $\mathbf{tp}_{\mathfrak{s}}(c, M^{\mathbf{d}}_{v,u}, M^{\mathbf{d}}_v)$ is q
(γ) $\mathbf{d}^*$ is explicitly regular (see Definition 12.8(5)).

2) Assume

(a) $\mathbf{m}$ is a stable I-system

(b) $J \subseteq I$ (no closure demands!) and for $u \in J, \mathbf{P}'_u = \{p \in \mathscr{S}^{\mathrm{bs}}_{\mathfrak{s}}(M^{\mathbf{m}}_u) : p \perp M^{\mathbf{s}}_{u,w}$ for every $w \subset u\}, \mathbf{P}''_u \subseteq \mathbf{P}'_u$ is auto-dense (see Definition 7.20(1)) and $((\mathbf{P}''_u)^{\perp})^{\perp} = \mathbf{P}'_u$.

Then there is an expanded stable I-system $\mathbf{d}^*$ such that

(α) $\mathbf{m}^{\mathbf{d}^*} = \mathbf{m}$

(β) if $u \in J, u \subset v \in I$ and $c \in \mathbf{J}^{\mathbf{d}^*}_{v,u}$ then $\mathbf{tp}_{\mathfrak{s}}(c, M^{\mathbf{d}^*}_{v,u}, M^{\mathbf{d}^*}_v)$ belongs to $\{h^{\mathbf{m}}_{v,u}(p) : p \in \mathbf{P}''_u\}$

(γ) if $u \in I \backslash J$ and $u \subset v \in I$ then $\mathbf{J}^{\mathbf{d}^*}_{v,u} = \mathbf{J}^{\mathbf{d}}_{v,u}$.

Proof. 1) Note that (γ) follows from (β). This holds by 12.11(6) and 10.16(5); easier to see when 12.3(3) holds.
2) Similarly. $\square_{12.12}$

12.13 Claim. *Assume $\ell \in \{1,2,3,4\}$ and*

(a) $\mathbf{d}_k$ *is an expanded stable (λ, I)-system for $k = 1, 2$*

(b) $\mathbf{m}^{\mathbf{d}_1} = \mathbf{m}^{\mathbf{d}_2}$.

Then $\mathbf{d}_1$ *is brimmed$^\ell$ iff* $\mathbf{d}^2$ *is brimmed$^\ell$.*

Proof. The least easy case is $\ell = 3$, see Definition 12.9(6), and the proof is similar to the proof of 12.11(4) or 12.12. $\square_{12.13}$

12.14 Definition. 1) For expanded stable I-systems $\mathbf{d}_0, \mathbf{d}_1$ let $\mathbf{d}_0 \leq_{\mathfrak{s}} \mathbf{d}_1$ or $\mathbf{d}_0 \leq^I_{\mathfrak{s}} \mathbf{d}_1$ mean[18] that:

(a) $M^{\mathbf{d}_0}_u \leq_{\mathfrak{s}} M^{\mathbf{d}_1}_u$ for $u \in I$ and $h^{\mathbf{d}_0}_{v,u} \subseteq h^{\mathbf{d}_1}_{v,u}$ for $u \subset u \in I$

(b) $\mathbf{J}^{\mathbf{d}_0}_{v,u} \subseteq \mathbf{J}^{\mathbf{d}_1}_{v,u}$ for $u \subseteq v \in I$

(c) if $c \in \mathbf{J}^{\mathbf{d}_0}_{v,u}$ then $\mathbf{tp}_{\mathfrak{s}}(c, M^{\mathbf{d}_1}_{v,u}, M^{\mathbf{d}_1}_v)$ does not fork over $M^{\mathbf{d}_0}_{v,u}$.

---

[18]in Definition 12.9's terms this means that $\langle \mathrm{id}_{M^{\mathbf{d}_0}_v} : v \in I \rangle$ embeds $\mathbf{d}_0$ into $\mathbf{d}_1$

2) We say that J is a successor of I <u>if</u> for some $t^* \notin \mathrm{Dom}(I)$ we have $J = I \cup \{u \cup \{t^*\} : u \in I\}$ and we call t^* the witness for J being a successor of I; so $\mathrm{Dom}(J) = \mathrm{Dom}(I) \cup \{t^*\}$.
3) For stable I-systems $\mathbf{m}_0, \mathbf{m}_1$ let[19] $\mathbf{m}_0 \le_{\mathfrak{s}} \mathbf{m}_1$ or $\mathbf{m}_0 \le^I_{\mathfrak{s}} \mathbf{m}_1$ mean that for some expansions $\mathbf{d}_0, \mathbf{d}_1$ of $\mathbf{m}_0, \mathbf{m}_1$ respectively we have $\mathbf{d}_0 \le^I_{\mathfrak{s}} \mathbf{d}_1$.
4) Assume that $\mathbf{d}_0 \le^I_{\mathfrak{s}} \mathbf{d}_1$ and J is a successor of I with the witness t^* and

(∗) if $u \subset v \in I$ and $c \in \mathbf{J}^{\mathbf{d}_1}_{v,u} \backslash \mathbf{J}^{\mathbf{d}_0}_{v,u}$ <u>then</u> $\mathbf{tp}_{\mathfrak{s}}(c, M^{\mathbf{d}_1}_{v,u}, M^{\mathbf{d}_1}_v)$ either does not fork over $M^{\mathbf{d}_0}_{v,u}$ or is orthogonal to $M^{\mathbf{d}_0}_{v,u}$.

<u>Then</u> we let $\mathbf{d} \approx \mathbf{d}_0 *_J \mathbf{d}_1$ mean that $\mathbf{d} = \langle M^{\mathbf{d}}_u, h^{\mathbf{d}}_{v,u}, \mathbf{J}^{\mathbf{d}}_{v,u} : u \subseteq v \in J\rangle$ (but note that $\mathbf{d}$ is not determined uniquely by $\mathbf{d}_0, \mathbf{d}_1, J$ as we have freedom concerning $\mathbf{J}^{\mathbf{d}}_{u \cup \{t^*\},u}$ for $u \in I$, still we will use $\mathbf{d}_0 *_J \mathbf{d}_1$ to denote such $\mathbf{d}$) where

(a) $M^{\mathbf{d}}_u$ is $M^{\mathbf{d}_0}_u$ <u>if</u> $u \in I$
(b) $\mathbf{J}^{\mathbf{d}}_{v,u} = \mathbf{J}^{\mathbf{d}_0}_{v,u}$ <u>if</u> $u \subseteq v \in I$
(c) $M^{\mathbf{d}}_u$ is $M^{\mathbf{d}_1}_{u \backslash \{t^*\}}$ <u>if</u> $u \in J \backslash I$
(d) $\mathbf{J}^{\mathbf{d}}_{v,u} = \{c \in \mathbf{J}^{\mathbf{d}_1}_{v \backslash \{t^*\},u} : c \notin \mathbf{J}^{\mathbf{d}_0}_{v \backslash \{t^*\},u}$ and $\mathbf{tp}_{\mathfrak{s}}(c, M^{\mathbf{d}_1}_{v,u}, M^{\mathbf{d}_1}_v)$ does not fork over $M^{\mathbf{d}_0}_{v \backslash \{t^*\},u}\}$ <u>if</u> $u \in I, v \in J \backslash I$ and $u \subset v \backslash \{t^*\}$
(e) $\mathbf{J}^{\mathbf{d}}_{v,u} = \{c \in \mathbf{J}^{\mathbf{d}_1}_{v \backslash \{t^*\},u \backslash \{t^*\}} : \mathbf{tp}_{\mathfrak{s}}(c, M^{\mathbf{d}_1}_{v \backslash \{t^*\},u \backslash \{t^*\}}, M^{\mathbf{d}_1}_v)$ is orthogonal to $M^{\mathbf{d}_0}_{v \backslash \{t^*\},u \backslash \{t^*\}}\}$ <u>if</u> $u \subset v$ are both from $J \backslash I$
(f) (α) $h^{\mathbf{d}}_{v,u}$ is $h^{\mathbf{d}_0}_{v,u}$ if $u \subseteq v \in I$
(β) $h^{\mathbf{d}}_{v,v}$ is $h^{\mathbf{d}_1}_{v \backslash \{t^*\},u \backslash \{t^*\}}$ if $t^* \in u \subseteq v \in I$ and
(γ) $h^{\mathbf{d}}_{v,u}$ is $h^{\mathbf{d}_0}_{v \backslash \{t(*)\},u}$ if $t(*) \in v \in J, u \subseteq v_1 \in I$
(g) $\mathbf{J}^{\mathbf{d}}_{u \cup \{t(*)\},u}$ is a maximal subset of

$$\{c \in M^{\mathbf{d}_1}_u : \mathbf{tp}_{\mathfrak{s}}(c, M^{\mathbf{d}_0}_u, M^{\mathbf{d}_1}_u) \in \mathscr{S}^{\mathrm{bs}}_{\mathfrak{s}}(M^{\mathbf{d}_0}_u)$$

[19] this relation, $\le_{\mathfrak{s}}$, is a two-place relation, we shall prove that it is a partial order.

is orthogonal to $M^{\mathbf{d}_0}_w$ for every $w \subset u\}$ which is independent in $(M^{\mathbf{d}_0}_u, M^{\mathbf{d}_1}_u)$ for every $u \in I$.

4A) Similarly for $\mathbf{m} = \mathbf{m}_0 *_J \mathbf{m}_1$ (but now $\mathbf{m}$ is uniquely determined).
5) If $\delta < \lambda^+_{\mathfrak{s}}$ and $\langle \mathbf{d}_\alpha : \alpha < \delta \rangle$ is a $\leq^I_{\mathfrak{s}}$-increasing sequence of expanded stable I-systems (see Claim 12.15 below) then we let $\mathbf{d} = \bigcup\limits_\alpha \mathbf{d}_\alpha$ be $\langle M^{\mathbf{d}}_u, h^{\mathbf{d}}_{v,u}, \mathbf{J}^{\mathbf{d}}_{v,u} : u \subseteq v \in I \rangle$ where $M^{\mathbf{d}}_u = \cup\{M^{\mathbf{d}_\alpha}_u : \alpha < \delta\}$ and $h^{\mathbf{d}}_{v,u} = \cup\{M^{\mathbf{d}_\alpha}_{v,u} : \alpha < \delta\}$ and $\mathbf{J}^{\mathbf{d}}_{v,u} = \cup\{\mathbf{J}^{\mathbf{d}_\alpha}_{v,u} : \alpha < \delta\}$. Similarly for $\langle \mathbf{m}_\alpha : \alpha < \delta \rangle$.

12.15 Claim. *1) $\leq^I_{\mathfrak{s}}$ is a partial order on the family of expanded stable I-systems.*
2) If $\delta < \lambda^+_{\mathfrak{s}}$ and $\langle \mathbf{d}_\alpha : \alpha < \delta \rangle$ is $\leq^I_{\mathfrak{s}}$-increasing sequence (of expanded stable I-systems) then $\mathbf{d} = \bigcup\limits_{\alpha<\delta} \mathbf{d}_\alpha$ is an expanded stable I-system and $\alpha < \delta \Rightarrow \mathbf{d}_\alpha \leq^I_{\mathfrak{s}} \mathbf{d}$.
2A) In part (2), for $\ell = 1, 2, 3$ if each $\mathbf{d}_\alpha$ is brimmed$^\ell$ for every $\alpha < \delta$ then $\mathbf{d}_\delta$ is brimmed$^\ell$.
3) If $\mathbf{d}_0 \leq^I_{\mathfrak{s}} \mathbf{d}_1$ and $u \subset v \in I$ then $\mathrm{NF}_{\mathfrak{s}}(M^{\mathbf{d}_0}_{v,u}, M^{\mathbf{d}_0}_v, M^{\mathbf{d}_1}_{v,u}, M^{\mathbf{d}_1}_v)$.
4) If $\mathbf{d}_0 \leq^I_{\mathfrak{s}} \mathbf{d}_1$ and $\mathbf{d}'_0$ is an expanded stable I-system satisfying $\mathbf{m}^{\mathbf{d}'_0} = \mathbf{m}^{\mathbf{d}_0}$ then we can find an expanded stable I-system $\mathbf{d}'_1$ such that $\mathbf{d}'_0 \leq^I_{\mathfrak{s}} \mathbf{d}'_1$ and $\mathbf{m}^{\mathbf{d}'_1} = \mathbf{m}^{\mathbf{d}_1}$.
5) The relation $\leq^I_{\mathfrak{s}}$ on the family of stable I-systems is a partial order.
6) In Definition 12.14(4), always there is $\mathbf{d}$ such that $\mathbf{d} \approx \mathbf{d}_0 *_J \mathbf{d}_1$ is an expanded stable (λ, J)-system.
7) For stable I-system $\mathbf{m}_1, \mathbf{m}_2$ we have $\mathbf{m}_1 \leq_{\mathfrak{s}} \mathbf{m}_2$ iff $\mathbf{m}_1 \leq \mathbf{m}_2$ and $u \subset v \in I \Rightarrow \mathrm{NF}_{\mathfrak{s}}(M^{\mathbf{m}_1}_u, M^{\mathbf{m}_1}_v, M^{\mathbf{m}_2}_u, M^{\mathbf{m}_2}_v)$; recall Definition 12.14(4) of $\leq$.

Proof. 1) Obvious. Check the definition.
2) The main point is why, for $u \subset v \in I$, the triple $(M^{\mathbf{d}}_{v,u}, M^{\mathbf{d}}_v, \mathbf{J}^{2,\mathbf{d}}_{v,u})$ belong to $K^{3,\mathrm{mx}}_{\mathfrak{s}}$. This is trivial by the definition of $K^{3,\mathrm{mx}}_{\mathfrak{s}}$, (have we used $K^{3,\mathrm{vq}}_{\mathfrak{s}}$ we should use e.g. 12.6).
2A) Easy.
3) As in the proof of 12.11(1).

4) As in the proof of 12.11(6).
5) Being partial order follows by (4) and (1).
6) For $u \in I$ we choose $\mathbf{J} = \mathbf{J}^{\mathbf{d}}_{u \cup \{t(*)\},u}$ such that

(a) $\mathbf{J} \subseteq \mathbf{I}_{M^{\mathbf{d}_0}_u, M^{\mathbf{d}_1}_u}$ is independent in $(M^{\mathbf{d}_0}_u, M^{\mathbf{d}_1}_u)$
(b) under (a), $\mathbf{J}$ is maximal.

Now we can check that all the demands in Definition 12.8(1) holds.
7) Easy by now (that is, let $\mathbf{d}_\ell$ be an expanded I-system with $\mathbf{m}[\mathbf{d}_\ell] = \mathbf{m}_\ell$ for $\ell = 1, 2$. Now we shall choose $\mathbf{J}'_{v,u}$ for $u \subset v \in \mathbf{I}$ such that $\mathbf{d}_1 \le^I_{\mathfrak{s}} \mathbf{d}'_2$ where $\mathbf{m}[\mathbf{d}'_v] = \mathbf{m}_2, \mathbf{J}^{\mathbf{d}'_2}_{v,u} = \mathbf{J}'_{v,u}$. We do this by induction on $|v|$, as in previous cases).

$\square_{12.15}$

12.16 Claim. *Assume*

(a) $\mathbf{m}_1$ *is a stable I-system*
(b) $\mathbf{m}_0$ *is an I-system*
(c) $M^{\mathbf{m}_0}_u \le_{\mathfrak{s}} M^{\mathbf{m}_1}_u$ *for $u \in I$ and $h^{\mathbf{m}_0}_{v,u} \subseteq h^{\mathbf{m}_1}_{v,u}$ for $u \subset v \in I$*
(d) *if $u \subset v \in I$ then* $\mathrm{NF}_{\mathfrak{s}}(M^{\mathbf{m}_0}_{v,u}, M^{\mathbf{m}_0}_v, M^{\mathbf{m}_1}_{v,u}, M^{\mathbf{m}_1}_v)$.

<u>Then</u> $\mathbf{m}_0$ *is a stable I-system and* $\mathbf{m}_0 \le^I_{\mathfrak{s}} \mathbf{m}_1$.

Proof. Let $\mathbf{d}_1$ be an expanded stable I-system such that $\mathbf{m}^{\mathbf{d}_1} = \mathbf{m}_1$. For each $u \subset v \in I$ we choose $\mathbf{I}_{v,u}$ as a maximal set such that

$\circledast_1$ (i) $\mathbf{I}_{v,u} \subseteq M^{\mathbf{m}_0}_v \backslash \cup \{M^{\mathbf{m}_0}_{v,w} : w \subset v\}$ and for any $c \in \mathbf{I}_{v,u}$ we have $\mathbf{tp}_{\mathfrak{s}}(c, M^{\mathbf{m}_0}_{v,u}, M^{\mathbf{m}_0}_v) \in \mathscr{S}^{\mathrm{bs}}_{\mathfrak{s}}(M^{\mathbf{m}_0}_{u,v})$ is orthogonal to $M^{\mathbf{m}_0}_{v,w}$ whenever $w \subset u$

(ii) $\mathbf{I}_{v,u} \cup \{\mathbf{J}^{\mathbf{m}_1}_{v,w,u} : u \subseteq w \subset v\}$ is independent in $(M^{\mathbf{m}_1}_{v,u}, M^{\mathbf{m}_1}_v)$.

Let $\mathbf{J}'_{v,u}$ be a maximal set such that

$\circledast_2$ (i) $\mathbf{J}'_{v,u} \subseteq M^{\mathbf{m}_1}_v \backslash \cup \{M^{\mathbf{m}_1}_{v,w} : w \subset v\}$

(ii) $\mathbf{J}'_{v,u} \cup \bigcup \{\mathbf{J}^{\mathbf{d}_1}_{v,w_1,w_0} : w_0 \subseteq w_1 \subset v, w_1 \not\subseteq u\}$ is independent in $(M^{\mathbf{m}_1}_{v,u}, M^{\mathbf{m}_1}_v)$

(iii) $\mathbf{I}_{v,u} \subseteq \mathbf{J}'_{v,u}$.

[Why $\mathbf{J}'_{v,u}$ exists? Because $\mathbf{I}_{v,u}$ satisfies clauses (i) + (ii) in the role of $\mathbf{J}'_{v,u}$, i.e., clause (i) because $u \subseteq v \in I \Rightarrow M^{\mathbf{m}_0}_{u \cap v} = M^{\mathbf{m}_1}_u \cap M^{\mathbf{m}_0}_v$ by clause (d) of the assumption and clause (ii) by clause (ii) of $\circledast_1$.]

So by 12.11(6), $\mathbf{d}'_1 := \langle M^{\mathbf{m}_1}_u, h^{\mathbf{m}_1}_{v,u}, \mathbf{J}'_{v,u} : u \subseteq v \in I \rangle$ is an expanded stable I-system. We could have chosen $\mathbf{d}_1, \langle (\mathbf{I}_{v,u}, \mathbf{J}'_{v,u}) : u \subset v \in I \rangle$ such that $I' = \{v \in I$: for every $u \subset v$ we have $\mathbf{J}^{\mathbf{d}_1}_{v,u} = \mathbf{J}'_{v,u}\}$ is maximal. By the proof so far, $I' = I$. Now for every $u \subset v \in I$, the set $\mathbf{I}^+_{v,u} := \cup \{h^{\mathbf{m}_0}_{v,w}(\mathbf{I}_{w,u}) : u \subset w \subseteq v\}$ is included in $\mathbf{J}'_{v,u} \cup \bigcup \{\mathbf{J}^{\mathbf{d}_1}_{v,w,u} : u \subset w \subset v\}$ hence is independent in $(M^{\mathbf{m}_1}_{v,u}, M^{\mathbf{m}_1}_v)$.

Now $\mathbf{I}^+_{v,u}$ is included in $M^{\mathbf{m}_0}_v$ and $\mathrm{NF}_{\mathfrak{s}}(M^{\mathbf{m}_0}_u, M^{\mathbf{m}_0}_v, M^{\mathbf{m}_1}_u, M^{\mathbf{m}_1}_v)$ holds by clause (d) of the assumption, hence $\mathbf{I}^+_{v,u}$ is independent in $(M^{\mathbf{m}_0}_{v,u}, M^{\mathbf{m}_1}_v)$.

Lastly, it is a maximal such set by the choice of $\mathbf{I}_{v,u}$ as maximal. Hence $\mathbf{d}_0 := \langle M^{\mathbf{m}_0}_u, h^{\mathbf{m}_0}_{v,u}; \mathbf{I}_{v,u} : u \subseteq v \in I \rangle$ is an expanded stable I-system. Now easily $\mathbf{m}_\ell = \mathbf{m}^{\mathbf{d}_\ell}$ and $\mathbf{d}_0 \le^I_{\mathfrak{s}} \mathbf{d}'_1$.
So we are done. $\square_{12.16}$

12.17 Claim. *Assume J is a successor of I with witness t^*.*
1) Assume $\mathbf{m}_0, \mathbf{m}_1$ are stable I-systems satisfying $\mathbf{m}_0 \le^I_{\mathfrak{s}} \mathbf{m}_1$. Then

(a) *for one and only one $\mathbf{m}, \mathbf{m} = \mathbf{m}_0 *_J \mathbf{m}_1$*

(b) *$\mathbf{m}$ is a stable J-system.*

2) Assume that $\delta < \lambda^+_{\mathfrak{s}}$ and $\langle \mathbf{m}_\alpha : \alpha < \delta \rangle$ is a $<^I_{\mathfrak{s}}$-increasing continuous sequence of stable I-systems and $\mathbf{m}_\alpha$ is brimmed$^\ell$ for $\alpha < \delta$ and $\ell \le 3$ then $\mathbf{m}_\delta = \cup \{\mathbf{m}_\alpha : \alpha < \delta\}$ is a stable I-system and is brimmed$^\ell$ and $\mathbf{m}_\alpha \le^I_{\mathfrak{s}} \mathbf{m}_\delta$ for $\alpha < \delta$.
3) Assume that $\mathfrak{s}$ is successful hence $\mathfrak{s}^+$ is as in 12.3 and $\ell = 3$. If $\langle \mathbf{m}_\alpha : \alpha < \lambda^+_{\mathfrak{s}} \rangle$ is a $<^I_{\mathfrak{s}}$-increasing sequence of stable I-systems and $\mathbf{m}_\alpha *_J \mathbf{m}_{\alpha+1}$ is brimmed$^\ell$ for each $\alpha < \lambda^+$ (or just for unboundedly many $\alpha < \lambda^+$) then $\mathbf{m} = \cup \{\mathbf{m}_\alpha : \alpha < \lambda^+_{\mathfrak{s}}\}$ is a brimmed$^\ell$ $(\lambda^+, I, \mathfrak{s})$-system.
4) In part (3), if $\ell \in \{0, 1, 2\}$ and $u \in I \Rightarrow M^{\mathbf{m}}_u \in K_{\mathfrak{s}(+)}$ then $\mathbf{m}$ is brimmed$^\ell$ $(I, \mathfrak{s}^+)$-system.

Remark. In 12.17(3),(4) the case $\ell = 4$ is ignored as we do not know to prove this.

Proof. 1) Clearly $\mathbf{m}$ is well defined.

Recall that $\mathbf{d}_0 \leq^I_{\mathfrak{s}} \mathbf{d}_1$ above does not imply that for a unique $\mathbf{d}, \mathbf{d} = \mathbf{d}_0 *_J \mathbf{d}_1$.

By Definition 12.14(3) there are stable expanded I-system $\mathbf{d}_1, \mathbf{d}_2$ such that $\mathbf{m}^{\mathbf{d}_\ell} = \mathbf{m}_\ell$ and $\mathbf{d}_0 \leq_{\mathfrak{s}} \mathbf{d}_1$. For each $u \in I$ let $\mathbf{P}^0_u$ be the set of $p \in \mathscr{S}^{\rm bs}_{\mathfrak{s}}(M^{\mathbf{m}_0}_u)$ orthogonal to $M^{\mathbf{m}_0}_{u,w}$ for every $w \subset u$ [if $\mathfrak{s}$ satisfies 12.3(3) you can use a maximal set of pairwise orthogonal regular types from $\mathscr{S}^{\rm bs}_{\mathfrak{s}}(M^{\mathbf{m}_0}_u)$ orthogonal to $M^{\mathbf{m}_0}_{u,w}$ for every $w \subset u$]. For each $u \in I$ let $\mathbf{P}^1_u$ be the set of $p \in \mathscr{S}^{\rm bs}_{\mathfrak{s}}(M^{\mathbf{m}_1}_u)$ such that $p \perp M^{\mathbf{m}_1}_{u,w}$ for $w \subset u$ and either p does not fork over $M^{\mathbf{m}_0}_u$ or is orthogonal to it [if e.g. 12.3(3), we can use a maximal set of pairwise orthogonal regular types $p \in \mathscr{S}^{\rm bs}_{\mathfrak{s}}(M^{\mathbf{m}_1}_u)$ orthogonal to $M^{\mathbf{m}_v}_w$ for every $w \subset u$ such that either p is a non-forking extension of some $q \in \mathbf{P}^0_u$ or $p \perp M^{\mathbf{m}_0}_u$]. By Claims 12.11(6), 12.12(2) (and see 12.15(4)), without loss of generality: if $i \in \{0,1\}, w \subset u \subset v \in I$ and $c \in \mathbf{J}^{\mathbf{d}_i}_{v,u}$ then $\mathbf{tp}_{\mathfrak{s}}(c, M^{\mathbf{d}_{i\ell}}_{v,u}, M^{\mathbf{d}_i}_{v,u}) \in \mathbf{P}^i_u$ and $\mathbf{J}^{\mathbf{d}_0}_{v,u} \subseteq \mathbf{J}^{\mathbf{d}_1}_{v,u}$.

So $\mathbf{d}_0, \mathbf{d}_1$ are as in Definition 12.14(4) (as $J, t(*)$ are given) hence there is a stable expanded I-system $\mathbf{d} \approx \mathbf{d}_0 *_J \mathbf{d}_1$. So $\mathbf{m} = \mathbf{m}^{\mathbf{d}}$ is as required.

2) If $\ell = 0$ we define $\mathbf{m}_\delta = \langle M^\delta_u, h^\delta_{u,v} : u \subseteq v \in I\rangle$ by $M^\delta_u := \cup\{M^{\mathbf{m}_\alpha}_u : \alpha < \delta\}$ and $h^\delta_{u,v} := \cup\{h^{\mathbf{m}_\alpha}_u : \alpha < \delta\}$. As $\mathfrak{K}_{\mathfrak{s}}$ is a λ-a.e.c. and the definition of I-systems, $\mathbf{m}_\delta$ is as required. So assume $\ell \geq 1$, i.e. we deal with stable systems, given a $\leq^I_{\mathfrak{s}}$-increasing continuous $\langle \mathbf{m}_\alpha : \alpha < \delta\rangle$, we choose by induction on $\alpha \leq \delta$ an expanded stable I-system $\mathbf{d}_\alpha$ such that $\mathbf{m}^{\mathbf{d}_\alpha} = \mathbf{m}_\alpha$ and $\langle \mathbf{d}_\beta : \beta \leq \alpha\rangle$ is $\leq^I_{\mathfrak{s}}$-increasing continuous. First for $\alpha = 0$ this is by the definition. Second, for $\alpha = \beta + 1$ by the definition of $\mathbf{m}_\beta \leq^I_{\mathfrak{s}} \mathbf{m}_\alpha$ there is a pair $(\mathbf{d}'_\beta, \mathbf{d}''_\beta)$ of expanded stable I-systems such that $\mathbf{d}'_\beta \leq^I_{\mathfrak{s}} \mathbf{d}''_\beta$ and $\mathbf{m}_\beta = \mathbf{m}^{\mathbf{d}'_\beta}$ and $\mathbf{m}_\alpha = \mathbf{m}^{\mathbf{d}''_\beta}$. But then by 12.15(4) there is an expanded stable I-system $\mathbf{d}_\alpha$ such that $\mathbf{d}_\beta \leq^I_{\mathfrak{s}} \mathbf{d}_\alpha$, but by 12.15(1) we have $\gamma < \alpha \Rightarrow \gamma \leq \beta \Rightarrow \mathbf{d}_\gamma \leq^I_{\mathfrak{s}} \mathbf{d}_\beta \leq^I_{\mathfrak{s}} \mathbf{d}_\alpha \Rightarrow \mathbf{d}_\gamma \leq^I_{\mathfrak{s}} \mathbf{d}_\alpha$ so we finish this case.

Lastly, for α limit ordinal $\leq \delta$ by 12.15(2) we have $\mathbf{d}_\alpha = \cup\{\mathbf{d}_\beta : \beta < \delta\}$ is an expanded stable I-system and $\gamma < \alpha \Rightarrow \mathbf{d}_\gamma \leq^I_{\mathfrak{s}} \mathbf{d}_\alpha$. Also if $\alpha < \delta$ then as $\langle \mathbf{m}_\beta : \beta \leq \alpha\rangle$ is increasing continuous, we have $\mathbf{m}^{\mathbf{d}_\alpha} = \mathbf{m}_\alpha$. For $\alpha = \delta, \mathbf{d}_\delta := \cup\{\mathbf{d}_\alpha : \alpha < \delta\}$ is an expanded stable I-system and $\mathbf{m}_\delta := \mathbf{m}^{\mathbf{d}_\alpha}$ is as required.

For $\ell = 1$ we are done.

For $\ell = 2$ note that the union of an $\le_{\mathfrak{s}}$-increasing continuous sequence of brimmed models is brimmed.

For $\ell = 3$, and for $\alpha < \delta$ above $\mathbf{d}_\alpha$ is brimmed$^\ell$ by 12.13 hence by 12.15(2A)) also $\mathbf{d}_\delta$ is brimmed$^\ell$ hence $\mathbf{m}_\delta$ is (by the definition).

3) Easy, too.

4) Easy, too. $\square_{12.17}$

12.18 Lemma. *1) Let $\ell \in \{0,2,3\}$. Assume that $\mathfrak{s}^+$ is successful (see 12.5(2)) and*

(a) *J a successor of I, in details $I \subseteq \mathscr{P}(u_*)$ is downward closed, u_* finite, $u_{**} = u_* \cup \{t^*\}, t^* \notin u_*$ and $J = I \cup \{u \cup \{t^*\} : u \in I\}$*

(b) *$\mathbf{m}$ is a brimmed$^\ell$ $(\lambda^+, I, \mathfrak{s}^+)$-system*

(c) *$\langle M^{\mathbf{m}}_{u,\alpha} : \alpha < \lambda^+\rangle$ is a $<_{\mathfrak{K}[\mathfrak{s}]}$-representation of $M^{\mathbf{m}}_u$*

(d) *for $\alpha < \lambda^+$ we try to define $(\lambda, I, \mathfrak{s})$-system $\mathbf{m}_\alpha$ by $M^{\mathbf{m}_\alpha}_u = M^{\mathbf{m}}_{u,\alpha}, h^{\mathbf{m}_\alpha}_{v,u} = h^{\mathbf{m}}_{v,u} \restriction M^{\mathbf{m}}_{u,\alpha}$*

(e) *for $\alpha < \beta < \lambda^+$ we try to define a $(\lambda, J, \mathfrak{s})$-system $\mathbf{m}_{\alpha,\beta}$ by*

$$M^{\mathbf{m}_{\alpha,\beta}}_u = M^{\mathbf{m}}_{u,\alpha} \text{ for } u \in I$$

$$M^{\mathbf{m}_{\alpha,\beta}}_{u\cup\{t^*\}} = M^{\mathbf{m}}_\beta$$

$$h^{\mathbf{m}_{\alpha,\beta}}_{v,u} = h^{\mathbf{m}}_{v,u} \restriction M^{\mathbf{m}_{\alpha,\beta}}_{u\cup\{t^*\}} \text{ if } u \in I, u \subseteq v \in J$$

$$h^{\mathbf{m}_{\alpha,\beta}}_{v,u\cup\{t^*\}} = h^{\mathbf{m}}_{v,u} \restriction M^{\mathbf{m}_{\alpha,\beta}}_u \text{ if } u \in I, u \cup \{t^*\} \subseteq v \in I.$$

Then for some club E of λ^+ we have

(α) *for every $\alpha \in E, \mathbf{m}_\alpha$ is an $(I,\mathfrak{s})$-system which is brimmed$^\ell$*

(β) *for every $\alpha < \beta$ from $E, \mathbf{m}_{\alpha,\beta}$ is a brimmed$^\ell$ $(J,\mathfrak{s})$-system*

(γ) *for every $\alpha < \beta$ from $E, \mathbf{m}_\alpha \le^I_{\mathfrak{s}} \mathbf{m}_\beta$ and $\mathbf{m}_{\alpha,\beta} = \mathbf{m}_\alpha *_J \mathbf{m}_\beta$.*

2) [Assume 12.3(3)] Assume that $\ell \in \{2,3\}$ and (a),(c),(d),(e) of part (1) holds for $\mathbf{m} = \mathbf{m}^{\mathbf{d}}$ and

(b)′ $\mathbf{d}$ is an explicitly regular brimmed$^\ell$ expanded stable $(\lambda^+, I, \mathfrak{s}^+)$-system.

Then for some club E of λ^+, we can define for $\alpha \in E, \mathbf{d}^\alpha$ as below and we can find $\mathbf{d}^{\alpha,\beta}$ for $\alpha < \beta$ from E as below such that

(α) $\mathbf{d}^\alpha$ is an explicitly regular expanded stable $(\lambda, I, \mathfrak{s})$-system

(β) $\mathbf{d}^{\alpha,\beta}$ is an explicitly regular brimmed$^\ell$ expanded stable $(\lambda^+, J, \mathfrak{s}^+)$-system

(γ) $\mathbf{m}[\mathbf{d}^{\alpha,\beta}] = \mathbf{m}^{\alpha,\beta}$ from part (1) and $\mathbf{m}[\mathbf{d}^\alpha] = \mathbf{m}^\alpha$ from part (1)

(δ) for $u \subset v \in I, \mathbf{J}^{\mathbf{d}^\alpha}_{v,u} = \mathbf{J}^{\mathbf{d}}_{v,u} \cap M^{\mathbf{d}^\alpha}_v$.

3) Assume that $\ell \in \{2,3\}$ and clauses (a),(b),(c),(d),(e) of parts (1) holds and $\mathbf{d}$ is an expanded stable $(\lambda^+, I, \mathfrak{s}^+)$-system with $\mathbf{m}^{\mathbf{d}} = \mathbf{m}$. Then we can find a club E of λ^+ and $\langle \mathbf{d}^\alpha : \alpha \in E\rangle$ such that $(\alpha), (\beta), (\gamma), (\delta)$ of part (2) holds provided that we omit the "explicitly regular".

Remark. In 12.18(1) we can add the case $\ell = 1$ as any brimmed$^1(\lambda^+$, I, $\mathfrak{s}^+)$-system is brimmed^2 as $M \in K_{\mathfrak{s}(+)}$ is brimmed.

Proof. 1) We leave the case $\ell = 0$ to the reader, so it is enough to prove parts (2),(3).
2) For each $u \in I$ let $\mathbf{P}_u := \{p \in \mathscr{S}^{\mathrm{bs}}_{\mathfrak{s}(+)}(M^{\mathbf{m}}_u) : p$ regular orthogonal to $M^{\mathbf{m}}_{u,w}$ for every $w \subset u\}$.

Recall that $\mathbf{d}$ is a stable expanded (λ, I)-system which is explicitly regular. Now

$\circledast_1$ if $p \in \mathbf{P}_u$ then for some $\alpha_0 = \alpha_0(p) \leq \alpha_1 = \alpha_1(p) < \lambda^+_{\mathfrak{s}}$ we have:

- (a) $M^{\mathbf{m}}_{v,\alpha_1}$ is a witness for p
- (b) $p \restriction M^{\mathbf{m}}_{u,\alpha_1}$ is orthogonal to $M^{\mathbf{m}}_{u,\gamma}$ iff $\gamma < \alpha_0$.

Easily

$\circledast_2$ for each $u \in I$ there is a club E_u of $\lambda^+_{\mathfrak{s}}$ such that if $p \in \mathbf{P}_u$ and $\alpha_0(p) \le \gamma \in E$ then $\alpha_1(p) \le \gamma$.

For each $p \in \mathbf{P}_u$ let $\mathbf{J}_p$ be a maximal subset of $\mathbf{I}_{M^{\mathbf{m}}_{u,\alpha_1(p)}, M^{\mathbf{m}}_u} = \cup\{\mathbf{I}_{M^{\mathbf{m}}_{u,\alpha_1(p)}, M^{\mathbf{m}}_{u,\beta}} : \beta \in [\alpha_1(p), \lambda^+_{\mathfrak{s}})\}$ of elements realizing $p \restriction M^{\mathbf{m}}_{u,\alpha_1(p)}$ in $M^{\mathbf{m}}_u$ which is independent in $(M^{\mathbf{m}}_{u,\alpha_1(p)}, M^{\mathbf{m}}_u)$.

We can find a club $E \subseteq \cap\{E_u : u \in I\}$ of $\lambda^+_{\mathfrak{s}}$ such that

$\circledast_3$ if $\delta \in E$ then

(a) for $u \subset v \in I$ we have $h^{\mathbf{d}}_{v,u}(M^{\mathbf{m}}_{u,\delta}) = M^{\mathbf{m}}_{v,u} \cap M^{\mathbf{m}}_{v,\delta}$

(b) for $u \subset v$ and $c \in \mathbf{J}^{\mathbf{d}}_{v,u}$:
if $c \in M^{\mathbf{m}}_{v,\delta}$ then $\delta \ge \alpha_1(\mathbf{tp}(c, M^{\mathbf{m}}_{v,u}, M^{\mathbf{m}}_v))$

(c) $(M^{\mathbf{m}}_{v,u} \cap M^{\mathbf{m}}_{v,\delta}, M^{\mathbf{m}}_{v,\delta}, \mathbf{J}^{2,\mathbf{d}}_{v,u} \cap M^{\mathbf{m}}_{v,\delta})$ belongs to $K^{3,\mathrm{mx}}_{\mathfrak{s}}$

(d) if $u \in I$ and $p \in \mathbf{P}_u$ and $\alpha_1(p) < \delta$ then $\mathbf{J}_p \cap M^{\mathbf{m}}_{u,\delta}$ is a maximal subset of $\mathbf{I}_{M^{\mathbf{m}}_{u,\alpha_1(p)}, M^{\mathbf{m}}_{u,\delta}}$ of elements realizing $p \restriction M^{\mathbf{m}}_{u,\alpha(p)}$ which is independent in $(M^{\mathbf{m}}_{u,\alpha_1(p)}, M^{\mathbf{m}}_{u,\delta})$

(e) if $u \subset v \in I$ and $\delta_1 < \delta_2$ are from E
then $\mathrm{NF}_{\mathfrak{s}}(M^{\mathbf{m}}_{u,\delta_1}, M^{\mathbf{m}}_{v,\delta_1}, M^{\mathbf{m}}_{u,\delta_2}, M^{\mathbf{m}}_{v,\delta_2})$

(f) if $u \subset v \in I, c \in \mathbf{J}^{\mathbf{m}}_{v,u}$ and $p = \mathbf{tp}_{\mathfrak{s}(+)}(c, M^{\mathbf{m}}_{v,u}, M^{\mathbf{m}}_v)$ and $\alpha_0(p) \le \delta$ <u>then</u> $\alpha_1(p) \le \delta$

(g) if $u \subset v \in I$ and $\delta \in E$ then $(h^{\mathbf{m}}_{v,u}(M^{\mathbf{m}}_{u,\delta}), M^{\mathbf{m}}_{v,\delta}, \mathbf{J}^{2,\mathbf{m}}_{v,u} \cap M^{\mathbf{m}}_{v,\delta}) \in K^{3,\mathrm{mx}}_{\mathfrak{s}}$.

For $\alpha \in E$ clearly $\mathbf{m}^\alpha$ as defined in clause (d) of part (1) is a $(\lambda, I, \mathfrak{s})$-system; and we expand it to $\mathbf{d}^\alpha$ by $\mathbf{J}^{\mathbf{d}^\alpha}_{v,u} = \mathbf{J}^{\mathbf{d}}_{v,\alpha} \cap M_u$ for $u \subset v \in I$ easily $\mathbf{d}^\alpha$ is an expanded stable $(I, \lambda, \mathfrak{s})$-system.

Let $\alpha < \beta$ be from E and let $\mathbf{m}^{\alpha,\beta}$ be as clause (e) of part (1) of the claim. We define an expanded (λ, J)-system $\mathbf{d}^{\alpha,\beta}$ by (essentially it is $\mathbf{d}^\alpha *_J \mathbf{d}^\beta$ recalling $\mathbf{d}$ is explicit)

$\circledast_4$(a) $\mathbf{m}^{\mathbf{d}^{\alpha,\beta}} = \mathbf{m}^{\alpha,\beta}$ (defined in part (1))

(b) if $u \subset v \in I$ then $\mathbf{J}^{\mathbf{d}^{\alpha,\beta}}_{u,v} = \mathbf{J}^{\mathbf{d}}_{u,v} \cap M^{\mathbf{m}}_{v,\alpha}$

(c) if $u \subset v \in J, u \in I, v = v_1 \cup \{t^*\}$ hence $u \subset v_1 \in I$ then $\mathbf{J}^{\mathbf{d}^{\alpha,\beta}}_{u,v} = \{c \in \mathbf{J}^{\mathbf{d}}_{v,u} : c \in M^{\mathbf{m}^{\alpha,\beta}}_v$ and $\alpha \geq \alpha_1(\mathbf{tp}_{\mathfrak{s}}(c, M^{\mathbf{d}}_{v,u}, M^{\mathbf{d}}_v))\}$

(d) if $u \subseteq v \in J, u = u_1 \cup \{t^*\}, v = v_1 \cup \{t^*\}, u \subset v \in I$ then $\mathbf{J}^{\mathbf{d}^{\alpha,\beta}}_{v,u} = \{c \in \mathbf{J}^{\mathbf{d}}_{v,u} : c \in \mathbf{J}^{\mathbf{d}}_{v,u}$ and $\alpha < \alpha_1(\mathbf{tp}(c, M^{\mathbf{d}}_{v,u}, M^{\mathbf{d}}_v))\}$

(e) if $u \in I, v = u \cup \{t^*\}$ then $\mathbf{J}^{\mathbf{d}^{\alpha,\beta}}_{v,u} =$ $\{c$: for some $p \in \mathbf{P}_u, \alpha_1(p) \leq \alpha$ and $c \in \mathbf{J}_p \cap M^{\mathbf{m}}_{u,\beta} \backslash M^{\mathbf{m}}_{u,\alpha}\}$.

Now check.

3) This is like part (2) but we are dealing with the general case: $\mathfrak{s}$ is not necessarily type-full or just has regulars. For each $u \in I$, there is a club E_u of $\lambda^+_{\mathfrak{s}}$ such that

$\circledast'_2$ (*a*) if $\alpha \in E_u$ then $M^{\mathbf{m}}_{u,\alpha} \in K_{\mathfrak{s}}$ is brimmed

(*b*) if $\alpha < \beta$ are from E_u then $M^{\mathbf{d}}_{u,\beta} \in K_{\mathfrak{s}}$ is brimmed over $M^{\mathbf{m}}_{u,\alpha}$.

Now for $u \in I$ we let $\mathbf{P}_u := \{p \in \mathscr{S}^{\mathrm{bs}}_{\mathfrak{s}(+)}(M^{\mathbf{m}}_u) : p$ is orthogonal to $M^{\mathbf{m}}_{u,w}$ for every $w \subset u\}$ and for some $\mathbf{P}'_u = \cup\{\mathbf{P}'_{u,\alpha} : \alpha \in E_u\}$ where for $u \in E_u$ we let $\mathbf{P}'_{u,\alpha} = \{p \in \mathbf{P}_u : M^{\mathbf{m}}_{u,\alpha}$ is a witness for p and $p \restriction M^{\mathbf{m}}_{u,\alpha} \in \mathscr{S}^{\mathrm{bs}}_{\mathfrak{s}}(M^{\mathbf{m}}_{u,\alpha})$ is orthogonal to $M^{\mathbf{m}}_{u,\beta}$ whenever $\beta \in E_u \cap \alpha\}$.

Now we have a choice: we can prove this through showing that $\mathbf{P}'_u$ is auto-dense (see Definition 7.20). But we also can use the brimmness of $\mathbf{m}^{\alpha,\beta}$'s; actually in the second we use weaker assumptions on $\mathfrak{s}$.

Let $\mathbf{d}'$ be an expanded stable $(\lambda^+, I, \mathfrak{s}^+)$-system and let $E \subseteq \{E_u : u \in I\}$ be a club of λ^+ satisfying clauses (a),(c),(e),(f) of $\circledast_3$ from the proof of part (2). Now for each $u \subset v \in I$ we choose $\mathbf{J}'_{v,u,\beta}$ by induciton on $\beta \in E$ such that

$\boxtimes$ (*a*) $\mathbf{J}'_{v,u,\beta} \subseteq \mathbf{I}_{M^{\mathbf{m}}_{v,u,\beta}, M^{\mathbf{m}}_{v,\beta}}$; moreover $\subseteq$ $\{c \in M^{\mathbf{m}}_{v,u,\beta} : \mathbf{tp}_{\mathfrak{s}(+)}(c, M^{\mathbf{m}}_{v,u}, M^{\mathbf{s}}_v) \in \cup\{\mathbf{P}'_{u,\gamma} : \gamma \leq \beta \wedge \gamma \in E\}$

(*b*) $\mathbf{J}'_{v,u,\beta}$ increasing with β

(*c*) $(M^{\mathbf{m}}_{v,u,\beta}, M^{\mathbf{m}}_{v,\beta}, \mathbf{J}'_{v,u,\beta}) \in K^{3,\mathrm{bs}}_{\mathfrak{s}}$

(*d*) under (a)+(c), $\mathbf{J}'_{v,u,\beta}$ is maximal.

Now define an expanded $(\lambda^+, I, \mathfrak{s}^+)$-system by $\mathbf{m}^{\mathbf{d}} = \mathbf{m}$ and $\mathbf{J}^{\mathbf{d}}_{v,u} = \cup\{\mathbf{J}'_{v,u,\beta} : \beta \in E\}$. By 7.17(3) it is as required. Now we continue as in part (2). $\square_{12.18}$

* * *

The following define the main notions of this section. Unfortunately they may seem too many, but for our inductive proofs they seem necessary. In those definitions the low cases ($n = 0, 1$ and sometimes $n = 2$) follows as proved later, still they are given their natural meaning and in the more general frameworks they will not be trivial.

12.19 Definition. Let $\ell \in \{1, 2, 3, 4\}$.
1) We say that $\mathfrak{s}$ satisfies (or has) the brimmed$^\ell$ weak (λ, n)-existence property <u>if</u>:

<u>Case 1</u>: $n = 0$.
There is a brimmed$^\ell$ model in $K_{\mathfrak{s}}$ (so always holds).

<u>Case 2</u>: <u>$n = 1$</u>.
If $M_\emptyset$ is brimmed$^\ell$ and $p_i \in \mathscr{S}^{\text{bs}}_{\mathfrak{s}}(M_\emptyset)$ for $i < i^* < \lambda^+_{\mathfrak{s}}$ then we can find $c_i (i < i^*)$ and $M_{\{\emptyset\}}$ such that $p_i = \mathbf{tp}_{\mathfrak{s}}(c_i, M_\emptyset, M_{\{\emptyset\}})$ and $(M_\emptyset, M_{\{\emptyset\}}, \{c_i : i < i^*\}) \in K^{3,\text{bs}}_{\mathfrak{s}}$ and $i < j \Rightarrow c_i \neq c_j$.

<u>Case 3</u>: $n \geq 2$.

Every brimmed$^\ell$ expanded stable $(\lambda, \mathscr{P}^-(n))$-system $\mathbf{d}$ can be completed to an expanded stable $(\lambda, \mathscr{P}(n))$-system $\mathbf{d}^+$, i.e. there is an expanded stable $\mathscr{P}(n)$-system $\mathbf{d}^+$ such that $\mathbf{d}^+ \restriction \mathscr{P}^-(n) = \mathbf{d}$; recall $\mathscr{P}^-(n) = \{u : u \in \{0, \ldots, n-1\}\} = \mathscr{P}(n) \backslash \{n\}$.
2) We say $\mathfrak{s}$ satisfies (or has) brimmed$^\ell$ weak (λ, n)-uniqueness property when:

<u>Case 1</u>: $n = 0$.
$\mathfrak{K}_{\mathfrak{s}}$ has the JEP (the joint embedding property).

<u>Case 2</u>: $n = 1$.
If $(M, N_k, \{a^k_i : i < i^*\}) \in K^{3,\text{mx}}_{\mathfrak{s}}$ for $k = 1, 2$ and $\mathbf{tp}_{\mathfrak{s}}(a^1_i, M, N_\ell) = \mathbf{tp}_{\mathfrak{s}}(a^2_i, M, N_\ell)$ <u>then</u> there is a $\leq_{\mathfrak{s}}$-embedding f of N_1 into some N'_2 such that $N_2 \leq_{\mathfrak{s}} N'_2$ and $f(a^1_i) = a^2_i$.

Case 3: $n \geq 2$.

If $\mathbf{d}_1, \mathbf{d}_2$ are brimmed$^\ell$ expanded stable $(\lambda, \mathscr{P}(n))$-systems and $\mathbf{d}^m = \mathbf{d}_m \restriction \mathscr{P}^-(n)$ and $\bar{f} = \langle f_u : u \in \mathscr{P}^-(n)\rangle$ is an isomorphism from $\mathbf{m}[\mathbf{d}^1]$ onto $\mathbf{m}[\mathbf{d}^2]$, then we can find a pair (f, N) such that:

(a) $M_n^{\mathbf{d}_2} \leq_{\mathfrak{s}} N$,

(b) f is a $\leq_{\mathfrak{s}}$-embedding of $M_n^{\mathbf{d}_1}$ into N

(c) for $u \subset n$ we have $f \circ h_{n,u}^{\mathbf{d}_1} = h_{n,u}^{\mathbf{d}_2} \circ f_u$.

3) We say that $\mathfrak{s}$ has the brimmed$^\ell$ strong (λ, n)-existence when:

Case 1: $n = 0$.

There is brimmed$^\ell$ model in $K_{\mathfrak{s}}$.

Case 2: $n = 1$.

As in part (1) but $(M_\emptyset, M_{\{\emptyset\}}, \{c_i : i < i^*\}) \in K_{\mathfrak{s}}^{3,\mathrm{mx}}$.

Case 3: $n \geq 2$.

For every brimmed$^\ell$ expanded stable $(\lambda, \mathscr{P}^-(n))$-system $\mathbf{d}$ we can find an expanded stable $(\lambda, \mathscr{P}(n))$-system $\mathbf{d}^+$ such that $\mathbf{d}^+ \restriction \mathscr{P}^-(n) = \mathbf{d}$ and $\mathbf{d}^+$ is reduced in n which means $u \subset n \Rightarrow \mathbf{J}_{n,u}^{\mathbf{d}^+} = \emptyset$ and if $\ell = 2,3,4$ then[20] $M_n^{\mathbf{d}^+}$ is brimmed (so if $\ell = 1,2$ then $\mathbf{d}^+$ is brimmed$^\ell$ but for $\ell = 3,4$ necessarily this fails).

4) We say that $\mathfrak{s}$ has the brimmed$^\ell$ strong (λ, n)-uniqueness property when:

Case 1: If $n = 0$, any two brimmed$^\ell$ models from $\mathfrak{K}_{\mathfrak{s}}$ are isomorphic.

Case 2: If $n = 1$, uniqueness for $K_{\mathfrak{s}}^{3,\mathrm{mx}}$, i.e., if $(M, N_k, \{a_\alpha^k : \alpha < \alpha^*\}) \in K_{\mathfrak{s}}^{3,\mathrm{mx}}$ so $\|N_k\| = \|M\|$ and M, N_k are brimmed$^\ell$ for $k = 1,2$ and $\mathbf{tp}_{\mathfrak{s}}(a_\alpha^1, M, N_1) = \mathbf{tp}_{\mathfrak{s}}(a_\alpha^2, M, N_2)$ for every $\alpha < \alpha^*$ then there is an isomorphism from N_1 onto N_2 over M mapping a_α^1 to a_α^2 for every $\alpha < \alpha^*$ (see 7.14(2) for sufficient conditions).

Case 3: $n \geq 2$ and $\ell \in \{3, 4\}$.

In part (2) we add $N = M_n^{\mathbf{d}_2}$ and f is onto N.

[20] if $K_{\mathfrak{s}}$ is categorical this is an empty demand

Case 4: $n \geq 2$ and[21] $\ell \in \{1,2\}$.

The conclusion of Case 3 holds[22] if we assume (what is said in part (2) and) that[23]:

> $\boxdot$ either $\mathbf{d}_1, \mathbf{d}_2$ are reduced in n or $\mathbf{d}_1, \mathbf{d}_2$ are brimmed$^\ell$ in n [on reduced in n see Case 3 of part 3; on brimmed$^\ell$ in n, see Definition 12.9(4)-(7), 12.10]. .

5) We say that $\mathfrak{s}$ has the brimmed$^\ell$ strong (λ, n)-primeness/weak (λ, n)-primeness property when: $n = 0, 1$ or for any brimmed$^\ell$ expanded stable $(\lambda, \mathscr{P}^-(n))$-system $\mathbf{d}_0$ and expanded stable $(\lambda, \mathscr{P}(n))$-system $\mathbf{d}_1$ such that $M_n^{\mathbf{d}_1}$ is brimmed$^\ell$ and $\mathbf{d}_1$ is reduced in n which means $u \subseteq n \Rightarrow \mathbf{J}_{n,u}^{\mathbf{d}_1} = \emptyset$ and $\mathbf{d}_1$ satisfy $\mathbf{d}_1 \restriction \mathscr{P}^-(n) = \mathbf{d}_0$ we have: $\mathbf{d}_1$ is strongly prime$^\ell$/weakly prime$^\ell$ over $\mathbf{d}_0$, see below.

5A) We say $\mathbf{d}_1$ is brimmed$^\ell$ strongly prime/weakly prime over $\mathbf{d}_0$ (and also say that $\mathbf{d}_1$ is weakly prime$^\ell$/strongly prime$^\ell$ over $\mathbf{d}_0$) when for some $n, \mathbf{d}_1$ is a stable expanded $(\lambda, \mathscr{P}(n))$-system, $\mathbf{d}_0 = \mathbf{d}_1 \restriction \mathscr{P}^-(n)$, if $\ell \geq 2$ then $M_n^{\mathbf{d}_1}$ is brimmed and:

> $(*)$ if $\mathbf{d}_2$ is a stable expanded $(\lambda, \mathscr{P}(n))$-system satisfying $\mathbf{d}_2 \restriction \mathscr{P}^-(n) = \mathbf{d}_0$ and $M_n^{\mathbf{d}_2}$ is brimmed$^\ell$, then there is an $\leq_{\mathfrak{s}}$-embedding f of $M_n^{\mathbf{d}_1}$ into $M_n^{\mathbf{d}_2}$ such that $u \in \mathscr{P}^-(n) \Rightarrow f \circ h_{n,u}^{\mathbf{d}_1} = h_{n,u}^{\mathbf{d}_2}$, but in the weak case, we further demand on $\mathbf{d}_2$ that it satisfies:
>
> $\odot_{\mathbf{d}_2}$ if $p \in \mathscr{S}_{\mathfrak{s}}^{\mathrm{bs}}(M_n^{\mathbf{d}_2})$ is orthogonal to $M_{n,u}^{\mathbf{d}_2}$ for every $u \subset n$

[21] those cases are problematic, in the sense of rarely holding but this does not concern us here

[22] in the case central for this section the $M_n^{\mathbf{d}_m}$ are brimmed so the only freedom left are about dimensions of types from $\mathscr{S}_{\mathfrak{s}}^{\mathrm{bs}}(M_{n,u}^{\mathbf{d}_m})$ for $u \subset n$. But for the general case, for $\mathfrak{s}$ which is not very "low", strong uniqueness holds for $\ell = 2$ in the beautiful case, but for $\ell = 1$ fails. However, we may have uniqueness of a prime model, see later.

[23] we could consider asking

> $\boxdot'$ if $u \in \mathscr{P}^-(n)$ and $p \in \mathscr{S}^{\mathrm{bs}}(M_{n,u}^{\mathbf{d}_1})$ then (here (and in part (5A)) as we concentrate on the case "$\mathfrak{K}_{\mathfrak{s}}$ is categorical in $\lambda_{\mathfrak{s}}$", the difference is minor; not so in subsequent works) the cardinality of $\{c \in \mathbf{J}_{n,u}^{\mathbf{d}_1} : \mathbf{tp}_{\mathfrak{s}}(c, M_{n,u}^{\mathbf{d}_1}, M_{un}^{\mathbf{d}_1}) \pm h_{n,u}^{\mathbf{d}_1}(p)\}$ is equal to the cardinality of $\{c \in \mathbf{J}_{n,u}^{\mathbf{d}_2} : \mathbf{tp}_{\mathfrak{s}}(c, M_{n,u}^{\mathbf{d}_2}, M_n^{\mathbf{d}_2}) \pm h_{n,u}^{\mathbf{d}_2}(f_u(p))\}$.

then for some $M \leq_{\mathfrak{s}} M_n^{\mathbf{d}_2}$ the type p does not fork over M and $\dim(p, M) = \lambda$.

6) Also in (1)-(5),(7) we may restrict ourselves to one brimmed$^\ell$ expanded stable $(\lambda, \mathscr{P}^-(n))$-system or $(\lambda, \mathscr{P}(n))$-system $\mathbf{d}$, i.e., consider the property as a property of $\mathbf{d}$; may omit brimmed$^\ell$ so in this case the brimmed$^\ell$ may refer to only $u = n$!
7) We say that $\mathfrak{s}$ has the brimmed$^\ell$ strong prime (λ, n)-existence/weak prime (λ, n)-existence if for every brimmed$^\ell$ expanded stable $(\lambda, \mathscr{P}^-(n))$-system $\mathbf{d}_1$ there is an expanded stable $(\lambda, \mathscr{P}(n))$-system $\mathbf{d}_2$ with $M_n^{\mathbf{d}_2}$ being brimmed when $\ell \geq 2$ and which is strongly prime$^\ell$/weakly prime$^\ell$ over $\mathbf{d}_1$, (note: $\mathbf{d}_2$ is only reduced in n).
8) Writing "$\ldots(< n)\ldots$ property" we mean "$\ldots m \ldots$ property for every $m < n$".

Our main aim is to show that when each $\mathfrak{s}^{+n}$ is successful and the sequence $\langle 2^{\lambda_{\mathfrak{s}}^{+n}} : n < \omega\rangle$ is increasing then every one of those properties is satisfied by $\mathfrak{s}^{+m}$ for $m < \omega$ large enough and we say $\mathfrak{s}^{+n}$ is k-beautiful when all these properties are satisfied for $m \leq k$.

12.20 Claim. *1) For $n = 0, 1, 2$, the frame $\mathfrak{s}$ has the brimmed$^\ell$ weak (λ, n)-existence property for $\ell = 1, 2, 3, 4$.*
2) For $n = 0, 1, 2$, the frame $\mathfrak{s}$ has the brimmed$^\ell$ weak (λ, n)-uniqueness property for $\ell = 1, 2, 3, 4$.
3) For $n = 0, 1$ the frame $\mathfrak{s}$ has the brimmed$^\ell$ strong (λ, n)-existence property for $\ell = 1, 2, 3, 4$.
4) For $n = 0, 1$, the frame $\mathfrak{s}$ has the brimmed$^\ell$/strong (λ, n)-primeness property for $\ell = 1, 2, 3, 4$.
5) For $n = 0, 1$ the frame $\mathfrak{s}$ has the brimmed$^\ell$/strong (λ, n)-prime existence property for $\ell = 1, 2, 3, 4$.
6) If $\mathfrak{s}$ has the brimmed$^\ell$ strong (λ, n)-existence property and the brimmed$^\ell$ strong (λ, n)-primeness/weak (λ, n)-primeness property, then it has the brimmed$^\ell$ strongly prime (λ, n)-existence/weak prime (λ, n)-existence property.

Proof. 1) If $n = 0$ this is trivial, and if $n = 1$ this holds by 5.8(6). Lastly, if $n = 2$ by II.6.20(3); "the existence of stable amalgamation"

we can find a $(\lambda, \mathscr{P}(n))$-system $\mathbf{m}^+$ such that $\mathbf{m}^+ \restriction \mathscr{P}^-(n) = \mathbf{m}^{\mathbf{d}}$ and $\mathrm{NF}_{\mathfrak{s}}(M^{\mathbf{m}^+}_{n,\emptyset}, M^{\mathbf{m}^+}_{n,\{0\}}, M^{\mathbf{m}^+}_{n,\{1\}}, M^{\mathbf{m}^+}_{n,\{0,1\}})$ hence we can extend $M^{\mathbf{m}^+}_n$ and choose $\mathbf{J}_{n,u}$ for $u \subset n$ to get $\mathbf{d}^+$ as required.
2) For $n = 2$, by the uniqueness of $\mathrm{NF}_{\mathfrak{s}}$-amalgamation, see II.6.22.
3) If $n = 0$, this is clear, for $n = 1$ this is the existence theorem for $K^{3,\mathrm{mx}}_{\mathfrak{s}}$, see 12.6(4).
4), 5) Easy, too.
6) Let $\mathbf{d}$ be a brimmed$^\ell$ $(\lambda, \mathscr{P}^-(n))$-system. By the brimmed$^\ell$ strong (λ, n)-existence property there is an expanded stable $(\lambda, \mathscr{P}(n))$-system $\mathbf{d}^*$, reduced in n such that $\mathbf{d}^* \restriction \mathscr{P}^-(n) = \mathbf{d}$ and $M^{\mathbf{d}^*}_n$ is brimmed$^\ell$. By the brimmed$^\ell$ weakly/strongly (λ, n)-primeness property $\mathbf{d}^*$ is weakly/strongly prime$^\ell$ so we are done.

$\square_{12.20}$

12.21 Claim. *Let* $\mathbf{d}$ *be an expanded stable* (λ, I)*-system.*

The following properties of $\mathbf{d}$ *are actually properties of* $\mathbf{m}$*, that is, their satisfaction depends just on* $\mathbf{m}^{\mathbf{d}}$

$(A)_{\bar f}$ $\bar f$ *is a stable* $\leq^I_{\mathfrak{s}}$*-embedding of* $\mathbf{d}$ *into* M

$(B)_{u,\ell}$ $\mathbf{d}$ *is brimmed*$^\ell$ *at* u *where* $\ell = 1,2,3,4$ *and* $u \in I$

(C) $\mathbf{d}$ *has the brimmed*$^\ell$ *weak/strong uniqueness/existence property*

(D) $\mathbf{d}$ *has the brimmed*$^\ell$ *strong prime/weak prime existence property*

(E) $\mathbf{d}$ *has the brimmed*$^\ell$ *weak primeness/strong primeness property.*

Proof. The least easy case is that replacing $\mathbf{J}^{\mathbf{d}}_{v,u}$ by similar $\mathbf{J}'_{v,u}$ does not make a difference which is proved as in 12.11(6). $\square_{12.21}$

<u>12.22 Discussion</u>: Why do we define the "weak primeness", "weak prime existence" properties?

The problem arises in 12.36. Assume $\mathbf{d}_0$ is a brimmed^3 expanded stable $(\lambda, \mathscr{P}(n))$-system, $\mathbf{d}_1$ is an expanded stable $(\lambda, \mathscr{P}(n))$-system, $\mathbf{d}_1 \restriction \mathscr{P}^-(n) = \mathbf{d}_0 \restriction \mathscr{P}^-(n)$, $\mathbf{d}_1$ is reduced in n, $M^{\mathbf{d}_1}_n \leq_{\mathfrak{s}} M^{\mathbf{d}_0}_n$ and $h^{\mathbf{d}_1}_{n,u} = h^{\mathbf{d}_0}_{n,u}$ for $u \subset n$. Is $M^{\mathbf{d}_0}$ really $(\lambda, *)$-brimmed over

$M_n^{\mathbf{d}_1}$? If $\mathfrak{s}$ has the NDOP (and $n \geq 2$) yes, but in general for $p \in \mathscr{S}_{\mathfrak{s}}^{\rm bs}(M_n^{\mathbf{d}_1})$ which is $\perp h_{n,u}^{\mathbf{d}_1}(M_u^{\mathbf{d}_1})$ for every $u \subset n$, we do not know that $\dim(p, M_n^{\mathbf{d}_0}) = \lambda_{\mathfrak{s}}$. This motivates the definition of weak primeness. But for $\ell = 3$, the weak and strong versions are equivalent.

12.23 Claim. *1) Let $\ell \in \{1,2,3,4\}$. Assume $I_1 \subseteq I_2$ and $\mathfrak{s}$ has the brimmed$^\ell$ weak $(\lambda, |u|)$-existence property whenever $u \in I_2 \backslash I_1$. Then for any brimmed$^\ell$ expanded stable (λ, I_1)-system $\mathbf{d}_1$ there is a brimmed$^\ell$ expanded stable $(\lambda, I_2, \mathfrak{s})$-system $\mathbf{d}_2$ satisfying $\mathbf{d}_2 \restriction I_1 = \mathbf{d}_1$.*
2) Let $\ell \in \{3,4\}$. Assume that for any $m < n, \mathfrak{s}$ has the brimmed$^\ell$ strong (λ, m)-uniqueness property. Then for any two brimmed$^\ell$ expanded stable $(\lambda, \mathscr{P}^-(n))$-systems $\mathbf{d}_1, \mathbf{d}_2$, the systems $\mathbf{m}[\mathbf{d}_1], \mathbf{m}[\mathbf{d}_2]$ are isomorphic. Similarly for (λ, I) if $u \in I \Rightarrow |u| < n$.
3) Let $\ell \in \{3,4\}$; if $\mathbf{d}_k$ is a brimmed$^\ell$ expanded stable (λ, I_2)-system, for $k = 1,2$ and $I_1 \subseteq I_2$ and $\mathfrak{s}$ has the brimmed$^\ell$ strong $(\lambda, |u|)$-uniqueness whenever $u \in I_2 \backslash I_1$ and $\bar{f} = \langle f_u : u \in I_1 \rangle$ is an isomorphism from $\mathbf{m}[\mathbf{d}_1 \restriction I_1]$ onto $\mathbf{s}[\mathbf{d}_2 \restriction I_1]$, then we can find $\bar{f}'$, an isomorphism from $\mathbf{m}[\mathbf{d}_1]$ onto $\mathbf{m}[\mathbf{d}_2]$ such that $\bar{f}' \restriction I_1 = \bar{f}$.

Proof. Natural (and part (2) is a special case of part (3) which is proved by induction on $|I_2 \backslash I_1|$). $\square_{12.23}$

Remark. If we like in 12.23(2) to deduce also for $\ell = 1, 2$ that $\mathbf{d}_1, \mathbf{d}_2$ are isomorphic, we should change Definition 12.19 accordingly.

12.24 Conclusion. Assume $\ell \in \{3,4\}$ and

(*a*) $\mathfrak{s}$ has the brimmed$^\ell$ strong $(\lambda, < n)$-uniqueness property
(*b*) there is a brimmed$^\ell$ stable $(\lambda, \mathscr{P}(n))$-system.

Then $\mathfrak{s}$ has the brimmed$^\ell$ weak (λ, n)-existence property.

Proof. The cases $n = 0, 1$ are trivial, see 12.20. Now by clause (b) there is a brimmed$^\ell$ expanded stable $(\lambda, \mathscr{P}(n))$-system $\mathbf{d}^*$. To prove the brimmed$^\ell$ weak (λ, n)-existence property, let $\mathbf{d}$ be a brimmed$^\ell$ expanded stable $(\lambda, \mathscr{P}^-(n))$-system. By assumption (a) and 12.23(2), the $(\lambda, \mathscr{P}^-(n))$-systems $\mathbf{m}[\mathbf{d}]$ and $\mathbf{m}[\mathbf{d}^* \restriction \mathscr{P}^-(n)]$ are isomorphic hence, by clause (c) of 12.21, without loss of generality they are equal, so $\mathbf{d}^*$ prove the existence. $\square_{12.24}$

12.25 Claim. *1) Let $\ell = 1, 2, 3, 4$. The brimmed$^\ell$ weak (λ, n)-existence property is equivalent to: for every brimmed$^\ell$ expanded stable $\mathscr{P}^-(n)$-system $\mathbf{d}$ there are $M \in \mathfrak{K}_\mathfrak{s}$ and $\bar{f} = \langle f_u : u \in \mathscr{P}^-(n)\rangle$, such that*

⊛ *$\bar{f}$ is an embedding of $\mathbf{d}$ into M which means*
- (a) *f_u is a $\leq_\mathfrak{s}$-embedding of $M^\mathbf{d}_u$ into M*
- (b) *if $u \subset v \in \mathscr{P}^-(n)$ then $f_u = f_v \circ f^\mathbf{d}_{v,u}$*
- (c) *for any $u \in \mathscr{P}^-(n)$, the set $\cup\{f_v(\mathbf{J}^\mathbf{d}_{v,u}) : v$ satisfies $u \subseteq v \in \mathscr{P}^-(n)\}$ is independent in $(f_u(M^\mathbf{d}_u), M)$, as an indexed set*[24].

1A) Similarly for "the brimmed$^\ell$ expanded stable $(\lambda, \mathscr{P}^-(n))$-system $\mathbf{d}$ has the brimmed$^\ell$ weak existence property".
2) Let $\ell = 1, 2, 3, 4$

- (a) *If $\mathfrak{s}$ has the brimmed$^\ell$ strong (λ, n)-existence property <u>then</u> $\mathfrak{s}$ has the brimmed$^\ell$ weak (λ, n)-existence property*
- (b) *if $\mathfrak{s}$ has the brimmed$^\ell$ strong (λ, n)-uniqueness property <u>then</u> $\mathfrak{s}$ has the brimmed$^\ell$ weak (λ, n)-uniqueness property*
- (c) *if $\mathfrak{s}$ has the brimmed$^\ell$ strong (λ, n)-primeness property <u>then</u> $\mathfrak{s}$ has the brimmed$^\ell$ weak (λ, n)-primeness property*
- (d) *if $\mathfrak{s}$ has the brimmed$^\ell$ strong prime (λ, n)-existence property <u>then</u> $\mathfrak{s}$ has the brimmed$^\ell$ weak prime (λ, n)-existence*
- (e) *in clauses (c),(d) we get equivalence <u>if</u> $\mathfrak{s}$ is categorical in λ.*

[24] "as index sets" this just means that $\langle\{f_v(c) : c \in \mathbf{J}^\mathbf{d}_{v,u}\} : u \subset v \in \mathscr{P}^-(n)\rangle$ is a sequence of pairwise disjoint sets

2A) Each of the clauses of (2) holds for every brimmed$^\ell$ $(\lambda, \mathscr{P}^-(n))$-system $\mathbf{d}$ *separately and*

(*f*) *if* $\mathbf{d}$ *is a brimmed$^\ell$ $(\lambda, \mathscr{P}(n))$-system and $M_n^{\mathbf{d}}$ is brimmed* <u>*then*</u> *the condition from $\odot_{\mathbf{d}}$ of 12.19(5A) holds.*

3) If $(\ell(1), \ell(2)) \in \{(1,2),(2,3),(3,4)\}$ and $\mathfrak{s}$ has the brimmed$^{\ell(1)}$ strong (λ, n)-existence property <u>*then*</u> *$\mathfrak{s}$ has the brimmed$^{\ell(2)}$ strong (λ, n)-existence property.*
3A) If $(\ell(1), \ell(2)) \in \{(2,3),(3,4)\}$ <u>*then*</u> *the same holds for the weak (λ, n)-existence property.*
4) If $(\ell(1), \ell(2)) \in \{(1,2),(2,3),(3,4))\}$ and $\mathfrak{s}$ has the brimmed$^{\ell(1)}$ weak uniqueness, <u>*then*</u> *$\mathfrak{s}$ has the brimmed$^{\ell(2)}$ weak uniqueness.*
*5)**

(*a*) *every expanded stable $(\lambda_{\mathfrak{s}}, I)$-system for $\mathfrak{s}$ is brimmed^2*

(*b*) *for each of the properties defined in Definition 12.19, the brimmed^2 version and the brimmed^1 one are equivalent.*

Proof of 12.25. 1) By part (1A).
1A) First assume that $\mathbf{d}$ has the brimmed$^\ell$ weak (λ, n)-existence property and we shall prove the condition $\circledast$ in 12.25(1). By the present assumption, we can find $\mathbf{d}'$ as in the definition 12.19(1), and we let $M = M_n^{\mathbf{d}'}, f_u = f_{n,u}^{\mathbf{d}'}$, clearly they are as required in $\circledast$.

Second assume that $\mathbf{d}$ satisfies the condition $\circledast$ from 12.25(1). So $\mathbf{d} = \mathbf{d}_0$, a brimmed$^\ell$ expanded stable $(\lambda, \mathscr{P}^-(n))$-system is given, so by our assumption there are $M, \langle f_u : u \in \mathscr{P}^-(n)\rangle$ as there. Now we define an expanded stable $(\lambda, \mathscr{P}(n))$-system $\mathbf{d}_1$ as follows: $\mathbf{d}_1 \restriction \mathscr{P}^-(n) = \mathbf{d}_0, M_n^{\mathbf{d}_1} = M$ for $u \subset n$ let $h_{n,u}^{\mathbf{d}_1} = f_u$ and let $\mathbf{J}_{n,u}^{\mathbf{d}_1}$ be a maximal subset $\mathbf{J}$ of $M \backslash \cup \{f_v(M_v) : v \subset n\}$ such that $\mathbf{J}_{v,u}^{*,\mathbf{d}_1} = \cup\{f_v(\mathbf{J}_{v,u}^{\mathbf{d}_0}) : v \text{ satisfies } u \subset v \subset n\} \cup \mathbf{J}$ is independent in $(f_u(M_u), M)$.

Note that $\langle f_{n,v}(\mathbf{J}_{v,u}^{\mathbf{d}_1}) : u \subset v \subseteq n\rangle$ are pairwise disjoint, i.e. $f_{v,n_1}(\mathbf{J}_{v_1,u_1}^{\mathbf{d}_1}), f_{n,v_2}(\mathbf{J}_{v_2,u_2}^{\mathbf{d}_1})$ are disjoint when $u_k \subset v_k \subseteq n$ for $k = 1, 2$ and $(v_1, u_1) \neq (u_2, v_2)$.
[Why? If $u_1 = u_2$ by an assumption so let $u_1 \neq u_2$, and then use super-orthogonality.]

So clearly $\mathbf{d}_1$ is as required except that in the brimmed^2 case we are missing "M is brimmed", and in the brimmed^3 case, we are missing the demand on $\mathbf{J}_{n,u}$ and in the brimmed^4 case the relevant condition. Choose M^* such that $M \leq_{\mathfrak{s}} M^*$ and M^* is $(\lambda,*)$-brimmed over M. Define the expanded $(\lambda, \mathscr{P}(n))$-system $\mathbf{d}_2$ as follows: $\mathbf{d}_2 \restriction \mathscr{P}^-(n) = \mathbf{d}_0, M_n^{\mathbf{d}_2} = M^*$ and $h_{n,u}^{\mathbf{d}_2} = h_{n,u}^{\mathbf{d}_1}$ for $u \subset n$ and lastly, for each $u \in \mathscr{P}^-(n)$ let $\mathbf{J}_{n,u}^{\mathbf{d}_2}$ be a maximal $\mathbf{J}$ such that:

- (α) $\mathbf{J}_{n,u}^{\mathbf{d}_1} \subseteq \mathbf{J} \subseteq M^* \cup \{f_u(\mathbf{J}_{v,u}^{\mathbf{d}}) : v \text{ satisfies } u \subset v \in \mathscr{P}^-(n)\}$
- (β) $c \in \mathbf{J} \Rightarrow \mathbf{tp}_{\mathfrak{s}}(c, f_u(M_u^{\mathbf{d}_0}), M^*) \in \mathscr{S}^{\text{bs}}(f_u(M_u^{\mathbf{d}_0}))$ is orthogonal to $f_w(M_w^{\mathbf{d}_0})$ for every $w \subset u$
- (γ) $\mathbf{J} \cup \bigcup\{f_{w_1}(\mathbf{J}_{w_1,u}^{\mathbf{d}}) : u \subset w_1 \in \mathscr{P}^-(n)\}$ is independent in $(f_u(M_u^{\mathbf{d}_\ell}), M^*)$
- (δ) $\mathbf{J}$ is maximal under $(\alpha) + (\beta) + (\gamma)$.

By 12.11(7) we are done.
2) By part (3).
2A) Clauses (a)-(d): Obvious.

Clause (e): Use clause (f).

Clause (f): Should be clear.
3), 4), 5) Left to the reader. $\square_{12.25}$

The following claim will give a crucial "saving" in our "spiralic going up".

12.26 Claim. *Assume that ($\ell = 3$ or just $\ell \in \{1, 2, 3, 4\}$ and $n \geq 2$ and)*

- (a) *$\mathfrak{s}$ has the brimmed$^\ell$ weak (λ, n)-uniqueness*
- (b) *$\mathbf{d}$ is an expanded stable $(\lambda, \mathscr{P}^-(n))$-system*
- (c) *$\mathbf{d} \restriction [n]^{<n-1}$ is a brimmed$^\ell$ system (but $\mathbf{d}$ not necessarily).*

Then the system $\mathbf{d}$ has weak uniqueness, i.e.

- ⊛ *if $\mathbf{d}_1, \mathbf{d}_2$ are expanded stable $(\lambda, \mathscr{P}(n))$-system satisfying $\mathbf{d}_1 \restriction \mathscr{P}^-(n) = \mathbf{d}_2 \restriction \mathscr{P}^-(n)$ then we can find (N, f) such that $M_n^{\mathbf{d}_2} \leq_{\mathfrak{s}} N$, f is a $\leq_{\mathfrak{s}}$-embedding of $M_n^{\mathbf{d}_1}$ into N and $u \subset n \Rightarrow h_{n,u}^{\mathbf{d}_2} = f \circ h_{n,u}^{\mathbf{d}_1}$.*

Proof. We prove this by induction on $k_{\mathbf{d}} = |\mathscr{P}_{\mathbf{d}}|$ where $\mathscr{P}_{\mathbf{d}} = \{v \in \mathscr{P}^-(n) : \mathbf{d}$ is not brimmed$^\ell$ in v (so $v \in [n]^{n-1}$)$\}$.

Clearly $k_{\mathbf{d}} \le \binom{n}{n-1} = n (\le |\mathscr{P}^-(n)| < 2^n)$. If $k_{\mathbf{d}} = 0$ the conclusion follows from assumption (a).

So assume that $k_{\mathbf{d}} > 0$ and choose $v_* \in \mathscr{P}_{\mathbf{d}}$. We can find M which is $(\lambda, *)$-brimmed$^\ell$ over $M^{\mathbf{d}}_{v_*}$. Next we define an expanded $(\lambda, \mathscr{P}^-(n))$-system $\mathbf{d}^+$:

$\circledast_1$ (*a*) $\mathbf{d}^+ \restriction (\mathscr{P}^-(n) \backslash \{v_*\}) = \mathbf{d} \restriction (\mathscr{P}^-(n) \backslash \{v_*\})$

(*b*) $h^{\mathbf{d}^+}_{v_*,u} = h^{\mathbf{d}}_{v_*,u}$ if $u \subset v_*$

(*c*) $M^{\mathbf{d}^+}_{v_*} = M$

(*d*) $\mathbf{J}^{\mathbf{d}^+}_{v,u} = \mathbf{J}^{\mathbf{d}}_{v,u}$ if $u \subset v \in \mathscr{P}^-(n) \backslash \{v_*\}$

(*e*) $\mathbf{J}^{\mathbf{d}^+}_{v_*,u}$, for $u \subset v_*$, is a maximal subset $\mathbf{J}$ of $\{c \in M \backslash \cup \{M^{\mathbf{d}}_{v_*,u} : u \subset v_*\} : \mathbf{tp}(c, M^{\mathbf{d}}_{v_*,u}, M) \in \mathscr{S}^{\mathrm{bs}}_{\mathfrak{s}}(M_{v,u})$ is orthogonal to $M_{v_*,w}$ when $w \subset u\}$ such that $\mathbf{J}$ is independent in $(M_{v_*,u}, M)$ and $\mathbf{J} \supseteq \mathbf{J}^{\mathbf{d}}_{v_*,u}$.

It is easy to check that

(∗) $\mathbf{d}^+$ is a stable $(\lambda, \mathscr{P}^-(n))$-system and $\mathbf{d}^+ \restriction [n]^{<n-1}$ is brimmed$^\ell$ and $k_{\mathbf{d}^+} = k_{\mathbf{d}} - 1$.

Now let $\mathbf{d}_1, \mathbf{d}_2$ be as in the assumption of $\circledast$. Next for $k = 1, 2$ by the existence of stable amalgamation there is a pair (N_k, f_k) such that $M^{\mathbf{d}_k}_n \le_{\mathfrak{s}} N_k$, f_k is a $\le_{\mathfrak{s}}$-embedding of $M = M^{\mathbf{d}^+}_{v_*}$ into N_k extending $f^{\mathbf{d}_k}_{n,v_*}$ and $\mathrm{NF}_{\mathfrak{s}}(M^{\mathbf{d}_k}_{n,v_*}, M^{\mathbf{d}_k}_n, f_k(M), N_k)$ holds.

By renaming without loss of generality $f_1 = f_2 = \mathrm{id}_M$. We now define an expanded stable $(\lambda, \mathscr{P}(n))$-system $\mathbf{d}^+_k$

$\circledast^2_k$ (*a*) $\mathbf{d}^+_k \restriction \mathscr{P}^-(n) = \mathbf{d}^+$

(*b*) $M^{\mathbf{d}^+_k}_n = N_k$.

(*c*) $h^{\mathbf{d}^+_k}_{n,u} = h^{\mathbf{d}_k}_{n,u}$ for $u \in \mathscr{P}^-(n) \backslash \{v_*\}$

(*d*) $h^{\mathbf{d}^+}_{n,v_*} = f_k$

(*e*) $\mathbf{J}^{\mathbf{d}^+_k}_{n,u}$ for $u \subset n$ are defined as in previous case, i.e. it is a maximal subset of $\{c \in N_k : \mathbf{tp}_{\mathfrak{s}}(c, M^{\mathbf{d}^+}_{n,u}, N_k)$ is orthogonal to $M^{\mathbf{d}^+}_{n,w}$ for every $w \subset u$ but $c \notin M^{\mathbf{d}^+}_{n,v}$ if $u \subset v \subset n\}$ such that $\mathbf{J}^{\mathbf{d}^+_k}_{n,u} \cup \bigcup\{\mathbf{J}^{\mathbf{d}^+}_{n,v,u} : v$ satisfies $u \subset v \subset n\}$ is independent in $(M^{\mathbf{d}^+}_{n,u}, N_k)$.

Easily

$\circledast^3$ $\mathbf{d}^+_k$ is a stable $(\lambda, \mathscr{P}(n))$-system and $\mathbf{d}^+_k \restriction \mathscr{P}^-(n) = \mathbf{d}^+$.

Now we use the induction hypothesis on $\mathbf{d}^+$, justified by $(*)$. So we can find (f', N') such that $N^{\mathbf{d}^+_2}_n \le_{\mathfrak{s}} N'$ and f' is a $\le_{\mathfrak{s}}$-embedding of $N^{\mathbf{d}_1}_n$ into N' such that $u \subset n \Rightarrow f^{\mathbf{d}^+_2}_{n,u} = f' \circ f^{\mathbf{d}^+_1}_{n,u}$.
So we are done. $\square_{12.26}$

12.27 Claim. *Assume that $I = \mathscr{P}^-(n)$ and (a) + (b) and: $(c)_1$ or $(c)_2$ where;*

(a) $\mathbf{m}^k$ *is a brimmed$^\ell$ stable (λ, I)-system for $k = 1, 2$*

(b) $\mathbf{m}^1 \restriction J = \mathbf{m}^2 \restriction J$ *where $J := \{u \in I : (\exists v \in I)(u \subset v)\}$*

$(c)_1$ *if $u \in I \backslash J$ then $\mathbf{m}^\ell \restriction \mathscr{P}^-(u)$ has the brimmed$^\ell$ weak uniqueness property*

$(c)_2$ *if $v \in I \backslash J$ then $u \subset v \Rightarrow h^{\mathbf{m}^1}_{v,u} = h^{\mathbf{m}^2}_{v,u}$ and $M^{\mathbf{m}^\ell}_v \le_{\mathfrak{s}} M^{\mathbf{m}^{3-\ell}}_v$ for some $\ell \in \{1, 2\}$.*

Then

(α) $\mathbf{m}^1$ *has the weak existence property iff $\mathbf{m}^2$ has the weak existence property*

(β) $\mathbf{m}^1$ *has the weak uniqueness property iff $\mathbf{m}^2$ has the weak uniqueness property.*

Proof. Similar to the proof of 12.26, we prove by induction on $k(\mathbf{m}^1, \mathbf{m}^2) = (\{u \in J : M^{\mathbf{m}^1}_u \ne M^{\mathbf{m}^2}_u\})$.

So without loss of generality $k(\mathbf{m}^1, \mathbf{m}^2) = 1$. When we use $(c)_2$ the proof is the same. When we use $(c)_1$ we have to take care of making the images of the $\mathbf{J}$ in $\mathbf{d}^\ell$ expanding $\mathbf{m}^\ell$ in the big model being independent using 12.11(6). $\square_{12.27}$

In 12.18 we have proved actually some things on $\mathfrak{s}^+$ concerning Definition 12.19. [Why have we ignored $\ell = 1$? As for $\mathfrak{s}^+$, brimmed^1 $\Rightarrow$ brimmed^2.]

12.28 Conclusion. Let $\ell \in \{0, 2, 3\}$ and assume that $\mathfrak{s}^+$ is successful hence $\mathfrak{s}^+$ satisfies the demands in 12.3 by 12.5(2). If for $\mathfrak{s}^+$ there is a brimmed$^\ell$ stable $\mathscr{P}(n)$-system, <u>then</u> for $\mathfrak{s}$ there is a brimmed$^\ell$ stable $\mathscr{P}(n+1)$-system.

Remark. For $\ell = 0, 2$ we can find trivial examples.

Proof. By 12.18. $\square_{12.28}$

12.29 Claim. *Assume that $\mathbf{d}_k$ is an expanded stable (λ, I)-system for $k = 1, 2$ and $\bar{f}$ is an embedding of $\mathbf{d}_1$ into $\mathbf{d}_2$, see Definition 12.9(1A).*
1) If $I = \mathscr{P}^-(n)$ and $\mathbf{d}_2$ has the weak existence property <u>then</u> $\mathbf{d}_1$ has the weak existence property.
2) If $I = \mathscr{P}^-(n)$ and $\bar{f} \restriction [n]^{<n-1}$ is an isomorphism from $\mathbf{d}_1 \restriction [n]^{<n-1}$ onto $\mathbf{d}_2 \restriction [n]^{<n-1}$ and $\mathbf{d}_2$ has the weak uniqueness property <u>then</u> $\mathbf{d}_1$ has weak uniqueness property.

Proof. 1) Easy (and was used inside the proof of 12.26).
2) Very similar to the proof of 12.26 (and we can use part (1)). $\square_{12.29}$

12.30 Lemma. *[$(f)^*$ of 12.1(2)]*
Let $\ell = 3$ and $n \geq 2$. Assume $2^\lambda < 2^{\lambda^+}$ and:

- (*a*) *$\mathfrak{s}^+$ is successful hence $\mathfrak{s}^+$ has the properties required in 12.3(1), (2)*
- (*b*) *$\mathfrak{s}$ has the brimmed$^\ell$ weak $(\lambda, \leq n+1)$-existence property*
- (*c*) *$\mathfrak{s}$ has the brimmed$^\ell$ strong $(\lambda, \leq n)$-uniqueness property*
- (*d*) *$\mathfrak{s}$ does not have the brimmed$^\ell$ weak $(\lambda, n+1)$-uniqueness property.*

Then $\mathfrak{s}^+$ *does not have the brimmed*$^\ell$ *strong* (λ^+, n)*-uniqueness property.*

Remark. Of course, it would be better to have "strong" in clause (d) of the assumption and it would be better to have weak in the conclusion. Still we can prove a slightly stronger claim.

A stronger variant of 12.30 is

12.31 Claim. *[(f)** of 12.1(2)]*
1) In 12.30 we can replace clause (c) by $(c)_1^- + (c)_2^-$ *(which obviously follows from it) where (recalling* $\ell = 3$*)*

- $(c)_1^-$ $\mathfrak{s}$ *has the brimmed*$^\ell$ *strong* $(\lambda, < n)$*-uniqueness property*
- $(c)_2^-$ $\mathfrak{s}$ *has the brimmed*$^\ell$ *weak* (λ, n)*-uniqueness property.*

2) We can strengthen the conclusion to: $\mathfrak{s}^+$ *fails the brimmed*$^\ell$ *weak* (λ^+, n)*-uniqueness property; used in the proof of* $(*)_4$ *in 12.37.*

Proof of 12.30. We by induction on $\alpha < \lambda^+$ choose $\mathbf{m}_\eta$ for every $\eta \in {}^\alpha 2$ such that:

- (α) $\mathbf{m}_\eta$ is a normal brimmed$^\ell$ stable $(\lambda, \mathscr{P}(n))$-system
- (β) the universe of $M_n^{\mathbf{m}_\eta}$ is the ordinal $\gamma_{\ell g(\eta)} = \gamma_\eta = \lambda \times (1 + \ell g(\eta)) < \lambda^+$
- (γ) the sequence $\langle M_u^{\mathbf{m}_{\eta \restriction \gamma}} : \gamma \leq \alpha \rangle$ is $\leq_{\mathfrak{s}}$-increasing continuous for $u \in \mathscr{P}(n)$
- (δ) if $\alpha = \beta + 1$ then $\mathbf{m}_\eta^* := \mathbf{m}_{\eta \restriction \beta} *_{\mathscr{P}(n+1)} \mathbf{m}_\eta$ is a brimmed$^\ell$ stable $(\lambda, \mathscr{P}(n+1))$-system
- (ε) if $\alpha = \beta + 1, \nu \in {}^\beta 2$, <u>then</u>: $\mathbf{m}^*_{\nu \char`\^ <0>} \restriction \mathscr{P}^-(n+1) = \mathbf{m}^*_{\nu \char`\^ <1>} \restriction \mathscr{P}^-(n+1)$
- (ζ) if $\alpha = \beta + 1$ and $\nu \in {}^\beta 2$ <u>then</u> for no f, N do we have: f is an $\leq_{\mathfrak{s}}$-embedding of $M_{n+1}^{\mathbf{m}^*_{\nu \char`\^ <0>}}$ into some $N$ for which $M_{n+1}^{\mathbf{s}^*_{\nu \char`\^ <1>}} \leq_{\mathfrak{s}} N$ and f is the identity on $M_u^{\mathbf{m}^*_{\nu \char`\^ <0>}} = M_u^{\mathbf{m}^*_{\nu \char`\^ <1>}}$ for $u \in \mathscr{P}^-(n+1)$

(η) if $\nu_1, \nu_2 \in {}^{\alpha}2$ then $\mathbf{m}_{\nu_1} \restriction \mathscr{P}^-(n) = \mathbf{m}_{\nu_2} \restriction \mathscr{P}^-(n)$ (this strengthens[25] clause (ε)).

Now

$(*)_1$ we can carry the induction
[Why? For $\alpha = 0$ trivial. For $\alpha = \beta + 1$ by clause (d) of the assumption there are normal brimmed$^\ell$ stable $\mathscr{P}(n+1)$-system $\mathbf{m}', \mathbf{m}''$ with $\mathbf{m}' \restriction \mathscr{P}^-(n+1) = \mathbf{m}'' \restriction \mathscr{P}^-(n+1)$ as in clause (ζ), i.e., for no (N, f) do we have $M_n^{\mathbf{m}''} \leq_{\mathfrak{s}} N$ and f is a $\leq_{\mathfrak{s}}$-embedding of $M_{n+1}^{\mathbf{m}'}$ into N which is the identity on $M_u^{\mathbf{m}'}$ for every $u \subset n+1$. Now as we are proving 12.30, by assumption (c) and claim 12.23(2) and renaming we have $\mathbf{m}_{\eta\char`^<0>}, \mathbf{m}_{\eta\char`^<1>}$ as required in clause (ε) and (ζ) of $\circledast$ and by renaming we have (β).

What about clause (η)? We redo the above. We can first choose $\bar{f} = \langle f_u : u \in \mathscr{P}^-(n)\rangle$, an isomorphism from $\mathbf{m}' \restriction \mathscr{P}^-(n) = \mathbf{m}'' \restriction \mathscr{P}^-(n)$ onto $\mathbf{m}_\eta \restriction \mathscr{P}^-(n)$ for every $\eta \in {}^{\beta}2$ recalling that clause (η) holds for β. Second, we choose f^*, a one to one mapping from $\cup\{M_u^{\mathbf{m}'} : u \in \mathscr{P}^-(n+1)\backslash\{n\}\}$ into $\gamma_\beta \cup \{\gamma_\beta + 2i : i < \lambda\}$. By f^* we define $\mathbf{m}_\eta \restriction (\mathscr{P}^-(n+1)\backslash\{n\})$ for $\eta \in {}^{\alpha}2$ which does not depend on η. So $\mathbf{m}_\eta \restriction \mathscr{P}^-(n+1)$ is defined for every $\eta \in {}^{\alpha}2$ and $\nu \in {}^{\beta}2 \Rightarrow \mathbf{m}_{\nu\char`^<0>} \restriction \mathscr{P}^-(n+1) = \mathbf{m}_{\nu\char`^<1>} \restriction \mathscr{P}^-(n+1)$. Now by the assumption (c) of 12.30, the system $\mathbf{m}_{\nu\char`^<0>} \restriction \mathscr{P}^-(n+1)$ fails the weak uniqueness property so we can choose $\mathbf{m}_{\nu\char`^<0>}, \mathbf{m}_{\nu\char`^<1>}$ as before.

Note that $M_v^{\mathbf{m}_\eta} \backslash \cup \{M_u^{\mathbf{m}_\eta} : u \subset v\}$ has cardinality λ in all relevant cases as the $J_{v,\emptyset}^{\mathbf{m}}$ witness if $v \neq \emptyset$ and trivially if $v = \emptyset$.

(For proving 12.31 more work will be needed).

For α limit $\mathbf{m}_\eta := \bigcup\limits_{\beta<\alpha} \mathbf{m}_{\eta\restriction\beta}$ is a stable $(\lambda, \mathscr{P}(n))$-system by Claim 12.17(2); moreover is brimmed$^\ell$ by 12.17(2).]

[25] By looking better at the weak diamond, by a less natural application we can avoid using clause (η).

For $\eta \in {}^{\lambda^+}2$ we define a normal $(\lambda^+, \mathscr{P}(n))$-system for $\mathfrak{s}^+$ called $\mathbf{m}_\eta$ by $M_u^{\mathbf{m}_\eta} = \cup\{M_u^{\mathbf{m}_{\eta\restriction\alpha}} : \alpha < \lambda^+\}$, clearly:

$(*)_2$ $\mathbf{m}_\eta$ is really a $(\lambda^+, \mathscr{P}(n))$-system for the frame $\mathfrak{s}^+$ (noting that $M_u^{\mathbf{m}_\eta} \in K_{\mathfrak{s}(+)}$ as it belongs to $K^{\mathfrak{s}}$ and is saturated over $\lambda_{\mathfrak{s}}$ because by clause (δ) by 12.11(10) as $(f)^{**}$ of 12.1 holds we know that $M_u^{\mathbf{m}_{\eta\restriction(\alpha+1)}}$ is $(\lambda, *)$-brimmed over $M_u^{\mathbf{m}_{\eta\restriction\alpha}}$ for every $\alpha < \lambda_{\mathfrak{s}}^+$; also $\leq_{\mathfrak{s}(+)} = \leq_{\mathfrak{K}[\mathfrak{s}]} \restriction K_{\mathfrak{s}(+}$ as $\mathfrak{s}$ is good$^+$ and successful)

$(*)_3$ $\mathbf{m}_\eta$ is a brimmed$^\ell$ stable $(\lambda^+, \mathscr{P}(n))$-system for $\mathfrak{s}(+)$
[why? by 12.17(3).]

$(*)_4$ $\mathbf{m}_\eta \restriction \mathscr{P}^-(n)$ is the same for all $\eta \in {}^{\lambda^+}2$ call it $\mathbf{m}$.
[Why? By Clause (η) of $\circledast$.]

Let $\rho \in {}^{\lambda^+}2$. To finish the proof of "$\mathfrak{s}^+$ fail the strong (λ, n)-uniqueness property" it is enough to find $\eta \in {}^{\lambda^+}2$ such that $h_* = \cup\{\mathrm{id}_{M_u^{\mathbf{t}}} : u \subset n\}$ cannot be extended to an isomorphism from $M_n^{\mathbf{m}_\rho}$ onto $M_n^{\mathbf{m}_\eta}$; toward contradiction assume that f_η is such an isomorphism for every $\eta \in {}^{\lambda^+}2$. By the weak diamond, (see I.0.5) for some $\eta_0, \eta_1 \in {}^{\lambda^+}2$ and $\delta < \lambda^+$ we have $\nu = \eta_\ell \restriction \delta, \nu\hat{}\langle \ell \rangle \triangleleft \eta_\ell$ and $f_{\eta_1} \restriction M_n^{\mathbf{m}_\nu} = f_{\eta_2} \restriction M_n^{\mathbf{m}_\nu}$. Clearly we get contradiction to clause (ζ) in the construction. $\square_{12.30}$

Proof of 12.31. 1) In the proof of 12.30 there one point in which the proofs differ. We are given the brimmed$^\ell$ stable $(\lambda, \mathscr{P}(n))$-systems $\mathbf{m}_\eta$ for $\eta \in {}^\beta 2$ and we know that there are normal brimmed$^\ell$ stable $\mathscr{P}(n+1)$-system $\mathbf{m}', \mathbf{m}''$ such that $\mathbf{m}' \restriction \mathscr{P}^-(n+1) = \mathbf{m}'' \restriction \mathscr{P}^-(n+1)$ but there is no $\leq_{\mathfrak{s}}$-embedding of $M_{n+1}^{\mathbf{m}'}$ into any $N, M_{n+1}^{\mathbf{m}''} \leq_{\mathfrak{s}} N$ over $\bigcup\limits_{u \subset n+1} M_u^{\mathbf{m}'}$. By the amount of uniqueness we have, i.e. by assumption $(c)_1^-$ without loss of generality $\mathbf{m}' \restriction \mathscr{P}^-(n) = \mathbf{m}_\eta \restriction \mathscr{P}^-(n)$ (for every $\eta \in {}^\beta 2$, hence also $\mathbf{m}'' \restriction \mathscr{P}^-(n) = \mathbf{m}_\eta$). Without loss of generality the universe of $M_{n+1}^{\mathbf{m}'}$ and of $M_{n+1}^{\mathbf{m}''}$ is $\gamma_\beta + \lambda$ (recall 12.11(7)) and, of course, the universe of $M_n^{\mathbf{m}'} = M_n^{\mathbf{m}''}$ is γ_η.

Now we define $\mathbf{m}_\eta^*$, a stable $(\lambda, \mathscr{P}^-(n+1))$-system by $\mathbf{m}_\eta^* \restriction (\mathscr{P}(n+1) \backslash \{n, n+1\}) = \mathbf{m}' \restriction (\mathscr{P}(n+1) \backslash \{n, n+1\}) = \mathbf{m}'' \restriction$

$(\mathscr{P}^-(n+1)\backslash\{n,n+1\})$ and $M_n^{\mathbf{m}^*_\eta} = M_n^{\mathbf{m}_\eta}$. Clearly $\mathbf{m}^*_\eta$ is stable and brimmed$^\ell$. Without loss of generality $\mathbf{m}^*_\eta$ is the same for all $\eta \in {}^\alpha 2$.
Now we can apply 12.27, clause (β), the version with $(c)_1$ there (because its assumption $(c)_1$ holds by $(c)_2^-$ here), so $\mathbf{m}' \restriction \mathscr{P}^-(n+1)$ has the brimmed$^\ell$ weak uniqueness iff $\mathbf{m}^*_\eta$ has it. So as $\mathbf{m}' \restriction \mathscr{P}^-(n+1)$ fails the brimmed$^\ell$ weak uniqueness, also $\mathbf{m}^*_\eta$ fails it. Hence we can find a stable $(\lambda, \mathscr{P}^*(n+1))$-systems $\mathbf{m}'_\eta, \mathbf{m}''_\eta$ witnessing it, so $\mathbf{m}'_\eta \restriction \mathscr{P}^-(n+1) = \mathbf{m}^*_\eta = \mathbf{m}''_\eta \restriction \mathscr{P}^-(n+1)$. By renaming we take care of clause (β) of $\circledast$ (we use freely 12.11).
2) We choose in addition to $\mathbf{m}_\eta$ also N_η such that

$\circledast$ $(\alpha), (\gamma) - (\eta)$ as in the proof

$(\beta)'$ N_η is brimmed over $M_\eta^{\mathbf{m}_n}$, the universe of N_η is the ordinal $\gamma_{\ell g(\eta)} < \lambda$ (instead of (β))

(θ) if $\nu \triangleleft \eta$ then $\mathrm{NF}_{\mathfrak{s}}(M_n^{\mathbf{m}_\nu}, N_\nu, M_n^{\mathbf{m}_\eta}, N_\eta)$ and N_η is brimmed over $M_n^{\mathbf{m}_\eta} \cup N_\nu$.

In the end for $\eta \in {}^{\lambda^+}2$ we define also $N_\eta = \cup\{N_{\eta\restriction\alpha} : \alpha < \lambda^+\}$ hence $M_n^{\mathbf{m}_\eta} \le_{\mathfrak{s}(+)} N_\eta$ and N_η is $(\lambda^+, *)$-brimmed over $M_n^{\mathbf{m}_\eta}$ by Definition II.7.4(1) and 1.25.
So assume that $\rho, \eta \in {}^{\lambda^+}2$, $M_n^{\mathbf{m}_\eta} \le_{\mathfrak{s}(+)} N$ and f is a $\le_{\mathfrak{s}(+)}$-embedding of $M_n^{\mathbf{m}_\rho}$ into N over $\cup\{M_u^{\mathbf{m}_\rho} : u \subset n\}$. Without loss of generality $N \le_{\mathfrak{s}} N_\eta$. We can find N'_η such that $N_\eta <_{\mathfrak{s}(+)} N'_\eta, N'_\eta$ is brimmed over N_η (for $\mathfrak{s}^+$). There is an isomorphism f_1 from N_ρ onto N'_η extending f and there is an isomorphism f_2 from N'_η onto N_η over $M_n^{\mathbf{m}_\eta}$. So $f_2 \circ f_1$ is an isomorphism from N_ρ onto N_η over $\cup\{M_u^{\mathbf{m}_\rho} : u \in \mathscr{P}^-(n)\}$. The rest should be clear. $\square_{12.31}$

12.32 Claim. *[(f)** of 12.3] 1) Assume $\ell = 3$ and $n \ge 1$ and*

(a) $\mathfrak{s}$ has the brimmed$^\ell$ strong $(\lambda, \le n)$-existence property

(b) $\mathfrak{s}$ has the brimmed$^\ell$ weak $(\lambda, \le n)$-primeness property

Then there is an expanded stable $\mathscr{P}(n+1)$-system $\mathbf{d}$ reduced at $n+1$ such that $\mathbf{d} \restriction \mathscr{P}^-(n)$ is brimmed$^\ell$.

2) If in addition clause (c) below holds then $\mathfrak{s}$ has the brimmed$^\ell$ strong $(\lambda, n+1)$-existence property where

(c) *$\mathfrak{s}$ has the brimmed$^\ell$ strong $(\lambda, \leq n)$-uniqueness property.*

Discussion: What is the aim of 12.32? Obviously it is to get the brimmed^3 strong $(\lambda, \leq n+1)$-existence property. As it happens that there is a unique brimmed^3 stable $(\lambda, \mathscr{P}^-(n+1))$-system $\mathbf{m}$, it is enough to find one expanded stable $(\lambda, \mathscr{P}(n+1))$-system $\mathbf{d}$ reduced at $n+1$ such that $\mathbf{d} \restriction \mathscr{P}^-(n+1)$ is brimmed^3. We construct a normal such $\mathbf{d}$ by first choosing $(M^{\mathbf{d}}_\emptyset, M^{\mathbf{d}}_{n+1})$, and then choosing $\mathbf{d} \restriction [n+1]^{\leq k}$ by induction on $k = 1, \ldots, n$. Note that for $k = n+1$ we have nothing to do in the proof. There are no $\mathbf{J}^{\mathbf{d}}_{n+1,u}$'s for $u \subset n+1$ because our intention is that $\mathbf{d}$ is reduced in $n+1$.

Of course we have various problems, The most transparent one is how to take care that for $u \in [n+1]^k$, on the one hand we will have large enough independent sets $\mathbf{J}_{u,w} = \mathbf{J}^{\mathbf{d}}_{u,w}$ when $w \subset u \subset [n+1]^k$ and, on the other hand, that when we finish, the set $\cup\{\mathbf{J}_{w,u} : u \subset w \subset n\}$ is a maximal subset of $\{c \in M^{\mathbf{d}}_{n+1} : \mathbf{tp}_{\mathfrak{s}}(c, M^{\mathbf{d}}_u, M^{\mathbf{d}}_{n+1}) \in \mathscr{S}^{\mathrm{bs}}_{\mathfrak{s}}(M^{\mathbf{d}}_u)$ is $\perp M^{\mathbf{d}}_w$ for $w \subset u\}$ which is independent in $(M^{\mathbf{d}}_u, M^{\mathbf{d}}_{n+1})$. The solution is to choose $\langle \mathbf{J}_{w,u} : u \subset w \subset n\rangle$ in the $k = |u|$-th stage.

But if the reader has glanced on/peeped into the proof, he may have noticed that we do not choose the $\langle \mathbf{J}_{w,u} : u \subset w \subset n\rangle$ but $\mathbf{J}^1_{w,u}$ and $\mathbf{J}^2_{w,u}$, and may feel lost from too many indexes. However, there is a real reason for them: dealing with $u \in [n+1]^k$, ($k \geq 2$, otherwise things are somewhat degenerated) we choose $N_u \leq_{\mathbf{m}} M^{\mathbf{d}}_{n+1}$ such that if we enrich $\mathbf{d} \restriction \mathscr{P}^-(u)$ to a stable $(\lambda, \mathscr{P}(u))$-system by N_u, it is stable and reduced in n, this is were assumption (a), i.e. $\mathfrak{s}$ has the brimmed^3 strong $(\lambda, \leq n)$-existence property is used to choose some such N_u (not necessarily $\leq_{\mathfrak{s}} M^{\mathbf{d}}_{n+1}$) and then assumption (b) says that $\mathfrak{s}$ has the brimmed$^\ell$ weak$(\lambda, \leq n)$-primeness property hence we can have $N_u \leq_{\mathfrak{s}} M^{\mathbf{d}}_{n+1}$. This does not hurt the "division of work" of the already defined $\mathbf{J}_{v,w}$'s. But choosing $M^{\mathbf{d}}_u$ (such that $N_u \leq_{\mathfrak{s}} M^{\mathbf{d}}_u \leq_{\mathfrak{s}} M^{\mathbf{d}}_{n+1}$) we should be careful if $p \in \mathscr{S}^{\mathrm{sb}}_{\mathfrak{s}}(N_u)$ is $\perp M^{\mathbf{d}}_w$ for $w \subset u$, maybe there is no $\mathbf{J} \subseteq \{c \in M^{\mathbf{d}}_{n+1} : c$ realizes $p\}$ independent in $(M^{\mathbf{d}}_{n+1}, M^{\mathbf{d}}_u, N_u)$ of cardinality λ. So we choose $\langle \mathbf{J}^1_{w,u} : w$ satisfies $u \subseteq w \subset n+1\rangle$, taking care of all such p's (yes! also for $w = u$). This is possible as $M^{\mathbf{d}}_{n+1}$ is brimmed over N_u by clause $(f)^*$ of 12.3. Now we choose $M^{\mathbf{d}}_u \leq_{\mathfrak{s}} M^{\mathbf{d}}_{n+1}$ such that $(N_u, M^{\mathbf{d}}_u, \mathbf{J}^1_{u,u}) \in K^{3,\mathrm{mx}}_{\mathfrak{s}}$ possible as $\mathfrak{s}$ has primes. So we can really carry the induction.

Proof. 1) Let $M_\emptyset \in K_{\mathfrak{s}}$ be brimmed$^\ell$. Let $M \in K_{\mathfrak{s}}$ be $(\lambda_{\mathfrak{s}}, *)$-brimmed over $M_\emptyset$. Let $\mathbf{P}^2_\emptyset = \mathscr{S}^{\mathrm{bs}}_{\mathfrak{s}}(M_0)$; (but if $\mathfrak{s}$ has regulars then we can let $\mathbf{P}^2_\emptyset \subseteq \{p \in \mathscr{S}^{\mathrm{bs}}_{\mathfrak{s}}(M_\emptyset) : p \text{ regular}\}$ be a maximal family of pairwise orthogonal types). Let $\mathbf{J}^2_\emptyset$ be a maximal subset of $\{c \in M : c$ realizes some $p \in \mathbf{P}^2_\emptyset$ in M over $M_\emptyset\}$ which is independent in $(M_\emptyset, M)$ so $|\mathbf{J}^2_\emptyset| = \lambda_{\mathfrak{s}}$ such that $p \in \mathbf{P}^2_\emptyset \Rightarrow \lambda = |\{c \in \mathbf{J}^2_\emptyset : \mathbf{tp}_{\mathfrak{s}}(c, M_\emptyset, N) = p\}|$. Next let $\{\mathbf{J}^2_{u,\emptyset} : \emptyset \subset u \subset n+1\}$ be a partition of $\mathbf{J}^2_\emptyset$ to sets each of cardinality $\lambda_{\mathfrak{s}}$ such that for every $p \in \mathbf{P}^2_\emptyset$ and $\emptyset \subset u \subset n+1$ the set $\{c \in \mathbf{J}^2_u : \mathbf{tp}_{\mathfrak{s}}(c, M_\emptyset, M) = p\}$ has cardinality $\lambda_{\mathfrak{s}}$. Let $I_k = \{u \subseteq n+1 : |u| \leq k\}$ for $k \leq n+1$.

We now choose by induction on $k \in \{1, 2, \ldots, n\}$, the objects $\mathbf{d}_k$ and $\langle N_u, \mathbf{P}^i_u : u \in I_k, |u| \geq 1\rangle, \langle \mathbf{J}^i_{u,v} : u \in I_k, |u| \geq 1$ and $u \subseteq v \subset n+1\rangle$ for $i = 1, 2$ such that

⊛(a) $\mathbf{d}_k$ is a normal brimmed$^\ell$ expanded stable (λ, I_k)-system embedded in M and $M^{\mathbf{d}_k}_\emptyset = M_\emptyset$ and for $u \in I_1 \backslash I_0$ we let $N_u = M_\emptyset$ and $\mathbf{J}^1_u = \emptyset$ and $\mathbf{J}^1_{v,u} = \emptyset$ when $u \subseteq v \subset n+1$ and also $\mathbf{J}^1_u = \emptyset, \mathbf{J}^1_{u,v} = \emptyset$ if $u \in I_0$, i.e. $u = \emptyset$ and $u \subseteq v \subset n+1$

(b) $1 \leq m < k \Rightarrow \mathbf{d}_m = \mathbf{d}_k \restriction I_m$

(c) for $u \in I_k \backslash I_1, N_u$ is such that $\cup\{M^{\mathbf{d}_k}_w : w \subset u\} \subseteq N_u \leq_{\mathfrak{s}} M$ and $\mathbf{d}^*_u$ is reduced in u where $\mathbf{d}^*_u$ is the normal brimmed^1 expanded stable $\mathscr{P}(u)$-system such that $\mathbf{d}^*_u \restriction \mathscr{P}^-(u) = \mathbf{d}_k \restriction \mathscr{P}^-(u), M^{\mathbf{d}^*_u}_u = N_u$ and $\mathbf{J}^{\mathbf{d}^*_u}_{u,w} = \emptyset$ for $w \subset u$

(d) for $u \in I_k \backslash I_1, \mathbf{P}^1_u = \{p \in \mathscr{S}^{\mathrm{bs}}_{\mathfrak{s}}(N_u) : p \perp M_w$ for $w \subset u\}$, (if $\mathfrak{s}$ truely has regulars, see 10.2, then we can use $\mathbf{P}^1_u$ is a maximal set of pairwise orthogonal types from $\{p \in \mathscr{S}^{\mathrm{bs}}_{\mathfrak{s}}(N_u) : p$ regular orthogonal to $M^{\mathbf{d}_k}_w$ for $w \subset u$ (hence to $M^{\mathbf{d}_k}_w$ for $w \in I_k$ such that $u \nsubseteq w$ by 10.10)$\}$)

(e) for $u \in I_k \backslash I_1$, the set $\mathbf{J}^1_u$ is a maximal subset of $\{c \in M : \mathbf{tp}_{\mathfrak{s}}(c, N_u, M) \in \mathbf{P}^1_u\}$ independent in (N_u, M) such that $p \in \mathbf{P}^1_u \Rightarrow \lambda = |\{c \in \mathbf{J}^1_u : \mathbf{tp}_{\mathfrak{s}}(c, N_u, M) = p\}|$

(*f*) if $u \in I_k \backslash I_1$, then $\langle \mathbf{J}^1_{v,u} : u \subseteq v \subset n+1 \rangle$ is a partition of $\mathbf{J}^1_u$ to sets each of cardinality $\lambda_{\mathfrak{s}}$ such that moreover, for every $p \in \mathbf{P}^1_u$ the set $\mathbf{J}^1_{v,u,p} := |\{c \in \mathbf{J}^1_{v,u}$: $\mathbf{tp}_{\mathfrak{s}}(c, N_u, M) = p\}|$ has cardinality $\lambda_{\mathfrak{s}}$

(*g*) $(N_u, M^{\mathbf{d}_k}_u, \bigcup\limits_{w \subseteq u} \mathbf{J}^1_{u,w} \cup \bigcup\limits_{w \subset u} \mathbf{J}^2_{u,w}) \in K^{3,\mathrm{mx}}_{\mathbf{m}}$ for $u \in I_k \backslash I_0 = I_k \backslash \{\emptyset\}$ recalling that we have $\mathbf{J}^1_{u,w} = \emptyset$ when $w \subseteq u, |w| \leq 1$

(*h*) $\mathbf{P}^2_u = \{p \in \mathscr{S}^{\mathrm{bs}}_{\mathfrak{s}}(M^{\mathbf{d}_1}_u) : p \perp N_u\}$ (or when $\mathfrak{s}$ truely has regulars is a maximal set of pairwise orthogonal types from $\{p \in \mathscr{S}^{\mathrm{bs}}_{\mathfrak{s}}(M^{\mathbf{d}_k}_u) : p$ orthogonal to N_u when $u \notin I_1$ and to $M_\emptyset$ if $u \in I_1 \backslash I_0\}$) when $u \in I_k \backslash I_0$ (if $u = \emptyset$ then $\mathbf{P}^2_u$ has already been chosen)

(*i*) $\mathbf{J}^2_u$ is a maximal subset of $\{c \in M : \mathbf{tp}_{\mathfrak{s}}(c, M^{\mathbf{d}_k}_u, M) \in \mathbf{P}^2_u\}$ independent in $(M^{\mathbf{d}_k}_u, M)$ such that for every $p \in \mathbf{P}^2_u$ the set $\{c \in \mathbf{J}^2_u : \mathbf{tp}_{\mathfrak{s}}(c, M^{\mathbf{d}_k}_u, M)$ is equal to $p\}$ has cardinality λ when $u \in I_k$, (if $u = \emptyset, \mathbf{J}^2_u$ has already been chosen), note that the $\mathbf{J}^2_{u,w}$ used is clause (g) has already been chosen)

(*j*) $\langle \mathbf{J}^2_{v,u} : u \subset v \in \mathscr{P}^-(n+1) \rangle$ is a partition of $\mathbf{J}^2_u$ such that for every $p \in \mathbf{P}^2_u$ and v such that $u \subset v \in \mathscr{P}^-(n+1)$ the set $\{c \in \mathbf{J}^2_{v,u} : \mathbf{tp}_{\mathfrak{s}}(c, M^{\mathbf{d}_k}_u, M) = p\}$ has cardinality λ when $u \in I_k$

(*k*) $\mathbf{J}^{\mathbf{d}_k}_{v,u} = \mathbf{J}^1_{v,u} \cup \mathbf{J}^2_{v,u}$ where $u \subset v \in I_k$

For $k = 1, \mathbf{d}_k$ is described by clauses (a) + (g) that is for $u \in I_1 \backslash I_0, N_u = M_\emptyset, \mathbf{J}^2_{\emptyset,u}$ has already been defined and $\mathbf{J}^1_{\emptyset,u} = \emptyset = \mathbf{J}^1_{u,u}$ so clauses (g) + (h) say that we should choose $M^{\mathbf{d}_1}_u \leq_{\mathfrak{s}} M$ such that $(M_\emptyset, M^{\mathbf{d}_1}_u, \mathbf{J}^2_{\emptyset,u}) \in K^{3,\mathrm{mx}}_{\mathfrak{s}}$ and if $|u| < n$ and $p \in \mathscr{S}^{\mathrm{bs}}_{\mathfrak{s}}(M^{\mathbf{d}}_u)$ is orthogonal to $M_\emptyset$ then $\dim(p, M) = \lambda$. Now if $|u| = n$ as $\mathfrak{s}$ has primes (see 12.3(1)(c), as $(M_\emptyset, M, \mathbf{J}^2) \in K^{3,\mathrm{bs}}_{\mathfrak{s}}$ there is $M^{\mathbf{d}}_u \leq_{\mathfrak{s}} M$ such that $(M_\emptyset, M, \mathbf{J}^2_{\emptyset,u}) \in K^{3,\mathrm{qr}}_{\mathfrak{s}}$ so by 12.6(3) it is as required. Now if $|u| < n$ we can find N'_u such that $M^{\mathbf{d}_1}_u \leq_{\mathfrak{s}} N'_u, (M_\emptyset, N'_u, \mathbf{J}^2_{\emptyset,u}) \in K^{3,\mathrm{mx}}_{\mathfrak{s}}$ and if $p \in \mathscr{S}^{\mathrm{bs}}_{\mathfrak{s}}(M^{\mathbf{d}_1}_u)$ is orthogonal to $M_\emptyset$ then $\dim(p, N'_u) = \lambda$. Why? Easy or see 12.33. So again by $(f)^{**}$ of 12.32 without loss of generality $N'_u \leq_{\mathfrak{s}} M$ and now also clause (h) will cause no problem.

Lastly, choose $\mathbf{P}^2_u, \mathbf{J}^2_u, \langle \mathbf{J}^2_{v,u} : v$ satisfies $u \subseteq v \subset n+1 \rangle$ as above for $u \in I_k \backslash I_0$ (i.e., $u = \{m\}, m < n+1$).

For $k = m+1 > 1$ for each $u \in I_k \backslash I_m$ clearly $\mathbf{d}_m \restriction \mathscr{P}^-(u)$ is a normal brimmed$^\ell$ expanded stable $\mathscr{P}^-(u)$-system hence by assumption (a) we can find $\mathbf{d}^*_u$ such that

$\odot_1$ (a) $\mathbf{d}^*_u$ is a normal brimmed^2 expanded stable $\mathscr{P}(u)$-system
(b) $\mathbf{d}^*_u \restriction \mathscr{P}^-(u) = \mathbf{d}_u \restriction \mathscr{P}^-(u)$
(c) $\mathbf{d}^*_u$ is reduced in u.

By claim 12.33 below without loss of generality there is N'_u such that

$\odot_2$ (d) $\cup \{M^{\mathbf{d}^*_u}_{u,w} : w \subset u\} \subseteq N'_u \leq_{\mathfrak{s}} N''_u \leq_{\mathfrak{s}} M^{\mathbf{d}^*_u}_u$
(f) if $p \in \mathscr{S}^{\rm bs}_{\mathfrak{s}}(N'_u)$ is orthogonal to $M^{\mathbf{d}_u}_{u,w}$ for every $w \subset u$ <u>then</u> $\lambda =$ $\dim(p, M^{\mathbf{d}^*_u}_u)$.

Note that

$\odot_3$ (g) if $p \in \mathscr{S}^{\rm bs}_{\mathfrak{s}}(N''_u)$ is orthogonal to $M^{\mathbf{d}_k}_w$ for $w \subset u$ then $\dim(p, M^{\mathbf{d}^*_u}_u) = \lambda$.

Now $\mathbf{d}^*_u$ is reduced in u hence $\mathbf{d}^*_u$ is prime over $\mathbf{d}_m \restriction \mathscr{P}^-(u)$, clause (b) of the assumption. We define a normal expanded $(\lambda, \mathscr{P}(u))$-system $\mathbf{d}'_u$ by $\mathbf{d}'_u \restriction \mathscr{P}^-(u) = \mathbf{d}_m \restriction \mathscr{P}^-(u)$ and $M^{\mathbf{d}'}_u = M$ and for $v \subset u$ we let $\mathbf{J}^{\mathbf{d}'}_{u,v}$ is a maximal subset of $M \backslash \cup \{M^{\mathbf{d}_m}_{w_1} : w_1 \subset u\}$ including $\mathbf{J}^1_{u,v} \cup \mathbf{J}^2_{u,v}$ such that $\mathbf{J}^{\mathbf{d}}_{u,v} \cup \bigcup \{\mathbf{J}^1_{u,w} : w \subseteq u\} \cup \bigcup \{\mathbf{J}^2_{u,w} : w \subset u\}$ is independent in $(M^{\mathbf{d}_m}_v, M)$.

So

$(*)_1$ $\mathbf{d}'_u$ is a normal expanded stable $(\lambda, \mathscr{P}(u))$-system
$(*)_2$ $\mathbf{d}'_u$ is brimmed$^\ell$

Now

$(*)_3$ $\mathbf{d}^*_u$ is weakly prime$^\ell$.

[Why? By assumption (b), primeness see Definition 12.19(5) because $\mathbf{d}^*_u$ is an expanded stable $(\lambda, \mathscr{P}^-(u))$-system, $\mathbf{d}^*_u \restriction \mathscr{P}^-(u) = \mathbf{d}_m \restriction \mathscr{P}^-(u)$ is brimmed$^\ell$ and $\mathbf{d}^*_u$ is reduced in u'.]

By the definition 12.19(6) of weakly prime, as $(*)_2 + (*)_3 + \mathbf{d}'_u \restriction \mathscr{P}^-(u) = \mathbf{d}^*_u \restriction \mathscr{P}^-(u)$ there is an $\leq_{\mathfrak{s}}$-embedding h of $M^{\mathbf{d}^*_u}_u$ into M

which is the identity on $\cup\{M_w^{\mathbf{d}_m} : w \subset u\}$. Hence without loss of generality $M_u^{\mathbf{d}_u^*} \leq_{\mathfrak{s}} M$. Define $N_u := N'_u$ and define $\mathbf{P}_u^1$ as in clause (d) of $\circledast$. Now.

$(*)_4$ if $p \in \mathscr{S}_{\mathfrak{s}}^{\rm bs}(N_u)$ then $\dim(p, M) = \lambda$.

[Why? Let $v(p) \subseteq u$ be such that $p \pm M_{u,v}^{\mathbf{d}_u^*}$ and $w \subset v(p) \Rightarrow p \perp M^{\mathbf{d}_u^*}$. First, if $v(p) \subset u$ we can use the set $\mathbf{J}^2_{u,v(p)}$. Second, if $v(p) = u$ then we use "$\lambda = \dim(p, M_u^{\mathbf{d}_u^*})$", which holds by our choice of $\mathbf{d}_u^*, N'_u$, i.e. clause (f) of $\circledast$.]

Now as $|u| > 1$ we can choose $\mathbf{J}_u^1$ and $\langle \mathbf{J}^1_{v,u} : v$ satisfies $u \subseteq v \subset n+1\rangle$ as required. Now let $M_u^{\mathbf{d}_k} \leq_{\mathfrak{s}} M$ be[26] such that:

(α) $M_u^{\mathbf{d}_k}$ is brimmed$^\ell$ over N_u

(β) $(N_u, M_u^{\mathbf{d}_k}, \cup\{\mathbf{J}^1_{u,w} \cup \mathbf{J}^2_{u,w} : w \subset u\} \cup \mathbf{J}^1_{u,u}) \in K_{\mathfrak{s}}^{3,\rm mx}$

(γ) if $p \in \mathscr{S}_{\mathfrak{s}}^{\rm bs}(M_u^{\mathbf{d}_k})$ is orthogonal to N_u then $|u| < n \Rightarrow \dim(p, M) = \lambda$ and $|u| = n \Rightarrow \dim(p, M) = 0$.

Concerning demands $(\alpha) + (\beta)$ this is possible as $\mathfrak{s}$ has primes and (f)* of 12.3.

If $|u| < n$ we first choose $M_u^{\mathbf{d}_k} = N''_u$. If $|u| = n$, we try to choose $N_{u,\alpha} \leq_{\mathfrak{s}} M$ by induction on $\alpha < \lambda^+$, which is $\leq_{\mathfrak{s}}$-increasing continuous, $N_{u,0} = M_u^{\mathbf{d}_k}$ and $(N_{u,\beta}, N_{u,\alpha}, a_\beta) \in K_{\mathfrak{s}}^{3,\rm pr}$ and $\mathbf{tp}_{\mathfrak{s}}(a_\beta, N_{u,\beta}, M) \perp M_w^{\mathbf{d}_k}$ for $w \subset u$ when $\alpha = \beta + 1$. For some α we are stuck. We then choose $\mathbf{P}_u^2, \mathbf{J}_u^2, \mathbf{J}^2_{v,u}$ $(u \subset v \subset n+1)$ as required; this is not in general possible but we could have chosen $M_u^{\mathbf{d}_u^k}$ such that clause (f) holds of $\circledast$.

Having carried the induction we define a normal $\mathscr{P}(n+1)$-system $\mathbf{d}_{n+1}$ by

$$\mathbf{d}_{n+1} \restriction \mathscr{P}^-(n+1) = \mathbf{d}_n$$

$$M_{n+1}^{\mathbf{d}_{n+1}} = M$$

---

[26]exists as for every $w \subset u$, the set $\mathbf{J}_{u,w}$ is independent in (N_u, M) because $(M_w, N_u, \cup\{\mathbf{J}_{v_1,u_1} : v_1 \subset u, u_1 \subseteq w, u_1 \subset v_1\}) \in K_{\mathfrak{s}}^{3,\rm mx}$

$$\mathbf{J}^{\mathbf{d}_{n+1}}_{n+1,u} = \emptyset \text{ for } u \subset n+1.$$

It is easy to check that $\mathbf{d}_{n+1}$ is as required.
2) Easy, by 12.23(2). $\square_{12.32}$

12.33 Claim. *Assume*

(a) $\mathbf{d}$ *is an expanded stable* $(\lambda, \mathscr{P}(n))$*-system*

(b) $\mathbf{d}$ *is reduced at* n.

Then we can find $\mathbf{d}'$ *such that*

(α) $\mathbf{d}'$ *is an expanded stable* $(\lambda, \mathscr{P}(n))$*-system reduced at* n

(β) $\mathbf{d}' \restriction \mathscr{P}^-(n) = \mathbf{d} \restriction \mathscr{P}^-(n)$

(γ) $h^{\mathbf{d}'}_{n,u} = h^{\mathbf{d}}_{n,u}$ *for* $u \subset n$

(δ) $M^{\mathbf{d}}_n \leq_{\mathfrak{s}} M^{\mathbf{d}'}_n$

(ε) *if* $p \in \mathscr{S}^{\mathrm{bs}}_{\mathfrak{s}}(M^{\mathbf{d}}_n)$ *is orthogonal*[27] *to* $M^{\mathbf{d}}_{n,u}$ *for every* $u \subset n$ *then* $\dim(p, M^{\mathbf{d}'}_n) = \lambda$.

Proof. Let $M^+ \in K_{\mathfrak{s}}$ be $(\lambda, *)$-brimmed over $M^{\mathbf{d}}_n$, let $\mathbf{P} = \{p \in \mathscr{S}^{\mathrm{bs}}_{\mathfrak{s}}(M^{\mathbf{d}}_n) : p \text{ regular} \perp M^{\mathbf{d}}_{n,u} \text{ for every } u \subset n\}$. Let $\mathbf{J} = \{c_{p,\alpha} : p \in \mathbf{P}$ and $\alpha < \lambda\}$ be such that

(*i*) $c_{p,\alpha} \in M^+$ realizes p

(*ii*) $p \in \mathbf{P} \wedge \alpha \neq \beta \Rightarrow c_{p,\alpha} \neq c_{p,\beta}$

(*iii*) $\mathbf{J}$ is independent in $(M^{\mathbf{d}}_n, M^+)$.

Now let $M \leq_{\mathfrak{s}} M^+$ be such that $(M^{\mathbf{d}}_n, M, \mathbf{J}) \in K^{3,\mathrm{mx}}_{\mathfrak{s}}$ and M is maximal under this condition and define $\mathbf{d}'$ such that $((\alpha), (\beta), (\gamma)$ above holds and) $M^{\mathbf{d}'}_n = M$. It is easy to check that $\mathbf{d}'$ is as required. $\square_{12.33}$

[27] when $\mathfrak{s}$ has regulars then enough if p is regular

12.34 Claim. *1) Assume $\ell = 3$ (and also $\ell = 1, 2$) and*

(a) *$\mathfrak{s}$ is successful hence $\mathfrak{s}^+$ satisfies the hypothesis 12.3*

(b) *$\mathfrak{s}$ has the brimmed$^\ell$ weak $(\lambda, \le n+1)$-uniqueness property [actually only the values $n+1$ and n are used]*

(c) *$\mathbf{m}$ is a brimmed$^\ell$ stable $(\mathscr{P}^-(n), \mathfrak{s}^+)$-system*

(d) *$\mathbf{m}^*$ is a stable $(\mathscr{P}(n), \mathfrak{s}^+)$-system reduced at n such that $\mathbf{m}^* \restriction \mathscr{P}^-(n) = \mathbf{m}$.*

Then $\mathbf{m}^*$ is weakly prime over $\mathbf{m}$ for $\mathfrak{s}^+$.
2) Moreover, $\mathbf{m}^*$ is strongly prime$^\ell$ over $\mathbf{m}$ for $\mathfrak{s}^+$.

Remark. This is similar to the proof of the existence of primes in $\mathfrak{s}^+$.

Proof. Without loss of generality $\mathbf{m}^*$ is normal. Let $\langle M_u^\alpha : \alpha < \lambda_\mathfrak{s}^+\rangle$ be $\le_\mathfrak{s}$-increasing continuous with union $M_u^{\mathbf{m}^*}$ and let E be a thin enough club of $\lambda_\mathfrak{s}^+$. By 12.18 for each $\alpha \in E, \mathbf{m}_\alpha^* = \langle M_u^\alpha : u \in \mathscr{P}(n)\rangle$ is a normal stable $\mathscr{P}(n)$-system reduced at n and letting $\mathscr{P} = \{u \subseteq n+1 : n \nsubseteq u\}$ for $\alpha < \beta$ from $E, \mathbf{m}_{\alpha,\beta}^* := \mathbf{m}_\alpha *_{\mathscr{P}(n+1)} \mathbf{m}_\beta$ is a stable $(\mathscr{P}(n+1), \mathfrak{s})$-system and $\mathbf{m}_{\alpha,\beta} \restriction \mathscr{P}$ is a normal brimmed^3 stable $(\mathscr{P}, \mathfrak{s})$-system. Suppose that $M \in K_{\mathfrak{s}(+)}$ and $\langle f_u : u \in \mathscr{P}^-(n)\rangle$ is a stable embedding of $\mathbf{m}$ into M (see Definition 12.9(3)).

As $\mathbf{m}$ is normal, we have $u \subseteq v \in \mathscr{P}^-(n) \Rightarrow f_u \subseteq f_v$. Let $\langle M_\alpha : \alpha < \lambda_\mathfrak{s}^+\rangle$ be $\le_\mathfrak{s}$-increasing continuous with union M and without loss of generality E is a thin enough club for this too so, e.g. $\alpha \in E$ & $u \subset n \Rightarrow f_u(M_u^\alpha) = f_u(M_u) \cap M_\alpha$; by renaming $E = \lambda_\mathfrak{s}^+$. Let $f_u^\alpha = f_u \restriction M_u^\alpha$, so $\bar{f}^\alpha = \langle f_u^\alpha : u \in \mathscr{P}^-(n)\rangle$ is an embedding of $\mathbf{m}_\alpha$ into $M_\alpha \le_{\mathfrak{K}[\mathfrak{s}]} M$. Now we choose f_n^α by induction on α such that

$\circledast$(i) f_n^α is a $\le_{\mathfrak{K}[\mathfrak{s}]}$-embedding of M_n^α into M (hence into $M_{\beta(\alpha)}$ for some $\beta(\alpha) < \lambda_\mathfrak{s}^+$)

(ii) f_n^α extends f_n^β for $\beta < \alpha$ and f_u^α for $u \in \mathscr{P}^-(n)$.

For $\alpha = 0, f_n^0$ exists as $\mathfrak{s}$ has the brimmed^3 weak (λ, n)-uniqueness property and M is λ^+-saturated above λ.
For α limit let $f_n^\alpha = \bigcup\limits_{\beta<\alpha} f_n^\beta$.

For $\alpha = \beta + 1$ let $\gamma < \lambda_{\mathfrak{s}}^+$ be such that M_γ is $(\lambda, *)$-brimmed over $\text{Rang}(f_n^\beta) \cup \bigcup\{\text{Rang}(f_u^\alpha : u \in \mathscr{P}^-(n)\}$. We shall show that there is a $\leq_{\mathfrak{s}}$-embedding of M_n^α into M and even into M_γ extending $f_n^\beta \cup \bigcup\{f_u^\alpha : u \in \mathscr{P}^-(n)\}$. We defined the normal $(\lambda, \mathscr{P}^-(n+1))$-system $\mathbf{m}'_{\beta,\alpha}$ defined by $\mathbf{m}' \restriction \{u \subset n+1 : n \notin u\} = \mathbf{m}^*_{\beta,\alpha}$ and $M_n^{\mathbf{m}'_{\alpha,\beta}} = M_n^\beta$. Clearly $\mathbf{m}'_{\beta,\alpha}$ is a normal stable $(\lambda, \mathscr{P}^-(n+1))$-system. Now we shall use "$\mathfrak{s}$ has the brimmed[3] weak $(\lambda, n+1)$-uniqueness property" defined in 12.19(2) for the $(\lambda_{\mathfrak{s}}, \mathscr{P}^-(n+1))$-system $\mathbf{m}^*_{\alpha,\beta}$. Now define $\bar{g}_\alpha = \langle g_u^\alpha : u \leq n\rangle$ by $g_u^\alpha = f_u^\alpha$ if $u \subseteq n$, $g_u^\alpha = f_{u\backslash\{n\}}^{\alpha+1}$ if $n \in u \subset n+1$; now $\bar{g}_\alpha$ is a stable embedding of $\mathbf{m}'_{\alpha,\beta}$ into $M_{\beta(\alpha)}$ (see Definition 12.9(3)) as M_n^α does not "contribute" (as $\mathbf{m}$ is reduced in n). But the assumption of "$\mathfrak{s}$ has brimmed$^\ell$ weak uniqueness" is not fully satisfied because for $\mathbf{m}'_{\alpha,\beta}$ the brimmed$^\ell$ demand does not (necessarily) holds for $u = n$, i.e., for $f_n^\alpha(M_n^\alpha)$; however, by Claim 12.26 this is overcomed. We get that there is a pair (f, N) such that $M_{\beta(\alpha)} \leq_{\mathfrak{s}} N$ and f a $\leq_{\mathfrak{s}}$-embedding M_n^α into N, but without loss of generality $N \leq_{\mathfrak{s}} M_{\beta(\alpha)+1}$ so we are done.

Now $f_n = \bigcup\limits_{\alpha < \lambda_{\mathfrak{s}}^+} f_n^\alpha$ is the required embedding. $\square_{12.34}$

12.35 Conclusion. Assume $\ell = 3$ and

(a) $\mathfrak{s}$ is successful hence $\mathfrak{s}^+$ satisfies 12.3

(b) $\mathfrak{s}$ has the brimmed$^\ell$ weak $(\lambda, \leq n+1)$-uniqueness property.

Then $\mathfrak{s}^+$ has the brimmed$^\ell$ weak (λ^+, n)-primeness$^\ell$ property.

Proof. By 12.34. $\square_{12.35}$

12.36 Claim. *[Assume $(f)^{**}$ of 12.3]*
1) Let $\ell = 3$ and

(a) $\mathfrak{s}$ has the brimmed$^\ell$ strong (λ, n)-existence property

(b) $\mathfrak{s}$ has the brimmed$^\ell$ weak (λ, n)-primeness property.

Then $\mathfrak{s}$ has the brimmed$^\ell$ strong (λ, n)-uniqueness property.
2) Let $\ell = 4$ and $\mathfrak{s}$ has the brimmed$^\ell$ weak (λ, n)-uniqueness property then $\mathfrak{s}$ has the brimmed$^\ell$ strong (λ, n)-uniqueness property.

Proof. 1) The cases $n = 0, 1$ are easy; so assume $n \geq 2$.

Assume $\mathbf{d}_1, \mathbf{d}_2$ are brimmed$^\ell$ stable $(\lambda, \mathscr{P}(n))$-system and $\bar{f} = \langle f_u : u \in \mathscr{P}^-(n) \rangle$ is an isomorphism from $\mathbf{d}_1 \restriction \mathscr{P}^-(n)$ onto $\mathbf{d}_2 \restriction \mathscr{P}^-(n)$. As $\mathfrak{s}$ has the brimmed$^\ell$ strong (λ, n)-existence property, (i.e., assumption (a)), clearly for $k = 1, 2$ there is an expanded stable $(\lambda, \mathscr{P}(n))$-system $\mathbf{d}^k, \mathbf{d}^k \restriction \mathscr{P}^-(n) = \mathbf{d}_k \restriction \mathscr{P}^-(n)$, $M_n^{\mathbf{d}^k}$ is brimmed$^\ell$ and $\mathbf{d}^k$ reduced in n, i.e., such that $u \subset n \Rightarrow \mathbf{J}_{n,u}^{\mathbf{d}^k} = \emptyset$).

Now we can find a pair $(N_k, \mathbf{J}_k)$ such that

$(*)_1$ (a) $(M_n^{\mathbf{d}^k}, N_k, \mathbf{J}_k) \in K_{\mathfrak{s}}^{3,\mathrm{mx}}$

(b) if $a \in \mathbf{J}_k$ then $\mathbf{tp}_{\mathfrak{s}}(a, M_n^{\mathbf{d}^k}, N_k)$ is orthogonal to $M_{n,u}^{\mathbf{d}^k}$ for $u \subset n$

(c) if $p \in \mathscr{S}_{\mathfrak{s}}^{\mathrm{bs}}(M_n^{\mathbf{d}^k})$ is orthogonal to $M_{n,u}^{\mathbf{d}^k}$ for $u \subset n$ then

$$\lambda = |\{a \in \mathbf{J}_k : p = \mathbf{tp}_{\mathfrak{s}}(a, M_n^{\mathbf{d}^k}, N_k)\}|.$$

By assumption (b), recalling that for $\ell \geq 2$, weak primeness and strong primeness are equivalent see the Definition 12.19 there is a $\leq_{\mathfrak{s}}$-embedding g_k of N_k into $M_n^{\mathbf{d}^k}$ over $\cup\{M_{n,u}^{\mathbf{d}^k} : u \subset n\}$.

Let $f'_u = f_u$ for $u \subset n$. Trivially, without loss of generality there is an isomorphism f'_n such that $\langle f'_u : u \in \mathscr{P}(n) \rangle$ is an isomorphism from $\mathbf{d}^1$ onto $\mathbf{d}^2$ and such that f'_n maps $g_1(N_1)$ onto $g_2(N_2)$ and it maps $N'_1 = g_1(M_n^{\mathbf{d}^1})$ onto $N'_2 = g_2(M_n^{\mathbf{d}^2})$.

By clause (b) of the assumption + $(*)$ of Definition 12.19(5A) without loss of generality $M_n^{\mathbf{d}^k} \leq_{\mathfrak{s}} M_n^{\mathbf{d}_k}$; recall $u \subset n \Rightarrow h_{n,u}^{\mathbf{d}^k} = h_{n,u}^{\mathbf{d}_k}$. For $u \subseteq n$ let

$$\mathbf{P}_u^k = \{p \in \mathscr{S}_{\mathfrak{s}}^{\mathrm{bs}}(N'_k) : p \perp M_{n,u,w}^{\mathbf{d}^k} \text{ if } w \subset u \text{ and } p \text{ does not fork over } M_{n,u}^{\mathbf{d}^k} \text{ if } u \subset n\}.$$

Lastly, let $\mathbf{P}_k = \cup\{\mathbf{P}_u^k : u \subseteq n\}$.

Now

$(*)_2$ if $k = 1, 2$ and $u \subset n$ and $p \in \mathbf{P}_u^k$ then $\dim(p, M_n^{\mathbf{d}}) = \lambda$.

[Why? We use "$\mathbf{d}$ is brimmed$^\ell$ $(\lambda, \mathscr{P}(n))$-system".]

Note that $(M^{\mathbf{d}^k}_{n,u}, M^{\mathbf{d}^k}_n, \mathbf{J}^{2,\mathbf{d}^k}_{n,u}) \in K^{3,\mathrm{mx}}_{\mathfrak{s}}$ hence $(M^{\mathbf{d}_k}_{n,u}, M^{\mathbf{d}_k}_n, \mathbf{J}^2_{n,u}) \in K^{3,\mathrm{mx}}_{\mathfrak{s}}$ hence $\mathbf{J}^{2,\mathbf{d}_k}_{n,u} \backslash \mathbf{J}^{2,\mathbf{d}^k}_{n,u}$ is independent in $(M^{\mathbf{d}_k}_{n,u}, M^{\mathbf{d}^k}_n, M^{\mathbf{d}_k}_n)$ hence by monotonicity is independent in $(M^{\mathbf{d}-k}_{n,u}, N'_k, M^{\mathbf{d}_k}_n)$ and it provides the necessary witnesses (the regular version is more transparent)

$(*)_3$ if $k = 1, 2$ and $p \in \mathbf{P}^k_n$ then $\dim(p, M^{\mathbf{d}^*_k}_n) = \lambda$.

[Why? Note that $p \in \mathscr{S}^{\mathrm{bs}}_{\mathfrak{s}}(N'_k)$. By the choice of g_k this means $\dim(g_k^{-1}(p), N_k) = \lambda$ which holds by the choice N_k.]

We can apply (d)(ii) + (f)* of 12.3 with $(M^{\mathbf{d}_k}_n, M^{\mathbf{d}^k}_n)$ here standing for (M, N) and $\langle M^{\mathbf{d}_k}_u : u \subset n\rangle$ there standing for $\langle M_\ell : \ell < n\rangle$ there, hence $M^{\mathbf{d}_k}_n$ is $(\lambda, *)$-brimmed over $M^{\mathbf{d}^k}_n$. As f'_n is an isomorphism from $M^{\mathbf{d}^2}_n$ onto $M^{\mathbf{d}^2}_n$ there is an isomorphism f_u from $M^{\mathbf{d}_1}_n$ onto $M^{\mathbf{d}_2}$ which extends f'_n. So clearly $\langle f_u : u \subseteq n\rangle$ is an isomorphism from $\mathbf{d}_1$ onto $\mathbf{d}_2$.

2) Left to the reader as an exercise (+ not used). $\square_{12.36}$

12.37 Theorem. $[2^{\lambda^{+n}} < 2^{\lambda^{+n+1}}$ *for* $n < \omega]$. *Let* $\ell = 3$. *Assume* $\mathfrak{s}$ *is* ω*-successful and* $\mathfrak{s}$ *satisfies 12.1(e)***,(f)*** *(starting with* $\mathfrak{s}^+$ *this follows).* *Then* $\mathfrak{s}^{+m}$ *is* $(m+2)$*-beautiful for every* $m < \omega$*; (see definition below, so* $\mathfrak{s}^{+m}$ *is* m'*-beautiful for* $m' = 0, \dots, m+2$*).*

12.38 Definition. 1) We say that $\mathfrak{s}$ is n-beautiful$^\ell$ or (n, ℓ)-beautiful if:

- (a) $\mathfrak{s}$ has the brimmed$^\ell$ strong $(\lambda, \leq n)$-existence property
- (b) $\mathfrak{s}$ has the brimmed$^\ell$ weak $(\lambda, \leq n)$-uniqueness property
- (c) $\mathfrak{s}$ has the brimmed$^\ell$ strong $(\lambda, < n)$-uniqueness property
- (d) $\mathfrak{s}$ has the brimmed$^\ell$ weak $(\lambda, < n)$-primeness property
- (e) $\mathfrak{s}$ has the brimmed$^\ell$ strong prime $(\lambda, < n)$-existence.

2) We say that $\mathfrak{s}$ is ω-beautiful$^\ell$ if $\mathfrak{s}$ is n-beautiful$^\ell$ for every n.

Remark. 1) In the Theorem we could restrict our demand to $n \leq n_*(< \omega)$ and get the appropriate conclusion, essentially $\mathfrak{s}^{+m}$ is n-excellent^3 if $\mathfrak{s}$ is $2n$-successful (and $\langle 2^{\lambda^{+\ell}} : \ell \leq 2n\rangle$ is increasing).

2) In clause (e) of 12.38 there is no difference between the strong and weak versions because 12.1(e)* holds.

Proof of 12.37. We know by 12.5(2) that

$(*)_1$ $\mathfrak{s}^{+m}$ satisfies the demands in 12.3(1)+(2) and is successful for $m < \omega$

We now prove by induction on $n \geq 2$ that

$\boxtimes_n$ $\mathfrak{s}^{+m}$ is n-beautiful$^\ell$ if $m \geq n-2$.

First we prove $\boxtimes_2$.

So we have to check in Definition 12.38(1) clauses (a),(b) for $n' \leq 2$ and clause (c),(d),(e) for $n' < 2$. First we deal with $n' = 0, 1$. Now clause (a) for $n' = 0$ holds trivially and for $n' = 1$ by the existence of primes (by 5.8(6) noting that being a brimmed^3 model is trivial as $\mathfrak{s}^{+n}$ is categorical by (e)* of 12.3). Clause (b), weak uniqueness holds by 12.20(2). Clause (c), strong uniqueness, for $n' = 0$ this means categoricity, i.e. (e)* of 12.3 which we have assumed and for $n' = 1$, this follows by 12.20(3).

Lastly, clause (d), weak primeness holds because of the same reason. Also clause (e) holds.

Now $\mathfrak{s}^{+m}$ has the brimmed$^\ell$ weak $(\lambda, 2)$-uniqueness as $\mathfrak{s}^{+m}$ is a weakly successful good frame (i.e., the uniqueness of $\mathrm{NF}_{\mathfrak{s}(+m)}$-amalgamation). Lastly, the brimmed$^\ell$ strong $(\lambda^{+m}, 2)$-existence holds by 12.32(2) for $n = 1$.

So let $n \geq 2$ and we assume $\boxtimes_n$ and we shall prove $\boxtimes_{n+1}$, this suffices. Now we should consider all $m \geq (n+1)-2 = n-1$; but the first two assertions below $(*)_2, (*)_3$ are shown even in more cases, i.e. for $m \geq n-2$

$(*)_2$ there is a brimmed$^\ell$ stable $(\mathscr{P}(n+1), \mathfrak{s}^{+m})$-system for $m \geq n-2$

[Why? By 12.28; why its assumptions hold? As for every $n' \leq n$, $\mathfrak{s}^{+m+1}$ has the brimmed$^\ell$ weak (λ^{+m+1}, n')-existence property by 12.23 there is a brimmed$^\ell$ expanded stable (λ^{+m+1}, n)-system for $\mathfrak{s}^{+m}$.]

$(*)_3$ $\mathfrak{s}^{+m}$ has the brimmed$^\ell$ weak $(\lambda^{+m}, n+1)$-existence property if $m \geq n-2$.

[Why? By $(*)_2$ there is brimmed$^\ell$ expanded stable $(\mathscr{P}(n+1), \mathfrak{s}^{+m})$-system call it $\mathbf{d}^*$. Let a brimmed$^\ell$ expanded stable $(\mathscr{P}^-(n+1), \mathfrak{s}^{+m})$-system $\mathbf{d}$ be given. By $\boxtimes_n$ we know that $\mathfrak{s}^{+m}$ has the strong $(\lambda^{+n}, < n)$-uniqueness property hence by 12.23(2) the systems $\mathbf{d}^* \restriction [n+1]^{<n}, \mathbf{d} \restriction [n+1]^{<n}$ are isomorphic so without loss of generality they are equal. Now we shall apply clause (α) of the conclusion of 12.27 to $\mathbf{d}$ and $\mathbf{d}^* \restriction \mathscr{P}^-(n+1)$, as the latter has the weak existence property (as $\mathbf{d}^*$ exemplify) it suffices to check the assumptions of 12.27. So here $I = \mathscr{P}^-(n+1)$ and $J = [n+1]^{<n}$, clause (a) of 12.27 is obvious, clause (b) was assumed above and clause $(c)_1$ follows from "$\mathfrak{s}^{+m}$ has the weak n-uniqueness property", which holds as we assume $\boxtimes_n$.]

$(*)_4$ $\mathfrak{s}^{+m}$ has the brimmed$^\ell$ weak $(\lambda^{+m}, n+1)$-uniqueness property if $m \geq n-2$.

[Why? We try to apply 12.31(2) hence implicitly 12.30 + 12.31(1) to $\mathfrak{s}^{+m}$ and n. Its conclusion fails by clause (b) of Definition 12.38, applied to $(\mathfrak{s}^{+m})^+ = \mathfrak{s}^{+m+1}$ for n which holds as we are assuming $\boxtimes_n$ so we have to check $m+1 \geq n-2$. Clause (a) from its assumptions, see 12.30, i.e. $(\mathfrak{s}^{+m})^+$ is successful, holds by $(*)_1$, clause (b) which says that $\mathfrak{s}^{+m}$ has the brimmed$^\ell$ weak $(\lambda^{+m}, \leq n+1)$-existence property, holds by $(*)_3$ for $n+1$ and by clause (a) of Definition 12.38 for $m \leq n$. Now clause $(c)_2^-$ which says that $\mathfrak{s}^{+m}$ has the brimmed$^\ell$ weak (λ^{+m}, n)-uniqueness property, see 12.31 holds by clause (b) of Definition 12.38 by $\boxtimes_n$ applied to $\mathfrak{s}^{+m}$ and clause $(c)_1^-$, which says that $\mathfrak{s}^{+m}$ has the brimmed$^\ell$ strong $(\lambda^{+m}, < n)$-uniqueness property, see 12.31 holds by clause (c) of Definition 12.38 by $\boxtimes_n$ applied to $\mathfrak{s}^{+m}$. So in 12.31 only the fourth assumption (d), may fail, so as its conclusion fails, (d) there fails.

Hence clause (d) from 12.30 has to fail which is the desired conclusion.]

$(*)_5$ $\mathfrak{s}^{+m}$ has the brimmed$^\ell$ weak (λ^{+m}, n)-primeness property if $m \geq n-1$.

[By 12.35 applied to $n' = n$ and $\mathfrak{s}' = \mathfrak{s}^{+(m-1)}$. It gives the desired conclusion. As for its assumption clause (a) there holds by $(*)_1$ and clause (b) there for $n+1$ by $(*)_4$ above (applicable as $m-1 \geq n-2$ because $m-1 \geq (n-1)-1 = n-2$ and for $0, \ldots, n$ by $\boxtimes_n$.]

$(*)_6$ $\mathfrak{s}^{+m}$ has the brimmed$^\ell$ strong (λ^{+m}, n)-uniqueness property if $m \geq n-1$.
[Why? By 12.36, assumption (a) there holds by clause (a) of the definition 12.38 of beautiful and $\boxtimes_n$ and clause (b) there holds by $(*)_5$ above.]

$(*)_7$ $\mathfrak{s}^{+m}$ has the brimmed$^\ell$ strong $(\lambda^{+m}, n+1)$-existence property for $m \geq n-1$.
[Why? We shall apply 12.32(2), its conclusion is what we need so we have to check its assumptions. First, assumption (a) there which says that $\mathfrak{s}^{+m}$ has the brimmed$^\ell$ strong $(\lambda^{+m}, \leq n)$-existence property, holds by $\boxtimes_n$. Assumption (b) there which says that $\mathfrak{s}^{+m}$ has the brimmed$^\ell$ weak $(\lambda^{+m}, \leq n)$-primeness property holds by $(*)_5$ and $\boxtimes_n$ (which says that).
Lastly, assumption (c) there which says that $\mathfrak{s}^{+m}$ has the brimmed$^\ell$ strong $(\lambda, \leq n)$-uniqueness property, holds by $(*)_6$ above $+\boxtimes_n$ so we are done.]

So $\boxtimes_{n+1}$ holds. Having carried the induction we are done. $\square_{12.37}$

We give here another cases of deriving a good λ-frame. It has a (limited) use. Recall that Chapter II has tried generalizing [Sh 87a], [Sh 87b] but through it give the parallel conclusions about each λ^{+n}, it does not say anything on $\mu \geq \lambda^{+\omega}$. In the claims (12.39), 12.41, 12.42, 12.43 below we derived the parallel of several of the further conclusions of [Sh 87a], [Sh 87b].
The aim of the following claim is to help proving for the case $\ell = 3$ that for non-uni-dimensional $\mathfrak{s}$, we can prove non-categoricity in higher cardinals (of course, we shall get better results when we prove beautifulness for $\ell = 1$ in [Sh 842]).

12.39 Claim. *1) Assume that*

(a) $\mathfrak{s}$ *is a successful good*$^+$ λ*-frame with primes*

(b) $M^* \in K_{\mathfrak{s}}$ *and* $\langle c_i : i < \lambda_{\mathfrak{s}} \rangle$ *list the elements of* M^*, $\mathbf{P} \subseteq \mathscr{S}^{\mathrm{bs}}_{\mathfrak{s}}(M^*)$ *is a non-empty set of types such that there is* $q \in \mathscr{S}^{\mathrm{bs}}_{\mathfrak{s}}(M^*)$ *orthogonal to* $\mathbf{P}$ *(so* $\mathfrak{s}$ *is not weakly uni-dimensional)*

(c) $\tau^* = \tau \cup \{c_i : c \in M_*\}$

(d) $K^* = \{M : M$ *is a* τ^**-model,* $M \restriction \tau \in K^{\mathfrak{s}}$ *and* $c_i \mapsto c_i^M$ *is a* $\leq_{\mathfrak{K}[\mathfrak{s}]}$*-embedding of* M^* *into* $M \restriction \tau$ *and if if* $M^* \leq_{\mathfrak{K}} M' \leq_{\mathfrak{K}} M \restriction \tau, M' \in K_\lambda$ *and* $p \in \mathbf{P}$ *<u>then</u>* p *has a unique extension in* $\mathscr{S}_{\mathfrak{s}}(M')\}$

(e) $\mathfrak{K}^* = (K^*, \leq_{\mathfrak{K}^*})$ *where* $M_1 \leq_{\mathfrak{K}^*} M_2$ *<u>iff</u>* $(M_1, M_2 \in K^*$ *and)* $M_1 \restriction \tau \leq_{\mathfrak{K}[\mathfrak{s}]} M_2 \restriction \tau$

(f) $\mathfrak{s}^* = (\mathfrak{K}^*, \mathscr{S}^{\mathrm{bs}}_*, \bigcup\limits_*)$ *where*

  (i) $\mathscr{S}^{\mathrm{bs}}_*(M_1)$ *is essentially* $\{\mathbf{tp}_{\mathfrak{K}^*}(a, M_1, M_2) : M_1 \leq_{\mathfrak{K}^*} M_2$ *both of cardinality* $\lambda_{\mathfrak{s}}$ *and* $\mathbf{tp}_{\mathfrak{s}}(a, M_1 \restriction \tau, M_2 \restriction \tau \in \mathscr{S}^{\mathrm{bs}}_{\mathfrak{s}}(M_1 \restriction \tau)\}$

  (ii) $\bigcup\limits_*$ *similarly, i.e.,* $\bigcup\limits_*(M_0, M_1, a, M_3)$ *<u>iff</u>* $M_0 \leq_{\mathfrak{K}^*} M_1 \leq_{\mathfrak{K}^*} M_3, a \in M_3$ *and* $\bigcup\limits_{\mathfrak{s}}(M_0, M_1, a, M_3)$.

<u>Then</u>

(α) $\mathfrak{s}^*$ *is a good* λ*-frame*

(β) $K^{\mathfrak{s}^*} \subseteq K^{\mathfrak{s}}, K^{\mathfrak{s}^*}_\lambda \neq \emptyset, K^{\mathfrak{s}^*}_{\lambda^+} \neq \emptyset$ *so* $\dot{I}(\mu, K^{\mathfrak{s}^*}) \leq \dot{I}(\mu, K^{\mathfrak{s}^*})$ *for* $\mu \geq \lambda$

(γ) *if* $\dot{I}(\lambda^{+n+1}, K^{\mathfrak{s}^*}) < \mu_{\mathrm{unif}}(\lambda^{+n+1}, 2^{\lambda^{+n}})$ *for* $n < \omega$, *<u>then</u>* $\mathfrak{s}^*$ *is* ω*-successful.*

Proof. <u>Clause (α)</u>.

Check

<u>Clause (β)</u>.

Trivial.

<u>Clause (γ)</u>.

By 12.37 applied to $\mathfrak{s}^*$. $\square_{12.39}$

12.40 Discussion: We are interested in the models of $K^{\mathfrak{s}}$ of cardinality $\geq \lambda_{\mathfrak{s}}^{+\omega}$. Of course, we assume that $\mathfrak{s}$ is ω-beautiful[3], which follows from $\langle 2^{\lambda_{\mathfrak{s}}^{+n}} : n < \omega \rangle$ is increasing and $\mathfrak{s}$ is ω-successful by Theorem 12.37 (otherwise we are stuck) and this follows from "$\mathfrak{s}$ is a good λ-frame and $\dot{I}(\lambda^{+n+1}, K^{\mathfrak{s}}) < \mu_{\text{unif}}(\lambda^{+n+1}, 2^{\lambda^{+n}})$ for each n. For simplicity we assume $\mathfrak{s}$ is good$^+$ (as $\mathfrak{s}^+$ is good$^+$ in any case) and without loss of generality $\mathfrak{s}$ is categorical (in λ).

First, let us look at $\mathfrak{s}$ which is weakly uni-dimensional (see §2 mainly, Definition 2.2, claims 2.9,2.11, 2.12) then $K^{\mathfrak{s}}$ is categorical in λ^+ hence $\mathfrak{K}_{\mathfrak{s}^+} = K^{\mathfrak{s}}_{\lambda_{\mathfrak{s}}^+}$ and in general $K_{\mathfrak{s}(+n)} = K^{\mathfrak{s}}_{\lambda^{+n}}$, so $K^{\mathfrak{s}}$ is categorical in λ^{+n} for each n. In this case we can prove that $K^{\mathfrak{s}}$ is categorical in μ for every μ and we can lift $\mathfrak{s}$ to an ω-beautiful $\mathfrak{s}[\mu]$ for every $\mu \geq \lambda$.

Second, we look at the case $\mathfrak{s}$ is not weakly uni-dimensional. We can naturally continue to define $\mathfrak{s}^{+\alpha} = \mathfrak{s}[\lambda^{+\alpha}]$ for $\alpha \geq \mu$ (restricting ourselves to superlimit models), again a good $\lambda^{+\alpha}$-frame categorical in $\lambda^{+\alpha}$ which is ω-beautiful[3], and relates naturally to $\mathfrak{s}^{+\beta}$ for $\beta < \alpha$. In particular, each $\mathfrak{s}^{+\alpha}$ is not weakly uni-dimensional (as $\mathfrak{s}$ is not weakly uni-dimensional, and we can lift this) hence $K^{\mathfrak{s}^{+\alpha}}$ is not categorical in $\lambda^{+\alpha+1}$. So $K^{\mathfrak{s}}$ has models in every cardinality $\mu \geq \lambda$ and is not categorical in every successor cardinality $\mu > \lambda$. This unfortunately leaves out the limit cardinals $\mu > \lambda$.

Thirdly, we can remedy this (take care of the limit cardinals) as follows. Without loss of generality $\mathfrak{s}$ has primes and is as in 12.3 (otherwise use $\mathfrak{s}^{+3}$, see 12.2). We can find $M^* \in K_{\mathfrak{s}}$ and $p, q \in \mathscr{S}^{\text{bs}}_{\mathfrak{s}}(M)$ which are orthogonal so we can define $\mathfrak{t}$ as in 12.39, so $\tau_{\mathfrak{K}[\mathfrak{t}]} = \tau_{\mathfrak{K}} \cup \{c_a : a \in M^*\}, N \in K^{\mathfrak{t}} \Rightarrow M^* \leq_{\mathfrak{K}[\mathfrak{s}]}, N \restriction \tau_{\mathfrak{K}}$. We still do not know that $\mathfrak{t}$ is ω-successful, as though $K^{\mathfrak{s}}_{\mu}, K^{\mathfrak{t}}_{\mu}$ are closely related for $\mu = \lambda_{\mathfrak{s}}$ and somewhat related for $\mu = \lambda_{\mathfrak{s}}^+$, we do not know about the relation later. Still if $n < \omega \Rightarrow \dot{I}(\lambda^{+n+1}, K^{\mathfrak{s}}) < \mu_{\text{unif}}(\lambda^{+n+1}, 2^{\lambda^{+n}})$ then also $n < \omega \Rightarrow \dot{I}(\lambda^{+n+1}, K^{\mathfrak{t}}) < \mu_{\text{unif}}(\lambda^{+n+1}, 2^{\lambda^{+n}})$ and so we can prove that $\mathfrak{t}$ is n-successful. From this by the above $\mu > \lambda \Rightarrow K^{\mathfrak{t}}_{\mu} \neq \emptyset$. But choose $M_1 \in K^{\mathfrak{s}}_{\mathfrak{s}[\mu]}, M_2 \in K_{\mathfrak{t}[\mu]}$ hence M_1 is $\lambda_{\mathfrak{s}}^+$-saturated above $\lambda_{\mathfrak{s}}$, whereas M_2 is not so $K^{\mathfrak{s}}$ is not categorical for each $\mu > \lambda$.

Fourth, though the assumption $\dot{I}(\lambda^{+n+1}, K^{\mathfrak{s}}) < \mu_{\text{unif}}(\lambda^{+n+1}, 2^{\lambda^{+n}})$ for $n < \omega$ is reasonable we may want to eliminate it. This involves a much more serious drawback of the above: it did not say much on $K^{\mathfrak{s}}$, $K^{\mathfrak{s}(+n)}$ and even on $\mathfrak{K}^{\mathfrak{s}(+\omega)}$. The last one can be remedied: working harder we can prove that $K^{\mathfrak{s}(+\omega)}$ is ω-beautiful[1].

This has strong consequences: in fact we can understand $\mathfrak{K}^{\mathfrak{s}(+\omega)}$, it has amalgamation and we can define a ω-beautiful[1], good$^+$, $\lambda^{+\alpha}$-frame $\mathfrak{s}(\lambda^{+\alpha})$ such that $\mathfrak{K}_{\mathfrak{s}(\lambda^{+\alpha})} = K^{\mathfrak{s}}_{\lambda^{+\alpha}}$ and it relates naturally to $\mathfrak{s}(\lambda^{+\beta})$ for $\beta < \alpha$.

Let us return to the consequences of "$\mathfrak{s}$ is not weakly uni-dimensional". Again $\mathfrak{s}^{+\omega}$ is not weakly uni-dimensional hence by ω-beautiful[1] (not proved here) for every $\mu > \lambda^{+\omega}$ we can build a model $M_\mu \in K^{\mathfrak{s}(+\omega)}_\mu$ which is not $\lambda^{+\omega+1}$-saturated for $\mathfrak{K}^{\mathfrak{s}(+\omega)}$. By the omitting type theorem for a.e.c. (see [Sh 394]) we can find $M'_\mu \in K^{\mathfrak{s}}_\mu$ which is not λ^+-saturated for $K^{\mathfrak{s}}$. As the $M \in \mathfrak{K}^{\mathfrak{s}[\mu]}$ is $\lambda^+_{\mathfrak{s}}$-saturated we are done.

12.41 Major Conclusion. Assume that $\mathfrak{s}$ satisfies the conclusion of 12.37 and let $\mathfrak{t} = \mathfrak{s}^{+\omega}$ (see Definition 0.4(4)) <u>then</u> we can define $\langle \mathfrak{s}_\mu : \mu \geq \lambda \rangle$ such that

(a) $\mathfrak{s}_{\lambda^{+\omega}} = \mathfrak{s}^{+\omega} = \mathfrak{s}(+\omega)$ is a good $\lambda^{+\omega}_{\mathfrak{s}}$-frame (recall that $\mathfrak{K}_{\mathfrak{s}(+\omega)}$ is $\cap\{\mathfrak{K}^{\mathfrak{s}(+n)}_{\lambda^{+\omega}} : n < \omega\}$)

(b) $\mathfrak{t} = \mathfrak{s}^{+\omega}$ is ω-beautiful[3]

(c) (α) if $\mu = \lambda^{+n}$ then $\mathfrak{s}_\mu = \mathfrak{s}^{+n}$

(β) if $\mu \geq \lambda^{+\omega}, \mathfrak{s}_\mu$ is a good μ-frame, which is beautiful and categorical in μ

(γ) $\mathfrak{s}_{\mu^+} = (\mathfrak{s}_\mu)^+$

(δ) if $\mu \geq \lambda^{+\omega}$ is a limit ordinal then

(i) $K_{\mathfrak{s}_\mu} = \cap\{K^{\mathfrak{s}_\theta} : \theta \in [\lambda, \mu)\}$

(ii) $\bigcup_{\mathfrak{s}_\mu}$ and $\mathscr{S}^{\rm bs}_{\mathfrak{s}_\mu}$ are defined as in II§2 from $\langle \mathfrak{s}_\theta : \theta \in [\lambda, \mu) \rangle$

(iii) similarly $\mathrm{NF}_{\mathfrak{s}_\mu}, K^{3,\mathrm{mx}}_{\mathfrak{s}}$

(d) if $\mathfrak{s}$ is weakly uni-dimensional and $\mu \geq \lambda$ <u>then</u> $\mathfrak{s}_\mu = \mathfrak{s}[\mu]$ and for $\mu \geq \lambda^{+\omega}, \mathfrak{t}[\mu] = \mathfrak{s}[\mu]$; hence $K^{\mathfrak{s}(+)}$ is categorical in μ for every $\mu \geq \lambda^+_{\mathfrak{s}}$ hence

$(d)'$ if in addition $K^{\mathfrak{s}}$ is categorical in $\lambda_{\mathfrak{s}}^{+}$ then $K^{\mathfrak{s}}$ is categorical in μ for every $\mu \geq \lambda_{\mathfrak{s}}$

(e) if $\mathfrak{s}$ is not weakly uni-dimensional <u>then</u> each $\mathfrak{s}(\mu)$ is not weakly uni-dimensional

(f) $\mathfrak{s}$ has NDOP iff $\mathfrak{t} = \mathfrak{s}^{+\omega}$ has NDOP iff $\mathfrak{s}(\mu)$ has NDOP (for any $\mu \geq \lambda_{\mathfrak{s}}$)

(g) $K_{\mu}^{\mathfrak{s}} \neq \emptyset$ for $\mu \geq \lambda, \mathfrak{s}[\mu], \mathfrak{s}(\mu)$ are well defined for $\mu \geq \lambda$ (on $\mathfrak{s}[\mu]$ see Definition 0.4(4)).

Before we prove note:

12.42 Conclusion. Assume $2^{\lambda^{+n}} < 2^{\lambda^{n+1}}$ for $n < \omega$ and

(a) $\mathfrak{s}$ is a good λ-frame not weakly uni-dimensional

(b) $\dot{I}(\lambda^{+n+1}, K^{\mathfrak{s}}) < \mu_{\text{unif}}(\lambda^{+n+1}, 2^{\lambda^{+n}})$ for $n < \omega$.

<u>Then</u> $K^{\mathfrak{s}}$ is not categorical in μ, for every $\mu > \lambda$.

Remark. We can add to 12.41 (but will be dealt with elsewhere)

(h) $\mathfrak{t}(\mu)$ is a beautiful[1] frame

(i) $\mathfrak{K}^{\mathfrak{t}}$ is essentially the class of $\lambda_{\mathfrak{s}}^{+\omega}$-saturated models from $\mathfrak{K}^{\mathfrak{s}}$, pednatically it is the class of $\cap\{K^{\mathfrak{s}(+n)} : n < \omega\}$.

This is done elsewhere mainly [Sh 842], where we separate the various aspects (in particular the existence).

Proof of 12.41. Should be clear if you arrive here.

We still give some details. The properties are defined such that they provably exist. E.g. a typical point of clause (a) is:

$(*)_1$ $\mathfrak{K}_{\mathfrak{t}}$ has the disjoint amalgamation property.

[Where? Let $\mu = \lambda^{+\omega}$. Assume that $M_{\emptyset} \leq_{\mathfrak{t}} M_{\{\ell\}}$ for $\ell = 0, 1$, hence $\|M_\ell\| = \mu$ for $\ell = 0, 1, 2$ and $M_1 \cap M_2 = M_0$. For $\alpha < \mu$ let $n(\alpha) = \text{Min}\{n \geq n : \lambda^{+n} \geq |\alpha|\}$. Let χ be large enough and we choose $\mathfrak{B}_\alpha \prec (\mathscr{H}(\chi), \in)$ for $\alpha < \mu$, increasing continuous with α

such that $\|\mathfrak{B}_\alpha\| = \lambda^{+1} + |\alpha|$ and $\mathfrak{s}, M_0, M_1, M_2$ belongs to $\mathfrak{B}_0$. Now let $M^u_\alpha := M_u \restriction \mathfrak{B}_\alpha$ for $\alpha \le \mu, u \subset \{0,1\}$; clearly $M^u_\alpha \in K_{\mathfrak{s}(+n(\alpha))}$ is $\le_{\mathfrak{K}}$-increasing continuous with union. Now we choose $M^{\{0,1\}}_\alpha$ by induction on α such that

⊛ (a) $M^{\{0,1\}}_\alpha \in \mathfrak{K}_{\mathfrak{s}(+n)}$ if $n = n(\alpha)$

(b) $M^{\{0,1\}} \in \mathfrak{K}_{\mathfrak{s}(+n)}$ if $n = n(\alpha)$

(c) $\mathbf{m}_\alpha = \langle M^u_\alpha : u \subseteq \{0,1\}\rangle$ is a brimmed[3] stable $((\langle_7^{+n(\alpha)}, \mathscr{P}(2), \mathfrak{s}^{+n(\alpha)})$-system

(d) $\mathbf{m}^\alpha = \mathbf{m}_\alpha *_{\mathscr{P}(3)} \mathbf{m}_{\alpha+1}$ is a brimmed[3] stable $(\lambda^{+n(\alpha)}, \mathscr{P}(3), \mathfrak{s}^{+n(\alpha)})$-system

(e) $M^{\{0,1\}}_\alpha \cap M^u_\alpha = M^u_\alpha$ for $u \subset 2$.

Note that easily $\mathbf{t}_\alpha$ is brimmed[3] stable $(\mathscr{P}^-(\{0,1,2,\})\backslash\{\{0,1\}\}, \mathfrak{s}^{n(\alpha)})$-system. The proof should be clear. $\square_{12.37}$

Proof of 12.42. For each $\mu > \lambda$, there is $M^1_\mu \in K_{\mathfrak{s}[\mu]}$ see 12.41. But we can define $\mathfrak{t}$ as $\mathfrak{s}^*[+\omega]$, where $\mathfrak{s}^*$ is as in 12.39 and apply it to 12.37, 12.41, and get $M^2_\mu \in K_{\mathfrak{t}[\mu]}$. Looking at $\le_{K[\mathfrak{s}]}$-submodels of M^1_μ, M^2_μ it is clear that $M^1_\mu \approx M^2_\mu$ so we are done. $\square_{12.42}$

We can sum up

12.43 Conclusion. Assume $2^{\lambda^{+n}} < 2^{\lambda^{+n+1}}$ for $n < \omega$. If an a.e.c. $\mathfrak{K}$ with $\mathrm{LS}(\mathfrak{K}) \le \lambda$, is categorical in λ, λ^+, $1 \le \dot{I}(\lambda^{++}, K)$ and $\dot{I}(\lambda^{+n+2}, K) < \mu_{\mathrm{unif}}(\lambda^{n+2}, 2^{\lambda^{+n+1}})$ for $n < \omega$ <u>then</u> $\mathfrak{K}$ is categorical in every $\mu \ge \lambda$.

12.44 Remark. 1) This through light on [MaSh 285], [KlSh 362], [Sh 472], [Sh 394], (see more in Chapter N) and see Theorem IV.7.12. In those works we start with an appropriate a.e.c. $\mathfrak{K}$ and assume that it is categorical in λ large enough then $\mathrm{LS}(\mathfrak{K})$ and prove that for some $\alpha_* < (2^{\mathrm{LS}(\mathfrak{K})})^+$ the class is categorical in every $\lambda' \in [\beth_{\alpha_*}, \lambda)$, but nothing is said about $\lambda' > \lambda$. However, if for some $\mu, \mu^{+\omega} \in (\beth_{\alpha_*}, \lambda]$ then by 12.43 we are done. This weak set theoretic assumption will

be eliminated in a sequel.
2) Moreover, we can eliminate the "λ successor" assumption.
3) We can say much more: ω-successful frames are very much like superstable first order classes and more. See on this mainly [Sh 842].

CATEGORICITY AND SOLVABILITY OF A.E.C., QUITE HIGHLY SH734

§0 Introduction

The hope which motivates this work is

0.1 Conjecture: If $\mathfrak{K}$ is an a.e.c. then either for every large enough cardinal $\mu, \mathfrak{K}$ is categorical in μ or for every large enough cardinal $\mu, \mathfrak{K}$ is not categorical in μ.
Why do we consider this a good dream? See Chapter N.

Our main result is 4.10, it says that if $\mathfrak{K}$ is categorical in μ (ignoring few exceptional μ's) and $\lambda \in [\mathrm{LS}(\mathfrak{K}), \mu)$ has countable cofinality and is a fix point of the sequence of the $\beth_\alpha$'s, (moreover a limit of such cardinals) then there is a superlimit $M \in K_\lambda$ for which $\mathfrak{K}_{[M]} = \mathfrak{K}_\lambda \restriction \{M' : M' \cong M\}$ has the amalgamation property (and a good λ-frame $\mathfrak{s}$ with $\mathfrak{K}_{\mathfrak{s}} = \mathfrak{K}_{[M]}$). Note that Chapter III seems to give a strong indication that finding good λ-frames is a significant advance. This may be considered an unsatisfactory evidence of an advance, being too much phrased in the work's own terms. So we prove in §5 - §7 that for a restrictive context we make a clear cut advance: assuming amalgamation and enough instances of $2^\lambda < 2^{\lambda^+}$ occurs, much more than the conjecture holds, see Chapter N on background.

Note that as we try to get results on $\lambda = \beth_\lambda > \mathrm{LS}(\mathfrak{K})$, clearly it does not particularly matter if for $\kappa \in (\mathrm{LS}(\mathfrak{K}), \lambda)$ we use, e.g. $\kappa_1 = \kappa^+$ or $\kappa_1 = \beth_{(2^\kappa)^+} (= \beth_{1,1}(\kappa))$ or even $\beth_{1,7}(\kappa)$.

After 4.10 the next natural step is to show that $\mathfrak{s}_\lambda$ has the better properties dealt with in Chapter II, Chapter III, see [Sh:F782]. Note that if we strengthen the assumption on μ in §4 (to $\mu = \mu^{<\lambda}$), then it relies on §1 only. Without this we need §2 (hence 5.1(1),(4)).

Originally we have used here categoricity assumptions but lately it seems desirable to use a weaker one: (variants of) solvability. About being solvable, see N§4(B), [Sh 842]. This seems better as it is a candidate for being an "outside" generalization of being superstable (rather than of being categorical).
Here we use solvable when it does not require much change; for more on it see [Sh 842], [Sh:F820] and on material delayed from here see [Sh:F782].
Note we can systematically use $K^{\mathrm{sc}(\theta)\text{-lin}}$, say with $\theta = \aleph_0$ or $\theta = \mathrm{LS}(\mathfrak{K})$ instead of K^{lin}; see Definition 0.14(8). In several respects this is better, but not enough to make us use it. Also working more it seemed we can get rid of "wide", "wide over", see Definition 0.14(1),(2),(3). If instead proving the existence of a good λ-frame it suffices for us to prove the existence of almost good λ-frame, then the assumption on λ can be somewhat weaker (fixed point instead limit of fix points of the sequence of the $\beth_\alpha$'s). In §7 we sometimes give alternative quotations in [Sh 394] but do not rely on it.
We thank Mor Doron, Esther Gruenhut, Aviv Tatarski and Alex Usvyatsov for their help in proofreading.
Basic knowledge on infinitary logics is assumed, see e.g. [Di]; though the reader may just read the definition here in N§5 and believe some quoted results.

0.2 Notation. Let $\beth_{0,\alpha}(\lambda) = \beth_\alpha(\lambda) := \lambda + \Sigma\{\beth_\beta(\lambda) : \beta < \alpha\}$. Let $\beth_{1,\alpha}(\lambda)$ be defined by induction on $\alpha : \beth_{1,0}(\lambda) = \lambda$, for limit β we let $\beth_{1,\beta} = \sum\limits_{\gamma<\beta} \beth_{1,\gamma}$ and $\beth_{1,\beta+1}(\lambda) = \beth_\mu$ where $\mu = (2^{\beth_{1,\beta}(\lambda)})^+$.

0.3 Remark. 1) For our purpose, usually $\beth_{1,\beta+1}(\lambda) = \beth_{\delta(\mu)}$ where $\mu = \beth_{1,\beta}(\lambda)$ suffice, see e.g. V.A§1 in particular on $\delta(-)$. Generally $\mu = (\beth_{1,\beta}(\lambda))^+$ is a more natural definition, but:

(*a*) the difference is not significant, e.g. for α limit we get the same value

(*b*) our use of omitting types makes our choice more natural.

2) We do not use but it is natural to define $\beth_{\gamma+1,0}(\lambda) = \lambda$, $\beth_{\gamma+1,\beta+1}(\lambda) = \beth_{\gamma,\mu}(\lambda)$ with $\mu = (2^{\beth_{\gamma+1,\beta}(\lambda)})^+$, $\beth_{\gamma+1,\delta}(\lambda) = \sum\limits_{\beta<\delta} \beth_{\gamma+1,\beta}(\lambda)$ and $\beth_{\delta,0}(\lambda) = \sup\{\beth_{\gamma,0}(\lambda) : \gamma < \delta\} = \lambda, \beth_{\delta,\beta+1}(\lambda) = \beth_{\delta,\beta}(\beth_{\delta,\beta}(\lambda))$, $\beth_{\delta,\delta_1} = \sup\{\beth_{\delta,\alpha}(\lambda) : \alpha < \delta_1\}$; this is used, e.g. in [Sh:g, ChV].

0.4 Definition. Assume M is a model, $\tau = \tau_M$ is its vocabulary and Δ is a language (or just a set of formulas) in some logic, in the vocabulary τ.

For any set $A \subseteq M$ and set Δ of formulas in the vocabulary τ_M, let $\mathrm{Sfr}^\alpha_\Delta(A, M)$ which we call the set of formal (Δ, α)-types over A in M, be the set of p such that

- (a) p a set of formulas of the form $\varphi(\bar{x}, \bar{a})$ where $\varphi(\bar{x}, \bar{y}) \in \Delta, \bar{x} = \langle x_i : i < \alpha\rangle$ and $\bar{a} \in {}^{\ell g(\bar{y})}A$
- (b) if Δ is closed under negation (which is the case we use here) then for any $\varphi(\bar{x}, \bar{y}) \in \Delta$ with $\bar{x}$ as above and $\bar{a} \in {}^{\ell g(\bar{y})}A$ we have $\varphi(\bar{x}, \bar{a}) \in p$ or $\neg\varphi(\bar{x}, \bar{a}) \in p$.

Recall

0.5 Definition. 1) For $\mathfrak{K}$ an a.e.c. we say $M \in \mathfrak{K}_\theta$ is a superlimit (model in $\mathfrak{K}$ or in $\mathfrak{K}_\theta$) when:

- (a) M is universal
- (b) if δ is a limit ordinal $< \theta^+$ and $\langle M_\alpha : \alpha \le \delta\rangle$ is $\le_{\mathfrak{K}_\theta}$-increasing continuous and $\alpha < \delta \Rightarrow M_\alpha \cong M$ then $M_\delta \cong M$ (equivalently, $\mathfrak{K}^{[M]}_\theta = \mathfrak{K} \restriction \{N : N \cong M\}$ is a θ-a.e.c.)
- (c) there is N such that $M <_{\mathfrak{K}} N \in \mathfrak{K}_\theta$ and N is isomorphic to M.

2) We say $M \in \mathfrak{K}_\theta$ is locally superlimit when we weaken clause (a) to

- $(a)^-$ if $N \in \mathfrak{K}_\theta$ is a $\le_{\mathfrak{K}}$-extension of M then N can be $\le_{\mathfrak{K}}$-embedded into M.

3) We say that M is pseudo superlimit when in part (1) clauses (b),(c) hold (but we omit clause (a)); see 0.6(7) below.
3A) For $M \in K_\lambda$ let $\mathfrak{K}_{[M]} = \mathfrak{K}_\lambda^{[M]}$ be $\mathfrak{K} \restriction \{N : N \cong M\}$.
4) In (1) we may say globally superlimit.

0.6 Observation. Assume ($\mathfrak{K}$ is an a.e.c. and) $\mathfrak{K}_\lambda \neq \emptyset$.
1) If $\mathfrak{K}$ is categorical in λ and there are $M <_{\mathfrak{K}_\lambda} N$ then every $M \in \mathfrak{K}_\lambda$ is superlimit.
2) If every/some $M \in \mathfrak{K}_\lambda$ is superlimit then every/some $M \in K_\lambda$ is locally superlimit.
3) If every/some $M \in \mathfrak{K}_\lambda$ is locally superlimit then every/some $M \in \mathfrak{K}_\lambda$ is pseudo superlimit.
4) If some $M \in \mathfrak{K}_\lambda$ is superlimit then every locally superlimit $M' \in \mathfrak{K}_\lambda$ is isomorphic to M.
5) If M is superlimit in $\mathfrak{K}$ then M is locally superlimit in $\mathfrak{K}$. If M is locally superlimit in $\mathfrak{K}$, then M is pseudo superlimit in $\mathfrak{K}$. If M is locally superlimit in $\mathfrak{K}_\theta$ then $\mathfrak{K}_\theta$ has the joint embedding property iff M is superlimit.
6) In Definition 0.5(1), clause (c) follows from

$(c)^-$ $\mathrm{LS}(\mathfrak{K}) \leq \theta$ and $K_{\geq \theta^+} \neq \emptyset$.

7) $M \in K_\lambda$ is pseudo-superlimit iff $\mathfrak{K}_{[M]}$ is a λ-a.e.c. and $\leq_{\mathfrak{K}_{[M]}}$ is not the equality. Also Definition 0.5(3A) is compatible with II.1.25.

0.7 Definition. For an a.e.c. $\mathfrak{K}$, let $\mathfrak{K}^{\mathrm{sl}}_\mu, \mathfrak{K}^{\mathrm{ls}}_\mu, \mathfrak{K}^{\mathrm{pl}}_\mu$ be the class of $M \in \mathfrak{K}_\mu$ which are superlimit, locally superlimit, pseudo superlimit respectively with the partial order $\leq_{\mathfrak{K}^{\mathrm{sl}}_\mu}, \leq_{\mathfrak{K}^{\mathrm{ls}}_\mu}, \leq_{\mathfrak{K}^{\mathrm{pl}}_\mu}$ being $\leq_{\mathfrak{K}} \restriction K^{\mathrm{sl}}_\mu, \leq_{\mathfrak{K}} \restriction K^{\mathrm{pl}}_\mu$ respectively.

0.8 Definition. 1) Φ is proper for linear orders when:

(a) for some vocabulary $\tau = \tau_\Phi = \tau(\Phi), \Phi$ is an ω-sequence, the n-th element a complete quantifier free n-type in the vocabulary τ

(b) for every linear order I there is a τ-model M denoted by $\mathrm{EM}(I, \Phi)$, generated by $\{a_t : t \in I\}$ such that $s \neq t \Rightarrow a_s \neq$

a_t for $s, t \in I$ and $\langle a_{t_0}, \ldots, a_{t_{n-1}} \rangle$ realizes the quantifier free n-type from clause (a) whenever $n < \omega$ and $t_0 <_I \ldots <_I t_{n-1}$; so really M is determined only up to isomorphism but we may ignore this and use $I_1 \subseteq J_1 \Rightarrow \mathrm{EM}(I_1, \Phi) \subseteq \mathrm{EM}(I_2, \Phi)$. We call $\langle a_t : t \in I \rangle$ "the" skeleton of M; of course again "the" is an abuse of notation as it is not necessarily unique.

1A) If $\tau \subseteq \tau(\Phi)$ then we let $\mathrm{EM}_\tau(I, \Phi)$ be the τ-reduct of $\mathrm{EM}(I, \Phi)$.
2) $\Upsilon^{\mathrm{or}}_\kappa[\mathfrak{K}]$ is the class of Φ proper for linear orders satisfying clauses $(a)(\alpha), (b), (c)$ of Claim 0.9(1) below and $|\tau(\Phi)| \leq \kappa$. The default value of κ is $\mathrm{LS}(\mathfrak{K})$ and then we may write $\Upsilon^{\mathrm{or}}_{\mathfrak{K}}$ or $\Upsilon^{\mathrm{or}}[\mathfrak{K}]$ and for simplicity always $\kappa \geq \mathrm{LS}(\mathfrak{K})$ (and so $\kappa \geq |\tau_{\mathfrak{K}}|$).
3) We define "Φ proper for K" similarly when in clause (b) of part (1) we demand $I \in K$, so K is a class of τ_K-models, i.e.

- (a) Φ is a function, giving for a quantifier free n-type in τ_K, a quantifier free n-type in τ_Φ
- $(b)'$ in clause (b) of part (1), the quantifier free type which $\langle a_{t_0}, \ldots, a_{t_{n-1}} \rangle$ realizes in M is $\Phi(\mathrm{tp}_{\mathrm{qf}}(\langle t_0, \ldots, t_{n-1} \rangle, \emptyset, M))$ for $n < \omega$, $t_0, \ldots, t_{n-1} \in I$.

0.9 Claim. *1) Let $\mathfrak{K}$ be an a.e.c. and $M \in K$ be of cardinality $\geq \beth_{1,1}(\mathrm{LS}(\mathfrak{K}))$ recalling we naturally assume $|\tau_{\mathfrak{K}}| \leq \mathrm{LS}(\mathfrak{K})$ as usual.* *Then there is a Φ such that Φ is proper for linear orders and:*

- (a) (α) $\tau_{\mathfrak{K}} \subseteq \tau_\Phi$,
  - (β) $|\tau_\Phi| = \mathrm{LS}(\mathfrak{K}) + |\tau_{\mathfrak{K}}|$
- (b) *for any linear order I the model $\mathrm{EM}(I, \Phi)$ has cardinality $|\tau(\Phi)| + |I|$ and we have $\mathrm{EM}_{\tau(\mathfrak{K})}(I, \Phi) \in K$*
- (c) *for any linear orders $I \subseteq J$ we have $\mathrm{EM}_{\tau(\mathfrak{K})}(I, \Phi) \leq_{\mathfrak{K}} \mathrm{EM}_{\tau(\mathfrak{K})}(J, \Phi)$*
- (d) *for every finite linear order I, the model $\mathrm{EM}_{\tau(\mathfrak{K})}(I, \Phi)$ can be $\leq_{\mathfrak{K}}$-embedded into M.*

2) If we allow $\mathrm{LS}(\mathfrak{K}) < |\tau_{\mathfrak{K}}|$ and there is $M \in \mathfrak{K}$ of cardinality $\geq \beth_{1,1}(\mathrm{LS}(\mathfrak{K}) + |\tau_{\mathfrak{K}}|)$, then there is $\Phi \in \Upsilon^{\mathrm{or}}_{\mathrm{LS}(\mathfrak{K})+|\tau(\Phi)|}[\mathfrak{K}]$ such that

$\mathrm{EM}(I,\Phi)$ *has cardinality* $\leq \mathrm{LS}(\mathfrak{K})$ *for* I *finite. Hence* $\mathscr{E}$ *has* $\leq 2^{\mathrm{LS}(\mathfrak{K})}$ *equivalence classes where* $\mathscr{E} = \{(P_1, P_2) : P_1, P_2 \in \tau_\Phi$ *and* $P_1^{\mathrm{EM}(I,\Phi)} = P_2^{\mathrm{EM}(I,\Phi)}$ *for every linear order* $I\}$.
3) Actually having a model of cardinality $\geq \beth_\alpha$ *for every* $\alpha < (2^{\mathrm{LS}(\mathfrak{K})+|\tau(\mathfrak{K})|})^+$ *suffice (in part (2)).*

Proof. Follows from the existence of a representation of $\mathfrak{K}$ as a $\mathrm{PC}_{\mu,2^\mu}$-class when $\mu = \mathrm{LS}(\mathfrak{K}) + |\tau(\mathfrak{K})|$ in I.1.4(3),(4),(5) and I.1.9 (or see [Sh 394, 0.6]). $\square_{0.9}$

0.10 Remark. Note that some of the definitions and claims below will be used only in remarks: $K_\theta^{\mathrm{sc}(\kappa)}$ from 0.14(8), in 1.7; and some only in §6,§7 (and part of §5 needed for it): $\Upsilon_\kappa^{\mathrm{lin}}[2]$ from 0.11(5) (and even less $\Upsilon_\kappa^{\mathrm{lin}}[\alpha(*)]$ from Definition 0.14(9)). Also the use of $\leq_\kappa^\otimes, \leq_\kappa^{\mathrm{ie}}, \leq_\kappa^\oplus$ is marginal.

0.11 Definition. We define partial orders $\leq_\kappa^\oplus, \leq_\kappa^{\mathrm{ie}}$ and $\leq_\kappa^\otimes$ on $\Upsilon_\kappa^{\mathrm{or}}[\mathfrak{K}]$ (for $\kappa \geq \mathrm{LS}(\mathfrak{K})$) as follows:
1) $\Psi_1 \leq_\kappa^\oplus \Psi_2$ <u>if</u> $\tau(\Psi_1) \subseteq \tau(\Psi_2)$ and $\mathrm{EM}_{\tau(\mathfrak{K})}(I,\Psi_1) \leq_{\mathfrak{K}} \mathrm{EM}_{\tau(\mathfrak{K})}(I,\Psi_2)$ and $\mathrm{EM}(I,\Psi_1) = \mathrm{EM}_{\tau(\Psi_1)}(I,\Psi_1) \subseteq \mathrm{EM}_{\tau(\Psi_1)}(I,\Psi_2)$ for any linear order I.
Again for $\kappa = \mathrm{LS}(\mathfrak{K})$ we may drop the κ.
2) For $\Phi_1, \Phi_2 \in \Upsilon_\kappa^{\mathrm{or}}[\mathfrak{K}]$, we say Φ_2 is an inessential extension of Φ_1 and write $\Phi_1 \leq_\kappa^{\mathrm{ie}} \Phi_2$ <u>if</u> $\Phi_1 \leq_\kappa^\oplus \Phi_2$ and for every linear order I, we have (note: there may be more function symbols in $\tau(\Phi_2)$!)

$$\mathrm{EM}_{\tau(\mathfrak{K})}(I,\Phi_1) = \mathrm{EM}_{\tau(\mathfrak{K})}(I,\Phi_2).$$

3) Let $\Upsilon_\kappa^{\mathrm{lin}}$ be the class of Ψ proper for linear order and (producing a linear order extending the original one, i.e.) such that:

- (a) $\tau(\Psi)$ has cardinality $\leq \kappa$ and the two-place predicate $<$ belongs to $\tau(\Psi)$
- (b) $\mathrm{EM}_{\{<\}}(I,\Psi)$ is a linear order which is an extension of I in the sense that $\mathrm{EM}(I,\Phi) \models$ "$a_s < a_t$" iff $I \models$ "$s < t$"; in fact we usually stipulate $[t \in I \Rightarrow a_t = t]$.

4) $\Phi_1 \leq_\kappa^\otimes \Phi_2$ <u>iff</u> there is Ψ such that

(a) $\Psi \in \Upsilon_\kappa^{\rm lin}$

(b) $\Phi_\ell \in \Upsilon_\kappa^{\rm or}[\mathfrak{K}]$ for $\ell = 1, 2$

(c) $\Phi'_2 \le_\kappa^{\rm ie} \Phi_2$ where $\Phi'_2 = \Psi \circ \Phi_1$, i.e. for every linear order I we have

$$\mathrm{EM}(I, \Phi'_2) = \mathrm{EM}(\mathrm{EM}_{\{<\}}(I, \Psi), \Phi_1).$$

5) $\Upsilon_\kappa^{\rm lin}[2]$ is the class of Ψ proper for $K_{\tau_2^*}^{\rm lin}$ and producing structures from $K_{\tau_2^*}^{\rm lin}$ extending the originals, i.e.

(a) $\tau_2^* = \{<, P_0, P_1\}$ where P_0, P_1 are unary predicates, $<$ a binary predicate

(b) $K_{\tau_2^*}^{\rm lin} = \{M : M$ a τ_2^*-model, $<^M$ a linear order, $\langle P_0^M, P_1^M\rangle$ a partition of $M\}$

(c) the two-place predicate $<$ and the one place predicates P_0, P_1 belong to $\tau(\Psi)$

(d) if $I \in K_{\tau_2^*}^{\rm lin}$ then $M = \mathrm{EM}_{\tau_2^*}(I, \Phi)$ belongs to $K_{\tau_2^*}^{\rm lin}$ and $<^M$ is a linear order and $I \models s < t \Rightarrow M \models a_s < a_t$ and $t \in P_\ell^I \Rightarrow a_\ell \in P_\ell^M$.

6) Similarly $\Upsilon_\kappa^{\rm lin}[\alpha(*)]$ using $K_{\tau_{\alpha(*)}^*}^{\rm lin}$ (see below in 0.14(9)).

0.12 Claim. *Assume $\Phi \in \Upsilon_{\mathfrak{K}}^{\rm or}$.*
1) If π is an isomorphism from the linear order I_1 onto the linear order I_2 then it induces a unique isomorphism $\hat\pi$ from $M_1 = \mathrm{EM}(I_1, \Phi)$ onto $M_2 = \mathrm{EM}(I_2, \Phi)$ such that:

(a) $\hat\pi(a_t) = a_{\pi(t)}$ *for $t \in I$*

(b) $\hat\pi(\sigma^{M_1}(a_{t_0}, \ldots, a_{t_{n-1}})) = \sigma^{M_2}(a_{\pi(t_0)}, \ldots, a_{\pi(t_{n-1})})$, *where $\sigma(x_0, \ldots, x_{n-1})$ is a τ_Φ-term and $t_0, \ldots, t_{n-1} \in I_1$.*

2) If π is an automorphism of the linear order I then it induces a unique automorphism $\hat\pi$ of $\mathrm{EM}(I, \Phi)$ (as above with $I_1 = I = I_2$).

0.13 Remark. 1) So in 0.11(2) we allow further expansion by functions definable from earlier ones (composition or even definition by

cases), as long as the number is $\leq \kappa$.
2) Of course, in 0.12 is true for trivial $\mathfrak{K}$.

So we may be interested in some classes of linear orders; below 0.14(1) is used much more than the others and also 0.14(5),(6) are used not so few times, in particular parts (8),(9) are not used till §5.

0.14 Definition. 1) A linear order I is κ-wide when for every $\theta < \kappa$ there is a monotonic sequence of lenth θ^+ in I.
2) A linear order I is κ-wider if $|I| \geq \beth_{1,1}(\kappa)$.
3) I_2 is κ-wide over I_1 if $I_1 \subseteq I_2$ and for every $\theta < \kappa$ there is a convex subset of I_2 disjoint to I_1 which is θ^+-wide. We say "I_2 is wide over I_1" if "I_2 is $|I_1|$-wide over I_2".
4) $K^{\text{lin}}[K^{\text{lin}}_\lambda]$ is the class of linear orders [of cardinality λ].
5) Let K^{flin} be the class of infinite linear order I such that every interval has cardinality $|I|$ and is with neither first nor last elements.
6) Let the two-place relation $\leq_{K^{\text{flin}}}$ on K^{flin} be defined by: $I \leq_{K^{\text{flin}}} J$ iff $I, J \in K^{\text{flin}}$ and $I \subseteq J$ and either $I = J$ or $J \backslash I$ is a dense subset of J and for every $t \in J \backslash I, I$ can be embedded into $J \restriction \{s \in J \backslash I : (\forall r \in I)(s <_J r \equiv t <_J r)\}$.
6A) Let the two-place relation $\leq^*_{K^{\text{flin}}}$ on K^{lin} be defined similarly omitting "$I \in K^{\text{flin}}$" (but not $J \in K^{\text{flin}}$).
7) $K^{\text{flin}}_\theta = \{I \in K^{\text{flin}} : |I| = \theta\}$ and $\leq_{K^{\text{flin}}_\theta} = \leq_{K^{\text{flin}}} \restriction K^{\text{flin}}_\theta$.
8) $K^{\text{sc}(\kappa)-\text{lin}}_\theta$ is the class of linear orders of cardinality θ which are the union of $\leq \kappa$ scattered linear orders (recalling I is scattered when there is no $J \subseteq I$ isomorphic to the rationals). If $\kappa = \aleph_0$ we may omit it (i.e. write $K^{\text{sc-lin}}_\theta$).
9) Let $\tau^*_{\alpha(*)} = \{<\} \cup \{P_i : i < \alpha(*)\}$, P_i a monadic predicate, $K^{\text{lin}}_{\tau^*_{\alpha(*)}} = \{I : I \text{ a } \tau^*_{\alpha(*)}\text{-model}, <^I \text{ a linear order and } \langle P^I_i : i < \alpha(*)\rangle \text{ a partition of } I\}$. If $\alpha(*) = 1$ we may omit P^I_0, so I is a linear order, so any ordinal can be treated as a member of $K^{\text{lin}}_{\tau^*_1}$.

0.15 Observation. 1) If $|I| > 2^\theta$ <u>then</u> I is θ^+-wide.
2) If $|I| \geq \lambda$ and λ is a strong limit cardinal <u>then</u> I is λ-wide.
3) $(K^{\text{flin}}_\theta, \leq_{K^{\text{flin}}_\theta})$ almost is a θ-a.e.c., only smoothness may fail.
4) If $I_1 \in K^{\text{lin}}$ <u>then</u> for some $I_2 \in K^{\text{flin}}$ we have: $|I_2| = |I_1| + \aleph_0$ and

$I_1 \leq^*_{K^{\rm flin}} I_2$; and $(\forall I_0)[I_0 \subseteq I_1 \wedge I_0 \in K^{\rm flin} \Rightarrow I_0 \leq_{K^{\rm flin}} I_2]$.
5) If I_1 is κ-wide and $I_1 <_{K^{\rm flin}} I_2$ <u>then</u> I_2 is κ-wide over I_2.

Remark. If in the definition of $\leq_{K^{\rm flin}}$ in 0.14(6) we can add "$(\forall t \in I)(\exists t' \in J)[t' <_J t \wedge (\forall s \in I)(s <_I t \rightarrow s <_J t')]$" (and its dual, i.e. inverting the order). So we can strengthen 0.14(6) by the demand above.

Proof. 1) By Erdös-Rado Theorem, i.e., by $(2^\theta)^+ \rightarrow (\theta^+)^2_2$.
2) Follows by part (1).
3),4),5) Easy. $\square_{0.15}$

0.16 Claim. *1) $(\Upsilon^{\rm or}_{\kappa[\mathfrak{K}]}, \leq^\otimes_\kappa), (\Upsilon^{\rm or}_\kappa[\mathfrak{K}], <^{\rm ie}_\kappa)$ and $(\Upsilon^{\rm or}_{\kappa[\mathfrak{K}]}, \leq^\oplus)$ are partial orders (and $\leq^\otimes_\kappa, \leq^{\rm ie}_\kappa \subseteq \leq^\oplus_\kappa$).*
2) If $\Phi_i \in \Upsilon^{\rm or}_\kappa[\mathfrak{K}]$ and the sequence $\langle \Phi_i : i < \delta \rangle$ is a $\leq^\otimes_\kappa$-increasing sequence, $\delta < \kappa^+$, <u>then</u> it has a $<^\otimes_\kappa$-l.u.b. $\Phi \in \Upsilon^{\rm or}_\kappa[\mathfrak{K}]$, and $\mathrm{EM}(I, \Phi) = \bigcup_{i<\delta} \mathrm{EM}(I, \Phi_i)$ for every linear order I, i.e. $\tau(\Phi) = \cup\{\tau(\Phi_i) : i < \delta\}$ and for every $j < \delta$ we have $\mathrm{EM}_{\tau(\Phi_j)}(I, \Phi) = \cup\{\mathrm{EM}_{\tau(\Phi_i)}(I, \Phi) : i \in [j, \delta)\}$.
3) Similarly for $<^\oplus_\kappa$ and $\leq^{\rm ie}_\kappa$.
4) If $\Phi \in \Upsilon^{\rm lin}_\kappa$ and $I \in K^{\rm lin}$ <u>then</u> $I \subseteq \mathrm{EM}_{\{<\}}(I, \Phi)$ as linear orders stipulating (as in 0.11(3)) that $a_t = t$.

Proof. Easy. $\square_{0.16}$

Recall various well known facts on $\mathbb{L}_{\infty,\theta}$.

0.17 Claim. *1) If M, N are τ-models of cardinality λ, $\mathrm{cf}(\lambda) = \aleph_0$ and $M \equiv_{\mathbb{L}_{\infty,\lambda}} N$ <u>then</u> $M \cong N$.*
2) If M, N are τ-models then $M \equiv_{\mathbb{L}_{\infty,\theta}} N$ <u>iff</u> there is $\mathscr{F}$ such that

$\circledast$(a) (α) *each $f \in \mathscr{F}$ is a partial isomorphism from M to N*
(β) $\mathscr{F} \neq \emptyset$
(γ) *if $f \in \mathscr{F}$ and $A \subseteq \mathrm{Dom}(f)$ then $f \restriction A \in \mathscr{F}$*
(b) *if $f \in \mathscr{F}$, $A \in [M]^{<\theta}$ and $B \in [N]^{<\theta}$ then for some $g \in \mathscr{F}$ we have $f \subseteq g$, $A \subseteq \mathrm{Dom}(g)$, $B \subseteq \mathrm{Rang}(g)$.*

2A) If $M \subseteq N$ *are* τ*-models,* <u>*then*</u> $M \prec_{\mathbb{L}_{\infty,\theta}} N$ *iff for some* $\mathscr{F}$ *clauses* $\circledast(a),(b)$ *hold together with*

(c) *if* $A \in [M]^{<\theta}$ *then for some* $f \in \mathscr{F}$ *we have* $\mathrm{id}_A \subseteq f$.

2B) In part (2) (and part (2A)), we can omit subclause (γ) *of clause (a), and if* $\mathscr{F}$ *satisfies* $(a)(\alpha),(\beta)+(b)$ *(and (c)),* <u>*then*</u> *also* $\mathscr{F}' = \{f \restriction A : f \in \mathscr{F}$ *and* $A \subseteq \mathrm{Dom}(f)\}$ *satisfies the demands.*
2C) Let M,N *be* τ*-models and define* $\mathscr{F} = \{f$*: for some* $\bar{a} \in {}^{\theta>}M$, f *is a function from* $\mathrm{Rang}(\bar{a})$ *to* N *such that* $(M,\bar{a}) \equiv_{\mathbb{L}_{\infty,\theta}} (N, f(\bar{a}))\}$ <u>*then*</u> $M \equiv_{\mathbb{L}_{\infty,\theta}} N$ <u>*iff*</u> $\mathscr{F} \neq \emptyset$ *iff* $\mathscr{F}$ *satisfies clauses (a),(b) of* $\circledast$.
3) If M *is a* τ*-model,* $\theta = \mathrm{cf}(\theta)$ *and* $\mu = \|M\|^{<\theta}$ <u>*then*</u> *for some* $\gamma < \mu^+$ *and* $\Delta \subseteq \mathbb{L}_{\mu^+,\theta}(\tau)$ *of cardinality* $\leq \mu$ *such that each* $\varphi(\bar{x}) \in \Delta$ *is of quantifier depth* $< \gamma$*, we have*

(a) *for* $\bar{a}, \bar{b} \in {}^{\theta>}M$ *we have* $(M,\bar{a}) \equiv_{\mathbb{L}_{\infty,\theta}} (M,\bar{b})$ *iff* $\mathrm{tp}_\Delta(\bar{a},\emptyset,M) = \mathrm{tp}_\Delta(\bar{a},\emptyset,M)$

(b) *for any* τ*-model* N *we have* $N \equiv_{\mathbb{L}_{\infty,\theta}} M$ <u>*iff*</u> $\{\mathrm{tp}_\Delta(\bar{a},\emptyset,N) : \bar{a} \in {}^{\theta>}N\} = \{\mathrm{tp}_\Delta(\bar{a},\emptyset,M) : \bar{a} \in {}^{\theta>}M\}$.

4) Assume $\chi > \mu = \mu^{<\kappa}$ *and* $x \in \mathscr{H}(\chi)$*. There is* $\mathfrak{B}$ *such that (in fact clauses (d)-(g) follow from clauses (a),(b),(c))*

(a) $\mathfrak{B} \prec (\mathscr{H}(\chi),\in)$ *has cardinality* μ,

(b) $\mu + 1 \subseteq \mathfrak{B}$ *and* $[\mathfrak{B}]^{<\kappa} \subseteq \mathfrak{B}$ *and* $x \in \mathfrak{B}$

(c) $\mathfrak{B} \prec_{\mathbb{L}_{\kappa,\kappa}} (\mathscr{H}(\chi),\in)$

(d) *if* $\mathfrak{K}$ *is an a.e.c. with* $\mathrm{LS}(\mathfrak{K}) + |\tau(\mathfrak{K})| \leq \mu$ *and* $\mathfrak{K} \in \mathfrak{B}$ *(which means* $\{(M,N) : M \leq_{\mathfrak{K}} N$ *has universes* $\subseteq \mathrm{LS}(\mathfrak{K})\} \in \mathfrak{B}$*)* <u>*then*</u>

  (α) $M \in \mathfrak{K} \cap \mathfrak{B} \Rightarrow M \restriction \mathfrak{B} := M \restriction (\mathfrak{B} \cap M) \leq_{\mathfrak{K}} M$

  (β) *if* $M \leq_{\mathfrak{K}} N$ *belongs to* $\mathfrak{B}$ *then* $M \restriction \mathfrak{B} \leq_{\mathfrak{K}} N \restriction \mathfrak{B}$

(e) *if* $\mathfrak{K}$ *is as in (d),* $\Phi \in \Upsilon^{\mathrm{or}}_{\leq\mu}[\mathfrak{K}] \cap \mathfrak{B}$ *and* $I \in \mathfrak{B}$ *is a linear order and so* $M = \mathrm{EM}(I,\Phi) \in \mathfrak{B}$ <u>*then*</u> $I' = I \restriction \mathfrak{B} \subseteq I$ *and* $M \restriction \mathfrak{B} = \mathrm{EM}(I',\Phi)$ *so* $(M \restriction \tau(\mathfrak{K})) \restriction \mathfrak{B} = \mathrm{EM}_{\tau(\mathfrak{K})}(I',\Phi) \leq_{\mathfrak{K}} M \restriction \tau(\mathfrak{K})$

(f) *if* $|\tau| \leq \mu, \tau \in \mathfrak{B}$ *and* $M,N \in \mathfrak{B}$ *are* τ*-models,* <u>*then*</u>

  (α) $M \restriction \mathfrak{B} \prec_{\mathbb{L}_{\kappa,\kappa}[\tau]} M$

(β) $M \not\equiv_{\mathbb{L}_{\infty,\kappa}[\tau]} N$ *iff* $(M \restriction \mathfrak{B}) \not\equiv_{\mathbb{L}_{\infty,\kappa}[\tau]} (N \restriction \mathfrak{B})$

(γ) *if* $M \subseteq N$ *then* $(M \prec_{\mathbb{L}_{\infty,\kappa}(\tau)} N)$ *iff* $(M \restriction \mathfrak{B}) \prec_{\mathbb{L}_{\infty,\kappa}(\tau)} (N \restriction \mathfrak{B})$; *this applies also to* $(M, \bar{a}), (N, \bar{a})$ *for* $\bar{a} \in {}^{\kappa>}M$

(g) *if* $I \in K^{\rm flin}$ *then* $I_1 \cap \mathfrak{B} \in K^{\rm flin}$ *and if* $I_1 <^*_{K^{\rm flin}} I_2$ *then* $(I_1 \cap \mathfrak{B}) <^*_{K^{\rm flin}} (I_2 \cap \mathfrak{B})$.

Proof. 1)-3) and 4)(a),(b),(c) Well known, e.g. see [Di].
4) Clauses (d),(e),(f): as in 0.9(1), i.e. by absoluteness. Also clause (g) should be clear. $\square_{0.17}$

0.18 Remark. 1) We will be able to add, in 0.17(4):

(g) if $\mathfrak{K}$ is as in clause (d) and $\tau = \tau_{\mathfrak{K}}$ then in clause (f) we can replace $\mathbb{L}_{\infty,\kappa}(\tau)$ by $\mathbb{L}_{\infty,\kappa}[\mathfrak{K}]$ and $\mathbb{L}_{\kappa,\kappa}(\tau)$ by $\mathbb{L}_{\kappa,\kappa}[\mathfrak{K}]$, see Definition 1.9 and Fact 1.10(5).

2) We use part (4) in 1.26(3).

0.19 Definition. For a model M and for a set Δ of formulas in the vocabulary of $M, \bar{x} = \langle x_i : i < \alpha \rangle, A \subseteq M$ and $\bar{a} \in {}^\alpha M$ let the Δ-type of $\bar{a}$ over A in M be $\mathrm{tp}_\Delta(\bar{a}, A, M) = \{\varphi(\bar{x}, \bar{b}) : M \models \varphi[\bar{a}, \bar{b}]$ where $\varphi = \varphi(\bar{x}, \bar{y}) \in \Delta$ and $\bar{b} \in {}^{\ell g(\bar{y})}A\}$.

§1 Amalgamation in K^*_λ

Our aim is to investigate what is implied by 1.3 below but instead of assuming it we shall shortly assume only some of its consequences. For our purpose here, for $\theta \in [\mathrm{LS}(\mathfrak{K}), \lambda), \lambda = \beth_\lambda$ it does not really matter if we use $\kappa = \beth_{1,1}(\theta)$ or $\kappa = \beth_{1,1}(\beth_n(\theta))$ or $\beth_{1,n}(\theta)$, as we are trying to analyze models in K_λ.

1.1 Remark. 1) We can in our claims use only $\Phi \in \Upsilon^{\rm or}_{\mathfrak{K}} = \Upsilon^{\rm or}_{\mathrm{LS}(\mathfrak{K})}[\mathfrak{K}]$ because for every $\theta \geq \mathrm{LS}(\mathfrak{K})$ we can replace $\mathfrak{K}$ by $\mathfrak{K}_{\geq\theta}$ as $\mathrm{LS}(\mathfrak{K}_{\geq\theta}) = \theta$ when $\mathfrak{K}_{\geq\theta} \neq \emptyset$, of course.

2) As usual we assume $|\tau_{\mathfrak{K}}| \le \mathrm{LS}(\mathfrak{K})$ just for convenience, otherwise we should just replace $\mathrm{LS}(\mathfrak{K})$ by $\mathrm{LS}(\mathfrak{K}) + |\tau_{\mathfrak{K}}|$.

1.2 Hypothesis.

- (*a*) $\mathfrak{K} = (K, \le_{\mathfrak{K}})$ is an a.e.c. with vocabulary $\tau = \tau(\mathfrak{K})$ (and we can assume $|\tau| \le \mathrm{LS}(\mathfrak{K})$ for notational simplicity)
- (*b*) $\mathfrak{K}$ has arbitrarily large models (equivalently has a model of cardinality $\ge \beth_{1,1}(\mathrm{LS}(\mathfrak{K}))$), not used, e.g. in 1.10, 1.11 but from 1.12 on it is used extensively.

1.3 Definition. We say (μ, λ) or really (μ, λ, Φ) is a weak/strong/ pseudo $\mathfrak{K}$-candidate when (weak is the default value):

- (*a*) $\mu > \lambda = \beth_\lambda > \mathrm{LS}(\mathfrak{K})$ (e.g. the first beth fix point $> \mathrm{LS}(\mathfrak{K})$, see 3.5; in the main case λ has cofinality $\aleph_0$)
- (*b*) $\mathfrak{K}$ categorical in μ and $\Phi \in \Upsilon^{\mathrm{or}}_{\mathfrak{K}}$ or just
- $(b)^-$ $\mathfrak{K}$ is weakly/strongly/pseudo solvable in μ and $\Phi \in \Upsilon^{\mathrm{or}}_{\mathfrak{K}}$ witnesses it; see below.

1.4 Definition. 1) We say $\mathfrak{K}$ is weakly (μ, κ)-solvable <u>when</u> $\mu \ge \kappa \ge \mathrm{LS}(\mathfrak{K})$ and there is $\Phi \in \Upsilon^{\mathrm{or}}_\kappa[\mathfrak{K}]$ witnessing it, which means that $\Phi \in \Upsilon^{\mathrm{or}}_\kappa[\mathfrak{K}]$ and $\mathrm{EM}_{\tau(\mathfrak{K})}(I, \Phi)$ is a locally superlimit member of $\mathfrak{K}_\mu$ for every linear order I of cardinality μ. We may say $(\mathfrak{K}, \Phi)$ is weakly (μ, κ)-solvable and we may say Φ witness that $\mathfrak{K}$ is weakly (μ, κ)-solvable.

If $\kappa = \mathrm{LS}(\mathfrak{K})$ we may omit it, saying $\mathfrak{K}$ or $(\mathfrak{K}, \Phi)$ is weakly μ-solvable in μ.
2) $\mathfrak{K}$ is strongly (μ, κ)-solvable <u>when</u> $\mu \ge \kappa \ge \mathrm{LS}(\mathfrak{K})$ and some $\Phi \in \Upsilon^{\mathrm{or}}_\kappa[\mathfrak{K}]$ witness it which means that if $I \in K^{\mathrm{lin}}_\lambda$ then $\mathrm{EM}_{\tau[\mathfrak{K}]}(I, \Phi)$ is superlimit (for $\mathfrak{K}_\lambda$). We use the conventions from part (1).
3) We say $\mathfrak{K}$ is pseudo (μ, κ)-solvable when $\mu \ge \kappa \ge \mathrm{LS}(\mathfrak{K})$ and there is $\Phi \in \Upsilon^{\mathrm{or}}_\kappa[\mathfrak{K}]$ witnessing it which means that for some μ-a.e.c. $\mathfrak{K}'$ with no $\le_{\mathfrak{K}'}$-maximal member, we have $M \in \mathfrak{K}'$ <u>iff</u> $M \cong \mathrm{EM}_{\tau(\mathfrak{K})}(I, \Phi)$ for some $I \in K^{\mathrm{lin}}_\mu$ iff $M \cong \mathrm{EM}_{\tau(\mathfrak{K})}(I, \Phi)$ for every

$I \in K^{\rm lin}_\mu$. We use the conventions from part (1).
4) Let (μ, κ)-solvable mean weakly (μ, κ)-solvable, etc., (including 1.3)

1.5 Claim. *1) In Definition 1.3, clause (b) implies clause $(b)^-$. Also in Definition 1.4 "$\mathfrak{K}$ is strongly (μ, κ)-solvable" implies "$\mathfrak{K}$ is weakly (μ, κ)-solvable" which implies "$\mathfrak{K}$ is pseudo (μ, κ)-solvable". Similarly for $(\mathfrak{K}, \Phi)$.*
2) Assume $\Phi \in \Upsilon^{\rm or}_\kappa[\mathfrak{K}]$; if clause $(b)^-$ of 1.3 or just $\dot{I}(\mu, \mathfrak{K}) < 2^\mu$, or just $2^\mu > \dot{I}(\mu, \{{\rm EM}_{\tau(\mathfrak{K})}(I, \Phi) : I \in K^{\rm lin}_\mu\})$ for some μ satisfying ${\rm LS}(\mathfrak{K}) < \kappa^+ < \mu$ then we can deduce that

(∗) *Φ, really $(\mathfrak{K}, \Phi)$ has the κ-non-order property, where the κ-non-order property means that:*
if I is a linear order of cardinality $\kappa, \bar{t}^1, \bar{t}^2 \in {}^\kappa I$ form a Δ-system pair (see below) and $\langle \sigma_i(\bar{x}) : i < \kappa\rangle$ lists the $\tau(\Phi)$-terms (with the sequence $\bar{x}$ of variables being $\langle x_i : i < \kappa\rangle$) and $\langle a_t : t \in I\rangle$ is "the" indiscernible sequence generating ${\rm EM}(I, \Phi)$ (i.e. as usual "$\langle a_t : t \in I\rangle$" is "the" skeleton of ${\rm EM}(I, \Phi)$, so generating it, see Definition 0.8) then for some $J \supseteq I$ there is an automorphism of ${\rm EM}_{\tau(\mathfrak{K})}(J, \Phi)$ which exchanges $\langle \sigma_i(\langle a_{t^1_i} : i < \kappa\rangle) : i < \kappa\rangle$ and $\langle \sigma_i(\langle a_{t^2_i} : i < \kappa\rangle) : i < \kappa\rangle$.
where

⊠ *$\bar{t}^1, \bar{t}^2 \in {}^\alpha I$ is a Δ-system pair when for some $J \supseteq I$ there are $\bar{t}^\zeta \in {}^\alpha J$ for $\zeta \in \kappa\backslash\{1, 2\}$ such that $\langle \bar{t}^\alpha : \alpha < \kappa\rangle$ is an indiscernible sequence for quantifier free formulas in the linear order J.*

Proof. 1) The first sentence holds by Claim 0.9(1) and Definition 0.8 (and Claim 0.6). The second and third sentences follows by 0.6.
2) Otherwise we get a contradiction by [Sh 300, Ch.III] or better [Sh:e, III]. $\square_{1.4}$

1.6 Definition. 1) If $\mathscr{M}'$ is a class of linear orders and $\Phi \in \Upsilon^{\rm or}_\kappa[\mathfrak{K}]$ then we let $K[\mathscr{M}', \Phi] = \{\mathrm{EM}_{\tau(\mathfrak{K})}(I, \Phi) : I \in \mathscr{M}'\}$.
2) Let $K^{u(\kappa)\text{-lin}}_\theta$ be the class of linear orders I of cardinality θ such that: for some scattered[1] linear order J and Φ proper for $K^{\rm lin}$ such that $<$ belongs to τ_Φ, $|\tau_\Phi| \le \kappa$ we have I is embeddable into $\mathrm{EM}_{\{<\}}(J, \Phi)$. If we omit κ we mean $\mathrm{LS}(\mathfrak{K})$. If $\kappa = \aleph_0$ we may omit it.

1.7 Remark. 1) Note that in Definition 1.4(1) we <u>can</u> restrict ourselves to $I \in K^{\mathrm{sc}(\theta)\text{-lin}}_\lambda$, see 0.14(8) and even $I \in K^{u(\theta)\text{-lin}}$ see 1.6(2), i.e., assume $2^\mu > \dot{I}(\mu, K[\mathscr{M}', \Phi])$, for $\mathscr{M}' = K^{\mathrm{sc}(\theta)\text{-lin}}_\lambda$ or $\mathscr{M}' = K^{u(\theta)\text{-lin}}_\lambda$ and restrict the conclusion $(*)$ to $I \in K^{\mathrm{sc}(\theta)\text{-lin}}$. A gain is that, if $\lambda > \theta$, every $I \in K^{\mathrm{sc}(\theta)\text{-lin}}_\lambda$ is λ-wide so later $K^* = K^{**}$, and being solvable is a weaker demand. But it is less natural. Anyhow we presently do not deal with this.
1A) Note that $K^{\mathrm{sc}(\theta)-\mathrm{lin}}_\lambda \supseteq K^{u(\theta)-\mathrm{lin}}_\lambda$.
2) An aim of 1.8 below is to show that: by changing Φ instead of assuming $I_1 \subset I_2 \wedge (I_2$ is κ-wide over $I_1)$ it suffices to assume $I_1 \subset I_2 \wedge (I_2$ is κ-wide).

1.8 Claim. *For every $\Phi_1 \in \Upsilon^{\rm or}_\kappa[\mathfrak{K}]$ there is Φ_2 such that*

- (a) $\Phi_2 \in \Upsilon^{\rm or}_\kappa[\mathfrak{K}]$ *and if Φ_1 witnesses $\mathfrak{K}$ is weakly/strongly/pseudo (λ, κ)-solvable then so does Φ_2*
- (b) $\tau_{\Phi_1} \subseteq \tau_{\Phi_2}$ *and* $|\tau_{\Phi_2}| = |\tau_{\Phi_1}| + \aleph_0$
- (c) *for any $I_2 \in K^{\rm lin}$ there are I_1 and h such that:*
  - (α) $I_1 \in K^{\rm lin}$ *and even* $I_1 \in K^{\rm flin}$, *see 0.14(5)*
  - (β) *h is an embedding of I_2 into I_1*
  - (γ) *there is an isomorphism f from $\mathrm{EM}_{\tau(\Phi_1)}(I_2, \Phi_2)$ onto $\mathrm{EM}(I_1, \Phi_1)$ such that $f(a_t) = a_{h(t)}$ for $t \in I_2$*
  - (δ) *if $J_1 = I_1 \restriction \mathrm{Rang}(h)$ and we let $\mathscr{E} = \{(t_1, t_2) : t_1, t_2 \in I_1 \backslash J_1$ and $(\forall s \in J_1)(s < t_1 \equiv s < t_2)\}$ <u>then</u>: $\mathscr{E}$ is*

[1] i.e. one into which the rational order cannot be embedded

an equivalence relation and each equivalence class has $\geq |I_2|$ *members and* $J_1 \leq_{K^{\rm flin}} I_1$*, see 0.14(6)*

(ε) *[not used] if* $\emptyset \neq J_2 \subseteq I_2$*,* $J_1 = \{t \in I_1$*: for some* $\tau(\Phi_2)$*-term* $\sigma(x_0, \ldots, x_{n-1})$ *and some* $t_0, \ldots, t_{n-1} \in J_2$ *we have* $f^{-1}(a_t) = \sigma^{{\rm EM}(I_2,\Phi_2)}(a_{t_0}, \ldots, a_{t_{n-1}})\}$ *and* $J'_1 \subseteq {\rm Rang}(h)\backslash J_1$ *and* $t \in J'_1$ *then* $\{s \in t/\mathscr{E} : f^{-1}(a_s)$ *belongs to the Skolem hull of* $\{f^{-1}(a_r) : r \in J'_1\}$ *in* ${\rm EM}(I_2, \Phi)\}$ *has cardinality* $\geq |J'_1|$ *and* J'_1 *and its inverse can be embedded into it; in fact,* I_1 *and its inverse are embeddable into any interval of* I_2*.*

Remark. 1) We can express it by $\leq^{\otimes}_{\mathfrak{K}}$, see 0.11(4). So for some Ψ proper for linear orders such that τ_Ψ is countable, the two-place predicate $<$ belongs to τ_Ψ and above ${\rm EM}_{\{<\}}(I_2, \Psi)$ is I_1.
2) In fact, $J_2 \subset I_2 \Rightarrow {\rm EM}_{\{<\}}(J_2, \Psi) <_{K^{\rm flin}} {\rm EM}_{\{<\}}(I_2, \Psi)$ and $I_2 <^*_{K^{\rm flin}} {\rm EM}_{\{<\}}(I_2, \Phi)$ when we identify $t \in I_2$ with a_t.

Proof. For $I_2 \in K^{\rm lin}$ let the set of elements of I_1 be $\{\eta : \eta$ is a finite sequence of elements from $(\mathbb{Z}\backslash\{0\}) \times I_2\}$. For $\eta \in I_1$ let $(\ell_{\eta,k}, t_{\eta,k})$ be $\eta(k)$ for $k < \ell g(\eta)$.

Lastly, I_1 is ordered by: $\eta_1 < \eta_2$ iff for some n one of the following occurs

$\circledast$(a) $\eta_1 \restriction n = \eta_2 \restriction n, \ell g(\eta_1) > n, \ell g(\eta_2) > n$ and $\ell_{\eta_1,n} < \ell_{\eta_2,n}$

(b) $\eta_1 \restriction n = \eta_2 \restriction n$, $\ell g(\eta_1) > n$, $\ell g(\eta_2) > n$, $\ell_{\eta_1,n} = \ell_{\eta_2,n} > 0$ and $t_{\eta_1,n} <_{I_2} t_{\eta_2,n}$

(c) $\eta_1 \restriction n = \eta_2 \restriction n$, $\ell g(\eta_1) > n$, $\ell g(\eta_2) > n$, $\ell_{\eta_1,n} = \ell_{\eta_2,n} < 0$ and $t_{\eta_2,n} <_{I_2} t_{\eta_1,n}$

(d) $\eta_1 \restriction n = \eta_2 \restriction n$, $\ell g(\eta_1) = n$, $\ell g(\eta_2) > n$ and $\ell_{\eta_2,n} > 0$

(e) $\eta_1 \restriction n = \eta_2 \restriction n$, $\ell g(\eta_1) > n$, $\ell g(\eta_2) = n$ and $\ell_{\eta_1,n} < 0$.

We identify $t \in I_1$ with the pair $(1, t)$. Now check. $\square_{1.8}$

1.9 Definition. 1) Let the language $\mathbb{L}_{\theta,\partial}[\mathfrak{K}]$ or $\mathbb{L}_{\theta,\partial,\mathfrak{K}}$ where $\theta \geq \partial \geq \aleph_0$ and θ is possibly ∞, be defined like the infinitary logic $\mathbb{L}_{\theta,\partial}(\tau_{\mathfrak{K}})$, except that we deal only with models from K and we add for $i^* < \partial$ the atomic formula "$\{x_i : i < i^*\}$ is the universe of a $\leq_{\mathfrak{K}}$-submodel", with obvious syntax and semantics. Of course, it is interesting normally only for $\partial >$ LS($\mathfrak{K}$) and recall that any formula has $< \partial$ free variables.
2) For M a $\tau_{\mathfrak{K}}$-model and $N \in K$ let $M \prec_{\mathbb{L}_{\theta,\partial}[\mathfrak{K}]} N$ means that $M \subseteq N$ and if $\varphi(\bar{x}, \bar{y})$ is a formula from $\mathbb{L}_{\theta,\partial}[\mathfrak{K}]$ and $N \models (\exists \bar{x})\varphi(\bar{x}, \bar{b})$ where $\bar{b} \in {}^{\ell g(\bar{y})}M$, <u>then</u> for some $\bar{a} \in {}^{\ell g(\bar{x})}M$ we have $N \models \varphi[\bar{a}, \bar{b}]$.

<u>1.10 Fact</u>: 1) If $\theta \geq \partial >$ LS($\mathfrak{K}$) and M, N are $\tau_{\mathfrak{K}}$-models and $N \in K$ and $M \prec_{\mathbb{L}_{\theta,\partial}[\mathfrak{K}]} N$, <u>then</u> $M \leq_{\mathfrak{K}} N$ and $M \in K$.
2) The relation $\prec_{\mathbb{L}_{\theta,\partial}[\mathfrak{K}]}$ can also be defined as usual: $M \prec_{\mathbb{L}_{\theta,\partial}[\mathfrak{K}]} N$ <u>iff</u> $M, N \in K, M \subseteq N$ and for every $\varphi(\bar{x}) \in \mathbb{L}_{\theta,\partial}[\mathfrak{K}]$ and $\bar{a} \in {}^{\ell g(\bar{x})}M$ we have $M \models \varphi[\bar{a}]$ iff $N \models \varphi[\bar{a}]$.
3) If $N \in \mathfrak{K}$ and M is a τ_K-model satisfying $M \prec_{\mathbb{L}_{\infty,\kappa}} N$ and $\kappa >$ LS($\mathfrak{K}$) <u>then</u> $M \in K, M \leq_{\mathfrak{K}} N$ and $M \prec_{\mathbb{L}_{\infty,\kappa}[\mathfrak{K}]} N$.
4) If $N \in K, M$ a τ_K-model and $M \equiv_{\mathbb{L}_{\infty,\kappa}} N$ where $\kappa >$ LS($\mathfrak{K}$) <u>then</u> $M \in K$ and $M \equiv_{\mathbb{L}_{\infty,\kappa}[\mathfrak{K}]} N$.
5) The parallel of 0.17(2) holds for $\mathbb{L}_{\infty,\kappa}[\mathfrak{K}]$, i.e. there is $\mathscr{F}$ satisfying clauses (a),(b) there and

- (d) if $f \in \mathscr{F}$ then
  - (α) $M \restriction \mathrm{Dom}(f) \leq_{\mathfrak{K}} M$
  - (β) $N \restriction \mathrm{Rang}(f) \leq_{\mathfrak{K}} M$.

6) Also the parallel of 0.17(2A) holds for $\mathbb{L}_{\infty,\kappa}[\mathfrak{K}]$.
7) The parallel of 0.17(4) holds for $\mathbb{L}_{\infty,\kappa}[\mathfrak{K}]$.

Proof. Part (1) is straight (knowing I§1 or [Sh 88, §1]). Part (2) is proved as in the Tarski-Vaught criterion and parts (5),(6),(7) are proved as in 0.17.

Toward proving parts (3),(4) we first assume just

$\boxtimes_1$ M, N are τ_K-models, $N \in K$ and $M \equiv_{\mathbb{L}_{\infty,\kappa}} N$ and $\kappa >$ LS($\mathfrak{K}$) and $\lambda \in [\mathrm{LS}(\mathfrak{K}), \kappa)$

and we define:

⊡(a) $I = I_\lambda = \{(f, M', N') : M' \subseteq M$ and $N' \subseteq N$ and f is an isomorphism from M' onto N' and $\|M'\| \leq \lambda$ and letting $\bar{a}$ list M' we have $(M, \bar{a}) \equiv_{\mathbb{L}_{\infty,\kappa}} (N, f(\bar{a}))\}$

(b) for $t \in I$ let $t = (f_t, M_t, N_t)$

(c) for $\ell = 0, 1, 2$ we define the two-place relation $\leq^\ell_I$ on I: let $s \leq^\ell_I t$ hold iff

(α) $\ell = 0$ and $M_s \subseteq M_t \wedge N_s \subseteq N_t$

(β) $\ell = 1$ and $M_s \leq_{\mathfrak{K}} M_t \wedge N_s \leq_{\mathfrak{K}} N_t$

(γ) $\ell = 2$ and $f_s \subseteq f_t$

(d) $I_1 = I^1_\lambda := \{t \in I_0 : N_t \leq_{\mathfrak{K}} N\}$ and let $\leq^\ell_{I_1} = \leq^\ell_I \restriction I_1$ for $\ell = 0, 1, 2$.

Now easily

$(*)_0$ (α) $I \neq \emptyset$ is partially ordered by $\leq^\ell_I$ for $\ell = 0, 1, 2$

(β) $s \leq^1_I t \Rightarrow s \leq^0_I t$

(γ) $s \leq^2_I t \Rightarrow s \leq^0_I t$.

[Why? Straight, e.g. $I \neq \emptyset$ by 0.17(1).]

$(*)_1$ if $t \in I_1$ then $M_t \in K_{\leq\lambda}$ and $N_t \in K_{\leq\lambda}$.

[Why? As $t \in I_1$ by the definition of I we have $N_t \in K_{\leq\lambda}$ (because $N_t \leq_{\mathfrak{K}} N$) and $M_t \in K_{\leq\lambda}$ as f_t is an isomorphism from M_t onto N_t.]

$(*)_2$ if $s \in I$, $A \in [M]^{\leq\lambda}$ and $B \in [N]^{<\lambda}$ then for some t we have $s \leq^2_I t$ and $A \subseteq M_t$ and $B \subseteq N_t$.

[Why? By the properties of $\equiv_{\mathbb{L}_{\infty,\kappa}}$, see 0.17(2C) as $\kappa > \lambda$, $M \equiv_{\mathbb{L}_{\infty,\kappa}}$ and the definition of I.]

$(*)_3$ if $s \leq^2_{I_1} t$ then $s \leq^1_I t$, i.e. $M_s \leq_{\mathfrak{K}} M_t$ and $N_s \leq_{\mathfrak{K}} N_t$.

[Why? As $s, t \in I_1$ we know that $N_s \leq_{\mathfrak{K}} N$ and $N_t \leq_{\mathfrak{K}} N$ and as $s \leq^2_I t$ we have $f_s \subseteq f_t$ hence $N_s \subseteq N_t$. By axiom V of a.e.c. it follows that $N_s \leq_{\mathfrak{K}} N_t$. Now $M_s \leq_{\mathfrak{K}} M_t$ as f_t is an isomorphism from

M_t onto N_t mapping M_s onto N_s (as it extends f_s by the definition of $\leq^2_I$) and $\leq_{\mathfrak{K}}$ is preserved by any isomorphism. So by the definition of $\leq^1_I$ we are done.]

$(*)_4$ if $s \in I$ <u>then</u> for some $t \in I_1$ we have $s \leq^2_I t$ (hence $I_1 \neq \emptyset$).

[Why? First choose $N' \leq_{\mathfrak{K}} N$ of cardinality $\leq \lambda$ such that $N_s \subseteq N'$, (possibly by the basic properties of a.e.c. (see I§1 or Chapter V.B)). Second we can find $t \in I$ such that $N_t = N' \wedge f_s \subseteq f_t$ by the characterization of $\equiv_{\mathbb{L}_{\infty,\kappa}}$ as in $(*)_2$. So $s \leq^2_I t$ by the definition of $\leq^2_I$ and $N_t = N' \leq_{\mathfrak{K}} N$ hence $t \in I_1$ as required. Lastly, $I_1 \neq \emptyset$ as by $(*)_0(\alpha)$ we know that $I \neq \emptyset$ and apply what we prove.]

$(*)_5$ if $s \leq^0_{I_1} t$ then $N_s \leq_{\mathfrak{K}} N_t$.

[Why? As in the proof of $(*)_3$ by AxV of a.e.c. we have $N_s \leq_{\mathfrak{K}} N_t$ (not the part on the M's!)]

$(*)_6$ if $s \in I_1, A \in [M]^{\leq\lambda}$ and $B \in [M]^{\leq\lambda}$ <u>then</u> for some t we have $s \leq^2_{I_1} t$ and $A \subseteq M_t, B \subseteq N_t$.

[Why? By $(*)_2$ there is t_1 such that $s \leq^2_I t_1, A \subseteq M_{t_1}$ and $B \subseteq N_{t_1}$. By $(*)_4$ there is $t \in I_1$ such that $t_1 \leq^2_I t$ hence by $(*)_0(\alpha)$ we have $s \leq^2_I t$. As $s, t \in I_1$ this implies $s \leq^2_{I_1} t$.]

Note that it is unreasonable to have "$(I_1, \leq^2_{I_1})$-directed" but

$(*)_7$ $(I_1, \leq^1_{I_1})$ is directed.

[Why? Let $s_1, s_2 \in I_1$. We now choose t_n by induction on $n < \omega$ such that

(a) $t_n \in I_1$
(b) M_{t_n} includes $\cup\{M_{t_k} : k < n\} \cup M_{s_1} \cup M_{s_2}$ if $n \geq 2$
(c) N_{t_n} includes $\cup\{N_{t_k} : k < n\} \cup N_{s_1} \cup N_{s_2}$ if $n \geq 2$
(d) $t_0 = s_1$
(e) $t_1 = s_2$
(f) if $n = m + 1 \geq 2$ then $t_m \leq^0_{I_1} t_n$
(g) if $n = m + 2$ then $t_m \leq^2_I t_n$ hence $t_m \leq^2_{I_1} t_n$.

For $n = 0, 1$ this is trivial. For $n = m + 2 \geq 2$, apply $(*)_6$ with $t_m, \cup\{M_{t_k} : k \leq m+1\}, \cup\{N_{t_k} : k \leq m+1\}$ here standing for s, A, B there getting t_n, so we get $t_n \in I_1$ in particular $t_m \leq^2_{I_1} t_n$, so clause (a) is satisfied by t_n. By the choice of t_n and as $s_1 = t_0, s_2 = t_1$, clauses (b) + (c) hold for t_n. By the choice of t_n, obviously also clause (g). Now why does clause (f) hold (i.e. $t_{m+1} \leq^0_I t_n$)? It follows from clauses (a),(b),(c), so t_n is as required. Hence we have carried the induction. Let $N^* = \cup\{N_{t_n} : 2 \leq n < \omega\}$, so clearly by $(*)_5$ and clause (f) we have $N_{t_n} \leq_{\mathfrak{K}} N_{t_{n+1}}$ for $n \geq 1$, and clearly $M_{t_n} \subseteq M_{t_{n+1}}$ for $n \geq 1$. Let $M^* = \cup\{M_{t_n} : 2 \leq n < \omega\}$. Note that by $(*)_3$ and clause (g) we have $M_{t_n} \leq_{\mathfrak{K}} M_{t_{n+2}}$, so $\langle M_{t_{n+2}} : n < \omega\rangle$ is $\subseteq$-increasing, and for $\ell = 0, 1$ the sequence $\langle M_{t_{2n+\ell}} : n < \omega\rangle$ is $\leq_{\mathfrak{K}}$-increasing with union M^*, hence by the basic properties of a.e.c. we have $M_{2n+\ell} \leq_{\mathfrak{K}} M^*$. So $M_{s_1} = M_{t_0} \leq_{\mathfrak{K}} M^*, M_{s_2} = M_{t_1} \leq_{\mathfrak{K}} M^*$. Now $M_{s_1}, M_{s_2} \subseteq M_{t_2} \leq_{\mathfrak{K}} M^*$ hence $M_{s_1}, M_{s_2} \leq_{\mathfrak{K}} M_{t_2}$, Recall that $N_{s_1} = N_{t_0} \leq_{\mathfrak{K}} N_{t_2}$ was proved above and $N_{s_2} = N_{t_1} \leq_{\mathfrak{K}} N_{t_2}$ was also proved above so t_2 is a common $\leq^1_I$-upper bound of s_1, s_2 as required.]

$(*)_8$ if $s \leq^0_{I_1} t$ then $s \leq^1_{I_1} t$.

[Why? By $(*)_7$ there is $t_1 \in I_1$ which is a common $\leq^1_{I_1}$-upper bound of s, t. So $M_s \subseteq M_t$ (as $s \leq^0_{I_1} t$) and $M_s \leq_{\mathfrak{K}} M_{t_1}$ (as $s \leq^1_{I_1} t_1$) and $M_t \leq_{\mathfrak{K}} M_{t_1}$ (as $t \leq^1_{I_1} t_1$). Together by axiom V of a.e.c. we get $M_s \leq_{\mathfrak{K}} M_t$ and by $(*)_5$ we have $N_s \leq_{\mathfrak{K}} N_t$. Together $s \leq^1_{I_1} t$ as required.]

$(*)_9$ $\langle M_s : s \in (I_1, \leq^1_{I_1})\rangle$ is $\leq_{\mathfrak{K}}$-increasing, $(I_1, \leq^1_{I_1})$ is directed and $\cup\{M_s : s \in I_1\} = M$.

[Why? The first phrase by the definition of $\leq^1_{I_1}$ in clause $(c)(\beta)$ of $\boxdot$, the second by $(*)_7$ and the third by $(*)_6 + (*)_4$.]

By the basic properties of a.e.c. (see I.1.6) we deduce

$\odot$ (a) $M \in K$

(b) $t \in I_1 \Rightarrow M_t \leq_{\mathfrak{K}} M$.

Now we strengthen the assumption $\boxtimes_1$ to

$\boxtimes_2$ the demands in $\boxtimes_1$ and $M \prec_{\mathbb{L}_{\infty,\kappa}[\tau_{\mathfrak{K}}]} N$.

We note

$\circledast_1$(a) if $\bar{a} \in {}^\alpha M, |\alpha| + \mathrm{LS}(\mathfrak{K}) \leq \lambda < \kappa$ then for some $t \in I_\lambda$, $f_t(\bar{a}) = \bar{a}$

(b) if $M' \subseteq M$ and $\|M\| \leq \lambda$ then $(\mathrm{id}_{M'}, M', M') \in I_\lambda$

(c) if $M_1 \subseteq N_1 \subseteq N$ and $M_1 \subseteq M$ and $\|N_1\| \leq \lambda$ then for some $t \in I$ we have $N_t = N_1$ and $\mathrm{id}_{M_1} \subseteq f_t$.

[Why? Clause (a) is a special case of clause (b) and clause (b) is a special case of clause (c). Lastly, clause (c) follows from the assumption $M \prec_{\mathbb{L}_{\infty,\kappa}[\tau_{\mathfrak{K}}]} N$ and 0.17(2A),(2B).]

We next shall prove

$\circledast_2$ $M \leq_{\mathfrak{K}} N$.

By I.1.6 and $(*)_9$ above for proving $\circledast_2$ it suffices to prove:

$\circledast_3$ if $s \in I_1$ then $M_s \leq_{\mathfrak{K}} N$.

[Why $\circledast_3$ holds? As $M \subseteq N$ there is $N_* \leq_{\mathfrak{K}} N$ of cardinality $\leq \lambda$ such that $M_s \cup N_s \subseteq N_*$. By $\circledast_1(c)$ there is $t \in I$ such that $N_t = N_*$ and $\mathrm{id}_{M_s} \subseteq f_t$. As $N_* \leq_{\mathfrak{K}} N$ it follows that $t \in I_1$. So by $\boxtimes_1 \Rightarrow \odot(b)$ applied to s and to t we can deduce $M_s \leq_{\mathfrak{K}} M$ and $M_t \leq_{\mathfrak{K}} M$. But as $\mathrm{id}_{M_s} \subseteq f_t$ it follows that $M_s \subseteq M_t$ hence by AxV of a.e.c. we know that $M_s \leq_{\mathfrak{K}} M_t$. But as $t \in I$ clearly f_t is an isomorphism from N_t onto M_t hence $f_t^{-1}(M_s) \leq_{\mathfrak{K}} N_t$, and as $\mathrm{id}_{M_s} \subseteq f_t$ this means that $M_s = f_t^{-1}(M_s) \leq_{\mathfrak{K}} N_t$. Recalling $N_t \leq_{\mathfrak{K}} N$ and $\leq_{\mathfrak{K}}$ is transitive it follows that $M_s \leq_{\mathfrak{K}} N$ as required.]

Let us check parts (3) and (4) of the Fact. Having proved $\boxtimes_1 \Rightarrow \odot(a)$, clearly in part (4) of the fact the first conclusion there, $M \in K$, holds. The second conclusion, $M \equiv_{\mathbb{L}_{\infty,\kappa}[\mathfrak{K}]} N$ holds by

$\circledast_4$ if $\varphi(\bar{x}) \in \mathbb{L}_{\infty,\kappa}[\mathfrak{K}]$ and $|\ell g(\bar{x})| + \mathrm{LS}(\mathfrak{K}) \leq \lambda < \kappa$ and $t \in I$ and $\bar{a} \in {}^{\ell g(\bar{x})}(M_t)$ then $M \models \varphi[\bar{a}] \Leftrightarrow N \models \varphi[f_t(\bar{a})]$.

[Why? Prove by induction on the depth of φ for all λ simultaneously. For $\alpha = 0$, first for the usual atomic formulas this should be clear. Second, by $(*)_4$ there is t_1 such that $t \leq_I^2 t_1 \in I_1$ hence by $\circledast_3$+ clause (d) of $\boxdot$+ clause (b) of $\odot$ we have $M_{t_1} \leq_{\mathfrak{K}} N \wedge N_{t_1} \leq_{\mathfrak{K}} N \wedge M_{t_1} \leq_{\mathfrak{K}} M$ respectively. So if $u \subseteq \ell g(\bar{x})$ then $M \restriction \mathrm{Rang}(\bar{a} \restriction u) \leq_{\mathfrak{K}} M \Leftrightarrow M \restriction$

$\text{Rang}(\bar{a} \restriction u) \le_{\mathfrak{K}} M_{t_1} \Leftrightarrow N \restriction \text{Rang}(f(\bar{a}) \restriction u) \le_{\mathfrak{K}} N_{t_1} \Leftrightarrow N \restriction \text{Rang}(f(\bar{a}) \restriction u) \le_{\mathfrak{K}} N$. So we have finished the case of atomic formulas, i.e. $\alpha = 0$. For $\varphi(\bar{x}) = (\exists \bar{y})\psi(\bar{x}, \bar{y})$ use $(*)_2$, the other cases are obvious.]

So part (4) holds. As for part (3), the first statement, "$M \in K$" holds by part (4), the second statement, $M \le_{\mathfrak{K}} N$, holds by $\circledast_2$ and the third statement, $M \prec_{\mathbb{L}_{\infty,\kappa}[\mathfrak{K}]} N$ follows by $\circledast_1(b) + \circledast_4$. As we have already noted parts (1),(2),(5),(6) and part (7) is proved as $\circledast_4$ is proved, we are done. $\square_{1.10}$

1.11 Claim. *For a limit cardinal $\kappa > \text{LS}(\mathfrak{K})$:*
1) $M \prec_{\mathbb{L}_{\infty,\kappa}[\mathfrak{K}]} N$ provided that

- (a) *if $\theta < \kappa$ and $\theta \in (\text{LS}(\mathfrak{K}), \kappa)$ then $M \prec_{\mathbb{L}_{\infty,\theta}[\mathfrak{K}]} N$*
- (b) *for every $\partial < \kappa$ for some $\theta \in (\partial, \kappa)$ we have: if $\bar{a}, \bar{b} \in {}^{\partial}M$ and $(M, \bar{a}) \equiv_{\mathbb{L}_{\infty,\theta}[\mathfrak{K}]} (M, \bar{b})$ then $(M, \bar{a}) \equiv_{\mathbb{L}_{\infty,\theta_1}[\mathfrak{K}]} (M, \bar{b})$ for every $\theta_1 \in [\theta, \kappa)$.*

1A) $M \equiv_{\mathbb{L}_{\infty,\kappa}[\mathfrak{K}]} N$ provided that

- (a) *if $\text{LS}(\mathfrak{K}) < \theta < \kappa$ then $M \equiv_{\mathbb{L}_{\infty,\theta}[\mathfrak{K}]} N$*
- (b) *as in part (1).*

2) In parts (1) and (1A) we can conclude

- $(b)^+$ *for every $\partial < \kappa$ for some $\theta \in (\partial, \kappa)$ we have: if $\bar{a}, \bar{b} \in {}^{\partial}M$ and $(M, \bar{a}) \equiv_{\mathbb{L}_{\infty,\theta}[\mathfrak{K}]} (M, \bar{b})$ then $(M, \bar{a}) \equiv_{\mathbb{L}_{\infty,\kappa}[\mathfrak{K}]} (M, \bar{b})$.*

3) If $\text{cf}(\kappa) = \aleph_0$ then $M \cong N$ when

- (a) *if $\theta < \kappa$ and $\theta \in (\text{LS}(\mathfrak{K}), \kappa)$ then $M \equiv_{\mathbb{L}_{\infty,\theta}[\mathfrak{K}]} N$*
- (b) *as in part (1), i.e., for every $\partial \in (\text{LS}(\mathfrak{K}), \kappa)$ for some $\theta \in (\partial, \kappa)$ we have: if $\bar{a} \in {}^{\partial}M$ and $\bar{b} \in {}^{\partial}N$ and $(M, \bar{a}) \equiv_{\mathbb{L}_{\infty,\theta}[\mathfrak{K}]} (N, \bar{b})$ then $(M, \bar{a}) \equiv_{\mathbb{L}_{\infty,\theta_1}[\mathfrak{K}]} (N, \bar{b})$ for every $\theta_1 \in (\theta, \kappa)$*
- (c) *M, N have cardinality κ.*

Proof. 1) By 1.10(3) it suffices to prove $M \prec_{\mathbb{L}_{\infty,\kappa}} N$, for this it suffices to apply the criterion from 0.17(2A).

Let $\mathscr{F}$ be the set of functions f such that:

$\odot$ (α) $\mathrm{Dom}(f) \subseteq M$ has cardinality $< \kappa$

(β) $\mathrm{Rang}(f) \subseteq N$

(γ) if $\bar{a}$ lists $\mathrm{Dom}(f)$ then for every $\theta \in (\ell g(\bar{a}), \kappa)$ we have $\mathrm{tp}_{\mathbb{L}_{\infty,\theta}[\mathfrak{K}]}(\bar{a}, \emptyset, M) = \mathrm{tp}_{\mathbb{L}_{\infty,\theta}[\mathfrak{K}]}(f(\bar{a}), \emptyset, N)$.

1A) Similarly.
2) Similarly to part (1) using 1.10(4) and 0.17(2) instead 1.10(3), 0.17(2A).
3) Recall 0.17(1). $\square_{1.11}$

1.12 Claim. *1) Assume 1.3(a) + (b), i.e. $\mathfrak{K}$ is categorical in $\mu > \mathrm{LS}(\mathfrak{K})$. If $\mu = \mu^{<\kappa}$ and $\kappa > \mathrm{LS}(\mathfrak{K})$ then for every $M \leq_{\mathfrak{K}} N$ from K_μ we have $M \prec_{\mathbb{L}_{\infty,\kappa}[\mathfrak{K}]} N$ (and there are such $M <_{\mathfrak{K}_\mu} N$).*
2) Assume $\mathfrak{K}$ is weakly or just pseudo μ-solvable as witnessed by Φ (see Definition 1.4 and Claim 1.5) and $M^* = \mathrm{EM}_{\tau(\mathfrak{K})}(\mu, \Phi)$. If $\mu = \mu^{<\kappa}$ and $\kappa > |\tau_\Phi|$ and $M \leq_{\mathfrak{K}} N$ are both isomorphic to M^* then $M \prec_{\mathbb{L}_{\infty,\kappa}[\mathfrak{K}]} N$.

Proof. 1) We prove by induction on γ that for any formula $\varphi(\bar{x})$ from $\mathbb{L}_{\infty,\kappa}[\mathfrak{K}]$ of quantifier depth $\leq \gamma$ (and necessarily $\ell g(\bar{x}) < \kappa$) we have

(*) if $M \leq_{\mathfrak{K}} N$ are from K_μ and $\bar{a} \in {}^{\ell g(\bar{x})}M$ then $M \models \varphi[\bar{a}] \Leftrightarrow N \models \varphi[\bar{a}]$.

If $\varphi(\bar{x})$ is atomic this is clear (for the "$\{x_i : i < i^*\}$ is the universe of a $\leq_{\mathfrak{K}}$-submodel", the implication $\Rightarrow$ holds as $\leq_{\mathfrak{K}}$ is transitive and the implication $\Leftarrow$ as $\mathfrak{K}$ satisfies AxV of a.e.c.).
If $\varphi(\bar{x})$ is a Boolean combination of formulas for which the assertion was proved, clearly it holds for $\varphi(\bar{x})$. So we are left with the case $\varphi(\bar{x}) = (\exists \bar{y})\psi(\bar{y}, \bar{x})$, so $\ell g(\bar{y}) < \kappa$. The implication $\Rightarrow$ is trivial by the induction hypothesis and so suppose that the other fails, say $N \models \psi[\bar{b}, \bar{a}]$ and $M \models \neg(\exists \bar{y})\psi(\bar{y}, \bar{a})$. We choose by induction on $i < \mu^+$ a model $M_i \in K_\mu$, $\leq_{\mathfrak{K}}$-increasing continuous, and for each i in addition we choose an isomorphism f_i from M onto M_i and if $i = j + 1$ we shall choose an isomorphism g_j from N onto M_{j+1} extending f_j. For $i = 0$, let $M_0 = M$, for i limit let $M_i = \bigcup\limits_{j<i} M_j$.

For any i, if M_i was chosen, f_i exists as $\mathfrak{K}$ is categorical in μ. Now if $i = j+1$ then M_j, f_j are well defined and clearly we can choose $M_i = M_{j+1}, g_j$ as required.

By Fodor lemma, as $\mu = \mu^{<\kappa}$ and the set $\{\delta < \mu^+ : \mathrm{cf}(\delta) \geq \kappa\}$ is stationary, clearly for some $\alpha < \beta < \mu^+$ we have $f_\alpha(\bar{a}) = f_\beta(\bar{a})$, now (by the choice of g_α) we have $M_{\alpha+1} \models \psi[g_\alpha(\bar{b}), g_\alpha(\bar{a})]$, hence by the induction hypothesis applied to the pair $(M_{\alpha+1}, M_\beta)$ we have $M_\beta \models \psi[g_\alpha(\bar{b}), g_\alpha(\bar{a})]$ so $M_\beta \models \varphi[g_\alpha(\bar{a})]$. But $g_\alpha(\bar{a}) = f_\alpha(\bar{a}) = f_\beta(\bar{a})$, contradiction to $M \models \neg\varphi[\bar{a}]$.

2) The same proof but we restrict ourselves to models in $K_{[M^*]}$ so, e.g. in $(*)$ we have $M, N \in K_{[M^*]}$ recalling that $\mathfrak{K}_{[M^*]}$ is a μ-a.e.c., see Definition 0.5(3A) and Claim 0.6(7). $\square_{1.12}$

1.13 Exercise: 1) For the proof (of 1.12(1)) it suffices to assume "$S \subseteq \{\delta < \mu^+ : \mathrm{cf}(\delta) \geq \kappa\}$ is a stationary subset of μ^+ and $M^* \in K_\mu$ is locally S-weakly limit (see I.3.3(5)).

2) Similarly we can weaken the demands "$M^* = \mathrm{EM}_{\tau(\mathfrak{K})}(\mu, \Phi)$ and (K, Φ) is pseudo solvable" to: for every $M \leq_{\mathfrak{K}} N$ isomorphic to M^* (which $\in K_\mu$) there is a $\leq_{\mathfrak{K}}$-increasing sequence $\langle M_\alpha : \alpha < \mu^+\rangle$ such that $\{\delta < \mu^+\colon \mathrm{cf}(\delta) \geq \kappa$ and $(M_\delta, M_{\delta+1})$ is isomorphic to (M, N) and $M_\delta = \cup\{M_\alpha : \alpha < \delta\}\}$ is a stationary subset of μ^+.

1.14 Claim. *Assume $\Phi \in \Upsilon^{\mathrm{or}}_{<\kappa}[\mathfrak{K}]$ satisfies the conclusion of 1.12(2) for (μ, κ) and $\mathrm{LS}(\mathfrak{K}) < \kappa \leq \mu$ and J, I_1, I_2 are linear orders and I_1, I_2 are κ-wide, see Definition 0.14(1). Then*

(a) *If $I_1 \subseteq I_2$ then $\mathrm{EM}_{\tau(\mathfrak{K})}(I_1, \Phi) \prec_{\mathbb{L}_{\infty,\kappa}[\mathfrak{K}]} \mathrm{EM}_{\tau(\mathfrak{K})}(I_2, \Phi)$*

(b) *Assume $J \subseteq I_1, J \subseteq I_2$; if $\varphi(\bar{x}) \in \mathbb{L}_{\infty,\kappa}[\mathfrak{K}]$ so $\ell g(\bar{x}) < \kappa$ and $\bar{a} \in {}^{\ell g(\bar{x})}(\mathrm{EM}(J, \Phi))$, then $\mathrm{EM}_{\tau(\mathfrak{K})}(I_1, \Phi) \models \varphi[\bar{a}] \Leftrightarrow \mathrm{EM}_{\tau(\mathfrak{K})}(I_2, \Phi) \models \varphi[\bar{a}]$*

(c) *Assume $\bar{\sigma} = \langle \sigma_i(\ldots, x_{\alpha(i,\ell)}, \ldots)_{\ell < \ell(i)} : i < i(*)\rangle$ where $i(*) < \kappa$, each σ_i is a $\tau(\Phi)$-term, $\alpha(i,\ell) < \alpha(*) < \kappa$. If $\bar{t}^\ell = \langle t^\ell_\alpha : \alpha < \alpha(*)\rangle$ is a sequence of members of I_ℓ for $\ell = 1, 2$ and $\bar{t}^1, \bar{t}^2$ realizes the same quantifier free type in I_1, I_2 respectively and $\bar{a}^\ell = \langle \sigma_i(\ldots, a_{t^\ell_{\alpha(i,j)}}, \ldots)_{j<j(i)} : i < i(*)\rangle$ for $\ell = 1, 2$ then $\bar{a}^1, \bar{a}^2$ realize the same $\mathbb{L}_{\infty,\kappa}[\mathfrak{K}]$ -type in $\mathrm{EM}_{\tau(\mathfrak{K})}(I_1, \Phi), \mathrm{EM}_{\tau(\mathfrak{K})}(I_2, \Phi)$ respectively.*

Proof of 1.14.

Clause (a): We prove that for $\varphi(\bar{x}) \in \mathbb{L}_{\infty,\kappa}[\mathfrak{K}]$ we have

$(*)_{\varphi(\bar{x})}$ if $I_1 \subseteq I_2$ are κ-wide linear orders of cardinality $\leq \mu$, and $\bar{a} \in {}^{\ell g(\bar{x})}(\mathrm{EM}_{\tau(\mathfrak{K})}(I, \Phi))$ then $\mathrm{EM}_{\tau(\mathfrak{K})}(I_1, \Phi) \models \varphi[\bar{a}] \Leftrightarrow \mathrm{EM}_{\tau(\mathfrak{K})}(I_2, \Phi) \models \varphi[\bar{a}]$.

This easily suffices as for any $I \in K^{\mathrm{lin}}$, the model $\mathrm{EM}_{\tau(\mathfrak{K})}(I, \Phi)$ is the direct limit of $\langle \mathrm{EM}(I', \Phi) : I' \subseteq I$ has cardinality $\leq \mu\rangle$, which is $\leq_{\mathfrak{K}}$-increasing and μ^+-directed and as we have:

$\odot$ $M^1 \prec_{\mathbb{L}_{\infty,\kappa}[\mathfrak{K}]} M^2$ when:

- (a) I is a κ-directed partial order
- (b) $\bar{M} = \langle M_t : t \in I\rangle$
- (c) $s <_I t \rightarrow M_s \prec_{\mathbb{L}_{\infty,\kappa}[\mathfrak{K}]} M_t$
- (d) $M^2 = \cup\{M_t : t \in I\}$
- (e) $M^1 \in \{M_t : t \in I\}$ or for some κ-directed $I' \subseteq I$ we have $M^1 = \cup\{M_t : t \in I'\}$.

We prove $(*)_{\varphi(\bar{x})}$ by induction on φ (as in the proof of 1.12 above). The only non-obvious case is $\varphi(\bar{x}) = (\exists \bar{y})\psi(\bar{y}, \bar{x})$, so let $I_1 \subseteq I_2$ be κ-wide linear orders of cardinality $\leq \mu$ and $\bar{a} \in {}^{\ell g(\bar{x})}(\mathrm{EM}_{\tau(\mathfrak{K})}(I_1, \Phi))$. Now if $\mathrm{EM}_{\tau(\mathfrak{K})}(I_1, \Phi) \models \varphi[\bar{a}]$ then for some $\bar{b} \in {}^{\ell g(\bar{y})}(\mathrm{EM}_{\tau(\mathfrak{K})}(I_1, \Phi))$ we have $\mathrm{EM}_{\tau(\mathfrak{K})}(I_1, \Phi) \models \psi[\bar{b}, \bar{a}]$ hence by the induction hypothesis $\mathrm{EM}_{\tau(\mathfrak{K})}(I_2, \Phi) \models \psi[\bar{b}, \bar{a}]$, hence by the satisfaction definition $\mathrm{EM}_{\tau(\mathfrak{K})}(I_2, \Phi) \models \psi[\bar{a}]$, so we have proved the implication $\Rightarrow$.

For the other implication assume that $\bar{b} \in {}^{\ell g(\bar{y})}(\mathrm{EM}_{\tau(\mathfrak{K})}(I_2, \Phi))$ and $\mathrm{EM}_{\tau(\mathfrak{K})}(I_2, \Phi) \models \psi[\bar{b}, \bar{a}]$. Let $\theta = |\ell g(\bar{a}{}^\frown\bar{b})| + \aleph_0$, so $\theta < \kappa$ and without loss of generality if κ is singular then $\theta \geq \mathrm{cf}(\kappa)$. Hence there is in I_1 a monotonic sequence $\bar{c} = \langle c_i : i < \theta^+\rangle$, without loss of generality it is increasing. Clearly there is I^* such that $\bar{a}{}^\frown\bar{b} \in {}^{\ell g(\bar{x}{}^\frown\bar{y})}(\mathrm{EM}(I^*, \Phi))$, $I^* \subseteq I_2$, $|I^*| \leq \theta$ and $\bar{a} \in {}^{\ell g(\bar{x})}(\mathrm{EM}(I^* \cap I_1, \Phi))$ and without loss of generality $i < \theta^+ \Rightarrow [c_0, c_i]_{I_2} \cap I^* = \emptyset$.

Similarly without loss of generality

$(*)$ $I_1 \backslash \cup\{[c_0, c_i)_{I_1} : i < \theta^+\}$ is κ-wide or $\kappa = \theta^+$.

Let $J_0 = I_2$; we can find J_1 such that $J_0 = I_2 \subseteq J_1$ and $J_1 \backslash I_2 = \{d_\alpha : \alpha < \mu \times \theta^+\}$ with d_α being $<_{J_1}$-increasing with α and $(\forall x \in I_2)(x <_{J_1} d_\alpha \equiv \bigvee\limits_{i<\theta^+} x <_{J_1} c_i)$.

As $\mathrm{EM}_{\tau(\mathfrak{K})}(I_2, \Phi) \models \psi[\bar{b}, \bar{a}]$ and $I_2 = J_0 \subseteq J_1, |J_1| \leq \mu$ and I_2 is κ-wide (and trivially J_1 is κ-wide), by the induction hypothesis $\mathrm{EM}_{\tau(\mathfrak{K})}(J_1, \Phi) \models \psi[\bar{b}, \bar{a}]$ hence $\mathrm{EM}_{\tau(\mathfrak{K})}(J_1, \Phi) \models \varphi[\bar{a}]$. Let $J_2 = J_1 \restriction \{x : x \in J_1 \backslash J_0$ or $x \in I_1 \backslash \cup \{[c_0, c_i]_{I_1} : i < \theta^+\}\}$. So $J_1 \supseteq J_2$, both linear orders have cardinality μ and are κ-wide as witnessed by $\langle d_\alpha : \alpha < \mu \times \theta^+ \rangle$ for both hence the conclusion of 1.12 holds, i.e. $\mathrm{EM}(J_2, \Phi) \prec_{\mathbb{L}_{\infty,\kappa}[\mathfrak{K}]} \mathrm{EM}(J_1, \Phi)$. Also $I^* \cap I_1 \subseteq J_2$ and recall that $\bar{a} \in {}^{\ell g(\bar{x})}(\mathrm{EM}(I^* \cap I_1, \Phi))$ hence $\bar{a} \in {}^{\ell g(\bar{x})}(\mathrm{EM}(J_2, \Phi))$. However, $\mathrm{EM}_{\tau(\mathfrak{K})}(J_1, \Phi) \models \varphi[\bar{a}]$, see above, hence by the last two sentences $\mathrm{EM}_{\tau(\mathfrak{K})}(J_2, \Phi) \models \varphi[\bar{a}]$.

So there is $\bar{b}^* \in {}^{\ell g(\bar{y})}(\mathrm{EM}_{\tau(\mathfrak{K})}(J_2, \Phi))$ such that $\mathrm{EM}_{\tau(\mathfrak{K})}(J_2, \Phi) \models \psi[\bar{b}^*, \bar{a}]$.
Let $J^* \subseteq J_2$ be of cardinality θ such that $\bar{b}^* \in {}^{\ell g(\bar{y})}(\mathrm{EM}_{\tau(\mathfrak{K})}(J^*, \Phi))$ and $I^* \cap I_1 \subseteq J^*$ recalling $I^* \cap [c_0, c_i)_{I_2} = \emptyset$ for $i < \theta^+$. Now let $u \subseteq \mu \times \theta^+$ be such that $J^* \backslash I_1 = \{d_\alpha : \alpha \in u\}$ so $|u| < \theta^+$. Let $J_3 = J_2 \restriction \{t : t \in J_2 \cap I_1$ or $t = d_\alpha \wedge \alpha > \sup(u)$ or $t = d_\alpha \wedge \alpha \in u\}$; as $\mathrm{cf}(\mu \times \theta^+) = \theta^+ > |u|$, clearly $\sup(u) < \mu \times \theta^+$ hence $|J_3| = \mu$ and J_3 is κ-wide. So by the conclusion of 1.12 (or by the induction hypothesis) also $\mathrm{EM}_{\tau(\mathfrak{K})}(J_3, \Phi) \models \psi[\bar{b}^*, \bar{a}]$. Let $w = \{\alpha < \mu \times \theta^+ : \alpha \in u$ or $\alpha > \sup(u) \wedge (\alpha - \sup(u) < \theta^+)\}$, so $\mathrm{otp}(w) = \theta^+$.

Let $J_4 = (J_3 \cap I_1) \cup \{d_\alpha : \alpha \in w\}$, so J_4 is κ-wide as witnessed by $I_1 \backslash \cup \{[c_0, c_i) : i < \theta^+\}$ or by $\{d_\alpha : \alpha \in w\}$ recalling $(*)$ above and $J_4 \subseteq J_3$ and $J^* \subseteq J_4$ hence $\bar{a}, \bar{b}^* \subseteq {}^{\kappa >}(\mathrm{EM}(J_4, \Phi))$ hence by the induction hypothesis $\mathrm{EM}_{\tau(\mathfrak{K})}(J_4, \Phi) \models \psi[\bar{b}^*, \bar{a}]$.

Let $J_5 = J_4 \cup \{c_i : i < \theta^+\} \backslash \{d_\alpha : \alpha \in w\}$ equivalently $J_5 = (J_3 \cap I_1) \cup \{c_\alpha : \alpha < \theta^+\} = (I_1 \backslash \cup \{[c_0, c_i)_{I_1} : i < \theta^+\}) \cup \{c_i : i < \theta^+\}$ so $J_5 \subseteq I_1$ and let $h : J_4 \to J_5$ be such that $h(d_\alpha) = c_{\mathrm{otp}(w \cap \alpha)}$ for $\alpha \in w$ and $h(t) = t$ for others, i.e. for $t \in J_3 \cap I_1$. So h is an isomorphism from J_4 onto J_5. Recalling 0.12 let $\hat{h}$ be the isomorphism from $\mathrm{EM}(J_4, \Phi)$ onto $\mathrm{EM}(J_5, \Phi)$ which h induces, so clearly $\hat{h}(\bar{a}) = \bar{a}$. Hence for some $\bar{b}^{**}$ we have $\bar{b}^{**} = \hat{h}(\bar{b}^*) \in {}^{\ell g(\bar{y})}(\mathrm{EM}_{\tau(\mathfrak{K})}(J_5, \Phi))$ and $\mathrm{EM}_{\tau(\mathfrak{K})}(J_5, \Phi) \models \psi[\bar{b}^{**}, \bar{a}]$. Note that by the choice of $\langle c_i : i < \theta^+\rangle$, (see (*) above), we know that J_5 is κ-wide. Also $J_5 \subseteq I_1$ so by the induction hypothesis applied to $\psi(\bar{y}, \bar{x}), J_5, I_1$ we have $\mathrm{EM}_{\tau(\mathfrak{K})}(I_1, \Phi) \models \psi[\bar{b}^{**}, \bar{a}]$ hence by the definition of satisfaction $\mathrm{EM}_{\tau(\mathfrak{K})}(I_1, \Phi) \models \varphi[\bar{a}]$, so we have finished proving the implication $\Leftarrow$ hence clause (a).

Clause (b): Without loss of generality for some linear order I we have $I_1 \subseteq I, I_2 \subseteq I$ and $\mathrm{EM}(I_\ell, \Phi) \subseteq \mathrm{EM}(I, \Phi)$ for $\ell = 1, 2$ and use clause (a) twice.

Clause (c): Easy by now, e.g. using a linear order I' extending I_1, I_2 which has an automorphism h such that $h(t^1_\alpha) = t^2_\alpha$ for $\alpha < \alpha(*)$. $\square_{1.14}$

1.15 Definition. Fixing $\Phi \in \Upsilon^{\mathrm{or}}_{\mathfrak{K}}$.
1) For $\theta \geq \mathrm{LS}(\mathfrak{K})$ let K^*_θ, [let K^{**}_θ] [let $K^{*,*}_\theta$] be the family of $M \in K_\theta$ isomorphic to some $\mathrm{EM}_{\tau(\mathfrak{K})}(I, \Phi)$ where I is a linear order of cardinality θ [which is θ-wide][which $\in K^{\mathrm{flin}}_\theta$]. More accurately we should write $K^*_{\Phi,\theta}, K^{**}_{\Phi,\theta}, K^{*,*}_{\Phi,\theta}$; similarly below.
2) Let K^* is the class $\cup\{K^*_\theta : \theta$ a cardinal $\geq \mathrm{LS}(\mathfrak{K})\}$, similarly $K^{*,*}, K^*_{\geq\lambda}, K^{**}_{\geq\lambda}$, etc.
3) Let $\mathfrak{K}^* = \mathfrak{K}^*_\Phi = (K^*, \leq_{\mathfrak{K}} \restriction K^*)$.
4) Let $\mathfrak{K}^*_\lambda = K^*_{\Phi,\lambda}$ be $(K^*_{\Phi,\lambda}, \leq_{\mathfrak{K}} \restriction K^*_{\Phi,\lambda})$.

1.16 Claim. *1) K^{**}_θ is categorical in θ if $\mathrm{LS}(\mathfrak{K}) < \theta \leq \mu$, $\mathrm{cf}(\theta) = \aleph_0$ and the conclusion of 1.12(2) hence of 1.14 holds for $\partial = \theta$ (and Φ), e.g. $\mathfrak{K}$ is pseudo solvable in μ as witnessed by Φ and $\mu = \mu^{<\theta}$.*
2) $K^{*,*}_\theta, K^{**}_\theta \subseteq K^*_\theta$.
3) If θ is strong limit $> \mathrm{LS}(\mathfrak{K})$ then $K^{**}_\theta = K^*_\theta$.

Proof. 1) By 1.14 and 0.17(1).
2) Read the definitions.
3) Recall 0.15(2). $\square_{1.16}$

1.17 Remark. 1) We will be specially interested in 1.16 in the case (μ, λ) is a $\mathfrak{K}$-candidate (see Definition .1.3) and $\theta = \lambda$.
2) Note that K^*_θ in general is not a θ-a.e.c.

3) If we strengthen 1.18(2) below, replacing (μ, λ) by (μ, λ^+) then categoricity of K^*_λ and in fact Claim 1.19(4) follows immediately from (or as in) Claim 1.16(1).

For the rest of this section we assume that the triple (μ, λ, Φ) is a pseudo $\mathfrak{K}$-candidate (see Definition 1.3) and rather than $\mu = \mu^\lambda$ we assume just the conclusion of 1.12, that is:

1.18 Hypothesis. 1) The pair (μ, λ) is a pseudo $\mathfrak{K}$-candidate and Φ witnesses this, so $|\tau_\Phi| \leq \mathrm{LS}(\mathfrak{K}) < \lambda = \beth_\lambda < \mu$ and $\Phi \in \Upsilon^{\mathrm{or}}_{\mathfrak{K}}$ is as in Definition 1.4 so $I \in K^{\mathrm{lin}}_\mu \Rightarrow \mathrm{EM}_{\tau(\mathfrak{K})}(I, \Phi) \in K^{\mathrm{pl}}_\mu$.
2) For every $\kappa \in (\mathrm{LS}(\mathfrak{K}), \lambda)$ the conclusion of 1.12(2) holds hence also of 1.14 (if $\mu = \mu^{<\lambda}$ this follows from (1) even for $\kappa = \lambda^+$ as $\mu^{<\kappa} = \mu^\lambda = \mu$ by cardinal arithmetic).

1.19 Claim. *1) If $M_1 \leq_{\mathfrak{K}} M_2$ are from K^*_λ or just $K^*_{\geq\lambda}$ and $\mathrm{LS}(\mathfrak{K}) < \theta < \lambda$ <u>then</u> $M_1 \prec_{\mathbb{L}_{\infty,\theta}[\mathfrak{K}]} M_2$; moreover $M_1 \prec_{\mathbb{L}_{\infty,\lambda}[\mathfrak{K}]} M_2$.*
2) If $M_1 \leq_{\mathfrak{K}} M_2$ are from K^* and $\|M_1\| \geq \kappa := \beth_{1,1}(\theta)$ (recall that this is $\beth_{(2^\theta)^+}$) and $\mu > \theta \geq \mathrm{LS}(\mathfrak{K})$ <u>then</u> $M_1 \prec_{\mathbb{L}_{\infty,\theta^+}[\mathfrak{K}]} M_2$.
3) Assume $\mathrm{LS}(\mathfrak{K}) < \theta < \kappa = \beth_{1,1}(\theta) \leq \chi < \mu, \chi_1 = \beth_{1,1}(\chi)$ and $M \in K^*_{\geq\chi_1}$ and $\bar{a}, \bar{b} \in {}^\gamma M$ where $\gamma < \theta^+$ and $(M, \bar{a}) \equiv_{\mathbb{L}_{\infty,\kappa}[\mathfrak{K}]} (M, \bar{b})$, i.e. $\varphi(\langle x_\beta : \beta < \gamma\rangle) \in \mathbb{L}_{\infty,\kappa^+}[\mathfrak{K}] \Rightarrow M \models \varphi[\bar{a}] \Leftrightarrow M \models \varphi[\bar{b}]$. <u>Then</u> $(M, \bar{a}) \equiv_{\mathbb{L}_{\infty,\chi}[\mathfrak{K}]} (M, \bar{b})$.
4) K^*_λ is categorical in λ provided that $\mathrm{cf}(\lambda) = \aleph_0$.

1.20 Remark. 1) What is the difference between say 1.19(3) and clause (a) of 1.14? Here there is no connection between the additional $\tau(\Phi)$-structures expanding M_1, M_2.
2) Note that Φ has the κ-non-order property (see 1.5(2)(*)) when $\kappa \geq \mathrm{LS}(\mathfrak{K}), \kappa^+ < \mu$ using 1.19(4).
3) Concerning 1.19(2), note that if $\|M_1\| \geq \mu$ it is easy to deduce this from 1.18(2), i.e, 1.12(2). But the whole point in this stage is to deduce something on cardinals $< \mu$.
4) Note that the proof of 1.19(2) gives:

> ⊛ assume $\mathrm{LS}(\mathfrak{K}) \leq \theta$ and $\delta(*) = \mathrm{Min}\{(2^\theta)^+, \delta(2^{\mathrm{LS}(\mathfrak{K})} + \theta)\}$ where on the function $\delta(-)$, see V.A.1.4,V.A.1.3, if $\beth_{\delta(*)} \leq \mu$

then for some $\alpha(*) < \delta(*)$ we have:

$\odot$ if $M_1 \le_{\mathfrak{K}} M_2$ are from K^* and $\|M_1\| \ge \beth_{\alpha(*)}$ then $M_1 \prec_{\mathbb{L}_{\infty,\theta^+}[\mathfrak{K}]} M_2$.

5) Similarly for 1.19(3) so we can weaken the demand $M \in K^*_{\ge \chi_1}$
6) We use "λ has countable cofinality, i.e. $\mathrm{cf}(\lambda) = \aleph_0$" in the proof of part (4) of 1.19, but not in the proof of the other parts.
7) Recall that for notational simplicity we assume $\mathrm{LS}(\mathfrak{K}) \ge |\tau_{\mathfrak{K}}|$ hence $\theta \ge |\tau_\Phi|$.
8) Note that for 1.19(2),(3) we can omit λ from Hypothesis 1.18.
9) Note that we shall use not only 1.19 but also its proof.

Proof of 1.19. 1) The first phrase holds by part (2) noting that $\kappa < \lambda$ if $\theta < \lambda$ as $\theta < \lambda = \beth_\lambda$. The second phrase holds by 1.11 as its assumption holds by parts (1) and (3).
2) We prove by induction on the ordinal γ that:

(*) if $M_1 \le_{\mathfrak{K}} M_2$ are from $K^*_{\ge \kappa}$ and the formula $\varphi(\bar{x}) \in \mathbb{L}_{\infty,\theta^+}[\mathfrak{K}]$ has depth $\le \gamma$ (so necessarily $\ell g(\bar{x}) < \theta^+$) and $\bar{a} \in {}^{\ell g(\bar{x})}(M_1)$ <u>then</u> $M_1 \models \varphi[\bar{a}] \Leftrightarrow M_2 \models \varphi[\bar{a}]$.

As in 1.12, the non-trivial case is to assume $\varphi(\bar{x}) = (\exists \bar{y})\psi(\bar{y}, \bar{x})$ where $\bar{a} \in {}^{\ell g(\bar{x})}(M_1)$ and $M_2 \models \varphi[\bar{a}]$ and we shall prove $M_1 \models \varphi[\bar{a}]$, so necessarily $\ell g(\bar{x}) + \ell g(\bar{y}) < \theta^+$ and we can choose $\bar{b} \in {}^{\ell g(\bar{y})}(M_2)$ such that $M_2 \models \psi[\bar{b}, \bar{a}]$. For $\ell = 1,2$ as $M_\ell \in K^*_{\ge \kappa}$ there is an isomorphism f_ℓ from $\mathrm{EM}_{\tau(\mathfrak{K})}(I_\ell, \Phi)$ onto M_ℓ for some linear order I_ℓ of cardinality $\ge \kappa$.

So we can find $J_\ell \subseteq I_\ell$ of cardinality θ for $\ell = 1,2$ such that $\bar{a} \subseteq M_1^-$ where $M_1^- = f_1(\mathrm{EM}_{\tau(\mathfrak{K})}(J_1, \Phi))$, and $\bar{a}\hat{\ }\bar{b} \subseteq M_2^-$ where $M_2^- = f_2(\mathrm{EM}_{\tau(\mathfrak{K})}(J_2, \Phi))$ and without loss of generality $M_1^- = M_2^- \cap M_1$. By 1.18(1), i.e. 0.9(1), clause (c) clearly $M_\ell^- \le_{\mathfrak{K}} M_\ell$ and so by AxV of a.e.c. (see Definition II.1.4), we have $M_1^- \le_{\mathfrak{K}} M_2^-$. First assume $\theta \ge 2^{\mathrm{LS}(\mathfrak{K})}$; in fact it is not a real loss to assume this. By renaming without loss of generality there is a transitive set B (in the set theoretic sense) of cardinality $\le \theta$ such that the following objects belong to it:

$\oplus(a)$ J_1, J_2

(b) Φ (i.e. τ_Φ and $\langle(\mathrm{EM}(n,\Phi), a_\ell)_{\ell<n} : n<\omega\rangle)$

(c) $\mathfrak{K}$, i.e., $\tau_{\mathfrak{K}}$ and $\{(M,N) : M \le_{\mathfrak{K}} N$ have universe included in $\mathrm{LS}(\mathfrak{K})\}$

(d) $\mathrm{EM}(J_\ell, \Phi)$ and $\langle a_t : t \in J_\ell\rangle$ for $\ell = 1, 2$.

Let χ be large enough, $\mathfrak{B} = (\mathscr{H}(\chi), \in, <^*_\chi)$ and $\mathfrak{B}^+$ be $\mathfrak{B}$ expanded by the individual constants $M^+_\ell = \mathrm{EM}(I_\ell, \Phi)$, $\langle a^\ell_t : t \in I_\ell\rangle$ the skeleton, M_ℓ, M^-_ℓ and f_ℓ (all for $\ell = 1, 2$), κ, B and x for each $x \in B$. By the assumption $\|M_1\| \ge \kappa = \beth_{1,1}(\theta)$, hence (see here V.A.1.3) there is $\mathfrak{C}$ such that

$\odot$ (a) $\mathfrak{C}$ is a $\tau(\mathfrak{B}^+)$-model elementarily equivalent to $\mathfrak{B}^+$ (that is, in first order logic)

(b) $\mathfrak{C}$ omits the type $\{x \ne b \ \& \ x \in B : b \in B\}$ but

(c) $|\{b : \mathfrak{C} \models \text{“}b \in \kappa^{\mathfrak{C}}\text{”}\}| = \mu = \|\mathfrak{C}\|$.

Without loss of generality $b \in B \Rightarrow b^{\mathfrak{C}} = b$.

Now

$\circledast_1$ if $\mathfrak{C} \models$ “$M \in K$”, so M is just a member of the model $\mathfrak{C}$ then we can define a $\tau_{\mathfrak{K}}$-model $M^{\mathfrak{C}} = M[\mathfrak{C}]$ as follows

(a) the set of elements of $M^{\mathfrak{C}}$ is $\{a : \mathfrak{C} \models$ “a is a member of the model M”$\}$

(b) if $R \in \tau_K$ is an n-place predicate then $R^{M[\mathfrak{C}]} = \{\langle a_\ell : \ell < n\rangle : \mathfrak{C} \models \text{“}\langle a_\ell : \ell < n\rangle \in R^M\text{”}\}$

(c) if $F \in \tau_K$ is an n-place function symbol, $F^{M[\mathfrak{C}]}$ is defined similarly.

$\circledast_2$(a) if $\mathfrak{C} \models$ “I is a linear order” then we define $I^{\mathfrak{C}}$ similarly

(b) similarly if $\mathfrak{C} \models$ “M is a $\tau(\Phi)$-model”

$\circledast_3$ if $\mathfrak{C} \models$ “I is a directed partial order, $\bar{M} = \langle M_s : s \in I\rangle$ satisfies $M_s \in K$ has cardinality $\mathrm{LS}(\mathfrak{K})$ and $s \le_I t \Rightarrow M_s \le_{\mathfrak{K}} M_t$” then also $\langle M^{\mathfrak{C}}_s : s \in I^{\mathfrak{C}}\rangle$ satisfies this.

By easy absoluteness (for clauses $(a)_1, (a)_2$ we use I.1.6, I.1.7 and $\circledast_3$):

$\boxtimes$(a)$_1$ if $\mathfrak{C} \models$ "$M \in K$" then $M^{\mathfrak{C}} \in K$

(a)$_2$ if $\mathfrak{C} \models$ "$M \leq_{\mathfrak{K}} N$" then $M^{\mathfrak{C}} \leq_{\mathfrak{K}} N^{\mathfrak{C}}$

(b)$_1$ if $\mathfrak{C} \models$ "I is a linear order" then $I^{\mathfrak{C}} = I[\mathfrak{C}]$ is a linear order

(b)$_2$ if $\mathfrak{C} \models$ "$I \subseteq J$ as linear orders" then $I^{\mathfrak{C}} \subseteq J^{\mathfrak{C}}$

(c) similarly for τ_Φ-models

(d)$_1$ if $\mathfrak{C} \models$ "$M = \text{EM}(I, \Phi)$" then there is a canonical isomorphism $f^{\mathfrak{C}}_I$ from $\text{EM}(I^{\mathfrak{C}}, \Phi)$ onto $M^{\mathfrak{C}}$ (hence it is also an isomorphism from $\text{EM}_{\tau(\mathfrak{K})}(I^{\mathfrak{C}}, \Phi)$ onto $M^{\mathfrak{C}} \restriction \tau(\mathfrak{K})$)

(d)$_2$ if $\mathfrak{C} \models$ "$I \subseteq J$ as linear orders" then $f^{\mathfrak{C}}_J$ extends $f^{\mathfrak{C}}_I$.

Now clearly $J^{\mathfrak{C}}_\ell = J_\ell$ and $I^{\mathfrak{C}}_\ell$ is a linear order of cardinality μ extending J_ℓ for $\ell = 1, 2$. Let $M^*_\ell = (M^-_\ell)^{\mathfrak{C}}$ for $\ell = 1, 2$.

So recalling clause (c) of $\odot$ we have: $M^{\mathfrak{C}}_1, M^{\mathfrak{C}}_2 \in K^*_\mu, M^{\mathfrak{C}}_1 \leq_{\mathfrak{K}} M^{\mathfrak{C}}_2, M^*_\ell \leq_{\mathfrak{K}} M^{\mathfrak{C}}_\ell, M^*_1 \leq_{\mathfrak{K}} M^*_2$ and $f^{\mathfrak{C}_0}_\ell, f^{\mathfrak{C}}_{I_\ell}$ are isomorphisms from $\text{EM}_{\tau(\mathfrak{K})}(I^{\mathfrak{C}}_\ell, \Phi)$ onto $M^{\mathfrak{C}}_\ell$, in fact, $f^{\mathfrak{C}}_{I_\ell}$ is the identity on $\text{EM}_{\tau(\mathfrak{K})}(J^{\mathfrak{C}}_\ell, \Phi) = \text{EM}_{\tau(\mathfrak{K})}(J_\ell, \Phi)$ and $f^{\mathfrak{C}}_\ell$ maps it onto M^*_ℓ for $\ell = 1, 2$.

Now $M_2 \models \psi[\bar{a}, \bar{b}]$, (why? assumed above) hence $M^{\mathfrak{C}}_2 \models \psi[\bar{a}, \bar{b}]$ (why? By 1.14, clause (b) or (c) and the situation recalling 1.18(2), of course noting that $I_2, I^{\mathfrak{C}}_2$ are of cardinality $\geq \kappa = \beth_{1,1}(\theta)$ hence are θ^+-wide), hence $M^{\mathfrak{C}}_2 \models \varphi[\bar{a}]$ (by definition of satisfaction), hence $M^{\mathfrak{C}}_1 \models \varphi[\bar{a}]$ (why? as $M^{\mathfrak{C}}_1, M^{\mathfrak{C}}_2 \in K^*_\mu$ hence $M^{\mathfrak{C}}_1 \prec_{\mathbb{L}_{\infty,\theta^+}[\mathfrak{K}]} M^{\mathfrak{C}}_2$ by $\boxtimes$ and 1.18(2) and recalling 1.12(2)) hence $M_1 \models \varphi[\bar{a}]$ (why? by clause (b) of 1.14 recalling 1.18(2)) as required in 1.19(2).]

So we are done except for a small debt: the case $\theta < 2^{\text{LS}(\mathfrak{K})}$ and $f^{\mathfrak{C}}_\ell$ is an isomorphism from $\text{EM}_{\tau(\mathfrak{K})}(I^{\mathfrak{C}}_\ell, \Phi)$.

In this case choose two sets B_1, B_2 such that $|B_1| = \theta, |B_2| = 2^{\text{LS}(\mathfrak{K})}$, $B_1 \subseteq B_2$ and concerning the demands in $\oplus$ above the objects from (a),(b),(d) and $\tau_{\mathfrak{K}}$ belong to B_1, the objects from (c) belong to B_2.

Again, without loss of generality B_1, B_2 are transitive sets and B_1, B_2 serve as individual constants of $\mathfrak{B}^+$ as well as each member of B_1. Now concerning $\mathfrak{C}$ we demand that it is elementarily equivalent to $\mathfrak{B}^+$; omit $\{x \in B_1 \wedge x \neq b : b \in B_1\}$ and for some $\mathfrak{B}^+_1 \prec \mathfrak{B}^+$ of

cardinality θ we have $\mathfrak{B}^+_1 \prec \mathfrak{C}$ and $\{b : \mathfrak{C} \models b \in B_2\} \subseteq \mathfrak{B}^+$. This influences just the proof of $\circledast_3$.
3) Without loss of generality $M = \mathrm{EM}_{\tau(\mathfrak{K})}(I, \Phi)$ and $I \in K^{\mathrm{lin}}_{\geq \chi_1}$. As $\gamma < \theta^+$ and $\bar{a}, \bar{b} \in {}^\gamma M$ there is $I_1 \subseteq I$ of cardinality θ such that $\bar{a}, \bar{b} \in {}^\gamma(M_1)$ where $M_1 = \mathrm{EM}_{\tau(\mathfrak{K})}(I_1, \Phi)$. As $(M, \bar{a}) \equiv_{\mathbb{L}_{\infty,\kappa^+}[\mathfrak{K}]} (M, \bar{b})$ necessarily there is $I_2 \subseteq I$ of cardinality κ and automorphism f of $M_2 = \mathrm{EM}_{\tau(\mathfrak{K})}(I_2, \Phi)$ mapping $\bar{a}$ to $\bar{b}$ such that $I_1 \subseteq I_2$. Why? Recalling 0.17(2), by the hence and forth argument as in the second part of the proof of 1.10(3).
Now as in the proof of part (2) there is a linear order I_3 extending I_1 of cardinality χ_1 and an automorphism g of $M_3 = \mathrm{EM}_{\tau(\mathfrak{K})}(I_3, \Phi)$ mapping $\bar{a}$ to $\bar{b}$. Without loss of generality for some linear order I_4 we have $I \subseteq I_4$ and $I_3 \subseteq I_4$.

Let $M_4 = \mathrm{EM}_{\tau(\mathfrak{K})}(I_4, \Phi)$, now $M \prec_{\mathbb{L}_{\infty,\chi^+}[\mathfrak{K}]} M_4$ by part (2), $M_3 \prec_{\mathbb{L}_{\infty,\chi^+}[\mathfrak{K}]} M_4$ by part (3) and $(M_3, \bar{a}) \equiv_{\mathbb{L}_{\infty,\chi^+}[\mathfrak{K}]} (M_3, \bar{b})$ by using the automorphism g of M_3 so together we are done.
4) So let $M, N \in K^*_\lambda$ (in fact, hence $\in K^{**}_\lambda$ recalling $K^*_\lambda = K^{**}_\lambda$ by 1.16(3) but not used). By parts (1),(3) the assumptions of 1.11(3) holds with λ here standing for κ there, hence its conclusion, i.e. $M \cong N$.

$\square_{1.19}$

Note: here the types below are sets of formulas.

1.21 Definition. Assume $M \in K, \mathbf{I} \subseteq {}^\gamma M$ and $\mathscr{L}, \mathscr{L}_1, \mathscr{L}_2$ are languages in the vocabulary $\tau_{\mathfrak{K}}$.
1) We say that $\mathbf{I}$ is $(\mathscr{L}, \partial, < \kappa)$-convergent in M, if: $|\mathbf{I}| \geq \partial$ and for every $\bar{b} \in {}^{\kappa>}M$, for some $\mathbf{J} \subseteq \mathbf{I}$ of cardinality $< \partial$ for some[2] p we have:

(*) for every $\bar{c} \in \mathbf{I} \backslash \mathbf{J}$, the $\mathscr{L}$-type of $\bar{c}\char`^\bar{b}$ in M is p.

2) Let $\mathrm{Av}_{\mathscr{L},\partial,<\kappa}(\mathbf{I}, M) = \{\varphi(\bar{x}, \bar{b}) : \varphi(\bar{x}, \bar{y})$ is an $\mathscr{L}$-formula, $\ell g(\bar{y}) < \kappa$ and $\bar{a} \in \mathbf{I} \Rightarrow \ell g(\bar{a}) = \ell g(\bar{x})$ and $\bar{b} \in {}^{\ell g(\bar{y})}M$ and for all but $< \partial$ of the sequences $\bar{c} \in \mathbf{I}$, the sequence $\bar{c}$ satisfies $\varphi(\bar{x}, \bar{b})$ in $M\}$. If ∂

[2] We could have demanded it for every single formula, here this distinction is not important

is missing, we mean $\partial = \kappa$. In parts (1) and (2) we may write "κ" instead of $< \kappa^+$; similarly below.
3) We say that $\mathbf{I}$ is $(\mathscr{L}_1, \mathscr{L}_2, \partial, < \kappa)$-based on A in M (if $\mathscr{L}_1 = \mathscr{L} = \mathscr{L}_2$ we may write only $\mathscr{L}$) <u>when</u>:

- (a) $A \subseteq M$
- (b) $\mathbf{I}$ is $(\mathscr{L}_1, \partial, < \kappa)$-convergent,
- (c) $\mathrm{Av}_{\mathscr{L}_1, \partial, <\kappa}(\mathbf{I}, M)$ does not $(\mathscr{L}_1, \mathscr{L}_2, < \kappa)$-split over A, see below.

4) We say that $p(\bar{x}) \in \mathrm{Sfr}^\alpha_{\mathscr{L}}(B, M)$ does not $(\mathscr{L}_1, \mathscr{L}_2, < \kappa)$-split over A <u>when</u>: if $\varphi(\bar{x}, \bar{y}) \in \mathscr{L}_1, \alpha = \ell g(\bar{x}) < \kappa, \ell g(\bar{y}) < \kappa$ and $\bar{b}, \bar{c} \in {}^{\ell g(\bar{y})}B$ realize the same $\mathscr{L}_2$-type in M over A <u>then</u> $\varphi(\bar{x}, \bar{b}) \in p \Leftrightarrow \varphi(\bar{x}, \bar{c}) \in p$; recalling that $\mathrm{Sfr}^\alpha_{\mathscr{L}}(A, M)$ is defined in 0.4 and normally $\mathscr{L}_1 = \mathscr{L}_2$ or at least $\mathscr{L}_1 \subseteq \mathscr{L}_2$.
5) Let $\mathrm{Av}_{<\kappa}(\mathbf{I}, M)$ be $\mathrm{Av}_{\mathbb{L}_{\infty,\kappa}[\mathfrak{K}]}(\mathbf{I}, M)$ and let $\mathrm{Av}_\kappa(\mathbf{I}, M)$ be $\mathrm{Av}_{\mathbb{L}_{\infty,\kappa^+}[\mathfrak{K}]}(\mathbf{I}, M)$.

1.22 Remark. 1) See definition of $\mathrm{Sav}^\alpha(M)$ in 1.34(2) below.
2) An alternative for clause (c) of 1.21(3) is:

- $(c)'$ the set $\{\mathrm{Av}_{\mathscr{L}, \partial, <\kappa}(f(\mathbf{I}), M) : f$ an automorphism of M over $A\}$ has cardinality $\leq \beth_{1,1}(\mathrm{LS}(\mathfrak{K}) + \theta + |A|) < \|M\|$.

1.23 Claim. *1) Assume that $M \in K, A \subseteq M, \mathbf{I} \subseteq {}^\theta M, |\mathbf{I}| \geq \partial = \mathrm{cf}(\partial) > \kappa \geq \theta + \mathrm{LS}(\mathfrak{K})$ and $\mathbf{I}$ is $(\mathscr{L}, \partial, \kappa)$-convergent. <u>Then</u> the type $p = \mathrm{Av}_{\mathscr{L}, \partial, \kappa}(\mathbf{I}, M)$ belongs to $\mathrm{Sfr}^\theta_{\mathscr{L}}(M)$, i.e., it is complete, recalling Definition 0.4 (no demand that it is realized in some $N, M \leq_{\mathfrak{K}} N$!).*
2) Also $\mathbf{I}$ is $(\mathscr{L}, \partial, \kappa)$-based on some set of cardinality $\leq \partial$, even on $\cup \mathbf{J}$, for any $\mathbf{J} \subseteq \mathbf{I}$ of cardinality $\geq \partial$.

Proof. 1) By the definition.
2) By the definitions: if $\bar{b} \in {}^{\kappa^+ >}M, \varphi = \varphi(\bar{x}, \bar{y}) \in \mathscr{L}$ and $\ell g(\bar{b}) = \ell g(\bar{y}), \ell g(\bar{x}) = \theta$, then by the convergence

$$\begin{aligned} \varphi(\bar{x}, \bar{b}) \in p \Leftrightarrow & \text{ for all but } < \partial \text{ members } \bar{a} \text{ of } \mathbf{I}, M \models \varphi[\bar{a}, \bar{b}] \Leftrightarrow \\ & \text{ for all but } < \partial \text{ members of } \mathbf{J}, M \models \varphi[\bar{a}, \bar{b}]. \end{aligned}$$

So only $\text{tp}_{\mathscr{L}}(\bar{b}, \cup \mathbf{J}, M)$ matters hence the non-splitting required in clause (c) of Definition 1.21(3). $\square_{1.23}$

As in V.A.1.12, we deduce non-splitting over a small set from non-order.

1.24 Claim. *Assume $M = \text{EM}_{\tau(\mathfrak{K})}(I, \Phi), \theta + \text{LS}(\mathfrak{K}) \leq \kappa < \lambda$ and $\beth_{1,1}(\partial) \leq |I|$ where $\partial = (2^{2^\kappa})^+$ or I is well ordered and $\partial = (2^\kappa)^+$. If $M \prec_{\mathbb{L}_{\infty,\partial}[\mathfrak{K}]} N$ <u>then</u> for every $\bar{a} \in {}^{\theta \geq} N$ there is $B \subseteq M$ of cardinality $< \partial$ such that $\text{tp}_{\mathbb{L}_{\infty,\kappa^+}[\mathfrak{K}]}(\bar{a}, M, N)$ does not $(\mathbb{L}_{\infty,\kappa^+}[\mathfrak{K}], \mathbb{L}_{\infty,\kappa^+}[\mathfrak{K}])$-split over B.*

Proof. Let $\bar{x} = \langle x_i : i < \ell g(\bar{a}) \rangle$.

We try to choose $B_\alpha, \gamma_\alpha, \bar{a}_\alpha, \bar{b}_\alpha, \bar{c}_\alpha, \varphi_\alpha(\bar{x}, \bar{y}_\alpha) \in \mathbb{L}_{\infty,\kappa^+}[\mathfrak{K}]$ by induction on $\alpha < \partial$ such that

⊛ (a) $B_\alpha = \cup\{\bar{a}_\beta : \beta < \alpha\}$

(b) $\bar{b}_\alpha, \bar{c}_\alpha \in {}^{\gamma_\alpha} M$ and $\gamma_\alpha < \kappa^+$

(c) $\varphi_\alpha(\bar{x}, \bar{y}_\alpha) \in \mathbb{L}_{\infty,\kappa^+}[\mathfrak{K}]$ such that $\ell g(\bar{y}_\alpha) = \gamma_\alpha$

(d) $N \models$ "$\varphi_\alpha[\bar{a}, \bar{b}_\alpha] \equiv \neg\varphi_\alpha[\bar{a}, \bar{c}_\alpha]$"

(e) $\bar{a}_\alpha \in {}^{\ell g(\bar{a})} M$ realizes $\{\varphi_\beta(\bar{x}, \bar{b}_\beta) \equiv \neg\varphi_\beta(\bar{x}, \bar{c}_\beta) : \beta < \alpha\}$ in M

(f) $M \models$ "$\varphi_\alpha[\bar{a}_\beta, \bar{b}_\alpha] \equiv \varphi_\alpha[\bar{a}_\beta, \bar{c}_\alpha]$" for $\beta \leq \alpha$.

If we are stuck at $\alpha(*) < \partial$ then we cannot choose $\gamma_\alpha, \bar{b}_\alpha, \bar{c}_\alpha, \varphi_\alpha(\bar{x}, \bar{y}_\alpha)$ clauses (b),(c),(d), because then $\bar{a}_\alpha$ as required in clauses (e),(f) exists because $M \prec_{\mathbb{L}_{\infty,\partial}[\mathfrak{K}]} N$. Hence $B := \cup\{\bar{a}_\alpha : \alpha < \alpha(*)\}$ is as required. So assume that we have carried the induction. As $\gamma_\alpha < \kappa^+ < \partial = \text{cf}(\partial)$ without loss of generality $\gamma_\alpha = \gamma < \kappa^+$ for every $\alpha < \partial$.

Let $\partial_1 = (2^\kappa)^+$.

Now by 1.25(5) below when I is not well ordered and by 1.25(4) below when I is well ordered (and part (1) of 1.25(1), recalling I is κ^+-wide as $\kappa < \partial$ and $\beth_{1,1}(\partial) \leq |I|$) clearly for some $S \subseteq \partial$ of order type ∂_1, the sequence $\langle \bar{a}_\alpha{}^\frown \bar{b}_\alpha{}^\frown \bar{c}_\alpha : \alpha \in S \rangle$ is $(\mathbb{L}_{\infty,\kappa^+}[\mathfrak{K}], \kappa^+, \kappa)$-convergent and $(\mathbb{L}_{\infty,\kappa^+}[\mathfrak{K}], < \omega)$-indiscernible in M hence without loss of generality $\alpha \in S \Rightarrow \varphi_\alpha = \varphi$. But as $\partial_1 > \kappa^+$ this contradicts (e) + (f) of ⊛ (if we use $\partial_1 = \kappa^+$, we can use a further conclusion of

1.25(1) stated in 1.25(2), i.e., $\langle \bar{a}_\alpha{}^\frown \bar{b}_\alpha{}^\frown \bar{c}_\alpha : \alpha \in S\rangle$ is a $(\mathbb{L}_{\infty,\kappa}[\mathfrak{K}], < \omega)$-indiscernible set not just a sequence, contradiction to (e) + (f) of $\circledast$). $\square_{1.24}$

1.25 Claim. *Assume* $M = \mathrm{EM}_{\tau(\mathfrak{K})}(I, \Phi), I$ *is* κ^+*-wide,* $\kappa < \lambda$ *and* $\mathrm{LS}(\mathfrak{K}) + \theta \le \kappa < \partial$.

1) Assume that $\mathscr{L} = \mathbb{L}_{\infty,\kappa^+}[\mathfrak{K}]$ *and* $\bar{a}_\alpha = \langle \sigma_i(\dots, a_{t(\alpha,i,\ell)}, \dots)_{\ell < n_i} : i < \theta\rangle$ *for* $\alpha < \partial$ *so* σ_i *is a* $\tau(\Phi)$*-term, and* $\mathrm{cf}(\partial) > \kappa$. *Assume further that letting* $\bar{t}_\alpha = \langle t(\alpha, i, \ell) : i < \theta, \ell < n_i\rangle$*, the sequence* $\langle \bar{t}_\alpha : \alpha < \partial\rangle$ *is indiscernible in* I *for quantifier free formulas (i.e. the truth values of* $t(\alpha_1, i_1, \ell_1) < t(\alpha_2, i_2, \ell_2)$ *depends only on* i_1, ℓ_1, i_2, ℓ_2 *and the truth value of* $\alpha_1 < \alpha_2, \alpha_1 = \alpha_2, \alpha_1 > \alpha_2$*).* <u>Then</u> $\langle \bar{a}_\alpha : \alpha < \partial\rangle$ *is* $(\mathscr{L}, \partial, \kappa)$*-convergent in the model* M.

2) In part (1), even dropping the assumption $\mathrm{cf}(\partial) > \kappa$*, moreover, the sequence* $\langle \bar{a}_\alpha : \alpha < \partial\rangle$ *is* $(\mathscr{L}, \kappa^+, \kappa)$*-convergent and* $(\mathscr{L}, < \omega)$*-indiscernible in* M.

3) In part (1) and in part (2), letting $J_0 = \{t(0, i, \ell) : t(0, i, \ell) = t(1, i, \ell)$ *and* $i < \theta, \ell < n_i\}$ *assume* $J_0 \subseteq J \subseteq I, J$ *is* κ^+*-wide (e.g.* $J = \{t(\alpha, i, \ell) : \alpha < \kappa^+, i < \theta, \ell < n_i\}$*) and* B *is the universe of* $\mathrm{EM}_{\tau(\mathfrak{K})}(J, \Phi)$ *and* $(i_1, i_2 < \theta, \ell_1 < n_{\ell_1}, \ell_2 < n_{i_2}$ *and* $[\alpha, \beta < \partial \Rightarrow t(\alpha, i_1, \ell_1) <_I t(\beta, i_2, \ell_2)] \Rightarrow \exists s \in J_0[\alpha, \beta < \partial \Rightarrow t(\alpha, i_1, \ell_1) <_I t <_I t(\beta, i_2, \ell_2)]$ <u>then</u> B *is a* (∂, κ)*-base of* $\{\bar{a}_\alpha : \alpha < \partial\}$.

4) If I *is well ordered (or just is* $\mathrm{EM}_{\{<\}}(J, \Psi), \Psi \in \Upsilon^{\mathrm{or}}, J$ *well ordered),* $\mathrm{LS}(\mathfrak{K}) + \theta \le \kappa, 2^\kappa < \partial, (\forall \alpha < \partial)[|\alpha|^\theta < \partial = \mathrm{cf}(\partial)]$ *and* $\bar{b}_\alpha \in {}^\theta M$ *for* $\alpha < \partial$, <u>then</u> *for some stationary* $S \subseteq \{\delta < \partial : \mathrm{cf}(\delta) \ge \theta^+\}$*, the sequence* $\langle \bar{b}_\alpha : \alpha \in S\rangle$ *is as in part (1) hence is* (κ^+, κ)*-convergent in* M*. Moreover, if* $S_0 \subseteq \{\delta < \partial : \mathrm{cf}(\delta) \ge \theta^+\}$ *is stationary we can demand* $S \subseteq S_0$.

5) If in (4) we omit the assumption "I is well ordered", and add $\partial \to (\partial_1)^2_{2^\kappa}$*, e.g.* $\partial_1 = (2^\kappa)^+, \partial = (2^{2^\kappa})^+$ <u>then</u> *we can find* $S \subseteq \partial, |S| = \partial_1$ *such that* $\langle \bar{a}_\alpha : \alpha \in S\rangle$ *is as in (1).*

Remark. In fact the well order case always applies at least if $\partial < \mu$.

Proof. 1) Let $\bar{b} \in {}^\kappa M$, so $\bar{b} = \langle \sigma^*_j(\dots, a_{s(j,\ell)}, \dots)_{\ell < m_j} : j < \kappa\rangle$ where σ^*_i is a $\tau(\Phi)$-term, $s(j, \ell) \in I$ and let $\bar{s} = \langle s(j, \ell) : \ell < m_j, j < \kappa\rangle$.

Now for each $i_1 < \theta, \ell_1 < n_{i_1}$ and $j_1 < \kappa, k_1 < m_{j_1}$ the sequence $\langle t(\alpha, i_1, \ell_1) : \alpha < \partial\rangle$ is monotonic (in I) hence there is $\alpha(i_1, \ell_1, j_1, k_1) < \partial$ such that

$(*)_1$ if $\beta, \gamma \in \partial\backslash\{\alpha(i_1, \ell_1, j_1, k_1)\}$ and $\beta < \alpha(i_1, \ell_1, j_1, k_1) \equiv \gamma < \alpha(i_1, \ell_1, j_1, k_1)$ then $\big(t(\beta, i_1, \ell_1) <_I s(j_1, k_1)\big) \equiv \big(t(\gamma, i_1, \ell_1) <_I s(j_1, k_1)\big)$ and $\big(t(\beta, i_1, \ell_1) >_I s(j_1, k_1)\big) \equiv \big(t(\gamma, i_1, \ell_1) >_I s(j_1, k_1)\big)$.

Let $u := \{\alpha(i_1, \ell_1, j_1, k_1) : i_1 < \theta, \ell_1 < n_{i_1}, j_1 < \kappa, k_1 < m_{j_1}\}$, it is a subset of ∂ of cardinality $\leq \theta + \kappa = \kappa$.
Hence

$(*)_2$ if $\beta, \gamma \in \partial\backslash u$ and $\beta \mathscr{E}_u \gamma$ which is defined by $(\forall \alpha \in u)(\alpha < \beta \equiv \alpha < \gamma)$ <u>then</u> $\bar{t}_\beta\hat{\ }\bar{s}, \bar{t}_\gamma\hat{\ }\bar{s}$ realizes the same quantifier free type in I

Now by clause (c) of 1.14 recalling I is κ^+-wide we have

$(*)_3$ if $\beta, \gamma \in \partial\backslash u$ and $\beta \mathscr{E}_u \gamma$ then $\bar{a}_\beta\hat{\ }\bar{b}, \bar{a}_\gamma\hat{\ }\bar{b}$ realizes the same $\mathbb{L}_{\infty,\kappa^+}[\mathfrak{K}]$-type in M.

As $\bar{b}$ was any member of ${}^\kappa M$ we have gotten

$(*)_4$ if $\bar{b} \in {}^{\kappa\geq}M$, then for some $u = u_{\bar{b}} \subseteq \partial$ of cardinality $\leq \kappa$ we have:
if $\beta, \gamma \in \partial\backslash u$ and $\beta \mathscr{E}_u \gamma$ then $\bar{a}_\beta\hat{\ }\bar{b}, \bar{a}_\gamma\hat{\ }\bar{b}$ realize the same $\mathbb{L}_{\infty,\kappa^+}[\mathfrak{K}]$-type in M.

As we are assuming $\mathrm{cf}(\partial) > \kappa (\geq \theta + \mathrm{LS}(\mathfrak{K}) \geq |\tau_\Phi|)$ we can conclude that

$(*)_5$ $\langle \bar{a}_\alpha : \alpha < \partial\rangle$ is $(\mathscr{L}, \partial, \kappa)$-convergent in M.

So we have proved 1.25(1).
2) We start as in the proof of part (1). However, after $(*)_3$ above letting for simplicity $u^+ = \{\alpha < \partial$: for some $\beta \in u \cap \alpha$ we have $\alpha + \kappa = \beta + \kappa\}$ we have

$(*)_6$ if $\beta, \gamma \in \partial\backslash u^+$ and $\beta < \gamma, \neg(\beta E_{u^+} \gamma)$
<u>then</u> we can find $(\mu^+, I^+, \bar{s}', \bar{b})$ such that
(α) $I \subseteq I^+ \in K^{\mathrm{lin}}$

(β) $M^+ = \mathrm{EM}_{\tau(\mathfrak{K})}(I, \Phi)$ hence $M \prec_{\mathbb{L}_{\infty,\kappa^+}[\mathfrak{K}]} N$

(γ) $\bar{s} = \langle s'(j,k) : k < m, j < \kappa \rangle$ a sequence of elements of I^+

(δ) $\bar{b}' = \langle \sigma^*_j(\dots, a_{s'(j,\ell)}, \dots)_{\ell < m_j} : j < \kappa \rangle \in {}^\kappa(M^+)$

(ε) $\bar{b}\char`^\bar{a}_\gamma, \bar{b}'\char`^\bar{a}_\gamma$ realize the same $\mathbb{L}_{\infty,\kappa^+}[\mathfrak{K}]$-types in M^+ as $\bar{b}\char`^\bar{a}_\gamma, \bar{b}\char`^\bar{a}_\beta$ respectively

(ζ) $\bar{s}\char`^\bar{t}_\beta, \bar{s}'\char`^\bar{t}_\beta$ form a Δ-system pair, i.e. are as in ⊠ from 1.5(2).

[Why?

Let $w^+ = \{(j,k) : k < m_j$ and $j < \kappa$ and for some $\ell < n_{i_1}, i_1 < \theta$ we have $\alpha(i_1, \ell_1, j, k) \in (\beta, \gamma)\}$

$$w^- := \{(j,k) : j < \kappa, k < m_j \text{ and } (j,\kappa) \notin w^+\}.$$

We choose I^+ extending I and $\bar{s}_\varepsilon = \langle s_i(j,k) : k < m_j, j < \kappa \rangle$ for $\varepsilon < \kappa$ such that

(a) the set of elements of I^+ is the disjoint union of I and $\{s_\varepsilon(j,k) : (j,k) \in w \text{ and } \varepsilon \in (0,\kappa)\}$

(b) $\bar{s}_\varepsilon, \bar{s}$ realize the same quantifier-free type in I^+

(c) if $\varepsilon, \zeta < \kappa$ then $\bar{t}_{\gamma+\varepsilon}\char`^\bar{s}_\zeta$ realizes in $I^+$ the quantifier-free type $\mathrm{tp}_{\mathrm{qf}}(\bar{t}_\beta\char`^\bar{s}, \emptyset, I)$ if $\varepsilon < \zeta$ and $\mathrm{tp}_q(\bar{t}_\gamma\char`^\bar{s}, \emptyset, I)$ if $\varepsilon \geq \zeta$

(d) $\langle \bar{t}_{\gamma+\varepsilon}\char`^\bar{s}_\varepsilon : \varepsilon < \kappa \rangle$ is indiscernible for quantifier-free formulas on I^+

(e) $\bar{s}_0 = \bar{s}$.

This is straight. Using $\bar{s}' = \bar{s}_1$ we are done.]

Now as Φ has the κ-non-order property (by Claim 1.5(2) which contains a definition, noting that the assumption of 1.5 holds by 1.18(1) and also 1.18(2)), repeating $(*)_4, (*)_5$ we get

$(*)_7$ for every $\bar{b} \in {}^{\kappa\geq}M$, for some $u = u^+_{\bar{b}} \in [\partial]^{\leq\kappa}$ if $\beta, \gamma \in \partial\backslash u^+$ then $\bar{a}_\beta\char`^\bar{b}, \bar{a}_\gamma\char`^\bar{b}$ realizes the same $\mathbb{L}_{\infty,\kappa^+}[\mathfrak{K}]$-type in M.

In other words

$(*)_8$ the sequence $\langle \bar a_\alpha : \alpha < \partial \rangle$ is $(\mathbb{L}_{\infty,\kappa^+}[\mathfrak{K}], \kappa^+)$-convergent.

The proof that it is a $(\mathbb{L}_{\infty,\kappa^+}[\mathfrak{K}], < \omega)$-indiscernible set is similar.
3) Not used; easy by 1.23(2) and convergence. [That is, note that we can find I^+ and $\bar a'_\alpha = \langle \sigma_i(\ldots, a_{t'(\alpha,i,\ell)}, \ldots)_{\ell_i < n_i} : i < \theta \rangle$ for $\alpha < \partial + \gamma$ such that:

(a) $I^+ \in K^{\rm lin}$ extend I

(b) $t'(\alpha, i, \ell) \in I^+$

(c) $\bar t'_\alpha = \langle t'(\alpha, i, \ell) : i < \theta, \ell < n_i \rangle$

(d) $\langle \bar t'_\alpha : \alpha < \partial + \gamma \rangle$ is indiscernible for quantifier-free formulas in I^+

(e) $\langle \bar t_\alpha : \alpha < \partial \rangle{}^\frown \langle \bar t'_\alpha : \alpha \in [\partial, \partial + \partial) \rangle$ is indiscernible for quantifier-free formulas in I'

(f) for each $i < \theta, \ell < n_i$ such that $t(0, i, \ell) = (j, i, t)$ the convex hull I_* of $\{t'(\alpha, i, \ell) : \alpha < \partial\}$ in I^+ is disjoint to I and if $s_1 <_I s_2$ and $(s_1, s_2)_{I^*} \cap I_* = \emptyset$ then $[s_1, s_2]_{I^*} \cap J_0 \neq \emptyset$.

So we can average over $\langle \bar a'_\alpha : \alpha < \partial \rangle$ instead averaging over $\langle \bar a_\alpha : \alpha < \partial \rangle$, and this implies the result. In fact we can weaken the assumption.]
4) Should be clear. [Still let $\bar t_\alpha = \langle t_{\alpha,i} : i < \theta \rangle$ be such that $\bar b_\alpha = \langle \sigma_{\alpha,j}(\ldots, a_{t_{\alpha,i(j,\alpha,\ell)}}, \ldots)_{\ell < n(\alpha,j)} : j < \theta \rangle$. So as $({\rm LS}(\mathfrak{K}) + |\tau_\Phi|)^\theta < \partial = {\rm cf}(\partial)$ for some stationary $S_1 \subseteq \{\delta < \partial\colon {\rm cf}(\delta) \geq \theta^+\}$ we have $\alpha \in S_1 \wedge j < \theta \Rightarrow \sigma_{\alpha,j} = \sigma_j$ (hence $j < \theta \Rightarrow n(\alpha, j) = n(j)$) and $\alpha \in S_1 \wedge j < \theta \wedge \ell < n(j) \Rightarrow i(j, \alpha, \ell) = i(j, \ell)$ and for every $i_1, i_2 < \theta$ we have $t_{\alpha,i_1} <_I t_{\alpha,i_2} \equiv (i_1, i_2) \in W$ for some sequence $\bar\sigma = \langle \sigma_j : j < \theta \rangle$ of τ_Φ-terms and $W \subseteq \kappa \times \kappa$ and sequence $\langle \langle i(j, \ell) : \ell < n(j) \rangle : j < \theta \rangle$.

If I is well ordered, for $\delta \in S_1$ let $\gamma_\delta = {\rm Min}\{\gamma\colon$ if $i < \theta$ and there are $\beta < \delta, j < \theta$ such that $t_{\delta,i} <_I t_{\beta,j}$ and then letting $(\beta_{\delta,i}, j_{\delta,i})$ be such a pair with $t_{\beta_{\delta,i}, j_{\delta,i}}$ being $<_I$-minimals, we have $\beta_{\delta,i} < \gamma\}$; clearly γ_δ is well defined and $< \delta$ so by Fodor lemma for some $\gamma_* < \partial$ the set $S_1 := \{\delta \in S_2 : \gamma_\delta = \gamma_*\}$ is stationary. As $|\gamma_*|^\theta < \partial$, for some $u \subseteq \theta$ and stationary $S_3 \subseteq S_2$ we have: if $\delta \in S_3$ then $j \in u \Leftrightarrow (\beta_{\delta,i}, j_{\delta,i})$ well defined and $j \in u \wedge \alpha \in S_3 \Rightarrow (\beta_{\delta,i}, j_{\delta,i}) = (\beta_i, j_i)$ and for each $i \in u$ the truth value of "$t_{\delta,i} = t_{\beta_i,j_i}$" is the same for all $\delta \in S_3$.

Now apply part (1) to $\langle \bar{b}_\alpha : \alpha \in S_3 \rangle$.]

5) By (1) and the definition of $\partial \to (\partial_1)^2_{2^\kappa}$. $\square_{1.25}$

1.26 Claim. *1) If $M \leq_{\mathfrak{K}} N$ are from K^*_λ and $\kappa \in [\mathrm{LS}(\mathfrak{K}), \lambda), \kappa^+ < \partial = \mathrm{cf}(\partial) < \lambda$ and moreover $\theta \leq \kappa$ and $\bar{a} \in {}^\theta N$ then there is a (κ^+, κ)-convergent set $\mathbf{I} \subseteq {}^\theta M$ of cardinality ∂ such that $\mathrm{Av}_\kappa(\mathbf{I}, M)$ is realized in N by $\bar{a}$.*

2) In fact we can weaken $M, N \in K^*_\lambda$ to $M, N \in K^*_{\geq \beth_{1,1}(\partial')}$ where, e.g. $\partial' = \beth_5(\kappa)^+$.

3) Assume $\theta \leq \kappa, \kappa \in [\mathrm{LS}(\mathfrak{K}), \lambda), \partial' = \beth_5(\kappa)^+$ and $M_1 \in K^*_{\geq \beth_{1,1}(\partial')}$. Assume further $M_1 \leq_{\mathfrak{K}} M_2 = \mathrm{EM}_{\tau(\mathfrak{K})}(I_2, \Phi), |\xi| = \theta$ and $\mathbf{I} \subseteq {}^\xi(M_1)$ is a (κ^+, κ)-convergent set (in M_1) of cardinality ∂'. If $I_2 <^*_{K^{\mathrm{flin}}} I_3$ (or just I_3 is κ^+-wide over I_2, which follows as $|I_2| \geq |\mathbf{I}| = \partial'$) and $M_3 = \mathrm{EM}_{\tau(\mathfrak{K})}(I_3, M_3)$ then

- *(a) we can find $\bar{d} \in {}^\xi(M_3)$ realizing $\mathrm{Av}_\kappa(M_2, \mathbf{I})$ so well defined*
- *(b) if $M_1 \leq_{\mathfrak{K}} N \in K^*$ and $\bar{d}^* \in {}^\xi N, |\xi| \leq \theta$ then we can find $\bar{d} \in {}^\xi(M_3)$ realizing $\mathrm{tp}_{\mathbb{L}_{\infty,\kappa^+}[\mathfrak{K}]}(\bar{d}^*, M_1, N)$ and $\mathrm{tp}_{\mathbb{L}_{\infty,\kappa^+}[\mathfrak{K}]}(\bar{d}, M_2, M_3)$ is the average of some (κ^+, κ)-convergent $\mathbf{I}' \subseteq {}^\alpha(M_1)$ of cardinality ∂'.*

Remark. The exact value of ∂' have no influences for our purpose.

Proof. 1) Without loss of generality $M = \mathrm{EM}_{\tau(\mathfrak{K})}(I, \Phi)$. Let $\partial_0 = \partial$ and $\partial_{\ell+1} = \beth_2(\partial_\ell)^+$ for $\ell = 0, 1$ so $\partial_\ell < \lambda$ and $\ell = 1, 2 \Rightarrow (\forall \alpha < \partial_\ell)(|\alpha|^{\kappa+\theta} < \partial_\ell = \mathrm{cf}(\partial_\ell) < \lambda)$ (if I is well ordered (which is O.K. by 1.19(4)) and $(\forall \alpha < \partial)(|\alpha|^\kappa < \partial)$ then we can use $\partial_\ell = \partial$).

By 1.24 there is $B_* \subseteq M$ of cardinality $< \partial_2$ (or just $\leq 2^{2^\kappa} < \partial_2$) such that $\mathrm{tp}_{\mathbb{L}_{\infty,\kappa^+}[\mathfrak{K}]}(\bar{a}, M, N)$ does not $(\mathbb{L}_{\infty,\kappa^+}[\mathfrak{K}], \mathbb{L}_{\infty,\kappa^+}[\mathfrak{K}])$-split over B_*.

Now by 1.19(1) for every $B \subseteq M, |B| < \partial_2$ there is $\bar{a}' \in {}^\theta M$ realizing in M, equivalently in N (with $\ell g(\bar{x}) = \theta$, of course), the type $\mathrm{tp}_{\mathbb{L}_{\infty,\kappa^+}[\mathfrak{K}]}(\bar{a}, B, N) = \{\varphi(\bar{x}, \bar{b}) : \bar{b} \in {}^{\kappa \geq} B, \varphi(\bar{x}, \bar{y}) \in \mathbb{L}_{\infty,\kappa^+}[\mathfrak{K}]$ and $N \models \varphi[\bar{a}, \bar{b}]\}$.

We can choose $J_\alpha, B_\alpha, \bar{a}_\alpha$ by induction on $\alpha < \partial_2$ such that B_α includes $\cup\{\bar{a}_\beta : \beta < \alpha\} \cup B_*$, B_α is the universe of $\mathrm{EM}(J_\alpha, \Phi)$, $J_\alpha \subseteq I$, $|J_\alpha| < \partial_2$, J_α increasing with α and J_α is quite closed (e.g. is $\mathfrak{B}_\alpha \cap I$ where $\mathfrak{B}_\alpha \prec_{\mathbb{L}_{\kappa^+,\kappa^+}} (\mathscr{H}(\chi), \in, <^*_\chi)$ with $M, N, \mathrm{EM}(I, \Phi), \mathfrak{K}, \langle \bar{a}_\beta : \beta < \alpha\rangle, \mathfrak{K}, \kappa, \theta$ belonging to $\mathfrak{B}_\alpha$ and $\mathfrak{B}_\alpha$ has cardinality $< \partial_2$ and $\mathfrak{B} \cap \partial_2 \in \partial_2$). Then choose $\bar{a}' = \bar{a}_\alpha$ as above, i.e. $\bar{a}_\alpha \in {}^\theta M$ realizes the same $\mathbb{L}_{\infty,\kappa^+}[\mathfrak{K}]$-type as $\bar{a}$ over $B_\alpha = M \cap \mathfrak{B}_\alpha = \mathrm{EM}_{\tau(\mathfrak{K})}(J_\alpha, \bar{a})$ in N; such $\bar{a}_\alpha$ exists by 1.19(1). So for some set $S_1 \subseteq \partial_2$ of order type ∂_1 the sequence $\mathbf{I} = \langle \bar{a}_\beta : \beta \in S_1\rangle$ is (κ^+, κ)-convergent (by 1.25(4),(5)).

It is enough to show that $\mathbf{I}$ is as required, toward contradiction assume that not. Then there is an appropriate formula $\varphi(\bar{x}, \bar{y})$ with $\ell g(\bar{x}) = \theta, \ell g(\bar{y}) = \kappa$ and $\bar{b} \in {}^\kappa M$ such that $N \models \varphi[\bar{a}, \bar{b}]$ but $u := \{\alpha \in S_1 : M \models \varphi[\bar{a}_\alpha, \bar{b}]\}$ has cardinality $< \kappa^+$. Now for $\alpha \in S_1$ as J_α was chosen "closed enough", there is $\bar{b}_\alpha \in {}^\kappa(\mathrm{EM}_{\tau(\mathfrak{K})}(J_\alpha, \Phi)) \subseteq {}^\kappa M$ realizing $\mathrm{tp}_{\mathbb{L}_{\infty,\kappa^+}[\mathfrak{K}]}(\bar{b}, B_*, M)$ such that $\beta \in S_1 \cap \alpha \Rightarrow M \models$ "$\varphi[\bar{a}_\beta, \bar{b}] \equiv \varphi[\bar{a}_\beta, \bar{b}_\alpha]$" (possible, e.g. as $|B_\alpha|^{|S \cap \alpha|} \le (2^{<\partial_1})^{<\partial_1} < \partial_2$).

So, again by 1.25(4),(5), for some $S_0 \subseteq S_1$ of order type $\partial = \partial_0$, the sequence $\langle \bar{a}_\alpha \hat{\ } \bar{b}_\alpha : \alpha \in S_0\rangle$ is $(\mathbb{L}_{\infty,\kappa^+}, \kappa^+, \kappa)$-convergent in M and $(\mathbb{L}_{\infty,\kappa^+}, < \omega)$-indiscernible. Let $\alpha \in S_0$ be such that $|S_0 \cap \alpha| > \kappa$, possible as $|S_0| = \partial_0 > \kappa_0$. So the set $\{\beta \in S_1 \cap \alpha : M \models \varphi[\bar{a}_\beta, \bar{b}_\alpha]\}$ has cardinality $\le \kappa$ (being equal to $\{\beta \in S_1 \cap \alpha : N \models \varphi[\bar{a}_\beta, \bar{b}]\}$) but $\alpha \in S_0 \subseteq S_1$ and $|S_0 \cap \alpha| > \kappa$, so for some $\beta < \alpha$ from $S_0, M \models \neg\varphi[a_\beta, \bar{b}_\alpha]$ hence by the indiscernibility $M \models \neg\varphi[\bar{a}_\beta, \bar{b}_\gamma]$ for every $\beta < \gamma$ from S_0.

On the other hand if $\alpha < \beta$ are from S_0 then by the choice of $\bar{b}_\alpha$ the sequences $\bar{b}, \bar{b}_\alpha$ realizes the same $\mathbb{L}_{\infty,\kappa^+}[\mathfrak{K}]$-type over B_*. Now $\mathrm{tp}_{\mathbb{L}_{\infty,\kappa^+}[\mathfrak{K}]}(\bar{a}, M, N)$ does not split over B_* by the choice of B_* so we have $N \models$ "$\varphi[\bar{a}, \bar{b}] \equiv \varphi[\bar{a}, \bar{b}_\alpha]$" but by the choice of $\bar{b}$ we have $N \models \varphi[\bar{a}, \bar{b}]$ hence $N \models \varphi[\bar{a}, \bar{b}_\alpha]$ hence $M \models \varphi[\bar{a}_\beta, \bar{b}_\alpha]$ by the choice of $\bar{a}_\beta$. Together this contradicts 1.5, i.e., 1.18(1).

2) Similarly (using 1.19(2) instead of 1.19(1).

3) Clause (a):

By 1.14 and the LS argument (i.e. by 0.17(4)) without loss of generality $M_1 \in K^*_{<\lambda}$. Let $\partial_\ell = \beth_\ell(\kappa)^+$ for $\ell \le 5$ so $\partial' = \partial_5$ and for notational simplicity assume $\theta \ge \aleph_0$.

Let $\{\bar{a}_\alpha : \alpha < \partial'\}$ list the members of $\mathbf{I}$, so for each $\alpha < \partial'$ there is $I_{2,\alpha} \subseteq I_2$ of cardinality θ such that $\bar{a}_\alpha$ is from $\mathrm{EM}_{\tau(\mathfrak{K})}(I_{2,\alpha}, \Phi)$.

For each $\alpha < \partial'$ let $\bar{t}^\alpha = \langle t_i^\alpha : i < \theta \rangle$ list $I_{2,\alpha}$ and so $\bar{a}_\alpha = \langle \sigma_{\alpha,\zeta}(\bar{t}^\alpha) : \zeta < \xi \rangle$ for some sequence $\langle \sigma_{\alpha,\zeta}(\bar{x}) : \zeta < \xi \rangle$ of τ_Φ-terms. We can find $S \subseteq \partial'$ of order type ∂_4 such that $\zeta < \xi \wedge \alpha \in S \Rightarrow \sigma_{\alpha,\zeta} = \sigma_\zeta$ and $\langle \bar{t}^\alpha : \alpha \in S \rangle$ is an indiscernible sequence (for quantifier free formulas, in I_2, of course).

By renaming $\kappa^+ \subseteq S$. We define a partition $\langle u_{-1}, u_0, u_1 \rangle$ of ξ by

$$u_0 = \{i < \theta : t_i^\alpha = t_i^\beta \text{ for } \alpha, \beta \in S\}$$

$$u_1 = \{i < \theta : t_i^\alpha <_{I_2} t_i^\beta \text{ for } \alpha < \beta \text{ from } S\}$$

$$u_{-1} = \{i < \theta : t_i^\beta <_{I_2} t_i^\alpha \text{ for } \alpha < \beta \text{ from } S\}.$$

We define an equivalence relation e on $u_{-1} \cup u_1$

$\odot$ $i_1 e i_2$ iff for some $\ell \in \{1, -1\}, i_1, i_2 \in u_\ell$ and $(t_{i_1}^\alpha <_I t_{i_2}^\beta) \equiv (t_{i_2}^\alpha <_I t_{i_1}^\beta)$ for every (equivalently some) $\alpha < \beta$ from S.

There is a natural set of representatives: $W = \{\zeta < \theta : \zeta \in u_{-1} \cup u_1$ and $\zeta = \min(\zeta/e)\}$.

We now define a linear order I_2^+; its set of elements is $\{t : t \in I_2\} \cup \{t_i^* : i \in u_{-1} \cup u_1\}$ where, of course, $t_i^* \in I_2^+$ are pairwise distinct and $\notin I_2$. The order is defined by (or see $\circledast_2$ and think)

$\circledast_1$ $s_1 <_{I_2^+} s_2$ iff

(a) $s_1, s_2 \in I_2$ and $s_1 <_{I_2} s_2$

(b) $s_1 \in I_2, s_2 = t_i^*$ and $s_1 <_{I_2} t_i^\alpha$ for every $\alpha < \kappa^+$ large enough

(c) $s_1 = t_i^*, s_2 \in I_2$ and $t_i^\alpha <_{I_2} s_2$ for every $\alpha < \kappa^+$ large enough

(d) $s_1 = t_i^*, s_2 = t_j^*$ and $t_i^\alpha <_I t_j^\alpha$ for every $\alpha < \kappa^+$.

Let $t^*_i = t^\alpha_i$ for $i \in u_0$ and any $\alpha < \kappa^+$. Let $M^+_2 = \mathrm{EM}_{\tau(\mathfrak{K})}(I^+_2, \Phi)$.

It is easy to check (by 1.14(a),(c)) that

$\circledast_2$ (a) $I_2 \subseteq I^+_2$

(b) $\bar{t}^* \in {}^\theta(I^+_2)$

(c) if $J \subseteq I_2$ has cardinality $\le \kappa$ <u>then</u> for every $\alpha < \kappa^+$ large enough, the sequences $\bar{t}^*, \bar{t}^\alpha$ realizes the same quantifier free type over J inside I^+_2.

Let

$\circledast_3$ $\bar{d} := \langle \sigma_\zeta(\bar{t}^*) : \zeta < \xi \rangle \in {}^\xi(M^+_2)$.

Recall that $\|M_2\| < \lambda$ hence $|I_2| < \lambda$ and I_2 is κ^+-wide having cardinality $\ge \partial' > 2^\kappa$.

Note

$\circledast_4$ $\bar{t}^*$ realizes $\mathrm{Av}_{\mathrm{qf}}(\{\bar{t}^\alpha : \alpha \in S\}, I_2)$ in the linear order I^+_2.

Without loss of generality $I^+_2 \cap I_3 = I_2$, so we can find a linear order I_4 of cardinality λ such that $I^+_2 \subseteq I_4 \wedge I_3 \subseteq I_4$. As I_3 is κ^+-wide over I_2 (see the assumption and Definition 0.14(6)+(3)), there is a convex subset I'_3 of I_3 disjoint to I_2 which contains a monotonic sequence $\langle s_\alpha : \alpha < \kappa^+ \rangle$. Without loss of generality there are elements s_α ($\alpha \in [\kappa^+, \lambda \times \kappa^+)$ in I_4 such that $\langle s_\alpha : \alpha < \lambda \times \kappa^+ \rangle$ is monotonic (in I_4), and its convex hull is disjoint to I_2. Let $I^-_3 = I_2 \cup \{s_\alpha : \alpha < \kappa^+\}$ and $I^\pm_3 = I_2 \cup \{s_\alpha : \alpha < \lambda \times \kappa^+\}$.

Now we use 1.14 several times. First, $\mathrm{EM}_{\tau(\mathfrak{K})}(I_2, \Phi) \prec_{\mathbb{L}_{\infty,\kappa^+}[\mathfrak{K}]} \mathrm{EM}_{\tau(\mathfrak{K})}(I^+_2, \Phi) \prec_{\mathbb{L}_{\infty,\kappa^+}[\mathfrak{K}]} \mathrm{EM}_{\tau(\mathfrak{K})}(I_4, \Phi)$ as $I_2 \subseteq I^+_2 \subseteq I_4$ are κ^+-wide, hence by $\circledast_4$ the sequence $\bar{d}$ realizes $q := \mathrm{Av}_\kappa(\{\langle \sigma_\zeta(\bar{t}^\alpha) : \zeta < \theta \rangle : \alpha < \kappa^+\}, M_2) = \mathrm{Av}(\{\bar{a}_\alpha : \alpha < \kappa^+\}, M_2) = \mathrm{Av}_\kappa(\mathbf{I}, M_2)$ in M^+_2 and also in $\mathrm{EM}_{\tau(\mathfrak{K})}(I_4, \Phi)$. Second, as $|I_2| < \lambda, I_2 \subseteq I^\pm_3 \subseteq I_4$ and $|I^\pm_3| = |I_4| = \lambda$, by 1.19(1) we have $\mathrm{EM}_{\tau(\mathfrak{K})}(I^\pm_3, \Phi) \prec_{\mathbb{L}_{\infty,\lambda}[\mathfrak{K}]} \mathrm{EM}_{\tau(\mathfrak{K})}(I_4, \Phi)$ so some $\bar{d}' \in {}^\xi(\mathrm{EM}_{\tau(\mathfrak{K})}(I^\pm_3, \Phi))$ realizes the type q in $\mathrm{EM}_{\tau(\mathfrak{K})}(I^\pm_3, \Phi)$. Let $w_1 \subseteq \lambda \times \kappa^+$ be of cardinality $\le \theta \le \kappa$ such that $\bar{d}'$ belongs to $\mathrm{EM}_{\tau(\mathfrak{K})}(I_2 \cup \{s_\alpha : \alpha \in w_1\}, \Phi)$. Choose $w_2 \subseteq \lambda \times \kappa^+$ of order type κ^+ including w_1, so $\mathrm{EM}_{\tau((\mathfrak{K})}(I_2 \cup \{s_\alpha : \alpha \in w_2\}, \Phi) \prec_{\mathbb{L}_{\infty,\kappa^+}[\mathfrak{K}]} \mathrm{EM}_{\tau(\mathfrak{K})}(I^\pm_3, \Phi)$ and $\bar{d}'$ belongs to the former hence realizes q in it. But there is an isomorphism h from $I_2 \cup \{s_\alpha : \alpha \in w_2\}$ onto I_3^- over I_2, hence it induces an isomorphism $\hat{h}$ from $\mathrm{EM}_{\tau(\mathfrak{K})}(I_2 \cup \{s_\alpha : \alpha \in w_2\}, \Phi)$ onto $\mathrm{EM}_{\tau(\mathfrak{K})}(I_3^-, \Phi)$ so $\hat{h}\,(\bar{d}')$ realizes q in the latter. But $I_3^- \subseteq I_3$ are both κ^+-wide hence by 1.14 the sequence $\hat{h}(\bar{d}')$ realizes q in $M_3 = \mathrm{EM}_{\tau(\mathfrak{K})}(I_3, \Phi)$ as required.

Clause (b):

By part (2) we can find appropriate **I** and then apply clause (a). $\square_{1.26}$

1.27 Remark. 1) In fact in 1.24, we can choose B of cardinality κ, hence similarly in the proof of 1.26(1).
2) Also using solvability to get well ordered I we can prove : if $A \subseteq M = \mathrm{EM}_{\tau(\mathfrak{K})}(\lambda, \Phi)$ and $|A| < \lambda$ then the set of $\mathbb{L}_{\infty,\kappa^+}[\mathfrak{K}]$-types realized in M over A is $\leq (|A| + 2)^\kappa$.

1.28 Claim. *1) If $M \in K^{**}_{\geq\kappa}$ and $\mathrm{LS}(\mathfrak{K}) \leq \theta$ and $\partial = \beth_{1,1}(\theta) \leq \kappa \leq \lambda$, then for $\bar{a}, \bar{b} \in {}^\theta M$ the following are equivalent: (the difference is using ∂ or κ)*

- *(a) $\bar{a}, \bar{b}$ realize the same $\mathbb{L}_{\infty,\partial}[\mathfrak{K}]$-type in M*
- *(b) $\bar{a}, \bar{b}$ realize the same $\mathbb{L}_{\infty,\kappa}[\mathfrak{K}]$-type in M.*

2) For $M, \theta, \partial, \kappa$ as above, the number of $\mathbb{L}_{\infty,\partial}[\mathfrak{K}]$-types of $\bar{a} \in {}^\theta M$ where $M = \mathrm{EM}_{\tau(\mathfrak{K})}(I, \Phi), |I| \geq \partial$ is $\leq 2^\theta$.

Remark. Part (1) improves 1.19(3).

Proof. 1) Clearly $(b) \Rightarrow (a)$, so assume clause (a) holds. As $M \in K^{**}_{\geq\kappa}$ without loss of generality there is a κ-wide linear order I such that $M = \mathrm{EM}_{\tau(\mathfrak{K})}(I, \Phi)$; hence for some $J \subseteq I, |J| = \theta$ we have $\bar{a}, \bar{b} \in {}^\theta(\mathrm{EM}_{\tau(\mathfrak{K})}(J, \Phi))$. So for every $\alpha < (2^\theta)^+$, by the hence and forth argument for $\mathbb{L}_{\infty,\beth^+_\alpha}[\mathfrak{K}]$ there are J_α, f_α such that $J \subseteq J_\alpha \subseteq I, |J_\alpha| = \beth_\alpha$ and f_α is an automorphism of $\mathrm{EM}_{\tau(\mathfrak{K})}(J_\alpha, \Phi)$ which maps $\bar{a}$ to $\bar{b}$. Hence as in the proof of 1.19 there is a linear order J^+ of cardinality μ extending J and an automorphism f of $M^+ = \mathrm{EM}_{\tau(\mathfrak{K})}(I^+, M)$

mapping $\bar a$ to $\bar b$. By clause (b) of Claim 1.14 we are done.
2) Easy by clause (c) of 1.14, i.e., by 1.18. $\square_{1.28}$

1.29 Claim. *Assume:*

(a) $I_1 \subseteq I_2, I_1 \neq I_2$, *moreover* $I_1 <_{K^{\rm flin}} I_2$, *see Definition 0.14(6)*
(b) $M_\ell = \mathrm{EM}_{\tau(\mathfrak{K})}(I_\ell, \Phi)$ *for* $\ell = 1, 2$
(c) $\bar b, \bar c \in {}^\alpha(M_2)$
(d) $\theta \geq |\alpha| + \mathrm{LS}(\mathfrak{K})$
(e) $\kappa = \beth_{1,1}(\theta_2) \leq \lambda$ *where* $\theta_1 = 2^\theta, \theta_2 = (2^{\theta_1})^+$
(f) $|I_1| \geq \kappa$
(g) $M_1 \leq_{\mathfrak{K}} M_2$, *follows from (a) + (b)*

1) Assume that for every $\bar a \in {}^{\kappa >}(M_1)$ *the sequences* $\bar a \hat{}\, \bar b, \bar a \hat{}\, \bar c$ *realize the same* $\mathbb{L}_{\infty,\kappa}[\mathfrak{K}]$*-type in* M_2. *Then there are* I_3, M_3 *and* f *such that* $I_2 \leq_{K^{\rm flin}} I_3 \in K^{\rm flin}_\lambda, M_3 = \mathrm{EM}_{\tau(\mathfrak{K})}(I_3, \Phi)$ *and* f *an automorphism of* M_3 *over* M_1 *mapping* $\bar b$ *to* $\bar c$.
2) Assume that for every $\bar a \in {}^{\kappa >}(M_1)$ *the sequences* $\bar a \hat{}\, \bar b, \bar a \hat{}\, \bar b$ *realize the same* $\mathbb{L}_{\infty,\kappa}[\mathfrak{K}]$*-type in* M_2 *(as in part (a)) and* $\beth_{1,1}(\partial) \leq |I_1|$ *and* $\partial < \lambda$. *Then for every* $\bar a \in {}^{\kappa >}(M_1)$, *the sequences* $\bar a \hat{}\, \bar b, \bar a \hat{}\, \bar c$ *realize the same* $\mathbb{L}_{\infty,\partial}[\mathfrak{K}]$*-type in* M_2.
3) Assume that $\mathrm{cf}(\lambda) = \aleph_0$ *and* $|I_1| = \lambda$ *and recall* $\lambda = \beth_\lambda > \mathrm{LS}(\mathfrak{K})$. *If* $M_1 \leq_{\mathfrak{K}} M^*_2 \in K^*_\lambda$ *then for some* I_3, *a linear order* $\leq_{K^{\rm flin}_\lambda}$*-extending* I_2 *the model* M^*_2 *can be* $\leq_{\mathfrak{K}}$*-embedded into* $M_3 := \mathrm{EM}_{\tau(\mathfrak{K})}(I_3, \Phi)$ *over* M_1.

Remark. 1) Under mild assumptions with somewhat more work in 1.29(1),(3) we can choose $I_3 = I_2$ (but for this has to be more careful with the linear orders). Recall that for $I \in K^{\rm lin}_\lambda$ like I_2 in 1.8(c) we have $\alpha < \lambda^+ \Rightarrow I \times \alpha$ can be embedded into I and 1.4(1)(d).

Proof. 1) There is $J_2 \subseteq I_2$ of cardinality $\leq \theta$ such that $\bar b, \bar c \in {}^\alpha(\mathrm{EM}_{\tau(\mathfrak{K})}(J_2, \Phi))$; let $J_1 = I_1 \cap J_2$.

We define a two-place relation $\mathscr{E}$ on $I_2 \backslash J_2 : s \mathscr{E} t$ iff $(\forall x \in J_2)(x <_{I_2} s \equiv x <_{I_2} t)$. Clearly $\mathscr{E}$ is an equivalence relation. As $I_1 <_{K^{\rm flin}} I_2$ clearly

$\odot_1$ (α) any interval of I_1 has cardinality $|I_1| \geq \kappa$

(β) for every $t \in I_2 \backslash J_2$ the equivalence class $t/\mathscr{E}$ is a singleton or has $|I_2| \geq \kappa$ members,

(γ) for every $t \in I_1 \backslash J_1$, $(t/\mathscr{E}) \cap I_1$ is a singleton or has $|I_1| \geq \kappa$ members

(δ) $I_1 \backslash J_2$ has at least κ elements

(ε) $\mathscr{E}$ has $\leq 2^{|J_2|} \leq 2^\theta$ equivalence classes

(ζ) we may $\leq_{K^{\rm flin}}$-increase I_2 so without loss of generality

$(*)_1$ $t \in I_2 \backslash J_2 \Rightarrow |t/\mathscr{E}| = |I_2|$

$(*)_2$ for every $t \in I_1$ for some $s_1, s_2 \in I_2$ we have $s_1 <_{I_2} t <_{I_2} s_2$ and $(s_1, t_{I_2}), (t, s_2)_{I_2}$ are disjoint to I_1.

Let $\langle \mathscr{U}_i : i < i(*) \rangle$ list the equivalence classes of $\mathscr{E}$, so without loss of generality $i(*) \leq 2^\theta$. For $\ell = 0, 1$ let $u_\ell = \{i < i(*) : \mathscr{U}_i \cap I_1$ has exactly ℓ members$\}$ and let $u_2 = i(*) \backslash u_0 \backslash u_1$, so by clause $\odot_1(\gamma)$, i.e. the definition of $I_1 \in K^{\rm flin}$ we have $i \in u_2 \Rightarrow |\mathscr{U}_i \cap I_1| = |I_1| \geq \kappa$. For $i \in u_1$ let t_i^* be the unique member of $\mathscr{U}_i \cap I_1$.

Without loss of generality $u_1 = \{i : i \in [j_0^*, j_1^*)\}$ for some $j_0^* \leq j_1^* \leq i(*)$ and let $i'(*) = i(*) + (j_1^* - j_0^*)$ and $u_1' = [i(*), i'(*))$ and define $\mathscr{U}_i'$ for $i < i'(*)$ by

$\odot_2$ (a) $\mathscr{U}_i' = \mathscr{U}_i$ if $i \in u_0 \cup u_2$

(b) $\mathscr{U}_i' = \{t \in \mathscr{U}_i : t < t_i^*\}$ if $i \in u_1$ and

(c) $\mathscr{U}_i' = \{t \in \mathscr{U}_\iota : t_\iota^* <_{I_2} t\}$ if $i \in [i(*), i'(*)], \iota \in (j_0^*, j_1^*)$ and $i - i(*) = \iota - j_0^*$.

For $i < i'(*)$ let $\langle t_{i,\alpha} : \alpha < \kappa \rangle$ be a sequence of pairwise distinct members of $\mathscr{U}_i'$ such that $i \in u_2 \Rightarrow t_{i,\alpha} \in I_1$ and $i \in u_0 \Rightarrow t_{i,\alpha} \notin I_1$, this actually follows. By $\odot_1(\zeta)$ and $\odot_1(\beta), (\gamma)$ we can find such $t_{i,\alpha}$'s.

For $\zeta < \theta_2$ (see clause (e) of the assumption so $\beth_\zeta < \kappa$) let $J_{1,\zeta} = \{t_{i,\alpha} : i \in u_2, \alpha < \beth_\zeta\} \cup J_1 \cup \{t_i^* : i \in u_1\}$. Now by the hence and forth argument (or see 0.17(2)) for each $\zeta < \theta_2$, there are $J_{2,\zeta}$ and f_ζ such that $J_{2,\zeta} \subseteq I_2$ is of cardinality $\beth_\zeta$, it includes $J_{1,\zeta} \cup J_2$ and also $\{t_{i,\alpha} : i < i'(*)$ and $\alpha < \beth_\zeta\}$ and f_ζ is an automorphism of $\mathrm{EM}_{\tau(\mathfrak{K})}(J_{2,\zeta}, \Phi)$ over $\mathrm{EM}_{\tau(\mathfrak{K})}(J_{1,\zeta}, \Phi)$ mapping $\bar{b}$ to $\bar{c}$.

(Why? Let $\bar a_0$ list $\mathrm{EM}(J_{1,\zeta},\Phi)$ so $\bar a_0 {}^\frown \bar b, \bar a_0 {}^\frown \bar c$ realize the same $\mathbb{L}_{\infty,\beth^+_\zeta}[\mathfrak{K}]$-type in M_2, and f be the mapping taking $\bar a_0 {}^\frown \bar b$ to $\bar a_0 {}^\frown \bar c$, etc.)
Now we shall immitate the proof of 1.19. By renaming without loss of generality there is a transitive set B (in the set theoretic sense) of cardinality $\le \theta_1 = 2^\theta$ which includes

$\oplus$(a) J_1, J_2

(b) Φ (i.e. τ_Φ and $\langle(\mathrm{EM}(n,\Phi), a_\ell)_{\ell<n} : n<\omega\rangle)$

(c) $\mathfrak{K}$, i.e., $\tau_{\mathfrak{K}}$ and $\{(M,N) : M \le_{\mathfrak{K}} N$ have universe included in $\mathrm{LS}(\mathfrak{K})\}$

(d) $\langle t^*_i : i\in u_1\rangle$ so each t^*_i for $i \in u_1$

(e) the ordinal $i(*)$.

Let χ be large enough, let $\mathfrak{B} = (\mathscr{H}(\chi), \in, <^*_\chi)$ and let $\mathfrak{B}^+_\zeta$ be $\mathfrak{B}$ expanded by

$\circledast_1$(a) $Q^{\mathfrak{B}_\zeta} = \{\alpha : \alpha < \beth_\zeta\}$

(b) $P^{\mathfrak{B}_\zeta}_i = J_{2,\zeta} \cap \mathscr{U}'_i$ for $i < i'(*)$

(c) $F^{\mathfrak{B}_\zeta}_2(t) = a_t$ for $t \in I_2$

(d) $H^{\mathfrak{B}_\zeta} = f_\zeta$ and $Q^{\mathfrak{B}_\zeta}_1 = J_{1,\zeta}, Q^{\mathfrak{B}_\zeta}_2 = J_{2,\zeta}$

(e) for $i < i'(*), H^{\mathfrak{B}_\zeta}_i$ is the function mapping $\alpha < \beth_\zeta$ to $t_{i,\alpha}$

(f) individual constants for B and for each $x \in B$, hence, e.g. for $t^*_i (i\in u_1), J_1, J_2, t$ for $t \in J_2$

(g) individual constants $J_{1,*}, J_{2,*}$ interpreted as the linear orders $J_{1,\zeta}, J_{2,\zeta}$ respectively and individual constants for $M^+_\ell = \mathrm{EM}(J_{0,\zeta},\Phi)$, and $\langle a_t : t \in I_\ell\rangle$ for $\ell = 1,2$.

As in the proof of 1.19 there is a $\tau(\mathfrak{B}^+)$-model $\mathfrak{C}$, such that

$\boxtimes$ (a) for some unbounded $S \subseteq \theta_2$

(α) $\mathfrak{C}$ is a first order elementarily equivalent to $\mathfrak{B}^+_\zeta$ for every $\zeta \in S$

(β) $\mathfrak{C}$ omits every type omitted by $\mathfrak{B}_\zeta$ for every $\zeta \in S$. In particular this gives

(γ) $\mathfrak{C}$ omits the type $\{x \neq b \wedge x \in B : b \in B\}$ so

(δ) without loss of generality $b \in B \Rightarrow b^{\mathfrak{C}} = b$

(b) $\mathfrak{C}$ is the Skolem hull of some infinite indiscernible sequence $\langle y_r : r \in I\rangle$, where I an infinite linear order and $y_r \in Q^{\mathfrak{C}}$ for $r \in I$.

Without loss of generality $I \in K^{\text{flin}}$ and I_2 can be $\leq_{K^{\text{flin}}}$-embedded into I say by the function g such that $(\forall t \in I_2)(\exists s_1, s_2 \in I)[s_1 <_I g(t) <_I s_2 \wedge (\forall t' \in I_2)(t' <_{I_2} t \rightarrow g(t') <_I s_1) \wedge (\forall t' \in I_2)(t <_{I_2} t' \rightarrow s_2 <_I g(t'))]$; and also $\|\mathfrak{C}\| = |I|$. Hence for each $i < i'(*)$ there is an embedding h_i of the linear order $\mathscr{U}'_i$, i.e., $I_2 \restriction \mathscr{U}'_i$ into $(P^{\mathfrak{C}}_i, (<_{I_2})^{\mathfrak{C}})$ such that $t \in \mathscr{U}'_i \Rightarrow (t \in I_1 \leftrightarrow h_i(t) \in Q^{\mathfrak{C}}_1)$.
[Why? Case 0: $i \in u_0$.

Trivial.

Case 1: $i \in u_1 \cup u'_1$.

Similar to Case 0 as $\mathscr{U}'_i \cap I_1 = \emptyset$, of course, we take care that $a = h_i(t) \wedge t \in \mathscr{U}'_i \wedge i \in u_1 \Rightarrow \mathfrak{C} \models$ "$a <_{I_2} t^*_i$" and similarly for u_{-1}.

Case 2: $i \in u_2$.

First approximation is $h'_i = (H^{\mathfrak{C}}_i \circ (g \restriction \mathscr{U}_i))$, so $t \in \mathscr{U}_i \Rightarrow h'_i(t) \in Q^{\mathfrak{C}}_1$. However by the choice of g we can find $\langle (s^-_t, s^+_t) : t \in \mathscr{U}_i\rangle$ such that:

(α) $s^-_t, s^+_t \in Q^{\mathfrak{C}}_2$

(β) $(s^-_t, s^+_t)_{I^{\mathfrak{C}}_2} \cap Q^{\mathfrak{C}}_2 = \{h'_i(t)\}$.

As I_2 is dense with no extremal members (being from K^{flin}) clearly $t_1 <_{I_2 \restriction \mathscr{U}'_i} t_2 \Rightarrow s^+_{t_1} <_{(I_2)^{\mathfrak{C}}} s^-_{t_2}$. Now choose h_i by: $h_i(t)$ is $h'_i(t)$ if $t \in I_1$ and is $s^+_{t_1}$ if $t \in I_1 \backslash I_2$.]

Hence there is an embedding h of the linear order I_2 into $J^{\mathfrak{C}}_{1,*}$ such that:

$\circledast_2$ $h(t)$ is:

(a) t if $t \in J_2 \cup \{t^*_i : i \in u_1\}$

(b) $h_i(t)$ if $t \in \mathscr{U}'_i$ and $i < i'(*)$.

Note

$\circledast_3$ for every $t \in I_2 \backslash J_2$ for some $i < i(*) \le \theta_1$ we have $(\forall s \in J_2)[s <_{I_2} t \equiv s <_{I_2} h_i(t_{i,0})]$

hence by the omitting type demand in $\boxtimes(a)(\beta)$:

$\circledast_3'$ for $t \in I_2^{\mathfrak{C}} \backslash J_2$ for some $i < i(*)$ we have $(\forall s \in J_2)[s <_{I_2^{\mathfrak{C}}} t \equiv s <_{I_2^{\mathfrak{C}}} (h_i(t_{i,0}))]$.

We can find a linear order $I_3, I_2 \subseteq I_3$ and an isomorphism h_* from I_3 onto $Q_2^{\mathfrak{C}}$ extending h, so clearly $I_3 \in K^{\text{flin}}$ and without loss of generality $h(I_2) <_{K^{\text{flin}}} I_3$. Now let $\hat{h}_*$ be the isomorphism which h_* induces from $\text{EM}_{\tau(\mathfrak{K})}(I_3, \Phi)$ onto $(\text{EM}_{\tau(\mathfrak{K})}(J_{2,*}^{\mathfrak{C}}, \Phi))^{\mathfrak{C}}$, so e.g., it maps for each $t \in I_2$, the member a_t of the skeleton to $F_2^{\mathfrak{C}}(h_*(t))$.

Note that h_* maps $\mathscr{U}_i \cap I_1$ into $Q_1^{\mathfrak{C}} \subseteq I_1^{\mathfrak{C}}$ when $\mathscr{U}_i \subseteq I_1$ and is the identity on $J_1 \cup \{t_i^* : i \in u_1\}$ so recalling $Q^{\mathfrak{B}_\zeta} = J_{1,\zeta} = \{t_{i,\alpha} : i \in u_2$ and $\alpha < \beth_\zeta\} \cup J_1 \cup \{t_i^* : i \in u_1\}$ hence it map I_1 into $Q_1^{\mathfrak{C}}$ but $\mathfrak{B}_\zeta \models$ "H is a unary function, an automorphism of $\text{EM}_{\tau(\mathfrak{K})}(J_{2,*}^{\mathfrak{C}}, \Phi)$ mapping $\bar{b}$ to $\bar{c}$ and is the identity on $\text{EM}_{\tau(\mathfrak{K})}(J_{1,*}^{\mathfrak{C}}, \Phi)$". Now $(\hat{h}_*)^{-1} H^{\mathfrak{C}}(\hat{h}_*)$ is an automorphism of $\text{EM}_{\tau(\mathfrak{K})}(I_3, \Phi)$ as required.

2) By part (1), i.e. choose I_3, M_3, f_3 as there; so as f is an automorphism of M_3 over M_1 mapping $\bar{b}$ to $\bar{c}$, clearly $\bar{b}, \bar{c}$ realize the same $\mathbb{L}_{\infty,\partial}[\mathfrak{K}]$-type over M_1 inside M_3. The desired result (the type inside M_2 rather than inside M_3) follows because $M_1 \prec_{\mathbb{L}_{\infty,\partial}[\mathfrak{K}]} M_2 \prec_{\mathbb{L}_{\infty,\partial}[\mathfrak{K}]} M_3$ by 1.14(a).

3) Let $M_2^* = \bigcup\limits_{n<\omega} M_{2,n}^*$ be such that $n < \omega \Rightarrow M_{2,n}^* \le_{\mathfrak{K}} M_{2,n+1}^*$ and $\|M_{2,n}^*\| < \lambda$. Let $\bar{c}_n$ list $M_{2,n}^*$ for $n < \omega$ (with no repetitions) and be such that $\bar{c}_n \triangleleft \bar{c}_{n+1}$. Let $\theta_n = \|M_{2,n}^*\| + \text{LS}(\mathfrak{K})$ so without loss of generality $\theta_n = \ell g(\bar{c}_n)$ and let $\theta_n' = \beth_3(\theta_n), \kappa_n = \beth_{1,1}(\theta_n')$, without loss of generality $\kappa_n < \theta_{n+1}$ and we choose for each $n < \omega$, a sequence $\bar{b}_n \in {}^{\ell g(\bar{c}_n)}(M_2)$ realizing $\text{tp}_{\mathbb{L}_{\infty,\kappa_n^+}[\mathfrak{K}]}(\bar{c}_n, M_1, M_2^*)$ in M_2. This is possible by 1.26(3) after possibly $<_{K^{\text{flin}}}$-increasing I_2.

Now we choose $(I_{3,n}, f_n, M_{3,n}, \bar{b}_n')$ by induction on n such that

$(*)$ (a) $I_{3,0} = I_2$ and $I_{3,n} \in K_\lambda^{\text{lin}}$

(b) $n = m + 1 \Rightarrow I_{3,m} <_{K^{\text{flin}}} I_{3,n}$

(c) $M_{3,n} = \mathrm{EM}_{\tau(\mathfrak{K})}(I_{3,n}, \Phi)$ (hence $n = m+1 \Rightarrow M_{3,m} \leq_{\mathfrak{K}_\lambda} M_{3,n}$)

(d) f_n is an automorphism of $M_{3,n}$ over M_1

(e) $\bar{b}'_n \in {}^{\ell g(\bar{b}_n)}(M_{3,n})$ realizes $\mathrm{tp}_{\mathbb{L}_{\infty,\kappa_n^+}[\mathfrak{K}]}(\bar{c}_n, M_1, M_2^*)$

(f) if $n = m+1$ then $\bar{b}'_m \trianglelefteq \bar{b}'_n$

(g) if $n = m+1$ then f_n maps $\bar{b}_{n+1} \restriction \ell g(\bar{b}_n)$ to $\bar{b}'_n$ and f_0 maps $\bar{b}_0$ to $\bar{b}'_0$.

For $n = 0, I_{3,0}, M_{3,0}$ are defined in clauses (a),(c) of (∗) and we let $f_0 = \mathrm{id}_{M_2} = \mathrm{id}_{M_{3,n}}, \bar{b}'_0 = \bar{b}_0$ this is trivially as required. For $n = m+1$ we apply part (1) with

⊡ $I_1, I_{3,m}, M_1, M_{3,m}, \bar{b}_{n+1} \restriction \ell g(\bar{c}_m), \bar{b}'_m, \theta_m, \kappa_m$ here standing for $I_1, I_2, M_1, M_2, \bar{b}, \bar{c}, \theta, \kappa$ there.

Why its assumptions holds? The main point is to check that for every $\bar{a} \in {}^{\kappa_m>}(M_1)$ the sequences $\bar{a}\hat{\ }(\bar{b}_{n+1} \restriction \theta_m), \bar{a}\hat{\ }\bar{b}'_m$ realize the same $\mathbb{L}_{\infty,\kappa_m}[\mathfrak{K}]$-type in $M_{3,m}$. Now $\bar{a}\hat{\ }(\bar{b}_{m+1} \restriction \theta_m), \bar{a}\hat{\ }\bar{b}'_m$ realize the same $\mathbb{L}_{\infty,\kappa_n}[\mathfrak{K}]$-type in $M_{3,m}$ by the induction hypothesis. Also the sequences $\bar{b}_{n+1} \restriction \theta_m, \bar{b}_{m+1} \restriction \theta_m$ satisfy for any $\bar{a} \in {}^{\kappa_m}(M_1)$ the sequences $\bar{a}\hat{\ }(\bar{b}_{n+1} \restriction \theta_m), \bar{a}\hat{\ }(\bar{b}_{m+1} \restriction \theta_m)$ realize the same $\mathbb{L}_{\infty,\kappa_m}[\mathfrak{K}]$-type in $M_{3,m}$ because the $\mathbb{L}_{\infty,\kappa_m}[\mathfrak{K}]$-type which $\bar{a}\hat{\ }(\bar{b}_{n+1} \restriction \theta_m)$ realizes in $M_{3,m}$ is the same as the $\mathbb{L}_{\infty,\kappa_m}[\mathfrak{K}]$-type it realizes in $M_2 = M_{3,0}$ which (by the choice of $\bar{b}_{n+1}$) is equal to the $\mathbb{L}_{\infty,\kappa_m}[\mathfrak{K}]$-type which $\bar{a}\hat{\ }(\bar{c}_{n+1} \restriction \theta_m)$ realizes in M_2^* which is the same as the $\mathbb{L}_{\infty,\kappa_m}[\mathfrak{K}]$-type which $\bar{a}\hat{\ }(\bar{c}_{m+1} \restriction \theta_m)$ realizes in M_2^* which is equal to the $\mathbb{L}_{\infty,\kappa_m}[\mathfrak{K}]$-type which $\bar{a}\hat{\ }(\bar{b}_{m+1} \restriction \theta_m)$ realizes in $M_{3,m}$.

By the last two sentences for every $\bar{a} \in {}^{\kappa_m>}(M_1)$ the sequences $\bar{a}\hat{\ }(\bar{b}_{n+1} \restriction \theta_m), \bar{a}\hat{\ }\bar{b}'_m$ realizes the same $\mathbb{L}_{\infty,\kappa_m}[\mathfrak{K}]$-type in $M_{3,m}$, so indeed the assumptions of part (1) holds for the case we are trying to apply it, see ⊡ above.

So we get the conclusion of part (1), i.e. we get $I_{3,n}, f_n$ here standing for I_3, f there so $I_{3,m} <_{K_\lambda^{\mathrm{flin}}} I_{3,n}$ and f_n is an automorphism of $M_{3,n} = \mathrm{EM}_{\tau(\mathfrak{K})}(I_{3,n}, \Phi)$ over M_1 mapping $\bar{b}_{n+1} \restriction \theta_m$ to $\bar{b}'_m$. Now we let $\bar{b}'_n = f_n(\bar{b}_{n+1} \restriction \theta_n)$ and can check all the clauses in (∗). Hence we have carried the induction. So we can satisfy (∗).

So $\bar b'_n$ satisfies the requirements on $\bar b_n$ and $\bar b'_n \triangleleft \bar b'_{n+1}$. Let $I_3 = \cup\{I_{3,n} : n < \omega\}$ and let $M_3 = \mathrm{EM}_{\tau(\mathfrak{K})}(I_3, \Phi)$ and let $g : M^*_2 \to M_3$ map $c_{n,i}$ to $b'_{n,i}$ for $i < \ell g(\bar c_n), n < \omega$, easily it is as required. That is, $g(c_{n,i})$ is well defined as $c_{n,i} \mapsto b'_{n,i}, (i < \ell g(\bar c_n))$ is a well defined mapping for each n and $i < \ell g(\bar c_n) \Rightarrow c_{n,i} = c_{n+1,i} \wedge b'_{n,i} = b'_{n+1,i}$. Also $g \restriction \{c_{n,i} : i < \ell g(\bar c_n)\}$ is a $\leq_{\mathfrak{K}}$-embedding of $M^*_{2,n}$ into M_3 and is the identity on $M^*_{2,n} \cap M_1$ as $\bar c_n$ list the elements of $M_{2,i}$ and $\mathrm{tp}_{\mathbb{L}_{\infty,\kappa^+_n}[\mathfrak{K}]}(\bar c_n, M_1, M^*_2) = \mathrm{tp}_{\mathbb{L}_{\infty,\kappa^+_n}[\mathfrak{K}]}(\bar b'_n, M_1, M_3)$ by clause (e) of $(*)$. But $\langle g \restriction M^*_{2,n} : n < \omega\rangle$ is $\subseteq$-increasing with union g so by Ax(V) of a.e.c. g is a $\leq_{\mathfrak{K}}$-embedding of M^*_2 into M_3. Lastly, obviously $g \supseteq \cup\{\mathrm{id}_{M^*_{2,n} \cap M_1} : n < \omega\} = \mathrm{id}_{M_1}$, so we are done. $\square_{1.29}$

We arrive to the crucial advance:

1.30 The Amalgamation Theorem. *If* $\mathrm{cf}(\lambda) = \aleph_0$, *then* $\mathfrak{K}^*_\lambda$, *i.e.,* $(K^*_\lambda, \leq_{\mathfrak{K}} \restriction K^*_\lambda)$ *has amalgamation, even disjoint one.*

Proof. So assume $M_0 \leq_{\mathfrak{K}^*_\lambda} M_\ell$ for $\ell = 1, 2$. Choose $I_0 \in K^{\mathrm{flin}}_\lambda$ so $M'_0 := \mathrm{EM}_{\tau(\mathfrak{K})}(I_0, \Phi) \in K^*_\lambda$ but K^*_λ is categorical (see 1.16 or 1.19(4)) hence $M'_0 \cong M_0$, so without loss of generality $M'_0 = M_0$. Choose $I_1 \in K^{\mathrm{flin}}_\lambda$ such that $I_0 <_{K^{\mathrm{flin}}} I_1$ and let $M'_1 = \mathrm{EM}_{\tau(\mathfrak{K})}(I_1, \Phi)$ so $M_0 \leq_{\mathfrak{K}} M'_1$. By applying 1.29(3) with I_0, I_1, M_0, M'_1, M_1 here standing for $I_1, I_2, M_1, M_2, M^*_2$ there, we can find a pair (I_2, f_1) such that $I_1 <_{K^{\mathrm{flin}}_\lambda} I_2$ and f_1 is a $\leq_{\mathfrak{K}}$-embedding of M_1 into $M'_2 := \mathrm{EM}_{\tau(\mathfrak{K})}(I_2, \Phi)$ over M_0.

Apply 1.29(3) again with $I_0, I_2, M_0, \mathrm{EM}_{\tau(\mathfrak{K})}(I_2, \Phi), M_2$ here standing for $I_1, I_2, M_1, M_2, M^*_2$ there. So there is a pair (I_3, f_2) such that $I_2 <_{K^{\mathrm{flin}}_\lambda} I_3$ and f_2 is $\leq_{\mathfrak{K}}$-embedding M_2 into $M_3 := \mathrm{EM}_{\tau(\mathfrak{K})}(I_3, \Phi)$ over $M_0 = \mathrm{EM}_{\tau(\mathfrak{K})}(I_0, \Phi)$. Of course, $M_3 \in K^*_\lambda$ and we are done proving the "has amalgamation".

Why disjoint? Let (I_4, h) be such that $I_3 <_{K^{\mathrm{flin}}_\lambda} I_4$ and h is a $\leq_{K^{\mathrm{flin}}}$-embedding of I_3 into I_4 over I_0 such that $h(I_3) \cap I_3 = I_0$. Now h induces an isomorphism $\hat h$ from $\mathrm{EM}_{\tau(\mathfrak{K})}(I_3, \Phi)$ onto $\mathrm{EM}_{\tau(\mathfrak{K})}(h(I_3), \Phi) \leq_{\mathfrak{K}} M_3$.

Lastly, by our assumptions on Φ if $J_1, J_2 \subseteq J$ and $I_1 \cap I_2$ is a dense linear order (in particular with neither first nor last member, e.g.

are from $K^{\rm flin}_\lambda$ as in our case) then $\mathrm{EM}_{\tau(\mathfrak{K})}(I_1,\Phi) \cap \mathrm{EM}_{\tau(\mathfrak{K})}(I_2,\Phi) = \mathrm{EM}_{\tau(\mathfrak{K})}(I_1 \cap I_2,\Phi)$. So in particular, above

$$\mathrm{EM}_{\tau(\mathfrak{K})}(I_3,\Phi) \cap \mathrm{EM}_{\tau(\mathfrak{K})}(\hat{h}(I_3,\Phi) = \mathrm{EM}_{\tau(\mathfrak{K})}(I_0,\Phi)$$

and $f_1, \hat{h} \circ f_2$ are $\le_{\mathfrak{K}}$-embeddings of M_1, M_2 respectively over $M_0 = \mathrm{EM}_{\tau(\mathfrak{K})}(I_0,\Phi)$ into $\mathrm{EM}_{\tau(\mathfrak{K})}(I_3,\Phi) \le_{\mathfrak{K}} \mathrm{EM}_{\tau(\mathfrak{K})}(I_4,\Phi)$ and $\mathrm{EM}_{\tau(\mathfrak{K})}(h(I_3),\Phi) \le_{\mathfrak{K}} \mathrm{EM}_{\tau(\mathfrak{K})}(I_4,\Phi)$, respectively, so we are done. $\square_{1.30}$

1.31 Claim. *Assume* $\mathrm{cf}(\lambda) = \aleph_0$. *If* $\delta < \lambda^+$, *the sequence* $\langle M_i : i < \delta\rangle$ *is* $\le_{\mathfrak{K}}$*-increasing continuous and* $M_i \in K^*_\lambda$ *for* $i < \delta$, *then* $M_\delta := \cup\{M_i : i < \delta\}$ *can be* $\le_{\mathfrak{K}}$*-embedded into some member of* K^*_λ.

Proof. We choose $I_i \in K^{\rm flin}_\lambda$ by induction on $i \le \delta$, which is $<_{K^{\rm flin}_\lambda}$-increasing continuous with i and a $\le_{\mathfrak{K}}$-embedding f_i of M_i into $N_i := \mathrm{EM}_{\tau(\mathfrak{K})}(I_i,\Phi)$, increasing continuous with i. For $i = 0$ choose $I_0 \in K^{\rm flin}_\lambda$, so $N_0 := \mathrm{EM}_{\tau(\mathfrak{K})}(I_0,M)$ is isomorphic to M_0 hence f_0 exists; for i limit use $I_i := \cup\{I_j : j < i\}$ and $f_i := \cup\{f_j : j < i\}$. So assume $i = j+1$. Now we can find M'_i, f'_i satisfying: f'_i is an isomorphism from M_i onto M'_i extending f_j such that $f_j(M_j) \le_{\mathfrak{K}} M'_i$ (actually this trivially follows) and $M'_i \cap N_j = f_j(M_j)$; so also M'_i belongs to K^*_λ. Now $f_j(M_j), \mathrm{EM}_{\tau(\mathfrak{K})}(I_j,\Phi), M'_i$ can be disjointly amalgamated (by 1.30) in $(K^*_\lambda, \le_{\mathfrak{K}})$, so there is $M^*_i \in K^*_\lambda$ such that $N_j = \mathrm{EM}_{\tau(\mathfrak{K})}(I_j,\Phi) \le_{\mathfrak{K}} M^*_i$ and $M'_i \le_{\mathfrak{K}} M^*_i$. Now by 1.29(3) there are I_i, g_i such that $I_j <_{K^{\rm flin}_\lambda} I_i$ and g_i is a $\le_{\mathfrak{K}}$-embedding of M^*_i into $N_i := \mathrm{EM}_{\tau(\mathfrak{K})}(I_i,\Phi)$ over $\mathrm{EM}_\tau(I_j,\Phi)$. Let $f_i = g_i \circ f'_i$, clearly it is as required. Having carried the induction, f_δ is a $\le_{\mathfrak{K}}$-embedding of M_δ into $\mathrm{EM}_{\tau(\mathfrak{K})}(\bigcup\limits_{j<\delta} I_j,\Phi)$, as promised. $\square_{1.31}$

1.32 Claim. *1) Assume* $\mathrm{cf}(\lambda) = \aleph_0$. *For every* $M_0 \in K^*_\lambda$ *there is a* $\le_{\mathfrak{K}}$*-extension* $M_1 \in K^*_\lambda$ *of* M_0 *such that: if* $M_0 \le_{\mathfrak{K}_\lambda} M_2 \in K^*_\lambda$ *and* $\bar{a} \in {}^{\lambda>}(M_2)$ *then for some* (M_3, f) *we have:*

> $M_1 \le_{\mathfrak{K}} M_3 \in K^*_\lambda$, *$f$ is a* $\le_{\mathfrak{K}}$*-embedding of* M_2 *into* M_3 *over* M_0 *and* $f(\bar{a}) \in {}^{\lambda>}(M_2)$.

2) Assume $\mathrm{cf}(\lambda) = \aleph_0$. *For every* $M_0 \in K^*_\lambda$ *there is a* $\le_{\mathfrak{K}}$*-extension* $M_1 \in K^*_\lambda$ *which is universal over* M_0 *for* $\le_{\mathfrak{K}_\lambda}$*-extensions.*
3) If (a) then (b) where

(a) $I_0 \le_{K^{\mathrm{flin}}_\lambda} I'_1 <_{K^{\mathrm{flin}}_\lambda} I_1$

(b) *if* $I_0 \subseteq I_2 \in K^{\mathrm{flin}}_\lambda$ *and* $\beta \le \gamma < \lambda, \bar b_1 \in {}^\beta(\mathrm{EM}_{\tau(\mathfrak{K})}(I'_1, \Phi))$ *and* $\bar c_2 \in {}^\gamma(\mathrm{EM}_{\tau(\mathfrak{K})}(I_2, \Phi))$ *and* $\bar b_2 = \bar c_2 \restriction \beta$ *and for every* $\kappa < \lambda$ *we have*

$$\mathrm{tp}_{\mathbb{L}_{\infty,\kappa}[\mathfrak{K}]}(\bar b_1, \mathrm{EM}_{\tau(\mathfrak{K})}(I_0, \Phi)), \mathrm{EM}_{\tau(\mathfrak{K})}(I_1, \Phi)) = $$
$$= \mathrm{tp}_{\mathbb{L}_{\infty,\kappa}[\mathfrak{K}]}(\bar b_2, \mathrm{EM}_{\tau(\mathfrak{K})}(I_0, \Phi), \mathrm{EM}_{\tau(\mathfrak{K})}(I_2, \Phi))$$

then for some (I_1^+, f) *we have* $I_1 \le_{K^{\mathrm{flin}}} I_1^+ \in K^{\mathrm{flin}}_\lambda$ *and* f *is a* $\le_{\mathfrak{K}}$*-embedding of* $\mathrm{EM}_{\tau(\mathfrak{K})}(I_2, \Phi)$ *into* $\mathrm{EM}_{\tau(\mathfrak{K})}(I_1^+, \Phi)$ *over* $\mathrm{EM}_{\tau(\mathfrak{K})}(I_0, \Phi)$ *mapping* $\bar b_2$ *to* $\bar b_1$ *and* $\bar c_2$ *into* $\mathrm{EM}_{\tau(\mathfrak{K})}(I_1, \Phi)$.

4) Assume $\mathrm{cf}(\lambda) = \aleph_0$. *If (c) then (d) and moreover* $(d)^+$ *when*

(c) $\langle J_\alpha : \alpha \le \omega \rangle$ *is* $<_{K^{\mathrm{flin}}_\lambda}$*-increasing,* $I_0 = J_0, I_1 = J_\omega$

(d) *if* $I_0 \subseteq I_2 \in K^{\mathrm{flin}}_\lambda$ *then some* f *is a* $\le_{\mathfrak{K}}$*-embedding of* $\mathrm{EM}_{\tau(\mathfrak{K})}(I_2, \Phi)$ *into* $\mathrm{EM}_{\tau(\mathfrak{K})}(I_1, \Phi)$ *over* $\mathrm{EM}_{\tau(\mathfrak{K})}(I_0, \Phi)$

$(d)^+$ $\mathrm{EM}_{\tau(\mathfrak{K})}(I_1, \Phi)$ *is* $\le_{\mathfrak{K}^*_\lambda}$*-universal over* $\mathrm{EM}_{\tau(\mathfrak{K})}(I_0, \Phi)$.

Proof. Note that by 1.29(3) clearly (3) $\Rightarrow$ (1) and (4) $\Rightarrow$ (2). So we shall prove (3) and (4).
3) First assume $\beta = 0, \gamma = 1$ so $\bar c_2 = \langle c \rangle$. Toward contradiction assume $I_0 \subseteq I_2 \in K^{\mathrm{lin}}_\lambda, a \in M_2 := \mathrm{EM}_{\tau(\mathfrak{K})}(I_2, \Phi)$ but there is no pair (I_1^+, f) as required in clause (b). Without loss of generality for some I_3 we have $I_0 \le_{K^{\mathrm{flin}}_\lambda} I_2 \le_{K^{\mathrm{flin}}_\lambda} I_3$ and $I_0 \le_{K^{\mathrm{flin}}_\lambda} I_1 \le_{K^{\mathrm{flin}}_\lambda} I_3$.

Let $\mathrm{EM}(I_2, \Phi) \models$ "$c_2 = \sigma(a_{t^2_0}, \dots, a_{t^2_{n-1}})$" where $\sigma(x_0, \dots, x_{n-1})$ a τ_Φ-term, $n < \omega$ and $I_2 \models$ "$t^2_0 < \dots < t^2_{n-1}$". Let $u = \{\ell < n : t^2_\ell \in I_0\}$. As $I_0 <_{K^{\mathrm{flin}}_\lambda} I_1$, we can find $\langle t^1_0, \dots, t^1_{n_1} \rangle$ such that:

⊛ (a) $t^1_\ell \in I_1$ for $\ell < n$

(b) $t^1_0 <_{I_1} \dots <_{I_1} t^1_{n-1}$

(c) if $\ell \in u$ then $t^2_\ell = t^1_\ell (\in I_0)$

(d) if $\ell < n \wedge \ell \notin u$ then $t^1_\ell \in I_1 \backslash I_0$

(e) if $\ell_1 \le \ell_2 < n$ and $[\ell_1, \ell_2] \cap u = \emptyset$ then $t^2_{\ell_2} <_{I_3} t^1_{\ell_1}$.

Let $M_\ell = \mathrm{EM}_{\tau(\mathfrak{K})}(I_\ell, \Phi)$ for $\ell = 0,1,2,3$ and let $c_2 = c$ and $c_1 = \sigma^{\mathrm{EM}(I_1,\Phi)}(a_{t^1_0}, \ldots, a_{t^1_{n-1}})$.

Let $\kappa < \lambda$ be large enough such that $\mathrm{tp}_{\mathbb{L}_{\infty,\kappa^+}[\mathfrak{K}]}(c_\ell, M_0, M_\ell)$ for $\ell = 1,2$ be distinct (exists by 1.29(1) because its conclusion fails by the "toward contradiction"). We easily get contradiction to the non-order property (see $(*)$ of 1.5(2)).

Note that if in addition $\langle I_{1,\alpha} : \alpha \le \lambda\rangle$ is $<_{K^{\mathrm{flin}}_\lambda}$-increasing continuous, $I_{1,0} = I'_1, I_{1,\lambda} = I_1$ then by what we have just proved and the proof of II.4.3 we can prove the general case (and part (4)). But we also give a direct proof.

In the general case, let $\theta = |\beta| + \aleph_0$, so we assume clause (a) and the assumptions of clause (b) and without loss of generality $I_1 \cap I_2 = I_0$ hence there is I_3 such that $I_\ell <_{K^{\mathrm{flin}}_\lambda} I_3$ for $\ell = 1,2$. Let $\kappa \in (\theta, \lambda)$ be large enough.

Hence

$$\mathrm{EM}_{\tau(\mathfrak{K})}(I_0, \Phi) \prec_{\mathbb{L}_{\infty,\lambda}[\mathfrak{K}]} \mathrm{EM}_{\tau(\mathfrak{K})}(I_\ell, \Phi) \prec_{\mathbb{L}_{\infty,\lambda}[\mathfrak{K}]} \mathrm{EM}_{\tau(\mathfrak{K})}(I_3, \Phi)$$

for $\ell = 1,2$. Applying 1.29(1) with $I_1, I_2, \bar{b}, \bar{c}$ there standing for $I_0, I_3, \bar{b}_1, \bar{b}_2$ here we can find a pair (I_4, f_4) such that $I_3 <_{K^{\mathrm{flin}}_\lambda} I_4$ and f_4 is an automorphism of $M_4 := \mathrm{EM}_{\tau(\mathfrak{K})}(I_4, \Phi)$ over $\mathrm{EM}_{\tau(\mathfrak{K})}(I_0, \Phi)$ mapping $\bar{b}_2$ to $\bar{b}_1$.
Clearly $M_3 := \mathrm{EM}_{\tau(\mathfrak{K})}(I_3, \Phi) \prec_{\mathbb{L}_{\infty,\lambda}[\mathfrak{K}]} \mathrm{EM}_{\tau(\mathfrak{K})}(I_4, \Phi)$. So $f_4(\bar{c}_2) \in {}^{\gamma}(M_4)$, hence we can apply clause (b) of Claim 1.26(3) with M_1, $M_2, I_2, N, \xi, \bar{d}^*$ there standing for $\mathrm{EM}_{\tau(\mathfrak{K})}(I'_1, \Phi)$, $\mathrm{EM}_{\tau(\mathfrak{K})}(I_1, \Phi), I_1$, $\mathrm{EM}_{\tau(\mathfrak{K})}(I_4, \Phi), \gamma, f_4(\bar{c}_2)$ here. Hence we can find $\bar{c}'_2 \in {}^{\gamma}(M_1)$ realizing in M_1 the type $\mathrm{tp}_{\mathbb{L}_{\infty,\kappa}[\mathfrak{K}]}(f_4(\bar{c}_2), \mathrm{EM}_{\tau(\mathfrak{K})}(I'_1, \Phi), \mathrm{EM}_{\tau(\mathfrak{K})}(I_1, \Phi))$.

Lastly, applying Claim 1.29(1) with $I_1, I_2, \bar{b}, \bar{c}$ there standing for $I'_1, I_4, f_4(\bar{c}_2), \bar{c}'_2$ here, clearly there is a pair (I_5, f_5) such that $I_4 <_{K^{\mathrm{flin}}_\lambda} I_5$ and f_5 is an automorphism of $\mathrm{EM}_{\tau(\mathfrak{K})}(I_5, \Phi)$ over $\mathrm{EM}(I'_1, \Phi)$ mapping to $f_4(\bar{c}_2)$ to $\bar{c}'_2$.
Let $I^+_1 := I_5, f = f'_5 \circ f'_4$ where $f'_5 = f_5 \restriction \mathrm{EM}_{\tau(\mathfrak{K})}(I_4, \Phi)), f'_4 = f_4 \restriction \mathrm{EM}_{\tau(\mathfrak{K})}(I_2, \Phi)$; now I^+_1, f are as required because $f_4(\bar{b}_2) = \bar{b}_1$ while $f_5(\bar{b}_1) = \bar{b}_1$.
4) Easy by part (3). First note that $(\mathrm{d})^+$ follows by (d) by 1.29(3), so we shall ignore clause $(\mathrm{d})^+$. Let $\mathrm{EM}_{\tau(\mathfrak{K})}(I_2, \Phi)$ be $\cup\{M_{2,n} : n < \omega\}$ where $M_{2,n} \in K_{<\lambda}$ and $n < \omega \Rightarrow M_{2,n} \le_{\mathfrak{K}} M_{2,n+1}$.

Let $\bar a_n$ list the elements of $M_{2,n}$ with no repetitions such that $\bar a_n \triangleleft \bar a_{n+1}$ for $n < \omega$. By induction on n, we choose $\bar b_n$ such that

$\circledast$ (a) $\bar b_n \in {}^{\ell g(\bar a_n)}(\mathrm{EM}_{\tau(\mathfrak{K})}(J_{n+1},\Phi)$

(b) if $n = m+1$ then $\bar b_m \triangleleft \bar b_n$

(c) for every $\kappa < \lambda$ the type
$\mathrm{tp}_{\mathbb{L}_{\infty,\kappa}[\mathfrak{K}]}(\bar b_n, \mathrm{EM}_{\tau(\mathfrak{K})}(I_0,\Phi), \mathrm{EM}_{\tau(\mathfrak{K})}(I_{n+1},\Phi)))$
is equal to the type
$\mathrm{tp}_{\mathbb{L}_{\infty,\kappa}[\mathfrak{K}]}(\bar a_n, \mathrm{EM}_{\tau(\mathfrak{K})}(I_0,\Phi), \mathrm{EM}_{\tau(\mathfrak{K})}(I_2,\Phi))$.

The induction step is by part (3). Let f_n be the unique function mapping $\bar a_n$ to $\bar b_n$ (with domain $\mathrm{Rang}(\bar a_n)$). So $f_n \subseteq f_{n+1}$ and f_n is a $\leq_{\mathfrak{K}}$-embedding of $M_{2,n}$ into $\mathrm{EM}_{\tau(\mathfrak{K})}(J_{n+1},\Phi)$ but $J_{n+1} \subseteq I_1$ hence into $\mathrm{EM}_{\tau(\mathfrak{K})}(I_1,\Phi)$. So $f := \cup\{f_n : n < \omega\}$ is a $\leq_{\mathfrak{K}}$-embedding of $\mathrm{EM}_{\tau(\mathfrak{K})}(I_2,\Phi)$ into $\mathrm{EM}_{\tau(\mathfrak{K})}(I_1,\Phi)$. Also f_n is the identity on $\mathrm{Rang}(\bar a_n) \cap \mathrm{EM}_{\tau(\mathfrak{K})}(I_0,\Phi)$ hence f is the identity on $\bigcup_n(\mathrm{Rang}(\bar a_n) \cap \mathrm{EM}_{\tau(\mathfrak{K})}(I_0,\Phi) = \mathrm{EM}_{\tau(\mathfrak{K})}(I_0,\Phi)$ so f is as required. $\square_{1.32}$

<u>1.33 Exercise</u>: 1) Assume $\mathfrak{K}_\lambda = (K_\lambda, \leq_{\mathfrak{K}_\lambda})$ satisfies axioms I,II (and 0, presented below) and amalgmation. Then $\mathbf{tp}(a,M,N)$ for $M \leq_{\mathfrak{K}_\lambda} N$ and $a \in N$ and $\mathscr{S}_{\mathfrak{K}_\lambda}(M)$ are well defined and has the basic properties of types from II§1.
2) If in addition $\mathfrak{K}_\lambda$ satisfies AxIII$^\odot$ below and $\mathfrak{K}_\lambda$ is stable (i.e. $|\mathscr{S}_{\mathfrak{K}_\lambda}(M)| \leq \lambda$ for $M \in K_\lambda$) <u>then</u> every $M \in \mathfrak{K}_\lambda$ has a $\leq_{\mathfrak{K}}$-universal extension N which means $M \leq_{\mathfrak{K}_\lambda} N$ and $(\forall N')(M \leq_{\mathfrak{K}_\lambda} N' \rightarrow (\exists f)[f$ is a $\leq_{\mathfrak{K}_\lambda}$-embedding of N' into N over $M])$.
3) AxIII (see II.1.4) implies AxIII$^\odot$
where:

Ax0: K is a class of $\tau_{\mathfrak{K}}$-models, $\leq_{\mathfrak{K}}$ a two place relation of K_λ, both preserved under isomorphisms

AxI: if $M \leq_{\mathfrak{K}_\lambda} N$ then $M \subseteq N$ (are $\tau(\mathfrak{K}_\lambda)$-models of cardinality λ

AxII: $\leq_{\mathfrak{K}_\lambda}$ is a partial order (so $M \leq_{\mathfrak{K}_\lambda} M$ for $M \in K_\lambda$)

AxIII$^\odot$: In following game the COM player has a winning strategy. A play last λ moves, they construct a $\leq_{\mathfrak{K}_\lambda}$-increasing continuous sequence $\langle M_\alpha : \alpha \leq \lambda\rangle$. In the α-th move M_α is chosen, by INC if

α is even by COM is α is odd. Now Com wins as long as INC has legal moves.

AxIV$^\odot$: For each $M \in K_\lambda$, in the following game, INC has no winning strategy: a play lasts $\lambda+1$ moves, in the α-th move $f_\alpha, M_\alpha, N_\alpha$ are chosen such that f_α is a $\le_{\mathfrak{K}}$-embedding of M_α into N_α, both are $\le_{\mathfrak{K}_\lambda}$-increasing continuous, f_α is $\subseteq$-increasing continuous, $M_0 = M$ and in the α-th move, M_α is chosen by INC, and the pair is chosen by the player INC if α is even and by the player COM if α is odd. The player COM wins if INC has always a legal move (the player COM always has: he can choose $N_\alpha = M_\alpha$)

1.34 Definition. 1) Let $<^*_\lambda = <^*_{\mathfrak{K}^*_\lambda}$ be the following two-place relation on K^*_λ (so $M \le^*_{\mathfrak{K}^*_\lambda} N$ mean $M = N \in \mathfrak{K}^*_\lambda$ or $M <^*_{\mathfrak{K}^*_\lambda} N$):

> $M_1 <^*_\lambda M_2$ iff $M_1 \le_{\mathfrak{K}_\lambda} M_2$ are from K^*_λ and M_2 is $\le_{\mathfrak{K}_\lambda}$-universal over M_1.

2) For $\alpha < \lambda, \kappa = \beth_{1,1}(|\alpha| + \mathrm{LS}(\mathfrak{K}))$ and $M \in K^*_\lambda$ let $\mathrm{Sav}^{\mathrm{bs},\alpha}(M)$ be the set of $\{\mathrm{Av}_\kappa(\mathbf{I}, M) : \mathbf{I}$ is a $((2^\kappa)^+, \kappa)$-convergent subset of ${}^\alpha M\}$. We define $\mathrm{tp}_*(\bar a, M, N)$ when $M \le_{\mathfrak{K}} N$ are from K^*_λ and $\bar a \in {}^\alpha N$, as $\mathrm{tp}_{\mathbb{L}_{\infty,\kappa}[\mathfrak{K}]}(\bar a, M, N) \in \mathrm{Sav}^{\mathrm{bs},\alpha}(M)$ naturally.
3) Let $\mathfrak{K}^*_\lambda = (K^*_\lambda, \le_{\mathfrak{K}} \restriction \mathfrak{K}^*_\lambda, \le^*_{\mathfrak{K}^*_\lambda})$, see 1.35 below but if $(K^*_\lambda, \le_{\mathfrak{K}} \restriction K^*_\lambda)$ is a λ-a.e.c. then we omit $\le^*_{\mathfrak{K}^*_\lambda}$.

1.35 Remark. 1) Note that the relation $<^*_\lambda = <^*_{\mathfrak{K}_\lambda}$ seemingly depends on the choice of Φ. However, assuming μ-solvability, by 1.37(2) below it does not depend.
2) The proof of 1.37 is like II.1.16(3).
3) So $\mathfrak{K}^*_\lambda$ is a semi-λ-a.e.c. (see Chapter N) but we do not use this notion here.

1.36 Claim. *Assume* $\mathrm{cf}(\lambda) = \aleph_0$.
0) If $M \in K^*_\lambda$ <u>then</u> for some $N, M <^*_{\mathfrak{K}^*_\lambda} N (\in K^*_\lambda)$.
1) If $M \le_{\mathfrak{K}} N$ are from $K^*_\lambda, \alpha < \lambda$ and $\bar a \in {}^\alpha N \backslash {}^\alpha M$ <u>then</u> $\bar a$ realizes some $p \in \mathrm{Sav}^{\mathrm{bs},\alpha}(M)$.

2) If $M_0 \le_{\mathfrak{K}} M_1 <^*_{\mathfrak{K}^*_\lambda} M_2 \le_{\mathfrak{K}} M_3$ and $M_\ell \in K^*_\lambda$ for $\ell < 4$, then $M_0 <^*_{\mathfrak{K}^*_\lambda} M_3$.

Proof. 0) As K^*_λ is categorical (by 1.16(1)) this follows by 1.32(2).
1) A proof of this is included in the proof of 1.29(2), i.e. by 1.26(1).
2) Easy recalling amalgamation. $\square_{1.36}$

1.37 Claim. *Assume* $\mathrm{cf}(\lambda) = \aleph_0$.
1) Assume $\langle M_i : i \le \delta\rangle$ is $\le_{\mathfrak{K}_\lambda}$-increasing continuous, $M_{2i+1} <^*_{\mathfrak{K}^*_\lambda} M_{2i+2}$ for $i < \delta$ then $M_\delta \in K^*_\lambda$.
2) Assume that $\langle M^\ell_i : i \le \delta\rangle$ is an $\le_{\mathfrak{K}^*_\lambda}$-increasing continuous sequence such that $M^\ell_{2i+1} <^*_{\mathfrak{K}^*_\lambda} M^\ell_{2i+2}$ for $i < \delta$ all for $\ell = 1, 2$. Any isomorphism f from M^1_0 onto M^2_0 (or just a $\le_{\mathfrak{K}_\lambda}$-embedding) can be extended to an isomorphism from M^1_δ onto M^2_δ.

Proof. 1) We prove this by induction on δ, hence without loss of generality $i < \delta \Rightarrow M_i \in K^*_\lambda$.

Let $M^1_\alpha = M_\alpha$ for $\alpha \le \delta$ and let $\langle I_\alpha : \alpha \le \delta\rangle$ be $<_{K^{\rm flin}_\lambda}$-increasing. Let $M^2_\alpha = \mathrm{EM}_{\tau(\mathfrak{K})}(I_\alpha, \Phi)$. Now there is an isomorphism f from M^1_0 onto M^2_0 as K^*_λ is categorical, so by part (2) there is an isomorphism g from M^1_α onto M^2_α, but $M^2_\alpha \in K^*_\lambda$ so we are done.
2) Note

$\boxtimes_2$ without loss of generality

  $\boxdot$ $M^2_i <^*_\lambda M^2_{i+1}$.

[Why? We can find $\langle M^3_i : i \le \delta\rangle$ which is $\le^*_{\mathfrak{K}^*_\lambda}$-increasing continuous and $M^3_0 = M^2_0$ and $M^3_i <^*_\lambda M^3_{i+1}$. Now apply the restricted version (i.e., with the assumption $\boxdot$) twice.]

By induction on $i \le \delta$ we choose (f_i, N^1_i, N^2_i) such that

(b) f_i is an isomorphism from N^1_i onto N^2_i
 (c) N^1_i, N^2_i, f_i are increasing continuous with i
 (d) for $i = 0, N^1_i = M^1_i, f_i = f$ and N^2_i is $f(M^1_i) = M^2_i$
 (e) if $i > 0$ is a limit ordinal then $N^1_i = M^1_i$ and $N^2_i = M^2_i$

(f) when $i = \omega\alpha + 2n < \delta$ we have

(α) $N^1_{\omega\alpha+2n+1} = M^1_{\omega\alpha+2n+1}$

(β) $N^2_{\omega\alpha+2n+1} \le_{\mathfrak{K}} M^2_{\omega\alpha+2n+1}$

(γ) $N^1_{\omega\alpha+2n+2} \le_{\mathfrak{K}} M^1_{\omega\alpha+2n+2}$

(δ) $N^2_{\omega\alpha+2n+2} = M^2_{\omega\alpha+2n+2}$.

Case 1: For $i = 0$ this is trivial by clause (d) and the assumption of the claim on f.

Case 2: $i = \omega\alpha + 2n + 1$.

Note that $N^2_{\omega\alpha+2n} = M^2_{\omega\alpha+2n}$. (Why? If $i = 0$ (i.e. $\alpha = 0 = n$) by $\circledast(d)$ and if i is a limit ordinal (i.e. $\alpha > 0 \wedge n = 0$) by clause (e) of $\circledast$ and if $n > 0$ by clause $((f)(\delta)$ of $\circledast)$.

Now we let $N^1_i = N^1_{\omega\alpha+2n+1} := M^1_{\omega\alpha+2n+1}$ and hence satisfying clause $(f)(\alpha)$ of $\circledast$. So $N^1_{i-1} = N^1_{\omega\alpha+2n} \le_{\mathfrak{K}} M^1_{\omega\alpha+2n} \le_{\mathfrak{K}} M^1_{\omega\alpha+2n+1} = N^1_{\omega\alpha+2n+1} = N^1_i$; and note that $N^2_{i-1} = N^2_{\omega\alpha+2n} <^*_\lambda M^2_{\omega\alpha+2n}$ by $\boxdot$ above hence we can apply Definition 1.34(1) and find an extension f_i of f_{i-1} to $\le_{\mathfrak{K}}$-embedding of $N^1_i = M^1_{\omega\alpha+2n+1}$ into $M^2_{\omega\alpha+2n+1}$ and let $N^2_i := f_i(N^1_i)$.

Case 3: $i = \omega\alpha + 2n + 2$.

Note that $N^1_{\omega\alpha+2n+1} = M^1_{\omega\alpha+2n+1}$ by clause $(f)(\alpha)$ of $\circledast$ hence by the assumption of the claim $N^1_{\omega\alpha+2n+1} <^*_{\mathfrak{K}^*_\lambda} M^1_{\omega\alpha+2n+2}$. We choose $N^2_{\omega\alpha+2n+2} := M^2_{\omega\alpha+2n+2}$ hence $N^2_{i-1} = N^2_{\omega\alpha+2n+1} \le_{\mathfrak{K}} M^2_{\omega\alpha+2n+1} \le_{\mathfrak{K}} M^2_{\omega\alpha+2n+2} = N^2_{\omega\alpha+2n+2} = N^2_i$. Now we apply Definition 1.34(1) to find a $\le_{\mathfrak{K}}$-embedding g_i of $N^2_{\omega\alpha+2n+2}$ into $M^1_{\omega\alpha+2n+2}$ extending f^{-1}_{i-1}.

Lastly, let $f_i = g^{-1}_i$ and $N^1_i = M^1_i \restriction \mathrm{Dom}(f_i)$. So we can carry the induction hence prove the claim. $\square_{1.37}$

Note that now we use more than in Hypothesis 1.18.

1.38 Claim. *Assume*

$\boxtimes$ (a) *$\langle \lambda_n : n < \omega \rangle$ is increasing, $\lambda = \lambda_\omega = \sum_{n<\omega} \lambda_n$ satisfying $\lambda_n = \beth_{\lambda_n} > \mathrm{LS}(\mathfrak{K})$ and $\mathrm{cf}(\lambda_n) = \aleph_0$ for $n < \omega$*

(b) *$\Phi \in \Upsilon^{\mathrm{or}}_{\mathfrak{K}}$ and for it each λ_n and $\lambda = \lambda_\omega$ is as in Hypothesis 1.18 or just satisfies all its conclusions so far.*

1) K^*_λ *is closed under unions* $\le_{\mathfrak{K}}$*-increasing chains (of length* $< \lambda^+$*).*
2) If $M_n \in K^*_{\lambda_n}, M_n \le_{\mathfrak{K}} M_{n+1}$ *and* $M = \bigcup_{n<\omega} M_n$ *then* $M \in K^*_\lambda$.
3) If $M \in K_\lambda$ *and* $\theta < \lambda \Rightarrow M \equiv_{\mathbb{L}_{\infty,\theta}[\mathfrak{K}]} \mathrm{EM}_{\tau(\mathfrak{K})}(\lambda, \Phi)$ *then* $M \in K^*_\lambda$.
4) K^*_λ *is categorical.*

Proof of 1.38. 1) We rely on part (2) which is proven below.

So let $\langle M_i : i < \delta \rangle$ be $\le_{\mathfrak{K}}$-increasing in K^*_λ with $\delta < \lambda^+$. Without loss of generality $\delta = \mathrm{cf}(\delta)$ hence $\delta < \lambda$ so call it θ and we prove this by induction on θ, so without loss of generality $\langle M_i : i < \theta \rangle$ is $\le_{\mathfrak{K}}$-increasing continuous such that $M_i \in K^*_\lambda$ for $i < \theta$, and let $M_\theta = \bigcup_{i<\theta} M_i$. By renaming without loss of generality $\theta < \lambda_0$.

Let I_n, I'_n be such that:

$\odot_1$ (a) I_n is a linear order of cardinality λ_n from K^{flin}

(b) I'_n is a linear order of cardinality 2^{λ_n} from K^{flin}

(c) I'_n is λ^+_n-saturated (which means that its cofinality is $> \lambda_n$, the cofinality of its inverse is $> \lambda_n$ and if $I'_n \models$ "$s_{\alpha_1} < s_{\beta_1} < t_{\beta_2} < t_{\alpha_2}$" where $\alpha_1 < \beta_1 < \gamma_1, \alpha_1 < \beta_2 < \gamma_2$ and $|\gamma_1| + |\gamma_2| < \lambda^+_n$ then for some r we have $I'_n \models$ "$s_{\alpha_1} < r < t_{\alpha_2}$" for $\alpha_1 < \gamma_1, \alpha_2 < \gamma_2$)

(d) $I_n <_{K^{\mathrm{flin}}} I'_n <_{K^{\mathrm{flin}}} I_{n+1}$ for $n < \omega$.

Let $I = \cup\{I_n : n < \omega\}$, so I is a universal member of K^{lin}_λ. Let $M^* = \mathrm{EM}_{\tau(\mathfrak{K})}(I, \Phi)$, so for every $i < \theta$ there is an isomorphism f_i from M^* onto M_i, exists as K^*_λ is categorical by 1.19(4) as $\mathrm{cf}(\lambda) = \aleph_0$.

Now

$\odot_2$ (a) every interval of I is universal in K^{lin}_λ

(b) if $n < \omega$, $J \subseteq I$, $\chi = |J| < \lambda$ and $\mathscr{E}_{J,I} = \{(t_1, t_2) : t_1, t_2 \in I \backslash J$ and $s \in J \Rightarrow s <_I t_1 \equiv s <_J t_2\}$ then for at most χ elements of t of $J \backslash I$ the set $t/\mathscr{E}_{J,I}$ is a singleton.

[Why? Clause (a) is obvious. For clause (b) assume $\langle t_\alpha : \alpha < \chi^+ \rangle$ are pairwise distinct members of $J \backslash I$ such that $t_\alpha/\mathscr{E}_{J,I}$ is a singleton for each $\alpha < \chi^+$. Without loss of generality for some $k < \omega$ we have $\alpha < \chi^+ \Rightarrow t_\alpha \in I_k$ hence $\chi \le \lambda_k$. For each $\alpha < \chi^+$ we can choose $s_\alpha \in I'_k$ such that $s_\alpha <_{I'_k} t_\alpha$ and $(s_\alpha, t_\alpha)_{I'_k} \cap J = \emptyset$. Clearly $\alpha < \beta < \chi^+ \Rightarrow (t_\alpha <_I s_\beta \vee t_\beta <_I s_\alpha)$ hence $\langle (s_\alpha, t_\alpha)_I : \alpha < \chi^+ \rangle$ are pairwise disjoint intervals of I, so for every $\alpha < \chi^+$ large enough, $(s_\alpha, t_\alpha)_I \cap J = \emptyset$, but then $(s_\alpha, t_\alpha)_I \subseteq t_\alpha/\mathscr{E}_{J,I}$, contradiction.]

Now by induction on $n < \omega$ and for each n by induction on $\varepsilon \leq \theta$ and for each $n < \omega$ and $\varepsilon \leq \theta$ for $i \leq \theta$, we choose $J_{n,\varepsilon,i} \in K^{\rm flin}_{\lambda_n}$ such that:

$\odot_3$ (a) $J_{n,\varepsilon,i} \subseteq I$

(b) $J_{n,\varepsilon,i}$ has cardinality λ_n

(c) $I_n <_{K^{\rm flin}} J_{n,0,i}$

(d) if $\zeta < \varepsilon \leq \theta$ and $i \leq \theta$ then $J_{n,\zeta,i} \subseteq J_{n,\varepsilon,i}$, moreover if for some $\xi, \zeta = 2\xi + 1$ and $\varepsilon = 2\xi + 2$ then there is a $<_{K^{\rm flin}_{\lambda_n}}$-increasing continuous sequence of length ω with first member $J_{n,\zeta,i}$ and union $J_{n,\varepsilon,i}$

(e) for ε limit, $J_{n,\varepsilon,i} = \bigcup\limits_{\zeta<\varepsilon} J_{n,\zeta,i}$

(f) if ε is odd and $i < j < \theta$ then $f_i(\mathrm{EM}_{\tau(\mathfrak{K})}(J_{n,\varepsilon,i}, \Phi)) = M_i \cap f_j(\mathrm{EM}_{\tau(\mathfrak{K})}(J_{n,\varepsilon,j}, \Phi))$

(g) $J_{n,\theta,i} \subseteq J_{n+1,0,i}$

(h) for every $k < \omega$ and $s <_I t$ from $J_{n,\varepsilon,i}$ if $[s,t]_I \cap I'_k \neq \emptyset$ then $[s,t]_I \cap I'_k \cap J_{n,\varepsilon,i} \neq \emptyset$

(i) if ζ is odd and $\varepsilon = \zeta + 1$ then $\mathrm{EM}_{\tau(\mathfrak{K})}(J_{n,\zeta,i}, \Phi) <^*_{\mathfrak{K}^*_{\lambda_n}} \mathrm{EM}_{\tau(\mathfrak{K})}(J_{n,\varepsilon,i}, \Phi)$.

There is no problem to carry the definition, for $\varepsilon = 2\xi + 2$ recalling $\odot_2$ above; the only non-trivial point is clause (i), which follows by 1.32(4) and clause (d) of $\odot_3$. Clearly $\langle J_{n,\varepsilon,i} : \varepsilon \leq \theta \rangle$ is $\subseteq$-increasing continuous by $\odot_3(d) + (e)$.
Let $M^*_{n,\varepsilon,i} = f_i(\mathrm{EM}_{\tau(\mathfrak{K})}(J_{n,\varepsilon,i}, \Phi))$ and $M^*_{n,\varepsilon} = M^*_{n,2\varepsilon,\varepsilon}$. So clearly $M^*_{n,\varepsilon,i} \in K^*_{\lambda_n}$ by $\odot_3(b)$ and the choice of $M^*_{n,\varepsilon,i}$ the sequence $\langle M^*_{n,\varepsilon} : \varepsilon < \theta \rangle$ is $\leq_{\mathfrak{K}}$-increasing continuous, all members in $K^*_{\lambda_n}$.

Now

$\odot_4$ $\langle M^*_{n,\varepsilon} : \varepsilon < \theta \rangle$ is $<^*_{\mathfrak{K}^*_{\lambda_n}}$-increasing.

[Why? As $\zeta < \varepsilon < \theta \Rightarrow M^*_{n,\zeta} = M_{n,2\zeta,\zeta} \leq_{\mathfrak{K}^*_{\lambda_n}} M_{n,2\zeta+1,\zeta} \leq_{\mathfrak{K}^*_{\lambda_n}} M_{n,2\zeta+1,\varepsilon} <^*_{\mathfrak{K}^*_{\lambda_n}} M_{n,2\zeta+2,\varepsilon} \leq_{\mathfrak{K}^*_{\lambda_n}} M_{n,2\varepsilon,\varepsilon} = M^*_{n,\varepsilon}$ by the choice of $M^*_{n,\zeta}$, by $\odot_3(d)$ and Ax(V) of a.e.c., by $\odot_3(f)$ and Ax(V) of a.e.c., by $\odot_3(i)$, by $\odot_3(d)$ + Ax(V) of a.e.c.(e), by the choice of $M^*_{n,\varepsilon}$ respectively). Now by 1.36(2) this argument shows that $\zeta < \varepsilon < \theta \Rightarrow M^*_{n,\zeta} <^*_{\mathfrak{K}^*_{\lambda_n}} M^*_{n,\varepsilon}$.]

We can conclude by using 1.37(1) for $\mathfrak{K}^*_{\lambda_n}$, that $M^*_n := \bigcup_{\varepsilon<\theta} M^*_{n,\varepsilon}$ belongs to $K^*_{\lambda_n}$. Also as $M^*_{n,\varepsilon} \leq_{\mathfrak{K}} M_\varepsilon \leq_{\mathfrak{K}} M_\delta$ for $\varepsilon < \theta = \delta$ by AxIV of a.e.c. we have $M^*_n \leq_{\mathfrak{K}} M_\delta$ and similarly $M^*_n \leq_{\mathfrak{K}} M^*_{n+1}$, and obviously for each $i < \theta$ we have $\bigcup_{n<\omega} M^*_n$ includes $\cup\{M^*_{n,\varepsilon} : n < \omega, \varepsilon < \theta\} = \cup\{M^*_{n,2,\varepsilon,\varepsilon} : n < \omega, \varepsilon < \theta\} = \cup\{M^*_{n,2\varepsilon,i} : n < \omega, i < \theta, \varepsilon < \theta\} = \bigcup_{n<\omega} M^*_{n,0,i}$ which recalling the choice of $M^*_{n,0,i}$ includes $\bigcup_n f_i(\mathrm{EM}_{\tau(\mathfrak{K})}(J_{n,0,i},\Phi)) \supseteq \bigcup_{n<\omega} f_i(\mathrm{EM}_{\tau(\mathfrak{K})}(I_n,\Phi)) = f_i(\mathrm{EM}_{\tau(\mathfrak{K})}(I,\Phi)) = M_i$. As this holds for every $i < \theta$ we get $\bigcup_{n<\omega} M^*_n = M_\delta$. So by part (2) we are done.

2) We choose I_n by induction on n such that:

$\odot_5$ (a) $I_n \in K^{\rm flin}_{\lambda_n}$
(b) $I_m <_{K^{\rm flin}} I_n$ if $n = m+1$.

Let $N_n = \mathrm{EM}_{\tau(\mathfrak{K})}(I_n,\Phi)$.

We now choose $(g_n, I'_n, I''_n, M'_n, M''_n, N'_n, N''_n)$ by induction on $n < \omega$ such that:

$\odot_6$ (a) g_n is an isomorphism from N''_n onto M''_n
(b) $I_n \subseteq I'_n \subseteq I''_n \subseteq I_{n+2}$ and $|I'_n| = \lambda_n, |I''_n| = \lambda_{n+1}$ and $I_{n+1} \subseteq I''_n$
(c) $N'_n = \mathrm{EM}_{\tau(\mathfrak{K})}(I'_n,\Phi)$ and $N''_n = \mathrm{EM}_{\tau(\mathfrak{K})}(I''_n,\Phi)$
(d) $M_n \leq_{\mathfrak{K}^*_{\lambda_n}} M'_n \leq_{\mathfrak{K}^*} M''_n \leq_{\mathfrak{K}^*} M_{n+2}$ and $M_{n+1} \leq_{\mathfrak{K}^*_{\lambda_{n+1}}} M''_n$
(e) g_n maps $N'_n = \mathrm{EM}_{\tau(\mathfrak{K})}(I'_n,\Phi)$ onto M'_n

(f) g_n extends $g_m \restriction N'_m$ if $n = m+1$

(g) $I'_n \subseteq I'_{n+1}$.

Case 1: For $n = 0$.

First, let $M''_n = M_1, I''_n = I_1$ so also N''_n is defined. Second, choose g_n satisfying (a) of $\odot_6$ by 1.16(1), i.e. 1.19(4), categoricity in $K^*_{\lambda_n}$. Third, choose $I^*_n \subseteq I''_n = I_1$ of cardinality λ_n such that $g_n(\mathrm{EM}_{\tau(\mathfrak{K})}(I^*_n, \Phi))$ includes M_0. Fourth, let $I'_n = I^*_n \cup I_n$ and $N'_n = \mathrm{EM}_{\tau(\mathfrak{K})}(I'_n, \Phi)$ and let $M'_n = g_n(N'_n)$.

Case 2: For $n = m+1$.

Let $k = n+2$, let $\bar{a} \in {}^{\lambda_m}(M'_m)$ list M'_m (with no repetitions). Now

$(*)_1$ If $\theta < \lambda_n$ then $\mathrm{tp}_{\mathbb{L}_{\infty,\theta}[\mathfrak{K}]}(\bar{a}, \emptyset, N_k) = \mathrm{tp}_{\mathbb{L}_{\infty,\theta}[\mathfrak{K}]}(\bar{a}, \emptyset, N''_m)$.

[Why? As $\mathrm{EM}_{\tau(\mathfrak{K})}(I''_m, \Phi) \prec_{\mathbb{L}_{\infty,\theta}[\mathfrak{K}]} \mathrm{EM}_{\tau(\mathfrak{K})}(I_k, \Phi)$ by 1.14(a) as $I''_m \subseteq I_k$.]

$(*)_2$ if $\theta < \lambda_n = \lambda_{m+1}$ then
$\mathrm{tp}_{\mathbb{L}_{\infty,\theta}}(\bar{a}, \emptyset, N''_m) = \mathrm{tp}_{\mathbb{L}_{\infty,\theta}}(g_m(\bar{a}), \emptyset, M''_m)$.

[Why? As g_m is an isomorphism from N''_m onto M''_m by $\odot_6(a)$, i.e. the induction hypothesis.]

$(*)_3$ if $\theta < \lambda_n$ then $\mathrm{tp}_{\mathbb{L}_{\infty,\theta}[\mathfrak{K}]}(g_m(\bar{a}), \emptyset, M''_m) = \mathrm{tp}_{\mathbb{L}_{\infty,\theta}[\mathfrak{K}]}(g_m(\bar{a}), \emptyset, M_k)$

[Why? This follows from $M'_m \prec_{\mathbb{L}_{\infty,\theta}[\mathfrak{K}]} M_k$ which we can deduce by 1.19(1) as $M''_m \in K^*_{\lambda_{m+1}} = K^*_{\lambda_n}$ by clause (d) of $\odot_6$, $M_k \in K^*_k$ by an assumption of the claim, $M''_m \leq_{\mathfrak{K}_\lambda} M_k$ by clause (d) of $\odot_6$.]

$(*)_4$ if $\theta < \lambda_n$ then $\mathrm{tp}_{\mathbb{L}_{\infty,\theta}[\mathfrak{K}]}(\bar{a}, \emptyset, N_k) = \mathrm{tp}_{\mathbb{L}_{\infty,\theta}[\mathfrak{K}]}(g_m(\bar{a}), \emptyset, M_k)$.

[Why? By $(*)_1 + (*)_2 + (*)_3$.]

$(*)_5$ $\mathrm{tp}_{\mathbb{L}_{\infty,\lambda^+_{n+1}}[\mathfrak{K}]}(\bar{a}, \emptyset, N_k) = \mathrm{tp}_{\mathbb{L}_{\infty,\lambda^+_{n+1}}[\mathfrak{K}]}(g_m(\bar{a}), \emptyset, M_k)$.

[Why? Clearly $N_k, M_k \in K^*_{\lambda_k}$ hence by 1.19(4) there is an isomorphism f_n from N_k onto M_k, so obviously $\mathrm{tp}_{\mathbb{L}_{\infty,\theta}[\mathfrak{K}]}(\bar{a}, \emptyset, N_k) = \mathrm{tp}_{\mathbb{L}_{\infty,\theta}[\mathfrak{K}]}(f_n(\bar{a}), \emptyset, N_k)$ so by $(*)_4$ we have $\mathrm{tp}_{\mathbb{L}_{\infty,\theta}[\mathfrak{K}]}(g_m(\bar{a}), \emptyset, M_k) = \mathrm{tp}_{\mathbb{L}_{\infty,\theta}[\mathfrak{K}]}(\bar{a}, \emptyset, N_k) = \mathrm{tp}_{\mathbb{L}_{\infty,\theta}[\mathfrak{K}]}(f_n(\bar{a}), \emptyset, M_k)$ so by 1.19(3) we have

$\mathrm{tp}_{\mathbb{L}_{\infty,\lambda^+_{n+1}}[\mathfrak{K}]}(g_n(\bar a),\emptyset,M_k) = \mathrm{tp}_{\mathbb{L}_{\infty,\lambda^+_{n+1}}[\mathfrak{K}]}(f_n(\bar a),\emptyset,M_k)$.
But as f_n is an isomorphism from N_k onto M_k and the previous sentence we get $\mathrm{tp}_{\mathbb{L}_{\infty,\lambda_{n+1}}[\mathfrak{K}]}(\bar a,\emptyset,N_k) = \mathrm{tp}_{\mathbb{L}_{\infty,\lambda^+_{n+1}}[\mathfrak{K}]}(f_n(\bar a),\emptyset,M_k) = \mathrm{tp}_{\mathbb{L}_{\infty,\lambda}}(g_n(\bar a),\emptyset,M_k)$ as required.]

$(*)_6$ there are g_n, I''_n, N''_n, M''_n as required in the relevant parts of $\odot_6$ (ignoring I'_n, N'_n, M'_n), i.e. clauses (a),(f) and the relevant parts of (b),(c),(d):

- $(b)'$ $I_n \subseteq I''_n \subseteq I_{n+2} = I_k$ and $|I''_n| = \lambda_{n+1}$ and $I_{n+1} \subseteq I'_n$
- $(c)'$ $N''_n = \mathrm{EM}_{\tau(\mathfrak{K})}(I''_n,\Phi)$
- $(d)'$ $M_n \le_{\mathfrak{K}^*} M''_n \le_{\mathfrak{K}^*} M_{n+2}$ and $M_{n+1} \le_{\mathfrak{K}^*_{\lambda_{n+2}}} M''_n$.

[Why? By the hence and forth argument, but let us elaborate.

First, let $\bar a'$ be a sequence of length λ_{n+1} listing (without repetitions) the set of elements of M_{n+1} and without loss of generality $g(\bar a) \triangleleft \bar a'$. Note that $\mathrm{Rang}(g_m) \subseteq M_{m+2} = M_{n+1}$.

Second, let g' be a function from $\mathrm{Rang}(\bar a')$ into N_k extending $(g_m{\restriction}N'_m)^{-1} = (g_m{\restriction}\mathrm{Rang}(\bar a))^{-1}$ such that $\mathrm{tp}_{\mathbb{L}_{\infty,\lambda^+_{n+1}}[\mathfrak{K}]}(g'(\bar a'),\emptyset,N_k) = \mathrm{tp}_{\mathbb{L}_{\infty,\lambda^+_{n+1}}[\mathfrak{K}]}(\bar a',\emptyset,M_k)$; it exists by $(*)_5$. Let $I''_n \subseteq I_k$ of cardinality λ_{n+1} be such that $\mathrm{Rang}(g') \subseteq \mathrm{EM}(I''_n,\Phi)$ and $I_{n+1} \subseteq I''_n$. Let $\bar a''$ list the elements of $\mathrm{EM}_{\tau(\mathfrak{K})}(I''_n,\Phi) \subseteq N_k$ and without loss of generality $g'(\bar a') \triangleleft \bar a''$ and let g_n be a function from $\mathrm{EM}_{\tau(\mathfrak{K})}(I''_n,\Phi)$ to M_k extending $(g')^{-1}$ such that $\mathrm{tp}_{\mathbb{L}_{\infty,\lambda^+_{n+1}}[\mathfrak{K}]}(\bar a'',\emptyset,N_k) = \mathrm{tp}_{\mathbb{L}_{\infty,\lambda^+_{n+1}}[\mathfrak{K}]}(g_n(\bar a''),\emptyset,M_k)$.

Lastly, let $N''_n = \mathrm{EM}_{\tau(\mathfrak{K})}(I''_n,\Phi)$ and $M''_n = g_n(N'_n)$ so we are done.]

$(*)_7$ there are I'_n, N'_n, M'_n as required.

[Why? By the LS argument we can choose I'_n and define N'_n, M'_n accordingly.]

So we can carry the induction. Now $N'_n \le_{\mathfrak{K}} N'_{n+1}$ (by clauses (g),(c) of $\odot_6$) and $g_n \restriction N'_n \subseteq g_{n+1} \restriction N'_{n+1}$ (by clause (f) + the previous statement). Hence $g = \cup\{g_n \restriction N'_n : n < \omega\}$ is an isomorphism from $\cup\{N'_n : n < \omega\}$ onto $\cup\{M'_n : n < \omega\}$. But $N = \cup\{N_n : n <$

$\omega\} \subseteq \cup\{N'_n : n < \omega\} \subseteq \mathrm{Dom}(g) \subseteq N$ and $M = \cup\{M_n : n < \omega\} \subseteq \cup\{M'_n : n < \omega\} \subseteq \mathrm{Rang}(g) \subseteq M$. Together g is an isomorphism from N onto M but obviously $N \in K^*_\lambda$ hence $M \in K^*_\lambda$ is as required.
3),4) Should be clear and depends just on 1.19(4). $\square_{1.38}$

1.39 Conclusion. Let λ be as in $\boxtimes$ of 1.38.
1) $\mathfrak{K}^*_\lambda$ is a λ-a.e.c. (with $\leq_{\mathfrak{K}} \restriction K^*_\lambda$) and it has amalgamation and is categorical.
2) $\mathfrak{K}^\oplus_{\geq\lambda}$ is an a.e.c., $\mathrm{LS}(\mathfrak{K}^\oplus_{\geq\lambda}) = \lambda$ and $(\mathfrak{K}^*_\lambda)^{\mathrm{up}} = K^\oplus_{\geq\lambda}$ and $(\mathfrak{K}^\oplus_{\geq\lambda})_\lambda = \mathfrak{K}^*_\lambda$, see Definition below.

1.40 Definition. Let $\mathfrak{K}^\oplus_{\geq\lambda} = \mathfrak{K} \restriction K^\oplus_{\geq\lambda}$ where $K^\oplus_{\geq\lambda} = \{M \in K_\lambda : M \equiv_{\mathbb{L}_{\infty,\lambda}[\mathfrak{K}]} \mathrm{EM}_{\tau(\mathfrak{K})}(\lambda, \Phi)\}$.

Proof. 1) It was clear defining $(K^*_\lambda, \leq_{\mathfrak{K}} \restriction K^*_\lambda)$ that it is of the right form and "$M \in K^*_\lambda$", "$M \leq_{\mathfrak{K}^*_\lambda} N$" are preserved by isomorphisms. Obviously "$\leq_{\mathfrak{K}} \restriction K^*_\lambda$ is a partial order", so AxI, AxII hold and obviously AxV holds (see II.1.4). The missing point was AxIII, about $\leq_{\mathfrak{K}}$-increasing union and it holds by 1.38(1). Then AxIV becomes easy by the definition of $\leq_{\mathfrak{K}^*_\lambda} = \leq_{\mathfrak{K}} \restriction K^*_\lambda$ and lastly the amalgamation holds by 1.30.
2) By II§1 we can "lift $\mathfrak{K}^*_\lambda$ up", the result is $\mathfrak{K}^\oplus_{\geq\lambda}$ (see II.1.23,II.1.24).
$\square_{1.39}$

Let us formulate a major conclusion in ways less buried inside our notation.

1.41 Conclusion. Assume $(\mathfrak{K}, \Phi)$ is pseudo solvable in μ, <u>then</u> $(\mathfrak{K}, \Phi)$ is pseudo solvable in λ provided that $\mathrm{LS}(\mathfrak{K}) < \lambda, \mu = \mu^{<\lambda}$ (or just the hypothesis 1.18 holds), $\mathrm{cf}(\lambda) = \aleph_0$ and λ is an accumulation point of the class of the fix point of the sequence of the $\beth$'s.

Proof. By 1.39(1). $\square_{1.41}$

Remark. About [weak] solvability, see [Sh:F782].

§2 Trying to Eliminate $\mu = \mu^{<\lambda}$

There was one point in §1 where we use $\mu = \mu^\lambda$ (i.e. in 1.12, more accurately in justifying hypothesis 1.18(1)). In this section we try to eliminate it. So we try to prove $M_1 \leq_{\mathfrak{K}_\mu} M_2 \Rightarrow M_1 \prec_{\mathbb{L}_{\infty,\theta}[\mathfrak{K}]} M_2$ for $\theta < \lambda$, hence we fix $\mathfrak{K}, \mu, \theta$. We succeed to do it with "few exceptions".

2.1 Hypothesis. (We shall mention $(b)_\mu$ or $(b)^-_\mu, (c), (d)$ when used! but not clause (a))

- (a) $\mathfrak{K}$ is an a.e.c. and $\Phi \in \Upsilon^{\rm or}_{\mathfrak{K}}$
- $(b)_\mu$ $\mathfrak{K}$ categorical in μ and $\Phi \in \Upsilon^{\rm or}_{\mathfrak{K}}$, or at least
- $(b)^-_\mu$ $\mathfrak{K}$ is pseudo μ-solvable as witnessed by $\Phi \in \Upsilon^{\rm or}_{\mathfrak{K}}$, see Definition 1.4 in particular $\mathrm{EM}_{\tau(\mathfrak{K})}(I, \mu)$ is pseudo superlimit for $I \in K^{\rm lin}_\lambda$,
- (c) $\mu \geq \beth_{1,1}(\mathrm{LS}(\mathfrak{K}))$
- (d) $\mu > \mathrm{LS}(\mathfrak{K})$.

2.2 Convention: $K^*_\lambda = K^*_{\Phi,\lambda}$, etc., see Definition 1.15.

2.3 Definition. Assume

$$\boxdot \ \mu \geq \chi \geq \theta > \mathrm{LS}(\mathfrak{K})$$

1) We let $K^1_{\mu,\chi} = \{(M,N) : N \leq_{\mathfrak{K}} M, N \in K_\chi, M \in K_\mu$ and $\mu = \chi \Rightarrow M = N\}$ and let $\leq_{\mathfrak{K}} = \leq_{\mathfrak{K},\mu,\chi}$ be the following partial order on $K_{\mu,\chi}$, $(M_0,N_0) \leq_{\mathfrak{K}} (M_1,N_1)$ iff $M_0 \leq_{\mathfrak{K}} M_1, N_0 \leq_{\mathfrak{K}} N_1$ (formally we should have written $\leq_{\mathfrak{K},\mu,\chi}$). Note that each pair $(M,N) \in K_{\mu,\chi}$ determine μ, χ. So if $\chi = \mu, K_{\mu,\chi}$ is essentially $\mathfrak{K}_\mu$. Let $K^1_\mu = K_\mu$ and let $\cup\{(M_i,N_i) : i < \delta\} = (\cup\{M_i : i < \delta\}, \cup\{N_i : i < \delta\})$ for any $\leq_{\mathfrak{K}}$-increasing sequence $\langle (M_i,N_i) : i < \delta \rangle$.
1A) Let $K_{\mu,\chi} = K^2_{\mu,\chi} = \{(M,N) \in K^1_{\mu,\chi} : M \in K^*_\mu\}$ and $K^2_\mu = K^*_\mu$ but we use them only when Φ witnesses $\mathfrak{K}$ is pseudo μ-solvable, i.e. $(b)^-_\mu$ from Hypothesis 2.1 holds.
2) For $k \in \{1,2\}$ a formula $\varphi(\bar{x}) \in \mathbb{L}_{\infty,\theta}[\mathfrak{K}]$ (so $\ell g(\bar{x}) < \theta$), cardinal

$\kappa \geq \theta$ the main case being $\kappa = \mu$; we may omit k if $k = 2$, and $M \in K^k_\kappa, \bar{a} \in {}^{\ell g(\bar{x})}M$ <u>we define</u> when $M \Vdash_k \varphi[\bar{a}]$ by induction on the depth of $\varphi(\bar{x}) \in \mathbb{L}_{\infty,\theta}[\mathfrak{K}]$, so the least obvious case is:

(∗) $M \Vdash_k (\exists \bar{y})\psi(\bar{y}, \bar{a})$ <u>when</u> for every $M_1 \in K^k_\kappa$ such that $M \leq_{\mathfrak{K}} M_1$ there is $M_2 \in K^k_\kappa$ satisfying $M_1 \leq_{\mathfrak{K}} M_2$ and $\bar{b} \in {}^{\ell g(\bar{y})}M_2$ such that $M_2 \Vdash_k \psi[\bar{b}, \bar{a}]$.

Of course

(α) for φ atomic, $M \Vdash_k \varphi[\bar{a}]$ <u>iff</u> $M \models \varphi[\bar{a}]$

(β) for $\varphi(\bar{x}) = \bigwedge_{i<\alpha} \varphi_i(\bar{x})$ let $M \Vdash_k \varphi[\bar{a}]$ iff $M \Vdash_k \varphi_i[\bar{a}]$ for each $i < \alpha$

(γ) $M \Vdash_k \neg\varphi[\bar{a}]$ iff for no N do we have $M \leq_{\mathfrak{K}} N \in K^k_\kappa$ and $N \Vdash_k \varphi[\bar{a}]$.

3) Let $k \in \{1, 2\}, \Lambda \subseteq \mathbb{L}_{\infty,\theta}[\mathfrak{K}]$ (each formula with $< \theta$ free variables, of course):

(a) Λ is downward closed if it is closed under subformulas

(b) Λ is (μ, χ)-model^k complete (when μ is clear from the context we may write χ-model^k complete) <u>if</u> $|\Lambda| < \mu$, and for every $(M_0, N_0) \in K^k_{\mu,\chi}$ we can find $(M, N) \in K^2_{\mu,\chi}$ above (M_0, N_0) which is Λ-generic, where:

(c) $(M, N) \in K^k_{\mu,\chi}$ is Λ-generic^k when:
if $\varphi(\bar{x}) \in \Lambda$ and $\bar{a} \in {}^{\ell g(\bar{x})}N$ then
$M \Vdash_k \varphi[\bar{a}] \Leftrightarrow N \models \varphi[\bar{a}]$ (yes! neither $(M, N) \Vdash_k \varphi[\bar{a}]$ which was not defined, nor "$M \models \varphi[\bar{a}]$")

(d) Λ is called $(\mu, < \mu)$-model^k complete when $|\Lambda| + \theta_\Lambda < \mu$ and for every χ: if $|\Lambda| + \theta_\Lambda \leq \chi < \mu$ then Λ is χ-model^k complete where $\theta_\Lambda := \min\{\partial : \partial > \mathrm{LS}(\mathfrak{K}) \text{ and } \Lambda \subseteq \mathbb{L}_{\infty,\partial}[\mathfrak{K}]\}$. We say Λ is model^k complete if it is $(\mu, < \mu)$-model^k complete and μ is understood from the context

(e) above if Φ or $(\mathfrak{K}, \Phi)$ is not clear from the context we may replace Λ by (Λ, Φ) or by $(\Lambda, \Phi, \mathfrak{K})$.

4) For $M \in K^k_\kappa, \bar{a} \in {}^{\theta>}M$ and $\Lambda \subseteq \mathbb{L}_{\infty,\theta}[\mathfrak{K}]$ let $\mathrm{gtp}^k_\Lambda(\bar{a}, \emptyset, M) = \{\varphi[\bar{a}] : M \Vdash_k \varphi[\bar{a}]\}$; if we write θ instead of Λ we mean $\mathbb{L}_{\infty,\theta}[\mathfrak{K}]$

(note: this type is not a priori complete) and we say that $\bar{a}$ materializes this type in M. To stress κ we may write $\text{gtp}^{\kappa,k}_{\Lambda}(\bar{a}, \emptyset, M)$ or $\text{gtp}^{\kappa,k}_{\theta}(\bar{a}, \emptyset, M)$ though M determines κ.
5) We say $M \in K_\kappa$ is Λ-generic^k <u>when</u> for every $\varphi(\bar{x}) \in \Lambda$ and $\bar{a} \in {}^{\ell g(\bar{x})}M$ we have $M \Vdash_k \varphi[\bar{a}] \Leftrightarrow M \models \varphi[\bar{a}]$. So $M \in K^k_\mu$ is Λ-generic^k iff $(M, M) \in K^k_{\mu,\mu}$ is Λ-generic^k. We say Λ is κ-model^k complete <u>when</u> every $M \in K^k_\kappa$ has a Λ-generic $\leq_{\mathfrak{K}}$-extension in K^k_κ (so depend on $\mathfrak{K}$ and if $k = 2$ also on Φ).
6) In all cases above, if $k = 2$ we may omit it.

2.4 Claim. *Assume that* $\text{LS}(\mathfrak{K}) < \theta \leq \chi < \mu$ *and* $\kappa > \theta$ *and* $k \in \{1, 2\}$ *so if* $k = 2$ *then 2.1(b)$^-_\mu$ holds, see 2.3(1A).*
1) $(K^k_{\mu,\chi}, \leq_{\mathfrak{K}})$ *is a partial order and chains of length* $\delta < \chi^+$ *of members has a* $\leq_{\mathfrak{K}}$*-lub, this is the union, see 2.3(1). If* $\text{EM}_{\tau(\mathfrak{K})}(\mu, \Phi)$ *is superlimit (not just pseudo superlimit)* <u>*then*</u> $K^2_{\mu,\chi}$ *is a dense subclass of* $K^1_{\mu,\chi}$ *under* $\leq_{\mathfrak{K}}$.
2) If $M_1 \Vdash_k \varphi(\bar{a})$ *and* $M_1 \leq_{\mathfrak{K}} M_2$ *are from* K^k_κ <u>*then*</u> $M_2 \Vdash_k \varphi[\bar{a}]$.
3) If $(M_\ell, N_\ell) \in K^k_{\mu,\chi}$ *are* Λ*-generic^k for* $\ell = 1, 2$ *and* $(M_1, N_1) \leq_{\mathfrak{K}} (M_2, N_2)$ <u>*then*</u> $N_1 \prec_\Lambda N_2$.
4) If $M_i \in K^k_\kappa$ *for* $i < \delta$ *is* $\leq_{\mathfrak{K}}$*-increasing,* $\delta < \kappa^+$, $\text{cf}(\delta) \geq \theta, \Lambda \subseteq \mathbb{L}_{\infty,\theta}[\mathfrak{K}]$ *and each* M_i *is* Λ*-generic^k,* <u>*then*</u> $M_\delta := \bigcup_{i<\delta} M_i$ *is* Λ*-generic^k and* $i < \delta \Rightarrow M_i \prec_\Lambda M_\delta$.
5) If $(M_i, N_i) \in K^k_{\mu,\chi}$ *for* $i < \delta$ *is* $\leq_{\mathfrak{K}}$*-increasing,* $\delta < \chi^+$, $\text{cf}(\delta) \geq \theta, \Lambda \subseteq \mathbb{L}_{\infty,\theta}[\mathfrak{K}]$ *and each* (M_i, N_i) *is* Λ*-generic^k,* <u>*then*</u> $(\bigcup_{i<\delta} M_i, \bigcup_{i<\delta} N_i)$ *is* Λ*-generic^k and* $N_j \prec_\Lambda \bigcup_{i<\delta} N_i$ *for each* $j < \delta$.

Proof. Should be clear; in part (1) for $k = 2$ we use clause $(b)^-_\mu$ of 2.1. In part (5) note that $\cup\{M_i : I < \delta\} \in K^*_\mu$ by Clause $(\text{b})^-_\mu$ of 2.1. $\square_{2.4}$

<u>2.5 Exercise</u>: If (M, N) is Λ-generic^k and $(M, N) \leq_{\mathfrak{K}} (M', N) \in K^k_{\mu,\chi}$ <u>then</u> (M', N) is Λ-generic^k.

2.6 Claim. *Assume that* $\mu \geq \chi \geq \theta > \mathrm{LS}(\mathfrak{K})$ *and* $k \in \{1,2\}$.
1) The set of quantifier free formulas in $\mathbb{L}_{\infty,\theta}[\mathfrak{K}]$ *is* (μ,χ)*-model*^{k} *complete.*
2) If $\Lambda_\varepsilon \subseteq \mathbb{L}_{\infty,\theta}(\tau_{\mathfrak{K}})$ *is downward closed,* (μ,χ)*-model*^{k} *complete for* $\varepsilon < \varepsilon^*$, *and* $\Lambda := \bigcup_{\varepsilon<\varepsilon^*} \Lambda_\varepsilon, \theta = \mathrm{cf}(\theta) \leq \chi \vee \theta < \chi, \varepsilon^* < \chi^+$ *(and* $\mu > \theta \vee \mu = \theta = \mathrm{cf}(\theta))$ *then* Λ *is* (μ,χ)*-model*^{k} *complete.*

Proof. 1) Easy.
2) Given $(M,N) \in K^k_{\mu,\chi}$ let θ_r be $\min\{\partial : \partial \geq \theta$ is regular$\}$. Clearly $\theta_r \leq \chi$ and we choose $(M_i,N_i) \in K^k_{\mu,\chi}$ for $i \leq \varepsilon^* \times \theta_r$ such that

⊛ (a) $\langle M_i : i \leq \varepsilon^* \times \theta_r\rangle$ is $\leq_{\mathfrak{K}}$-increasing continuous
(b) $\langle N_i : i \leq \varepsilon^* \times \theta_r\rangle$ is $\leq_{\mathfrak{K}}$-increasing continuous
(c) if $i = \varepsilon^* \times \gamma + \varepsilon$ and $\varepsilon < \varepsilon^*$ then (M_{i+1}, N_{i+1}) is Λ_ε-generic^k
(d) $(M_0,N_0) = (M,N)$.

There is no problem to do this.

Now for each $\varepsilon < \varepsilon^*$ the sequence $\langle (M_{\varepsilon^*\times\gamma+\varepsilon+1}, N_{\varepsilon^*\times\gamma+\varepsilon+1}) : \gamma < \theta_r\rangle$ is $\leq_{\mathfrak{K},\mu,\chi}$-increasing with union $(M_{\varepsilon^*\times\theta_r}, N_{\varepsilon^*\times\theta_r})$, and each member of the sequence is Λ_ε-generic^k hence by 2.4(5) we know that the pair $(M_{\varepsilon^*\times\theta_r}, N_{\varepsilon^*\times\theta_r})$ is Λ_ε-generic^k. As this holds for each Λ_ε it holds for Λ so $(M_{\varepsilon^*\times\theta_r}, N_{\varepsilon^*\times\theta_r})$ is as required. $\square_{2.6}$

From now on in this section

2.7 Hypothesis. We assume (a) + $(\mathrm{b})^-_\mu$ of 2.1 and we omit k using Definition 2.3 meaning $k = 2$.

2.8 Claim. *1) For* $M \in K^*_\mu$ *and* $\mathrm{LS}(\mathfrak{K}) < \theta < \mu$ *the number of complete* $\mathbb{L}_{\infty,\theta}[\mathfrak{K}]$*-types realized by sequences from* ${}^{\theta>}M$ *is* $\leq 2^{<\theta}$, *moreover, the relation* $\mathscr{E}^{<\theta}_M := \{(\bar a,\bar b) : \bar a,\bar b \in {}^{\theta>}M$ *and some automorphism of* M *maps* $\bar a$ *to* $\bar b\}$ *is an equivalence relation with* $\leq 2^{<\theta}$ *equivalence classes.*
2) Hence there is a set $\Lambda_* = \Lambda^*_\theta = \Lambda^*_{\mathfrak{K},\Phi,\mu,\theta} \subseteq \mathbb{L}_{\infty,\theta}[\mathfrak{K}]$ *such that:*

(a) $|\Lambda_*| \leq 2^{<\theta}$ *and* $\Lambda_* \subseteq \mathbb{L}_{(2^{<\theta})^+,\theta}[\mathfrak{K}]$

(*b*) Λ_* *is closed under sub-formulas and finitary operations*

(*c*) *each* $\varphi(\bar{x}) \in \Lambda_*$ *has quantifier depth* $< \gamma^*$ *for some* $\gamma^* < (2^{<\theta})^+$

(*d*) *for* $\alpha < \theta, M \in K^*_\mu$ *and* $\bar{a} \in {}^\alpha M$, *the* Λ_**-type which* $\bar{a}$ *realizes in* M *determines the* $\mathbb{L}_{\infty,\theta}[\mathfrak{K}]$*-type which* $\bar{a}$ *realizes in* M, *moreover one formula in the type determine it*

(*e*) *similarly for materialize in* $M \in K^*_\mu$, *see Definition 2.3(4)*

(*f*) *if* $\mathrm{LS}(\mathfrak{K}) \leq \chi < \mu$ *and* $(M, N) \in K_{\mu,\chi}$ *is* Λ_**-generic* *<u>then</u> it is* $\mathbb{L}_{\infty,\theta}[\mathfrak{K}]$*-generic*

(*g*) *if* $M \in K^2_\mu$ *is* Λ_**-generic* <u>*then*</u> *it is* $\mathbb{L}_{\infty,\theta}[\mathfrak{K}]$*-generic.*

Remark. Part (1) can also be proved using just $(\lambda + 1) \times I_*$ with I_* a θ-saturated dense linear order with neither first nor last element, but this is not clear for 2.11(1).

Proof. 1) By 5.1(1) and categoricity of K^*_λ.
2) Follows but we elaborate.

Let $\{\bar{a}_\alpha : \alpha < \alpha^* \leq 2^{<\theta}\}$ be a set of representatives of the $\mathscr{E}^{<\theta}_M$-equivalence classes. For each $\alpha \neq \beta$ such that $\ell g(\bar{a}_n) = \ell g(\bar{a}_\beta)$, let $\bar{x}_\alpha = \langle x_i : i < \ell g(\bar{a}_\alpha)\rangle$ and choose $\varphi_{\alpha,\beta}(\bar{x}_\alpha), \psi_{\alpha,\beta}(\bar{x}_\alpha) \in \mathbb{L}_{(2^{<\theta})^+,\theta}[\mathfrak{K}]$ such that, if possible we have $M \models \varphi_{\alpha,\beta}[\bar{a}_\alpha] \wedge \neg\varphi_{\alpha,\beta}[\bar{a}_\beta]$ and, under this, if possible $M \Vdash$ "$\psi_{\alpha,\beta}(\bar{a}_\alpha) \wedge \neg\psi_{\alpha,\beta}(\bar{a}_\beta)$ but in any case $M \models \varphi_{\alpha,\beta}[\bar{a}_\alpha]$ and $M \Vdash \psi_{\alpha,\beta}[\bar{a}_\alpha]$. Let $\varphi_\alpha(\bar{x}) = \wedge\{\varphi_{\alpha,\beta}(\bar{x}_\alpha) : \beta < \alpha^*, \beta \neq \alpha$ and $\ell g(\bar{a}_\beta) = \ell g(\bar{a}_\alpha)\}$ and similarly $\psi_\alpha(\bar{x}_\alpha)$. Let Λ_* be the closure of $\{\varphi_{\alpha,\beta}, \psi_{\alpha,\beta}, \varphi_\alpha, \psi_\alpha : \alpha \neq \beta < \alpha^*\}$ under subformulas and finitary operations. Obviously, clauses (a), (b) hold hence the existence of $\gamma^* < (2^{<\theta})^+$ as required in clause (c) follows. Clause (d) holds as $\bar{a}\mathscr{E}^{<\theta}_M\bar{b} \Rightarrow \mathrm{tp}_{\mathbb{L}_{\infty,\theta}[\mathfrak{K}]}(\bar{a}, \emptyset, M) = \mathrm{tp}_{\mathbb{L}_{\infty,\theta}[\mathfrak{K}]}(\bar{b}, \emptyset, M)$ using the automorphisms and for $\alpha, \beta < \alpha_*$ such that $\ell g(\bar{a}_\alpha) = \ell g(\bar{a}_\beta)$ we have $M \models (\forall \bar{x}_\alpha)(\varphi_\alpha(\bar{x}_\alpha) = \varphi_\beta(\bar{x}_\beta)$ implies $\mathrm{tp}_{\mathbb{L}_{(2^{<\theta})^+,\theta}[\mathfrak{K}]}(\bar{a}_\alpha, \emptyset, M) = \mathrm{tp}_{\mathbb{L}_{(2^{<\theta})^+,\theta}[\mathfrak{K}]}(\bar{a}_\beta, \emptyset, M)$ and even $\mathrm{tp}_{\mathbb{L}_{\infty,\theta}[\mathfrak{K}]}(\bar{a}_\alpha, \emptyset, M) = \mathrm{tp}_{\mathbb{L}_{\infty,\theta}[\mathfrak{K}]}(\bar{a}_\beta, \emptyset, M)$ recalling the choice of the $\varphi_{\alpha,\beta}$'s.

Clause (e) holds similarly by the choice of the $\psi_{\alpha,\beta}$'s. Clauses (f),(g) should also be clear. (The proof is similar to the proof of the classical 0.17(3).) $\square_{2.8}$

2.9 Observation. Assume $(2.1(\mathrm{b})^-_\mu$ of course and) $\Lambda \subseteq \mathbb{L}_{\infty,\theta}[\mathfrak{K}]$ and $\mu > 2^{<\theta}$ and $\theta >$ LS$(\mathfrak{K})$.
1) The number of complete $\mathbb{L}_{\infty,\theta}[\mathfrak{K}]$-types realized in some $M \in K^*_\mu$, by a sequence of length $< \theta$ of course, is $\leq 2^{<\theta}$. Hence every formula in $\mathbb{L}_{\infty,\theta}[\mathfrak{K}]$ is equivalent, for models from K^*_μ to a formula of quantifier depth $< (2^{<\theta})^+$, even from $\Lambda_* \subseteq \mathbb{L}_{(2^{<\theta})^+,\theta}[\mathfrak{K}]$ where Λ_* is in 2.8(2).
2) Assume that $I_1 \subseteq I_2$ are well ordered, cf(I_1), cf$(I_2) > 2^{<\theta}$ and $t \in I_2 \backslash I_1 \Rightarrow 2^{<\theta} <$ cf$(I_1 \restriction \{s \in I_1 : s <_{I_2} t\})$ and $t \in I_2 \backslash I_1 \Rightarrow 2^{<\theta} <$ cf$(I_2 \restriction \{s \in I_2 : (\forall r \in I_1)(r <_{I_2} t \equiv r <_{I_2} s)\})$. Then $\mathrm{EM}_{\tau(\mathfrak{K})}(I_1, \Phi) \prec_{\mathbb{L}_{\infty,\theta}[\mathfrak{K}]} \mathrm{EM}_{\tau(\mathfrak{K})}(I_2, \Phi)$.
3) If $M = \mathrm{EM}_{\tau(\mathfrak{K})}(I, \Phi), |I| = \mu, I$ well ordered of cofinality $> 2^{<\theta}, \bar{a} \in {}^\alpha M$ where $\alpha < \theta$ and $a_i = \sigma_i(\ldots, a_{t_{i,\ell}}, \ldots)_{\ell < n(i)}$ for $i < \alpha$ then $\mathrm{tp}_{\Lambda_*}(\bar{a}, \emptyset, M)$ is determined by $\langle \sigma_i(x_0, \ldots, x_{n(\ell)-1}) : i < \ell g(\bar{a})\rangle$ and the essential θ-type of $\langle t_{i,\ell} : i < \ell g(\bar{a}), \ell < n(i)\rangle$, see Definition 2.10 below.

Before proving 2.9

2.10 Definition. 1) For $\bar{t} = \langle t_i : i < \alpha\rangle \in {}^\alpha I, I$ well ordered, let the essential θ-type of $\bar{t}$ in I be the essential $(\theta, (2^{<\theta})^+)$-type where for an ordinal γ we let the essential (θ,γ)-type of $\bar{t}$ in I, $\mathrm{estp}_{\theta,\gamma}(\bar{t}, \emptyset, I)$ be the following information stipulating $t_\alpha = \infty$:

- (a) the truth value of $t_i < t_j$ (for $i, j < \alpha$)
- (b) $\mathrm{otp}([r_i, t_i)_I)$ for $i < \alpha$ where for $i \leq \alpha$ we let r_i be the minimal member r of I such that $\mathrm{otp}([r, t_i)_I) < \theta \times \gamma$ and $r \leq_I t_i$ and $j < \alpha \wedge t_j < t_i \Rightarrow t_j \leq r$
- (c) $\mathrm{Min}\{\theta \times \gamma, \mathrm{otp}[s_i, r_i)_I\}$ for $i \leq \alpha$ where we let s_i be the minimal member of I such that $(\forall j < \alpha)[t_j <_I t_i \Rightarrow t_j <_I s_i]$
- (d) $\mathrm{Min}\{\theta, \mathrm{cf}(I \restriction \{s : s <_I r_i\})\}$ for $i \leq \alpha$ which may be zero.

2) Let the function implicit in 2.9(3) be called $\mathbf{t}^\mu_\Lambda = \mathbf{t}^\mu_{\mathfrak{K},\Lambda} = \mathbf{t}^\mu_{\mathfrak{K},\Phi,\Lambda}$, i.e., $\mathbf{t}^\mu_\Lambda(\mathbf{s}, \bar{\sigma}) = \mathrm{tp}_\Lambda(\bar{a}, \emptyset, M)$ when $\bar{a} = \langle \sigma_i(\ldots, a_{t_{\beta(i,\ell)}}, \ldots)_{\ell < n_i} : i < \ell g(\bar{a})\rangle, \bar{\sigma} = \langle \sigma_i(\ldots, x_{\beta(i,\ell)}, \ldots)_{\ell < n}; i < \ell g(\bar{a})\rangle$ and $\mathbf{s}$ is the essential θ-type of $\langle t_{i,\ell} : i < \ell g(\bar{a}), \ell < n_i\rangle$ in I.
If $\Lambda = \mathbb{L}_{\infty,\theta}[\mathfrak{K}]$ we may write just θ.

Proof of 2.9. 1) By 2.8(1) this holds for each $M \in K^*_\mu$.
2) It is known by Kino [Kin66] that $I_1 \prec_{\mathscr{L}} I_2$ if $\mathscr{L} \subseteq \{\varphi \in \mathbb{L}_{\infty,\theta}(\{<\}) : \varphi$ has quantifier depth $< (2^{<\theta})^+\}$. From this the result follows by part (1).

More fully let θ_r be the first regular cardinal $\geq \theta$, and we say that the pair (I_1, I_2) is γ-suitable when we replace in the assumptions "of cofinality $> 2^{<\theta}$" by "of cofinality $\geq \theta$ and of order type divisible by $\theta \times \gamma$". Now we prove by induction on γ that

$\odot_1$ assume that for $\alpha < \theta$ and for $\ell = 1, 2$ we have: I_ℓ is a well ordering, $\bar{t}^\ell = \langle t^\ell_i : i < \alpha\rangle$ is $<_{I_\ell}$-increasing, t^ℓ_0 is the first element of I_ℓ, we stipulate $t^\ell_\alpha = \infty$ and $\text{otp}([t^\ell_i, t^\ell_{i+1})_{I_0}) = \theta_r\gamma\alpha^\ell_i + \beta_i$ where $\beta_i < \theta\gamma$ and $(\text{cf}(\alpha^1_i) = \text{cf}(\alpha^1_i)) \vee (\text{cf}(\alpha^1_i) \geq \theta \wedge \text{cf}(\alpha^2_i) \geq \theta)$.
Then for any formula $\varphi(\langle x_i : i < \alpha\rangle) \in \mathbb{L}_{\infty,\theta}(\{<\})$ of quantifier depth $\leq \gamma$ we have $I_1 \models \varphi[\bar{t}^1] \Leftrightarrow I_2 \models \varphi[\bar{t}^2]$.

Hence

$\odot_2$ if $\vartheta(\bar{x}) \in \mathbb{L}_{\infty,\theta}(\{<\})$ has quantifier depth $< \gamma$ and (I_1, I_2) is γ-suitable and $\bar{t} \in {}^{\ell g(\bar{x})}(I_1)$ then $I_1 \models \varphi[\bar{t}] \Leftrightarrow I_2 \models \theta[\bar{t}]$.

3) Follows by part (2). $\square_{2.9}$

2.11 Claim. *Assume*

$\boxdot$ (a) (a) $M \in K^*_\mu$

(b) $\Lambda \subseteq \mathbb{L}_{\infty,\theta}[\mathfrak{K}]$ *is downward closed,* $|\Lambda| \leq \chi$, $\text{LS}(\mathfrak{K}) < \theta \leq \chi < \mu$ *and* $2^{<\theta} \leq \chi$ *and* $\theta = \text{cf}(\theta) \vee \theta < \chi$ *so* $\Lambda = \Lambda_*$ *from 2.8 is O.K.*

(c) *in part (3),(4),(5) we assume* $(\chi^{<\theta} \leq \mu) \vee (\text{cf}(\mu) \geq \theta)$

(d) *for part (6) we assume* $\text{cf}(\mu) \geq \theta$ *(hence the demand in clause (c) holds).*

1) If $M \in K^*_\mu$ *then* $\{\text{gtp}_\Lambda(\bar{a}, \emptyset, M) : \bar{a} \in {}^{\theta>}M\}$ *has cardinality* $\leq 2^{<\theta}$.
2) If $(M, N) \in K_{\mu,\chi}$ *then we can find* N', $(M, N) \leq_{\mathfrak{K}} (M, N') \in K_{\mu,\chi}$ *such that*

(∗) *if* $\alpha < \theta$ *and* $\bar{b} \in {}^{\alpha}M$ *and* $\Lambda \subseteq \mathbb{L}_{\infty,\theta}[\mathfrak{K}]$ *then for some* $\bar{b}' \in {}^{\alpha}(N')$ *we have: for every* $\bar{a} \in {}^{\theta>}N$, $\mathrm{gtp}_\Lambda(\bar{a}\hat{\ }\bar{b}, \emptyset, M) = \mathrm{gtp}_\Lambda(\bar{a}\hat{\ }\bar{b}', \emptyset, M)$.

3) If $(M, N) \in K_{\mu,\chi}$, *then we can find* (M_1, N_1) *such that* $(M, N) \leq_{\mathfrak{K}} (M_1, N_1) \in K_{\mu,\chi}$ *and (note that* $\bar{y}$ *may be the empty sequence)*

(∗) *if* $\exists \bar{y}\varphi(\bar{y}, \bar{x}) \in \Lambda$ *and* $\bar{a} \in {}^{\ell g(\bar{x})}N$ *then* $M_1 \Vdash \neg\exists\bar{y}\varphi(\bar{y}, \bar{x})$ *or for some*
$\bar{b} \in {}^{\ell g(\bar{y})}(N_1)$ *we have* $M_1 \Vdash \varphi[\bar{b}, \bar{a}]$.

4) In part (3) we can demand

$(*)^+$ *if* $\exists\bar{y}\varphi(\bar{y}, \bar{x}) \in \Lambda$ *and* $\bar{a} \in {}^{\ell g(\bar{x})}(N_1)$ *then* $M_1 \Vdash \neg(\exists\bar{y})\varphi(\bar{y}, \bar{x})$ *or for some* $\bar{b} \in {}^{\ell g(\bar{y})}(N_1)$ *we have* $M_1 \models \varphi[\bar{b}, \bar{a}]$.

5) In part (4) it follows that the pair (M_1, N_1) *is* Λ*-generic (most interesting for* Λ_*, *see 2.8).*
6) If $M_1 \in K^*_\mu$ *then it is* Λ*-generic.*

Proof. 1) Proved just like 2.8(1).
2) First assume θ is a successor cardinal. As $M \in K^*_\mu$ without loss of generality $M = \mathrm{EM}_{\tau(\mathfrak{K})}(I, \Phi)$ for some linear order I of cardinality μ as in 5.1(1),(4) with $\theta^-, \theta, \chi^+, \mu$ here standing for $\mu, \theta_1, \theta_2, \lambda$ there. It follows that for some $J \subseteq I$ of cardinality χ we have $N \subseteq \mathrm{EM}_{\tau(\mathfrak{K})}(J, \Phi)$, and let $J^+ \subseteq I$ be such that $J \subseteq J^+, |J^+ \backslash J| = \chi$ and for every $\bar{t} \in {}^{\theta>}I$ there is an automorphism f of I over J which maps $\bar{t}$ to some member of ${}^{\ell g(\bar{t})}(J^+)$.

Lastly, let $N' = \mathrm{EM}_{\tau(\mathfrak{K})}(J^+, \Phi)$, it is easy to check (see 1.4) that (∗) holds. If θ is a limit ordinal it is enough to prove for each $\partial < \theta$, a version of (∗) with $\alpha < \partial$; and this gives N'_∂. Now we choose N' such that $\partial < \theta \Rightarrow N'_\partial \leq_{\mathfrak{K}} N'$ and $(M, N') \in K_{\mu,\chi}$.
3),4),5),6) We prove by induction on γ that if we let Λ_γ be $\{\varphi(\bar{x}) : \varphi(\bar{x}) \in \Lambda$ has quantifier depth $< 1 + \gamma\}$ then parts (3),(4),(5),(6) holds for Λ_γ. For all four parts, $|\Lambda| \leq \chi$ hence $|\Lambda_\gamma| \leq \chi$ and it suffices to consider $\gamma < \chi^+$. For $\gamma = 0$ they are trivial and for γ limit also easy (let θ_r be the first regular $\geq \theta$ and extend $|\gamma|^+ \times \theta_r$ times taking care of Λ_β in stage $\gamma \times \zeta + \beta$ for each $\beta < \gamma$). So let $\gamma = \beta + 1$.

We first prove (3), but we have two cases (see clause (c)) of the assumption. If $\chi^{<\theta} \le \mu$ this is straight by bookkeeping. So assume $\mathrm{cf}(\mu) \ge \theta$. Given $(M,N) \in K_{\mu,\chi}$ we try to choose by induction on $i < \chi^+$ a pair (M_i, N_i) and for i odd also $\psi_i(\bar{y}_i, \bar{x}_i), \bar{a}_i, \bar{b}_i$ such that

$\circledast_1$ (a) $(M_0, N_0) = (M, N)$

(b) $(M_i, N_i) \in K_{\mu,\chi}$ is $\le_{\mathfrak{K}}$-increasing continuous

(c) M_{i+1} is Λ_β-generic for i even

(d) for i odd $\psi_i(\bar{y}_i, \bar{x}_i) \in \Lambda_\beta$ and $\bar{a}_i \in {}^{\theta>}N$ and $\bar{b}_i \in {}^{\theta>}(N_{i+1})$ are such that $\ell g(\bar{a}_i) = \ell g(\bar{x}_i), \ell g(\bar{b}_i) = \ell g(\bar{y}_i)$ and

(α) $\bar{b} \in {}^{\ell g(\bar{y}_i)}(M_i) \Rightarrow M_i \nVdash \psi_i[\bar{b}_i, \bar{a}]$ but

(β) $M_{i+1} \Vdash \psi_i[\bar{b}_i, \bar{a}_i]$

(γ) for every $\bar{b} \in {}^{\theta>}(M_{i+1})$ there is an automorphism of M_{i+1} over N_i mapping $\bar{b}$ into N_{i+1}.

If we succeed, by part (2) applied to the pair of models $(\bigcup_{i<\chi^+} M_i, N)$ as $\chi^+ \le \mu$ this pair belongs to $K_{\mu,\chi}$ we get N' as there, hence for some odd $i < \chi^+, N' \subseteq M_i$, let $\zeta = i+2$ and this gives a contradiction to the choice of $(\psi_\zeta, \bar{a}_\zeta, \bar{b}_\zeta)$.
[Why? There is an automorphism f of $M := \cup\{M_j : j < \chi^+\}$ over N mapping $\bar{b}_\zeta$ into N' hence into M_i hence $f(\bar{b}_\zeta) \in {}^{\theta>}(M_\zeta)$. We know (by clause $(d)(\beta)$ above) that $M_{\zeta+1} \Vdash \psi_\zeta[\bar{b}_\zeta, \bar{a}_\zeta]$ but $M_{\zeta+1} \le_{\mathfrak{K}_\mu} M$ hence $M \Vdash \psi_\zeta[\bar{a}_\zeta, \bar{b}_\zeta]$. Recall that f is an automorphism of M over N hence $M \Vdash \psi_\zeta[f(\bar{b}_\zeta), f(\bar{a}_\zeta)]$, but $\bar{a}_\zeta \in {}^{\theta>}N$ so $f(\bar{a}_\zeta) = \bar{a}_\zeta$ hence $M \Vdash \psi_\zeta[\bar{b}_\zeta, f(\bar{a}_\zeta)]$ but $M_\zeta \le_{\mathfrak{K}_\mu} M$ and $\bar{a}, f(\bar{b}_\zeta)$ are from M_ζ hence $M_\zeta \nVdash \neg\psi_\zeta[f(\bar{b}_\zeta), (\bar{a}_\zeta)]$. However by clause (d)($\alpha$) of $\circledast_1$ we have $M_\zeta \nVdash \psi_\zeta[f(\bar{b}_\zeta), \bar{a}_\zeta]$. But as i is an odd ordinal the last two sentences contradicts clause (c) of $\circledast_1$ applied to $i+1$.]
Hence we are stuck for some $i < \chi^+$. Now for $i = 0$ clause $\circledast(a)$ gives a permissible value and for i limit take unions noting that clauses (c),(d) required nothing. So $i = j+1$; if j is even we apply the induction hypothesis to part (6) for the pair (M_i, N_i). Hence j is odd so we cannot choose $\psi_j(\bar{y}, \bar{x}), \bar{a}_j, \bar{b}_j$, recalling part (2) so the pair (M_j, N_j) is as required thus proving (3) (for Λ_γ).

Second, we prove part (4). We can now again try to choose by induction on $i < \chi^+$ a pair (M_i, N_i) satisfying

$\circledast_2$ (a) $(M_0, N_0) = (M, N)$

(b) $(M_i, N_i) \in K_{\mu,\chi}$ is $\leq_{\mathfrak{K}}$-increasing continuous

(c) if $i = 2j + 1$, <u>then</u> (M_{i+1}, N_{i+1}) is as in part (3) for Λ_γ with $(M_i, N_i), (M_{i+1}, N_{i+1})$ here standing for (M, N), (M_1, N_1) there

(d) if $i = 2j$ <u>then</u> for some $\psi_i(\bar{y}_i, \bar{x}_i) \in \Lambda_\beta$ and $\bar{a}_i \in {}^{(\ell g(\bar{x}_i))}(N_i)$ and $\bar{b}_i \in {}^{(\ell g(\bar{y}_i))}(N_{i+1})$ we have $M_{i+1} \Vdash \psi_i(\bar{b}_i, \bar{a}_i)$ but $\bar{b} \in {}^{\ell g(\bar{y}_i)}(M_i) \Rightarrow M_i \nVdash \psi_i[\bar{b}, \bar{a}_i]$.

If we succeed, let $S_0 = \{\delta < \chi^+ : \text{cf}(\delta) \geq \theta\}$, so by an assumption S is a stationary subset of χ^+, i.e. as by clause $\boxdot(b)$ we have $\theta = \text{cf}(\theta) \leq \chi \vee \theta < \chi$; also for $\delta \in S_0$, as $\langle N_i : i < \delta \rangle$ is increasing with union N_δ, and $\delta = 2\delta$ clearly $\bar{a}_\delta$ is well defined, so for some $i(\delta) < \delta$ we have $\bar{a}_\delta \in {}^{\theta>}(N_{i(\delta)})$ and without loss of generality $i(\delta) = 2j(\delta)+1$ for some $j(\delta)$ hence by clause (c) of $\circledast_2$ the pair $(M_{i(\delta)+1}, N_{i(\delta)+1})$ is as required there contradiction as in the proof for part (3). Hence for some i we cannot choose (M_i, N_i).

For $i = 0$ let $(M_i, N_i) = (M, N)$ so only clauses (a) + (b) of $\circledast_2$ apply and are satisfied. For i limit take unions. So $i = j + 1$. If $j = 1 \bmod 2$, clause (d) of $\circledast_2$ is relevant and we use part (3) for Λ_β which holds as we have just proved it.

Lastly, if $j = 2 \bmod 2$ and we are stuck then the pair (M_j, N_j) is as required.

Third, Part (5) should be clear but we elaborate.

We prove by induction on γ' that if $\varphi(\bar{x}) \in \Lambda_\gamma$ has quantifier depth $< 1 + \gamma'$ then for every $\bar{a} \in {}^{\ell g(\bar{x})}(N_1)$ we have $M_1 \models \varphi[\bar{a}] \Leftrightarrow N_1 \models \varphi[\bar{a}]$. For atomic φ this is obvious and for $\varphi = \bigwedge\limits_{i<\alpha} \varphi_i$ should be clear. If $\varphi(\bar{x}) = \neg\psi(\bar{x})$ note that in $(*)^+$ of part (4) we can use empty $\bar{y}$ so $\neg(\exists \bar{y})\psi(\bar{x}) = \neg\psi(\bar{x})$. Also for $\varphi(\bar{x}) = (\exists \bar{y})\varphi'(\bar{y}, \bar{x})$ we apply part (4).

Fourth, we deal with part (6), so (see clause (d) of the assumption) we have $\text{cf}(\mu) \geq \theta$. Let $\chi = \langle \chi_i : i < \text{cf}(\mu) \rangle$ be constantly μ^- (so $\mu = \chi_i^+$) if μ is a successor cardinal, and be increasing continuous with limit μ, $2^{<\theta} < \chi_i < \mu$ if μ is a limit cardinal recalling $2^{<\theta} < \mu$ by $\boxdot(b)$. Consider $K_{\mu,\bar{\chi}} = \{\bar{M} : \bar{M} = \langle M_i : i \le \text{cf}(\mu)\rangle$ is $\le_{\mathfrak{K}}$-increasing continuous, $M_{\text{cf}(\mu)} \in K^*_\mu$ and $M_i \in K_{\chi_i}$ for $i < \text{cf}(\mu)\}$ ordered by $\bar{M}^1 \le_{\mathfrak{K}} \bar{M}^2$ iff $i \le \text{cf}(\mu) \Rightarrow M^1_i \le_{\mathfrak{K}} M^2_i$.

By 2.11 and part (5) for Λ_γ which we proved we can easily find $\bar{M} \in K_{\mu,\bar{\chi}}$ such that $i < \text{cf}(\mu) \Rightarrow (M_{\text{cf}(\mu)}, M_{i+1})$ is Λ_γ-generic; such $\bar{M}$ we call Λ_*-generic. Next

> $\boxtimes$ if $\varphi(\bar{x}) \in \Lambda_\gamma$ and $\bar{M}$ is Λ_γ-generic, $\bar{a} \in {}^{\theta>}(M_i)$, i successor, $\varphi(\bar{x}) \in \mathbb{L}_{\infty,\theta}[\mathfrak{K}]$ and $\ell g(\bar{x}) = \ell g(\bar{a})$ then $M_{\text{cf}(\mu)} \models \varphi[\bar{a}] \Leftrightarrow M_{\text{cf}(\mu)} \Vdash \varphi[\bar{a}]$.

[Why? Recalling $\text{cf}(\mu) \ge \theta$, we prove this by induction on the quantifier depth of φ.]

By the definition of "M is Λ-generic" and categoricity of K^*_μ we are done. $\square_{2.11}$

2.12 Conclusion. If $\mu \ge (2^{<\theta})^+, \theta >$ LS($\mathfrak{K}$) and $\text{cf}(\mu) \ge \theta >$ LS($\mathfrak{K}$) then every $M \in K^*_\mu$ is $\mathbb{L}_{\infty,\theta}[\mathfrak{K}]$-generic, hence if $M_1 \le_{\mathfrak{K}} M_2$ are from K^*_μ then $M_1 \prec_{\mathbb{L}_{\infty,\theta}[\mathfrak{K}]} M_2$.

Remark. 1) With a little more care, if $\mu = \mu_0^+$ also $\theta = \mu$ is O.K. but here this is prepheral.
2) $\theta \le$ LS($\mathfrak{K}$) is not problematic, we just ignore it.
3) So 2.12 improve 1.12, i.e. we need $\text{cf}(\mu) \ge \lambda(>$ LS($\mathfrak{K}$)) instead $\mu = \mu^{<\lambda}$ but still there is a class of μ which are not covered.

Proof. Let Λ_* be as in 2.8(2) so in particular $|\Lambda_*| \le 2^{<\theta}$. Now 2.11(6) and clause (g) of 2.8 proves the first assertion in 2.12. For the second assume that $M_1 \le_{\mathfrak{K}_\mu} M_2$ and we shall prove that $M_1 \prec_{\mathbb{L}_{\infty,\theta}[\mathfrak{K}]} M_2$.

By the categoricity of $\mathfrak{K}$ in μ or clause $(b)^-_\mu$ of Hypothesis 2.1, K^* is categorical in μ hence $M_1, M_2 \in K^*_\mu$ are Λ_*-generic. Suppose $\bar{a} \in {}^{(\ell g(\bar{x}))}(M_1), \varphi(\bar{x}) \in \Lambda_*$, so by M'_1 being Λ_*-generic (or $\boxtimes$ from the end of the proof of 2.11 applied to $\bar{M}^2$) we have

$$(*)_1 \;\; M_1 \models \varphi[\bar{a}] \Rightarrow M_1 \Vdash \varphi[\bar{a}] \Rightarrow M_1 \models \varphi[\bar{a}]$$

and by M_2 being Λ_*-generic (or ⊠ from the end of the proof of 2.11 applied to $\bar{M}^2$) we have

$(*)_2$ $M_2 \models \varphi[\bar{a}] \Rightarrow M_2 \Vdash \varphi[\bar{a}] \Rightarrow M_2 \models \varphi[\bar{a}]$

and by the definition of "$M \Vdash \varphi[\bar{a}]$" recalling $M_1 \leq_{\mathfrak{K}_\mu} M_2$,

$(*)_3$ if $M_1 \Vdash \varphi'[\bar{a}]$ then $M_2 \Vdash \varphi'[\bar{a}]$ for $\varphi'(\bar{x}) \in \{\varphi(\bar{x}), \neg\varphi(\bar{x})\}$.

So both M_1 and M_2 satisfy $\varphi[\bar{a}]$ if M_1 satisfy it, but this applies to $\neg\varphi[\bar{a}]$ too; so we are done. $\square_{2.12}$

2.13 Claim. *If K is categorical also in μ^* or just Hypothesis 2.7 apply also to μ^*, too, (with the same Φ) and $\mu^* \geq \mu^{<\theta} > \mu > \theta >$ $\mathrm{LS}(\mathfrak{K})$ and $(*)$ below, then every $M \in K^*_\mu$ is $\mathbb{L}_{\infty,\theta}[\mathfrak{K}]$-generic and $M_1 \in K^*_\mu \wedge M_2 \in K^*_\mu \wedge M_1 \leq_{\mathfrak{K}_\mu} M_2 \Rightarrow M_1 \prec_{\mathbb{L}_{\infty,\theta}[\mathfrak{K}]} M_2$, i.e. the conclusions of 1.12, 2.12 hold where*

> $(*)$ *if $M \in K^*_{\mu^*}$ and $A \in [M]^\mu$ then we can find $N \leq_{\mathfrak{K}} M$ such that $A \subseteq N \in K^*_\mu$ and for every $\varphi(\bar{x}) \in \mathbb{L}_{\infty,\theta}[\mathfrak{K}]$ and $\bar{a} \in {}^{\ell g(\bar{x})}N$ we have $M \Vdash \varphi[\bar{a}] \Leftrightarrow N \Vdash \varphi[\bar{a}]$.*

Proof. We shall choose $(M_i, N_i) \in K_{\mu^*,\mu}$ by induction on $i \leq \theta^+$ such that not only $M_i \in K^*_{\mu^*}$ (see the definition of $K_{\mu^*,\mu}$) but also $N_i \in K^*_\mu$ and this sequence of pairs is $\leq_{\mathfrak{K}}$-increasing continuous. For $i = 0$ use any pair, e.g. $M_0 = \mathrm{EM}_{\tau(\mathfrak{K})}(\mu^*, \Phi)$ and $N_0 = \mathrm{EM}_{\tau(\mathfrak{K})}(\mu, \Phi)$.

For i limit take unions, recalling M_j, N_j are pseudo superlimit for $j < i$.

For $i = j + 1$, let $N^+_j \leq_{\mathfrak{K}} M_j$ be such that $N_j \subseteq N^+_j \in K_\mu$ and (M_j, N^+_j) satisfies $(*)$ of the claim (standing for (M, N)). Let Λ_* be as in 2.8 for μ^*. Then by 2.11(5) with (μ^*, μ, θ) here standing for (μ, χ, θ) there noting that in $\boxdot(c)$ there we use the case $\chi^{<\theta} \leq \mu$ which here means $\mu = \mu^{<\theta}$, we can choose a Λ_*-generic pair $(M_i, N_i) \in K_{\mu^*,\mu}$ above (M_j, N^+_j) hence by 2.8(2)(g) also it is a $\mathbb{L}_{\infty,\theta}[\mathfrak{K}]$-generic pair. Now for $j < \theta^+$, for $\bar{a} \in {}^{\theta>}(N_j)$, we can read $\mathrm{gtp}^{\mu^*}_\theta(\bar{a}, \emptyset, M_{j+1})$ and it is complete, but as by our use of $(*)$ it is the same as $\mathrm{gtp}^\mu_\theta(\bar{a}, \emptyset, N^+_{j+1})$. So $\mathrm{gtp}^\mu_\theta(\bar{a}, \emptyset, N^+_{j+1})$ is complete for every $\bar{a} \in {}^{\theta>}(N_j)$, so also $\mathrm{gtp}^\mu(\bar{a}, \emptyset, N_{\theta^+})$ is complete by monotonicity.

Now if $\bar{a} \in {}^{\theta>}(N_{\theta+})$ then for some $j < \theta^+$ we have $\bar{a} \in {}^{\theta>}(N_j)$, so by the above $p_{\bar{a}} := \mathrm{gpt}^{\mu^*}_\theta(\bar{a}, \emptyset, M_{j+1}) = \mathrm{gtp}^\mu_\theta(\bar{a}, \emptyset, N^+_{j+1}) = \mathrm{gtp}^\mu_\theta(\bar{a}, \emptyset, N_{\theta+})$ is complete and does not depend on j as long as j is large enough.

Now we prove that if $\bar{a} \in {}^{\theta>}(N_{\theta+})$ then $\varphi(\bar{x}) \in p_{\bar{a}} \Rightarrow N_{\theta+} \models \varphi[\bar{a}]$; and we prove this by induction on the quantifier depth of $\varphi(\bar{x})$; as usual the real case is $\varphi(\bar{x}) = (\exists \bar{y})\varphi(\bar{y}, \bar{x})$. Let $j < \theta^+$ be such that $\bar{a} \in {}^{\ell g(\bar{x})}(N_j)$, so $p_{\bar{a}} = \mathrm{gtp}^{\mu^*}_\theta(\bar{a}, M_{j+1})$ so $M_{j+1} \Vdash \varphi[\bar{a}]$ and by the choice of (M_{j+1}, N_{j+1}) it follows that $N_{j+1} \models \varphi[\bar{a}]$ hence for some $\bar{b} \in {}^{\ell g(\bar{y})}(N_{j+1})$ we have $N_{j+1} \models \psi[\bar{b}, \bar{a}]$ hence $M_{j+1} \Vdash \psi(\bar{b}, \bar{a})$, hence $\psi(\bar{y}, \bar{x}) \in p_{\bar{b}\hat{\ }\bar{a}}$ hence by the induction hypothesis $N_{\theta+} \models \psi[\bar{b}, \bar{a}]$ hence $N_{\theta+} \models \varphi[\bar{a}]$.

$\square_{2.13}$

2.14 Conclusion. 1) For each $\theta \geq \mathrm{LS}(\mathfrak{K})$ the family of $\mu > 2^{<\theta}$ in which K is categorical but some (equivalent every) $M \in K_\mu$ is not $\mathbb{L}_{\infty,\theta}[\mathfrak{K}]$-generic is $\subseteq \{[\mu_i, \mu_i^{<\theta}] : i < 2^{2^\theta}\}$ for some sequence $\langle \mu_i : i < 2^{2^\theta} \rangle$ of cardinals.
2) Similarly for pseudo solvable, i.e. for each $\theta \geq \mathrm{LS}(\mathfrak{K})$ and $\Phi \in \Upsilon^{\mathrm{or}}_\theta$ for at most $\beth_2(\theta)$ cardinals $\mu > 2^{<\theta}$ we have $(\forall \alpha < \mu)(|\alpha|)^{<\theta} < \mu)$ and for some $\mu^* \in [\mu, \mu^{<\theta}]$ the pair $(\mathfrak{K}, \Phi)$ is pseudo μ^*-solvable but some $\equiv$ every $M \in K^*_{\Phi,\mu^*}$ is not $\mathbb{L}_{\infty,\theta^+}[\mathfrak{K}]$-generic.

Proof. Straight. Note that it is enough to prove this for each Φ separately.

Toward contradiction assume $\langle \mu_\varepsilon : \varepsilon < (\beth_2(\theta))^+ \rangle$ is an increasing sequence of such cardinals, satisfying $(\mu_\varepsilon)^{<\theta} < \mu_{\varepsilon+1}$ and choose $I_\varepsilon \times \mu_\varepsilon \times (2^{<\theta})^+$, hence $\langle I_\varepsilon : \varepsilon < (\beth_2(\theta))^+ \rangle$ is an increasing sequence of linear orders as in 2.9, in particular, well ordered. Let $\Lambda_\varepsilon = \Lambda^*_{\mathfrak{K},\Phi,\mu_\varepsilon,\theta}$ be from 2.8(2) applied to $I = I_\varepsilon$ hence to any ordinal $< \mu_\varepsilon^+$ of cofinality $> 2^{<\theta}$. Now the number of functions $\mathbf{t}^{\mu_\varepsilon}_{\mathfrak{K},\Lambda_\varepsilon}$ (see Observation 2.9(3) and Definition 2.10(2)) is at most $\beth_2(\theta)$, so for some $\varepsilon < \zeta < (\beth_2(\theta))^+$ we have $\mathbf{t}^{\mu_\varepsilon}_{\mathfrak{K},\Lambda_\varepsilon} = \mathbf{t}^{\mu_\zeta}_{\mathfrak{K},\Lambda_\zeta}$.

Now apply 2.13 with $(\mu_\zeta, \mu_\varepsilon)$ here standing for (μ^*, μ) there, $(*)$ there holds easily by 2.9(3) so we get a contradiction. $\square_{2.14}$

* * *

For the rest of this section we note some basic facts on the dependency on Φ (not used here).

2.15 Definition. 1) We define a two-place relation $\mathscr{E}_\kappa = \mathscr{E}^{\rm or}_\kappa[\mathfrak{K}]$ on $\Upsilon^{\rm or}_\kappa[\mathfrak{K}]$, so $\kappa \geq$ LS$(\mathfrak{K})$: $\Phi_1 \mathscr{E}_\kappa \Phi_2$ iff for every linear orders I_1, I_2 there are linear orders J_1, J_2 extending I_1, I_2 respectively such that $\mathrm{EM}_{\tau(\mathfrak{K})}(J_1, \Phi)$, $\mathrm{EM}_{\tau(\mathfrak{K})}(J_2, \Phi)$ are isomorphic.
2) We define $\leq^{\rm or}_\kappa = \leq^{\rm or}_\kappa [\mathfrak{K}]$, a two-place relation on $\Upsilon^{\rm or}_\kappa[\mathfrak{K}]$ as in part (1) only in the end $\mathrm{EM}_{\tau(\mathfrak{K})}(J_1, \Phi_1)$ can be $\leq_{\mathfrak{K}}$-embedded into $\mathrm{EM}_{\tau(\mathfrak{K})}(J_2, \Phi_2)$.

2.16 Claim. *1) The following conditions on $\Phi_1, \Phi_2 \in \Upsilon^{\rm or}_\kappa[\mathfrak{K}]$ are equivalent*

- (a) $\Phi_1 \mathscr{E}_\kappa \Phi_2$
- (b) *there are $I_1, I_2 \in K^{\rm lin}$ of cardinality $\geq \beth_{1,1}(\kappa)$ such that $\mathrm{EM}_{\tau(\mathfrak{K})}(I_1, \Phi_1)$, $\mathrm{EM}_{\tau(\mathfrak{K})}(I_2, \Phi)$ are isomorphic*
- (c) *there are Φ'_1, Φ'_2 satisfying $\Phi_\ell \leq^\otimes \Phi'_\ell \in \Upsilon^{\rm or}_\kappa[\mathfrak{K}]$ for $\ell = 1, 2$ such that Φ'_1, Φ'_2 are essentially equal (see Definition 2.17 below).*

2) The following conditions are equivalent

- (a) $\Phi_1 \leq^{\rm or}_\kappa \Phi_2$ *recall* $\leq_\kappa = \leq^{\rm or}_\kappa [\mathfrak{K}]$
- (b) *there are $I_1, I_2 \in K^{\rm lin}$ of cardinality $\geq \beth_{1,1}(\kappa)$ such that $\mathrm{EM}_{\tau(\mathfrak{K})}(I_1, \Phi_1)$ can be $\leq_{\mathfrak{K}}$-embedded into $\mathrm{EM}_{\tau(\mathfrak{K})}(I_2, \Phi_2)$*
- (c) *for every $I_1 \in K^{\rm lin}$ there is $I_2 \in K^{\rm lin}$ such that $\mathrm{EM}_{\tau(\mathfrak{K})}(I_1, \Phi_1)$ can be $\leq_{\mathfrak{K}}$-embedded into $\mathrm{EM}_{\tau(\mathfrak{K})}(I_2, \Phi_2)$.*

2.17 Definition. $\Phi_1, \Phi_2 \in \Upsilon^{\rm or}_\kappa[\mathfrak{K}]$ are essentially equal when for every linear order I there is an isomorphism f from $\mathrm{EM}_{\tau(\mathfrak{K})}(I, \Phi_1)$ onto $\mathrm{EM}_{\tau(\mathfrak{K})}(I, \Phi_2)$ such that for any τ_{Φ_1}-term $\sigma_1(x_0, \ldots, x_{n-1})$ there is a τ_{Φ_2}-term $\sigma_2(x_0, \ldots, x_{n-1})$ such that: $t_0 <_I \ldots <_I t_{n-1} \Rightarrow f(a_1) = a_2$, where a_ℓ is $\sigma_\ell(a_{t_0}, \ldots, a_{t_{n-1}})$ as computed in $\mathrm{EM}(I, \Phi_\ell)$ for $\ell = 1, 2$.

Proof of 2.16. Straight (particularly recalling such proof in 1.29(1)).
$\square_{2.17}$

2.18 Claim. *1)* $\mathscr{E}_\kappa = \mathscr{E}^{\rm or}_\kappa[\mathfrak{K}]$ *is an equivalence relation, and* $\Phi_1 \mathscr{E}^{\rm or}_\kappa[\mathfrak{K}]\Phi_2 \Rightarrow \Phi_1 \le^{\rm or}_\kappa [\mathfrak{K}]\Phi_2$.
1A) In fact if $\langle \Phi_\varepsilon : \varepsilon < \varepsilon(*)\rangle$ *are pairwise* $\mathscr{E}_\kappa$*-equivalent and* $\varepsilon(*) \le \kappa$ *then we can find* $\langle \Phi'_\varepsilon : \varepsilon < \kappa\rangle$ *satisfying* $\Phi'_\varepsilon \le^\otimes \Phi'_\varepsilon$ *for* $\varepsilon < \varepsilon(*)$ *such that the* Φ'_ε *for* $\varepsilon < \varepsilon(*)$ *are pairwise essentially equal.*
2) $\le^{\rm or}_\kappa$ *is a partial order.*
3) If $\Phi_1, \Phi_2 \in \Upsilon^{\rm or}_\kappa[\mathfrak{K}]$ *are essentially equal then* $(\mathfrak{K}, \Phi_1)$ *is pseudo/ weakly/strongly* (μ, κ)*-solvable iff* $(\mathfrak{K}, \Phi_2)$ *is pseudo/weakly/strongly* (μ, κ)*-solvable.*
4) If $\Phi_1 \in \Upsilon^{\rm or}_\kappa[\mathfrak{K}]$ *is strongly* (μ, κ)*-solvable and* Φ_2 *exemplifies* $\mathfrak{K}$ *is* (μ, κ)*-solvable then* $\Phi_1 \mathscr{E}_\kappa \Phi_2$.
5) If $\mathfrak{K}$ *is categorical in* μ *and* $\mu > \kappa \ge {\rm LS}(\mathfrak{K})$ *then every* $\Phi \in \Upsilon^{\rm or}_\kappa[\mathfrak{K}]$ *is strongly* (μ, κ)*-solvable.*
6) Assume $(\mathfrak{K}, \Phi_\ell)$ *is pseudo* (μ, κ)*-solvable and* $\mu \ge \beth_{1,1}(\kappa)$ *for* $\ell = 1, 2$. *Then* $\Phi_1 \mathscr{E}_\kappa \Phi_2$ *iff* $\Phi_1 \le^{\rm or}_\kappa [\mathfrak{K}]\Phi_2 \wedge \Phi_2 \le^{\rm or}_\kappa [\mathfrak{K}]\Phi_1$.
7) If $\Phi_1 \le^{\rm or}_\kappa \Phi_2$ *and* Φ_1 *is strongly* (μ, κ)*-solvable or just pseudo* (μ, κ)*-solvable then* Φ_1, Φ_2 *are* $\mathscr{E}^{\rm or}_\kappa[\mathfrak{K}]$*-equivalent.*

Proof. Easy, use 1.29(1) and its proof. $\square_{2.18}$

§3 Categoricity for cardinals on a club

We draw here an easy conclusion from §2, getting that on a closed unbounded class of cardinals which is $\aleph_0$-closed we get a constant answer to being categorical. This is, of course, considerably weaker than conjecture 0.1 but still is a progress, e.g. it shows that the categoricity spectrum is not totally chaotic.

We concentrate on the case the results of §1 holds (e.g. $\mu = \mu^\lambda$) for the λ's with which we deal. To eliminate this extra assumption we need §2. This section is not used later. Note that 3.4 is continued (and improved) in [Sh:F820] and Exercise 3.8, [Sh:F782] improve 3.6; similarly 3.7.

In the claims below we concentrate on fix points of the sequence of $\beth_\alpha$'s.

3.1 Hypothesis. As in Hypothesis 1.2, (i.e. $\mathfrak{K}$ is an a.e.c. with models of arbitrarily large cardinality).

3.2 Definition. 1) Let $\mathrm{Cat}_{\mathfrak{K}}$ be the class of cardinals in which $\mathfrak{K}$ is categorical.
1A) Let $\mathrm{Sol} = \mathrm{Sol}_{\mathfrak{K},\Phi} = \mathrm{Sol}^1_{\mathfrak{K},\Phi}$ be the class of $\mu > \mathrm{LS}[\mathfrak{K}]$ such that $(\mathfrak{K},\Phi)$ is pseudo μ-solvable. Let $\mathrm{Sol}^2_{\mathfrak{K},\Phi}[\mathrm{Sol}^3_{\mathfrak{K},\Phi}]$ be the class of $\mu > \mathrm{LS}(\mathfrak{K})$ such that $(\mathfrak{K},\Phi)$ is weakly [strongly] μ-solvable.
2) Let $\text{mod-com}_{\mathfrak{K},\Phi}$ be the class of pairs (μ,θ) such that: $\mu > \theta \geq \mathrm{LS}(\mathfrak{K})$ and $\mathbb{L}_{\infty,\theta^+}[\mathfrak{K}]$ is μ-model complete (on $K^*_{\Phi,\mu}$, see Definition 2.3(3)(b), 2.3(5)).
3) Let $\mathrm{Cat}'_{\mathfrak{K}}$ be the class of $\mu \in \mathrm{Cat}_{\mathfrak{K}}$ such that: $\mu \geq \beth_{1,1}(\mathrm{LS}(\mathfrak{K}))$ and if $\mathrm{LS}(\mathfrak{K}) \leq \theta$ and $\beth_{1,1}(\theta) \leq \mu$ <u>then</u> $\mathbb{L}_{\infty,\theta^+}[\mathfrak{K}]$ is μ-model complete.
3A) For $\Phi \in \Upsilon^{\mathrm{or}}_{\mathfrak{K}}$ let $\mathrm{Sol}^{k,*}_{\mathfrak{K},\Phi}$ be the class of $\mu \in \mathrm{Sol}^k_{\mathfrak{K},\Phi}$ such that $\mu \geq \beth_{1,1}(\mathrm{LS}(\mathfrak{K}))$ and: if $\mathrm{LS}(\mathfrak{K}) \leq \theta$ and $\beth_{1,1}(\theta) \leq \mu$ then the pair $(\mathbb{L}_{\infty,\theta^+}[\mathfrak{K}],\Phi)$ is μ-model complete.

Let $\mathrm{Sol}^{\ell,<\theta}_{\mathfrak{K},\Phi}$ be the class of $\lambda \in \mathrm{Sol}^\ell_{\mathfrak{K},\Phi}$ such that $\mathbb{L}_{\infty,\theta}[\mathfrak{K}]$ is μ-model complete (see §2).

Let $\mathrm{Sol}'_{\mathfrak{K},\Phi} = \mathrm{Sol}^{1,*}_{\mathfrak{K},\Phi}$. Instead $k,*$ we may write $3+k$.
4) Let $\mathbf{C} = \{\lambda : \lambda = \beth_\lambda \text{ and } \mathrm{cf}(\lambda) = \aleph_0\}$.

<u>3.3 Exercise</u>: 1) The conclusion of 1.12(1) equivalently 1.12(2) means that $\theta \leq \lambda \Rightarrow (\mu,\theta) \in \text{mod-com}_{\mathfrak{K},\Phi}$.
2) Write down the obvious implications.

3.4 Claim. *If* $\mu > \lambda = \beth_\lambda > \kappa \geq \mathrm{LS}(\mathfrak{K})$ *and* $\Phi \in \Upsilon^{\mathrm{or}}_\kappa[\mathfrak{K}]$, $\mathrm{cf}(\lambda) = \aleph_0$ <u>*then*</u> $\mu = \mu^{<\lambda} \Rightarrow \mu \in \mathrm{Sol}'_{\mathfrak{K},\Phi} \Rightarrow \lambda \in \mathrm{Sol}'_{\mathfrak{K},\Phi}$.

Proof. The first implication holds by 1.12(2) and 3.3. The second implication, its assumption implies Hypothesis 1.18, see 3.3(1) hence its conclusion holds by 1.41.

$\square_{3.4}$

3.5 Observation. K_λ is categorical in λ (hence Hypothesis 1.18 holds), <u>if</u>:

$\circledast_\lambda$ $\lambda = \beth_\lambda = \sup(\lambda \cap \mathrm{Cat}'_{\mathfrak{K}}) > \mathrm{LS}(\mathfrak{K})$ and $\aleph_0 = \mathrm{cf}(\lambda)$.

Proof. Fix $\Phi \in \Upsilon^{\rm or}_{\mathfrak{K}}$, now clearly $\mathrm{Sol}'_{\mathfrak{K},\Phi} \supseteq \mathrm{Cat}'_{\mathfrak{K}}$ by their definitions.

By the assumptions we can find $\langle \mu_n : n < \omega \rangle$ such that $\lambda = \Sigma\{\mu_n : n < \omega\}$, $\mathrm{LS}(\mathfrak{K}) < \mu_n \in \mathrm{Cat}'_{\mathfrak{K}}$ and $\beth_{1,1}(\mu'_n) < \mu_{n+1}$ where $\mu'_n = \beth_{1,1}(\mu_n)$. As every $M \in K_{\mu_{n+1}}$ is $\mathbb{L}_{\infty,\mu'_n}[\mathfrak{K}]$-generic (as $K_{\mu_{n+1}} \subseteq K_{\Phi,\mu_{n+1}}$ and $\mu_{n+1} \in \mathrm{Cat}'_{\mathfrak{K}}$) easily

$(*)_0$ if $M \le_{\mathfrak{K}} N$ are from $K^*_{\Phi,\ge\mu_{n+1}}$ then $M \prec_{\mathbb{L}_{\infty,\mu'_n}[\mathfrak{K}]} N$.

Let $M^\ell \in K_\lambda$, for $\ell \in \{1,2\}$; so we can find a $\le_{\mathfrak{K}}$-increasing sequence $\langle M^\ell_n : n < \omega \rangle$ such that $M^\ell_n \in K_{\mu_n}, M^\ell_n \le_{\mathfrak{K}} M^\ell_{n+1} \le_{\mathfrak{K}} M^\ell$ and $M^\ell = \cup\{M^\ell_n : n < \omega\}$. Now

$(*)_1$ $M^\ell_n \in K^*_{\Phi,\mu_n}$.

[Why? As $\mathfrak{K}$ is categorical in $\mu_n = \|M^\ell_n\|$.]

$(*)_2$ if $\alpha \le \mu_n, n < m < k$ and $\bar a, \bar b \in {}^\alpha(M^\ell_m)$ <u>then</u>:

(a) $\mathrm{tp}_{\mathbb{L}_{\infty,\mu'_n}[\mathfrak{K}]}(\bar a, \emptyset, M^\ell_m) = \mathrm{tp}_{\mathbb{L}_{\infty,\mu'_n}}(\bar b, \emptyset, M^\ell_m)$ <u>iff</u> $\mathrm{tp}_{\mathbb{L}_{\infty,\mu'_n}[\mathfrak{K}]}(\bar a, \emptyset, M^\ell_k) = \mathrm{tp}_{\mathbb{L}_{\infty,\mu'_n}}(\bar b, \emptyset, M^\ell_k)$.

(b) if $\mathrm{tp}_{\mathbb{L}_{\infty,\mu'_n}[\mathfrak{K}]}(\bar a, \emptyset, M^\ell_k) = \mathrm{tp}_{\mathbb{L}_{\infty,\mu'_n}[\mathfrak{K}]}(\bar b, \emptyset, M^\ell_k)$ <u>then</u> $\mathrm{tp}_{\mathbb{L}_{\infty,\mu'_m}[\mathfrak{K}]}(\bar a, \emptyset, M^\ell_k) = \mathrm{tp}_{\mathbb{L}_{\infty,\mu'_m}[\mathfrak{K}]}(\bar b, \emptyset, M^\ell_k)$.

[Why? Clause (a) by $(*)_0$, clause (b) by 1.19(3).]

$(*)_3$ $M^1_n \cong M^2_n$.

[Why? As $\mathfrak{K}$ is categorical in μ_n.]

We now proceed as in the proof of 1.38.

Let $\mathscr{F}_n = \{f$: for some $\bar a_1, \bar a_2$ and $\alpha < \mu_n$ we have $\bar a_\ell \in {}^\alpha(M^\ell_{n+2})$ for $\ell = 1,2$, $\mathrm{tp}_{\mathbb{L}_{\infty,\mu_{n+1}}[\mathfrak{K}]}(\bar a_1, \emptyset, M^1_{n+2}) = \mathrm{tp}_{\mathbb{L}_{\infty,\mu_{n+1}}[\mathfrak{K}]}(\bar a_2, \emptyset, M^2_{n+1})$ and f is the function which maps $\bar a_1$ into $\bar a_2\}$, (actually can use $\alpha = \mu_n$).

By the hence and forth argument we can find $f_n \in \mathscr{F}_n$ by induction on $n < \omega$ such that $M^1_n \subseteq \mathrm{Dom}(f_{2n+2}), M^2_n \subseteq \mathrm{Rang}(f_{2n+2})$ and $f_n \subseteq f_{n+1}$; hence $\cup\{f_n : n < \omega\}$ is an isomorphism from M^1 onto M^1. $\square_{3.4}$

3.6 Claim. *$\mathfrak{K}$ is categorical in λ when:*

$\circledast^+_\lambda$ $\lambda = \beth_\lambda > \mathrm{LS}(\mathfrak{K})$ *and* $\lambda = \mathrm{otp}(\mathrm{Cat}_{\mathfrak{K}} \cap \lambda \cap \mathbf{C})$ *and* $\mathrm{cf}(\lambda) = \aleph_0$.

Proof. Fix Φ as in the proof of 3.4. Let $\langle \theta_n : n < \omega \rangle$ be increasing such that $\lambda = \Sigma\{\theta_n : n < \omega\}$ and $\mathrm{LS}(\mathfrak{K}) < \theta_0$. For each n, by 2.14 we know $\{\mu \in \mathrm{Cat}_{\mathfrak{K}} : \mu > \theta_n$ and the $M \in K_\mu$ is not $\mathbb{L}_{\infty,\theta_n^+}$-generic$\}$ is "not too large", i.e. is included in the union of at most $\beth_2(\theta_n)$ intervals of the form $[\chi, \chi^{\theta_n}]$. Now we choose $(n(\ell), \mu_\ell)$ by induction on $\ell < \omega$ such that

$\circledast$ (a) $n(\ell) < \omega$ and $\mu_\ell \in \mathrm{Cat}_{\mathfrak{K}} \cap \lambda$

(b) if $\ell = k+1$ then $n(\ell) > n(k), \theta_{n(\ell)} > \mu_k, \mu_\ell \in \mathrm{Cat}_{\mathfrak{K}} \cap \lambda \backslash \theta^+_{n(\ell)}$ and the $M \in K_{\mu_\ell}$ is $\mathbb{L}_{\infty,\theta_{n(\ell)}}[\mathfrak{K}]$-generic (hence $\mathbb{L}_{\infty,\mu_k^+}[\mathfrak{K}]$-generic).

This is easy and then continue as in 3.5. $\square_{3.6}$

We have essentially proved

3.7 Theorem. *In 3.5, 3.6 we can use* $\mathrm{Sol}_{\mathfrak{K},\Phi}$, $\mathrm{Sol}'_{\mathfrak{K},\Phi}$ *instead of* $\mathrm{Cat}_{\mathfrak{K}}$, $\mathrm{Cat}'_{\mathfrak{K}}$.

<u>3.8 Exercise</u>: For Claim 1.38(2), Hypothesis 1.18 suffice.
[Hint: The proof is similar to the existing one using 1.19.]

§4 Good Frames

Here comes the main result of Chapter : from categoricity (or solvability) assumptions we derive the existence of good λ-frames.

Our assumption is such that we can apply §1.

4.1 Hypothesis. 1)

(*a*) $\mathfrak{K}$ is an a.e.c.

(*b*) $\mu > \lambda = \beth_\lambda >$ $\mathrm{LS}(\mathfrak{K})$ and $\mathrm{cf}(\lambda) = \aleph_0$;

(*c*) $\Phi \in \Upsilon^{\mathrm{or}}_{\mathfrak{K}}$

(d) $\mathfrak{K}$ is categorical in μ or just

$(d)^-$ $(\mathfrak{K}, \Phi)$ is pseudo superlimit in μ (this means $\Phi \in \mathrm{Sol}^1_{\mathfrak{K},\Phi}$; so 1.18(1) holds)

(e) also 1.18(2)(a) holds, i.e. the conclusion of 1.12(2) holds.

2) In addition we may use some of the following but then we mention them and (we add superscript $*$ when used; note that $(g) \Rightarrow (f)$ by 1.39)

(f) K^*_λ is closed under $\leq_{\mathfrak{K}}$-increasing unions (justified by 1.38)

(g) $\langle \lambda_n : n < \omega \rangle$ is increasing, $\lambda_0 > \mathrm{LS}(\mathfrak{K})$, $\lambda = \Sigma\{\lambda_n : n < \omega\}$ and the assumptions of 1.38 holds.

4.2 Observation. 1) $\mathfrak{K}^*_\lambda$ is categorical.
2) $\mathfrak{K}^*_\lambda$ has amalgamation.
3)* (We assume (f) of 4.1(2)). $\mathfrak{K}_\lambda$ is a λ-a.e.c.

Proof. 1) By 1.16(1) or 1.19(4) as $\mathrm{cf}(\lambda) = \aleph_0$.
2) By 1.30(1).
3) As in 1.39, (i.e. as $\leq_{\mathfrak{K}^*_\lambda} = \leq_{\mathfrak{K}} \restriction \mathfrak{K}$, closure under unions of $\leq_{\mathfrak{K}}$-increasing chains is the only problematic point and it holds by (f) of 4.1(2)). $\square_{4.2}$

4.3 Remark. 1) Why do we not assume 4.1(1),(2) all the time? The main reason is that for proving some of the results assuming 4.1(1),(2) we use some such results on smaller cardinals on which we use 4.1(1) only.
2) Note that it is not clear whether improvement by using 4.1(1) only will have any affect when (or should we say if) we succeed to have the parallel of III§12.

4.4 Claim. *1) Assume $M_0 \leq_{\mathfrak{K}^*_\lambda} M_\ell, \alpha < \lambda$ and $\bar{a}_\ell \in {}^\alpha(M_\ell)$ for $\ell = 1, 2$ and $\kappa := \beth_{1,1}(\beth_2(\theta)^+)$ where $\theta := |\alpha| + \mathrm{LS}(\mathfrak{K})$ so $\kappa < \lambda$. If $\mathrm{tp}_{\mathbb{L}_{\infty,\kappa}[\mathfrak{K}]}(\bar{a}_1, M_0, M_1) = \mathrm{tp}_{\mathbb{L}_{\infty,\kappa}[\mathfrak{K}]}(\bar{a}_2, M_0, M_2)$ then $\mathbf{tp}_{\mathfrak{K}^*_\lambda}(\bar{a}_1, M_0, M_1) = \mathbf{tp}_{\mathfrak{K}^*_\lambda}(\bar{a}_2, M_0, M_2)$.*

2) If $M_1 \leq_{\mathfrak{K}^*_\lambda} M_2$ <u>then</u> $M_1 \prec_{\mathbb{L}_{\infty,\theta}[\mathfrak{K}]} M_2$ for every $\theta < \lambda$, and moreover $M_1 \prec_{\mathbb{L}_{\infty,\lambda}[\mathfrak{K}]} M_2$.
2A) If $M_0 \leq_{\mathfrak{K}^*_\lambda} M_\ell$ for $\ell = 1,2$ and $\mathbf{tp}_{\mathfrak{K}^*_\lambda}(\bar{a}_1, M_0, M_1) = \mathbf{tp}_{\mathfrak{K}^*_\lambda}(\bar{a}_2, M_0, M_2)$ and $\bar{a}_\ell \in {}^\alpha(M_0), \alpha < \kappa \leq \lambda$ <u>then</u> $\text{tp}_{\mathbb{L}_{\infty,\kappa}[\mathfrak{K}]}(\bar{a}_1), M_0, M_1) = \text{tp}_{\mathbb{L}_{\infty,\kappa}[\mathfrak{K}]}(\bar{a}_2, M_0, M_2)$.
2B) In part (1), if $M_\ell \leq_{\mathfrak{K}^*_\lambda} M'_\ell$ for $\ell = 1,2$ <u>then</u> $\text{tp}_{\mathbb{L}_{\infty,\kappa}[\mathfrak{K}]}(\bar{a}_1, M, M'_1) = \text{tp}_{\mathbb{L}_{\infty,\kappa}[\mathfrak{K}]}(\bar{a}_2, M, M'_2)$.
3) Assume that $M_0 \leq_{\mathfrak{K}^*_\lambda} M_1 \leq_{\mathfrak{K}^*_\lambda} M_2 \leq_{\mathfrak{K}^*_\lambda} M_3, \bar{a} \in {}^\alpha(M_2), \alpha < \lambda$ and $\kappa = \beth_{1,1}(|\alpha| + \text{LS}(\mathfrak{K})) < \theta < \lambda$. <u>Then</u>

- (a) *from $\text{tp}_{\mathbb{L}_{\infty,\kappa}[\mathfrak{K}]}(\bar{a}, M_1, M_2)$ we can compute $\text{tp}_{\mathbb{L}_{\infty,\theta}[\mathfrak{K}]}(\bar{a}, M_1, M_2)$ and $\text{tp}_{\mathbb{L}_{\infty,\lambda}[\mathfrak{K}]}(\bar{a}, M_0, M_3)$*
- (b) *from $\text{tp}_{\mathbb{L}_{\infty,\kappa}[\mathfrak{K}]}(\bar{a}, \emptyset, M_2)$ we can compute $\text{tp}_{\mathbb{L}_{\infty,\theta}[\mathfrak{K}]}(\bar{a}, \emptyset, M_2)$ and even $\text{tp}_{\mathbb{L}_{\infty,\lambda}[\mathfrak{K}]}(\bar{a}, \emptyset, M_2)$*
- (c) *from $\mathbf{tp}_{\mathfrak{K}^*_\lambda}(\bar{a}, M_1, M_2)$ we can compute $\text{tp}_{\mathbb{L}_{\infty,\lambda}[\mathfrak{K}]}(\bar{a}, M_1, M_2)$ and $\mathbf{tp}_{\mathfrak{K}^*_\lambda}(\bar{a}, M_0, M_3)$.*

4) If $M_1 \leq_{\mathfrak{K}^*_\lambda} M_2$ and $\alpha < \kappa^* < \lambda, \mathbf{I}_\ell \subseteq {}^\alpha(M_1), |\mathbf{I}_\ell| > \kappa, \mathbf{I}_\ell$ is $(\mathbb{L}_{\infty,\theta}[\mathfrak{K}], \kappa^*)$- convergent in M_1 for $\ell = 1,2$ and $\text{Av}_{<\kappa}(\mathbf{I}_1, M_1) = \text{Av}_{<\kappa}(\mathbf{I}_1, M_1)$ <u>then</u> $\mathbf{I}_\ell$ is $(\mathbb{L}_{\infty,\kappa}[\mathfrak{K}], \kappa^*)$-convergent in M_ℓ for $\ell = 1,2$ and $\text{Av}_{<\kappa}(\mathbf{I}_1, M_\ell) = \text{Av}_{<\kappa}(\mathbf{I}_1, M_2)$.

Proof. 1) Without loss of generality $M_0 = \text{EM}_{\tau(\mathfrak{K})}(I_0, \Phi)$ and $I_0 \in K^{\text{flin}}_\lambda$. By 1.29(3) for $\ell = 1,2$ there is a pair (I_ℓ, f_ℓ) such that $I_0 \leq_{K^{\text{flin}}} I_\ell \in K^{\text{flin}}_\lambda$ and f_ℓ is a $\leq_{\mathfrak{K}}$-embedding of M_ℓ into $M'_\ell = \text{EM}_{\tau(\mathfrak{K})}(I_\ell, \Phi)$ over M_0. By renaming without loss of generality f_ℓ is the identity on M_ℓ hence $M_\ell \leq_{\mathfrak{K}} M'_\ell$. By 1.19(1) we know that $M_\ell \prec_{\mathbb{L}_{\infty,\kappa}[\mathfrak{K}]} M'_\ell$ hence $\text{tp}_{\mathbb{L}_{\infty,\kappa}[\mathfrak{K}]}(\bar{a}_1, M_0, M'_1) = \text{tp}_{\mathbb{L}_{\infty,\kappa}[\mathfrak{K}]}(\bar{a}_1, M_0, M_1) = \text{tp}_{\mathbb{L}_{\infty,\kappa}[\mathfrak{K}]}(\bar{a}_2, M_0, M_2) = \text{tp}_{\mathbb{L}_{\infty,\kappa}[\mathfrak{K}]}(\bar{a}_2, M_0, M'_2)$.

By 1.29(1) we can find (I_3, g_1, g_2, h) such that $I_0 \leq_{K^{\text{flin}}} I_3 \in K^{\text{flin}}_\lambda, g_\ell$ is a $\leq_{\mathfrak{K}}$-embedding of M'_ℓ into $M_4 := \text{EM}_{\tau(\mathfrak{K})}(I_3, \Phi)$ over M_0 for $\ell = 1,2$ and h is an automorphism of M_4 over M_0 mapping $g_1(\bar{a}_1)$ to $g_2(\bar{a}_2)$. By the definition of orbital types, this gives $\mathbf{tp}_{\mathfrak{K}^*_\lambda}(\bar{a}_1, M_0, M_1) = \mathbf{tp}_{\mathfrak{K}^*_\lambda}(\bar{a}_2, M_0, M_2)$ as required.
2) This holds by 1.19(1) for $\theta \in (\text{LS}(\mathfrak{K}), \lambda)$, hence by 1.11(1) also for $\theta = \lambda$ (the assumptions of 1.11 hold as clause (a) there holds by the case above $\theta < \lambda$ and clause (b) there holds by 1.28(1)).
2A) Should be clear:

(a) by part (2) this holds if $\bar{a}_1 = \bar{a}_2$ and $M_1 \leq_{\mathfrak{K}} M_2$

(b) trivially it holds if there is an isomorphism from M_1 onto M_2 over M_0 mapping $\bar{a}_1$ to $\bar{a}_2$

(c) by the definition of **tp** we are done.

2B) Should be clear by part (2).

3) Clause (a):
By parts (1) + (2).

Clause (b): By 1.28(1).

Clause (c): By part (2A) and the definition of **tp**.

4) Easy, too. $\square_{4.4}$

4.5 Definition. Assume $M_0 \leq_{\mathfrak{K}^*_\lambda} M_1 \leq_{\mathfrak{K}^*_\lambda} M_2, \alpha < \lambda$ and $\bar{a} \in {}^\alpha(M_2)$ and $p = \mathbf{tp}_{\mathfrak{K}^*_\lambda}(\bar{a}, M_1, M_2)$. We say that p does not fork over M_0 (for $\mathfrak{K}^*_\lambda$) when, letting $\theta_0 = |\alpha| + \text{LS}(\mathfrak{K}), \theta_1 = \beth_{1,1}(\beth_2(\theta_0)^+), \theta_2 = 2^{\theta_1}, \theta_2 = \beth_2(\theta_1)$ we have:

(∗) for some $N \leq_{\mathfrak{K}^*} M_0$ satisfying $\|N\| \leq \theta_2$ we have $\text{tp}_{\mathbb{L}_{\infty,\theta_1}[\mathfrak{K}]}(\bar{a}, M_1, M_2)$ does not split over N.

We now would like to show that there is $\mathfrak{s}_\lambda$ which fits Chapter II and Chapter III and $\mathfrak{K}_{\mathfrak{s}_\lambda} = \mathfrak{K}^*_\lambda$.

4.6 Observation. Assume that $M_0 \leq_{\mathfrak{K}^*_\lambda} M_1 \leq_{\mathfrak{K}^*_\lambda} M_2, \bar{a} \in {}^\alpha(M_2), \alpha < \lambda, \lambda > \kappa_0 \geq |\alpha| + \text{LS}(\mathfrak{K}), \kappa_1 = \beth_{1,1}(\beth_2(\kappa_0)^+)$ and $\kappa_2 = \beth_2(\kappa_1)$. Then the following conditions are equivalent

(a) $\mathbf{tp}_{\mathfrak{K}^*_\lambda}(\bar{a}, M_1, M_2)$ does not fork over M_0

(b) for some (κ_1^+, κ_1)-convergent $\mathbf{I} \subseteq {}^\alpha(M_0)$ of cardinality $> \kappa_2$ we have
$\text{tp}_{\mathbb{L}_{\infty,\kappa_1}[\mathfrak{K}]}(\bar{a}, M_1, M_2) = \text{Av}_{<\kappa_1}(\mathbf{I}, M_1)$ hence this type does not split over $\cup\mathbf{I}'$ for any $\mathbf{I}' \subseteq \mathbf{I}$ of cardinality $> \kappa_1$

(c) for every $N \leq_{\mathfrak{K}} M_0$ of cardinality $\leq \kappa_2$, if $\text{tp}_{\mathbb{L}_{\infty,\kappa_1}[\mathfrak{K}]}(\bar{a}, M_0, M_2)$ does not split over N then the type $\text{tp}_{\mathbb{L}_{\infty,\kappa_1}[\mathfrak{K}]}(\bar{a}, M_1, M_2)$ does not split over N.

4.7 Remark. 1) See verification of axiom (E)(c) in the proof of Theorem 4.10.
2) Note that have we used $\beth_7(\kappa_1)^+$ instead of κ_1 in 4.5, 4.6, the difference would be small.
3) We could in clause (c) of 4.6 use "for some $N \leq_{\mathfrak{K}} M_0$ of cardinality $< \kappa_1, \mathrm{tp}_{\mathbb{L}_{\infty,\kappa_1}[\mathfrak{K}]}$ ..." The proof is the same.
4) We can allow below $M_0 \leq_{\mathfrak{K}} M_1$ if $M_0 \in K_{\geq \kappa_2}$.

Proof. (a) ⇒ (b)

Let $\theta_0, \theta_1, \theta_2$ be as in Definition 4.5. By Definition 4.5 there is $N \leq_{\mathfrak{K}} M_0$ of cardinality $\leq \theta_2$ such that

$(*)_1$ the type $\mathrm{tp}_{\mathbb{L}_{\infty,\theta_1}[\mathfrak{K}]}(\bar{a}, M_1, M_2)$ does not split over N.

By Claim 1.26(1) there is a (κ_1^+, κ_1)-convergent set $\mathbf{I} \subseteq {}^{\alpha}(M_0)$ of cardinality κ_2^+ (convergence in M_0, of course) such that $\mathrm{tp}_{\mathbb{L}_{\infty,\kappa_1}[\mathfrak{K}]}(\bar{a}, M_0, M_2) = \mathrm{Av}_{<\kappa_1}(\mathbf{I}, M_0)$. So as $M_0 \prec_{\mathbb{L}_{\infty,\lambda}[\mathfrak{K}]} M_1 \prec_{\mathbb{L}_{\infty,\lambda}[\mathfrak{K}]} M_2$, by Claim 4.4(2), clearly $\mathbf{I}$ is (κ_1^+, κ_1)-convergent also in M_1 and in M_2 hence $\mathrm{Av}_{<\kappa_1}(\mathbf{I}, M_1)$ is well defined. Hence, by Claims 1.23(2), 1.21(3) the type $\mathrm{Av}_{<\kappa_1}(\mathbf{I}, M_1)$ does not split over $\cup\mathbf{I}$ but $\theta_2 \leq \kappa_2$ and $\cup\mathbf{I} \subseteq \cup\mathbf{I} \cup N$ hence

$(*)_2$ $\mathrm{Av}_{<\theta_1}(\mathbf{I}, M_1)$ does not split over $\cup\mathbf{I} \cup N$.

But also

$(*)_3$ $\mathrm{tp}_{\mathbb{L}_{\infty,\theta_1}[\mathfrak{K}]}(\bar{a}, M_1, M_2)$ does not split over N (by the choice of N) hence over $\cup\mathbf{I} \cup N$.

As $M_0 \prec_{\mathbb{L}_{\infty,\lambda}[\mathfrak{K}]} M_1$ and $|\cup\mathbf{I} \cup N| < \lambda$ and $\mathrm{tp}_{\mathbb{L}_{\infty,\theta_1}[\mathfrak{K}]}(\bar{a}, M_0, M_2) = \mathrm{Av}_{<\theta_1}(\mathbf{I}, M_0)$ clearly, by $(*)_2 + (*)_3$ we have $\mathrm{tp}_{\mathbb{L}_{\infty,\theta_1}[\mathfrak{K}]}(\bar{a}, M_1, M_2) = \mathrm{Av}_{<\theta_1}(\mathbf{I}, M_1)$.
Now there is a pair $(M_2', \bar{a}')$ satisfying that $M_1 \leq_{\mathfrak{K}} M_2' \in K_\lambda^*$ and $\bar{a}' \in {}^{\alpha}(M_2')$ such that $\mathrm{tp}_{\mathbb{L}_{\infty,\theta_1}[\mathfrak{K}]}(\bar{a}', M_1, M_2') = \mathrm{Av}_{<\theta_1}(\mathbf{I}, M_1)$ hence by the previous sentence $\mathrm{tp}_{\mathbb{L}_{\infty,\theta_1}[\mathfrak{K}]}(\bar{a}', M_1, M_2') = \mathrm{tp}_{\mathbb{L}_{\infty,\theta_1}[\mathfrak{K}]}(\bar{a}, M_1, M_2)$. Now by 4.4(1) and then 4.4(2A) it follows that $\mathrm{tp}_{\mathbb{L}_{\infty,\kappa_1}[\mathfrak{K}]}(\bar{a}, M_1, M_0) = \mathrm{Av}_{<\kappa_1}(\mathbf{I}, M_1)$ as required.

(b) ⇒ (c)

Let $\mathbf{I}$ be as in clause (b), so $\mathbf{I}$ is (κ_1^+, κ_1)-convergence in M_0 and is of cardinality $> \kappa_1$. We know that $M_0 \prec_{\mathbb{L}_{\infty,\lambda}[\mathfrak{K}]} M_1$, so by the previous sentence, $\mathbf{I}$ is (κ_1^+, κ_1)-convergent in M_1. To prove clause (c) assume that $N \leq_{\mathfrak{K}} M_0$ is of cardinality κ_2 and $\mathrm{tp}_{\mathbb{L}_{\infty,\kappa_1}[\mathfrak{K}]}(\bar{a}, M_0, M_2)$ does not split over N. Hence $\mathrm{Av}_{<\kappa_1}(\mathbf{I}, M_0) = \mathrm{tp}_{\mathbb{L}_{\infty,\kappa_1}[\mathfrak{K}]}(\bar{a}, M_0, M_2)$ does not split over N. Again as $M_0 \prec_{\mathbb{L}_{\infty,\lambda}[\mathfrak{K}]} M_1$ we can deduce that $\mathrm{Av}_{<\kappa_1}(\mathbf{I}, M_1)$ does not split over N but by the choice of $\mathbf{I}$ it is equal to $\mathrm{tp}_{\mathbb{L}_{\infty,\kappa_1}[\mathfrak{K}]}(\bar{a}, M_1, M_2)$, so we are done.

(c) ⇒ (a)

By Claim 1.24 there is $B \subseteq M_0$ of cardinality $\leq \kappa_2$ such that $\mathrm{tp}_{\mathbb{L}_{\infty,\kappa_1}[\mathfrak{K}]}(\bar{a}, M_0, M_2)$ does not split over B.

As we can increase B as long as we preserve "of cardinality $\leq \kappa_2$", without loss of generality $B = |N|$ where $N \leq_{\mathfrak{K}} M_0$. So the antecedent of clause (c) holds, but we are assuming clause (c) so the conclusion of clause (c) holds, that is $\mathrm{tp}_{\mathbb{L}_{\infty,\kappa_1}[\mathfrak{K}]}(\bar{a}, M_1, M_2)$ does not split over N.

Also by 1.26(1) there is $\mathbf{I}_1 \subseteq {}^{\alpha}(M_0)$ of cardinality κ_2^+ which is (κ_1^+, κ_1)-convergent and $\mathrm{Av}_{<\kappa_1}(\mathbf{I}_1, M_0) = \mathrm{tp}_{\mathbb{L}_{\infty,\kappa_1}[\mathfrak{K}]}(\bar{a}, M_0, M_1)$. Clearly $\kappa_1 \geq \theta_1$ hence $\kappa_2 = (\kappa_2)^{\theta_1}$. Now as K_λ^* is categorical clearly $M_0 \cong \mathrm{EM}_{\tau(\mathfrak{K})}(\lambda, \Phi)$ hence applying 1.25(4) we can find $\mathbf{I}_2 \subseteq \mathbf{I}_1$ of cardinality κ_2^+ which is (θ_1^+, θ_1)-convergent. As above $M_0 \prec_{\mathbb{L}_{\infty,\kappa_1}[\mathfrak{K}]} M_1$ so we deduce that $\mathbf{I}_2$ is (θ_1^+, θ_1)-convergent and (κ_1^+, κ_1)-convergent also in M_1.

As above we have $M_0 \prec_{\mathbb{L}_{\infty,\kappa_1}[\mathfrak{K}]} M_1$ by 1.19(1) hence $\mathrm{Av}_{<\kappa_1}(\mathbf{I}_2, M_1)$ is well defined and does not split over N hence is equal to $\mathrm{tp}_{\mathbb{L}_{\infty,\kappa_1}[\mathfrak{K}]}(\bar{a}, M_1, M_2)$. This implies that $\mathrm{Av}_{<\theta_1}(\mathbf{I}_2, M_1) = \mathrm{tp}_{\mathbb{L}_{\infty,\theta_1}[\mathfrak{K}]}(\bar{a}, M_1, M_2)$.

Now choose $\mathbf{I}_3 \subseteq \mathbf{I}_2 \subseteq M_0$ of cardinality θ_2 and $N_3 \leq_{\mathfrak{K}} M_0$ of cardinality θ_2 such that $\mathbf{I}_3 \subseteq {}^{\alpha}(N_3)$. Now by 1.23(2) we know that $\mathrm{tp}_{\mathbb{L}_{\infty,\theta_1}[\mathfrak{K}]}(\bar{a}, M_1, M_2)$ does not split over $\mathbf{I}_3$ hence it does not split over N_3, so N_3 witnesses clause (a). $\square_{4.6}$

4.8 Definition. We define a pre-frame $\mathfrak{s}_\lambda = (\mathfrak{K}_{\mathfrak{s}_\lambda}, \bigcup_{\mathfrak{s}_\lambda}, \mathscr{S}^{\mathrm{bs}}_{\mathfrak{s}_\lambda})$ as follows:

(a) $\mathfrak{K}_{\mathfrak{s}_\lambda} = \mathfrak{K}^*_\lambda$

(b) $\mathscr{S}^{\rm bs}_{\mathfrak{s}_\lambda}$ is defined by $\mathscr{S}^{\rm bs}_{\mathfrak{s},\lambda}(M) := \{\mathbf{tp}_{\mathfrak{K}^*_\lambda}(a,M,N) : M \le_{\mathfrak{K}^*_\lambda} N, a \in N \backslash M\}$,

(c) $\underset{\mathfrak{s}_\lambda}{\bigcup} = \{(M_0, M_1, a, M_3) : M_0 \le_{\mathfrak{K}^*_\lambda} M_1 \le_{\mathfrak{K}^*_\lambda} M_2$ and $\mathbf{tp}_{\mathfrak{K}^*_\lambda}(a, M_1, M_3)$ does not fork over $M_0\}$, see Definition 4.5.

4.9 Remark. 1) Recall $\le_{\mathfrak{s}_\lambda} = \le_{\mathfrak{K}} \restriction K_{\mathfrak{s}_\lambda} = \le_{\mathfrak{K}^*_\lambda}$.
2) Concerning the proof of 4.10 below we mention a variant which the reader may ignore. This variant, from weaker assumptions gets weaker conclusions. In detail, define the weak versions $(f)^-$ of (f) of 4.1(2); see Definition 1.34 and Claim 1.37(1)

$(f)^-$ if $\langle M_\alpha : \alpha \le \delta \rangle$ is $\le_{\mathfrak{K}}$-increasing continuous and $\alpha < \delta \Rightarrow M_{2\alpha+1} <^*_{\mathfrak{K}^*_\lambda} M_{2\alpha+2}$ (e.g. $M_{2\alpha+2}$ is $\le_{\mathfrak{K}^*_\lambda}$-universal over $M_{2\alpha+1}$) hence both are from K^*_λ then $M_\delta \in K^*_\lambda$.

Assuming only 4.1(1) + (f)$^-$ we do not know whether $\mathfrak{K}^*_\lambda$ is a λ-a.e.c. but still $(K^*_\lambda, \le_{\mathfrak{K}} \restriction K^*_\lambda, <^*_{\mathfrak{K}^*_\lambda})$, see Definition 1.34, is a so called semi λ-a.e.c., see Chapter N.

If clause (f) from 4.1(2) holds (i.e., $K_{\mathfrak{s}_\lambda}$ is closed under unions), we can omit "$<^*_{\mathfrak{s}_\lambda}$".
3) It will be less good but not a disaster if we have assumed below $\lambda = \sup(\mathrm{Cat}'_{\mathfrak{K}} \cap \lambda)$.
4) It will be better to have $\mathfrak{K}_{\mathfrak{s}_\lambda} = K_\lambda$; of courses, this follows from categoricity so by §3 is not unreasonable for conjecture 0.1.
5) But we can ask only for $M \in K_{\mathfrak{s}_\lambda}$ to be universal in $\mathfrak{K}_\lambda$,
6) We can ask that for every $\mu > \lambda$ large enough, for every $M \in K_\mu$ for a club of $N \in K_\lambda$ satisfying $N \le_{\mathfrak{K}} M$ we have $N \in K_{\mathfrak{s}_\lambda}$.

4.10 Theorem*. *(Assume 4.1(2),(g) hence (f)).*
$\mathfrak{s}_\lambda$ is a good λ-frame categorical in λ and is full.

Proof. We check the clauses in the definition II.2.1.

Clause (A):
By observation 4.2(3), [in the weak version using (f)$^-$ from 4.9(1)].

Clause (B):

Categoricity holds by 1.16 (or 4.2(1)) and this implies "there is a superlimit model", the non-maximality by $\leq_{\mathfrak{K}^*_\lambda}$ holds by the choice of Φ.

Clause (C):

Observation 4.2(2) guarantee amalgamation, categoricity (of $\mathfrak{K}^*_\lambda$ by 4.2(1)) implies the JEP and "no-maximal model" holds by clause (B).

Clause $(D)(a), (b)$:

Obvious by the definition.

(D) (c) (density).

Assume $M <_{\mathfrak{K}^*_\lambda} N$, then there are $a \in N \backslash M$ and for any such a the type $\mathbf{tp}_{\mathfrak{K}^*_\lambda}(a, M, N)$ belongs to $\mathscr{S}^{\mathrm{bs}}_{\mathfrak{s}_\lambda}(M)$. In fact

⊛ $\mathfrak{s}_\lambda$ is type-full

(D) (d) (bs-stability).

The demand means $M \in K^*_\lambda \Rightarrow |\mathscr{S}^1_{\mathfrak{K}^*_\lambda}(M)| \leq \lambda$.

This holds by 1.32(2) (and amalgamation).

$(E)(a), (b)$. By the definition.

$(E)(c)$ (local character)

This says that if $\langle M_i : i \leq \delta + 1 \rangle$ is $\leq_{\mathfrak{s}_\lambda}$-increasing continuous and $p = \mathbf{tp}_{\mathfrak{s}_\lambda}(a, M_\delta, M_{\delta+1}) \in \mathscr{S}^{\mathrm{bs}}_{\mathfrak{s}_\lambda}(M_\delta)$ then for some $i < \delta$ the type p does not fork over M_i (for $\mathfrak{s}_\lambda$).

From now on (in the proof of 4.10) we use 4.6 freely and let (noting $\mathrm{cf}(\delta) < \lambda$ as λ is singular)

$$\odot \ \kappa_0 = \ \mathrm{LS}(\mathfrak{K}) + \ \mathrm{cf}(\delta), \kappa_1 = \beth_{1,1}(\beth_2(\kappa_0))^+, \kappa_2 = \beth_2(\kappa_1).$$

Now by 4.6 there is a (κ_1^+, κ_1)-convergent $\mathbf{I} \subseteq M_\delta$ with $\mathrm{Av}_{<\kappa_1}(\mathbf{I}, M_\delta) = \mathrm{tp}_{\mathbb{L}_{\infty,\kappa_1}[\mathfrak{K}]}(a, M_\delta, M_{\delta+1})$ such that $\mathbf{I}$ is of cardinality $> \kappa_2$. For some $i(*) < \delta, |\mathbf{I} \cap M_{i(*)}| > \kappa_2$, so without loss of generality $\mathbf{I} \subseteq M_{i(*)}$, so by 4.6 we are done.

$(E)(d)$ Transitivity of non-forking

We are given $M_0 \leq_{\mathfrak{s}_\lambda} M_1 \leq_{\mathfrak{s}_\lambda} M_2 \leq_{\mathfrak{K}_\mathfrak{s}} M_3$ and $a \in M_3$ such that $\mathbf{tp}_{\mathfrak{s}_\lambda}(a, M_{\ell+1}, M_3)$ does not fork over M_ℓ for $\ell = 0, 1$. So for

$\ell = 0, 1$ there is $\mathbf{I}_\ell \subseteq M_\ell$ which is (κ_1^+, κ_1)-convergent in $M_{\ell+1}$ of cardinality κ_2^+ such that $\mathrm{Av}_{<\kappa_1}(\mathbf{I}_\ell, M_{\ell+1}) = \mathrm{tp}_{\mathbb{L}_{\infty,\kappa_1}[\mathfrak{K}]}(a, M_{\ell+1}, M_3)$. As $\mathrm{Av}_{<\kappa_1}(\mathbf{I}_0, M_1) = \mathrm{Av}_{<\kappa_1}(\mathbf{I}_1, M_1)$ (being both realized by a) because $M_1 \prec_{\mathbb{L}_{\infty,\lambda}[\mathfrak{K}]} M_2$ by 4.4(4) clearly we have $\mathrm{Av}_{<\kappa_1}(\mathbf{I}_0, M_2) = \mathrm{Av}_{<\kappa_1}(\mathbf{I}_1, M_2) = \mathrm{tp}_{\mathbb{L}_{\infty,\kappa_1}[\mathfrak{K}]}(a, M_2, M_3)$ all well defined. So $\mathbf{I}_0$ witness by 4.6 that $\mathrm{tp}_{\mathbb{L}_{\infty,\kappa_1}[\mathfrak{K}]}(a, M_2, M_3)$ does not fork over M_0, which means that $\mathbf{tp}_{\mathfrak{K}^*_\lambda}(a, M_2, M_3)$ does not fork over M_0 as required.

$(E)(e)$ Uniqueness.

Recalling 4.4(1), the proof is similar to $(E)(d)$; the two witnesses are now in M_0.

$(E)(f)$ Symmetry

Towards a contradiction, recalling II.2.19 assume $M_0 \leq_{\mathfrak{K}^*_\lambda} M_1 \leq_{\mathfrak{K}^*_\lambda} M_2 \leq_{\mathfrak{K}^*_\lambda} M_3$ and $a_\ell \in M_{\ell+1} \backslash M_\ell$ for $\ell = 0, 1, 2$ are such that $p_\ell = \mathbf{tp}_{\mathfrak{K}^*_\lambda}(a_\ell, M_\ell, M_{\ell+1})$ does not fork over M_0 for $\ell = 0, 1, 2$ and $\mathbf{tp}_{\mathfrak{K}^*_\lambda}(a_0, M_0, M_1) = \mathbf{tp}_{\mathfrak{K}^*_\lambda}(a_2, M_0, M_3)$ but $\mathbf{tp}_{\mathfrak{K}^*_\lambda}(\langle a_0, a_1\rangle, M_0, M_3) \neq \mathbf{tp}_{\mathfrak{K}^*_\lambda}(\langle a_2, a_1\rangle, M_0, M_3)$.

By 4.6 we can deal with $p_\ell = \mathrm{tp}_{\mathbb{L}_{\infty,\kappa_1}[\mathfrak{K}]}(a_\ell, M_\ell, M_{\ell+1})$ for $\ell = 0, 1, 2$. For each $\ell \leq 2$, we can find convergent $\mathbf{I}_\ell = \{a^\ell_\alpha : \alpha < \kappa_2^+\} \subseteq M_0$ which is (κ_1^+, κ_1)-convergent such that $\mathrm{Av}_{<\kappa_1}(\mathbf{I}_\ell, M_\ell) = p_\ell$.

So as $M_0 \prec_{\mathbb{L}_{\infty,\kappa_1}[\mathfrak{K}]} M_k$ we deduce the set $\mathbf{I}_\ell$ is (κ_1^+, κ_1)-convergent in M_k for $\ell, k = 0, 1, 2$, also $\mathrm{Av}_{<\kappa_1}(\mathbf{I}_0, M_0) = \mathrm{Av}_{<\kappa_1}(\mathbf{I}_2, M_0)$ hence $\mathrm{Av}_{<\kappa_1}(\mathbf{I}_0, M_2) = \mathrm{Av}_{<\kappa_1}(\mathbf{I}_2, M_2)$ so without loss of generality $\mathbf{I}_0 = \mathbf{I}_2$.

Now use the non-order property to get symmetry.

$(E)(g)$ Existence

So assume $M \leq_{\mathfrak{s}_\lambda} N$ and $p \in \mathscr{S}^{\mathrm{bs}}_{\mathfrak{s}_\lambda}(M)$. So we can find a pair (M', a) such that $M \leq_{\mathfrak{s}_\lambda} M', a \in M_1$ and $p = \mathbf{tp}_{\mathfrak{s}_\lambda}(a, M, M')$. By 1.26(1) there is a (κ_1^+, κ_1)- convergent $\mathbf{I} \subseteq M$ of cardinality κ_2^+ such that $\mathrm{Av}_{<\kappa_1}(M, \mathbf{I}) = \mathrm{tp}_{\mathbb{L}_{\infty,\kappa_1}[\mathfrak{K}]}(a, M, M')$. By 1.26(3) + 4.6 there is a pair (N', a') such that $N \leq_{\mathfrak{s}_\lambda} N', a' \in N'$ and $\mathrm{tp}_{\mathbb{L}_{\infty,\kappa_1}}(a', N, N') = \mathrm{Av}_{<\kappa_1}(\mathbf{I}, N)$. So by 4.6 the type $\mathbf{tp}_{\mathfrak{s}_\lambda}(a', N, N')$ easily $\in \mathscr{S}^{\mathrm{bs}}_{\mathfrak{s}_\lambda}(N)$, does not fork over N and extend p, as required.

$(E)(h)$ Continuity

Follow by II.2.17. Alternatively assume $\langle M_i : i \leq \delta + 1\rangle$ is $\leq_{\mathfrak{s}_\lambda}$-increasing continuous, and $a \in M_{\delta+1} \backslash M_\delta$ and $\mathbf{tp}_{\mathfrak{s}_\lambda}(a, M_i, M_{\delta+1})$

does not fork over M_0 for $i < \delta$. So there is a convergent $\mathbf{I}_i \subseteq M_0$ such that $i < \delta \Rightarrow \mathrm{tp}_{\mathbb{L}_{\infty,\kappa}[\mathfrak{K}]}(a, M_i, M_{\delta+1}) = \mathrm{Av}_\kappa(\mathbf{I}, M_i)$.
As above, without loss of generality $\mathbf{I}_i = \mathbf{I}_0$. We can find a convergent $\mathbf{I} \subseteq M_\delta$ of cardinality $> \mathrm{cf}(\delta) + \kappa$ (recall $\mathrm{cf}(\delta) < \lambda$!) such that $\mathrm{tp}_{\mathbb{L}_{\infty,\kappa}[\mathfrak{K}]}(a, M_0, M_{\delta+1}) = \mathrm{Av}_\kappa(\mathbf{I}, M_\delta)$. So for some $i(*) < \delta, |\mathbf{I} \cap M_{i(*)}| > \kappa$ so without loss of generality (by equivalence) $\mathbf{I} \subseteq M_{i(*)}$. We finish as in $(E)(f)$.

Axiom (E)(i):
Follows by II.2.16. $\square_{4.10}$

4.11 Exercise: Replace above $\mathrm{Av}_{<\kappa_1}(\mathbf{I}, M)$ by $\cup\{\mathrm{Av}_{\beth_\zeta(\kappa_0)}(\mathbf{I}, M) : \zeta < (2^{\kappa_0})^+\}$.

§5 Homogeneous enough linear orders

5.1 Claim. *Assume $\mu^+ = \theta_1 = \mathrm{cf}(\theta_1) < \theta_2 = \mathrm{cf}(\theta_2) < \lambda$.*
1) Then there is a linear order I of cardinality λ such that: the following equivalence relation $\mathscr{E} = \mathscr{E}^{\mathrm{aut}}_{I,\mu}$ on ${}^\mu I$ has $\le 2^\mu$ equivalence classes, where
$\eta_1 \mathscr{E} \eta_2$ iff there is an automorphism of I mapping η_1 to η_2.
2) Moreover if $I' \subseteq I$ has cardinality $< \theta_2$ and $n < \omega$ then the following equivalence relation $\mathscr{E}$ on ${}^n I$ has $\le \mu + |I'|$ equivalence classes: $\bar{s} \mathscr{E} \bar{t}$ iff there is an automorphism h of I over I' mapping $\bar{s}$ to $\bar{t}$.
3) Moreover, there is Ψ proper for $K^{\mathrm{lin}}_{\tau_2^*}$ (i.e. $\Psi \in \Upsilon^{\mathrm{lin}}_{\aleph_0}[2]$, see Definitions 0.11(5) and 0.14(9)) with $\tau(\Psi)$ countable such that $I = \mathrm{EM}_{\{<\}}(I^{\mathrm{lin}}_{\theta_2,\lambda\times\theta_2}, \Phi)$ where $I^{\mathrm{lin}}_{\theta_2,\zeta} = (\zeta, <, P_0, P_1), P_\ell = \{\alpha < \zeta : (\mathrm{cf}(\alpha) < \theta_2) \equiv (\ell = 0)\}$.
4) If $I^*_0 \subseteq I$ has cardinality $< \theta_2$ then for some $I^*_1 \subseteq I$ of cardinality $\le \mu^+ + |I^*_0|$ for every $J \subseteq I$ of cardinality $\le \mu$ there is an automorphism of I over I^*_0 mapping J into I^*_1.
5) If $I^*_1, I^*_2 \subseteq I^{\mathrm{lin}}_{\mu,\lambda\times\mu^+}$ has cardinality $\le \mu$ and h is an isomorphism from I^*_1 onto I^*_2 then there is an automorphism $\hat{h}$ of the linear order $I = \mathrm{EM}_{\{<\}}(I^{\mathrm{lin}}_{\theta,\lambda}, \Psi)$ extending the natural isomorphism $\check{h}$ from $\mathrm{EM}_{\{<\}}(I^*_1, \Psi)$ onto $\mathrm{EM}_{\{<\}}(I^*_2, \Psi)$.

Remark. 1) Of course, if $\lambda = \lambda^{<\theta_2}$ and I is a dense linear order of cardinality λ which is θ-strongly saturated (hence θ-homogeneous) then the demand in 5.1(1) is satisfied (and in part (2) of 5.1 the number (of $\mathscr{E}$ equivalence classes) is $\le 2^\chi$ for every $\chi \in [\aleph_0, \theta_2)$). Also if $\lambda = \sum_{i<\delta} \lambda_i, \delta < \theta_2$ and $i < \delta \Rightarrow \lambda_i^{<\theta_2} = \lambda$ we have such order.

2) Laver [Lv71, §2] deals with related linear orders but for his aims I_1, I_2 are equivalent if each is embeddable into the other; see more in [Sh:e, AP,§2]. For a cardinal ∂ and linear order I let $\Theta_{I,\partial} = \{\text{cf}(J)$: for some $<_I$-decreasing sequence $\langle t_i : i < \partial \rangle$ we have $J = I \restriction \{t \in I : t <_I t_i$ for every $i < \partial\}\}$. So if $\partial \le \mu$ then $({}^\mu I)/E^{\text{aut}}_{I,\mu}$ has $\ge |\Theta_{I,\partial}|$. So we have to be careful to make $\Theta_{I,\partial}$ small. We choose a very concrete construction which leads quickly to defining I and the checking is straight so we thought it would be easy but a posteriori the checking is lengthy; [Sh:e, AP,§2] is an anti-thetical approach.

3) We can replace $\theta_1 = \mu^+$ by $\theta_1 = \text{cf}(\theta_1) > \aleph_0$ and "of cardinality $\le \mu$" by "of cardinality $< \theta_1$".

4) In 2.8(1), 2.11(2) we use parts (1),(1)+(4) respectively. Also we use 5.1 in the proof of 7.8.

5) The case $2^\mu \ge \lambda$ in 5.1(1) says nothing, in fact if $2^\mu \ge \lambda$ then $2^\mu = \lambda^\mu = ({}^\mu M)/\mathscr{E}^{\text{aut}}_{I,\mu}$ for any model M of cardinality $\le 2^\mu$ but ≥ 2, for any vocabulary τ_M.

6) Claim 5.1(1),(2) holds also if we replace μ by $\chi \in [\mu, \theta_2)$.

Proof. 1) Fix an ordinal $\zeta, \lambda \le \zeta < \lambda^+$ such that $\text{cf}(\zeta) = \theta_2$, e.g., $\zeta = \lambda \times \theta_2$ (almost always $\text{cf}(\zeta) \ge \theta_2$ suffice).

Let I_1 be the following linear order, its set of elements is $\{(\ell, \alpha) : \ell \in \{-2, -1, 1, 2\}, \alpha < \zeta + \omega\}$ ordered by $(\ell_1, \alpha_1) <_{I_1} (\ell_2, \alpha_2)$ iff $\ell_1 < \ell_2$ or $\ell_1 = \ell_2 \in \{-1, 2\} \wedge \alpha_1 < \alpha_2$ or $\ell_1 = \ell_2 \in \{-2, 1\} \wedge \alpha_1 > \alpha_2$.

For $t \in I_1$ let $t = (\ell^t, \alpha^t)$.

Let I_2^* be the set $\{\eta : \eta$ is a finite sequence of members of $I_1\}$ ordered by $\eta_1 <_{I_2} \eta_2$ <u>iff</u> $(\exists n)(n < \ell g(\eta_1) \wedge n < \ell g(\eta_2) \wedge \eta_1 \restriction n = \eta_1 \restriction n$ & $\eta_1(n) <_{I_1} \eta_2(n))$ or $\eta_1 \triangleleft \eta_2 \wedge \ell^{\eta_2(\ell g(\eta_1))} \in \{1, 2\}$ or $\eta_2 \triangleleft \eta_1 \wedge \ell^{\eta_1(\ell g(\eta_2))} \in \{-2, -1\}$.

Let I_2 be I_2^* restricted to the set of $\eta \in I_2^*$ satisfying $\circledast$ where

$\circledast$ for no $n < \omega$ do we have:

- (a) $\ell g(\eta) > n + 1$
- (b) $\alpha^{\eta(n)}$ is a limit ordinal of cofinality $\ge \theta_1$

(c) $\alpha^{\eta(n+1)} \geq \zeta$

(d) $\ell^{\eta(n)} \in \{-1, 2\}$, $\ell^{\eta(n+1)} = -2$ or $\ell^{\eta(n)} \in \{-2, 1\}$, $\ell^{\eta(n+1)} = 2$.

Let M_0 be the following ordered field:

$(*)_1$ (a) M_0 as a field, is $\mathbb{Q}(a_t : t \in I_2)$, the field of rational functions with $\{a_t : t \in I_2\}$ algebraically independent

(b) the order of M_0 is determined by

(α) if $t \in I_2, n < \omega$ then $M_0 \models n < a_t$

(β) if $s <_{I_2} t$ and $n < \omega$ then $M_0 \models$ "$(a_s)^n < a_t$".

(c) let M be the real[3] (algebraic) closure of M_0 (i.e. the elements algebraic over M_0 in the closure by adding elements realizing any Dedekind cut of M_0).

Now we shall prove that I, which is M as a linear order, is as requested.

$\boxtimes_1$ each of I_1, I_2^* and I_2 is anti-isomorphic to itself.

[Why? Let $g : I_1 \to I_1$ be $g(t) = (-\ell^t, \alpha^t)$, clearly it is an anti-isomorphism of I_1. Let $\hat{g} : I_2^* \to I_2^*$ be defined by $\hat{g}(\eta) = \langle g(\eta(m)) : m < \ell g(\eta)\rangle$, it is an anti-isomorphism of I_2^*. Lastly $\hat{g}$ maps I_2 onto itself, in particular by the character of clause (d) of ⊛, i.e. the two cases are interchanged by $\hat{g}$]

$\boxtimes_2$ (a) I_1, I_2^*, I_2 have cofinality $\aleph_0$.

(b) if $t \in I_2$ then $I_{2,<t} := I_2 \restriction \{s : s <_{I_2} t\}$ has cofinality $\aleph_0$.

[Why? For clause (a), $\{(2, \lambda + n) : n < \omega\}$ is a cofinal subset of I_1 of order type ω and $\{< t >: t \in I_1\}$ is a cofinal subset of I_2^* and of I_2 of order type being the same as I_1. For clause (b) for $\eta \in I_2$ the set $\{\eta\hat{\ }\langle(-1, \lambda + n)\rangle : n < \omega\}$ is a cofinal subset of $I_{2,<\eta}$ of order type ω by ⊡ below.]

---

[3] in fact, we could just use M_0

Now

$\boxdot$ if η satisfies $\circledast$ and $\ell \in \{1, -1\}$ then also $\eta\hat{\ }\langle(\ell, \alpha)\rangle$ satisfies $\circledast$ for any $\alpha < \lambda + \omega$.

[Why? By clause (d) of $\circledast$ as the only value of n there which is not obvious is $n = \ell g(\eta) - 1$, but to be problematic we should have $\ell(\eta\hat{\ }<(\ell,\alpha)>)(n+1) \in \{-2, 2\}$ whereas $\ell = -1$.]

$\boxtimes_3$ if $\partial = \text{cf}(\partial)$ so ∂ is $0, 1$ or an infinite regular cardinal and $\bar{\eta} = \langle \eta_i : i < \partial \rangle$ is a $<_{I_2}$-decreasing sequence and we let $J_{\bar{\eta}} = \{s \in I_2 : s <_{I_2} \eta_i$ for every $i < \partial\}$ <u>then</u> (clearly exactly one of the following clauses applies)

(a) if $J_{\bar{\eta}} = \emptyset$ then $\partial = \aleph_0$

(b) if $\text{cf}(J_{\bar{\eta}}) = 1$ then $\partial = \aleph_0$

(c) if $\text{cf}(J_{\bar{\eta}}) = \aleph_0$ then $\partial < \theta_1$

(d) if $\aleph_1 \leq \text{cf}(J_{\bar{\eta}}) < \theta_1$ then $\partial = \aleph_0$ and for some $\ell \in \{-1, 2\}, \nu \in I_2$ and ordinal $\delta < \zeta$ of cofinality $\text{cf}(J_{\bar{\eta}})$ the set $\langle \nu\hat{\ }\langle(\ell, \alpha)\rangle : \alpha < \delta \rangle$ is an unbounded subset of $J_{\bar{\eta}}$

(e) if $\theta_1 \leq \text{cf}(J_{\bar{\eta}})$ then $\partial \geq \theta_1$ and moreover $\partial = \theta_2 \vee \text{cf}(J_{\bar{\eta}}) = \theta_2$.

[Why does $\boxtimes_3$ hold? The proof is split into cases and finishing a case we can then assume it does not occur.

Clearly we can replace $\bar{\eta}$ by $\langle \eta_i : i \in u \rangle$ for any unbounded subset u of ∂ and by $\langle \nu_i : i \in u \rangle$ if $\eta_{\zeta_{2i+1}} \leq_{I_2} \nu_i \leq_{I_2} \eta_{\zeta_{2i}}$ and $\langle \zeta_i : i < \partial \rangle$ an increasing sequence of ordinals $< \partial$. We shall use this freely.

<u>Case 0</u>: $\partial = 0$ or $\partial = 1$.

By $\boxtimes_2$ clearly clause (c) of $\boxtimes_3$ holds.

<u>Case 1</u>: $\partial = \aleph_0$ and there is $\nu \in {}^\omega(I_1)$ such that $(\forall n < \omega)(\exists i < \partial)(\eta_i \restriction n \triangleleft \nu)$.

Let $n_i = \ell g(\eta_i \cap \nu)$, it is impossible that $\{i : n_i = k\}$ is infinite for some k, so without loss of generality $\langle n_i : i < \omega \rangle$ is an increasing sequence and $n_0 > 0$.

For every $i < \omega$ we have $\nu \restriction (n_i+1) \trianglelefteq \eta_{i+1}$ and $\eta_{i+1} <_{I_2} \eta_i$, so by the definition of $<_{I_2}$ also $\nu \restriction (n_i+1) <_{I_2} \eta_i$, and we choose $\beta_{n_i} < \zeta+\omega$ so that $(-2, \beta_{n_i}) <_{I_1} \nu(n_i)$ hence letting $\rho_i = \nu \restriction n_i\hat{\ }\langle(-2, \beta_{n_i})\rangle$ we have $\rho_i \in I_2$. This can be done, e.g. because we can choose β_{n_i} such that $\beta_{n_i} = \alpha^{\nu(n_i)} + 1$ if $\ell^{\nu(n_i)} = -2$ and $\beta_{n_i} = 0$ otherwise.

For every $i, j < \omega$ we have $\rho_i <_{I_2} \rho_{i+1} <_{I_2} \eta_{i+1} <_{I_2} \eta_i$, so if $i \leq j$ then $\rho_i <_{I_2} \rho_j <_{I_2} < \eta_j$, and if $i > j$ then $\rho_i <_{I_2} \eta_i <_{I_2} \eta_j$, so $\rho_i \in J_{\bar{\eta}}$.

Now $\langle \rho_i : i < \omega \rangle$ is $<_{I_2}$-increasing also it is cofinal in $J_{\bar{\eta}}$, for if $\rho \in J_{\bar{\eta}}$ let $n = \ell g(\rho \cap \nu)$, so for $i < \omega$ such that $n_i \leq n < n_{i+1}$ we have $\rho <_{I_2} \eta_{i+1}$ so $\rho(n) <_{I_1} \eta_{i+1}(n) = \rho_{i+1}(n)$ and as $\rho \restriction n = \nu \restriction n = \rho_{i+1} \restriction n$ we have $\rho <_{I_2} \rho_{i+1}$.

As $\langle \rho_i : i < \omega \rangle$ is of order type ω clearly $\mathrm{cf}(J_{\bar{\eta}}) = \aleph_0 = \partial$ hence clause (c) of $\boxtimes_3$ applies, and we are done.

So from now on assume that case 1 fails.

As $\ell g(\eta_i) < \omega$ and as not Case 1 without loss of generality for some n, we have $i < \partial \Rightarrow \ell g(\eta_i) = n$. Similarly without loss of generality for some m and $\nu \in I_2$ we have $i < \partial \Rightarrow \eta_i \restriction m = \nu$ and $\langle \eta_i(m) : i < \partial \rangle$ with no repetitions so $m < n$. Without loss of generality $i < \partial \Rightarrow \ell^{\eta_i(m)} = \ell^*$ and so $\langle \alpha^{\eta_i(m)} : i < \partial \rangle$ is with no repetitions; and without loss of generality is monotonic hence, as $\partial \geq \aleph_0$ is an increasing sequence of ordinals. As $\bar{\eta}$ is $<_{I_2}$-decreasing necessarily $\ell^* \in \{-2, 1\}$ and let $\delta = \cup\{\alpha^{\eta_i(m)} : i < \partial\}$, so clearly $\mathrm{cf}(\delta) = \partial$ and δ is a limit ordinal $\leq \zeta + \omega$. Now those ℓ^*, δ will be used till the end of the proof of $\boxtimes_3$. So for the rest of the proof we are assuming

$\odot$ (a) $i < \partial \Rightarrow \eta_i \restriction m = \nu$

(b) $\langle \eta_i(m) : i < \partial \rangle$ is (strictly) increasing with limit δ

(c) $\ell^{\eta_i(m)} = \ell^* \in \{-2, 1\}$

(d) $\mathrm{cf}(\delta) = \partial, \delta \leq \zeta + \omega$.

Also note by $\circledast$ that $\nu\hat{\ }\langle(\ell^*, \delta)\rangle \notin I_2 \Rightarrow \delta \in \{\zeta + \omega, \zeta\}$ and if $\delta = \zeta \wedge \nu\hat{\ }\langle(\ell^*, \delta)\rangle \notin I_2$ then $\ell g(\nu) > 0$ and the ordinal $\alpha^{\nu(\ell g(\nu)-1)}$ is limit of cofinality $\geq \theta_1$ (and more).

<u>Case 2</u>: $J_{\bar{\eta}} = \emptyset$.

Clearly $m = 0 \wedge \ell^* = -2 \wedge \delta = \zeta + \omega$ hence $\partial = \aleph_0$ so clause (a) of $\boxtimes_3$ holds.

Case 3: $\ell^* = 1$ and $\nu\hat{}\langle(\ell^*, \delta)\rangle \notin I_2$.

As $\ell^* = 1$ clearly we cannot have $\delta = \zeta$ by clause (d) of $\circledast$ so $\delta = \zeta + \omega$ and recalling $\partial = \text{cf}(\delta)$ we have $\partial = \aleph_0$. Now clearly $J_{\bar{\eta}}$ has a last element, ν, so case (b) of $\boxtimes_3$ applies.

Case 4: $\ell^* = -2, \partial = \aleph_0$ and $\nu\hat{}\langle(\ell^*, \delta)\rangle \notin I_2$.

Again $\delta = \zeta + \omega$ as $\aleph_0 = \partial = \text{cf}(\delta)$ and $\text{cf}(\zeta) = \theta_2 > \mu \geq \aleph_0$ making $\delta = \zeta$ impossible; now $\ell g(\nu) > 0$ (as we have discarded the case $J_{\bar{\eta}} = \emptyset$, i.e. Case 2); and let $k = \ell g(\nu) - 1$. Now we prove case 4 by splitting to several subcases.

Subcase 4A: $\ell^{\nu(k)} \in \{-2, 1\}$.

Let $\nu_1 = (\nu \restriction k)\hat{}\langle(\ell^{\nu(k)}, \alpha^{\nu(k)} + 1)\rangle$, note that $\nu_1 \in I_2$ as $\nu \in I_2 \wedge (\alpha^{\nu(k)} < \zeta \equiv \alpha^{\nu(k)} + 1 < \zeta)$ and (as $\ell^{\nu(k)} \in \{-2, 1\}$) clearly $\{\rho : \nu_1 \trianglelefteq \rho \in I_2\}$ is a cofinal subset of $J_{\bar{\eta}}$ even an end segment. Now for $n < \omega$ we have $\nu_1\hat{}\langle(2, \zeta + n)\rangle \in I_2^*$ and it satisfies $\circledast$. (Why? As $\nu_1 \in I_2$, only $n = k$ may be problematic, but $\alpha^{\nu(k)} + 1 = \alpha^{\nu_1(k)}$ here stands for $\alpha^{\eta(n)}$ there hence clause (b) of $\circledast$ does not apply), so by the definition of I_2, clearly $\{\nu_1\hat{}\langle(2, \zeta + n)\rangle : n < \omega\}$ is $\subseteq I_2$ and is a cofinal subset of $J_{\bar{\eta}}$ so $\partial = \aleph_0 = \text{cf}(J_{\bar{\eta}})$ and clause (c) of $\boxtimes_3$ holds.

Subcase 4B: $\ell^{\nu(k)} \in \{-1, 2\}$ and $\alpha^{\nu(k)}$ is a successor ordinal.

Let $\nu_1 = (\nu \restriction k)\hat{}\langle(\ell^{\nu(k)}, \alpha^{\nu(k)} - 1)\rangle$, of course $\nu_1 \in I_2^*$ and as $\nu \in I_2$ clearly $\nu_1 \in I_2$ so the set $\{\rho : \nu_1 \trianglelefteq \rho \in I_2\}$ is an end segment of $J_{\bar{\eta}}$ and has cofinality $\aleph_0$ because $n < \omega \Rightarrow \nu_1\hat{}\langle(2, \zeta + n)\rangle \in I_2$. (Why? It $\in I_2^*$ and as $\nu_1 \in I_2$ checking $\circledast$ only $n = k$ may be problematic, but $(\ell^{\nu(k)}, 2)$ here stand for $(\ell^{\eta(n)}, \ell^{\eta(n+1)})$ there but presently $\ell^{\nu(k)} \in \{-1, 2\}$ contradicting clause (d) of $\circledast$). So clause (c) of $\boxtimes_3$.

Subcase 4C: $\ell^{\nu(k)} \in \{-1, 2\}$ and $\alpha^{\nu(k)} = 0$.

Then let $\nu_1 = (\nu \restriction k)\hat{}\langle(\ell^{\nu(k)} - 1, 0)\rangle$. Now $\nu_1 \in I_2$ as $\nu \restriction k \in I_2$ and for $n = k - 1$ clause (c) of $\circledast$ fails and $\nu_1\hat{}\langle(2, \zeta + n)\rangle \in I_2$ because of $\nu_1 \in I_2$ and for $n = k$ the failure of clause (b) of $\circledast$ so continue as in Subcase 4B above.

Lastly,
Subcase 4D: $\ell^{\nu(k)} \in \{-1, 2\}$ and $\alpha^{\nu(k)}$ is a limit ordinal.

Then $\{(\nu \restriction k)\hat{}\langle(\ell^{\nu(k)}, \alpha)\rangle : \alpha < \alpha^{\nu(k)}\}$ is $\subseteq I_2$ and is an unbounded subset of $J_{\bar{\eta}}$ hence $\mathrm{cf}(J_{\bar{\eta}}) = \mathrm{cf}(\alpha^{\nu(k)})$. If $\mathrm{cf}(\alpha^{\nu(k)}) = \aleph_0$, then clause (c) in $\boxtimes_3$ holds, and if $\mathrm{cf}(\alpha^{\nu(k)}) \in [\aleph_1, \theta_1)$ then necessarily $\alpha^{\nu(k)} \neq \zeta$ so being a limit ordinal $< \zeta + \omega$ clearly $\alpha^{\nu(k)} < \zeta$ so clause (d) from $\boxtimes_3$ holds. To finish this subcase note that $\mathrm{cf}(\alpha^{\nu(k)}) \geq \theta_1$ is impossible.
[Why "impossible"? Clearly for large enough $i < \partial$ we have $\eta_i(m) \geq \zeta$ (because $\delta = \zeta + \omega$ as said in the beginning of the case) and recall $\nu \triangleleft \eta_i \in I_2$. We now show that clauses (a)-(d) of $\circledast$ hold with η_i, k here standing for η, n there. For clause (a) recall $\ell g(\eta_i) \geq \ell g(\nu) + 1$ and $m = \ell g(\nu) = k+1$. Now $\ell^{\eta_i(k+1)} = \ell^{\eta_i(m)} = \ell^* = -2$ as $\ell^* = -2$ is part of the case, $\ell^{\eta_i(k)} = \ell^{\nu(k)} \in \{-1, 2\}$ in this subcase, so clause (d) of $\circledast$ holds. Also $\alpha^{\eta_i(k+1)} = \alpha^{\eta_i(m)} \geq \zeta$ as said above so clause (c) of $\circledast$ holds and $\mathrm{cf}(\alpha^{\eta_i(k)}) = \mathrm{cf}(\alpha^{\nu(k)}) \geq \theta_1$ (as we are trying to prove "impossible"), so clause (b) of $\circledast$ holds. Together we have proved (a)-(d) of $\circledast$. But $\eta_i \in I_2$, contradiction.]

Now subcases 4A,4B,4C,4D cover all the possibilities hence we are done with case 4.

Case 5: $\ell^* = -2, \partial > \aleph_0$ and $\nu\hat{}\langle(\ell^*, \delta)\rangle \notin I_2$.

Recalling δ is the limit of the increasing sequence $\langle\alpha^{\eta_i(m)} : i < \partial\rangle$ hence $\mathrm{cf}(\delta) = \partial > \aleph_0$ and $\nu\hat{}\langle(-2, \delta)\rangle \notin I_2$, necessarily $\delta = \zeta$ so $\partial = \theta_2$. As $\nu\hat{}\langle(-2, \delta)\rangle \notin I_2$ necessarily clauses (a) - (d) of $\circledast$ hold for some n and as $\nu \in I_2$, clearly $n = \ell g(\nu) - 1$ (see clause (a) of $\circledast$) so we have $\ell g(\nu) > 0$, and letting $k = \ell g(\nu) - 1$, by clause (d) of $\circledast$ the $\ell^{\eta(n+1)}$ there stands for $\ell^* = -2$ here so we have $\ell^{\nu(k)} \in \{-1, 2\}$ and by clause (b) of $\circledast$ we have $\mathrm{cf}(\alpha^{\nu(k)}) \geq \theta_1$. Hence $\{(\nu \restriction k)\hat{}\langle(\ell^{\nu(k)}, \beta)\rangle : \beta < \alpha^{\nu(k)}\}$ is cofinal in $J_{\bar{\eta}}$ and its cofinality is $\mathrm{cf}(\alpha^{\nu(k)})$ as $(\nu \restriction k)\hat{}\langle(\ell^{\nu(k)}, \beta)\rangle$ increase (by $\leq_{I_2}$) with β as $\ell^{\nu(k)} \in \{-1, 2\}$. But $\mathrm{cf}(\alpha^{\nu(k)}) \geq \theta_1$ and $\partial = \theta_2$ (see first sentence of the present case), so clause (e) of $\boxtimes_3$ holds.

Case 6: $\nu\hat{}\langle(\ell^*, \delta)\rangle \in I_2$.

Subcase 6A: $\nu\hat{}\langle(\ell^*, \delta), (2, \zeta)\rangle \in I_2$.

Note that for $m = \ell g(\nu)$ and the pair $(\nu\hat{}\langle(\ell^*,\delta),(2,\zeta)\rangle, m)$ standing for (η, n) in $\circledast$, clauses (a),(c),(d) of $\circledast$ hold (recall $\ell^* \in \{-2,1\}$, see the discussion after case 1) so necessarily clause (b) of $\circledast$ fails hence $\mathrm{cf}(\delta) < \theta_1$ but $\partial = \mathrm{cf}(\delta)$ so $\partial < \theta_1$. Now as $\nu\hat{}\langle(\ell^*,\delta),(2,\zeta)\rangle \in I_2$ clearly if $\ell < \omega$, then $\nu\hat{}\langle(\ell^*,\delta),(2,\zeta+\ell)\rangle$ belongs to I_2 hence $\{\nu\hat{}\langle(\ell^*,\delta),(2,\zeta+\ell)\rangle : \ell < \omega\}$ is a cofinal subset of $J_{\bar\eta}$ by the choice of I_2 hence $\mathrm{cf}(J_{\bar\eta}) = \aleph_0$ so clause (c) of $\boxtimes_3$ applies.

Subcase 6B: $\nu\hat{}\langle(\ell^*,\delta),(2,\zeta)\rangle \notin I_2$.

As $\nu\hat{}\langle(\ell^*,\delta)\rangle \in I_2$, necessarily clauses (a)-(d) of $\circledast$ hold with $(\nu\hat{}\langle(\ell^*,\delta),(2,\zeta)\rangle, m)$ here standing for (η, n) there, recalling $m = \ell g(\nu)$ so by clause (b) of $\circledast$ we know that $\mathrm{cf}(\delta) \geq \theta_1$ but $\partial = \mathrm{cf}(\delta)$ hence $\partial \geq \theta_1$. Also $\{\nu\hat{}\langle(\ell^*,\delta),(2,\alpha)\rangle : \alpha < \zeta\}$ is a subset of I_2 and cofinal in $J_{\bar\eta}$ and is increasing with α so $\mathrm{cf}(J_{\bar\eta}) = \theta_2$ so clause (e) of $\boxtimes_3$ applies.

As the two subcases 6A,6B are complimentary case 6 is done.

Finishing the proof of $\boxtimes_3$:

It is easy to check that our cases cover all the possibilities (as after discarding cases 0,1, if not case (6) then $\nu\hat{}\langle(\ell^*,\delta)\rangle \notin I_2$, as not case (3), $\ell^* \neq 1$ but (see clause $\odot$(c) before case 2), $\ell^* \in \{-2,1\}$ so necessarily $\ell^* = -2$, so case (4),(5) cover the rest). Together we have proved $\boxtimes_3$.]

- $\boxtimes_4$ recall $\aleph_0 \leq \mu < \theta_1 < \theta_2$; if $X \subseteq I_2, |X| < \theta_2$ then we can find Y such that $X \subseteq Y \subseteq I_2, |Y| = \mu + |X|, Y$ is unbounded in I_2 from below and from above and for every $\nu \in I_2 \backslash Y$ the following linear orders have cofinality $\aleph_0$:
  - (a) $J^2_{Y,\nu} = I_2 \restriction \{\eta \in I_2 \backslash Y : (\forall \rho \in Y)(\rho <_{I_2} \nu \equiv \rho <_{I_2} \eta)\}$
  - (b) the inverse of $J^2_{Y,\nu}$
  - (c) $J^-_{Y,\nu} = I_2 \restriction \{\eta \in I_2 : (\forall \rho \in J^2_{Y,\nu})(\eta <_{I_2} \rho)\}$
  - (d) the inverse of $J^+_{Y,\nu} := I_2 \restriction \{\eta \in I_2 : (\forall \rho \in J^2_{Y,\nu})(\rho <_{I_2} \eta)\}$.

[Why? Let $\mathscr{U} = \{\alpha^{\eta(\ell)} : \eta \in X \text{ and } \ell < \ell g(\eta)\}$.
We choose W_n by induction on $n < \omega$ such that

$\boxdot_1$ (a) $d\mathscr{U} \subseteq W_n \subseteq \zeta + \omega$

(b) W_n has cardinality $\mu + |\mathscr{U}| = \mu + |X|$ and $m < n \Rightarrow W_m \subseteq W_n$

(c) $\mu \subseteq W_0$ and $\zeta + n \in W_0$ for $n < \omega$

(d) $\alpha \in W_n \Rightarrow \alpha + 1 \in W_{n+1}$

(e) $\alpha + 1 \in W_n \Rightarrow \alpha \in W_{n+1}$

(f) if $\delta \in W_n$ is a limit ordinal of cofinality $< \theta_1$ then $\delta = \sup(\delta \cap W_{n+1})$

(g) if $\delta \in W_n$ and $\text{cf}(\delta) \geq \theta_1$ (or just $\text{cf}(\delta) \leq \mu + |X|$) then $\sup(\delta \cap W_n) + 1 \in W_{n+1}$.

This is straight. Let $W = \cup\{W_n : n < \omega\}$, so

$\boxdot_2$ $\mathscr{U} \subseteq W$ and $|W| = \mu + |X|$ and W satisfies

(a) $W \subseteq \zeta + \omega$

(b) $|W| < \theta_2$

(c) $0 \in W$ and $\{\zeta + m : m < \omega\} \subseteq W$

(d) $\alpha \in W \Leftrightarrow \alpha + 1 \in W$

(e) if $\delta \in W$ and $\aleph_0 < \text{cf}(\delta) \leq \mu$ <u>then</u> $\delta = \sup(W \cap \delta)$

(f) if $\delta \in W$ and $\text{cf}(\delta) \geq \theta_1$ or $\text{cf}(\delta) = \aleph_0$ then $\text{cf}(\text{otp}(W \cap \delta))) = \aleph_0$.

Let $Y = \{\eta \in I_2 : \alpha^{\eta(\ell)} \in W$ for every $\ell < \ell g(\eta)\}$. Clearly $X \subseteq Y$ and $|Y| = \aleph_0 + |W| = \mu + |\mathscr{U}| < \theta_2$. It suffices to check that Y is as required in $\boxtimes_4$. From now on we shall use only the choice of Y and clauses (a)-(f) of $\boxdot_2$. By $\boxdot_2(c)$ and the choice of Y clearly Y is unbounded in I_2 from above and from below.

So let $\nu \in I_2 \backslash Y$, as $\nu \restriction 0 \in Y$ there is $n < \ell g(\nu)$ such that $\nu \restriction n \in Y, \nu \restriction (n+1) \notin Y$, so $\alpha^{\nu(n)} < \zeta + \omega$, and $\alpha^{\nu(n)} \notin W$, but by clause (c) of $\boxdot_2$ we have $\{\zeta + m : m < \omega\} \subseteq W$ hence $\alpha^{\nu(n)} < \zeta$ and so $\alpha_1 := \text{Min}(W \backslash \alpha^{\nu(n)})$ is well defined, is $\leq \zeta$ and $> \alpha^{\nu(n)}$. As clearly $0 \in W, \beta \in W \Leftrightarrow \beta + 1 \in W$ by the choice of W, obviously α_1 is a limit ordinal. By clause (e) of $\boxdot_2$ clearly α_1 is of cofinality $\aleph_0$ or $\geq \theta_1 = \mu^+$. So clearly $\alpha_0 := \sup(W \cap \alpha^{\nu(n)}) = \sup(W \cap \alpha_1) =$

$\min\{\alpha : W \cap \alpha = W \cap \alpha^{\nu(n)}\}$ is a limit ordinal $\leq \alpha^{\nu(n)}$ and $\alpha_0 \notin W$ so $\mathrm{cf}(\alpha_0) \leq |W| < \theta_2$ but by the assumption on W, (see clause (f) of $\boxdot_2$) we have $\mathrm{cf}(\alpha_0) = \aleph_0$. So $(\nu \restriction n)\hat{}\langle(\ell^{\nu(n)}, \alpha_0)\rangle \in J^2_{Y,\nu}$; moreover

$\boxdot_3$ $\rho \in J^2_{Y,\nu}$ iff $\rho \in I_2$ satisfies one of the following:

(a) (i) $\nu \restriction n = \rho \restriction n$, and $\ell^{\nu(n)} = \ell^{\rho(n)}$,

(ii) $\alpha^{\rho(n)} \in [\alpha_0, \alpha_1)$

(b) (i) $\nu \restriction n = \rho \restriction n$, and $\ell^{\nu(n)} = \ell^{\rho(n)}$,

(ii) $\alpha^{\rho(n)} = \alpha_1$ and $\alpha^{\rho(n+1)} \in [\sup(W \cap \zeta), \zeta)$

(iii) $(\ell^{\rho(n+1)}, \ell^{\rho(n)}) = (\ell^{\rho(n_1)}, \ell^{\nu(n)}) \in$

$$\{(2,-2),(2,1),(-2,-1),(-2,2)\}$$

(c) (i) $\alpha_1 = \zeta$ and $n > \theta$ and $(\nu \restriction n)\hat{}(\ell^{\nu(n)}, \alpha_1) \notin I_2$

(ii) $(\ell^{\nu(n)}, \ell^{\nu(n-1)}) \in \{(2,-2),(2,1),(-2,2),(-2,-1)\}$

(iii) $\mathrm{cf}(\nu(n)) \geq \theta_1$ and $\nu(n) > \sup(W \cap \nu(n))$

(iv) $\rho \restriction (n-1) = \nu \restriction (n-1), \ell^{\rho(n-1)} = \ell^{\nu(n-1)}$

(v) $\alpha^{\rho(n-1)} \in [\sup(\nu(n-1) \cap W), \nu(n-1))$.

[Why? First note that if $\rho \in J^2_{Y,\nu}$ and $\rho \restriction k = \nu \restriction k$, $\rho(k) \neq \nu(k)$, and $k \leq n$ then necessarily $k = n \wedge \ell^{\rho(k)} = \ell^{\nu(k)}$. We now proceed to check "if". Let $f : \{-2,-1,1,2\} \to \{2,-2\}$ so that $f^{-1}(\{2\}) = \{-2,1\}$ and $f^{-1}(\{-2\}) = \{-1,2\}$. Case (a) is obvious. In case (b) in order for $\eta \in Y$ to separate between ν and ρ it is necessary that $\eta \restriction (n+1) = \rho \restriction (n+1)$, $\ell^{\eta(n+1)} = \ell^{\rho(n+1)} = f(\ell^{\rho(n)})$ and that $\alpha^{\eta(n+1)} \geq \zeta$, but then $\eta \notin I_2$. In case (c) in order to separate between ρ and ν by $\eta \in Y$ there are two possibilities. Either $\eta \restriction n = \nu \restriction n$ and then $\ell^{\eta(n)} = \ell^{\nu(n)} = f(\ell^{\nu(n-1)})$ (recall that $\nu \restriction n\hat{}\langle(\ell^{\nu(n)}, \alpha_1)\rangle \notin I_2$), and $\alpha^{\eta(n)} \geq \zeta$, but then also $\eta \notin I_2$. The other possibility is that $\eta \restriction (n-1) = \nu \restriction (n-1), \ell^{\eta(n-1)} = \ell^{\nu(n-1)}$ and $\alpha = \alpha^{\eta(n-1)}$ is such that $\alpha \in W$ and $\alpha^{\rho(n-1)} < \alpha < \alpha^{\nu(n-1)}$ which is also impossible by the choice of $\alpha^{\rho(n-1)}$. Showing that these are the only cases (the "only if" direction) is similar and is actually done below.]

Now we proceed to check that clauses of $\boxtimes_4$ hold.

Clause (a):

First assume $\ell^{\nu(n)} \in \{-2,1\}$, and let $J = \{\nu \restriction n\char`\^\langle(\ell^{\nu(n)},\alpha_0),(2,\zeta+m) : m < \omega\}$. Now $J \subseteq I_2$ [why? clearly if $\rho \in J$ then $\rho \restriction (n+1) \in I_2$ so we only need to check ⊛ for n, recall that $\mathrm{cf}(\alpha_0) = \aleph_0 < \theta_1$, hence clause (b) of ⊛ fails]. Now by clause (a) of $\boxdot_3$ we have that $J \subseteq J^2_{Y,\nu}$, and we claim that it is also cofinal in it. [Why? Note that as $\ell^{\nu(n)} \in \{-2,1\}$ then $\nu \restriction n\char`\^\langle(\ell^{\nu(n)},\alpha_0) <_{I_2} \nu \restriction (n+1)$, and if $\rho \in J^2_{Y,\nu}$ is as in clauses $(a)$ or $(b)$ of $\boxdot_3$ then for every $m$ large enough $\rho <_{I_2} \nu \restriction n\char`\^\langle(\ell^{\nu(n)},\alpha_0),(2,\zeta+m)\rangle$. If $\rho \in J^2_{Y,\nu}$ is as in clause (c) of $\boxdot_3$ then $\ell^{\nu(n)} \in \{-2,2\}$ by (ii) there, and as in this case $\ell^{\nu(n)} \in \{-2,1\}$, necessarily $\ell^{\nu(n)} = -2$ and so by (ii) of (c) of $\boxdot_3$ we have $\ell^{\nu(n-1)} \in \{-1,2\}$, but then $\rho <_{I_2} \nu$ and so it is below every element in J.]

Second, assume $\ell^{\nu(n)} \in \{-1,2\}$ and $\nu \restriction n\char`\^\langle(\ell^{\nu(n)},\alpha_1)\rangle \in I_2$; let $\delta^* = \sup(W \cap \zeta)$, so as above $\delta^* \notin W$, and has cofinality $\aleph_0$ (which is less than θ_1), recall also that $\mathrm{cf}(\alpha_1) \geq \theta_1$. So (for $\ell \in \{-2,-1,1,2\}$) by ⊛ we have $(\nu \restriction n)\char`\^\langle(\ell^{\nu(n)},\alpha_1),(\ell,\beta)\rangle \in I_2$ iff $\beta < \zeta \wedge \ell \in \{-2,-1,1,2\}$ or $(\zeta \leq \beta < \zeta+\omega \wedge \ell \neq -2)$. Hence we have $(\nu \restriction n)\char`\^\langle(\ell^{\nu(n)},\alpha_1),(-2,\beta)) \in I_2 \Leftrightarrow \beta < \zeta$. Also $(\nu \restriction n)\char`\^\langle(\ell^{\nu(n)},\alpha_1),(-2,\beta)\rangle \in Y \Leftrightarrow \beta \in W$, and as $\nu(n) < \alpha_1 \wedge \ell^{\nu(n)} \in \{-1,2\}$ clearly $\nu <_{I_2} (\nu \restriction n)\char`\^\langle(\ell^{\nu(n)},\alpha_1),(-2,\beta)\rangle$. Easily $\{(\nu \restriction n)\char`\^\langle(\ell^{\nu(n)},\alpha_1),(-2,\varepsilon)\rangle : \varepsilon \in W \cap \zeta)\}$ is a subset of $\{\eta \in Y : \nu <_{I_2} \eta\}$ unbounded from below in it.

So $\{(\nu \restriction n)\char`\^\langle(\ell^{\nu(n)},\alpha_1),(-2,\delta^*),(2,\alpha)\rangle : \zeta < \alpha < \zeta+\omega\}$ is included in I_2 (recalling clause (b) of ⊛ as $\mathrm{cf}(\delta^*) = \aleph_0$) and moreover is a cofinal subset of $J^2_{Y,\nu}$ of order type ω, so $\mathrm{cf}(J^2_{Y,\nu}) = \aleph_0$ as required.

Third, assume $\rho^{\nu(n)} \in \{-1,2\}$ and $(\nu \restriction n)\char`\^\langle(\ell^{\nu(n)},\alpha_1)\rangle \in I_2$ and $\mathrm{cf}(\alpha_1) < \theta_1$, equivalently $\mathrm{cf}(\alpha_1) = \aleph_0$ by clause (e) of $\boxdot_2$. In this case $\{(\nu \restriction n)\char`\^\langle(\ell^{\nu(n)},\alpha)(-2,\beta)\rangle : \zeta \leq \beta < \zeta+\omega\}$ is included in I_2 (recalling clause (b) of ⊛) and in Y, hence recalling $\boxdot_3(a)$ the set $\{(\nu \restriction n)\char`\^\langle(\ell^{\nu(n)},\alpha)\rangle : \alpha \in [\alpha_0,\alpha_1)\}$ is a cofinal subset of $J^2_{Y,\nu}$ hence its cofinality is $\mathrm{cf}(\alpha_1) = \aleph_0$ as required.

Fourth, we are left with the case that $\ell^{\nu(n)} \in \{-1,2\}$ and $(\nu \restriction n)\char`\^\langle(\ell^{\nu(n)},\alpha_1)\rangle \notin I_2$ so necessarily $n > 0$ and clauses (a)-(d) of ⊛ hold for it for $n-1$; then by clause (c) of ⊛ (recalling $\alpha_1 \leq \zeta$ as shown before $\boxdot_3$) necessarily $\alpha_1 = \zeta$. Clearly $k := n-1 \geq 0$ and as clause

(d) of ⊛ holds and it says there "$\ell^{\eta(n+1)} \in \{2,-2\}$" which means here $\ell^{\nu(n)} \in \{2,-2\}$ but we are assuming presently $\ell^{\nu(n)} \in \{-1,2\}$ hence $\ell^{\nu(n)} = \ell^{\nu(k+1)} = 2$ so using clause (d) of ⊛, see above, it follows that $\ell^{\nu(k)} \in \{-2,1\}$ and by clause (b) of ⊛ we have $\mathrm{cf}(\alpha^{\nu(k)}) \geq \theta_1$. Let $\delta_* = \sup(W \cap \alpha^{\nu(k)})$. Now if $\delta_* < \alpha^{\nu(k)}$ then by clause (f) of $\boxdot_2$ we know $\mathrm{cf}(\delta_*) = \aleph_0$ and $\{(\nu \restriction k)\hat{}\langle(\ell^{\nu(k)},\delta_*)(2,\zeta+m)\rangle : m < \omega\}$ is included in I_2 (as $\nu \in I_2$ and $\delta_* \leq \alpha^{\nu(k)}$ we have to check in ⊛ only with $k+1$ here standing for n there, but $\mathrm{cf}(\delta_*) = \aleph_0$ so clause (b) there fails) and so recalling $\boxdot_3(c)$ this set is a cofinal subset of $J^2_{Y,\nu}$ exemplifying that its cofinality is $\aleph_0$.

Lastly, if $\delta_* = \alpha^{\nu(k)}$ then $\langle(\nu \restriction n)\hat{}\langle(\ell^{\nu(n)},\alpha)\rangle : \alpha \in W \cap \zeta\rangle$ is $<_{I_2}$-increasing with α, all members in Y, and in $J^2_{Y,\nu}$, cofinal in it and has order type $\mathrm{otp}(W \cap \zeta)$ which has cofinality $\aleph_0$ so also $J^2_{Y,\nu}$ has cofinality $\aleph_0$ as required.

<u>Clause (b)</u>: What about the cofinality of the inverse? Recall that I_2 is isomorphic to its inverse by the mapping $(\ell,\beta) \mapsto (-\ell,\beta)$, but this isomorphism maps Y onto itself hence it maps $J^2_{Y,\nu}$ onto $J^2_{Y,\nu'}$ for some $\nu' \in I_2 \backslash Y$, but clause (a) was proved also for ν', so this follows.

<u>Clause (c)</u>: As Y is unbounded from below in I_2 (containing $\{\langle(-2,\zeta+n)\rangle : n < \omega\}$) it follows that $J^-_{Y,\nu}$ is non-empty, hence $\mathrm{cf}(J^-_{Y,\nu}) \neq 0$, but what is $\mathrm{cf}(J^-_{Y,\nu})$?

First, if $\ell^{\nu(n)} \in \{-1,2\}$ then $\{(\nu \restriction n)\hat{}\langle(\ell^{\nu(n)},\alpha)\rangle : \alpha < \alpha_0\}$ is an unbounded subset of $J^-_{Y,\nu}$ of order type α_0 hence $\mathrm{cf}(J^-_{Y,\nu}) = \mathrm{cf}(\alpha_0) = \aleph_0$ (see the assumption on W and the choice of α_0).

Second, if $\ell^{\nu(n)} = \{-2,1\}$ and $(\nu \restriction n)\hat{}\langle(\ell^{\nu(n)},\alpha_1)\rangle \in I_2$ and $\mathrm{cf}(\alpha_1) \geq \theta_1$ then as in the proof of clause (a) we have $\{(\nu \restriction n)\hat{}\langle(\ell^{\nu(n)},\alpha_1),(2,\zeta+m)\rangle \notin I_2$ for $m < \omega$ and again letting $\delta^* = \sup(W \cap \zeta)$ we have $\{(\nu \restriction n)\hat{}\langle(\ell^{\nu(n)},\alpha_1),(2,\beta)\rangle : \beta \in W \cap \zeta\}$ is included in I_2 and in $J^-_{Y,\nu}$ and even is an unbounded subset of $J^-_{Y,\nu}$ of order type $\mathrm{otp}(W \cap \delta^*)$ which has the same cofinality as δ^* which is $\aleph_0$.

Third, if $\ell^{\nu(n)} \in \{-2,1\}$ and $(\nu \restriction n)\hat{}\langle(\ell^{\nu(n)},\alpha_1)\rangle \in I_2$ and $\mathrm{cf}(\alpha_1) < \theta_1$, equivalently $\mathrm{cf}(\alpha_1) = \aleph_0$, <u>then</u> $\{(\nu \restriction n)\hat{}\langle(\ell^{\nu(n)},\alpha_1),(2,\zeta+$

$m)\rangle : m < \omega\}$ is a subset of I_2 (as $\mathrm{cf}(\alpha_1) = \aleph_0$) is included in $J^-_{Y,\nu}$, unbounded in it and has cofinality $\aleph_0$, so we are done.

Fourth and lastly, if $\ell^{\nu(n)} \in \{-2, 1\}$ and $(\nu \restriction n)\hat{}\langle(\ell^{\nu(n)}, \alpha_1)\rangle \notin I_2$ then as in the proof of clause (a) we have $\alpha_1 = \zeta$ and again letting $\delta^* = \sup(W \cap \zeta)$ we have $\mathrm{cf}(\delta^*) = \aleph_0$ and $(\nu \restriction n)\hat{}\langle(\ell^{\nu(n)}, \delta^*)\rangle \in I_2$ and $\{(\nu \restriction n)\hat{}\langle(\ell^{\nu(n)}, \delta^*), (2, \zeta + m)\rangle : m < \omega\}$ is a subset of I_2, moreover a subset of $J^-_{Y,\nu}$ unbounded in it and $(\nu \restriction n)\hat{}\langle(\ell^{\nu(n)}, \delta^*), (2, \zeta + m)\rangle$ is $<_{I_2}$-increasing with m. So indeed $J^-_{Y,\nu}$ has cofinality $\aleph_0$.

Clause (d): As in clause (b) we use the anti-isomorphism.
So $\boxtimes_4$ holds.]

$\boxtimes_5$ if $I' \subseteq I_2$ then the number of cuts of I' induced by members of $I_2 \backslash I'$, that is $\{\{s \in I' : s <_{I_2} t\} : t \in I_2 \backslash I'\}$ is $\leq |I'| + 1$.
[Why? Let $\mathscr{U} := \{\alpha^{\eta(\ell)} : \ell < \ell g(\eta) \text{ and } \eta \in I'\}$, it belongs to $[\zeta + \omega]^{\leq \mu}$. Now (by inspection) $\eta_1, \eta_2 \in I_2 \backslash I'$ realizes the same cut of I' when:

- (a) $\ell g(\eta_1) = \ell g(\eta_2)$
- (b) $\ell^{\eta_1(n)} = \ell^{\eta_2(n)}$ for $n < \ell g(\eta_1)$
- (c) $\alpha^{\eta_1(n)} \in \mathscr{U} \Leftrightarrow \alpha^{\eta_2(n)} \in \mathscr{U} \Rightarrow \alpha^{\eta_1(n)} = \alpha^{\eta_2(n)}$ for $n < \omega$
- (d) $\beta < \alpha^{\eta_1(n)} \equiv \beta < \alpha^{\eta_2(n)}$ for $\beta \in \mathscr{U}$ and $n < \omega$

[Why? Now clauses (a)-(d) define an equivalence relation on $I_2 \backslash I'$ which refines "inducing the same cut" and has $\leq |\mathscr{U}| + \aleph_0 = |I'| + \aleph_0$ equivalence classes. As the case I' is finite is trivial, we are done proving $\boxtimes_5$.]

$\boxtimes_6$ if ∂ is regular uncountable, $n^* < \omega$ and $t_{\varepsilon,\ell} \in I_2$ for $\varepsilon < \partial, \ell < n^*$ and $t_{\varepsilon,0} <_{I_2} \cdots <_{I_2} t_{\varepsilon,n^*-1}$ for $\varepsilon < \partial$ then for some unbounded (and even stationary) set $S \subseteq \partial, m \leq n^*$ and $0 = k_0 < k_1 < \ldots < k_m = n^*$ stipulating $t_{\varepsilon,k_m} = \infty$ and letting $\varepsilon(*) = \mathrm{Min}(S)$ we have:

(a) for each $i < m$:

(α) if $\varepsilon < \xi$ are from S and $\ell_1, \ell_2 \in [k_i, k_{i+1})$ then $t_{\varepsilon,\ell_1} <_{I_2} t_{\xi,\ell_2}$ or

(β) if $\varepsilon < \xi$ are from S and $\ell_1, \ell_2 \in [k_i, k_{i+1})$ then $t_{\xi,\ell_2} <_{I_2} t_{\varepsilon,\ell_1}$ <u>or</u>

(γ) $k_{i+1} = k_i + 1$ and for every $\varepsilon \in S$ we have $t_{\varepsilon,k_i} = t_{\varepsilon(*),k_i}$

(b) there is a sequence $\langle s_i^-, s_i^+ : i < m \rangle$ such that

(α) $i < m \Rightarrow s_i^- <_{I_2} s_i^+$

(β) if $i < m-1$ then $s_i^+ < s_{i+1}^-$ except possibly when $\langle t_{\varepsilon,k_i} : \varepsilon < \partial \rangle$ is $<_{I_2}$-decreasing and there is no $t \in I_2$ such that $\varepsilon < \partial \Rightarrow t_{\varepsilon,k_i} <_{I_2} t <_{I_2} t_{\varepsilon,k_{i+1}}$, hence (by $\boxtimes_3$) we have $\partial \geq \theta_2$

(γ) for each $i < m$ the set $\{t_{\varepsilon,\ell} : \varepsilon \in S$ and $\ell \in [k_i, k_{i+1})\}$ is included in the interval $(s_i^-, s_i^+)_{I_2}$.

[Why? Straight. For some stationary $S_1 \subseteq \partial$ and $\langle n_k : k < n^* \rangle$ we have $\varepsilon \in S_1 \wedge k < n^* \Rightarrow \ell g(t_{\varepsilon,k}) = n_k$. Without loss of generality also $\langle \ell^{t_{\varepsilon,k}(i)} : i < n_k \rangle$ does not depend on $\varepsilon \in S_1$. By $\sum\limits_{k<n^*} n_k$ application of $\partial \rightarrow (\partial, \omega)^2$, without loss of generality for each $k < n^*$ and $i < n_k$ the sequence $\langle \alpha^{t_{\varepsilon,k}(i)} : \varepsilon \in S_1 \rangle$ is constant or increasing. Cleaning a little more we are done.
So $\boxtimes_6$ holds.]

Lastly, recall that we chose I to be $(|M|, <^M)$, where M was the real closure of M_0 and (see $(*)_1$), M_0 the ordered field generated over $\mathbb{Q}$ by $\{a_t : t \in I_2\}$ as described in $(*)_1$ above and for every $u \subseteq \zeta$ let:

$(*)_2$ (a) $I_u^1 = \{(\ell, \beta) \in I_1 : \beta \in u$ or $\beta \in [\zeta, \zeta + \omega)\}$

(b) $I_u^{*,2} = \{\eta \in I_2^* : \alpha^{\eta(\ell)} \in I_u^1$ for every $\ell < \ell g(\eta)\}$

(c) $I_u^2 = \{\eta \in I_2 : \alpha^{\eta(\ell)} \in I_u^1$ for every $\ell < \ell g(\eta)\}$

(d) $I_u =$ the real closure of $\mathbb{Q}(a_t : t \in I_u^2)$ in M

(e) for $t \in I_2 \backslash I_u^2$ let $I_{u,t}^2 = I_2 \restriction \{s \in I_2 : s \notin I_u^2$ and for every $r \in I_u^2$ we have $r <_{I_2} t \equiv r <_{I_2} s\}$

(f) for $x \in I \backslash I_u$ let $I_{u,x} = I \restriction \{y \in I : y \notin I_u$ and $(\forall a \in I_u)(a <_I y \equiv a <_I x)\}$

(g) let $\hat{I}_u$ be the set $I_u \cup \{I_{u,a} : a \in I \backslash I_u\}$ ordered by: $x <_{\hat{I}_u} y$ iff
one of the following holds:

(α) $x, y \in I_u$ and $x <_{I_u} y$

(β) $x \in I_u, y = I_{u,b}$ and $x <_{I_u} b$

(γ) $x = I_{u,a}, y \in I_u$ and $a <_{I_u} y$

(δ) $x = I_{u,a}, y = I_{u,b}$ and $a <_{I_u} b$ (can use it more!)
(note that by $\boxtimes_5$, $|u| \leq \mu \Rightarrow |\hat{I}_u| \leq \mu$).

Now observe

$(*)_3$ for $u \subseteq \zeta$, I^2_u is unbounded in I_2 from below and from above.

We define

$(*)_4$ we say[4] that u is μ-reasonable if:

(a) $u \subseteq \zeta, |u| < \theta_2$ and $\mu \subseteq u$

(b) $\alpha \in u \equiv \alpha + 1 \in u$ for every α

(c) if $\delta \in u$ and $\aleph_0 \leq \text{cf}(\delta) \leq \mu$ then $\delta = \sup(u \cap \delta)$

(d) if $\delta \leq \zeta$ and $\text{cf}(\delta) > \mu$ then $\text{cf}(\text{otp}(\delta \cap u)) = \aleph_0$.

Now we note

$(*)_5$ if $X \subseteq I$ has cardinality $< \theta_2$ and $u_* \subseteq \zeta$ has cardinality $< \theta_2$ then we can find a μ-reasonable u such that $X \subseteq I_u$ and $u_* \subseteq u$ and $|u| = \mu + |X| + |u_*|$.

[Why? By the proof of $\boxtimes_4$.]

$(*)_6$ if u is μ-reasonable then $Y := I^2_u$ satisfies the conclusions of $\boxtimes_4$.

---

[4] we may in clauses (e) + (c) replace μ by $\mu + |\mathscr{U}|$, no harm and it makes (c)(β) of $(*)_1$, redundant

[Why? By the proof of $\boxtimes_4$, that is if $u^+ := u \cup \{\zeta + n : n < \omega\}$ then Y as defined in the proof there using u^+ for W, is I^2_u from $(*)_2(c)$, and it satisfies demands (a)-(f) from $\boxdot_2$ so the proof there applies.]

$(*)_7$ if u is μ-reasonable and $x \in I \backslash I_u$ then $\mathrm{cf}(I_{u,x}) \leq \aleph_0$.

Why? The proof takes awhile. Toward contradiction assume $\partial = \mathrm{cf}(I_{u,x})$ is $> \aleph_0$ and let $\langle b_\varepsilon : \varepsilon < \partial\rangle$ be an increasing sequence of members of $I_{u,x}$ unbounded in it. So for each $\varepsilon < \partial$ there is a definable function $f_\varepsilon(x_0, \ldots, x_{n(\varepsilon)-1})$ where definable of course means in the theory of real closed fields and $t_{\varepsilon,0} <_{I_2} t_{\varepsilon,1} <_{I_2} \cdots <_{I_2} t_{\varepsilon,n(\varepsilon)-1}$ from I_2 such that $M \models$ "$b_\varepsilon = f_\varepsilon(a_{t_{\varepsilon,0}}, \ldots, a_{t_{\varepsilon,n(\varepsilon)-1}})$" and $n(\varepsilon)$ is minimal. As $\mathrm{Th}(\mathbb{R})$ is countable and $\aleph_0 < \partial = \mathrm{cf}(\partial)$, without loss of generality $\varepsilon < \partial \Rightarrow f_\varepsilon = f_*$ so $\varepsilon < \partial \Rightarrow n(\varepsilon) = n(*)$.

Apply $\boxtimes_6$ to $\langle \bar{t}^\varepsilon = \langle t_{\varepsilon,\ell} : \ell < n(*)\rangle : \varepsilon < \partial\rangle$ and get $S \subseteq \partial$ and $0 = k_0 < k_1 < \ldots < k_m = n(*)$ and $\langle (s^-_i, s^+_i) : i < m\rangle$ and $\varepsilon(*) = \mathrm{Min}(S)$ as there. Without loss of generality the truth value of "$t_{\varepsilon,\ell} \in I^2_u$" for $\varepsilon \in S$, depends just on ℓ. Let $w_1 = \{i < m : (\forall \varepsilon \in S)(t_{\varepsilon,k_i} = t_{\varepsilon(*),k_i})\}$, $w_2 = \{\ell < n(*) : t_{\varepsilon(*),\ell} \in I^2_u\}$; clearly for every $\ell < n(*)$ we have $(\forall \varepsilon \in S)(t_{\varepsilon,\ell} = t_{\varepsilon(*),\ell}) \Leftrightarrow \ell \in \{k_i : i \in w_1\}$ and $i \in w_1 \Rightarrow k_i + 1 = k_{i+1}$.

Let $t^*_{k_i} = t_{\varepsilon,k_i}$ for ($\varepsilon < \partial$ and $i \in w_1$). Renaming without loss of generality $S = \partial$ and $\varepsilon(*) = 0$.

We have some free choice in choosing $\langle b_\varepsilon : \varepsilon < \partial\rangle$ (as long as it is cofinal in $I_{u,x}$), so without loss of generality we choose it such that $n(*)$ is minimal and then $|w_1|$ is maximal and then $|w_2|$ is maximal.

Now does the exceptional cae in $(b)(\beta)$ of $\boxtimes_6$ occurs? This is an easier case and we delay it to the end.

As I_2 and $I_{2,<t}$ for $t \in I_2$ have cofinality $\aleph_0$ (see $\boxtimes_2(a),(b)$) and $\boxtimes_3$ and this holds for the inverse of I_2, too, while $\partial = \mathrm{cf}(\partial) > \aleph_0$ and we can replace $\langle b_\varepsilon : \varepsilon < \partial\rangle$ by $\langle b_{n(*)+\varepsilon} : \varepsilon < \partial\rangle$ we can find $t_{\partial,\ell}$ ($\ell < n(*)$) such that

$\odot$ (a) $t_{\partial,0} <_{I_2} t_{\partial,1} <_{I_2} \cdots <_{I_2} t_{\partial,n(*)-1}$

(b) if $\varepsilon < \xi < \partial$ and $\ell_1, \ell_2 < n(*)$ then $(t_{\varepsilon,\ell_1} <_{I_2} t_{\partial,\ell_2}) \equiv (t_{\varepsilon,\ell_1} <_{I_2} t_{\xi,\ell_2})$ and $(t_{\partial,\ell_1} <_{I_2} t_{\varepsilon,\ell_2}) \equiv (t_{\xi,\ell_1} < t_{\varepsilon,\ell_2})$

(c) if $\ell \in [k_i, k_{i+1})$ then $t_{\partial,\ell} \in (s^-_i, s^+_i)_{I_2}$.

Case 0: $\{0, \ldots, m-1\} = w_1$.

This implies $i < m \Rightarrow k_i + 1 = k_{i+1}$ hence $m = n$ hence $\ell < n \Rightarrow t_{\xi,\ell} = t^*_\ell$ and so contradicts "$\langle b_\varepsilon : \varepsilon < \partial \rangle$ is increasing" (as it becomes constant).

Case 1: $[0, m) \backslash w_1$ is not a singleton.

It cannot be empty by "not case 1". Choose $i(*) \in \{0, \ldots, m-1\} \backslash w_1$ and for $\varepsilon, \xi < \partial$ let $\bar{t}^{\varepsilon,\xi} = \langle t^{\varepsilon,\xi}_\ell : \ell < n(*) \rangle$ be defined by: $t^{\varepsilon,\xi}_\ell$ is $t_{\varepsilon,\ell}$ if $\ell \in [k_{i(*)}, k_{i(*)+1})$ and $t_{\xi,\ell}$ otherwise. Let $b_{\varepsilon,\xi} = f_*(a_{t^{\varepsilon,\xi}_0}, \ldots, a_{t^{\varepsilon,\xi}_{n(*)-1}}) \in M$.

Clearly

$\circledast_0$ for any $\varepsilon_1, \varepsilon_2, \xi_1, \xi_2 \leq \partial$ the truth value of $b_{\varepsilon_1,\xi_1} < b_{\varepsilon_2,\xi_2}$ depend just on the inequalities which $\langle \varepsilon_1, \varepsilon_2, \xi_1, \xi_2 \rangle$ satisfies and even just on the inequalities which the $t_{\varepsilon_1,\ell}, t_{\varepsilon_2,\ell}, t_{\xi_1,\ell}, t_{\xi_2,\ell}$ ($\ell < n(*)$) satisfy.

[Why? Recall $\langle \langle t_{\varepsilon,\ell} : \ell < n(*) \rangle : \varepsilon \in S \rangle$ is an indiscernible sequence in the linear order I_2 (for quantifier free formulas) and M has elimination of quantifiers.]

$$\circledast_1 \quad \bigwedge_{\ell=1,2} \varepsilon(0) < \varepsilon_\ell < \varepsilon(1) < \partial \Rightarrow b_{\varepsilon(0)} <_I b_{\varepsilon_1,\varepsilon_2} <_I b_{\varepsilon(1)}.$$

[Why? By $\circledast_0$ the desire statement, $b_{\varepsilon(0)} <_I b_{\varepsilon_1,\varepsilon_2} <_I b_{\varepsilon(1)}$ is equivalent to $b_{\varepsilon(0)} < b_{\varepsilon_1,\varepsilon_1} < b_{\varepsilon(1)}$ which means $b_{\varepsilon(0)} < b_{\varepsilon_1} < b_{\varepsilon(1)}$ which holds.]

$\circledast_2$ $b_{0,2} <_I b_1$.

[Why? Otherwise $b_1 \leq_I b_{0,2}$ hence $\varepsilon \in (0, \partial) \Rightarrow b_\varepsilon <_I b_{0,\varepsilon+1} <_I b_{\varepsilon+2}$ (by $\circledast_0 + \circledast_1$) so $\langle b_{0,\varepsilon} : \varepsilon \in (1, \partial) \rangle$ is also an increasing sequence unbounded in $I_{u,x}$ contradiction to "w_1 maximal".]

$\circledast_3$ $b_{0,2} < b_{1,2}$.

[Why? By $\circledast_0 + \circledast_2$ we have $b_{0,4} < b_1$ and by $\circledast_1$ we have $b_1 < b_{2,4}$ together $b_{0,4} < b_{2,4}$ so by $\circledast_0$ we have $b_{0,2} < b_{1,2}$.]

But then $\langle b_{\varepsilon,\partial} : \varepsilon < \partial \rangle$ increases (by $\circledast_3 + \circledast_0$) and $\varepsilon < \partial \Rightarrow b_\varepsilon = b_{\varepsilon,\varepsilon} < b_{\varepsilon+1,\partial} < b_{\varepsilon+2}$ (by $\circledast_1$ and $\circledast_2$ respectively) hence is an unbounded subset of $I_{u,x}$ contradiction to the maximality of $|w_1|$.

Case 2: $m \backslash w_1 = \{0, \ldots, m-1\} \backslash w_1$ is $\{i(*)\}$.

Subcase 2A: For some $i < m, i \neq i(*)$ and $j := k_i \notin w_2$.

Choose such i with $|i - i(*)|$ maximal. For any s let $t_{\varepsilon,\ell,s}$ be $t_{\varepsilon,\ell}$ if $\ell \neq j$ and be s if $\ell = j$.

Let $I' = \{s \in I^2_{u,t_{\varepsilon(*),j}} : s, t_{\varepsilon(*),j}$ realize the same cut of $\{t_{\varepsilon,\ell} : \varepsilon < \partial, \ell \neq j\}\}$, note that $k_{j+1} = k_j + 1$. Recalling $\boxtimes_2(b)$, the cofinality of $I_{2,<t_{\varepsilon(*),j}}$ is $\aleph_0$ and also the cofinality of the inverse of $I_{2,>t_{\varepsilon(*),j}}$ is $\aleph_0$ recalling the choice of $\langle (s^-_\iota, s^+_\iota) : \iota < m \rangle$ there is an open interval[5] of I_2 around $t_{\varepsilon(*),j}$ which is $\subseteq I'$. Note that I' is dense in itself and has neither first nor last member by $\boxtimes_2 + \boxtimes_4(a), (b)$.

As f_* is definable, by the choice of M_0 and M and of $I' \subseteq I^2_{u,t_{\varepsilon(*),j}}$ we have: if $\varepsilon < \partial \wedge s \in I'$ then $t_{\varepsilon(*),j}$ and s realize the same cut of $I^2_u \cup \{t_{\varepsilon,\ell} : \varepsilon < \partial, j \neq \ell\}$ hence $f^M_*(\ldots, a_{t_{\varepsilon,\ell,s}}, \ldots)_{\ell<n}, b_\varepsilon$ realize the same cut of I_u which means that $f_*(\ldots, a_{t_{\varepsilon,\ell,s}}, \ldots)_{\ell<n} \in I_{u,x}$ hence by the choice of $\langle b_\varepsilon : \varepsilon < \partial \rangle$ we have $(\exists \xi < \partial)(f_*(\ldots, a_{t_{\varepsilon,\ell,s}}, \ldots) < b_\xi)$.

So again by the definability (and indiscernibility)

$$\circledast_4 \quad \varepsilon < \partial \wedge s \in I' \Rightarrow f^M_*(\ldots, a_{t_{\varepsilon,\ell,s}}, \ldots) < b_{\varepsilon+1}.$$

As I' is dense in itself, what we say on the pair $(s, t_{\varepsilon(*),j})$ when $s \in I' \wedge s <_{I_2} t_{\varepsilon(*),j}$ holds for the pair $(t_{\varepsilon(*),j}, s)$ when $s \in I' \wedge t_{\varepsilon(*),j} <_I s$ so

$$\circledast_5 \quad \varepsilon < \partial \wedge s \in I' \Rightarrow b_\varepsilon < f^M_*(\ldots, a_{t_{\varepsilon+1,\ell,s}}, \ldots)$$

(more fully let $s_1 <_{I_2} t_{\varepsilon(*),j} <_{I_2} s_2$ and $s_1, s_2 \in I'$ then the sequences $\langle t_{\varepsilon,\ell} : \ell \neq j, \ell < n(*) \rangle \hat{}\langle s_1 \rangle \hat{}\langle t_{\varepsilon+1,\ell} : \ell \neq j, \ell < n(*) \rangle \hat{}\langle t_{\varepsilon(*),j} \rangle$ and $\langle t_{\varepsilon,\ell} : \ell \neq j, \ell < n(*) \rangle \hat{}\langle t_{\varepsilon(*),j} \rangle \hat{}\langle t_{\varepsilon+1,\ell} : \ell \neq j, \ell < n(*) \rangle \hat{}\langle s_2 \rangle$ realizes the same quantifier free type in I_2, (recalling $t_{\varepsilon,j} = t_{\varepsilon(*),j}$).

By $\circledast_4 + \circledast_5$ and indiscernibiity we can replace $t_{\varepsilon(*),j}$ by any $t' \in I'$ which realizes the same cut as $t_{\varepsilon(*),j}$ of $\{t_{\varepsilon,\ell} : \varepsilon < \partial, \ell \neq j\}$. But if $j > i(*)$ then $\{t^*_{j+1}, \ldots, t^*_{n(*)-1}\} \subseteq I^2_u$ by the choice of j, and the set $I'' = \{t \in I_2$: if $\varepsilon < \partial, \ell \neq j$ then $t \neq t_{\varepsilon,\ell}$ and $t_{\varepsilon,\ell} <_{I_2} t \equiv t_{\varepsilon,\ell} <_{I_2} t^*_j\}$ include an initial segment of $J^+_{I^2_u, t_{\varepsilon(*),j}}$,

[5] if we allow $+\infty, -\infty$ as end points

see $\boxtimes_4(d)$, i.e. $(*)_6$ so its inverse has cofinality $\aleph_0$, say $\langle s^*_n : n < \omega \rangle$ exemplifies this, so $n < \omega \Rightarrow s^*_{n+1} <_{I_2} s^*_n$. So for every $\varepsilon < \partial$ for some $n < \omega, f^M_*(\ldots, a_{t_{\varepsilon+1,\ell}, s^*_n}, \ldots) \in (b_\varepsilon, b_{\varepsilon+1})_I$. So for some $n_* < \omega$ this holds for unboundedly many $\varepsilon < \partial$, contradictory to "$|w_2|$ is maximal". Similarly if $j < i(*)$.

<u>Subcase 2B</u>: For every $\varepsilon < \partial$ for some $\xi \in (\varepsilon, \partial)$, the interval of I_2 which is defined by $t_{\varepsilon, k_{i(*)}}, t_{\xi, k_{i(*)}}$ is not disjoint to I^2_u [so without loss of generality has $\geq k_{i(*)+1} - k_{i(*)}$ members of I^2_u].

In this case as in case 1, without loss of generality $\{k_{i(*)}, \ldots, k_{i(*)+1}\} \subseteq w_2$ so as $|w_2|$ is maximal this holds. So as not subcase 2A, $\{t_{\varepsilon,\ell} : \varepsilon < \partial, \ell < n\} \subseteq I^2_u$ hence $\{b_\varepsilon : \varepsilon < \partial\} \subseteq I_u$, contradiction.

<u>Subcase 2C</u>: None of the above.

As not subcase(2B), without loss of generality $\{t_{\varepsilon,\ell} : \varepsilon < \partial$ and $\ell \in [k_{i(*)}, k_{i(*)+1})\} \subseteq I^2_{u, t_{\varepsilon(*), k_{i(*)}}}$. Then as in subcase(2A) the sequence $\langle t_{\varepsilon, k_{i(*)}} : \varepsilon < \partial \rangle$ is increasing/decreasing and is unbounded from above/below in $I^2_{u, t_{\varepsilon(*), k_{i(*)}}}$ contradiction to $(*)_6$.

In more detail, so $I' := I^2_{u, t_{0, k_{i(*)}}}$ includes all $\{t_{\varepsilon,\ell} : \varepsilon < \partial$ and $\ell \in [k_{i(*)}, k_{i(*)+1})\}$. Also I' and its inverse are of cofinality $\aleph_0$ by $(*)_6$ hence without loss of generality we can find (new) $\langle t_{\partial,\ell} : \ell \in [k_{i(*)}, k_{i(*)+1}) \rangle$ such that $t_{\partial,\ell} <_{I_2} t_{\partial,\ell+1}, t_{\partial,\ell} \in (s^-_{i(*)}, s^+_{i(*)})_{I_2}$ and $\varepsilon < \partial \Rightarrow t_{\varepsilon,\ell_1} <_{I_2} t_{\partial,\ell} \equiv t_{\varepsilon,\ell_1} < t_{\varepsilon+1,\ell_2}$ and the convex hull in I_2 of $\{t_{\zeta,\ell} : \zeta \leq \partial$ and $\ell \in [k_{i(*)}, k_{i(*)+1}]\}$ is disjoint to I^2_u. Let $t_{\partial,\ell} = t_{\partial,\ell}$ for $\ell \notin [k_{i(*)}, k_{i(*)+1}], \ell < m, b_\partial = f_*(a_{t_{\partial,0}}, \ldots, a_{t_{\partial,n-1}})$.

Easily $\varepsilon < \partial \Rightarrow b_\varepsilon <_I b_\partial$. As $\varepsilon < \xi < \partial \Rightarrow (b_\varepsilon, b_\xi)_{I_2} \cap u = \emptyset$ easily $\varepsilon < \partial \Rightarrow (b_\varepsilon, b_\partial)_{I_2} \cap u = 0$, contradiction to $\langle b_\varepsilon : \varepsilon < \partial \rangle$ being cofinal in $I_{u,x}$.

To finish proving $(*)_7$, we have to consider the possibility that applying $\boxtimes_6$, the exceptional case in $(b)(\beta)$ of $\boxtimes_6$ occurs for some $i < m$ say for $i(*)$; see before $\odot$.

Also without loss of generality as $\partial \geq \theta_2$ then without loss of generality $\ell \in w_2 \Rightarrow t_{\varepsilon,\ell} = t_{\varepsilon(*),\ell}$ and for each $\ell < n(*)$ we have $(\forall \varepsilon, \zeta < \partial)(\forall s \in I^2_u)(s <_{I_2} t_{\varepsilon,\ell} \equiv s <_{I_2} t_{\zeta,\ell})$.

Now we can define $\bar{t}^{\varepsilon,\xi} = \langle t^{\varepsilon,\xi}_\ell : \ell < n(*) \rangle$ as in case 1 and prove $\circledast_0 - \circledast_3$ there.

Clearly all members of $\{t_{\varepsilon,\ell} : \varepsilon < \partial, \ell \in [k_{i(*)}, k_{i(*)+2})\}$ realize the same cut of I^2_u and we get easy contradiction.

As we can use only $\langle t_{n(*),\varepsilon} : \varepsilon < \partial\rangle$ and add to f_* dummy variables, without loss of generality $k_{i(*)+1} - k_{i(*)} = k_{i(*)+2} - k_{i(*)+1}$. Let J be $\{1,-1\} \times \partial$ ordered by $(\ell_1, \varepsilon_1) <_J (\ell_2, \varepsilon_2)$ iff $\ell_1 = 1 \wedge \ell_2 = -1$ or $\ell_1 = 1 = \ell_2 \wedge \varepsilon_1 < \varepsilon_2$ or $\ell_1 = -1 = \ell_2 \wedge \varepsilon_1 > \varepsilon_2$.

For $\iota \in J$ let $\iota = (\ell^\iota, \varepsilon^\iota) = (\ell[\iota], \varepsilon[\iota])$. For $\zeta < \partial$ and $\iota_1, \iota_2 \in J$ we define $\bar t_{\zeta,\iota_1,\iota_2} = \langle t_{\zeta,\iota_1,\iota_2,n} : n < n(*)\rangle$ by $t_{\zeta,\iota_1,\iota_2,n}$ is $t_{\varepsilon[\iota_1],n}$ if $n \in [k_{i(*)}, k_{i(*)+1})$, $t_{\varepsilon[\iota_2],n}$ if $n \in [k_{i(*)+1}, k_{i(*)+2})$ and $t_{\zeta,n}$ otherwise. Now letting $b_{\zeta,\iota_1,\iota_2} := f_*(\bar t_{\zeta,\iota_1,\iota_2})$

$\circledast_6$ all $b_{\zeta,\iota_1,\iota_2}$ realize the ame cut of I^2_u.

Now

$\circledast_7$ indiscernibility as in $\circledast_0$ holds

$\circledast_8$ $\neg(b_{\zeta,(1,\varepsilon),(1,\varepsilon+1)} \le_{I_*} b_{\zeta,(1,\varepsilon+2),(1+\varepsilon+3)})$.

[Why? Otherwise by indiscernibility, if $\zeta \in (6,\partial)$ then $b_{\zeta,(1,\zeta),(-1,3)} <_I b_{\zeta,(-1,5),(-1,4)}$. Hence $\langle b_{\zeta,(-1,5),(-1,4)} : \zeta \in (6,\partial)\rangle$ is monotonic in I_*, all members realizing the fix cut of I^2_u and is unbounded in it (by the inequality above) so contradiction to maximality of $|w_j|$.]

$\circledast_9$ $\neg(b_{\zeta,(1,\varepsilon+2),(1,\varepsilon+3)} <_I b_{\zeta,(1,\varepsilon),(1,\varepsilon+1)})$.

[Similarly, as otherwise if $\zeta \in (6,\partial)$ then $b_{\zeta,(1,\zeta),(-1,\zeta)} <_I b_{\zeta,(1,4),(1,5)})$. Hence $\langle b_{\zeta,(1,4),(1,5)} : \zeta \in (6,\partial)\rangle$ contradict the maximality of (w_1).]

So we have proved $(*)_7$

$(*)_8$ if u is μ-reasonable, $x \in I \backslash I_u$ then $\mathrm{cf}(I_{u,x}) = \aleph_0$.

[Otherwise by $(*)_7$ it has a last element say $b = f_*(a_{t_0}, \dots, a_{t_{n-1}})$ where $t_0, \dots, t_{n-1} \in I_2$ and f_* a definable function, without loss of generality with n minimal hence $\{a_{t_0}, \dots, a_{t_{n-1}}\}$ is transcendentally independent and with no repetitions and b is not algebraic over $\{a_{t_0}, \dots, a_{t_{n-1}}\} \backslash \{a_{t_\ell}\}$ for $\ell < n$. So $\{t_0, \dots, t_{n-1}\} \nsubseteq I^2_u$ and let $\ell < n$ be such that $t_\ell \notin I^2_u$ hence there are $s_0 <_{I_2} s_1$ such that $t_\ell \in (s_0, s_1)_{I_2}$ and $(s_0, s_1)_{I_2} \cap I^2_u = \emptyset$ (recall $\boxtimes_4$(a), (b) and $(*)_6$ about cofinality $\aleph_0$ and I_2 being dense).

Also without loss of generality $\{t_0, \dots, t_{n-1}\} \cap (s_0, s_1)_{I_2} = \{t_\ell\}$, now the function $c \mapsto f^M_*(a_{t_0}, \dots, a_{t_{\ell-1}}, c, a_{t_{\ell+1}}, \dots, a_{t_{n-1}})$ for $c \in$

$(a_{s_0}, a_{s_1})_I$ is increasing or decreasing (cannot be constant by the minimality on n and the elimination of quantifiers for real closed fields and the transcendental independence of $\{t_0, \ldots, t_{n-1}\}$). So we can find s'_0, s'_1 such that $s_0 <_{I_2} s'_0 <_{I_2} t_\ell <_{I_2} s'_1 <_{I_2} s_1$ such that $X := \{f^M_*(a_{t_0}, \ldots, a_{t_{\ell-1}}, c, a_{t_{\ell+1}}, \ldots, a_{t_{n-1}}) : c \in (a_{s'_0}, a_{s'_1})_I\}$ is included in $I_{u,x}$. Again as the function defined above is monotonic on $(a_{s'_0}, a_{s'_1})_I$ so for some value $b' \in (a_{s'_0}, a_{s'_1})$ we have $b <_I b'$. But b is last in $I_{u,x}$ by our assumption toward contradiction hence $(b, b')_{I_u} \cap I_u = \emptyset$. But this is impossible as all members of $\{f(a_{t_0}, \ldots, a_{t_{\ell-1}}, c, a_{t_{\ell+1}}, \ldots, a_{t_{n-1}}) : c \in (a_{s'_1}, a_{s'_2})_I\}$ realize the same cut of I_u so $(*)_8$ holds.]

$(*)_9$ if u is μ-reasonable, $x \in I \backslash I_u$ then also the inverse of $I_{u,x}$ has cofinality $\aleph_0$.

[Why? Similarly to the proof of $(*)_7 + (*)_8$ or note that the mapping $y \mapsto -y$ (defined in M) maps I_u onto itself and is an isomorphism from I onto its inverse.]

$(*)_{10}$ if u is μ-reasonable, then I_u is unbounded in I from below and from above.

[Why? Easy.]

$(*)_{11}$ if h, u_1, u_2 are as in clauses (a),(b),(c) below <u>then</u> the function h_4 defined below is (well defined and) is, recalling $(*)_2(g)$, an order preserving function from $\hat{I}_{u_1}$ onto $\hat{I}_{u_2}$ mapping u_1 onto u_2 and also the functions $h_0, h_1, h^*_2, h_2, h_3$ are as stated where

- (a) $u_1, u_2 \subseteq \zeta$ are μ-reasonable
- (b) h is an order preserving function from u_1 onto u_2
- (c) (α) for $\alpha \in u_1$, we have $\mathrm{cf}(\alpha) \geq \theta_1 \Leftrightarrow \mathrm{cf}(h(\alpha)) \geq \theta_1$
  - (β) if $\gamma \in u_1$ then $(\forall \alpha < \gamma)(\exists \beta \in u_1)(\alpha \leq \beta < \gamma)$ iff $(\forall \alpha < h(\gamma))(\exists \beta \in u_2)(\alpha \leq \beta < h(\gamma)))$
- (d) (α) h_1 is the induced order preserving function from $I^1_{u_1}$ onto $I^1_{u_2}$, i.e., $h_1((\ell, \beta')) = (\ell, \beta'')$ when $h(\beta') = \beta'' < \zeta$ or $\beta' = \beta'' \in [\zeta, \zeta + \omega)$;
  - (β) let h_0 be the partial function from $\zeta + \omega$ into $\zeta + \omega$ such that $h_0(\alpha) = \beta \Leftrightarrow (\exists \ell)[h_1((\ell, \alpha)) = (\ell, \beta)]$

(e) h_2^* is the order preserving function from $I_{u_1}^{*,2}$ onto $I_{u_2}^{*,2}$ defined by: for $\eta \in I_{\eta_1}^{*,2}, h_2^*(\eta) = \langle h_1(\eta(\ell)) : \ell < \ell g(\eta)\rangle = \langle (\ell^{\eta(\ell)}, h_0(\alpha^{\eta(\ell)})) : \ell < \ell g(\eta)\rangle$, recalling (d)

(f) $h_2 = h_2^* \restriction I_{u_1}^2$ is an order preserving function from $I_{u_1}^2$ onto $I_{u_2}^2$

(g) h_3 is the unique isomorphism from the real closed field $M_{I_{u_1}^2}$ onto the real closed field $M_{I_{u_2}^2}$ mapping a_t to $a_{h_2(t)}$ for $t \in I_{u_1}^2$, where for $I' \subseteq I_2$ we let $M_{I'} \subseteq M$ be the real closure of $\{a_t : t \in I'\}$ inside M

(h) h_4 is the map defined by:
$h_4(x) = y$ iff

(α) $x \in I_{u_1} \wedge y = h_3(x)$ or

(β) for some $a \in I \backslash I_{u_1}, b \in I \backslash I_{u_2}$ we have $x = I_{u,a}, y \in I_{u,b}$ and $(\forall c \in I_u)(c <_I a \equiv h_3(c) <_I b)$

(i) $\hat{I}_{u_1} = \mathrm{Dom}(h_4)$ and $\hat{I}_{u_2} = \mathrm{Rang}(h_4)$ ordered naturally.

[Why? Trivially h_1 is an order preserving function from $I_{u_1}^1$ onto $I_{u_2}^1$. Recall $I_{u_\ell}^{2,*} = \{\eta \in I_2^* : \eta(\ell) \in I_{u_\ell}^1$ for $\ell < \ell g(\eta)\}$. So obviously h_2^* is an order preserving function from $I_{u_1}^{*,2}$ onto $I_{u_2}^{*,2}$. Now $h_2 = h_2^* \restriction I_{u_1}^2$, but does it map $I_{u_1}^2$ onto $I_{u_2}^2$? we have excluded some members of $I_{u_2}^{*,2}$ by ⊛ above. But by clauses (c) and (d)(α) of the assumption being excluded/not excluded is preserved by the natural mapping, i.e., h_2^* maps $I_{u_1}^2$ onto $I_{u_2}^2$ hence $h_2 = h_2^* \restriction I_{u_1}^1$ is an isomorphism from $I_{u_1}^1$ onto $I_{u_2}^1$. Also by $(*)_1$ being the real closure of the ordered field M_0, and the uniqueness of "the real closure" h_3 is the unique isomorphism from the real closed field $M_{I_{u_1}^2}$ onto $M_{I_{u_2}^2}$ mapping a_t to $a_{h_2(t)}$ for $t \in I_{u_1}^2$.

Let $\langle (\mathscr{U}_\varepsilon^1, \mathscr{U}_\varepsilon^2) : \varepsilon < \varepsilon^* \rangle$ list the pairs $(\mathscr{U}_1, \mathscr{U}_2)$ such that:

$\circledast_{10}$ (a) $\mathscr{U}_\ell$ has the form $I_{u_\ell, x}$ for some $x \in I \backslash I_{u_\ell}$ for $\ell = 1, 2$

(b) for every $a \in I_{u_1}$,
$(\exists y \in \mathscr{U}_1)(a <_I y) \Leftrightarrow (\exists y \in \mathscr{U}_2)(h_2(a) <_I y)$.

Now

$\circledast_{11}$ $\langle \mathscr{U}^\ell_\varepsilon : \varepsilon < \varepsilon^* \rangle$ is a partition of $I \backslash I_{u_\ell}$ for $\ell = 1, 2$.

[Why? First, note the parallel claim for I_1. For this note that $h_1((\ell, 0)) = (\ell, 0)$ as $0 \in u_1 \cap u_2$ as u_1, u_2 are μ-reasonable, see clause (e) of $(*)_4$ and $h_1((\ell, \alpha)) = (\ell, \beta) \Leftrightarrow h_1((\ell, \alpha + 1)) = (\ell, \beta + 1)$, by clause (b) of $(*)_4$ and if $h((\ell, \delta_1)) = (\ell, \delta_2), \delta_1$ is a limit (equivalently δ_2 is limit) then

$$\delta_1 = \sup\{\alpha < \delta : (\ell, \alpha) \in I^1_{u_1}\} \Leftrightarrow \delta_2 = \sup\{\alpha < \delta : (\ell, \alpha) \in I^1_{u_2}\}.$$

Second, note the parallel claim for $h_2, I^{*,2}_{u_\ell}, h^*_2$.

Third, note the parallel claim for $I^2_{u_\ell}, h_2$.

Fourth, note the parallel claim for I_{u_ℓ}, h_3 (which is the required one).]

So it follows that

$\circledast_{12}$ h_4 is as promised.

So we are done proving $(*)_{11}$.
[Why? By clauses (b),(c) of $(*)_{11}$.]

$(*)_{12}$ if u_1, u_2 are μ-reasonable, h is an order preserving mapping from $\hat{I}_{u_1}$ onto $\hat{I}_{u_2}$ which maps I_{u_1} onto I_{u_2} then there is an automorphism h^+ of the linear order I extending $h \restriction I_{u_1}$.

[Why? Let $\langle \mathscr{U}^1_\varepsilon : \varepsilon < \varepsilon^* \rangle$ list $\hat{I}_{u_1} \backslash I_{u_1}$ and $\mathscr{U}^2_\varepsilon = h(\mathscr{U}^1_\varepsilon)$. Now for every ε we choose $\langle a^\ell_{\varepsilon,n} : n \in \mathbb{Z} \rangle$ such that

$\circledast_{13}$ (a) $a^\ell_{\varepsilon,n} \in \mathscr{U}^\ell_\varepsilon$

(b) $a^\ell_{\varepsilon,n} <_I a^\ell_{\varepsilon,n+1}$ for $n \in \mathbb{Z}$

(c) $\{a^\ell_{\varepsilon,n} : n \in \mathbb{Z}, n \geq 0\}$ is unbounded from above in $\mathscr{U}^\ell_\varepsilon$

(d) $\{a^\ell_{\varepsilon,n} : n \in \mathbb{Z}, n < 0\}$ is unbounded from below in $\mathscr{U}^\ell_\varepsilon$.

This is justified by u_ℓ being μ-reasonable by $(*)_6$, $\boxtimes_4$. Now define $h_5 : I \to I$ by:

$h_5(x) = h_4(x)$ <u>if</u> $x \in I_{u_1}$ and otherwise
$h_5(x) = a^2_{\varepsilon,n} + (a^2_{\varepsilon,n+1} - a^2_{\varepsilon,n})(x - a^1_{\varepsilon,n})/(a^1_{\varepsilon,n+1} - a^1_{\varepsilon,n})$

<u>if</u> $a^1_{\varepsilon,n} \leq_{I_2} x < a^1_{\varepsilon,n+1}$ and $n \in \mathbb{Z}$.
Now check using linear algebra.]

$(*)_{13}$ $({}^{\mu}I)/\mathscr{E}^{\rm aut}_{I,\mu}$ has $\leq 2^\mu$ members recalling that $f_1\ \mathscr{E}^{\rm aut}_{I,h}\ f_2$ <u>iff</u> f_1, f_2 are functions from μ into I and for some automorphism h of I we have $(\forall \alpha < \mu)(h(f_1(\alpha)) = f_2(\alpha))$ [Why? Should be clear recalling $|I^1_u| \leq \mu$, recalling $(*)_5, (*)_{11}, (*)_{12}$.]

So we have finished proving part (1) of 5.1.
2) Really the proof is included in the proof of part (1). That is, given $I' \subseteq I$ of cardinality $< \theta_2$ by $(*)_5$ there is a μ-reasonable $u \subseteq \zeta$ such that $I' \subseteq I_u$ and $|u| = \mu + |I'|$. Now clearly

$(*)_{14}$ for μ-reasonable $u \subseteq \zeta$, the family $\{I^2_{u,x} : x \in I_2 \backslash I^2_u\}$ has $\leq \mu + |u|$ members.

[Why? By $\boxtimes_5$.]

$(*)_{15}$ for a μ-reasonable $u \subseteq \zeta$, the family $\{I_{u,x} : x \in I \backslash I_u\}$ has $\leq \mu$ members.

[Why? By $(*)_{16}$ below.]

$(*)_{16}$ if u is μ-reasonable then $I_{u,b_1} = I_{u,b_2}$ when
- (a) $b_k = f(a_{t_{k,0}}, \ldots, a_{t_{k,n-1}})$ for $k = 1, 2$
- (b) f a definable function in M
- (c) $t_{k,0} <_{I_2} \cdots <_{I_2} t_{k,n-1}$ for $k = 1, 2$
- (d) $t_{1,\ell} \in I^2_u \vee t_{2,\ell} \in I^2_u \Rightarrow t_{1,\ell} = t_{2,\ell}$
- (e) if $t_{1,\ell} \notin I^2_u$ then $I^2_{u,t_{1,\ell}} = I^2_{u,t_{2,\ell}}$ for $\ell = 0, \ldots, n-1$.

[Why? Use the proof of $(*)_{11}$, for $u_1 = u = u_2, h = \mathrm{id}_{u_2}$ so $\mathscr{U}^1_\varepsilon = \mathscr{U}^2_\varepsilon$ for $\varepsilon < \varepsilon^*$.

By the assumptions for each ℓ there is ε such that $a_{t_{\varepsilon,1,\ell}}, a_{t_{2,\ell}} \in \mathscr{U}^1_\varepsilon = \mathscr{U}^2_\varepsilon$. Now for each $\varepsilon < \varepsilon^*$ there is an automorphism π_ε of $\mathscr{U}^1_\varepsilon$

as a linear order mapping $t_{1,\ell}$ to $t_{2,\ell}$ if $t_{1,\ell} \in \mathscr{U}^1_\varepsilon$. Let $\pi = \cup\{\pi_\varepsilon : \varepsilon < \varepsilon^*\} \cup \mathrm{id}_{I_u}$.]

$(*)_{17}$ if $n < \omega, t^\ell_0 <_I t^\ell_1 <_I \ldots <_I t^\ell_{n-1}$ for $\ell = 1, 2, I_{u,t^1_k} = I_{u,t^2_k}$ for $k = 0, 1, \ldots, n-1$ then for some automorphism g of I over I_u we have $k < n \Rightarrow g(t^1_k) = t^2_k$.

[Why? We shall use g such that $g \restriction I_u = \mathrm{id}_{I_u}$ and $g \restriction I_{u,x}$ is an automorphism of $I_{u,x}$ for each $x \in I \backslash I_u$. Clearly it suffices to deal with the case $\{t^\ell_k : \ell < n \text{ and } \ell \in \{1, n\}\} \subseteq I_{u,x}$ for one $x \in I \backslash I_u$. We choose $s_1 < s_2$ from $I_{u,x}$ such that $s_1 <_I t^\ell_k < s_2$ for $\ell = 1, 2$. We choose $g \restriction I_{u,x}$ such that it is the identity on $\{s \in I_{u,x} : s \leq_I s_1$ or $s_2 \leq_I s$, now stipulates $t_{-1} = s_1, t_n = s_2$ and maps $(t^1_k, t^1_{k+1})_I$ onto $(t^2_k, t^2_{k+1})_I$ for $k = -1, 0, \ldots, n-1$ as in the definition above.]
So we have completed the proof of part (2) of 5.1.
3) Obvious from the Definition (0.14(9)) and the construction.
4) First

$\odot_1$ there is $J^*_1 \subseteq I$ of cardinality μ^+ such that: for every $J^*_2 \subseteq I$ of cardinality $\leq \mu$ there is an automorphism π of I which maps J^*_2 into J^*_1.

[Why? Let $u = \mu^+ \times \mu^+ \subseteq \zeta$ and let $J^*_1 = I_u$. Clearly u has cardinality μ^+ and so does $J^*_1 = I_u$. So suppose $J^*_2 \subseteq I$ has cardinality $\leq \mu$. There is $u_2 \subseteq \zeta$ of cardinality μ such that $J^*_2 \subseteq I_{u_2}$ and without loss of generality u_2 is reasonable. We define an increasing function h from u_2 into u_1, by defining $h(\alpha)$ by induction on α:

$(*)_{17}$ if $\mathrm{cf}(\alpha) \leq \mu$ then $h(\alpha) = \cup\{h(\beta) + 1 : \beta \in u_2 \cap \alpha\}$

$(*)_{18}$ if $\mathrm{cf}(\alpha) > \mu$ then $h(\alpha) = \cup\{h(\beta) + 1 : \beta \in u_2 \cap \alpha\} + \mu^+$.

Let $u_1 := \{h(\alpha) : \alpha \in u_2\}$ so $u_1 \subseteq u$. Now h, u_1, u_2 satisfies clauses (a),(b),(c) of $(*)_{11}$ hence $h_1, h^*_2, h_2, h_3, h_4, \hat{I}_{u_1}, \hat{I}_{u_2}$ are as there.

By $(*)_{12}$ there is an isomorphism h^+ of I which extends h_4; now does h^+ map J^*_2 into J^*_1? Yes, as $J^*_2 \subseteq I_{u_2}$ and $h^+ \restriction I_{u_2}$ is an isomorphism from I_{u_2} onto I_{u_1} but $I_{u_1} \subseteq I_u, I_u = J^*_1$, so we are done proving $\odot_1$.]

Finally

$\odot_2$ part (4) of 5.1 holds, i.e. if $I_0^* \subseteq I, |I_0^*| < \theta_2$ then for some $I_1^* \subseteq I$ of cardinality $\leq \mu^+ + |I_0^*|$ we have: for every $J \subseteq I$ of cardinality $\leq \mu$ there is an automorphism of I over I_0^* mapping J into I_1^*.

Why? Given $I_0^* \subseteq I$ of cardinality $< \theta_2$ we can find $u_1 \subseteq \zeta$ of cardinality $\mu + |I_0^*|$ such that $I_0^* \subseteq I_{u_1}$. By $(*)_5$ we can find a μ-reasonable set $u_2 \subseteq \zeta$ of cardinality $\mu + |u_1|$ such that $u_1 \subseteq u_2$.

Let $\langle \mathscr{U}_\varepsilon : \varepsilon < \varepsilon^* \rangle$ list the sets of the form $I_{u_2,x}, x \in I_2 \backslash I_{u_1}$, so by $(\boxdot_5)$ $\varepsilon^* \leq \mu + |I_0^*|$. For each ε we choose $\langle a_{\varepsilon,n} : n \in \mathbb{Z} \rangle$ as in $\circledast_{13}$ from the proof of $(*)_{12}$. For each $\varepsilon < \varepsilon^*$ and $n \in \mathbb{Z}$ let $\pi_{\varepsilon,n}$ be an isomorphism from I onto $(a_{\varepsilon,n}, a_{\varepsilon,n+1})_I$, exists by the properties of ordered fields. Let $J_1^* \subseteq I$ be as in $\odot_1$ above and let $I_2^* = I_1^* \cup \{a_{\varepsilon,n} : \varepsilon < \varepsilon^* \text{ and } n < \omega\} \cup \{\pi_{\varepsilon,n}(J_1^*) : \varepsilon < \varepsilon^* \text{ and } n \in \mathbb{Z}\}$. Easily, I_2^* is as required.

5) By 0.12. $\square_{5.1}$

Remark. Concerning $(*)_{11}$, we could have used more time

$(*)'_{11}$ h_2 is an order preserving function from $I^2_{u_1}$ onto $I^2_{u_2}$ and h_3 is an isomorphism from I_{u_1} onto I_{u_2} and h_1 is an order preserving mapping from $\hat{I}_{u_2}$ onto $\hat{I}_{u_2}$.

§6 Linear orders and equivalence relations

This section deals with a relative of the stability spectrum. We ask: what can be the number of equivalence classes in ${}^\mu I$ for an equivalence realtion on ${}^\mu I$ which is so called "invariant", in fact definable (essentially by a quantifier free infinitary formula, mainly for well ordered I).

It is done in a very restricted context, but via EM-models has useful conclusions, for a.e.c. and also for a.e.c. with amalgamation; i.e. it is used in 7.8.

There are two versions; one for well ordering and one for the class of linear orders both expanded by unary relations.

On $\tau^*_{\alpha(*)}, K^{\text{lin}}_{\tau^*_{\alpha(*)}}$ see 0.14(4). We may replace sequences, i.e. $\text{inc}_J(I)$ by subsets of I of cardinality $|J|$, this may help to eliminate $2^{|J|}$

later, but at present it seems not to help in the final bounds in §7. We do here only enough for §7.

6.1 Context. We fix $\alpha(*), \bar{u}^* = (u^-, u^+)$ such that

(a) $\alpha(*)$ is an ordinal ≥ 1
(b) $u^- \subseteq \alpha(*)$
(c) $u^+ \subseteq \alpha(*)$.

6.2 Remark. 1) The main cases are

(A) $\alpha(*) = 1$, so $K^{\rm lin}_{\tau^*_{\alpha(*)}}$ is the class of linear orders
(B) $\alpha(*) = 2, u^+ = \emptyset, u^- = \{0\}$.

2) Usually the choice of the parameters does not matter.

6.3 Definition. 1) For $I, J \in K^{\rm lin}_{\tau^*_{\alpha(*)}}$, i.e. both linear orders expanded by a partition $P_\alpha(\alpha < \alpha(*))$, pedantically the interpretation of the P_α's, let $\mathrm{inc}'_J(I)$ be the set of embedding of J into I; see below, we denote members by h.
2) Recalling $\bar{u}^* = (u^-, u^+)$ where $u^- \cup u^+ \subseteq \alpha(*)$ let $\mathrm{inc}^{\bar{u}^*}_J(I)$ be the set of h such that

(a) h is an embedding of J into I, i.e. one-to-one, order preserving function mapping P^J_α into P^I_α for $\alpha < \alpha(*)$
(b) if $\alpha \in u^-$ and $t \in P^J_\alpha$ and $s <_I h(t)$ <u>then</u> for some $t_1 <_J t$ we have $s \leq_I h(t_1)$
(c) if $\alpha \in u^+$ and $t \in P^J_\alpha$ and $h(t) <_I s$ <u>then</u> for some t_1 we have $t <_J t_1$ and $h(t_1) \leq_I s$.

Concerning $\bar{u}^*$

6.4 Observation. 1) For any $h \in \mathrm{inc}^{\bar{u}^*}_J(I)$

(a) if t is the successor of s in J (i.e. $s <_J t$ and $(s,t)_J = \emptyset$) and $t \in P^J_\alpha, \alpha \in u^-$ <u>then</u> $h(t)$ is the successor of $h(s)$ in I

(b) if $\langle t_i : i < \delta \rangle$ is $<_J$-increasing with limit $t_\delta \in J$ (i.e. $i < \delta \Rightarrow t_i <_J t_\delta$ and $\emptyset = \cap\{(t_i, t_\delta)_J : i < \delta\}$) and $t_\delta \in P^J_\alpha, \alpha \in u^-$ <u>then</u> $\langle h(t_i) : i < \delta \rangle$ is $<_I$-increasing with limit $h(t_\delta)$ in I

(c) if t is the first member of J and $t \in P^J_\alpha, \alpha \in u^-$ <u>then</u> $h(t)$ is the first member of I.

2) If $h_1, h_2 \in \operatorname{inc}^{\bar{u}^*}_J(I)$ <u>then</u>

(a) if t is the successor of s in J and $t \in P^J_\alpha, \alpha \in u^-$ <u>then</u> $h_1(s) = h_2(s) \Leftrightarrow h_1(t) = h_2(t)$ and $h_1(s) <_I h_2(s) \Leftrightarrow h_1(t) <_I h_2(t)$ and $h_1(s) >_I h_2(s) \Leftrightarrow h_1(t) >_I h_2(t)$

(b) if $\langle t_i : i < \delta \rangle$ is $<_J$-increasing with limit t_δ and $t_\delta \in P^J_\alpha, \alpha \in u^-$, <u>then</u> $(\forall i < \delta)(h_1(t_i) = h_2(t_i)) \Rightarrow h_1(t_\delta) = h_2(t_\delta)$ moreover $(\forall i < \delta)(\exists j < \delta)(h_1(t_i) <_I h_2(t_j) \wedge h_2(t_i) <_I h_1(t_j)) \Rightarrow h_1(t_\delta) = h_2(t_\delta)$ and also $(\exists j < \delta)(\forall i < \delta)(h_1(t_i) <_I h_2(t_j)) \Rightarrow h_1(t_\delta) <_I h_2(t_\delta)$.

3) Similar to parts (1) + (2) for $\alpha \in u^+$ (inverting the orders of course).
4) $\operatorname{inc}'_I(J) = \operatorname{inc}^{(\emptyset,\emptyset)}_I(J)$.

Proof. Straight (and see the proof of 6.7). $\square_{6.4}$

6.5 Convention. 1) $\alpha(*), \bar{u}^*$ will be constant so usually we shall not mention them, e.g. write $\operatorname{inc}_J(I)$ for $\operatorname{inc}^{\bar{u}^*}_I(I)$ and pedantically below we should have written $\mathbf{e}^{\bar{u}^*}(J, I), \mathbf{e}^{\bar{u}^*}_*(J)$ and also in notions like reasonable and wide in Definition 6.10 mention $\bar{u}^*$.
2) I, J denote members of $K^{\text{lin}}_{\tau^*_{\alpha(*)}}$.

Below we use mainly "e-pairs" (and weak e-pairs and the reasonable case).

6.6 Definition. 1) let $\mathbf{e}(J)$ be the set of equivalence relations on some subset of J such that each equivalence class is a convex subset of J.
2) For $h_1, h_2 \in \operatorname{inc}_J(I)$ we say that (h_1, h_2) is a strict e-pair (for (I, J)) <u>when</u> $e \in \mathbf{e}(J)$ and (h_1, h_2) satisfies

(a) $s \in J \setminus \mathrm{Dom}(e)$ iff $h_1(s) = h_2(s)$

(b) if $s <_J t$ and $s/e \neq t/e$ (so $s, t \in \mathrm{Dom}(e)$) <u>then</u> $h_1(s) <_I h_2(t)$ and $h_2(s) <_I h_1(t)$

(c) if $s <_J t$ and $s/e = t/e$ (so $s, t \in \mathrm{Dom}(e)$) <u>then</u> $h_1(t) <_I h_2(s)$.

2A) We say that (h_1, h_2) is a strict $(e, \mathscr{Y})$-pair where $e \in \mathbf{e}(J)$ and $\mathscr{Y} \subseteq \mathrm{Dom}(e)/e$ <u>when</u> clauses (a)+(b) from part (2) hold and

(c)′ if $s <_J t$ and $s/e = t/e$ (so $s, t \in \mathrm{Dom}(e)$) <u>then</u> $(h_1(t) <_I h_2(s)) \equiv (s/e \in \mathscr{Y}) \equiv (h_1(s) < h_2(t))$.

2B) We say that (h_1, h_2) is an e-pair when (h_1, h_2) is a strict $(e, \mathscr{Y})$-pair for some $\mathscr{Y}$ (this relation is symmetric, see below).
3) We say that (h_1, h_2) is a weak e-pair where $h_1, h_2 \in \mathrm{inc}_J(I)$ <u>when</u> clauses (a),(b) hold (this, too, is symmetric!)
4) For $h_1, h_2 \in \mathrm{inc}_J(I)$, let $e = \mathbf{e}(h_1, h_2)$ be the (unique) $e \in \mathbf{e}(J)$ such that (see 6.8(1) below)

(a) $\mathrm{Dom}(e) = \{s \in J : h_1(s) \neq h_2(s)\}$

(b) (h_1, h_2) is a weak e-pair

(c) if $e' \in \mathbf{e}(J)$ and (h_1, h_2) is a weak e'-pair <u>then</u> $\mathrm{Dom}(e) \subseteq \mathrm{Dom}(e')$ and e refines $e' \restriction \mathrm{Dom}(e)$.

5) If $e \in \mathbf{e}(J)$ and $\mathscr{Y} \subseteq \mathrm{Dom}(e)/e$ <u>then</u> we let $\mathrm{set}(\mathscr{Y}) = \{s \in J : s/e \in \mathscr{Y}\}$ and $e \restriction \mathscr{Y} = e \restriction \mathrm{set}(\mathscr{Y})$.
6) Let $\mathbf{e}(J, I)$ be the set of $e \in \mathbf{e}(J)$ such that there is an e-pair.
7) Let $\mathbf{e}_*(J) = \cup\{\mathbf{e}(J, I) : I \in K^{\mathrm{lin}}_{\tau^*_{\alpha(*)}}\}$.

Concerning $\bar{u}^*$

6.7 Observation. Assume that $e \in \mathbf{e}(J, I)$.
0)

(a) If t is the first member of J and $t \in P^J_\alpha, \alpha \in u^-$ <u>then</u> $t \notin \mathrm{Dom}(e)$.

(b) If $t \in \mathrm{Dom}(e)$ and t is the first member of t/e and $t \in P^J_\alpha$ <u>then</u> $\alpha \notin u^-$.

1) If t is the $<_J$-successor of s and $t \in P^J_\alpha, \alpha \in u^-$ <u>then</u> $s \in \mathrm{Dom}(e) \Leftrightarrow t \in \mathrm{Dom}(e)$ and $s \in \mathrm{Dom}(e) \Rightarrow s \in t/e$.
2) If $\langle t_i : i < \delta \rangle$ is $<_J$-increasing with limit t_δ and $t_\delta \in P^J_\alpha$ and $\alpha \in u^-$ <u>then</u>:

(a) if $(\forall i < \delta)(t_i \notin \mathrm{Dom}(e))$ <u>then</u> $t_\delta \notin \mathrm{Dom}(e)$
(b) if $(\forall i < \delta)(\neg t_i e t_{i+1})$ or just $(\forall i < \delta)(\exists j < \delta)(i < j \wedge \neg t_i e t_j)$ <u>then</u> $t_\delta \notin \mathrm{Dom}(e)$
(c) if $(\forall i < \delta)(t_i \in t_0/e)$ <u>then</u> $t_\delta \in t_0/e$.

3) Similar to parts (0),(1),(2) when $\alpha \in u^+$ (inverting the order, of course).
4) $\mathbf{e}_*(J)$ is the family of $e \in \mathbf{e}(J)$ satisfying the requirements in parts (0),(1),(2),(3) above so if $\bar{u}^* = (\emptyset, \emptyset)$ then $\mathbf{e}_*(J) = \mathbf{e}(J)$.

Proof. Easy by 6.4, e.g.

<u>Part (1)</u>: We are assuming $e \in \mathbf{e}(J, I)$ hence by Definition 6.6 there is an e-pair (h_1, h_2) where $h_1, h_2 \in \mathrm{inc}_J(I)$. Now for $\ell = 1, 2$, clearly $h_\ell(s), h_\ell(t) \in I$ and as $s <_J t$ we have $h_\ell(s) < h_\ell(t)$. Now if $h_\ell(t)$ is not the $<_I$-successor of $h_\ell(s)$ then there is $s'_\ell \in (h_\ell(s), h_\ell(t))_I$ hence by clause (b) of Definition 6.3(2) there is $s^*_\ell \in [s, t)_J$ such that $s'_\ell \leq_I h_\ell(s^*_\ell) <_I h_\ell(t)$ so as $h_\ell(s) <_I s'_\ell$ we have $h_\ell(s) <_I h_\ell(s^*_\ell) <_I h_\ell(t)$ hence $s <_I s^*_\ell <_J t$, contradiction to the assumption "t is the successor of s in J". So indeed $h_\ell(t)$ is the successor of $h_\ell(s)$ in I.

As this holds for $\ell = 1, 2$, clearly $h_1(s) = h_2(s) \Leftrightarrow h_1(t) = h_2(t)$ but by Definition 6.3(2) we know $s \in \mathrm{Dom}(e) \Leftrightarrow (h_1(s) \neq h_2(s))$ and similarly for t hence $s \in \mathrm{Dom}(e) \Leftrightarrow t \in \mathrm{Dom}(e)$. Lastly, assume $s, t \in \mathrm{Dom}(e)$, but s, t are nor e-equivalent so by Definition 6.6(2) clause (b) we have $h_1(s) <_I h_2(t) \wedge h_2(s) <_I h_1(t)$ clear contradiction.

<u>Part 2</u>: We leave clauses (a),(b) to the reader.

For clause (c) of part (2), if $t_\delta \notin t_0/e$ then choose $h_1, h_2 \in \mathrm{inc}^{\bar{u}^*}_J(I)$ such that (h_1, h_2) is an e-pair, hence an $(e, \mathscr{Y})$-pair for some $\mathscr{Y} \subseteq \mathrm{Dom}(e)/e$. If $(t_0/e) \in \mathscr{Y}$ then $h_2(t_0)$ is above $\{h_1(t_i) : i < \delta\}$ by $<_I$ so we have $h_1(t_\delta) \leq_I h_2(t_0)$ but if $t_\delta \notin t_0/e$ this contradicts clause (b) in Definition 6.6(2),(2A). The proof when $t_0/e \notin \mathscr{Y}$ is similar. $\square_{6.7}$

6.8 Observation. Let $h_1, h_2 \in \operatorname{inc}_J(I)$ and $e \in \mathbf{e}(J)$.
1) $\mathbf{e}(h_1, h_2)$ is well defined.
2) (h_1, h_2) is a strict $(e, \mathscr{Y}_1)$-pair iff (h_2, h_1) is a strict $(e, \mathscr{Y}_2)$-pair when $(\mathscr{Y}_1, \mathscr{Y}_2)$ is a partition of $\mathrm{Dom}(e)/e$.
3) (h_1, h_2) is a strict e-pair iff (h_2, h_1) is a strict $(e, \emptyset)$-pair.
4) (h_1, h_2) is an e-pair iff (h_2, h_1) is an e-pair.
5) (h_1, h_2) is a weak e-pair iff (h_2, h_1) is a weak e-pair.
6) If (h_1, h_2) is a strict e-pair then (h_1, h_2) is an e-pair which implies (h_1, h_2) being a weak e-pair.
7) If $e_\alpha \in \mathbf{e}(J)$ for $\alpha < \alpha^*$, then $e := \cap\{e_\alpha : \alpha < \alpha^*\} = \{(s,t) : s, t$ are e_α-equivalent for every $\alpha < \alpha^*\}$ belongs to $\mathbf{e}(J)$ with $\mathrm{Dom}(e) = \cap\{\mathrm{Dom}(e_\alpha) : \alpha < \alpha^*\}$.
8) If $e \in \mathbf{e}(J, I)$ then for every $\mathscr{Y} \subseteq \mathrm{Dom}(e)/e$ also $e \restriction \mathrm{set}(\mathscr{Y})$ belongs to $\mathbf{e}(J, I)$ and there is a strict $(e \restriction \mathrm{set}(\mathscr{Y}))$-pair (h_1', h_2'); moreover, for every $\mathscr{Y}_1 \subseteq \mathscr{Y}$ there is a strict $(e \restriction \mathrm{set}(\mathscr{Y}), \mathscr{Y}_1)$-pair. $\square_{6.8}$

Proof. Easy, e.g.:
1) Let

$$\begin{aligned} e = \{(s_1, s_2) : h_1(s_\ell) \neq h_2(s_\ell) &\text{ for } \ell = 1,2 \text{ and if } s_1 \neq s_2 \text{ then} \\ &\text{for some } t_1 <_J t_2 \text{ we have } \{s_1, s_2\} = \{t_1, t_2\} \\ &\text{and there is no initial segment } J' \text{ of } J \text{ such that} \\ &J' \cap \{t_1, t_2\} = \{t_1\} \text{ and} \\ &(\forall t' \in J')(\forall t'' \in J \backslash J')[h_1(t') <_I h_2(t'') \wedge h_2(t') <_I h_1(t'')]\}. \end{aligned}$$

Clearly e is an equivalence relation on $\{t \in J : h_1(t) \neq h_2(t)\}$ and each equivalence class is convex hence $e_1 \in \mathbf{e}(J)$, so clauses (a),(b) of 6.6(1),(4) holds. Easily e is as required.
8) Let (h_1, h_2) be an e-pair and $\mathscr{Y}_1, \mathscr{Y}_2, \mathscr{Y}_3$ be a partition of $\mathrm{Dom}(e)/e$. We define $h_1', h_2' \in \operatorname{inc}_J(I)$ as follows, for $\ell \in \{1, 2\}$

(a) if $t \in J \backslash \mathrm{Dom}(e)$ then $h_\ell'(t) = h_1(t)\ (= h_2(t))$

(b) if $t \in \mathrm{set}(\mathscr{Y}_1)$ then $h_\ell'(t) = h_1(t)$

(c) if $t \in \operatorname{set}(\mathscr{Y}_2)$ then $h'_\ell(t)$ is $\min\{h_1(t), h_2(t)\}$ if $\ell = 1$, and is $\max\{h_1(t), h_2(t)\}$ if $\ell = 2$

(d) if $t \in \operatorname{set}(\mathscr{Y}_3)$ then $h'_\ell(t)$ is $\max\{h_1(t), h_2(t)\}$ if $\ell = 1$ and is $\min\{h_1(t), h_2(t)\}$ if $\ell = 2$.

Now (h'_1, h'_2) is a strict $(e \restriction (\operatorname{set}(\mathscr{Y}_2) \cup \operatorname{set}(\mathscr{Y}_3)), \mathscr{Y}_2)$-pair, so we are done. $\square_{6.8}$

6.9 Definition. 1) For a subset u of $J \in K^{\text{lin}}_{\tau^*_{\alpha(*)}}$ we define $e = e_{J,u} \in \mathbf{e}(J)$ on $J \backslash u$ as follows:

$$s_1 e s_2 \text{ iff } (\forall t \in u)(t <_J s_1 \equiv t <_J s_2).$$

2) For $I, J \in K^{\text{lin}}_{\alpha(*)}$, we say that the pair (I, J) is non-trivial <u>when</u>: $\mathbf{e}(J, I) \neq \emptyset$.

6.10 Definition. 1) For $h_0, \dots, h_{n-1} \in \operatorname{inc}_J(I)$ let

$$\operatorname{tp_{qf}}^J(\langle h_0, \dots, h_{n-1}\rangle, I) = \{(\ell, m, s, t) : s, t \in J \text{ and } h_\ell(s) < h_m(t)\}.$$

We may write $\operatorname{tp_{qf}}^J(h_0, \dots, h_{n-1}; I)$ and we usually omit J as it is clear from the context.

2) For $h_1, h_2 \in \operatorname{inc}_J(I)$ let $\operatorname{eq}(h_1, h_2) = \{s \in J : h_1(s) = h_2(s)\}$.

3) We say that the pair (I, J) is a reasonable $(\mu, \alpha(*)))$-base <u>when</u>:

(a) $I, J \in K^{\text{lin}}_{\tau^*_{\alpha(*)}}$, $|J| \leq \mu$ and the pair (I, J) is non-trivial

(b) if $e \in \mathbf{e}(J, I)$ and $h_1, h_2 \in \operatorname{inc}_J(I)$ and (h_1, h_2) is an e-pair then we can find $h'_1, h'_2, h'_3 \in \operatorname{inc}_J(I)$ and $\mathscr{Y} \subseteq \operatorname{Dom}(e)/e$ such that

 (α) $\operatorname{tp_{qf}}((h'_1, h'_2), I) = \operatorname{tp_{qf}}((h_1, h_2), I)$

 (β) (h'_1, h'_3) and (h'_2, h'_3) are strict $(e, \mathscr{Y})$-pairs.

4) We say that the pair (I, J) is a wide $(\lambda, \mu, \alpha(*))$-base <u>when</u>:

(a) $I, J \in K^{\text{lin}}_{\tau^*_{\alpha(*)}}$, $|J| \leq \mu$ and the pair (I, J) is non-trivial

(b) for every $e \in \mathbf{e}(J, I)$ there is a sequence $\bar{h} = \langle h_\alpha : \alpha < \lambda \rangle$ such that

(α) h_α is an embedding of J into I

(β) if $\alpha < \beta < \lambda$ then (h_α, h_β) is an e-pair.

5) We say that the pair (I, J) is a strongly wide $(\lambda, \mu, \alpha(*))$-base <u>when</u>:

(a) $I, J \in K^{\text{lin}}_{\tau^*_{\alpha(*)}}$, the pair (I, J) is non-trivial and J has cardinality $\leq \mu$

(b) for every $e \in \mathbf{e}(J, I)$ and $\mathscr{Y} \subseteq \text{Dom}(e)/e$ there is $\bar{h} = \langle h_\alpha : \alpha < \lambda \rangle$ such that

(α) $h_\alpha \in \text{inc}_J(I)$

(β) if $\alpha < \beta$ then (h_α, h_β) is a strict $(e, \mathscr{Y})$-pair.

6) Above we may omit μ meaning $\mu = |J|$ and we may omit $\alpha(*)$, as it is determined by J (and by I), and then may omit "base" so in part (3) we say (I, J) is reasonable and in part (4) we say λ-wide and in part (5) say strongly λ-wide.

6.11 Observation. 1) If (I, J) is a reasonable $(\mu, \alpha(*))$-base <u>then</u> (I, J) is a reasonable $(\mu', \alpha(*))$-base for $\mu' \geq \mu$.
2) If (I, J) is a wide $(\lambda, \mu, \alpha(*))$-base and $\lambda' \leq \lambda, \mu' \geq \mu$ <u>then</u> (I, J) is a wide $(\lambda', \mu', \alpha(*))$-base.
3) If (I, J) is a strongly wide $(\lambda, \mu, \alpha(*))$-base, <u>then</u> (I, J) is a wide $(\lambda, \mu, \alpha(*))$-base.

Proof. Obvious. $\square_{6.11}$

6.12 Claim. *1) If $\alpha(*) = 1$ and $\mu \leq \zeta(*) < \mu^+ \leq \lambda$, <u>then</u> the pair $(\lambda \times \zeta(*), \zeta(*))$ is a reasonable $(\mu, \alpha(*))$-based which is a wide $(\lambda, \mu, \alpha(*))$-base.*
2) If $\alpha(*) = 2$ and $\bar{u}^* = (\{0\}, \emptyset)$ as in 6.2 and $\mu \leq \zeta(*) < \mu^+ <$

λ and $\zeta'(*) = \zeta(*) \times 3$ and $w \subseteq \zeta(*), w \neq \zeta(*)$ then the pair $(I^{\rm lin}_{\mu,\lambda\times\zeta(*)}, I^{\rm lin}_{\mu,\zeta(*),w})$ is a reasonable $(\mu,\alpha(*))$-base which is a wide $(\lambda,\mu,\alpha(*))$-base where

- $(*)$ *for any ordinal β and $w \subseteq \beta$ we define $I = I^{\rm lin}_{\mu,\beta,w}$, a $\tau^*_{\alpha(*)}$-model (if $w = \emptyset$ we may omit it)*
  - (α) *its universe is β*
  - (β) *the order is the usual one*
  - (γ) *$P^I_1 = \{\alpha < \beta : {\rm cf}(\alpha) > \mu$ or $\alpha \in w\}$, (if we write $I^{\rm lin}_{\geq\mu,\beta,w}$ we mean here ${\rm cf}(\alpha) \geq \mu$).*

Proof. 1) First: $(I,J) = \underline{(\lambda \times \zeta(*), \zeta(*))}$ is a wide $\underline{(\lambda,\mu,\alpha(*))}$-base

Easily $\mathbf{e}(J,I) \neq \emptyset, |J| \leq \mu$ and $I,J \in K^{\rm lin}_{\tau^*_{\alpha(*)}}$ so clause (a) of Definition 6.10(4) holds (recalling Definition 6.9(2)), so it suffices to deal with clause (b).

Let $e \in \mathbf{e}(J,I)$ and define

$$u = \{\zeta < \zeta(*) : \zeta \in {\rm Dom}(e) \text{ is minimal in } \zeta/e \text{ or } \zeta \in \zeta(*) \setminus {\rm Dom}(e)\}.$$

Now for every $\alpha < \lambda$ we define $h_\alpha \in {\rm inc}_J(I)$ as follows:

- (a) if $\zeta \in \zeta(*) \setminus {\rm Dom}(e)$ then $h_\alpha(\zeta) = \lambda \times \zeta$
- (b) if $\zeta \in {\rm Dom}(e)$ and $\varepsilon = \min(\zeta/e)$ then $h_\alpha(\zeta) = \lambda \times \varepsilon + \zeta(*) \times \alpha + \zeta$.

Second: $(I,J) = \underline{(\lambda \times \zeta(*), \zeta(*))}$ is a reasonable $\underline{(\mu,\alpha(*))}$-base

Again clause (a) of Definition 6.10(3) holds so we deal with clause (b).

So assume $e \in \mathbf{e}(J,I)$ and $h_1,h_2 \in {\rm inc}_J(I)$ and (h_1,h_2) is just a weak e-pair and $\mathscr{Y} \subseteq {\rm Dom}(e)/e$. Let $u = {\rm Rang}(h_1) \cup {\rm Rang}(h_2)$. For $\ell = 1,2$ let $h^*_\ell \in {\rm inc}_J(I)$ be $h^*_\ell(\zeta) = {\rm otp}(u \cap h_\ell(\zeta))$, so ${\rm Rang}(h^*_\ell) \subseteq \xi(*) := {\rm otp}(u) \leq \zeta(*) \times 3$.

[Why? If $\zeta(*)$ is finite this is trivial, so assume $\zeta(*) \geq \omega$. Let $n < \omega$ and α be such that $\omega^\alpha n \leq \zeta(*) < \omega^\alpha(n+1)$, so $\alpha \geq 1, n \geq 1$. As

ω^α is additively indecomposable $\text{otp}(u) \le \omega^\alpha(2n+1)$, alternatively use natural sums [MiRa65] which gives a better bound $\zeta(*) \oplus \zeta(*)$, [actually $< \mu^+$ sufices using $\zeta(*) < \mu^+$ large enough below, still.]

For $\ell = 1, 2, 3$ we define $h'_\ell \in \text{inc}_J(I)$ as follows:

(a) if $\zeta \in \zeta(*) \setminus \text{Dom}(e)$ then $h'_\ell(\zeta) = (\zeta(*) \times 4) \times \zeta$

(b) if $\zeta \in \text{Dom}(e)$ and $\varepsilon = \min(\zeta/e)$ and $\zeta/e \in \mathscr{Y}$ then

  (α) if $\ell = 3$ then $h'_\ell(\zeta) = (\zeta(*) \times 4) \times \varepsilon + \zeta(*) \times 3 + \zeta$

  (β) if $\ell = 1, 2$ then $h'_\ell(\zeta) = (\zeta(*) \times 4) \times \varepsilon + h^*_\ell(\zeta)$

(c) if $\zeta \in \text{Dom}(e)$ and $\varepsilon = \min(\zeta/e)$ and $\zeta/e \notin \mathscr{Y}$ then

  (α) if $\ell = 3$ then $h'_\ell(\zeta) = (\zeta(*) \times 4) \times \varepsilon + \zeta$

  (β) if $\ell = 1, 2$ then $h'_\ell(\zeta) = (\zeta(*) \times 4) \times \varepsilon + \zeta(*) + h^*_\ell(\zeta)$.

Now check.

2) First: $(I, J) = (I^{\text{lin}}_{\mu,\lambda\times\zeta(*)}, I^{\text{lin}}_{\mu,\zeta(*),w})$ is a wide $(\lambda, \mu, \alpha(*))$-base.

Note that $P^J_1 = w$ because $\zeta(*) < \mu^+$ and $P^I_1 = \{\alpha \in I : \text{cf}(\alpha) > \mu\}$. As above clause (a) of the Definition 6.10 holds so we deal with clause (b).

Let

$$u = \{\zeta < \zeta(*) : \zeta \in \text{Dom}(e) \text{ is minimal in } \zeta/e \text{ or } \zeta \in \zeta(*) \setminus \text{Dom}(e)\}.$$

Clearly u is a closed subset of $\zeta(*)$ and $0 \in u$.

Given $\zeta < \zeta(*)$ let $\varepsilon_\zeta := \max(u \cap (\zeta + 1))$, clearly well defined by the choice of u and $\varepsilon_\zeta \le \zeta$.

For every $\alpha < \lambda$ we define $h_\alpha \in \text{inc}_J(I)$ as follows:

We define $h_\alpha(\zeta)$ by induction on $\zeta < \zeta(*)$ such that $h_\alpha(\zeta) < \lambda \times (\varepsilon_\zeta + 1)$.

Case A: for $\zeta \in \zeta(*) \setminus \text{Dom}(e)$

Subcase A1: $\zeta \in P^J_1$

Let $h_\alpha(\zeta)$ be $\lambda \times \varepsilon_\zeta + \mu^+$.

Subcase A2: $\zeta \in P^J_0$ and $\zeta = 0$

Let $h_\alpha(\zeta) = 0$.

Subcase A3: $\zeta \in P_0^J, \zeta = \xi + 1$
Let $h_\alpha(\zeta) = h_\alpha(\xi) + 1$.

Subcase A4: $\zeta \in P_0^J, \zeta$ is a limit ordinal, $\zeta = \sup(u \cap \zeta)$
Let $h_\alpha(\zeta) = \lambda \times \varepsilon_\zeta$ which is equal to $\cup\{h_\alpha(\zeta') : \zeta' < \zeta\}$.

Subcase A5: $\zeta \in P_0^J, \zeta$ is a limit ordinal and $\xi = \sup(u \cap \zeta) < \zeta$.
So $(\xi + 1)/e$ is an end-segment of ζ, but this is impossible by 6.7(2)(c).

Case B: $\zeta \in \mathrm{Dom}(e)$:

Subcase A1: $\zeta = \min(\zeta/e)$ hence $\zeta \in P_1^J$ (see 6.7(0)(b))
Let $h_\alpha(\zeta) = \lambda \times \varepsilon_\zeta + \mu^+ \times \zeta(*) \times \alpha + \mu^+$.

Subcase A2: $\zeta \in P_0^J$ hence $\zeta > \min(\zeta/e)$
Let $h_\alpha(\zeta) = \cup\{h_\alpha(\zeta') + 1 : \zeta' < \zeta\}$.

Subcase A3: $\zeta \in P_1^J$ and $\zeta > \min(\zeta/e)$
Let $h_\alpha(\zeta) = \cup\{h_\alpha(\zeta') : \zeta' < \zeta\} + \mu^+$.

So clearly we can show by induction on $\zeta < \zeta(*)$ that:

$$h_\alpha(\zeta) < \lambda \times \varepsilon_\zeta + \mu^+ \times \zeta(*) \times (\alpha 2 + 2).$$

Now check.

Also recalling $\mu^+ < \lambda$ clearly for $\alpha < \lambda, \zeta < \zeta(*)$ we have $h_\alpha(\zeta) < \lambda \times \varepsilon_\zeta + \lambda$.

Now check.

Second $(I^{\mathrm{lin}}_{\mu,\lambda\times\zeta(*)}, I^{\mathrm{lin}}_{\mu,\zeta(*),w})$ is a reasonable $(\mu, \alpha(*))$-base

Combine the proof of "first" with the parallel proof in part (1). $\square_{6.12}$

6.13 Definition. 1) Let $I, J \in K^{\mathrm{lin}}_{\tau^*_{\alpha(*)}}$. We say that $\mathscr{E}$ is an invariant (I, J)-equivalence relation when:

- (a) $\mathscr{E}$ is an equivalence relation on $\mathrm{inc}_J(I)$, so $\mathscr{E}$ determines I and J
- (b) if $h_1, h_2, h_3, h_4 \in \mathrm{inc}_J(I)$ and $\mathrm{tp}_{\mathrm{qf}}(h_1, h_2; I) = \mathrm{tp}_{\mathrm{qf}}(h_3, h_4; I)$ then $h_1 \mathscr{E} h_2 \Leftrightarrow h_3 \mathscr{E} h_4$.

2) We add non-trivial <u>when</u>:

(c) if $\mathrm{eq}(h_1, h_2) = \{t \in J : h_1(t) = h_2(t)\}$ is co-finite then $h_1 \mathscr{E} h_2$
(d) there are $h_1, h_2 \in \mathrm{inc}_J(I)$ such that $\neg(h_1 \mathscr{E} h_2)$.

3) Let $J, I_1, I_2 \in K^{\mathrm{lin}}_{\tau^*_{\alpha(*)}}$. <u>Then</u> $I_1 \leq^1_J I_2$ means that:

(a) $I_1 \subseteq I_2$
(b) for every $h_1, h_2, h_3 \in \mathrm{inc}_J(I_2)$ we can find $h'_1, h'_2, h'_3 \in \mathrm{inc}_J(I_1)$ such that $\mathrm{tp}_{\mathrm{qf}}(h'_1, h'_2, h'_3; I_1) = \mathrm{tp}_{\mathrm{qf}}(h_1, h_2, h_3; I_2)$.

6.14 Claim. *Assume* $J, I_1, I_2 \in K^{\mathrm{lin}}_{\tau^*_{\alpha(*)}}$.
1) If $I_1 \subseteq I_2, \mathscr{E}$ *is an invariant* (I_2, J)*-equivalence relation* <u>*then*</u> $\mathscr{E} \restriction \mathrm{inc}_J(I_1)$ *is an invariant* (I_1, J)*-equivalence relation.*
2) If $I_1 <^1_J I_2$ *and* $\mathscr{E}_1$ *is an invariant* (I_1, J)*-equivalence relation* <u>*then*</u> *there is one and only one invariant* (I_2, J)*-equivalence relation* $\mathscr{E}_2$ *such that* $\mathscr{E}_2 \restriction \mathrm{inc}_J(I_1) = \mathscr{E}_1$.
3) Assume $e \in \mathbf{e}(J)$ *and* $\mathscr{Y} \subseteq \mathrm{Dom}(e)/e$. *If* (h'_1, h'_2) *is a strict* $(e, \mathscr{Y})$*-pair for* (I_1, J) *and* (h''_1, h''_2) *is a strict* $(e, \mathscr{Y})$*-pair for* (I_2, J) <u>*then*</u> $\mathrm{tp}_{\mathrm{qf}}(h'_1, h'_2; I_1) = \mathrm{tp}_{\mathrm{qf}}(h''_1, h''_2; I_2)$.
4) Assume $\alpha(*) = 1, J = \zeta(*), I_\ell = \beta_\ell$ *with the usual order (for* $\ell = 1, 2$*),* $\mu \leq \zeta(*) < \mu^+$ *and* $\mu^+ \leq \beta_1 \leq \beta_2$. <u>*Then*</u> $I_1 <^1_J I_2$ *(see Definition 6.13(3)).*
5) Assume $\alpha(*) = 2, J = I^{\mathrm{lin}}_{\mu,\zeta(*),w}, I_\ell = I^{\mathrm{lin}}_{\mu,\beta_\ell}$ *for* $\ell = 1, 2$ *and* $\mu^{++} \leq \beta_1 \leq \beta_2$. <u>*Then*</u> $I_1 <^1_J I_2$ *(see Definition 6.13(3)).*

Proof. 1) Obvious.
2) We define

$$\mathscr{E}^*_2 = \{(h_1, h_2) : h_1, h_2 \in \mathrm{inc}_J(I_2) \text{ and for some } h'_1, h'_2 \in \mathrm{inc}_J(I_1) \text{ we have } \mathrm{tp}_{\mathrm{qf}}(h'_1, h'_2; I_1) = \mathrm{tp}_{\mathrm{qf}}(h_1, h_2; I_2) \text{ and } h'_1 \mathscr{E}_1 h'_2\}.$$

Now

$(*)_1$ $\mathscr{E}^*_2$ is a set of pairs of members of $\mathrm{inc}_J(I_2)$.

[Why? By its definition]

$(*)_2$ $h_1 \mathscr{E}_2^* h_1$ if $h_1 \in \mathrm{inc}_J(I_2)$.

[Why? Let $h' \in \mathrm{inc}_J(I_1)$ so clearly $h' \mathscr{E}_1 h'$ and $\mathrm{tp_{qf}}(h', h'; I_1) = \mathrm{tp_{qf}}(h, h; I_2)$]

$(*)_3$ $\mathscr{E}_2^*$ is symmetric.

[Why? As $\mathscr{E}_1$ is.]

$(*)_4$ $\mathscr{E}_2^*$ is transitive.

[Why? Assume $h_1 \mathscr{E}_2^* h_2$ and $h_2 \mathscr{E}_2^* h_3$ and let $h'_1, h'_2 \in \mathrm{inc}_J(I_1)$ witness $h_1 \mathscr{E}_2^* h_2$ and $h''_2, h''_3 \in \mathrm{inc}_J(I_1)$ witness $h_2 \mathscr{E}_2^* h_3$.

Apply clause (b) of part (3) of Definition 6.13 to (h_1, h_2, h_3) so there are $g_1, g_2, g_3 \in \mathrm{inc}_J(I_1)$ such that $\mathrm{tp_{qf}}(g_1, g_2, g_3; I_1) = \mathrm{tp_{qf}}(h_1, h_2, h_3; I_2)$. Now $h'_1 \mathscr{E}_1 h'_2$ by the choice of (h'_1, h'_2), and $\mathrm{tp_{qf}}(g_1, g_2; I_1) = \mathrm{tp_{qf}}(h_1, h_2; I_2) = \mathrm{tp_{qf}}(h'_1, h'_2; I_1)$ so as $\mathscr{E}_1$ is invariant we get $g_1 \mathscr{E}_1 g_2$. Similarly $g_2 \mathscr{E}_1 g_3$, so as $\mathscr{E}_1$ is transitive we have $g_1 \mathscr{E}_1 g_3$. But clearly $\mathrm{tp_{qf}}(g_1, g_3; I_1) = \mathrm{tp_{qf}}(h_1, h_3; I_2)$ hence g_1, g_2 witness that $h_1 \mathscr{E}_2 h_3$ is as required.]

$(*)_5$ $\mathscr{E}_2^*$ is invariant.

[Why? See its definition.]

$(*)_6$ $\mathscr{E}_2^* \restriction \mathrm{inc}_I(I_1) = \mathscr{E}_1$.

[Why? By the way $\mathscr{E}_2^*$ is defined and $\mathscr{E}_1$ being invariant.]

So together $\mathscr{E}_2^*$ is as required. The uniqueness (i.e. if $\mathscr{E}_2$ is an invariant equivalent relation on $\mathrm{inc}_J(I)$ such that $\mathscr{E}_2 \restriction \mathrm{inc}_J(I_1) = \mathscr{E}_1$ then $\mathscr{E}_2 = \mathscr{E}_2^*$) is also easy.

3) Straight.

4) See[6] the proof of "Second" in the proof of 6.12(1).

5) Combine[7] the proof of part (4) and of "First" in the proof of 6.12(2). $\square_{6.14}$

Below mostly it suffices to consider $\mathscr{D}_{\mathscr{E},e}$.

---

[6]Actually instead "$\mu^+ \leq \beta_1$" it suffice to have $\zeta(*) \times 4 \leq \beta_1$ because if $\zeta(*) = \sum\limits_{i<\gamma} \zeta_i$ then $\sum\limits_{i<\gamma} \zeta_i \times 4 \leq \zeta(*) \times 4$ or just the natural sum $\zeta(*) \oplus \zeta(*) \oplus \zeta(*)$.

[7]Here $(\mu^+ + 1) \times (\zeta(*) \times 4)$ will suffice.

6.15 Definition. 1) Let $\mathscr{E}$ be an invariant (I, J)-equivalence relation; we define

$$\mathscr{D}_{\mathscr{E}} = \{u \subseteq J : \text{if } h_1, h_2 \in \text{inc}_J(I) \text{ satisfies eq}(h_1, h_2) \supseteq u \\ \text{then } h_1 \mathscr{E} h_2\}$$

recalling

$$\text{eq}(h_1, h_2) := \{t \in J : h_1(t) = h_2(t)\}.$$

2) If in addition $e \in \mathbf{e}(J, I)$ then we let

$$\mathscr{D}_{\mathscr{E},e} = \{u \subseteq \text{Dom}(e)/e : \text{if } h_1, h_2 \in \text{inc}_J(I) \text{ and } (h_1, h_2) \text{ is an} \\ (e \restriction (\text{Dom}(e) \backslash \text{set}(u)))\text{-pair then } h_1 \mathscr{E} h_2\}.$$

6.16 Claim. *Assume $I, J \in K^{\text{lin}}_{\mathcal{T}^*_{\alpha(*)}}$ and (I, J) is reasonable (see Definition 6.10(3),(6)) and $\mathscr{E}$ is an invariant (I, J)-equivalence relation.*
1) For $u \subseteq J$ such that $e_{J,u} \in \mathbf{e}(J, I)$ we have: $u \in \mathscr{D}_{\mathscr{E}}$ iff $h_1 \mathscr{E} h_2$ for every $e_{J,u}$-pair (h_1, h_2) iff $h_1 \mathscr{E} h_2$ for some $e_{J,u}$-pair (h_1, h_2); see Definition 6.9(1).
2) Assume $e \in \mathbf{e}(J, I)$, then for any $u \subseteq \text{Dom}(e)/e$ we have: $u \in \mathscr{D}_{\mathscr{E},e}$ iff $h_1 \mathscr{E} h_2$ for any $(e \restriction \text{set}(u))$-pair iff $h_1 \mathscr{E} h_2$ for some $(e \restriction \text{set}(u))$-pair.
3) If $e \in \mathbf{e}(J, I)$ and $u_1, u_2 \subseteq \text{Dom}(e)/e$ then we can find $h_1, h_2, h_3 \in \text{inc}_J(t)$ such that (h_1, h_2) is a strict $(e \restriction \text{set}(u_1))$-pair, (h_2, h_3) is a strict $(e \restriction \text{set}(u_2))$ pair and (h_1, h_3) is a strict $(e \restriction (\text{set}(u_1 \cup u_2))$-pair.
4) Assume $e \in \mathbf{e}(J, I)$ and that in clause (b) of Definition 6.10(3) we allow (h_1, h_2) to be a weak e-pair, then for any $u \subseteq \text{Dom}(e)/e$ we have: $\text{Dom}(e) \backslash u \in \mathscr{D}_{\mathscr{E},e}$ iff $h_1 \mathscr{E} h_2$ for every weak e-pair (h_1, h_2).

Proof. 1) Like part (2).
2) In short, by transitivity of equivalence and the definitions + mixing, but we elaborate.
The "first implies the second" holds by Definition 6.15(2) and "the second implies the third" holds trivially as there is such a pair (h_1, h_2)

by the assumption $e \in \mathbf{e}(J, I)$. So it is enough to prove "the third implies the first"; hence suppose that $g_1 \mathscr{E} g_2$, where (g_1, g_2) is an $e_1 := e \restriction \text{set}(u)$-pair (recalling that $e_1 \in \mathbf{e}(J, I)$ by 6.8(8)), and let (h_1, h_2) be an e_1-pair, we need to show that $h_1 \mathscr{E} e_2$. By Definition 6.6(2B) for some sets $\mathscr{Y}_g, \mathscr{Y}_h \subseteq \text{Dom}(e_1)/e_1$ the pair (g_1, g_2) is a strict $(e_1, \mathscr{Y}_g)$-pair and the pair (h_1, h_2) is a strict $(e_1, \mathscr{Y}_h)$-pair. Recalling clause (b) of 6.10(3) there are g'_1, g'_2, g'_3 and $\mathscr{Y}$ such that:

$(*)_1$ (a) $g'_\ell \in \text{inc}_J(I)$ for $\ell = 1, 2, 3$
(b) $\text{tp}_{\text{qf}}(g_1, g_2) = \text{tp}_{\text{qf}}(g'_1, g'_2)$
(c) $\mathscr{Y} \subseteq \text{Dom}(e_1)/e_1$
(d) (g'_1, g'_3) and (g'_2, g'_3) are strict $(e_1, \mathscr{Y})$-pairs.

Now for each $s \in \text{Dom}(e_1)$, we can find a permutation $\bar{\ell}_s = (\ell_{s,1}, \ell_{s,2}, \ell_{s,3})$ of $\{1, 2, 3\}$ such that $I \models g'_{\ell_{s,1}}(s) < g'_{\ell_{s,2}}(s) < g'_{\ell_{s,3}}(s)$. By $(*)_1(d)$ and $(*)_1(b)$ and (g_1, g) being an e_1-pair, clearly $\bar{\ell}_s$ depends only on s/e_1 and every member of $\{(g'_{\ell_{s,1}}(t) : t \in s/e_1\}$ is below every member of $\{g'_{\ell_{s,2}}(t) : t \in s/e_1\}$ and similarly for the pair $(g'_{\ell_{s,2}}, g'_{\ell_{s,3}})$. Now we can find (g''_1, g''_2, g''_3) such that:

$(*)_2$ (a) $g''_\ell \in \text{inc}_J(I)$ for $\ell = 1, 2, 3$
(b) (g''_1, g''_2) is a strict $(e_1, \mathscr{Y}_h)$
(c) (g''_1, g''_3) and (g''_2, g''_3) are strict $(e_1, \mathscr{Y}_g)$-pairs

[Why? We do the choice for each s/e_1 separately such that: $\{g''_1 \restriction (s/e_1), g''_2 \restriction (s/e_1), g''_3 \restriction (s/e_1)\} = \{g'_1 \restriction (s/e_1), g'_2 \restriction (s/e_1), g'_3 \restriction (s/e_1)\}$.]

Clearly $\text{tp}_{\text{qf}}(g''_1, g''_3; I) = \text{tp}_{\text{qf}}(g_1, g_2; I) = \text{tp}_{\text{qf}}(g''_2, g''_3; I)$ so as $\mathscr{E}$ is invariant and $g_1 \mathscr{E} g_2$ clearly $g''_1 \mathscr{E} g''_3 \wedge g''_2 \mathscr{E} g''_3$ which implies $g''_1 \mathscr{E} g''_2$. For $\mathscr{Y}' = \mathscr{Y}_h$ by clause (b) of $(*)_2$ we conclude that $\text{tp}_{\text{qf}}(g''_1, g''_2; I) = \text{tp}_{\text{qf}}(h_1, h_2; I)$ so as $\mathscr{E}$ is invariant we are done.
3),4) Similarly. $\square_{6.16}$

6.17 Claim. *Assume $I, J \in K^{\text{lin}}_{\tau^*_{\alpha(*)}}$ and $\mathscr{E}$ is an invariant (I, J)-equivalence relation.*
0) If $\mathrm{e} \in \mathbf{e}(J, I)$ and $\mathscr{E}$ is non-trivial <u>then</u> $\mathscr{D}_{\mathscr{E},\mathrm{e}}$ contains all co-finite subsets of $\text{Dom}(\mathrm{e})/\mathrm{e}$.

1) If the pair (I, J) is reasonable and $e \in \mathbf{e}(I, J)$ then $\mathscr{D}_{\mathscr{E},e}$ is a filter on $\mathrm{Dom}(e)/e$ *but possibly* $\emptyset \in \mathscr{D}_{\mathscr{E},e}$.
2) (a) $\mathscr{D}_{\mathscr{E}}$ is a filter on J

(b) if $\mathscr{E}$ is non-trivial then all cofinite subsets of J belongs to $\mathscr{D}_{\mathscr{E}}$ but $\emptyset \notin \mathscr{D}_{\mathscr{E}}$.

Proof. 0) Easy, see Definition 6.13(2).
1) By 6.16(2) and 6.16(3).
2) Trivial by Definition 6.15(1). $\square_{6.17}$

6.18 Main Claim. *Assume*

(a) $I, J \in K^{\mathrm{lin}}_{\tau^*_{\alpha(*)}}$
(b) *$\mathscr{E}$ is an invariant (I, J)-equivalence relation*
(c) *(I, J) is a reasonable $(\mu, \alpha(*))$-base which is a wide $(\lambda, \mu, \alpha(*))$-base*
(d) $e \in \mathbf{e}(J, I)$
(e) *g is a function from* $\mathrm{Dom}(e)/e$ *into some cardinal θ*
(f) *$\mathscr{D}^* = \{Y \subseteq \theta : g^{-1}(Y) \in \mathscr{D}_{\mathscr{E},e}\}$ is a filter, i.e., $\emptyset \notin \mathscr{D}^*$.*

Then $\mathscr{E}$ has at least $\chi := \lambda^\theta/\mathscr{D}^*$ equivalence classes.

Proof. Let $\langle f_\alpha : \alpha < \chi\rangle$ be a set of functions from θ to λ exemplifying $\chi := \lambda^\theta/\mathscr{D}^*$ so $\alpha \neq \beta \Rightarrow \{i < \theta : f_\alpha(i) = f_\beta(i)\} \notin \mathscr{D}^*$.

Let $\langle h_\zeta : \zeta < \lambda\rangle$ exemplify the pair (I, J) being a wide $(\lambda, \mu, \alpha(*))$-base, see Definition 6.10(4), so $h_\zeta \in \mathrm{inc}_J(I)$.

Lastly for each $\alpha < \chi$ we define $h^\alpha \in \mathrm{inc}_J(I)$ as follows:
$h^\alpha(t)$ is: $h_0(t)$ if $t \in J\backslash\mathrm{Dom}(e)$
$h_{f_\alpha(g(t/e))}(t)$ if $t \in \mathrm{Dom}(e)$.

Now

$(*)_1$ h^α is a function from J to I.

[Why? Trivially recalling each h_ζ is.]

$(*)_2$ h^α is increasing.

[Why? Let $s <_J t$ and we split the proof to cases.
If $s, t \in J \backslash \mathrm{Dom}(e)$ use "$h_0 \in \mathrm{inc}_J(I)$".
If $s \in J \backslash \mathrm{Dom}(e)$ and $t \in \mathrm{Dom}(e)$, then $h^\alpha(t) = h_{f_\alpha(g(t/e))}(t), h^\alpha(s) = h_0(s) = h_{f_\alpha(g(t/e))}(s)$ because $\langle h_\alpha \restriction (J \backslash \mathrm{Dom}(e)) : \alpha < \lambda \rangle$ is constant (recalling (h_0, h_α) is an e-pair (for $\alpha > 0$)), so as $h_{f_\alpha(g(t/e))} \in \mathrm{inc}_J(I)$ we are done.
If $s \in \mathrm{Dom}(e), t \in J \backslash \mathrm{Dom}(e)$, the proof is similar.
If $s, t \in \mathrm{Dom}(e), s/e \neq t/e$, we again use Definition 6.6(2B), clause (b)(β) of Definition 6.10(4).

Lastly, if $s, t \in \mathrm{Dom}(e), s/e = t/e$ we get $g(s/e) = g(t/e)$ hence $f_\alpha(g(s/e)) = f_\alpha(g(t/e))$ call it γ so $h^\alpha(s) = h_\gamma(s), h^\alpha(t) = h_\gamma(t)$ and of course $h_\gamma \in \mathrm{inc}_J(I)$ hence $h_\gamma(s) <_I h_\gamma(t)$ so necessarily $h^\alpha(s) <_I h^\alpha(t)$ as required. So $(*)_2$ holds.]

$(*)_3$ $h^\alpha \in \mathrm{inc}_J(I)$.

[Why? Clearly if $i < \alpha(*)$ and $t \in P_i^J$ then $(\forall \beta < \lambda) h_\beta(t) \in P_i^J$ hence $\alpha < \chi \Rightarrow h_{f_\alpha(g(t/e))}(t) \in P_i^J$ which means $\alpha < \chi \Rightarrow h^\alpha(t) \in P_i^J$; so recalling $(*)_2$, clause (a) of Definition 6.3(2) holds. We should check clauses (b),(c) of Definition 6.3(2) which is done as in the proof of 6.7 and of $(*)_2$ above.]

$(*)_4$ if $\alpha < \beta$ and we let $u = u_{\alpha,\beta} := \cup\{g^{-1}(\zeta) : \zeta < \theta$ and $f_\alpha(\zeta) \neq f_\beta(\zeta)\}$ so $u \subseteq \mathrm{Dom}(e)/e$ <u>then</u> (h^α, h^β) is a $(e \restriction \mathrm{set}(u))$-pair.

[Why? <u>Case 1</u>: If $s \in J \backslash \mathrm{Dom}(e)$ then $h^\alpha(s) = h_0(s) = h^\beta(s)$.

<u>Case 2</u>: If $s \in \mathrm{Dom}(e) \backslash \mathrm{set}(u)$ then $h^\alpha(s) = h_{f_\alpha(g(s/e))}(s) =$
$= h_{f_\beta(g(s/e))}(s) = h^\beta(s)$.

<u>Case 3</u>: If $s, t \in \mathrm{set}(u), s/e \neq t/e, s <_J t$ <u>then</u> $h^\alpha(s) <_I h^\beta(t) \wedge h^\beta(s) <_I h^\alpha(t)$ because

<u>Subcase 3A</u>: If $f_\alpha(g(s/e)) = f_\beta(g(t/e))$ we use $h_{f_\alpha(g(t/e))} \in \mathrm{inc}_J(I)$ hence

$$h^\alpha(s) = h_{f_\alpha(g(s/e))}(s) <_I h_{f_\alpha(g(s/e))}(t) = h_{f_\beta(g(t/e))}(t) = h^\beta(t)$$

and similarly $h^\beta(s) <_I h^\alpha(t)$.

Subcase 3B: $f_\alpha(g(s/e)) \neq f_\beta(g(t/e))$ we use "$(h_{f_\alpha(g(s/e))}, h_{f_\beta(g(t/e))})$ is an e-pair".

Case 4: And lastly, if $s, t \in \text{set}(u), s/e = t/e$ and $s <_J t$ then $h^\alpha(t) <_I h^\beta(s) \equiv (s/e \in u) \equiv h^\alpha(s) <_I h^\beta(t)$.

Why? Recalling $f_\alpha(g(s/e)) \neq f_\beta(g(t/e))$ as $s, t \in \text{set}(u)$ by the definition of u, see $(*)_4$ and we just use "$(h_{f_\alpha(g(s/e))}, h_{f_\beta(g(s/e))})$" is an e-pair and clause (c)$'$ of Definition 6.6.]

$(*)_5$ if $\alpha < \beta$ then $u_{\alpha,\beta} \neq \emptyset \mod \mathscr{D}_{\mathscr{E},e}$.

[Why? By the choice of $\langle f_\alpha : \alpha < \lambda \rangle$.]

$(*)_6$ if $\alpha < \beta$ then h^α, h^β are not $\mathscr{E}$-equivalent.

[Why? By $(*)_4 + (*)_5$ and 6.16(2).]

Together we are done. $\square_{6.18}$

6.19 Claim. *Assume $\mathscr{E}$ is an invariant (I, J)-equivalence relation, I, J are well ordered and $|\text{inc}_J(I)/\mathscr{E}| \geq \lambda = \text{cf}(\lambda) > \mu = |I| > |2 + \alpha(*)|^{|J|}$. Then for some $e \in \mathbf{e}(I, J)$ there is an ultrafilter $\mathscr{D}$ on $\text{Dom}(e)/e$ extending $\mathscr{D}_{\mathscr{E},e}$ which is not principal.*

Remark. This is close to [Sh 620, §7].

Proof. Without loss of generality as linear orders, J is $\zeta(*)$ and I is $\xi(*) \in [\mu, \mu^+)$.

Toward contradiction assume the conclusion fails. Let g be a one-to-one function from μ onto $[\xi(*)]^{<\aleph_0}$ and χ be large enough and $\kappa = |J|$ and $\partial = |2 + \alpha(*)|^{|J|}$ so $\partial^\kappa = \partial$.

We now choose $\langle N_\eta : \eta \in {}^n\mu \rangle$ by induction on $n < \omega$ such that

$\circledast_1$ (a) $N_\eta \prec (\mathscr{H}(\chi), \in)$

(b) $\|N_\eta\| = \partial$ and $\partial + 1 \subseteq N_\eta$

(c) $A \subseteq N_\eta \wedge |A| \leq \kappa \Rightarrow A \in N_\eta$

(d) I, J and g as well as η belong to N_η

(e) $\nu \triangleleft \eta \Rightarrow N_\nu \in N_\eta$ (hence $N_\nu \subseteq N_\eta$ so $N_\nu \prec N_\eta$).

There is no problem to do this. Now it suffices to prove that for every $h \in \text{inc}_J(I)$, for some $h' \in \cup\{N_\eta : \eta \in {}^{\omega>}\mu\} \cap \text{inc}_J(I)$ we have $h\mathscr{E}h'$.

Fix $h_* \in \text{inc}_J(I)$ such that $h_* \notin \cup\{h/\mathscr{E} : h \in \text{inc}_J(I) \cap N_\eta$ for some $\eta \in {}^{\omega>}\mu\}$ and for each $\eta \in {}^{\omega>}\mu$ we define $\bar{\alpha}_\eta, e_\eta$ as follows:

$\circledast_2$ (a) $\bar{\alpha}_\eta = \langle \alpha_{\eta,t} : t \in J\rangle$

(b) $\alpha_{\eta,t} = \min((\xi(*)+1) \cap N_\eta \backslash h_*(t))$

(c) $e_\eta := \{(s,t) : s,t \in J$ and $\alpha_{\eta,s} = \alpha_{\eta,t}$ and $\alpha_{\eta,s} > h_*(s)$ and $\alpha_{\eta,t} > h_*(t)\}$

(d) for $\alpha \in N_\eta$ let $X_{\eta,\alpha} := \{t \in J : \alpha_{\eta,t} = \alpha > h_*(t)\}$.

Note

$(*)_1$ $\bar{\alpha}_\eta \in N_\eta$.

[Why? As $[N_\eta]^{\leq\kappa} \subseteq N_\eta$ and $|J| = \kappa$ and $\alpha_{\eta,t} \in N_\eta$ for every $t \in J$.]

$(*)_2$ *(a)* $e_\eta \in \mathbf{e}(J)$, i.e. e_η is an equivalence relation on some subset of J with each equivalence class a convex subset of J, see Definition 6.6(1)

(b) $\langle X_{\eta,\alpha} : \alpha \in \{\alpha_{\eta,t} : t \in \text{Dom}(e)\}$ hence $X_{\eta,\alpha} \neq \emptyset\rangle$ list the e_η-equivalence classes.

[Why? Think.]

$(*)_3$ $h_\eta := h_* \restriction (J \backslash \text{Dom}(e_\eta)) \in N_\eta$.

[Why? By the definition of e_η we have $t \in J \wedge t \notin \text{Dom}(e_\eta) \Rightarrow h_*(t) \in N_\eta$ and recall $[N_\eta]^{\leq\kappa} \subseteq N_\eta$.]

$(*)_4$ if $t \in \text{Dom}(e_\eta)$ then $\text{cf}(\alpha_{\eta,t}) > \partial$.

[Why? As $\alpha_{\eta,t} \in N_\eta \prec (\mathscr{H}(\chi), \in)$ if $\text{cf}(\alpha_{\eta,t}) = \theta \leq \partial$ then there is a cofinal set B of $\alpha_{\eta,t}$ of cardinality θ in N_η but $\theta \leq \partial + 1 \subseteq N_\eta$ therefore $B \subseteq N_\eta$. In particular as $h_*(t) < \alpha_{\eta,t}$ there is $\beta \in B$ so that $h_*(t) < \beta$, but this contradicts the choice of $\alpha_{\eta,t}$.]

$(*)_5$ $e_\eta \in \mathbf{e}(J, I)$.

[Why? Choose $h' \in \mathrm{inc}_J(I) \cap N_\eta$ similar enough to h_*, specifically: $t \in J \setminus \mathrm{Dom}(e_\eta) \Rightarrow h'(t) = h_*(t)$ and $t \in \mathrm{Dom}(e_\eta) \Rightarrow \sup\{\alpha_{\eta,s} : s \in J, s <_J t \text{ and } s \notin t/e_\eta\} < h'(t) < \alpha_{\eta,t}$. The point being that $\sup\{\alpha_{\eta,s} : s \in J, s <_J t \text{ and } s \notin t/e_n\} \in N_\eta$. Now (h', h_*) is a strict e-pair.]

$(*)_6$ there is $\ell_\eta < \omega$ and a finite sequence $\langle \beta_{\eta,\ell} : \ell < \ell_\eta \rangle$ of members of $\mathrm{Rang}(\bar{\alpha}_\eta \restriction \mathrm{Dom}(e_\eta))$ so $X_{\eta,\beta_{\eta,\ell}} \in \mathrm{Dom}(e_\eta)/e_\eta$ for $\ell < \ell_\eta$ such that $\cup\{X_{\eta,\beta_{\eta,\ell}} : \ell < \ell_\eta\} \in \mathscr{D}_{\mathscr{E},e_\eta}$.

[Why? Otherwise there is an ultrafilter as desired, but toward contradiction we have assumed this does not occur; in trying to get generalizations we should act differently.]

Now we choose (η_n, h_n) by induction on $n < \omega$ such that

$\boxdot$(a) (a) $\eta_n \in {}^n\mu$

(b) if $n = m + 1$ then $\eta_m = \eta_n \restriction m$

(c) $h_n \in \mathrm{inc}_J(I)$

(d) $h_0 = h_*$

(e) if $n = m + 1$ then:

(α) $h_n \mathscr{E} h_m$ hence $h_n \mathscr{E} h_*$ and $\mathrm{Dom}(e_{\eta_n}) \subseteq \mathrm{Dom}(e_{\eta_m})$

(β) $h_m \restriction (J \setminus \mathrm{Dom}(e_{\eta_m})) \subseteq h_n$

(γ) $(h_m \restriction \cup\{X_{\eta_m,\beta_{\eta_m,\ell}} : \ell < \ell_{\eta_m}\}) \subseteq h_n$

(δ) $h_n \restriction (\mathrm{Dom}(e_{\eta_m}) \setminus \cup\{X_{\eta_m,\beta_{\eta_m,\ell}} : \ell < \ell_{\eta_m}\})$ belongs to N_{η_m}

(ε) moreover $t \in \mathrm{Dom}(e_{\eta_m}) \setminus \cup\{X_{\eta_m,\beta_{\eta_m,\ell}} : \ell < \ell_{\eta_m}\}$ implies $h_n(t) < h_m(t)$

(ζ) $\ell_{\eta_m} > 0$

(f) $Y_{m+1} \subseteq Y_m$ where $Y_m := \cup\{X_{\eta_m,\beta_{\eta_m,\ell}} : \ell < \ell_\eta\}$.

Why can we carry out the construction? For $n = 0$ we obviously can (choose $h_0 = h_*$). For $n = m + 1$ first choose $h'_m \in N_{\eta_m}$ as we choose in the proof of $(*)_5$. Now recalling $\langle X_{\eta_m}, \beta_{\eta_m,\ell} : \ell < \ell_{\eta_m} \rangle$ was chosen in $(*)_6$, and define h_n by $h_n \restriction (\mathrm{Dom}(e_{\eta_m}) \setminus \cup\{X_{\eta_m,\beta_{\eta_m,\ell}} : \ell < \ell_{\eta_m}\}) = h'_m \restriction (\mathrm{Dom}(e_{\eta_m}) \setminus \cup\{X_{\eta_m,\beta_{\eta_m,\ell}} : \ell < \ell_\eta\})$ and $h_n \restriction$

$(J\backslash \mathrm{Dom}(e_{\eta_m})) = h_m \restriction (J\backslash \mathrm{Dom}(e_{\eta_m})$ and $h_n \restriction (\cup\{X_{\eta_m,\beta_{\eta_m,\ell}} : \ell < n_{\eta_m}\}) = h_m \restriction (\cup\{X_{\eta_m,\beta_{\eta_m,\ell}} : \ell < \ell_{\eta_m}\})$. Why $h_n \mathscr{E} h_m$? Because

- (i) as in the proof of $(*)_5$, (h_n, h_m) form a strict ℓ_η-pair
- (ii) they agree on $\cup\{X_{\eta_m,\beta_{\eta_m,\ell}} : \ell < \ell_\eta\}$
- (iii) $\{X_{\eta_m,\beta_{\eta_m,\ell}} : \ell < n\} \in \mathscr{D}_{\mathscr{E},\mathbf{e}_\eta}$.

Lastly, choose $\eta_n = \eta_m \hat{\ } \langle \gamma_m \rangle$ where γ_m is chosen such that $g(\gamma_m) = \{\sup(\beta_{\eta_m,\ell} \backslash \sup\{h_m(t) : t \in X_{\beta_{\eta_m,\ell}}\}) : \ell < \ell_{\eta_m}\}$ recalling that g is a function from μ onto $[\xi(*)]^{<\aleph_0} = [I]^{<\aleph_0}$.

Now check that η_n, h_n are as required.

Note that this induction never stops in the sense that $h_n \notin N_{\eta_n}$ recalling the choice of h_* and $h_n \mathscr{E} h_*$. Now $\mathscr{U}_n := \{\beta_{\eta_m,\ell} : \ell < n_\eta\}$ is a finite non-empty set of ordinals, and if $n = m+1$, then easily $(\forall \ell < \ell_{\eta_n})(\exists k < \ell_{\eta_m})(\beta_{\eta_n,\ell} < \beta_{\eta_m,k})$ because for $\ell < \ell_{\eta_n}$ letting $t \in X_{\eta_n,\ell}$ we know that for some $k \le \ell_{\eta_m}$ we have $t \in X_{\eta_m,k}$ and $\eta_n(m)$ was chosen above such that as γ_m, now $h_*(t) \le \gamma_n \in N_{\eta_n}, \gamma_m \le \alpha_{\eta_m,t}$ and the inequality is strict as $\mathrm{cf}(\alpha_{\eta_m,t}) > 0$. So $\langle \max(\mathscr{U}_n) : n < \omega \rangle$ is a decreasing sequence of ordinals, contradiction, so we are done. $\square_{6.19}$

<u>6.20 Example</u>: For $e \in \mathbf{e}(J, I)$, $J \in K^{\mathrm{lin}}_{\tau^*_{\alpha(*)}}$ and $I \in K^{\mathrm{lin}}_{\tau^*_{\alpha(*)}}$ we define $\mathscr{E}^*_e = \mathscr{E}^*_{e,I}$; it is an invariant equivalent relation on $\mathrm{inc}_J(I)$, by: $h_1 \mathscr{E}^*_{e,I} h_2$ iff:

- (a) if $t \in J\backslash\mathrm{Dom}(e)$ then $h_1(t) = h_2(t)$
- (b) if $t \in \mathrm{Dom}(e)$ then $\mathrm{cnv}_{I,h_1}(t) = \mathrm{cnv}_{I,h_2}(t)$ where $\mathrm{cnv}_{I,h}(s) :=$ the convex hull (in I) of the set $\{h_1(s)\} \cup \bigcup\{[h_1(s), h_1(t)]_I : s <_J t \text{ and } t \in s/e\} \cup \bigcup\{[h(t), h(s)]_I : t <_J s \text{ and } t \in s/e\}$.

1) If $J, I \in K^{\mathrm{lin}}_{\tau^*_{\alpha(*)}}$ are well ordered and $e = J \times J$ then $\mathscr{E}^*_{e,I}$ from part (1) has $\le |I| + \aleph_0$ equivalence classes.
2) If $J \in K^{\mathrm{lin}}_{\tau^*_{\alpha(*)}}$ and e as in part (2), $\theta = \mathrm{cf}(J)$ and $|J| < \lambda = \lambda^{<\theta} < \lambda^\theta$ <u>then</u> there is $I \in K^{\mathrm{lin}}_{\tau^*_{\alpha(*)}}$ of cardinality λ such that $\mathscr{E}^*_{e,I}$ has λ^θ equivalence classes.

Remark. We can define the stability spectrum for some classes, essentially this is done in §7, generally we intend to look at it in [Sh:F782].

§7 Categoricity for a.e.c. with bounded amalgamation

Recall that 4.10 is the main result of this chapter; we think that it will lead to understanding the categoricity spectrum of an a.e.c. In particular we hope eventually to prove that this spectrum contains or is disjoint to some end segments of the class of cardinals. Still here we like to show that what we have is enough at least for restricted enough families of a.e.c. $\mathfrak{K}$'s, those definable by $\mathbb{L}_{\kappa,\omega}$, κ a measurable cardinal or with enough amalgamation (concerning them and earlier results see Chapter N). We could have relied on[8] [Sh 394], but though we mention connections, we do not rely on it, preferring self-containment.

We can say much even if we replace categoricity by strong solvability, but do this only when it is cheap; we can work even with weak and even pseudo-solvability but not here.

7.1 Hypothesis. 1) $\mathfrak{K}$ is an a.e.c., so $\mathscr{S}(M) = \mathscr{S}_{\mathfrak{K}_\lambda}(M)$ for $M \in K_\lambda$, see II.1.9.
2) Let K^{x}_{μ} be the class K_μ if K is categorical in μ and the class of superlimit models in $\mathfrak{K}_\mu$ if there is one, (the two definitions are compatible).

The following is a crucial claim because lack of locality is the problem in [Sh 394].

7.2 Claim. *Assume*

- (a) $\mathrm{cf}(\mu) > \kappa \geq \mathrm{LS}(\mathfrak{K})$
- (b) $\mathfrak{K}_{<\mu}$ *has amalgamation*
- (c) $\Phi \in \Upsilon^{\mathrm{or}}_{\kappa}[\mathfrak{K}]$ *satisfies: if I is θ-wide and $\theta \in (\kappa, \mu)$ <u>then</u> $\mathrm{EM}_{\tau(\mathfrak{K})}(I, M)$ is θ-saturated (see 0.14(1), II.1.13(2) and II.1.14).*

<u>*Then*</u>

- (α) *for some $\mu_* < \mu$, the class $\{M \in K_{<\mu} : M$ is saturated$\}$ is $[\mu_*, \mu)$-local, see Definition 7.4(3) below*

[8]In the references to [Sh 394], e.g. 1.6tex is to 1.6 in the published version and 1.8 is in the e-version.

$(\alpha)^+$ *this applies not only to* $\mathscr{S}(M) = \mathscr{S}^1(M)$ *but also for* $\mathscr{S}^\partial(M)$ *if* $\mathrm{cf}(\mu) > \kappa^\partial$.

Recall

7.3 Definition. $\mathfrak{K}$ is μ-stable if $\mu \geq \mathrm{LS}(\mathfrak{K})$ and $M \in K_{\leq\mu} \Rightarrow |\mathscr{S}(M)| \leq \mu$.

Recall ([Sh 394, Def.1.8=1.6tex](1),(2).

7.4 Definition. 1) For $M \in \mathfrak{K}, \mu \geq \mathrm{LS}(\mathfrak{K})$, satisfying $\mu \leq \|M\|$ and α, let $\mathbb{E}_{M,\mu,\alpha}$ be the following equivalence relation on $\mathscr{S}^\alpha(M)$: $p_1 \mathbb{E}_{M,\mu,\alpha} p_2$ iff for every $N \leq_{\mathfrak{K}} M$ of cardinality μ we have $p_1 \restriction N = p_2 \restriction N$. We may suppress α if it is 1, similarly below; let $\mathbb{E}_{\mu,\alpha}$ be $\bigcup\{\mathbb{E}_{M,\mu,\alpha} : M \in K\}$ and so $\mathbb{E}_\mu = \mathbb{E}_{\mu,1}$.
2) We say that $M \in \mathfrak{K}$ is $\mu-\alpha$-local when $\mathbb{E}_{M,\mu,\alpha}$ is the equality; we say that $p \in \mathscr{S}^\alpha(M)$ is μ-local if $p/\mathbb{E}_{M,\mu,\alpha}$ is a singleton and we say e.g. $K' \subseteq \mathfrak{K}$ is $\mu-\alpha$-local (in $\mathfrak{K}$, if not clear from the context) when every $M \in K'$ is.
3) We say $K' \subseteq \mathfrak{K}$ is $[\mu_*, \mu) - \alpha$-local if every $M \in K' \cap \mathfrak{K}_{[\mu^*,\mu)}$ is $\mu_* - \alpha$-local.
4) We say that $\bar{a} \in N$ realizes $\mathbf{p} \in \mathscr{S}^\alpha_{\mathfrak{K}}(M)/\mathbb{E}_{\mu,\alpha}$ if $M \leq_{\mathfrak{K}} N$ and for every $M' \leq_{\mathfrak{K}} N$ of cardinality μ the sequence $\bar{a}$ realizes $\mathbf{p} \restriction M'$ in N or pedantically realizes $q \restriction M'$ for some, equivalently every $q \in \mathbf{p}$.

Remark. If $M \in \mathfrak{K}_\mu$, then M is $\mu - \alpha$-local.

Proof of 7.2. Recall $\Phi \in \Upsilon^{\mathrm{or}}_\kappa[\mathfrak{K}]$, see Definition 0.8(2) and Claim 0.9. Easily there is $\langle I_\theta : \theta \in [\kappa, \mu)\rangle$, an increasing sequence of wide linear orders which are strongly $\aleph_0$-homogeneous (that is dense with neither first nor last element such that if $n < \omega$ and $\bar{s}, \bar{t} \in {}^n(I_\theta)$ are $<_I$-increasing then some automorphism of I_θ maps $\bar{s}$ to $\bar{t}$, e.g. the order of any real closed field/or just ordered field) satisfying $|I_\theta| = \theta$.

Recalling $\mathbb{Q}$ here is the rational order, we let $J_\theta = \mathbb{Q} + I_\theta, M_\theta = \mathrm{EM}_{\tau(\mathfrak{K})}(I_\theta, \Phi)$ and $N_\theta = \mathrm{EM}_{\tau(\mathfrak{K})}(J_\theta, \Phi)$. So

$\circledast$(a) $M_\theta \leq_{\mathfrak{K}_\theta} N_\theta$

(b) $M_{\theta_1} \leq_{\mathfrak{K}} M_{\theta_2}$ and $N_{\theta_1} \leq_{\mathfrak{K}} N_{\theta_2}$ when $\kappa \leq \theta_1 < \theta_2 < \mu$

(c) M_θ is saturated (for $\mathfrak{K}$, of course) when $\theta > \kappa$

(d) every type from $\mathscr{S}(M_\theta)$ is realized in N_θ

(e) if $n < \omega, \bar{a} \in {}^n(N_\theta)$ then for some $\bar{a}' \in {}^n(N_\kappa)$ and automorphism π of $N_\theta, \pi(\bar{a}) = \bar{a}'$ and π maps M_θ onto itself.

[Why? Clauses (a),(b) holds by clause (c) of Claim 0.9(1) recalling Definition 0.8(2).
Clause (c) holds by Clause (c) of the assumption of 7.2; you may note [Sh 394, 6.7=6.4tex](2).
Clause (d) holds as $\mathrm{EM}_{\tau(\mathfrak{K})}(\theta^+ + J_\theta, \Phi) \in \mathfrak{K}_{\theta^+}$ is saturated, and use the definition of a type (or like the proof of claue (e) below using appropriate $I' + I_\theta$ instead $\theta^+ + J_\theta$); you may note [Sh 394, 6.8=6.5tex].
Clause (e) holds as for every finite sequence $\bar{t}$ from J_θ there is an automorphism π of J_θ such that: π is the identity on $\mathbb{Q}$, it maps I_θ onto itself and it maps $\bar{t}$ to a sequence from $J_\kappa = \mathbb{Q} + I_\kappa$, such π exists as I_θ is strongly $\aleph_0$-homogeneous and $I_\kappa \subseteq I_\theta$ is infinite.]
For any $a \neq b$ from N_κ let

$$\begin{aligned}\mu(a,b) = \ &\mathrm{Min}\{\theta : \theta \geq \kappa \text{ and if } \theta < \mu \\ &\qquad \text{then } \mathbf{tp}_{\mathfrak{K}}(a, M_\theta, N_\theta) \neq \mathbf{tp}_{\mathfrak{K}}(b, M_\theta, N_\theta)\}.\end{aligned}$$

So $\mu(a,b) \leq \mu$. Let

$$\mu_* = \sup\{\mu(a,b) : a, b \in N_\kappa \text{ and } \mu(a,b) < \mu\}.$$

So μ_* is defined as the supremum on a set of $\leq \kappa \times \kappa$ cardinals $< \mu$, which is a cardinal of cofinality $\mathrm{cf}(\mu) > \kappa$, hence clearly $\mu_* < \mu$. Also $\mu_* \geq \kappa$ as there are $a \neq b$ from M_κ hence $\mu(a,b) = \kappa$. Now suppose that $\theta \in [\mu_*, \mu), M \in \mathfrak{K}_\theta$ is saturated and $p_1 \neq p_2 \in \mathscr{S}(M)$ and we shall find $M' \leq_{\mathfrak{K}} M, M' \in \mathfrak{K}_{\mu_*}$ such that $p_1 \restriction M' \neq p_2 \restriction M'$, this suffice.

Clearly $M_\theta \in K_\theta$ is saturated (by clause (c) of ⊛) hence the models M, M_θ are isomorphic so without loss of generality $M = M_\theta$. But by clause (d) of ⊛ every type from $\mathscr{S}(M_\theta)$ is realized in N_θ, so let b_ℓ be such that $p_\ell = \mathbf{tp}_{\mathfrak{K}}(b_\ell, M_\theta, N_\theta)$ for $\ell = 1, 2$. Now there is an automorphism π of N_θ which maps M_θ onto itself and maps b_1, b_2 into N_κ (by clause (e) of $\circledast$) and let $a_\ell = \pi(b_\ell)$ for $\ell = 1, 2$, so $a_1, a_2 \in N_\kappa$.

Now

$$\mathbf{tp}(a_1, M_\theta, N_\theta) = \mathbf{tp}(\pi(b_1), \pi(M_\theta), \pi(N_\theta)) = \pi(\mathbf{tp}(b_1, M_\theta, N_\theta)) \neq$$
$$\neq \pi(\mathbf{tp}(b_2, M_\theta, N_\theta)) = \mathbf{tp}(\pi(b_2), \pi(M_\theta), \pi(N_\theta)) = \mathbf{tp}(a_2, M_\theta, N_\theta).$$

Hence by the definition of $\mu(a_1, a_2)$ we have $\mu(a_1, a_2) \leq \theta < \mu$. Hence by the definition of μ_* we have $\mu(a_1, a_2) \leq \mu_*$ which implies that $\mathbf{tp}_{\mathfrak{K}}(a_1, M_{\mu_*}, N_{\mu_*}) \neq \mathbf{tp}_{\mathfrak{K}}(a_2, M_{\mu_*}, N_{\mu_*})$.
As π is an automorphism of N_θ and $M_{\mu^*} \leq_{\mathfrak{K}} M_\theta$ it follows that

$$\mathbf{tp}_{\mathfrak{K}}(\pi^{-1}(a_1), \pi^{-1}(M_{\mu^*}), \pi^{-1}(N_\theta)) \neq$$
$$\neq \mathbf{tp}_{\mathfrak{K}}(\pi^{-1}(a_2), \pi^{-1}(M_{\mu^*}), \pi^{-1}(N_\theta))$$

which means
$\mathbf{tp}_{\mathfrak{K}}(b_1, \pi^{-1}(M_{\mu^*}), N_\theta) \neq \mathbf{tp}_{\mathfrak{K}}(b_2, \pi^{-1}(M_{\mu^*}), N_\theta)$, but $\pi^{-1}(M_{\mu^*}) \leq_{\mathfrak{K}} M_\theta$ as π maps M_θ onto itself and recall that $p_\ell = \mathbf{tp}_{\mathfrak{K}}(b_\ell, M_\theta, N_\theta)$ so $p_\ell \restriction \pi^{-1}(M_{\mu^*})$ is well defined for $\ell = 1, 2$. Hence $p_1 \restriction \pi^{-1}(M_{\mu^*}) \neq p_2 \restriction \pi^{-1}(M_{\mu^*})$ and clearly $\pi^{-1}(M_{\mu^*})$ has cardinality μ^* and is $\leq_{\mathfrak{K}} M_\theta$, so we are done proving clause (α). The proof of clause $(\alpha)^+$ is the same except that

- $(*)_1$ if $\theta \in [\kappa, \mu), \bar{t} \in {}^{\partial}(I_\theta)$ then some automorphism π of I_θ maps $\bar{t}$ to some $\bar{t}' \in {}^{\partial}(I_\kappa)$, justified by 5.1
- $(*)_2$ we replace $\mathbb{Q}$ by ∂^+
- $(*)_3$ ${}^{\partial}(N_\kappa)$ has cardinality $\leq (\partial^+ + \kappa)^{\partial} \leq \kappa^{\partial} < \operatorname{cf}(\mu)$.

$\square_{7.2}$

Implicit in non-μ-splitting is

7.5 Definition. Assume $\alpha < \mu^+, N \in K_{\leq \mu}, N \leq_{\mathfrak{K}} M$ and $p \in \mathscr{S}^\alpha(M)$ does not μ-split over N, see Definition III.2.18(1). The scheme of the non-μ-splitting, $\mathfrak{p} = \operatorname{sch}_\mu(p, N)$ is $\{(N'', c, \bar{b})_{c \in N}/\cong$: we have $N \leq_{\mathfrak{K}} N' \leq_{\mathfrak{K}} M$ and $N' \leq_{\mathfrak{K}} N'', \{N', N''\} \subseteq K_\mu$ and the sequence $\bar{b}$ realizes $p \restriction N'$ in the model $N''\}$.

7.6 Definition. For a cardinal μ and model M let
1)

$$\begin{aligned} \text{ps} - \mathscr{S}_\mu(M) = \mathscr{S}_{\mathfrak{K},\mu}(M) = \{\mathbf{p} :& \mathbf{p} \text{ is a function with domain} \\ & \{N \in K_\mu : N \le_{\mathfrak{K}} M\} \\ & \text{such that } \mathbf{p}(N) \in \mathscr{S}(N) \\ & \text{and } N_1 \le_{\mathfrak{K}} N_2 \in \text{Dom}(\mathbf{p}) \\ & \Rightarrow \mathbf{p}(N_1) = \mathbf{p}(N_2) \restriction N_1\}. \end{aligned}$$

2) For $p \in \mathscr{S}(M)$ let $p \restriction (\le \mu)$ be the function $\mathbf{p}$ with domain $\{N \in K_\mu : N \le_{\mathfrak{K}} M\}$ such that $\mathbf{p}(N) = p \restriction N$.

7.7 Observation. 1) The function $p \mapsto p \restriction (\le \mu)$ is a function from $\mathscr{S}(M)$ into ps-$\mathscr{S}_\mu(M)$ such that for $p_1, p_2 \in \mathscr{S}(M)$ we have $p_1 \restriction (\le \mu) = p_2 \restriction (\le \mu) \Leftrightarrow p_1 \mathbb{E}_\mu p_2$.
2) The subset $\{p \restriction (\le \mu) : p \in \mathscr{S}(M)\}$ of ps-$\mathscr{S}_\mu(M)$ has cardinality $|\mathscr{S}(M)/\mathbb{E}_\mu|$.

Proof. Should be clear. $\square_{7.7}$

7.8 Claim. *Every (equivalently some) $M \in K^x_\mu$ is λ^+-saturated* <u>*when*</u>*:*

(a) (α) $\mathfrak{K}$ is categorical in μ
or just
(β) $\mathfrak{K}$ is strongly solvable in μ

(b) $\text{LS}(\mathfrak{K}) \le \lambda < \chi \le \mu$ and $2^{2^\lambda} \le \mu$ (actually $2^\lambda \le \mu$ suffice)

(c) (α) $\aleph_{\lambda+4} = \lambda^{+\lambda^{+4}} \le \chi$
<u>*or*</u> *at least*
(β) if $\theta = \text{cf}(\theta) \le \lambda$ is $\aleph_0$ or a measurable cardinal <u>*then*</u> *for some $\partial \in (\lambda, \chi)$ we have: $\partial = \partial^{<\theta} < \partial^\theta$* <u>*or*</u> *at least $\partial^{<\theta>_{\text{tr}}} > \partial$ (i.e. there is a tree $\mathscr{T}$ with θ levels, ∂ nodes and the number of θ-branches of $\mathscr{T}$ is $> \chi$, see [Sh 589])*

(d) $\mathfrak{K}_{\geq \partial} \neq \emptyset$ *for every* ∂, *equivalently* $K_{\geq \theta} \neq \emptyset$ *for arbitrarily large* $\theta < \beth_{1,1}(\mathrm{LS}(\mathfrak{K}))$

(e) (α) $\mathfrak{K}_{<\mu}$ *has amalgamation and* JEP
or just

(β) *if* $\mathrm{LS}(\mathfrak{K}) \leq \partial < \chi$ *then*

(i) $\mathfrak{K}_\partial$ *has amalgamation and* JEP *and*

(ii) $\mathfrak{K}$ *has* $(\partial, \leq \partial^+, \mu)$*-amalgamation*[9] *(see I.2.7(2)) hence*[10]

(iii) *every* $M \in K_{\partial+}$ *has a* $\leq_{\mathfrak{K}}$*-extension in* K^x_μ *(actually (i) + (iii) suffices).*

Remark. 1) M is λ^+-saturated is well defined as $\mathfrak{K}_{\leq\lambda}$ has amalgamation.
2) We assume $2^{2^\lambda} \leq \mu$ because the proof is simpler with not much loss (at least as long as other parts of the analysis are not much tighter).
3) We can weaken the assumptions. In particular using solvability instead categoricity, but for non-essential reasons this is delayed; similarly in 7.12.
4) If $\mu = \mu^\lambda$ the claim is easy (as in §1).

Proof. Note that by [Sh:g, IX,§2], [Sh:g, II,3.1] if clause $(c)(\alpha)$ holds then clause $(c)(\beta)$ holds, hence we can assume $(c)(\beta)$.

Let $\Phi \in \Upsilon^{\mathrm{or}}_{\mathfrak{K}}$ see Definition 0.8(2), exist by 0.9 and clause (d) of the assumption and $I \in K^{\mathrm{lin}}_\mu \Rightarrow \mathrm{EM}_{\tau(\mathfrak{K})}(I, \Phi) \in K^x_\mu$ (trivially if K is categorical in μ, otherwise by the definition of solvable).

Clearly

$(*)_0$ if $\partial \in [\mathrm{LS}(\mathfrak{K}), \chi)$ <u>then</u> $\mathfrak{K}$ is stable in ∂.

---

[9]It suffices to have: if $M_0 \leq_{\mathfrak{K}} M_1 \in K_{\partial+}, M_1 \leq_{\mathfrak{K}} M_2 \in K^x_\mu$ and $M_0 \in K_\partial$ then M_1 can be $\leq_{\mathfrak{K}}$-embedded into some $M_3 \in K^x_\mu$. Similarly in 7.12.

[10]Why? Assume $M \in K_{\partial+}$ let $M_2 \in K^x_\mu$, let $M_0 \leq_{\mathfrak{K}} M_2$ be of cardinality ∂, let $M_1 \in K_{\partial+}$ be a $\leq_{\mathfrak{K}}$-extension of M_0 which there is an $\leq_{\mathfrak{K}}$-embedding f of M into M_1 (exists as $\mathfrak{K}_\partial$ has amalgamation and JEP). Lastly, use "$\mathfrak{K}$ has $(\partial, \leq \partial^+, \mu)$-amalgamation

[Why? We prove assuming clause $(e)(\beta)$, as the case of clause $(e)(\alpha)$ is easier. Otherwise as $\mathfrak{K}_\partial$ has amalgamation there are $M_0 \leq_{\mathfrak{K}} M_1$ such that $M_0 \in K_\partial, M_1 \in K_{\partial^+}$ and $\{\mathbf{tp}_{\mathfrak{K}}(a, M_0, M_1) : a \in M_1\}$ has cardinality ∂^+. By assumption $(e)(\beta)(iii)$ there is N_1 such that $M_1 \leq_{\mathfrak{K}} N_1 \in \mathfrak{K}_\mu$ and without loss of generality $N_1 \in K^x_\mu$. Let I be as in 5.1 with $(\lambda, \theta_2, \theta_1, \mu)$ there standing for $(\mu, \partial^{++}, \partial^+, \partial)$ here and $N_2 := \mathrm{EM}_{\tau(\mathfrak{K})}(I, \Phi)$. Now by 5.1(2), $N_1 \not\cong N_2$, contradiction to "$\mathfrak{K}$ categorical in μ". Or you may see [Sh 394, 1.7=1.5tex].]

The proof now splits to two cases.
Case 1: For every $M \in K^x_\mu$ we have $\mu \geq |\mathscr{S}(M)/\mathbb{E}_\lambda|$.

For every $M \in K^x_\mu$ there is M' such that: $M \leq_{\mathfrak{K}} M' \in K_\mu$ and for every $\mathbf{p} \in \mathscr{S}(M)/\mathbb{E}_\lambda$ either $\mathbf{p}$ is realized in M' or there are no M'', a such that $M' \leq_{\mathfrak{K}} M'' \in K_\mu$ and $a \in M''$ realizes p in M''.
[Why? Let $\langle p_i/\mathbb{E}_\lambda : i < \mu\rangle$ list $\mathscr{S}(M)/\mathbb{E}_\lambda$, exists by the assumptions and choose M_i for $i \leq \mu$, $\leq_{\mathfrak{K}_\mu}$-increasing continuous such that M_{i+1} satisfies the demand for $\mathbf{p} = p_i/\mathbb{E}_\lambda$, possibly no $p \in p_i/\mathbb{E}_\lambda$ has an extension in $\mathscr{S}(M_{i+1})$ (hence is not realized in it), so then the desired demand holds trivially; note that it is not unreasonable to assume $\mathfrak{K}_\mu$ has amalgamation and it clarifies but it is not necessary.]

Also without loss of generality $M' \in K^x_\mu$ as any model M from K_μ has a $\leq_{\mathfrak{K}}$-extension in K^x_μ (at least if M does $\leq_{\mathfrak{K}}$-extend some $M' \in K^x_\mu$).

Now we can choose by induction on $i \leq \lambda^+$ a model $M_i \in K^x_\mu$, $\leq_{\mathfrak{K}}$-increasing continuous with i, such that for every $p \in \mathscr{S}(M_i)$ either there is $q \in \mathscr{S}(M_i)$ realized in M_{i+1} which is $\mathbb{E}_\lambda$-equivalent to p or there is no $\leq_{\mathfrak{K}}$-extension of M_{i+1} satisfying this. Now we shall prove that M_{λ^+} is λ^+-saturated recalling Definition II.1.13. Now if $N \leq_{\mathfrak{K}} M_{\lambda^+}, \|N\| \leq \lambda$ and $p \in \mathscr{S}(N)$ then there is $i < \lambda^+$ such that $N \leq_{\mathfrak{K}} M_i$ and we can find $p' \in \mathscr{S}(M_{\lambda^+})$ extending p. (Why? If clause $(e)(\alpha)$ holds then this follows by $\mathfrak{K}_{<\mu}$ having amalgamation, see I.2.12. If clause $(e)(\beta)$ holds, use "$\mathfrak{K}$ has the $(\lambda, \leq \lambda^+, \mu)$-amalgamation property" recalling $\mathrm{LS}(\mathfrak{K}) \leq \lambda < \chi$.) Hence there is $a \in M_{i+1}$ such that $\mathbf{tp}(a, M_i, M_{i+1})\mathbb{E}_\lambda(p' \restriction M_i)$, hence a realizes p in M_{i+1} hence in M_{λ^+}.

Case 2: Not Case 1.

Let I be as in 5.1 with $(\lambda, \theta_2, \theta_1, \mu)$ there standing for $(\mu, \lambda^{++}, \lambda^+, \lambda)$ here, so $|I| = \mu$. Let $M = \mathrm{EM}_{\tau(\mathfrak{K})}(I, \Phi)$, so by not Case 1 we can find $p_i \in \mathscr{S}(M)$ for $i < \mu^+$ pairwise non-$\mathbb{E}_\lambda$-equivalent. As $\mathfrak{K}_\lambda$ is a λ-a.e.c. with amalgamation and is stable in λ (by $(*)_0$) we can deduce, see III.2.21(2), that: if $p \in \mathscr{S}(M)$ then for some $N \leq_{\mathfrak{K}} M$ of cardinality λ the type p does not λ-split over N (or see [Sh 394, 3.2 = 3.2tex](1)). For each i choose $N_i \leq_{\mathfrak{K}} M$ of cardinality λ such that p_i does not μ-split over N_i. As there is no loss in increasing N_i (as long as it is $\leq_{\mathfrak{K}} M$ and has cardinality λ) without loss of generality

$(*)_1$ $N_i = \mathrm{EM}_{\tau(\mathfrak{K})}(I_i, \Phi)$ where $I_i \subseteq I$ and $|I_i| = \lambda$ and let $\bar{t}_i = \langle t^i_\varepsilon : \varepsilon < \lambda \rangle$ list I_i with no repetitions.

As $2^\lambda \leq \mu$ without loss of generality the I_i's are pairwise isomorphic, so without loss of generality for $i, j < \mu^+$, the mapping $t^i_\varepsilon \mapsto t^j_\varepsilon$ is such an isomorphism. Moreover, without loss of generality

$(*)_2$ for every $i, j < \mu^+$ there is an automorphism $\pi_{i,j}$ of I mapping t^i_ε to t^j_ε for $\varepsilon < \lambda$.

[Why? By 5.1(1) as we can replace $\langle p_i : i < \mu^+ \rangle$ by $\langle p_i : i \in \mathscr{U} \rangle$ for every unbounded $\mathscr{U} \subseteq \mu^+$.]

Let $\mathfrak{p}_i$ be the non-λ-splitting scheme of p over N_i (see Definition 7.5). Without loss of generality:

$(*)_3$ for $i, j < \mu^+$, the isomorphism $h_{i,j}$ from $N_j = \mathrm{EM}_{\tau(\mathfrak{K})}(I_j, \Phi)$ onto $N_i = \mathrm{EM}_{\tau(\mathfrak{K})}(I_i, \Phi)$ induced by the mapping $t^j_\zeta \mapsto t^i_\zeta$ (for $\zeta < \lambda$) satisfies

- (i) it is an isomorphism from N_j onto N_i
- (ii) it maps $\mathfrak{p}_j$ to $\mathfrak{p}_i$.

[Why? For (i) this holds by the definition of $\mathrm{EM}(I_i, \Phi)$. For (ii) let $h_{i,0}$ map $\mathfrak{p}_i$ to $\mathfrak{p}'_i$. The number of schemes is $\leq 2^{2^\lambda}$; so if $\mu \geq 2^{2^\lambda}$ then without loss of generality $i < \mu^+ \Rightarrow \mathfrak{p}'_i = \mathfrak{p}'_1$ hence we are done (with no real loss). If we weaken the assumption $\mu \geq 2^{2^\lambda}$ to $\mu \geq 2^\lambda$ (or even $\mu > \lambda$ so waive $(*)_2$) using 5.1(4) we can find I^+_i such that $I_i \subseteq I^+_i \subseteq I, |I^+_i| \leq \lambda^+$ and for every $J \subseteq I$ of cardinality $\leq \lambda$

there is an automorphism of I over I_i mapping J into I_i^+. So only $\langle \mathfrak{p}'_i((\mathrm{EM}_{\tau(\mathfrak{K})}(I_0^+,\Phi),c,\bar b))_{c\in \mathrm{EM}_{\tau(\mathfrak{K})}(I_0,\Phi)}/\cong) : \bar b \in {}^{\lambda}(\mathrm{EM}_{\tau(\mathfrak{K})}(I_0^+,\Phi))\rangle$ matters (an overkill) but this is determiend by $p_i \restriction \mathrm{EM}_{\tau(\mathfrak{K})}(I_0^+,\Phi))$ which $\in \mathscr{S}(\mathrm{EM}_{\tau(\mathfrak{K})}(I_0^+,\Phi))$ by $(*)_0$ and as $\mathfrak{K}$ is stable in λ^+ without loss of generality $\mathfrak{p}'_{1+i} = \mathfrak{p}'_1$ and we are done.]

Now we translate our problem to one on expanded (by unary predicates) linear orders which was treated in §6. Recall that by 5.1(3), we can use $I = \mathrm{EM}_{\{<\}}(I^*,\Psi)$ where $\Psi \in \Upsilon^{\mathrm{lin}}_{\aleph_0}[2]$, see Definition 0.11(5), and $I^* = I^{\mathrm{lin}}_{\lambda,\mu\times\lambda^+}$ from 6.12(2) with $\alpha(*) = 2$. Recall that $I^* = I^{\mathrm{lin}}_{\lambda,\mu\times\lambda^{++}}$ is $\mu\times\lambda^{++}$ expanded by $P_1 = \{\alpha \in I^* : \mathrm{cf}(\alpha) \geq \lambda^+\}, P_0 = I_*\backslash P_0$ so I^* is a well ordered τ_2^*-model, i.e. $\in K^{\mathrm{lin}}_{\tau_2^*}$, see Definition 0.11(5). Without loss of generality $I_i = \mathrm{EM}_{\{<\}}(I_i^*,\Psi)$ where $I_i^* \subseteq I^*$ has cardinality λ and the pair (I^*,I_i^*) is a reasonable $(\lambda,\alpha(*))$-base which is a wide $(\mu,\lambda,\alpha(*))$-base, see Definition 6.10(3)(4), Claim 6.12(2). Without loss of generality for every $i<\mu^+$ there is h_i, an isomorphism from I_0^* onto I_i^* such that (see below) the induced function $h_1^{[1]}$ maps $\bar t_0$ to $\bar t_i$. Let $J^* = I_0^*$ and $J = I_0$. We like to apply §6 for J^*, I^* fixing $\alpha(*) = 2, \bar u^* = (u^-,u^+) = (\{0\},\emptyset)$. So recalling Definition 6.3(2) for every $h \in \mathrm{inc}^{\bar u^*}_{J^*}(I^*)$ we can naturally define the function $h^{[1]}$ by $h^{[1]}(\sigma^{\mathrm{EM}(J^*,\Psi)}(t_0,\dots,t_{n-1})) = \sigma^{\mathrm{EM}(J^*,\Psi)}(a_{h(t_0)},\dots,a_{h(t_{n-1})})$ whenever $\sigma(x_0,\dots,x_{n-1})$ is a $\tau(\Psi)$-term and $J^* \models$ "$t_0 < \dots < t_{n-1}$" so it is an isomorphism from $\mathrm{EM}_{\{<\}}(J^*,\Psi)$ onto $\mathrm{EM}_{\{<\}}(I^* \restriction \mathrm{Rang}(h),\Psi)$ so as $J^* \subseteq I^*$ by 5.1(5) there is an automorphism $h^{[2]}$ of I extending $h^{[1]}$ and so there is an automorphism $h^{[3]}$ of $\mathrm{EM}(I,\Phi)$ such that $h^{[3]}(a_t) = a_{h^{[2]}(t)}$ for $t\in I$ and $h^{[3]}(\sigma^{\mathrm{EM}(I,\Phi)}(a_{t_0},\dots,a_{t_{n-1}})) = \sigma^{\mathrm{EM}(I,\Phi)}(a_{h^{[2]}(t_0)},\dots,a_{h^{[2]}(t_n)})$ where $t_0 <_I \dots <_I t_{n-1}$ and $\sigma(x_0,\dots,x_{n-1})$ is a $\tau(\Phi)$-term.

Note that

$(*)_4$ if h',h'' are automorphisms of $\mathrm{EM}_{\tau[\mathfrak{K}]}(I,\Phi)$ extending $h^{[3]} \restriction \mathrm{EM}_{\tau[\mathfrak{K}]}(I_0)$ then $h'(p_0/\mathbb{E}_\lambda) = h''(p_0/\mathbb{E}_\lambda)$.

[Why? Because p_0 does not λ-split over $\mathrm{EM}_{\tau[\mathfrak{K}]}(I_0,\Phi)$.]

We define a two-place relation $\mathscr{E}$ on $\mathrm{inc}_{J^*}(I^*)$ by: $h_1 \mathscr{E} h_2$ if $h_1^{[3]}(p_0/\mathbb{E}_\lambda) = h_2^{[3]}(p_0/\mathbb{E}_\lambda)$. (Note that $h \mapsto h^{[3]}$ is a function so this is well defined and $h^{[3]}$ is an automorphism of $\mathrm{EM}_{\tau(\mathfrak{K})}(I,\Phi)$). By

$(*)_4$ clearly $\mathscr{E}$ is an invariant equivalence relation on $\mathrm{inc}^{\bar{u}^*}_{J^*}(I^*)$ with $> \mu$ equivalence classes as exemplified by $\langle h_i : i < \mu^+ \rangle$.

By 6.19 there is $e \in \mathbf{e}(J^*, I^*)$ such that (recalling Definition 6.16) the filter $\mathscr{D}_{\mathscr{E},e}$ has an extension to a non-principal ultrafilter $\mathscr{D}$ so for some regular $\theta \le \lambda$ there is a function g from $\mathrm{Dom}(\mathbf{e})/e$ onto θ which maps $\mathscr{D}$ to a uniform ultrafilter $g(\mathscr{D})$ on θ, so $\partial^{<\theta>_{\mathrm{tr}}} \le \partial^{\mathrm{Dom}(e)/e}/\mathscr{D}_{\mathscr{E},e}$ for every cardinal ∂. Choose such a pair (g, θ) with minimal θ so $\mathscr{D}$ is θ-complete hence $\theta = \aleph_0$ or θ is a measurable cardinal $\le \lambda$. By clause $(c)(\beta)$ of our assumption justified in the beginning of the proof there is $\partial \in (\lambda^+, \chi)$ such that $\partial < \partial^{<\theta>_{\mathrm{tr}}}$ hence $\partial^+ \le \partial^{<\theta>_{\mathrm{tr}}} \le \partial^{\mathrm{Dom}(e)/e}/\mathscr{D}_{\mathscr{E},e}$. So letting $I^0_\partial = I^{\mathrm{lin}}_{\lambda,\partial\times\lambda^{++}} \subseteq I^*$ the set $\{\bar{t}/\mathscr{E} : \bar{t} \in \mathrm{incr}_{J^*}(I^*)$ and $\mathrm{Rang}(\bar{t}) \subseteq I^0_\partial\}$ has cardinality $> \partial$. Now for each $\bar{t} \in \mathrm{inc}_{J^*}{}^{\bar{u}^*}(I^*)$ let $\pi_{\bar{t}} \in \mathrm{Aut}(I)$ be such that $\pi_{\bar{t}}(\bar{t}_0) = \bar{t}$ and let $\hat{\pi}_{\bar{t}}$ be the automorphism of $\mathrm{EM}_{\tau(\mathfrak{K})}(I, \Phi)$ which $\pi_{\bar{t}}$ induce, and let $p_t = \hat{\pi}_{\bar{t}}(p_0) \in \mathscr{S}(M)$. Hence $\{\hat{\pi}_{\bar{t}}(p_0) \restriction \mathrm{EM}_{\tau(\mathfrak{K})}(I^{\mathrm{lin}}_{\lambda,\partial\times\lambda^+}, \Phi) : \bar{t} \in \mathrm{inc}_{J^*}{}^{\bar{u}^*}(I^*)$ and $\mathrm{Rang}(\bar{t}) \subseteq I^{\mathrm{lin}}_{\lambda,\partial\times\lambda^{++}}\}$ is of cardinality $> \partial$, contradicting "$\mathfrak{K}$ stable in ∂" from $(*)_0$. $\square_{7.8}$

Note but we shall not use

7.9 Conclusion. 1) Under the assumptions of 7.8 we have $\kappa(\mathfrak{K}_\mu) = \aleph_0$, see below.
2) Moreover, $\kappa_{\mathrm{st}}(\mathfrak{K}_\mu) = \emptyset$.

Recall

7.10 Definition. If $\mathfrak{K}_\mu$ is an μ-a.e.c. with amalgamation which is stable, <u>then</u>:

- (a) $\kappa(\mathfrak{K}_\mu) = \aleph_0 + \sup\{\kappa^+ : \kappa$ regular $\le \mu$ and there is an $\le_{\mathfrak{K}_\mu}$-increasing continuous sequence $\langle M_i : i \le \kappa \rangle$ and $p \in \mathscr{S}(M_\kappa)$ such that M_{2i+2} is universal over M_{2i+1} and $p \restriction M_{2i+2}$ does μ-split over $M_{2i+1}\}$
- (b) $\kappa_{\mathrm{sp}}(\mathfrak{K}_\mu) := \{\kappa : \kappa$ regular $\le \mu$ and there is an $\le_{\mathfrak{K}_\mu}$-increasing continuous sequence $\langle M_i : i \le \kappa \rangle$ and $p \in \mathscr{S}(M_\kappa)$ which μ-splits over M_i for each $i < \kappa$ and M_{2i+2} is universal over $M_{2i+1}\}$.

Proof of 7.9. By playing with $\mathrm{EM}(I, \Phi)$, (or see Claim [Sh 394, 5.7=5.7tex] and Definition [Sh 394, 4.9=4.4tex]). $\square_{7.9}$

7.11 Question: Can we omit assumption 7.8(c) (see below so $\chi = \mathrm{LS}(\mathfrak{K})$)?

7.12 Theorem. *For some cardinal $\lambda_* < \chi$ and a cardinal $\lambda_{**} < \beth_{1,1}(\lambda_*^{+\omega})$ above λ_*, $\mathfrak{K}$ is categorical in every cardinal $\lambda \geq \lambda_{**}$ but in no $\lambda \in (\lambda_*, \lambda_{**})$ provided that:*

$\circledast^{\mu,\chi}_{\mathfrak{K}}$ (a) *K is an a.e.c. cateogorical in μ*

(b) *$\mathfrak{K}$ has amalgamation and* JEP *in every $\lambda < \aleph_\chi, \lambda \geq \mathrm{LS}(\mathfrak{K})$*

(c) *χ is a limit cardinal,* $\mathrm{cf}(\chi) > \mathrm{LS}(\mathfrak{K})$, *and for arbitrarily large $\lambda < \chi$ the sequence $\langle 2^{\lambda^{+n}} : n < \omega \rangle$ is increasing*

(d) *$\mu > \beth_{1,1}(\lambda)$ for every $\lambda < \chi$ hence $\mu \geq \aleph_\chi$*

(e) *every $M \in K_{<\aleph_\chi}$ has a $\leq_{\mathfrak{K}}$-extension in K_μ.*

Remark. 1) Concerning [Sh 394] note

(a) there the central case was $\mathfrak{K}$ with full amalgamation (not just below $\chi \ll \mu$!), trying to concentrate on the difficulty of lack of localness,

(b) when we use clause (e) this is just to get the "$M \in K_\mu$ is λ-saturated", this is where we use 7.8

(c) we demand "$\mathrm{cf}(\chi) > \mathrm{LS}(\mathfrak{K})$" to prove locality.

2) We rely on Chapter II and Chapter III in the end.
3) The assumption (e) of 7.12 follows if $\mathfrak{K}$ has amalgamation in every $\lambda' \leq \beth_{1,1}(\lambda)$ for $\lambda < \chi$ which is a reasonable assumption.
4) Most of the proof works even if we weaken the assumption (a) to "$\mathfrak{K}$ is strongly solvable in μ" and even weakly solvable, i.e. up to $\boxdot_7$, we continue in and see more [Sh:F782].
5) Theorem 7.12 also continue Kolman-Shelah [KlSh 362], [Sh 472], as its assumptions are proved there.

Proof. Let $\kappa = \mathrm{LS}(\mathfrak{K})$ and let $\Phi \in \Upsilon^{\mathrm{or}}_{\kappa}[\mathfrak{K}]$ be as guaranteed by 0.9(1) hence

$(*)_1$ if $I \in K^{\mathrm{lin}}_{\lambda}$ then $\mathrm{EM}_{\tau(\mathfrak{K})}(I,\Phi)$ belongs to K_λ for $\lambda \geq \mathrm{LS}(\mathfrak{K})$ (and in the strongly solvable case, $I \in K^{\mathrm{lin}}_{\mu} \Rightarrow \mathrm{EM}_{\tau(\mathfrak{K})}(I,\Phi) \in K^{x}_{\mu}$)

and

$(*)_2$ if $I \subseteq J$ are from K^{lin} then $\mathrm{EM}_{\tau(\mathfrak{K})}(I,\Phi) \leq_{\mathfrak{K}} \mathrm{EM}_{\tau(\mathfrak{K})}(J,\Phi)$.

Also

$(*)_3$ $\langle \mathscr{S}_{\mathfrak{K}}(M) : M \in \mathfrak{K}_{<\aleph_\chi}\rangle$ has the reasonable basic properties.

[Why? See II.1.9 and II.1.11 because $\mathfrak{K}_{<\aleph_\chi}$ has the amalgamation property by clause (b) of the assumption $\circledast^{\mu,\chi}_{\mathfrak{K}}$).]

$(*)_4$ if $M \in K_\mu$ then M is χ-saturated (hence χ-model homogeneous).

[Why? We shall prove that: if $\mathrm{LS}(\mathfrak{K}) \leq \lambda < \chi$ and $M \in K^{x}_{\mu}$ <u>then</u> M is λ^+-saturated. We shall show that all the assumptions of 7.8 with (μ,χ,λ) there standing for $(\mu,\aleph_\chi,\lambda)$ here hold. Let us check; clause (a) of 7.8 means "$\mathfrak{K}$ is categorical in μ" (or is strongly solvable) which holds by clause (a) of $\circledast^{\mu,\chi}_{\mathfrak{K}}$. Clause (b) of 7.8 says that $\mathrm{LS}(\mathfrak{K}) \leq \lambda < \aleph_\chi \leq \mu$ and $2^{2^\lambda} \leq \mu$; the first holds because of the way λ was chosen above and the second holds as clause (d) of $\circledast^{\mu,\chi}_{\mathfrak{K}}$ says that $\mu > \beth_{1,1}(\lambda)$ and $\mu \geq \aleph_\chi$. Clause (c)(α) of 7.8 holds as $\lambda^{+\lambda^{+4}} < \aleph_{\lambda+5}$ which is $< \aleph_\chi$ as χ is a limit cardinal and $\aleph_\chi$ here plays the role of χ there. Clause (d) of 7.8 says $\mathfrak{K}_{\geq\partial} \neq \emptyset$ for every cardinal ∂, holds by $(*)_1$ above. Lastly, clause (e) of 7.8 holds more exactly clauses (e)(β)(i) + (iii) hold by clauses (b) + (e) of $\circledast^{\mu,\chi}_{\mathfrak{K}}$ and they suffice.

We have shown that all the assumptions of 7.8 holds, hence its conclusion, which says, as $M \in K_\mu$, that M is λ^+-saturated. The "χ-model homogeneous" holds by II.1.14.]

$(*)_5$ if $M \leq_{\mathfrak{K}} N$ are from K^{x}_{μ} then $M \prec_{\mathbb{L}_{\infty,\chi}[\mathfrak{K}]} N$.

[Why? Obvious by $(*)_4$.]

$(*)_6$ if $\lambda \in (\kappa, \chi)$ and $I \in K^{\rm lin}_{\geq\lambda}$ is λ-wide then $\mathrm{EM}_{\tau(\mathfrak{K})}(I, \Phi)$ is λ-saturated; moreover, if $I^+ \in K^{\rm lin}_\lambda$ is wide over I then every $p \in \mathscr{S}(\mathrm{EM}_{\tau(K)}(I, \Phi))$ is realized in $\mathrm{EM}_{\tau(\mathfrak{K})}(I^+, \Phi)$.

[Why? By 1.14, its assumption "Φ satisfies the conclusion of 1.12" holds by $(*)_5$, (or as in [Sh 394, 6.8=6.5tex]). The "moreover" is immediate by $(*)_4$ as in the proof of $\circledast(d)$ inside the proof of 7.2 above or see the proof of $(*)_{10}$ below.]

$(*)_7$ $\mathfrak{K}$ is stable in λ when $\kappa \leq \lambda < \chi$.

[Why? Recalling clause (e) of the assumption of 7.12, by Claim 7.8 or more accurately $(*)_0$ in its proof as we have proved (in the proof of $(*)_4$) that the assumptions of 7.8 holds with (μ, χ, λ) there standing for $(\mu, \aleph_\chi, \lambda)$ here.]

$(*)_8$ if $\lambda \in [\kappa, \chi)$ and $M \in K^x_\lambda$ then there is $N \in \mathfrak{K}_\lambda$ which is $(\lambda, \aleph_0)$-brimmed over M

[Why? By $(*)_7$ and II.1.16(1)(b) remembering the amalgamation, clause (b) of the assumption of the theorem.]

$(*)_9$ if $\langle M_\alpha : \alpha \leq \lambda\rangle$ is $\leq_{\mathfrak{K}}$-increasing continuous, $\kappa \leq \|M_\lambda\| \leq \lambda < \chi$, then no $p \in \mathscr{S}_{\mathfrak{K}}(M_\lambda)$ satisfies $p \restriction M_{i+1}$ does λ-split over M_i for every $i < \lambda$.

[Why? Otherwise we get contradiction to stability in λ, i.e. $(*)_7$, see in III.2.21(1B), using amalgamation (using the tree ${}^{\theta>}2$ when $\theta = \min\{\partial : 2^\partial > \lambda\}$; also we can prove it as in the proof of case 2 inside the proof of 7.8.]

We could use more

$(*)_{10}$ if I_1, I_2 are wide linear orders of cardinality $\lambda \in (\kappa, \chi)$ and I_2 is wide over I_1 so $I_1 \subseteq I_2$ and $M_\ell = \mathrm{EM}_{\tau(\mathfrak{K})}(I_\ell, \Phi)$, then M_2 is universal over M_1 and even brimmed over I_1, even (λ, ∂)-brimmed for any regular $\partial < \lambda$.

[Why? As I_2 is wide over I_1, we can find a sequence $\langle J_\gamma : \gamma < \lambda\rangle$ of pairwise disjoint subsets of $I_2 \backslash I_1$ such that each J_γ is a convex subset of I_2 and in J_γ there is a monotonic sequence $\langle t_{\gamma,n} : n < \omega\rangle$ of members. Let $\langle \gamma_\varepsilon : \varepsilon < \lambda \times \partial\rangle$ list λ, and let $I_{2,0} = I_1$ and

$I_{2,1+\varepsilon} = I_2 \backslash \cup \{J_{\gamma_\zeta} : \zeta \in [1+\varepsilon, \lambda \times \partial)\}$ and $M'_\zeta = \mathrm{EM}_{\tau(\mathfrak{K})}(I_{2,\varepsilon}, \Phi)$. So $\langle M'_\zeta : \zeta \leq \lambda \times \partial \rangle$ is $\leq_{\mathfrak{K}}$-increasing continuous sequence of members of K_λ; first member M_1, last member M_2.

By II.1.16(4)(b) it is enough to prove that if $\varepsilon < \lambda \times \partial$ and $p \in \mathscr{S}(M_\varepsilon)$ then p is realized in $M_{\varepsilon+1}$. As I_1 is wide of cardinality λ so is $I_{2,\varepsilon}$ hence M'_ε is saturated. Also for each ε we can find a linear order $I^+_{2,\varepsilon}$ of cardinality λ such that $I_{2,\varepsilon+1} \subseteq I^+_{2,\varepsilon}$ and $J^+_\varepsilon = I^+_{2,\varepsilon+1} \backslash I_{2,\varepsilon}$ is a convex subset of $I^+_{2,\varepsilon+1}$ and is a wide linear order of cardinality λ which is strongly $\aleph_0$-homogeneous, (recall $J_{\gamma_\varepsilon} \subseteq J^+_{\gamma_\varepsilon}$ is infinite). So in $M^+_{\varepsilon+1} = \mathrm{EM}_{\tau(\mathfrak{K})}(I^+_{2,\varepsilon+2}, \Phi)$ every $p \in \mathscr{S}(M^1_\varepsilon)$ is realized (as $I^+_{2,\varepsilon+1}$ is wide over $I_{2,\varepsilon}$ as J^+_ε is wide of cardinality λ), moreover realized in $M'_{\varepsilon+1}$ (why? by the strong $\aleph_0$-homogeneous every element and even finite sequence from $M^+_{\varepsilon+1}$ can be mapped by some automorphism of $M^+_{\varepsilon+1}$ over M_ε into $M_{\varepsilon+1}$). As said above, this suffices.]

> $\circledast_1$ χ_* is well defined $\in (\kappa, \chi)$ where
>
> $$\chi_* = \mathrm{Min}\{\theta : \kappa < \theta < \chi \text{ and for every saturated } M \in \mathfrak{K}, \text{ if } \theta \leq \|M\| < \chi, \text{ every } p \in \mathscr{S}(M) \text{ is } \theta\text{-local, see Definition 7.4(2)}\}.$$

[Why? By 7.2 which we apply with (μ, κ) there standing for (χ, κ) here recalling $\kappa = \mathrm{LS}(\mathfrak{K})$; this is O.K. as: clause (a) in 7.2 holds by clause (c) of the assumption here, clause (b) in 7.2 holds by clause (b) of the assumption here as $\chi \leq \aleph_\chi$. Lastly, clause (c) in 7.2 easily follows by $(*)_6$ above.]

> $\circledast_2$ if $\lambda \in (\kappa, \chi)$ and $\langle M_i : i \leq \delta \rangle$ is $\leq_{\mathfrak{K}_\lambda}$-increasing continuous, M_{i+1} is $\leq_{\mathfrak{K}}$-universal over M_i for $i < \delta$ <u>then</u> M_δ is saturated and moreover every $p \in \mathscr{S}(M_\delta)$ does not λ-split over M_α for some $\alpha < \delta$.

[Why? For $i \leq \delta$ let I_i be the linear order $\lambda \times \lambda \times (1+i)$ and $M'_i = \mathrm{EM}_{\tau(\mathfrak{K})}(I_i, \Phi)$. So $\langle M'_i : i \leq \delta \rangle$ is $\leq_{\mathfrak{K}_\lambda}$-increasing continuous. Also for $i \leq \delta, \zeta \leq \lambda$ let $I_{i,\zeta} = \lambda \times \lambda \times (1+i) + \lambda \times \zeta$ and $M'_{i,\zeta} = \mathrm{EM}_{\tau(\mathfrak{K})}(I_{i,\zeta}, \Phi)$, so for each $i < \delta$ the sequence $\langle M'_{i,\zeta} : \zeta \leq \lambda \rangle$ is $\leq_{\mathfrak{K}_\lambda}$-increasing continuous, $M'_{i,0} = M'_i, M'_{i,\lambda} = M'_{i+1}$. Now for

$i < \delta, \zeta < \lambda$ every $p \in \mathscr{S}(M_{i,\zeta})$ is realized in $M'_{i,\zeta+1}$ by $(*)_6$ and the definition of type, varying the linear order. By II.1.16(4)(b) the model M'_{i+1} is $\leq_{\mathfrak{K}_\lambda}$-universal over M'_i and by Definition II.1.15 the models M'_δ and M_δ are $(\lambda, \mathrm{cf}(\delta))$-brimmed hence by II.1.16(3) are isomorphic. But M'_δ is saturated by $(*)_6$, hence M_δ is saturated.

What about the "moreover"? (Note that if $\lambda = \lambda^{\mathrm{cf}(\delta)}$ then $(*)_9$ does not cover it.) We can find easily $\langle I''_\alpha : \alpha \leq \lambda \times \delta + 1\rangle$ such that:

- (a) I''_α is a linear order of cardinality λ into which λ can be embedded
- (b) I''_α is increasing continuous with α
- (c) I''_α is an initial segment of I''_β for $\alpha < \beta \leq \delta + 1$
- (d) $I''_{\alpha+1}$ has a subset of order types $\lambda \times \lambda$ whose convex hull is disjoint to I''_α
- (e) if $\alpha \leq \beta < \lambda \times \delta$ and $s \in I''_{\lambda\times\delta+1} \backslash I''_{\lambda\times\delta}$ then there is an automorphism $\pi_{\alpha,\beta,s}$ of $I''_{\lambda\times\delta+1}$ mapping $I''_{\beta+1}$ onto $I''_{\lambda\times\delta}$ and is over $I''_\alpha \cup \{t \in I''_{\lambda\times\delta+1} : s \leq_{I''_{\lambda\times\delta+1}} t\}$.

Let $M''_\alpha = \mathrm{EM}_{\tau(\mathfrak{K})}(I''_\alpha, \Phi)$, so $\langle M''_{\lambda\times\alpha} : \alpha \leq \delta\rangle$ has the properties of $\langle M'_\alpha : \alpha \leq \delta\rangle$, i.e. every $p \in \mathscr{S}(M''_\alpha)$ is realized in $M''_{\alpha+1}$ hence $M''_{\alpha+\lambda}$ is $\leq_{\mathfrak{K}_\lambda}$-universal over M''_α. So (easily or see II.1.16,II.1.15) there is an isomorphism f from M_δ onto $M''_{\lambda\times\delta}$ such that $M''_{\lambda\alpha} \leq_{\mathfrak{K}} f(M_{\alpha+1}) \leq M''_{\lambda\alpha+2}$. So it suffices to prove the "moreover" for $\langle M''_{\lambda\times\alpha} : \alpha \leq \delta\rangle$, equivalently for $\langle M''_\alpha : \alpha \leq \lambda \times \delta\rangle$. Let $p \in \mathscr{S}(M''_{\lambda\times\delta})$ so some $a \in M''_{\lambda\times\delta+1}$ realizes it, hence for some $t_0 < \ldots < t_{n-1}$ from $I''_{\lambda\times\delta+1}$ and τ_Φ-term $\sigma(x_0, \ldots, x_{n-1})$ we have $a = \sigma^{\mathrm{EM}(I''_{\lambda\times\delta+1},\Phi)}(a_{t_0}, \ldots, a_{t_{n-1}})$, it follows that for some $m \leq n$ we have $t_\ell \in I''_{\lambda\times\delta} \Leftrightarrow \ell < m$ and let $\alpha < \lambda \times \delta$ be such that $\{t_\ell : \ell < m\} \subseteq I''_\alpha$; if $m = n$ choose any $t_n \in I''_{\lambda\times\delta+1} \backslash I''_{\lambda\times\delta}$. If $\beta \in (\alpha, \lambda \times \delta)$ and $\mathbf{tp}_{\mathfrak{K}}(a, M''_\delta, M''_{\delta+1})$ does λ-split over M''_β then $\pi' := \pi_{\beta,\beta,t_m}$ is an automorphism of $I''_{\lambda\times\delta+1}$ mapping $I''_{\beta+1}$ onto $I''_{\lambda\times\delta}$ and is over $I''_\beta \cup \{s \in I''_{\lambda\times\delta+1} : t_m \leq_{I''_{\lambda\times\delta+1}} s\}$ hence it is the identity on the set $\{t_\ell : \ell < n\}$; now π' induces an automorphism $\hat{\pi}'$ of $\mathrm{EM}_{\tau(\mathfrak{K})}(I''_{\lambda\times\delta+1}, \Phi)$, so clearly it maps a to itself and maps $\mathbf{tp}_{\mathfrak{K}}(a, M''_{\beta+1}, M''_{\lambda\times\delta+1})$ to $\mathbf{tp}_{\mathfrak{K}}(a, M''_{\lambda\times\delta}, M''_{\lambda\times\delta+1})$ and it maps M''_β onto itself, hence also $\mathbf{tp}_{\mathfrak{K}}(a, M''_{\beta+1}, M''_{\delta+1})$ does λ-split over M''_β. So if for some $\beta \in (\alpha, \lambda \times \delta)$, the type $\mathbf{tp}_{\mathfrak{K}}(a, M''_\delta, M''_{\delta+1})$

does not λ-split over M''_β we get the desired conclusion, but otherwise this contradicts $(*)_9$.]

$\circledast_3$ If $\lambda \in [\chi_*, \chi)$ and $M \in K_\lambda$ is saturated and $p \in \mathscr{S}(M)$ <u>then</u> for some N we have:

- (a) $N \leq_{\mathfrak{K}} M$
- (b) $N \in K_{\chi_*}$ is saturated
- (c) p does not χ_*-split over N
- (d) p does not λ-split over N (follows by (a),(b),(c)).

[Why $\circledast_3$ holds? For clauses (a),(b),(c) use $\circledast_2$ or just $(*)_9$; for clause (d) use localness, i.e. recall $\circledast_1$ and Definition 7.4.]

$\circledast_4$ Assume $\lambda \in [\kappa, \chi)$ and $M_1 \leq_{\mathfrak{K}} M_2 \leq_{\mathfrak{K}} M_3$ are members of K, M_2 is λ^+-saturated and $p \in \mathscr{S}(M_3)$. If $N_\ell \leq_{\mathfrak{K}} M_\ell$ is from $K_{\leq\lambda}$ and $p \restriction M_{\ell+1}$ does not λ-split over N_ℓ for $\ell = 1, 2$ <u>then</u> p does not λ-split over N_1.

[Why? Easy manipulations. Without loss of generality $N_1 \leq_{\mathfrak{K}} N_2$ as we can increase N_2. So for some pair (M_4, a) we have $M_3 \leq_{\mathfrak{K}} M_4, a \in M_4$ and $p = \mathbf{tp}_{\mathfrak{K}}(a, M_3, M_4)$. Assume $\alpha < \lambda^+$ and let $\bar{b}, \bar{c} \in {}^\alpha(M_3)$ be such that $\mathbf{tp}_{\mathfrak{K}}(\bar{b}, N_1, M_3) = \mathbf{tp}_{\mathfrak{K}}(\bar{c}, N_1, M_3)$. As M_2 is λ^+-saturated and $N_2 \leq_{\mathfrak{K}} M_2 \leq_{\mathfrak{K}} M_3$ we can find $\bar{b}', \bar{c}' \in {}^\alpha(M_2)$ such that $\mathbf{tp}_{\mathfrak{K}}(\bar{b}'\char`\^\bar{c}', N_2, M_3) = \mathbf{tp}_{\mathfrak{K}}(\bar{b}\char`\^\bar{c}, N_2, M_3)$ using II.1.14. Hence $\mathbf{tp}_{\mathfrak{K}}(\bar{b}', N_1, M_3) = \mathbf{tp}_{\mathfrak{K}}(\bar{b}, N_1, M_3) = \mathbf{tp}_{\mathfrak{K}}(\bar{c}, N_1, M_3) = \mathbf{tp}_{\mathfrak{K}}(\bar{c}', N_1, M_3)$.

By the choice of (M_4, a) and the assumption on N_1 that $p \restriction M_2$ does not λ-split over N_1 we get

$$\mathbf{tp}_{\mathfrak{K}}(\langle a\rangle\char`\^\bar{b}', N_1, M_4) = \mathbf{tp}_{\mathfrak{K}}(\langle a\rangle\char`\^\bar{c}', N_1, M_4).$$

Clearly $\mathbf{tp}_{\mathfrak{K}}(\bar{b}', N_2, M_3) = \mathbf{tp}_{\mathfrak{K}}(\bar{b}, N_2, M_3)$ hence by the choice of (M_4, a) and the assumption on N_2 that p does not λ-split over N_2 we have $\mathbf{tp}_{\mathfrak{K}}(\langle a\rangle\char`\^\bar{b}', N_2, M_4) = \mathbf{tp}(\langle a\rangle\char`\^\bar{b}, N_2, M_4)$ hence by monotonicity

$$\mathbf{tp}_{\mathfrak{K}}(\langle a\rangle\char`\^\bar{b}', N_1, M_4) = \mathbf{tp}_{\mathfrak{K}}(\langle a\rangle\char`\^\bar{b}, N_1, M_4).$$

Similarly

$$\mathbf{tp}_{\mathfrak{K}}(\langle a\rangle\hat{\ }\bar{c}', N_1, M_4) = \mathbf{tp}_{\mathfrak{K}}(\langle a\rangle\hat{\ }\bar{c}, N_1, M_4).$$

As equality of types is transitive, we have $\mathbf{tp}_{\mathfrak{K}}(\langle a\rangle\hat{\ }\bar{c}, N_1, M_4) =$ $= \mathbf{tp}_{\mathfrak{K}}(\langle a\rangle\hat{\ }\bar{c}', N_1, M_4) = \mathbf{tp}_{\mathfrak{K}}(\langle a\rangle\hat{\ }\bar{b}', N_1, M_4) = \mathbf{tp}_{\mathfrak{K}}(\langle a\rangle\hat{\ }\bar{b}, N_1, M_4)$ as required.]

$\circledast_5$ Assume $I_3 = I_0 + I_1' + I_2'$ are wide linear orders of cardinality λ where $\chi > \lambda > \kappa$ and let $I_\ell = I_0 + I_\ell'$ for $\ell = 1, 2$ and $M_\ell = \mathrm{EM}_{\tau(\mathfrak{K})}(I_\ell, \Phi)$ for $\ell = 0, 1, 2, 3$. If $\ell \in \{1, 2\}$ and $\bar{a} \in {}^{\lambda>}(M_\ell)$ <u>then</u> $\mathbf{tp}_{\mathfrak{K}_\lambda}(\bar{a}, M_{3-\ell}, M_3)$ does not λ-split over M_0, (moreover if $\mathbf{tp}_{\mathfrak{K}_\lambda}(\bar{a}, M_0, M_3)$ does not λ-split over $N \in K_{\leq\lambda}$ then also $\mathbf{tp}_{\mathfrak{K}_\lambda}(\bar{a}, M_{3-\ell}, M_3)$ does not λ-split over N).

[Why? For $\ell = 2$, if the desired conclusion fails we get a contradiction as in the proof of $\circledast_2$ so for $\ell = 2$ we get the conclusion. For $\ell = 1$ if the desired conclusion fails (but it holds for $\ell = 2$) we get a contradiction to categoricity in μ by the order property (by 1.5).]

$\circledast_6$ If $\lambda \in (\chi_*, \chi), \delta < \lambda^+, \langle M_i : i \leq \delta\rangle$ is $\leq_{\mathfrak{K}_\lambda}$-increasing continuous and $i < \delta \Rightarrow M_i$ saturated <u>then</u> M_δ is saturated.

[Why? Let $N \leq_{\mathfrak{K}} M_\delta, \|N\| < \lambda$ and $p \in \mathscr{S}(N)$. If $\mathrm{cf}(\delta) > \|N\|$ this is easy so assume $\mathrm{cf}(\delta) \leq \|N\|$ hence $\mathrm{cf}(\delta) < \lambda$ and without loss of generality $\delta = \mathrm{cf}(\delta)$ and choose a cardinal θ such that $\mathrm{LS}(\mathfrak{K}) < \chi_* + |\mathrm{cf}(\delta)| + \|N\| \leq \theta < \lambda$ and $\|N\|^+ < \lambda \Rightarrow \|N\| < \theta$ and let $q \in \mathscr{S}(M_\delta)$ extend p, exist as $\mathfrak{K}_{\leq\lambda}$ has amalgamation.

Now for every $X \subseteq M_\delta$ of cardinality $\leq \theta$ we can choose $N_i \leq_{\mathfrak{K}} M_i$ by induction on $i \leq \delta$ such that $N_i \in K_\theta$ is saturated, is $\leq_{\mathfrak{K}}$-increasing continuous with i and N_i is $\leq_{\mathfrak{K}}$-universal over N_j and includes $(X \cup N) \cap M_i$ when $i = j+1$. So by $\circledast_2$ (we justify the choice of N_i for limit i and) the model N_δ is saturated, so if $\|N\|^+ < \lambda$ then $N \leq_{\mathfrak{K}} N_\delta, N_\delta$ is saturated of cardinality $\theta > \|N\|$ so we are done as $N_\delta \leq_{\mathfrak{K}} M_\delta$, so without loss of generality $\lambda = \|N\|^+$ hence $\lambda = \theta^+$.

Also for some $\alpha_* < \delta$ and $N_* \leq_{\mathfrak{K}} M_{\alpha_*}$ of cardinality θ, the type q does not θ-split over N_*. [Why? Otherwise we choose (N_i, N_i^+) by induction on $i \leq \delta$ such that $N_i \leq_{\mathfrak{K}} N_i^+$ are from $K_\theta, N_i \leq_{\mathfrak{K}} M_i, N_i^+ \leq_{\mathfrak{K}} M_\delta, N_i$ is $\leq_{\mathfrak{K}}$-increasing continuous, N_i is $\leq_{\mathfrak{K}}$-universal over N_j if $i = j+1$ and $q \restriction N_i^+$ does θ-split over N_i and $\cup\{N_j^+ \cap M_i : j < i\} \subseteq N_i$. In the end we get a contradiction to $\circledast_2$.]

We can find $N' \leq_{\mathfrak{K}} M_{\alpha_*}$ from K_{χ_*} such that $q \restriction M_{\alpha_*}$ does not θ-split over N', (why? by $\circledast_3$) and without loss of generality $N' \leq_{\mathfrak{K}} N_*$ and $N' \leq_{\mathfrak{K}} N$. Also q does not θ-split over N' (why? by applying $\circledast_4$, with $\theta, N_*, M_{\alpha_*}, M_\delta$ here standing for $\lambda, M_1, M_2, M_3, N_1, N_2$ there; or use $N' = N_*$).

By $(*)_6$ as M_{α_*} is saturated without loss of generality $M_{\alpha_*} = \mathrm{EM}_{\tau(\mathfrak{K})}(\lambda, \Phi)$ and for $\varepsilon < \lambda$ let $M_{\alpha_*,\varepsilon} = \mathrm{EM}_{\tau(\mathfrak{K})}(\theta \times \theta \times (1+\varepsilon), \Phi)$, so $M_{\alpha_*,\varepsilon} \in K_\theta$ is saturated and is brimmed over $M_{\alpha^*,\zeta}$ when $\varepsilon = \zeta + 1$ by $(*)_{10}$. So for each $\varepsilon < \lambda$ there is $a_\varepsilon \in M_{\alpha^*,\varepsilon+1}$ realizing $q \restriction M_{\alpha_*,\varepsilon}$. Also without loss of generality $M_\delta \leq_{\mathfrak{K}} \mathrm{EM}_{\tau(\mathfrak{K})}(\lambda + \lambda, \Phi)$ as in the proof of $\circledast_2$ or by $(*)_{10}$, now for some $\varepsilon(*) < \lambda$ we have $N \leq_{\mathfrak{K}} \mathrm{EM}_{\tau(\mathfrak{K})}(I_2, \Phi)$ and $N_* \leq_{\mathfrak{K}} \mathrm{EM}_{\tau(\mathfrak{K})}(I_0, \Phi)$ where $I_0 = \theta \times \theta \times (1 + \varepsilon(*))$ and $I_2 = [\lambda, \lambda + \varepsilon(*)) \cup I_0$. Let $I_1 = \theta \times \theta \times \zeta(*)$ where $\zeta(*) \in (\varepsilon(*), \lambda)$ is large enough such that $a_{\varepsilon(*)} \in \mathrm{EM}_{\tau(\mathfrak{K})}(I_1, \Phi)$, e.g. $\zeta(*) = 1 + \varepsilon(*) + 1$ and let $I_3 = I_1 \cup I_2 \subseteq \lambda + \lambda$. Let $M'_\ell = \mathrm{EM}_{\tau(\mathfrak{K})}(I_\ell, \Phi)$ for $\ell = 0, 1, 2, 3$.

Now we apply $\circledast_5$, the "moreover" with θ, I_0, I_1, I_2, $I_1 \backslash I_0$, $I_2 \backslash I_0$, $a_{\varepsilon(*)}$, N' here standing for λ, I_0, I_1, I_2, I'_1, I'_2, $\bar{a}$, N there and we conclude that $\mathbf{tp}_{\mathfrak{K}_\lambda}(a_{\varepsilon(*)}, M'_2, M'_3)$ does not θ-split over N'.

As $N' \leq_{\mathfrak{K}} M'_0 \leq_{\mathfrak{K}} M'_2$ also the type $q' := \mathbf{tp}_{\mathfrak{K}_\lambda}(a_{\varepsilon(*)}, M'_2, M'_3)$ does not θ-split over N'. Let us sum up: $q \restriction M'_2, q'$ belong to $\mathscr{S}_{\mathfrak{K}_\lambda}(M'_2)$, does not θ-split over N', $N' \in K_{\chi_*}$ and $\chi_* \leq \theta$. Also $N' \leq_{\mathfrak{K}_*} M'_0 \leq_{\mathfrak{K}_*} M'_2$, the model M'_0 is θ-saturated and $q \restriction M_{\alpha_*} = q' \restriction M_{\alpha_*}$. By the last two sentences obviously $q = q'$ (it may be more transparent to consider $q \restriction (\leq \chi_*) = q' \restriction (\leq \chi_*)$), so we are done proving $\circledast_6$.]

$\circledast_7$ If $\lambda \in (\chi_*, \chi)$ <u>then</u> the saturated $M \in \mathfrak{K}_\lambda$ is superlimit.

[Why? By $\circledast_6$, (existence by $(*)_6$, the non-maximality by $(*)_6$+ uniqueness; you may look at [Sh 394, 6.7=6.4tex](1).]

Now we have arrived to the main point

$\odot_1$ If $\lambda \in (\chi_*, \chi)$ <u>then</u> $\mathfrak{s}_\lambda$ is a full good λ-frame, $K_{\mathfrak{s}_\lambda}$ categorical where $\mathfrak{s}_\lambda$ is defined by

(a) $\mathfrak{K}_{\mathfrak{s}_\lambda} = \mathfrak{K}_\lambda \restriction \{M \in K_\lambda : M \text{ saturated}\}$

(b) $\mathscr{S}^{\mathrm{bs}}_{\mathfrak{s}_\lambda}(M) = \mathscr{S}^{\mathrm{na}}_{\mathfrak{s}_\lambda}(M) := \{\mathbf{tp}_{\mathfrak{s}}(a, M, N) : M \leq_{\mathfrak{K}_\lambda} N$ and $a \in N \backslash M\}$ for $M \in K_{\mathfrak{s}_\lambda}$

(c) $p \in \mathscr{S}^{\rm bs}_{\mathfrak{s}_\lambda}(M_2)$ does not fork over M_1 when $M_1 \le_{\mathfrak{s}_\lambda} M_2$ and for some $M \le_{\mathfrak{K}} M_1$ of cardinality χ_*, the type p does not χ_*-split over N.

[Why? We check the clauses of Definition II.2.1.

$K_{\mathfrak{s}_\lambda}$ is categorical:
By II.1.26(1) and $\circledast_7$.

Clause (A),Clause (B): By $\circledast_7$ recalling that there is a saturated $M \in K_{\mathfrak{s}_\lambda}$ (and it is not $<_{\mathfrak{s}_\lambda}$-maximal) by $(*)_6$ and trivially recalling II.1.26, of course.

Clause (C): By categoricity and $(*)_6$ clearly no $M \in K_{\mathfrak{s}_\lambda}$ is maximal; amalgamation and JEP holds by clause (b) of the assumption of the claim.

Clause (D)(a),(b): By the definition.

Clause (D)(c): Density is obvious; in fact $\mathfrak{s}_\lambda$ is full.

Clause (D)(d): (bs - stability).
Easily $\mathscr{S}_{\mathfrak{s}_\lambda}(M) = \mathscr{S}_{\mathfrak{K}_\lambda}(M)$ which has cardinality $\le \lambda$ by the moreover in $(*)_6$.

Clause (E)(a): By the definition.

Clause (E)(b): Monotonicity (of non-forking).
By the definition of "does not χ_*-split".

Clause (E)(c): Local character.
Why? Let $\langle M_\alpha : \alpha \le \delta \rangle$ be $\le_{\mathfrak{s}_\lambda}$-increasing continuous, $\delta < \lambda^+$ and $q \in \mathscr{S}^{\rm bs}_{\mathfrak{s}_\lambda}(M_\delta)$. Using the third paragraph of the proof of $\circledast_6$ for $\theta = \chi_*$, for some $\alpha_* < \delta$ and $N_* \le_{\mathfrak{s}_\lambda} M_{\alpha_*}$ of cardinality θ the type q does not θ-split over N_*. So clearly q does not fork over M_{α_*} (for $\mathfrak{s}_\lambda$), as required.

Clause (E)(d): Transitivity of non-forking.
By $\circledast_4$.

Clause (E)(e): Uniqueness.
Holds by the choice of χ_*, i.e. by $\circledast_1$.

Clause (E)(f): Symmetry.

Why? Let M_ℓ for $\ell \le 3$ and a_0, a_1, a_2 be as in (E)(f)$'$ in II.2.19. We can find a $\le_{\mathfrak{K}}$-increasing continuous sequence $\langle M_{0,\alpha} : \alpha \le \lambda^+\rangle$ such that $M_{0,0} = M_0, M_{0,\alpha+1}$ is $\le_{\mathfrak{s}_\lambda}$-universal over $M_{0,\alpha}$ and without loss of generality $M_{0,\alpha} = \mathrm{EM}_{\tau(\mathfrak{K})}(\gamma_\alpha, \Phi)$ so is $\le_{\mathfrak{K}}$-increasing continuous, and λ divides γ_α.

By (E)(g) proved below we can find $a^\ell_\alpha \in M_{0,\alpha+1}$ realizing $\mathbf{tp}_{\mathfrak{s}_\lambda}(a_\ell, M_0, M_{\ell+1})$ such that $\mathbf{tp}_{\mathfrak{s}_\lambda}(a^\ell_\alpha, M_{0,\alpha}, M_{0,\alpha+1})$ does not fork over $M_0 = M_{0,0}$, for $\ell = 1, 2$. We can find $N_* \le_{\mathfrak{K}} M_0$ of cardinality χ_* such that $\mathbf{tp}_{\mathfrak{s}_\lambda}(\langle a_1, a_2\rangle, M_0, M_3)$ does not χ_*-split over N_* so $N_* \le_{\mathfrak{K}} M_{0,0}$.

Then as in 1.5 we get a contradiction (recalling II.2.19).

<u>Clause (E)(g)</u>: Extension existence.

If $M \le_{\mathfrak{s}_\lambda} N$ and $p \in \mathscr{S}^{\mathrm{bs}}_{\mathfrak{s}_\lambda}(M) = \mathscr{S}^{\mathrm{na}}_{\mathfrak{K}}(M)$, then p does not χ_*-split over M_* for some $M_* \le_{\mathfrak{K}} M$ of cardinality χ_* by $\circledast_3$. Let $M^* \in K_{\chi^*}$ be such that $M_* \le_{\mathfrak{K}} M^* \le_{\mathfrak{K}} M$ and M^* is $\le_{\mathfrak{K}}$-universal over M_*. As $M, N \in K_{\mathfrak{s}_\lambda} \subseteq K_\lambda$ are saturated there is an isomorphism π from M onto N over M^* and let $q = \pi(p)^+$.

Now $q \restriction M = p$ by $\circledast_1$ as both are from $\mathscr{S}^{\mathrm{na}}_{\mathfrak{K}}(M)$, does not χ_*-split over M_* and has the same restriction to M^*.

<u>Clause (E)(h)</u>: Follows by II.2.17(3),(4) recalling $\mathfrak{s}_\lambda$ is full.

<u>Clause (E)(i)</u>: Follows by II.2.16.

So we have finished proving "$\mathfrak{s}_\lambda$ is a good λ-frame.]

$\odot_2$ If $\lambda \in (\chi_*, \chi)$ then $\mathfrak{K}^{\mathfrak{s}_\lambda}$ is $\mathfrak{K} \restriction \{M : M \text{ is } \lambda\text{-saturated}\}$.

[Why? Should be clear.]

$\odot_3$ λ_* is well defined where
$\lambda_* = \mathrm{Min}\{\lambda : \chi_* < \lambda < \chi \text{ and } 2^{\lambda^{+n}} < 2^{\lambda^{+n+1}} \text{ for every } n < \omega\}$.

[Why? By clause (c) of the assumption.]

Let $\Theta = \{\lambda_*^{+n} : n < \omega\}$.

$\odot_4$ $\mathfrak{s}_\lambda$ is weakly succsesful for $\lambda \in \Theta$.

[Why? Recalling that "$\mathfrak{s}_\lambda$ categorical", by Definition III.1.1, Definition II.5.2 and Observation II.5.8(b) this means that if $(M, N, a) \in K^{3,\mathrm{bs}}_{\mathfrak{s}_\lambda}$ then for some $(M_1, N_1, a) \in K^{3,\mathrm{uq}}_{\mathfrak{s}_\lambda}$ we have $(M, N, a) \le^{\mathrm{bs}}_{\mathfrak{s}_\lambda}$

(M_1, N_1, a) (see Definition II.5.3). Toward contradiction, assume that this fails. Let $\langle M_\alpha : \alpha < \lambda^+ \rangle$ be $\leq_{\mathfrak{s}_\lambda}$-increasing continuous, $M_{\alpha+1}$ is brimmed over M_α for $\alpha < \lambda^+$ such that $M_0 = M$. Now directly by the definitions (as in II§5, see more in Chapter VII) we can find $\langle M_\eta, f_\eta : \eta \in {}^{\lambda^+>}2 \rangle$ such that:

(a) if $\eta \triangleleft \nu \in {}^{\lambda^+>}2$ then $M_\eta \leq_{\mathfrak{s}_\lambda} M_\nu$

(b) if $\eta \in {}^{\lambda^+>}2$ then f_η is a one-to-one function from $M_{\ell g(\eta)}$ to M_η over $M_0 = M$ such that $\rho \triangleleft \eta \Rightarrow f_\rho \subseteq f_\eta$ and $f_\eta(M_{\ell g(\eta)}) \leq_{\mathfrak{s}_\lambda} M_\eta$ in fact $f_0 = \mathrm{id}_M$ and $(M, N, a) \leq^{\mathrm{bs}}_{\mathfrak{s}_\lambda} (f_\eta(M_{\ell g(\eta)}), M_\eta, a) \in K^{\mathrm{bs}}_{\mathfrak{s}}$

(c) if $\nu = \eta\char`\^\langle \ell \rangle \in {}^{\lambda>}2$ then M_ν is brimmed over M_η

(d) if $\eta \in {}^{\lambda^+>}2$ then $f_{\eta\char`\^<0>}(M_{\ell g(\eta)+1}) = f_{\eta\char`\^<1>}(M_{\ell g(\eta)+1})$

(e) if $\eta \in {}^{\lambda>}2$ then there is no triple (N, f_0, f_1) such that $f_{\eta\char`\^\langle 1 \rangle}(M_{\ell g(\eta)+1}) \leq_{\mathfrak{s}} N$, and $f_\ell$ is a $\leq_{\mathfrak{s}_\lambda}$-embedding of $M_{\eta\char`\^<\ell>}$ into N over $f_{\eta\char`\^<\ell>}(M_{\ell g(\eta)+1})$ for $\ell = 0, 1$ and $f_0 \restriction M_\eta = f_1 \restriction M_\eta$.

Having carried the induction by renaming without loss of generality $\eta \in {}^{\lambda^+>}2 \Rightarrow f_\eta = \mathrm{id}_{M_{\ell g(\eta)}}$. Now $M_* := \cup\{M_\alpha : \alpha < \lambda^+\}$; it belongs to $\mathfrak{s}_{\lambda^+}$ and is saturated and for $\eta \in {}^{\lambda^+}2$ let $M_\eta := \cup\{M_{\eta \restriction \alpha} : \alpha < \lambda^+\}$ so $M_* \leq_{\mathfrak{s}_{\lambda^+}} M_\eta \in K_{\mathfrak{s}_{\lambda^+}}$. But χ is a limit cardinal so also $\lambda^+ \in (\kappa, \chi)$ so let $N_* \in K_{\mathfrak{s}_{\lambda^+}}$ be $\leq_{\mathfrak{s}_{\lambda^+}}$-universal over M_*, so for every $\eta \in {}^{\lambda^+}2$ there is an $\leq_{\mathfrak{s}^+}$-embedding h_η of M_η into N_* over M_*. But $2^\lambda < 2^{\lambda^+}$ by the choice of λ_* so by I.0.5 we get a contradiction to clause (e).]

$\odot_5$ for $\lambda \in \Theta$, if $M \in K^{\mathfrak{s}_\lambda}_{\lambda^+}$ is saturated above λ for $K^{\mathfrak{s}_\lambda}$, <u>then</u> M is saturated for $\mathfrak{K}$.

[Why? Should be clear and implicitly was proved above.]

$\boxdot_1$ $\mathrm{NF}_{\mathfrak{s}_\lambda}$ is well defined and is a non-forking relation on $\mathfrak{K}_{\mathfrak{s}_\lambda}$ respecting $\mathfrak{s}_\lambda$ (for $\lambda \in \Theta$).

[Why? By II§6 as $\mathfrak{s}_\lambda$ is a weakly successful good λ frame.]

$\boxdot_2$ $\mathfrak{s}_\lambda$ is a good$^+$ λ-frame (for $\lambda \in \Theta$).

[Recalling Definition III.1.3, assume that this fails so there are $\langle M_i, N_i : i < \lambda^+ \rangle$ and $\langle a_{i+1} : i < \lambda^+ \rangle$, as there, i.e. $a_{i+1} \in M_{i+2} \backslash M_{i+1}$, $\mathbf{tp}_{\mathfrak{s}_\lambda}(a_{i+1}, M_{i+1}, M_{i+2})$ does not fork over M_0 for $\mathfrak{s}_\lambda$, but $\mathbf{tp}_{\mathfrak{s}_\lambda}(a_{i+1}, N_0, M_{i+1})$ forks over M_0. Also, recalling Definition III.1.3 the model $M = \cup\{M_i : i < \lambda^+\}$ is saturated for $\mathfrak{K}^{\mathfrak{s}_\lambda}_{\lambda^+}$ hence by $\odot_5$ for $\mathfrak{K}$, so it belongs to $K_{\mathfrak{s}_{\lambda^+}}$.
We can find an isomorphism f_0 from M onto $\mathrm{EM}_{\tau(\mathfrak{K})}(\lambda^+, \Phi)$, by $(*)_6$. By the "moreover" from $(*)_6$, more exactly by $(*)_{10}$ we can find a $\leq_{\mathfrak{K}}$-embedding f_1 of $N =: \cup\{N_i : i < \lambda^+\}$ into $\mathrm{EM}_{\tau(\mathfrak{K})}(\lambda \times \lambda, \Phi)$ extending f_0. As we can increase the N_i's without loss of generality f_1 is onto $\mathrm{EM}_{\tau(\mathfrak{K})}(\lambda \times \lambda, \Phi)$. We can find $\delta < \lambda^+$ such that $N_\delta = \mathrm{EM}_{\tau(\mathfrak{K})}(u, \Phi)$ where $u = \{\lambda\alpha + \beta : \alpha, \beta < \delta\}$. By $a_{\delta+1}$ we get a contradiction to $\circledast_5$.]

$\boxdot_3$ Let $\lambda \in \Theta$

(α) $\leq^*_{\mathfrak{s}_\lambda}$ is a partial order on $K^{\mathrm{nice}}_{\lambda^+}[\mathfrak{s}_\lambda] = K_{\mathfrak{s}_{\lambda^+}}$ and $(K_{\mathfrak{s}_{\lambda^+}}, \leq^*_{\mathfrak{s}_\lambda})$ satisfies the demands on a.e.c. except possibly smoothness, see II§7

(β) if $M \in K_{\lambda^+}$ is saturated and $p \in \mathscr{S}_{\mathfrak{K}}(M)$ then for some pair (N, a) we have $M \leq^*_{\mathfrak{s}_\lambda} N$ and $a \in N$ realizes p

(γ) if $M \in K_{\lambda^+}$ is saturated then some N satisfies:

- (a) $N \in K_{\lambda^+}$ is saturated
- (b) N is $\leq_{\mathfrak{K}}$-universal over M
- (c) $M \leq^*_{\mathfrak{s}_\lambda} N$

(δ) $\mathfrak{s}_\lambda$ is successful.

[Why? Clause (α):

We know that both $K^{\mathrm{nice}}_{\lambda^+}[\mathfrak{s}_\lambda]$ and $K_{\mathfrak{s}_{\lambda^+}}$ are the class of saturated $M \in K_\lambda$. The rest holds by II§7,§8.

Clause (β):

By $\circledast_3$ we can find $M_* \leq_{\mathfrak{K}} M$ of cardinality χ_* such that p does not χ_*-split over it (equivalently does not λ^+-split over it).

Let $\langle M_\alpha : \alpha < \lambda^+ \rangle$ be $\leq_{\mathfrak{s}_\lambda}$-increasing continuous such that $M_{\alpha+1}$ is brimmed over M_α for $\mathfrak{s}_\lambda$ for every $\alpha < \lambda^+$ and $M_* \leq_{\mathfrak{K}} M_0$ (so

$\|M_*\| < \|M_0\|$ otherwise we would require M_0 is brimmed over M_*). Hence $\cup\{M_\alpha : \alpha < \lambda^+\} \in K_{\lambda^+}$ is saturated (by $\odot_5$) so without loss of generality is equal to M. We can choose $a_*, N_\alpha(\alpha < \lambda)$ such that $\langle N_\alpha : \alpha < \lambda^+\rangle$ is $\leq_{\mathfrak{s}_\lambda}$-increasing continuous, $M_\alpha \leq_{\mathfrak{s}_\lambda} M_\alpha$, $\mathrm{NF}_{\mathfrak{s}_\lambda}(M_\alpha, N_\alpha, M_\beta, M_\beta)$ for $\alpha < \beta < \lambda^+, N_{\alpha+1}$ is brimmed over $M_{\alpha+1} \cup N_\alpha$ and $\mathbf{tp}_{\mathfrak{s}_\lambda}(a, N_0, M_0) = p \restriction M_0$ so $a \in N_0$. Let $N = \cup\{N_\alpha : \alpha < \lambda^+\}$ so again $N \in K_{\lambda^+}$ is saturated (equivalently $N \in K^{\mathrm{nice}}_{\lambda^+}[\mathfrak{s}_\lambda]$) and $M \leq_{\mathfrak{K}} N$ and even $M \leq^*_{\mathfrak{s}_\lambda} N$ (by the definition of $\leq^*_{\mathfrak{s}_\lambda}$). For each $\alpha < \lambda^+$ we have $\mathrm{NF}_{\mathfrak{s}_\lambda}(M_0, N_0, M_\alpha, N_\alpha)$ but $\mathrm{NF}_{\mathfrak{s}_\lambda}$ respect $\mathfrak{s}_\lambda$ hence $\mathbf{tp}_{\mathfrak{s}_\lambda}(a, M_\alpha, N_\alpha)$ does not fork over M_0 hence by the definition of $\mathfrak{s}_\lambda$ the type $\mathbf{tp}_{\mathfrak{s}_\lambda}(a, M_\alpha, N_\alpha)$ does not λ-split over M_* hence $\mathbf{tp}_{\mathfrak{s}_\lambda}(a, M_\alpha, N_\alpha) = p \restriction M_\alpha$. As this holds for every $\alpha < \lambda^+$, by the choice of χ_*, i.e. by $\circledast_1$ clearly a realizes p.

Clause (γ):

By clause (β) as in the proofs in II§4; that is, we choose $N \in K_{\lambda^+}$ which is $\leq_{\mathfrak{K}_\lambda}$-universal over M. We now try to choose $(M_\alpha, f_\alpha, N_\alpha)$ by induction on $\alpha < \lambda^+$ such that: $M_0 = M, N_0 = N, f_0 = \mathrm{id}_M, M_\alpha$ is $\leq^*_{\mathfrak{s}_\lambda}$-increasing continuous, N_α is $\leq_{\mathfrak{K}}$-increasing continuous, f_α is a $\leq_{\mathfrak{K}}$-embedding of M_α into N_α, f_α is $\subseteq$-increasing continuous with α and $\alpha = \beta + 1 \Rightarrow f_\alpha(M_\alpha) \cap N_\beta \neq f_\beta(M_\beta)$.

For $\alpha = 0, \alpha$ limit no problems. If $\alpha = \beta + 1$ and $f_\alpha(M_\alpha) = N_\alpha$ we are done and otherwise use clause (β). But by Fodor lemma we cannot carry the induction for every $\alpha < \lambda^+$, so we are done proving (γ).

Clause (δ):

We should verify the conditions in Definition III.1.1. Now clause (a) there, being weakly successful, holds by $\odot_4$. As for clause (b) there, it suffices to prove that if $M_1, M_2 \in K^{\mathrm{nice}}_{\lambda^+}[\mathfrak{s}_\lambda] = K_{\mathfrak{s}^+_\lambda}$ and $M_1 \leq_{\mathfrak{K}} M_2$ then $M_1 \leq^*_{\mathfrak{s}_\lambda} M_2$ which means: if $\langle M^\ell_\alpha : \alpha < \lambda^+\rangle$ is $\leq_{\mathfrak{s}_\lambda}$-increasing continuous, $M^\ell_{\alpha+1}$ is brimmed over M^ℓ_α with $M_\ell = \cup\{M^\ell_\alpha : \alpha < \lambda^+\}$, then for some club E of λ^+ for every $\alpha < \beta$ from E, $\mathrm{NF}_{\mathfrak{s}_\lambda}(M^1_\alpha, M^2_\alpha, M^1_\beta, M^2_\delta)$.

By clause (γ) there is $N \in K_{\mathfrak{s}^+_\lambda}$ such that $M_1 \leq^*_{\mathfrak{s}_{\lambda^+}} N$ (hence $M_1 \leq_{\mathfrak{K}} N$) and N is $\leq_{\mathfrak{K}^{\mathfrak{s}_\lambda}}$-universal over M_1. So without loss of generality $M_2 \leq_{\mathfrak{K}} N$ but by II.7.4(3) all this implies $M_1 \leq^*_{\lambda^+} M_2$. So we are done proving $\boxdot_3$.

$\boxdot_4$ $\mathfrak{s}_{\lambda^+}$ is the successor of $\mathfrak{s}_\lambda$ for $\lambda \in \Theta$.

[Why? Now by $\boxdot_3$ the good frame $\mathfrak{s}_\lambda$ is successful; by III.1.6 we know that $\mathfrak{s}^+_\lambda$ is a well defined good λ^+-frame. Clearly $K_{\mathfrak{s}_\lambda(+)}$ is the class of saturated $M \in \mathfrak{K}_{\lambda^+}$, by $\odot_5$, see the definitions in II.7.2, II.8.7(5). But $\mathfrak{s}_\lambda$ is good$^+$ by $\boxdot_2$ so by III.1.8 we know that $\leq_{\mathfrak{s}_\lambda(+)} = <^*_{\lambda^+} [\mathfrak{s}_\lambda]$ is equal to $\leq_{\mathfrak{K}} \restriction K_{\mathfrak{s}_\lambda(+)}$, so $\mathfrak{K}_{\mathfrak{s}_\lambda(+)} = \mathfrak{K}_{\mathfrak{s}_{\lambda^+}}$. As both $\mathfrak{s}_\lambda(+)$ and $\mathfrak{s}_{\lambda^+}$ are full, clearly $\mathscr{S}^{\rm bs}_{\mathfrak{s}_\lambda(+)} = \mathscr{S}^{\rm bs}_{\mathfrak{s}_{\lambda^+}}$. For $M_1 \leq_{\mathfrak{s}_\lambda(+)} M_2 \leq_{\mathfrak{s}_\lambda(+)} M_3$ and $a \in M_3 \backslash M_2$, comparing the two definitions of "$\mathbf{tp}_{\mathfrak{K}_{\mathfrak{s}_\lambda(+)}}(a, M_2, M_1)$ does not fork over M_1" they are the same. So we are done.]

$\boxdot_5$ $\mathfrak{s}_{\lambda_*^{+\omega}}$ is the limit of $\langle \mathfrak{s}^{+n}_{\lambda_*} : n < \omega \rangle$.

[Why? Should be clear.]

$\boxdot_6$ $\mathfrak{s}_\lambda$ satisfies the hypothesis III.12.3 of III§12 if $\lambda \in \Theta \backslash \lambda^{+3}_*$ holds.

[Why? By $\boxdot_2, \boxdot_3, \boxdot_4$ and III.12.2.]

Hence

$\boxdot_7$ $\mathfrak{s}_{\lambda_*}$ is beautiful $\lambda^{+\omega}_*$-frame.

[Why? By III.12.37 and III.12.41.]

$\boxdot_8$ $K[\mathfrak{s}_{\lambda^{+\omega}_*}]$ is categorical in one $\chi > \lambda^{+\omega}_*$ iff it is categorical in every $\chi > \lambda^{+\omega}$.

[Why? By III.12.41(d),(e).]

$\boxdot_9$ if $\lambda \geq \beth_{1,1}(\lambda^{+\omega}_*)$ then $\mathfrak{K}_\lambda = \mathfrak{K}_\lambda[\mathfrak{s}_{\lambda^{+\omega}_*}]$.

[Why? The conclusion $\supseteq$ is obvious. For the other inclusion let $M \in K_\lambda$, now by the definition of class in the left, it is enough to prove that M is $(\lambda^{+\omega}_*)^+$-saturated. But otherwise by the omitting type theorem for a.e.c., i.e. by 0.9(1),(d), (or see [Sh 394, 8.6=X1.3A]) there is such a model $M' \in K_\mu$, contradiction to $(*)_4$.]

By $\boxdot_8 + \boxdot_9$ we are done. $\square_{7.12}$

BIBLIOGRAPHY FOR *UNIVERSAL CLASSES*

[Bal88] John Baldwin. *Fundamentals of Stability Theory.* Perspectives in Mathematical Logic. Springer-Verlag, Berlin, 1988.

[Bal0x] John Baldwin. *Categoricity,* volume to appear. 200x.

[BlSh 862] John Baldwin and Saharon Shelah. Examples of non-locality. *Journal of Symbolic Logic,* **accepted**.

[Bl85] John T. Baldwin. Definable second order quantifiers. In J. Barwise and S. Feferman, editors, *Model Theoretic Logics,* Perspectives in Mathematical Logic, chapter XII, pages 445–477. Springer-Verlag, New York Berlin Heidelberg Tokyo, 1985.

[BKV0x] John T. Baldwin, David W. Kueker, and Monica VanDieren. Upward Stability Transfer Theorem for Tame Abstract Elementary Classes. *Preprint,* 2004.

[BLSh 464] John T. Baldwin, Michael C. Laskowski, and Saharon Shelah. Forcing Isomorphism. *Journal of Symbolic Logic,* **58**:1291–1301, 1993. math.LO/9301208.

[BlSh 156] John T. Baldwin and Saharon Shelah. Second-order quantifiers and the complexity of theories. *Notre Dame Journal of Formal Logic,* **26**:229–303, 1985. Proceedings of the 1980/1 Jerusalem Model Theory year.

[BlSh 330] John T. Baldwin and Saharon Shelah. The primal framework. I. *Annals of Pure and Applied Logic,* **46**:235–264, 1990. math.LO/9201241.

[BlSh 360] John T. Baldwin and Saharon Shelah. The primal framework. II. Smoothness. *Annals of Pure and Applied Logic,* **55**:1–34, 1991. Note: See also 360a below. math.LO/9201246.

[BlSh 393] John T. Baldwin and Saharon Shelah. Abstract classes with few models have 'homogeneous-universal' models. *Journal of Symbolic Logic,* **60**:246–265, 1995. math.LO/9502231.

[BaFe85] Jon Barwise and Solomon Feferman (editors). *Model-theoretic logics.* Perspectives in Mathematical Logic. Springer Verlag, Heidelberg-New York, 1985.

[BKM78] Jon Barwise, Matt Kaufmann, and Michael Makkai. Stationary logic. *Annals of Mathematical Logic,* **13**:171–224, 1978.

Typeset by $\mathcal{AMS}$-TEX

[BY0y] Itay Ben-Yaacov. Uncountable dense categoricity in cats. *J. Symbolic Logic*, **70**:829–860, 2005.

[BeUs0x] Itay Ben-Yaacov and Alex Usvyatsov. Logic of metric spaces and Hausdorff CATs. *In preparation.*

[BoNe94] Alexandre Borovik and Ali Nesin. *Groups of finite Morley rank*, volume 26 of *Oxford Logic Guide*. The Clarendon Press, Oxford University Press, New York, 1994.

[Bg] John P. Burgess. Equivalences generated by families of borel sets. *Proceedings of the AMS*, **69**:323–326, 1978.

[ChKe66] Chen-Chung Chang and Jerome H. Keisler. *Continuous Model Theory*, volume 58 of *Annals of Mathematics Studies*. Princeton University Press, Princeton, NJ, 1966.

[ChKe62] Chen chung Chang and Jerome H. Keisler. Model theories with truth values in a uniform space. *Bulletin of the American Mathematical Society*, **68**:107–109, 1962.

[CoSh:919] Moran Cohen and Saharon Shelah. Stable theories and Representation over sets. *preprint*.

[DvSh 65] Keith J. Devlin and Saharon Shelah. A weak version of $\diamondsuit$ which follows from $2^{\aleph_0} < 2^{\aleph_1}$. *Israel Journal of Mathematics*, **29**:239–247, 1978.

[Di] M. A. Dickman. Larger infinitary languages. In J. Barwise and S. Feferman, editors, *Model Theoretic Logics*, Perspectives in Mathematical Logic, chapter IX, pages 317–364. Springer-Verlag, New York Berlin Heidelberg Tokyo, 1985.

[Eh57] Andrzej Ehrenfeucht. On theories categorical in power. *Fundamenta Mathematicae*, **44**:241–248, 1957.

[Fr75] Harvey Friedman. One hundred and two problems in mathematical logic. *Journal of Symbolic Logic*, **40**:113–129, 1975.

[GiSh 577] Moti Gitik and Saharon Shelah. Less saturated ideals. *Proceedings of the American Mathematical Society*, **125**:1523–1530, 1997. math.LO/9503203.

[GbTl06] Rüdiger Göbel and Jan Trlifaj. *Approximations and endomorphism algebras of modules*, volume 41 of *de Gruyter Expositions in Mathematics*. Walter de Gruyter, Berlin, 2006.

[Gr91] Rami Grossberg. On chains of relatively saturated submodels of a model without the order property. *Journal of Symbolic Logic*, **56**:124–128, 1991.

[GrHa89] Rami Grossberg and Bradd Hart. The classification of excellent classes. *Journal of Symbolic Logic*, **54**:1359–1381, 1989.

[GIL02] Rami Grossberg, Jose Iovino, and Olivier Lessmann. A primer of simple theories. *Archive for Mathematical Logic*, **41**:541–580, 2002.

[GrLe0x] Rami Grossberg and Olivier Lessmann. The main gap for totally transcendental diagrams and abstract decomposition theorems. *Preprint.*

[GrLe00a] Rami Grossberg and Olivier Lessmann. Dependence relation in pregeometries. *Algebra Universalis*, **44**:199–216, 2000.

[GrLe02] Rami Grossberg and Olivier Lessmann. Shelah's stability spectrum and homogeneity spectrum in finite diagrams. *Archive for Mathematical Logic*, **41**:1–31, 2002.

[GrSh 259] Rami Grossberg and Saharon Shelah. On Hanf numbers of the infinitary order property. *Mathematica Japonica*, **submitted**. math.LO/9809196.

[GrSh 174] Rami Grossberg and Saharon Shelah. On universal locally finite groups. *Israel Journal of Mathematics*, **44**:289–302, 1983.

[GrSh 238] Rami Grossberg and Saharon Shelah. A nonstructure theorem for an infinitary theory which has the unsuperstability property. *Illinois Journal of Mathematics*, **30**:364–390, 1986. Volume dedicated to the memory of W.W. Boone; ed. Appel, K., Higman, G., Robinson, D. and Jockush, C.

[GrSh 222] Rami Grossberg and Saharon Shelah. On the number of nonisomorphic models of an infinitary theory which has the infinitary order property. I. *The Journal of Symbolic Logic*, **51**:302–322, 1986.

[GrVa0xa] Rami Grossberg and Monica VanDieren. Galois-stbility for Tame Abstract Elementary Classes. *submitted.*

[GrVa0xb] Rami Grossberg and Monica VanDieren. Upward Categoricity Transfer Theorem for Tame Abstract Elementary Classes. *submitted.*

[Ha61] Andras Hajnal. Proof of a conjecture of S.Ruziewicz. *Fundamenta Mathematicae*, **50**:123–128, 1961/1962.

[HHL00] Bradd Hart, Ehud Hrushovski, and Michael C. Laskowski. The uncountable spectra of countable theories. *Annals of Mathematics*, **152**:207–257, 2000.

[HaSh 323] Bradd Hart and Saharon Shelah. Categoricity over P for first order T or categoricity for $\phi \in \mathrm{L}_{\omega_1\omega}$ can stop at $\aleph_k$ while holding for $\aleph_0, \cdots, \aleph_{k-1}$. *Israel Journal of Mathematics*, **70**:219–235, 1990. math.LO/9201240.

[He74] C. Ward Henson. The isomorphism property in nonstandard analysis and its use in the theory of Banach spaces. *Journal of Symbolic Logic*, **39**:717–731, 1974.

[HeIo02] C. Ward Henson and Jose Iovino. Ultraproducts in analysis. In *Analysis and logic (Mons, 1997)*, volume 262 of *London Math. Soc. Lecture Note Ser.*, pages 1–110. Cambridge Univ. Press, Cambridge, 2002.

[He92] A. Hernandez. *On ω_1-saturated models of stable theories.* PhD thesis, Univ. of Calif. Berkeley, 1992. Advisor: Leo Harrington.

[HuSh 342] Ehud Hrushovski and Saharon Shelah. A dichotomy theorem for regular types. *Annals of Pure and Applied Logic*, **45**:157–169, 1989.

[Hy98] Tapani Hyttinen. Generalizing Morley's theorem. *Mathematical Logic Quarterly*, **44**:176–184, 1998.

[HySh 474] Tapani Hyttinen and Saharon Shelah. Constructing strongly equivalent nonisomorphic models for unsuperstable theories, Part A. *Journal of Symbolic Logic*, **59**:984–996, 1994. math.LO/0406587.

[HySh 529] Tapani Hyttinen and Saharon Shelah. Constructing strongly equivalent nonisomorphic models for unsuperstable theories. Part B. *Journal of Symbolic Logic*, **60**:1260–1272, 1995. math.LO/9202205.

[HySh 632] Tapani Hyttinen and Saharon Shelah. On the Number of Elementary Submodels of an Unsuperstable Homogeneous Structure. *Mathematical Logic Quarterly*, **44**:354–358, 1998. math.LO/9702228.

[HySh 602] Tapani Hyttinen and Saharon Shelah. Constructing strongly equivalent nonisomorphic models for unsuperstable theories, Part C. *Journal of Symbolic Logic*, **64**:634–642, 1999. math.LO/9709229.

[HySh 629] Tapani Hyttinen and Saharon Shelah. Strong splitting in stable homogeneous models. *Annals of Pure and Applied Logic*, **103**:201–228, 2000. math.LO/9911229.

[HySh 676] Tapani Hyttinen and Saharon Shelah. Main gap for locally saturated elementary submodels of a homogeneous structure. *Journal of Symbolic Logic*, **66**:1286–1302, 2001, no.3. math.LO/9804157.

[HShT 428] Tapani Hyttinen, Saharon Shelah, and Heikki Tuuri. Remarks on Strong Nonstructure Theorems. *Notre Dame Journal of Formal Logic*, **34**:157–168, 1993.

[HyTu91] Tapani Hyttinen and Heikki Tuuri. Constructing strongly equivalent nonisomorphic models for unstable theories. *Annals Pure and Applied Logic*, **52**:203–248, 1991.

[JrSh 875] Adi Jarden and Saharon Shelah. Good frames minus stability. *Preprint.*

[J] Thomas Jech. *Set theory.* Springer Monographs in Mathematics. Springer-Verlag, Berlin, 2003. The third millennium edition, revised and expanded.

[Jn56] Bjarni Jónsson. Universal relational systems. *Mathematica Scandinavica*, **4**:193–208, 1956.

[Jo56] Bjarni Jónsson. Universal relational systems. *Mathematica Scandinavica*, **4**:193–208, 1956.

[Jn60] Bjarni Jónsson. Homogeneous universal relational systems. *Mathematica Scandinavica*, **8**:137–142, 1960.

[Jo60] Bjarni Jónsson. Homogeneous universal relational systems. *Mathematica Scandinavica*, **8**:137–142, 1960.

[KM67] H. Jerome Keisler and Michael D. Morley. On the number of homogeneous models of a given power. *Israel Journal of Mathematics*, **5**:73–78, 1967.

[Ke70] Jerome H. Keisler. Logic with the quantifier "there exist uncountably many". *Annals of Mathematical Logic*, **1**:1–93, 1970.

[Ke71] Jerome H. Keisler. *Model theory for infinitary logic. Logic with countable conjunctions and finite quantifiers*, volume 62 of *Studies in Logic and the Foundations of Mathematics.* North–Holland Publishing Co., Amsterdam–London, 1971.

[KiPi98] Byunghan Kim and Anand Pillay. From stability to simplicity. *Bull. Symbolic Logic*, **4**:17–36, 1998.

[Kin66] Akiko Kino. On definability of ordinals in logic with infinitely long expressions. *Journal of Symbolic Logic*, **31**:365–375, 1966.

[KjSh 409] Menachem Kojman and Saharon Shelah. Non-existence of Universal Orders in Many Cardinals. *Journal of Symbolic Logic*, **57**:875–891, 1992. math.LO/9209201.

[KlSh 362] Oren Kolman and Saharon Shelah. Categoricity of Theories in $L_{\kappa,\omega}$, when κ is a measurable cardinal. Part 1. *Fundamenta Mathematicae*, **151**:209–240, 1996. math.LO/9602216.

[KoSh 796] Péter Komjáth and Saharon Shelah. A partition theorem for scattered order types. *Combinatorics Probability and Computing*, **12**:621–626, 2003, no.5-6. Special issue on Ramsey theory. math.LO/0212022.

[Las88] Michael C. Laskowski. Uncountable theories that are categorical in a higher power. *The Journal of Symbolic Logic*, **53**:512–530, 1988.

[LwSh 871] Michael C. Laskowski and Saharon Shelah. Karp height of models of stable theories. 0711.3043.

[LwSh 489] Michael C. Laskowski and Saharon Shelah. On the existence of atomic models. *Journal of Symbolic Logic*, **58**:1189–1194, 1993. math.LO/9301210.

[LwSh 560] Michael C. Laskowski and Saharon Shelah. The Karp complexity of unstable classes. *Archive for Mathematical Logic*, **40**:69–88, 2001. math.LO/0011167.

[LwSh 687] Michael C. Laskowski and Saharon Shelah. Karp complexity and classes with the independence property. *Annals of Pure and Applied Logic*, **120**:263–283, 2003. math.LO/0303345.

[Lv71] Richard Laver. On Fraissé's order type conjecture. *Annals of Mathematics*, **93**:89–111, 1971.

[Le0x] Olivier Lessmann. Abstract group configuration. *Preprint.*

[Le0y] Olivier Lessmann. Pregeometries in finite diagrams. *Preprint.*

[McSh 55] Angus Macintyre and Saharon Shelah. Uncountable universal locally finite groups. *Journal of Algebra*, **43**:168–175, 1976.

[MaSh 285] Michael Makkai and Saharon Shelah. Categoricity of theories in $L_{\kappa\omega}$, with κ a compact cardinal. *Annals of Pure and Applied Logic*, **47**:41–97, 1990.

[Mw85a] Johann A. Makowsky. Abstract embedding relations. In J. Barwise and S. Feferman, editors, *Model-Theoretic Logics*, pages 747–791. Springer-Verlag, 1985.

[Mw85] Johann A. Makowsky. Compactnes, embeddings and definability. In J. Barwise and S. Feferman, editors, *Model-Theoretic Logics*, pages 645–716. Springer-Verlag, 1985.

[MkSh 366] Alan H. Mekler and Saharon Shelah. Almost free algebras . *Israel Journal of Mathematics*, **89**:237–259, 1995. math.LO/9408213.

[MiRa65] Eric Milner and Richard Rado. The pigeon-hole principle for ordinal numbers. *Proc. London Math. Soc.*, **15**:750–768, 1965.

[MoVa62] M. D. Morley and R. L. Vaught. Homogeneous and universal models. *Mathematica Scandinavica*, **11**:37–57, 1962.

[Mo65] Michael Morley. Categoricity in power. *Transaction of the American Mathematical Society*, **114**:514–538, 1965.

[Mo70] Michael D. Morley. The number of countable models. *Journal of Symbolic Logic*, **35**:14–18, 1970.

[Pi0x] Anand Pillay. Forking in the category of existentially closed structures. In *Connections between model theory and algebraic and analytic geometry*, volume 6 of *Quad. Mat.*, pages 23–42. Dept. Math., Seconda Univ. Napoli, Caserta, 2000.

[RuSh 117] Matatyahu Rubin and Saharon Shelah. Combinatorial problems on trees: partitions, Δ-systems and large free subtrees. *Annals of Pure and Applied Logic*, **33**:43–81, 1987.

[Sc76] James H. Schmerl. On κ-like structures which embed stationary and closed unbounded subsets. *Annals of Mathematical Logic*, **10**:289–314, 1976.

[Sh 88r] Saharon Shelah. *Abstract elementary classes near* $\aleph_1$. Chapter I. 0705.4137.

[Sh:E12] Saharon Shelah. Analytical Guide and Corrections to [Sh:g]. math.LO/9906022.

[Sh:F888] Saharon Shelah. Categoricity in λ and a superlimit in λ^+.

[Sh 322] Saharon Shelah. Classification over a predicate. *preprint.*

[Sh 482] Saharon Shelah. Compactness in ZFC of the Quantifier on "Complete embedding of BA's". In *Non structure theory, Ch XI*, accepted. Oxford University Press.

[Sh:922] Saharon Shelah. Diamonds. *Proceedings of the American Mathematical Society*, **submitted**. 0711.3030.

[Sh:F820] Saharon Shelah. From solvability of an aec in μ to large extensions in λ.

[Sh 832] Saharon Shelah. Incompactness in singular cardinals. *Preprint.*

[Sh 840] Saharon Shelah. Model theory without choice: Categoricity. *Journal of Symbolic Logic*, **submitted**. math.LO/0504196.

[Sh:e] Saharon Shelah. *Non–structure theory*, accepted. Oxford University Press.

[Sh:F782] Saharon Shelah. On categorical a.e.c. II.

[Sh 800] Saharon Shelah. On complicated models. *Preprint.*

[Sh:F841] Saharon Shelah. On h-almost good λ-frames: More on [SH:838].

[Sh:F735] Saharon Shelah. Revisiting 705.

[Sh 842] Saharon Shelah. Solvability and Categoricity spectrum of a.e.c. with amalgamation. *Preprint.*

[Sh 839] Saharon Shelah. Stable Frames and weight. *Preprint.*

[Sh 705] Saharon Shelah. Toward classification theory of good λ frames and abstract elementary classes.

[Sh 868] Saharon Shelah. When first order T has limit models. *Notre Dame Journal of Formal Logic*, **submitted**. math.LO/0603651.

[Sh 1] Saharon Shelah. Stable theories. *Israel Journal of Mathematics*, **7**:187–202, 1969.

[Sh 3] Saharon Shelah. Finite diagrams stable in power. *Annals of Mathematical Logic*, **2**:69–118, 1970.

[Sh 10] Saharon Shelah. Stability, the f.c.p., and superstability; model theoretic properties of formulas in first order theory. *Annals of Mathematical Logic*, **3**:271–362, 1971.

[Sh 16] Saharon Shelah. A combinatorial problem; stability and order for models and theories in infinitary languages. *Pacific Journal of Mathematics*, **41**:247–261, 1972.

[Sh 31] Saharon Shelah. Categoricity of uncountable theories. In *Proceedings of the Tarski Symposium (Univ. of California, Berkeley, Calif., 1971)*, volume XXV of *Proc. Sympos. Pure Math.*, pages 187–203. Amer. Math. Soc., Providence, R.I, 1974.

[Sh 48] Saharon Shelah. Categoricity in $\aleph_1$ of sentences in $L_{\omega_1,\omega}(Q)$. *Israel Journal of Mathematics*, **20**:127–148, 1975.

[Sh 46] Saharon Shelah. Colouring without triangles and partition relation. *Israel Journal of Mathematics*, **20**:1–12, 1975.

[Sh 43] Saharon Shelah. Generalized quantifiers and compact logic. *Transactions of the American Mathematical Society*, **204**:342–364, 1975.

[Sh 54] Saharon Shelah. The lazy model-theoretician's guide to stability. *Logique et Analyse*, **18**:241–308, 1975.

[Sh 56] Saharon Shelah. Refuting Ehrenfeucht conjecture on rigid models. *Israel Journal of Mathematics*, **25**:273–286, 1976. A special volume, Proceedings of the Symposium in memory of A. Robinson, Yale, 1975.

[Sh:a] Saharon Shelah. *Classification theory and the number of nonisomorphic models*, volume 92 of *Studies in Logic and the Foundations of Mathematics*. North-Holland Publishing Co., Amsterdam-New York, xvi+544 pp, $62.25, 1978.

[Sh 108] Saharon Shelah. On successors of singular cardinals. In *Logic Colloquium '78 (Mons, 1978)*, volume 97 of *Stud. Logic Foundations Math*, pages 357–380. North-Holland, Amsterdam-New York, 1979.

[Sh:93] Saharon Shelah. Simple unstable theories. *Annals of Mathematical Logic*, **19**:177–203, 1980.

[Sh:b] Saharon Shelah. *Proper forcing*, volume 940 of *Lecture Notes in Mathematics*. Springer-Verlag, Berlin-New York, xxix+496 pp, 1982.

[Sh 132] Saharon Shelah. The spectrum problem. II. Totally transcendental and infinite depth. *Israel Journal of Mathematics*, **43**:357–364, 1982.

[Sh 87a] Saharon Shelah. Classification theory for nonelementary classes, I. The number of uncountable models of $\psi \in L_{\omega_1,\omega}$. Part A. *Israel Journal of Mathematics*, **46**:212–240, 1983.

[Sh 87b] Saharon Shelah. Classification theory for nonelementary classes, I. The number of uncountable models of $\psi \in L_{\omega_1,\omega}$. Part B. *Israel Journal of Mathematics*, **46**:241–273, 1983.

[Sh 202] Saharon Shelah. On co-κ-Souslin relations. *Israel Journal of Mathematics*, **47**:139–153, 1984.

[Sh 200] Saharon Shelah. Classification of first order theories which have a structure theorem. *American Mathematical Society. Bulletin. New Series*, **12**:227–232, 1985.

[Sh 205] Saharon Shelah. Monadic logic and Lowenheim numbers. *Annals of Pure and Applied Logic*, **28**:203–216, 1985.

[Sh 197] Saharon Shelah. Monadic logic: Hanf numbers. In *Around classification theory of models*, volume 1182 of *Lecture Notes in Mathematics*, pages 203–223. Springer, Berlin, 1986.

[Sh 155] Saharon Shelah. The spectrum problem. III. Universal theories. *Israel Journal of Mathematics*, **55**:229–256, 1986.

[Sh 88a] Saharon Shelah. Appendix: on stationary sets (in "Classification of nonelementary classes. II. Abstract elementary classes"). In *Classification theory (Chicago, IL, 1985)*, volume 1292 of *Lecture Notes in Mathematics*, pages 483–495. Springer, Berlin, 1987. Proceedings of the USA–Israel Conference on Classification Theory, Chicago, December 1985; ed. Baldwin, J.T.

[Sh 88] Saharon Shelah. Classification of nonelementary classes. II. Abstract elementary classes. In *Classification theory (Chicago, IL, 1985)*, volume 1292 of *Lecture Notes in Mathematics*, pages 419–497. Springer, Berlin, 1987. Proceedings of the USA–Israel Conference on Classification Theory, Chicago, December 1985; ed. Baldwin, J.T.

[Sh 220] Saharon Shelah. Existence of many $L_{\infty,\lambda}$-equivalent, nonisomorphic models of T of power λ. *Annals of Pure and Applied Logic*, **34**:291–310, 1987. Proceedings of the Model Theory Conference, Trento, June 1986.

[Sh 225] Saharon Shelah. On the number of strongly $\aleph_\epsilon$-saturated models of power λ. *Annals of Pure and Applied Logic*, **36**:279–287, 1987. See also [Sh:225a].

[Sh 300] Saharon Shelah. Universal classes. In *Classification theory (Chicago, IL, 1985)*, volume 1292 of *Lecture Notes in Mathematics*, pages 264–418. Springer, Berlin, 1987. Proceedings of the USA–Israel Conference on Classification Theory, Chicago, December 1985; ed. Baldwin, J.T.

[Sh 225a] Saharon Shelah. Number of strongly $\aleph_\epsilon$ saturated models—an addition. *Annals of Pure and Applied Logic*, **40**:89–91, 1988.

[Sh:c] Saharon Shelah. *Classification theory and the number of nonisomorphic models*, volume 92 of *Studies in Logic and the Foundations of Mathematics*. North-Holland Publishing Co., Amsterdam, xxxiv+705 pp, 1990.

[Sh 284c] Saharon Shelah. More on monadic logic. Part C. Monadically interpreting in stable unsuperstable **T** and the monadic theory of ${}^{\omega}\lambda$. *Israel Journal of Mathematics*, **70**:353–364, 1990.

[Sh 429] Saharon Shelah. Multi-dimensionality. *Israel Journal of Mathematics*, **74**:281–288, 1991.

[Sh 351] Saharon Shelah. Reflecting stationary sets and successors of singular cardinals. *Archive for Mathematical Logic*, **31**:25–53, 1991.

[Sh 420] Saharon Shelah. Advances in Cardinal Arithmetic. In *Finite and Infinite Combinatorics in Sets and Logic*, pages 355–383. Kluwer Academic Publishers, 1993. N.W. Sauer et al (eds.). 0708.1979.

[Sh:g] Saharon Shelah. *Cardinal Arithmetic*, volume 29 of *Oxford Logic Guides*. Oxford University Press, 1994.

[Sh 430] Saharon Shelah. Further cardinal arithmetic. *Israel Journal of Mathematics*, **95**:61–114, 1996. math.LO/9610226.

[Sh:f] Saharon Shelah. *Proper and improper forcing*. Perspectives in Mathematical Logic. Springer, 1998.

[Sh 394] Saharon Shelah. Categoricity for abstract classes with amalgamation. *Annals of Pure and Applied Logic*, **98**:261–294, 1999. math.LO/9809197.

[Sh 620] Saharon Shelah. Special Subsets of ${}^{\mathrm{cf}(\mu)}\mu$, Boolean Algebras and Maharam measure Algebras. *Topology and its Applications*, **99**:135–235, 1999. 8th Prague Topological Symposium on General Topology and its Relations to Modern Analysis and Algebra, Part II (1996). math.LO/9804156.

[Sh 589] Saharon Shelah. Applications of PCF theory. *Journal of Symbolic Logic*, **65**:1624–1674, 2000.

[Sh 460] Saharon Shelah. The Generalized Continuum Hypothesis revisited. *Israel Journal of Mathematics*, **116**:285–321, 2000. math.LO/9809200.

[Sh 576] Saharon Shelah. Categoricity of an abstract elementary class in two successive cardinals. *Israel Journal of Mathematics*, **126**:29–128, 2001. math.LO/9805146.

[Sh 472] Saharon Shelah. Categoricity of Theories in $L_{\kappa^*\omega}$, when κ^* is a measurable cardinal. Part II. *Fundamenta Mathematicae*, **170**:165–196, 2001. math.LO/9604241.

[Sh 603] Saharon Shelah. Few non minimal types and non-structure. In *Proceedings of the 11 International Congress of Logic, Methodology and Philosophy of Science, Krakow August'99; In the Scope of Logic, Methodology and Philosophy of Science*, volume 1, pages 29–53. Kluwer Academic Publishers, 2002. math.LO/9906023.

[Sh 715] Saharon Shelah. Classification theory for elementary classes with the dependence property - a modest beginning. *Scientiae Mathematicae Japonicae*, **59, No. 2; (special issue: e9, 503–544)**:265–316, 2004. math.LO/0009056.

[Sh 829] Saharon Shelah. More on the Revised GCH and the Black Box. *Annals of Pure and Applied Logic*, **140**:133–160, 2006. math.LO/0406482.

[ShHM 158] Saharon Shelah, Leo Harrington, and Michael Makkai. A proof of Vaught's conjecture for ω-stable theories. *Israel Journal of Mathematics*, **49**:259–280, 1984. Proceedings of the 1980/1 Jerusalem Model Theory year.

[ShUs 837] Saharon Shelah and Alex Usvyatsov. Model theoretic stability and categoricity for complete metric spaces. *Israel Journal of Mathematics*, **submitted**. math.LO/0612350.

[ShVi 648] Saharon Shelah and Andrés Villaveces. Categoricity may fail late. *Journal of Symbolic Logic*, **submitted**. math.LO/0404258.

[ShVi 635] Saharon Shelah and Andrés Villaveces. Toward Categoricity for Classes with no Maximal Models. *Annals of Pure and Applied Logic*, **97**:1–25, 1999. math.LO/9707227.

[Sh:E45] Shelah, Saharon. Basic non-structure for a.e.c.

[Sh:E54] Shelah, Saharon. Comments to Universal Classes.

[Sh:E56] Shelah, Saharon. Density is at most the spread of the square. 0708.1984.

[Sh:F709] Shelah, Saharon. Good* λ-frames.

[Sh:E36] Shelah, Saharon. Good Frames.

[Str76] Jacques Stern. Some applications of model theory in Banach space theory. *Annals of Mathematical Logic*, **9**:49–121, 1976.

[Va02] Monica M. VanDieren. *Categoricity and Stability in Abstract Elementary Classes.* PhD thesis, Carnegie Melon University, Pittsburgh, PA, 2002.

[Zi0xa] B.I. Zilber. Dimensions and homogeneity in mathematical structures. preprint, 2000.

[Zi0xb] B.I. Zilber. A categoricity theorem for quasiminimal excellent classes. preprint, 2002.

www.ingramcontent.com/pod-product-compliance
Lightning Source LLC
LaVergne TN
LVHW012325060326
832902LV00011B/1727